Readings
in Mammalian
Cell Culture

Readings in Mammalian Cell Culture

SECOND EDITION

Completely Revised
and Expanded

Edited by

ROBERT POLLACK

Department of Biological Sciences
Columbia University

Cold Spring Harbor Laboratory
1981

Readings in Mammalian Cell Culture
© 1981 by Cold Spring Harbor Laboratory
Second edition 1981
All rights reserved
Printed in the United States of America
Book design by Emily Harste
Cover design by Emily Harste and Michael Verderame

Library of Congress Cataloging in Publication Data
Main entry under title:

Readings in mammalian cell culture.

   Includes bibliographies.
   1. Cell culture--Addresses, essays,
lectures.   2. Tissue culture--Addresses,
essays, lectures.   3. Mammals--Cytology--
Addresses, essays, lectures.   I. Pollack,
Robert, 1940–   [DNLM: 1. Cells,
Cultured--Essays.   2. Tissue culture--
Essays.   QH 585 R287]
QH585.R43   1981   599.08'7'0724   81-67728
ISBN 0-87969-137-9   AACR2

Front cover: Fluorescence photomicrograph of two mouse cells soon after division. Actin in the cytoskeleton (seen here in green) is visualized with fluorescein isothiocynate-phalloidin, a chemical reagent specific for polymerized actin (Chapter 11). The nuclear tumor antigen and DNA-binding protein SV40 large T (seen here in yellow) is visualized with a double immune stain. The cells were first stained with monoclonal mouse antibody produced by a hybridoma cell line (Chapter 2), then with rhodamine-conjugated antibody to mouse immunoglobulin G. Fluorescein isothiocynate-phalloidin was generously supplied by Dr. T. Wieland, Max Planck Institute. Monoclonal antibody to SV40 large T antigen was generously supplied by Dr. E. Gurney, U. Utah, Salt Lake City.
Photo by Michael Verderame, Columbia University

Back cover: Connective tissue and endothelial (?) cells liberated from cultures of the heart muscle and abdominal muscle of a 3-day-old rat. Mononuclear cells from the blood of the same animal. Wright's stain. This is one of the first studies on the dissociation of live cells with trypsin. (Rous and Jones, *J. Exp. Med. 23:* 549 [1916])

All Cold Spring Harbor Laboratory publications are available through booksellers or may be ordered directly from Cold Spring Harbor Laboratory, Box 100, Cold Spring Harbor, New York 11724.

Contents

Chapter 4   Mechanisms of Growth Control   243

Chapter 11   Cytoplasmic Organization   651

Introduction to the Second Edition

The first edition of this book was revised in 1975 by the simple addition of 12 papers. In that revision I made the safe bet that Cell Biology would someday yield entirely new concepts. After only five years, it has already done so, and cultured cells have been the matrix from which many of these concepts have sprung.

In this new, second edition I have taken apart and reassembled the entire book. To give full space to some major developments in biology that have occurred in the past five years through studies on cultured mammalian cells, I have contracted the space given to certain systems heavily represented in the first edition. For example, some papers on heterokaryons and somatic cell hybrids have been deleted to make room for ones that describe powerful new techniques to clone, transfer, and localize genes. The format and the size of the book remains otherwise the same. It has served well in many courses, including my own graduate course at Columbia.

Five subjects are given their own chapters for the first time: Tumorigenicity and Metastasis (Chapter 5), Somatic Cell Genetics (Chapter 8), Transfer of Chromosomes and Genes (Chapter 9), Cell Membrane Organization and Function (Chapter 10), and The Cytoskeleton (Chapter 11).

All chapters now contain papers reflecting some of the newest developments in the subject being considered, as well as those older papers that I have found serve best as teaching aids. For example, in many cases a complex, but defined, mixture of hormones and inorganic ions can replace serum (Chapter 1). Differentiated cells that grow clonally in culture now include normal keratinocytes, adipocytes, blood vessel cells, and B and T lymphoblasts (Chapter 2). Although a clear mechanism for the maintenance of the transformed state is still lacking (Chapter 3), viral genes and their products have been localized and purified in many transformation systems. Growth control of normal fibroblasts is the combined response to signals of cell shape as well as to the presence of hormones (Chapter 4). Metastatic capacity is clearly a cellular property in clones derived from tumors, and anchorage independence seems to be a prerequisite to tumorigenicity (Chapter 5). Although a controversy remains over the existence of a stochastic period in the normal cell cycle of DNA replication (S) and transcription (G1), it is clear that regulation of the cycle is seriously deranged in transformed cells (Chapter 6). The nucleosome, the unit of eukaryotic chromatin, is well-understood, and the structures of mitotic, replicating, and transcribing cell DNA are becoming clearer (Chapters 7 and 8). As genes can now be put into eukaryotic chromatin at will, it appears that gene position can in some cases determine gene function (Chapter 9). The cell membrane as a transducer of information about the extracellular world may be seen as a highly organized collection of transmembrane and intramembranous proteins in the lipid bilayer, and protein-protein interactions are at least as important for function as are protein-lipid interactions (Chapter 10). The cytoplasm is not a soup in which organelles float, but rather a mesh of cytoskeletal proteins in large macromolecular arrays, capable of interaction with all cytoplasmic organelles, the nucleus, ribosomes, and the cell membrane (Chapter 11). In keeping with these changes, the Supplemental Readings lists, which follow each chapter and in which the readings are grouped historically by decade, are now up-to-date as of 1980.

In preparing this book I have been struck by the tyranny that time extends over practicing scientists. A paper has a right time to appear: too early and it makes no kink in the flow of argument; too late and it is obvious and dry. In picking the papers to appear, I was often drawn to ones that had the most influence, and these were not always the initial papers. I find it very hard to understand the element, other than precedence, that makes a paper influential, but I can guess that the techniques used in it can play a surprisingly important role.

Often techniques define ideas. Although we would like to think the inverse is more reasonable; in fact, an idea that cannot be tested will have little or no influence at all. Therefore, ideas flow rapidly once a new way of testing them appears. This edition is freighted with new techniques. I hope that those who use it will learn from them how to extract the maximum amount of physiologically relevant information from the lovely simple systems of cell culture.

Acknowledgments

Assembling this book was a lot of work and required the assistance and forebearance of many individuals. My own laboratory at Columbia was my mainstay, and all the people in it helped in countless ways. Marisa Bolognese kept track of thousands of papers with amazing ease. Mike Verderame patiently helped me to teach in order to learn and prepared the cover photograph. Nancy Ford, Annette Kirk, Liz Ritcey, and Elizabeth Jacobellis of the Cold Spring Harbor Publications office were unflagging in enthusiasm and provided first-rate editorial assistance. Among the many colleagues who critically reviewed my efforts to describe the various fields covered in this book, my special thanks go to Eric Holtzman, Harvey Lodish, Carol Prives, Larry Chasin, Sherman Beychok, Harvey Ozer, Seung-Il Shin, and Dan Rifkin.

It is my pleasure to acknowledge the permission granted by publishers and authors to reproduce the papers in this volume. Citations to the original source are included with each paper, and I encourage readers to make a habit of glancing through current issues of these journals.

Thanks also go to Amy, Marya, the Dicksteins, and the Greenbergs, who all exhibited enormous patience at certain moments in the summer of 1980. Finally, this book would never have existed without Jim Watson's unique intensity and style as Director of Cold Spring Harbor Laboratory.

R.P.
Sag Harbor
Summer 1980

Readings
in Mammalian
Cell Culture

Chapter 1 Growth of Cells in Culture

Rous, P. and F.S. Jones. 1916. A method for obtaining suspensions of living cells from the fixed tissues, and for the plating out of individual cells. *J. Exp. Med. 23:* 549–555.

Gey, G.O., W.D. Coffman, and M.T. Kubicek. 1952. Tissue culture studies of the proliferative capacity of cervical carcinoma and normal epithelium. *Cancer Res. 12:* 264–265.

Eagle, H. 1955. Nutrition needs of mammalian cells in tissue culture. *Science 122:* 501–504.

Nowell, P.C. 1960. Phytohemagglutinin: An initiator of mitosis in cultures of normal human leukocytes. *Cancer Res. 20:* 462–466.

Hayflick, L. and P. Moorhead. 1961. The serial cultivation of human diploid cell strains. *Exp. Cell Res. 25:* 585–621.

Todaro, G.J. and H. Green. 1963. Quantitative studies of the growth of mouse embryo cells in culture and their development into established lines. *J. Cell Biol. 17:* 299–313.

Hayashi, I. and G.H. Sato. 1976. Replacement of serum by hormones permits growth of cells in a defined medium. *Nature 259:* 132–134.

Hamilton, W.G. and R.G. Ham. 1977. Clonal growth of Chinese hamster cell lines in protein-free media. *In Vitro 13:* 537–547.

Before cell culture could be carried out successfully, two problems had to be overcome. Populations of cells had to be grown from single cells, and these populations had to be maintained, alive, for many generations. The papers in this chapter report the discovery and refinement of in vitro conditions for the cultivation of cells.

Antibiotics created a great divide in the history of cell culture because they suppressed the growth of bacteria and permitted long-term culture without fear of catastrophic contamination. All publications until 1945 describe work carried out under laboratory conditions made to resemble the surgical amphitheater. As a result, they now have a rather awkward and even comical air. The paper by **Rous** and **Jones** shows that work under such difficult conditions could still be done with care and precision. Their paper presents the first modern cell culture study, for they handled single chick cells as if they were simple microorganisms.

This modern approach was all the more remarkable considering the culture system of that day—the growth of bits of tissue under droplets of clotted blood plasma. A clot begins to form when the plasma protein fibrinogen is converted to fibrin by thrombin. As fibrin forms an insoluble meshwork, blood platelets touching it break open to release a contractile complex of actomyosin and a mixture of soluble molecules. Thus, a clot is a mixture of blood components suspended in a mesh of fibrin and actomyosin.

In Rous's day, tissue cells survived outside the body best when bits of tissue were imbedded in a clot, and there were few reports of cell suspensions retaining viability. Rous and Jones dispersed cells by digesting the cultured bit of tissue with the proteolytic enzyme trypsin. Dispersed cells grew in a new clot and could be resuspended with trypsin and replated again in a fresh clot. Theirs is the first report of serial passage of vertebrate tissue cells in culture. Rous was quite struck by the way growing cells changed shape: Rounding occurred both during cell division and as a result of trypsin treatment. Such changes in shape are still under intense study today (see Chapters 4 and 11).

The next two papers show the rapid development of cell culture with the advent of antibiotics. Most tumors were thought to be unculturable. The note from **Gey, Coffman, and Kubicek** changed this. A surgeon, Gey had attempted to culture cells from normal and cancerous cervical biopsies. For the most part, these cells failed to grow. However, from one cervical cancer he obtained a rapidly growing cell population, HeLa. This line is so vigorous that to this day it periodically turns up as an unwanted overgrower of other cultures in laboratories carrying it in their collections. Notice the medium Gey et al. used to grow HeLa—chicken plasma, bovine embryo extract, and human placental cord serum—just about as chemically complex and as poorly defined a medium as one can imagine.

Eagle, working with HeLa and mouse L cells, found that the need for undefined complex body fluids could be satisfied by as little as 1% of dialyzed horse serum in an otherwise defined mixture of nutrients. Other human tumors grew in this medium, and viruses such as polio grew on HeLa cells in the absence of serum entirely. This result suggests that media such as Eagle's potentially might be useful for the production of attenuated virus vaccines in the total absence of unwanted serum proteins.

Nucleated cells (leukocytes) found in samples of blood from healthy people are rarely, if ever, dividing. Their progenitors in bone and spleen are dividing rapidly. Is the absence of mitotic lymphoid cells in peripheral blood a permanent situation, or can such lymphoid cells be recovered and induced to divide in culture? **Nowell** describes a way to get normal lymphoid cells to divide. He incubated them with phytohemagglutinin (PHA), a glycoprotein capable of sticking cells to each other through linkage to material at the surfaces of the cells (see Chapter 10). Division occurred only under conditions where the agglutinin PHA was added. Note that Nowell concluded that an agglutinin can make lymphoid cells divide, even though his initial intention in adding PHA was merely to agglutinate, and thereby remove, red cells from the lymphoid cell suspension.

Not all normal human cells need such stimulation in order to divide. Like HeLa, the normal scar-forming cells, or fibroblasts, of most human tissues are quite capable of growing in a dish in Eagle's or other media, provided that serum is present as well. Antibiotics permitted **Hayflick** and **Moorhead** to determine the eventual fate of normal human fibroblast cells subjected to continued serial passage by trypsinization. Remarkably, the fate of normal human cells is to die after a few dozen passages. Death comes only after a very large increase in cell mass. From one primary culture of perhaps 10 million cells, the total accumulated progeny would add up to a cell mass equivalent to about 100 million people. Thus, precrisis human cell strains are useful when large numbers of normal cells are needed, as in vaccine production. Since cells can be frozen at any passage and thawed at any later time, the vast potential cell number really can be harnessed for practical uses. The eventual death of a culture of normal human fibroblasts follows a period of poor cell growth, called "crisis." Cells of other species, in particular rodent cells, can survive crisis to generate permanent cell lines (see Chapter 2).

Hayflick and Moorhead show that the cells are free of contaminating microorganisms, are unable to grow into tumors upon injection into animals and even humans, and are capable of supporting the growth of clinically interesting viruses such as rabies and polio. The inoculation of cells into terminally ill people seems a bit raw and probably would not be permitted today. In a small footnote we find an early reference to contamination of monkey kidney cells used for vaccine development by an adventitious virus called SV40. SV40 is, of course, now one of the most-well-studied tumor-causing viruses (see Chapters 3, 4, and 5).

Why do human cells die in a crisis after about 50 passages? If death is due to dilution of an essential cellular molecule, that molecule must be constantly made by the cells or else it would be gone well before the 50th division. Perhaps a plasmid carrying essential genes replicates in normal cells, albeit less well than does the host genome. Loss of some function necessary for growth would then occur with some fixed probability at each division. It is not even necessary to imagine loss by dilution, since functional change could occur simply by transposition of a DNA sequence in the genome. At the time of Hayflick and Moorhead's paper, autonomous bits of DNA were not thought of as part of the genomes of human cells, but we now know that DNA does indeed colonize and move around in eukaryotic cells (see Chapter 8).

How common is the death of a culture as that described by Hayflick and Moorhead? The crisis of normal fibroblasts is a universal phenomenon, but events subsequent to it vary with species. Rodent fibroblasts, for instance, survive crisis to become postcrisis cell lines. These cells are not the same as any cells that come out of a normal tissue. Their chromosome number is abnormal, and sometimes they are even able to initiate tumors when injected into susceptible animals (see Chapter 5). In a very careful analysis of the process of establishment of cell lines, **Todaro** and **Green** passaged mouse embryo cells under various sets of conditions and found that cell-cell contact during crisis can determine the properties of postcrisis cells. Most of their established cell lines grew into multilayered, dense cell sheets when continuously fed without passaging. However, the 3T3 line, derived from a culture that had never been permitted to grow dense during the passage through crisis, ceased to grow at a low final density. Apparently, dense growth during establishment selected for cells insensitive to the growth-inhibiting effects of cell-cell contact (see Chapter 4).

Why is serum necessary for growth of cells in the otherwise defined medium of Eagle? The papers by **Hayashi** and **Sato** and **Hamilton** and **Ham** approach this problem in two different ways. Hayashi and Sato show that cells known to require a hormone for growth will grow without serum in the presence of that hormone and a series of other hormones. They conclude that a set of hormones, some defined and some not yet purified, are the active molecules of serum. Hamilton and Ham reworked a variation of Eagle's procedure, asking whether mixes of known nutrients will promote colony growth from single cells. The Chinese hamster line CHO formed colonies without serum provided that Ham's medium was supplemented with a large number of trace metals. The conclusion of this study was that serum provides trace metals. The apparent paradox of these two different lines of work would be resolved if the hormone preparations of Hayashi and Sato were shown to contain trace metals, or if trace metals were shown to mimic the effects of hormones. It would also be useful to pool both procedures and use Ham's medium with normal cells to determine whether colony growth of normal cells requires hormones (see Chapter 4).

Reprinted from Journal of Experimental Medicine
Vol. 23, pp. 549–555. 1916

A METHOD FOR OBTAINING SUSPENSIONS OF LIVING CELLS FROM THE FIXED TISSUES, AND FOR THE PLATING OUT OF INDIVIDUAL CELLS.

By PEYTON ROUS, M.D., AND F. S. JONES, V.M.D.

(*From the Laboratories of The Rockefeller Institute for Medical Research.*)

PLATES 84 TO 87.

(Received for publication, January 15, 1916.)

The only cells of the mammalian body which lend themselves as individuals to accurate experimentation *in vitro* while yet alive are the blood cells, the cells of exudates, and the spermatozoa. In saying this we do not overlook the usefulness of tissue cultivation or of experiments with living tissue fragments, of the transplantable tumors for instance. But both means of study involve, not individual cells, but complexes of different cells, which can be standardized only roughly, and which cannot be broken up into their component elements or protected from confusing factors, such, for example, as are introduced by death and autolysis of the central tissue portions. These difficulties have led us to work out a method whereby living tissue cells can be obtained as individuals in suspension, and, if desired, can be plated out in a culture medium (plasma) just as are bacteria. After growth the cells can be liberated again, and again plated successfully.

The method consists, in brief, in the growth of tissue in plasma, according to Carrel and Burrow's modification of Harrison's technique, and the liberation of the new cells by digestion of the clot with trypsin. We had noted that if the serum of a growing tissue culture is replaced with Locke's solution at room temperature the cells of the growing strands that extend out into the medium sometimes contract into spheres, which may be separate or, when growth has been dense, loosely attached, side by side. The general outline of the culture is maintained because the cells are held in place by the fibrin network; and if serum is added and incubation renewed they again put forth processes, and, joining each other, again form strands. The problem

549

5

has been to cause the cells to contract and then to liberate them from the fibrin network. This is readily done with trypsin in Locke's solution (Fig. 1); and the resulting suspension can be freed by filtration of all but individual cells.

Method.

We have used the trypsin powders of Merck, Grübler, and Kahlbaum. It is necessary to free them as far as possible from the ammonium sulphate which constitutes the greater part of their bulk. According to Kirchheim,[1] the trypsin of Merck does not contain ammonium sulphate; but we have found it present in as great amount as in the other preparations mentioned. It should be got rid of by Kirchheim's method. The trypsin powder is shaken briefly in absolute alcohol and allowed to stand while the heavy sulphate settles out. The supernatant flocculus is collected on a filter, rapidly washed with ether, dried in the air, and dissolved in Locke's solution (Locke's modification of Ringer's solution, but without sugar). The yield from 2 gm. of the unpurified trypsin is dissolved in 98 cc. of Locke's solution. The cloudy, yellowish fluid is filtered, first through paper, then through a Berkefeld cylinder (N) to sterilize it, and is distributed in test-tubes and kept in the ice box. It loses very slowly its ability to digest and can still be used after 2 months. 3 per cent trypsin digests plasma clots more rapidly and does not harm most cells; but 5 per cent kills cells. Unpurified trypsin powders can be employed but the results are not so good.

The tissue from which cells are to be obtained should be cultivated preferably in plasma diluted with Locke's solution in order that the fibrin network to be digested shall be slight. A mixture of one part of plasma with three of Locke's solution is a medium suitable for most tissues. If there is need for a thick suspension of cells many bits of tissue should be grown. It is convenient to flood them in small Petri dishes with a thin layer of the dilute plasma. After clotting has taken place each dish is sealed to prevent evaporation, and placed in the incubator. A stout cord dipped in hot, sterile paraffin and thrust between the outer and inner rim of the dish, with one end

[1] Kirchheim, L., Arch. f. exper. Path. u. Pharm., 1911, lxvi, 352.

left free, is useful for sealing. A pull on the free end will release the top of the dish.

When growth is established the trypsin solution, warmed to 37°C., is poured on, filling the dish above the plasma, and incubation is continued. In a few minutes some of the tissue fragments are free, and within about an hour the clot has disappeared and there remains a clear fluid containing numerous tissue particles. This is taken up with a pipette, stirred to break up any loose aggregations of cells, diluted with Locke's solution, filtered through sterile gauze, and centrifugalized. The fine, powdery, yellowish gray sediment will consist of discrete cells, nearly all of them alive. They can be washed repeatedly if need be. We prefer for this purpose the "gelatin-Locke's,"—Locke's solution containing $\frac{1}{8}$ per cent of gelatin,—which, as Rous and Turner showed,[2] protects fragile cells against mechanical injury. If the cells are to be plated again in plasma they need not be washed, but after centrifugalization can be suspended in the Locke's solution used to dilute the plasma. Plating is done, as before, in Petri dishes.

Results.

The cells liberated as individuals by trypsin are those which grow out into the medium in strands or a meshwork, or which wander out separately (connective tissue cells, endothelium (?), choroid, sarcoma, and splenic tissue cells). Thus far we have used successfully the tissue of rat and chick embryos, of rat and chicken tumors, and the normal tissue of young rats. Sheets of growing cells (epithelium) are not readily broken up. Whether individual epithelial cells can be liberated in this way is as yet uncertain. But small groups of epithelial cells are obtained, and bits of striated muscle which live for a brief period when plated again.

The individual cells become approximately spherical when in suspension and the nuclei also tend to, though less perfectly. The change in form is especially noteworthy in the case of elements which, when growing in culture, are stellate or of an attenuated spindle shape with an elongated nucleus. When freed, suspended in

[2] Rous, P., and Turner, J. R., *Jour. Exper. Med.*, 1916, xxiii, 219.

serum, and stained, such cells show no trace of the long protoplasmic processes which they had while growing. With Wright's stain certain of them derived from connective tissue and probably of fibroblastic and endothelial origin have a resemblance to the mononuclear series of the blood (Fig. 2). Their cytoplasm is basophilic. Other cells from the same source are three or four times the diameter of any blood element. These morphological features will be taken up in a later paper.

The freed cells, distributed in plasma as separate individuals and incubated, soon put forth processes and assume their original form. Bits of striped muscle from the embryo may round at the ends, thus gaining a leech shape, and put out short processes (Fig. 4). We have not observed them to proliferate. But the spindle-shaped and stellate cells of connective tissue, sarcoma, and the choroid coat of the eye multiply rather rapidly. If the cells are numerous the plate will show at the end of 24 hours a thick mesh- or feltwork con- sisting of elements once separate which have reached out and joined each other by means of attenuated processes (Fig. 3). The tendency of scattered cells thus to connect with each other again is striking. At the end of 48 hours the number of growing elements is greatly increased, not only by proliferation but by the "waking up" of cells previously spherical. If small masses of cells are present in the culture, as the result of incorrect filtration, growth from them may be almost explosive, each mass resolving itself into elements that radiate in every direction.

The Replating of Cultures.

The limits of the method have not yet been reached. The freed and plated cells can be liberated anew after growth and successfully plated again in fresh plasma. To judge from our results, the process can be repeated indefinitely. Isolated cells of the chick's choroid continue to form pigment after they have been twice liberated with trypsin and twice replated (Fig. 5).

Cells that have been growing in tissue cultures for more than 24 hours when freed and examined in suspension show, as a rule, fat droplets, and corresponding vacuoles when fixed and stained in the

8

spherical state. Fat droplets have often been noted in tissue cultures and their source is to some extent known.[3] But they are much less prominent in the culture with its extended cells than in the freed, contracted elements. We wish to emphasize the fact that they develop very early, even when growth is taking place in a dilute plasma medium. Only during the first 24, or rarely the first 48 hours, do the cells appear absolutely normal. Later the culture consists for the most part of abnormal elements. This is true also of the freed and plated cells. It follows that replating should be carried out at least every 48 hours.

Technical Difficulties.

The initial cultures must be free from bacteria if the cells are to be replated after their liberation. For the tryptic digestion liberates not only tissue cells but bacterial colonies, and a single one of these latter can by its dispersion ruin all of the new plates. For this reason it is best to cut up the tissue to be grown, in a sterile, glass-sided box, closed with pieces of rubber dam at the ends, through apertures in which the instruments and tissue are introduced, and the hands thrust, encased in sterile, rubber gloves. A small, glass hood with cloth sides will do nearly as well, and it is useful for the replating of cultures. Needless to say a single contamination at any time will ruin a sequence of plates. If the cells are to be used in suspension it is of less importance.

The centrifugalization to bring down tissue cells brings down also fine débris such as bits of cotton, particles of dust, etc., from the fluid. By the time cultures have been twice digested and plated, enough of this will have been collected to mar their appearance, unless special care is taken. Such care consists in the use of well filtered fluids, and centrifuge tubes closed with corks instead of cotton or gauze stoppers. Much time can be saved if the corks are hollowed to fit over the end of the tube, but with a central core to prevent dislodgement (Text-fig. 1). They may be boiled or autoclaved. The central core should be rather short in order that it may remain uncontaminated when the cork is placed on an unsterile surface.

[3] Lambert, R. A., *Jour. Exper. Med.*, 1914, xix, 398.

TEXT-FIG. 1. Centrifuge tube closed with an easily removable cork designed to keep the contents sterile.

SUMMARY.

Individual, living, tissue cells can be obtained in suspension by digesting with trypsin the clot of growing tissue cultures. Under these circumstances the living cells assume a spherical form. When washed and plated in fresh plasma they put out processes and proliferate. After growth in the new plates has occurred the digestion and plating can be repeated. The limits of the method have not yet been reached. We are at work on a number of the problems which it has opened up.

EXPLANATION OF PLATES.

PLATE 84.

FIG. 1. Edge of a culture undergoing digestion with trypsin. The cells have begun to contract into spheres. (Chick embryo.)

PLATE 85.

FIG. 2. Connective tissue and endothelial (?) cells liberated from cultures of the heart muscle and abdominal muscle of a 3 day old rat. Mononuclear cells from the blood of the same animal. Wright's stain. All the cells are drawn to the same magnification.

Fig. 1.

(Rous and Jones: Living Cells from Fixed Tissues.)

Blood   Heart   Muscle
Mononuclears   Small Series   Large Series   Large Series

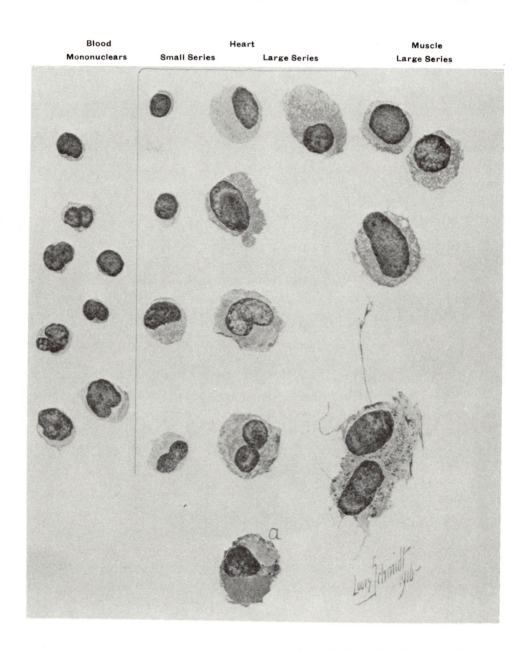

(Rous and Jones: Living Cells from Fixed Tissues.)

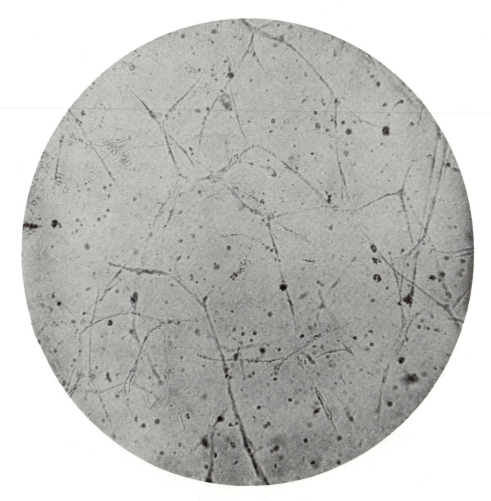

Fig. 3

(Rous and Jones: Living Cells from Fixed Tissues.)

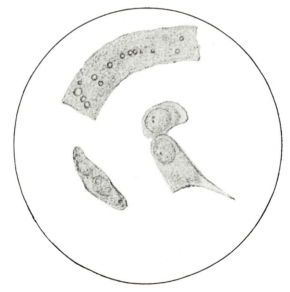

FIG. 4.

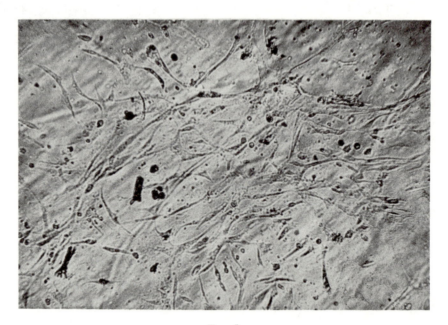

FIG. 5.

(Rous and Jones: Living Cells from Fixed Tissues.)

The cell marked *a* has ingested two red cells. One cell of the muscle series shows vacuoles resulting from a fatty change, and another has attached to it undigested fibrin threads.

PLATE 86.

FIG. 3. Meshwork formed by the anastomosis of connective tissue cells liberated by trypsin and plated as separate individuals. (Chick embryo.)

PLATE 87.

FIG. 4. Striped muscle from a culture incubated 24 hours after liberation by trypsin and replating.

One fragment of muscle, with sharp-cut ends has not grown and has undergone fatty change. But the others give evidence of life, as shown by their change in form, and one has put forth a process. (Rat embryo.)

FIG. 5. Cells from the chick's choroid growing after two liberations with trypsin and two replatings. The formation of pigment is going on actively.

Reprinted from Cancer Research
Vol. 12, pp. 264–265. 1952

TISSUE CULTURE STUDIES OF THE PRO-
LIFERATIVE CAPACITY OF CERVICAL
CARCINOMA AND NORMAL EPITHE-
LIUM. George O. Gey, Ward D. Coffman,*
and Mary T. Kubicek.* (Departments of
Surgery and Gynecology, Johns Hopkins
Hospital and University, Baltimore 5, Md.)

This is a report of an evaluation *in vitro* of the
growth potential of normal, early intra-epithelial,
and invasive carcinoma from a series of cases of
cervical carcinoma. Comparable cytological and
tissue culture studies were actually carried out on
selected biopsies of normal and neoplastic areas of
the same cervix. Thus far, only one strain of epidermoid carcinoma has been established and
grown in continuous roller tube cultures for almost a year. It grows well in a composite medium
of chicken plasma, bovine embryo extract, and
human placental cord serum. The autologous normal prototype is most difficult to maintain under
comparable cultural conditions. Most of the tissue
from other cases showed rapid keratinization of
the cells grown in cultures whether from normal
or neoplastic areas. Some of the hormonal aspects
of the problem will be discussed.

16 September 1955, Volume 122, Number 3168

SCIENCE

Nutrition Needs of Mammalian Cells in Tissue Culture

Harry Eagle

Although cells grown in tissue culture are usually imbedded in a supporting structure such as a plasma or fibrinogen clot, a number of cell lines have been propagated that do not require this support but, instead, adhere to the surface of the glass container. As they multiply, they spread out horizontally on the surface of the glass to form a thin adherent sheet. Several cell lines, notably the single-cell line of mouse fibroblasts (strain L) isolated by Sanford, Earle, and Likely (1) and a human uterine carcinoma cell (strain HeLa) isolated by Gey (2), have been propagated in this manner for years. The media ordinarily employed consist of a "balanced" salt solution enriched with serum, embryonic tissue extracts, and ultrafiltrates of these materials, in varying combination. Such complex systems do not, however, lend themselves to the identification of the specific requirements for growth.

The problem was simplified with the finding (3, 4) that these two cell lines could be propagated in a medium consisting of an arbitrary mixture of amino acids, vitamins, cofactors, carbohydrates, and salts, supplemented with a small amount of serum protein, the latter supplied as *dialyzed* horse or human serum. In such a system the omission of a single essential component resulted in the early death of the culture. It thus proved possible to determine most of the specific nutrients that are essential for the

growth and multiplication of mammalian cells in tissue culture, to produce specific nutritional deficiencies, to study the microscopic lesions thereby produced, and to "cure" these deficiencies by the restoration of the missing component.

With the optimum medium as defined for the HeLa cell, another human carcinoma has been cultured directly onto glass (5), without intervening culture in a plasma clot; and a number of human cell lines cultured by other workers (6–8) have been propagated in the same medium for many months, with an average generation time of 40 to 60 hours. It has become possible to compare the growth requirements of normal and malignant human cells, to approach the problem of the specific metabolic requirements for the propagation of virus in such cells, and to study the incorporation of various nutrilites into cellular protein and nucleic acid. The results obtained to date are summarized in this report (9).

Amino Acid Requirements of a Mouse Fibroblast and of a Human Carcinoma Cell

These two cell lines proved remarkably similar with respect to their amino acid requirement. For both, 13 amino acids proved to be essential (arginine, cyst(e)ine, glutamine, histidine, isoleucine, leucine, lysine, methionine, phenylalanine, threonine, tryptophan, tyrosine, and valine). Only the L-amino acids were active. The D-enantiomorphs, although inactive, did not inhibit the growth-promoting action of the L-isomers. For the mouse fibroblast, the requirements for optimal growth varied

from 0.005 millimolar (mM) in the case of L-tryptophan, to 0.1 to 0.2 millimolar for L-isoleucine (3), and 0.2 to 0.5 millimolar for L-glutamine (10). The optimal concentrations for the growth of the HeLa cell (4, 10) were from 1 to 3 times those required by the mouse fibroblast.

Both cell species have been found to have active transaminating systems (11); and the carbon sources used for the synthesis of the six nonessential amino acids are under study.

Particular interest attaches to the glutamine requirement. The optimal requirement for growth proved to be 0.2 to 0.5 millimolar for the L cell and 1 to 2 millimolar for the HeLa cell. An unexpected finding was the fact that glutamic acid at any concentration, even supplemented by NH_4^+ and adenosinetriphosphate (ATP), failed to permit the growth of the L-fibroblast. This was not due to the impermeability of the cell, for isotopically labeled glutamic acid could be shown to be actively incorporated into protein, even from concentrations as low as 0.01 millimolar. With the HeLa cell, although glutamic acid did permit growth, approximately 10 to 20 times as much was required as of glutamine; and at the optimal level of 20 to 30mM, there was regularly less growth than with glutamine at 1mM. In this case also it is difficult to ascribe the relative inactivity of glutamic acid to the impermeability of the cell, since it was shown to be actively incorporated into protein from low concentrations. These data strongly suggest that glutamine plays an essential metabolic role which glutamic acid, NH_4^+, and ATP cannot fulfill in the case of the L cell; and with the HeLa cell glutamic acid may be active only by virtue of the fact that, at high concentrations, it can be transformed to glutamine in amounts adequate for growth (10).

Proline, ornithine, α-ketoglutaric acid, and asparagine, in any concentration tested, with or without NH_4^+ and ATP, failed to substitute for glutamine in either cell (10). Attempts to demonstrate glutamine synthetase activity, either with intact cells or with cell-free extracts, have to date been unsuccessful (11).

The parallelism in the relative amounts of the 13 amino acids required by the two cell lines is seen in

Dr. Eagle is chief of the section on experimental therapeutics, Laboratory of Infectious Diseases, National Microbiological Institute, National Institutes of Health, U.S. Public Health Service, Bethesda, Md. This article is based on a paper given at the AAAS Gordon Research Conference on Vitamins and Metabolism, 19 Aug. 1955, New London, N.H.

Table 1. In both animal species, the concentrations present in the serum (12, 13), and thus available for distribution to the body fluids, were in most instances significantly in excess of those that suffice for the maximal growth of the cell line *in vitro*. However, with a few amino acids (arginine and isoleucine in the case of the mouse fibroblast; methionine, leucine, and isoleucine in the case of the HeLa cell), the requirement for optimum growth was of the same order of magnitude as the reported values for the concentration in the serum. If the requirements of these cells were the same *in vivo* as are now found after prolonged cultivation *in vitro*, then the availability of these compounds in the body fluids could have been a growth-limiting factor.

With both cell lines, a number of amino acids caused partial inhibition of growth in concentrations 2 to 5 times the maximally effective level. With both cell lines also, dipeptides were just as effective as the component amino acids in promoting growth (14). The degree to which these cells can use precursors of the essential amino acids is under study.

On the omission of a single amino acid from the medium, microscopic changes indicative of cell injury developed within 2 to 3 days, and the cells eventually died (Figs. 1 and 2). These changes differed significantly with the individual amino acids, perhaps reflecting their varying metabolic functions or differences in the amino acid composition and turnover rate of individual proteins. It is perhaps significant that, with many of the amino acid deficiencies, the cytopathogenic changes closely resembled those that result from viral infection.

Fig. 1. Illustrative amino acid and vitamin deficiencies produced in the HeLa cell by the omission of a single nutrilite from the complete medium of Table 4. Left to right: normal control, 6 days; phenylalanine, 7 days; riboflavin, 8 days; glutamine, 5 days.

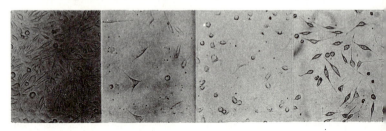

Fig. 2. Illustrative amino acid, vitamin, and salt deficiencies produced in the mouse fibroblast by the omission of a single nutrilite from the complete medium of Table 4. Left to right: normal control, 8 days; tyrosine, 9 days; choline, 4 days; Mg^{++}, 6 days.

In their early stages, the cytopathogenic effects of amino acid deficiencies were reversible. Cells that had been exposed to a medium deficient in a single amino acid and had largely degenerated in consequence could be revived on the restoration of the missing component (3, 4). The details of the microscopic changes produced by specific amino acid deficiencies and their reversal by the addition of the missing compound are now under study by phase and electron microscopy (15).

Minimum Vitamin Requirements

To date, seven vitamins have proved demonstrably essential for the growth of both the mouse fibroplast and the HeLa cell (choline, folic acid, nicotinamide, pantothenate, pyridoxal, riboflavin, and thiamine) (16). On the omission of any one of these from the medium, degenerative changes developed after 5 to 15 days, and the culture eventually died (Figs. 1 and 2). In their early stages, these specific vitamin deficiencies could be "cured" by the addition of the missing vitamin component to the medium. The minimum amounts of the individual vitamins and of the corresponding conjugates required for the maximal growth of the L-fibroblast are shown in Table 2. Nicotinic acid and nicotinamide proved interchangeable, as did pyridoxine, pyridoxamine, and pyridoxal; and vitamin conjugates regularly proved capable of substituting for the corresponding

vitamin—for example, flavin mononucleotide (FMN) or flavin adenine dinucleotide (FAD) for riboflavin; diphosphopyridine nucleotide (DPN) or triphosphopyridine nucleotide (TPN) for nicotinamide; coenzyme-A for pantothenic acid; cocarboxylase for thiamine).

It must be emphasized that the seven vitamins so far found to be essential are not necessarily the total vitamin requirement of these two cell lines. It is possible that a number of other vitamins are essential, but probably more prolonged cultivation in an appropriately deficient medium would be necessary in order to produce the specific deficiency. Further, a number of essential vitamins may well be present as trace contaminants in the other components of the medium and would, therefore, appear to be nonessential under the conditions of the present experiments.

Salt and Glucose Requirements

The ions demonstrably essential for the survival and growth of both the mouse fibroblast and the HeLa cell were Na^+, K^+, Mg^{++}, Ca^{++}, Cl^-, and $H_2PO_4^-$ (17). The concentration of each ionic species required for the optimal growth of both cell lines is shown in Table 3. No information is as yet available with respect to the need for trace elements.

Both the L-fibroblast and the HeLa cell grew well in a medium containing glucose as the only carbon source over and above the essential amino acids. A

Table 1. Amino acid requirements of a mouse fibroblast (strain L) and a human carcinoma cell (strain HeLa)

L-Amino acid	Optimum for growth of HeLa cell in vitro, (mM)	Optimum for growth of L-fibroblast in vitro, (mM)
Tryptophan	0.01	0.005
Histidine	0.02	0.01
Cystine	0.03	0.01
Tyrosine	0.03	0.02–0.05
Methionine	0.03	0.02–0.05
Phenylalanine	0.05	0.02
Arginine	0.05	0.05
Leucine	0.1	0.05
Threonine	0.1	0.05
Valine	0.1	0.05
Lysine	0.1	0.05
Isoleucine	0.1	0.1–0.2
Glutamine	1.0	0.2–0.5

502

number of carbohydrates could substitute for glucose. Some (galactose, mannose, and maltose) were almost as active as glucose, mole for mole; a few were only slightly less active; and a number were weakly effective in high concentrations. The degree to which these varying activities reflect (i) the varying permeability of the cell, or (ii) the varying ability of the cell to transform them to compounds that can then enter into the normal metabolic pathways remains to be determined.

Serum Protein

A medium containing the 13 amino acids and the seven vitamins found to be essential for the growth of the L and HeLa cells, each at the optimum concentration, and appropriately supplemented with glucose did not permit growth unless a small amount of serum protein was added, conveniently supplied as dialyzed serum. For the L cell, the concentration of protein that permitted maximal growth was 1 part in 1500, and 1 part in 5000 sufficed for limited growth. Approximately 3 to 4 times these concentrations were required for the HeLa cell.

The function of the serum protein is not yet clear. It is obviously not supplying the amino acids and vitamins already shown to be essential, at least in concentrations sufficient for survival and growth. Some of the serum fractions obtained by alcohol-salt precipitation proved inert (I, II, and III), while others (IV, V) were only weakly active, separately or in combination (9). Exhaustively dialyzed serum was similarly inactive. On the other hand, fractions obtained by simple salting out with $(NH_4)_2SO_4$, followed by 24-hour dialysis, were all more or less equally active. The possibility that the protein contributes trace elements or vitamins that are bound to the protein but slowly dissociate in the culture medium is under study.

Applications

Chemically defined medium. The initial objective of these studies was the identification of the specific metabolites required for the growth of various cell types rather than the development of a chemically defined medium for the cultivation of mammalian cells. Obviously, however, such studies may ultimately result in the provision of a completely defined medium in which every component has been shown to be essential for the growth of a specific cell, and each component is present in the concentration optimal for that cell.

To date, as is indicated in the foregoing sections, 27 factors have been identified as essential for the growth of a mouse fibroblast and a human carcinoma cell. The factors are listed in Table 4, together with the concentrations of each used for the propagation of the two cell lines under present consideration. There remains to be determined the function of the small amounts of serum protein that must be added over and above these essential components, and in the absence of which the cells degenerate and die.

Isolation of new cell lines in tissue culture. The striking similarity of the nutritional requirements of the HeLa cell and of the mouse fibroblast suggested that the optimum medium as so far defined for the growth of the HeLa cell might be similarly effective for other human cells. The medium of Table 4, containing the essential growth factors, all at the concentrations found optimal for the HeLa cell, and supplemented with 10-percent whole human serum was used for this purpose. A human epidermoid carcinoma of the floor of the mouth (strain KB) was cultured (5) by implantation of the tumor cells directly onto the glass surface, overlaid with the medium of Table 4; and initially promising results have been obtained with several other tumors.

Strain KB grows in this medium at a rapid rate, the generation time in the logarithmic phase of growth averaging 30 hours. Human liver and kidney cells cultured by Chang (6) and Henle (7),

a human embryonic intestinal cell cultured by Henle (7), and a leukemia cell cultured by Osgood (8) have also been propagated in this medium, with average generation times of approximately 40 to 60 hours.

Nutritional deficiencies; antimetabolite assays. The fact that both cell types here studied degenerate and die on the omission of a single essential growth factor, whether that factor is a single vitamin, a single amino acid, or glucose, makes it possible to follow the cytopathogenic and biochemical changes that develop as a result of specific nutritional deficiencies and to follow also their reversal when the deficiency is cured by the restoration of the missing component to the medium (15).

It becomes possible also to determine by direct assay the growth-inhibitory activity of various antimetabolites on specific cell lines, and to determine also whether these inhibitors are competitive with normal metabolites.

The nutritional requirements for viral synthesis. With the identification of the specific metabolites required for the survival and growth of mammalian cells, it became possible to identify the components of the medium that are essential for the intracellular propagation of viruses. The amount of poliomyelitis virus released into the medium by HeLa cells was found to be quantitatively unaffected by the omission of serum protein, amino acids, or vitamins from the medium of Table 4 (18). In such a deficient medium the protein components of the virus are necessarily built up entirely at the expense either of the host cell protein or of its amino acid and peptide pool; and if cofactors are required for that synthesis, those already in the cell, or their precursors, suffice. On the other hand, the omission of both glucose and glutamine from the medium resulted in a marked decrease in virus production. The omission of each singly either had no effect or caused only a partial reduction.

Table 2. Minimum vitamin requirements of a mouse fibroblast

Vitamin, precursor, or conjugate	Optimal concentration range (mM)
Choline	10^{-2}
Acetylcholine	10^{-2}
Folic acid	10^{-5}
Citrovorum factor	$10^{-6} \pm$
Nicotinamide	10^{-3}
Nicotinic acid	10^{-3}
DPN	10^{-3}
TPN	$10^{-3}*$
Pantothenic acid	10^{-5}
Coenzyme-A	10^{-4}
Pyridoxine	10^{-5} to 10^{-6}
Pyridoxal	10^{-5}
Pyridoxamine	10^{-4}
Pyridoxal phosphate	10^{-3}
Riboflavin	10^{-6}
Flavin mononucleotide	10^{-6}
Flavin adenine dinucleotide	10^{-5}
Thiamine	10^{-5}
Thiamine phosphate	$10^{-5}*$
Cocarboxylase	10^{-5}

* Significantly less growth than with vitamin.

Table 3. Minimum electrolyte requirements of a mouse fibroblast and a human carcinoma cell

Ionic species	Concn. (mM) permitting maximal growth* of	
	HeLa cell	L-fibroblast
Na +	100	120
K +	1	1
Ca ++		1
Mg ++		0.2
H$_2$PO$_4^-$		0.2 to 0.5

* In medium of Table 4, supplemented with dialyzed serum as there indicated.

Protein synthesis, turnover, and amino acid exchange in growing and resting cells. Given seven essential vitamins, the unidentified growth factors present in serum protein, and the necessary salts, both cell lines here studied could obtain their total requirements for energy and growth from 13 amino acids and glucose. On the omission of any one of the amino acids from the medium, the cells stopped multiplying, and the amount of cell protein then usually remained unchanged during a period of 48 to 72 hours or decreased slightly as the cells began to degenerate. There was, however, a continuing incorporation of labeled amino acids into cell protein during this entire period.

The amount of the individual amino acid incorporated was the same, whether one or six amino acids had been omitted from the medium, and in the case of L-phenylalanine approached as a limiting value approximately 35 percent of the total amount of that amino acid in the cell protein. When several labeled amino acids were used in conjunction, the amounts incorporated were additive.

This incorporation of amino acids into protein from a medium in which there is no net synthesis could reflect protein turnover—that is, protein degradation and resynthesis—with total recapture of the essential amino acids that were not available from the medium. Alternatively, there may be an exchange of amino acid residues between the intact protein and the amino acid pool, similar to that described for bacteria by Gale and Folkes (19). These two possibilities are presently under study.

Studies are in progress on the metabolic pathways involved in the synthesis of nucleic acid and of the six nonessential amino acids from the 13 essential amino acids and glucose.

Table 4. Basal media for cultivation of the HeLa cell and mouse fibroblast (10)

L-Amino acids* (mM)		Vitamins‡ (mM)		Miscellaneous	
Arginine	0.1	Biotin	10^{-3}	Glucose	5mM§
Cystine	0.05 (0.02)†	Choline	10^{-3}	Penicillin	0.005%#
Glutamine	2.0 (1.0)‖	Folic acid	10^{-3}	Streptomycin	0.005%#
Histidine	0.05 (0.02)†	Nicotinamide	10^{-3}	Phenol red	0.0005%#
Isoleucine	0.2	Pantothenic acid	10^{-3}		
Leucine	0.2 (0.1)†	Pyridoxal	10^{-3}	For studies of cell nutrition	
Lysine	0.2 (0.1)†	Thiamine	10^{-3}	Dialyzed horse serum, 1%†	
Methionine	0.05	Riboflavin	10^{-4}	Dialyzed human serum, 5%	
Phenylalanine	0.1 (0.05)†				
Threonine	0.2 (0.1)†	Salts§ (mM)			
Tryptophan	0.02 (0.01)†			For stock cultures	
Tyrosine	0.1			Whole horse serum, 5%†	
Valine	0.2 (0.1)†	NaCl	100	Whole human serum, 10%	
		KCl	5		
		NaH$_2$PO$_4$·H$_2$O	1		
		NaHCO$_3$	20		
		CaCl$_2$	1		
		MgCl$_2$	0.5		

* Conveniently stored in the refrigerator as a single stock solution containing 20 times the indicated concentration of each amino acid.
† For mouse fibroblast.
‡ Conveniently stored as a single stock solution containing 100 or 1000 times the indicated concentration of each vitamin; kept frozen.
§ Conveniently stored in the refrigerator in two stock solutions, one containing NaCl, KCl, NaH$_2$PO$_4$, NaHCO$_3$, and glucose at 10 times the indicated concentration of each, and the second containing CaCl$_2$ and MgCl$_2$ at 20 times the indicated concentration.
‖ Conveniently stored as a 100mM stock solution; frozen when not in use.
Conveniently stored as a single stock solution containing 100 times the indicated concentrations of penicillin, streptomycin, and phenol red.

References and Notes

1. K. K. Sanford, W. R. Earle, G. D. Likely, *J. Natl. Cancer Inst.* 9, 229 (1948).
2. W. F. Scherer, J. T. Syverton, G. O. Gey, *J. Exptl. Med.* 97, 695 (1953).
3. H. Eagle, *J. Biol. Chem.* 214, 839 (1955).
4. ———, *J. Exptl. Med.* 102, 37 (1955).
5. ———, *Proc. Soc. Exptl. Biol. Med.* 89, 362 (1955).
6. R. S.-M. Chang, *Proc. Soc. Exptl. Biol. Med.* 87, 440 (1954).
7. G. Henle, personal communication (1955).
8. E. E. Osgood, personal communication (1955).
9. The essential and able assistance of Ralph Fleischman, Clara L. Horton, Mina Levy, and Vance I. Oyama in the conduct of these experiments is gratefully acknowledged. The courtesy of J. L. Oncley of the Harvard Medical School who supplied generous samples of freshly prepared human plasma fractions I, II, III, IV, and V is also gratefully acknowledged.
10. H. Eagle *et al.*, *J. Biol. Chem.*, in press.
11. S. Barban and H. Eagle, in preparation.
12. E. C. Albritton, *Standard Values in Blood* (Saunders, Philadelphia, 1953), p. 99.
13. W. H. Stein and S. Moore, *J. Biol. Chem.* 211, 915 (1954).
14. H. Eagle, *Proc. Soc. Exptl. Biol. Med.* 89, 96 (1955).
15. A. J. Dalton and H. Eagle, in preparation.
16. H. Eagle, *J. Exptl. Med.*, in press.
17. ———, in preparation.
18. H. Eagle and K. Habel, in preparation.
19. E. F. Gale and J. P. Folks, 55, 721, 730 (1953); 59, 661, 675 (1955).

"Fact," as I intend the term, can only be defined ostensively. Everything that there is in the world I call a "fact." The sun is a fact; Caesar's crossing of the Rubicon was a fact; if I have toothache, my toothache is a fact. If I make a statement, my making it is a fact, and if it is true there is a further fact in virtue of which it is true, but not if it is false. The butcher says: "I'm sold out, and that's a fact"; immediately afterwards, a favored customer arrives and gets a nice piece of lamb from under the counter. So the butcher told two lies, one in saying he was sold out, and the other in saying that his being sold out was a fact. Facts are what make statements true or false. I should like to confine the word "fact" to the minimum of what must be known in order that the truth or falsehood of any statement may follow analytically from those asserting that minimum. For example, if "Brutus was a Roman" and "Cassius was a Roman" each assert a fact, I should not say that "Brutus and Cassius were Romans" asserted a new fact. We have seen that the questions whether there are negative facts and general facts raise difficulties. These niceties, however, are largely linguistic.—BERTRAND RUSSELL, *Human Knowledge*, 1948.

Reprinted from Cancer Research, Vol. 20, pp. 462–466. 1960

Phytohemagglutinin: An Initiator of Mitosis in Cultures of Normal Human Leukocytes[*]

PETER C. NOWELL

(Dept. of Pathology, University of Pennsylvania School of Medicine, Philadelphia 4, Pennsylvania)

SUMMARY

Possible factors responsible for the initiation of mitotic activity in "gradient" cultures of leukocytes from normal human blood were investigated. Variations of temperature, pH, oxygen tension, carbon dioxide tension, plasma and cell concentrations, as well as the amount of agitation, over at least as wide a range as might be encountered *in vivo*, produced only moderate quantitative changes in mitotic activity.

The mucoprotein plant extract, phytohemagglutinin (PHA), employed originally as a means of separating the leukocytes from whole blood in preparing the cultures, was found to be a specific initiator of mitotic activity: in its presence, cell division occurred; in its absence, no mitoses appeared. The studies suggest that the mitogenic action of PHA does not involve mitosis *per se* but rather the stage preceding mitosis—the alteration of circulating monocytes and large lymphocytes to a state wherein they are capable of division. The relationship of this mitogenic action of PHA to mitotic and premitotic processes in the body remains to be investigated.

In a recent paper (8) we have described the differentiation, in short-term tissue culture, of both normal and leukemic human leukocytes obtained from peripheral blood. In the course of this study, considerable mitotic activity was observed in the cultures of normal leukocytes as well as in the cultures of leukemic "blasts." The dividing cells in the normal cultures were tentatively identified as monocytes and large lymphocytes which had become mitotically active *in vitro* after a 2-day latent period. Ordinarily, normal leukocytes do not divide in the peripheral blood. There is evidence, however, to suggest that some circulating white cells do have mitotic potentialities in the body (2). Experiments were, therefore, designed to attempt to determine what specific factor or condition in our culture system was responsible for activating this latent mitotic potential of circulating mononuclear cells. Physiologic factors such as oxygen tension, carbon dioxide tension, and plasma concentration were varied individually but proved to have only minor quantitative effects

on mitotic activity in the cultures. Instead, an apparently nonphysiologic mitogenic agent was uncovered: the plant extract, phytohemagglutinin (PHA) (11). This substance was originally employed for its erythrocyte-agglutinating ability in obtaining leukocytes from whole blood. The present studies, however, indicate that it also has the ability to initiate mitosis among these leukocytes, apparently by stimulating the alteration of monocytes and large lymphocytes to a state wherein they are capable of division.

MATERIALS AND METHODS

Standard cultures.—The "gradient" technic for culturing leukocytes has been described in detail elsewhere (4, 8, 10). In preparing the "standard" cultures which were used as a base line for the present studies, phytohemagglutinin (PHA)[1] was used to separate the leukocytes from heparinized peripheral blood (10). To each 10 ml. of blood, 0.2

[*] This investigation was supported by Senior Research Fellowship SF-4 from the U.S. Public Health Service and by Grants C-3562 and C-4659 from the National Cancer Institute of the National Institutes of Health, U.S. Public Health Service.

Received for publication October 12, 1959.

[1] The phytohemagglutinin used in these studies was obtained from Difco Laboratories, Detroit 1, Michigan (Bacto-Phytohemagglutinin-Code 0528). It is a partially purified mucoprotein hemagglutinin prepared from *Phaseolus vulgaris* (red kidney beans) by a modification of the method of Rigas and Osgood (11). For use it is rehydrated with Bacto-Hemagglutination Buffer (Difco). Each ml. of the rehydrated solution contains 10 mg. of PHA. The buffer alone has no mitogenic activity.

ml. of PHA was added. After mixing, the blood was allowed to stand for 45 min. at 4° C. and was then centrifuged at 350 r.p.m. for 10 min. to remove nearly all of the agglutinated erythrocytes. The supernatant plasma, and the leukocytes remaining suspended in it, were then utilized in setting up the cultures. No additional PHA was added. The cells were incubated at 37° C. in 60 × 22-mm. screw-top bottles, in a mixture of commercial tissue culture medium (TC-199, Difco) and plasma, with penicillin and streptomycin added. Room air served as the gas phase, and the bottle tops were not tightened. An initial inoculum of 1.5–2.0 × 10⁶ cells/ml was employed in a total culture volume of 8–10 ml. (15–20 per cent plasma). This gave a culture depth of approximately 20 mm. A piece of standard microscope slide (30 × 10 mm.) was placed in each bottle at an angle. The cultures were incubated without agitation, and the cells settled out and grew on the slide as well as on the bottom of the bottle. To examine the cells or to estimate mitotic activity in the cultures, the slide was removed and stained with Giemsa (8); or the entire culture was sacrificed and squash preparations made of the harvested cells after they were fixed and stained in acetic-orcein (4). Repeated examination of the same culture was made possible by simply replacing each removed slide with a new one and depositing cells on the new slide by brief agitation of the culture. Ordinarily, no changes of medium were made during the life of these cultures.

"Spinner" cultures of whole blood.—To investigate a number of variables simultaneously, cultures were set up consisting of 10 ml. of heparinized whole blood. PHA (0.2 ml.) was added to each of these cultures, and penicillin and streptomycin were also included. The cultures were incubated with constant agitation by means of a magnetic stirrer. The vigorous agitation not only maintained all the cells in suspension but also prevented any agglutination of the erythrocytes, despite the presence of PHA. To maintain the pH at physiologic levels, it was necessary to seal the bottles with paraffin to prevent loss of carbon dioxide.

Other factors in our standard culture system which were considered to be of possible critical importance in initiating mitosis were tested as follows:

Gas phase.—The effect of very high and very low concentrations of oxygen were tested by equilibrating standard PHA-containing cultures with 50 per cent oxygen, 100 per cent oxygen, or 100 per cent nitrogen at the time of planting and then sealing them with paraffin. Equilibration was obtained by gently bubbling the gas through the culture for 10–15 min. and then layering the gas over the surface. The effect of increased carbon dioxide tension was similarly tested by equilibrating cultures with varying concentrations of carbon dioxide gas up to 100 per cent (i.e., 5, 15, 50, and 100 per cent). During equilibration, the pH was continually adjusted by addition of 0.1 N sodium hydroxide. Perfusion with carbon dioxide was continued until no further pH adjustment was required, and the cultures were then sealed. To test low carbon dioxide levels, bicarbonate-free TC-199 (Difco) plus 10 per cent plasma was employed as the tissue culture medium. Cultures were planted in this medium and equilibrated with air in order to drive off as much carbon dioxide from the plasma component as possible. These cultures were then left unsealed and continuously agitated by a magnetic stirrer to prevent reaccumulation of carbon dioxide. In these cultures, the pH during the first day tended to rise and had to be adjusted with 0.1 N hydrochloric acid; thereafter, the pH remained constant.

Conditioned medium.—The importance of conditioned medium in initiating mitosis was tested on leukocytes which had been separated from whole blood by the usual PHA method. These cells were planted in cultures in which the liquid phase was replaced, completely or in part, by medium removed from 3-day-old standard cultures of actively dividing leukocytes.

Heparin and antibiotics.—The significance of these substances with respect to mitotic activity was investigated by setting up cultures free of both. The leukocytes were separated from oxalated blood with PHA and then planted in medium which contained heparin-free serum instead of plasma, and no antibiotics.

Phytohemagglutinin (PHA).—A variety of cultures were set up containing leukocytes separated from whole blood by some means other than PHA. In most of these, the cells were separated with dextran (12). Cells were also obtained by the fibrinogen method (12), as well as by simply centrifuging heparinized blood rapidly so as to yield leukocytes in the form of a "buffy coat." When it was observed that no mitotic activity occurred in any of these cultures, additional studies on the action of PHA were undertaken by means of the following culture systems:

1. Ten-ml. cultures planted with dextran-separated leukocytes, and hence free of PHA, were incubated at high and low levels of oxygen and carbon dioxide as described above. Similar cultures, to which 0.2 ml. of PHA was added at the time of planting, were employed as controls.

2. Ten-ml. spinner cultures of whole blood were set up exactly as described above, except that the PHA was omitted.

3. Ten-ml. cultures were planted with dextran-separated leukocytes, and 0.2 ml. of PHA was then added, either immediately, 1 day, or 2 days after planting.

4. Ten-ml. cultures were set up containing leukocytes which had been separated from whole blood with PHA as previously described but which were then washed 3 times and planted in PHA-free medium.

5. Cultures were set up containing leukemic "blasts" from the blood of patients with acute leukemia or containing normal human or rat bone marrow cells (7, 8). The cells in these cultures were not exposed to PHA. Control cultures of similar cells, but containing PHA (0.2 ml/10 ml culture), were run simultaneously or at different times.

RESULTS

Standard cultures.—After incubation for 3 days, slides from the standard cultures, containing PHA, typically showed clumps of cells consisting of large mononuclear forms interspersed with small lymphocytes. Very few granulocytes remained. The majority of the large mononuclear cells had a uniformly round nucleus with one or more prominent nucleoli and finely dispersed chromatin. The cytoplasm was intensely basophilic and agranular, often with a pale area adjacent to one side of the nucleus. In all our experiments, such "large mononuclears" were frequent only in cultures in which mitoses were present.

In the standard cultures, after 3 days, the mitotic index on the slides and among the cells from the bottom of the culture was in the range of 0.5–1 per cent (number of mitoses/100 nucleated cells). The addition of colchicine (1×10^{-7}M) 18 hours prior to sacrifice, in order to accumulate mitoses, usually resulted in indices of 5–10 per cent (Fig. 1). These data are based on more than 50 standard cultures examined so far. Mitotic activity was generally uniform over the entire length of the slide. However, the cells harvested from the bottom of the culture often showed a slightly lower mitotic index than did those on the slide.

By the 3d day, in five standard cultures on which cell counts were done, the total cell number had generally decreased to 30–50 per cent of the original inoculum. From the 3d to the 5th day, during the period of maximal mitotic activity, the total cell number remained nearly constant or in-

creased slightly. Thereafter, in cultures followed into the 2d week, the cell population gradually decreased.

Cultures of the dimensions noted above ordinarily required no pH adjustment. Good mitotic activity was observed over a pH range of 6.9–7.7, with occasional mitoses occurring at pH as high as 8.0. Temperature variation of 35°–39° C. also had little effect on mitotic activity; higher temperatures produced cell death.

"Spinner" cultures of whole blood.—In smears of the cells remaining in these cultures after incubation for three days, the red cells were uniformly dispersed as single cells and were nearly all intact. Although some leukocytes were degenerating, many more polymorphonuclear forms remained than in the standard cultures. Lymphocytes and occasional monocytes were also present, as well as many large mononuclear forms similar to those previously described. The leukocytes showed little clumping, although, occasionally, groups of three or four cells appeared together. Mitoses were present both in these small clumps (Fig. 2) and among the individually dispersed cells. In all of the four spinner cultures examined, the mitotic index, after colchicine, on the 3d or 4th day, was in the range of 0.5–1.0 per cent. This definite decrease in mitotic activity, as compared with the standard cultures (mitotic index=5–10 per cent), is only partially accounted for by the greater number of mitotically inactive granulocytes surviving in the spinner cultures.

Gas phase.—Over a surprisingly wide range, variation of oxygen or carbon dioxide concentration had little effect on mitotic activity. At least two cultures were tested under each set of conditions. Standard cultures, containing PHA, incubated for 3 days in oxygen concentrations of 50 per cent and near zero (i.e., equilibrated with 100 per cent nitrogen) showed essentially the same mitotic index as standard cultures incubated in air (mitotic index = 5–10 per cent, after colchicine). Similarly, carbon dioxide concentrations of 5, 15, and 50 per cent had no effect on the mitotic index. However, the two extreme carbon dioxide concentrations tested, near 100 per cent and near zero, did reduce, although they definitely did not abolish, mitotic activity. Mitotic indices of only 1–2 per cent, after colchicine, were observed in these latter cultures.

Cultures incubated in 100 per cent oxygen showed no mitotic activity at all at the end of 3 days despite the presence of PHA. The cells in these cultures appeared healthy, but the number of large mononuclears was definitely lower than in standard 3-day-old cultures grown with air as the gas phase. To investigate further this effect of 100 per cent oxygen, cultures of actively dividing cells which had been grown for 3 days in air were then placed in an atmosphere of 100 per cent oxygen. Mitotic activity in these cultures decreased only slightly over the subsequent 48 hours. On the other hand, when 3-day-old cultures which had been grown in 100 per cent oxygen, and hence showed no mitotic activity and few large mononuclears, were exposed to an atmosphere of room air, only occasional mitoses appeared in these cultures on the subsequent 2 days. However, after these cultures had been in an air atmosphere for 3 or 4 days, mitotic indices of 5–10 per cent, after colchicine, were observed. Apparently, the inhibitory effect of 100 per cent oxygen in our leukocyte cultures is directed more toward preventing the cells from undergoing the transition to a state capable of mitosis than toward the mitotic process *per se*.

Conditioned medium.—Leukocytes separated with PHA and then planted in conditioned medium (i.e., medium removed from mitotically active leukocyte cultures) showed, on the 3d day, mitotic activity equal to that of standard cultures. However, no mitoses were observed during the first 2 culture days. In fact, none of the wide variety of experimental conditions employed in our studies of leukocytes from normal peripheral blood ever resulted in mitotic activity on the 1st or 2d culture day.

Heparin and antibiotics.—Two leukocyte cultures which contained PHA but did not contain either heparin or antibiotics showed mitotic activity after 3 days, comparable to that seen in standard cultures.

Phytohemagglutinin (PHA).—As indicated above, cultures containing leukocytes which had been obtained from whole blood by any method other than that using PHA uniformly failed to show mitotic activity. Generally, the cells in these cultures looked healthy and underwent the same population changes over the first few days as did those in the PHA-containing cultures, except that typical large mononuclears appeared only in very low numbers. Mitotic figures were not observed, even after colchicine, in two PHA-free cultures of fibrinogen-separated cells, or in three cultures of leukocytes obtained from PHA-free "buffy coats." In 23 of 25 cultures of cells separated with dextran, not a single mitosis was found on the 3d day on any of the culture slides despite careful search of the many thousands of cells on each slide. In the remaining two cultures, a single mitosis was found on each slide.

The various PHA-free cultures of dextran-separated cells incubated at high and low oxygen and carbon dioxide concentrations, as well as the spinner cultures of whole blood from which the PHA was omitted, all failed to show any mitotic activity. The control cultures, to which PHA was added at the time of planting, showed mitotic indices comparable to those previously observed under similar conditions in standard cultures of cells separated with PHA. At least two experimental and two control cultures were tested under each set of conditions.

Addition of PHA to a culture of dextran-separated cells either immediately after planting, or on the 1st or 2d day following planting, resulted in the appearance of mitoses, in normal numbers, on the 3d, 4th, or 5th culture day, respectively. Only very rarely was a mitosis observed in these cultures earlier than the 3d day following the addition of PHA.

Two cultures of PHA-separated, washed leukocytes planted in PHA-free medium showed definite, though somewhat decreased, mitotic activity on the 3d day. Cultures of both leukemic "blasts" and normal bone marrow cells regularly showed considerable mitotic activity, first appearing on the 1st or 2d culture day. Mitotic indices were the same whether or not the leukemic or marrow cells had ever been exposed to PHA.

Although no detailed quantitative studies were made, it appeared that in the cultures of dextran-separated normal leukocytes, as well as in the spinner cultures of whole blood, a minimal PHA concentration of approximately 0.2 ml. in a 10-ml. culture was required for mitotic activity to be initiated. Heating PHA at 100° C. for 30 min. completely abolished its mitogenic action as well as its ability to agglutinate erythrocytes; heating at 56° C. for 30 min. did not affect either activity.

DISCUSSION

The mechanism by which the mitogenic effect of PHA is exerted is not clear, although it would appear to be a direct action on the leukocytes themselves. The fact that PHA initiates mitosis in spinner cultures of whole blood rules out the possibility that its mitogenic action is an indirect one resulting from the removal of red cells through agglutination. In these spinner cultures, the red cells were not removed but remained present in normal numbers. PHA agglutinates erythrocytes by linkage of the euglobulin portion of the mucoprotein PHA molecule with a polysaccharide on the red cell surface (11). Perhaps the action of PHA on leukocytes also involves the cell surface. Possibly, it alters the cell membrane to permit entrance of some substance from the culture medium which, in turn, initiates the mitotic process. Our studies have not ruled out the possibility that some such essential factor, either serum component or cellular product, does exist in our cultures, although the fact remains that it alone, in the absence of PHA, does not initiate mitosis.

The present findings further suggest that PHA acts on leukocytes not to stimulate mitosis *per se* but rather to initiate the conversion of monocytes and large lymphocytes to a state capable of division. Certainly the cells as they come from normal peripheral blood are not mitotically active. Neither conditioned medium nor any other variation of the culture conditions, with or without PHA, was capable of stimulating these normal leukocytes to divide immediately. On the other hand, bone marrow cells and leukemic blasts from peripheral blood, under the same culture conditions, showed mitotic activity within a few hours. This difference may be the result of what Swann has recently called "long-latent-period" and "short-latent-period" responses to mitogenic stimuli (13). Tissues which normally have a relatively high mitotic index *in vivo* (e.g., bone marrow) and thus always contain many cells capable of division, respond to stimulation within a few hours; cells which normally show very low mitotic activity (e.g., liver) respond only after a latent period of several days, during which time differentiated cells of the involved organ are "switching over" from their usual specific functions to the synthesis of mitotic protein and other materials for division.

The consistent failure to observe mitoses in cultures of normal leukocytes before the 3d day *in vitro* would seem to fit this concept of a "long-latent-period" response; and the studies with PHA suggest that it is the substance, in our culture system, which initiates this "switching-over" process. The change to a mitotically active state of dextran-separated cells in our PHA-free cultures did not begin until PHA was added, as indicated by the fact that mitoses never appeared until 3 days after the addition of PHA regardless of the total age of the culture. Leukocytes in a state already capable of division, on the other hand, (i.e., marrow cells, leukemic blasts) did not require the presence of PHA for the mitotic process itself.

There must be a number of other factors which can operate both *in vivo* and *in vitro* to stimulate or inhibit the mitotic potential of circulating leuko-

cytes (5, 9). Not only have previous workers obtained some mitotic activity in leukocyte cultures, in the absence of PHA (1, 3), but there is also at least indirect evidence that mononuclear leukocytes from the circulating blood can undergo mitosis in the body (e.g., at sites of inflammation) (6). However, our present attempts, *in vitro*, to uncover evidence of a *physiologic* control mechanism have been unsuccessful. In the absence of PHA, none of the variety of conditions studied resulted in the appearance of mitoses in our cultures; and, with PHA present, variation of these conditions, over at least as wide a range as might be encountered *in vivo*, uniformly failed to prevent the initiation of mitotic activity.

ACKNOWLEDGMENTS

The technical assistance of Elizabeth Krohnert is gratefully acknowledged. Photomicrographs are by William Fore.

REFERENCES

1. Bloom, W. Tissue Culture of Blood and Blood-Forming Organs, pp. 1471–1585. *In:* H. Downey (ed.), Handbook of Hematology. New York: P. N. Hoeber, 1938.
2. Bond, V. P.; Fliedner, T. M.; Cronkite, E. P.; Rubini, J. R.; Brecher, G.; and Schork, P. K. Proliferative Potentials of Bone Marrow and Blood Cells Studied by in Vitro Uptake of H3-Thymidine. Acta Haematol., 21:1–15, 1959.
3. Chrustschoff, G. K., and Berlin, E. A. Cytological Investigation on Cultures of Normal Human Blood. J. Genetics, 31:243–51, 1935.
4. Hungerford, D. A.; Donnelly, A. J.; Nowell, P. C.; and Beck, S. The Chromosome Constitution of a Human Phenotypic Intersex. Am. J. Human Genetics, 11:215–36, 1959.
5. Loomis, W. F. pCO₂ Inhibition of Normal and Malignant Growth. J. Nat. Cancer Inst., 22:207–17, 1959.
6. McCutcheon, M. Inflammation, pp. 13–61. *In:* W. Anderson (ed.), Pathology, 3d ed. St. Louis: Mosby, 1957.
7. Nowell, P. C. Stimulation of Mitosis in Rat Marrow Cultures by Serum from Infected Rats. Proc. Soc. Exper. Biol. & Med., 101:347–50, 1959.
8. ———. Differentiation of Human Leukemic Leukocytes in Tissue Culture. Exper. Cell Research (in press).
9. Osgood, E. E. A Unifying Concept of the Etiology of the Leukemias, Lymphomas, and Cancers. J. Nat. Cancer Inst., 18:155–66, 1957.
10. Osgood, E. E., and Krippaehne, M. L. The Gradient Tissue Culture Method. Exper. Cell Research, 9:116–27, 1955.
11. Rigas, D. A., and Osgood, E. E. Purification and Properties of the Phytohemagglutinin of *Phaseolus vulgaris*. J. Biol. Chem., 212:607–9, 1955.
12. Skoog, W. A., and Beck, W. S. Studies on the Fibrinogen, Dextran, and Phytohemagglutinin Methods of Isolating Leukocytes. Blood, 11:436–54, 1956.
13. Swann, M. M. The Control of Cell Division: *A Review.* II. Special Mechanisms. Cancer Research, 18:1118–60, 1958.

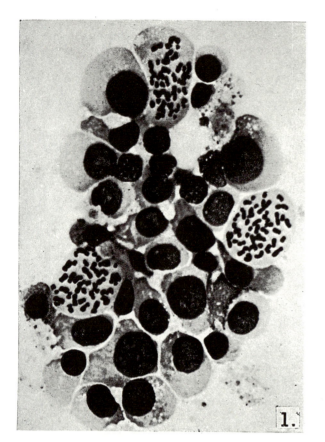

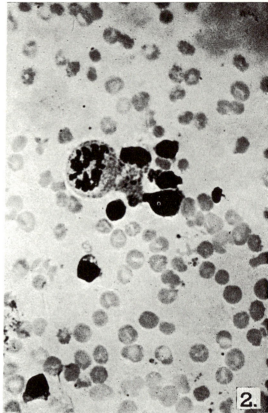

Fig. 1.—Three mitoses in clump of "large mononuclears" and lymphocytes on slide from 4-day "standard" culture of normal human leukocytes, containing phytohemagglutinin. Cells arrested in metaphase by colchicine treatment (18 hours). Swelling of cytoplasm of all cells results from water rinse before air drying and staining with Giemsa. ×720.

Fig. 2.—Mitotic figure in smear from 4-day "spinner" culture of whole blood. Apparently healthy lymphocytes are present, as well as degenerating forms. Erythrocytes are mainly intact and not agglutinated, despite presence of phytohemagglutinin. Colchicine treatment—18 hours. Wright's stain. ×720.

Experimental Cell Research **25**, *585–621 (1961)*

THE SERIAL CULTIVATION OF HUMAN DIPLOID
CELL STRAINS[1]

L. HAYFLICK and P. S. MOORHEAD

Wistar Institute of Anatomy and Biology, Philadelphia, Pa., U.S.A.

Received May 15, 1961

Oɴʟʏ limited success has been obtained in developing strains of human cells that can be cultivated for long periods of time *in vitro* and that still preserve the diploid chromosomal configuration [41, 47, 48, 58, 59]. Indeed, heteroploidy may be a necessary corollary or even the cause of the alteration of primary or diploid cells *in vitro* to the status of a cell line. Such changes in chromosome number appear to be independent of the type of primary tissue since they have been observed in cells derived from both normal and malignant tissue [4, 22, 23, 31].

These cell lines, of which over two hundred have been reported in the literature, have serious limitations for many kinds of biological studies. Chief among these is the exclusion of their use for the production of human virus vaccines. This limitation is based on the supposition that such heteroploid cell lines, whether of normal or malignant origin, share many of the properties of malignant cells [29, 30, 37]. This objection would be even more important if viruses played a role in human neoplasia. In general, if strains of human cells could be kept continuously under conditions of rapid growth for extended periods of time with the retention of the diploid configuration these objections would not apply.

Furthermore, diploid cell strains would parallel more closely the biology of cells *in vivo*. Although characterizations of heteroploid cell lines are often stated in terms of a modal chromosomal number, this should not obscure the fact that extensive pleiomorphism is present [21]. The cells comprising the modal class in heteroploid cell lines are found to be heterogeneous if chromosomal analysis is extended beyond a simple enumeration [49]. This genomic variability constitutes an important consideration in experiments using cell lines for the study of metabolic or other phenotypic cell markers. The use of cloning as a means of reducing this variability in heteroploid cell lines is unfortunately limited by the rapid re-emergence of a range of chromosomal types among the progeny of the clone [5, 49, 50].

[1] This investigation was supported by a research grant (C-4534) from the National Cancer Institute, National Institutes of Health, United States Public Health Service.

The results to be presented stress, incidentally, the need for a clarification of certain terms used by tissue culturists to describe a number of phenomena. Consideration of eleven important criteria show the term "cell line" to be inapplicable to the type of cells described in this report. Precedence and usage confine the term "cell line" to only those cells that have been grown *in vitro* for extended periods of time (years). This period of time presumes potential "immortality" of the cells when serially cultivated *in vitro*. (The situation is analogous to transplantable tumors which are also apparently "immortal" in the sense that serial subcultivation in proper hosts guarantees the growth of the tumor for an indefinite period of time.) The diploid cell strains presently described are assumed to lack this characteristic of potential immortality. In addition, all mammalian cell lines examined to date vary from the diploid chromosome number. This fact alone should exclude the diploid cells from being termed "cell lines" and we have chosen to refer to them as "cell strains".

A *cell strain*, therefore, is a population of cells derived from animal tissue, subcultivated more than once *in vitro*, and lacking the property of indefinite serial passage while preserving the chromosomal karyotype characterizing the tissue of origin. Conversely, a *cell line* is a population of cells derived from animal tissue and grown *in vitro* by serial subcultivations for indefinite periods of time with a departure from the chromosome number characterizing its source.

It is also possible, when observed, to include in the definition of a *cell line* the characteristic of cell "alteration" as described by Parker, Castor and McCulloch [44], and by Hayflick [17]. No such alterations have been found in the cell strains which are the subject of this report. It is proposed that the term "cell transformation", which has been used interchangeably with "cell alteration" [23], be excluded on grounds that the latter term has precedence and also that "transformation" has specific implications in the allied field of bacteriology which do not, as yet, apply to cell culture. The terms "established cell lines" and "stable cell lines" should likewise be avoided as the former is redundant and the latter implies that changes have ceased to take place or are no longer possible. The term *primary cells* should indicate those cells obtained from the original tissue that have been cultivated *in vitro* for the first time. If subsequent passages of these cells are made, it is assumed that such cells can properly be called a *cell strain* until they are either lost through further subcultivations or alter to the heteroploid state, in which case they could properly be referred to as a *cell line* (Text-Fig. 1).

It is the subject of this study to describe and characterize the development of 25 strains of human cells derived from fetal tissue which retain true diploidy for extended periods of cultivation without alteration. Specific attention will be given to those cell characteristics which serve to distinguish cell lines from cell strains.

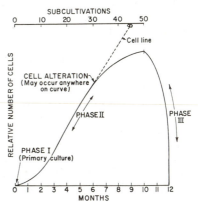

Text-Fig. 1.—Diagrammatic representation of the history of cell strains and the phenomenon of cell alteration. Phase I, or the primary culture, terminates with the formation of the first confluent sheet. Phase II is characterized by luxuriant growth necessitating many subcultivations. Cells in this phase are termed "cell strains". An alteration may occur at any time giving rise to a "cell line" whose potential life is infinite. Conversely, cell strains characteristically enter Phase III and are lost after a finite period of time.

MATERIALS AND METHODS

Media.—The growth medium (GM) used was Eagle's Medium in Earle's Balanced Salt Solution [8] supplemented with 10 per cent calf serum.[1] Twenty-five ml of 5.6 per cent $NaHCO_3$, 10^5 units of penicillin and 10^5 μg of streptomycin, were added per liter. The final pH of the medium was 7.3, and before use it was brought to 37°C. Phosphate buffered saline (PBS) was prepared as described by Dulbecco and Vogt [7]. Difco trypsin (1:250) was prepared as a 0.25 per cent solution in PBS and supplemented after filtration with the antibiotics described above.

Isolation of primary cells.—Two methods of cell cultivation from primary tissue were employed in this study with identical qualitative results. The use of trypsin yielded far more cells initially than cultures prepared from fragmented or minced tissue. Since high cell yields were not required from the starting tissue, most cultures were started from fragmented or minced tissue. Such preparations gave fewer cells initially than could have been obtained from tissue treated with the enzyme preparation. Minced preparations were obtained by cutting the tissue in a Petri dish containing GM with paired scalpels or a scissors until the size of each piece approximated 1–4 mm³. Fragmented preparations were obtained by tearing apart the tissue with two pairs of forceps in a Petri dish containing GM until the pieces could no longer conveniently be grasped and shredded. The entire contents of the dish were emptied into one or more Pyrex Blake bottles (surface area 100 cm²), depending on the size of the original starting tissue. The fragmented lungs, for example, from a three-month-old human fetus were usually placed in four Blake bottles. Treatment of tissue with trypsin was done, in general, according to the method of Fernandes [11].

---

[1] Obtained from Microbiological Associates, Inc., Bethesda, Maryland.

Experimental Cell Research 25

Initiation of cultures.—If the fetal tissue was viable when received, cells could be found in bottles planted by any one of the methods described after about three days of incubation at 36°C. When growth was first observed the cultures were refed. The spent medium and any fragments present were discarded. If additional bottles were required these fragments could be replanted in a new bottle. Fresh GM was added and as soon as the cells formed a confluent sheet the cultures were subcultivated. This normally occurred in about 10 days. Periodic feeding of the cultures was done when a sharp drop in the pH of the medium made it necessary.

In the beginning of this study attempts were made to minimize the period of time elapsing between the receipt of the fetus or fetal tissue and its cultivation *in vitro.* It was subsequently found that if either was viable upon receipt it could be kept for at least 5 days at room temperature, or 5°C, without apparent loss of viability. Minced tissue, kept in a minimal amount of GM has been found to be viable for periods of time up to 3 weeks, either at room temperature or 5°C.

Subcultivation of confluent cultures.—As soon as cell cultures were fully sheeted they were put on a strict schedule of subcultivations, which were done alternately every third and fourth day. The spent GM was discarded and trypsin solution was added to each bottle. After incubation at 37°C, or room temperature, for 15 min, the enzyme solution containing the dislodged cells was centrifuged for 10 min at 600 r.p.m. in an International Size 2 Model V Centrifuge. The trypsin solution was decanted after centrifugation and the cells were resuspended in a small amount of GM, aspirated with a 5 ml pipette, and evenly distributed to two Blake bottles. Sufficient fresh medium was added to each bottle to cover the surface adequately. This was called a 2:1 split. In the early part of this study split ratios of 3:1 were used with equal success. Incubation was carried out at 36°C.

Preservation of cells by freezing.—After trypsinization of a mature culture and resuspension of the centrifuged cells in a few ml of GM, the cell concentration was adjusted with GM to 1.5–2.0×10^6 cells per ml. Sterile glycerol was added to give a final concentration of 10 per cent and the suspension was dispensed in 2 ml portions in 5 ml ampules. The ampules were then sealed and held at 5°C overnight. The next day the ampules were placed directly at -70°C.

Recovery of frozen cells.—Ampules to be reconstituted were removed from the dry ice chest and placed quickly in a 37°C water bath. After the contents had thawed, the suspension was placed in a milk dilution bottle (surface area 40 cm²) and sufficient fresh GM added to cover the surface of the bottle adequately. After incubation at 36°C for one day the medium was completely changed. Periodic feedings of the culture were made until the cell sheet was confluent at which time the culture was manipulated as described above. Reconstituted cells frozen for up to one year invariably yielded viable cultures if these conditions were met. Although quantitative recovery of the frozen cells was not achieved, the fraction of the frozen population that did survive was always sufficient to recover the culture.

Chromosome analysis.—Thirteen of the strains were studied for purposes of chromosome analysis. Actively dividing cultures (usually 48 hr after seeding a Blake bottle) of these strains were sacrificed for chromosome studies of cells arrested in metaphase by colchicine treatment. Following 6 hr subjection to a concentration of 2×10^{-6} M colchicine in the medium, the cells were trypsinized free. Suspended cells were then processed for spreading on glass slides according to an air-drying technique [38] based upon that of Rothfels and Siminovitch [51]. Aceto-orcein stained chromosome preparations were scanned under low power optics (20 × objective) for adequately spread metaphase cells showing a minimum of scattering of the chromosomes. Selected metaphases were then studied under oil immersion optics for simple counts and for detailed analysis of karyotype, where warranted.

From 250 metaphases per determination the proportions of polyploid and diploid cell classes were obtained by rough chromosome estimates (approximate accuracy ±6 chromosomes).

Sex chromatin.—The scoring of nuclei for the presence or absence of the sex chromatin, or Barr body [3], was done upon the same slide preparations made for the study of hypotonically spread metaphases. Interphase nuclei which were not overstained and which did not have numerous chromocentres were scored for sex chromatin determinations. These considerations required the elimination of 30–40 per cent of the interphase nuclei from the scoring. This ambiguous class of nuclei was considerably reduced by using less intense staining with aceto-orcein as follows.

Coverslips bearing sheets of fibroblasts were rinsed briefly in PBS, fixed in methyl alcohol for 1 min, air-dried, aceto-orcein stained for 10–30 sec, rinsed twice briefly in each of 95 per cent ethyl alcohol, absolute ethyl alcohol, xylol, and then mounted in Permount.

These provided excellent detail of the sex chromatin body with very few over-stained nuclei. Only those nuclei which formed numerous chromocentres, especially common in non-dividing cultures, comprised the ambiguous class, while 80–90 per cent of the remaining female nuclei clearly displayed a single sex chromatin body. In male material lightly stained in this way 80–90 per cent of the nuclei (excluding those with many chromocentres) clearly lacked any such stained body. A minimum of 200 nuclei were scored per determination.

Implantation of diploid cells into hamster cheek pouches.—Cells of diploid strain WI-25 at the 14th passage (61 days *in vitro*) were selected for inoculation into the cheek pouches of unconditioned hamsters. In general, the technique used was that described by Handler and Foley [12, 15]. Trypsin dispersed cells were centrifuged and resuspended so that the desired inoculum was contained in 0.1 ml of GM. Inoculum sizes ranged from 10^3 to 10^5 cells per 0.1 ml. Five hamsters were inoculated with

Text-Fig. 2.—Diagrammatic representation of the history of the WI-1 strain of human diploid cells. Series A represents the continuous subcultivation of the strain through a period of 50 passages during which time surplus cultures from each passage were committed to storage at −70°C. Other series denote those cultures of the strain that were removed from storage at −70°C after periods of time represented by the vertical dash line. Surplus cultures from each series were committed to storage, the origin of the dash line denotes that passage which was thawed to start a new series. ——, series; − − −, months; +, still in cultivation as of February 28, 1961.

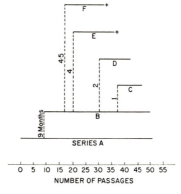

Experimental Cell Research 25

each cell concentration selected. For control purposes, five hamsters were also in-oculated with 10^5 HeLa cells per 0.1 ml.

Homotransplantation of WI-1 cells into terminal cancer patients.—Two pools of strain WI-1 cells were used to inoculate six terminal cancer patients. The first pool consisted of those cells that had been grown serially for nine passages, held for 9 months at $-70°C$, restored, carried for an additional 21 passages, stored again at $-70°C$ for 2 months, restored and subsequently carried for another seven passages (Series D in Text-Fig. 2). Thus this first pool of cells represented a total of 37 sub-cultivations. The second pool of WI-1 cells consisted of cells that had been grown serially for nine passages, held for 9 months at $-70°C$, restored, and subsequently carried for 36 more passages (Series B in Text-Fig. 2). Thus the second pool of cells represented a total of 45 serial subcultivations *in vitro*.

The cells of both pools were grown as indicated previously, harvested with trypsin, resuspended in PBS and adjusted to a concentration of 6×10^5 cells per ml. Cell counts were made in trypan blue and only viable cells were counted; dead cells constituted less than 3 per cent of the total harvest. One-half ml of this suspension was inoculated subcutaneously on the flexor surface of the forearm with a tuberculin syringe fitted with a No. 20 needle. The area was tattooed for subsequent identifica-tion when taking biopsies. The six patients used in this study were also skin tested with GM alone prior to cell inoculation in order to ascertain their sensitivity to calf protein. These patients had advanced incurable cancer and a very short life ex-pectancy.

EXPERIMENTAL RESULTS

Establishment of diploid cultures.—In all cases where the original human fetal tissue was viable, indicated by cell growth in the first culture, the strains from the various organs were kept in serial cultivation as shown in Table I. When a particular fetus was found to yield non-viable cells from one organ, invariably cultures made from other organs were also found to be non-viable. It was found, therefore, that if a primary culture was obtained from tissue, the cell strain could be cultivated serially for periods of time up to 11 months. With the exception of strains WI-6 and WI-22 derived from heart tissue, all strains could be carried for at least 25 passages, extending over a period of 5 months. The maximum number of subcultivations obtained is exempli-fied by the WI-23 strain derived from lung, which lasted for 8 months during which 55 subcultivations were made. Without exception, all of the human strains were of the fibroblast cell type from about the 5th subcultivation. The kidney strains began as epithelial cell cultures with scattered nests of fibroblasts. The least successful cultures were those obtained from liver in which fully sheeted primary cultures were rarely obtained.

Although the subject of this report is confined to experiments involving human fetal cells, adult human cells have also been carried for similarly extensive periods of time with retention of the diploid configuration. Other workers [41, 59] have reported similar results with adult human diploid cells.

Morphology of diploid human fibroblast cell strains.—Figs. 1 and 2 represent fibroblasts of the WI-1 strain of diploid human fetal lung cells after 35 subcultivations and 9 months *in vitro*. Characteristically these cells are extremely elongated fibroblasts averaging about 185 $\mu \times 15$ μ. The single nucleus contains from 1 to 4 nucleoli which vary from oval to branching bodies. Individual cells are markedly transparent with characteristically

TABLE I. *History of human diploid cell strains.*

Strain designation	Fetus no.	Tissue of origin	Months in serial cultivation[a]	No. of subcultivations
WI-1	1	Lung	11	51
WI-2		Skin and muscle	6[b]	20[b]
WI-3	2	Lung	5	35
WI-4	3	Kidney	6	29
WI-5		Muscle	7	33
WI-6	4	Heart	2.5	10
WI-7		Thymus and thyroid	5	25
WI-8	5	Skin	8.5	32
WI-9		Kidney	8.5	29
WI-10	6	Kidney	5[c]	32[c]
WI-11	7	Lung	5[c]	30[c]
WI-12	8	Skin and muscle	8	41
WI-13		Kidney	8	40
WI-14	9	Skin	8	43
WI-15	10	Kidney	7.5	28
WI-16	11	Lung	8	44
WI-17		Liver	5[b]	24[b]
WI-18	12	Lung	8	53
WI-19	13	Lung	8	50
WI-20	14	Skin and muscle	5[b]	25[b]
WI-21	15	Heart	5	26
WI-22	16	Heart	1	5
WI-23	17	Lung	8	55
WI-24	18	Lung	7	39
WI-25	19	Lung	6[c]	38[c]

[a] Continuously passaged cells, never reconstituted from frozen stock (Series A).
[b] Serial cultivation of strain lost through bacterial contamination but cells from previous passages stored at $-70°C$.
[c] Still in culture as of February 28, 1961.

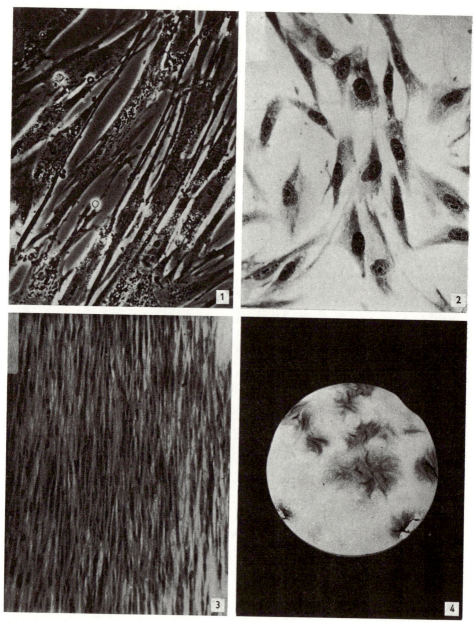

Fig. 1.—Strain WI-1. Diploid human fetal lung cells after 35 subcultivations and 9 months *in vitro*. Phase contrast. ×360.

Fig. 2.—Strain WI-1. Diploid human fetal lung cells after 35 subcultivations and 9 months *in vitro*. Stained with May-Grünwald Giemsa. ×300.

fine cytoplasmic granularity. The cell membrane is filamentous at either end of the long axis of the cell and undulating membranes are common. Multinucleated cells are rarely found. The cells adhere firmly to the glass surface about 1 hr after subcultivation. Within 48 hr many highly polarized fibroblasts oriented in different directions can be seen. At this time, under ideal conditions of growth, almost every field (150 ×) contains from two to eight rounded, highly refractile cells which represent fibroblasts in various stages of mitosis. As the cell sheet becomes confluent the sheet takes on a quilt-like appearance when the bottle is examined macroscopically with incident light. Upon microscopic examination, the sheet is seen to be composed of numerous swirls of fibroblasts, each swirl containing elongated cells oriented in a common direction (Fig. 3). The direction of cell orientation resulting from parallel alignment varies from area to area. It is this growth pattern that gives to the bottle its characteristic "sheen" when the cell sheet is viewed with incident light macroscopically. Puck [46] has described the appearance of such fibroblasts, growing in Petri dishes for studies of plating efficiency (Fig. 4), as "rough, hairy, colonies". The concept of contact inhibition as described by Abercrombie and Heaysman [1] seems to explain adequately this growth pattern. When the movement of a fibroblast in a particular direction brings it into contact with another fibroblast so that an adhesion forms, further movement in that direction is inhibited. All these characteristics apply to each of the human diploid cell strains throughout their periods of active proliferation.

Chromosome analysis.—To determine if heteroploid or aneuploid classes had emerged within the growing population of these strains the level of tetraploidy was determined, exact chromosome counts made and karyotypic analyses of the better metaphases prepared (Table II). Exact chromosome counts of 28 or more metaphases per determination revealed primarily diploid and few tetraploid cells[1] (Table II). The level of tetraploidy at various passages was found to be low, exceeding 3 per cent in the dividing population only in strain WI-25 (Table II). There was no evidence of alteration in number or morphology of the somatic cell chromosomal comple-

---

[1] Exceptions with respect to available metaphases for exact counts being strains WI-15 and WI-1 of the 39th passage in which mitoses were scanty (Table II).

Fig. 3.—Strain WI-1. Directional orientation of cells after 3 days in culture. Stained with May-Grünwald Giemsa. × 100.

Fig. 4.—Strain WI-1. Colony development from single cells planted in a 50 mm Petri dish. Photographed after 35 days in culture. Colonies are rough and hairy in appearance. May-Grünwald Giemsa.

ment. Karyotypic analyses of 6–12 excellent metaphases for each determination revealed all 23 chromosome pairs present and recognizable either individually or according to their group: #1, #2, #3, #4-5, X-6-12, #13-15, #16, #17, #18, #19-20, #21-22-Y as defined by the Denver Conference of 1960.[1,2]

TABLE II. *Karyology of human cell strains.*

Strain	Passage no.			Chromosome counts (exact counts only)									Other counts	Tetraploids per 250 metaphases	Sex chromosomes
		40	41	42	43	44	45	46	47	48	91	92			
WI-1 (Ser. A)	9						1	56						3 (1.2 %)	XY
WI-1 (Ser. B)	21[a]			1	—	1	1	24			1	1	59±0, 88±0	4 (1.6 %)	XY
WI-1 (Ser. D) (Phase III)	39			1	1	—	—	14					38±0	4	XY
WI-12	32	1	1	—	1	—	3	26						2 (0.8 %)	XX
WI-12	40					1	1	31						2	XX
WI-13	25						1	25						3	XX
WI-14	29				1	1	7	25	1	2	—	1	57±0, 54±0	3	XY
WI-15	21				1	—	—	7						—	XY
WI-16	22				1	1	4	25			—	1		3	XX
WI-17	17				1	1	2	27						3	XX
WI-18	21				2	1	1	26					40±0, 52±0	6 (2.4 %)	XY
WI-19	21			1	1	2	—	24						2	XY
WI-21	17					1	4	26						1	XY
WI-23	22			1	—	1	2	25						3	XY
WI 24	22			1	—	—	2	27			—	1		4	XX
WI-25	30			1	1	—	1	24			—	1	68±0, 62±0	16 (6.4 %)	XX

[a] Period of 9 months in frozen state at 9th passage. See Text-Fig. 2.

Those counts falling below the 2n number of 46 (Table II) are regarded as the result of excessive spreading and occasional loss of one or a few chromosomes from a complement. This is supported by the general "left-skewness" of such counts and the lack of consistency with respect to the identity of the

[1] Representative karyotypes of male and female diploid strains are shown in Figs. 5–8.
[2] Report of a study group: Am. J. Human Genet. **12**, 384 (1960).

Experimental Cell Research **25**

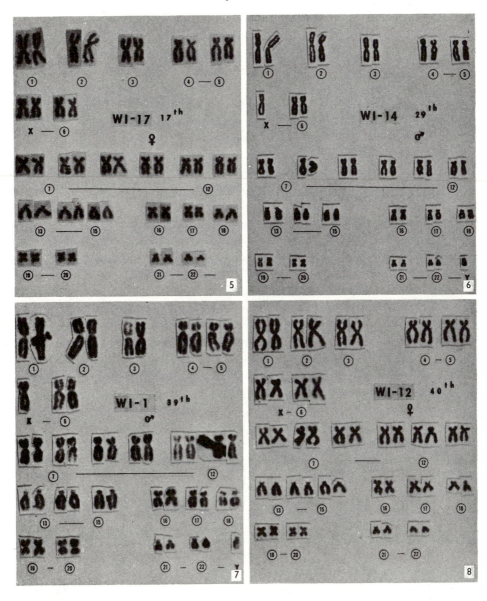

Fig. 5.—Strain WI-17, 17th passage. Classic female diploid karyotype of metaphase cell.

Fig. 6.—Strain WI-14, 29th passage. Classic male diploid karyotype of metaphase cell.

FIG. 7.—Strain WI-1, 39th passage. Classic male diploid karyotype of metaphase cell. Note that one member of pair #1 overlies a member of pair #11.

Fig. 8.—Strain WI-12, 40th passage. Classic female diploid karyotype of metaphase cell.

missing chromosome in each case. Approximately half of these seeming aneuploid counts were analyzable and in only one case (WI-14) was the missing chromosome from the same group. Thus four of seven counts at $2n-1$ lacked a chromosome from group X–12. The usually rare $2n+1$ and $2n+2$ counts here might be regarded as further evidence for suspicion of this strain. On the other hand, this particular material was excessively spread and broken metaphase groups were common.

Sex chromatin.—The retention of a single sex chromatin body in the interphase nucleus was noted in those strains having a female (XX) karyotype throughout all passages examined (Table II). Each of the four female strains was readily detected from a coded group of slide preparations which included seven male strain representatives. Since this material was prepared primarily for chromosome studies, 30–40 per cent of the nuclei were not included in the scoring. Of the suitable nuclei remaining 57–66 per cent had a single chromatin mass (Figs. 9 and 10), usually peripheral, such as described by Barr [3]. In male material 10–32 per cent of the scorable nuclei possessed smaller chromatic bodies which might be considered as sex chromatin. However, these bodies did not permit interpretation as typical bipartite sex chromatin. Male nuclei regarded as lacking sex chromatin are shown in Figs. 11 and 12. Presumptive tetraploid nuclei (size class) in female cells usually displayed two distinct sex chromatin bodies (Fig. 13). This finding is consistent with the extensive work of Klinger and Schwartzacher [28], who used human amnion in studies also involving spectrophotometric measurements of sex chromatin. Other examples of female strain nuclei with sex chromatin are presented in Figs. 14 and 15. Examination of female cell strains at high passages revealed no reduction in the proportion of nuclei with the typical body. (Strain WI-16 in the 44th passage showed 82 per cent with single body, 9 per cent with two bodies, less than 2 per cent with three bodies and 8 per cent with no visible body. Total: 237 nuclei.)

Fig. 9.—Strain WI-16, 22nd passage. Single sex chromatin bodies (arrows) in nuclei from female cells. (Chromosome preparation: hypotonically pretreated, 15 min in orcein.) Note dense nuclei. × 1050.

Fig. 10.—Same material as in Fig. 9. × 1050.

Fig. 11.—Strain WI-1, 9th passage. Absence of sex chromatin in a nucleus from male cells. Chromosome preparation. × 1050.

Fig. 12.—Strain WI-1, 22nd passage. Absence of sex chromatin in a nucleus from male cells. Chromosome preparation. × 1050.

Fig. 13.—Strain WI-25, 16th passage. Single sex chromatin bodies (arrows) in diploid nuclei from female cells. Note 2 sex chromatin bodies in the large presumptive tetraploid nucleus. (Coverslip grown preparation: 10 seconds orcein.) × 1000.

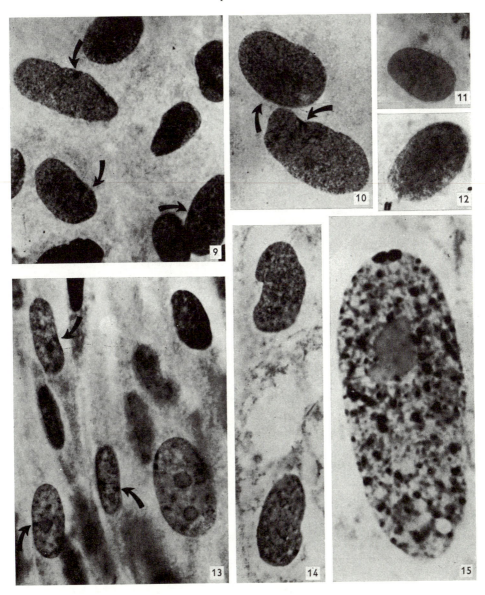

Fig. 14.—Strain WI-16, 44th passage. Single sex chromatin bodies in nuclei from female diploid cells. (Coverslip grown preparation: 10 seconds orcein.) ×1200.

Fig. 15.—Strain WI-16, 44th passage. Presumptive tetraploid nucleus with two adjacent sex chromatin bodies located peripherally. Female strain. (Coverslip grown preparation: 10 seconds orcein.) ×2800.

39 – 61173267 *Experimental Cell Research* 25

General growth characteristics of human diploid cell strains.—The diploid cell strains were, in all cases, established without recourse to special procedures. Once the cultures have become confluent cell sheets, the cells are very active metabolically, as shown by the fact that the GM becomes acid faster than in cultures of heteroploid cell lines inoculated with the same number of cells. In general, all diploid strains were subcultivated with a split ratio of 2:1, twice a week. This split ratio and rigid semi-weekly schedule of subcultivation was maintained until the last passage, when the cultures failed to show the numerous mitotic figures characteristic of good growth and started to accumulate debris. These degenerative changes did not appear suddenly, but took place over a period of from 1 to 3 months (Figs. 16 and 17). The fine debris appears to be the first indication that the strain is waning. The debris aggregates about the cell surfaces and presumably consists of protoplasmic material resulting from cell degeneration. In addition to the loss of mitotic activity and accumulation of debris, the cells become much less polarized, more spread out, and lose contact inhibition. Although mitotic activity eventually ceases in the culture, acid production continues; and as dictated by drops in the pH, cultures can be fed for a few months until all acid production ceases and the culture is observed to have completely degenerated.

At the first signs of these changes, various attempts were made to recover the strain. Changes in media components, further subcultivations and attempts to crowd into one culture a number of cultures of degenerating cells all failed to accomplish the intended result. It is apparent, however, that media composition plays no role in explaining the cause of this degeneration, since portions of the same large pool of medium can be used to support the luxuriant growth of "young" strains while older strains are observed to degenerate in the same pool of medium. Even strains derived from different organs of the same fetus support this contention. A lung strain may be degenerating while a skin strain is luxuriating in the same pool of medium (which is prepared in large enough batches to last a number of weeks). Even cells which are

---

Fig. 16.—Strain WI-1, 50th passage. Degenerative changes characteristic of Phase III. Mitosis ceases and granular debris accumulates. Unstained. × 140.

Fig. 17.—Same as Fig. 16.

Fig. 18.—Strain WI-1, 30th passage. Degeneration of cells in the presence of Poliovirus Strain Koprowski-Chat Type 1 (attenuated virus). × 140.

Fig. 19.—Strain WI-1, 30th passage. Plaque formation in the presence of Poliovirus Strain Koprowski-Chat Type 1. 50 mm Petri dish.

Experimental Cell Research **25**

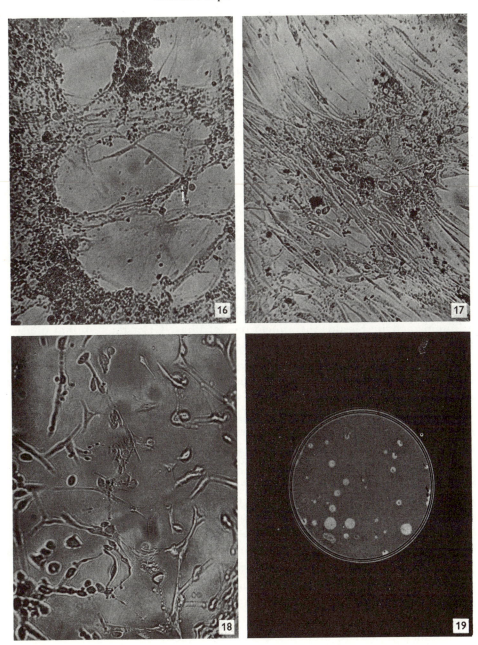

41

reconstituted (e.g.) from the frozen 9th or 37th passage flourish in the same medium that was used on the identical strain undergoing degeneration in the 50th passage (Text-Fig. 2). Such observations were confirmed many times whenever a diploid strain started to degenerate.

The general history of a diploid cell strain may, therefore, be divided into three distinct phases (Text-Fig. 1). Phase I, or early growth phase, constitutes that period when the cells have been freed from the intact tissue and are just establishing themselves on glass (primary culture). In general this phase lasts from 1 to 3 weeks. Phase I ends with the formation of the first confluent sheet, at which time the culture is ready for its first subcultivation and entry into Phase II. Phase II is characterized by rapid cell multiplication and great acid production. During this phase the diploid strains must be subcultured at least twice a week with split ratios of 2 or 3:1. This coincides with the formation of confluent sheets. Apparently, success in keeping the diploid strains in serial cultivation for long periods of time depends on their being kept in the log phase of growth as much as possible during their history. Phase II lasts from 2 to 10 months, at the end of which time cell degeneration begins to take place. This degeneration and lessening of mitotic activity heralds the appearance of Phase III or the terminal phase. It is characterized by the appearance of debris, as illustrated in Figs. 16 and 17, reduction of mitotic activity and a consequently longer period of time for the development of confluent sheets. The frequency of tetraploids among the few dividing cells of this degenerative phase was as low as it was at all passage levels (Table II, WI-1 Series D). However, in interphase cells of Phase III, bizarre nuclear forms and sizes become more frequent and the appearance of these nuclei is reminiscent of irradiated cultures. Our attempts to reverse Phase III have been uniformly unsuccessful.

Absence of pleuropneumonia-like organisms (PPLO) and latent viruses in human diploid cell strains.—Periodic monitoring of all diploid cell strains herein reported has never revealed the presence of PPLO as a contaminant. The techniques used were those described previously by Hayflick and Stinebring [19]. These results were predictable on the basis of evidence recently presented describing the probable origin of this type of contaminant [18].

Numerous attempts were made to detect the presence of latent viruses in all the diploid cell strains. Spent medium with and without cells was passed into many types of tissue cultures and laboratory animals with consistently negative results. Special attention was given to spent fluids and cells from cultures in Phase III. In no case was cytopathology (or hemadsorption) observed in any of the cell strains before or during Phase III that could be attributed to the unmasking of latent viruses, as is the case, for example, with primary monkey kidney cells [24] or primary human tonsil and adenoid tissue [52, 53]. Cells obtained from Phase II and Phase III were stained by various methods and observed for possible virus inclusions with negative results.

In order to investigate this point more critically, a culture of strain WI-1 (known to be male) at the 49th passage was selected for the following experiment. Since this strain was in Phase III it had not formed a confluent sheet but consisted of numerous scattered cells that were still lowering the pH when fed. This culture was seeded directly with a suspension of female strain WI-25 in Phase II which was then at the 13th passage; sister cultures of WI-25 were carried simultaneously. The mixed culture (WI-1 plus WI-25) began to proliferate as expected and has now been subcultured 17 times, which represents the identical number of passages of the WI-25 sister culture from the date of the experiment. In addition, cells examined at this passage of the mixed culture have been characterized as female by the presence of sex chromatin and also by karyotype analysis (Table II). Single sex chromatin body: 79 per cent; double: 10 per cent; triple: less than 1 per cent; none: 10 per cent (418 nuclei). If a latent virus had been responsible for Phase III in the WI-1 strain it seems unlikely that it would have been able to discriminate between male and female cells.

Although it cannot be said with certainty that any cell line or cell strain is absolutely free of contaminating viruses our evidence does not suggest the presence of such viruses.

Growth rate of human diploid cell strain WI-1.—WI-1 cells at the 25th subcultivation were used to determine the rate of growth. These had been in continuous cultivation for 5 months. The growth curve of these cells is shown in Text-Fig. 3 and was in general derived according to the method of Fernandes [11]. Each point is the average of four separate tube cultures, inoculated initially with 2.5×10^4 cells per ml per tube. The counts show viable cells only, ascertained by trypan blue staining. The generation time of 24 hr obtained during log phase does not differ significantly from that obtained for HeLa [20]. An initial lag of 24 hr was found, after which time the cells entered the log phase for 96 hr. After approximately 9 days the cell numbers remained constant with viability remaining at 90 per cent for one month.

Size of initial cell concentration as a function of population survival.—Early in these studies it was found that, under the conditions described, both a minimum concentration of 10^4 cells per ml and a period of 13 days were necessary to obtain a confluent culture in bottles or tubes as indicated in

Text-Fig. 4. Cell concentrations below 10⁴ per ml did not result in confluent sheet development but in isolated islands of cell colonies (Fig. 4).

It has been consistently observed that per unit area of glass surface about 25 per cent fewer diploid cells are found than cells of heteroploid cell lines (e.g. HeLa) in confluent cultures.

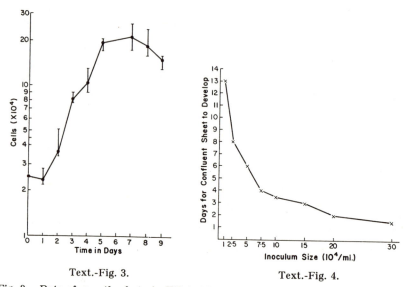

Text.-Fig. 3.   Text.-Fig. 4.

Text-Fig. 3.—Rate of growth of strain WI-1, 25th passage. Four separate tube cultures were used for each point on the curve. The starting inoculum was 2.5 × 10⁴ cells per ml per tube.

Text-Fig. 4.—Period of time necessary for confluent sheet formation starting with different inoculum sizes. Strain WI-1, 27th passage.

Long-term maintenance of tube cultures.—Various media were investigated in order to determine the best medium for maintaining tube cultures of diploid cells for long periods of time. Pyrex test tubes were inoculated with 5 × 10⁴ cells each of strain WI-25 (14th passage) in 2 ml each of the medium to be tested. The media used were GM with 1, 5, and 10 per cent calf or horse serum; medium 199 [39]; and Eagle's medium supplemented with 50 per cent bovine amniotic fluid. One set of tubes was not refed, another set was refed once a week and the third set twice a week. No pH changes were recorded for those sets that were refed. The results indicated that those cultures in which 1 per cent serum was used maintained themselves in the best condition for 4 weeks whether or not fluid changes were made. Horse serum in concentrations greater than 1 per cent sometimes caused cell granularity.

Experimental Cell Research 25

A trial was then made of the various media containing no serum. Strain WI-11 (8th passage) was used after being washed once with medium 199. Those cultures refed once a week, regardless of the medium used, survived most successfully for 4 weeks with no added serum.

In order to restrict pH reduction by tube cultures being maintained on serum-containing media, cultures of WI-11 were maintained with 1, 0.5, 0.25, and 0.1 per cent calf serum, in Eagle's Minimum Essential Medium [9] containing 0.01 M final concentration of tris buffer. Although the cells metabolized well without addition of tris buffer, the pH was more easily maintained in its presence. After 4 weeks with no refeeding, excellent tube cultures were obtained with all serum concentrations in tris buffered Eagle's Minimum Essential Medium. The pH had dropped from 7.8 to 7.1.

The diploid strains are therefore unusually well-suited for virus isolations where single cultures may have to remain in a viable state and metabolically active for long periods of time without fluid changes. The fact that excellent maintenance was obtained with serumless media also recommends them as important hosts for many virus studies.

Lack of growth of human diploid cell strains in suspended cultures.—Many attempts were made to grow cells from diploid human strains in suspended cultures involving cells from various passage levels. In spite of the changes made in such parameters as inoculum size, medium components, speed of agitation, and the condition of the harvested cells for suspended culture inoculation, success was never achieved. A steady decline of the viable cells inoculated was observed in every case and within three days the entire suspended culture was dead. It appears, therefore, that human diploid cell strains share with primary cells an apparent failure to multiply from individually suspended cells in agitated culture. Under conditions which have repeatedly resulted in successful suspended cultures of HeLa and WISH [17] (both of which are heteroploid cell lines), diploid human cell strains failed to grow.

Restoration of diploid human cell strains from frozen stock.—Although present evidence indicates that human diploid cell strains cannot be carried continuously for an indefinite period of time as is usual with cell lines, it is possible to have a diploid cell strain *continuously available*. This is achieved by storing diploid cells at $-70°C$ from any or all of the excess cultures that are made when the strains are subcultured and before they enter Phase III. Thus it was possible, for example, in the case of the WI-1 strain, to have stored at $-70°C$ ampules of cells derived from almost every passage up to the 50th, at which time the strain entered Phase III and excess cells were unobtainable. In this case where the WI-1 cells were lost after the 51st sub-cultivation and 11 months in continuous cultivation, it was possible to restore from frozen stock those cells that were derived from, for example, the 9th passage of this strain. These cells had been frozen for 9 months and the culture successfully restored. The WI-1 cells restored from frozen 9th passage cells and representing Series B of this strain were cultivated for an additional 41 passages over 5 months. Excess cells derived from each of the 41 passages of Series B were also frozen at $-70°C$ and restoration achieved. The cells from Series B entered Phase III at the 41st passage after restoration of the 9th passage cells, or a total period of 50 passages. As indicated in Text-Fig. 2, restoration of other series of WI-1 cells from frozen stock yielded cultures that, regardless of the passage frozen, entered Phase III at total cumulative passages of 42 or better.

It is apparent, therefore, that by freezing cells at each subcultivation or every few subcultivations one could have cells available at any given time and in almost limitless numbers.[1] If it is assumed that the cell strains have a "passage potential" of 50 passages, as in the WI-1 strain, almost unlimited numbers of cells can be obtained by restoring cells frozen from surplus cultures at each passage and committing the excess cultures of this second series of cells to the freezer. This pattern can presumably be repeated until the theoretical progeny limit of 10^{22} cells has been achieved (Text-Fig. 2). Diploid cell strains restored from the frozen state still retain all those characteristics investigated for cell strains that have never been stored at $-70°C$, including their diploid karyotype.

Susceptibility of diploid human cell strains to viruses.—The WI-1 and WI-10 strains of diploid human fibroblasts were used, for the most part, in deter-mining virus susceptibility. The virus titrations were made from the 17th to the 35th subcultivation of these cells. The qualitative results are tabulated in Table III. Tubes were prepared each with 5×10^4 cells per ml in GM. The tubes were refed 24 hr after planting and titrations were performed using three tubes for each tenfold dilution of virus. The final titrations yielded values consistently less than those obtained with the system in which the virus pool had originally been grown. WI-1 cells reconstituted from the frozen state did not vary in virus susceptibility. The end points of titrations made in WI-1 cells were read on the 6th day post inoculation. Subsequent production of pools of Polio Strain Koprowski-Chat in strain WI-1 yielded

---

[1] Assuming a 2:1 split ratio, the theoretical maximum cell yield for 50 generations is 10^{22} cells where 1×10^7 cells can be obtained from one Blake bottle. This total potential yield is equal to 2×10^7 metric tons of cells based on 5×10^8 cells equivalent to 1 g wet weight.

TABLE III. *Virus susceptibility of human diploid cell strains.*

Cell strain	Virus	Cytopathogenic effect
	Measles	+
	COE	÷
	Adenovirus Type 2	+
	Adenovirus Type 12	+
	Coxsackie A 9[a]	+
	Coxsackie A 13	+
	ECHO 9	+
	Reovirus Type 1[a]	+
	Poliovirus Koprowski–Chat Type 1 (attenuated)	+
	Poliovirus Strain TN Type 2 (attenuated)	+
WI-1	Poliovirus Koprowski–Fox Type 3 (attenuated)	+
(Passages	Poliovirus Type 1 (Mahoney)	+
20–35)	ECHO 21	+
	Varicella[b]	+
	Rabies Strain CVS	+ (see text)
	Mengo	+
	Endomyocarditis virus (EMC)	+
	Herpes simplex	+
	Vaccinia	+
	Yellow Fever, Strain 17D	+
	Influenza Type A, Strain jap	+
	Influenza Type B, Strain lee	+
	Influenza Type C, Strain colindale	−
	Coxsackie B 1	−
	Coxsackie A 1	−
	Western Equine Encephalitis	−
	Polyoma	−−
	ECHO 11[a]	+
	ECHO 20[a]	+
	ECHO 22[a]	+
	ECHO 28[a]	+
	Respiratory Syncytial (Strain Long)[a]	+
	HE virus (Pett. strain)[a]	+
WI-10	Salisbury strain H.G.P.[a]	+
(Passages	Salisbury strain F.E.B.[a]	+
17–25)	Parainfluenza 1[a, c]	−
	Parainfluenza 2[a, c]	−
	Parainfluenza 3[a, c]	+

[a] Tested by Drs. Vincent V. Hamparian, Albert Ketler, and Maurice R. Hilleman of the Division of Virus and Tissue Culture Research, Merck Institute for Therapeutic Research, West Point, Pa.   [b] Tested by Dr. Eugene Rosanoff, Wyeth Laboratories, Radnor, Pa.
[c] Detected by hemadsorption.

titers of $10^{-7.2}$, indicating that with the passage of other viruses in the diploid strains, higher titers may be achieved. WI-1 cell degeneration in the presence of Polio Strain Koprowski-Chat is illustrated in Fig. 18.

The growth of the CVS 24 strain of rabies fixed virus in the WI-1 strain was determined by mouse inoculations, since a variable cytopathogenic effect was observed. A complete cytopathogenic effect was observed with most of the viruses listed in Table III. A culture of WI-1 cells inoculated with this strain of rabies virus continued to replicate virus for periods up to one month after periodic complete medium changes as measured *in vivo*. The cultures continued to metabolize during this time. When the medium became acid every 4–5 days, the sheet was washed with PBS and refed. Intracerebral inoculations of 3 to 4-week-old Swiss mice with aliquots of spent medium taken 1 and 4 weeks post inoculation of the WI-1 culture and fluorescent antibody staining of cell sheets at $2\frac{1}{2}$ weeks showed the presence of rabies virus [25, 26]. Recent reports [10, 27] indicate similar growth of rabies virus in primary hamster kidney cells with no concomitant cytopathology. The experiments with this virus in WI-1 cells indicate that rabies virus can now be grown in nonneural human cells *in vitro*.

Plaque formation was also readily obtained with WI-1 cells inoculated with Koprowski-Chat poliovirus Type 1 as indicated in Fig. 19 and with Poliovirus strain Mahoney as indicated in Fig. 20. Characteristically smaller plaques were obtained with Chat than with Mahoney.

It is also of interest that Coxsackie A9 can be grown in passaged WI-1 cells since it has been reported that this virus can only be grown on primary primate cells [32]. The Salisbury strains [60], which are closely identified with the common cold, were also observed to give an unmistakable cytopathogenic effect in high passaged human kidney and lung strains. Varicella reacted similarly in high passaged human lung strains.

Implantation of diploid human cell strains into hamster cheek pouches.— Since it is known that heteroploid cell lines of malignant origin will form tumors which develop progressively when implanted into the hamster cheek

---

Fig. 20.—Strain WI-1, 30th passage. Plaque formation in the presence of Poliovirus Type 1, strain Mahoney. 50 mm Petri dish.

Fig. 21.—Strain WI-1, 35th passage. Multilayer growth after prolonged incubation. Each cell plane is oriented in a different direction. May-Grünwald Giemsa stain. × 140.

Fig. 22.—Strain WI-1, 35th passage. Incubation for one month in a 50 mm Petri dish. Note membrane is beginning to curl away from the edges of the dish. May-Grünwald Giemsa stain.

Fig. 23.—Strain WI-1, 39th passage. Membrane produced in a Blake bottle and curled up in the medium. × 12.

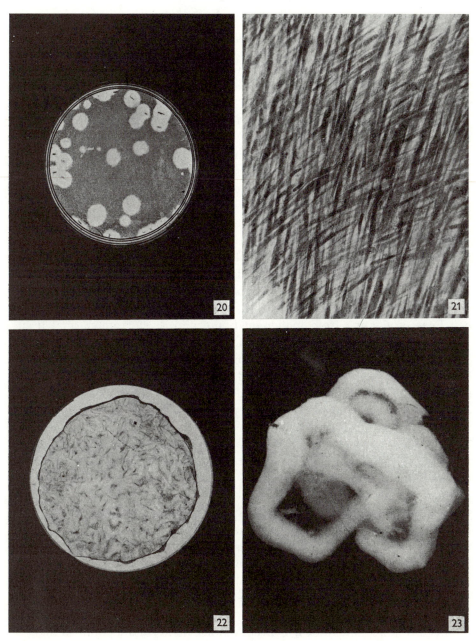

pouch [12, 13, 15], the degree of heterotransplantability of human diploid cell strains was investigated. After 7 days, three of the five hamsters receiving 10^6 WI-25 cells in the 14th passage showed small nodules. Animals receiving less than this amount never revealed nodules. The 5 hamsters receiving 10^5 HeLa cells all showed nodule formation. After 14 days the nodules in three of the five animals receiving 10^6 WI-25 cells had regressed considerably, while those receiving HeLa cells had enlarged. After 21 days only one very small nodule remained in one hamster receiving the WI-25 cells. This nodule was biopsied and regarded as a hemorrhagic area with an inflammatory response. The nodules of two hamsters receiving HeLa cells were also biopsied at this time and it was determined that there was progressive cellular growth with vascularization.

As pointed out by Foley and Handler [13], cell lines derived from neoplastic tissue are frankly invasive and exhibit a greater growth potential than those derived from non-malignant tissue when titrated in unconditioned hamsters. Accordingly all cell lines of malignant origin examined by these investigators formed tumors in cell concentrations of 10^6 per hamster. Lower cell concentrations and the fate of the nodule served to distinguish between cell lines originating from malignant tissue and those from normal tissue. It is therefore of interest that the diploid human cell strain WI-25, which was inoculated in concentrations as high as 10^6 cells per animal, caused no tumor formation, and it is concluded that even after 14 passages *in vitro* strain WI-25 had not acquired the degree of malignancy associated with those cell lines investigated by Foley and Handler.

Homotransplantation of WI-1 cells into terminal cancer patients.—The WI-1 strain was selected for studies paralleling those performed by Southam, Moore and Rhoads [56]. These workers showed that when cells from heteroploid human cell lines (derived from both normal and malignant tissue) were inoculated into terminal cancer patients, they multiplied in most of the recipients, as indicated by the formation of a palpable nodule at the site of implantation; and that upon biopsy healthy cancer cells with active mitosis were found. The only "normal" cells utilized by these investigators were human embryonic fibroblasts with "normal cytology". In this case examination of recipient cancer patients revealed that no growth took place at the site of inoculation but that cell line cells inoculated simultaneously into the same patients did grow.

It is presumed that the five preparations of normal human embryonic fibroblasts used by these workers were primary cells and therefore diploid [36, 56]. Our objective was to determine whether WI-1 cells that had been subcultivated up to 45 times *in vitro*, and that appeared to be normal by all other criteria, would react in the same way as the primary fetal fibroblasts of Southam *et al.* [56] and unlike their heteroploid cell lines.

The results of this study are shown in Table IV. As pointed out by Southam *et al.* [56]: "Slight local induration and erythema frequently followed inoculations but subsided completely by the third day." Although our results indicated that nodules could be found in some cases up to the seventh day, biopsies of these nodules did not reveal anything suggestive of the results

TABLE IV. *Results of homotransplantation of WI-1 cells into terminal cancer patients.*

Patient	Diagnosis	Therapy	Pool inoculated	Results	Biopsy results
A.L.	Diffuse abdominal carcinomatosis	Symptomatic	1	Nodule developed and disappeared on 6th day.	Not done.
M.G.	Metastatic breast carcinoma	Prednisone for 8 weeks prior to and after homotransplantation.	1	5 mm nodule formed on 6th day. Disappeared by 10th day. Patient died on 14th day.	Not done.
W.H.	Metastatic bronchogenic carcinoma	Pronounced leukopenia from chemotherapy (nitrogen mustard)	2	3 mm nodule formed on 6th day. Biopsied.	Negative
R.M.	Metastatic carcinoma of the colon	5-fluorouracil given 1 week prior to homotransplatation (no leukopenia)	2	No nodule developed up to 9 days.	Not done.
S.J.	Metastatic breast carcinoma	Pancytopenia following Cytoxan given 2 days before homotransplantation.	2	Minimum induration after 7 days. Biopsied. Patient died on 8th day.	Negative
P.M.	Metastatic breast carcinoma	Mild leukopenia following Cytoxan administration 9 days prior to homotransplantation.	2	No nodule developed. Biopsied on 7th day.	Negative

obtained by Southam *et al.* [56] when they inoculated heteroploid cell lines. None of the biopsies revealed anything of an abnormal character and the presence or growth of the inoculated WI-1 cells could not be detected. In two of our patients (A.L. and M.G.) where nodules developed, subsequent total regression was observed. In two other patients (W.H. and S.J.) where nodules developed and were biopsied, nothing abnormal was found. The two remaining patients (R.M. and P.M.) developed no nodules.

These results indicate that the WI-1 strain of embryonic, diploid, human fibroblasts subcultivated up to 45 passages does not produce the kind of growth obtained when heteroploid cell lines are inoculated into human cancer patients. The activity of the WI-1 cells closely parallels the results obtained when inoculating primary human fetal cells under similar conditions. It is therefore concluded that although the WI-1 strain was propagated for 45 passages over a period of 10 months, with intervening storage at $-70\,^{\circ}$C, the malignant characteristics exhibited by heteroploid cell lines were not acquired as demonstrated by inoculation into terminal cancer patients.

Histotypic growth of human diploid cell strains.—When a confluent culture of any diploid cell strain is kept at 36°C for approximately one month, with 1 or 2 pH adjustments, it is found that in almost all cases the cell sheet formed is not a monolayer of fibroblasts, but is instead a multilayer cell sheet, or membrane. The membrane formed is such that each layer of fibroblasts is oriented in a different direction as illustrated in Fig. 21 for the WI-1 strain. After about one month the entire membrane will start to peel from the glass surface at the edges (Fig. 22) and can often be found floating free in the medium. Agitation of the culture vessel will cause the membrane to roll up as illustrated in Fig. 23. Slight agitation of the culture vessel may help to lift the membrane from the glass surface and such membranes can be spread out on a glass slide, stained, and examined microscopically (Fig. 24). The rolled up membranes can also be fixed, imbedded, sectioned, and stained using routine histological methods.

Hemotoxylin and eosin stains of a representative number of such membranes were performed, some of which are illustrated in Figs. 25 and 26. They illustrate the appearance of a strain WI-1 membrane from the 36th

---

Fig. 24.—Strain WI-1, 36th passage. Membrane spread out on the surface of a glass slide, fixed and stained with May-Grünwald Giemsa. $\times 1$.

Fig. 25.—Strain WI-1, 36th passage. Microscopic appearance of a membrane. Hematoxylin–eosin. Note multilayer cellular organization and intercellular matrix. $\times 90$.

Fig. 26.—Same as Fig. 25. Higher power magnification. $\times 410$.

Experimental Cell Research 25

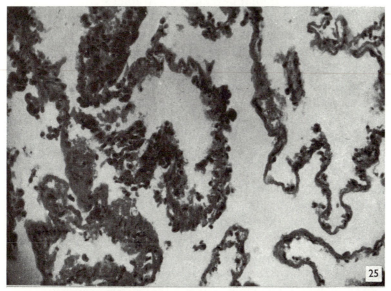

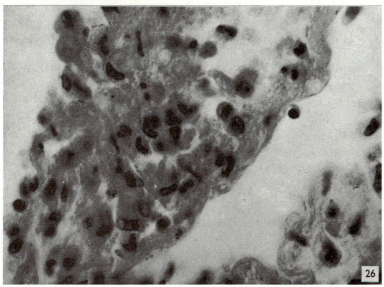

subcultivation representing a period of time in cultivation of 7 months. Many of the membranes that were detached from glass were fixed in Teilyesniezky's fixative, imbedded in paraffin and sectioned as routine histological material. Consecutive slides were stained with hematoxylin–eosin, Weigert–Van Gieson for the identification of elastic and connective tissue, and Rocque's chromotrope and toluidine blue with alcohol differentiation. At microscopic examination the membranes appeared to be constituted of an inner portion of amorphous eosinophilic material, lined on both sides by one or two cell layers. The cells were essentially round and occasionally a cell or group of cells of cylindrical shape had one border attached to the membrane. The amorphous material stained a pale red with the Van Gieson, blue with the chromotrope, and did not show any metachromasia with the toluidine blue. No fibrillar structure was observed with light microscopy. Several groups of pyknotic nuclei at different stages of degeneration were observed in some fields. Similar membrane formation with the same histological characteristics were observed with WI-8, WI-9 and WI-23 strains. The amorphous matrix does not contain acid mucopolysaccharide and the staining reactions seem to suggest the presence of collagen. The ultrastructure of this matrix is now being studied with the electron microscope.

The results are compatible with an interpretation of primitive histotypical differentiation characterized by the ability of the cells to lay down an intercellular matrix and to organize into a multilayered membrane, suggesting a retention of some functional capacities.

DISCUSSION

The experiments herein described illustrate the simplicity with which strains of human fibroblasts can be kept in serial cultivation for long periods of time with maintenance of the integrity of the diploid karyotype. It is apparent that the exacting conditions described by others [41, 47] are not critical in order to achieve this result.

Chromosome cytology and sex chromatin.—The observed conformity between sex chromatin characterization and chromosomal sex exhibited through at least 44 passages fulfills expectations for primary cells in culture [34, 35]. Others report reduction or loss of sex chromatin with continued subcultivations [43] and of variations in number per nucleus in cell lines [6].

The absence of evidence of heteroploid alteration in the 13 strains studied implies a "stability" of the diploid cell in culture in spite of repeated trypsinization which is known to increase division anomalies in mouse cells [31].

Experimental Cell Research 25

However, since alterations to heteroploidy occur haphazardly and under influences poorly understood, chromosome monitoring is necessary to establish confidence in any particular strain. While chromosome counts alone may rule out heteroploidy and aneuploidy, monitoring must be extended to include karyotype analysis for those changes which are less obvious, such as quasi-diploidy [14].

Although certain human chromosomes are specifically identifiable, some rearrangements between and within those chromosomes recognizable only as members of a group would probably pass unnoticed. Such "cryptostructural" changes in relation to cancer hypotheses have been stressed by Hauschka and Levan [16]. An appreciation of the limits of detection of visible chromosome aberrations is afforded by the recent finding made by Nowell and Hungerford [42], and independently by others [2]. These workers found a chromosomal change in blood cells consistently associated with chronic granulocytic leukemia which involved a reduction in length of one of the smallest chromosomes comprising less than 0.8 per cent of the total haploid chromosome length. Aside from "cryptostructural" limitations, it should be stated that however large the sample of the dividing population, absolute certainty as to the exclusion of chromosomally altered cells is impossible. However, considerations concerning "overgrowth" or replacement of the original cell population by the altered cell and sampling at various subculture levels should ensure detection of heteroploid alterations.

Obtaining nearly limitless numbers of cells of a uniform genetic constitution has marked practical advantages over using the pleiomorphic population comprising the cell line. This is analogous to the use of inbred as compared with randomly bred animals for biological study. *In vitro* studies of the diploid somatic cell as well as specific types of trisomics, monosomics, etc. should prove valuable. A cell strain has been obtained from a mongol (now in the 25th passage) which remains trisomic for chromosome #21 as is characteristic of the tissue of origin [33].

The Phase III phenomenon.—The characteristic degeneration typical of Phase III in the history of diploid cell strains, wherein the strain loses its growth potential and is itself gradually lost, has important implications. Of three possible explanations of this phenomenon that are considered, only one appears to be acceptable.

The fact that Phase III occurs in one strain of cells while other strains from the same or from different fetuses are luxuriating in Phase II in the same medium pool apparently excludes lack of metabolites or presence of toxic materials in the medium as an explanation for this phenomenon.

Furthermore, if cells from Phase II are restored from the frozen state in the same pool of medium that is being used on the continuous passage cells of that strain in Phase III, luxurious growth invariably results. It appears, therefore, that the growth medium used for these experiments is entirely adequate; and inasmuch as about 15 separate pools of calf serum have been employed throughout this work the impression is that no "toxic" sera were encountered. No attempt was made to pretest the medium. Swim and Parker [57] arrived at a similar conclusion, with cell strains assumed by us to be diploid.

A second explanation for the eventual loss of the diploid strains, or their entry into Phase III, is the possibility that some pool of essential metabolites originally present in the primary cell population and not synthesized during *in vitro* cultivation is gradually depleted. In the case of the WI-1 strain, such a possibility would require a hypothetical pool originally present in the cell population large enough to impart at least one molecule to those cells of the 50th passage. This possibility can be rejected on purely mathematical grounds.

The 50th passage represents more than 2^{50} generations or 10^{15} progeny. Each cell isolated from the fetal lung tissue must then contain a pool of the hypothetical metabolite in a concentration of 10^{15} molecules, assuming equal distribution to daughter cells. A pool of 10^{15} hydrogen molecules would weight 3.3×10^{-9} g. Since one WI-1 cell weighs approximately 2×10^{-9} g (based on 5×10^8 cells equivalent to 1 g wet weight) this hypothesis is invalid.

The third possible explanation for entry into Phase III may bear directly upon problems of ageing, or more precisely, "senescence". This concept, although vague at the level of the whole organism, may have some validity in explaining the phenomenon at the cellular level, at least as an operational concept.

This may be explained by postulating a factor, necessary for cell survival, whose rate of duplication is less than that of the cell (asynchronous). It is possible to conceive of two separate self-duplicating systems one of which (the hypothetical) is contained within the other (the cell). A slight reduction in replication rate of the hypothetical system could possibly lead to a gradual depletion of this factor to a critical or threshold level within 40–50 generations. Conversely the rate of synthesis itself may be unchanged, but a slightly higher rate of loss (through some unknown *in vitro* condition) would eventually yield the same result. This may be interpreted as a cumulative effect.

A parallel situation exists in biology in a number of protozoa and has been well investigated in certain species of Paramecium which contain the Kappa factor [55]. This factor is a self-duplicating system within the Paramecium and multiplies at a rate that can be influenced by the environment independently of the rate of multiplication of the Paramecium. Thus such Kappa containing Paramecium or killer organisms can be depleted of their Kappa by manipulating environmental conditions in such a way that the Paramecium can "outgrow" the Kappa factor and become a non-killer. As pointed out by Sonneborn [55]: "The maintenance of Kappa appears to be precarious. It does not necessarily divide synchronously with the Paramecia. It is distributed, not precisely, but randomly to the products of fission. It often cannot keep pace with the reproduction of the Paramecia; this can result in total loss which is irreversible."

This concept appears to explain best the fact that in our experiments and those of others [41, 47, 48, 58, 59] no one has succeeded in carrying diploid human cells for periods much over one year. It is not implied that any exact number of passages are required or that a definite period of time is necessary to predispose the cells to Phase III, but simply that there does exist a finite limit to the cultivation period of diploid cell strains (Text-Fig. 1). The fact that heteroploid cell lines can apparently be carried for indefinite periods of time may be compared directly with transplantable tumors. Both can be cultivated indefinitely and both are heteroploid. It is interesting to speculate on the possibility of constructing an analogous situation with the diploid cell strains. Could mammalian diploid strains having an appropriate marker (e.g. sex chromatin) be carried indefinitely *in vivo* in isologous hosts which would indicate an escape by the marked cells from the phenomenon of senescence?

Although no abnormalities were noted in the dividing population just prior to Phase III, the bizarre interphase nuclei seen may represent mitotically incompetent cells. Sax and Passano [54] have shown that "culture age is a major factor in the incidence of spontaneous chromatid aberrations" seen at anaphase.

Histotypical differentiation.—Diploid human strains cultivated under the conditions described form membranes consisting of a multilayer of cells with the concomitant formation of an interstitial matrix. This membrane formation has been observed, when last attempted, as late as the 36th subcultivation (7 months *in vitro*) with the WI-1 strain. It has long been felt by organ culturists that in order to retain tissue organization *in vitro* conditions must be provided other than those required for rapid multiplication. By adjusting the pH and thus depriving the cells of any new source of nutrients (glucose?) which are presumed to be exhausted, membrane formation has been achieved in these diploid strains.

The fact that only a very primitive type of membrane develops may be related to the observation that all of the human strains studied in this way were composed exclusively of fibroblasts (beyond the 5th passage) without detection of any epithelial elements. The use of such membranes for wound repair is under investigation. It may be surmised that the type of histotypical development observed is restricted to that type of organization possible with stromal elements only. It may be postulated, therefore, that if methods could be developed to retain all types of cells present in the primary tissue, a greater degree of differentiation could be obtained. Inclusion of all cellular elements, which is apparently the case when reaggregation of primary cells occurs *in vitro*, is demonstrable when no serial cultivations intervene. This results in the histotypical differentiation reported by Moscona [40] and the more spectacular morphological organization reported by Weiss [61] when dissociated chick embryo cells are grown on the chorioallantoic membrane of 8-day chick embryos. It is important, therefore, that human diploid fibroblasts can, after repeated subcultivations, exposures to trypsin, and passages of considerable lengths of time, form membranes with some degree of histotypical differentiation. Such differentiation, primitive as it is, serves to underline the concepts of Weiss [61] which stress the role of internal "self organization" as opposed to the more classic views, which emphasize the role of "inducers" in supplying cellular "information" from without.

Virus sensitivity of diploid human cell strains.—The cytopathogenic effect observed when a broad spectrum of viruses was titrated on the WI-1 and WI-10 strains indicates the high degree of susceptibility of this type of cell to virus infection. The variations in diploid human cell strain susceptibility to the Koprowski-Chat Type 1 strain of poliovirus had more of a quantitative than a qualitative difference (Table V). It is our impression that the consistently lower titers of all viruses titrated in the WI-1 strain as opposed to titers obtained in the optimum cell system may be a result of titration in the WI-1 strain directly from virus pools grown in the optimum system. Where the Koprowski-Chat Type 1 poliovirus was first grown in the WI-1 cells and then titrated in these cells, the titer obtained was *not* appreciably lower than titrations made in primary monkey kidney (Table V).

This raises the question of the use of diploid human cell strains for the production of killed or attenuated human virus vaccines; and in particular poliovirus vaccines. The objections raised against using heteroploid cell lines in the production of human virus vaccines have been pointed out by Westwood [62]: "It is the fear of malignancy more than any other single factor which precludes the cell lines at present available from use in the production of virus vaccines." In view of the filtration procedures used in making oral polio vaccines the question of feeding live cells can be discounted. As pointed out further by Westwood [62]: "The risk lies in the possibility of inducing malignant changes in the cells of the human subject by the introduction of an, as yet hypothetical, virus or non-living transforming principle analogous to that inducing change of type in the pneumococcus."

TABLE V. *Comparative titrations of poliovirus Koprowski-Chat Type 1 in diploid human cell strains.*

Cells	No. of passages	TCID$_{50}$/ml.
Monkey kidney	Primary	$10^{-7.5}$
HeLa	Cell line	$10^{-6.5}$
WI-1[a]	35	$10^{-7.2}$
WI-13	26	$10^{-6.5}$
WI-5	27	$10^{-6.5}$
WI-9	24	$10^{-5.7}$
WI-15	23	$10^{-5.7}$
WI-11	23	$10^{-5.5}$
WI-14	30	$10^{-3.0}$
WI-12	32	$10^{-3.0}$

[a] Virus pool inoculated, previously grown in WI-1. All other cells titrated with a virus pool grown in primary monkey kidney.

In view of these objections human poliovirus vaccines (attenuated and killed) are now acceptable in this country only when grown in primary monkey kidney, as such tissue is presumed to have no malignant properties.[1]

Serious objections, however, may be raised even to the use of this tissue. It is well known that primary monkey kidney has a very high content of latent simian viruses. Indeed, at least 18 such viruses have now been isolated from primary monkey kidney [24]. One of these latent viruses (B virus) is even known to have caused fatalities in man [45]. It is now becoming apparent that all primary monkey kidney may contain one or more latent viruses whose characteristic cytopathology becomes evident when such tissue is cultured *in vitro*. The appearance of such viruses and the associated cytopathology is probably a reasonable explanation of the fact that such

[1] *Note added in proof.*—A recent report (Eddy, B. E. *et al.*, *Proc. Soc. Exptl. Biol. Med.* **107**, 191 (1961)) indicates that extracts of pools of monkey kidney cells are capable of inducing neoplasms in hamsters. Krooth, R. S. and Tjio, J. H. (*Virology* **14**, 289 (1961)) have independently reported on the growth of poliovirus strain mahoney in human diploid cell strains.

tissue cannot be subcultured successfully beyond about the 5th passage. This excludes those rare cases [44] in which alterations to heteroploidy occurred. This degenerative phenomenon has also been observed when certain members of the adenovirus group are unmasked in tonsil and adenoid tissue cultivated *in vitro* [52, 53].

The isolation and characterization of human diploid cell strains from fetal tissue make this type of cell available as a substrate for the production of live virus vaccines. Other than the economical advantages, such strains, in contrast to the heteroploid cell lines, exhibit those characteristics usually reserved for "normal" or "primary" cells (Table IV) and therefore make the consideration of their use in the production of human virus vaccines a distinct possibility.

TABLE VI. *Differential characteristics for human cell lines and cell strains.*

Character	Cell lines	Cell strains
1. Chromosome number	Heteroploid	Diploid
2. Sex chromatin	Not retained or variable	Retained
3. Histotypical differentiation	Not retained	Partially retained
4. Growth in suspended culture	Generally successful	Unsuccessful
5. Pathological criteria for malignancy as determined on biopsies of cells inoculated into hamsters or human terminal cancer patients	Positive	Negative
6. Limitation of cell multiplication (life of strain or line)	Unlimited	Limited
7. Virus spectrum compared to corresponding primary tissue	Often different	Same
8. Cell morphology compared to corresponding primary tissue	Characteristically different	Same
9. Acid production	Less than that produced by equal number of cell strain cells	More than that produced by equal number of cell line cells
10. Retention of Coxsackie A 9 receptor substance	Lost	Retained
11. Ease of establishment	Difficult (not predictable)	Usually successful

The question of the presence of latent viruses in any cellular material is one that can probably never be answered with absolute certainty; yet it is possible to perform exhaustive studies with techniques now available (e.g. irradiation) to rule out effectively the presence of latent viruses in one strain of diploid cells so that attention can be concentrated on the use of such a "clean" strain for the production of live human virus vaccines.

It would not be necessary to test large numbers of such strains for latent virus content. Even though these strains do degenerate as late as the 50th passage (strain WI-1), if all the surplus cells from each subcultivation were stored in the frozen state a potential yield of 20 metric tons of cells could be obtained from any single strain if its "passage potential" was even as low as 30 subcultivations. Clearly, the potential "senescence" of any diploid strain should not detract from its usefulness, since the potential cell yield is abundant, if not inexhaustible, for all practical purposes.

SUMMARY

The isolation and characterization of 25 strains of human diploid fibroblasts derived from fetuses are described. Routine tissue culture techniques were employed. Other than maintenance of the diploid karyotype, ten other criteria serve to distinguish these strains from heteroploid cell lines. These include retention of sex chromatin, histotypical differentiation, inadaptability to suspended culture, non-malignant characteristics *in vivo*, finite limit of cultivation, similar virus spectrum to primary tissue, similar cell morphology to primary tissue, increased acid production compared to cell lines, retention of Coxsackie A9 receptor substance, and ease with which strains can be developed.

Survival of cell strains at $-70°C$ with retention of all characteristics insures an almost unlimited supply of any strain regardless of the fact that they degenerate after about 50 subcultivations and one year in culture. A consideration of the cause of the eventual degeneration of these strains leads to the hypothesis that non-cumulative external factors are excluded and that the phenomenon is attributable to intrinsic factors which are expressed as senescence at the cellular level.

With these characteristics and their extremely broad virus spectrum, the use of diploid human cell strains for human virus vaccine production is suggested. In view of these observations a number of terms used by cell culturists are redefined.

Experimental Cell Research 25

The authors are indebted to Dr. Sven Gard of the Karolinska Institutet Medical School, Stockholm, Sweden, and to Dr. Jacob Gershon-Cohen, Department of Radiology, Einstein Hospital, Northern Division, Philadelphia, Pa., for much valuable assistance. Drs. Richard Carp, Stanley Plotkin, Eberhardt Wecker, and Mrs. Barbara Cohen of the Wistar Institute, Philadelphia, Pa., participated in several of the virus studies, and Dr. Vittorio Defendi of the same Institute undertook the histological examinations. Dr. Anthony J. Girardi of the Merck Institute for Therapeutic Research, West Point, Pa., performed the studies on maintenance media and hamster inoculations. We are also indebted to Dr. Robert G. Ravdin and Dr. William Elkins of the Harrison Department of Research Surgery, Medical School, University of Pennsylvania, Philadelphia, Pa., who undertook the studies in human subjects. Acknowledgment is also made of the excellent technical assistance of Mr. Fred Jacks.

REFERENCES

1. ABERCROMBIE, M. and HEAYSMAN, J. E. M., *Exptl. Cell Research* **6**, 293 (1954).
2. BAIKIE, A. G., COURT-BROWN, W. M., BUCKTON, K. E., HARNDEN, D. G., JACOBS, P. A. and TOUGH, I. M., *Nature* **188**, 1165 (1960).
3. BARR, M. L., *Am. J. Human Genet.* **12**, 118 (1960).
4. BERMAN, L., STULBERG, C. S. and RUDDLE, F. H., *Cancer Research* **17**, 668 (1957).
5. CHU, E. H. Y. and GILES, N. H., *J. Natl. Cancer Inst.*, **20**, 383 (1958).
6. DeWITT, S. H., RABSON, A. S., LEGALLAIS, F. Y., DEL VECCHIO, P. R. and MALMGREN, R. A., *J. Natl. Cancer Inst.* **23**, 1089 (1959).
7. DULBECCO, R. and VOGT, M., *J. Exptl. Med.* **99**, 167 (1954).
8. EAGLE, H., *J. Exptl. Med.* **102**, 595 (1955).
9. EAGLE, H., *Science* **130**, 432 (1959).
10. FENJE, P., *Can. J. Microbiol.* **6**, 479 (1960).
11. FERNANDES, M. V., *Texas Repts. Biol. Med.* **16**, 48 (1958).
12. FOLEY, G. E. and HANDLER, A. H., *Proc. Soc. Exptl. Biol. Med.* **94**, 661 (1957).
13. —— *Ann. N.Y. Acad. Sci.* **76**, 506 (1958).
14. FORD, D. K. and YERGANIAN, G., *J. Natl. Cancer Inst.* **21**, 393 (1958).
15. HANDLER, A. H. and FOLEY, G. E., *Proc. Soc. Exptl. Biol. Med.* **91**, 237 (1956).
16. HAUSCHKA, T. S. and LEVAN, A., *J. Natl. Cancer Inst.* **21**, 77 (1958).
17. HAYFLICK, L., *Exptl. Cell Research* **23**, 14 (1961).
18. —— *Nature* **185**, 783 (1960).
19. HAYFLICK, L. and STINEBRING, W. R., *Ann. N.Y. Acad. Sci.* **79**, 433 (1960).
20. HSU, T. C., *Texas Repts. Biol. Med.* **18**, 31 (1960).
21. HSU, T. C. and KLATT, O., *J. Natl. Cancer Inst.* **21**, 437 (1958).
22. HSU, T. C. and MOORHEAD, P. S., *J. Natl. Cancer Inst.* **18**, 463 (1957).
23. HSU, T. C., POMERAT, C. M. and MOORHEAD, P. S., *J. Natl. Cancer Inst.* **19**, 867 (1957).
24. HULL, R. N., MINNER, J. R. and MASCOLI, C. C., *Am. J. Hyg.* **68**, 31 (1958).
25. KAPLAN, M. M., (personal communication).
26. KAPLAN, M. M., FORSEK, Z. and KOPROWSKI, H., *Bull. World Health Organization* **22**, 434 (1960).
27. KISSLING, R. E., *Proc. Soc. Exptl. Biol. Med.* **98**, 223 (1958).
28. KLINGER, H. P. and SCHWARZACHER, H. G., *J. Biophys. Biochem. Cytol.* **8**, 345 (1960).
29. LEIGHTON, J., KLINE, I., BELKIN, M., LEGALLAIS, F. and ORR, H. C., *Cancer Research* **17**, 359 (1957).
30. LEVAN, A., *Cancer* **9**, 648 (1956).
31. LEVAN, A. and BIESELE, J. J., *Ann. N.Y. Acad. Sci.* **71**, 1022 (1958).
32. McLAREN, C. L., HOLLAND, J. J. and SYVERTON, J. T., *J. Exptl. Med.* **112**, 581 (1960).
33. MELLMAN, W. J., HAYFLICK, L. and MOORHEAD, P. S., Unpublished observations.
34. MILES, C. P., *Nature* **184**, 477 (1959).
35. —— *Cancer* **12**, 299 (1959).

36. MOORE, A. E., *Ann. N.Y. Acad. Sci.* **76**, 497 (1958).
37. MOORE, A. E., SOUTHAM, C. M. and STERNBERG, S. S., *Science* **124**, 127 (1956).
38. MOORHEAD, P. S., NOWELL, P. C., MELLMAN, W. J., BATTIPS, D. M. and HUNGERFORD, D. A., *Exptl. Cell Research* **20**, 613 (1960).
39. MORGAN, J. F., MORTON, H. J. and PARKER, R. C., *Proc. Soc. Exptl. Biol. Med.* **73**, 1 (1950).
40. MOSCONA, A., *Proc. Natl. Acad. Sci., U.S.* **43**, 184 (1957).
41. MOSER, H., *Experientia*, **16**, 385 (1960).
42. NOWELL, P. C. and HUNGERFORD, D. A., *Science* **132**, 1497 (1960).
43. ORSI, E. V., WALLACE, R. E. and RITTER, H. B., *Science* **133**, 43 (1961).
44. PARKER, R. C., CASTOR, L. N. and McCULLOCH, E. A., *Spec. Publ. N.Y. Acad. Sci.* **5**, 303 (1957).
45. PIERCE, E. C., PIERCE, J. D. and HULL, R. N., *Am. J. Hyg.* **68**, 242 (1958).
46. PUCK, T. T., *in* Perspectives in Virology. John Wiley and Sons, Inc., New York, 1959.
47. PUCK, T. T., CIECIURA, S. J. and FISHER, H. W., *J. Exptl. Med.* **106**, 145 (1957)
48. PUCK, T. T., CIECIURA, S. J. and ROBINSON, A., *J. Exptl. Med.* **108**, 945 (1958).
49. ROTHFELS, K. H., AXELRAD, A. A., SIMINOVITCH, L., McCULLOCH, E. A. and PARKER, R. C., *in* Proc. 3rd Canad. Cancer Research Conf., New York, p. 189. Academic Press, Inc., 1959.
50. ROTHFELS, K. H. and PARKER, R. C., *J. Exptl. Zool.* **142**, 507 (1959).
51. ROTHFELS, K. H. and SIMINOVITCH, L., *Stain Technol.* **33**, 73 (1958).
52. ROWE, W. P., HUEBNER, R. J., GILMORE, L. K., PARROTT, R. H. and WARD, T. G., *Proc. Soc. Exptl. Biol. Med.* **84**, 570 (1953).
53. ROWE, W. P., HUEBNER, R. J., HARTLEY, J. W., WARD, T. G. and PARROTT, R. H., *Am. J. Hyg.* **61**, 197 (1955).
54. SAX, H. J. and PASSANO, K. N., *Am. Naturalist* **95**, 97 (1961).
55. SONNEBORN, T. M., *Advances in Virus Research* **6**, 229 (1959).
56. SOUTHAM, C. M., MOORE, A. E. and RHOADS, C. P., *Science* **125**, 158 (1957).
57. SWIM, H. E. and PARKER, R. F., *Am. J. Hyg.* **66**, 235 (1957).
58. SYVERTON, J. T., *Spec. Pub. N.Y. Acad. Sci.* **5**, 331 (1957).
59. TJIO, J. H. and PUCK, T. T., *J. Exptl. Med.* **108**, 259 (1958).
60. TYRRELL, D. A. J. and PARSONS, R., *Lancet* **1**, 239 (1960).
61. WEISS, P. and TAYLOR, A. C., *Proc. Natl. Acad. Sci. U.S.* **46**, 1177 (1960).
62. WESTWOOD, J. C. N., MACPHERSON, I. A. and TITMUSS, D. H. J., *Brit. J. Exptl. Pathol.* **38**, 138 (1957).

QUANTITATIVE STUDIES OF THE
GROWTH OF MOUSE EMBRYO CELLS IN
CULTURE AND THEIR DEVELOPMENT
INTO ESTABLISHED LINES

GEORGE J. TODARO and HOWARD GREEN, M.D.

From the Department of Pathology, New York University School of Medicine, New York

ABSTRACT

Disaggregated mouse embryo cells, grown in monolayers, underwent a progressive decline in growth rate upon successive transfer, the rapidity of the decline depending, among other things, on the inoculation density. Nevertheless, nearly all cultures developed into established lines within 3 months of culture. The first sign of the emergence of an established line was the ability of the cells to maintain a constant or rising potential growth rate. This occurred while the cultures were morphologically unchanged. The growth rate continued to increase until it equaled or exceeded that of the original culture. The early established cells showed an increasing metabolic autonomy, as indicated by decreasing dependence on cell-to-cell feeding. It is suggested that the process of establishment involves an alteration in cell permeability properties. Chromosome studies indicated that the cells responsible for the upturn in growth rate were diploid, but later the population shifted to the tetraploid range, often very rapidly. Still later, marker chromosomes appeared. Different lines acquired different properties, depending on the culture conditions employed; one line developed which is extremely sensitive to contact inhibition.

INTRODUCTION

When mammalian cells are placed in culture they grow rapidly, often at a rate substantially exceeding that in the intact animal. However, this growth does not continue indefinitely. Most frequently, after a variable interval, for reasons as yet unclear, the cells die (1, 2). In some cases, usually thought to be uncommon, changes in the cell population are observed to occur, culminating in the development of an established line having a variety of properties which distinguish it from the strain of origin (2). One of these properties is the ability to produce tumors when injected into suitable hosts (3–5), so that the mechanism by which the normal cell is converted into an established cell may have a bearing on the problem of carcinogenesis.

In the following experiments the growth properties of mouse embryonic fibroblasts were closely studied from the time they were placed in culture, and especially during their conversion to established lines. The changes in growth properties were related to chromosomal and morphological changes, and certain criteria were set up for the established condition.

MATERIALS AND METHODS

Media

The medium used was Dulbecco's modification of Eagle's medium (6), containing an approximately fourfold higher concentration of the amino acids

Reprinted from THE JOURNAL OF CELL BIOLOGY, 1963, Vol. 17, No. 2, pp. 299–313

and vitamins described, plus serine and glycine, and 10 per cent calf serum (Colorado Serum Co., Denver). This medium has a high bicarbonate concentration, and the pH was kept at 7.2 by equilibration with 10 per cent CO_2 in air. In the earlier experiments 10 per cent tryptose phosphate was supplemented. All cultures were maintained in 50 mm diameter plastic Petri dishes and were transferred by trypsinization in phosphate-buffered saline (PBS) (7).

Primary Cultures

Cultures of 17 to 19 day old Swiss mouse embryos were prepared by fine mincing of the whole embryos

detached, in 10 to 15 minutes, aliquots of the suspension were taken from each plate for counting in a hemocytometer chamber under phase microscopy. At least 250 cells, and in most cases 500 to 1000 cells, were counted. Appropriate aliquots were then taken for inoculation. Similar counts were made on duplicate cultures 24 hours after inoculation.

The growth in the interval between transfers was calculated by dividing the cell number at the end of the growth interval by the number at 24 hours after inoculation (N/N_0). With fast growing cultures N_0 exceeded the number of cells inoculated, while in slowly growing cultures it was slightly less than the

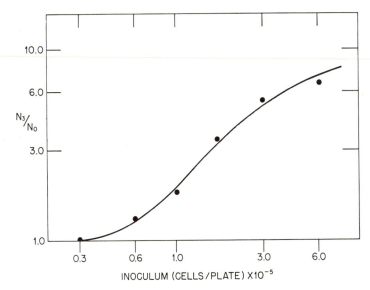

FIGURE 1

Growth of secondary cultures of embryonic fibroblasts at different inoculation densities.

and disaggregating with 0.25 per cent trypsin. The trypsin was removed by centrifugation and the cells were resuspended in medium. The cells were counted and plated at 3×10^6 cells per plate. After 2 to 3 days, confluent monolayers formed. The growth experiments to be described begin with the first transfer thereafter.

Subculture Schedules

All cultures were put on a rigid transfer schedule, being transferred either every 3 days or every 6 days, and inoculated always at the same cell density. The medium was changed on the 1st day for the 3-day transfer regime, and on the 1st, 3rd, and 5th days for the 6-day transfer regime.

At each transfer, duplicate cultures were washed once with dilute trypsin or PBS, and then trypsinized in 2 ml of 0.1 per cent trypsin. After the cells had

number of cells inoculated, in the range between 80 per cent and 90 per cent. As any growth during the first 24 hours after inoculation is neglected in all calculations of the total number of generations through which a cell population has grown, these are minimal estimates and might, in some cases, be as much as 20 per cent low.

In all experiments, growth over a 2- or 3-day interval is denoted as N_2/N_0 or N_3/N_0, respectively.

Growth of Cells under Agar

Agar casts were prepared by pouring a 4 per cent agar solution in diluted growth medium into Petri dishes and allowing it to harden. The casts were then transferred with a spatula onto monolayers on the 1st day after transfer, and liquid medium was added to the Petri dishes above the agar. Growth of the cells continued under the agar for the subsequent

5 days. The liquid medium was changed at the usual times. On the day the cells were transferred, the agar casts were removed and the cells treated as described above.

Chromosome Preparations

Actively growing cultures were arrested in metaphase for 4 to 6 hours with vincaleukoblastine (Velban) at a concentration of 0.1 $\mu g/ml$. Cells were then treated according to a modification of the procedure of Hastings et al. (8). Preparations were stained with 2 per cent aceto-orcein. The number of chromosomes was estimated in each metaphase seen, and assigned to a diploid or non-diploid category. This was done without selection of cells in order to avoid prejudice created because the metaphases with higher chromosome numbers are more difficult to count exactly. Exact counts were performed wherever possible, and Table II is based on both kinds of data.

RESULTS

The Growth Rate of Normal Mouse Embryo Cells upon Successive Transfer

Growth in culture of a variety of mammalian cell types, both normal and established, has been known to be prevented when the cell density falls below a certain critical level (9, 10). This is thought to be due to the loss of labile substrates or intermediates by leakage from the cells and is compensated at high cell densities or on irradiated feeder cells by mutual cell feeding (11, 12). This behavior is shown in the growth of disaggregated mouse embryo cells. In addition, there is a range of cell concentrations where growth occurs but the rate of growth depends upon the inoculation size. Fig. 1 shows the results of a representative experiment to demonstrate this effect on the short term growth of normal secondary cultures. Cells from a healthy subconfluent primary culture were plated at varying dilutions on replicate plates. At 24 hours duplicate samples were counted. The medium was changed on the remaining plates, and the cells were allowed to grow for a further 3 days and were then trypsinized and counted. Growth during the 3-day interval, N_3/N_0, is plotted against inoculation density in Fig. 1. The results show that below a cell concentration of 3×10^4 there is no net growth in the population.[1] At somewhat higher

densities, growth is a function of the cell density and rises to a maximum at 6×10^5 cells per plate. At concentrations higher than 6×10^5 cells per plate, growth is again reduced owing to the effect of crowding and the relatively slow rate of growth of normal cells out of the plane of the monolayer.

When cells are transferred successively in culture, the growth rate and ultimate fate of the cultures depend, among other things, on the inoculation density. To show the result of repeated transfer of normal cells at slightly suboptimal densities, cells from a confluent primary culture were plated at densities of 1×10^5, 3×10^5, and 6×10^5, grown for 3 days, counted, and transferred, always at the original cell density. Fig. 2 shows the results of such an experiment. The cells carried at 1×10^5, a markedly suboptimal density, grew through several transfers, but at a decreasing rate, until after 8 generations they had lost all ability to divide in vitro and eventually died. The cells transferred at 3×10^5 and 6×10^5 (the latter representing the density which permitted maximal growth in the short term experiments) grew considerably better and had doubling times of 60 hours and 30 hours, respectively, at a time when those maintained at 1×10^5 had ceased to divide. The growth rate of cells at higher densities, however, did decline later to a very low level. When the growth rate of these cultures during each transfer is plotted as a function of the number of generations in vitro (Fig. 3), it becomes much more obvious that not only the cells inoculated at 1×10^5 but also those at the other cell densities show a decline in their growth rates beginning virtually as soon as they are put into culture. The rate of this decline, however, is much slower for cells maintained at the higher densities. Between 10 and 20 generations after being put into culture the doubling time for all these cultures exceeded 70 hours.

At concentrations of 12×10^5 cells per plate (Fig. 4, curve C), the initial growth rate of the secondary cultures is somewhat lower than that at 6×10^5 cells per plate because of cell crowding

---

[1] A small minority of these cells can grow at densities below 3×10^4 cells per plate and would form colonies in sufficient number to permit the

calculation of a plating efficiency, which would, of course, be quite low. These cells would constitute too small a proportion of the total population to be detected as an over-all increase in cell number during a 3-day interval. The low plating efficiency reflects the difficulty the cells have in growing in the absence of cell-to-cell feeding and would therefore be a test of their ability to function as independent organisms rather than of their intrinsic ability to grow.

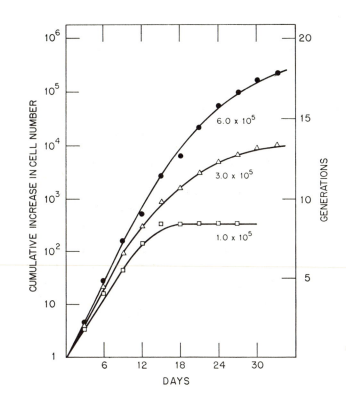

FIGURE 2

Growth of embryonic fibroblasts upon successive transfer at different inoculation densities.

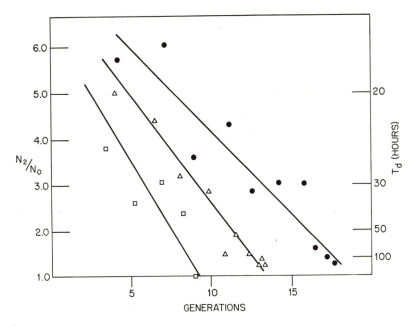

FIGURE 3

Decline in growth rate upon successive transfer at different inoculation densities. Solid circles, inoculum 6×10^5 cells; triangles, inoculum 3×10^5 cells; squares, inoculum 1×10^5 cells. T_d, doubling time.

before the end of the growth interval, but the rate of decline on successive transfer is also slower. Nevertheless the doubling time of this culture, too, rises to over 70 hours by 18 generations. Increasing the cell concentration affects the slope of the decline of the growth rate, but no cell density tested enabled these cells to maintain their growth potential during this period. Since the maximum number of cells per plate attainable (the saturation density) is 5 to 6 $\times$ 10^6 for secondary cultures and even less for later subcultures, clearly it is not possible to raise the inoculation density much higher and still permit appreciable cell division.

The Process of Establishment

The results obtained from a long term study of the cultures just described and other embryo cultures of independent origin are shown in Figs. 4 and 5. For each of these cultures, transferred at 3-day intervals, it can be seen that after a variable time, from 15 to 30 generations, or from 45 to 75 days after the beginning of culture, the growth rate began to rise again and soon reached a value similar to that at the beginning, with a T_d of from 14 to 24 hours. These cultures may now be said to be established lines; some of them have been carried in culture for over a hundred generations with either a constant or a rising growth rate and have never shown any indication of dying out. Out of a total of 11 secondary cultures, 9 have led to the production of established lines. It appears that higher inoculation size favors establishment, but it will be noted that one experiment maintaining cells at only 3 $\times$ 10^5 also led to establishment (Fig. 4,

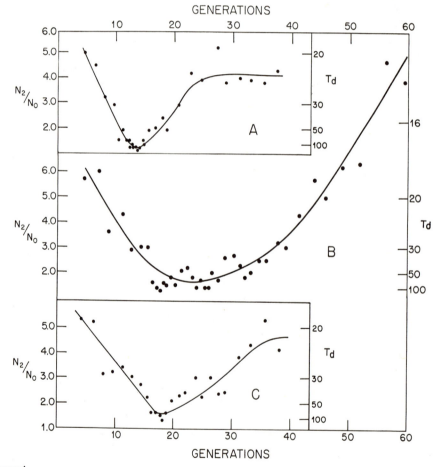

FIGURE 4

Growth rates of fibroblasts upon successive transfer. A, 3T3 (3-day transfer, inoculum 3 $\times$ 10^5 cells); B, 3T6; C, 3T12. T_d, doubling time.

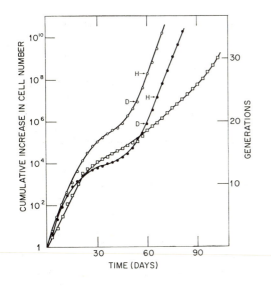

FIGURE 5

Formation of established lines from cells on 3-day transfer regime. Open circles, 3T12; solid circles, 3T3; squares, 3T12A. D, population essentially diploid; H, population essentially heteroploid.

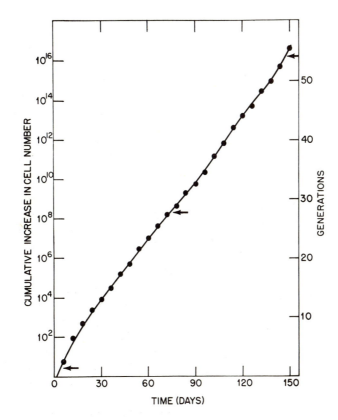

FIGURE 6

Growth of 6T6 upon successive transfer. Arrows indicate times of potential growth rate measurements (see text and Table I).

curve A). In some cases the upward turn in growth rate occurred sharply (Fig. 4, curves A and C) and in others more slowly (curve B).

Cells maintained on the 6-day transfer regime also gave rise to established cell lines. While in the case of the 3-day transfers the phases of falling and rising growth rate are obvious from the curves described, the phenomenon is not so clearly shown by the cultures maintained on 6-day transfer (Fig. 6). The growth rate of 6T6 declined only slightly following the beginning of culture, the doubling time (T_d) rising from 45 hours to 63 hours at 28 generations, and this difference might

TABLE I
Actual vs. Potential Growth Rate

Strain	No. of generations in culture	Actual		Potential*	
		N_2/N_0	T_d(hr.)	N_2/N_0	T_d(hr.)
6T6	2	2.1‡	45	5.1	20
	28	1.7‡	63	1.9	52
	54	2.1‡	45	3.8	25
	116	2.4‡	38	6.3	18
3T12A	2	4.0	24	5.1	19
	23	1.4	99	1.3	128
	35	1.7	63	2.0	48
	51	2.0	48	2.8	33
	70	3.6	26	4.6	22

* Inoculation density, 6×10^5 cells.
‡ Calculated from the total growth over the 5-day interval.

be regarded as of doubtful significance. However, during most of the time these cells were studied, their growth rate was reduced by the cell crowding which occurred during the last half of each transfer interval. At a cell inoculation density of 6×10^5 per plate, the maximum rate of cell growth occurs during the 2nd and 3rd days after inoculation. The growth rate during this interval represents the potential growth rate of which the cells are capable under optimal conditions, and is appreciably higher than the average rate of growth over longer intervals or at higher inoculation densities.

If one examines the potential growth rate of cells taken from the 6T6 line at different times, it is clear (Table I) that while the actual growth rate declined only slightly between the 2nd and 28th generation, the potential growth rate declined markedly, with the T_d rising from 20 hours to 52 hours at 28 generations. At the latter time, the actual and potential rates virtually coincided, so that the cells were growing at close to their maximal rate through the entire 5-day growth interval. By 54 generations (144 days) the potential growth rate was found to have increased considerably ($T_d = 25$ hours) while the actual growth rate rose relatively little ($T_d = 45$ hours). It is therefore clear that while the cells on the slower transfer regime might appear to be maintaining an almost constant growth rate, in fact they do lose growth potential, but this is concealed by the transfer regime, which does not allow its expression. The subsequent rise in potential growth rate indicates establishment of the line. The potential and actual growth rates of 3T12A (Fig. 5), a line maintained on a 3-day transfer regime, are also shown in Table I. The pattern of development is the same for the two transfer schedules.

Growth Properties of the Established Lines

ABILITY TO GROW AT A LOW INOCULATION DENSITY

The established cells have a much greater ability to grow at low cell density. Reduced dependence on cell feeding is characteristic of all cell lines tested and serves to separate them as a class from normal cells. Fig. 7 shows a representative example of 3T12A, tested at different times: as a secondary culture, during the phase of slowest potential growth, and then again after establishment at 67 and 107 generations. The cells at 24 generations grew more slowly at all densities than did cells at 2 generations and were similarly unable to grow substantially at inoculation densities of 5×10^4 cells per plate or less. These were as yet unestablished diploid cells. The established cells that subsequently emerged showed a progressively improving capacity to function as independent organisms by acquiring ever greater ability to grow in the absence of cell-to-cell feeding. The 3T12A cells, in their entire *in vitro* life, had never been at a density less than 10^6 cells per plate and were therefore not selected for growth at low density; yet, when tested after 67 generations *in vitro*, they could grow at almost 50 per cent of the maximal rate at 3×10^4 cells per plate, and after 107 generations they could grow maximally at this density.

Maximum Cell Number in Crowded Culture

Secondary mouse cultures allowed to get as crowded as possible reached densities of 5 to 6×10^6 cells per plate, but on subsequent transfers this saturation density would fall to 2 to 3×10^6, only to rise again after establishment. The established cells showed a progressive increase in saturation density; some established lines reached densities of 10^7 cells per plate in cultures maintained for 10 to 14 days with frequent medium changes. The increased saturation density of the cell lines reflects both the increased growth rate and the ability of the cells to grow over one another and form multilayers, the latter suggesting changes in cell surface properties that are known to be associated with malignant properties *in vivo* (13).

A striking exception to this pattern of growth is shown by line 3T3, established by other criteria, but virtually unable to grow once a confluent state is reached. Fig. 8 shows the results of an experiment in which 3T3 and 3T12 cells were plated on replicate plates at 3×10^5 and 12×10^5 cells per

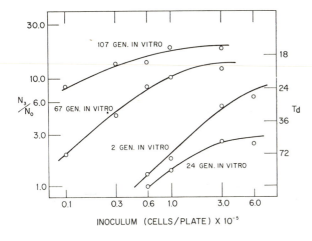

FIGURE 7

Relation between inoculation size and growth rate of mouse fibroblasts at different times in the course of formation of the established cell line 3T12A. T_d, doubling time.

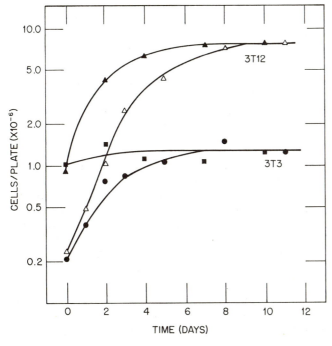

FIGURE 8

Growth of two established lines, 3T12 and 3T3. Solid triangles and squares, inoculum 1.2×10^6 cells; open triangles and solid circles, inoculum 0.3×10^6 cells. The saturation density is independent of the inoculum size for both lines.

plate and the medium changed the following day, and every other day thereafter. Duplicate plates of each line were counted for each time point. 3T3 cells and 3T12 cells when plated at 3×10^5 grew equally well at first, with a doubling time of about 24 hours. When 3T3 reached a density of close to 10^6 cells per plate all growth ceased, while 3T12 continued to grow to a saturation density of about 7×10^6 cells, over 6 times higher than that of 3T3. With larger inocula

trypsinized suspensions was 16.8 μ for 3T3, and 17.9 μ for 3T12. It appears, therefore, that the growth properties (and morphology, see below) of the emerging established line may be very different according to the culture conditions employed during and after the process of establishment.

GROWTH OF CELLS UNDER AGAR

The rate of normal embryo cell growth under agar is reduced, since the application of the agar in-

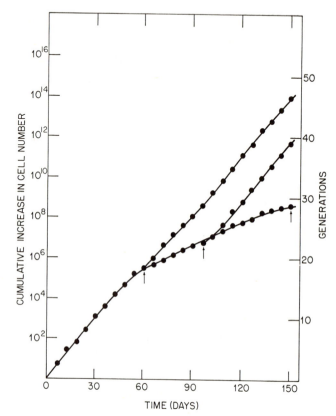

FIGURE 9

Growth of 6T6 under agar (6-day transfer, inoculum 6×10^5 cells per plate). Time of release of cells from agar inhibition indicated by arrows.

$(12 \times 10^5$ cells) the final saturation densities were not changed. In the case of 3T3, over 10^6 cells attached but no growth occurred, as the saturation density was already attained. These cells could be kept with frequent medium change for a month with no increase in cell number; yet they appeared fully viable on replating at low densities. 3T12 inoculated at 12×10^5 cells began to grow at maximal rate, and reached the same density as previously. These differences in the behavior of the two cell lines are not to be accounted for by differences in cell volume. Mean cell diameter of

hibits growth out of the plane of the monolayer (14). Fig. 9 shows the growth curve for cells maintained on a 6-day transfer regime under agar. Their growth rate was reduced throughout the entire period, and especially after 60 days under agar. The application of agar was terminated in some cultures at 60 days, in others at 96 days, and in still others as late as 150 days, by which time cell growth was practically arrested ($N_2/N_0 = 1.3$). In spite of the repression of growth by the agar, the first two of these lines had become established by 150 days of culture, as indicated by study of their

G. J. TODARO AND H. GREEN *Mouse Embryo Cells in Culture* 307 potential growth rate. The cells maintained under agar for the entire interval were not established at this time, but subsequently they too became established.

Morphology and Growth Pattern of the Established Lines

After the upturn in growth rate the cells in most experiments could be seen to become slightly more

TABLE II

Chromosome Alterations during the Process of Establishment

| Cell strain | Transfer | Gene-ration | Diploid cells | | N_2/N_0 |
			Counted	Percentage	
3T3	2	2	93/100	(93)	5.0
	5	10	23/25	(92)	2.9
	10	12	—	—	1.2
	15	14	—	—	1.2
	18	16	—	—	1.9
	20	18	79/100	(79)	2.4
	22	20	3/72	(4)	2.9
	25	24	0/50	(0)	3.8
3T6	5	11	77/81	(95)	4.3
	16	21	46/50	(92)	2.1
	21	25	—	—	1.4
	24	28	50/56	(89)	1.7
	27	31	46/52	(87)	2.3
	30	35	24/50	(48)	2.5
	32	39	4/74	(6)	4.0
3T12	2	2	93/100	(93)	5.3
	5	10	45/50	(90)	3.2
	7	13	67/82	(82)	3.0
	12	18	—	—	1.5
	14	19	—	—	1.6
	16	21	41/50	(82)	2.3
	18	23	89/100	(89)	3.0
	20	26	11/50	(22)	3.0
	22	29	0/67	(0)	3.8

refractile and less firmly attached to the Petri dish substrate. They tended to appear less fusiform in shape, and in crowded cultures there was less parallel orientation of the cells and more cell interlacing and overgrowth, suggestive of the loss to some degree of contact inhibition (20). Nevertheless layered membrane formation did occur in cultures left without frequent transfer. As the growth rate increased, the cell size, as seen in the counting chamber, became notably smaller. All the changes that occurred were gradual and were not readily apparent until long after the establish-

ment was detected by the upturn in growth rate. In no case did morphological variants arise as a locus of cells differing markedly from the parents.

The cells of line 3T3 differed in appearance from the other lines. In sparse culture, they also looked fibroblastic but grew considerably flatter, appeared finely granular, and were more difficult to trypsinize. In confluent cultures cell borders were obscured and a thin syncytium-like sheet formed with no tendency toward multilayering.

Cells maintained under agar also became very flat and epithelium-like. This morphology persisted for months after the removal of agar, but later gradually changed toward a more fibroblastic type.

Changes in Chromosome Constitution and the Relation of the Alterations in Karyotype to the Process of Establishment

Levan and Biesele (5) and Rothfels and Parker (15) have demonstrated a relation between changes in karyotype and the development of established mouse lines. Since we were able to detect an established line quite early in its development, it was possible to examine this relation in greater detail.

In Table II it is seen that as early as the second transfer in strains 3T3 and 3T12 there is a significant polyploid element in the population. An average of 8.7 per cent of 527 cells counted in secondary cultures were in the tetraploid range. During the phase of declining growth rate the percentage of tetraploids stays constant or rises only slightly, the population remaining mainly diploid for from 20 generations (3T3) to over 40 generations (6T6). The subsequent increase in the growth rate is accompanied in all cases by a change from a primarily diploid population to one with few, if any, diploid cells. All the lines, after establishment, had the great majority of cells in the hypo-tetraploid region with considerable variation around a modal number in the seventies. The mode of 6T6 at 45 generations (135 days) was 76, and 3T12 at 34 generations (69 days) had a mode of 78. The early established cells contained no grossly abnormal chromosomes; no metacentrics were seen at this time. At a later time abnormal chromosomes did appear and the number of metacentrics and minute chromosomes progressively increased in the population. 3T12 at 92 generations had in 30 per cent of its cells from one to three metacentrics. A significant minority also had minute chromosomes.

The mode in this case had also shifted from the hypo- to the hypertetraploid region.

Examination of the data indicates that in at least two of the three lines, 3T3 and 3T12, and probably 3T6 as well, the period of rapid change in ploidy occurred *after* the upturn in growth rate. 3T3, for example, after a long period of virtually no growth had, by the 20th transfer, increased its growth rate ($N_2/N_0 = 2.4$); yet the population was still predominantly diploid. Two transfers later, however, well after the start of the upturn in growth rate (Fig. 4, curve A; Fig. 5) the population had become almost entirely heteroploid. Similarly, 3T12 at the 18th transfer (23rd generation), after N_2/N_0 had risen from a low of 1.5 to 3.0, was still 89 per cent diploid. At this time, exact counts on 50 cells in the diploid range showed the great majority of cells to have exactly 40 chromosomes, and the remainder to have the same

small variation in chromosome number that is seen in secondary cultures. Two transfers (3 generations) later, the population was largely in the tetraploid range. In 3T3 the change in ploidy occurred between the 18th and the 20th generation, and in 3T6 between the 31st and the 39th generation. This brief period when the mass of the population develops an altered karyotype is not associated with any striking further increase in growth rate.

The change in the cases of 3T3 and 3T12 occurred rapidly enough, within as little as 2 or 3 generations, to make it unlikely that it was the result of selection of a rare heteroploid variant. Rather, it is suggestive of conversion of a large fraction of the population to the heteroploid state.

Origin of the Established Line

The time of appearance of the cell type with improved growth properties under the 3-day transfer

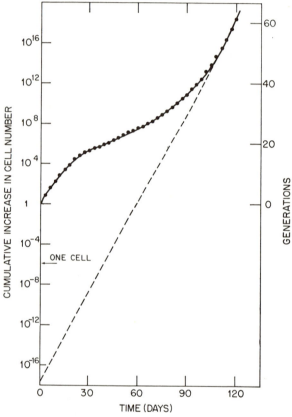

FIGURE 10

The formation of established cell line 3T6.

regime varies from roughly 45 to 90 days. This population could represent the outgrowth of a small minority of unaltered cells present in the original population and requiring this period of time to become predominant. Alternatively, they could be cells that acquired new properties during the time *in vitro*, giving them selective advantage. The simple case of the first alternative can be eliminated by extrapolating the final growth curve of 3T6 back to zero time (Fig. 10), whereupon one sees that even if just one cell present in the original population were growing at the same rapid rate from the beginning, it would have taken over the population at a much earlier time.[2]

The growth rates of the established cells, after they have emerged, progressively increase and continue to do so long after the diploid elements are no longer present in the population. 3T6, after 96 days, is almost completely non-diploid. In the subsequent 8 transfers the generation time decreased from 29 to 15 hours and morphological and karyotypic changes also continued to occur. The cells of the long established line are different genetically and metabolically from both the original euploid cells and the early established cells.

DISCUSSION

The development in cell culture of established lines from normal cell populations has in most cases been considered a rather infrequent and unpredictable occurrence (1, 2), perhaps like the development of a "spontaneous" tumor in the animal. The results presented here show that it is possible to produce established lines from mouse embryo cells with a high degree of probability (9 out of 11 cultures initiated), most of them within 3 months of culture.

Established lines and normal cell strains have been distinguished on the basis of a number of

---

[2] It could be argued that the cells which finally take over the population were indeed the unchanged progeny of a small number of cells present from the beginning, but that their growth rate was repressed by the other cells while the latter constituted the majority of the population. Experiments in which the cells of the established 3T6 were mixed with 10^2 or 10^4 times their number of secondary embryo cells showed, however, that there was no inhibition at all of the growth of the 3T6 cells under these conditions.

criteria by Hayflick and Moorhead (2). Perhaps the most essential of these is the capacity for unlimited growth, possessed by established cell lines, but not by cell strains. This difference, while quite significant as a matter of experience, is difficult to employ practically as a distinguishing criterion. The assumption is usually made that if cells have been in continuous culture for many months or more than a year, they probably constitute an established line (16, 17). Billen and Debrunner (18) have routinely maintained mouse bone marrow cells in serial transfer from 6 months to a year and assumed that they were established. On the other hand, it is possible for cells to go through over 50 generations and still not be established (2, 19).

The experiments presented here provide a relatively precise criterion for establishment, in terms of growth capacity. In all cases when cells were placed in culture they showed at first a progressively declining potential growth rate, with the doubling time rising to over 60 hours. Thereafter, the ability of the cells to maintain a rising growth rate, *under conditions where the maximum growth rate could be expressed*, constituted the first indication of establishment. This occurred as early as 30 to 45 days of culture in some of the 3-day transfer experiments (Fig. 4) but required a longer time in the 6-day transfer experiments. This may be due, at least in part, to the more rapid rate of growth under the 3-day regime. Cell lines which were found to be established by this criterion never died out subsequently, and always began a series of evolutionary changes expressed in a variety of new properties.

Under the conditions of our experiments, a change involving the development of new properties appears to be necessary in order that the cells continue to grow indefinitely *in vitro*. That this is not simply a population change resulting from selective overgrowth of cell types whose properties were fixed since the beginning of the culture is strongly suggested by the experiment of Fig. 10. New growth properties must have been developed by some cells during the first 15 to 30 generations of culture. Whether a large or only small proportion of the original population was capable of undergoing this change cannot be decided from these experiments.

Following establishment, a further rise in growth rate always occurred, the doubling time reaching values equal to or less than that of the initial cell

75

population, usually by 45 generations of *in vitro* life. In some cases, evolution continued until doubling times of less than 14 hours were reached (line 3T6).

Cell morphology of the early established lines was virtually indistinguishable from that of normal fibroblasts. In crowded cultures (with the exception of 3T3) the cells became progressively more able to form multilayers and interlace with one another. With time, saturation density increased, indicating a progressive loss of the contact inhibition characteristic of normal cells.

The particular conditions under which cells are cultivated influence the properties of the emerging cell type, as seen from examining line 3T3 and, to a lesser degree, those cell lines maintained under agar for many months. The cells of line 3T3 were always plated at a relatively low cell density and were transferred frequently so that they never were allowed to become confluent. There was little or no cell-to-cell contact prior to establishment. The resultant line, established by all other criteria and grossly abnormal in karyotype, remained extremely sensitive to homologous contact inhibition and ceased growing completely as the culture reached confluence. Its final saturation density was less than one-fifth that of the other established lines. Contact inhibition (20) is a property of normal cells lost to some degree both by malignant and by most established cells. The strikingly different behavior of 3T3, as compared with the other established lines, may reflect the peculiar conditions of its establishment. The loss of contact inhibition and, perhaps, the malignant properties of many established lines may be the result of the selective processes usually operating in cell culture and not related to the process of establishment *per se*.

The karyotypes which develop when mouse cells become established in culture have been studied in detail by Levan and Biesele (5), by Rothfels and Parker (15), and by Hsu *et al.* (21), and the changes found in our experiments were similar to those described by these investigators: first a change of chromosome number to the tetraploid and hypotetraploid range, and, later, the gradual development of grossly abnormal chromosomes. However, in our experiments, wherever the relation between growth rate and chromosomal changes was closely examined it was found that the population began to grow more rapidly before the chromosome number shifted. It therefore appears that the drastic karyotypic alterations characteristic of established mouse lines are not essential for establishment, or at least its initial phase. When the shift to heteroploidy did occur it took place very rapidly, during an interval in which the growth rate was not changing very much, suggesting that the shift was not produced by selection of a very small rapidly growing heteroploid minority of cells, but rather by a conversion of a large fraction of the diploid or quasidiploid population.

In the mouse, long established cell lines have been found invariably to be markedly heteroploid (22). In other species, however, such as the hamster (23–25) and the pig (26), establishment need not be followed by such drastic changes in the chromosome constitution. The mouse, it has been suggested, may have an inherently unstable karyotype. To determine whether this instability makes establishment of lines easier than for cells of other species would require systematic efforts to produce established lines in a variety of species.

One of the cellular properties that emerges very soon and very strongly after the initial phase of establishment is the ability to grow at low inoculation density. The relative inability of freshly cultured cells to do so is an important factor in the progressive reduction in their growth rate upon successive transfer (Fig. 2). No improvement takes place unless establishment occurs, whereupon, as the potential growth rate rises, the ability to grow at low density also improves. This represents an increased ability to function as independent organisms (a highly unnatural condition for normal mammalian cells). Eagle and Piez (27) have demonstrated that even established lines, at lowered population density, demonstrate growth factor requirements for substances the cells are capable of making. The reason for this is believed to be that the synthetic processes of the cells are unable to keep up with the leakage of these substances to the medium. However incomplete the autonomy of established cells, it is considerably more developed than that of non-established strains.

Trypsinization, because it increases the leakiness of cells to small molecules (28, 29), may be expected to aggravate this condition. It may be that the process of establishment involves a reduction of the normal leakiness of the cell for metabolites to a level where cell growth can continue, under the given culture conditions. Further improvement in this capacity can result, by mutation-selection or by some form of adaptation. The end result, at any rate, is a cell which can grow at very low inoculation density and therefore gives a relatively high plating efficiency when tested by conventional methods. Experiments (24) comparing the relative capacity of established mouse lines and normal mouse strains to act as feeders for normal cells suggest that part of the adjustment made by the established cell is, in fact, an increased ability to control the rate of loss of intermediates. (See also reference 30.)

Inability to become established seems to be more common in the case of cells of certain species other than the mouse. Normal human fibroblasts, for example, are thought to undergo establishment very rarely, if at all, and Hayflick and Moorhead (2) have suggested that the finite *in vitro* lifetime of such cells may reflect an aging process expressed at the cellular level, because of the loss of some factor necessary for cell survival, or because its rate of duplication is less than that of the cell. All cell densities used in culture are orders of magnitude below those *in vivo* (27) and it might be that all cell densities technically feasible in culture will be suboptimal with respect to cell-to-cell feeding, and lead to the depletion of substances needed for cell division. In our experiments, the rate of decline of potential growth rate and total number of generations the cells went through before establishment occurred was rather constant for the particular cell type maintained under a particular set of conditions. However, the fact that the inoculation density used, where all other conditions are identical, directly affects the decline of growth rate and the number of generations grown shows that the latter is not a fixed property of these cells but can be greatly modified by culture conditions. Whether the ever increasing generation time displayed by cells which fail to establish in culture reflects a fundamental property of the normal cell, or rather these inherent difficulties in the *in vitro* system, has not been resolved.

If methods were devised to make the cell : medium ratio comparable to that *in vivo*, or to reduce the rate of loss of metabolic intermediates, it might well be possible to maintain diploid cells at their maximal potential growth rate for much longer periods of time, perhaps indefinitely. In this case the modifications of cell properties which lead to establishment, if adaptive in nature, need not occur; if mutational, they need not be selected.

Some of the results described in this paper are in close agreement with those of similar experiments recently reported by Rothfels *et al.* (30). In addition, these investigators showed that under their conditions most established lines developed neoplastic properties over the course of long term serial culture. Various factors involved in cell feeding were also extensively investigated, and differences were found in the capacity of different cell types to act as feeders for normal embryo cells.

This work was aided by grants from the United States Public Health Service.

Received for publication, October 8, 1962.

BIBLIOGRAPHY

1. HAFF, R. F., and SWIM, H. E., Serial propagation of 3 strains of rabbit fibroblasts, *Proc. Soc. Exp. Biol. and Med.*, 1956, **93**, 200.

2. HAYFLICK, L., and MOORHEAD, P. S., The serial cultivation of human diploid cell strains, *Exp. Cell Research*, 1961, **25**, 585.

3. EARLE, W. R., and NETTLESHIP, A., Production of malignancy *in vitro*. V. Results of injections of cultures into mice, *J. Nat. Cancer Inst.*, 1943, **4**, 213.

4. GEY, G. O., Cytological and cultural observations on transplantable rat sarcomata produced by inoculation of altered normal cells maintained in continuous culture, *Cancer Research*, 1941, **1**, 737.

5. LEVAN, A., and BIESELE, J. J., Role of chromosomes in cancerogenesis, as studied in serial tissue culture of mammalian cells, *Ann. New York Acad. Sc.*, 1958, **71**, 1022.

6. EAGLE, H., OYAMA, V. I., LEVY, M., and FREEMAN, A. E., Myo-inositol as an essential growth factor for normal and malignant human cells in tissue culture, *J. Biol. Chem.*, 1957, **226**, 191.

7. DULBECCO, R., and VOGT, M., Plaque formation and isolation of pure lines with poliomyelitis virus, *J. Exp. Med.*, 1954, **99**, 167.

8. HASTINGS, J., FREEDMAN, S., RENDOW, O., COOPER, H. L., and HIRSCHHORN, K., Culture of human white cells using differential leucocyte separation, *Nature*, 1961, **192**, 1214.

9. SANFORD, K. K., EARLE, W. R., and LIKELY, G. D., The growth *in vitro* of single isolated tissue cells, *J. Nat. Cancer Inst.*, 1948, **9**, 229.

10. EARLE, W. R., SANFORD, K. K., EVANS, V. J.,

WALTZ, H. K., and SHANNON, J. E., Influence of inoculum size on proliferation in tissue cultures, *J. Nat. Cancer Inst.*, 1951, **12**, 133.

11. FISCHER, H. W., and PUCK, T. T., On the functions of x-irradiated "feeder" cells in supporting single mammalian cell growth, *Proc. Nat. Acad. Sc.*, 1956, **42**, 900.

12. MOSER, H., Modern approaches to the study of mammalian cells in culture, *Experientia*, 1960, **16**, 385.

13. VOGT, M., and DULBECCO, R., Virus-cell interaction with a tumor producing virus, *Proc. Nat. Acad. Sc.*, 1960, **46**, 365.

14. GREEN, H., and NILAUSEN, K., Repression of growth of mammalian cells under agar, *Nature*, 1962, **194**, 406.

15. ROTHFELS, K., and PARKER, R. C., The karyotypes of cell lines recently established from normal mouse tissue, *J. Exp. Zool.*, 1959, **142**, 507.

16. CLAUSEN, J. J., and SYVERTON, J. T., Comparative chromosomal study of 31 cultured mammalian cell lines, *J. Nat. Cancer Inst.*, 1962, **28**, 1, 117.

17. FOLEY, G. E., DROLET, B. P., McCARTHY, R. E., GOULET, K. A., DOKOS, J. M., and FILLER, D. A., Isolation and serial propagation of malignant and normal cells in semi-defined media, *Cancer Research*, 1960, **20**, 930.

18. BILLEN, D., and DEBRUNNER, G. A., Continuously propagating cells derived from the normal mouse bone marrow, *J. Nat. Cancer Inst.*, 1960, **25**, 1127.

19. PUCK, T. T., CIECIURA, S. J., and ROBINSON, A., Genetics of somatic mammalian cells. III. Long-term cultivation of euploid cells from human and animal subjects, *J. Exp. Med.*, 1958, **108**, 945.

20. ABERCROMBIE, M., The bases of the locomotory behavior of fibroblasts, *Exp. Cell Research*, 1961, Suppl. 8, 188.

21. HSU, T. C., BILLEN, D., and LEVAN, A., Mammalian chromosomes *in vitro*. XV. Patterns of transformation, *J. Nat. Cancer Inst.*, 1961, **27**, 515.

22. HSU, T. C., Chromosomal evolution in cell populations, *Internat. Rev. Cytol.*, 1961, **12**, 69.

23. YERGANIAN, G., and LEONARD, M. J., Maintenance of normal *in situ* chromosomal features in long-term tissue cultures, *Science*, 1961, **133**, 1600.

24. MACPHERSON, I., and STOKER, M., Polyoma transformation of hamster cell clones—An investigation of genetic factors affecting cell competence, *Virology*, 1962, **16**, 147.

25. TODARO, G. J., and GREEN, H., unpublished observations.

26. RUDDLE, F. H., Chromosome variation in cell populations derived from pig kidney, *Cancer Research*, 1961, **21**, 885.

27. EAGLE, H., and PIEZ, K., Population dependent requirements by cultured mammalian cells for metabolites they can synthesize, *J. Exp. Med.*, 1962, **116**, 29.

28. PUCK, T. T., MARCUS, P. I., and CIECIURA, S. J., Clonal growth of mammalian cells *in vitro*: Growth characteristics of colonies from single HeLa cells with and without a "feeder" layer, *J. Exp. Med.*, 1956, **103**, 273.

29. LEVINE, S., Effect of manipulation on P^{32} loss from tissue culture cells, *Exp. Cell Research*, 1960, **19**, 220.

30. ROTHFELS, K. H., KUPELWIESER, E. B., and PARKER, R. C., Effects of x-irradiated feeder layers on mitotic activity and development of aneuploidy in mouse embryo cells in *vitro*, in Fifth Canadian Cancer Congress (Honey Harbor), New York, Academic Press, Inc., 1963.

78

Reprinted from Nature, Vol. 259, No. 5539, pp. 132–134. 1976. © 1976 Macmillan Journals, Ltd.

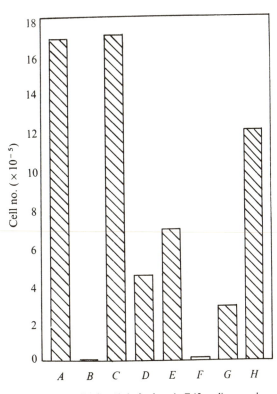

Fig. 1 GH$_3$ cells after 10 d of culture in F-12 medium supplemented with 8% foetal calf serum (8% FCS F-12), serum-free F-12 medium (F-12 SF), and F-12 SF supplemented with hormones. The cells from stock plates (in Dulbecco's modified Eagle's medium supplemented with 12.5% horse serum and 2.5% foetal calf serum) were trypsinised into single cells, treated with trypsin inhibitor, suspended in F-12 SF, and then inoculated at 1×10^5 cells per 60-mm plate. The hormones were added at the time of plating. The final concentrations of hormones were 3×10^{-10} M T$_3$, 1×10^{-9} M TRH, 50 mU ml^{-1} somatomedin preparation, 5 μg ml^{-1} transferrin and 0.5 ng ml^{-1} PTH. At the end of the experiments, both the floating and the attached cells were collected by trypsinisation and centrifugation, treated with Trypan blue, and the live cells counted by haemocytometer. *A*, 8% FCS F-12; *B*, F-12 SF; *C*, F-12 SF + hormones (T$_3$ + TRH + somatomedin A + transferrin + PTH); *D*, F-12 SF + hormones minus T$_3$, *E*, F-12 SF + hormones minus TRH; *F*, F-12 SF + hormones minus somatomedin A, *G*, F-12 SF + hormones minus transferrin; *H*, F-12 SF + hormones minus PTH. The first-day cell counts for *A* and *C* were identical (0.9×10^5 cells per plate). T$_3$ and transferrin (human) were obtained from Sigma, TRH is a synthetic peptide and PTH is the synthetic peptide of the first 30 amino acids of human PTH.

Replacement of serum by hormones permits growth of cells in a defined medium

MOST cell cultures require the addition of serum to synthetic media for their maintenance and growth, and we believe that the primary role of the serum is to provide hormones[1]. We have been led to this hypothesis by a series of experiments showing that serum depleted of certain hormones no longer supports growth of cells, unless the medium is supplemented with the hormones that were removed[2-4]. Clear evidence for the validity of this hypothesis has not yet been obtained because it is difficult to grow cells in the absence of serum. Recently, however, we have succeeded in growing an established rat pituitary cell line, GH$_3$, in a defined serum-free medium supplemented with physiological concentrations of four hormones together with the iron transport protein, transferrin. Preliminary investigation shows that serum-free medium supplemented with hormone will also support the growth of several other cell lines.

GH$_3$ is a functional rat pituitary cell line which produces growth hormone and prolactin[5,6], and depends on thyroid hormones for growth[4]. Initially, the basal medium used in our growth experiments consisted of 8% charcoal-extracted foetal calf serum. At this concentration of serum, the stimulation by triiodothyronine (T$_3$) of the growth of these cells was readily demonstrated. As the serum concentration was decreased in a stepwise manner, we were able to find which additional hormones were required; eventually it was

found possible to eliminate serum entirely if the medium was supplemented with T$_3$, thyrotropin-releasing hormone (TRH), transferrin, the biologically active peptide of parathyroid hormone (PTH), and a partially purified somatomedin preparation (5,000-fold purification from serum)[7]. The effect of these hormones on cell growth is shown in Figs 1 and 2. In the control plates, where neither serum nor hormones were added, there were no live cells after 3 d of incubation. When the five components are present in the serum-free medium, the growth is ~60–100% that in the rich medium (8% foetal calf serum in F-12). If any of the components is not added to the medium, growth either does not occur or is severely depressed. This deficiency is not overcome by substituting other pituitary hormones, steroids, hypothalamic releasing hormones, caeruloplasmin, insulin, glucagon, glycyl-histidyl-lysyl acetate, calcitonin, or prostaglandin E$_1$.

The results obtained with GH$_3$ cells encouraged us to

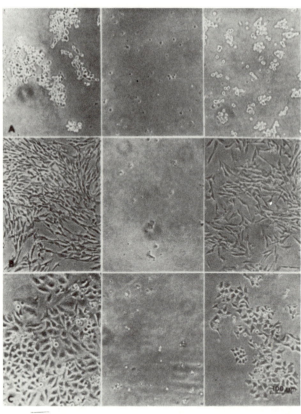

Fig. 2 Three cell lines in various media after 7 d in culture. Cells: A, GH₃; B, BHK; C, HeLa. Media from left to right, 8% FCS F-12, F-12 SF, F-12 SF supplemented with hormones. Hormone additions to GH₃ and BHK are described in the text and in Fig. 3. The final concentrations of hormones added to HeLa cultures were: 1×10^{-9} M TRH, 10 ng ml^{-1} SRIF and LRH, 2×10^{-9} M hydrocortisone acetate, 17 β-oestradiol and testosterone, 2×10^{-8} M progesterone, 3×10^{-10} M T₃, 5 μg ml^{-1} caeruloplasmin and transferrin, 0.2 μg ml^{-1} glycyl-histidyl-lysyl acetate, 50 mU ml^{-1} somatomedin A, 50 ng ml^{-1} insulin and glucagon 0.5 ng ml^{-1} PTH and 0.1 ng ml^{-1} calcitonin.

study the growth of other established cell culture lines in a defined serum-free medium supplemented with hormones. Initial experiments showed that BHK and HeLa cells do not survive in serum-free medium, but can grow in serum-free medium supplemented with a mixture of 25 hormones. The growth in the latter medium is comparable to that in the rich medium. The growth response of these cells to hormones is shown in Figs 2 and 3. Determinations of the specific hormones which support the growth of these cell lines is in progress.

There have been several reports of the growth of cells in completely defined medium[8–10]. Most if not all of these cases are, however, examples of selection or adaptation of cells to defined culture media. By contrast, our data show that the medium supplemented with hormone substitutes for serum without altering the characteristics of the individual cells or the overall population. Our experiments indicate that a combination of hormones and other specific factors, such as transferrin, can substitute completely for serum. Though the possibility of serial propagation of GH₃ is still under investigation, the results obtained with BHK (Fig. 3) strongly suggest that serum-free medium supplemented with hormones will also support the long term growth of GH₃ cells in a completely defined medium. It has already been shown that hormones can partially substitute for some functions of serum such as the overcoming of density-inhibited growth on changing from serum-containing to serum-free medium[11].

In these experiments, there is a possibility that residual serum factors are still present and active even after extensive washing. We eliminate this possibility in our experiments in which the cells are transferred from serum-containing medium by trypsinisation into defined medium,

and the growth studied entirely in the defined, serum-free medium. It is important to note that there is no delay in the initiation of growth, suggesting that there is neither selection of cells capable of growth in the serum-free medium, nor extensive adaptation to the growth conditions. Furthermore, growth, in the case of GH₃ cells, is not a nonspecific response to the addition of protein to the serum-free medium. The total amount of protein added (transferrin and the somatomedin preparation) is about 8 μg ml^{-1}. Serum protein equivalent to this protein concentration (0.016% foetal calf serum), when added to the serum-free medium, does not support the growth of these cells, even when supplemented with T₃. We attribute the growth-promoting activity of the transferrin and somatomedin preparations to these two specific factors, although we recognise that other components within these preparations might also be responsible for stimulation of growth. We are continuing to investigate this as more highly purified preparations become available.

The results presented here strongly suggest that it is possible to eliminate serum from culture medium, and that the main function of serum in cell culture is to furnish hormones. We expect that any cell culture line showing a requirement for serum can be grown in synthetic medium supplemented with a combination of hormones and a few factors such as transferrin. Comparison of the hormonal requirements of GH₃, BHK and HeLa already indicates that the specific hormones will vary with the cell type, although a few hormones may be common to all cell types. We believe that the elucidation of these requirements will be useful in advancing our understanding of integrated physiology. The possibility of culturing cells in serum-free medium supplemented with hormones will enhance the importance

Nature Vol. 259 January 15 1976

2 Armelin, H. A., *Proc. natn. Acad. Sci. U.S.A.*, **70**, 2702–2706 (1973).
3 Nishikawa, K., Armelin, H. A., and Sato, G., *Proc. natn. Acad. Sci. U.S.A.*, **72**, 483–487 (1975).
4 Samuels, H. H., Tsai, J. S., and Cintron, R., *Science*, **181**, 1253–1256 (1973).
5 Tashjian, A. H., Yasumura, Y., Levine, L., Sato, G. H., and Parker, M. L., *Endocrinology*, **82**, 342–452 (1968).
6 Tashjian, A. H., Bankroft, F. C., and Levine, L., *J. Cell Biol.*, **47**, 61–70 (1970).
7 Uthne, K., *Acta Endocrinol. Suppl.*, **175**, 1–35 (1973).
8 Waymouth, C., *J. natn. Cancer Inst.*, **22**, 1003–1018 (1956).
9 Takaoka, T., and Katsuta, H., *Expl Cell Res.*, **67**, 295–304 (1971).
10 Donta, S. T., *Expl Cell Res.*, **82**, 119–124 (1975).
11 Armelin, H. A., and Armelin, M. C. S., *Biochem. biophys. Res. Commun.*, **62**, 260–267 (1975).

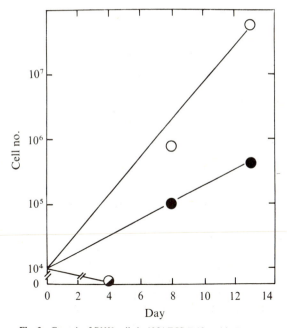

Fig. 3 Growth of BHK cells in 10% FCS F-12 and in F-12 SF supplemented with hormones. There were no live cells in F-12 SF only, 4 d after the initiation of culture. The initial plating procedure is as described in Fig. 1. The cells were inoculated at 1×10^4 cells per 35-mm plate, and the medium was changed on day 5. The cells, both in the rich medium and in the serum-free medium supplemented with hormones were subcultured on day 8 into the respective media. For subcultures, both trypsin and trypsin inhibitor were also used. On day 8 and on day 13 the cells were washed with buffer, trypsinised, and counted with a Coulter counter. The final concentrations of hormones and factors used were: 0.5 µg ml^{-1} LH, 10 mU ml^{-1} FSH, 5 µU ml^{-1} ACTH, 5 ng ml^{-1} GH, 10 ng ml^{-1} prolactin, 5 ng ml^{-1} TSH, 10 ng ml^{-1} prostaglandin E$_1$, 5 µg ml^{-1} caeruloplasmin and transferrin, 0.2 µg ml^{-1} glycyl-histidyl-lysyl acetate, 50 mU ml^{-1} somatomedin A, 50 ng ml^{-1} insulin and glucagon, 0.5 ng ml^{-1} PTH, 0.1 ng ml^{-1} calcitonin, 1×10^{-9} M TRH, 10 ng ml^{-1} SRIF (somatotropin releasing inhibitory factor) and LRH (LH releasing hormone). ○, 8% FCS F-12; ●, F-12 SF; ◐, F-12 SF supplemented with hormones. LH, FSH, GLH, prolactin (bovine) and TSH were supplied by NIAMD Program of National Institutes of Health; ACTH was obtained from ARmour Pharmaceutical Co., Chicago Illinois; caeruloplasmin, transferrin and insulin were from Sigma (Bovine) and glycyl-histidyl-lysyl acetate and glucagon were from Calbiochem. Prostaglandin E was from Dr J. Pike, Upjohn Co.

of cell culture as an experimental tool, especially in overcoming the present difficulties associated with obtaining primary cultures.

We are grateful for the generous gift of somatomedin preparations from Dr Knut Uthne, Ab Kabi, Stockholm, Sweden, of TRH, SRIF and LRH from Dr Roger Guillemin, Salk Institute, La Jolla, California, and of PTH and calcitonin from Dr Leonard Deftos of the Veterans Administration Hospital, La Jolla, California. We also thank Dr Walter Desmond for comments. This work was inspired by a conversation with the late Gordon Tomkins, and was supported by US Public Health Service grants.

IZUMI HAYASHI
GORDON H. SATO

Department of Biology,
John Muir College,
University of California, San Diego,
La Jolla, California 92093

Received September 5; accepted October 28, 1975.

1 Sato, G. H., in *Biochemical Actions of Hormones* (edit. by Litwack, G.) (Academic, New York, in the press).

In Vitro
Volume 13, No. 9, 1977
All rights reserved ©

CLONAL GROWTH OF CHINESE HAMSTER CELL LINES IN PROTEIN-FREE MEDIA

W. GREGORY HAMILTON[1] AND RICHARD G. HAM[2]

Department of Molecular, Cellular and Developmental Biology, University of Colorado, Boulder, Colorado 80309

SUMMARY

A protein-free synthetic medium, MCDB 301, has been developed for clonal growth of Chinese hamster ovary cell lines. Medium F12 was developed originally for that purpose, but later failed to support good growth without small amounts of serum protein. Growth was restored by the addition of nonphysiological amounts of commercially prepared thyroxine or smaller amounts of the trace element selenium. The thyroxine preparation was shown to contain sufficient selenium to account for all of its growth-promoting activity. MCDB 301 contains increased concentrations of calcium chloride and glutamine, and a smaller amount of cysteine than medium F12. It also has been supplemented with 19 inorganic ions, in addition to selenium and those in medium F12, in order to insure against possible future deficiencies as chemicals are purified further. A Chinese hamster lung line which will not grow in MCDB 301 alone will grow when the medium is supplemented either with methylcellulose or with insulin. The growth-promoting activity is thought to be an impurity shared in common by both substances. The probable "essential" role of impurities in cellular growth in most synthetic media and the problems involved in attempting to develop a truly "defined" medium are discussed.

Key words: defined medium; Chinese hamster cells; clonal growth; thyroxine; selenium; trace elements.

INTRODUCTION

Medium F12 (1) was developed to support clonal growth of Chinese hamster cell lines without the addition of serum or other undefined supplements. For reasons which were never completely clear, medium F12 prepared in our current laboratory in Boulder never supported good clonal growth of the Chinese hamster cells without protein supplementation. Recent results make it appear likely that contaminants in the chemicals or water used to prepare the original medium were probably essential for cellular growth. This paper describes the development of an improved synthetic medium, MCDB 301, which will support clonal growth of Chinese hamster ovary lines without protein supplementation and of a Chinese hamster lung line when supplemented only with methylcellulose.

---

[1]Present address: Clayton Foundation Biochemical Institute, University of Texas, Austin, Tex. 78712.

[2]To whom requests for reprints should be sent.

MATERIALS AND METHODS

Cells. Three sublines of Chinese hamster ovary cells (CHO) have been used. All are proline dependent, and all are derived from the original CHBOC1 clone isolated by Dr. T. T. Puck and coworkers at the University of Colorado Medical Center (2-4). Most of the studies were done with a line designated CHOP which was obtained from Dr. D. M. Prescott (University of Colorado, Boulder), who in turn had obtained it from Dr. D. F. Peterson (Los Alamos Scientific Laboratories, Los Alamos, N. Mex.). The Los Alamos strain was obtained originally from the laboratory of Dr. Puck in 1962. (In a previous paper (5), we referred to these cells simply as CHO.)

A second culture of CHO cells was obtained directly from Dr. Peterson and carried his laboratory designation C14. It is a presumably unaltered descendant of the culture originally obtained from Dr. Puck and is representative of the widely studied Los Alamos CHO cultures. The third subline is clone $CHOK_1$, which was isolated from ear-

537

TABLE 1

Composition of Medium MCDB 301[a]

Component	Δ[b]	Source[c]	Mg per l	Mol per l
Amino acids (20)				
L-Alanine	U	S	8.909	1.0×10^{-4}
L-Arginine·HCl	U	S	210.7	1.0×10^{-3}
L-Asparagine·H₂O	U	S	15.01	1.0×10^{-4}
L-Aspartic acid	U	S	13.31	1.0×10^{-4}
L-Cysteine·HCl·H₂O	L	S	17.56	1.0×10^{-4}
L-Glutamic acid	U	S	14.71	1.0×10^{-4}
L-Glutamine	H	S	438.6	3.0×10^{-3}
Glycine	U	S	7.507	1.0×10^{-4}
L-Histidine·HCl·H₂O	U	S	20.96	1.0×10^{-4}
L-Isoleucine	U	S	3.936	3.0×10^{-5}
L-Leucine	U	S	13.12	1.0×10^{-4}
L-Lysine·HCl	U	S	36.54	2.0×10^{-4}
L-Methionine	U	S	4.476	3.0×10^{-5}
L-Phenylalanine	U	S	4.956	3.0×10^{-5}
L-Proline	U	S	34.53	3.0×10^{-4}
L-Serine	U	S	10.51	1.0×10^{-4}
L-Threonine	U	S	11.91	1.0×10^{-4}
L-Tryptophan	U	S	2.042	1.0×10^{-5}
L-Tyrosine	U	S	5.436	3.0×10^{-5}
L-Valine	U	S	11.71	1.0×10^{-4}
Vitamins (9)				
D-Biotin	U	S	0.007329	3.0×10^{-8}
Folic acid	U	S	1.324	3.0×10^{-6}
DL-αLipoic acid (DL-6,8-Thioctic acid)	U	S	0.2063	1.0×10^{-6}
Niacinamide	U	S	0.03663	3.0×10^{-7}
D-Pantothenic acid (hemi-calcium salt)	U	S	0.2383	1.0×10^{-6d}
Pyridoxine·HCl	U	S	0.06171	3.0×10^{-7}
Riboflavin	U	S	0.03764	1.0×10^{-7}
Thiamine·HCl	U	S	0.3373	1.0×10^{-6}
Vitamin B₁₂	U	S	1.357	1.0×10^{-6}
Other organic compounds (8)				
Choline chloride	U	S	13.96	1.0×10^{-4}
D-Glucose	U	S	1801.6	1.0×10^{-2}
Hypoxanthine	U	S	4.083	3.0×10^{-5}
i-Inositol	U	S	18.02	1.0×10^{-4}
Linoleic acid	U	S	0.08412	3.0×10^{-7}
Putrescine·2HCl	U	S	0.01611	1.0×10^{-6}
Sodium pyruvate	U	S	110.1	1.0×10^{-3}
Thymidine	U	S	0.7266	3.0×10^{-6}
Major inorganic salts (5)				
CaCl₂·H₂O	H	F	88.22	6.0×10^{-4}
KCl	U	F	223.65	3.0×10^{-3}
MgCl₂·6H₂O	U	F	122.0	6.0×10^{-4}
NaCl	U	F	7599.	1.3×10^{-1}
Na₂HPO₄·7H₂O	U	F	268.1	1.0×10^{-3}

[a] The alphabetical listing and groupings in this table do not reflect the sequences or groupings used to prepare the medium. Except for altered concentrations of cysteine, glutamine and calcium chloride, and the addition of trace elements not in medium F12, the medium is prepared exactly as described for medium F12 by Ham (9, *Appendix A*).

[b] This column indicates changes from medium F12 (1): U = unchanged; H = higher concentration than in F12; L = lower concentration than in F12; N = new component not in F12.

[c] This column indicates the source of chemicals: A = Baker and Adamson, Allied Chemicals, Morristown, N.J.; B = J.T. Baker Chemical Co., Phillipsburg, N.J.; F = Fisher Scientific Co., Pittsburg, Pa.; G = B.D.H. Chemicals, Gallard-Schlesinger Chemicals Mfg. Co., Larle Place, N.Y.; J = Johnson Matthey Co., London, England (obtained in USA through Jarrell-Ash Division, Fisher Scientific); M = Matheson, Coleman and Bell, Los Angeles, Calif.; S = Sigma Chemical Co., St. Louis, Mo.

[d] Molar concentration is for pantothenate ion.

TABLE 1 (continued)

Component	Δ^b	Source^c	Mg per l	Mol per l
Trace elements (23)^e				
AlCl$_3$·6H$_2$O	N	A	0.001207	5.0×10^{-9}
AgNO$_3$	N	F	0.0001699	1.0×10^{-9}
Ba(C$_2$H$_3$O$_2$)$_2$	N	A	0.002554	1.0×10^{-8}
KBr	N	F	0.000119	1.0×10^{-9}
CdCl$_2$·2½H$_2$O	N	M	0.002284	1.0×10^{-8}
CoCl$_2$·6H$_2$O	N	A	0.002379	1.0×10^{-8}
Cr(SO$_4$)$_3$·15H$_2$O	N	G	0.0006624	1.0×10^{-9}
CuSO$_4$·5H$_2$O^e	U	J	0.002497	1.0×10^{-8}
NaF	N	A	0.004199	1.0×10^{-8}
FeSO$_4$·7H$_2$O^e	U	J	0.834	3.0×10^{-6}
GeO$_2$	N	F	0.000529	5.0×10^{-9}
KI	N	B	0.0001660	1.0×10^{-9}
MnSO$_4$·5H$_2$O^e	N	J	0.000241	1.0×10^{-9}
(NH$_4$)$_6$Mo$_7$O$_{24}$·4H$_2$O^e	N	J	0.01236	1.0×10^{-8h}
NiCl$_2$·6H$_2$O	N	J	0.0002377	1.0×10^{-9}
RbCl	N	F	0.001209	1.0×10^{-8}
H$_2$SeO$_3$^{e}	N	J	0.001290	1.0×10^{-8}
Na$_2$SiO$_3$·9H$_2$O	N	F	0.0002841	1.0×10^{-9}
SnCl$_2$·2H$_2$O	N	J	0.002256	1.0×10^{-8}
TiCl$_4$	N	F	0.0009486^f	5.0×10^{-9}
NH$_4$VO$_3$^{e}	N	J	0.001170	1.0×10^{-8}
ZnSO$_4$·7H$_2$O^e	U	J	0.8626	3.0×10^{-6}
ZrOCl$_2$·8H$_2$O	N	F	0.003222	1.0×10^{-8}
Buffers and indicators				
NaHCO$_3$^{g}	U	F	1176.	1.4×10^{-2}
Phenol red (sodium salt)	U	S	1.242	3.3×10^{-6}
Final pH of medium 7.4^g	U			

^{e} Medium MCDB 302 contains only seven trace elements: copper, iron, manganese, molybdenum, selenium, vanadium and zinc. Their concentrations and molecular forms are the same as shown in the table for MCDB 301.

^{f} TiCl$_4$ is sold as a liquid. The volume added is 0.55 μl per l of medium.

^{g} The pH is adjusted to 7.4 *before* the sodium bicarbonate is added. The amount of bicarbonate specified in the table is correct for use with 5% CO$_2$. After the bicarbonate has been added, the medium will be alkaline when equilibrated with air, but should have a pH of 7.4 when equilibrated with 5% CO$_2$. The use of HEPES buffer in this medium is not recommended due to slight toxicity in serum-free culture (*unpublished results*).

^{h} The molar concentration of molybdenum is 7.0×10^{-8}.

lier CHO clones in Dr. Puck's laboratory by Dr. F. T. Kao and has been used widely in studies of cellular genetics.

All three of the CHO lines are related closely to the proline-dependent Chinese hamster ovary lines utilized in the original development of media F7, F10 and F12 (1, 4, 6). Detailed Giemsa-banded karyotypes have been published for the Los Alamos cultures (7). The C14 and CHOP cells both have 21 chromosomes, whereas the K$_1$ clone has only 20 (8).

The Chinese hamster lung line, designated CHL, was derived originally from the lung of a newborn male Chinese hamster. Current cultures are related closely to the CHL cells used in the development of medium F12 (1) although they have undergone an increase in chromosome number and are now subtetraploid.

Media. The starting point for the development of new media was medium F12 (1). Detailed pro-

cedures for the preparation of medium F12 have been published (9, *Appendix 1*). These procedures have been followed with appropriate modifications for all media used in the current studies. Medium MCDB 301 (Table 1) is a modification of medium F12 containing more glutamine and calcium chloride, less cysteine, and a supplement of 20 inorganic elements which have known or suspected biological roles or tend to accumulate in biological systems. All of the compounds in the supplement have been tested for toxicity and are present in MCDB 301 at concentrations at least 100-fold below the minimum toxic level under the experimental conditions employed.

Medium MCDB 301, or an earlier version which lacked manganese, was employed routinely in most of the studies reported in this paper in order to avoid possible marginal deficiencies of inorganic nutrients. In all cases where the medium without manganese was used, it has been shown subsequently that the addition of manganese does not change the results. A second medium, MCDB 302, which contains only those trace elements which can be shown to be beneficial to one or more of the cell culture systems currently under study in this laboratory (5, 10, 11; *unpublished results*), also is described in Table 1. No antibiotics are used with either medium.

Spectrographically analyzed "specpure" grade selenious acid (Johnson Matthey Chemicals Ltd., London, England) has been used as the source of selenium in all experiments to be certain that effects ascribed to selenium are not due to contaminants. The sources of all chemicals used in preparation of media are indicated in Table 1.

Culture procedures. Monolayer stock cultures are grown in Corning No. 25100 plastic tissue culture flasks (25-cm²) with 5 ml of medium containing 2.0% (v/v) whole fetal bovine serum (Flow Laboratories, Inglewood, Calif.). Initially, medium F12 (1) was used, but more recently, it has been replaced with medium MCDB 301 (Table 1).

A low-temperature harvesting procedure, similar to that described by McKeehan (12), is utilized. The culture flasks are chilled on ice and the medium is removed. The monolayers then are rinsed twice with 2.0 ml aliquots of a chilled saline solution (130 mM NaCl, 3.0 mM KCl, 1.0 mM NaH$_2$PO$_4 \cdot$7H$_2$O, 10 mM glucose, 0.0033 mM phenol red, 30 mM HEPES buffer, pH adjusted to 7.4 with 4 N NaOH). After the second rinse, 2.0 ml of 500 μg per ml (0.05% w/v) 3X crystalline trypsin (Sigma Chemical Co.), prepared in the saline solution described above, is added to the flask. The trypsin solution is gently swirled over the cells and immediately removed. The monolayer, covered only by a residual film of trypsin solution, is kept on ice for 8 min. After the digestion period, 2.0 ml of chilled medium MCDB 301, neutralized to pH 7.4 with 1.0 N HCl, is added to the flask. The flask then is shaken gently to release the cells from the culture surface, and the cell suspension is pipetted gently to break up clumps of cells. The resultant suspension of single cells is used to inoculate new monolayer cultures or is diluted as described below for clonal growth experiments.

Clonal growth assay. The clonal growth technique (9, 13) is used to assay growth and nutrient responses. Experimental media are prepared in advance, and 5.0 ml aliquots in Corning No. 25010 60-mm tissue culture plastic Petri dishes are equilibrated in the cell culture CO$_2$-incubator while the cell suspension is being prepared. Cells harvested as described above are counted with a hemacytometer and serially diluted with chilled and neutralized medium to a concentration of 1000 to 2500 cells per ml, depending on the desired inoculum. Each Petri dish is inoculated with 0.1 ml of a suspension containing 100 to 150 CHOP cells, or 200 to 250 K$_1$, C14 or CHL cells.

Duplicate Petri dishes are incubated for 10 days in an atmosphere of 5% CO$_2$ and 95% air, saturated with water vapor at 37°C. All work has been done in a laboratory at an altitude of 5400 ft (1650 meters), with an average barometric pressure of 625 mm Hg.

At the end of the 10-day incubation period, the cultures are fixed with 10% (v/v) formalin for 5 min and stained with 0.1% (w/v) crystal violet for 5 min. A photometric measurement is made of "colony size," using equipment previously described (6, 11). The measurement is sensitive both to the area and the density of the colony and is approximately proportional to the number of cells per colony. Five typical colonies from each plate are measured and the values are averaged. In addition, each experiment is examined visually to determine which dishes subjectively appear to have the "best" growth, and significant results are recorded photographically.

Thyroxine. L-Thyroxine, DL-thyroxine and 3,3′,5-triodo-L-thyronine were obtained from Sigma Chemical Co., and 5.0 mM stock solutions were prepared in absolute ethanol, adjusted to pH 10.0 with 4.0 N NaOH. The stock solutions were diluted serially with absolute ethanol as needed and normally were added to culture media in amounts of less than 1% (v/v) to yield desired final concentrations.

Determination of selenium. The amount of selenium present as a contaminant in commercial preparations of thyroxine was determined by use of an assay based on catalysis by selenium of the reduction of methylene blue by sodium sulfide (14, 15). The procedure is as follows: Dissolve thyroxine in ethanol at a concentration of 5 mM and add 1.0 ml of the thyroxine solution to 9.0 ml of water in an 18- by 150-mm test tube (a precipitate may form, but it will redissolve when the alkaline sulfide solution is added). Prepare reference standards of "specpure" grade selenious acid (Johnson, Matthey Chemicals Ltd.) in 10% ethanol in water at concentrations from 10 to 300 nMol per l and place 10-ml aliquots in test tubes. Add 5.0 ml of "conditioner" solution (25 g EDTA, 0.5 g FeCl$_3$ and 50 ml triethanolamine, diluted to 1 l with water) to each 10-ml test or reference sample and mix thoroughly. Add 5.0 ml of concentrated formaldehyde solution (Fisher Scientific, 37.6%) and mix. Add 1.0 ml of alkaline sulfide solution (2.4 g of sodium sulfite, 2.4 g of sodium sulfide and 4.0 g of sodium hydroxide in 100 ml of water) and mix. As rapidly as possible, add 2 drops of 0.05% (w/v) methylene blue to each tube, shake to mix and start a stopwatch. Record the time required for complete decolorization of the methylene blue in each tube. Plot the reciprocal of time (min^{-1}) against the concentration of selenium to produce a standard curve and read the selenium content of the test samples from the standard curve. Confirmatory determinations of selenium content of thyroxine preparations based on atomic absorption were obtained from the laboratory of Dr. Forrest Nielsen, U.S. Department of Agriculture, Human Nutrition Laboratory, Grand Forks, N. Dak.

Insulin and reduced insulin. Crystalline insulin (Sigma) is dissolved in water at a concentration of 1 mg per ml (pH 4.0) and is stored at 4°C. Reduced insulin, which should be hormonally inactive (16, 17), is prepared as follows: Incubate 10 ml of 1.0 mg per ml insulin with 0.5 M β-mercaptoethanol at pH 7.5 for 15 min at 37°C, then freeze the mixture and lyophilize to remove the β-mercaptoethanol. Weigh the dry powder and redissolve at 100 μg per ml in distilled water.

Methylcellulose. Methylcellulose (Fisher Scientific, 15 cP) is dissolved in distilled water by stirring 15 min at 0°C. The solution normally is prepared at a concentration of 100 μg per ml so that it can be sterilized easily with a Millipore type GS (0.22-μm) filter membrane. More concentrated solutions can be prepared and sterilized by autoclaving, which does not alter the growth-promoting activity of the methylcellulose solutions.

RESULTS

Thyroxine. High concentrations of commercially prepared L-thyroxine were found to be effective in promoting the growth of CHOP cells in medium F12 without protein supplementation (Fig. 1). However, optimum growth occurred at concentrations between 10^{-6} and 10^{-5} M, which is far above the physiological level.

At the same time that we were studying this phenomenon, other research in the laboratory revealed that the trace element selenium is essential for clonal growth of human diploid fibroblasts in media containing small amounts of serum protein (5). In the presence of "optimum" amounts of L-

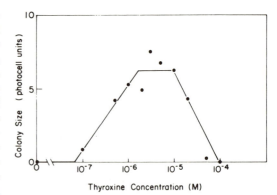

FIG. 1. Growth response of CHOP cells to L-thyroxine. CHOP cells (100 per 60-mm Petri dish) were grown for 10 days in medium F12 supplemented with the indicated amounts of L-thyroxine. Colony size was measured photometrically as described in Materials and Methods.

thyroxine, selenium has no effect on the growth of CHOP cells. However, in media without L-thyroxine, small amounts of selenium support good growth of CHOP cells (5). The effects of selenium, thyroxine and a combination of selenium plus thyroxine on colony formation by CHOP cells in medium F12 are shown in Fig. 2. Selenium at a concentration of 30 pMol per 1 has approximately the same effect as thyroxine at a concentration of 1.0 μMol per 1.

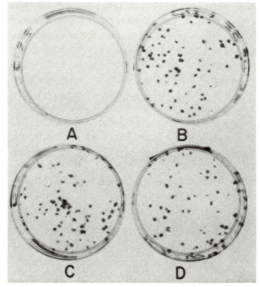

FIG. 2. Growth response of CHOP cells to selenium and thyroxine. CHOP cells (100 per 60-mm Petri dish) were grown for 10 days in medium F12 supplemented as indicated: *A*, no supplement; *B*, 30 pMol per 1 H$_2$SeO$_3$; *C*, 1.0 μMol per 1 L-thyroxine; *D*, 3.0 nMol per 1 H$_2$SeO$_3$ plus 1.0 μMol per 1 L-thyroxine.

Since the concentration of thyroxine required to produce growth was much higher than normal physiological levels of thyroxine in serum (18, 19), and since selenium was active at much lower levels, we suspected that selenium might be present as a contaminant in the thyroxine. Analysis for selenium by catalysis of reduction of methylene blue, as described in Materials and Methods, yielded an average of 212 pMol of selenium per μMol of L-thyroxine or approximately 21 ppm on a weight basis in four separate determinations (Fig. 3). This amount is adequate to account for all of the growth-promoting activity of thyroxine for CHOP cells in medium F12.

An independent determination in the laboratory of Dr. Forrest Nielsen (USDA Human Nutrition Laboratory, Grand Forks, N. Dak.) revealed a concentration of 25 ppm selenium in the L-thyroxine. Commercial preparations of L-triiodothyronine and DL-thyroxine, which also possess growth-promoting activity, contain similar amounts of selenium.

Quantitative adjustment of nutrient concentrations. After growth of CHOP cells in medium F12 without added protein was obtained through the addition of thyroxine, a systematic study was made to determine if the nutrients in the medium were at optimum concentrations. Each nutrient was omitted and added back at a series of different concentrations to determine its optimum range. Similar titrations also were performed with CHL cells in media containing insulin (described

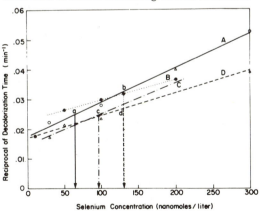

FIG. 3. Assay for selenium in commercial L-thyroxine. Four replicate tests were performed as described in Materials and Methods. Lines A, B, C and D are standard curves derived from samples containing known concentrations of H_2SeO_3. Points a, b, c and d represent the values obtained for test samples each containing 0.5 mM L-thyroxine (a = 66 nM Se, b = 130 nM Se, c = 98 nM Se and d = 130 nM Se). The average of the four assays is 106 nM selenium in 0.5 mM thyroxine.

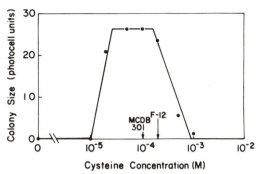

FIG. 4. Growth response of CHOP cells to L-cysteine. CHOP cells (100 per 60-mm Petri dish) were grown for 10 days in medium F12 prepared without cysteine and supplemented with 2.0 μM L-thyroxine plus the indicated amounts of L-cysteine HCl (neutralized to pH 7.0). The amounts of cysteine in media F12 and MCDB 301 are indicated (*arrows*).

below). The concentration of cysteine in medium F12 (2.0×10^{-4} M) was found to be near the edge of the toxic range and therefore was reduced to 1.0×10^{-4} M (Fig. 4). A small increase in colony size for CHL cells was achieved by increasing the concentration of glutamine to 3.0×10^{-3} M. Increasing the concentration of calcium chloride to 6.0×10^{-4} M was also beneficial for CHL cells. These three changes, doubling the calcium chloride concentration, tripling the glutamine concentration and reducing the cysteine concentration to one-half, have been incorporated into medium MCDB 301. All three of these changes are similar to changes which have been found beneficial for clonal growth of WI-38 cells (11, 20).

Trace elements. Selenium appears to be an absolute requirement for clonal growth of CHOP cells in the current medium. CHL cells also require selenium and need significantly higher concentrations than CHOP cells for optimum growth (Fig. 5). Marginal growth stimulation for CHOP cells has been obtained with concentrations of selenium as low as 1.0 pMol per l, whereas comparable stimulation of CHL cells required approximately 100 pMol per l.

Addition of the other 19 "trace elements" in medium MCDB 301 results in some stimulation of growth of CHOP cells, but clearly defined responses to individual elements have not yet been obtained. Reasons for adding all of these elements to medium MCDB 301 will be presented in the Discussion section of this paper.

Growth of CHOP cells in MCDB 301. Single CHOP cells consistently develop into large well-formed colonies with a plating efficiency of approximately 60% in medium MCDB 301 in the

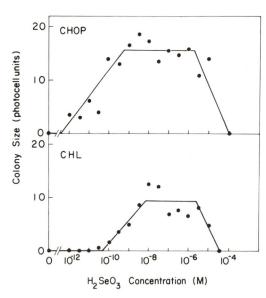

FIG. 5. Growth response of CHOP and CHL cells to selenium. *A*, CHOP cells (100 per 60-mm Petri dish) were grown for 10 days in medium F12 supplemented with the indicated amount of selenium (added as a dilute solution of H_2SeO_3, neutralized with sodium hydroxide). *B*, CHL cells (200 per 60-mm Petri dish) were grown in MCDB 301 prepared without selenium and supplemented with 1.0 μg per ml insulin plus the indicated amount of neutralized selenious acid.

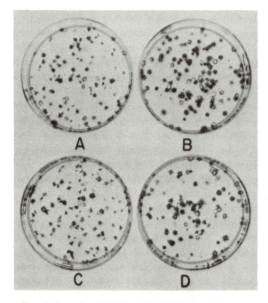

FIG. 6. Growth of CHOP cells in MCDB 301 with and without supplements. CHOP cells (150 per 60-mm Petri dish) were grown for 10 days in MCDB 301 supplemented as indicated: *A*, no supplement; *B*, 1.0% (v/v) fetal bovine serum; *C*, 1.0 μg per ml methylcellulose; *D*, 1.0 μg per ml insulin.

total absence of added proteins or other undefined supplements (Fig. 6*A*). The addition of 1.0% fetal bovine serum increases growth rate slightly and results in somewhat larger colonies at the end of the 10-day growth period (Fig. 6*B*). There is also an increased tendency to form satellite colonies when the dishes are disturbed during the incubation period. In MCDB 301 without serum, cellular attachment to the tissue culture plastic surface is very tight, and satellite colonies are seen rarely, even if the dishes are handled repeatedly during the incubation period. As will be shown below, either methylcellulose or insulin must be added to MCDB 301 to obtain clonal growth of CHL cells. However, methylcellulose has no effect on clonal growth of CHOP cells (Fig. 6*C*), and insulin causes only a slight increase in colony size (Fig. 6*D*).

Growth of C14 and K₁ cells in protein-free media. Medium MCDB 301 also supports clonal growth of the K_1 and C_{14} strains of CHO cells without protein supplementation, but with a somewhat slower growth rate. Both strains require selenium for growth, and both show positive responses to the adjusted concentrations of nutrients in medium MCDB 301.

Limited data suggest that cells which have been carried in stock cultures with 2.0% serum for extended periods of time will grow better in the protein-free media than those grown with higher levels of serum. A stock of K_1 cells carried continuously for 13 months in media containing 2.0% serum was compared with freshly thawed samples of K_1 cells which were frozen soon after being received in our laboratory and previously had been grown with higher concentrations of serum. Growth rate (as determined by colony size after 10 days) was distinctly higher in protein-free medium MCDB 301 for the continuous culture than for the recently thawed cells.

Methylcellulose. Medium MCDB 301 will not support clonal growth of CHL cells without supplementation (Fig. 7*A*). However, the addition of 1.0 μg per ml of methylcellulose to medium MCDB 301 results in good clonal growth of CHL cells (Fig. 7*B*). Autoclaving the methylcellulose does not affect its activity. Preliminary attempts to digest the methylcellulose with cellulase for 24 hr at 39°C have not destroyed the biological activity and have made a portion of it capable of passing through an Amicon UM-2 Diaflo filter. Reduction with β-mercaptoethanol does not destroy the activity. The activity is not extractable with ether, and cellobiose (Sigma) will not replace the

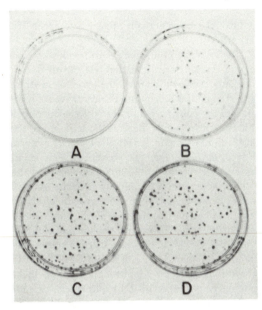

FIG. 7. Growth response of CHL cells to methylcellulose and insulin. CHL cells (200 per 60-mm Petri dish) were grown for 10 days in MCDB 301 supplemented as indicated: *A*, no supplement; *B*, 1.0 µg per ml methylcellulose; *C*, 1.0 µg per ml insulin; *D*, 1.0 µg per ml methylcellulose plus 1.0 µg per ml insulin.

methylcellulose. The minimum amount of methylcellulose required for growth stimulation is 0.5 µg per ml, with an optimum between 1.0 and 2.0 µg per ml and toxicity beginning at 1000 µg per ml.

Insulin. Insulin at 1.0 µg per ml also will support good clonal growth of CHL cells in medium MCDB 301 (Fig. 7*C*). Growth-promoting activity of the insulin is not destroyed by autoclaving for 20 min, boiling in 1.0 *N* HCl for 15 min, or incubating in 1.0 *N* NaOH at 37°C for 3 hr, all of which have been reported to destroy the hormonal activity in standard assays. However, reduction with β-mercaptoethanol does destroy the growth-promoting activity. The minimum amount of insulin which will support growth is 0.5 µg per ml, and the optimum range is between 2.0 and 10 µg per ml. Little or no additive effect is obtained by mixing insulin and methycellulose (Fig. 7*D*).

DISCUSSION

At the time that medium F12 was developed, cell lines closely related to the current CHOP and CHL cultures were grown routinely in it for extended periods without protein supplementation (1; *unpublished observations*). However, in our present laboratory in Boulder, with new sources of water and chemicals, comparable growth could

never be achieved until the results described in this paper were obtained. It now seems probable that selenium was present as a contaminant in the chemicals (or in something else, such as water or sterilizing filters) used in the preparation of the original medium F12.

The results reported here are probably just the tip of an iceberg. A variety of "defined" media have been reported in the literature to support monolayer or suspension culture growth of many different established cell lines (21, 22), and several investigators have succeeded in establishing lines directly in defined media (23-25). None of these media contain all of the inorganic trace elements which are now known or strongly suspected to be essential for whole animal nutrition (26, 27). As chemical purity is gradually improved, it can be anticipated that one after another of these media will "mysteriously" fail to support cellular growth if specific steps are not taken to supply a complete set of trace elements.

With this in mind, we have included a supplement of 19 trace elements (in addition to selenium and those in medium F12) in the new medium, MCDB 301, despite the fact that we cannot demonstrate actual requirements of CHOP cells for them at this time. The supplement includes elements known or suspected to have specific biological roles, plus a number which accumulate in biological systems and possibly could be shown to have biological roles in the future. All have been added at levels which are not toxic for CHOP cells.

It is likely that some of the elements in the supplement ultimately will be shown not to have biological activity. It is also true that some of the elements, such as cadmium, are highly toxic, except when balanced by other elements such as zinc. The inorganic supplement in MCDB 302 contains only those elements which have been shown to be beneficial to WI-38 cells (11). It is much more "reasonable" by current standards, but there is a real risk that it ultimately will prove to be inadequate when highly purified medium components become available. Elements such as chromium and arsenic, which were viewed only as toxic a few years ago, already are beginning to be classified as probably essential for living systems (28, 29).

The greater risk may well be that MCDB 301, even with its extensive supplement, still does not contain all of the "essential" elements. For example, arsenic was excluded from the formula because of severe toxicity in the protein-free system.

If current data in the literature concerning suspected biological roles of arsenic (29) are correct, it may prove necessary to find a way to introduce arsenic into future media with less toxicity, either through improved balance or by supplying it in a bound form.

At first glance, it might seem that clonal growth of CHOP cells in a completely synthetic medium such as MCDB 301 would provide a very sensitive system for studying trace-element requirements. However, two problems are involved. First, on the basis of studies with human diploid fibroblasts grown with varying amounts of serum protein (11), and on the basis of chemical analysis of synthetic media, it appears that the chemicals used in preparation of media are a more significant source of trace-element contamination than serum proteins, and that protein-free media therefore offer little advantage over media supplemented with small amounts of serum protein. Second, comparative studies in this laboratory indicate that CHOP cells utilize selenium much more efficiently than other cells which have been studied (5; and Fig. 6). If the same is true for other elements, as we suspect to be the case, it will be very difficult to obtain background media which are clean enough to demonstrate responses of CHOP cells to copper, manganese, molybdenum or vanadium, whereas human diploid fibroblasts do exhibit marginal responses in similar media, even in the presence of small amounts of serum protein (11).

We have avoided deliberate attempts to adapt cells to media with progressively lower concentrations of serum protein, since our goal is to develop synthetic media which will replace fully the functions of serum rather than to develop new cell lines which do not need serum. Stock cultures routinely have been grown with 2% (v/v) serum, and the growth assay has demanded that cells taken directly from those cultures form colonies in a protein-free medium without any appreciable lag or period of adaptation.

It now appears that despite our efforts to avoid it, some adaptation occurs during prolonged culture with 2% serum, since K_1 and C14 cells which have been grown for long periods with 2% serum form better colonies in the protein-free medium than those recently transferred from media with more serum. The only evidence for such adaptation, however, is the improved growth in the protein-free medium. The shift from 5% or 10% serum to 2% serum is achieved with no obvious stress to the cells, and, in most cases, monolayer and clonal growth actually appear better with 2% serum.

The large multiplication factor involved in colony formation largely eliminates growth due to nutrients carried over from the previous medium with the cellular inoculum. The carryover of trypsin into the cultures is minimized by using only a thin film of trypsin to disperse the cells and by the large dilution factor involved in clonal culture. We cannot rule out the possibility that small amounts of trypsin may remain adsorbed to the cell surfaces, but it is unlikely that the amount would be sufficient to support growth of single cells into colonies containing more than a thousand cells each. The nutritional adequacy of medium MCDB 301 has been demonstrated by maintaining monolayer cultures of CHOP cells in it without protein supplementation for eight monolayer pasages with 1:10 split ratios, or a total multiplication factor of 10^8 (approximately 26.5 population doublings), with no loss of growth potential.

The fact that CHL cells do not grow well in MCDB 301 indicates that we still do not have a complete understanding of their nutritional requirements. Since CHL cells require more selenium than CHOP cells, they also may require larger amounts of the trace contaminants which are present in sufficient amounts in current preparations of MCDB 301 to satisfy the needs of CHOP cells. Alternately, they may require one or more additional growth factors which are not needed at all by CHOP cells, just as CHOP cells require proline, which is not needed by CHL cells.

The fact that either methylcellulose or insulin will support growth of CHL cells in MCDB 301 suggests a possible requirement for a contaminant shared in common by both substances. The failure to affect growth-promoting activity by treatments with heat and mild hydrolysis, which are expected to destroy the hormonal activity of insulin (16, 17), suggests that something other than hormonal function may be involved. Likewise, the failure of cellulase digestion to inactivate methylcellulose suggests that the activity may consist of something else.

Methylcellulose is a frequent component of "defined" media, supposedly because of its effect on physiochemical properties of solutions (30, 31). At least twice in the past, a requirement for methylcellulose has been replaced by the addition of iron to a culture medium (32, 33). Such a replacement is unlikely in the current system, however, since a substantial amount of iron is present in medium MCDB 301 and growth of CHOP cells in that medium has been shown to be dependent on the presence of that iron.

The realization that a substance such as thyroxine added to a culture medium at a concentration of 1 μMol per l can carry enough of a contaminant to have a major effect on growth is sobering. There are 41 components in medium MCDB 301 at concentrations of 1 μm or higher, including 20 amino acids, five vitamins, seven other organic compounds, eight salts and the phenol red indicator. Any one of these could be carrying a contaminant which is essential for cellular growth. Hopefully, the trace-element supplement will take care of most future inorganic requirements, but there is no reason to believe that organic contaminants with vitamin-like or hormone-like properties may not play equally important roles in clonal growth of CHOP and related cell lines in medium MCDB 301.

It is tempting to refer to MCDB 301 as a "chemically defined" medium for CHOP cells. However, until chemicals of higher purity than are currently available have been prepared and combined into a truly defined medium, and that medium has been shown to support cellular growth, it is very risky to use the term "defined" to refer to any cell culture medium.

REFERENCES

1. Ham, R. G. 1965. Clonal growth of mammalian cells in a chemically defined, synthetic medium. Proc. Natl. Acad. Sci. U.S.A. 53: 288–293.
2. Puck, T. T., S. J. Cieciura, and A. Robinson. 1958. Genetics of somatic mammalian cells. III. Long-term cultivation of euploid cells from human and animal subjects. J. Exp. Med. 108: 945–955.
3. Tjio, J. H., and T. T. Puck. 1958. Genetics of somatic mammalian cells. II. Chromosomal constitution of cells in tissue culture. J. Exp. Med. 108: 259–268.
4. Ham, R. G. 1962. Clonal growth of diploid Chinese hamster cells in a synthetic medium supplemented with purified protein fractions. Exp. Cell Res. 28: 489–500.
5. McKeehan, W. L., W. G. Hamilton, and R. G. Ham. 1976. Selenium is an essential trace nutrient for growth of WI-38 diploid human fibroblasts. Proc. Natl. Acad. Sci. U.S.A. 73: 2023–2027.
6. Ham, R. G. 1963. An improved nutrient solution for diploid Chinese hamster and human cell lines. Exp. Cell Res. 29: 515–526.
7. Deaven, L. L., and D. F. Petersen. 1973. The chromosomes of CHO, an aneuploid Chinese hamster cell line: G-band, C-band, and autoradiographic analyses. Chromosoma 41: 129–144.
8. Kao, F. T., and T. T. Puck. 1968. Genetics of somatic mammalian cells. VII. Induction and isolation of nutritional mutants in Chinese hamster cells. Proc. Natl. Acad. Sci. U.S.A. 60: 1275–1281.
9. Ham, R. G. 1972. Cloning of mammalian cells. Methods Cell Physiol. 5: 37–74.
10. McKeehan, W. L., and R. G. Ham. 1976. Stimulation of clonal growth of normal fibroblasts with substrata coated with basic polymers. J. Cell Biol. 71: 727–734.
11. McKeehan, W. L., K. A. McKeehan, S. L. Hammond, and R. G. Ham. 1977. Improved medium for clonal growth of human diploid cells with limiting concentrations of serum protein. In Vitro 13: 399–416.
12. McKeehan, W. L. 1977. The effect of temperature during trypsin treatment on viability and multiplication of single normal and chicken fibroblasts. Cell Biol. Int. Reports 1: 335–343.
13. Puck, T. T., P. I. Marcus, and S. J. Cieciura. 1956. Clonal growth of mammalian cells in vitro. Growth characteristics of colonies from single HeLa cells with and without a "feeder" layer. J. Exp. Med. 103: 273–284.
14. West, P. W., and T. V. Ramakrishna. 1968. A catalytic method for determining traces of selenium. Anal. Chem. 40: 966–968.
15. Shendrikar, A. D. 1974. Critical evaluation of analytical methods for the determination of selenium in air, water and biological materials. Sci. Total Environ. 3: 155–168.
16. Lens, J., and J. Neutelings. 1950. The reduction of insulin. Biochim. Biophys. Acta 4: 501–508.
17. Kutsky, R. J. 1973. *Handbook of Vitamins and Hormones*. Van Nostrand Reinhold Co., New York, pp. 190–195.
18. Audhya, T. K., and K. D. Gibson. 1975. Enhancement of somatomedin titers of normal and hypopituitary sera by addition of L-triiodothyronine in vitro at physiological concentrations. Proc. Natl. Acad. Sci. U.S.A. 72: 604–608.
19. Sterling, K., and M. A. Brenner. 1966. Free thyroxine in human serum: simplified measurement with the aid of magnesium precipitation. J. Clin. Invest. 45: 155–163.
20. Ham, R. G., S. L. Hammond, and L. L. Miller. 1977. Critical adjustment of cysteine and glutamine concentrations for improved clonal growth of WI-38 cells. In Vitro 13: 1–10.
21. Higuchi, K. 1973. Cultivation of animal cells in chemically defined media, a review. Adv. Appl. Microbiol. 16: 111–136.
22. Katsuta, H., and T. Takaoka. 1973. Cultivation of cells in protein and lipid-free synthetic media. Methods Cell Biol. 6: 1–42.
23. Waymouth, C. 1965. The cultivation of cells in chemically defined media and the malignant transformation of cells in vitro. In: C. V. Ramakrishnan (Ed.), *Tissue Culture*. Dr. W. Junk, Hague, pp. 168–179.
24. Andresen, W. F., F. M. Price, J. L. Jackson, T. B. Dunn, and V. J. Evans. 1967. Characterization and spontaneous neoplastic transformation of mouse embryo cells isolated and

continuously cultivated in vitro in chemically defined medium NCTC 135. J. Natl. Cancer Inst. 38: 169-183.

25. Jackson, J. L., K. K. Sanford, and T. B. Dunn. 1970. Neoplastic conversion and chromosomal characteristics of rat embryo cells in vitro. J. Natl. Cancer Inst. 45: 11-23.

26. Underwood, E. J. 1971. *Trace Elements in Human and Animal Nutrition*. Third edition. Academic Press, New York.

27. Miller, W. J. (Ed.). 1974. Symposium on newer candidates for essential trace elements. Fed. Proc. 33: 1747-1775.

28. Mertz, W., E. W. Toepfer, E. E. Roginski, and M. M. Polansky. 1974. Present knowledge of the role of chromium. Fed. Proc. 33: 2275-2280.

29. Nielsen, F. H., S. H. Givand, and D. R. Myron. 1975. Evidence of a possible requirement for arsenic by the rat (abstr. 3987). Fed. Proc. 34: 923.

30. Higuchi, K., and R. C. Robinson. 1973. Studies on the cultivation of mammalian cell lines in a serum-free, chemically defined medium. In Vitro 9: 114-121.

31. Pollack, R., R. Risser, S. Conlon, and D. Rifkin. 1974. Plasminogen activator production accompanies loss of anchorage regulation in transformation of primary rat embryo cells by Simian virus 40. Proc. Natl. Acad. Sci. U.S.A. 71: 4792-4796.

32. Thomas, J. A., and M. J. Johnson. 1967. Trace-metal requirements of NCTC clone 929 strain L cells. J. Natl. Cancer Inst. 39: 337-345.

33. Birch, J. R., and S. J. Pirt. 1970. Improvements in a chemically defined medium for the growth of mouse cells (strain LS) in suspension. J. Cell Sci. 7: 661-670.

This research was supported by Grant No. CA15305 from the National Cancer Institute. We thank Karen Brown, Catherine Verhulst and Dick Carter for assistance in preparing the manuscript and illustrations.

Supplementary Readings

Before 1961

Carrel, A. 1912. On the permanent life of tissues outside of the organism. *J. Exp. Med. 15:* 516–528.

Eagle, H. 1960. The sustained growth of human and animal cells in a protein-free environment. *Proc. Natl. Acad. Sci. 46:* 427–432.

Earle, W. 1943. Production of malignancy in vitro. IV. Mouse fibroblast cultures and changes seen in living cells. *J. Natl. Cancer Inst. 4:* 165–212.

Fisher, H.W., T.T. Puck, and G. Sato. 1958. Molecular growth requirements of single mammalian cells: The action of fetuin in promoting cell attachment to glass. *Proc. Natl. Acad. Sci. 44:* 4–10.

Gey, G. 1954. Some aspects of the constitution and behavior of normal and malignant cells maintained in continuous culture. *Harvey Lectures 50:* 154–229.

McLimmans, W.F., E.V. Davis, F.L. Glover, and G.W. Rake. 1957. The submerged culture of mammalian cells: The spinner culture. *J. Immunol. 79:* 428–433.

Puck, T.T. and P.I. Marcus. 1955. A rapid method for viable cell titration and clone production with HeLa cells in tissue culture: The use of X-irradiated cells to study conditioning factors. *Proc. Natl. Acad. Sci. 41:* 432–437.

Rinaldini, L. 1959. An improved method for the isolation and quantitative cultivation of embryonic cells. *Exp. Cell Res. 16:* 477–505.

Sanford, K.K., W.R. Earle, and G.D. Likely. 1948. The growth in vitro of single isolated tissue cells. *J. Natl. Cancer Inst. 9:* 229–246.

Shannon, J., W. Earle, and H. Walts. 1952. Massive tissue cultures prepared from chick embryos plated as a cell suspension on glass substrate. *J. Natl. Cancer Inst. 13:* 349–365.

1961–1970

Broder, S.W., P.R. Glade, and K. Hirschhorn. 1970. Establishment of long-term lines from small aliquots of normal lymphocytes. *Blood 35:* 539–542.

Choi, K.W. and A.D. Bloom. 1970. Cloning human lymphocytes in vitro. *Nature 227:* 171–172.

Eagle, H. and K. Piez. 1962. The population-dependent requirement by cultured mammalian cells for metabolites which they can synthesize. *J. Exp. Med. 116:* 29–43.

Martin, G.M., C.A. Sprague, and C.J. Epstein. 1970. Replicative life-span of cultivated human cells. Effects of donor's age, tissue, and genotype. *Lab. Invest. 23:* 86–92.

Merz, G. and J. Ross. 1969. Viability of human diploid cells as a function of in vitro age. *J. Cell. Physiol. 74:* 219–222.

Petursson, G., J.I. Coughlin, and C. Meylan. 1964. Long-term cultivation of diploid rat cells. *Exp. Cell Res. 33:* 60–67.

1971–1980

Birch, J. and S. Pirt. 1971. The quantitative glucose and mineral nutrient requirement of mouse LS (suspension) cells in chemically defined medium. *J. Cell Sci. 8:* 693–700.

Cassiman, J.J. and M.R. Bernfield. 1974. Morphogenetic properties of human embryonic cells: Aggregation of dissociated cells and histogenesis in cultured aggregates. *Pediatr. Res. 8:* 184–192.

Dykhuizen, D. 1974. Evolution of cell senescence, atherosclerosis and benign tumors. *Nature 251:* 616–618.

Harley, C.B. and S. Goldstein. 1980. Retesting the commitment theory of cellular aging. *Science 207:* 191–193.

Harley, C.B., J.W. Pollard, J.W. Chamberlain, C.P. Stanners, and S. Goldstein. 1980. Protein synthetic errors do not increase during aging of cultured human fibroblasts. *Proc. Natl. Acad. Sci. 77:* 1885–1889.

Kirkwood, T.B.L. and R. Holliday. 1975. Commitment to senescence: A model for the finite and infinite growth of diploid and transformed human fibroblasts in culture. *J. Theor. Biol. 53:* 481–486.

Martin, G.M., C.A. Sprague, T.H. Norwood, and W.R. Pendergrass. 1974. Clonal selection, attenuation and differentiation in an in vitro model of hyperplasia. *Amer. J. Pathol. 74:* 137–153.

Norwood, T.H., W.R. Pendergrass, C.A. Sprague, and G.M. Martin. 1974. Dominance of the senescent phenotype in heterokaryons between replicative and post-replicative human fibroblast-like cells. *Proc. Natl. Acad. Sci. 71:* 2231–2235.

Pienkowski, M., D. Solter, and H. Koprowski. 1974. Early mouse embryos: Growth and differentiation in vitro. *Exp. Cell Res. 85:* 424–428.

Rafferty, K.A., Jr. 1973. Morphology and longevity of cells cultured from primate, carnivore and rodent tissues. *Differentiation 1:* 363–372.

Shechter, Y. and S.J.D. Karlish. 1980. Insulin-like stimulation of glucose oxidation in rat adipocytes by vanadyl (IV) ions. *Nature 284:* 556–558.

Chapter 2 Retention of Differentiated Properties

Harrison, R.G. 1907. Observations on the living developing nerve fiber. *Proc. Soc. Exp. Biol. Med.* **4:** 140–143.

Buonassisi, V., G. Sato, and A.I. Cohen. 1962. Hormone-producing cultures of adrenal and pituitary tumor origin. *Proc. Natl. Acad. Sci.* **48:** 1184–1190.

Thompson, E.B., G.M. Tomkins, and J.F. Curran. 1966. Induction of tyrosine α-ketoglutarate transaminase by steroid hormones in a newly established tissue culture cell line. *Proc. Natl. Acad. Sci.* **56:** 296–303.

Yaffe, D. 1968. Retention of differentiation potentialities during prolonged cultivation of myogenic cells. *Proc. Natl. Acad. Sci.* **61:** 477–483.

Augusti-Tocco, G. and G. Sato. 1969. Establishment of functional clonal lines of neurons from mouse neuroblastoma. *Proc. Natl. Acad. Sci.* **64:** 311–315.

Friend, C., W. Scher, J.G. Holland, and T. Sato. 1971. Hemoglobin synthesis in murine virus-induced leukemic cells *in vitro*: Stimulation of erythroid differentiation by dimethyl sulfoxide. *Proc. Natl. Acad. Sci.* **68:** 378–382.

Green, H. and M. Meuth. 1974. An established preadipose cell line and its differentiation in culture. *Cell* **3:** 127–133.

Köhler, G. and C. Milstein. 1975. Continuous cultures of fused cells secreting antibody of predefined specificity. *Nature* **256:** 495–497.

Christian, C.N., P.G. Nelson, J. Peacock, and M. Nirenberg. 1977. Synapse formation between two clonal cell lines. *Science* **196:** 995–998.

Green, H., J.G. Rheinwald, and T.-T. Sun. 1977. Properties of an epithelial cell type in culture: The epidermal keratinocyte and its dependence on products of the fibroblast. *Prog. Clin. Biol. Res.* **17:** 493–500.

Jones, P.A. 1979. Construction of an artificial blood vessel wall from cultured endothelial and smooth muscle cells. *Proc. Natl. Acad. Sci.* **76:** 1882–1886.

Bennett, D.C. 1980. Morphogenesis of branching tubules in cultures of cloned mammary epithelial cells. *Nature* **285:** 657–659.

The growth of cells in culture is usually a poor imitation of the finely tuned growth of cells in any tissue. As cells in tissues grow, they become highly specialized in function, appearance, and presumably internal biochemistry. Retention by cells in culture of such specializations is hard to achieve but not impossible. In the papers in this chapter many different ways are given to preserve differentiated properties of cells from various tissues while retaining their capacity to divide in culture.

The early paper by **Harrison** examines the question of whether a nerve fiber grows from distant central cell bodies. Here the organism is the developing frog embryo, and the means for study is the classic lymph, or serum, clot. Harrison carefully notes that nerve axons can grow from chunks of neuronal tissue placed in these clots. Although totally descriptive, this paper quite convincingly established that developing nerve fibers can originate from central cells and, therefore, that single nerve cell processes can be extremely long.

By the 1960s it had become clear that most other differentiated cells from mammalian tissue were not easy to grow, because fibroblasts would overgrow the culture. Indeed, for many decades differentiated epithelial cells were simply thought to be doomed in culture. The paper by **Buonassisi, Sato,** and **Cohen** shows a way to retain differentiated functions in cultured epithelial cells. Many epithelial tumors are made of cells that differentiate after the fashion of the normal cells they arise from. Such tumor cells remain differentiated in culture, as well, until they are overgrown by fibroblasts. The method of Buonassisi et al. was to place tumor cells in culture and then, after a while, to reimplant the cultured cells into a new animal. Only tumor cells will be progenitors of a tumor in the new animal, and such tumor cells must also have been cells that had survived and grew in culture. Multiple passages alternating between culture and animal therefore select for differentiated cell lines that can be cloned in vitro. This frees them from the background of fibroblasts. Differentiation is measured here as steroid production in a cultured adrenal cortex tumor line in response to stimulation by ACTH. The ACTH was made by another successful tumor cell culture of pituitary origin.

Certain chemicals fed to rats cause their livers to develop nodules that subsequently become malignant tumors. These tumors, called hepatomas, can be excised from the liver and placed in culture. **Thompson, Tompkins,** and **Curran** describe the differentiated sensitivity of one such hepatoma cell line to steroid hormones. Steroids induce the enzyme tyrosine aminotransferase in the cultured cells, as they do in normal liver. Note the paradoxical stimulation of enzyme synthesis when actinomycin D is added after steroid stimulation. This result implies that an unstable inhibitor may be made during steroid stimulation, but a final explanation is not yet in.

When a differentiated property is absolutely incompatible with cell division, the only possible way to get a population of cells to display it in culture is to grow the parental or stem cells of the differentiation pathway. These stem cells divide, but their daughters differentiate, so the tactic is to find conditions under which stem cells in culture will cease to divide as they begin to specialize. In the next papers in this chapter we see cell populations that can be stimulated to differentiate but that otherwise are maintained as growing, undifferentiated cultures. In Chapters 7 and 8 we consider the molecular implications of a cell division in which one daughter remains a stem cell while the other differentiates.

As myoblast cells differentiate to become striated muscle, they fill with actomyosin and fuse to each other to produce the mature, multinuclear contractile myotube. Since myotubes cannot divide, culture of striated muscle requires that myoblasts be able first to divide and then to differentiate and fuse in culture. By chemical transformation (see Chapter 3), **Yaffe** generated a population of myoblast stem cells that grow indefinitely, so long as they are maintained in sparse culture. However, Yaffe's transformants included cells that grew in sparse culture but began to differentiate into myotubes once they had grown into a colony. Note that the myoblasts were grown in a mixed cell culture containing a feeder of irradiated precrisis fibroblasts. The feeders were necessary to initiate growth of sparse myoblast cultures. In a similar study, **Augusti-Tocco** and **Sato** applied the technique of alternate animal and culture passage to the mouse neuroblastoma and recovered a clonal population of cells that grow in an undifferentiated state but can be induced by starvation to send out axonal processes.

The stem cells of the lymphoid system reside in the marrow. Branches arise from the earliest stem cell to become macrophages, T cells, B cells, and the cells that make erythrocytes. The mouse tumor virus murine leukemia virus (MLV) (see Chapter 3) can induce a leukemia in which differentiating precursors of erythrocytes are found dividing in the blood. These precursors can be grown as established cell lines in culture. They will express red blood cell functions when plated as separate cells. Under normal culture conditions, however, no more than 1–2% of the cells differentiate sufficiently to be stained by benzidine, a probe for hemoglobin. The paper by **Friend, Scher, Holland,** and **Sato** shows that the drug dimethylsulfoxide (DMSO) is capable of inducing 90% of these tumor cells to undergo a more complete erythroid differentiation, elevating hemoglobin synthesis by almost 100-fold. The population induced in this fashion still retains the capacity to make tumors. Final differentiation is not achieved, and both stem and differentiated cells continue to exist in the presence of DMSO. How DMSO works is unknown, but the paper by Friend et al. offers the tantalizing possibility that drugs may be designed to reverse the course of a malignancy by forcing tumor cells to switch from the stem to the terminally differentiated state.

The two papers in this chapter from the laboratory of Howard Green describe differentiations that can be carried out in culture by nontumorigenic normal stem cells. In the first, **Green** and **Meuth** begin with the nontumorigenic, anchorage-dependent mouse fibroblast cell line 3T3 (see Chapters 1, 3, and 4). In a dish that has been permitted to sit at confluence for some days, not all 3T3 cells look alike. Some regions of the dish fill with adipose cells. When such fatty cells are cloned, cell lines can be recovered that stably express the capacity to become full of fat. One such cell line, 3T3L1, continues to make collagen and therefore might be thought of as a fibroblast. However, at confluence, 3T3L1 cells respond to high concentrations of serum by filling with fat as well. Bromodeoxyuridine (BrdU) stops this adipocyte differentiation completely, while not affecting collagen synthesis at all. The second differentiation, of keratinocytes, requires a mixed culture and is discussed as a mixed-culture system later in this chapter.

Antibodies are secreted into the blood by lymphoid B cells. The initial specificity of this immune reaction is determined by regions or portions of inducing foreign molecules called antigens. **Köhler** and **Milstein** provide a major technological breakthrough that permits the development in culture of cell lines that secrete immunoglobulins directed against specific antigens. How is it possible to get lymphoid lines to secrete antibodies to order? To begin with, Köhler and Milstein use two different B-cell tumors (mouse myelomas) as parents in a cell-fusion experiment. Each parent line secretes a different immunoglobulin. The result of cell fusion is the appearance of a new hybrid cell line (see Chapter 8) that continues to secrete the two parental types of immunoglobulin molecules.

The spleen contains many normal B cells. Having established a capacity to fuse myeloma cells with each other, they then were able to fuse myeloma cells with normal spleen cells and to grow the hybrids in culture. Spleen cells can be obtained from a mouse previously immunized by any antigen of interest. Köhler and Milstein used sheep red blood cells as antigen, and hybrid myeloma-spleen clones were recovered in a screen based on their capacity to secrete immunoglobulin directed against sheep red blood cells. These lines are now called monoclonal hybridomas. The entire process is done with cells from inbred BALB mice, and, as a result, the hybrid cell lines obtained can be reinjected into new mice, where they grow as tumors. Surrounding each tumor in the ascites cavity of the mouse is a fluid that becomes extremely rich in the monoclonal antibody molecules produced by the hybrid cell lines.

The last papers in this chapter study mixed cultures. There is no killing of one type of cell to make a feeder, nor is there any intentional cell fusion. Instead, cell-to-cell interaction of a specific sort occurs. **Christian, Nelson, Peacock,** and **Nirenberg** culture neuronal cells together with cultured muscle cells. After many weeks, connections are established between some neuronal cells and some multinuclear myotubes. They then impale the nerve and muscle cells of these joined sets with sets of electrodes. Depolarizing the nerve cell causes it to produce an action potential. The muscle response to the nerve is a depolarization very reminiscent of the standard response of a neuromuscular junction to a neuronal action potential. This system might be useful in the analysis of the specificity of normal neuronal connections.

Epithelial cells normally only divide when they are on the basement membrane, an acellular material that separates epithelial cells from the mesodermal or fibroblastic part of a tissue. In skin, for example, epithelial stem cells form a single cell layer on the basement membrane. As they divide, one daughter differentiates into a keratinocyte as it is pushed from the basement membrane, eventually to be shed off as dandruff. The other daughter of each stem cell division remains at the basement membrane and can divide again. Epithelial skin cells in the body thus maintain a geometric arrangement of proximity to fibroblasts but are separated from them by a basement membrane. **Green, Rheinwald,** and **Sun** carried this arrangement of keratinocytes and fibroblasts into culture. For success, keratinocyte stem cells had to be grown on dishes that had previously been plated with a sparse culture of irradiated 3T3 cells. The 3T3 feeder apparently secreted into the medium and onto the dish materials that were necessary for the satisfactory attachment and proliferation of keratinocytes. Tumor and normal keratinocytes grown this way were surprisingly similar in culture.

The topological separation by a basement membrane of epithelium from blood vessels, fibroblasts, and other mesodermal cell types is not restricted to the skin. Blood vessels are constructed in the same fashion. In the blood vessel, an epithelium lies on the lumen side of a basement membrane and smooth muscle cells lie on the other side. This arrangement is essential to normal blood vessel function: Penetration of the blood vessel basement membrane by tumor cells is a necessary step in the development of metastases (see Chapter 5), and rips of the epithelial monolayer of major arteries may be an early event in the development of atherosclerotic plaques. **Jones** reassembled the cell layers of a blood vessel in their proper topology, using endothelial cells obtained from a bovine aorta and smooth muscle cells obtained from rabbit heart. He shows that both cell types contribute to the production in culture of a basement membrane that resembles that of a blood vessel. The binary cell culture is stable for many months and therefore should be a good tool for the study of atherosclerosis, metastatic invasion, and perhaps other diseases, such as diabetes, where connections between blood vessel cells are abnormally weak.

Certainly, it is a daunting challenge to get cells to form shapes typical of developing tissues. Morphogenesis in three dimensions would seem to be impossible to obtain in a dish. **Bennett** shows that branching structures, typical of developing glands such as the breast, grow from the descendants of a single cell. Bennett's model to explain this requires that a sheet that grows from a single cell have two different surfaces and that one of these surfaces be collapsed to a minimum. Interestingly, one surface of the in vivo developing epithelium of mammary tissue is very rich in cytoskeletal actomyosin along the one side. Perhaps the directed movement of cell sheets in morphogeneis is the consequence of intracellular localization of actomyosin (see Chapter 11).

Reprinted from Proceedings of the Society for Experimental Biology and Medicine
Vol. 4, pp. 140–143. 1907

Observations on the living developing nerve fiber.

By Ross G. Harrison.

[*From the Anatomical Laboratory of the Johns Hopkins University.*]

The immediate object of the following experiments was to obtain a method by which the end of a growing nerve could be brought under direct observation while alive, in order that a correct conception might be had regarding what takes place as the fiber extends during embryonic development from the nerve center out to the periphery.

The method employed was to isolate pieces of embryonic tissue known to give rise to nerve fibers, as for example, the whole or fragments of the medullary tube, or ectoderm from the branchial region, and to observe their further development. The pieces were taken from frog embryos about 3 mm. long, at which stage, *i. e.*, shortly after the closure of the medullary folds, there is no visible differentiation of the nerve elements. After carefully dissecting it out the piece of tissue is removed by a fine pipette to a cover slip upon which is a drop of lymph freshly drawn from one of the lymph sacs of an adult frog. The lymph clots very quickly, holding the tissue in a fixed position. The cover slip is then inverted over a hollow slide and the rim sealed with paraffine. When reasonable aseptic precautions are taken, tissues will live under these conditions for a week and in some cases specimens have been kept alive for nearly four weeks. Such specimens may be readily observed from day to day under highly magnifying powers.

While the cell aggregates, which make up the different organs and organ complexes of the embryo, do not undergo normal transformation in form, owing no doubt in part to the abnormal conditions of mechanical tension to which they are subjected, nevertheless the individual tissue elements do differentiate characteristically. Groups of epidermis cells round themselves off into little spheres or stretch out into long bands, their cilia remain active for a week or more and a typical cuticular border develops. Masses of cells taken from the myotomes differentiate into muscle fibers showing fibrillæ with typical striations. When portions of myotomes are left attached to a piece of the medullary cord the muscle fibers which develop will, after two or three days, exhibit frequent contractions. In pieces of nervous tissue numerous fibers are formed, though owing to the fact that they are developed largely within the mass of transplanted tissue itself, their mode of development cannot always be followed. However, in a large number of cases fibers were observed which left the mass of nerve tissue and extended out into the surrounding lymph clot.   It is these structures which concern us at the present time.

In the majority of cases the fibers were not observed until they had almost completed their development, having been found usually two, occasionally three and once or twice four days after isolation of the tissue.   They consist of an almost hyaline protoplasm, entirely devoid of the yolk granules, with which the cell-bodies are gorged.   Within this protoplasm there is no definiteness of structure ; though a faint fibrillation may sometimes be observed and faintly defined granules are discernible.   The fibers are about $1.5–3\,\mu$ thick and their contours show here and there irregular varicosities.   The most remarkable feature of the fiber is its enlarged end, from which extend numerous fine simple or branched filaments.   The end swelling bears a resemblance to certain rhizopods and close observation reveals a continual change in form, especially as regards the origin and branching of the filaments. In fact the changes are so rapid that it is difficult to draw the details accurately.   It is clear we have before us a mass of protoplasm undergoing amœboid movements.   If we examine sections of young normal embryos shortly after the first nerves have developed, we find exactly similar structures at the end of the developing nerve fibers.   This is especially so in the case of the fibers which are connected with the giant cells described by Rohon and Beard.

Still more instructive are the cases in which the fiber is brought under observation before it has completed its growth.   Then it is found that the end is very active and that its movement results in the drawing out and lengthening of the fiber to which it is attached.   One fiber was observed to lengthen almost $20\,\mu$ in 25 minutes, another over $25\,\mu$ in 50 minutes.   The longest fibers observed were 0.2 mm. in length.

When the placodal thickenings of the branchial region are isolated, similar fibres are formed and in several of these cases they have been seen to arise from individual cells.   On the other hand, other tissues of the embryo such as myotomes, yolk endoderm, notochord and indifferent ectoderm from the abdominal region do not give rise to structures of this kind.   There can therefore be no doubt that we are dealing with a specific characteristic of nervous tissue.

It has not yet been found possible to make permanent specimens which show the isolated nerve fibers completely intact. The structures are so delicate that the mere immersion in the preserving fluid is sufficient to cause violent tearing and this very frequently results in the tearing away of the tissue in its entirety from the clot. Nevertheless, sections have been cut of some of the specimens and nerves have been traced from the walls of the medullary tube, but they were in all cases broken off short.

In view of this difficulty an effort, which resulted successfully, was made to obtain permanent specimens in a somewhat different way. A piece of medullary cord about four or five segments long was excised from an embryo and this was replaced by a cylindrical clot of proper length and caliber, which was obtained by allowing blood or lymph of an adult frog to clot in a capillary tube. No difficulty was experienced in healing the clot into the embryo in proper position. After two, three or four days the specimens were preserved and examined in serial sections. It was found that the funicular fibers from the brain and anterior part of the cord, consisting of naked axones without sheath cells, had grown for a considerable distance into the clot.

These observations show beyond question that the nerve fiber develops by the outflowing of protoplasm from the central cells. This protoplasm retains its amœboid activity at its distal end, the result being that it is drawn out into a long thread which becomes the axis cylinder. No other cells or living structures take part in this process. The development of the nerve fiber is thus brought about by means of one of the very primitive properties of living protoplasm, ambœoid movement, which, though probably common to some extent to all the cells of the embryo, is especially accentuated in the nerve cells at this period of development.

The possibility becomes apparent of applying the above method to the study of the influences which act upon a growing nerve. While at present it seems certain that the mere outgrowth of the fibers is largely independent of external stimuli, it is of course probable that in the body of the embryo there are many influences which guide the moving end and bring about contact with the proper end structure. The method here employed may be of value in analyzing these factors.

Reprinted from the Proceedings of the NATIONAL ACADEMY OF SCIENCES
Vol. 48, No. 7, pp. 1184–1190.   July, 1962.

HORMONE-PRODUCING CULTURES OF ADRENAL AND PITUITARY TUMOR ORIGIN*

By VINCENZO BUONASSISI,† GORDON SATO, AND ARTHUR I. COHEN

GRADUATE DEPARTMENT OF BIOCHEMISTRY, BRANDEIS UNIVERSITY, AND THE ARTHUR G. ROTCH
LABORATORY OF THE BOSTON DISPENSARY

Communicated by Theodore T. Puck, May 11, 1962

Animal cell cultures rarely perform the differentiated functions of the tissue of origin for any practical length of time.   This has been especially true of cultures of endocrine tissue although a few cultures of endocrine origin have been reported to maintain their function.[1]   The present work deals with a systematic procedure which may prove to be generally applicable for developing cultures of physiologically specialized cells which maintain these specialized functions in culture.   To this end, hormone-producing mouse tumors were put into culture, and after various times in culture, the cultures were injected into mice to obtain new tumors.   Tumors arising from cultures were checked for hormonal activity and put back into culture, and the whole process was repeated.   It was hoped that this process would selectively enrich the tumors for cells better able to withstand the conditions of culture.   Furthermore, only the cancerous endocrine cells would benefit from this selective process because stromal elements which survive the culture period should not participate in the process of new tumor formation.   The experiments indicate that successive passages of these tumor cells through culture and animal give rise to cells with increased growth capacity and enhanced hormonal activity in culture.

103

Materials and Methods.—The functional adrenocortical and ACTH-secreting mouse tumors used in this study have been developed and extensively investigated by Furth and his associates.[2-7] The culture techniques have been previously described.[8,9] The only modification adopted here was that the disaggregated tissue was incubated in 1% Viokase‡ solutions in physiological saline for a few minutes at room temperature.

The assay of delta 4-3-ketosteroid secretion by the adrenal tumor cultures was performed in a manner similar to that previously described for tumor slices.[5] The medium was aspirated from the cultures and replaced with 2.0 ml of incubation medium per petri dish, containing a determined amount of ACTH. The incubation medium was either Krebs-Ringer bicarbonate buffer fortified with 200 mg per cent glucose or medium 199 plus 5% horse serum. The two incubation media gave equivalent results for short incubation periods (0–4 hr), but for long (4–24 hr) incubations, medium 199 plus horse serum was superior. The cultures were incubated at 37°C in a humidified atmosphere of 5% CO_2–95% air for a period ranging from 2 to 24 hr. The rate of steroid secretion was found to be constant over the 24-hr period. The incubation medium was extracted with a two-fold volume of methylene chloride (spectrophotometric grade), evaporated to dryness under nitrogen, and dissolved in 95% ethanol. The optical densities of the alcoholic solutions were then measured on the Cary spectrophotometer between 300 and 200 millimicrons. A blank was prepared in the same way as the sample from fresh incubation medium. Quantitation of steroid secretion was obtained by comparing the optical density of the samples at the characteristic peak absorption (242 millimicrons) with those obtained with standard solutions of cortisol. The ACTH used in these studies was obtained from the Armour Company (40 I.U./vial). U.S.P. reference standard ACTH gave equivalent results for short incubation periods but seemed to be toxic over long periods. Steroid production by these cultures was found to be proportional to the logarithm of the dose of ACTH. Estimates of the amount of ACTH in pituitary tumor cultures were made by stimulating adrenal tumor cultures with the pituitary tumor culture media and comparing the steroid production to that obtained with known amounts of Armour ACTH.

Antisera were prepared as previously described.[8] Antisera against normal mouse adrenal cortex were absorbed with homogenates of mouse lung, kidney, liver, and abdominal muscle to remove nonspecific antibodies.

Protein content of the tumors and culture cells were determined by the method of Lowry.[10]

Results.—(1) *Effect of previous culture passage on growth of tumors in culture:* The growth capacity of adrenal tumors put in culture for the first time was compared to that of tumors arising from previous cultures (Fig. 1). The two adrenal tumors were plated at a level of 10^5 cells per petri dish. Under these conditions, almost 100 per cent of the inoculated cells become attached and stretched on the petri dish surface. It is seen that the original adrenal tumor, when put into culture, gives rise to a cell population which does not increase in number with time, but in fact, slowly declines. On the other hand, the adrenal tumor which had been previously passed through culture increases by a factor of ten in about ten days of culture.

The pituitary tumor also exhibits this behavior. When dispersed cells from a pituitary tumor which had never been in culture previously are plated, less than 0.1

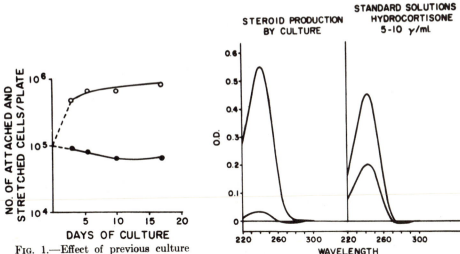

Fig. 1.—Effect of previous culture passage on growth of tumors in culture. 10⁵ cells of the original adrenal tumor were plated per petri dish, and the number of attached and stretched cells per petri dish was measured as a function of time in culture (lower curve). The upper curve represents the growth obtained when the same number of cells from an adrenal tumor which had been previously cultured, in this case once for thirty days and subsequently for three days, was plated.

Fig. 2.—Steroid production by adrenal tumor cultures. Absorption spectra of steroidal material obtained from adrenal tumor cultures are presented on the left. The upper curve was obtained upon maximal stimulation of the cultures with ACTH and the lower curve in the absence of ACTH. The curves on the right represent the absorption spectra of hydrocortisone solutions. The adrenal tumor cultures used in this experiment were derived from a tumor which had been in culture twice previously (once for 45 days and then for 27 days).

per cent of the inoculated cells attach, stretch, and initiate growth; for pituitary tumors arising from injection of cultures, this figure is close to 100 per cent.

(2) *Hormonal activity of tumors arising from cultures:* After the first passage through culture, slices of adrenal tumor usually have a greater capacity (1.3–1.8 gamma steroids/mg protein/2 hr) for steroid production than slices of the original tumor (about 1.1 gamma steroids/mg protein/2 hr). Pituitary tumors also have enhanced hormonal activity after the first passage through culture. The original pituitary line has an ACTH content of 25–50 mu ACTH/mg tissue, while a tumor arising from culture has a content greater than 400 mu ACTH/mg tissue.

(3) *Immunologic characterization of culture populations:* In the early stages of this work, the production of steroids in culture by the original adrenal tumor was barely and not constantly detectable. If only a small fraction of the cells in culture were of adrenal cortex origin, the low production of hormones by the cultures could be accounted for. To get a more precise description of the cell population of these cultures, four-week-old cultures of adrenal tumor and 4-week-old control cultures of mouse lung and pituitary tumor were treated with anti-adrenal cortex antiserum as shown in Table 1. It is seen that mouse lung and mouse pituitary tumor cultures are relatively unaffected over the range of antiserum used, while mouse adrenal tumor cultures are destroyed even at the lowest concentration of antiserum used. It appears that the adrenal tumor cultures are composed almost exclusively of cells containing antigens specific to adrenal cortex, and therefore, population hetero-

105

TABLE 1

TREATMENT OF CULTURES WITH ANTI-ADRENAL CORTEXT ANTISERUM

Treatment	4-week-old culture of mouse lung	4-week-old culture of pituitary tumor	4-week-old culture of adrenal tumor
	(Number of attached and stretched cells per petri dish)		
No antiserum	10,020	2,033	7,014
6% anti-adrenal cortex antiserum	9,202	1,980	211
10% anti-adrenal cortex antiserum	16,664	—	3
25% anti-adrenal cortex antiserum	9,208	522	1
6% anti-adrenal cortex antiserum (unabsorbed)	57	20	0

Cultures were trypsinized and plated in the presence of various concentrations of anti-adrenal cortex antiserum and 1 to 2 C'H$_{50}$ units of guinea pig complement/ml. Antisera were absorbed with homogenates of mouse liver, lung, kidney, and abdominal muscle. After 12 hr of incubation, plates were fixed, stained, and scored for number of attached and stretched cells/plate.

geneity can probably be excluded as an explanation for the low production of steroids by tumors in culture for the first time.

(4) *Steroid production by adrenal tumor cultures:* Figure 2 illustrates the procedure used to detect the production of steroids by cultures of adrenal tumors. Cultures were incubated with and without ACTH as described in the section on *Methods.* The typical absorption spectrum for adrenal delta-4-3-ketosteroids is obtained when the cultures are stimulated with ACTH, while a smaller but still distinct peak representing basal secretion is obtained in the absence of added ACTH. Extraction of the culture cells themselves from replicate cultures prior to ACTH stimulation does not yield any steroid-like material. The steroids found in the incubation buffer must, therefore, represent synthesis rather than release of preformed material.

The steroids secreted by these cultures as analyzed by paper chromatography yielded a spectrum similar to that observed earlier in the case of tumor slice secretion,[7] with the exception that 11-OH-androstene-3,17-dione, formerly a major steroid component was now definitely absent. An unidentified steroid (Compound VI), which was produced in increasing quantities by tumor slices following successive tumor transplantations, was now the major steroid component of the adrenal cell culture secretion. On the basis of paper chromatographic characteristics, this compound is probably a hydroxylated progesterone derivative with two hydroxyl groups on the molecule, but without hydroxylation at C_{21}. Three minor steroids were also observed on the paper chromatogram as was the case with adrenal tumor slice incubation extracts.

(5) *Culture growth and steroid production:* In order to know how the specific activity of steroid production is maintained as the culture grows, the experiment presented in Figure 3 was performed with an adrenal tumor arising from injection of a 45-day old culture. An unexpected finding was that the specific activity of steroid production (measured as gammas steroid/mg protein/2 hr) of cultures was six times as great as the specific activity of slices of tumor from which the culture was derived. It is seen that the cells can multiply with maintenance of the specific activity of steroid production. However, as the plates become overgrown, the specific activity begins to decline, in this case by a factor of two. Upon subculture, steroidogenesis becomes undetectable and reappears after some time to an ex-

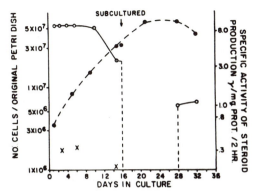

FIG. 3.—Culture growth and steroid production. Growth (solid circles) and the rate of steroid production under maximal ACTH stimulation (open circles) were followed for 34 days for one adrenal tumor culture derived from a tumor previously passaged through culture. After subculture, steroid production is not detectable for a variable length of time as indicated by the dashed line. The X's represent the specific activity of steroid production of an adrenal tumor which had never been in culture previously.

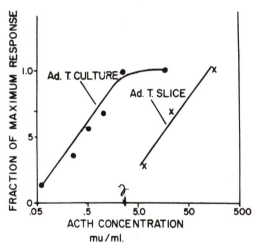

FIG. 4.—ACTH dose-response relationship. Adrenal tumor cultures and adrenal tumor slices were stimulated with graded doses of ACTH and the steroid response measured. The cultures and the tissue slices were derived from the same tumor. The relationship between response and ACTH dose is the same for slices and cultures but higher ACTH doses are required for stimulation of the slices.

tent which is variable. This result is a constant feature of our experiments and is found whether the plates are harvested for subculture by means of trypsin, scraping, or versene. The initial high level of steroid production has been maintained for up to 90 days, but usually the level of steroid production begins to decline after the first month of culture. Cultures which no longer produce any detectable steroids, upon injection into mice, produce tumors with fully restored hormonal activity. Figure 3 also shows that tumors in culture for the first time possess a low specific activity of steroid production.

(6) *ACTH-dose-response relationship:* In the course of the previous experiments, it was found that adrenal tumor cultures are maximally stimulated by a concentration of ACTH as low as 5 mu/plate. On the other hand, it was known that tumor slices require about 200 mu ACTH/100 mg tissue for maximal stimulation. The response of both tumor slices and tumor cultures to graded doses of ACTH is presented in Figure 4. It is seen that the same relationship between response and ACTH concentration holds for both cultures and slices; namely, the response approaches a maximum level at high ACTH concentrations and is proportional to the logarithm of the ACTH concentration at limiting concentrations, but the slices require about 40 times as much ACTH as the cultures to obtain the same rate of steroid synthesis. The response of these cultures to ACTH is apparently specific, since in preliminary experiments, tissue extracts, horse serum, culture media from cultures other than pituitary tumor cultures, and pituitary hormones other than ACTH do not elicit this response.

(7) *Detection and quantitation of ACTH produced by pituitary tumor cultures:* The production of steroids by adrenal tumor cultures upon ACTH stimulation sug-

gested that the system could be used to detect ACTH production by cultures of pituitary tumors. In Figure 5 are presented the results obtained when adrenal tumor cultures are stimulated with graded doses of Armour ACTH and graded doses of pituitary tumor culture medium. Although the maximum response obtained with pituitary tumor culture medium is always higher than the maximum response obtained with Armour ACTH, the same relationship between response and dosage is obtained. This enables us to estimate the amount of ACTH in the culture medium. For this purpose, we assume that the same fractional degree of stimulation is obtained with the same number of units of ACTH. It is not yet known whether the difference between Armour ACTH and mouse pituitary tumor culture medium is due to a species difference between hog and mouse ACTH or to accessory potentiating factors in the culture medium. Unlike pituitary tumor cells *in vivo* which are capable of storing up to 400 mu ACTH/mg tissue, cells in culture seem incapable of storing any ACTH, since ACTH activity has only been found in the supernatant fluid and not in the cells themselves.

(8) *Growth of pituitary tumor cultures and ACTH synthesis:* Figure 6 illustrates the growth curve of a culture of a pituitary tumor which had previously been in culture for six days and the amounts of ACTH produced per petri dish during various 24-hr periods. As the number

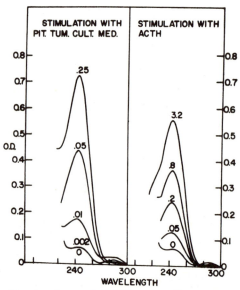

FIG. 5.—Assay of ACTH produced by pituitary tumor cultures. The family of curves on the right represent the absorption spectra obtained when adrenal cultures are stimulated with 0, 0.05, 0.2, 0.8, and 3.2 mμ ACTH per ml incubation medium. The curves on the left were obtained when the adrenal tumor cultures were stimulated with 0, 0.002, 0.01, 0.05, and 0.25 ml of pituitary tumor culture medium per ml incubation medium.

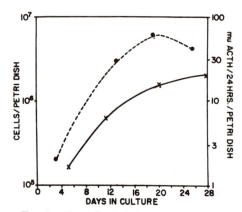

FIG. 6.—Growth of pituitary tumor cultures and ACTH synthesis. The solid curve represents the growth of a pituitary tumor in culture. The dashed curve represents the amount of ACTH produced per petri dish over a 24-hr period.

of cells increases, the rate of ACTH synthesis increases in a parallel manner. At about the thirtieth day of culture, ACTH production usually begins to decline. However, cultures which have ceased producing ACTH are capable of giving rise to tumors with renewed capacity for ACTH synthesis.

Discussion.—Since the adrenal tumor cells multiply and maintain steroid secretion for as long as ninety days in culture, it would seem that in their present state of development, the cultures provide a sensitive system both for the study of the mechanism of ACTH action and for the assay of ACTH. The pituitary tumor cultures, which maintain function for several weeks, also provide a convenient system for the study of the mechanism controlling ACTH secretion. Both cultures offer possibilities for study of the factors involved in the loss of specialized function in culture. The adrenal tumor cultures are especially suited for this purpose because considerable knowledge is available on the enzymatic steps involved in steroidogenesis, and the loss of the enzymatic steps can be followed individually.

Although cultures of tumors which no longer show any detectable specific function are still able to give rise to tumors with restored hormonal activity, it is not clear whether cells which have lost this activity regain it *in vivo*, or whether non-producing cultures still contain a few cells which continue to produce hormones and which are responsible for generating the new tumor cell population. It is hoped that cloning experiments will resolve this question.

Summary.—Cultures of adrenal and pituitary tumor have been obtained by alternate passaging of tumors through culture and animal. This procedure enhances the hormonal activity and growth capacity of these cells in culture. Cultures ultimately lose their ability to produce hormones but retain the ability to give rise to tumors with restored activity.

The steroid analyses were kindly performed by Eric Bloch of Albert Einstein Medical College, Yeshiva University. We are deeply appreciative of the excellent technical assistance of Jeanne Thivierge and Maria Braeats. We thank Victor Z. Stollar and Stanley E. Mills for their helpful suggestions.

* This is publication no. 163 of the Graduate Department of Biochemistry, Brandeis University, Waltham, Massachusetts. Supported by grants from the National Institutes of Health and the National Science Foundation.

† Supported by a National Institutes of Health Training Grant (2B-891).

‡ Purchased from the Viobin Corporation, Monticello, Illinois.

[1] Levintow, L., and H. Eagle, *Ann. Rev. of Biochem.*, **30**, 605 (1961).

[2] Furth, J., in *Recent Progress in Hormone Research*, ed. Gregory Pincus (New York, Academic Press, 1955), vol. 2, p. 22.

[3] Furth, J. in *Hormone Production in Endocrine Tumours*, Ciba Foundation Colloquia on Endocrinology, vol. 12, ed. G. E. W. Wolstenholme and Maeve O'Connor (Boston: Little, Brown and Company), p. 3.

[4] Cohen, A. I., J. Furth, and R. F. Buffet, *Am. J. Path.*, **23**, 631 (1957).

[5] Cohen, A. I., E. Bloch, and E. Celozzi, *Proc. Soc. Exp. Biol. Med.*, **95**, 304 (1957).

[6] Cohen, A. I., and J. Furth, *Cancer Res.*, **19**, 72 (1959).

[7] Bloch, E., and A. I. Cohen, with introduction by J. Furth, *J. National Cancer Inst.*, **24**, 97 (1960).

[8] Sato, G., L. Zaroff, and S. E. Mills, these PROCEEDINGS, **46**, 963 (1960).

[9] Zaroff, L., G. Sato, and S. E. Mills, *Exp. Cell Res.*, **23**, 565 (1961).

[10] Lowry, O. H., N. F. Rosebrough, A. L. Farr, and R. J. Randall, *J. Biol. Chem.*, **193**, 265 (1951).

Reprinted from the Proceedings of the National Academy of Sciences
Vol. 56, No. 1, pp. 296–303. July, 1966.

INDUCTION OF TYROSINE α-KETOGLUTARATE TRANSAMINASE BY STEROID HORMONES IN A NEWLY ESTABLISHED TISSUE CULTURE CELL LINE

By E. Brad Thompson, Gordon M. Tomkins, and Jean F. Curran

NATIONAL INSTITUTE OF ARTHRITIS AND METABOLIC DISEASES,
NATIONAL INSTITUTES OF HEALTH, BETHESDA, MARYLAND

Communicated by C. B. Anfinsen, May 4, 1966

Our attention was drawn to the possibility of studying enzyme induction in tissue culture by a brief report from Pitot et al.[1] that the Reuber hepatoma in tissue culture showed an increase in tyrosine transaminase activity in response to treatment with hydrocortisone.[1] Through the courtesy of Dr. H. P. Morris of the National Cancer Institute, we obtained primary cultures from rats containing two lines of hepatomas in the ascites form. Each of these resulted in a permanent tissue culture line. In this paper we describe the characteristics of one of these lines in which glucocorticoids induce a rapid, substantial increase in the activity of tyrosine α-ketoglutarate transaminase (L-tyrosine:2-oxoglutarate aminotransferase, EC 2.6.1.5). While being carried in serial transfer for over a year, this cell line (designated HTC for hepatoma tissue culture) has remained stable as to growth and inducibility. Furthermore, from single cells of the original line, clones have been isolated which continue to show the same characteristics.

Materials and Methods.—Sera were obtained from Microbiological Associates; the NIH medie unit prepared Swim's medium. Merck, Inc., very kindly supplied the dexamethasone phosphata (Dx). Porcine kidney p-hydroxyphenyl pyruvate keto-enol tautomerase (4.3 K units/ml) was purchased from Sigma Chemical Co., centrifuged at 20,000 g for 10 min, the pellet discarded, and the soluble fraction stored frozen.

The cell line described here came from an ascites tumor[2] which in turn had been derived from a solid hepatoma (#7288c) originally induced by feeding male Buffalo rats a diet containing 0.04% N,N′-2,7-fluorenylenebis-2,2,2-trifluoroacetamide for 12.4 months.[3, 4] Primary culture was carried out by sterile peritoneal puncture and withdrawal of 0.1 ml ascitic fluid which was placed in a T30 culture flask to which 5 ml growth medium was at once added. After an initial lag of a few days, a layer of epithelioid cells grew out. For the first 8 months, growth was maintained in tightly stoppered bottles in a standard laboratory incubator, but since then a humidified CO_2 incubator running with 3% CO_2—97% air has been used with the bottles stoppered loosely. The growth medium was Swim's medium 77 (S77). S77 has the same composition as S103 described by Swim and Barker,[5] except that hydroxyproline was omitted, the serine concentration was 0.2 mM, and choline bitartrate was substituted for choline chloride. S77, when used to support growth, was supplemented with 20% bovine serum and 5% fetal bovine serum. Penicillin G, 100 units/ml, streptomycin sulfate, 12.5 μg/ml, were added for routine culturing; however, periodically they have been omitted for several days and the medium was then cultured for bacterial contamination. Checks for PPLO contamination also have been carried out by plating on "mycoplasma agar" as described by Hayflick.[6] No mycoplasmas have been found.

Periodically, cells have been frozen in 5% glycerol—95% growth medium by standard techniques[7] and stored in liquid nitrogen. Upon thawing after as much as a year of such storage, HTC cells exhibited the same growth and induction as the original line. Chromosome preparations, stained with Giemsa, were carried out by a slight modification of the method of Tjio and Puck.[8]

For enzyme induction studies, growth medium was replaced by serum-free S77; then the cells were gently shaken free, pooled, and apportioned as needed for a given experiment.

Tyrosine transaminase was assayed as follows: Each aliquot of cells was centrifuged from the induction medium at 600 g for 5 min at 0°, twice washed and recentrifuged with aliquots of 0.15 M sodium phosphate, pH 7.9 at 0–4°, and the resulting pellet frozen. To the frozen pellet 0.5 ml

296

of the buffer was added and the cells were ruptured with a chilled probe sonicator, using two 10-sec bursts at 2 amp in an ice bath. The broken cell suspensions were centrifuged at 20,000 *g* for 10 min and the supernatant solution was assayed for tyrosine transaminase activity by the following modification of the method of Lin and Knox.[9] A total volume of 1.0 ml buffered with a final concentration of 0.5 *M* sodium borate, pH 7.8, contained 0–0.25 ml of centrifuged cell extract in phosphate buffer, 5×10^{-3} mmoles of tyrosine, 0.010 ml porcine p-hydroxyphenylpyruvate-enol-keto-tautomerase, 2×10^{-4} mmoles of pyridoxal phosphate, 0.01 mmoles of α-ketoglutarate, and 0.15 *M* sodium phosphate, pH 7.8, to volume. Enzyme activity was followed by the rate of increase of absorption of light at 310 mμ on a Gilford recording spectrophotometer. One unit of transaminase activity was defined as the quantity of enzyme which catalyzed the formation of 1 mμmole of the p-hydroxyphenylpyruvate-enol borate complex per minute. The molar absorbancy of the complex was taken to be 10,700 at 310 mμ.[10] Specific activity was expressed as units per mg protein. Proteins were estimated by the method of Lowry *et al.*[11] using bovine serum albumin as a standard.

Partial purification of tyrosine transaminase with negligible loss of activity was obtained by centrifuging cell sonicates as described above and submitting the supernatant solution to 60°C for 1 hr in the presence of 8×10^{-4} *M* pyridoxal phosphate and 1.3×10^{-2} *M* α-ketoglutarate. Denatured proteins were then removed by centrifugation at 12,000 *g* for 10 min.

Disc electrophoresis was carried out on 7% lower gel in a standard Canalco apparatus by the method of Ornstein and Davis.[12] The experiments were run for about 2 hr at a constant current of 3 ma per tube in pH 8.3 Tris-glycine buffer. To the upper tank were added 1% tracking dye, 1.56×10^{-2} *M* α-ketoglutarate and 1.2×10^{-2} *M* pyridoxal phosphate. Gels were stained either for proteins (Amido Schwartz) or for tyrosine transaminase activity. The latter stain was carried out by linking the tyrosine transaminase reaction to the reduction of a tetrazolium dye to its insoluble red formazan. The reaction mixture contained per ml: 0.05 ml of 0.33 *M* α-ketoglutarate, 0.2 ml of 0.025 *M* tyrosine, 0.005 ml tautomerase (4.3–5.0 K units/ml), 0.005 ml of 0.02 *M* pyridoxal phosphate, 0.0025 ml of a suspension of 20 mg/ml crystalline glutamate dehydrogenase, 0.25 ml of iodonitrotetrazolium 3.2 mg/ml, 0.1 ml of DPN 10 mg/ml, 0.05 ml of 0.4 mg/ml phenazine methosulfate, and 0.15 *M* sodium phosphate pH 7.9 to make 1.0 ml. This solution was added directly to the unstained disc gel which was observed while the reaction took place (10–30 min); then the reaction was stopped by the addition of 7% acetic acid.

Results and Discussion.—General characteristics of HTC cells: This cell line was originally cultured in October 1964, and then carried in an unbroken series of 59 transfers over 12 months. Since then, cells frozen at passage 33 have been used and have been carried another 35 transfers. Like other tumor lines in culture, these cells form multilayered confluent sheets on glass surfaces[13] and exhibit logarithmic growth with a doubling time of approximately 24 hr (Fig. 1). Histologically they have the characteristics of "epithelioid" cells (Fig. 2) showing irregular cytoplasmic projections when growing in contact with glass and isolated from other cells, but becoming more rounded as intercellular contact is established. No blood elements or fibroblasts were seen. Compared to normal rats, which have chromosome number of 42,[14–16] a count of 100

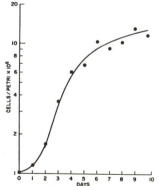

FIG. 1.—Growth curve of uncloned HTC cell line. Aliquots of 10^6 cells each were placed in a series of Petri dishes to each of which 10 ml of fresh growth medium was added. Each day the cells in one Petri were trypsinized and counted, and the medium was renewed by replacing half with fresh. Each point represents the average of at least two cell counts of at least 200 cells.

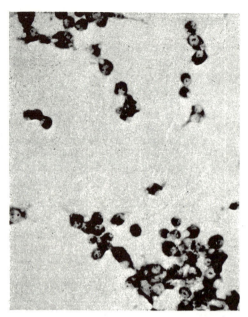

Fig. 2.—HTC cells grown on a coverslip and stained *in situ*. Giemsa stain, ×260 magnification. Area chosen to show both grouped and individually growing cells.

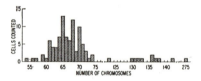

Fig. 3.—Distribution of chromosome content in uncloned HTC line, 100 cells in mitosis counted.

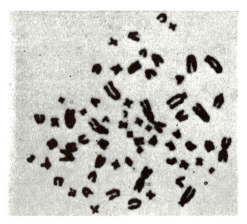

Fig. 4.—Typical set of chromosomes from cell of uncloned HTC line.

mitotic figures of HTC cells at the 46th transfer revealed a hypotetraploid number with a mean around 66, and 5% 2s and 1% 4s figures (Fig. 3). Of special interest, since they are not seen in normal rat cells, were the several metacentric chromosomes observed in all mitotic figures studied from HTC cells (Fig. 4). Tjio and Levan have described similar chromosomes in the Yoshida rat sarcoma.[14]

Cloning: Where approximately 10^5 cells/ml were used, there was a plating efficiency of 95 per cent. In contrast to the Reubner hepatoma cell line cultured by Pitot *et al.*,[1] HTC cells form clones quite easily either by serial dilution with growth on glass or in agar or by direct manual isolation of individual cells.[14-19] The cloning efficiency of dilute cell suspensions (10 cells/ml) was approximately 60 per cent.

Induction of tyrosine transaminase: Serum, although essential for growth, was not required for the induction of tyrosine transaminase. In serum-containing medium, the enzyme was inducible during either the logarithmic or the stationary phase of the growth curve. In order to provide better-defined conditions, the induction experiments described in this paper were performed in the serum-free medium, S77. In S77, cell division was not observed during a 4-day period, but when serum was added, growth promptly resumed.

After the addition of the synthetic steroid hormone, dexamethasone phosphate (Dx) 10^{-5} M, to a culture of HTC cells in S77 (Fig. 5), there was a 2-hr lag and then enzyme activity rose for 5–8 hr to a new plateau about 10 times the baseline level where it was maintained or slightly increased for the next 30 hr. Figure 6 shows the rate of enzyme increase at various concentrations of Dx, and in separate experi-

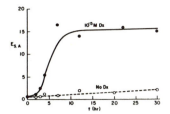

Fig. 5.—Kinetics of induction of tyrosine transaminase activity in HTC cells at 37°. At 0 time pooled cells were divided into two Petri dishes, one of which contained $10^{-5} M$ Dx. At times shown, aliquots were assayed for tyrosine transaminase as described in *Materials and Methods*. $E_{S.A.}$ refers to tyrosine transaminase specific activity.

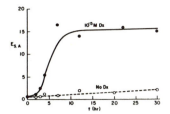

ments, Dx concentrations higher than $10^{-5} M$ produced no further increase in induction rate. Among the hormones tested, nonsteroid as well as steroid, glucocorticoids were the most efficient inducers of tyrosine transaminase activity (Table 1). Deoxycorticosterone, with its relatively weak glucocorticoid action, induced erratically and to a lower level. Aldosterone at $3 \times 10^{-8} M$ was not active but at $10^{-5} M$ induced to about one half the level achieved with glucocorticoids.

Steroid must be continuously present in order to maintain an induced level of enzyme, for if maximally induced cells were gently washed with induction medium containing no steroid and allowed to incubate further, the enzyme activity returned in a few hours to its uninduced level as the following experiment demonstrates: Several bottles of HTC cells were induced with $10^{-5} M$ Dx for 16 hr, the original induction medium was decanted and reserved, and the cells were pooled and washed twice with steroid-free S77. Duplicate samples were taken for enzyme activity determination and the remaining cells were divided into 5 aliquots, each of which was resuspended in one of the following media: (A) fresh S77 containing $10^{-5} M$ Dx, (B) the original induction medium, (C) conditioned medium containing $10^{-5} M$ Dx, (D) steroid-free conditioned medium, or (E) steroid-free fresh medium. (Conditioned medium refers to hormone-free S77 left 16 hr in contact with a fully grown layer of cells and then freed of cells by centrifugation prior to use.) As can be seen (Fig. 7), with conditions A, B, and C the induced enzyme level was maintained and even increased somewhat during the 10 hr of the experiment, whereas under conditions D and E, the enzyme after the first hour rapidly fell to basal levels. These data suggest that steroid was required for the maintenance of the induced level of enzyme activity. Kenney has mentioned that a very similar phenomenon occurs in cells of the line cultured from the Reuber hepatoma.[20]

Characteristics of tyrosine transaminase from HTC cells: As is found with liver tyrosine transaminase,[21] the enzyme from HTC cells was relatively heat-stable and was even more so in the presence of α-ketoglutarate. The heat stability of the steroid-induced enzyme from HTC cells was investigated further under various conditions as follows: a centrifuged cell extract containing tyrosine transaminase was passed through a column of Sephadex

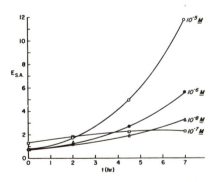

Fig. 6.—Relation of rate of tyrosine transaminase induction to concentration of Dx. A pool of cells was divided into Petri dishes in S77 at 37° with Dx at the concentrations indicated. Aliquots were assayed for enzyme at times shown. *Ordinate:* tyrosine transaminase specific activity as in Fig 5.

TABLE 1

RESPONSE OF TYROSINE TRANSMINASE ACTIVITY IN
HTC CELLS TO VARIOUS STEROID HORMONES

Treatment	Tyrosine transaminase specific activity, units/mg protein
Zero time, uninduced	2.8
No additions	1.6
0.1% ethanol (95%)	2.4
Dx, 10^{-4} M	15.0
Dx, 10^{-4} M, + 0.1% ethanol	12.8
Triamcinolone, 10^{-4} M, + 0.1% ethanol	10.5
Hydrocortisone hemisuccinate, 10^{-5} M, + 0.1% ethanol	14.3
Deoxycorticosterone, 10^{-5} M, + 0.1% ethanol	7.7
Aldosterone, 10^{-5} M, + 0.1% ethanol	6.7
Aldosterone, 3×10^{-8} M, + 0.1% ethanol	1.8
Stilbesterol diphosphate, 10^{-5} M, + 0.1% ethanol	1.4
17 β-Estradiol, 10^{-5} M, + 0.1% ethanol	2.2
Testosterone, 10^{-5} M, + 0.1% ethanol	2.4
Progesterone, 10^{-5} M, + 0.1% ethanol	3.6

After sampling in duplicate for zero time, uninduced enzyme a pool of cells was divided into Petri dishes in 10 ml S77 pre-warmed to 37°. To duplicate Petri dishes, steroid and/or ethanol was added to give the final concentrations indicated. After 15 hr in the CO_2 incubator, cells were harvested, washed, and assayed for enzyme as before. Tyrosine transaminase specific activity expressed as in Fig. 5. Figures given represent the average of duplicate samples.

G-25. The gel-filtered extract was then heated in a water bath at 56° and under these conditions about 50 per cent of the enzyme activity remained after 22 min. Either α-ketoglutarate, 1.3×10^{-2} M, or pyridoxal phosphate, 8.0×10^{-4} M, or both almost completely protected the activity. Interestingly, however, pyridoxal, 2.0×10^{-3} M, destabilized the enzyme so that there was complete loss of activity on heating for 20 min. Pyridoxal phosphate protected the enzyme against the loss of activity incurred by heating in the presence of pyridoxal but did not restore activity to the inactivated protein. Neither L-tyrosine at 1.4×10^{-3} M nor Dx at 1.25×10^{-3} M influenced the heat stability of the transaminase.

Using heat denaturation, the tyrosine transaminase activities from induced and uninduced cells have been compared as illustrated in Figure 8. Figure 8a shows that basal and induced enzymes when heated for 5 min in the presence of pyridoxal phosphate and α-ketoglutarate demonstrated identical temperature stability curves. In Figure 8b, the kinetics of heat denaturation at 76° are compared and again the induced and basal enzymes behave identically. Disc gel electrophoresis also suggested identity of induced and uninduced enzymes. When cell sonicates, partially purified by heating at 60° for 1 hr in the presence of pyridoxal phosphate and α-ketoglutarate, were electrophoresed and stained for tyrosine transaminase activity, basal and induced enzyme were found at a single, identical position. Therefore, the basal enzyme was indistinguishable from the steroid-induced enzyme. With a gel-filtered extract of steroid-induced HTC cells, the K_m for α-ketoglutarate was 6.65×10^{-4}; for tyrosine, 1.82×10^{-3}; and the pH optimum was 7.8.

Inhibition of induction: In contrast to experiments in intact animals, there was no detectable stimulation of over-all amino acid incorporation into protein during induction (Table 2).

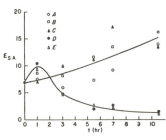

FIG. 7.—Effect of removing Dx on tyrosine transaminase activity. See text for experimental details. *Ordinate:* tyrosine transaminase specific activity as before.

TABLE 2

INFLUENCE OF INHIBITORS OF PROTEIN SYNTHESIS ON THE INCREASE IN TYROSINE TRANSAMINASE ACTIVITY INDUCED BY DEXAMETHASONE

0 hr			1 hr	5 hr	
Tyrosine transaminase (units/mg prot)	Leucine-C14 incorporation (cpm/mg prot)	Additions	Leucine-C14 incorporation (cpm/mg prot)	Tyrosine transaminase (units/mg prot)	Leucine-C14 incorporation (cpm/mg prot)
		None	257	3.2	1000
		Dexamethasone phosphate, 10^{-5} M	236	10.9	1018
3.0 (pooled cells)	12	Dexamethasone + puromycin, 4.2×10^{-4} M	48	1.7	68
		Dexamethasone + cycloheximide, 10^{-4} M	33	1.8	53
		Dexamethasone + chloramphenicol, 10^{-2} M	30	3.0	33
		Dexamethasone + progesterone, 10^{-4} M	80	2.0	273

Cells from five bottles were pooled in 100 ml of S77 at 37°. Two µc of C14-leucine (Schwarz, 5 µc/µmole, 2 µmoles/ml) was added and duplicate aliquots were immediately taken for 0 hr enzyme activity and radioactivity estimation. The remaining pooled cells were then divided among prewarmed Petri dishes containing various additions and replaced in the CO2 incubator. After 1 hr and 5 hr, aliquots were removed for estimation of radioactivity. Each aliquot was pipetted into an equal volume of 10% trichloroacetic acid (TCA), allowed to stand at 4° overnight, heated for 25 min at 85°, and chilled in an ice bath for 15 min. The precipitates were separated by slow centrifugation, washed twice in 5% TCA at 25°, drained, and dissolved in a small volume of 0.1 N NaOH. Aliquots from this were taken for protein estimation and for counting in Bray's solution in a Nuclear-Chicago scintillation counter. Five-hr aliquots were also taken for estimation of enzyme activity.

However, as is seen in liver, both perfused and *in vivo*,[22–25] induction of tyrosine transaminase in HTC cells in culture was blocked by the inhibitors of RNA synthesis, actinomycin D, and mitomycin C, as well as by compounds which more specifically prevent protein synthesis. Mitomycin C at the levels used in these experiments inhibits RNA as well as DNA synthesis. Table 2 demonstrates that puromycin, cycloheximide, and chloramphenicol, sufficient to inhibit C14-leucine incorporation into TCA-precipitable material by at least 90 per cent, also completely inhibited induction. Table 3 summarizes three other experiments in which induction was followed by a longer period. In these experiments, actinomycin D, mitomycin C, or cycloheximide prevented the rise in induced enzyme activity. The variety of inhibitors of protein synthesis which also interfere with enzyme induction strongly suggests that in this case, just as in liver, enzyme induction is due to a more rapid *de novo* synthesis of enzyme molecules.[2] Many recent studies indicate that actinomycin D and mitomycin C have numerous

FIG. 8.—A pool of cells was divided in half and incubated in S77 overnight, one half with and one half without 10^{-5} M Dx. Cell extracts were obtained by sonication and α-ketoglutarate and pyridoxal phosphate added to 1.3×10^{-2} M and 8×10^{-4} M, respectively. Aliquots were assayed for enzyme activity, and from the remaining extracts aliquots were heated for various times and temperatures in a controlled-temperature water bath. After heating, any precipitate was removed by centrifugation at 12,000 × g for 10 min, and the supernate assayed for transminase activity. (a) Each point average of duplicate assays. (b) Sonicates heated for various times at 76°. Each induced enzyme point was the average of three experiments in duplicate, and each basal enzyme point the average of one experiment in duplicate.

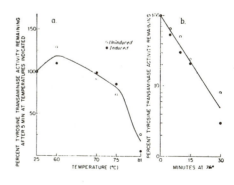

TABLE 3

INFLUENCE OF ACTINOMYCIN D, MITOMYCIN C, AND CYCLOHEXIMIDE ON
DEXAMETHASONE INDUCTION OF TYROSINE TRANSAMINASE

Additions	Fraction of Initial Tyrosine Transaminase Specific Activity Time (hr)						
	0	2	5	6	8	12	24
(Expt. 1)							
None	1.0	0.7	0.9	—	0.8	0.7	0.6
Dexamethasone phosphate, 10^{-5} M	1.0	0.8	1.5	—	5.2	4.2	5.2
Actinomycin D, 5 μg/ml	1.0	1.0	1.0	—	1.7	0.9	1.0
Actinomycin D + dexamethasone phosphate, 10^{-5} M	1.0	0.9	1.0	—	1.1	1.1	0.8
(Expt. 2)							
None	1.0	1.5	1.1	—	1.0	—	0.7
Dexamethasone phosphate, 10^{-5} M	1.0	3.1	6.9	—	13.2	—	7.5
Dexamethasone + cycloheximide, 10^{-4} M	1.0	0.8	0.6	—	0.9	—	0.7
(Expt. 3)							
Dexamethasone phosphate, 10^{-5} M	1.0	—	—	6.1	—	—	7.3
Dexamethasone + mitomycin C, 15 μg/ml	1.0	—	—	3.3	—	—	1.9

Three experiments are shown. In each, cells were pooled in S77 and, after taking aliquots for basal enzyme estimation, subdivided into appropriate groups which were treated as indicated. At the times shown, aliquots were removed for enzyme determination. Experiment 3, using mitomycin C, was carried out by Dr. Beverly Peterkofsky, who kindly permitted us to include these data. All activities are expressed in fractions relative to basal tyrosine transaminase specific activity, which varied slightly, as discussed in the text.

actions in addition to their effect on DNA-directed RNA synthesis;[25-31] therefore, interference with induction by these compounds can be taken only as suggestive evidence that RNA synthesis specifically is required for induction.

In contrast to the inhibition of induction seen when actinomycin D was added simultaneously with steroid, a stimulatory action of the inhibitor was seen when 5 μg/ml was added to previously induced cells in which tyrosine transaminase had reached plateau levels. The enzyme level did not fall as might be expected, but rose instead, and the rise was inhibited by puromycin (Fig. 9). This paradoxical effect of actinomycin seems analogous to that described *in vivo* in the rat liver[25] and will be presented in detail in a separate communication.

Summary.—Tyrosine α-ketoglutarate transaminase can be induced by steroid hormones in a newly established line of tissue culture cells, derived from primary culture of the ascites form of an experimental rat hepatoma. The isolation, growth, and morphology of the tissue culture cells are described. Heat inactivation studies and disc gel electrophoresis suggest that induced and basal tyrosine transaminase are the same. The enzyme activity is stabilized to heat by pyridoxal phosphate or α-ketoglutarate, but made unstable to heat by pyridoxal. Kinetics of tyrosine transaminase induction reveal a 2-hr lag after addition of steroid, followed by a rapid

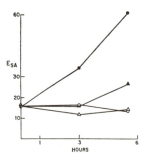

FIG. 9.—Several bottles of cells induced by incubating 40 hr in S77 containing 2 × 10^{-5} M Dx were pooled, washed twice with steroid-free S77, and suspended in S77 at 37°. After taking duplicate samples for zero time tyrosine transaminase levels, the remaining cells were divided into Petri dishes. Dx at 2 × 10^{-5} M was added to all dishes, and then groups of four received, respectively, actinomycin D 5 μg/ml (●), puromycin 8 × 10^{-4} M (○), or actinomycin plus puromycin (Δ). At 3 hr and 5½ hr, aliquots were taken for estimation of enzyme activity. Each point represents the average of quadruplicate samples. Controls received only Dx (▲). $E_{S.A.}$ signifies tyrosine transaminase specific activity as in previous figures.

rise in enzyme for 5–8 hr to a plateau level about 10 times the basal activity. The induced level is maintained as long as steroid is present. The induction is blocked by puromycin, cycloheximide, chloramphenicol, progesterone, actinomycin D, and mitomycin C. After induction by steroid has taken place, actinomycin D produces a further increase in enzyme activity.

The authors would like to express their gratitude to Drs. Pitot, Morse, and Potter for very kindly supplying details of their culture methodology; to Dr. H. P. Morris for allowing cultures to be obtained from his lines of *in vivo* hepatomas; and to Drs. J. H. Tjio and J. Whang for their instruction (and friendly encouragement) in the techniques of chromosomal analysis.

Dr. Shin-ichi Hayashi in our laboratory has crystallized rat liver tyrosine transaminase and has obtained rabbit antibody to this preparation. Preliminary experiments indicate that the basal and induced enzyme are immunologically identical in liver and HTC cells. Furthermore, transaminase induction in the tissue culture cells, as well as in the liver, is accompanied by an increase in immunologically reactive protein. These experiments will be presented in detail in subsequent communications.

[1] Pitot, H. C., C. Peraino, P. A. Morse, Jr., and Van R. Potter, *Natl. Cancer Inst. Monograph*, **13**, 229 (1964).

[2] Odashima, S., and H. P. Morris, personal communication, article in press.

[3] Morris, H. P., and B. P. Wagner, personal communication, article submitted for publication.

[4] Morris, H. P., *Advan. Cancer Res.*, **9**, 227 (1965).

[5] Swim, H. E., and R. F. Barker, *J. Lab. Clin. Med.*, **52**, 309 (1958).

[6] Hayflick, L., *Texas Rept. Biol. Med.*, **23**, 285 (1965).

[7] Peterson, W. D., Jr., and C. S. Stulberg, *Cryobiol.*, **1**, 80 (1964).

[8] Tjio, J. H., and T. T. Puck, *J. Exptl. Med.*, **108**, 259 (1958).

[9] Lin, E. C. C., and W. E. Knox, *Biochim. Biophys. Acta*, **26**, 85 (1957).

[10] Jacoby, G. A., and B. N. LaDu, *J. Biol. Chem.*, **239**, 419 (1964).

[11] Lowry, O., N. Rosebrough, A. Farr, and R. Randall, *J. Biol. Chem.*, **193**, 265 (1951).

[12] Ornstein, L., and B. J. Davis, preprint, Canal Industrial Corp., Bethesda, Md. (1961).

[13] Eagle, H., *Science*, **143**, 42 (1965).

[14] Tjio, J. H., and A. Levan, *Hereditas*, **42**, 218 (1956).

[15] Fitzgerald, P. H., *Exptl. Cell Res.*, **25**, 191 (1961).

[16] Krooth, R. S., M. J. Shaw, and B. K. Campbell, *J. Natl. Cancer Inst.*, **32**, 1031 (1964).

[17] Ham, R. G., and T. T. Puck, in *Methods in Enzymology*, ed. S. P. Colowick and N. O. Kaplan (New York: Academic Press, 1962), vol. 5, p. 90.

[18] Puck, T. T., P. I. Marcus, and S. J. Cieciura, *J. Exptl. Med.*, **103**, 273 (1956).

[19] MacPherson, I., and L. Montagnier, *Virology*, **23**, 291 (1964).

[20] Kenney, F. T., *J. Cellular Comp. Physiol.*, **66**, 141 (1965).

[21] Kenney, F. T., *J. Biol. Chem.*, **237**, 1605 (1962).

[22] Barnabei, O., and F. Sereni, *Biochim. Biophys. Acta*, **91**, 239 (1964).

[23] Greengard, O., and G. Acs, *Biochim. Biophys. Acta*, **61**, 652 (1962).

[24] Greengard, O., M. A. Smith, and G. Acs, *J. Biol. Chem.*, **238**, 1548 (1963).

[25] Garren, L. D., R. R. Howell, G. M. Tomkins, and R. M. Crocco, these PROCEEDINGS, **52**, 1121 (1964).

[26] Revel, M., H. H. Hiatt, and J. Revel, *Science*, **146**, 1311 (1964).

[27] Wiesner, R., G. Acs, E. Reich, and A. Shafig, *J. Cell Biol.*, **27**, 47 (1965).

[28] Paul, J., and M. G. Struthers, *Biochem. Biophys. Res. Commun.*, **11**, 135 (1963).

[29] Lazlo, J., D. S. Miller, K. S. McCarty, and P. Hochstein, *Science*, **151**, 1007 (1966).

[30] Smith-Kielland, I., *Biochim. Biophys. Acta*, **91**, 360 (1964).

[31] Lipsett, M. N., and A. Weissbach, *Biochemistry*, **4**, 206 (1965).

Reprinted from Proceedings of the National Academy of Sciences, Vol. 61, pp. 477–483. 1968

RETENTION OF DIFFERENTIATION POTENTIALITIES DURING PROLONGED CULTIVATION OF MYOGENIC CELLS*

By David Yaffe

DEPARTMENT OF CELL BIOLOGY, THE WEIZMANN INSTITUTE OF SCIENCE, REHOVOTH, ISRAEL

Communicated by J. Brachet, June 21, 1968

Recent studies on the regulation of macromolecule synthesis during embryogenesis indicated that in several differentiating systems, proteins are formed on a stable form of messenger RNA (mRNA).[1-4] These observations may lead to speculation as to the role of stable informational macromolecules in the determination of the irreversible nature of the differentiation process, especially with relevance to systems in which a precursor cell is differentiating, after a limited number of cell divisions, into a postmitotic one (nerve, muscle, etc.). Thus, a pre-existent information store may be responsible for the "committed" development of such cells.

In the present communication, the influence of prolonged multiplication *in vitro* on the capacity of specific precursor cells to differentiate has been studied. It was assumed that if the differentiation capacity depends on preformed informational molecules, continuous multiplication will result in gradual loss of this potentiality.

Cultures prepared from trypsin-suspended embryonic or newborn skeletal muscle cells are at first a heterogeneous population of mononucleated cells, the majority of which are spindle-shaped myoblasts. The myoblasts, which are the mononucleated precursor cells of muscle fibers, divide in culture two to three times and then begin to aggregate and fuse into postmitotic multinucleated muscle fibers.[5, 6] Further differentiation is associated with the appearance of cross-striation and contractility.

The present investigation was designed to test whether the fusion and differentiation of myoblasts into multinucleated cells is a rigidly progressive phenomenon which must occur within a definite number of cell divisions, or whether the capacity of myoblasts to fuse and differentiate can be retained over extended periods of multiplication *in vitro*. Experiments showed that, under appropriate culture conditions, it is possible to maintain myogenic cell lines for many months in a continuous state of replication without loss of differentiation potentialities.

Materials and Methods.—*Primary skeletal muscle cultures* were prepared from thigh muscle of newborn rats as described previously.[6]

Feeder layer (FL) was prepared from secondary muscle cultures exposed to 6,000 r and then plated at a cell density of 1×10^5 cells/dish.

Cloning was carried out by seeding 25–200 cells onto a feeder layer contained in collagen-coated[7] plates.

Growth media: Primary cultures were grown in a mixture of 25% medium 199 (GIBCO, powder media) and 75% Eagle's essential medium (EM), supplemented with 10% horse serum (HS) and 3% embryo extract. Clones and cell lines were cultivated in medium 199 supplemented with 2% embryo extract, 0.5% bovine serum albumin (Armor fraction V), and 10% horse or fetal calf serum.

Embryo extract was prepared from 10-day-old chick embryos according to the technique of Hauschka and Konigsberg,[7] but omitting the ultracentrifugation step.

Autoradiography and polyoma virus infection were performed as previously.[6, 8, 9]

20(3)Methylcholanthrene (MC) pellets were prepared by dissolving 6% w/w MC in melted paraffin. MC was supplied to the cultures by adding one to two pellets (2 mm in diameter) to the growth medium.

Results.—Establishment of myogenic cell lines: Attempts were made to maintain myogenic cells *in vitro* by conventional serial cell passages. However, in all cases, loss of myoblasts by their fusion into fibers, by overgrowth of other cell types, or by various degenerative phenomena occurred. Hence, we performed experiments that were aimed at developing cultivation methods which would (1) promote multiplication, (2) prevent cell fusion, and (3) select myoblasts out of the heterogeneous population of cells in the primary cultures.

In preliminary experiments, we attempted to obtain pure populations of myoblasts, using the differential trypsinization method described by Kaighn *et al.*[10] However, it was later observed that the reattachment of myoblasts to the plastic plate after trypsinization takes place much more slowly than that of the other cell types present in primary cultures. Therefore, a short time after a heterogeneous cell suspension is plated on a Petri dish, most of the fibroblastic and epithelial cells begin to attach to the surface while most of the myoblasts remain floating in the medium. Accordingly, cultures were trypsinized, the trypsin was removed by centrifugation, and the cell pellet was resuspended in growth medium. The cells were then plated in plastic 60-mm Petri dishes and incubated at 37°C. After 40 minutes the medium was recollected and the cells which remained floating were counted and plated at a concentration of 1.5×10^5 cells/dish, on collagen-coated dishes containing 1×10^5 irradiated feeder layer cells. When these cells were seeded under cloning conditions, 94 per cent of the colonies formed muscle fibers (MFC) as compared to 74 per cent MFC obtained when unfractionated cell suspensions of primary cultures were cloned.

This method was applied to thigh muscle cells serially passaged in tissue culture. In each generation, cells obtained by trypsinization of one to three cultures were plated in a Petri dish and those floating in the medium after 40 minutes were recollected and seeded as described. The cells multiplied rapidly under these conditions (with a generation time, calculated from growth curves, of 18–20 hr), and after three to four days, the cultures consisted of a network of myoblasts with very few nonmyoblastic cells. The cultures were examined daily and were repassaged in the same manner at the onset of fusion. It has repeatedly been reported that cells of primary cultures can ordinarily not be propagated by serial passages in cultures unless transformed either spontaneously or intentionally.[11, 12] Therefore, in some experiments a pellet of MC dissolved in paraffin was introduced into the culture medium during the first two cell passages, to promote "transformation" of myoblasts. Seven experiments for the serial passage of myoblasts were initiated, in three of which MC treatment was employed. Two lines (one MC-treated, one untreated) were lost by accidental bacterial contamination and three were lost after four to six passages by gradual cessation of cell multiplication, whereas two cell lines from cultures exposed to MC in the first two passages are still being maintained by serial passage without any loss in their capacity to replicate and differentiate. One of them (designated L_6) has now been maintained *in vitro* for more than 18 months.   In the course of the first few passages, the latter line was dependent on collagen and feeder layer for good growth and differentiation; however, sublines obtained by cloning now grow readily (with a generation time of about 18 hr) and differentiate into a very dense network of contracting fibers, even in the absence of collagen and feeder layer (Figs. 1–3).   The second line, L_{41}, is now eight months in culture and has characteristics indistinguishable from the L_6 line.   Cells of both lines can also be kept in a frozen state (at $-80°C$) for extended periods of time without loss of viability and of differentiation potentialities.

Fig. 1.—Muscle fiber formation by a myogenic cell line of rat origin. 1×10^5 cells of the line L_6 plated in an untreated 60-mm Petri dish.   Fixed at day 8 (Giemsa, mag. $\times 9$).

Cessation of DNA synthesis at fusion: Fusion of myoblasts into multinucleated muscle fibers is associated with the cessation of DNA synthesis.[5, 6]   Therefore, the effect on this property of prolonged cultivation of myogenic cells *in vitro* was investigated.   Cultures of the line L_6 (350 days in culture) were allowed to differentiate.   At different stages after the onset of fusion, the cells were exposed to a five-hour pulse of thymidine-H^3, then fixed and processed for autoradiography.   In no case was thymidine-H^3 incorporated into the nuclei of multinucleated fibers, thus demonstrating that cells of the myogenic lines stop DNA synthesis at fusion as myoblasts of primary cultures do.   As in primary cultures,[13] infection of the cultures with polyoma virus resulted in the induction of DNA synthesis within multinucleated fibers.

Inheritance of differentiation potentialities by single cells: To test the homo-

Fig. 2.—Same as Fig. 1, magnification ×52.

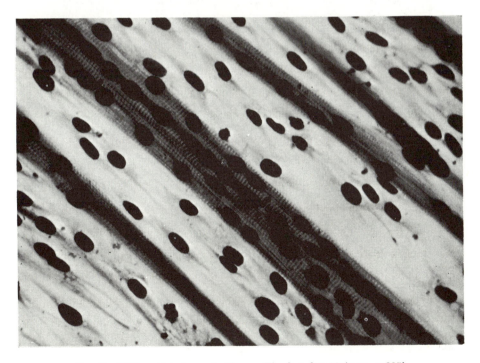

Fig. 3.—Differentiated muscle fibers. Fixed at day 12 (mag. ×325).

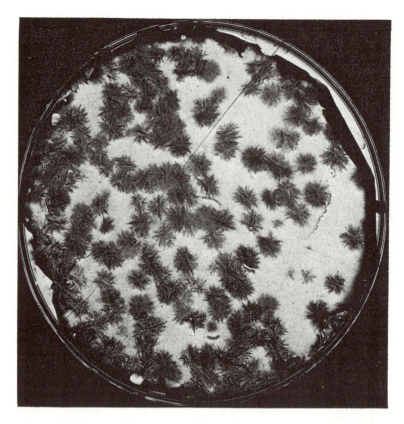

Fig. 4.—Muscle-forming colonies developed from single cells.   150 cells of L_6 plated in 60-mm collagen-coated Petri dishes, preseeded with 1×10^5 irradiated FL cells.   Fixed at day 13 (mag. $\times 1.86$).

geneity and stability of the differentiation properties of the cell population, an experiment was designed to analyze the pattern of inheritance by daughter cells of the differentiation potentialities of single cells.   Cultures of the line L_6 (300 days in culture) were trypsinized and plated at cell densities of 25–150 cells/plate on collagen-coated plates containing a feeder layer.   After 10–13 days, the cultures were fixed and the number of clones which differentiated into fibers was counted.   The plating efficiency under these conditions ranged between 65 and 80 per cent, and more than 98 per cent (in most cases, 100%) of the colonies formed muscle fibers (Figs. 4 and 5).   Single clones were isolated from these cultures and replated under cloning conditions.   All colonies developed in these cultures formed muscle fibers.

These results indicate that cells maintained in exponential growth in tissue culture for many months pass on to virtually all their progeny the capacity to fuse and differentiate into muscle fibers.   The possibility that the cloning conditions (feeder layer and collagen) were selecting out only muscle-forming cells seems unlikely, since under identical culture conditions cells of primary cultures gave rise to colonies which did not form fibers.

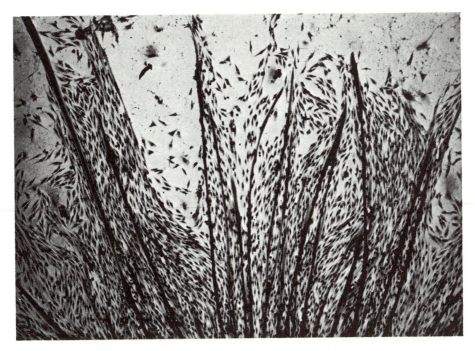

Fig. 5.—Part of a muscle-forming colony (mag. ×52).

Discussion.—Replication *in vitro* of differentiated cells often resulted in their "dedifferentiation" after a relatively short period of cultivation.[14, 15] Recently, however, Coon and Cahn have demonstrated that both retina pigment cells and cartilage cells which would dedifferentiate under "conventional" culture conditions redifferentiated when grown under appropriate cloning conditions.[16, 17] Konigsberg[18] has shown that chick thigh muscle cells can be cloned *in vitro* and differentiate into muscle-forming colonies. The present study extends this observation in demonstrating that myogenic cells can be maintained by serial passage and can multiply for at least several months in tissue culture and still retain their capacity to differentiate. Unlike retina pigment cells and cartilage cells, the myogenic cells differentiate readily both in mass culture and under cloning conditions.

The muscle system differs from other systems employed in similar studies in one important feature: differentiation of myoblasts results in the fusion of the cells into postmitotic multinucleated fibers. Therefore, throughout the experiments, cells which continued to differentiate were automatically excluded, and the line was maintained only by the *precursor cells* which did not phenotypically express their differentiation potentiality beyond the myoblastic level. Thus, these experiments clearly demonstrated that the ability to transmit differentiation potentialities to progeny cells for many generations is not dependent on the morphological manifestation of these traits. Cloning analysis has shown that these traits are transmitted in an exact manner to most or all cell progeny.

This is reminiscent of a rather common situation for many tissues *in vivo*, in which differentiated cells are "dead ends" and the continuous renewal of the tissue is carried out only by primitive, morphologically undifferentiated cells which multiply continuously (skin, hemopoietic tissues, etc.). While the possibility that the stem cells *in situ* are continuously induced to differentiate cannot be excluded, the results of the *in vitro* experiments indicate that in several differentiating cell types, the determined state or the potentiality to differentiate is maintained within the stem cell itself, without interaction with other cell types. The transmission of this state to nearly all progeny, in spite of continuous multiplication, indicates the existence within the cells of a mechanism for the *continuous reproduction* of this property.

While it is impossible to predict whether the cells of the established lines will be capable of replicating indefinitely in culture, the possibility of freeze storage makes the use of this cell system convenient for investigations of various aspects of cellular differentiation. Nevertheless, the factors which determine whether an attempt to isolate a myogenic cell line will fail or not and the nature of the changes in the cells during the establishment of the line are still obscure and need further investigation. It is of obvious importance to elucidate whether treatment with the carcinogen MC had any effect on the two lines described here. However, in a second series of experiments[19] three myogenic cell lines have been isolated from rat muscle cultures and one of mouse origin. None of them have been exposed to carcinogens. One of the rat lines is now more than seven months in culture with no decline in viability and in differentiation potentialities. It can therefore be stated that exposure of the cells to a carcinogen was *not* essential for the establishment of myogenic lines.

Summary.—Myoblasts maintained *in vitro* for many months in a state of continuous multiplication retained their capacity to fuse and differentiate into postmitotic multinucleated muscle fibers. Cloning experiments showed that the potentiality to differentiate is inherited in these lines by virtually all the cells.

I wish to thank Professor Michael Feldman, Head of the Department of Cell Biology, for his continued interest and for reviewing the manuscript. The skillful assistance of Mrs. Sara Neuman is gratefully acknowledged.

* Supported by a grant no. C-6165 from the National Institutes of Health, USPHS, Bethesda, Maryland, and by a grant from the French Government through Fonds de la Recherche Scientifique et Technique.

[1] Gross, P. R., L. I. Malkin, and W. A. Moyer, these PROCEEDINGS, **51**, 407 (1964).
[2] De Bellis, R. H., N. Gluck, and P. A. Marks, *J. Clin. Invest.*, **43**, 1329 (1964).
[3] Reeder, R., and E. Bell, *Science*, **150**, 71 (1965).
[4] Humphreys, T., S. Penman, and E. Bell, *Biochem. Biophys. Res. Commun.*, **17**, 618 (1964).
[5] Stockdale, F. K., and H. Holtzer, *Exptl. Cell Res.*, **24**, 508 (1961).
[6] Yaffe, D., and M. Feldman, *Develop. Biol.*, **11**, 300 (1965).
[7] Hauschka, S. P., and I. R. Konigsberg, these PROCEEDINGS, **55**, 119 (1966).
[8] Yaffe, D., and S. Fuchs, *Develop. Biol.*, **15**, 33 (1967).
[9] Yaffe, D., and D. Gershon, *Nature*, **215**, 421 (1967).
[10] Kaighn, M. E., J. D. Ebert, and P. M. Stott, these PROCEEDINGS, **56**, 133 (1966).
[11] Hayflick, L., and P. S. Moorhead, *Exptl. Cell Res.*, **25**, 585 (1961).
[12] Berwald, Y., and L. Sachs, *J. Natl. Cancer Inst.*, **35**, 641 (1965).
[13] Yaffe, D., and D. Gershon, *Israel J. Med. Sci.*, **3**, 329 (1967).
[14] Davidson, E. H., *Advan. Genet.*, **12**, 143 (1964).
[15] Eagle, H., *Science*, **148**, 42 (1965).
[16] Cahn, R. P., and M. B. Cahn, these PROCEEDINGS, **55**, 106 (1966).
[17] Coon, H. G., these PROCEEDINGS, **55**, 66 (1966).
[18] Konigsberg, I. R., these PROCEEDINGS, **47**, 1868 (1961).
[19] Yaffe, D., and C. Revivi, manuscript in preparation.

Reprinted from the Proceedings of the National Academy of Sciences
Vol. 64, No. 1, pp. 311–315. September, 1969.

ESTABLISHMENT OF FUNCTIONAL CLONAL LINES OF NEURONS FROM MOUSE NEUROBLASTOMA*

By Gabriella Augusti-Tocco† and Gordon Sato

GRADUATE DEPARTMENT OF BIOCHEMISTRY, BRANDEIS UNIVERSITY,
WALTHAM, MASSACHUSETTS

Communicated by Marshall Nirenberg, June 27, 1969

Abstract.—Clonal lines of neurons were obtained in culture from a mouse neuroblastoma. The neuroblastoma cells were adapted to culture growth by the animal-culture alternate passage technique and cloned after single-cell plating. The clonal lines retained the ability to form tumors when injected back into mice. A striking morphological change was observed in the cells adapted to culture growth; they appeared as mature neurons, while the cells of the tumor appeared as immature neuroblasts.

Acetylcholinesterase and the enzymes for the synthesis of neurotransmitters, cholineacetylase and tyrosine hydroxylase were assayed in the tumor and compared with brain levels; tyrosine hydroxylase was found to be particularly high, as described previously in human neuroblastomas. The three enzymes were found in the clonal cultures at levels comparable to those found in the tumors. Similarly, there were no remarkable differences between the three clones examined.

The difficulty in separating glial cells and neurons has proved to be a major obstacle in the biochemical characterization of the components of the nervous system. Methods of separation, which yield homogeneous populations of cells, are limited by the low amount of cells obtainable.[1, 2] On the other hand, the methods described for large scale preparation[3-5] give highly heterogeneous fractions and produce a large amount of cell damage.[6] Clonal cell lines of the components of nervous tissue would, therefore, provide a useful tool for the study of neurobiology. Recently, a glial cell line which retains in culture the ability to synthesize the brain specific protein S-100 has been developed from a rat brain tumor.[7] This cell line was obtained using the technique of alternate animal-culture passage described by Buonassisi et al.[8] to obtain in culture functional cell lines from functional tumors.

Previous work on human neuroblastomas indicates that this is a functional tumor and can adapt to growth in culture. Short-term cultures of human neuroblastoma explants have been reported to metabolize norepinephrine as "*in vivo*";[9] also long-term cultures of human neuroblastoma have been described to be able to rapidly metabolize norepinephrine.[10] Fast degradation of norepinephrine[11-12] and high tyrosine hydroxylase content[13] have been reported as biochemical features of neuroblastoma.

The availability of a transplantable mouse neuroblastoma offered the opportunity to apply the alternate animal-culture passages technique to select clonal lines which are more adaptable to the culture conditions and at the same time maintain their function. Choline acetylase, acetylcholinesterase, and tyrosine hydroxylase activities were measured as a test of the functional capacity of the derived cell cultures.

Materials and Methods.—The mouse neuroblastoma, C 1300, was obtained from Jackson Laboratory, Bar Harbor, Maine. C-14 acetyl CoA, spec. act. 60 mc/mM, was obtained from New England Nuclear Corp., Boston, Mass.; 3,5 H-3 L-tyrosine, 36 c/mM, from Amersham/Searle Corp., Des Plaines, Ill.; Triton X-100, L-tyrosine and butyrylthiocholine from Mann Research Laboratories, New York, N.Y.; acetyl CoA, choline iodide, acetylthiocholine iodide, neostigmine methyl sulfate, 5,5'-dithio-bis-(2-nitrobenzoic acid) (DTNB), 6-7-dimethyl-5,6,7,8-tetrahydropterine hydrochloride from Calbiochem, Los Angeles, Calif.; Dowex 50 X8, 100–200 mesh, and Dowex 1 X10, 200–400 mesh from Baker, Phillipsburg, N.J.

Tissue culture methods: The technique of alternate passage animal-culture[8] was followed to adapt the neuroblastoma cells to growth in monolayer culture. The tumor tissue was dissociated by viokase treatment and the single cells plated in Falcon plastic plates or flasks pretreated with 5% gelatin solution. The medium used was Ham's F 10[14] supplemented with 15% horse serum and 2.5% fetal calf serum.

Clonal lines were isolated following the single-cell plating technique described by Puck *et al.*[15] from cultures at the second or third passage *in vitro.* Tumors grown from clonal cultures injected into host animal will be referred to as clonal tumors and designated with the lettering of the clone of origin.

Enzyme assays: Tyrosine hydroxylase and choline acetylase were assayed as described by Wilson *et al.*[16] Acetylcholinesterase was assayed according to the method of Ellman,[17] with 1 ml as final volume of the reaction mixture. The assays were run using as substrate both acetylthiocholine ($0.5 \times 10^{-3} M$) and butyrylthiocholine ($1 \times 10^{-3} M$) to ascertain that true acetylcholinesterase activity was measured. Cholinesterases nonspecific for acetylcholine hydrolyze butyrylcholine at a faster rate than acetylcholine.[18, 19] The tumors were finely minced with scissors and then homogenized in the appropriate buffer in a glass homogenizer with a motor-driven Teflon pestle. Cultures were washed twice with phosphate buffered saline solution (PBS: NaCl 8 gm, KCl 0.2 gm, Na_2HPO_4 1.15 gm, KH_2PO_4 0.2 gm, $MgCl_2$ 0.1 gm, $CaCl_2$ 0.1 gm per liter at pH 7) and scraped with a rubber policeman directly in the required volume of buffer for homogenization. Alternatively, the cells were scraped in 2–3 ml of PBS, centrifuged down at low speed, and then homogenized in the required volume of buffer; this procedure was followed when several flasks were needed for an assay.

Proteins were determined by the Lowry method,[20] and all enzyme activities expressed as μmole or mμmoles of substrate converted in 10 min per milligram protein.

Results.—*Morphology:* Neuroblastomas have been described as highly undifferentiated tumors. The C 1300 Jackson tumor, described as a spontaneous tumor of the region of spinal cord, showed the usual morphology of neuroblastomas. Histological section of the tumor revealed the presence of only round cells. Fibers were absent.

When placed into culture, the cells undergo a striking change in morphology. The most striking characteristic of these cells is the large number of elongated processes which emanate from the cell body. These processes begin development soon after subculture or the initiation of primary culture, and within a few days form a complex network. A typical colony of a clonal line is shown in Figure 1. In each colony the cells remain rather sparse. After a few days in culture, round cells appear on the colonies (Fig. 2). They pile up on the colonies and form clumps, which tend to float away, while the cells with long processes remain attached to the plates.

Enzyme assay: The clonal line NB42B and the cultures obtained from the

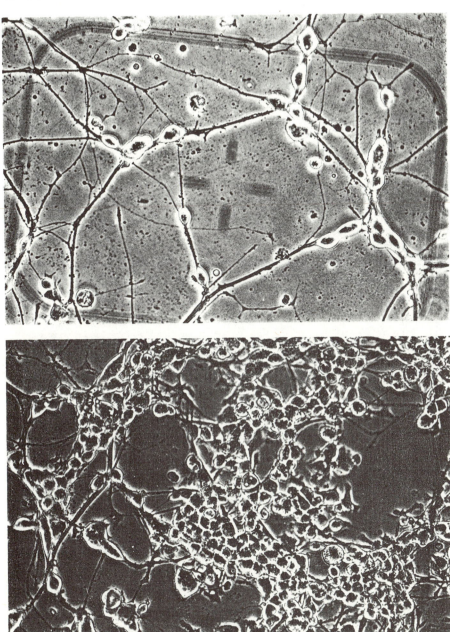

Figs. 1–2.—Phase contrast photomicrograph (×370). 4-month-old continuous culture of clone NB42B.

clonal tumor NB41A and NB41B were assayed for acetylcholinesterase, choline acetylase, and tyrosine hydroxylase activities. These results are reported in Table 1. The three enzymes were present in the original tumor. As compared with the brain, the tumor showed a higher content of tyrosine hydroxylase and a lower content of choline acetylase and acetylcholinesterase. The three clones studied, both as tumors and in culture, did not show striking differences in their enzyme contents. The differences observed between tumor and brain were much greater. Some variations, however, seem to occur going from tumor to culture. Choline acetylase activity is lower and possibly tyrosine hydroxylase activity is higher in culture than in the tumor (NB41A and NB41B). Similarly, the clonal tumors NB41A and NB41B seem to have higher choline acetylase activity and lower tyrosine hydroxylase activity than the original tumor. Acetylcholinesterase content appears to be more constant in all the conditions.

Choline acetylase, tyrosine hydroxylase, and specific acetylcholinesterase were not detected in control culture of mouse fibroblasts (AF1—Table 1).

Discussion.—Morphological observations indicate that the neuroblastoma cells can adapt to culture conditions and, moreover, that the culture conditions stimulate the immature neuroblasts present in the tumor to complete (or at least to proceed further in) their maturation. The same observations had been described a few years ago by Goldstein,[21–22] culturing explants of human neuroblastoma.

TABLE 1. *Enzyme activities of neuroblastoma cell lines and tumors.*

| | Choline Acetylase | | Tyrosine Hydroxylase | | Acetylcholinesterase | | | |
| | | | | | Acetyl | | Butyryl | |
	Tumor	Culture	Tumor	Culture	Tumor	Culture	Tumor	Culture
Brain	8.1		0.003		1.16		0.031	
Neuroblastoma	0.206	...	0.0541	...	0.125	...	0.045	...
NB41A	0.790	0.126	0.0041	...	0.107	0.144	0.024	...
NB41B	0.840	0.275	0.0082	0.0140	0.340	0.192	0.026	...
NB42B	...	0.210	...	0.0145	...	0.247	...	n.d.
AF 1	...	n.d.	...	n.d.	...	0.034	...	0.066

Choline acetylase and tyrosine hydroxylase activities are expressed in mμmoles/10 min/mg protein; acetylcholinesterase in μmoles/10 min/mg protein; n.d. indicates that enzyme activity was not detectable at homogenate concentration at least as high as that of neuroblastoma cells. NB41A and NB41B, cultures obtained from clonal tumors of mouse neuroblastoma; NB42B, clonal line of mouse neuroblastoma; AF1, clonal line of mouse fibroblast.

The data reported in Table 1 show that the neuroblastoma cells, adapted to growth in monolayer, keep the ability of the tumor of origin to synthesize the key enzymes for the transmission of the nerve impulse. At present, it is not possible to evaluate the significance of the observed variations of enzyme activities. In fact, specific activity of enzymes in the various tumors could be affected by the variable extent of necrotic areas; on the other hand, it has not yet been determined whether the enzyme activity of the cultured cells may vary during the growth cycle and therefore, whether the measured enzyme activity is the maximal one the cells can reach under the conditions of culture. Further work is in progress to ascertain this.

Our findings of the enzymes for the synthesis of norepinephrine and acetylcholine and acetylcholinesterase in clonal cell lines is in agreement with the Burn and Rand hypothesis for the transmission of impulse in adrenergic fibers[23] and demonstrate that the two neurotransmitters are present in the same neurons.

Summary.—Clonal lines of neurons were obtained from a mouse neuroblastoma by alternate animal-culture passage technique. Choline acetylase, acetylcholinesterase, and tyrosine hydroxylase activities were assayed.   Enzyme activities of cultures and tumors were of the same order of magnitude.

The authors acknowledge the expert technical assistance of Mrs. Maria Brasats.

* This is publication no. 667 of the Graduate Department of Biochemistry, Brandeis University, Waltham, Massachusetts 02154.   Supported by grants from the National Science Foundation (GB 7104) and the National Institutes of Health (CA 4123).

† On leave of absence from International Laboratory of Genetics and Biophysics (CNR), Naples, Italy.

[1] Hyden, H., *Nature*, **184**, 433 (1959).

[2] Roots, B. I., and P. V. Johnston, *J. Ultrastruct. Res.*, **10**, 350 (1964).

[3] Rose, S. P. R., *Biochem. J.*, **102**, 33 (1967).

[4] Satake, M., and S. Abe, *J. Biochem.*, **59**, 72 (1966).

[5] Bocci, V., *Nature*, **212**, 826 (1964).

[6] Cremer, J. E., P. V. Johnston, B. I. Roots, and A. J. Trevor, *J. Neurochem.*, **15**, 1361 (1968).

[7] Benda, P., J. Lightbody, G. Sato, L. Levine, and W. Sweet, *Science*, **161**, 370 (1968).

[8] Buonassisi, V., G. Sato, and A. I. Cohen, these Proceedings, **48**, 1184 (1962).

[9] La Brosse, E. H., J. Belehradek, G. Barski, C. Bohuon, and O. Schweisguth, *Nature*, **203**, 195 (1964).

[10] Goldstein, M., B. Anagnoste, and M. N. Goldstein, *Science*, **160**, 768 (1968).

[11] Robinson, R., in *Neuroblastomas Biochemical Studies*, ed. C. Bohuon (New York: Springer-Verlag, 1966), p. 37.

[12] Bohuon, C., E. H. La Brosse, M. Assicot, and A. Amar-Costesec, in *Neuroblastomas Biochemical Studies*, ed. C. Bohuon (New York: Springer-Verlag, 1966), p. 16.

[13] Goldstein, M., in *Neuroblastomas Biochemical Studies*, ed. C. Bohuon (New York: Springer-Verlag, 1966), p. 66.

[14] Ham, R. G., *Exptl. Cell Res.*, **29**, 515 (1963).

[15] Puck, T. T., P. I. Marcus, and S. J. Cieciura, *J. Exptl. Med.*, **103**, 273 (1956).

[16] Wilson, S., J. Farber, and M. Nirenberg, personal communication.

[17] Ellman, J. L., K. D. Courtney, V. Andres, Jr., and R. M. Featherstone, *Biochem. Pharmacol.*, **1**, 88 (1961).

[18] Augustinsson, K.-B., *Arch. Biochem.*, **23**, 111 (1969).

[19] Nachmansohn, D., and M. A. Rothenberg, *J. Biol. Chem.*, **158**, 653 (1945).

[20] Lowry, O. H., N. F. Rosebrough, A. L. Farr, and R. J. Randall, *J. Biol. Chem.*, **193**, 265 (1951).

[21] Goldstein, M. N., and D. Pinkel, *J. Natl. Canc. Inst.*, **20**, 675 (1958).

[22] Goldstein, M. N., J. A. Burdman, and L. J. Journey, *J. Natl. Canc. Inst.*, **32**, 165 (1964).

[23] Burn, J. H., and M. J. Rand, *Ann. Rev. Pharmacol.*, **5**, 163 (1965).

Proceedings of the National Academy of Sciences
Vol. 68, No. 2, pp. 378–382, February 1971

Hemoglobin Synthesis in Murine Virus-Induced Leukemic Cells *In Vitro*: Stimulation of Erythroid Differentiation by Dimethyl Sulfoxide

CHARLOTTE FRIEND, WILLIAM SCHER, J. G. HOLLAND, AND TORU SATO

Center for Experimental Cell Biology, Mollie B. Roth Laboratory, The Mount Sinai School of Medicine of the City University of New York, New York, N.Y. 10029, and Division of Cytology, the Sloan-Kettering Institute for Cancer Research, New York, N.Y. 10021

Communicated by S. E. Luria, November 30, 1970

ABSTRACT Cells of a cloned line of murine virus-induced erythroleukemia were stimulated to differentiate along the erythroid pathway by dimethyl sulfoxide at concentrations that did not inhibit growth. A rise in the number of benzidine-positive normoblasts was accompanied by increased synthesis of heme and hemoglobin and a decrease in the malignancy of the cells. This action of dimethyl sulfoxide, which was reversible, may represent the derepression of leukemic cells to permit their maturation.

The opportunity to explore the possibility that leukemia is a disease resulting from a block in the process of maturation of hematopoietic cells has been provided by established tissue culture lines of murine virus-induced erythroleukemic cells (1–3). These cells, which grow in suspension, have continued to exhibit a limited degree of differentiation along the erythroid line throughout their 4-year serial passage history. Although they synthesize hemoglobin (4, 5), they are malignant as tested by bioassay in syngeneic hosts. They produce virus which, although low in leukemogenic activity, is a highly effective immunizing agent (6).

During the course of studies to determine the effect of superinfecting these cells with Friend leukemia virus, dimethyl sulfoxide (DMSO) was added to the medium. DMSO had been demonstrated to enhance infectivity of both poliovirus RNA (7) and mengovirus RNA (8), as well as transformation by polyoma virus (9). It is also known to stabilize enveloped viruses (10). The wide range of biological activities of DMSO has been described (11). The present report describes an effect of DMSO on the differentiation of established lines of murine virus-induced leukemic cells and illustrates still another property of this compound.

In the dose–response experiments initially set up to determine the toxicity of DMSO on the leukemic cell lines, a striking effect on the differentiation of these cells was noted. Of the cells allowed to grow in medium containing 2% DMSO for 4 days, a majority of the erythroblasts had matured to normoblasts which stained benzidine-positive (B+). The increase in the number of cells maturing along the erythroid series in DMSO-containing medium was accompanied by an increase in the amount of hemoglobin synthesized.

Abbreviation: DMSO, dimethyl sulfoxide.
This study is dedicated to the memory of Dr. Austin S. Weisberger.

MATERIALS AND METHODS

The origin of the cell lines of murine Friend leukemia virus-induced leukemic cells and their clonal derivatives has been described previously (1, 2). Methods for maintaining the cell culture, cloning on semisoft agar, and the medium were also detailed in these earlier reports.

The present experiments were done on a clone designated 707, in which 1–2% of the cells are B+ (5). Cells were generally seeded at a concentration of 10^5 per ml (except where otherwise indicated) either in 32-oz. (900 ml) prescription bottles containing 60 ml of medium or in Falcon plastic Petri dishes (60 × 15 mm) containing 10 ml of medium. Dehydrated Eagle's basal medium diluted with Earle's balanced salt solution and supplemented with 15% fetal calf serum was used. The cultures were grown in a humidified incubator containing 5% CO_2 in air.

Certified reagent grade DMSO (Fisher) was stored in brown bottles and just prior to use was added, unsterilized, to the medium in the concentrations cited in the text on a volume-per-volume (v/v) basis.

For the dose–response experiments, replicate Petri dish cultures were incubated with 10^6 cells in 10 ml of medium with or without DMSO. At designated intervals the number of cells in each of two replicate cultures was determined by counting in a hemocytometer. Cell counts were done with trypan blue added to estimate the ratio of living to dead cells. Smears were prepared with Ralph's hemoglobin stain (12), then counterstained with Leischman, and the percent of benzidine-positive (B+) cells was determined.

For malignancy tests, the cells were inoculated subcutaneously into 8-week-old DBA/2J SPF mice of either sex (Jackson Laboratory).

The techniques for studying iron metabolism in these cells are described elsewhere (5). In brief, cells grown in medium containing ferrous (^{59}Fe) citrate (0.15 μCi of ^{59}Fe per ml) at an iron concentration of 6.8×10^{-9} M were washed in saline, lysed by freezing and thawing, and analyzed for iron as: intracellular iron (radioactivity of the total lysate), heme iron (cyclohexanone-extractable radioactivity after acid dissociation of heme and globin), and hemoglobin iron (radioactivity of B+ bands, separated by discontinuous acrylamide gel electrophoresis, of hemoglobin precipitated with ($NH_4)_2SO_4$ between 60 and 85% saturation).

RESULTS

Effect of DMSO on differentiation

After we had observed that the number of differentiating normoblasts increased among the leukemic cells grown in DMSO-containing medium, we studied the effect of several concentrations of the compound on cell growth. The dose–response curve obtained from a representative experiment of 4 days duration is shown in Fig. 1.

Cells grown in media containing 0.5 or 1% DMSO multiplied at approximately the same rate as those of the control (untreated) cultures. When the concentration was increased to 2%, there was a lag in the rate of growth during the first 48 hr, but by the 96th hour the number of cells approximated that of the controls. At a concentration of 3% DMSO, cell growth was inhibited, the cultures remaining almost stationary, and at 5% the compound was cytocidal, no living cells remaining after 72 hr.

When the cells were centrifuged for staining, it was noted that the cell pellets of the cultures exposed to DMSO became increasingly tinted with time. By the 4th day, the pellets were pink to red because of the presence of hemoglobin, whereas the control pellets were almost colorless.

In the cultures treated with 2% DMSO no clear-cut morphologic changes were detected for the first 48 hr. At 72 hr there was a slight but significant rise in the number of differentiating B+ cells and by the 96th hour a striking increase was apparent. Daily analysis of a representative experiment over a period of 5 days yielded consistently low frequencies of B+ cells in control cultures (0, 0.2, 0.2, 1.0, and 0.2%) and increasing frequencies in DMSO-treated cultures (0.4, 0.4, 2.8, 66.0, and 95.6%). The cells survived in 2% DMSO-containing medium as long as 7 days, at which time most of the living cells had reached the stage of orthochromatophilic normoblast. The appearance of the cells grown in the absence or presence of DMSO for 6 days is shown in Fig. 2. The blast cells characteristic of this leukemic cell line predominated in the untreated culture (Fig. 2A), whereas normoblasts were numerous in the treated cultures (Fig. 2B). These maturing cells were smaller, had condensed nuclei, and had a lower nuclear to cytoplasmic ratio than the cells from the control cultures.

Electron microscopic studies (manuscript in preparation) of the control and DMSO-treated cells revealed a marked difference in the number and association of ribosomes. Single ribosomes were numerous in the control cells, whereas fewer were seen in the treated cells, whose ribosomes often appeared

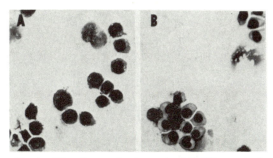

FIG. 2. Microphotographs of control and DMSO-treated cells of clone 707. (*A*) Control cells show pleomorphism and relatively large, basophilic erythroblasts. (*B*) Treated cells show many orthochromatophilic normoblasts with small, dense nuclei and more prominent cytoplasm. Both photographs were taken after 6 days of growth and are the same final magnification (×1250).

in clusters of 3 or 4. However, growth of the cells in DMSO does not eliminate the production of virus, since virus particles could be detected budding from the cell membranes and in the tissue-culture supernatant fluid.

The response of the cells to lower concentrations of DMSO, although less dramatic, also showed an increase in the number of B+ cells. Similarly, short exposures to the compound were sufficient to stimulate the mechanism involved in differentiation. The cells were seeded in replicate cultures in medium containing 2% DMSO and incubated for 1 and 24 hr, after which times aliquots of each were transferred to DMSO-free medium for an additional 4 days of incubation. As compared to the untreated cells, in which the frequency of B+ cells was 1.6% at the end of the observation period, the percentage in cultures exposed to DMSO for 1 hr had increased to 7.7 and in those exposed for 24 hr to 16.

In the experiments described, DMSO did not affect all of the cells since there were a few stem cells and early erythroblasts remaining in the cultures. Serially transferred in DMSO-free medium, these cultures gave rise to cell populations that were morphologically identical to those of the untreated cultures.

Thus far, the cells have not developed resistance to the treatment. When alternate passages were made in DMSO-containing and DMSO-free medium, there was no apparent change in the sensitivity of the cells to stimulation. The leukemic cells of the mass cultures, as well as of all clones tested, have demonstrated the ability to differentiate in the presence of DMSO.

Attempts to passage the cells serially in DMSO were also made. The cells could be maintained in medium containing 1% DMSO, but not in 2%, where few viable cells remained after two consecutive passages.

Bioassay for malignancy

To determine whether any change in the malignant properties of the differentiating cells had occurred, various concentrations of control or 2% DMSO-treated cells were inoculated subcutaneously into groups of syngeneic DBA/2J mice. Five mice were used for each concentration. There was little variation in the time of appearance of tumors in the mice receiving untreated cells or cells treated with DMSO for up to 48 hr. Mice inoculated with cells exposed to DMSO

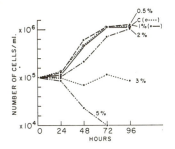

FIG. 1. Dose–response curves of leukemic cells of clone 707 in medium containing different concentrations of DMSO. Control: O- - -O; DMSO: 0.5% ●— —●; 1% ●———●; 2% ●— — —●; 3% ●- - -●; 5% ●- - - - - -●.

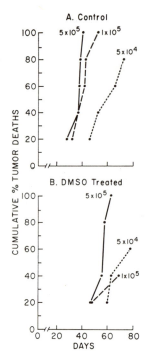

FIG. 3. Comparison of the survival times of mice inoculated with 5×10^5, 1×10^5, or 5×10^4 cells of clone 707 from control (*A*) and 2% DMSO-treated (*B*) cultures grown for 96 hr. Cumulative percentage of deaths is plotted against number of days after inoculation.

for 72 hr, however, lived longer and had slower-growing tumors than mice injected with untreated cells. This was even more evident with cells treated for 96 hr, as shown in Fig. 3. The groups of mice receiving the treated cells survived significantly longer than those inoculated with the control cells. Among the mice receiving the highest concentration of cells (5×10^5), the mean survival time of the control group was 38 days as compared to 56 days for those that received the treated cells. This difference is being further investigated with inocula containing smaller numbers of cells, in the hope of determining the number of tumorigenic cells that resist the effect of DMSO.

Iron metabolism
The effects of DMSO on the rates of cellular iron accumulation and of iron incorporation into heme and hemoglobin are shown in Fig. 4. As compared to the control cells, the accumulation of iron in the DMSO-treated cells was lower than in the controls for the first 72 hr, after which it was greatly and reproducibly stimulated. The stimulatory effect on heme and hemoglobin synthesis occurred earlier: it was apparent by the 48th hour for heme and by the 72nd hour for hemoglobin. By the 72nd and 96th hours respectively, there was a 20- and 60-fold stimulation in heme and a 12- and 40-fold stimulation of hemoglobin synthesis in the treated cells as compared to the untreated cells.

Fig. 5 illustrates the absorption spectra of the hemoglobin extracted from cells that had been grown in DMSO-supplemented medium. The spectra obtained from these cell ly-

sates were identical with those of hemolysates of normal adult DBA/2J mouse cells.

Influence of other factors
While DMSO has thus far proved to be the most efficient inducer of erythroid differentiation in our leukemic cell lines, other factors that might influence erythropoiesis have been explored. Some effect was observed by adding steroids to the medium or by altering the conditions of growth. For example, in a series of 2-step experiments in which the cells were first grown in serumless medium (13) for 4 days and then transferred to the complete medium with 15% fetal calf serum for an additional 4 days of growth, differentiation was induced. Heme production was stimulated 2–3 times and hemoglobin synthesis 10 times over the control cells during the second step.

DISCUSSION
The murine virus-induced leukemia under study is always associated with an erythropoietic response *in vivo* and has been thought to resemble the erythremic myelosis seen in diGuglielmo's disease in man (3). The cells of the mass culture derived from the leukemic tissues, as well as the cells of cloned lines, are capable of performing specific functions characteristic of their normal erythrocytic counterparts, i.e., they synthesize heme, which is utilized to produce measurable amounts of hemoglobin. The hemoglobin present in these cells has been identified by its histochemical staining properties, peroxidase activity, incorporation of ^{59}Fe, absorption spectrum, and characteristic migration pattern during discontinuous electrophoresis on polyacrylamide gels. The electrophoretic pattern of hemoglobin bands appears to be

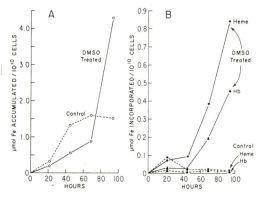

FIG. 4. Effect of DMSO on the rate of cellular iron accumulation (*A*) and the rates of incorporation of iron into heme and hemoglobin (*B*). Clone 707 cells were grown in the absence or presence of 2% DMSO in medium supplemented with ferrous (^{59}Fe) citrate (Squibb) to yield a final concentration of 1.2×10^4 cpm/μmol Fe. The ^{59}Fe solution was first incubated with fetal calf serum. Cells harvested on the first or second day of incubation had been seeded at 2.0×10^5 cells/ml and on the third or fourth day at 1.0×10^5 cells/ml. The cells from 20 ml of medium were harvested for each determination of cellular iron accumulation and iron incorporation into heme. The cells from 80 ml were harvested for each duplicate hemoglobin determination. The methods for the harvesting of the cells and for these determinations are described in a forthcoming paper (5) and briefly in *Methods*.

similar, if not identical, to that of hemolysates of adult DBA/2J mice, the strain from which the cells originated (4, 5).

In the present report, DMSO, at concentrations that did not appreciably inhibit cell replication, has been shown to induce a high percentage of the cells to differentiate along the erythroid pathway *in vitro*. Exposure of the cel s to DMSO for 1 hr followed by transfer to DMSO-free medium was sufficient to "trigger" the synthesis of hemoglobin. In the treated cultures, the increase in the number of benzidine-positive cells was always correlated with an increase in the amount of hemoglobin synthesis. The effect—as measured by morphologic and histochemical alteration, by increased synthesis of heme and hemoglobin, and by a change in the growth pattern when assayed in syngeneic mice—was clearly discernible after 96 hr and correlates with data on the kinetics of maturation of erythroid cells (14, 15).

Electron-microscopic examination of the cells revealed that both control and DMSO-treated cells had viruses budding from the membrane surfaces. Whether the properties of virus from the treated cells are altered remains to be determined. Single ribosomes appear to predominate in the control cells and clustered ribosomes in the form of triads and tetrads are abundant in the treated cells. Similar differences between ribosomes of maturing erythroid cells before and during hemoglobin synthesis have been described (16, 17).

The survival curves of mice receiving treated or untreated cells suggest that the normoblasts in the DMSO-treated cultures may have lost their malignancy. Approximately 15% of the population of the DMSO-treated cells bioassayed remained at the blastic benzidine-negative stage and these are presumed to have given rise to the observed tumors. This is indicated by the finding that the survival time of the mice inoculated with 5×10^4 control cells was similar to that of the mice inoculated with 5×10^5 treated cells.

Dimethyl sulfoxide may stimulate differentiation of the leukemic cells directly or indirectly through one or several of its known biological activities by affecting synthesis of nucleic acids and proteins (18–26). It could presumably act also by altering the intracellular concentration of low molecular weight compounds (27–30) and by affecting the secondary and tertiary structure of macromolecules (31–35).

Of particular interest was the report of Schrek *et al.* on the difference in cytocidal effect of DMSO on normal and leukemic lymphocytes (36). DMSO at a concentration of 2% was more cytotoxic *in vitro* to human lymphocytes from patients with chronic lymphocytic leukemia and from AKR leukemic mice than to normal control lymphocytes.

In studies with fibroblasts of strain L-929 exposed to DMSO, Berliner and Ruhman (37) observed a sensitivity pattern somewhat analogous to that described here. The fibroblasts in 1% DMSO grew at a rate similar to that of the controls, but proliferation was suppressed at 3 or 5%. Cells transferred from medium containing growth-inhibiting concentrations of DMSO to control medium grew normally and showed no evidence of increased resistance to a second exposure to DMSO. This reversibility of inhibition parallels the findings in the leukemic cell system in the present study.

Dimethyl sulfoxide has also been observed to affect maturation and pigmentation of fungi. Sproston and Setlow (38) reported that the UV requirement for sporulation of fungi could be replaced by DMSO (1–10%). They suggested that DMSO acted by releasing ergosterol, the sterol necessary for

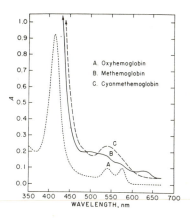

Fig. 5. Absorption spectra of extracts obtained from cells of clone 707 grown in medium containing 2% DMSO for 70 hr. A hemoglobin-rich fraction obtained by $(NH_4)_2SO_4$ precipitation (see *Methods*) was diluted with 0.1 M KPO_4 buffer pH 6.6. The readings were taken on a Cary model 14 recording spectrophotometer with a 1.0-cm light path. *A*, Cell extract, oxyhemoglobin. *B*, Oxidation of the extract (1.0 ml) with 5 μl of 5% $K_3Fe(CN)_6$ to form methemoglobin. *C*, Solution corresponding to Curve *B* treated with 5 μl of 1% KCN to form cyanmethemoglobin.

conidial production. Other investigators (39–41) described the blockage of pigment formation in various species of fungi by concentrations over 1%. Growth rate and sporulation were unaffected until concentrations of DMSO well in excess of those required for pigment inhibition were used.

The observation that differentiation can be induced among the leukemic cells by transfer to complete medium after an initial period of "starvation" in serumless medium cannot be interpreted at present. It is interesting to note that axon outgrowth on neuroblastoma cells occurs after the removal of serum from the culture (42) and differentiation of slime molds can be induced by starvation of the plasmodium (43).

The possibility that the regulatory mechanisms that control differentiation and malignancy are similar has been considered (44–46). Differentiation of human as well as murine neuroblastoma cells *in vitro* has been observed by a number of investigators (47–49). Kleinsmith and Pierce have shown that not only are mouse teratoma cells multipotential but a clone derived from a single cell can differentiate into as many as 11 different tissues (50). Other reports suggest that human leukemia cells may also possess the ability to differentiate (51–53). Finally, 5-bromodeoxyuridine has been shown to effect the differentiation of malignant cells (54, 55) and of embryonic cells (56).

It is evident that established lines of leukemic cells can be used for the investigation of problems related to the molecular control of differentiation and oncogenesis. Perhaps an interference in the normal flow of information from DNA into cell proteins results in malignancy. It may now be possible to ascertain if agents such as DMSO influence part(s) of the genome of the malignant cell to permit it to mature fully to the stage of its normal counterpart.

This work was supported in part by NCI grants CA 10,000 and CA 08,748, and by the Health Research Council of the City of New York, contract U-1840. The skillful technical assistance of Mrs. Ella Poran is gratefully acknowledged.

1. Friend, C., M. C. Patuleia, and E. de Harven, *Nat. Cancer Inst. Monogr.*, **22**, 505 (1966).

2. Patuleia, M. C., and C. Friend, *Cancer Res.*, **27**, 726 (1967).

3. Friend, C., and G. B. Rossi, in *Can. Cancer Conf.*, **8**, ed. J. F. Morgan (Pergamon Press, Toronto, 1969), pp. 171–182.

4. Friend, C., W. Scher, and G. B. Rossi, in *The Biology of Large RNA Viruses*, ed. B. W. J. Mahy and R. Barry (Academic Press, London, 1970), in press.

5. Scher, W., J. G. Holland, and C. Friend, *Blood*, in press.

6. Friend, C., and G. B. Rossi, *Int. J. Cancer*, **3**, 523 (1968).

7. Amstey, M. S., and P. D. Parkman, *Proc. Soc. Exp. Biol. Med.*, **123**, 438 (1966).

8. Tovell, D. R., and J. Colter, *Virology*, **37**, 624 (1969).

9. Kisch, A. L., *Virology*, **37**, 32 (1969).

10. Wallis, C., and J. L. Melnick, *J. Virol.*, **2**, 953 (1968).

11. *Ann.* pp. 1–671 *N.Y. Acad. Sci.*, **141**, 1 (1967).

12. LoBue, J., B. S. Dornfest, A. S. Gordon, J. Hurst, and H. Quastler, *Proc. Soc. Exp. Biol. Med.*, **112**, 1058 (1963).

13. Neuman, R. E., and A. A. Tytell, *Proc. Soc. Exp. Biol. Med.*, **104**, 252 (1960).

14. Granick, S., and R. D. Levere, in *Progress in Hematology*, ed. C. V. Moore and E. B. Brown (Grune and Stratton, New York, 1964), vol. 4, pp. 1–46.

15. Marks, P. A., and J. S. Kovick, in *Current Topics in Developmental Biology*, ed. A. A. Moscana and A. Monroy (Academic Press, New York, 1966), vol. 1, pp. 213–252.

16. Rifkind, R. A., D. Danon, and P. A. Marks, *J. Cell Biol.*, **22**, 599 (1964).

17. Warner, J. R., A. Rich, and C. E. Hall, *Science*, **138**, 1399 (1962).

18. Hellman, A., J. G. Farrelly, and D. H. Martin, *Nature*, **213**, 982 (1967).

19. Hagemann, R. F., and T. C. Evans, *Proc. Soc. Exp. Biol. Med.*, **128**, 648 (1967).

20. Hagemann, R. F., *Experientia*, **15**, 1298 (1969).

21. Gerhards, E., and H. Gibian, *Ann. N.Y. Acad. Sci.*, **141**, 65 (1967).

22. Gerhards, E., and M. J. Ashwood-Smith, *Ann. N.Y. Acad. Sci.*, **141**, 45 (1967).

23. Stock, B. A., A. R. Hansen, and J. R. Fouts, *Toxicol. Appl. Pharmacol.*, **16**, 728 (1970).

24. Ritter, P. O., F. J. Kull, and K. B. Jacobson, *J. Biol. Chem.*, **245**, 2114 (1970).

25. Rammler, D. H., *J. Biol. Chem.*, **245**, 291 (1970).

26. Monder, C., *J. Biol. Chem.*, **245**, 300 (1970).

27. Kligman, A. M., *J. Amer. Med. Ass.*, **193**, 140 (1965).

28. Steinberg, A., *Ann. N.Y. Acad. Sci.*, **141**, 532 (1967).

29. Leonard, C. D., *Ann. N.Y. Acad. Sci.*, **141**, 148 (1967).

30. Franz, T. J., and J. T. van Bruggen, *Ann. N.Y. Acad. Sci.*, **141**, 302 (1967).

31. Katz, L., and S. Penman, *Biochem. Biophys. Res. Commun.*, **23**, 557 (1966).

32. Strauss, J. H., Jr., R. B. Kelly, and R. L. Sinsheimer, *Biopolymers*, **6**, 793 (1968).

33. Duesberg, P. H., and R. D. Cardiff, *Virology*, **36**, 697 (1968).

34. Montagnier, L., A. Golde, and P. Vigier, *J. Gen. Virol.*, **4**, 449 (1969).

35. Rammler, D. H., and A. Zaffaroni, *Ann. N.Y. Acad. Sci.*, **141**, 13 (1967).

36. Schrek, R., L. M. Elrod, and K. Vir Batra, *Ann. N.Y. Acad. Sci.*, **141**, 202 (1967).

37. Berliner, D. L., and A. G. Ruhman, *Ann. N.Y. Acad. Sci.*, **141**, 159 (1967).

38. Sproston, T., and R. B. Setlow, *Mycologia*, **60**, 104 (1968).

39. Carley, H. E., R. D. Watson, and D. M. Huber, *Can. J. Bot.*, **45**, 1451 (1967).

40. Bean, G. A., G. W. Rambo, and W. L. Klarman, *Life Sci.*, **8**, 1185 (1969).

41. Tillman, R. W., and G. A. Bean, *Mycologia*, **62**, 428 (1970).

42. Seeds, N. W., A. G. Gilman, T. Amano, and M. W. Nirenberg, *Proc. Nat. Acad. Sci. USA*, **66**, 160 (1970).

43. Rusch, H. P., in *Advances in Cell Biology*, ed. D. M. Prescott, L. Goldstein, and E. McConkey (Appleton-Century-Crofts, New York, 1970), pp. 297–327.

44. Braun, A. C., in *The Cancer Problem: A Critical Analysis and Modern Synthesis* (Columbia University Press, New York and London, 1969), 209 pp.

45. Pierce, G. B., *Fed. Proc.*, **29**, 1248 (1970).

46. Markert, C. L., *Cancer Res.*, **28**, 1908 (1968).

47. Goldstein, M. N., J. A. Burdman, and L. J. Journey, *J. Nat. Cancer Inst.*, **32**, 165 (1964).

48. Schubert, D., S. Humphreys, C. Baroni, and M. Cohn, *Biochemistry*, **64**, 316 (1969).

49. Tumilowicz, J. J., W. W. Nichols, J. J. Cholon, and A. E. Greene, *Cancer Res.*, **30**, 2110 (1970).

50. Kleinsmith, L. J., and G. B. Pierce, *Cancer Res.*, **24**, 1544 (1964).

51. Nowell, P. C., *Exp. Cell Res.*, **19**, 26 (1960).

52. Farnes, P., and F. E. Trobaugh, Jr., *J. Lab. Clin. Med.*, **57**, 568 (1961).

53. Clarkson, B., A. Strife, and E. de Harven, *Cancer*, **21**, 926 (1967).

54. Silagi, S., and S. A. Bruce, *Proc. Nat. Acad. Sci. USA*, **66**, 72 (1970).

55. Schubert, D., and F. Jacob, *Proc. Nat. Acad. Sci. USA*, **67**, 247 (1970).

56. Coleman, A. W., J. R. Coleman, D. Kankel, and I. Werner, *Exp. Cell Res.*, **59**, 319 (1970).

Cell, Vol. 3, 127–133, October 1974, Copyright © 1974 by MIT

An Established Pre-Adipose Cell Line and its Differentiation in Culture

Howard Green and Mark Meuth
Department of Biology
Massachusetts Institute of Technology
Cambridge, Massachusetts 02139

Summary

The established cloned line, 3T3–L1, is a pre-adipose line. When the cells enter a resting state, either in monolayers or in suspension culture stabilized with methyl cellulose, they accumulate triglyceride fat and become adipose cells. A high serum concentration in the culture medium increases the rapidity and extent of the fat accumulation. The adipose conversion can be delayed indefinitely in surface cultures by keeping the cells in a growing state.

3T3–L1 is also specialized for collagen synthesis; prior to its adipose conversion, it makes about as much collagen as other 3T3 cells. We may therefore regard 3T3–L1 as a fibroblast line with an additional form of specialization.

After 3T3–L1 cells are grown to confluence in the presence of low concentrations of bromodeoxyuridine, their rate of collagen synthesis is not affected, but their conversion to adipose cells is completely prevented. If the cells are then permitted to grow in medium free of bromodeoxyuridine, their ability to convert to adipose cells is regained. The conversion of 3T3–L1 from pre-adipose to adipose cells therefore involves a process of differentiation which can be studied under cell culture conditions.

Introduction

Beginning with disaggregated Swiss mouse embryo cells, the established mouse fibroblast line 3T3 was evolved under defined culture conditions (Todaro and Green, 1963). The cell population at the end of this period ($\sim$ 100 cell generations) was quite homogeneous with regard to sensitivity to contact inhibition (high serum requirement for growth) and ability to remain in a resting state for long periods. The line had never been cloned up to the time these properties were established, and though many clones were subsequently isolated, we have preserved the original uncloned stock (usually designated 3T3–M) in the frozen state.

A recent communication (Green and Kehinde, 1974) described the isolation from 3T3–M of clones (3T3–L1 and 3T3–L2) which accumulated a great deal of lipid when they entered a resting state. This lipid consisted of triglycerides, suggesting a relation between these fatty 3T3 cells and adipose cells. Since the fat was divided among multiple droplets

in a still abundant cytoplasm, and the nucleus remained central, the resemblance was greatest to brown adipose cells or to developing white adipose cells. No cells were seen to have the large central fat droplet and eccentric nucleus characteristic of the mature white adipose cell. We show here that the adipose conversion of these cells is a process of differentiation that can go virtually to completion in cell culture.

Results

The Adipose Conversion of 3T3–L1

In order to permit 3T3–L1 cells to differentiate more fully into adipose cells, we found it important to increase the concentration of calf serum in the medium to 20–30%. The serum stimulated cell multiplication to a greater saturation density before the resting state was attained, and then the accumulation of fat took place much more rapidly and to a greater extent than in medium supplemented with 10% serum.

For electron microscopy, we prepared monolayer cultures in medium containing 20% calf serum. Several weeks later, when lipid accumulation was advanced, the cells were removed by trypsinization. Upon centrifugation, cells with the most fat rose to the top and were removed with a spatula as in creaming; they were then fixed, embedded, and sectioned. Figure 1 shows a group of low power electron micrographs of cells with features typical of maturing adipose cells — the large central fat droplet and the thin rim of cytoplasm rich in mitochondria. Two of the micrographs show a slightly compressed peripheral nucleus in the plane of section. The largest cells had diameters of 30–35 microns.

The resting state is not attained in monolayer culture until the cells have grown to confluence, but growth can be arrested at any time by trypsinizing a culture and inoculating the cells as suspensions in medium containing methyl cellulose (Stoker, 1968). Under these conditions the cells do not grow into colonies; instead they accumulate lipid. The presence of calf serum at a final concentration of 30% was highly stimulatory, and within a few days many cells were considerably enlarged by lipid deposits. Some of the cells could be seen, especially by phase microscopy, to possess the typical signet-ring appearance of white adipose cells.

The Fibroblastic Nature of 3T3–L1

Though many cell types synthesize very small amounts of collagen in culture, a high rate of synthesis is a property distinctive of fibroblasts and related cell types (Green and Goldberg, 1968). The mouse lines 3T3, 3T6, 3T12, etc. are known to be

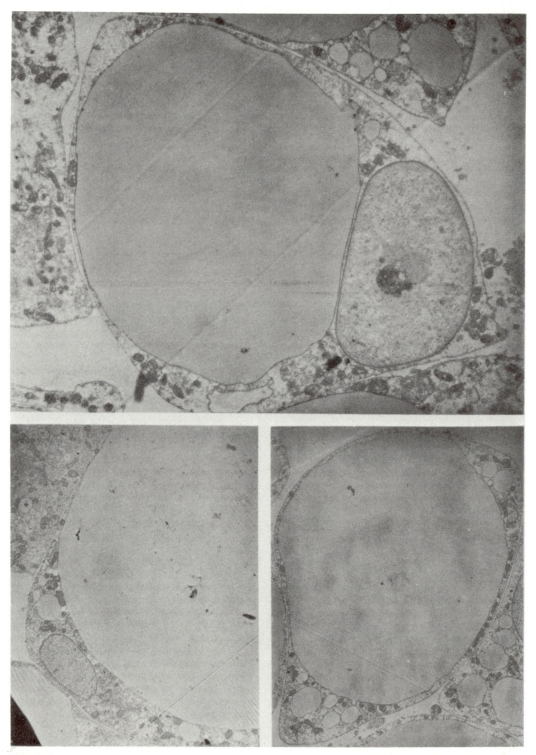

Figure 1. Maturation of the Adipose Cells

A surface culture of 3T3–L1 was allowed to remain at confluence for about a month while it was fed three times weekly with medium containing 20% calf serum. Electron micrographs (courtesy of Elaine Lenk) made from cells isolated by flotation show the large central fat droplet, thin cytoplasmic rim, and flattened eccentric nucleus characteristic of adipose cells. Magnification: (upper photo) 7,500 ×, (lower photos) 4,000 ×.

fibroblasts since they are specialized for the synthesis of fibrous collagen (Goldberg, Green, and Todaro, 1963; Green and Goldberg, 1965) and hyaluronic acid (Hamerman, Todaro, and Green, 1965, Saito and Uzman, 1971). Since cells with a marked tendency to accumulate lipid were distributed clonally in 3T3 cultures (Green and Kehinde, 1974), it was conceivable that fat-forming cells of an independent lineage unrelated to fibroblasts could have been carried along through the selection process used to establish the 3T3 line. Clone 3T3–L1 therefore was examined for its ability to synthesize collagen.

The most accurate and sensitive measure of this function is the incorporation of labeled proline into hydroxyproline residues of protein. Cultures of 3T3–L1 and of 3T3–M2 (a subline of uncloned 3T3 in which very few cells are able to accumulate fat) were allowed to grow to saturation density, and were tested before an appreciable number of fatty cells accumulated in the 3T3–L1 cultures. ^{14}C-labeled proline and 50 μg/ml of ascorbate were added in fresh medium; after 20 hr, the medium and cell layer were harvested, and the amounts of labeled hydroxyproline and proline incorporated into protein were determined.

Cell types not specialized for collagen synthesis, if they synthesize collagen at all, do so in amounts of less than 0.2% of total protein synthesis, whereas fibroblasts make 10–60 fold more (Green and Goldberg, 1968). Table 1 shows that for 3T3–M3 the ratio of collagen synthesized to total protein synthesized, derived from the HyPro/Pro ratio, was slightly under 2%, somewhat lower than values reported earlier, though within the range for fibroblasts. The corresponding value for the fat forming 3T3–L1 cells was 3.25%. It is therefore clear that the pre-adipose cell is at least as specialized for collagen synthesis as the nonfat-accumulating cells.

Electron microscopy provided morphologic evidence for collagen synthesis by 3T3–L1. A culture which had accumulated a large number of adipose cells was fixed and embedded in the petri dish. Two

classes of periodic fibers were seen in the extracellular spaces (Figure 2). The first consisted of typical but slender collagen fibers which, in suitable regions, could be seen to have the characteristic major period of 700 A. The second more abundant class contained thicker fibers similar to those observed earlier in 3T6 cultures (Goldberg and Green, 1964) and in rat odontoblasts and predentin (Weinstock and LeBlond, 1974). The appearance of these fibers is consistent with fibrous long spacing forms of collagen or procollagen.

We do not know from these experiments how long the ability to synthesize collagen persists once the adipose conversion has begun. Since not all the pre-adipose cells undergo the conversion, it would be necessary to isolate the adipose cells from the others. This separation could be made on the basis of their different buoyant density.

Potential for Adipose Cell Formation by Individual Cells of the Pre-Adipose Line

Since 3T3–L1 is a recent clonal isolate, genetic divergence of the population should be small. Yet even after cultures had been in the resting state for several weeks in medium supplemented with 20% calf serum, only 10–50% of the cells appeared to convert to adipose cells. This raised the question of whether some cells of the clone had lost the capacity to do so.

This was tested in two ways. First, we isolated 10 subclones of 3T3–L1. After growth to confluence, some cells of every subclone accumulated lipid, indicating that all cells of 3T3–L1 have some potential to convert to adipose cells; but the rapidity and intensity of lipid deposition, as well as the number of cells participating, did vary considerably in the different subclones. None gave rise to 100% adipose cells. This may in part be due to the fact that in the adipose conversion the cells become more spherical, more self-associating, and less spread out on the vessel surface, thereby making room for a few nonfatty cells to multiply and keep the monolayer complete.

Table 1. Collagen Synthesis by Preadipose and Non-Preadipose 3T3 Cells

	$\frac{HyPro}{Pro}$ (%)	Collagen Synthesized / Total Cell Protein Synthesized (%)	Formation of Adipose Cells in Resting Culture
3T3–M2	5.5	1.9	virtually none
3T3–L1	9.6	3.2	abundant
	9.9	3.3	
3T3–L1 grown in presence of BUdR (10^{-5} M) and deoxycytidine (4×10^{-5} M)	10.5	3.5	none
	10.8	3.6	

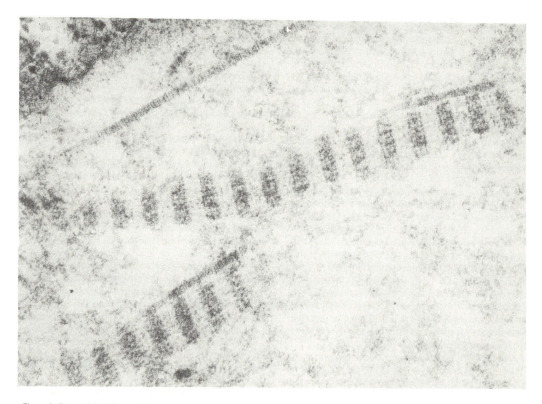

Figure 2. Extracellular Fibers of Collagen in Cultures of 3T3–L1
(Magnification 90,000 ×)

In a second test, 3T3–L1 cells were inoculated
at very low density (50 and 100 cells per 60 mm
petri dish). Ten days later, some of the cultures
were fixed and stained with hematoxylin for colony
counts; others were allowed to grow to confluence,
and three weeks later were fixed and stained with
Oil Red O for the counting of fat centers (Figure
3). Though the proportion of cells that underwent
the adipose conversion was quite variable in the different centers, each center that could be identified
was counted, regardless of the density of adipose
cells in it. From a total of 700 cells plated, 175 fat
centers resulted. The same number of cells plated
produced 158 cell colonies. By this criterion, it must
be concluded that all the cells of clone 3T3–L1 are
potentially able to convert to adipose cells.

Prevention of the Adipose Conversion by Bromodeoxyuridine

Perhaps the most striking single common feature
of differentiating systems is their failure to develop
subsequent to exposure of the growing precursor
cells to bromodeoxyuridine (Wessells, 1964; Stock-

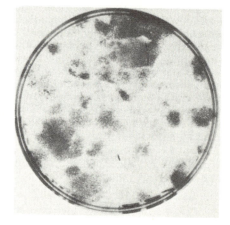

Figure 3. Culture of 3T3–L1 Showing Fat-Forming Centers
One hundred cells were inoculated into a petri dish and fed three
times weekly thereafter. Three weeks after the colonies became
confluent the culture was fixed and stained with Oil Red O. The
red stain shows as black on the photograph. There is marked variation in size and staining intensity of the fat-forming centers. This
is due to differences in the proportion of the cells undergoing the
adipose conversion in the different centers. Each center is presumably formed within a different colony.

dale et al., 1964; for a review see Rutter, Pictet, and Morris, 1973).

3T3–L1 cells were tested for their ability to convert to adipose cells after growth in the presence of bromodeoxyuridine at different concentrations. 3×10^4 cells were inoculated into 60 mm petri dishes and grown to confluence ($\sim$ 5 cell generations) in the presence of the drug. The cells then entered a resting state, and after a period of 3 weeks were fixed and stained with Oil Red O. Figure 4 shows that at a concentration of 5×10^{-6} Molar or higher, virtually no fat accumulation was visible. At lower concentrations the effectiveness of the bromodeoxyuridine diminished, and below 5×10^{-8} M the drug had no effect. Some effects of bromodeoxyuridine are known to be due to the induction of a deoxycytidineless state and consequent interference with DNA synthesis (Meuth and

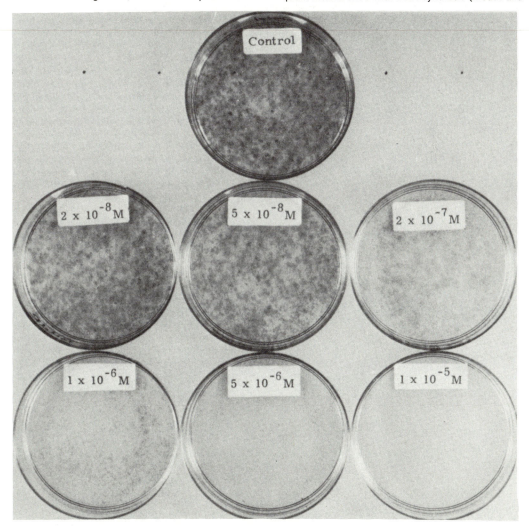

Figure 4. Effect of Bromodeoxyuridine on Differentiation of Pre-Adipose Cells

A number of petri dishes were inoculated with approximately 3×10^4 growing 3T3–L1 cells, and the cultures were fed with medium containing 10% calf serum and different concentrations of bromodeoxyuridine. The cultures were refed 3 times weekly with the same medium. All cultures reached confluence within about a week and were allowed to remain in the confluent state for another 3 weeks. They were then fixed and stained with Oil Red O. The red stain shows as black on the photograph. The presence of deoxycytidine at 4×10^{-5} Molar did not affect the result, though it did enable the cells in higher bromodeoxyuridine concentrations to reach confluence somewhat more rapidly.

Green, 1974). The addition of deoxycytidine improved the growth of the pre-adipose cells at the highest bromodeoxyuridine concentrations, but did not affect the suppression of the adipose conversion at any bromodeoxyuridine concentration.

The effect of bromodeoxyuridine was examined quantitatively as shown in Table 2. Growth in the presence of 10^{-5} M bromodeoxyuridine for 5 generations lowered the number of adipose cells by 70–1000 fold, and those few fatty cells seen had much smaller amounts of fat than the control cells. The suppression of the adipose conversion was reversible, since growth for 5 generations after removal of the bromodeoxyuridine restored the ability of confluent monolayers to accumulate lipid. If, on the other hand, the bromodeoxyuridine was removed a few days *after* the cells reached confluence when virtually no growth was taking place, the adipose conversion was still prevented (Experiment 2).

In contrast to its effect on the adipose conversion, bromodeoxyuridine had no effect on the rate of collagen synthesis. After growth in 10^{-5} M bromodeoxyuridine, 3T3–L1 cultures synthesized collagen in the same proportion of total protein synthesis as in its absence (Table 1).

Discussion

The embryonic development of adipose cells has been extensively studied in the tissues of animals and in organ cultures. The early histologists regarded the adipose cell as a mesenchymal cell distinct from the fibroblast, and possibly even of reticuloendothelial origin (for references see Barrnett, 1962). From light and electron microscopic studies of developing fat, it was later concluded that adipose cells could arise from fibroblasts (Clark and Clark, 1940; Napolitano, 1963). All studies on this

question have been based on morphological correlations. The more rigorous examination made possible by our cloned material shows that cells of the pre-adipose line 3T3–L1 must be considered modified fibroblasts. During the early stages of their maturation they resemble brown fat cells (Fawcett, 1952; Sidman, 1956); but ultimately their appearance is that of nearly mature white adipose cells. The demonstration that fibroblasts can convert to adipose cells does not, of course, exclude the possibility that in animal tissues other cell types might also be able to give rise to adipose cells.

Under the culture conditions used to establish the 3T3 line (Todaro and Green, 1963), the cells were maintained at low density and were prevented from entering a resting state. These conditions would have prevented the development of nonmultiplying mature adipose cells from pre-adipose cells and so preserved the latter in the population. It would be of interest to know what fraction of the established 3T3 population has the ability to undergo the adipose conversion. Experiments in progress indicate that most of the cells have this capacity to some degree, though only very few convert with a probability at all comparable to 3T3–L1.

The conversion of pre-adipose to adipose cells involves a process of differentiation which is prevented by previous incorporation of bromodeoxyuridine. On the other hand, bromodeoxyuridine does not affect the ability of the pre-adipose cell to carry out another differentiated function—the synthesis of collagen. Evidently, the drug is able to prevent the change in program necessary for the adipose conversion, while the previously established program of collagen synthesis continues without hindrance.

Though the number of adipose cells in the adipose tissues of adult animals is thought to re-

Table 2. Effect of Growth in Bromodeoxyuridine on Subsequent Lipid Accumulation by 3T3–L1

Experiment Number	Culture Conditions	Frequency of Fatty Cells	% of Control
1	Control	0.224	100
	After growth in BUdR (10^{-5} M) for 5 generations	0.0002	0.09
	After growth in BUdR (10^{-5} M) + deoxycytidine (4×10^{-5} M)	0.0003	0.14
	After growth in BUdR for 5 generations and growth in absence of BUdR for 5 generations	0.11	49
2	Control	0.124	100
	BUdR (10^{-5} M) throughout	0.0018	1.4
	BUdR (10^{-5} M) during growth, removed for 2 weeks while cells rested at confluence	0.002	1.6

main constant (Hirsch and Han, 1969), pre-adipose cells appear to be present as well and are capable of some growth and differentiation in culture (Clark and Clark, 1940; Smith, 1970; Poznanski, Waheed, and Van, 1973). These cells must therefore be capable of maintaining themselves in tissues without converting to adipose cells. The pre-adipose line 3T3–L1 seems to have a similar property, since even in essentially resting cultures, a significant fraction of the cells does not convert to adipose cells. As noted earlier, the rounded shape and self-associating nature of the adipose cell may have some bearing on this behavior.

The expression of adipocyte character (accumulation of triglycerides) becomes evident when the growth of the cells is arrested, either in methocel suspension culture, in which 3T3 cells cannot multiply, or in monolayer culture at saturation density. It seems likely that the resting condition is accompanied by important metabolic changes. The lipid metabolism of cultured mammalian cells has been examined (for a review see Rothblat, 1969), but the cell types employed were quite different from the pre-adipose line we describe here. Studies in progress show that when 3T3–L1 accumulates lipid, the rates of triglyceride synthesis from fatty acids, acetate, and glucose are all increased. As the accumulation rate is subject to chemical influence (Green and Kehinde, 1974), the line can be used to study this aspect of the adipose conversion as well as the differentiation process itself.

Experimental Procedures

Culture Conditions

All cultures were grown in fortified Eagle's medium supplemented with 10–30% calf serum. The cells were maintained in the pre-adipose state by keeping them in exponential growth. Fixation and staining for lipid were carried out as described earlier (Green and Kehinde, 1974).

Rate of Collagen Synthesis

The rate of collagen synthesis was determined as previously described (Green and Goldberg, 1965). After incubation of resting cultures with proline uniformly labeled with ^{14}C (7 μC, 100 mC/ mMole) and ascorbate (50 μg/ml), metabolic activity was stopped with 5% trichloracetic acid. After addition of unlabeled proline and hydroxyproline, the entire mixture was dialyzed exhaustively and hydrolyzed in 6 N HCl. A portion of the hydrolysate was subjected to ion exchange chromatography on Bio-Rad Aminex Q-150S. The proline and hydroxyproline peaks were isolated, and their radioactivity was determined. The proportion of collagen synthesized to total cell protein synthesized was calculated from this ratio (Green and Goldberg, 1965).

Electron Microscopy

For examination of the adipose cells, surface cultures were trypsinized, and the cells were centrifuged at 2000 rpm. The pellicle of cells which floated to the top was removed with a spatula and fixed for 2 hr in a 1% solution of osmium tetroxide in veronal acetate buffer (pH 7.5). The cells were embedded in an epon–araldite mixture. 300 Å sections were cut, post-stained in Reynold's lead for 5 min, and examined at low magnification in the JEOL 100 B microscope.

For the demonstration of extracellular fibers, a several week-old surface culture was fixed and embedded in the plastic petri dish as described by Goldberg and Green (1964).

Acknowledgment

We thank Dr. B. Goldberg for valuable advice and assistance in the course of these studies, Mrs. E. Lenk for the electron microscopy, and Mr. O. Kehinde and Miss J. Thomas for technical assistance. These investigations were aided by grants from the National Cancer Institute.

Received June 13, 1974; revised July 10, 1974

References

Barrnett, R. J. (1972). In Adipose Tissue as an Organ. L. W. Kinsell, ed. (Springfield, Ill.: Charles C. Thomas), p. 3.

Clark, E. R., and Clark, E. L. (1940). Am. J. Anat. 67, 255.

Fawcett, D. W. (1952). J. Morph. 90, 363.

Goldberg, B., and Green, H. (1964). J. Cell Biol. 22, 227.

Goldberg, B., Green, H., and Todaro, G. J. (1963). Exptl. Cell Res. 31, 444.

Green, H., and Goldberg, B. (1965). Proc. Nat. Acad. Sci. USA 53, 1360.

Green, H., and Goldberg, B. (1968). In Sym. Intern. Soc. for Cell Biol., 7, (New York: Academic Press), p. 123.

Green, H. and Kehinde, O. (1974). Cell 1, 113.

Hamerman, D., Todaro, G. J., and Green, H. (1965). Biochim. Biophys. Acta 101, 343.

Hirsch, J., and Han, P. W. (1969). J. Lipid Res. 10, 77.

Meuth, M., and Green, H. (1974). Cell 2, 109.

Napolitano, L. (1963). J. Cell Biol. 18, 663.

Poznanski, W. J., Waheed, I., and Van, R. (1973). Lab. Invest. 29, 570.

Rothblat, G. H. (1969). Adv. in Lipid Res. 7, 135.

Rutter, W. J., Pictet, R. L., and Morris, P. W. (1973). Ann. Review Biochem. 42, 601.

Saito, H., and Uzman, B. G. (1971) Biochem. Biophys. Res. Commun. 43, 723.

Sidman, R. L. (1956). Anat. Rec. 124, 723.

Smith, U. (1970). Anat. Rec. 169, 97.

Stockdale, F., Okazaki, K., Nameroff, M., and Holtzer, H. (1964). Science 146, 533.

Stoker, M. (1968). Nature 218, 234.

Todaro, G. J., and Green, H. (1963). J. Cell Biol. 17, 299.

Wessells, N. K. (1964). J. Cell. Biol. 20, 415.

Weinstock, M., and LeBlond, C. P. (1974). J. Cell Biol. 60, 92.

Reprinted from Nature, Vol. 256, No. 5517, pp. 495–497. 1975. © 1975 Macmillan Journals, Ltd.

Continuous cultures of fused cells secreting antibody of predefined specificity

THE manufacture of predefined specific antibodies by means of permanent tissue culture cell lines is of general interest. There are at present a considerable number of permanent cultures of myeloma cells[1,2] and screening procedures have been used to reveal antibody activity in some of them. This, however, is not a satisfactory source of monoclonal antibodies of predefined specificity. We describe here the derivation of a number of tissue culture cell lines which secrete anti-sheep red blood cell (SRBC) antibodies. The cell lines are made by fusion of a mouse myeloma and mouse spleen cells from an immunised donor. To understand the expression and interactions of the Ig chains from the parental lines, fusion experiments between two known mouse myeloma lines were carried out.

Each immunoglobulin chain results from the integrated expression of one of several *V* and *C* genes coding respectively for its variable and constant sections. Each cell expresses only one of the two possible alleles (allelic exclusion; reviewed in ref. 3). When two antibody-producing cells are fused, the products of both parental lines are expressed[4,5], and although the light and heavy chains of both parental lines are randomly joined, no evidence of scrambling of *V* and *C* sections is observed[4]. These results, obtained in an heterologous system involving cells of rat and mouse origin, have now been confirmed by fusing two myeloma cells of the same mouse strain,

The protein secreted (MOPC 21) is an IgG1 (κ) which has been fully sequenced[7,8]. Equal numbers of cells from each parental line were fused using inactivated Sendai virus[9] and samples contining 2×10^5 cells were grown in selective medium in separate dishes. Four out of ten dishes showed growth in selective medium and these were taken as independent hybrid lines, probably derived from single fusion events. The karyotype of the hybrid cells after 5 months in culture was just under the sum of the two parental lines (Table 1). Figure 1 shows the isoelectric focusing[10] (IEF) pattern of the secreted products of different lines. The hybrid cells (samples *c–h* in Fig. 1) give a much more complex pattern than either parent (*a* and *b*) or a mixture of the parental lines (*m*). The important feature of the new pattern is the presence of extra bands (Fig. 1, arrows). These new bands, however, do not seem to be the result of differences in primary structure; this is indicated by the IEF pattern of the products after reduction to separate the heavy and light chains (Fig. 1*B*). The IEF pattern of chains of the hybrid clones (Fig. 1*B*, *g*) is equivalent to the sum of the IEF pattern (*a* and *b*) of chains of the parental clones with no evidence of extra products. We conclude that, as previously shown with interspecies hybrids[4,5], new Ig molecules are produced as a result of mixed association between heavy and light chains from the two parents. This process is intracellular as a mixed cell population does not give rise to such hybrid molecules (compare *m* and *g*, Fig. 1*A*). The individual cells must therefore be able to express both isotypes. This result shows that in hybrid cells the expression of one isotype and idiotype does not exclude the expression of another: both heavy chain

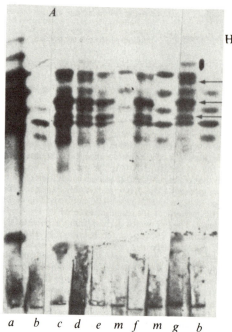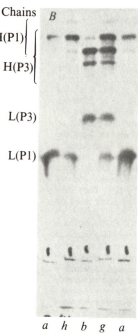

Chains

H(P1)
H(P3)
L(P3)
L(P1)

a   b   c   d   e   m   f   m   g   h   a   h   b   g   a

Fig. 1 Autoradiograph of labelled components secreted by the parental and hybrid cell lines analysed by IEF before (*A*) and after reduction (*B*). Cells were incubated in the presence of ^{14}C-lysine[14] and the supernatant applied on polyacrylamide slabs. *A*, pH range 6.0 (bottom) to 8.0 (top) in 4 M urea. *B*, pH range 5.0 (bottom) to 9.0 (top) in 6 M urea; the supernatant was incubated for 20 min at 37 °C in the presence of 8 M urea, 1.5 M mercaptoethanol and 0.1 M potassium phosphate pH 8.0 before being applied to the right slab. Supernatants from parental cell lines in: *a*, P1Bul; *b*, P3-X67Ag8; and, *m*, mixture of equal number of P1Bul and P3-X67Ag8 cells. Supernatants from two independently derived hybrid lines are shown: *c–f*, four subclones from Hy-3; *g* and *h*, two subclones from Hy-B. Fusion was carried out[4,9] using 10^6 cells of each parental line and 4,000 haemagglutination units inactivated Sendai virus (Searle). Cells were divided into ten equal samples and grown separately in selective medium (HAT medium, ref. 6). Medium was changed every 3 d. Successful hybrid lines were obtained in four of the cultures, and all gave similar IEF patterns. Hy-B and Hy-3 were further cloned in soft agar[14]. L, Light; H, heavy.

and provide the background for the derivation and understanding of antibody-secreting hybrid lines in which one of the parental cells is an antibody-producing spleen cell.

Two myeloma cell lines of BALB/c origin were used. P1Bul is resistant to 5-bromo-2′-deoxyuridine[4], does not grow in selective medium (HAT, ref. 6) and secretes a myeloma protein, Adj PC5, which is an IgG2A (κ), (ref. 1). Synthesis is not balanced and free light chains are also secreted. The second cell line, P3-X63Ag8, prepared from P3 cells[2], is resistant to 20 μg ml^{-1} 8-azaguanine and does not grow in HAT medium.

isotypes (γ1 and γ2a) and both V_H and both V_L regions (idiotypes) are expressed. There are no allotypic markers for the C_K region to provide direct proof for the expression of both parental C_K regions. But this is indicated by the phenotypic link between the *V* and *C* regions.

Figure 1*A* shows that clones derived from different hybridisation experiments and from subclones of one line are indistinguishable. This has also been observed in other experiments (data not shown). Variants were, however, found in a survey of 100 subclones. The difference is often associated with changes

Nature Vol. 256 August 7 1975

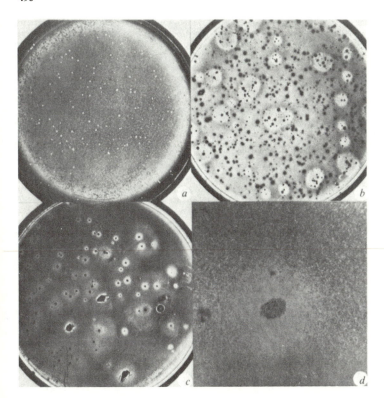

Fig. 2 Isolation of an anti-SRBC antibody-secreting cell clone. Activity was revealed by a halo of haemolysed SRBC. Direct plaques given by: *a*, 6,000 hybrid cells Sp-1; *b*, clones grown in soft agar from an inoculum of 2,000 Sp-1 cells; *c*, recloning of one of the positive clones Sp-1/7; *d*, higher magnification of a positive clone. Myeloma cells (10⁷ P3-X67A g8) were fused to 10⁸ spleen cells from an immunised BALB/c mouse. Mice were immunised by intraperitoneal injection of 0.2 ml packed SRBC diluted 1:10, boosted after 1 month and the spleens collected 4 d later. After fusion, cells (Sp-I) were grown for 8 d in HAT medium, changed at 1–3 d intervals. Cells were then grown in Dulbecco modified Eagle's medium, supplemented for 2 weeks with hypoxanthine and thymidine. Forty days after fusion the presence of anti-SRBC activity was revealed as shown in *a*. The ratio of plaque forming cells/total number of hybrid cells was 1/30. This hybrid cell population was cloned in soft agar (50% cloning efficiency). A modified plaque assay was used to reveal positive clones shown in *b–d* as follows. When cell clones had reached a suitable size, they were overlaid in sterile conditions with 2 ml 0.6% agarose in phosphate-buffered saline containing 25 µl packed SRBC and 0.2 ml fresh guinea pig serum (absorbed with SRBC) as source of complement. *b*, Taken after overnight incubation at 37 °C. The ratio of positive/total number of clones was 1/33. A suitable positive clone was picked out and grown in suspension. This clone was called Sp-1/7, and was recloned as shown in *c*; over 90% of the clones gave positive lysis. A second experiment in which 10⁶ P3-X67Ag8 cells were fused with 10⁸ spleen cells was the source of a clone giving rise to indirect plaques (clone Sp-2/3-3). Indirect plaques were produced by the addition of 1:20 sheep anti-MOPC 21 antibody to the agarose overlay.

in the ratios of the different chains and occasionally with the total disappearance of one or other of the chains. Such events are best visualised on IEF analysis of the separated chains (for example, Fig. 1*h*, in which the heavy chain of P3 is no longer observed). The important point that no new chains are detected by IEF complements a previous study[4] of a rat–mouse hybrid line in which scrambling of V and C regions from the light chains of rat and mouse was not observed. In this study, both light chains have identical C_κ regions and therefore scrambled V_L–C_L molecules would be undetected. On the other hand, the heavy chains are of different subclasses and we expect scrambled V_H–C_H to be detectable by IEF. They were not observed in the clones studied and if they occur must do so at a lower frequency. We conclude that in syngeneic cell hybrids (as well as in interspecies cell hybrids) V–C integration is not the result of cytoplasmic events. Integration as a result of DNA translocation or rearrangement during transcription is also suggested by the presence of integrated mRNA molecules[11] and by the existence of defective heavy chains in which a deletion of V and C sections seems to take place in already committed cells[12].

The cell line P3-X63Ag8 described above dies when exposed to HAT medium. Spleen cells from an immunised mouse also die in growth medium. When both cells are fused by Sendai virus and the resulting mixture is grown in HAT medium, surviving clones can be observed to grow and become established after a few weeks. We have used SRBC as immunogen, which enabled us, after culturing the fused lines, to determine the presence of specific antibody-producing cells by a plaque assay technique[13] (Fig. 2*a*). The hybrid cells were cloned in soft agar[14] and clones producing antibody were easily detected by an overlay of SRBC and complement (Fig. 2*b*). Individual clones were isolated and shown to retain their phenotype as almost all the clones of the derived purified line are capable of lysing SRBC (Fig. 2*c*). The clones were visible to the naked eye (for example, Fig. 2*d*). Both direct and indirect plaque

assays[13] have been used to detect specific clones and representative clones of both types have been characterised and studied.

The derived lines (Sp hybrids) are hybrid cell lines for the following reasons. They grow in selective medium. Their karyotype after 4 months in culture (Table 1) is a little smaller than the sum of the two parental lines but more than twice the chromosome number of normal BALB/c cells, indicating that the lines are not the result of fusion between spleen cells. In addition the lines contain a metacentric chromosome also present in the parental P3-X67Ag8. Finally, the secreted immunoglobulins contain MOPC 21 protein in addition to new, unknown components. The latter presumably represent the chains derived from the specific anti-SRBC antibody. Figure 3*A* shows the IEF pattern of the material secreted by two such Sp hybrid clones. The IEF bands derived from the parental P3 line are visible in the pattern of the hybrid cells, although obscured by the presence of a number of new bands. The pattern is very complex, but the complexity of hybrids of this type is likely to result from the random recombination of chains (see above, Fig. 1). Indeed, IEF patterns of the reduced material secreted by the spleen–P3 hybrid clones gave a simpler pattern of Ig chains. The heavy and light chains of the P3 parental line became prominent, and new bands were apparent.

The hybrid Sp-1 gave direct plaques and this suggested that it produces an IgM antibody. This is confirmed in Fig. 4 which shows the inhibition of SRBC lysis by a specific anti-IgM

Table 1 Number of chromosomes in parental and hybrid cell lines

Cell line	Number of chromosomes per cell	Mean
P3-X67Ag8	66,65,65,65,65	65
P1Bu1	Ref. 4	55
Mouse spleen cells	—	40
Hy-B (P1–P3)	112,110,104,104,102	106
Sp-1/7-2	93,90,89,89,87	90
Sp-2/3-3	97,98,96,96,94,88	95

antibody. IEF techniques usually do not reveal 19S IgM molecules. IgM is therefore unlikely to be present in the unreduced sample *a* (Fig. 3*B*) but μ chains should contribute to the pattern obtained after reduction (sample *a*, Fig. 3*A*).

The above results show that cell fusion techniques are a powerful tool to produce specific antibody directed against a predetermined antigen. It further shows that it is possible to isolate hybrid lines producing different antibodies directed against the same antigen and carrying different effector functions (direct and indirect plaque).

The uncloned population of P3–spleen hybrid cells seems quite heterogeneous. Using suitable detection procedures it should be possible to isolate tissue culture cell lines making different classes of antibody. To facilitate our studies we have used a myeloma parental line which itself produced an Ig. Variants in which one of the parental chains is no longer expressed seem fairly common in the case of P1–P3 hybrids (Fig. 1*h*). Therefore selection of lines in which only the specific antibody chains are expressed seems reasonably simple. Alternatively, non-producing variants of myeloma lines could be used for fusion.

We used SRBC as antigen. Three different fusion experiments were successful in producing a large number of antibody-producing cells. Three weeks after the initial fusion, 33/1,086

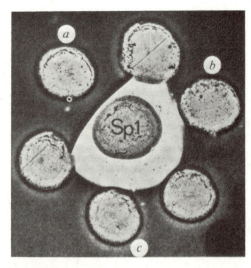

Fig. 4 Inhibition of haemolysis by antibody secreted by hybrid clone Sp-1/7-2. The reaction was in a 9-cm Petri dish with a layer of 5 ml 0.6% agarose in phosphate-buffered saline containing 1/80 (v/v) SRBC. Centre well contains 2.5 µl 20 times concentrated culture medium of clone Sp-1/7-2 and 2.5 µl mouse serum. *a*, Sheep specific anti-mouse macroglobulin (MOPC 104E, Dr Feinstein); *b*, sheep anti-MOPC 21 (P3) IgG1 absorbed with Adj PC-5; *c*, sheep anti-Adj PC-5 (IgG2a) absorbed with MOPC 21. After overnight incubation at room temperature the plate was developed with guinea pig serum diluted 1:10 in Dulbecco's medium without serum.

clones (3%) were positive by the direct plaque assay. The cloning efficiency in the experiment was 50%. In another experiment, however, the proportion of positive clones was considerably lower (about 0.2%). In a third experiment the hybrid population was studied by limiting dilution analysis. From 157 independent hybrids, as many as 15 had anti-SRBC activity. The proportion of positive over negative clones is remarkably high. It is possible that spleen cells which have been triggered during immunisation are particularly successful in giving rise to viable hybrids. It remains to be seen whether similar results can be obtained using other antigens.

The cells used in this study are all of BALB/c origin and the hybrid clones can be injected into BALB/c mice to produce solid tumours and serum having anti-SRBC activity. It is possible to hybridise antibody-producing cells from different origins[4,5]. Such cells can be grown *in vitro* in massive cultures to provide specific antibody. Such cultures could be valuable for medical and industrial use.

G. KÖHLER
C. MILSTEIN

MRC Laboratory of Molecular Biology,
Hills Road, Cambridge CB2 2QH, UK

Received May 14; accepted June 26, 1975.

1 Potter, M., *Physiol. Rev.*, **52**, 631-719 (1972).
2 Horibata, K., and Harris, A. W., *Expl Cell Res.*, **60**, 61-70 (1970).
3 Milstein, C., and Munro, A. J., in *Defence and Recognition* (edit. by Porter, R. R.), 199-228 (MTP Int. Rev. Sci., Butterworth, London, 1973).
4 Cotton, R. G. H., and Milstein, C., *Nature*, **244**, 42-43 (1973).
5 Schwaber, J., and Cohen, E. P., *Proc. natn. Acad. Sci. U.S.A.*, **71**, 2203-2207 (1974).
6 Littlefield, J. W., *Science*, **145**, 709 (1964).
7 Svasti, J., and Milstein, C., *Biochem. J.*, **128**, 427-444 (1972).
8 Milstein, C., Adetugbo, K., Cowan, N. J., and Secher, D. S., *Progress in Immunology*, II, 1 (edit. by Brent, L., and Holborow, J.), 157-168 (North-Holland, Amsterdam, 1974).
9 Harris, H., and Watkins, J. F., *Nature*, **205**, 640-646 (1965).
10 Awdeh, A. L., Williamson, A. R., and Askonas, B. A., *Nature*, **219**, 66-67 (1968).
11 Milstein, C., Brownlee, G. G., Cartwright, E. M., Jarvis, J. M., and Proudfoot, N. J., *Nature*, **252**, 354-359 (1974).
12 Frangione, B., and Milstein, C., *Nature*, **244**, 597-599 (1969).
13 Jerne, N. K., and Nordin, A. A., *Science*, **140**, 405 (1963).
14 Cotton, R. G. H., Secher, D. S., and Milstein, C., *Eur. J. Immun.*, **3**, 135-140 (1973).

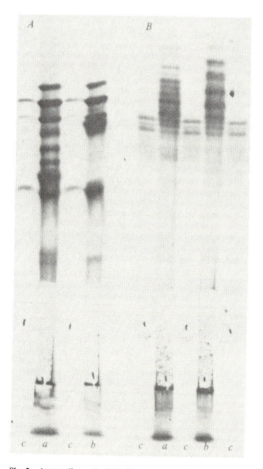

Fig. 3 Autoradiograph of labelled components secreted by anti-SRBC specific hybrid lines. Fractionation before (*B*) and after (*A*) reduction was by IEF. *p*H gradient was 5.0 (bottom) to 9.0 (top) in the presence of 6 M urea. Other conditions as in Fig. 1. Supernatants from: *a*, hybrid clone Sp-1/7-2; *b*, hybrid clone Sp-2/3-3; *c*, myeloma line P3-X67Ag8.

Reprinted from Science, Vol. 196, pp. 995–998. 1977. © 1977 AAAS

Synapse Formation Between Two Clonal Cell Lines

Abstract. *Clonal neuroblastoma × glioma hybrid cells frequently formed synapses with clonal mouse striated muscle cells. Clonal myotubes were similar to cultured mouse embryo myotubes with respect to acetylcholine sensitivity and other membrane properties examined. However, acetylcholine sensitivity measurements indicate that acetylcholine receptors of clonal myotubes are distributed more uniformly over the cell surface than the receptors of cultured mouse embryo myotubes.*

A primary reason for the use of cell tissue culture is the possibility of analyzing a relatively simple system which displays the relevant phenomenon to be investigated (*1*). Certain clonal cell lines provide a large and readily manipulated population of homogenous cells with neural or muscle characteristics and recently have been demonstrated to be synaptically competent. Kidokoro and Heinemann (*2*) found that striated myotubes formed from clonal myogenic lines can be innervated by normal neurons in explants of embryonic spinal cord. We have previously shown that clonal neuroblastoma × glioma hybrid NG108-15 cells form functional synapses with cultured striated myotubes from the embryonic mouse (*3*). In this report, we describe the establishment from mouse muscle cultures of a myogenic clonal line, G-8, and show that the clonal hybrid cells form synapses with clonal striated myotubes.

The clonal line, G-8, was obtained from M114, an uncloned myogenic cell line which arose spontaneously in a culture of embryonic spinal cord. We have previously viously shown that clonal neuroblastoma × glioma hybrid NG108-15 cells form functional synapses with cultured striated myotubes from the embryonic mouse (*3*). In this report, we describe the establishment from mouse muscle cultures of a myogenic clonal line, G-8, and show that the clonal hybrid cells form synapses with clonal striated myotubes.

cells dissociated from Swiss Webster mouse hindlimb muscle (*4*). Whereas the diploid number of chromosomes in the mouse is 40, the modal chromosome number of G-8 cells is 75. When confluent, most G-8 cells form parallel arrays of spindle-shaped mononucleated cells which fuse to form multinucleated myotubes up to 500 μm in length; multinucleated spherical cells also are found in some older cultures. The myotubes attach to the plastic surface of the petri dish and to neighboring cells. Well-differentiated G-8 myotubes possess striations and closely resemble normal mouse myotubes in morphology. Many G-8 myotubes contract spontaneously.

Neuroblastoma × glioma NG108-15 cells generate action potentials in response to electrical or chemical stimuli and synthesize, store, and release acetylcholine (*3*). When added to plates con-

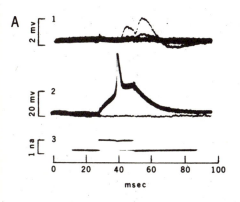

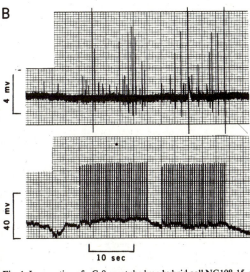

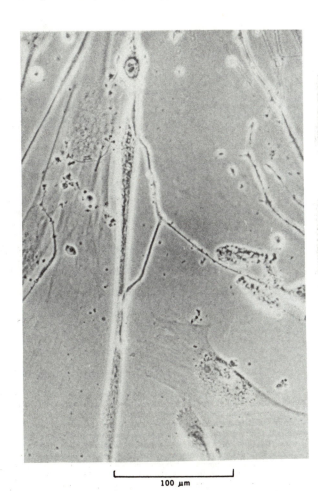

Fig. 1. Innervation of a G-8 myotube by a hybrid cell NG108-15. (A) Multiple oscilloscope sweeps demonstrating a functional connection between a hybrid and myotube. 1, Myotube recording; resting membrane potential is 45 mv. 2, Hybrid recording. 3, Current through hybrid recording electrode. (B) Chart record of same connection showing dependence of myotube responses (upper trace) on spikes elicited in hybrid (lower trace) by depolarizing currents passed through hybrid recording pipette. (C) Appearance of a coculture plated 31 days previously with myotubes and 18 days previously with hybrid cells.

taining G-8 myotubes (5), the hybrid cells attach readily to the plastic substratum and to neighboring hybrid or muscle cells and then extend long processes (Fig. 1C). The formation of synapses between NG108-15 hybrid cells and G-8 muscle cells was assessed after the cells had been cocultured for 3, 5, and 9 days. During the synapse assay the concentration of horse serum in the medium was reduced to 3 percent and the concentrations of choline chloride and calcium chloride were increased to 124 μM and 3.8 mM, respectively. The NG108-15 hybrid and G-8 muscle cells in physical contact were impaled with separate microelectrodes; the hybrid cell was stimulated electrically to elicit action potentials and the muscle responses were recorded. In 8 of 23 pairs tested, NG108-15 cell action potentials evoked depolarizing responses in myo-

tubes. An example is shown in Fig. 1, A and B. The amplitude of the myotube responses to NG108-15 action potentials ranged from 0.3 mv, the electrical noise level, to 8 mv, and their latency varied from 1 to 40 msec. The efficiency of transynaptic communication varied from one synapse to another; but in most cases 10 to 50 percent of the NG108-15 hybrid action potentials evoked muscle responses.

As shown in Fig. 2B, myotube responses to NG108-15 cell action potentials were mimicked by the iontophoretic application of acetylcholine. Responses in G-8 myotubes produced by NG108-15 cell innervation were reversibly blocked by $3 \times 10^{-5}M$ d-tubocurarine released from a nearby blunt micropipette, as previously demonstrated (3, 6) in normal mouse myotubes innervated by NG108-

15 cells. The NG108-15 cells thus appear to form chemical synapses with either normal mouse or G-8 myotubes, utilizing acetylcholine as a neurotransmitter.

In the present study, synapses were found with 34 percent of the NG108-15 hybrid cell and G-8 muscle cell pairs tested. Thus, these clonal cell lines form synapses with each other with high frequency. The hybrid cells form synapses with normal mouse embryo myotubes with the same high frequency (3). The formation of synaptic connections between clonal hybrid cells and clonal myotubes demonstrates that synapse formation is not dependent upon Schwann cells or other cell types.

To determine whether G-8 myotubes could be innervated by normal spinal cord neurons as well as by the hybrid cells, myotubes were examined 12 to 19 days after the addition of spinal cord cells. Spontaneously occurring postsynaptic potentials were recorded from 13 of 26 myotubes. Muscle response amplitudes were similar to those found in the biclonal system, as was the muscle resting membrane potential, 49 mv ($\pm$ 2.9) for innervated compared to 47 mv ($\pm$ 2.2) for noninnervated myotubes. When added to the recording medium, d-tubocurarine produced a complete block of spontaneous postsynaptic potentials at $10^{-8}M$ and a partial block at $10^{-9}M$ concentrations.

The membrane properties of G-8 myotubes were compared to those of normal cultured mouse muscle cells (7). Both G-8 and normal mouse embryo muscle cells were cocultured with NG108-15 hybrid cells, but the passive properties were analyzed without regard to whether individual myotubes were innervated. Table 1 shows that clonal and mouse embryo myotubes differed primarily in the resting membrane potential and in the variation in sensitivity to acetylcholine at different sites on the membrane surface.

The average measured resting membrane potential of cultured normal mouse myotubes was 11 mv greater than found in clonal G-8 myotubes. This may indicate that normal myotubes mature more quickly in vitro than do clonal myotubes. The measured resting membrane potential of mouse (8), rat (9), and chick (10) myotubes has been shown to increase with age as the cultured myotubes mature in vitro. At the time of testing, normal myotubes appear to be more mature than G-8 myotubes with respect to striations and hypolemmal nuclei.

Many of the myotubes in both types of cultures contracted spontaneously, an effect accentuated by the presence of di butyryl 3',5'-adenosine monophosphate

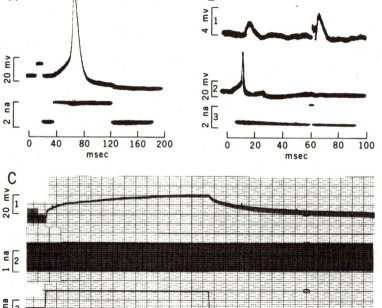

Fig. 2. Characteristics of the G-8 myotube and its innervation by the NG108-15 hybrid. All records are from tissue culture dishes of fused myotubes, cocultured for at least 3 days with hybrid cells. (A) An oscilloscope trace of an action potential (upper trace) in a myotube elicited by anodal break stimulation. A constant hyperpolarizing current (lower trace) was passed through a second electrode in the myotube, producing a membrane potential of 65 mv, then briefly terminated to produce an action potential. (B) An oscilloscope trace demonstrating the similarity of a myotube depolarization produced by an action potential in a nearby hybrid and by direct focal application of acetylcholine from an iontophoretic pipette. 1, Intracellular recording from a G-8 myotube. 2, Intracellular record of contiguous hybrid cell caused to spike by depolarizing current passed through recording pipette. 3, Current passed through an acetylcholine iontophoretic pipette placed close to the myotube membrane. (C) Chart record demonstrating a maintained myotube depolarization and conductance increase caused by continued application of acetylcholine. 1, Myotube membrane potential; resting potential 48 mv. 2, Depolarizing pulses (one per second) passed through a second intracellular pipette. 3, Current passed through an acetylcholine iontophoretic pipette. Baseline represents the -2 na holding current, which was turned off for approximately 3 minutes to release the drug.

996

(dibutyryl cyclic AMP). In all G-8 myotubes tested, action potentials could be elicited by intracellular depolarizing pulses, although as with rat muscle clone L-6 cells [11], it was often first necessary to hyperpolarize the G-8 myotube membrane potential (Fig. 2A). Most G-8 myotubes contracted when adequately stimulated.

All G-8 myotubes tested responded to short pulses of iontophoretically applied acetylcholine with brief monophasic depolarizations (Fig 2B); longer applications of acetylcholine were accompanied by an approximately threefold decrease in membrane resistance (Fig. 2C). The polarity of the response reversed when the membrane potential was adjusted to a more positive value than −5 mv. These results suggest that the response of the G-8 cells to acetylcholine involves an increase in permeability to sodium and potassium ions, as reported for other cultured myotubes [12].

The G-8 muscle cells were highly sensitive to acetylcholine, although the average maximum sensitivities of G-8 myotubes were somewhat lower than those of primary mouse myotubes (875 as opposed to 1474 mv/ncoulomb). This may be due to the lower resting membrane potential and membrane resistance of the G-8 myotubes, rather than to differences in the maximum acetylcholine receptor concentrations on the two types of myotubes. Whereas the acetylcholine sensitivity of normal myotubes at sites other than the site of maximum sensitivity often was not detected with short iontophoretic acetylcholine pulses, the responsiveness of G-8 myotubes was uniformly high over the entire surface of the cell (at least 400 mv/ncoulomb in a number of myotubes). Such uniformly high sensitivity to acetylcholine was not observed with clonal L-6 rat muscle cells [13] or normal cultured myotubes [14].

To compare the efficiencies of the synapses formed by the hybrid cell on the clonal and normal cultured myotubes, we determined the peak postsynaptic conductance change caused by the presumed release of a single quantum of acetylcholine. Because synapses on both types of myotubes had high failure rates (> 0.5), and the wide variability of response latencies indicated that more than one quantum released on any trial would be detected as two separate events, each muscle response elicited by hybrid stimulation was taken to result from the release of the contents of a single vesicle. The average highest rate of voltage increase of synaptic responses in a number of myotubes was used to compute the average conductance change [15]. As shown in

Table 1. A comparison of passive membrane properties, acetylcholine sensitivity, and postsynaptic parameters of normal mouse and G-8 myotubes. Electrical membrane constants were determined in normal mouse myotubes 22 days in culture, and G-8 myotubes 23 days after plating. Both plates were cocultured with hybrid cells and x-irradiated. For each myotube, foci of maximal and minimal acetylcholine sensitivity were determined by iontophoretic application of the drug. Some normal mouse myotubes had areas of membrane which failed to yield detectable responses when up to 10 ncoulombs of current were passed through the iontophoretic pipette; they were assigned a response amplitude of 0.5 mv. Maximum rates of voltage increase of evoked responses were determined visually from records of G-8 myotube connections obtained in the present study, and from records of mouse myotube connections previously described [3]. For each myotube, response current and conductance change were calculated from the average maximum rate of voltage increase of its synaptic responses [15] with the assumption of a reversal potential of −5 mv and a fixed capacitance given in this table. The data are expressed as means ± standard errors.

Parameter	Unit	Mouse	N	G-8	N
Passive electrical properties					
Resting potential	Millivolt	59.9 ± 2.8	7	49.0 ± 2.2	6
Input resistance	Megaohm	10.2 ± 2.4	5	7.5 ± 1.7	6
Time constant	Millisecond	5.8 ± 1.3	5	8.7 ± 1.0	6
Specific resistance	Ohm-cm²	1855 ± 531	5	2555 ± 467	6
Specific capacitance	Microfarad/cm²	4.3 ± 1.4	5	4.1 ± 0.9	6
Sensitivity to acetylcholine					
Maximum sensitivity	Millivolt/ncoulomb	1474 ± 295	6	875 ± 137	6
Minimum sensitivity	Millivolt/ncoulomb	< 61 ± 22	4	299 ± 47	4
Ratio		> 24		2.9	
Quantal postsynaptic response parameters					
Maximum rate of rise	Volt/second	1.03 ± .09	7	0.50 ± .03	3
Maximum current	Nanoamp	0.58 ± .05	7	0.57 ± .03	3
Conductance change	Nanomhos	13.0 ± 1.8	7	13.8 ± 1.2	3

Table 1, the conductance increase associated with the release of a single quantum from a hybrid cell was 1.3×10^{-8} mhos in primary mouse cells and 1.38×10^{-8} mhos in G-8 myotubes [10 to 25 percent of the conductance changes found at frog or snake neuromuscular junctions [16, 17]]. Spontaneous responses in a G-8 myotube cocultured with normal mouse spinal cord cells had an average conductance change of 1.2×10^{-8} mhos. These conductance changes would correspond to 5 to 10×10^6 Na^+ ions passing across the muscle membrane during a synaptic event.

If one assumes that a similar number of acetylcholine molecules are released by the hybrid at synapses with both normal and G-8 myotubes, and that the response magnitude is a function of acetylcholine receptor concentration, then the similar magnitudes of the postsynaptic conductance changes imply that at regions of synaptic contact, normal and G-8 myotubes may have approximately the same acetylcholine receptor concentration and the same access to released transmitter.

This would suggest that synaptic sites on normal mouse myotubes are restricted to the relatively small (< 10 percent) portion of the normal myotube surface which has acetylcholine sensitivity comparable to at least the lower range of sensitivity of the G-8 myotubes (greater than 200 mv/ncoulomb). The results indicate that NG108-15 hybrid cells interact with normal myotubes so that the acetylcholine release sites of hybrid neurites are closely

apposed to areas on the normal myotube surface membrane which have relatively high concentrations of nicotinic acetylcholine receptors. Evidence compatible with the interpretation is furnished by immunohistochemical data which show areas of normal mouse myotubes with high concentrations of nicotinic acetylcholine receptors in apposition to NG108-15 hybrid neurites known to innervate the myotube [18]. Whether specific interactions between molecules on the hybrid and myotube membranes are needed to juxtapose acetylcholine release sites and acetylcholine receptors remains to be determined.

If we assume that the activation of one acetylcholine receptor results in a change in conductance of 50×10^{-12} mhos [19], then the conductance change produced by a hybrid cell quantum represents the action of at least 240 acetylcholine molecules. Although somewhat lower than the estimates of the minimal number of acetylcholine molecules found in a quantum at neuromuscular junctions [17], this number indicates that the formation and release of acetylcholine from the hybrid cells and the response of the myotubes to quanta are well developed at biclonal synapses. The most prominent functional difference between the NG108-15 hybrid synapse and the mature neuromuscular synapse is that the hybrid cell releases fewer quanta than a spinal cord motorneuron.

The synapses between the NG108-15 hybrid cell and myotubes resemble normal immature, neuromuscular synapses in this regard. The normal maturation of the neuromuscular junction involves a hundredfold increase in the number of quanta released by a nerve cell action potential. It is an attractive possibility that the number of quanta released by the hybrid cell may be regulated and that its regulatory mechanisms are the same as in the normal maturational process. Relatively little is known about these regulatory mechanisms at the molecular level. The demonstration that clonal cells form synapses provides a system that can be used to explore both presynaptic and postsynaptic regulatory events.

C. N. Christian
P. G. Nelson

Behavioral Biology Branch, National Institute of Child Health and *Human Development, National Institutes of Health, Bethesda, Maryland 20014*

J. Peacock

Department of Neurology, Stanford Medical School, Stanford, California

M. Nirenberg

Laboratory of Biochemical Genetics, National Heart, Lung, and Blood Institute, National Institutes of Health

References and Notes

1. P. G. Nelson, *Physiol. Rev.* **55**, 1 (1975).
2. Y. Kidokoro and S. Heinemann, *Nature (London)* **252**, 593 (1974). Synapse formation between NG108-15 hybrid cells and L-6 myotubes has also been shown by Heinemann and Kidokoro (personal communication).
3. P. G. Nelson, C. N. Christian, M. Nirenberg, *Proc. Natl. Acad. Sci. U.S.A.* **73**, 123 (1976).
4. Proliferating M114 cells were grown in Dulbecco's modified Eagle's medium (DMEM) supplemented with 10 percent horse serum (HS) and 10 percent fetal calf serum (FCS) in plastic flasks; in the third and subsequent passages flasks were coated with collagen. After approximately six generations, the M114 cells were dissociated and plated into collagen-coated cloning wells at an average of less than one cell per well. A myotube-forming clone, G-8, was subcultured by three serial passages in flasks and frozen [J. Peacock, in preparation]. The G-8 cells have been subcultured 15 times (an estimated 50 cell divisions) without loss of the ability to form myotubes. The G-8 cells were grown in DMEM, 10 percent fetal bovine serum, 10 percent horse serum, penicillin (50 unit/ml), sodium salt, and streptomycin sulfate (50 μg/ml; Microbiological Associates) in 250-ml Falcon plastic flasks, without collagen. When cultures became confluent, but before cell fusion occurred, they were dissociated with a solution of 0.125 percent crude trypsin (Microbiological Associates) in Puck's balanced salt solution D_1, adjusted with NaCl to 340 mosmole per kilogram of H_2O and subcultured at a 20-fold lower cell concentration. Myotube cultures were prepared by plating 10^5 dissociated G-8 cells in a 35-mm Falcon plastic culture dish, in growth medium supplemented with 50 μg/ml of acid soluble collagen (Calbiochem). Fusion was promoted by feeding the culture infrequently, that is, only as needed to maintain the pH at 7.0 to 7.4, thus allowing myoblasts to "condition" the medium [I. R. Konigsberg, *Dev. Biol.* **26**, 133 (1971)].
5. NG108-15 hybrid cells were grown as previously described, and "predifferentiated" for at least 1 week by the addition of 1 mM N^6,O^2-dibutyryl cyclic AMP to the growth medium. Seven days after the G-8 cells were plated, cocultures were established by the addition of 3×10^4 NG108-15 cells in 90 percent DMEM, 10 percent HS plus 1 mM dibutyryl cyclic AMP, 0.1 mM hypoxanthine, and 0.016 mM thymidine. To compare G-8 myotubes with normal myotubes, cultures were prepared from dissociated cells of the hindlimbs of 18- to 21-day-old C57B1/6N mouse embryos [E. L. Giller *et al.*, *Science* **182**, 588 (1973)] and plated with NG108-15 cells after 11 days. Occasionally, both types of muscle plates were x-irradiated (4000 rad at 274 rad/minute) which stopped the proliferation of both clonal and primary cells. The x-irradiation did not inhibit synapse formation between NG108-15 cells and normal mouse muscle cells (unpublished observations). To compare the synaptogenesis of G-8 and normal spinal cord neurons to the biclonal condition, cultures were prepared by the addition of dissociated cells from spinal cords of 12- to 14-day-old G-8 cultures. 5-Fluorodeoxyuridine and uridine were used to inhibit proliferation of nonneuronal cells derived from the spinal cord (E. L. Giller *et al.*, *J. Cell Biol.*, in press.)
6. P. G. Nelson *et al.*, in preparation.
7. Each myotube was penetrated with two micropipettes filled with 3M potassium acetate, one electrode recording the cell's response to intracellular current passed through the second electrode. Input resistance was calculated from the amount of current producing a hyperpolarizing response of from 20 to 30 mv. The time constant of the membrane was taken to be the time necessary to reach 66 percent of the maximal hyperpolarizing response, and membrane area was approximated by two rectangles of the same dimension as the myotube. After determining membrane constants, each myotube was tested for sensitivity to acetylcholine by ionotophoresis of the drug from high-resistance pipettes. Pulses lasting less than 4 msec were used to eject acetylcholine onto the surface of the myotube, and various areas were assayed until its maximal sensitivity was determined. Back currents of less than 2 na were sufficient to prevent desensitization.
8. J. A. Powell and D. M. Fambrough, *J. Cell Physiol.* **82**, 21 (1973).
9. D. Fambrough and J. E. Rash, *Dev. Biol.* **26**, 55 (1971).
10. G. D. Fischbach, M. Nameroff, P. G. Nelson, *J. Cell Physiol.* **78**, 289 (1971).
11. Y. Kidokoro, *Nature (London) New Biol.* **241**, 158 (1973).
12. J. H. Steinbach, *J. Physiol. (London)* **247**, 393 (1975); A. H. Ritchie and D. M. Fambrough, *J. Gen. Physiol.* **66**, 327 (1975).
13. J. H. Steinbach, A. J. Harris, J. Patrick, D. Schubert, S. Heinemann, *J. Gen. Physiol.* **62**, 255 (1973).
14. G. D. Fischbach and S. A. Cohen, *Dev. Biol.* **31**, 147 (1973).
15. J. G. Blackman, B. L. Ginsberg, C. Ray, *J. Physiol. (London)* **167**, 389 (1963); C. C. Hunt and P. G. Nelson, *ibid.* **177**, 1 (1965).
16. A. Takeuchi and N. Takeuchi, *J. Neurophysiol.* **23**, 397 (1960); P. W. Gage and R. N. McBurney, *J. Physiol. (London)* **226**, 79 (1972).
17. C. H. Hartzell, S. W. Kuffler, D. Yoshikami, *J. Physiol. (London)* **251**, 427 (1975).
18. M. Daniels, unpublished data.
19. F. Sachs and H. Lecar, *Nature (London) New Biol.* **246**, 214 (1973); C. R. Anderson and C. F. Stevens, *J. Physiol. (London)* **235**, 655 (1973).
20. We thank E. Godfrey and D. Rush for their contributions to this study. This work was supported by PHS grant FO2 HD 55299-01 to C.N.C. and NSF grant 6B 43526 and NIH grant NS 12151 to J.P.

3 August 1976; revised 9 November 1976

Properties of an Epithelial Cell Type in Culture: The Epidermal Keratinocyte and Its Dependence on Products of the Fibroblast

Howard Green, James G. Rheinwald, and Tung-Tien Sun

Department of Biology, Massachusetts Institute of Technology, Cambridge, Massachusetts 02139

Keratinocytes of stratified squamous epithelium can be grown serially in culture and retain the various markers typical of their form of differentiation. In order to form colonies at each transfer, the keratinocytes must be suitably supported by fibroblasts. Established keratinocyte lines of teratomal origin show this dependence, as do diploid strains of finite culture life derived from human skin. For at least some keratinocyte lines, this requirement can be satisfied by soluble products elaborated by the fibroblasts.

It is suggested that epithelial cells in general may not be independent cell types and that their poor cultivability may be due to failure to provide suitable fibroblast support. The existence of a number of established lines of epithelial origin that can grow without such support and of lines of fibroblastic origin which cannot support keratinocytes suggests that both epithelial dependence and the fibroblast supporting function can sometimes be lost in established cell lines.

INTRODUCTION

Mammalian cell culture began with the development of established cell lines that could be grown serially (1, 2). As well-brought-up microbiologists, subsequent investigators insisted on the purity of the cultured cell types to be studied; any evidence of diversity in the culture was dealt with by cloning.

The desire to work only with pure cell types was quite appropriate for studies of basic cell functions, for the field of virology, for somatic cell hybridization, etc., but when using cell culture for the study of differentiation, it is necessary to consider that very few tissues of the mammalian organism consist of a pure cell type. While most animals are clonally pure, nearly every part of the body consists of a mixture of cell types. Moreover, the development of a new cell type in embryogenesis usually results from the interaction of already differing cell types. It seems obvious that earlier studies on embryonic induction based on interactions of developing organ rudiments in culture dealt more realistically with this aspect of differentiation than do modern studies on pure cell types.

In what follows, we wish to emphasize the importance of such interactions insofar as they concern the growth and differentiation of an epithelial cell type—the keratinocyte. This cell type requires products synthesized by fibroblasts in order to grow and function

493

properly. The existence of such products was suggested by earlier studies on organ culture recombinations of dermal and epidermal tissues, and attempts were made to purify mesenchymal products necessary for keratinocytes (3, 4) and other epithelial cell types (5).

There are indications that more than a single product of fibroblasts may be involved. For example, the role of mesenchyme in the differentiation of embryonic epithelia may be either determining (instructive) or permissive (6, 7). The dermis instructs the morphogenetic behavior of the epidermis, even in adult animals (8). The effects of fibroblasts on the keratinocytes we have studied thus far in cell culture may be mainly of a permissive nature, although a search for instructive effects demonstrable at this level seems clearly indicated.

SERIAL CULTIVATION OF THE KERATINOCYTE

Fibroblasts are the most easily cultivable of mammalian cells. Serial cultivation of disaggregated cells of a piece of tissue from almost any location soon leads to predominance of the fibroblasts and the emergence of strains of homogeneous appearance. Other cell types of mesenchymal origin have been obtained from tissues and grown serially in culture, e.g., the myoblast precursor of striated muscle (9), the cartilage cell (10), and the smooth muscle cell (11). Like the fibroblast, these cell types grow in pure culture, though myoblasts show some dependence on fibroblast products (12).

In contrast to the mesenchymal cell types, it has been difficult or impossible to reproducibly cultivate epithelial cell types serially out of any epithelium. Established lines of epithelial origin have been developed mostly from tumors, but even tumors do not give rise to serially cultivable epithelial lines with a high probability. By employing a principle not hitherto utilized for the purpose, we have been successful in serially cultivating mammalian keratinocytes of 2 types:

1) Those originating from a serially transplanted mouse teratoma. One such keratinocyte line (XB) studied in detail is capable of indefinite serial propagation in culture (13). Although the cells share many properties in common with skin keratinocytes, they can easily be distinguished from them by their cellular and colonial morphology and their movements, as seen in time-lapse cinematography.

2) Those originating from normal human skin. Cultures of diploid human epidermal keratinocytes can be started routinely and serially transferred, but they possess a finite lifetime measured in total number of cell generations (14).

Both classes of keratinocytes are strongly dependent on fibroblasts or fibroblast products. They survive with difficulty or not at all without this form of support. For example, when human epidermal keratinocytes are inoculated alone, colony formation is abortive, but when inoculated together with lethally irradiated 3T3 cells at a density of $2 \times 10^4/cm^2$ they form macroscopic colonies. The keratinocytes make contact with the dish surface and, in growing, push the surrounding 3T3 cells back at the expanding colony perimeter. Figures 1 and 2 show colonies of human epidermal keratinocytes grown in this way.

GENERAL PROPERTIES OF CULTURED KERATINOCYTES

1) Colonies originate from single cells.

2) Cell stratification takes place so that the colonies ultimately consist of multiple cell layers (Fig. 3).

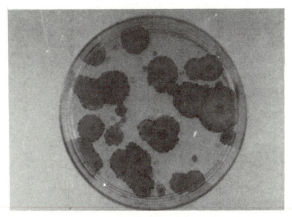

Fig. 1. Primary Culture of Human Epidermal Keratinocytes. Primary culture of human epidermal keratinocytes (12-yr-old abdominal skin). Fixed and stained 28 days after inoculation. The Rhodanile Blue stains keratinocyte colonies red and the 3T3 feeder cells blue. Each colony has grown from a single cell.

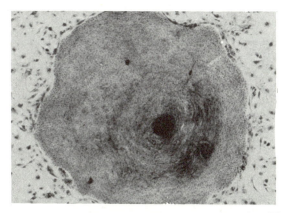

Fig. 2. Colony of Human Epidermal Keratinocytes. This is a low-power view (30X) of a 10-day-old colony grown from a single cell in a secondary culture of keratinocytes originating from foreskin of a 3-year-old child. The colony consists of about 5,000 cells. The cells have completed a total of about 23 cell generations in culture (hematoxylin and eosin).

3) The colonies stain red with Rhodanile Blue, a property of keratinizing epithelia (15).

4) The basal cell layer contains the dividing cells of the colony.

5) The cells of the superficial layers differentiate. They increase in size, their envelopes become cornified and the cells convert to flattened squames, which are ultimately shed into the medium.

6) The cells have the ultrastructure of keratinocytes; they contain abundant tonofilaments and are linked by numerous desmosomes.

7) The cells contain abundant keratin-type polypeptides. About 30–40% of cell protein consists of a group of polypeptides of molecular weight 45,000–60,000 similar to those which account for nearly all the protein present in human stratum corneum (Sun and Green, in preparation). The basal cells as well as the superficial cells contain these proteins.

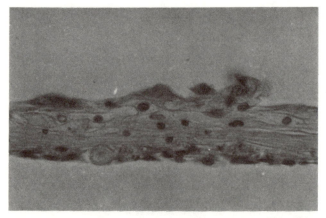

Fig. 3. Section Through a Colony of Mouse XB Teratomal Keratinocytes. The cell layers are less regular and the superficial layers less flattened than in the stratified structure produced by human epidermal keratinocytes. In the upper layers, the cell boundaries are very distinct probably because of the cornified envelopes, and the nuclei appear to be more widely separated; the cytoplasmic volume of keratinocytes increases during their differentiation (T.-T. Sun and H. Green, in preparation). Magnification 200 X.

SPECIFIC PROPERTIES OF DIFFERENT CULTURED KERATINOCYTE STRAINS

Keratinocyte strains derived from foreskin of different newborn humans display only small variations in their growth rate and their tendency to stratify and terminally differentiate. On the other hand, the behavior of teratomal keratinocyte line XB is quite different from that of human epidermal keratinocytes. Derived from the mouse teratoma 69891 of Dr. Leroy Stevens, XB was the first established keratinocyte line we identified (13). We have obtained colonies of similar type routinely from primary platings of the same strain of teratoma, but not from another teratoma, PS-A4 (16); this teratoma gives rise to keratinocytes more like those derived from human skin. The properties which line XB shares with the other classes of keratinocyte have been summarized above. Those properties which distinguish XB from the others may be considered in relation to the possibility that XB represents a more primitive type of keratinocyte.

1) The growth rate of XB cells is much more rapid. Their doubling time is about 18 hr, compared with about 24—30 hr for human epidermal keratinocytes.

2) Medium harvested from 3T3 cultures can substitute effectively for the 3T3 cells themselves in supporting the growth of XB colonies. Such substitution is in general much less satisfactory for the growth of human epidermal cells; under special conditions strains of human epidermal keratinocytes are able to make colonies in medium conditioned by 3T3, but at a lower frequency than in the presence of 3T3 cells themselves.

3) Even in small colonies, human epidermal cells pack tightly together and adhere to each other. Stratification and Rhodanile Blue staining begin early. In contrast, the cells of small XB colonies retain a loose, noncohesive structure. Only when the colony becomes large do the cells begin to adhere tightly to each other, stratify and stain red with Rhodanile blue. Observations made by time-lapse cinematography show that in young XB colonies (less than 8 days) the cells move individually and very quickly in random pattern with respect to each other. Only later do the cells begin to adhere to each other

and presumably begin to form the numerous desmosomes that are seen in older colonies (13). On the other hand, human epidermal cells even in small colonies move as if they are highly adherent to each other. Movement within the colonies takes place, not of single-cell units moving randomly with respect to each other, but of a group of cells flowing as a unit past other groups of cells. Perhaps even this type of movement involves breaking and reformation of desmosomes, but if so, its extent would likely be much less than that required to permit independent movement of single cells.

4) The number of terminally differentiating cells in epidermal colonies can be obtained by counting the cornified cells, defined as those whose envelopes are insoluble in solutions of sodium dodecyl sulfate and mercaptoethanol (Sun and Green, in preparation). In small colonies of human epidermal cells, the proportion of cornified cells is not much different from that in large colonies (5–10% in both). The proportion of cornified cells in small XB colonies is extremely low (less than 0.1%). Only when the colonies reach a size at which the cells associate and adhere (~ 8 days) does the proportion of cornified cells begin to become appreciable. The cornification process is initiated much later by XB cells, possibly owing to the same delay in cell-cell interaction cited above.

5) XB cells have a much higher colony-forming efficiency on transfer than human epidermal cells. Though a large fraction of the basal cells of human epidermal colonies is growing, most cells do not succeed in initiating colony formation on transfer.

6) Large colonies of human epidermal cells growing with 3T3 support respond to epidermal growth factor (17) by increasing their average growth rate. The colony-forming efficiency of the cells is increased and the proportion of cells in a state of terminal differentiation is reduced (Rheinwald and Green, in preparation). Though EGF has minor effects on the morphology of XB cells, it does not have these effects on growth rate or differentiation.

7) Human epidermal cells show improved growth when the culture medium is supplemented with low concentrations of hydrocortisone (> 0.03 γ/ml). When XB cells grow in the presence of 3T3 cells the addition of hydrocortisone is highly inhibitory to differentiation. No such effect occurs when hydrocortisone is added to cultures of human epidermal cells at any concentration.

DEPENDENCE OF KERATINOCYTES ON FIBROBLASTS OR THEIR PRODUCTS

The support of keratinocytes does not require contact with 3T3 cells, for medium harvested from 3T3 cultures can substitute for the 3T3 cells in supporting colony formation by XB teratomal keratinocytes. Medium harvested after 24-hr contact with a confluent layer of 3T3 cells is effective in supporting growth, stratification and differentiation of XB cells (Fig. 3).

This effect of 3T3 cells is one with considerable specificity and is not analogous to feeding effects described earlier in cell culture work. The use of lethally irradiated feeder cells of the same type as the viable cells was introduced by Puck and Marcus (18) for the cultivation of cells that grow with difficulty at low density. The feeder cells provide metabolites lost to the medium by the viable cells (19). The feeder cells are unnecessary if the viable cells are present at a sufficiently high density, for the viable cells can synthesize everything provided by the feeders.

As might be expected from earlier work (20), the ability to support keratinocytes is not a property possessed by any cell type. Human epidermal cells inoculated even at quite high density will not multiply without fibroblast support. XB cells cannot sustain multiplication even if they are inoculated at high density, and the cells do not stratify or stain red with Rhodanile Blue. Medium conditioned by different cell lines can be tested for ability to substitute for 3T3 cells in supporting the growth of XB colonies. Figure 4 shows that in the presence of medium harvested from cultures of some nonfibroblast lines, XB cells failed to undergo more than a few cell divisions. Medium from other nonfibroblast lines supported very slow progressive growth, but the colonies did not stratify or stain red with Rhodanile Blue. Even medium harvested from some established fibroblast lines (3T6 or a viral transformant of 3T3) was unable to support the formation of large stratifying colonies. The same was true for medium harvested from L cells or V79 cells. On the other hand, medium of cultures of 4 independent strains of human diploid fibroblasts (2 originating from newborn skin, and 2 from embryonic lung) and medium from cultures of mouse embryonic diploid fibroblasts were able to support keratinocyte colonies about as well as medium from 3T3 cultures. The evidence thus far indicates that when fibroblasts convert to established lines or are transformed by oncogenic viruses they may lose this fibroblast function. It may be relevant that 3T3 cells, the only established fibroblast line so far demonstrated to condition medium effectively, are also the most sensitive to contact inhibition of growth.

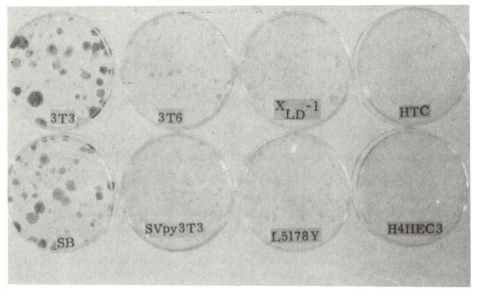

Fig. 4. Support of XB Cells by Medium Harvested From Cultures of Different Cell Lines. Medium harvested from cultures of rat hepatoma lines HTC and H_4II EC_3 permitted virtually no colony formation. That harvested from a lymphoma line (L5178y), an epithelial teratoma line (X_{LD}-1), and mouse fibroblast lines 3T6 and SV Py 3T3 supported colony formation, but the colonies grew very slowly, and did not stratify or stain red with Rhodanile Blue. Medium harvested from 3T3 cultures, or cultures of human diploid fibroblasts (SB) supported colony formation, rapid growth, stratification, and red staining by Rhodanile Blue (data from [13]).

LIFE (BIRTH TO DEATH) OF THE EPIDERMAL KERATINOCYTE IN CELL CULTURE

There now exist numerous serially cultivable lines of teratoma stem cells that, under suitable culture conditions, differentiate into defined somatic cell types. Some of these give rise to keratin-forming cells in culture (16, 21). Like epidermal keratinocytes, the teratomal keratinocytes stratify and give rise to cornified cells. The end product of epidermal differentiation in culture, as in vivo, is the squame which, in its mature form, lacks a nucleus. The entire life history of the keratinocyte can be studied in culture as follows: The keratinocyte is created by differentiation of teratomal stem cells. Growth and terminal differentiation may be followed either in the teratomal or the skin keratinocytes. At least some stages of this life history depend absolutely on mesenchymal support in the form of fibroblasts or their products, and it is now possible to study this dependence in detail.

GENERAL IMPORTANCE OF FIBROBLAST SUPPORT FOR MULTIPLICATION AND DIFFERENTIATED FUNCTION OF DEPENDENT CELL TYPES

As we pointed out at the beginning of this article, the easy cultivability of fibroblasts and other mesenchymal cell types stands in sharp contrast to the inability of epithelial cell types, when pure, to undergo more than a small number of cell doublings. For the keratinocyte the reason for this is clear – it is a cell type dependent on fibroblast products. When these are provided, multiplication has been shown to continue for over 50 cell generations (14); when conditions are further improved, the number exceeds 120 (Rheinwald and Green, in preparation). Are other epithelial cell types similarly dependent on fibroblast products? The older literature on organ culture strongly suggests an affirmative answer. Pituitary rudiments cannot grow in culture after removal of the connective tissue (22), and the same is true for epithelial cells of mouse mammary gland (23) and the thymus epithelium of mouse embryos (24).

To be sure, some established lines of epithelial origin are indefinitely cultivable in the absence of fibroblasts, but these lines have usually been derived from tumors and may have been aided in evolving independence of mesenchymal support by their initial neoplastic state. It seems clear that ideas about the cultivability of any epithelial cell type (and possibly others as well) will have to be reexamined with respect to a role of connective tissue cells in their support. One must distinguish a dependent cell type – one which requires fibroblast support – from an independent cell type which can grow without such support (25). The evolution of culture lines of epithelial cells independent of fibroblasts means that it is possible for a dependent cell type to become independent. The character of dependence may be of importance in normal growth control, and its loss a factor in neoplasia.

The nature of the fibroblast chosen for use in supporting epithelial growth may be an extremely important one. First, one must consider possible specificity in the relation of the fibroblast to the epithelial cell. Evidence for such specificity is clear from the work of Billingham and Silvers (8) and McLoughlin (7). Second, most of the established fibroblast lines we have tested do not have the ability to support keratinocyte growth and must be presumed to have lost it during their life in culture. Even short-term serial cultivation of the mesenchyme of submandibular gland rudiments was observed to lead to loss of its ability to influence the behavior of the epithelial components (26). The retention of the capacity of 3T3 cells to support keratinocytes may be regarded as additional evidence that 3T3 cells resemble fibroblasts as they exist in vivo more than do most other established fibroblast lines.

ACKNOWLEDGMENTS

This investigation was aided by grants from the National Cancer Institute. We are greatly indebted to Mr. Peter Yanover for the time-lapse microcinematography.

REFERENCES

1. Sanford, K. K., Earle, W. R., and Likely, G. D., J. Nat. Cancer Inst. 9:229 (1948).
2. Gey, G. O., Coffman, W. D., and Kubicek, M. T., Cancer Res. 12:264 (1952).
3. Wessells, N. K., Proc. Nat. Acad. Sci. USA 52:252 (1964).
4. Melbye, S. W., and Karasek, M. A., Exptl. Cell Res. 79:279 (1973).
5. Ronzio, R. A., and Rutter, W. J., Devel. Biol. 30:307 (1973).
6. Grobstein, C., Science 118:52 (1953).
7. McLoughlin, C. B., J. Embryol. Exp. Morphol. 9:385 (1961).
8. Billingham, R. E., and Silvers, W. K., J. Exptl. Med. 125:429 (1967).
9. Yaffe, D., Proc. Nat. Acad. Sci. USA 61:477 (1968).
10. Horwitz, A. L., and Dorfman, A., J. Cell Biol. 45:434 (1970).
11. Ross, R., Glomset, J., Kariya, B., and Harker, L., Proc. Nat. Acad. Sci. USA 71:1207 (1974).
12. Hauschka, S. D., and Konigsberg, I. R., Proc. Nat. Acad. Sci. USA 55:119 (1966).
13. Rheinwald, J. G., and Green, H., Cell 6:317 (1975).
14. Rheinwald, J. G., and Green, H., Cell 6:331 (1975).
15. MacConnaill, M. A., and Gurr, E., Irish J. Med. Sci., June p. 243, (1964).
16. Martin, G. R., and Evans, M. J., Cell 6:467 (1975).
17. Cohen, S., Carpenter, G., and Lembach, K. J., in "Advances in Metabolic Disorders," R. Lust and K. Hall (Eds.). Academic Press, New York, vol. 8, p. 265 (1975).
18. Puck, T. T., and Marcus, P. I., Proc. Nat. Acad. Sci. USA 41:432 (1955).
19. Eagle, H., and Piez, K., J. Exptl. Med. 116:29 (1962).
20. Wessells, N. K., New Eng. J. Med. 277:21 (1967).
21. Nicolas, J. F., Dubois, P., Jakob, H., Gaillard, J., and Jacob, F., Ann. Microbiol. (Inst. Pasteur) 126A:3 (1975).
22. Sobel, H., J. Embryol. Exp. Morphol. 6:518 (1958).
23. Lasfargues, E. Y., Murray, M. R., and Moore, D. H., in "Symposium on Phenomena of the Tumor Viruses," Nat. Cancer Inst. Monogr. No. 4, p. 151 (1960).
24. Auerbach, R., Devel. Biol. 2:271 (1960).
25. McLoughlin, C. B., Symp. Soc. Exp. Biol. 17:359 (1963).
26. Grobstein, C., J. Exptl. Zool. 124:383 (1953).

Reprinted from

Proc. Natl. Acad. Sci. USA
Vol. 76, No. 4, pp. 1882–1886, April 1979
Cell Biology

Construction of an artificial blood vessel wall from cultured endothelial and smooth muscle cells

(mixed cultures/basal lamina/growth kinetics)

PETER A. JONES

Departments of Pediatrics and Biochemistry, University of Southern California School of Medicine; and Department of Pediatrics, Division of Hematology–Oncology, Childrens Hospital of Los Angeles, Los Angeles, California 90027

Communicated by Charles Heidelberger, January 25, 1979

ABSTRACT   Cloned bovine endothelial cells were grown on a preformed layer of cultured rat smooth muscle cells that contained large amounts of connective tissue proteins. The successful growth of the endothelial cells was dependent upon the addition of more than 2.5×10^4 cells per cm^2, and the final density reached was approximately 2.5 times higher than that obtained for the same cells growing on plastic. The endothelial cells anchored more firmly to the smooth muscle cells than to plastic, and electron microscopy showed the existence of an irregular, dense, basal lamina-like structure between the two cell types. Biochemical analysis of the lamina produced by the endothelial cells in isolation demonstrated the presence of collagen and two fucosylated glycoproteins. The structure produced, which has some of the characteristics of a blood vessel wall, was stable for several months in culture and has many potential applications.

The current interest in the culture of endothelial cells (1–5) stems, in part, from the important role that blood vessels play in many diseases. Human umbilical cord and bovine aortic endothelial cells have been successfully cultured (1, 3) and bovine aortic endothelial cells have been cloned (5). Endothelial cells synthesize several endothelial-specific products *in vitro*, including Factor VIII antigen (4), and Weibel–Palade bodies are present in the cells (1, 2), although these are much less common in bovine than in human cells (6). They also synthesize type IV collagen (7, 8) and fibronectin (5, 9).

The endothelium and basement membrane rest on layers of smooth muscle cells in arterioles and large vessels *in vivo*. Since rat smooth muscle cells may also be easily cultured and form a multilayer *in vitro* (10), it was realized that a preformed smooth muscle layer could be used as a substrate for bovine endothelial cells. The present studies demonstrate that cloned bovine endothelial cells attach to rat smooth muscle cells and grow to form a structure possessing some of the morphological characteristics of a blood vessel wall. The characteristics of this system are described together with its potential applications.

MATERIALS AND METHODS

Isolation of Cells. The pulmonary artery obtained from a 13-inch (32.5-cm) bovine fetus was cut longitudinally to expose the inside wall. Endothelial cells were gently scraped off the artery with a scalpel blade and cultured in Eagle's minimal essential medium buffered with Earle's salts (GIBCO) containing 10% fetal calf serum (Irvine Scientific, Irvine, CA), 2% tryptose phosphate broth (Difco), and penicillin and strepto-

mycin (100 μg/ml). Three weeks after explantation, an area of the dish containing cells growing with a flat "cobblestone" appearance was isolated with a glass ring and the cells were trypsinized. These cells were seeded at 50–100 cells per 60-mm dish; discrete colonies, containing vigorously growing cells, were retrypsinized by the ring technique. The clone used in these studies was called Al$_4$Cl-1.

The rat smooth muscle cells used were the R22 strain obtained from newborn rat hearts (10). The majority of the cells in this strain have the phenotype of smooth muscle, although it is not known whether they are derived from the coronary vessels or from the endocardium.

Immunofluorescence. Cells were grown on coverslips, fixed in cold acetone, and processed exactly as described (3). Rabbit antiserum to bovine Factor VIII was a kind gift from E. Kirby (Temple University) and was used at a 1:40 dilution. Goat anti-rabbit immunoglobulin coupled to fluorescein was purchased from Behring.

Growth of Endothelial Cells on Preformed Smooth Muscle Layer. The rat cells (R22) (passage 6–10) were seeded into 35-mm dishes (Falcon or Corning) at $1–2 \times 10^5$ cells per dish or into 16-mm wells (Costar, Cambridge, MA) at 0.5×10^5 cells per well. The cultures were given ascorbic acid (50 μg/ml) the next day and every day thereafter and the medium was changed twice weekly. The cell layer contained visible amounts of extracellular matrix proteins 2 weeks after seeding (10), at which time the endothelial cells were added to the layer. For the standard production of mixed cultures, 5×10^4 or 2×10^5 endothelial cells were used per well or 35-mm dish, respectively, and daily ascorbic acid treatment was continued.

Growth Kinetics. Although a variety of techniques were tested to remove selectively the endothelial layer from the smooth muscle layer, none was successful. The number of endothelial cells in mixed cultures containing both cell types was therefore determined by subtraction of the number of cells in duplicate cultures containing smooth muscle cells only. Trypsin was inadequate to dissociate the layered smooth muscle cells because of the large amounts of elaborated elastin and collagen. Cultures were washed once with phosphate-buffered saline and then treated with 0.25% Viokase (Viobin Corp., Monticello, IL) containing 0.05% bacterial collagenase (Worthington Type CLS) at 37°C for 30 min. The suspension was vigorously aspirated with a pasteur pipette and returned to the incubator for another 60 min, followed by further aspiration. Samples of these cell suspensions were counted in a Coulter Counter.

Ultrastructural Procedures. Cultures were washed and the cell layer was either fixed *in situ* or carefully loosened from the dish with a flat rubber policeman and a pair of fine forceps. The layer was made into a roll and transferred directly to 10% for-

Cell Biology: Jones

Proc. Natl. Acad. Sci. USA 76 (1979) 1883

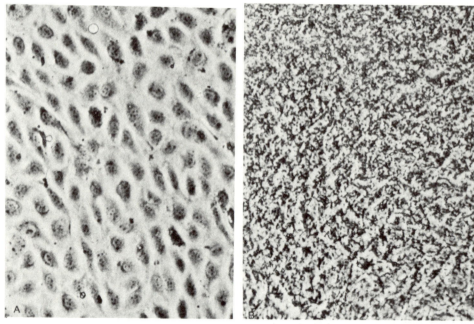

FIG. 1. Phase contrast micrographs of living endothelial (*A*) and smooth muscle (*B*) cells growing as separate cultures on plastic. (×285.)

malin or to 2% glutaraldehyde in phosphate buffer for electron microscopy. Formalin-fixed membranes were embedded in paraffin, cut at right angles to the long axis of the roll, and stained with hematoxylin/eosin. Glutaraldehyde and osmium-fixed specimens were embedded in Epon/Araldite (50:50) stained with uranyl acetate and lead citrate, and sections were examined in a Siemens Elmiskop A electron microscope.

Basal Lamina Characterization. The matrix proteins produced by the smooth muscle cells were prepared on 35-mm dishes by treating 2-week-old cultures of R22 cells with 0.25 M NH_4OH, which removes the cells but leaves the extracellular matrix intact (11). Endothelial cells were then grown on these matrix proteins in the presence of 1 μCi (1 Ci = 3.7 × 10^{10} becquerels) of L-[3,4-^{3}H(N)]proline (44 Ci/mmol) or L-[1-^{3}H]fucose (2.8 Ci/mmol) (New England Nuclear) per ml and received daily additions of ascorbic acid (50 $\mu g/ml$). The cultures were lysed for 30 min with 0.25 M NH_4OH 12 days after seeding to obtain the insoluble radioactive proteins produced by the endothelial cells. The percentages of radioactivity solubilized in sequential 3-hr treatments by trypsin (Sigma Type III pretreated with 5 mg of bovine elastin per ml to remove contaminating elastase), elastase (Worthington, ESFF), and collagenase (Worthington, CLSPA) were determined as described (10). All enzymes were used at a concentration of 10 $\mu g/ml$ in 0.1 M Tris·HCl, pH 7.6/10 mM $CaCl_2$.

RESULTS

Growth of Individual Cell Types on Plastic. The bovine cells (Al₄Cl-1) grew with the characteristic "cobblestone" morphology of endothelial cells (Fig. 1*A*). The endothelial cells produced immunologically detectable Factor VIII antigen, whereas the R22 cells or human skin fibroblasts were negative under identical conditions. The borders between the cells stained with silver nitrate and they exhibited a high level of

fibrinolytic activity in contrast to smooth muscle cells isolated from the same vessel (not shown). Although the morphology of the cells was not changed by the daily addition of ascorbate, the monolayer tended to strip off the culture dish and could not be successfully maintained for more than 2 weeks under these conditions.

The rat cells (R22) contained many of the ultrastructural characteristics of smooth muscle cells (see Fig. 4) and secreted prodigious amounts of glycoprotein(s), elastin, and collagen (10). The large amount of extracellular material secreted by these cells can be seen in Fig. 1*B*. They could be maintained for many months in this condition.

Growth of Endothelial Cells on Preformed Smooth Muscle Layers. The Al₄Cl-1 cells attached more quickly to plastic than to the smooth muscle layer. Almost all of the cells had attached to plastic 1 hr after plating, whereas the majority of cells were still floating in dishes containing smooth muscle layers. The cells grown on plastic were well spread 2 hr after seeding, but those grown on smooth muscle layers were still rounded up. The endothelial cells were well spread on the muscle layer 18–24 hr later, but tended to be slightly clumped.

It was subsequently difficult to observe the endothelial cells with the inverted microscope because of the thickness of the smooth muscle multilayer and associated extracellular matrix (Fig. 1*B*). An opaque layer due to the endothelial cells was, however, clearly visible to the naked eye several weeks after seeding. These mixed cultures were stable, and some have been maintained for up to 6 months without stripping off the dish. As mentioned above, the endothelial layer peeled off plastic dishes after 2 weeks when given daily ascorbate, showing that the adhesiveness of the cells to plastic was less than that for the smooth muscle layer.

The growth kinetics of the Al₄Cl-1 cells grown in plastic dishes and in dishes containing a smooth muscle cell layer are shown in Fig. 2. The endothelial cells rapidly reached a monolayer containing approximately 4 × 10^5 cells per well when

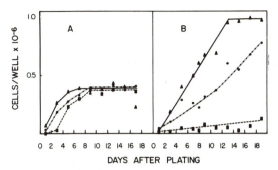

FIG. 2. Growth curves for endothelial cells on plastic (*A*) or preformed smooth muscle cell layers (*B*). Cells (Al₄Cl-1) were seeded onto plastic or preformed muscle layers on day 0 and the number of endothelial cells was determined on the days indicated. Seeding levels were 10^4 (■), 5×10^4 (●), or 10^5 (▲) cells per 16-mm well.

Table 1.   Sequential enzyme digestion of endothelial basal lamina

Label	% total radioactivity solubilized by		
	Trypsin	Elastase	Collagenase
[³H]Proline	67	4	28
[³H]Fucose	83	6	4

Endothelial cells (Al₄Cl-1) were grown on unlabeled matrix proteins in the presence of [³H]fucose or [³H]proline for 12 days and were given ascorbate daily (50 µg/ml). The cells were then lysed with 0.25 M NH₄OH. The basal lamina produced by the endothelial cells was analyzed by sequential enzyme digestion with trypsin, elastase, and collagenase. The percentage of the total radioactivity on the dish solubilized by each enzyme was then calculated. Total fucose radioactivity was 31,000 cpm; proline radioactivity was 310,000 cpm.

plated onto plastic (Fig. 2*A*) and the final density reached was the same for the three different seeding densities used.

The growth kinetics of cells plated on a smooth muscle layer were markedly different from those plated on plastic (Fig. 2*B*). The endothelial cells seeded with 10^5 cells per well grew at the same rate as those seeded on plastic, but the saturation density reached was approximately 2.5-fold higher. Higher saturation densities were also observed for endothelial cells growing on the smooth muscle matrix proteins (not shown), suggesting that the method used to derive the number of endothelial cells in the mixed cultures was a reasonable approximation of their actual number. The growth of cultures containing 5×10^4 cells was slightly retarded, but the final density reached was again increased when compared to plastic. In contrast, cultures initiated with 10^4 endothelial cells grew slowly and only a small increase in cell number was observed even after 19 days of growth. This result indicates that a threshold number of endothelial cells must be plated onto the smooth muscle layer for successful endothelial growth to occur.

Microscopic Examination. The smooth muscle cells and their extracellular matrix were easily removed from culture dishes as a discrete structure by gently scraping with a rubber policeman. Fig. 3*A* shows the appearance of such a structure after formalin fixation and staining with hematoxylin/eosin. The section, which was rolled for ease of processing, has the appearance of a "tissue" since it contains both cells and matrix and is several cell layers thick. When endothelial cells were grown on the smooth muscle layer, they could be seen as a layer

of more darkly staining cells (Fig. 3*B*). The layer appeared to be continuous and, in general, to be one cell thick.

Cells with many of the fine structural characteristics of smooth muscle were seen in cultures that had been fixed *in situ* with glutaraldehyde before removal from culture dishes and subsequently cut at right angles to the plane of the dish (Fig. 4). The cells contained many myofibrils, frequent plasmalemmal vesicles, an external lamina, and fusiform densities. In addition to the external lamina, extracellular material with morphological characteristics of elastin was present and fibrillar collagen was often apparent, confirming biochemical evidence for the presence of these proteins (10).

The endothelial cells growing on the smooth muscle multilayer contained an extensive granular endoplasmic reticulum with well-developed Golgi apparatus (Fig. 4). It was, however, not always easy to distinguish between two cell types based purely on ultrastructural criteria. The distinction between the two cell types could be more easily made if the structure was removed from the culture dish prior to fixation so that the cells contracted to a rounder shape when placed in glutaraldehyde (Fig. 5*A*). Distinct cell junctions were often apparent between adjacent endothelial cells, and darkly stained material was present between the endothelial cell layer and the smooth muscle cells (Fig. 5*A*). This material had the appearance of modified basal lamina (Fig. 5*B*) and, although not continuous, was always evident between the two cell layers but was only seen very occasionally on the upper surfaces of the endothelial cells.

Characterization of Endothelial Cell Basal Lamina. When endothelial cells were grown on plastic with daily additions of ascorbate and subsequently lysed with 0.25 M NH₄OH, a delicate membranous structure remained, which floated into the

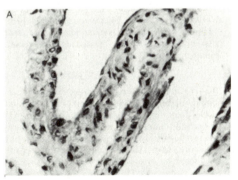
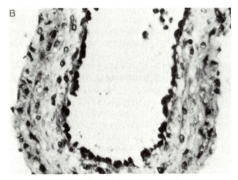

FIG. 3.   Hematoxylin/eosin-stained sections of rat smooth muscle layers (*A*) and endothelial cells growing on a preformed smooth muscle layer (*B*). (×300.) The endothelial cells can be seen as the darkly staining monolayer on the inside of the section in *B*.

Cell Biology: Jones

Proc. Natl. Acad. Sci. USA 76 (1979) 1885

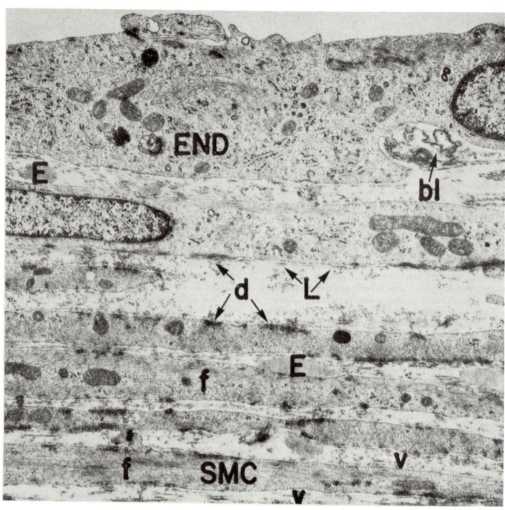

FIG. 4. Electron micrograph of a section of endothelial cells growing on a smooth muscle multilayer fixed *in situ* before removal from the culture dish. The smooth muscle cells (SMC) contain many myofilaments (f) cut both longitudinally and in cross section, with fusiform densities (d) particularly at the periphery, and also numerous plasmalemmal vesicles (v). Extracellularly, there is evidence for an external lamina (L), elastin (E), and collagen cut in cross section. The endothelial cell (END) at the top of the section appears to be separated from the underlying multilayer by the dense material, which has the appearance of a modified basal lamina (bl). (×18,000.)

solution. If, however, the endothelial cells were cultured on the insoluble matrix proteins that had been produced by the smooth muscle cells, the basal lamina remained anchored to the matrix proteins and could therefore be analyzed. Endothelial cells were grown on the unlabeled matrix proteins in the presence of [^{3}H]proline to label all of the membrane proteins or [^{3}H]fucose, which specifically labels glycoproteins. The basal lamina proteins were then analyzed by sequential enzyme digestion (Table 1). Trypsin solubilized 67% of the proline radioactivity, and subsequent treatment with pancreatic elastase released only a further 4% of this radioactivity. Since elastin is digested by elastase but not by trypsin (12), the lamina contained little or no elastin. The remainder of the proline radioactivity (28%) was solubilized by purified collagenase, demonstrating the presence of collagen. The basal lamina preparation could also be labeled with [^{3}H]fucose, and in this case the majority of the radioactivity (83%) was trypsin sensitive (Table

1). Two major [^{3}H]fucose-containing proteins with apparent molecular weights of 140,000 and 270,000 were seen when sodium dodecyl sulfate/polyacrylamide gels were run with detergent extracts of the lamina under reducing conditions (results not shown).

DISCUSSION

The interactions between different cell types are of fundamental importance to the structure of tissues. The system described here demonstrates that an endothelial layer can be successfully established on cultured smooth muscle cells to create an artificial vessel wall. Although the use of bovine endothelium with a rat smooth muscle layer is possibly nonoptimal, there is no apparent reason why a homologous system could not be constructed. The rat cells were used because of their uniform growth, rapid production of extracellular matrix, and, most importantly, their ability to remain firmly anchored to

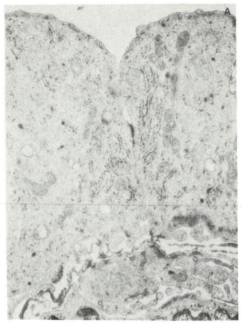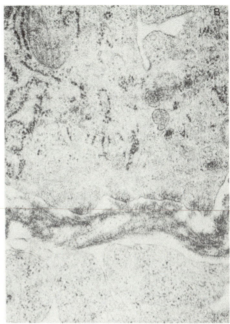

FIG. 5.   Electron micrographs of a section of endothelial cells growing on a smooth muscle layer and fixed *after* removal of the structure from the culture dish. (*A*) Two endothelial cells growing on the preformed layer showing a cell junction and membranous material between the cell layers. (×10,500). (*B*) Membranous material between the two layers. (×35,000.)

culture dishes for extended periods. Although the R22 cells appear to consist predominantly of smooth muscle cells, other cell types are probably present in the cultures (10). Sublines of smooth muscle cells have, however, been isolated from the parent cultures (10), so that purer cultures could be used. The bovine endothelial cells are readily available in cloned form and have been well characterized (3, 5).

The attachment of the endothelial cells to the smooth muscle layer appeared to be stronger than to plastic even though the attachment time was longer. Similarly, the endothelial basal lamina was more firmly attached to the matrix proteins than to plastic. These observations suggest that the adhesion mechanisms used in attachment to plastic and to smooth muscle layers may be different. Also, the different growth kinetics indicate that the substratum can have marked effects on cell behavior, particularly since the endothelial cells did not grow well when seeded below a certain density.

Ultrastructural studies have described the presence of elastin in the basal lamina of the vessel wall (8, 13) although it is not clear which of the two adjoining cell types is responsible for its production. Smooth muscle cells are, however, known to produce elastin (14, 15), and the rat cells used here produce large amounts of this protein in culture (10). Jaffe *et al.* (8) identified elastin in micrographs of the proteins produced by cultured human umbilical cord endothelial cells, suggesting that these cells are capable of elastin production. The endothelial cells (Al₄Cl-1) produced little or no elastin, although the basal lamina contained collagen and two fucosylated glycoproteins. The larger of the glycoproteins, which had an apparent molecular weight of 270,000, was probably fibronectin (5, 9).

The structure produced by culturing endothelial cells on a smooth muscle substrate, therefore, does approximate a blood vessel wall. The artificial vessel wall should be useful for studies on thrombosis, atherosclerosis, blood cell migration, and tumor invasion. Preliminary experiments with human tumor cells have

already suggested that the endothelial layer markedly increases the resistance of the structure to tumor cell invasion.

The author thanks Dr. H. Issacs and Dr. H. Neustein for their help in preparing and examining the histological sections and also for their encouragement. This work was supported by Contract N01-CP-55641 from the National Cancer Institute, and the electron microscopy was performed in the Michael J. Connell Foundation Electron Microscope Laboratory of Childrens Hospital of Los Angeles.

1. Jaffe, E. A., Nachman, R. L., Becker, C. G. & Minick, C. R. (1973) *J. Clin. Invest.* **52**, 2745–2756.
2. Gimbrone, M. A., Cotran, R. S. & Folkman, J. (1974) *J. Cell Biol.* **60**, 673–684.
3. Booyse, F. M., Sedlak, B. J. & Rafelson, M. E. (1975) *Thromb. Diathes. Haemorrh.* **34**, 825–839.
4. Jaffe, E. A., Hoyer, L. W. & Nachman, R. L. (1973) *J. Clin. Invest.* **52**, 2757–2764.
5. Gospodarowicz, D., Greenburg, G., Bialecki, H. & Zetter, B. R. (1978) *In Vitro* **14**, 85–118.
6. Gospodarowicz, D., Moran, J., Braun, D. & Birdwell, C. (1976) *Proc. Natl. Acad. Sci. USA* **73**, 4120–4124.
7. Howard, B. V., Macarak, E. J., Gunson, D. & Kefalides, N. A. (1976) *Proc. Natl. Acad. Sci. USA* **73**, 2361–2364.
8. Jaffe, E. A., Minick, C. R., Adelman, B., Becker, C. G. & Nachman, R. (1976) *J. Exp. Med.* **144**, 209–225.
9. Jaffe, E. A. & Mosher, D. F. (1978) *J. Exp. Med.* **147**, 1779–1791.
10. Jones, P. A., Scott-Burden, T. & Gevers, W. (1979) *Proc. Natl. Acad. Sci. USA* **76**, 353–357.
11. Jones, P. A. & Scott-Burden, T. (1979) *Biochem. Biophys. Res. Commun.* **76**, 71–77.
12. Hartley, B. S. & Shotton, D. M. (1971) in *The Enzymes,* ed. Boyer, P. D. (Academic, New York), Vol 3, pp. 323–373.
13. Porter, K. R. & Bonneville, M. A. (1973) *Fine Structure of Cells and Tissue,* (Lea & Febiger, Philadelphia), 4th Ed., pp. 169–170.
14. Ross, R. (1971) *J. Cell Biol.* **50**, 172–186.
15. Foster, J. A., Mecham, R. P., Rich, C. B., Cronin, M. F., Levine, A., Imberman, M. & Salcedo, L. L. (1978) *J. Biol. Chem.* **253**, 2797–2803.

Reprinted from Nature, Vol. 285, No. 5767, pp. 657–659. 1980.

Morphogenesis of branching tubules in cultures of cloned mammary epithelial cells

D. C. Bennett

The Salk Institute, PO Box 85800, San Diego, California 92138

Morphogenesis (the development of biological form, usually of multicellular organisms or their parts) is generally studied in simple organisms like the slime mould *Dictyostelium*[1], for in mammals even single tissues like the mammary epithelium discussed here appear complex. Mammary epithelium, supported by mesenchymal tissue, forms a system of branching, tubular ducts. During phases of rapid growth these ducts end in solid, swollen 'end-buds', and when mature in globular 'alveoli'[2]. The mesenchyme influences the morphogenesis of the epithelium and may be essential for this process[3,4]. An unknown number of cell types are present in both the epithelium and the mesenchyme. One step towards a better-defined 'model gland' was taken by Yang *et al.*[5], who recently described three-dimensional, solid, tumour-like outgrowths from clumps of mouse mammary tumour cells cultured in floating 'collagen gel'[6] instead of on a plastic surface. I now describe behaviour retaining some elements of natural morphogenesis, in a cloned line of epithelial cells and thus in the unequivocal absence of mammary mesenchyme—unless mesenchyme can arise from epithelium. On floating collagen gel Rama 25 cells (derived from a rat mammary tumour[7]) could generate three-dimensional structures which, although often disorganized and tumour-like, included branching, hollow tubules, sometimes with bulbous ends. Thus all the information to specify such organization resided in a single cell type and survived cloning. This raises the possibility of a simple mammalian system in which morphogenetic mechanisms, and their relation to cell differentiation, can be studied as readily as in *Dictyostelium*.

Cultures of Rama 25 'cuboidal' cells were maintained as described previously[7]. These cells behave in culture as 'stem cells': despite repeated cloning they form other types of cells including 'elongated' cells, believed to be mammary myoepithelial cells[7–9]. Elongated cells show epithelial behaviour during early passages, forming cohesive colonies and resisting detachment by trypsin, but they share some structural and antigenic features with mesenchymal cells[7,9]. Studies of Rama 25 cells on floating collagen gel were undertaken when Emerman *et al.*[10] reported exceptional functional differentiation in normal, mouse mammary epithelial cells maintained on this substrate. 'Fixed' collagen gels were prepared as described in Fig. 1 legend, or with minor differences in protocol whose effects are analysed below. Rama 25 cells were plated on top. The cells attached and proliferated less quickly than on plastic, but formed a confluent cell layer in 4–9 days. The layer was very dense but predominantly one cell thick[7]. Occasional foci of elongated cells were present beneath the monolayer, but usually no other departures from the monolayer were present at this stage.

The gels were now released to float[6]. Over the subsequent 3–4 weeks, three-dimensional structures of several types appeared, with marked variation between different gels (see below). The most striking type, often observed, consisted of blunt projections into the gel. These were seen from about day 3–7 after floating and were usually in clusters, of which between 5 and 30 generally developed per gel. Branching of these outgrowths was observed (Fig. 1), sometimes up to five times in succession. Sometimes the tips of the branches were bulbous (Fig. 1*a*), like mammary 'end-buds'[2]. Some such outgrowths resembled early mammary rudiments[3]. Histological studies showed that these structures were

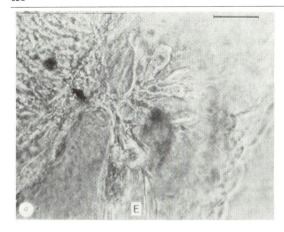

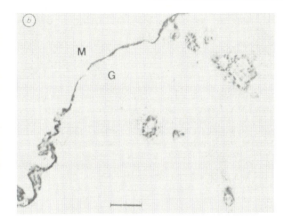

generally hollow tubules (Fig. 1*b*). These tubules resembled tubular carcinoma more closely than normal mammary epithelium: their walls were usually one cell thick (sometimes up to three); the cells were small and somewhat pleomorphic; mitotic figures were seen quite frequently; pyknotic cells were scattered both in and outside some tubules, and the outer layer of myoepithelial cells present in normal mammary ducts[2] was not observed in these.

Elongated (myoepithelial-like) cells were present focally in all cultures, but they usually migrated into the gel in densely branching chains and spikes instead of forming a second cell layer. Cloned elongated cells of the line Rama 29 (ref. 7) produced similar formations, although these cells also grew preferentially on the gel surface. Mammary cells invading collagen gel have previously been assumed to be mesenchymal[5,10]. However, these elongated cells arose from cloned epithelial cells and so are probably myoepithelial (see above). Perhaps this behaviour was associated with the failure of Rama 25 cells on collagen gel to produce a continuous basal lamina, as shown by electron microscopy[7]. Some otherwise similar spiky outgrowths had thicker, apparently solid branches and were reminiscent of the predominant structures described by Yang *et al.*[5]. Many small, apparently solid 'lumps' a few cells in diameter appeared from days 1–4. Broader areas with a convoluted surface like that of brain coral arose later, probably by coalescence of the lumps. In section these areas resembled papillary hyperplasia (Fig. 1*b*). Finally, groups of globules or spheres arose from days 16–22, in a few cultures only, either directly from the monolayer or from the tubular outgrowths (Fig. 2). Preliminary histological sections showed, among scattered elongated cells, circular cavities lined by one layer of small, flat cells. Some mitoses were seen, while pyknotic cells were again present, Thus, despite a superficial resemblance to mammary alveoli, these outgrowths were again more like some form of dysplasia.

The variation in the numbers and types of structures produced, between gels and between experiments, seemed largely due to small variations in the preparation of the gel. For example, sterilization of the gels under UV light as generally used[6,10] proved to cause drying of the gels. When gels were deliberately dried at 37 °C for different times (16–48 h) before use, the longer-dried gels were thin and flaccid when released. Cells initially attached and proliferated very well on these dehydrated gels, but after floating tended to form

Fig. 1 Gland-like outgrowths from Rama 25 cells on floating collagen gels. Collagen solution was prepared by acetic acid extraction of rat-tail tendons[6] in sterile conditions then centrifuged for 1 h at 10,000*g* and the pellet discarded. 1 M NaOH solution was mixed with 2× concentrated DEM (Dulbecco's modification of Eagle's medium) to give 0.11 M NaOH, and the precipitate removed using a 0.22 μm filter (Nalge). This medium and the collagen solution were kept at 4 °C. To make each of four gels, 1.5 ml of the alkaline medium were mixed at 4 °C with 5 ml of collagen solution and 1.5 ml were dispensed quickly to each 33-mm culture dish (Falcon). When set, each gel was overlaid with 2 ml of culture medium (DEM with 7% or 10% fetal calf serum), and left for 1 day or more in a humid incubator at 37 °C, gassed with 10% v/v CO_2 in air. A suspension of Rama 25 cells was prepared[7] and plated on the gels in a fresh 2 ml of the same medium at 2×10^5 cells per dish. Incubation was resumed and the medium was renewed every 2–3 days until the cultures were confluent. Each gel was now released to float[6] and transferred to an 85-mm, bacteriological grade dish (Falcon) with 10 ml of medium to support the many cells now present. Incubation was resumed, with fresh medium every 5 days or less. *a*, Living culture seen by phase contrast microscopy, 2 weeks after floating, showing outgrowth from cell layer at left, branching about four times and with bulbous ends. E denotes chain of elongated cells. *b*, 5 μm paraffin section of similar culture, fixed in neutrally buffered formalin (haematoxylin and eosin stain; transmission optics). G denotes gel; M, surface of cell sheet which faced culture medium. Several tubules are sectioned, showing branch-points and mostly single-layered epithelium; at left cell layer forms papillae. Scale bars represent 100 μm.

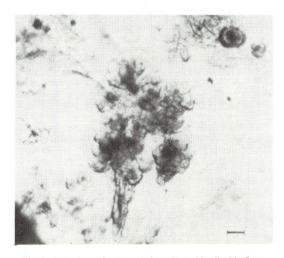

Fig. 2 Later form of outgrowth from Rama 25 cells. The figure shows living culture, prepared as for Fig. 1 and photographed on day 17. Duct-like outgrowths present in this gel up to day 14 had given place to clusters of spheres as shown (only part of cluster in focus). Transmission optics. Scale bar, 100 μm.

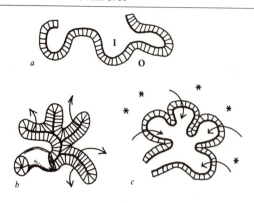

Fig. 3 Hypothetical set of properties allowing cells of a single type to form tubules and alveoli. *a*: (1) Cells cohere to form a membrane. (2) The membrane has limited thickness, for example, a single cell layer with an upper limit to cell height. (3) Cell volume has a lower limit. (4) Cells are readily distorted. (5) The membrane remains continuous (cannot fragment). (6) 'Inside' (I) and 'outside' (O) surfaces are different. *b*: (7) Processes exist which keep the inside surface area at a minimum (zero). The interior is thus one-dimensional: a point for very small membranes or a line, which can branch. The membrane then forms an occluded sphere or (branching) tube respectively. It can grow while maintaining this form. (7a) (optional) if the membrane is relatively impermeable and separates two fluid compartments, a strong force contributing to (7) can be the net outward transport of fluid by cells (arrows). The resulting pressure difference tends to collapse the interior. *c*: (8) Cells respond to an environmental signal (*) by transporting fluid inwards, causing inflation into an alveolar form.

Rama 25 cuboidal cells can produce all the cell types required.

I thank R. Hallowes for valuable advice and discussion; H. Durbin, K. Miller and B. Armstrong for assistance; R. Dulbecco, H. Battifora and D. Schubert for helpful criticism of the manuscript and B. Lang for typing it. This work was supported by the Imperial Cancer Research Fund, London and by grant DRG 254-F from the Damon Runyon Cancer Fund.

Received 29 October 1979; accepted 8 April 1980.

1. Kay, R. R., Town, C. D. & Gross, J. D. *Differentiation* **13**, 7–14 (1979).
2. Mayer, G. & Klein, M. in *Milk: the Mammary Gland and its Secretion* 1 (eds Kon, S. K. & Cowie, A. T.) 47–126 (Academic, New York, 1961).
3. Kratochwil, K. *Devl Biol.* **20**, 46–71 (1969).
4. Sakakura, T., Nishizuka, Y. & Dawe, C. J. *Science* **194**, 1439–1441 (1976).
5. Yang, J. *et al. Proc. natn. Acad. Sci. U.S.A.* **76**, 3401–3405 (1979).
6. Michalopoulos, G. & Pitot, H. C. *Expl Cell Res.* **94**, 70–78 (1975).
7. Bennett, D. C., Peachey, L. A., Durbin, H. & Rudland, P. S. *Cell* **15**, 283–298 (1978).
8. Lennon, V. A., Unger, M. & Dulbecco, R. *Proc. natn. Acad. Sci. U.S.A.* **75**, 6093–6097 (1978).
9. Rudland, P. S., Bennett, D. C. & Warburton, M. J. *Cold Spring Harb. Symp. Cell Proliferation* **6**, 677–699 (1979).
10. Emerman, J. T., Enami, J., Pitelka, D. R. & Nandi, S. *Proc. natn. Acad. Sci. U.S.A.* **74**, 4466–4470 (1977).
11. Gierer, A. *Q. Rev. Biophys.* **10**, 529–593 (1977).
12. Puchtler, H., Waldrop, F. S., Carter, M. G. & Valentine, L. S. *Histochemistry* **40**, 281–289 (1974).
13. Hollman, K. H. in *Lactation: A Comprehensive Treatise* 1 (eds Larson, B. L. & Smith, V. R.) 3–95 (Academic, New York, 1974).
14. McGrath, C. M. *Am. Zool.* **15**, 231–236 (1975).
15. Cereijido, M., Robbins, E. S., Dolan, W. J., Rotunno, C. A. & Sabatini, D. D. *J. Cell Biol.* **77**, 853–880 (1978).

only lumps and 'brain coral' instead of the more penetrating structures. The depth of the gel was significant, as was the mixing procedure. The initiation of tubular outgrowths seemed in general to be promoted by irregularities in the cell monolayer, including lumps and especially foci of elongated cells.

It is the formation of hollow, branching tubules in these cultures which seems particularly interesting. Mesenchyme may control the initial determination of epithelium as mammary tissue, and can certainly modulate the epithelial branching pattern[3,4], but this study shows that the basic 'glandular' form of branching tubes can be generated by isolated, cloned epithelial cells. (The same form is seen in developing salivary gland, pancreas and lung.) This is not too surprising, for Gierer has shown that in theory only one cell type is needed to form tubes, among other structures[11]. Figure 3 illustrates this with a combination of properties which could plausibly be possessed by the lining epithelial cells of the breast, and also by Rama 25 cuboidal cells. Property (7) may be the least plausible, but glandular epithelial cells contain at their apical or inner face a 'terminal web' of microfilaments which contain myosin-like protein[12] and which may thus be contractile. Mammary ducts and ductules are not always occluded[13], but morphogenesis can depend on transient processes[11]. Property (7a) would explain the production of 'domes' or multicellular blisters in cultured sheets of mammary epithelial cells[7,14], for domes seem to result from transport of ions and water from the apical to the basal ('outer') cell surface[14,15]. Milk secretion could provide property (8). The example in Fig. 3 is not meant as a claim that glands contain only one cell type, nor to account for all of mammary morphogenesis. For example it does not explain the branching of ducts. The essential point is that both ducts and alveoli have simple, easily defined forms: forms which do not require multiple cell types.

In conclusion, either the morphogenesis of branching tubules can be achieved by a single epithelial cell population, or

Supplementary Readings

Before 1961

Sato, G., L. Zaroff, and S. Mills. 1960. Tissue culture populations and their relation to the tissue of origin. *Proc. Natl. Acad. Sci. 46:* 963–972.

1961–1970

Bischoff, R. and H. Holtzer. 1968. The effect of mitotic inhibitors on myogenesis in vitro. *J. Cell Biol. 36:* 111–127

Davidson, R. and K. Yamamoto. 1968. Regulation of melanin synthesis in mammalian cells as studied by somatic hybridization. *Proc. Natl. Acad. Sci. 60:* 894–901.

de La Haba, G., G.W. Cooper, and V. Elting. 1966. Hormonal requirements for myogenesis of striated muscle in vitro. Insulin and somatotropin. *Proc. Natl. Acad. Sci. 56:* 1719–1723.

Fahey, J.L., I. Finegold, A.S. Rabson, and R.A. Manaker. 1966. Immunoglobulin synthesis in vitro by established human cell lines. *Science 152:* 1259–1261.

Green, H., B. Ephrussi, M. Yoshida, and D. Hamerman. 1966. Synthesis of collagen and hyaluronic acid by fibroblast hybrids. *Proc. Natl. Acad. Sci. 55:* 41–44.

Gross, W.O., E. Schopf-Ebner, and O.M. Bucher. 1968. Technique for the preparation of homogeneous cultures of isolated heart muscle cells. *Exp. Cell Res. 53:* 1–10.

Hauschka, S. and I.R. Konigsberg. 1966. The influence of collagen on the development of muscle clones. *Proc. Natl. Acad. Sci. 55:* 119–126.

Konigsberg, I.R. 1963. Clonal analysis of myogenesis. *Science 140:* 1273–1284.

Laskov, R. and M. Scharff. 1970. Synthesis, assembly and secretion of gamma globulin by mouse myeloma cells. *J. Exp. Med. 131:* 515–541.

Nameroff, M. and H. Holtzer. 1969. Contact-mediated reversible suppression of myogenesis. *Dev. Biol. 19:* 380–396.

Nelson, P., W. Ruffner, and M. Nirenberg. 1969. Neuronal tumor cells with excitable membranes grown in vitro. *Proc. Natl. Acad. Sci. 64:* 1004–1010.

Pluznik, D.H. and L. Sachs. 1965. The cloning of normal "mast" cells in tissue culture. *J. Cell Comp. Physiol. 66:* 319–324.

Richardson, U., A. Tashjian, and L. Levine. 1969. Establishment of a clonal strain of hepatoma cells which secrete albumin. *J. Cell Biol. 40:* 236–247.

Schimmer, B.P. 1969. Phenotypically variant adrenal tumor cell cultures with biochemical lesions in the ACTH-stimulated steroidogenic pathway. *J. Cell. Physiol. 74:* 115–122.

Schubert, D. and F. Jacob. 1970. 5-Bromodeoxy-uridine-induced differentiation of a neuroblastoma. *Proc. Natl. Acad. Sci. 67:* 247–254.

Shainberg, A., G. Yagil, and D. Yaffee. 1969. Control of myogenesis in vitro by Ca^{2+} concentration in nutritional medium. *Exp. Cell Res. 58:* 163–167.

1971–1980

Bloom, A.D. and F.T. Nakamura. 1974. Establishment of a tetraploid, immunoglobulin-producing cell line from the hybridization of two human lymphocyte lines. *Proc. Natl. Acad. Sci. 71:* 2689–2692.

Bodick, N. and C. Levinthal. 1980. Growing optic nerve fibers follow neighbors during embryogenesis. *Proc. Natl. Acad. Sci. 77:* 4374–4378.

Civerchia-Perez, L., B. Faris, G. LaPointe, J. Beldekas, H. Leibowitz, and C. Franzblau. 1980. Use of collagen-hydroxyethylmethacrylate hydrogels for cell growth. *Proc. Natl. Acad. Sci. 77:* 2064–2068.

Clark, J.L., K.L. Jones, D. Gospodarowicz, and G.H. Sato. 1972. Growth response to hormones by a new rat ovary cell line. *Nature New Biol. 236:* 180–181.

Cooke, J. 1980. Compartments, "switches," "programs" and vertebrate development. *Nature 284:* 216–218.

Coffino, P., B. Knowles, S. Nathenson, and M. Scharff. 1971. Suppression of immunoglobulin synthesis by cellular hybridization. *Nature New Biol. 231:* 87–90.

Darlington, G.J., H.P. Bernhard, and F.H. Ruddle. 1974. Human serum albumin phenotype activation in mouse hepatoma-human leukocyte cell hybrids. *Science 185:* 859–861.

DeLain, D., M.C. Meienhofer, D. Proux, and F. Schapira. 1973. Studies on myogenesis in vitro: Changes of creatine kinase, phosphorylase and phosphofructokinase isozymes. *Differentiation 1:* 349–354.

Elliot, B.J. and D.G.F. Harriman. 1974. Growth of human muscle spindles in vitro. *Nature 251:* 622–624.

Eyre, D.R. 1980. Collagen: Molecular diversity in the body's protein scaffold. *Science 207:* 1315–1322.

Franks, L.M. and T.W. Cooper. 1972. The origin of human embryo lung cells in culture: A comment on cell differentiation, in vitro growth and neoplasia. *Int. J. Cancer 9:* 464–468.

Gimbrone, M. 1976. Culture of vascular endothelium. *Prog. Hemostasis Thromb. 13:* 1–28.

Goudie, R.B., J.C. Spence, and R.J. Scothorne. 1980. Do vascular clones determine developmental patterns? *Lancet,* March 15, i: 570–572.

Green, H. 1977. Terminal differentiation of cultured human epidermal cells. *Cell 11:* 405–416.

Green, H. and O. Kehinde. 1976. Spontaneous heritable changes leading to increased adipose conversion in 3T3 cells. *Cell 17:* 105–113.

Harris, A.J., S. Heinemann, D. Schubert, and H. Tarakis. 1971. Trophic interaction between cloned tissue culture lines of nerve and muscle. *Nature 231:* 296–301.

Koch, K. and H.L. Leffert. 1974. Growth control of differentiated fetal rat hepatocytes in primary monolayer culture. VI. Studies with conditioned medium and its functional interactions with serum factors. *J. Cell Biol. 62:* 780–791.

Kuri-Harcuch, W. and H. Green. 1978. Adipose conversion of 3T3 cells depends on a serum factor. *Proc. Natl. Acad. Sci. 75:* 6107–6109.

Kuri-Harcuch, W., L.S. Wise, and H. Green. 1978. Interruption of the adipose conversion of 3T3 cells by biotin deficiency: Differentiation without triglyceride accumulation. *Cell 14:* 53–59.

Leffert, H.L. 1974. Growth control of differentiated fetal rat hepatocytes in primary monolayer culture. V. Occurrence in dialyzed fetal bovine serum of macromolecules having both positive and negative growth regulatory function. *J. Cell Biol. 62:* 767–779.

Lennon, V.A. and E.H. Lambert. 1980. Myasthenia gravis induced by monoclonal antibodies to acetylcholine receptors. *Nature 285:* 238–240.

Lewis, L.J., J.C. Hoak, R.D. Maca, and G.L. Fry. 1973. Replication of human endothelial cells in culture. *Science 181:* 453–454.

Lin, H.-S. and C.C. Stewart. 1974. Peritoneal exudate cells. I. Growth requirement of cells capable of forming colonies in soft agar. *J. Cell. Physiol. 83:* 369–378.

Mandel, J.-L. and M.L. Pearson, 1974. Insulin stimulates myogenesis in a rat myoblast line. *Nature 251:* 618–620.

McCredie, K.B., E.M. Hersh, and E.J. Freireich. 1971. Cells capable of colony formation in the peripheral blood of man. *Science 171:* 293–294.

Minna, J., P. Nelson, J. Peacock, D. Glazer, and M. Niremberg. 1971. Genes for neuronal properties expressed in neuroblastoma × L cell hybrids. *Proc. Natl. Acad. Sci. 68:* 234–239.

Mohit, B. and K. Fan. 1971. Hybrid cell line from a clonal immunoglobulin producing mouse myeloma and a nonproducing mouse lymphoma. *Science 171:* 75–77.

Parker, K.K., M.D. Norenberg, and A. Vernadakis. 1980. "Transdifferentiation" of C6 glial cells in culture. *Science 208:* 179–181.

Peterson, J. and M. Weiss. 1972. Depression of differentiated functions in hepatoma cell hybrids. *Proc. Natl. Acad. Sci. 69:* 571–575.

Rheinwald, J.G. and H. Green. 1975. Formation of a keratinizing epithelium in a culture by a cloned cell line derived from a teratoma. *Cell 16:* 317–330.

Rutishauser, U., P. D'Eustachio, and G.M. Edelman. 1973. Immunological function of lymphocytes fractionated with antigen-derivatized fibers. *Proc. Natl. Acad. Sci. 70:* 293–294.

Schied, M.P., M.K. Hoffmann, K. Komuro, U. Hämmerling, J. Abbott, E.A. Boyse, G.H. Cohen, J.A. Hooper, R.S. Schulof, and A.L. Goldstein. 1973. Differentiation of T cells induced by preparations from thymus and by nonthymic agents. *J. Exp. Med. 138:* 1027–1032.

Schubert, D., A. Harris, D.E. Devine, and S. Heinemann. 1974. Characterization of a unique muscle cell-line. *J. Cell Biol. 61:* 398–413.

Schubert, D., S. Humphreys, F. de Vitry, and F. Jacob. 1971. Induced differentiation of a neuroblastoma. *Dev. Biol. 25:* 514–546.

Singer, D., M. Cooper, G.M. Maniatis, P.A. Marks, and R.A. Rifkind. 1974. Erythropoietic differentiation in colonies of cells transformed by Friend virus. *Proc. Natl. Acad. Sci. 71:* 2668–2670.

Spiegel, F.W. and E.C. Cox. 1980. A one-dimensional pattern in the cellular slime mould *Polysphondylium pallidum. Nature 286:* 806–807.

Weiss, R.E. and A.H. Reddi. 1980. Synthesis and localization of fibronectin during collagenous matrix-mesenchymal cell interaction and differentiation of cartilage and bone *in vivo. Proc. Natl. Acad. Sci. 77:* 2074–2078.

Zucker-Franklin, D., G. Grusky, and P. L'Esperance. 1974. Granulocyte colonies derived from lymphocyte fractions of normal human peripheral blood. *Proc. Natl. Acad. Sci. 71:* 2711–2714.

Chapter 3   Transformations

Temin, H.M. and H. Rubin. 1958. Characteristics of an assay for Rous sarcoma virus and Rous sarcoma cells in tissue culture. *Virology* 6: 669–688.

Todaro, G.J. and H. Green. 1964. An assay for cellular transformation by SV40. *Virology* 23: 117–119.

Macpherson, I. and L. Montagnier. 1964. Agar suspension culture for the selective assay of cells transformed by polyoma virus. *Virology* 23: 291–294.

Freeman, A.E., P.H. Black, R. Wolford, and R.J. Huebner. 1967. Adenovirus type 12-rat embryo transformation system. *J. Virol.* 1: 362–367.

Smith, H.S., C.D. Scher, and G.J. Todaro. 1971. Induction of cell division in medium lacking serum growth factor by SV40. *Virology* 44: 359–370.

Risser, R. and R. Pollack. 1974. A nonselective analysis of SV40 transformation of mouse 3T3 cells. *Virology* 59: 477–489.

Sutherland, B.M., J.S. Cimino, N. Delihas, A.G. Shih, and R.P. Oliver. 1980. Ultraviolet light-induced transformation of human cells to anchorage-independent growth. *Cancer Res.* 40: 1934–1939.

Kennedy, A.R., M. Fox, G. Murphy, and J.B. Little. 1980. Relationship between X-ray exposure and malignant transformation in C3H 10T½ cells. *Proc. Natl. Acad. Sci.* 77: 7262–7266.

As assemblages of differentiated cell types, we survive because our cell types show growth control. Most of the cells in a normal body show either one or the other of two types of growth control. In one type, cells are dividing all the time, but one daughter differentiates and dies. In the other type, cells normally do not divide. As a result of these restrictions, our net cell mass is constant.

For the fibroblast, growth control operates in the body by the cessation of cell division in the absence of wounding or scar formation. Cell culture analogs of this state exist, since there are a reasonably large number of conditions in which normal fibroblasts remain viable but do not divide. Any agent that can cause a normal fibroblast to lose one or more of these growth controls in a heritable fashion will show itself in the appropriate restrictive assay as an inducer of a growing clone of cells, which is seen against a background of normal cell quiescence. Such clones are said to be transformed.

Because agents that overthrow one or the other type of growth control may cause tumors, transformation assays are of great interest in the study of cancer. Indeed, the manner in which these different transformations are linked to tumorigenic growth in vivo remains an open question (see Chapter 5).

The first four papers in this chapter describe four different growth controls of fibroblasts and the capacity of certain tumor viruses and X rays to generate transformed clones able to grow despite them. As early as the first decade of the century, Rous had shown that a virus causes a fibroblastic tumor or sarcoma in chickens and that chicken sarcoma cells in culture are rounder than normal fibroblasts. In their careful and elegant paper, **Temin** and **Rubin** quantitate this viral transformation. They infected confluent precrisis cultures of chick fibroblasts with Rous sarcoma virus. By imbedding the monolayer of cells under agar, they were able to detect and count discrete, morphologically changed cells whose shapes had become rounder than normal. Eventually, such cells gave rise to discrete foci of rounded, poorly spread cells. Presumably, then, one normal growth control involves the maintenance of a particular cell shape when spread. Temin and Rubin showed that the number of foci is in strict inverse proportion to the dilution of virus stock used for infection. That is, a single particle (virion) was capable of initiating a transformed focus. Are Rous-transformed cells stable in their phenotype (Chapter 1)? They can be passaged for 20 passages, but then crisis afflicts them as completely as it does untransformed chick cells, and the cultures die. Apparently, transformation need not accomplish establishment.

Some established lines, however, can subsequently be transformed by viruses. 3T3 mouse cells (Chapter 1) show a different growth control—the ability to cease division when confluent. **Todaro** and **Green** infected 3T3 cells with the monkey virus SV40. This virus had already been shown to cause tumors in rodents (see footnote in Hayflick and Moorhead, Chapter 1). Infection resulted in dense-foci colonies of cells overgrowing the 3T3 monolayer. As with Rous sarcoma virus in chick cells, dense transformed 3T3 colonies arose with low frequency, but linearly with virus dose.

Some transformed cells are less adherent and rounder than normal. In the extreme case, they grow even when they cannot adhere at all to the surface on which they are plated. Working with the hamster cell line BHK21, **Macpherson** and **Montagnier** constructed an assay in which growth requires a total loss of any anchorage requirement. After infection by polyoma virus, cells were imbedded in agar. Unable to spread on agar, the cells initially remained suspended as single spheres. Uninfected BHK cells did not grow when prevented from spreading, but a few days after polyoma infection, many large, multicellular, spherical colonies appeared in the agar. When these were picked and grown on a dish, most, but not all, gave rise to dense, morphologically altered colonies.

The concentration of calcium ion in plasma is about 1–2 mM. Some human adenoviruses can cause tumors when injected into hamsters. Cells from these tumors cannot be cultured in media containing 2 mM calcium, but they will grow in a medium carrying a fraction of the normal amount of calcium ions. **Freeman, Black, Wolford,** and **Huebner** demonstrate that reduction of the calcium concentration in medium is doubly selective for adenovirus-transformed clones, since normal cells die in low calcium, whereas transformed colonies grow. In this assay, once again, transformation is linear with virus dose but very inefficient. The relationships between a low calcium requirement and freedom from anchorage or density inhibitions are not yet well understood.

Serum is necessary for the growth of normal fibroblasts. Some of the factors normal fibroblasts require can be removed from serum by chemical treatments (Chapter 1). Cells themselves exhaust an aliquot of serum of its growth factors as they grow in it. In their study of 3T3 cells and their SV40-transformed derivatives, **Smith, Scher,** and **Todaro** used serum depleted by chemical removal of gamma globulins and by preincubation with monolayers of 3T3 cells. 3T3 cells cannot grow in depleted serum. SV40-transformed clones isolated in the density assay grow in depleted serum, suggesting that they had lost a growth requirement for a limiting factor in serum. Some clones selected for growth in depleted medium did not grow to high densities in medium with fresh serum, which shows that loss of contact inhibition cannot be the only phenotypic difference generated by SV40. Loss of a serum requirement is another autonomous phenotypic alteration.

Are these different transformed phenotypes the reflection of a single heritable change in a cell? Can a single transforming agent generate two different colonies with two different stable phenotypes? **Risser** and **Pollack** infected 3T3 mouse cells with SV40 and permitted clones to grow up without any selective restraints. Many clones were picked, recloned, and assayed separately for their ability to grow in some of the various restrictive transformation assay conditions described above. Viral gene expression was monitored as the presence of a viral antigen (T antigen) in the nuclei of cells in each clone. The various clones recovered were not identical to one another; rather, they exhibited a spectrum of phenotypes. Loss of the requirement for serum was most common, and acquisition of anchorage independence most rare. All anchorage-transformed colonies were serum-independent as well, and viral gene expression was coordinate with growth in the density and anchorage assays. About a third of the transformed clones were variegated for T antigen and for growth in agar, a feature not seen before in transformed cells.

In a formal sense, viral transformations of cells are mutations; the rate-limiting step in such transformations seems to be the integration and expression of at least one new viral gene. X rays and ultraviolet light can also transform cells. It is tempting to imagine that they too transform by mutagenesis, but this is difficult to prove. In the last two papers in this chapter, we see that whereas DNA damage by one physical agent (ultraviolet light) is a step in transformation, X rays cannot be transforming fibroblasts through immediate operation of mutations of the ordinary sort.

Sutherland, Cimino, Delihas, Shih, and **Oliver** transformed human fibroblasts by exposing them to multiple short doses of UV light. This is much less toxic than a single large dose, and among the survivors of the split UV doses are cells capable of continued growth in agar. DNA damage of a specific sort was implicated in this transformation assay, because Sutherland et al. were able to block transformation by irradiating the cells with visible light immediately after UV irradiation. The action spectrum of this sparing process matched that for photoreactivating repair of UV-induced pyrimidine dimers, suggesting that pyrimidine dimers play a role in UV induction of anchorage transformants. Whether mutagenesis directly follows the appearance of UV dimers or whether the dimers lead to transformation by some other set of events is not yet known.

Kennedy, Fox, Murphy, and **Little** used X rays to show that cells of the mouse line 10T½ are unlikely to be transformed by direct mutation. After X-ray treatment, the cells must be left at confluence for weeks before the transformed foci appear. Remarkably, the number of transformed foci per confluent dish was proportional to X-ray dose, but not to the number of irradiated cells initially plated per dish. Apparently, all irradiated cells, and all their progeny, were identically unable to make a transformed colony so long as they were growing, whereas all cells also became identically able to yield transformants with the same low frequency once they reached confluence. For structural genes, X-ray-induced mutagenesis generates rare mutant clones directly, but in the paper by Kennedy et al. transformation seems to be at least a two-step process, and the first step seems unlikely to be a classic mutation.

VIROLOGY **6,** 669–688 (1958)

Characteristics of an Assay for Rous Sarcoma Virus and Rous Sarcoma Cells in Tissue Culture[1]

Howard M. Temin[2] and Harry Rubin

Division of Biology, California Institute of Technology, Pasadena, California

Accepted July 30, 1958

An accurate tissue culture assay for Rous sarcoma virus (RSV) and Rous sarcoma cells is described. The Rous sarcoma virus changes a chick fibroblast into a morphologically new and stable cell type with the same chromosomal complement as ordinary chick embryo cells. One virus particle is enough to change one cell, but at any one time 90% of the cells in a culture are not affected by RSV. The cellular resistance is the same in a clonal population. The physiological state of the cell is of some importance in deciding whether or not it is competent to be infected by RSV but so far attempts to infect all the cells in a chick embryo culture by altering the physiological condition have failed.

INTRODUCTION

The customary technique for assaying the Rous sarcoma virus (RSV) has been the production of tumors in chickens (Rous, 1911; Bryan, 1946). More recently, the virus has been assayed by infecting the chorioallantoic membrane of the developing chicken embryo (Keogh, 1938; Rubin, 1955; Prince, 1957). Although these techniques are satisfactory for many experiments, they do not provide the accuracy nor the ease of manipulation required for a diversified study of virus-host interactions at the level of the individual cell. An assay technique comparable to the methods used for assaying bacteriophages and cytopathogenic animal viruses has now been developed and is described in the present paper. The assay depends upon the morphological change caused by RSV in chick embryo cells.

The first clear demonstration that chick embryo cells grown in tissue culture were changed into Rous sarcoma cells by infection with RSV was

[1] Supported by grants from the American Cancer Society and the United States Public Health Service, grant E 1531.

[2] United States Public Health Service predoctoral fellow.

given by Halberstaedter *et al.* (1941), who exposed chick fibroblast cultures to pieces of intensely irradiated Rous sarcoma cultures and observed gross changes in the culture and cytological changes similar to those present in cultures of Rous sarcoma cells derived from a chicken sarcoma. Lo and associates (1955) succeeded in changing fibroblasts *in vitro* with partially purified RSV. They also showed that the culture containing changed cells produced virus for several months. Following this lead Manaker and Groupé (1956) succeeded in producing discrete foci of changed cells upon treatment of monolayer cultures of chick embryo cells with RSV. In the range tested the number of foci per culture was proportional to the concentration of virus added.

In the assay method to be described here an agar overlay has been introduced to decrease the chance of formation of additional foci by virus liberated from infected cells and by detachment and reattachment of infected cells. In addition, a technique will be described for determining the number of infected cells in a mixed population. The properties and nature of the changed cells are also discussed.

MATERIALS AND METHODS

Solutions and Media

"Standard medium":
    8 parts Eagle's medium (Eagle, 1955) with double concentration of
        amino acids and vitamins and 4.2 g NaHCO$_3$ per liter.
    1 part Difco Bacto-Tryptose phosphate.
    1 part serum (usually 0.8 calf and 0.2 chicken serum).
Trypsin:
    0.25 % Bacto-Trypsin in tris buffer.
Eagle's minus:
    Eagle's medium in which the salts of Ca^{++} and Mg^{++} have been
    omitted.

Incubation

All cultures were made in petri dishes. They were incubated at 37° in a water-saturated atmosphere. The pH was regulated at about 7.3 by a bicarbonate buffer controlled by CO$_2$ injected into the incubator.

Examination of Cultures

Cultures were examined at 25- and 100-fold magnifications of a Zeiss plankton (inverted) microscope.

Chick Embryo Cultures

Primary cultures of 9- to 13-day-old chicken embryo cells were made by the method of Dulbecco as modified by Rubin (1957). Four million cells in 10 ml of standard medium were placed in a 100-mm petri dish. After 3–5 days of incubation secondary cultures were made. The primary cultures were washed twice with Eagle's minus, and 2 ml of 0.05 % trypsin in Eagle's minus were added. After 15 minutes' incubation 2 ml of standard medium were added to stop the action of the trypsin. The cells were pipetted twice and centrifuged at 1500 rpm for 1 minute. The pellet was resuspended in standard medium and the cells counted. Two or three hundred thousand cells in 3 or 5 ml of standard medium were placed in 50-mm petri dishes. After incubation overnight these plates were used for the assay of Rous sarcoma virus and Rous sarcoma cells. The cultures consisted chiefly of fibroblasts.

Assay for Virus

The secondary culture was washed once with Eagle's medium and then virus in a volume varying from 0.1 to 0.8 ml was added. After an adsorption period of from 10 minutes to 1 hour 5 ml of standard medium containing 0.6 % agar was added. (The concentration of agar is not critical for focus development.) Three days later the culture was fed by adding 2 ml of agar medium on top of the first agar layer. Five to seven days after infection 3 ml of a 1/20,000 solution of neutral red in Eagle's medium was added to allow easier counting of the Rous sarcoma foci. After 2 hours' incubation the neutral red was removed and the plate was placed on a piece of glass with a rectangular grid of 2-mm squares and was scanned for foci at a 25-fold magnification of the inverted microscope.

Assay for Rous Sarcoma Cells or Foci Formers

The cells to be assayed were trypsinized, counted, diluted, and added to a secondary plate. After incubation for 8–16 hours to allow attachment of the cells, the medium was removed and standard medium with 0.6 % agar was added. Three days later the culture was fed by adding 2 ml of agar medium on top of the first agar layer. After 5–7 days 3 ml of a 1/20,000 solution of neutral red in Eagle's medium was added and the plate scanned for foci at a 25-fold magnification.

Virus Stocks

The original virus used in this work was generously supplied by Dr. Bryan of the National Cancer Institute. This preparation contained about 2×10^6 focus-forming units (FFU) per milliliter and was derived from tumors in chickens. Tissue culture virus was obtained by infecting secondary chick embryo cultures with 10^5 FFU of RSV and growing the cultures in standard medium without chicken serum for 2 weeks with four transfers. The supernatants were then collected. They contained between 10^5 and 10^6 FFU per milliliter.

Chromosome Counts

Cells were transferred to dishes containing sterile 22×22-mm No. 0 cover slips and incubated in standard medium. Eighteen hours later the cover slips were put through the following modification of the procedure of Hsu and co-workers (1957):

1 part Eagle's:9 parts distilled water	15 minutes
Dry rapidly under blower	
100 % Methyl alcohol	1 hour
Aceto-orcein	2 hours
Ethyl alcohol (absolute)	Rinse
Ethyl alcohol (absolute)	10 minutes
Mount in Euparal	

Drawings of well-spread chromosome figures were made with a camera lucida at 1350 magnification. The actual counts were made from the camera lucida drawings.

Cloning Technique

Cells were cloned by the feeder layer technique of Puck *et al.* (1957). Twenty-four hours before cloning cultures containing 5×10^4 HeLa cells or chick fibroblasts were X-rayed with a dose of 5000 r. The cells to be cloned were dispersed with trypsin, counted, and 2×10^3 cells added to each plate. At sixteen hours 5 ml of standard medium containing 0.3 % agar was added. At 8 days and every 4 days thereafter the cultures were fed by adding agar medium on top of the first agar layer. When the clones were large enough to be picked, the agar was removed and trypsin added. Under a dissecting microscope the clones were picked with a micropipette and transferred to a new dish.

X-Irradiation

Cultures were X-irradiated from a Machlett OEG.60 tube with tungsten target and beryllium end window operated at 50 kilovolts and

174

30 milliamperes. The outlet of the tube was covered with a 0.38-mm Al filter. The irradiation of the cells was carried out at a distance of 6.6 cm from the target in a covered petri dish at an intensity of 2640 roentgens per minute.

Ultraviolet Irradiation

Ultraviolet irradiation was given from a Westinghouse germicidal lamp. The virus was in 1 ml of medium, 11¾ inches from the lamp.

Abbreviations

RSV—Rous sarcoma virus
FFU—Focus-forming unit.

EXPERIMENTAL

Morphological Changes in Chick Embryo Cells following Infection with Rous Sarcoma Virus

One or two days after addition of Rous sarcoma virus to a chick embryo culture rounded refractile cells appear. As single cells these can be confused with cells in mitosis. However, these cells and their progeny retain their changed appearance throughout interphase, giving rise after 2 or 3 days to groups of rounded refractile cells. These groups or foci can easily be distinguished from the background of fibroblasts (Plate I, Fig. A). The cells differ markedly in their colonial characteristics from the fibroblasts. They grow in a grapelike cluster only loosely attached to one another and to the glass. Often a focus becomes multilayered and the rounded cells migrate out on top of the other cells (Plate I, Fig. B). The cell sheet may also tear and retract in the region of a focus leaving a hole surrounded by the round, refractile cells (Plate I, Fig. C). As long as the culture is fed every 3 days, the number of cells in a focus doubles about every 18 hours (Fig. 1). If the culture is not fed, the cells in a focus lose their refractility and become difficult to distinguish from the background of fibroblasts. Upon addition of fresh medium the cells in a focus may regain their refractility.

Mechanism of Growth of a Focus

There are two processes which could be concerned in the growth of a focus: division of the cells in a focus; and infection of normal cells by the virus released from cells in the focus. Several experiments were carried out to determine whether both of these processes actually occur. Cells

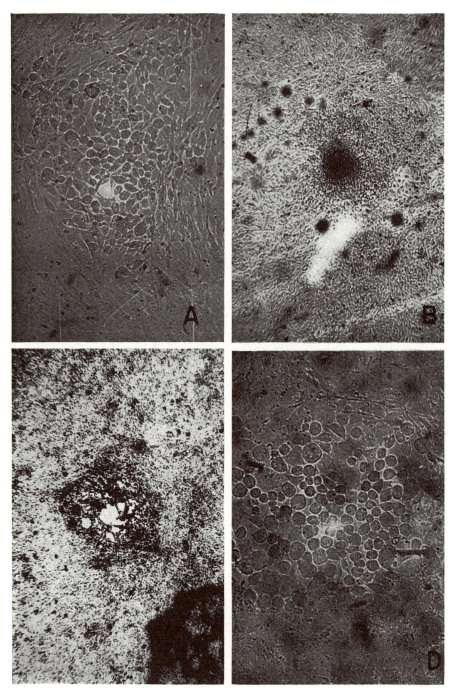

PLATE I
674

176

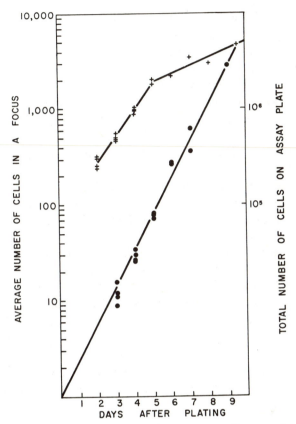

FIG. 1. Increase in number of cells in a focus. Rous cells were plated on assay plates. Twenty hours later agar was added to the plates. Every day the number of cells in 10–20 foci was counted. The total number of cells on the plate was also counted: first, by directly counting a fraction of the cells on the plate; and after the third day, by removing the agar, trypsinizing the cells, and counting in a hemocytometer. ● = Average numer of cells in a focus. + = Number of cells on an assay plate.

FIG. A. Focus of Rous sarcoma cells formed on a secondary chick embryo culture. Unstained; magnification: × 80.

FIG. B. Focus of Rous sarcoma cells formed on a secondary chick embryo culture. Lightly stained with neutral red; magnification: × 20.

FIG. C. Focus of Rous sarcoma cells formed on a secondary chick embryo culture. Heavily stained with neutral red; magnification: × 20.

FIG. D. Focus of Rous sarcoma cells formed on a clonal population of chick embryo fibroblasts. Unstained; magnification: × 80.

which had been infected more than 2 days before with RSV were plated in the following ways:

1. The cells were plated as described under methods for assay of Rous sarcoma cells.
2. The cells were plated as in (1) but on a secondary culture of cells 1/40 as sensitive to virus infection as the strain regularly used.
3. The cells were plated as in (1), but the secondary culture was X-rayed previously.
4. The cells were X-rayed to prevent their multiplication and plated as in (1).

Under (1) both reinfection and division can occur. Under (2) reinfection is 1/40 as likely and under (3) impossible. (Previous work had shown that X-rayed fibroblasts cannot be infected by RSV.) Under (4) infection would be necessary for initiation of a focus.

The relative number of foci resulting from the four methods of plating are presented in Table 1. The table shows that the number of foci was the same in all cases. It is concluded that a focus can be formed by dividing Rous sarcoma cells in the absence of reinfection but that nondividing Rous sarcoma cells can initiate formation of a focus, showing that virus released from a cell on the glass can infect the surrounding cells. Since there was little difference in the size of the foci obtained in the different ways even when reinfection was impossible, (3), the major role in the development of the foci must be assigned to division of changed cells.

TABLE 1

INITIATION OF A FOCUS[a]

Number of foci after standard plating of cells (1)	Treatment	Number of foci after designated treatment
105	2	111
23, 44, 79	3	40, 46, 66
197, 212	4	180, 204

[a] Rous sarcoma cells were plated in the following ways: (1) The cells were plated on a secondary plate as described under methods for assaying Rous sarcoma cells. (2) The cells were plated on a secondary plate composed of chick embryo cells 1/40 as sensitive to RSV as the strain usually used. (3) The cells were plated on a X-rayed feeder layer. (4) The Rous sarcoma cells were X-rayed and then plated as in (1). The number of foci resulting by each procedure are given in relation to (1).

Properties and Nature of Rous Sarcoma Cells

After the agar overlay is removed from an infected culture, the cells can be transferred and grown in fluid medium. Chick serum is omitted to allow further infection. After 2 weeks the culture becomes composed chiefly of round, refractile cells. These cells are producing RSV. They resemble the basophilic, round cells of Doljanski and Tenenbaum (1943).

The Rous sarcoma cell is a stable cell type which can be grown in pure culture. Rous sarcoma cells were cloned and grown for twenty generations in the absence of ordinary chick embryo cells. After more prolonged culture the cells became giant and diverse in appearance.

To check whether or not the morphological change was associated with a gross change in the chromosome complement, chromosome counts were done on Rous sarcoma cells and on chick embryo fibroblasts. The number of chromosomes found in less than 1-month-old Rous sarcoma cells and in normal fibroblasts was the same (Fig. 2). The mode was in the middle seventies for both types of cells and the spread was similar.

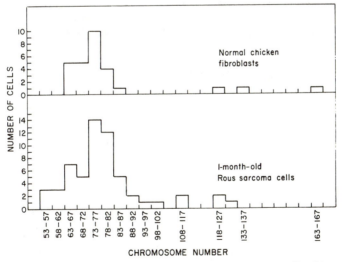

FIG. 2. Chromosome number of Rous sarcoma cells and of fibroblasts. Rous cells or normal cells were transferred to cover slips and fixed and stained by a modification of the procedure of Hsu. Drawings of well-spread chromosome figures were made with a camera lucida at 1350 magnifications, and counts were made from the drawings.

179

After more prolonged culture of the Rous sarcoma cells at a time when many giants were present, over half of the cells were polyploid.

The nature of the chick embryo cell which is changed by RSV to a Rous sarcoma cell has been controversial since Carrel's claim that only macrophages could be infected (Carrel, 1924, 1925). Several workers have disputed this contention, claiming the fibroblast as the cell of origin of the Rous sarcoma cell (quoted in Lo *et al.*, 1955). To determine unequivocally whether the fibroblast, which is the major cell type of the secondary chick embryo culture, is indeed the cell which is changed by infection with RSV, clonal lines of fibroblasts were exposed to the virus. The foci which resulted were similar in number and appearance to those found on the usual assay plates (Plate I, D).

Quantitative Features of the Assay

Adsorption volume. The following experiment was carried out to determine the relative efficiency of virus adsorption as a function of the volume of virus inoculum placed on the secondary culture. Virus was diluted in Eagle's medium. A constant amount of virus was suspended in different volumes of fluid and placed on separate assay plates. After 30 minutes the supernatant was removed and an agar overlay added. The

TABLE 2

ADSORPTION VOLUME[a]

Volume of original virus suspension	Volume of inoculum	Number of foci
1/5000 ml	0.1 ml	177
	.2	200
	.4	135
	.8	55
1/20,000	.1	41
	.2	40
	.4	23
	.8	11
1/80,000	.1	11
	.2	9
	.4	6
	.8	3

[a] Virus was diluted in Eagle's medium. A constant amount of virus was suspended in different volumes of fluid and placed on separate assay plates. After 30 minutes the supernatant was removed and an agar overlay added.

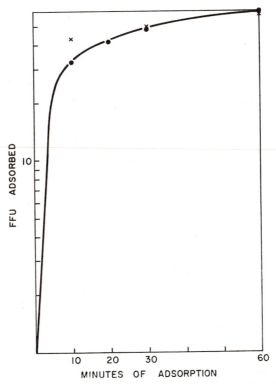

FIG. 3. Kinetics of adsorption. An inoculum of virus containing 50 FFU in 0.2 ml was added to each of several secondary cultures. After adsorption for varying lengths of time the supernatant was removed and an agar overlay added. Two separate experiments are shown. Each point is the average of two assay plates.

results are in Table 2. The number of foci per plate for the same amount of virus doubles as the volume of the inoculum was lowered from 0.8 ml to 0.4 ml and from 0.4 ml to 0.2 ml. No difference was found between 0.2 ml and 0.1 ml. Two-tenths of a milliliter is therefore used as a standard inoculum.

Kinetics of adsorption. An inoculum of virus containing 50 FFU in 0.2 ml was added to each of several cultures. After adsorption for varying lengths of time the supernatant was removed and an agar overlay added. The number of foci per plate is plotted as a function of the time of adsorption (Fig. 3). For practical reasons 30 minutes has been used as a standard adsorption time.

Relationship between virus concentration and number of foci. A series of twofold dilutions between 1/80 and 1/2560 of a virus stock containing 3.5×10^5 FFU/ml were placed on secondary culture for 30 minutes and agar added. The number of foci per plate was found to be proportional to the virus concentration over a thousand fold range (Fig. 4). This linear dose response confirms earlier results, using *in vivo* assay techniques, which showed that one virus particle is sufficient to cause an infection (Keogh, 1938; Rubin, 1955; Prince, 1957).

Relationship between virus concentration and number of infected cells. A series of twofold dilutions from undiluted virus to 1/512 of a virus stock

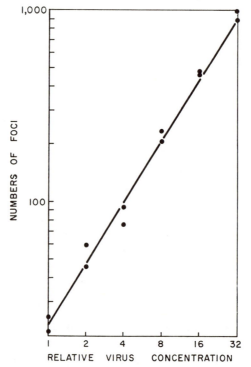

FIG. 4. Relationship between virus concentration and number of foci. A series of twofold virus dilutions between 1/80 and 1/2560 of a virus stock containing 7×10^4 FFU/ml were placed on secondary cultures for 30 minutes and agar added. To enable counting of as many as 1000 foci per plate the cultures were counted on the fifth day after infection.

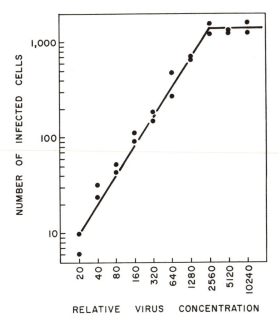

RELATIVE   VIRUS   CONCENTRATION

FIG. 5. Relationship between virus concentration and number of infected cells. Virus dilutions from undiluted to 1/512 of a stock containing 10^5 FFU/ml were placed on secondary cultures for 30 minutes. The cultures were washed, trypsinized, and various dilutions of the infected cell suspension were placed on each of two secondary cultures. Sixteen hours later agar was added. The plates were counted on the sixth day after infection.

containing 10^5 FFU/ml were placed on secondary cultures for 30 minutes. The cultures were washed, trypsinized, and various dilutions of the infected cell suspension were placed on each of two secondary cultures. Sixteen hours later agar was added. The number of foci per plate on the sixth day is plotted in Fig. 5. The number of Rous sarcoma cells per plate was a linear function of virus concentration until 0.5 % of the cells were infected. Further increases in virus concentration failed to increase the fraction of cells infected. In other experiments the maximal fraction of cells infected usually varied from 1.0 to 10 %.

The efficiency of the technique for detecting the number of infected cells by plating them in suspension as focus-formers was tested in the following manner. Three plates were infected with 100–600 FFU of RSV. After a 30-minute adsorption period the plates were washed. Agar was added to two plates and they were incubated as in the standard virus assay. The number of foci showed the number of cells infected. The cells on the third plate were suspended with trypsin and plated on secondary cultures. The results of several experiments are presented in Table 3. The efficiency of the technique to determine the number of infected cells by plating them in suspension varied between 10 and 100 % when compared with the number of foci in the direct assay. This variation in efficiency of plating is partly responsible for the fluctuation in the maximum fraction of cells infected measured above, but does not explain why no more than 10 % of the cells can be infected in *any* experiment.

Factors limiting the fraction of cells which can be infected at any given time. There are several possibilities to explain the restriction in the proportion of cells infected at a given time by RSV. The culture may be heterogeneous with regard to the genetic or embryological origin of the cells; with regard to their physiological state, either in general or in reference to the division cycle; or there may be interfering substances, such as an inactive virus, in the inoculum. Investigation of these possibilities was carried out.

1. Variation in genetic or embryological origin of the cells: Two lines of clonal fibroblasts were obtained. They were infected with serial three-fold dilutions at high virus concentrations and the number of infected cells determined as above. The results of such an experiment are presented in Fig. 6. A plateau in the number of infected cells was found at

TABLE 3

EFFICIENCY OF PLATING INFECTED CELLS IN SUSPENSION[a]

Number of foci resulting		Efficiency of plating infected cells in suspension (%)
From direct assay	From plating of infected cell in suspensions	
555	48, 54	9.2
280, 210	80	33
135, 123	55, 75	50
855, 815	645, 510	69
101, 102	84, 104	93
595, 610	653, 662	109

[a] Three plates were infected with from 100 to 600 FFU of RSV. After a 30-minute adsorption period the plates were washed. Agar was added to two plates and they were incubated as in the standard virus assay. The number of foci showed the number of cells infected. The cells on the third plate were suspended with trypsin and plated at a dilution of 1/1 to 1/4. A number of different experiments are listed.

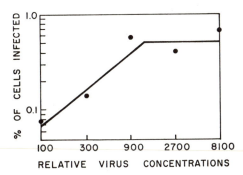

Fig. 6. Relationship between virus concentration and number of infected cells in a cloned population. Cultures of 2×10^5 cells were made from a clonal line of chick embryo fibroblasts. They were infected with serial threefold dilutions at high virus concentrations and the number of infected cells determined.

the same level as was found with secondary chick embryo cultures of varied cell origin. Since the clonal population was from a single cell, the plateau could not be due to heterogeneity in the origin of the cell.

2. Variation in physiological susceptibility: The fraction of infectible cells on secondary cultures was determined under different conditions as a function of the time which had elapsed since the plates were made. One group of cultures was kept in standard conditions; a second was placed in Eagle's medium without glutamine or glucose for 24 hours (Vogt, personal communication); a third group was kept at 14° for 24 hours in an attempt at obtaining synchronization (Chèvremont et al., 1957). At 3-hour intervals after the cells were returned to standard conditions, different cultures were infected with 10^5 FFU of RSV and the number of infected cells determined as above. The results presented in Fig. 7 show there is a considerable fluctuation in the fraction of susceptible cells in a culture with time and under different environmental conditions.

3. Interference: The possible occurrence of interference by inactive virus particles was tested by the following experiment. Stock virus was killed by irradiation with ultraviolet-light or by incubation at 37°. Two-tenths milliliter of killed virus was placed on a secondary plate. After 30 minutes the supernatant was removed and live virus added as in the direct assay. The results presented in Table 4 indicate that such killed virus preparations interfere with infection by live virus.

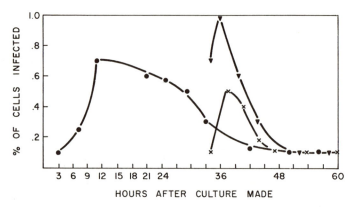

FIG. 7. Physiological state and per cent cells infected. Secondary cultures were made at time 0. One group of cultures was incubated in the usual manner; a second group was placed in Eagle's medium without glucose or glutamine for 24 hours and then standard medium was replaced; a third group in sealed dishes was placed at 14° for 24 hours and then returned to 37°. At the indicated times the cultures were infected with a high concentration of virus. After adsorption they were trypsinized, the cells counted, and 10^4 cells placed on each of two secondary cultures. Each point represents an average of two plates. ● = Control; ✕ = starved for 24 hours; ▼ = 14° for 24 hours.

TABLE 4

INTERFERENCE[a]

Treatment	Untreated virus	Treated virus (diluted 1:10)	Treated virus (undiluted) followed by untreated
UV 10 min	533	11	121, 120
37° 72 hours	1005, 1100	0, 1	85, 110

[a] Stock virus was inactivated by UV irradiation or by prolonged incubation at 37°C. The undiluted inactivated virus was placed on secondary cultures for thirty minutes. The supernatants were then removed and live virus at a concentration of about 1/500 was added.

Since the standard virus stocks would be expected to contain significant amounts of 37°-inactivated virus, such virus could cause autointerference and restrict the number of cells infected. To minimize the amount of inactivated virus, fluid was collected from a culture of Rous sarcoma cells 6 hours after washing. The maximum number of infectible cells, using this virus stock for infection, was only slightly higher than when

TABLE 5

INFECTION WITH 6-HOUR VIRUS[a]

Virus stock (hours after washing)	Titer (FFU/ml)	Per cent cells infected
6	5×10^5	8.5
48	2.5×10^5	5.6

[a] Fluid was collected from a culture of Rous sarcoma cells 6 hours after washing. Secondary cultures were infected with this virus stock. The percentage of cells infected is compared with the percentage of cells infected using a standard virus stock.

standard virus was used for infection (Table 5). This finding indicates that autointerference by 37°-inactivated virus plays a minor role in limiting the number of cells infected.

4. Repeated infection with RSV: The results in the previous sections suggest that the physiological condition of the cells is an important factor in limiting their susceptibility to infection. An experiment was carried out to determine whether or not the physiological variation of chick embryo cells to RSV infection was due to a transient period of competence in individual cells. Several cultures were infected with a saturating dose (10^5 FFU) of RSV for 1 hour. Supernatants were removed from the cultures and medium added. Each hour thereafter for 4 hours, virus was added to a different culture for 1 hour to determine whether any more cells had become susceptible in the interval. One culture was not re-exposed to virus. At the end of 5 hours the number of infected cells in each culture was determined. The results of two such experiments are given in Fig. 8. The number of infected cells in a culture could be increased beyond the usual plateau level by re-exposing the cells to virus at a time after the initial exposure. The maximum increase was obtained when the interval between the first and second inocula was 2–3 hours; longer intervals did not increase it further, suggesting that cells undergo a 2–3-hour transient period of competence.

Relative efficiency of in vitro assay and chorioallantoic membrane assay. Comparative titrations of a virus stock sent by Bryan were carried out on the chorioallantoic membrane of the developing chick embryo and by the standard tissue culture technique. The titer on various occasions varied from 1 to 4 $\times 10^6$ FFU/ml in both assays.

187

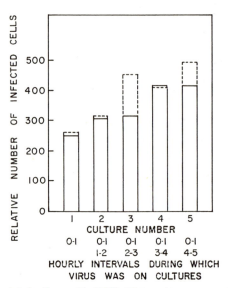

Fig. 8. Repeated infection with RSV. Five cultures were infected with 10^5 FFU of RSV for 1 hour. Supernatants were removed from the cultures and medium added. Each hour thereafter for 4 hours virus was added to one culture for 1 hour. One culture was not re-treated. At the end of 5 hours the number of infected cells in each culture was determined. Each line is the average of two plates. Two separate experiments are shown.

DISCUSSION

Use of the tissue culture assay for RSV which has been described here has shown that at a given time over 90 % of the cells in at chick embryo culture are not suscetpible to being infected by RSV. This resistance is not due to the diverse genetic or embryological nature of the cells. There is some interference by material in the virus stock which influences the number of infectible cells but does not appear to be the major limiting factor. The physiological state of the cells does affect their susceptibility and experiments with repeated infection suggest that the cells pass through a transient state of competence to infection and change, similar to that described by Hotchkiss (1954) for transformation of pneumococci. Among temperate phage, the physiological condition of the cells has also been shown to play an important role in determining the proportion of cells which will undergo the lysogenic cycle (Bertani, 1957).

Attempts to synchronize cultures of chick embryo cells with prolonged starvation has given results indicative of cyclical changes in competence (Temin, unpublished). At the time of maximum competence, however, 90 % of the cells still are not infected.

It is possible that methods of synchronization which require less drastic treatment of the cells will enable all cells to be infected with RSV simultaneously.

REFERENCES

BERTANI, L. E. (1957). The effect of the inhibition of protein synthesis on the establishment of lysogeny. *Virology* **4,** 53–71.

BRYAN, W. R. (1946). Quantitative studies on the latent period of tumors induced with subcutaneous injections of the agent of chicken tumor. I. 1. Curve relating dosage of agent and chicken response. *J. Natl. Cancer Inst.* **6,** 225–237.

CARREL, A. (1924). Action de l'extrait filtre du sarcome de Rous sur les macrophages du sang. *Compt. rend. soc. biol.* **91,** 1069–1071.

CARREL, A. (1925). Effets de l'extrait de sarcomes fusocellulaires sur des cultures pures de fibroblastes. *Compt. rend. soc. biol.* **92,** 477–479.

CHÈVREMONT, S., FIRKET, H., CHÈVREMONT, M., and FREDERIC, J. (1957). Contribution à l'étude de la préparation à la mitose. *Acta Anat.* **30,** 175–193.

DOLJANSKI, L., and TENENBAUM, E. (1943). Studies on Rous sarcoma cells cultivated *in vitro.* 2. Morphologic properties of Rous cells. *Cancer Research* **3,** 585–603.

EAGLE, H. (1955). The specific amino acid requirements of a human carcinoma (cell strain HeLa) in tissue culture. *J. Exptl. Med.* **102,** 37–48.

HALBERSTAEDTER, L., DOLJANSKI, L., and TENENBAUM, E. (1941). Experiments on the cancerization of cells *in vitro* by means of Rous sarcoma agent. *Brit. J. Exptl. Pathol.* **22,** 179–187.

HOTCHKISS, R. D. (1954). Cyclical behaviour in pneumococcal growth and transformability occasioned by environmental changes. *Proc. Natl. Acad. Sci. U. S.* **40,** 49–54.

HSU, T. C., POMERAT, C. M., and MOORHEAD, P. S. (1957). Mammalian chromosomes *in vitro.* VIII. Heteroploid transformation in the human cell strain Majes. *J. Natl. Cancer Inst.* **19,** 867–872.

KEOGH, E. V. (1938). Ectodermal lesions produced by the virus of Rous sarcoma. *Brit. J. Exptl. Pathol.* **19,** 1–8.

LO, W. H. Y., GEY, G. O., and SHAPRAS, P. (1955). The cytopathogenic effect of the Rous sarcoma virus on chicken fibroblasts in tissue cultures. *Bull. Johns Hopkins Hosp.* **97,** 248–256.

MANAKER, R. A., and GROUPÉ, V. (1956). Discrete foci of altered chicken embryo cells associated with Rous sarcoma virus in tissue culture. *Virology* **2,** 838–840.

PRINCE, A. M. (1957). Quantitative studies in Rous sarcoma virus. I. The titration of Rous sarcoma virus on the chorioallantoic membrane of the chick embryo. *J. Natl. Cancer Inst.* **20,** 147–158.

Puck, T. T., Cieciura, S. J., and Fisher, H. W. (1957). Clonal growth *in vitro* of human cells with fibroblastic morphology. *J. Exptl. Med.* **106,** 145–158.

Rous, P. (1911). A sarcoma of the fowl transmissible by an agent separable from the tumor cells. *J. Exptl. Med.* **13,** 397–411.

Rubin, H. (1955). Quantitative relations between causative virus and cell in the Rous No. 1 chicken sarcoma. *Virology* **1,** 445–473.

Rubin, H. (1957). Interactions between Newcastle disease virus (NDV), antibody and cell. *Virology* **4,** 533–562.

Reprinted from Virology, Vol. 23, pp. 117–119. 1964

An Assay for Cellular Transformation by SV40[1]

SV40 has been demonstrated to transform *in vitro* the properties of human and hamster fibroblasts (*1–6*) and, more recently, cells from a variety of other species, including the mouse (*7*). In most cases the transformed cells become evident only some weeks after infection although under certain conditions they can be detected quite early (*8*). The cytopathic effect of the virus in human cell cultures, and the low cloning efficiency of fibroblast strains in general, make quantitative study of the transformation difficult.

An established mouse cell line, 3T3, whose evolution has already been described (*9*), possesses properties making it particularly well suited to analysis of *in vitro* transformation by SV40. The cells have a cloning efficiency of 30–50 % in ordinary medium and grow with an 18-hour doubling time. However, when the cells reach confluence, growth is sharply arrested and no growth occurs out of the plane of the monolayer, the cells appearing, in spite of their aneuploid karyotype, to possess an unusually high degree of contact inhibition of cell division. Infection with SV40 leads to the production of cells able to grow out of the plane of the monolayer as discrete colonies against a background of the monolayer. This communication describes an assay for the transforming effect of SV40 using this cell line. Other changes in cellular properties induced by the virus are described elsewhere (*10*).

Cultures were grown in 50-mm plastic petri dishes in Dulbecco and Vogt's modi-

[1] Aided by grants from the United States Public Health Service and an Institutional Grant from the American Cancer Society.

fication of Eagle's medium, supplemented with 10% calf serum. The cell line was maintained by 1:1000 dilution of trypsinized cultures. Subconfluent monolayers containing 2×10^5 to 5×10^5 cells per plate were exposed for 3 hours to 0.5 ml of a stock of SV40 strain 776 (*11*) containing $10^{5.6}$ to $10^{8.4}$ $TCID_{50}$ per milliliter, kindly provided by Dr. John Easton. The plates were then washed three times with phosphate-buffered saline (PBS), pH 7.2, and fresh medium was added. The following day the monolayers were trypsinized and plates were inoculated with 100 and 500 cells. Ten to fourteen days later, the cells were fixed with 10% formalin in PBS and stained for 10 minutes with 1% hematoxylin.

The cloning efficiency of line 3T3 varied from 30 to 50%, and there was no decrease when the cells were infected with even the highest viral concentrations. No detectable cytocidal effects of SV40 were observed microscopically.

Two weeks after plating, the untransformed cells formed a confluent monolayer with no nuclear overlapping while transformed colonies were multilayered and readily detected with the unaided eye or under low power magnification by their deep hematoxylin staining. Figure 1 shows a large transformed colony against a background of untransformed cells; the deeper staining and extensive multilayering are evident. Vertical sections through such an area show nuclear overlapping from as many as 5–6 cells, while in areas of untransformed cells, nuclei do not overlap (*10*).

Table 1 shows frequency of transformation of 3T3 cells as a function of viral titer. Within the limits tested it is seen that the transformation rate is nearly proportional to the input of virus. Infection with the most potent virus available ($10^{8.4}$ $TCID_{50}$ per milliliter) resulted in the production of transformed colonies from 0.75% of the cells plated. Since the transformed cells, like 3T3, have a plating efficiency of 40%, it is assumed that somewhat less than 2% of the cells infected were transformed. The overall efficiency of transformation, i.e., the ratio of transformed cells to infectious

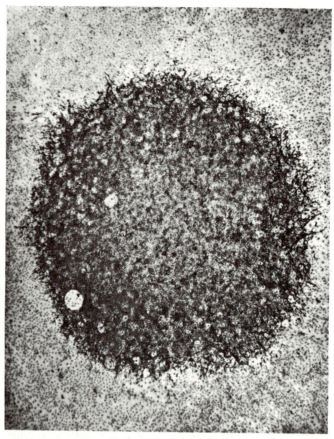

FIG. 1. A colony of SV40-transformed 3T3 cells against a background of untransformed cells 14 days after plating 500 virus-infected cells. Hematoxylin stain. Magnification: × 10.

TABLE 1

FREQUENCY OF TRANSFORMATION OF CELL LINE 3T3 BY SV40[a]

Viral titer (log $TCID_{50}$ per ml)	Total cells plated	Number of transformed colonies	Frequency of transformed colonies (% of total cells plated)
5.6	14,500	0	0.00
6.6	10,500	2	0.02
7.4	24,000	20	0.08
8.4	28,000	209	0.75

[a] Subconfluent monolayers were infected with 0.5 ml of virus and transferred the following day at an inoculation density of 100 and 500 cells per plate. Transformed colonies were scored 14 days later.

particles, in this system is of the order of 10^{-4} to 10^{-5}, a range similar to that reported for DNA containing tumor viruses in other systems (12, 13).

Smaller groups of transformed cells were sometimes seen in the vicinity of a large colony; these were presumed to arise by dissemination of transformed cells from an already developed colony during medium changes, and were not counted. However, in a few cases, well isolated colonies were seen, consisting primarily of unaltered 3T3 cells but containing what appeared to be a sector of typical transformed cells. These rare cases may be examples of delayed transformation described by Stoker (14) for polyoma virus transformation. Transformed 3T3 cells, once cloned, uniformly gave rise to transformed colonies.

This quantitative system is similar to

that described by MacPherson and Stoker (*15*) for polyoma virus on the BHK21 hamster kidney cell line in that the cytocidal actions of the viruses are minimal in both cases. Polyoma virus also transforms 3T3 cells (*10*), but the cytocidal effect produced by that virus makes quantitative studies less satisfactory.

Strains of polyoma virus have been shown by Medina and Sachs (*16*) to differ widely in their ability to transform mouse embryo cultures, and Black and Rowe (*17*) have presented evidence that a similar situation exists with SV40 strains. The system reported here using cell line 3T3 should allow more precise analysis of these differences and of other factors involved in the transformation by SV40.

REFERENCES

1. SHEIN, H. M., and ENDERS, J. F., *Proc. Natl. Acad. Sci. U. S.* **48**, 1164–1172 (1962).
2. KOPROWSKI, H., PONTÉN, J. A., JENSEN, F., RAVDIN, R. G., MOORHEAD, P., and SAKSELA, E., *J. Cellular Comp. Physiol.* **59**, 281–292 (1962).
3. PONTÉN, J. A., JENSEN, F., and KOPROWSKI, H., *J. Cellular Comp. Physiol.* **61**, 145–154 (1963).
4. RABSON, A. S., and KIRSCHSTEIN, R. L., *Proc. Soc. Exptl. Biol. Med.* **111**, 323–328 (1962).
5. BLACK, P. H., and ROWE, W. P., *Virology* **19**, 107–109 (1963).
6. ASHKENAZI, A., and MELNICK, J. L., *J. Natl. Cancer Inst.*, **30**, 1227–1269 (1963).
7. BLACK, P. H., and ROWE, W. P., *Proc. Soc. Exptl. Biol. Med.* **114**, 721–727 (1963).
8. TODARO, G. J., WOLMAN, S., and GREEN, H., *J. Cellular Comp. Physiol.* **62**, 257–266 (1963).
9. TODARO, G. J., and GREEN, H., *J. Cell Biol.* **17**, 299–313 (1963).
10. TODARO, G. J., GREEN, H., and GOLDBERG, B. D., *Proc. Natl. Acad. Sci. U. S.* **51**, 66–73 (1964).
11. SWEET, B. H., and HILLEMAN, M. R., *Proc. Soc. Exptl. Biol. Med.* **105**, 420–427 (1960).
12. STOKER, M., and ABEL, P., *Cold Spring Harbor Symp. Quant. Biol.* **27**, 375–385 (1962).
13. DULBECCO, R., *Science* **142**, 932–936 (1963).
14. STOKER, M., *Virology* **20**, 366–371 (1963).
15. MACPHERSON, I., and STOKER, M., *Virology* **16**, 147–151 (1962).
16. MEDINA, D., and SACHS, L., *Virology* **19**, 127–139 (1963).
17. BLACK, P., and ROWE, W. H., *Proc. Natl. Acad. Sci. U. S.* **50**, 606–613 (1963).

GEORGE J. TODARO
HOWARD GREEN

Department of Pathology
New York University School of Medicine
New York, New York
Accepted March 2, 1964

Reprinted from Virology, Volume 23, No. 2, June 1964
Copyright © 1964 by Academic Press Inc. *Printed in U.S.A.*

Virology **23,** ·291–294 (1964)

Agar Suspension Culture for the Selective Assay of Cells Transformed by Polyoma Virus

When cell suspensions of the cloned hamster line BHK21/13 are infected with polyoma virus and cultured at low cell density on glass, colonies of two distinct morphologic types develop (*1*). Most of the colonies are small monolayers in which the cells are orientated parallel to each other. However, after infection of the cells with a high multiplicity of virus, 1–5% of the colonies are "transformed" and consist of cells piled in disarray. Not more than 100–150 separate colonies can be accommodated in a 60-mm petri dish culture, and since the proportion of transformed colonies is low, many replicate cultures are necessary to obtain sufficient numbers for accurate assays. Clearly a technique that suppresses the growth of normal cells without impairing colony formation by transformed cells would be very advantageous, particularly when small differences in the transformation rate are expected or when the rate might be very low, for example, following infection with polyoma virus DNA.

Methods of enrichment for transformed cells in cultures of hamster embryo cells have been reported by Stanners *et al.* (*2*), but no completely selective assay has been described despite the known differences between transformed and untransformed cells (*3*).

Sanders and Burford (*4*) found that after a polyoma-transformed clone derived from BHK21 cells had been adapted to grow as an ascites tumor the cells were capable of forming colonies in semisolid agar medium. Wildy (*5*) has found that some established lines of transformed BHK21 cells are also capable of growing in agar medium without adaptation as ascites cells. These observations prompted us to investigate the possibility that freshly transformed cells acquire the ability to form colonies in agar medium and that this ability may be used as the basis of a selective assay. For this to be successful it would be necessary to adjust the conditions so that the plating of large numbers of infected cells resulted in the formation of colonies only by transformed cells. These conditions have been realized by plating cells in soft agar medium (*6*). The applications of this technique are the subject of this report.

Cultures were prepared as follows. Base layers of 7 ml of medium (*1*) containing 0.5% Difco Bacto agar were set in 60-mm petri dishes before the addition of a second layer with 10^3 to 5×10^5 cells in 1.5 ml of medium containing 0.3% agar. The plates were incubated for 7–10 days at 37° in a humidified atmosphere of 5% CO_2 in air. A single stock of small-plaque polyoma virus was used for all the experiments. Cells were infected in suspension for 1 hour at 37° before they were incorporated in the top layer. Colony counts were made on unstained cultures with the aid of a low power microscope. In cultures containing many colonies an estimate of the total was obtained by counting those in a known fraction of the dish.

Uninfected BHK21/13 cells and hamster embryo cells derived from primary to fifth passage cultures did not form colonies when 10^3 to 5×10^5 cells were plated in agar medium. The ability of several established lines of transformed cells to form colonies under these conditions is shown in Table 1. Reconstruction experiments in which mix-

TABLE 1

THE GROWTH OF CELLS IN AGAR MEDIUM

Cells	Multiplicity of infection (PFU/cell)	Plated on glass (G) or in agar (A)	Number of cells plated (*a*)	No. of Transformed Colonies (*b*)	$\frac{b}{a}$ %
BHK21/13/PyX[a]	None	A	5×10^3	227	4.5
BHK21/13/PyY[a]	None	A	5×10^3	275	5.5
BHK21/13/Py6[a]	None	A	5×10^3	893	17.8
BHK21/13/PyC[a]	None	A	5×10^3	2027	40.5
BHK21/13	None	A	5×10^5	No colonies	
BHK21/13	10^3	G	1.8×10^3	14	0.8
BHK21/13	10^3	A	10^5	4076	4.0
Hamster embryo (primary)	None	A	5×10^5	No colonies	
	10^3	G	2×10^5	12	0.006
	10^3	A	10^6	66	0.007
Hamster embryo (fifth passage)	10^3	G	4×10^5	14	0.004
	10^3	A	10^6	30	0.003

[a] Polyoma transformed lines.

tures of BHK21/13 cells and transformed cells were cultured in agar showed that transformed cells in the presence of a great excess of BHK21/13 cells could be recovered with an undiminished plating efficiency.

When freshly infected BH21/13 cells and hamster embryo cells were plated in agar medium, colonies developed in both cases. The results of these experiments are given in Table 1 together with parallel platings of the infected cell suspension on glass. The transformation rate obtained with BHK21/13 cells by the agar assay (AA) is 3–5 times higher than that obtained on glass. Compared with similarly infected hamster embryo cells, the transformation rate in BHK21/13 cells by AA is 700 times higher. These results show that AA is selective, sensitive, and economical. One AA plate seeded with 10^5 infected cells is equivalent to 1000 petri dishes using the previous method. A typical colony that developed in a culture of infected BHK21/13 cells is shown in Fig. 1. The colonies are smaller and denser than those obtained by Sanders and Burford (4) with the ascitic variant. Cells that do not form colonies remain as single cells or, in plates seeded with large numbers of cells, may undergo 1–3 divisions and form minute colonies. As

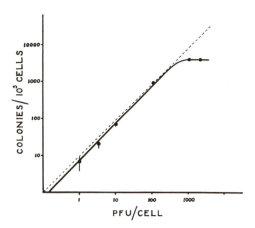

FIG. 2. Colonies developing in agar medium 10 days after seeding cultures with BHK21/13 cells that had been infected with different amounts of polyoma virus. The vertical line through each point represents the standard error of the estimate. The broken line is a slope of 1.

shown by their ability to absorb neutral red, these cells remain viable for more than 10 days. Experiments in which infected cells were mixed with uninfected cells and then plated in the presence of antiserum showed that the nondividing cells acted as feeders for the developing transformed colonies. Other experiments in which mixtures of BHK21/13 cells and transformed cells were plated on glass and then overlaid with agar medium indicated that this medium was not selectively toxic for untransformed cells.

Transformed colonies were easily removed from the agar with finely drawn pipettes. On transfer to fluid medium they rapidly grew out on the glass. When the cells from 3 transformed colonies in agar were grown on glass and replated in agar they had plating efficiencies of 40, 70, and 70%. These are higher than the plating efficiencies of transformed lines that have been repeatedly passaged on glass (Table 1). Of 125 colonies selected at random from different plates and cultured on glass 121 grew with the characteristic appearance of transformed cells. The other 4 cultures gave rise to "normal" monolayers. These cultures are being studied further, and investigations are also in progress to determine whether the transformed colonies obtained, respectively, on glass and in agar represent

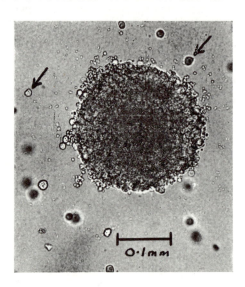

FIG. 1. A colony of transformed cells in agar medium 7 days after seeding the culture with BHK21/13 cells infected with polyoma virus. Single nondividing cells are arrowed.

populations of transformed cells with different properties or whether they are cells with the same original characteristics that have diverged owing to differences in the way they were isolated and subcultured. Whatever the outcome of these particular studies, it is clear that a highly significant correlation exists between transformation of cells by polyoma virus and their ability to form colonies in agar.

A study of the transformation dose response curve using the agar assay showed that, over a range of multiplicities from less than 1 up to 500 plaque-forming units (PFU) per cell, the curve is linear and has a slope of 1 on a log log plot (Fig. 2). These are the characteristics of a "one hit" response to the transforming agent. At multiplicities of 500 PFU/cell and higher no further increase in the transformation rate was obtained. This "plateau effect" has been demonstrated before, utilizing the colony forming assay on glass (7). No satisfactory explanation of this effect has been obtained. The transformation dose response curve obtained by AA can be measured with a much higher degree of precision than was previously possible by colony formation assays on glass. For the stock virus used, one cell-transforming "dose" is equivalent to 10^4 PFU.

It should be noted that growth in agar medium is not peculiar to polyoma-transformed hamster cells. A variant of BHK21/13 cells that is capable of growing in suspension (8) and has a high transplantability comparable with that of polyoma-transformed lines also forms colonies in agar medium with a plating efficiency of 13.8%. Another variant of BHK21/13 cells that

forms colonies in agar (9) is also highly transplantable, a property that is enhanced when the cells are transformed by polyoma virus. HeLa, Hep 2, and L strain of mouse fibroblasts also form colonies in agar.

Further studies are in progress to elucidate the mechanism of the selective action of agar medium and to determine whether it has a more general application for the study of cells transformed by other oncogenic viruses and for neoplastic cells from other sources.

REFERENCES

1. Macpherson, I., and Stoker, M., *Virology* **16**, 147–151 (1962).
2. Stanners, C. P., Till, J. E., Siminovitch, L., *Virology* **21**, 448–463 (1963).
3. Macpherson, I., *J. Natl. Cancer Inst.* **30**, 795–815 (1963).
4. Sanders, F. K., and Burford, B. O., *Nature* **201**, 786–789 (1964).
5. Wildy, P., personal communication.
6. Montagnier, L., and Macpherson, I., *Compt. Rend. Acad. Sci.* in press.
7. Stoker, M., and Abel, P., *Cold Spring Harbor Symp. Quant. Biol.* **27**, 375–385 (1962).
8. Capstick, P. B., Telling, R. C., Chapman, W. G., and Stewart, D. L., *Nature* **195**, 1163–1164 (1962).
9. Montagnier, L., Macpherson, I., and Jarrett, O., to be published.

Ian Macpherson
Luc Montagnier[1]
M. R. C. Unit for Experimental Virus Research
Institute of Virology
University of Glasgow
Scotland

Accepted April 8, 1964

[1] Present address: Institut du Radium, 26 rue d'Ulm, Paris.

JOURNAL OF VIROLOGY, Apr. 1967, p. 362–367
Copyright © 1967 American Society for Microbiology

Vol. 1, No. 2
Printed in U.S.A.

Adenovirus Type 12-Rat Embryo Transformation System

AARON E. FREEMAN, PAUL H. BLACK, RONALD WOLFORD, AND ROBERT J. HUEBNER

Department of Virus Research, Microbiological Associates, Inc., and National Institute of Allergy and Infectious Diseases, National Institutes of Health, Bethesda, Maryland 20014

Received for publication 23 December 1966

Adenovirus type 12 (Huie) inoculated into cultures of primary whole rat embryo produced foci of morphologically altered cells. The number and identification of these transformed areas was dependent upon the calcium concentration of the medium; more foci appeared in 0.1 mM than in 1.8 mM calcium. Cell lines derived from these inoculated cultures did not yield infectious virus, and also were similar to cell lines derived from adenovirus type 12-induced tumors with respect to morphology, presence of virus-specific tumor antigen, and oncogenicity. Dose-response curves revealed that transformation of rat embryo cells by adenovirus type 12 followed one-hit kinetics, and that approximately 7×10^5 infectious virus particles were required for one transformation event. Our results indicate that the transformation system described for adenovirus type 12 is reproducible, and that previous difficulties experienced in developing such a system may well be explained by the higher calcium concentration of the tissue culture media used.

There have been several reports of hamster cells being transformed in vitro from normal to malignant states by adenovirus type 12 (10, 13). Unlike the polyoma (11, 14) or simian virus 40 (1, 15) transformation systems, however, there has been little quantitation of adenovirus transformation. Not only has transformation been a relatively rare event, but the derivation of cell lines has been accomplished with difficulty. It has been suggested that, because of more consistent transformation and cell line derivation, rat cells may be superior to hamster or rabbit cells for adenovirus type 12 transformation studies (J. D. Levinthal and W. Peterson, Federation Proc. **24:**174, 1965).

It has been found that cell lines derived from in vivo adenovirus-induced tumors have a characteristic sensitivity to calcium at the 1.8 mM concentration found in tissue culture media based on Earle's balanced salt solution (4). At this calcium concentration, the tumor cells formed aggregates and came off the glass; initiation and propagation of adenovirus tumor cell lines could be more easily achieved in medium containing low concentrations of calcium (5, 6). The present study was undertaken to investigate whether a reliable adenovirus type 12 transformation system could be developed by use of rat cells grown in a medium containing an optimal calcium concentration for cells derived from adenovirus-induced tumors.

MATERIALS AND METHODS

Virus. Adenovirus type 12, strain Huie, was obtained from the American Type Culture Collection and was passed three times in KB cell cultures and once in human embryonic kidney (HEK) cultures. The HEK-grown pool was subdivided into ampoules and stored in the vapor phase above a liquid nitrogen reservoir. This pool, which was used for all quantitative experiments, was titered in HEK before and after the course of these experiments. The mean titer of $10^{8.2}$ TCID$_{50}$ per ml did not decrease during the storage period. A sample of this pool was found to be free from adenovirus-associated viruses types I, II, and III.

Cell cultures. Embryos, delivered by caesarian section from near-term inbred Fisher rats, were minced, washed, trypsinized, and planted at 2×10^5 cells per milliliter in Eagle's basal medium (2) with 10% fetal bovine serum, 2 mM glutamine, and penicillin and streptomycin in concentrations of 100 units and 100 µg/ml, respectively. These cultures, which were incubated at 37 C under 5% CO_2 and 95% air, were fed on the 3rd day and used when confluent, usually 5 days after seeding.

Media. Transformation studies were carried out in Eagle's minimal essential medium (3) formulated without calcium, and supplemented with 5% dialyzed calf serum, 2% fetal bovine serum, 2 mM glutamine, 0.1 mM nonessential amino acids (3), and antibiotics. From a 0.5 M stock of calcium chloride, calcium was added to the media at a final concentration of 0.1 or 1.8 mM.

Quantitation studies. Dilutions of the virus were made in the appropriate growth medium, and, when

362

confluent, cultures were drained and inoculated with 0.1 ml per tube or 0.4 ml per 4-oz flask. The virus was adsorbed at 37 C for 2 to 4 hr, with manipulation of the culture every 15 min to assure optimal virus contact with the cells. After the adsorption period, the cultures were fed by adding 1 ml per tube or 10 ml per flask of the appropriate growth medium. The cultures were maintained for periods up to 60 days by feeding the cells every other day with the growth medium. At 3- to 4-day intervals, the tubes were carefully scanned, and the number of foci was recorded. The count used for quantitation was taken approximately 7 weeks after infection.

Testing for virion. Transformed cell lines were tested for infectious virus by plating 2×10^5 viable cells on confluent tube cultures of HEK. In addition, a 20% extract of thrice frozen and thawed cells was prepared from each cell line. An equivalent of 2×10^6 cells was added in extract form to each of nine tube cultures of HEK. All tubes were examined for 21 days, at which time supernatant fluids were passed into new HEK cultures. These latter cultures were also observed for 21 days.

Oncogenicity. To determine the tumor-producing potential of the transformed cells, flask cultures were trypsinized and the viable cell count was determined with trypan blue as a vital stain. Weanling Fisher rats and weanling Syrian hamsters were inoculated subcutaneously with 10^5 to 10^8 cells per 0.5 ml.

Calcium sensitivity tests. Prescription bottles (4-oz) were seeded with 2×10^5 cells per milliliter in 10 ml of Eagle's minimal essential medium with 0.1 mM calcium, 5% dialyzed calf serum, 2% fetal bovine serum, 0.2 mM nonessential amino acids, and antibiotics. When confluent, replicate cultures were fed with media containing 0.1, 1.8, 5, or 7.5 mM calcium. The cultures were observed daily for 6 days, with a feeding on the 3rd day. Cultures were considered calcium-sensitive, or "positive," if retracting or clumping, or both, occurred at any calcium concentration as compared with the 0.1 mM control.

RESULTS

Response of rat cells to adenovirus type 12. Inoculation of rat embryo cells with as much as $10^{8.2}$ TCID$_{50}$ of adenovirus type 12 produced no cytopathogenic effect. From 18 to 36 days after the time of inoculation, there appeared morphologically altered, apparently transformed foci, composed of small, tightly packed epithelioid cells which were several layers thick.

Effect of calcium on morphological transformation and establishment of cell lines. The appearance of clearly identifiable morphologically transformed foci was greatly enhanced by using a medium with a low calcium concentration. As seen in Fig. 1A, control cells grown in Eagle's medium (1.8 mM calcium) were fully confluent after 48 days. If there were transformed foci in the inoculated cultures, they were difficult to identify (Fig. 1B). In a medium containing 0.1

mM calcium (Fig. 1C), however, the control cultures were not confluent, and in the inoculated cultures the transformed foci were readily identifiable (Fig. 1D). As described by McBride, the transformed foci looked very much like "sombreros" viewed from above (10). In the six cultures grown in medium containing 0.1 mM calcium, there were 56 foci, as compared with no foci in those cultures grown in the 1.8 mM calcium medium (Table 1). After 48 days, the calcium concentration in three flasks was increased from 0.1 to 1.8 mM. In those flasks kept in 0.1 mM calcium, the number of transformed foci continued to increase, whereas in those flasks changed to 1.8 mM calcium the number of foci decreased. After an additional 25 days, there were approximately 10-fold more foci in the 0.1 mM calcium cultures as compared with those with the increased calcium concentration. Similarly, three flasks from the 1.8 mM calcium group were changed to 0.1 mM calcium, 48 days after inoculation. Those cultures kept in the higher calcium concentration continued to contain no identifiable foci; those changed to 0.1 mM calcium developed six foci after an additional 25 days, which indicated that transformation had occurred but could not be expressed in a medium containing 1.8 mM calcium. In a second experiment, confluent cultures were exposed to 0.1 or 1.8 mM calcium during the 4-hr adsorption period (Table 1, B). Immediately thereafter, half of the tubes from each group were changed to the contrasting calcium concentration. Regardless of the calcium concentration during adsorption, clearly identifiable transformed foci appeared only in those cells maintained in 0.1 mM calcium. The results of this experiment indicate that adsorption occurs equally well in either calcium concentration.

Cultures containing transformed foci were subdivided in a 0.1 mM calcium medium. The daughter culture bottles appeared to be pocked with colonies of transformed cells (Fig. 2A). If kept in 0.1 mM calcium, the colonies grew (Fig. 2B), but, if changed to 1.8 mM calcium, the colonies appeared to become walled-in and no obvious growth occurred for weeks (Fig. 2C, D). Obviously, cell lines were derived much more readily from the 0.1 mM calcium cultures than the 1.8 mM calcium cultures. In each of 18 attempts, a cell line was derived by use of Eagle's minimal essential medium containing 0.1 mM calcium instead of the usual 1.8 mM calcium concentration.

Characteristics of the transformed cell lines. Each of the 18 transformed lines was made up of

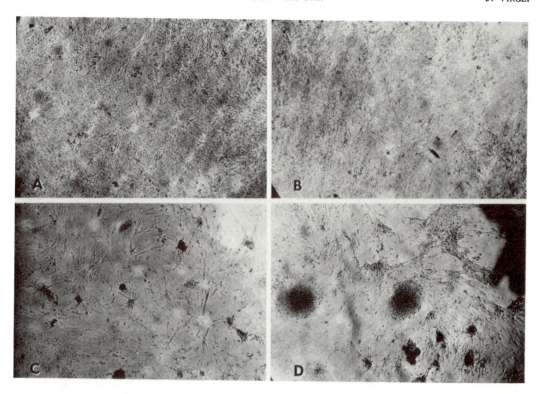

FIG. 1. *Effect of calcium concentration on the appearance of adenovirus type 12 transformed foci. (A) Uninoculated control, rat embryo cells after 48 days, 1.8 mM calcium. (B) Inoculated culture, rat embryo cells after 48 days, 1.8 mM calcium. (C) Uninoculated control, rat embryo cells after 48 days, 0.1 mM calcium. (D) Inoculated culture, rat embryo cells after 48 days, 0.1 mM calcium.* × 41.

cells with an epithelioid morphology. There was loss of contact inhibition, in that the cells grew in multilayered foci. Of four lines tested for calcium sensitivity, each demonstrated the calcium effect at concentrations of 1.8 or 5 mM calcium, or at both concentrations. Of six cell lines tested for infectious adenovirus 12, all were negative. However, in these cell lines, approximately 95% of the cells were positive for adenovirus type 12 tumor antigen by the indirect fluorescent-antibody (FA) technique (13) when tested with a serum pool from hamsters bearing adenovirus 12-induced tumors. Each cell line was also positive for adenovirus type 12 tumor antigen by the complement-fixation (CF) test (7) when tested with 4 to 8 units of antibody from another serum pool similar to the one used for FA tests; the average CF titer at passage 20 to 24 was greater than 1:16. None of the lines tested had demonstrable simian virus 40 "T" antigen by either FA or CF test. One of the cell lines at passage 19 was compared by chromosome analysis with passage 2 normal rat cells. There appeared to be no obvious difference in the percentage of diploid cells. Of the metaphases from 111 transformed cells, 70% were diploid, as compared with 61% diploid in the 61 normal metaphases counted. Further, there appeared to be no chromosomal markers or aberrations in the transformed cells.

Tumorigenicity of cell lines derived from transformed cultures. Each of three of the cell lines was inoculated subcutaneously into 18 weanling Fisher rats, but only a single rat developed a tumor (10^6 cells, 170 days). This tumor had a characteristic adenovirus histology (12) and contained the adenovirus 12 "T" antigen by CF test. In addition, the serum from this animal had a titer of 1:80 when tested against another adenovirus 12 hamster tumor antigen. When inoculated subcutaneously into hamsters, three of four lines tested produced progressively growing (10^6 cells, 28 days) tumors. These tumors were pathologically and serologically similar to adenovirus-induced tumors, but analysis of the chromosomes indicated that the cells which were growing were rat, rather than hamster, cells.

Dose-response relationship. Attempts were

made to quantitate the transformation of whole rat embryo cultures by adenovirus type 12, by use of a medium containing 0.1 mM calcium. During the course of these experiments, the number of transformed foci continuously increased in cultures inoculated with $10^{6.2}$ or $10^{7.2}$ infectious units of virus. Approximately 7 weeks after the time of infection, however, the number

TABLE 1. *Effect of calcium on the number of identifiable transformed foci in the adenovirus type 12-rat embryo transformation system*

Expt	Virus dose (TCID50)/culture	Multiplicity of infection	Initial Ca++ concn of medium (mM)	Time at which some cultures were changed to a contrasting calcium concn	Distribution of foci at end of time period in initial calcium concn	New Ca++ concn (mM)	No. of foci after additional 25 days
(A)	$10^{6.2}$	0.1	0.1	48 days	31[a]	0.1	120[a]
					25[a]	1.8	15[a]
			1.8	48 days	0[a]	0.1	6[a]
					0[a]	1.8	0[a]
(B)	$10^{7.2}$	40	0.1	4 hr	0[b]	0.1	35[b,c]
					0[b]	1.8	1[b]
			1.8	4 hr	0[b]	0.1	24[b]
					0[b]	1.8	0[b]

[a] Total in three 4-oz prescription bottles.
[b] Total in nine tubes.
[c] Reading at 50 days = 194/8 tubes.

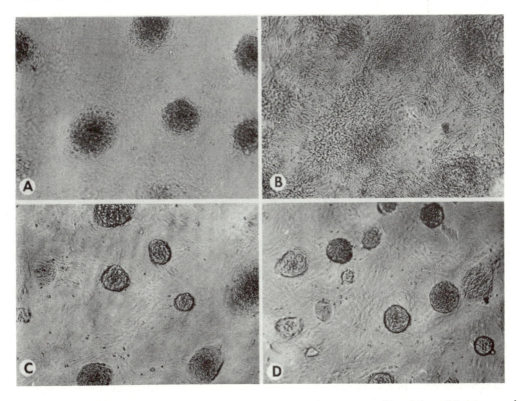

FIG. 2. *Calcium-dependent inhibition of transformed foci. (A) Adenovirus type 12 transformed foci in normal whole rat embryo cells grown in 0.1 mM calcium medium. (B) Replicate culture after 3 weeks in 0.1 mM calcium medium. (C) Replicate culture after 1 week in 1.8 mM calcium medium. (D) Replicate culture after 3 weeks in 1.8 mM calcium medium. × 72.*

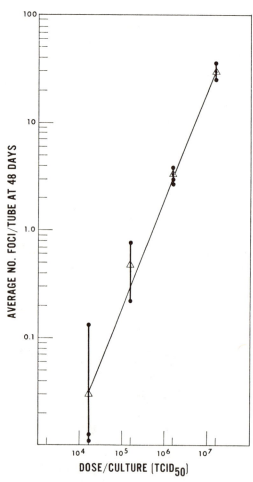

FIG. 3. *Relationship of dose of adenovirus type 12 to the number of transformed foci. Symbols:* △, *average of all experiments;* ●, *average of nine tubes representing one point in each individual dose-response experiment.*

of foci appeared to remain constant or decrease somewhat due to coalescence. Metastatic colony formation did not appear to influence these counts, since cultures inoculated with $10^{5.2}$ infectious units of virus developed single foci which remained localized for at least 3 to 4 weeks.

Results tabulated from four separate dose-response experiments indicated that there was a one-hit relationship between the number of transformed foci and the virus dose, with one transformation occurring for each 7×10^5 infectious units (Fig. 3). In cultures exposed to $10^{7.2}$ infectious units of adenovirus type 12, which was the most concentrated virus used, an average of 35 transformed foci per tube were present when counts were determined 48 days after inoculation.

Since each tube culture contained approximately 4×10^5 cells when confluent, the transformation frequency was approximately 1 per 10^4 cells or 0.01%.

DISCUSSION

A number of cell lines derived from rat embryo cultures infected with adenovirus type 12 exhibited altered morphology, loss of contact inhibition, the presence of adenovirus type 12 specific tumor antigen, and oncogenicity in rats and hamsters. Since these cell lines derived from apparently transformed foci had the general characteristics of adenovirus-induced tumor cells (8), there appears to be justification for considering the morphological alterations as an indication of transformation.

By using the number of morphologically altered areas as an index of transformation, it became evident that a calcium concentration of 1.8 mM not only reduced the number of transformed foci, but also inhibited the growth of these foci once they were detectable. We consider that the effect of calcium on the rate of transformation depends upon the double selective property of the medium. The double selection can be broken down into (i) the selective force for the growth of normal cells and against the growth of adenovirus transformed cells in a medium containing 1.8 mM calcium and (ii) the selective force for the growth of adenovirus transformed cells and against the growth of normal cells in a medium containing 0.1 mM calcium. That transformation occurs, but cannot be expressed, in a medium containing 1.8 mM calcium was evident from the switch experiments (Table 1), and is consistent with previous reports (5, 6) regarding the sensitivity to calcium of cell lines derived from adenovirus-induced tumors. It should be stressed that 1.8 mM calcium is the concentration found in all tissue culture media formulated with Earle's balanced salt solution (4).

When the low calcium medium was used, transformation of primary whole rat embryo cells by adenovirus type 12 became so consistent that an attempt was made to quantitate the transformation event by counting the number of transformed foci. Such focus counting in a liquid medium is subject to error resulting from metastasis from the primary focus. Although it may have occurred, we do not believe metastasis significantly affected the increase in the number of foci which was observed in cultures inoculated with relatively high doses of virus, since dissemination from a single transformed focus was not observed in cultures inoculated with lower concentrations of virus. Further, if the increase in the number of foci was due to metastasis, the ultimate number of foci should have been the same regardless of the virus input, but this was not observed. Rather, there was a linear dose-response relationship as shown in Fig. 3.

Although the dose-response curves were linear at the concentrations of virus used, it may be that, with a more concentrated virus inoculum, there would be a plateau in the transformation rate, similar to that reported for polyoma virus (14) and simian virus 40 (1). On the basis of the data presented here, however, adenovirus type 12 was a much less efficient transforming agent than either of these two viruses, since 7×10^5 infectious units were required for one transformation event, as compared with 10^4 and 10^3 infectious units for polyoma virus and simian virus 40, respectively. The frequency of transformation by adenovirus type 12 was 1 per 10^4 cells or 0.01%, as compared with the 1 per 10^4 to 1 per 10^5 transformation frequency in an adenovirus type 12-hamster embryo system reported by W. A. Strohl, H. C. Rouse, and R. W. Schlesinger (Bacteriol. Proc., p. 136, 1966). In the hamster system, however, the number of foci was determined 3 weeks after infection, and the dose-response curves were not linear. Furthermore, in the transformation system reported here, no end point was reached with the virus concentrations used and an increase in virus titer may well have increased the frequency of transformation.

A more recently derived adenovirus type 12 transformed rat embryo line caused tumors in 9 of 11 newborn Fisher rats in 30 to 33 days when 10^6 cells were inoculated subcutaneously.

ACKNOWLEDGMENTS

We gratefully acknowledge the contributions of the following: John Lehman of the Wistar Institute for the chromosome analysis of the adenovirus type 12 transformed rat cells growing as a subcutaneous tumor in hamsters, Horace C. Turner of the National Institute of Allergy and Infectious Diseases for supervising the CF tests, Eustace Vanderpool and Inez Archie for assisting with the transformation studies, Howard Igel for pathology evaluations.

This investigation was supported by Public Health Service contract PH-43-63-81 to Microbiological Associates, Inc.

LITERATURE CITED

1. BLACK, P. H. 1966. Transformation of mouse cell line 3T3 by SV_{40}: dose response relationship and correlation with SV_{40} tumor antigen production. Virology 28:760–763.
2. EAGLE, H. 1955. Nutrition needs of mammalian cells in tissue culture. Science 122:501–504.
3. EAGLE, H., V. I. OYAMA, M. LEVY, AND A. E. FREEMAN. 1957. myo-Inositol as an essential growth factor for normal and malignant human cells in tissue culture. J. Biol. Chem. 226:191–207.
4. EARLE, W. R. 1943. Production of malignancy "in vitro." IV. The mouse fibroblast cultures and changes seen in the living cells. J. Natl. Cancer Inst. 4:165–212.
5. FREEMAN, A. E., C. H. CALISHER, P. J. PRICE, H. C. TURNER, AND R. J. HUEBNER. 1966. Calcium sensitivity of cell cultures derived from adenovirus-induced tumors. Proc. Soc. Exptl. Biol. Med. 122:835–840.
6. FREEMAN, A. E., S. HOLLINGER, P. J. PRICE, AND C. H. CALISHER. 1965. The effect of calcium on cell lines derived from adenovirus type 12-induced hamster tumors. Exptl. Cell Res. 39:259–264.
7. HUEBNER, R. J., W. P. ROWE, H. C. TURNER, AND W. T. LANE. 1963. Specific adenovirus complement-fixing antigens in virus-free hamster and rat tumors. Proc. Natl. Acad. Sci. U.S. 50:379.
8. KITAMURA, I., G. VAN HOOSIER, JR., L. SAMPER, G. TAYLOR, AND J. J. TRENTIN. 1964. Characteristics of human adenovirus type 12 induced hamster tumor cells in tissue culture. Proc. Soc. Exptl. Biol. Med. 116:563–568.
9. LOCKART, R. Z., AND H. EAGLE. 1959. Requirements for growth of single human cells. Science 129:252–254.
10. McBRIDE, W. D., AND A. WIENER. 1964. In vitro transformation of hamster kidney cells by human adenovirus type 12. Proc. Soc Exptl. Biol. Med. 115:870–874.
11. MACPHERSON, I., AND L. MONTAGNIER. 1964. Agar suspension culture for the selective assay of cells transformed by polyoma virus. Virology 23:291–294.
12. OGAWA, K., A. TSUTSUMI, K. IWATA, Y. FUJII, M. OHMORI, K. TAGUCHI, AND Y. YABE. 1966. Histogenesis of malignant neoplasm induced by adenovirus type 12. Gann 57:43.
13. POPE, J. H., AND W. P. ROWE. 1964. Immunofluorescent studies of adenovirus 12 tumors and of cells transformed or infected by adenoviruses. J. Exptl. Med. 120:577–588.
14. STOKER, M., AND P. ABEL. 1962. Conditions affecting transformation by polyoma virus. Cold Spring Harbor Symp. Quant. Biol. 27:375–386.
15. TODARO, G. J., AND H. GREEN. 1966. High frequency of SV_{40} transformation of mouse cell line 3T3. Virology 28:756–759.

Reprinted from VIROLOGY, Volume 44, No. 2, May 1971
Copyright © 1971 by Academic Press, Inc. *Printed in U.S.A.*

VIROLOGY **44**, 359–370 (1971)

Induction of Cell Division in Medium Lacking Serum Growth Factor By SV40

HELENE S. SMITH,[1] CHARLES D. SCHER, AND GEORGE J. TODARO

National Cancer Institute, Bethesda, Maryland, and Meloy Laboratories, Springfield, Virginia

Accepted December 21, 1970

Infection with SV40 induces many cells to synthesize DNA and divide in a medium lacking serum protein growth factor(s) essential for division of uninfected cells. Virus infection can induce the cells to go through several rounds of cell division since colonies containing more than 100 cells are formed. A functioning virus genome is necessary to induce cell division since UV-irradiation of the virus destroys this ca-.pacity. The total number of cell divisions induced is proportional to the input multiplicity. Most of the colonies induced to divide show no evidence of permanent functioning virus information: they no longer have cells with T-antigen, and most do not grow to high saturation density when shifted to complete medium.

The ability of a cell to grow in factor-free medium is not the same property as its loss of contact inhibition; a transformed clone has been selected which still is strongly contact inhibited in spite of producing SV40 T-antigen and having rescuable virus information. This transformant would not have been recognized in the standard transformation assay. It has gained one of the growth properties associated with virus-transformation, the ability to grow in factor-free medium.

INTRODUCTION

Methods used to study SV40 transformation of the established mouse cell lines, 3T3, and BALB/3T3, have been described (Todaro, 1969). The fully transformed cells differ from their normal counterparts in a number of properties (Black, 1968; Dulbecco, 1969; Eckhart, 1969). They are characterized by the ability to form multiple cell layers under conditions where uninfected cells remain confined to a monolayer. Cells permanently transformed by SV40 contain the genetic information of the virus detectable by nucleic acid hybridization (Benjamin, 1966; Sambrook *et al.*, 1968), synthesize virus specific tumor (T) and transplantation antigens (Black *et al.*, 1963; Habel and Eddy, 1963), and release infectious virus upon fusion with permissive lines (Gerber, 1966; Koprowski *et al.*, 1967; Watkins and Dulbecco, 1967).

[1] Present address: Bionetics Research Laboratory, 7300 Pearl Street, Bethesda, Maryland.

SV40-transformed BALB/3T3 cells grow in medium lacking serum growth factor(s) (Jainchill and Todaro, 1970). Growth in this medium appears to be a characteristic of the virus-transformed state, since 3T3 and BALB/3T3 cells transformed by SV40 (Holley and Kiernan, 1968), polyoma virus, murine sarcoma virus (Jainchill and Todaro, 1970), and Rous sarcoma virus (Smith and Scher, 1971) all are able to grow in this selective medium. A similar change in requirement for multiplication stimulating factors in serum has been shown for chick embryo fibroblasts infected with avian sarcoma virus (Temin, 1967), and hamster BHK cells transformed by polyoma virus (Burk, 1966).

The present report describes the use of medium lacking serum growth factor(s) to study the early events in the transformation process. Cells plated sparsely in growth factor-free medium are induced to divide and form colonies after infection with SV40.

The total number of cell divisions induced by SV40 is proportional to the multiplicity used. Many of the infected cells only transiently gain the ability to grow in factor-free medium without becoming permanently transformed by the virus. In addition, growth in factor-free medium allows the recognition of transformants that remain contact inhibited and, therefore, would not be recognized by the standard transformation assay. A preliminary report of this work has been presented (Smith et al., 1970).

MATERIALS AND METHODS

Cell lines and culture. BALB/3T3 clone A31 (Aaronson and Todaro, 1968) cells were grown in Dulbecco's modification of Eagle's medium (DM) supplemented with 10% calf serum (Colorado Serum Co.); this will be referred to as complete medium. Cells were routinely maintained by frequent passage at low cell concentrations (100–1000 cells/plate). Since long-term passage of BALB/3T3 clones results in the emergence of variants with reduced dependence on serum factor(s), the clone used in these experiments was not carried for more than 3 months; after this time stocks frozen at a relatively early passage (125 cell generations in culture) were thawed, grown up, and used.

Virus stocks. The small plaque (SV-S) mutant of SV40 (Takemoto et al., 1966) was prepared in primary African green monkey cells (GMK) growing in DM containing 5% fetal calf serum. The GMK was inoculated with 10^{-3} PFU/cell to limit the number of defective particles (Uchida et al., 1968). Virus was harvested at 10–14 days, 3 days after the last medium change. The cell debris was removed by low speed centrifugation and the stocks were used without further purification. Virus stocks were titered on GMK using the agar-overlay technique (Takemoto et al., 1966).

Ultraviolet irradiation of SV40 was carried out using a 15-W GE germicidal lamp. A virus suspension of 1.0 ml was irradiated for 10 min in a 50-mm diameter petri dish, while being shaken, at an incident dose of 22 ergs/mm/sec. The dose rate was determined with a Blak-Ray ultraviolet intensity meter.

SV40 was recovered from transformed lines by fusion with permissive African green monkey kidney cells (GMK) using UV-inactivated Sendai virus (Takemoto et al., 1968).

Growth factor-free medium. Prepared agamma newborn calf serum, made γ-globulin-free by a modified Cohn alcohol precipitation, was obtained from North American Biological, Inc. (North Miami, Florida). This was added to DM to a final concentration of 10% or 20%. Growth factor-free medium was prepared by depleting medium containing this prepared agamma calf serum on confluent BALB/3T3 (30 ml of medium per 32-ounce bottle) for 3–4 days. Cell debris was removed by low speed centrifugation and the medium stored at $-20°$. The medium containing 20% calf serum was diluted with DM to a final concentration of 10% prior to use.

A similar and more direct way of preparing growth factor-free serum involves heating prepared agamma calf serum. Heating serum to 70° for 30 min destroys its ability to support cell division of sparse cultures of BALB/3T3; heating to 65° partially inactivates the growth-promoting activity. To prevent gel-formation, the 70° heated prepared agamma serum was added to DM before cooling, to a final concentration of 10%. Heating calf serum, that has not been made γ-globulin-free by Cohn fractionation, to 70° results in extensive gel formation. Growth factor-free medium, prepared by either of the two methods described, supports the growth of SV40-transformed cells. BALB/3T3 grows when whole calf serum or the precipitate from the Cohn fractionation is added back to factor-free medium.

SV40 infection of BALB/3T3. To allow the cessation of serum-induced DNA synthesis, confluent cultures of BALB/3T3 that had not had their medium changed for 3 days were used. After removal of the medium, the cells were infected with 0.5 ml of virus suspension for 3 hr. Control plates were incubated with DM with 5% fetal calf serum depleted for 3 days on confluent GMK, or with disrupted preparations of GMK in DM with 5% fetal calf serum. All virus dilutions were made in these media. After infection, the cells were trypsinized, counted in a hemacytometer, and plated into factor-free medium at 5000 and 1500 cells per plate. After 4 days, some of the plates at each cell concentration were shifted to complete medium to determine the efficiency of plating. Parallel plates were left in factor-free medium and scanned with an inverted microscope using a vernier holder at a magnification of 31 ×. The number of cells per colony was counted at various times after infection. The scoring was done with the observers unaware of the previous treatment to the individual plates. For plates in which fewer than 300 cells survived until day 4, two cells were considered to be in a single colony if they were less than 0.8 mm apart. When BALB/3T3 was grown in complete medium, or SV40-transformed BALB/3T3 was grown in either complete or factor-free medium, individual colonies maintained these standards.

In other experiments, BALB/3T3 was plated in factor-free medium; after 3 days the cells were infected as described. Duplicate plates were shifted to complete medium after the adsorption period to determine the number of viable cells. The plates remaining in factor-free medium were scanned at intervals to determine both the number of colonies per plate and the number of cells per colony.

Autoradiography. Cells were inoculated onto coverslips in factor-free medium at 15,000 cells per plate. After 3 days the cells were infected with SV40 and returned to factor-free medium or shifted to complete medium containing ^{3}H-TdR (New England Nuclear Corp., Boston, Massachusetts), at concentrations of 1 μCi/ml. Twenty-four hours later, fresh ^{3}H-TdR was added to the cells (final isotope concentrations; 2 μCi/ml). At various times cover slips were removed, and incubated for 2 hr in isotope-free medium. The cover slips were washed with phosphate buffered saline, fixed in cold acetone and dried. The slides were covered with emulsion (Eastman NTB2), stored in vacuo at 25° for 3 days and developed. The cells were stained through the stripping film with 1% hematoxylin.

Detection of SV40 T-antigen. Cells were inoculated onto cover slips in factor-free medium at various concentrations. After remaining in factor-free medium for 3 days, the cells were infected with SV40 as described, and fresh factor-free medium was added. Forty-eight hours after infection, cover slips from plates containing 1×10^5 cells/plate were fixed and assayed for T-antigen by indirect immunofluorescence (Aaronson and Todaro, 1968). Ten to 12 days after infection, cover slips from sparsely seeded plates containing individual colonies were similarly treated. Antiserum from hamsters bearing virus-free SV40-induced tumors was kindly supplied by Dr. S. Aaronson; goat antihamster fluorescein conjugated globulin was obtained from the Resources and Logistics Segment (NCI). The presence of T-antigen in the clonal lines was determined by immunofluorescence and in certain cases, checked by complement fixation (kindly performed by Dr. Harvey Ozer, NCI).

Relative plating efficiency in growth factor-free medium. Exponentially growing cultures of the cell lines to be tested were trypsinized and plated at various concentrations in complete medium and in factor-free medium. Ten to fifteen days later the cells were fixed with 10% formalin in phosphate-buffered saline, stained with 1% hematoxylin, and counted.

RESULTS

SV40-Induced Cell Division in Growth Factor-Free Medium

Table 1 shows that SV40 can induce several rounds of cell division in factor-free medium. BALB/3T3 that had been plated sparsely and had remained in factor-free medium for 3 days was infected with SV40. Cell division induced by the virus could be seen as early as 3 days after infection. By 5 days, colonies containing 8 or more cells were observed (see Table 1). The size of the colonies in infected cultures increased throughout the observation period. After infection with 5×10^7 PFU/ml, 32–40% of viable cells present at the time of infection formed colonies containing 16 or more cells. When 5×10^5 PFU of SV40 was used, a small but significant effect was still noted in this and in other experiments.

TABLE 1

Induction of Cell Division in Growth Factor-Free Medium by SV40[a]

Sample	Days after infection	Number of colonies[b]							Total colonies
		1–2	3–4	5–7	8–15	16–31	32–63	≧64	
					(cells/colony)				
SV40	0	85	0	0	0	0	0	0	85
5 × 10⁷	5	25	8	3	3	1	0	0	40
PFU	12	6	9	3	6	5	7	2	38
SV40	0	82	0	0	0	0	0	0	82
5 × 10⁵	5	39	0	0	0	0	0	0	39
PFU	12	30	3	2	0	0	0	0	35
Control	0	86	0	0	0	0	0	0	86
	5	35	2	0	0	0	0	0	86
	12	17	0	0	0	0	0	0	17

[a] Sparse cultures in growth factor-free medium for 3 days were infected with SV40.

[b] There were 80–100 viable cells on day 0.

Effect of Virus Multiplicity on the Induction of Cell Division in Growth Factor-Free Medium

In the preceding experiment, sparse cells were induced to undergo cell division after infection with SV40. Although the inoculum of virus was comparable to that used in transformation studies (Aaronson and Todaro, 1968), the ratio of virus to cells (input multiplicity) was considerably larger, being $\geq 10^4$. To study the effect of low multiplicities of SV40 on the induction of cell division in factor-free medium, confluent BALB/3T3 (1×10^6 cells) was infected with 5×10^6 PFU (input multiplicity, 5) and 5×10^7 PFU (input multiplicity, 50) of virus. The cells were trypsinized, and plated in growth factor-free medium. Four days later duplicate plates were shifted to complete medium to determine the number of viable cells. The transformation frequency of the cells shifted to complete medium at this time was proportional to the inoculum size. Eleven days after infection, the plates remaining in factor-free medium were scored for colony formation. Colony formation was observed with as few as 5 PFU/cell (Fig. 1). Nearly the same number of cells was induced to divide at this multiplicity as at the higher multiplicity (50 PFU/cell); in the latter case, however, the average colony size was considerably larger. At both multiplicities colonies were

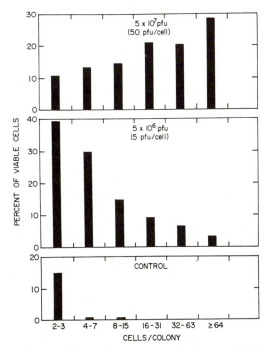

Fig. 1. Effect of the input multiplicity of SV40 on the colony size of BALB/3T3 in growth factor-free medium. Confluent cultures (1×10^6 cells) were infected with 5×10^7 PFU (input multiplicity, 50) and 5×10^6 PFU (input multiplicity, 5) of virus; after the adsorption period, 5000 cells were plated in growth factor-free medium. The number of viable cells was determined by shifting parallel plates to complete medium 4 days after infection. The number of viable cells/plate was: 5×10^7 PFU, 108; 5×10^6 PFU, 177; control, 98.

observed that contained more than 100 cells.

The transformation frequency of BALB/3T3 is proportional to the multiplicity of virus used (Aaronson and Todaro, 1968). To determine whether the total number of cell divisions induced is also proportional to the multiplicity, the number of colonies as well as the number of cells per colony was scored 10–11 days after infection. Since each colony arose from a single cell, the number of cell divisions per colony was taken as one less than the colony size. Thus, a colony which has grown from one to 16 cells (four generations) has undergone fifteen independent cell divisions. This method of expressing the data takes into account both the number of cells induced to divide and the number of cells per colony. Figure 2 shows that the number of divisions induced per viable cell is directly proportional to the virus inoculum. The total divisions induced was greater when cells were infected prior to plating than when infected after plating; however, in each experiment it was proportional to the inoculum size. A significant induction of cell division was observed at an input multiplicity as low as 0.5 PFU/cell.

Effect of UV-irradiation of SV40 on the Induction of Cell Division in Growth Factor-Free Medium

To determine whether a functioning virus genome was necessary to induce cell division in factor-free medium, ultraviolet light was used to inactivate the virus. A dose of ultraviolet light that caused a 200-fold loss of infectivity (PFU) and approximately a 10-fold loss of transforming efficiency decreased the ability of the virus to induce growth in factor-free medium (Table 2). This suggests that the virus genome rather than virus protein (coat or internal) is responsible for the cell division-inducing capacity. Induction of cell division by residual serum proteins in the virus preparation is unlikely, since they too would not be efficiently inactivated by UV-irradiation.

Induction of Cellular DNA Synthesis in Growth Factor-Free Medium by SV40

The preceding experiments showed that SV40 induces BALB/3T3 cells to undergo

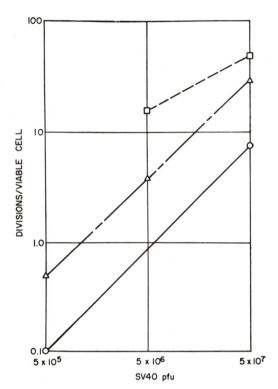

FIG. 2. Effect of the virus inoculum on the number of cell divisions induced in growth factor-free medium. Cultures were infected with 0.5 ml of virus suspension; cell divisions were scored 10–11 days after infection. Three separate experiments are shown. In each, control values from mock infected cultures were subtracted from the total cell divisions. □, confluent cultures (1×10^6 cells) that had not had their medium changed for 3 days were infected with SV40 and plated in factor-free medium; control cultures had 0.3 divisions per viable cell. △, confluent cultures (1×10^6 cells) that had not had their medium changed for 3 days were infected with SV40 and plated in factor-free medium; control cultures had 0.7 divisions per viable cell. ○, sparse cultures (1×10^2 cells) already in factor-free medium were infected with SV40; control cultures had 0.02 divisions per viable cell.

several rounds of cell division in factor-free medium. However, one could not conclude that the initial round of DNA synthesis was induced by SV40. It is possible that colonies arose from cells which were synthesizing DNA at the time of infection. In the control cultures, DNA synthesis by a few cells coupled with the death of others would not have been detected. To rule out

TABLE 2

UV IRRADIATION OF SV40: EFFECT ON THE INDUCTION OF CELL DIVISION IN GROWTH FACTOR-FREE MEDIUM[a]

| Sample[b] | Number of colonies/plate[c] | | | | | | Total colonies |
| | 1–3 | 4–7 | 8–15 | 16–31 | 32–63 | ≥ 64 | |
			(cells/colony)				
SV40	18	9	11	9	6	2	55
UV-treated SV40	57	6	1	1	0	0	65
Control	116	4	1	0	0	0	121

[a] Sparse cultures in growth factor-free medium for 3 days were infected with SV40.

[b] UV-inactivated virus had a titer of 6×10^5 PFU/ml. The unirradiated stock contained 1×10^8 PFU/ml.

[c] Plates were scored for colony formation 10 days after infection.

this possibility, DNA synthesis was measured autoradiographically; the percent of cells which incorporated ^{3}H-TdR into their nuclei was determined at intervals after infection.

Table 3 shows that only 3% of the uninfected cells synthesized DNA within 24 hr, and only 8% within 48 hr. After infection with SV40, 51% of the cells synthesized DNA in the first 48 hr, indicating that the initial round of DNA synthesis was also induced by the virus. Under similar conditions, 88% of the cells of an SV40-transformed cloned line (SV10-A-7) underwent DNA synthesis within 48 hr.

Shifting BALB/3T3 to complete medium results in 40% of the cells synthesizing DNA within the first 24 hr; approximately 90% had labeled nuclei after 48 hr. The fraction of cells synthesizing DNA during this interval was the same in infected and uninfected cultures. Thus, BALB/3T3 that has remained in growth factor-free medium for 72 hr is capable of rapidly resuming DNA synthesis when returned to complete medium.

Synthesis of T-antigen by Cells in Factor Free Medium after Infection by SV40

Cells that are permanently transformed by SV40 synthesize T-antigen. If all the colonies induced to grow in factor-free medium were permanently transformed, one would expect them to also synthesize T-antigen. Before testing whether these colonies were positive for T-antigen, it was necessary to determine whether T-antigen could

TABLE 3

SV40-INDUCED DNA SYNTHESIS IN GROWTH FACTOR-FREE MEDIUM[a]

| Cell line | SV40 (PFU) | Percent of nuclei with grains[b] | | | |
| | | Factor-free medium | | Complete medium | |
		24 hr	48 hr	24 hr	48 hr
BALB/3T3	5×10^6	10	51	38	87
	0	3	8	42	93
SV10 A-7	0	40	88	ND	ND

[a] Sparse cultures in growth factor-free medium for 3 days were infected with SV40. Cells were left in factor-free medium or shifted to complete medium after the adsorption period.

[b] Two hundred cells were examined for each point. ND = not done.

be synthesized by cells maintained in factor-free medium. Fully transformed cells plated in factor-free medium were positive when stained for T-antigen. Furthermore, SV40 induced T-antigen synthesis 2 days after infection when the cells were infected and maintained in factor-free medium (Table 4). Cells infected in factor-free medium behave similarly to those infected in complete medium (Aaronson and Todaro, 1968); with both conditions the fraction of T-antigen positive cells is proportional to viral dose.

To determine whether the colonies induced to grow in factor-free medium were positive for T-antigen, cells were plated relatively sparsely and infected with 5 ×

10^7 PFU of SV40. Ten to twelve days later the cover slips were assayed for T-antigen by indirect immunofluorescence. No colonies containing more than 4 cells were found in the mock-infected control. Table 5 shows that 36 of the 57 colonies induced to divide by SV40 were negative for T-antigen; these included colonies that had grown to over 50 cells. In other colonies (12% of the total) every cell in the colony was positive for T-antigen. One of these positive colonies contained only 9 cells. In other colonies,

TABLE 4

SV40 T-Antigen Induction in BALB/3T3 Maintained in Growth Factor-Free Medium

SV40 (PFU)	Percent of cells positive 48 hr after infection[a]
5×10^7	16
5×10^6	2.0
5×10^5	0.2
0	0.0

[a] 1000 cells were scored for each experimental point.

TABLE 5

SV40 T-Antigen Synthesis in Colonies Induced to Divide in Factor-Free Medium

Size of colony (cells/colony)[a]	Total colonies examined	Number of colonies[b]			Percent positive
		Negative	Mixed	Positive	
5–20	22	20	1	1	5
21–50	15	8	4	3	20
>50	20	8	9	3	15
Total	57	36	14	7	—
Average	—	—	—	—	12

[a] Cover slips were fixed and stained 11–12 days after infection with 5×10^7 PFU of SV40. In control cultures which were mock infected with medium containing 5% fetal calf serum that had been depleted by 3 days of growth on GMK cells, no colonies containing more than 4 cells were observed.

[b] "Negative" means every cell in the colony was T-antigen negative; "Mixed" means the colony had both negative and positive cells; in the "positive" colonies every nucleus showed clear SV40 T-antigen.

some of the cells were positive for T-antigen, while other cells in the same colony were clearly negative. The data indicate that many of the cells induced to divide in factor-free medium did not demonstrate at least one virus function, the T-antigen.

Saturation Density of the Colonies Induced to Grow in Factor-Free Medium by SV40

The ability to grow to a high saturation density is also characteristic of SV40-transformed cells (Aaronson and Todaro, 1968). An experiment was performed to determine whether cells induced to form colonies by SV40 in factor-free medium have also acquired this property. Confluent (1.0×10^6 cells) BALB/3T3 cells were infected with 5×10^7 or 5×10^6 PFU and plated after the absorption period at various dilutions in factor-free medium. Eleven days after infection, the location on the dish of all the colonies present and the number of cells per colony was determined for each plate; the cells were then shifted to complete medium and allowed to grow until they reached confluence. The plates were stained as in a standard transformation assay (Aaronson and Todaro, 1968). Only 10–15% of the colonies that were induced to grow in factor free medium showed the typical SV40-transformed morphology (Table 6). This was true when cells were infected with either 5×10^7 or 5×10^6 PFU. These colonies were categorized according to size attained in factor-free medium 11 days after infection. Table 7 shows that the colony size in factor-free medium was not related to the ability of the cells to grow to high saturation density. Many colonies that contained more than 64 cells showed a normal rather than a transformed morphology, while others that had grown to only 8–15 cells showed the typical transformed morphology when put in complete medium. The percentage of morphologically transformed colonies agrees well with the percent of colonies growing in factor-free medium in which all the cells have T-antigen (see Table 5). With the lower virus inoculum (5×10^6 PFU), there was some suggestion that the smaller colonies were less likely to be transformed. However, with both viral doses and at all

TABLE 6

RELATIONSHIP BETWEEN VIRUS-INDUCED GROWTH IN FACTOR-FREE MEDIUM AND MORPHOLOGICAL TRANSFORMATION

Virus inoculum (PFU)	Total colonies with >7 cells/colony in factor-free medium	Number of "transformed" colonies[b]	"Transformed" colonies / total colonies (%)
5×10^7	300	47	15.6
5×10^6	231	24	10.4

[a] Scored 11 days after infection. In a mock-infected culture, 2 colonies containing >7 cells were observed.

[b] Total number of colonies that grew to high saturation density and appeared morphologically transformed.

TABLE 7

RELATIONSHIP BETWEEN COLONY SIZE IN FACTOR-FREE MEDIUM AND MORPHOLOGICAL TRANSFORMATION

Virus inoculum (PFU)	Size of colony[a] (cells/colony)	No. of colonies scored	No. of "transformed" colonies	No. of "transformed" colonies / No. of colonies scored (%)
5×10^7	8–15	50	7	14.0
	16–63	152	20	13.2
	≥ 64	98	20	20.4
5×10^6	8–15	94	3	3.2
	6–63	113	17	15.0
	≥ 64	24	4	16.6
Control[b]	8–15	2	0	0
	16–63	0	—	—
	≥ 64	0	—	—

[a] Scored 11 days after infection.

[b] Control cells were mock infected with medium containing 5% fetal calf serum that had been depleted by 3 days of growth on GMK cells. This medium was similar to that in which the virus was prepared.

colony sizes, the great majority of the cells induced to form colonies in factor-free medium did not lose contact inhibition.

Characteristics of Clones Isolated after Growth in Factor-Free Medium

The previous experiments showed that only 10–20% of the colonies induced by SV40 to grow in factor-free medium lost contact inhibition of cell division and con-

tained cells that were positive for T-antigen. This result would be expected if SV40 only temporarily changed the growth properties of the cells. Alternatively, the selection procedure may have allowed recognition of transformed cells which had permanently associated SV40, but which, nevertheless, were still subject to contact inhibition. To distinguish between these two possibilities, 12 colonies that grew in factor-free medium after SV40 infection were isolated and grown up to mass cultures. When the progeny cells were tested for properties associated with permanent virus transformation, it was clear that both situations existed.

Table 8 shows that different classes of colonies were found. The first class (5 of the 12 clones) had the properties associated with the transformed state. They all had SV40 T-antigen and their saturation densities ranged from 6.0 to 12.5 $\times 10^6$ cells/plate. The colonies in this class all retained the ability to grow in factor-free medium. Their plating efficiency in this medium ranged from 5 to 64% of the plating efficiency in complete medium, values that are comparable to those found for SV40 transformants obtained by standard methods (Smith and Scher, 1971). Two of the 4 clones tested, however, failed to yield infectious virus after fusion with permissive cells. The second class (5 of the 12 clones) had none of the properties of transformed cells. Their saturation densities in complete medium were similar to normal BALB/3T3 cells. They did not grow in factor-free medium and did not have T-antigen. None of these colonies yielded virus after fusion with permissive cells.

Two of the 12 colonies did not fall into either class. One of these clones, 10 A-7, grew to the same saturation density as normal cells, but was still able to grow in factor-free medium, although more slowly than the other transformants. It also had T-antigen, and, after fusion with GMK, yielded virus. This transformed colony would not have been detected by the standard transformation assay. Another colony, 10 A-4, was able to slowly grow in factor-free medium, although it had the same saturation density as normal BALB/3T3 cells. In this clone, however, no other evi-

TABLE 8

PROPERTIES OF CLONES ISOLATED FROM FACTOR-FREE MEDIUM

Clone	H story			Saturation density in complete medium ($\times 10^6$/plate)	Relative plating efficiency in factor-free medium[b]	T-Antigen		Recovery of virus by fusion
	Virus inoculum (PFU)	Approx. No. of cells per colony in factor-free medium[a]	Allowed to grow in complete medium before cloning			Immunofluorescence	Complement fixation	
Class I								
11-A-4	5×10^7	100	−	12.5	42.0	+	NT[d]	NT
11-A-6	5×10^7	500	−	9.0	78.0	+	NT	+
11-A-6	5×10^7	300	−	8.4	64.2	+	NT	−
10-A-6	5×10^7	300	+	6.0	5.4	+	NT	−
11-A-1	5×10^7	300	−	6.0	28.0	−	+	+
Class II								
12-A-3	5×10^7	200	−	2.0	<1.0	−	−	−
12-A-2	5×10^7	200	−	1.6	<0.1	−	−	−
12-A-1	5×10^7	200	−	1.6	<0.1	−	−	−
12-A-4	5×10^7	200	−	1.2	<0.4	−	−	−
10-B-3	5×10^5	34	+	0.8	<0.2	−	−	−
Class III								
10-A-7	5×10^7	200	+	0.9	34[c]	+	+	+
10-A-4	5×10^7	94	+	0.9	10[c]	−	−	−
Control								
10-C-1	0	7	+	1.0	NT	−	−	NT
10-C-3	0	6	+	1.0	<0.5	−	−	−

[a] Observed 2–3 weeks after infection.

[b] Colonies growing in factor-free medium × 100/colonies growing in complete medium.

[c] Colonies grew more slowly in factor-free medium than did those in Class I.

[d] NT = Not tried.

dence of persisting virus information has yet been found; it was negative for T-antigen, and did not yield virus after fusion.

DISCUSSION

SV40 infection of BALB/3T3 cells in growth factor-free medium induces several rounds of mitosis. The ability of the virus to induce cell division with the formation of colonies can be detected at input multiplicities as low as 0.5. Dividing cells can be recognized a few days after infection. Many of the cells induced to divide do not have permanently altered growth properties.

Several rounds of cell division are needed for the full expression of loss of contact inhibition (Todaro and Green, 1966), a characteristic of the transformed state. One explanation for the delay in the expression of transformation is that normal cell components (e.g., membrane structures) must be diluted out before cells gain the ability to grow to a high saturation density. The present studies demonstrate that the ability of cells to grow in the absence of serum growth factor(s), an independent criterion of virus transformation (Jainchill and Todaro, 1970), occurs without a need for dilution of cell structures.

Several virus functions are known that are directly proportional to the input multiplicity. These include infectivity, transforming ability and T-antigen inducing ability. From the present studies it appears that the ability of the virus to induce cell division is also proportional to the multiplicity. At high multiplicities the cells are induced to undergo many rounds of cell division, although the majority are not permanently transformed. Although the nature of the experiments make precise determinations difficult, it would appear that the division inducing capacity is of comparable magnitude to the infectivity. One possible explanation is that each infectious unit can induce one round of cell division. Cells that are multiply infected then would be able to divide several times. If, however, the viral genome successfully integrates, cellular growth properties are permanently altered.

The induction of cell division by SV40 appears to depend upon the presence of a functioning virus genome since UV-irradiation inactivates it efficiently. It is reasonable to assume that this function is caused by the DNA of the virus. No attempt has been made to determine the "target size" of the division-inducing capacity of SV40, since repair of UV damaged virus genetic functions can be extensive (Aaronson, 1970).

In this paper it is shown that the small-plaque mutant of SV40 induces cell division of BALB/3T3 in growth factor-free medium. 3T3 behaves similarly to BALB/3T3 and can be induced to divide in factor-free medium after infection with SV40. The large-plaque, temperature-sensitive mutant of SV40 (SV-L) also induces cell division under these conditions, although, as with transformation (Todaro and Takemoto, 1969), it is considerably less efficient per infectious unit than the small-plaque mutant (unpublished observations).

Many of the cells induced to divide in factor-free medium by SV40 are not permanently altered in their growth properties. Only 10–20% of the colonies that grew in factor-free medium showed the typical morphology of SV40-transformed cells, many colonies induced to grow in factor-free medium had no cells with detectable T-antigen when observed as early as 10–12 days after infection; some of the colonies that were cloned after growing in factor-free medium were still contact inhibited, could no longer grow in factor-free medium and showed no evidence of the continued presence of functioning virus information. A transient change in growth properties (abortive transformation) has also been observed after polyoma virus infection of hamster cells (Stoker, 1968; Stoker et al., 1968); after infection with polyoma virus, many cells transiently acquired the ability to grow in agar suspension.

Growth in factor-free medium provides a method for selecting virus-transformed cells that is independent of the methods commonly used to select SV40-transformants; the latter rely on a recognizable morphological alteration and the loss of contact inhibition of cell division. By isolating colonies that grew in factor-free medium, one virus-transformed line, 10 A-7, has been obtained that is still subject to contact inhibition. The line is similar in phenotype to the "flat revertants" (Pollack et al., 1968) selected by using FUdR to kill dividing cells. A limitation of the FUdR selection procedure is that it may select a secondary alteration in the already fully transformed population. Since there was no secondary selection involved in the development of 10 A-7, it is likely that the virus induced only one growth change, the ability to divide in factor-free medium. This line indicates that the property of growth in factor-free medium is not identical to the property of loss of contact inhibition. Another colony with unexpected growth properties was isolated after SV40-induced growth in factor-free medium. This colony, 10 A-4, showed no evidence of permanent virus association; it was negative when tested for T-antigen, and did not yield virus after fusion. Although the cells were still subject to contact inhibition, they had gained the ability to grow in factor-free medium. This line may represent a "spontaneous" selection for this growth property unrelated to the virus infection. However, repeated attempts to obtain such a variant from uninfected BALB/3T3 in the absence of SV40 have not been successful. Alternately, the ability of the cells to grow in factor-free medium may be the only property associated with the input virus genome that is expressed in this clone. Nucleic acid hybridization methods may resolve whether the virus genome, or a portion of it, is still present in these cells.

To limit cell division of sparse cultures, serum was used that had most growth promoting factor(s) removed by Cohn fractionation. Medium made with such serum was further depleted on confluent BALB/3T3 to remove any remaining growth promoting activity. One of the problems associated with the use of medium prepared in such a manner is that the depletion of essential nutrients or the accumulation of toxic substances by the cells may be responsible for the inability of normal BALB/3T3 to divide. That this is not the explanation for the selective effect of factor-free medium is clearly shown by heating the serum and destroying its growth promoting activity. Whether this treatment removes one or many growth stimulating factors is not known. A factor(s) in serum which stimulates growth of contact inhibited cells has been described (Todaro et al., 1967; Holley and Kiernan, 1968; Cunningham and Pardee, 1969). It is possible that the same factor(s) is necessary for the growth of sparse cultures, and that the heating procedure described destroys this growth-stimulating factor(s). By heating the serum, growth factor-free medium can be prepared in a direct and uniform way.

The ability of SV40 to rapidly, although transiently, remove the requirement for a serum factor(s) from BALB/3T3 suggests that the virus itself may be inducing this activity within the cells it infects. Recently, Rubin has described a nonviral factor produced by Rous sarcoma virus-transformed cells which stimulates cellular overgrowth in crowded chick embryo cultures (Rubin, 1970). It is not clear whether similar factors can be isolated from SV40-transformed cells.

ACKNOWLEDGMENT

We would like to acknowledge the excellent technical assistance of Miss Kay Reynolds and Mrs. Kay Gottesman. We would also like to thank Dr. Mark Nameroff for teaching us the technique of autoradiography.

REFERENCES

AARONSON, S. A. (1970). Effect of ultraviolet irradiation on the survival of Simian virus 40 functions on human and mouse cells. J. Virol. 6, 393–399.

AARONSON, S. A., and TODARO, G. J. (1968). Development of 3T3-like lines from Balb/c mouse embryo cultures: Transformation susceptibility to SV40. J. Cell Physiol. 72, 141–148.

BENJAMIN, T. L. (1966). Virus-specific RNA in cells productively infected or transformed by polyoma virus. J. Mol. Biol. 16, 359–373.

BLACK, P. H. (1968). The oncogenic DNA viruses: A review of in vitro transformation studies. Annu. Rev. Microbiol. 22, 391–426.

BLACK, P. H., ROWE, W. P., TURNER, H. C., and

HUEBNER, R. J. (1963). A specific complement-fixing antigen present in SV40 tumor and transformed cells. Proc. Nat. Acad. Sci. U. S. 50, 1148–1156.

BURK, R. R. (1966). Growth inhibitor of hamster fibroblast cells. Nature (London) 212, 1261–1262.

CUNNINGHAM, D. D., and PARDEE, A. B. (1969). Transport changes rapidly initiated by serum addition to "contact inhibited" 3T3 cells. Proc. Nat. Acad. Sci. U. S. 64, 1049–1056.

DULBECCO, R. (1969). Cell transformation by viruses. Science 166, 962–968.

ECKHART, W. (1969). Cell transformation by polyoma virus and SV40. Nature (London) 224, 1069–1071.

GERBER, P. (1966). Studies on the transfer of subviral infectivity from SV40-induced hamster tumor cells to indicator cells. Virology 28, 501–509.

HABEL, K., and EDDY, B. E. (1963). Specificity of resistance to tumor challenge of polyoma and SV40 virus-immune hamsters. Proc. Soc. Exp. Biol. Med. 113, 1–16.

HOLLEY, R. W., and KIERNAN, J. A. (1968). "Contact inhibition" of cell division in 3T3 cells. Proc. Nat. Acad. Sci. U. S. 60, 300–304.

JAINCHILL, J., and TODARO, G. J. (1970). Stimulation of cell growth in vitro by serum with and without growth factor. Exp. Cell Res. 59, 137–146.

KOPROWSKI, H., JENSEN, F. C., and STEPLEWSKI, Z. (1967). Activation of production of infectious tumor virus SV40 in heterokaryon cultures. Proc. Nat. Acad. Sci. U. S. 58, 127–133.

POLLACK, R. E., GREEN, H., and TODARO, G. J. (1968). Growth control in cultured cells: Selection of sublines with increased sensitivity to contact inhibition and decreased tumor-producing ability. Proc. Nat. Acad. Sci. U. S. 60, 126–133.

RUBIN, H. (1970). Overgrowth stimulating factor released from Rous sarcoma cells. Science 167, 1271–1272.

SAMBROOK, J., WESTPHAL, H., SRINIVASAN, P. R., and DULBECCO, R. (1968). The integrated state of viral DNA in SV40-transformed cells. Proc. Nat. Acad. Sci. U. S. 60, 1288–1295.

SMITH, H. S., and SCHER, C. D. (1971). Cell division in medium lacking serum growth factor: comparison of lines transformed by different agents. Nature (London). In press.

SMITH, H. S., SCHER, C. D., and TODARO, G. J. (1970). Abortive transformation of Balb/3T3 by Simian virus 40. Bacteriol. Proc. 187.

STOKER, M. (1968). Abortive transformation by polyoma virus. Nature (London) 218, 234–238.

STOKER, M., O'NEILL, C., BERRYMAN, S., and WAXMAN, V. (1968). Anchorage and growth regulation in normal and virus-transformed cells. *Int. J. Cancer* **3**, 683–693.

TAKEMOTO, K. K., KIRSCHSTEIN, R. L., and HABEL, K. (1966). Mutants of Simian virus 40 differing in plaque size, oncogenicity, and heat sensitivity. *J. Bacteriol.* **92**, 990–994.

TAKEMOTO, K. K., TODARO, G. J., and HABEL, K. (1968). Recovery of SV40 virus with genetic markers of original inducing virus from SV40-transformed mouse cells. *Virology* **35**, 1–8.

TEMIN, H. M. (1967). Control by factors in serum of multiplication of uninfected cells and cells infected and converted by avian sarcoma virus. *In* "Growth Regulating Substances for Animal Cells in Culture" (V. Defendi and M. Stoker, eds.), pp. 103–116. Wistar Inst. Press, Philadelphia, Pennsylvania.

TODARO, G. J. (1969). Transformation assay using cell line 3T3. *In* "Fundamental Techniques in Virology" (K. Habel and N. Salzman, eds.), pp. 220–228. Academic Press, New York.

TODARO, G. J., and GREEN, H. (1966). Cell growth and the initiation of transformation by SV40. *Proc. Nat. Acad. Sci. U. S.* **55**, 302–306.

TODARO, G. J., and TAKEMOTO, K. K. (1969). "Rescued" SV40: Increased transforming efficiency in mouse and human cells. *Proc. Nat. Acad. Sci. U. S.* **62**, 1031–1037.

TODARO, G. J., MATSUYA, Y., BLOOM, S., ROBBINS, A., and GREEN, H. (1967). Stimulation of RNA synthesis and cell division in resting cells by a factor present in serum. *In* "Growth Regulating Substances for Animal Cells in Culture" (V. Defendi and M. Stoker, eds.), pp. 87–101. Wistar Inst. Press, Philadelphia, Pennsylvania.

UCHIDA, S., YOSHIIKE, K., WATANABE, S., and FURANO, A. (1968). Antigen-forming defective viruses of Simian virus 40. *Virology* **34**, 1–8.

WATKINS, J. F., and DULBECCO, R. (1967). Production of SV40 virus in heterokaryons of transformed and susceptible cells. *Proc. Nat. Acad. Sci. U. S.* **58**, 1396–1403.

VIROLOGY 59, 477–489 (1974)

A Nonselective Analysis of SV40 Transformation of Mouse 3T3 Cells

REX RISSER AND ROBERT POLLACK

Cold Spring Harbor Laboratory, Cold Spring Harbor, New York 11724

Accepted February 5, 1974

Mouse cells transformed by simian virus 40 show many alterations in their growth properties *in vitro*. In order to investigate the coordinate nature of these changes, we have analyzed the growth properties of 40 randomly selected colonies arising after SV40 infection of 3T3 cells. Clones of cells, established from these colonies, were characterized as to saturation density and doubling time in 10% and 1% calf serum, growth in methyl cellulose suspension, colony formation on monolayers of normal cells, and presence of viral antigens. This analysis revealed that only 5 of the clones were indistinguishable from 3T3 cells; the remaining 35 clones differed from 3T3 cells in that they grew as rapidly in 1% calf serum as standard SV40 transformed cells. Of these 35 clones, ten corresponded to standard transformants previously described. Another ten showed other growth properties intermediate between 3T3 cells and standard transformants. These intermediate clones had lower levels of viral T-antigen than standard transformants and showed considerable heterogeneity in staining from cell to cell. The remaining 15 clones were T-antigen negative and had saturation densities slightly higher than that of 3T3 cells. These changes in cellular behavior are stable on recloning.

INTRODUCTION

Simian virus 40 can radically alter the *in vitro* behavior of fibroblastic cells. This transformation of cellular growth properties into patterns similar to those seen in tumor cells is detected in a minority of the infected cells by selective assays. Various approaches have been used to analyze the role of the virus in this process; in particular, viral mutants temperature-sensitive for lytic growth have been isolated and some have been found to affect transformation (Tegtmeyer, 1972; Kimura and Dulbecco, 1973; Robb *et al.*, 1972). Also cells reverted to a more normal cellular behavior have been selected from transformed populations, and SV40 recovered from them has been analyzed (Pollack *et al.*, 1968; Renger and Basilico, 1972; Ozanne, 1973; Vogel and Pollack, 1973; Vogel *et al.*, 1973; Culp and Black, 1972). Though many of these experiments indicate that viral functions are required for the transformation process, the relationship of a viral function(s) to the observed alterations in cellular behavior is not at all clear. In this study we have approached the problem of virally induced growth alterations somewhat differently. Rather than subject infected cells directly to a selective transformation assay, we have obtained random clones of cells arising after SV40 infection and then tested them in a number of transformation assays. Such an analysis should detect changes in one or a few growth properties if such changes occur and thus define more completely the physiological effect of SV40 on most of the cells it infects.

The numerous assays which have been used to monitor *in vitro* transformation include growth to high density (Todaro *et al.*, 1964), growth in agar or methyl cellulose suspension (Macpherson and Montagnier, 1964; Stoker *et al.*, 1968), focus formation on monolayers of normal cells (Temin and Rubin, 1958), growth in depleted or low concentrations of serum (Smith *et al.*, 1971;

477

Holley and Kiernan, 1968) and morphological changes (Stoker and Abel, 1962). Additionally, a virus-specific nuclear antigen (T-antigen) has been found in cells acutely infected with or transformed by SV40 (Black et al., 1963; Pope and Rowe, 1964).

Several observations suggest that each assay of transformation reflects a different change in cellular physiology. In particular, the frequency of cellular transformation induced by SV40 depends upon the particular assay used (Black, 1966). Also, it is possible to obtain cells which, though transformed in their ability to grow in depleted serum, are normal in their saturation density (Smith et al., 1971; Scher and Nelson-Rees, 1971). The results presented here extend these observations, and demonstrate that after SV40 infection most clones show a reduced growth requirement for serum components. Furthermore, a high proportion of SV40 transformants show only intermediate changes in other cellular properties associated with oncogenic transformation. It is concluded from the data presented here that SV40 infection of 3T3 cells can induce several stages of transformed behavior. A preliminary report of these results has appeared elsewhere (Risser and Pollack, 1974).

MATERIALS AND METHODS

Cell cultures. Cell lines and clones were maintained in Dulbecco's modified Eagle's medium (DME) (Gibco-H21) containing 10% calf serum (Colorado Serum Company) with 50 μg/ml Gentamicin. All cells were transferred weekly at an inoculation density of about 5×10^2 cells/cm^2 on Falcon plastic dishes. A recently recloned line of mouse 3T3 cells (Todaro and Green, 1963) and a SV40 transformed 3T3 cell line, SV101 (Todaro et al., 1964), were used routinely in each series of assays.

Virus. SV40 (strain 776, originally from NIH) was serially plaque purified three times on BSC-1 cells, a line of African green monkey kidney cells. The lysate obtained from a single plaque was passaged twice at low multiplicities of infection (<0.1 PFU cell), and virus was then purified according to the method of Black et al. (1964) and twice banded to equilibrium in CsCl. All plaque assays were done on BSC-1 cells in

medium containing 2% FCS, 98% DME mycostatin (50 units/ml) and 0.9% agar (Difco-Bacto). Plaques were detected at 10 days and counted at 14 days.

Random cloning of SV40 cells. A 60 mm plate containing 5×10^5 3T3 cells was infected with 0.2 ml DME containing 10^9 PFU of purified SV40 or mock infected with 0.2 ml DME for 2 hr at 37°. Medium was then replaced, and the next day cells were trypsinized, counted, and replated at densities of $5 \times 10^4, 5 \times 10^3, 5 \times 10^2, 5 \times 10^1$, and 5 cells per plate. Plates were cultured for 2 weeks, and then most plates were stained with Harris hemotoxylin. The plating efficiency (EOP) was scored as the number of colonies formed per cells plated and the transformation efficiency as the number of dense colonies per total colonies. From unfixed duplicate plates containing 10–20 colonies, all well isolated colonies were circled, classified as morphologically normal or transformed, and cloned using a steel cloning cylinder. Colonies were scored as morphologically transformed if they showed several layers of cell growth and the tightly packed random orientation seen in isolated colonies of SV101 cells.

To clone, colonies were picked without preference for a particular morphology; 50–75% of the colonies on a given plate were cloned. Each clone was passaged for 1 month, and then tested in each assay or frozen for future testing.

Viral antigens. Assays for the presence of viral antigens were done by indirect immunofluorescence. Cells were plated on coverslips at a density of 2×10^4 cells/cm^2. Four to twelve hours before staining, medium was changed to 1% calf serum-DME. Coverslips were rinsed in phosphate-buffered saline medium (PBS) (at 4°), fixed 5–10 min in acetone (at −20°), and stained for 1 hr at 37° with direct antibody. Coverslips were rinsed and stained with rhodamine-conjugated bovine serum albumin and fluorescein-conjugated counterstain for 1 hr at 37°. After rinsing in PBS, coverslips were examined at 400× magnification under dark-field ultraviolet illumination (Zeiss). Hamster anti-T antibody (Flow Laboratories), fluorescein-conjugated rabbit anti-hamster γ-globulins (Antibodies, Inc.) and rhodamine-conjugated bovine serum albumin (Huntingdon Research Center) were used in PBS at dilutions of $\frac{1}{5}$, $\frac{1}{10}$ and $\frac{1}{30}$, respectively. Complement fixation assays were performed according to the method of Osborn and Weber (manuscript in preparation).

Growth in medium containing 10% or 1% calf serum. 2×10^4 cells in 2 ml of medium containing 10% or 1% calf serum were seeded onto 35 mm Falcon dishes. Medium was changed every third day and cell counts taken daily using a Coulter counter. Saturation densities are the equilibrium densities which the cells maintain for the last 3 to 4 days of the growth experiment. Doubling times were calculated from the slope of the initial logarithmic growth curve before density-dependent inhibition of cell increase was seen.

Growth in medium containing methyl cellulose. Cells were plated in 4 ml of medium containing 10% calf serum, 1.2% Methocel (Dow Chemical Company, 4000 cps), 90% DME at densities of 10^5, 10^4, 10^3, or 10^2 cells/60 mm dish over a layer of 0.9% agar (Difco-Bacto), 10% calf serum, 90% DME. Cultures were fed with an additional 4 ml of Methocel medium at 1 and 2 weeks and scored and measured at 3 weeks. Colonies $\geq$ 100 cells were scored as positive colonies in plating efficiency experiments. Measurements of colony size were done on cells and colonies by use of an eyepiece reticle. The reticle was calibrated using corn, pecan, ragweed, and mulberry pollen grains of diameters 87.5, 47.5, 19.5 and 13.5 μm, respectively. From the diameter of the colony the volume was calculated. Volume increments were assumed to be proportional to increases in cell number, and the cell number was inferred from the increase in volume.

Growth on normal monolayers. 100 or 1000 cells were plated on confluent 3T3 monolayers and on plastic dishes. Medium was changed every third day and dishes fixed with formalin-PBS and stained with Harrishemotoxylin at day 10. Colonies sufficiently dense to be seen without the aid of a microscope were scored. A single medium preparation was used throughout a plating experiment. The variation observed among the absolute plating efficiencies of different clones (Tables 1 and 2) is largely due to differences in serum preparations; a 2-fold difference in plating efficiencies is not considered significant.

RESULTS

Random Selection of Clones of SV40 Cells

Subconfluent 3T3 cells were infected with SV40 and plated at various dilutions as in Materials and Methods. After 2 weeks the transformation efficiency was determined and 40 randomly selected colonies were cloned from plates which received infected cells. The transformation efficiency of 10.5% in this experiment (Table 1) is comparable to that found by others for similar doses of virus. It is apparent from the low number of morphological transformants picked that no preference was shown in picking clones (Table 1). To serve as control normal lines, four clones were picked from plates which received mock infected cells. To serve as control transformed clones, four dense foci were picked from plates which received 10^3 infected cells. SV101 cells and the parental 3T3 cells served as additional controls. All

TABLE 1
RANDOM SELECTION OF SV40 CLONES

Protocol	Virus Multiplicity (PFU/cell)	Transformation[a] efficiency $\left(\dfrac{\text{transformed colonies} \times 100}{\text{total colonies}}\right)$	Efficiency of plating $\left(\dfrac{\text{colonies}}{\text{cells plated}} \times 100\right)$	Clones picked	Clones morphologically transformed (%)
Virus Infected	2×10^3	10.5	83	40	15
Mock Infected	0.0	≤ 0.001	91	4	0

[a] As in Materials and Methods.

TABLE 2
Properties of SV40 Clones

SVR clone number	T-anti-gen	10% Calf serum		1% Calf serum		EOP[b] in methyl cellulose	EOP[b] on plastic	EOP[b] on 3T3 monolayer	EOP on monolayer/ EOP on plastic
		Satura-tion density[a]	Doubling time (hr)	Satura-tion density[a]	Doubling time (hr)				
Experimental clones									
11	I	19.5	19	3.8	34	14.6	41.5	≤0.05	≤0.001
12	I	27.0	18	3.6	28	3.7	42.0	0.5	0.012
13	I	18.0	17	3.3	33	0.2	47.3	0.2	0.004
14	I	22.0	17	4.0	26	6.9	36.0	≤0.05	≤0.001
15	+	40.0	24	4.0	45	11.8	7.0	5.5	0.79
16	I	22.0	22	3.2	29	0.5	70.0	≤0.05	≤0.001
17	I	29.0	18	3.9	32	2.2	59.7	0.1	0.002
18	I	31.0	19	5.7	32	1.3	34.5	0.15	0.004
21	+	46.0	16	8.8	45	30.1	35.7	30.3	0.85
22	I	45.0	22	3.6	41	3.5	52.0	13.0	0.25
23	+	60.0	18	7.0	44	36.7	47.3	28.3	0.60
25	+	60.0	16	10.0	31	29.0	58.0	31.7	0.55
26	+	46.0	17	3.0	34	53.0	48.5	4.7	0.10
32	I	18.5	24	2.5	38	0.7	25.5	1.4	≤.055
33	–	8.5	26	0.9	84	≤0.001	22.0	≤0.03	≤0.001
34	–	8.5	28	1.0	88	≤0.001	27.3	≤0.03	≤0.001
35	–	7.6	26	0.9	100	≤0.001	25.0	≤0.03	≤0.001
36	–	8.0	24	0.6	63	≤0.001	7.8	≤0.03	≤0.001
41	–	9.5	19	2.0	48	≤0.001	38.5	≤0.03	≤0.001
42	–	12.5	15	2.4	37	≤0.001	40.5	≤0.03	≤0.001
43	–	8.0	20	2.5	32	≤0.001	44.7	≤0.03	≤0.001
44	–	11.0	24	1.8	43	≤0.001	36.0	≤0.03	≤0.001
45	–	12.5	22	1.8	38	≤0.001	17.3	≤0.03	≤0.001
47	–	11.5	24	1.8	26	≤0.001	32.0	≤0.03	≤0.001
51	–	16.5	17	1.8	36	≤0.001	27.0	≤0.03	≤0.001
54	–	15.0	17	1.9	29	≤0.001	30.7	≤0.03	≤0.001
55	–	13.0	22	2.3	46	≤0.001	42.0	≤0.03	≤0.001
56	–	16.0	19	3.0	49	≤0.001	51.8	≤0.03	≤0.001
57	–	16.0	17	3.0	31	≤0.001	58.0	≤0.03	≤0.001
58	–	11.5	22	0.7	55	≤0.001	31.0	≤0.03	≤0'001
63	I	15.0	15	2.5	29	≤0.001	43.0	≤0.03	≤0.001
81	+	50.0	17	10.5	24	10.7	53.0	52.0	1.0
82	–	12.0	20	3.0	29	≤0.001	37.3	≤0.03	≤0.001
84	+	45.0	16	11.0	27	10.6	15.3	40.0	2.6
85	+	43.0	14	7.6	22	44.0	29.0	23.5	0.81
86	+	46.0	17	8.0	34	58.0	36.5	39.5	1.1
87	+	35.0	19	9.3	34	38.6	27.7	28.3	1.0
94	–	11.0	22	3.4	26	≤0.001	51.0	≤0.03	≤0.001
95	–	13.0	24	2.9	31	≤0.001	37.3	≤0.03	≤0.001
97	–	10.7	19	3.2	48	≤0.001	31.7	≤0.03	≤0.001
Control mock clones									
101	–	9.0	20	1.0	67	≤0.001	36.0	≤0.03	≤0.001
103	–	7.0	22	1.3	66	≤0.001	32.7	≤0.03	≤0.001
104	–	7.0	26	0.8	70	≤0.001	35.5	≤0.03	≤0.001
106	–	8.5	19	1.4	86	≤0.001	26.5	≤0.03	≤0.001
3T3	–	7.6	23	0.9	66	≤0.001	39.0	≤0.03	≤0.001
Control transformed clones									
111	+	60	18	16.0	36	16.1	33.0	24.3	0.74
114	+	45	17	11.5	19	26.0	37.0	19.5	0.53
122	+	50	24	10.0	43	28.5	39.3	10.0	0.25
123	+	45	21	9.0	41	45.0	25.5	17.0	0.67
SV101	+	57	16	11.0	26	27.0	38.0	41.3	1.1

[a] Units are cells/cm² × 10⁴.
[b] Units are colonies/100 cells plated.

clones were passaged a minimum of 1 month before they were tested for transformed properties.

Viral Antigens

Though not a selective assay, the persistence of SV40 specific antigens has served as an indication of the presence of a functioning viral genome. For that reason all clones were stained for viral T-antigen by indirect immunofluorescence. Three patterns of staining were seen: positive, negative, and intermediate; examples are shown in the photographs in Fig. 6. Intermediate clones showed 10–20 % bright positive nuclei, 50–80 % weakly fluorescent nuclei and 10–30 % dark nonfluorescent nuclei. Control transformants stained uniformly positive while negative clones did not show nuclear fluorescence. These staining patterns can be used to classify clones, and thus clones will be referred to as T-antigen negative, positive or intermediate (Table 2).

The staining method was checked by staining mixtures of SV101 and 3T3 cells at various ratios. When SV101 cells were mixed with 3T3 cells in ratios of $\frac{1}{5}$ or $\frac{1}{100}$, $\frac{215}{1000}$ or $\frac{11}{1000}$ cells, respectively, showed bright nuclear fluorescence. When a mixture of SV101 and 3T3 cells (ratio 1:4) were plated very sparsely and allowed to form colonies on coverslips, $\frac{7}{25}$ colonies were uniformly T-antigen positive. No intermediate or mixed colonies were seen. These results suggest that T-antigen intermediate clones seen among the experimental clones are not due to artifacts introduced by the staining or cloning procedures used.

Levels of complement-fixing T-antigen were determined on equal numbers of cells from various clones. As can be seen in Table 3, indirect immunofluorescence and complement fixation show close agreement. Those clones designated intermediate by immunofluorescence have approximately a 10-fold reduction in complement-fixing titers. Clones designated negative by immunofluorescence show background complement fixing titers, comparable to that of 3T3.

V-antigen production was also determined by indirect immunofluorescence. Of the 25 clones tested, none were V-antigen positive, whereas BSC-1 cells infected 48 hr earlier with SV40 fluoresced brightly in the nucleus.

TABLE 3

T-Antigen Determination on SV40 Clones

SVR clone number	Immuno-fluorescene	Complement fixation
13	I	10.0
16	I	3.0
18	I	10.0
22	I	3.0
35	−	0.2
43	−	0.8
63	I	10.0
85	+	100.0
86	+	25.0
3T3	−	0.1
SV101	+	100.0

[a] Complement fixation assays were performed on frozen-thawed lysates of approximately 5×10^6 cells resuspended in about 100 μl of PBS according to the procedure of Osborn and Weber (manuscript in preparation). Values are reported as the percentage of the SV101 complement-fixing titer.

Growth in Medium Containing 10% and 1% Calf Serum

Growth experiments were carried out on all clones in 10 % and 1 % calf serum (Table 2). As can be seen in Fig. 1, such experiments differentiated 3T3 cells and SV101 cells on the basis of saturation density in either serum concentration. As is graphically shown in the histograms in Fig. 2, a wide range of saturation densities was found among experimental clones in either 10 % calf serum (Fig. 2A) or 1 % calf serum (Fig. 2B). In contrast, control transformants or control mock-infected clones showed a relatively narrow distribution of saturation densities. The level of T-antigen in a clone is rather well correlated with its saturation density (Fig. 2).

The doubling times of all clones in 10 % calf serum ranged from 16 to 24 hr, providing little distinction between normal and transformed cells. The doubling time in 1 % calf serum, however, differed considerably between normal and transformed lines; 3T3 cells doubled in about 80 hr in 1 % calf serum, whereas SV101 cells doubled in only 30 hr (Table 2, Fig. 1). Thirty-five of the experimental clones showed doubling times comparable to those of transformed cells in 1 % calf serum (Fig. 3). As can be seen many of these clones lacked viral T-antigen.

Growth in Medium Containing Methyl Cellulose

Each clone was plated in suspension culture containing 1.2% methyl cellulose. Again, experimental clones showed a spectrum of plating efficiencies (Fig. 4). In this assay, however, less than one in 10^5 cells from any T-antigen negative clone plated, a result which did not distinguish them from mock-infected clones. T-antigen intermediate clones ranged from about 0.1% to 10% in plating efficiencies, whereas T-antigen positive clones plated efficiencies of 10–50%.

To determine whether cells from T-antigen negative clones were dividing in methyl cellulose, yet not reaching the size necessary for visual scores, measurements were made on individual cells or colonies. As can be seen in Fig. 5a, 3T3 cells underwent few if any divisions during the 3-week course of the assay, whereas SV101 cells underwent many divisions to finally give a mean colony size of 250 cells (Fig. 5b). This represents a significant lengthening of the doubling time of SV101 cells as compared to growth in medium lacking methyl cellulose (62 vs 16 hr), suggesting this is a more restrictive growth assay. T-antigen negative clones SVR82,

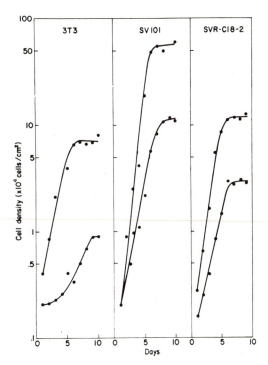

FIG. 1. Typical growth experiments of 3T3, SV101, SVR82. Cells were seeded as in Materials and Methods in 10% calf serum (upper curves) or 1% calf serum (lower curves), and cell counts taken daily. Medium was changed every third day.

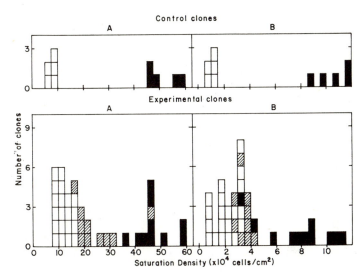

FIG. 2. Histograms of saturation densities of SV40 clones. Growth experiments were carried out in (A) 10% calf serum and (B) 1% calf serum as in Materials and Methods. □, T-antigen negative clones; ▨, T-antigen intermediate clones; ■, T-antigen positive clones

SVR34, and SVR42 and T-antigen intermediate clone SVR63 showed few divisions
during the 3-week assay (Figs. 5k, 5h, 5i, 5j,
respectively). The behavior of SVR63 is

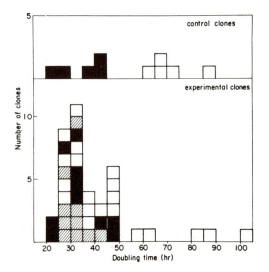

Fig. 3. Histogram of doubling times of SV40
clones in 1% calf serum. Doubling times were
determined during the exponential growth phase
before density-dependent inhibition of cell increase was seen. □, T-antigen negative clones;
▨, T-antigen intermediate clones; ■, T-antigen
positive clones.

unique, since the other five T-antigen intermediate clones measured showed some
limited growth in methyl cellulose suspension. Cells from T-antigen intermediate
clones SVR13 and SVR32 continuously
divided during the three week assay (Figs.
5c and d; and Fig. 5f and g, respectively).
In their continuous though slow division in
methyl cellulose they differ from the abortive
transformants described by Stoker (1968)
which underwent 3 to 4 divisions and then
ceased growing in suspension culture. The
great majority of cells in these intermediate
clones do not divide as rapidly as SV101
cells or other T-antigen-positive transformed
cells. Most colonies from these clones are
much smaller than SV101 colonies, and thus
are not detected when plating efficiency is
scored without a microscope.

Growth on Normal Monolayers

When clones were plated on confluent 3T3
monolayers, a range of plating efficiencies
was seen (Fig. 6). T-antigen negative clones
form dense colonies on normal monolayers
with efficiencies of $\leq 10^{-3}$ as compared to
plastic dishes, whereas T-antigen positive
clones and control transformants plate as
efficiently on normal monolayers as they do

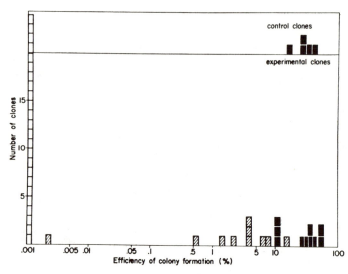

Fig. 4. Histogram of plating efficiencies of SV40 clones in methyl cellulose suspension. EOP was determined as in Materials and Methods. □, T-antigen negative clones; ▨, T-antigen intermediate clones;
■, T-antigen positive clones.

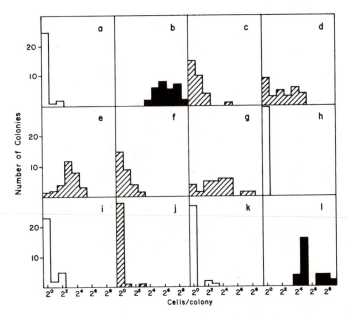

Fig. 5. Size of colonies arising in methyl cellulose suspension. Colony size was determined as in Materials and Methods. (a) 3T3—3 weeks, (b) SV101—3 weeks, (c) SVR13—1 week, (d) SVR13—3 weeks, (e) SVR17—3 weeks, (f) SVR32—1 week, (g) SVR32—3 weeks, (h) SVR34—3 weeks, (i) SVR42—3 weeks, (j) SVR63—3 weeks, (k) SVR82—3 weeks, (l) SVR85—3 weeks.

on plastic. T-antigen intermediate clones form dense colonies with varying efficiencies ranging from ≤0.1 to 10% of the plating efficiency on plastic.

Stability of Growth Properties

The stability of these growth properties was tested by culturing 6 representative clones for 4 months and then recloning each. Both parental and subclone were then retested in each assay. The data in Table 4 demonstrate that the properties of a given clone are quite stable. The greatest variation is seen in the doubling time of clone SVR42 and subclone SVR42-1 in 1% calf serum. The difference seen between initial testing and later testing of parent and recloned population is no greater than that seen among different control transformants, however.

T-antigen intermediate clones were more intensively investigated for the stability of the heterogeneous T-antigen staining pattern. Thirty-six subclones were generated from 4 different T-antigen intermediate lines, and each was tested for T-antigen by indirect immunofluorescence. Sixteen of these subclones showed the same heterogeneous pattern described earlier. Twenty of the subclones were homogeneous from cell to cell; however, the intensity of fluorescence was less than that in control transformants. No T-antigen negative clones were observed, though they would have been expected from the frequency of cells which showed no nuclear fluorescence in the original population of these clones of intermediate cells (about 20%).

With low frequency, T-antigen intermediate clones give rise to cells which are scored as transformants in methyl cellulose suspension and monolayer plating assays. Since plating in methyl cellulose suspension offers a simple means of obtaining clones of such cells, a number of colonies of the > 200 cell size were picked and analyzed in terms of viral T-antigen, Methocel plating and colony formation on normal monolayers (Table 5). These are referred to as methyl cellulose or MC subclones. For comparison recloned populations were also generated by random picking of colonies on plastic dishes. Randomly selected subclones give rise to visible colonies in methyl cellulose suspen-

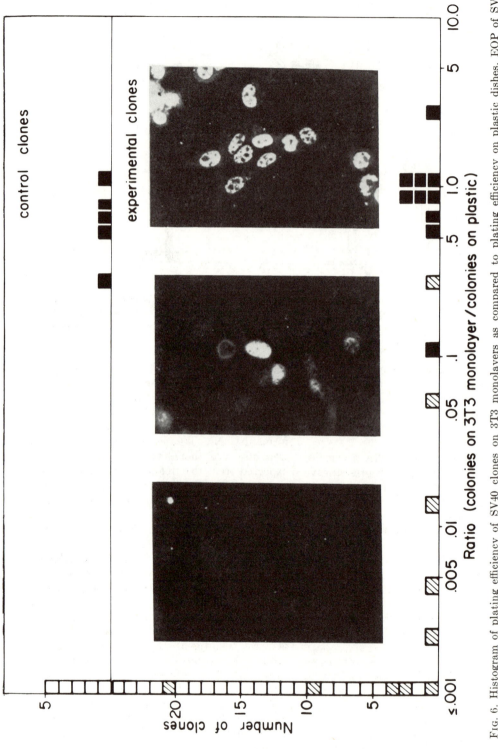

FIG. 6. Histogram of plating efficiency of SV40 clones on 3T3 monolayers as compared to plating efficiency on plastic dishes. EOP of SV40 clones was determined on plastic and on 3T3 monolayers as in Materials and Methods and the monolayer EOP divided by the EOP on plastic to give the values reported here. □, T-antigen negative clones; ▨, T-antigen intermediate clones; ■, T-antigen positive clones. The inserted photographs are examples of T-antigen staining patterns and are from left to right T-antigen negative 3T3 cells, T-antigen intermediate SVR16 cells, and T-antigen positive SV101 cells.

TABLE 4

Stability of Growth Characteristics

SVR clone No.	T-antigen			Saturation density in 10% CS			Doubling time in 1% CS			EOP in methyl cellulose			EOP on monolayers / EOP on plastic		
	1 mo.	5 mo.	5 mo. Reclone	1 mo.	5 mo.	5 mo. Reclone	1 mo.	5 mo.	5 mo. Reclone	1 mo.	5 mo.	5 mo. Reclone	1 mo.	5 mo.	5 mo. Reclone
42	−	−	−	8	12.5	10.0	24	37	47	0.001	NT[b]	0.001	0.01	NT	0.01
57	−	−	−	15	16	15	40	31	31	0.001	NT	0.001	0.01	NT	0.01
63	I*	I*	I*	9	15	15	33	29	29	0.01	0.001	0.001	0.01	NT	0.01
81	+	+	+	60	50	70	20	24	22	11	NT	20	1.0	NT	0.7
82	−	−	−	10.5	12	13.5	40	29	29	0.001	NT	0.001	0.01	NT	0.01
84	+	+	+	50	45	60	32	27	30	11	NT	28.5	2.6	NT	2.0

[a] I* Intermediate.

[b] NT, not tested.

TABLE 5

Properties of Subcloned Lines from Intermediate Clones

Clone	T-antigen	EOP in methyl cellulose (%)	EOP on monolayer / EOP on plastic
Parental clones			
13	I	0.2	0.004
32	I	2.2	0.055
Random subclones			
13–1	I	0.01	NT
13–2	I	0.11	NT
13–3	I	0.58	NT
Methyl cellulose subclones			
13MC1	+	65.0	1.2
13MC2	+	66.5	0.3
13MC3	I	102	0.5
13MC4	+	22.0	0.9
32MC1	+	16.0	1.5
32MC2	I	26.0	NT
32MC3	+	24.0	1.1
32MC4	NT	29.0	1.0

[a] NT, not tested.

sion with low frequencies, comparable to those of the parental clone. When methyl cellulose selected subclones were picked and tested, however, they plated with high efficiencies comparable to those of standard transformants. In general, these MC subclones also plated better on normal monolayers. The T-antigen staining pattern of such clones also differs from that of parental lines in that methyl cellulose subclones show more intense nuclear fluorescence. In some cases, e.g., 13MC 2, 13MC 4, heterogeneity is still seen from cell to cell. It is apparent, then, that cells which show fully transformed properties can be isolated from intermediate clones. Once such cells are obtained, i.e., by cloning in methyl cellulose, they continue to behave like standard transformants.

DISCUSSION

The transformation of mouse 3T3 cells by SV40 is a low-frequency event requiring input multiplicities of 10^4 PFU/cell to achieve 50% transformation in a standard assay (Black, 1966; Todaro and Green, 1966). In the present study we observed 10% transformation at a multiplicity of 2×10^3 PFU/cell. When random clones of cells arising out of this infection were tested in a number of transformation assays, 87.5% were detectably different from 3T3 cells. Thus while the application of a selective assay detected only a small fraction of the cells as being affected by virus, most cells were in fact stably altered by SV40.

The cell types obtained from this experiment may be classed into four general categories (Table 6). The first two are those expected from the assay, namely clones indistinguishable from 3T3 cells and clones which behave like standard transformants. The percentage of clones which are classed as standard transformants is higher than that seen when clones were first morphologically scored (10 of 40 clones vs 6 of 40 clones).

TABLE 6

CLASSES OF SV40 CLONES[a]

Class	Percent of clones	T-antigen	Saturation density in 10% CS ($\times 10^5$ cells/cm²)	Doubling time in 1% CS (hr)	EOP in methyl cellulose (%)	Colony formation on normal cells[b]
Normal	12.5	−	8.5 (7.6–9.5)	78 (55–100)	≤0.001	≤0.001
Transformed	25	+	47 (35–60)	34 (24–45)	32 (11–58)	0.94 (0.1–2.6)
Minimal-transformed	37.5	−	15 (9.5–16.5)	34 (26–49)	≤0.001	≤0.001
Intermediate-transformed	25	±	25 (15–45)	32 (29–41)	4 (0.5–14.6)	0.03 (0.001–0.20)
Control						
Normal		−	7.8 (7.9)	70 (60–86)	≤0.001	≤ .05
Transformed		+	53 (47–60)	37 (26–43)	25 (16–30)	0.7 (0.25–1.1)

[a] The mean numerical value for each parameter of transformation is reported for each class of transformant, and in parentheses the range of values seen among different clones in each class is given.

[b] Expressed as the ratio of EOP on 3T3 monolayers to EOP on plastic dishes.

Delayed transformations have previously been observed (Stoker, 1963; Todaro and Green, 1966), and could well account for the increase in transformed clones. This is quite reasonable since only the center of the colony was observed when colonies were morphologically scored and thus a small transformed sector might have been missed.

The third class of cells obtained differ from 3T3 cells in their lower doubling time in 1% calf serum and their slightly higher saturation densities. They are, however, T-antigen negative. Two facts strongly suggest that these cells are significantly different from 3T3 cells. The serum dependence of such cells is stable on recloning, and no minimal transformed cells were detected among mock controls. Also, 3T3 cells continuously passaged during the course of this experiment retained their high serum requirement. In a preliminary report (Risser and Pollack, 1974) this class of cells was referred to as serum transformants; the term "minimal transformants" better describes their behavior and will be used in this and all subsequent reports.

The last category of cell types is actually a collection of cell lines showing growth properties intermediate between those of normal cells, e.g., 3T3, and transformed cells, e.g., SV101. All intermediate lines grow well in 1% calf serum and have a lower titer of complement fixing T-antigen as well as considerable intercellular variation in antibody staining. The plating efficiency on 3T3 monolayers of T-antigen intermediate lines is approximately 30-fold reduced from that observed among standard transformants. The saturation density of T-antigen intermediate lines is about 50% of that of standard transformants, however. In their relative inability to plate on normal cells, SV40 intermediate lines resemble polyoma-transformed BHK cells (Stoker, 1964). When intermediate clones are plated in methyl cellulose suspension a small minority of the cells form large colonies (>200 cells) by the end of 3 weeks. Such segregated colonies behave in all assays like standard transformants and have increased levels of viral T-antigen.

Some lines found in this study are in some ways similar to those seen previously. Smith et al. (1971) and Scher and Nelson-Rees (1971) have described cell lines which grow to low or intermediate densities, grow in medium containing serum depleted of γ-globulin fractions, and contain viral T-antigen. Some of their lines may correspond to intermediate clones described here, though closer comparison is necessary to establish this. In addition we have detected a class of minimal transformants which lack viral T-antigen. It must be pointed out that the assay used by Smith et al. (1971) and Scher and Nelson-Rees (1971) differs from the assay used here in that in their assay cells are plated sparsely in depleted medium and the plating efficiencies measured. In our assay cells are plated at $\frac{1}{100}$ confluence in 1% calf serum and the growth rates measured. The former assay is probably more complicated in that cells are required to overcome the effects of depletion of medium on Balb/3T3 monolayers as well as removal of serum γ-globulin fractions.

It could be argued that the wide variation in transformed properties observed in this study resulted from the heteroploid condition of the 3T3 parental cell. This is probably not the reason for such a spectrum of transformed lines since SV40 infection of secondary embryonic rat cells gives rise to clones of cells showing a similar diversity of transformed expression (Pollack, Risser, Arelt, and Rifkin, unpublished observations). After SV40 infection we have obtained clones of established rat cells which grow to the same density as secondary rat embryo fibroblasts and do not plate in methyl cellulose suspension, clones which plate quite well in suspension and grow to high saturation densities, and clones intermediate in both these properties. All of these clones grow well in 1% serum.

The nonselective nature of this experiment should allow one to correlate given changes in growth behavior induced by SV40 without the preselection of a particular cell type. In general, this study shows quite good correlation of transformed properties; cells that grow to high saturation densities plate well on normal monolayers and in methyl cellulose suspension, and have high levels of SV40 specific T-antigen. Cell lines intermediate in densities plate with intermediate efficiencies on monolayers and in methyl cellulose and show intermediate levels of T-antigen. Furthermore, when transformed segregants are selected from such intermediate clones, they show coordinate increases in plating in suspension culture, plating on monolayers and levels of T-antigen.

There are, however, some significant exceptions to the coordinate nature of these changes. The great majority of the infected clones grow rapidly in 1% calf serum, regardless of the level of viral T-antigen expression. In 40% of the experimental lines this rapid growth in low serum concentrations is the major difference from 3T3 cells. A somewhat analogous situation has been seen in various revertant lines obtained by negative selection of variants from transformed cell populations (Pollack et al., 1968; Vogel and Pollack, 1973; Vogel et al., 1973; Ozanne, 1973). When revertant cells are selected for the loss of transformed properties other than growth in low serum concentration, they generally retain the serum dependence of transformed cells. When they are selected for the inability to grow in 1% serum, they also revert in many other transformed properties (Vogel and Pollack, 1973). Taken together, these results suggest the primary physiological effect of SV40 infection on mouse 3T3 cells is a decrease in the cellular requirement for serum components. This change does not require continued viral T-antigen expression, and in revertant cells is retained even though other growth properties have returned to a more normal state.

This study was undertaken to better define what is meant by SV40 "transformation" of cellular growth properties. It is apparent from our data that several patterns of transformed behavior can be induced by SV40. These may correspond to different stages of transformed development as has been previously described for polyoma infected hamster cells in vitro (Vogt and Dulbecco, 1963). We do not, however, have any data that relates the stages described here sequentially one to another. It is entirely possible these different types of cellular behavior are the result of different cellular responses to viral infection. Either explanation of this diversity of transformed behavior suggests the process of cellular transformation by SV40 is the result of several complicated interactions of cellular and viral genes and not simply the result of a single viral gene acting to directly transform a cell. Previous in vitro transformation studies have emphasized the most "transformed" cells, e.g., cells resembling SV101, in attempting to correlate virus expression with transformed cellular behavior. It appears to us, however, that less "transformed" stages, e.g., minimal transformants or intermediate transformants, may more directly relate virus expression to cellular growth alterations.

ACKNOWLEDGMENTS

We thank Mary Osborn for performing comple- ment fixation assays, Nancy Hopkins and Art Vogel for their helpful discussions, and Carole Thomason and Sue Arelt for their excellent tech- nical assistance. This work was supported by a Public Health Service predoctoral fellowship to R.R. (4-F01-GM-49503) and a grant from the National Cancer Institute (CA13106).

REFERENCES

BLACK, P. H. (1966). Transformation of mouse cell line 3T3 by SV40: Dose response relationship and correlation with SV40 tumor antigen pro- duction. *Virology* **28,** 760–763.

BLACK, P. H., CRAWFORD, E. M., and CRAWFORD, L. V. (1964). The purification of simian Virus 40. *Virology* **24,** 381–387.

BLACK, P. H., ROWE, W. P., TURNER, H. C., and HUEBNER, R. J. (1963). A specific complement- fixing antigen present in SV40 tumor and trans- formed cells. *Proc. Nat. Acad. Sci. U.S.* **50,** 1148– 1156.

CULP, L. AND BLACK, P. (1972). Contact inhibited revertant cell lines isolated from Simian Virus 40 transformed cells. III. Concanavalin A- selected revertant cells. *J. Virol.* **9,** 611–620.

HOLLEY, R. W., and KIERNAN, J. W. (1968). "Contact inhibition" of cell division in 3T3 cells. *Proc. Nat. Acad. Sci. U.S.* **60,** 300–304.

KIMURA, G., and DULBECCO, R. (1973). A tempera- ture-sensitive mutant of simian virus 40 affect- ing transforming ability. *Virology* **52,** 529–534.

MACPHERSON, I., and MONTAGNIER, L. (1964). Agar suspension culture for the selective assay of cells transformed by polyoma virus. *Virology* **23,** 291–294.

OZANNE, B. (1973). Variants of simian virus 40- transformed 3Y3 cells that are resistant to concanavalin A. *J. Virol.* **12,** 79–89.

POPE, J. H. and ROWE, W. P. (1964). Detection of specific antigen in SV40-transformed cells by immunofluorescence. *J. Exp. Med.* **120,** 121–128.

POLLACK, R., GREEN, H., and TODARO, G. (1968). Growth control in cultured cells: Selection of sublines with increased sensitivity to contact inhibition and decreased tumor-producing capacity. *Proc. Nat. Acad. Sci. U.S.* **60,** 126–133.

RENGER, H. C., and BASILICO, C. (1972). Mutation causing temperature-sensitive expression of cell transformation by a tumor virus. *Proc. Nat. Acad. Sci. U.S.* **69,** 109–114.

RISSER, R., and POLLACK, R. (1974). Biological analysis of clones of SV40-infected mouse 3T3 cells. *In* "Control of Proliferation in Animal Cells" (B. Clarkson and R. Baserga, eds.) Cold Spring Harbor Laboratory, Cold Spring Harbor, New York, in press.

ROBB, J. A., SMITH, H. S., and SCHER, C. D. (1972). Genetic analysis of simian virus 40. IV. In- hibited transformation of Balb/3T3 cells by a temperature-sensitive early mutant. *J. Virol.* **9,** 969–972.

SCHER, C. D., and NELSON-REES, W. A. (1971). Direct isolation and characterization of "flat" SV40-transformed cells. *Nature (London) New Biol.* **223,** 263–265.

SMITH, H. S., SCHER, C. D., and TODARO, G. (1971). Induction of cell division in medium lacking serum growth factor by SV40. *Virology* **44,** 359–370.

STOKER, M. (1963). Delayed transformation by polyoma virus. *Virology* **20,** 366–371.

STOKER, M. (1964). Regulation of growth and orientation in hamster cells transformed by polyoma virus. *Virology* **24,** 165–174.

STOKER, M. (1968). Abortive transformation by polyoma virus. *Nature (London)* **218,** 234–238.

STOKER, M., and ABEL, P. (1962). Conditions affecting transformation by polyoma virus. *Cold Spring Harbor Symp. Quant. Biol.* **27,** 375– 385.

STOKER, M., O'NEILL, C., BERRYMAN, S., and WAXMAN, V. (1968). Anchorage and growth regulation in normal and virus-transformed cells. *Int. J. Cancer* **3,** 683–693.

TEGTMEYER, P. (1972). Simian virus 40 deoxyribo- nucleic acid synthesis: The viral replicon. *J. Virol.* **10,** 591–598.

TEMIN, H. M., and RUBIN, H. (1958). Charac- teristics of an assay for Rous sarcoma virus and Rous sarcoma cells in tissue culture. *Virology* **6,** 669–688.

TODARO, G. J., and GREEN, H. (1963). Quantita- tive studies of the growth of mouse embryo cells in culture and their development into estab- lished lines. *J. Cell Biol.* **17,** 299–313.

TODARO, G. J., and GREEN, H. (1966). High fre- quency of SV40 transformation of mouse cell line 3T3. *Virology* **28,** 756–759.

TODARO, G. J., GREEN, H., and GOLDBERG, B. D. (1964). Transformation of properties of an established cell line by SV40 and polyoma virus. *Proc. Nat. Acad. Sci. U.S.* **51,** 66–73.

VOGEL, A., and POLLACK, R. (1973). Isolation and characterization of revertant cell lines. IV. Direct selection of serum-revertant sublines of SV40-transformed 3T3 mouse cells. *J. Cell. Physiol.* **82,** 189–198.

VOGEL, A., RISSER, R., and POLLACK, R. (1973). Isolation and characterization of revertant cell lines. III. Isolation of density-revertants of SV40-transformed 3T3 cells using colchicine. *J. Cell. Physiol.* **82,** 181–188.

VOGT, M., and DULBECCO, R. (1963). Steps in the neoplastic transformation of hamster embryo cells by polyoma virus. *Proc. Nat. Acad. Sci. U.S.* **49,** 171–179.

[CANCER RESEARCH 40, 1934–1939, June 1980]
0008-5472/80/0040-0000$02.00

Ultraviolet Light-induced Transformation of Human Cells to Anchorage-independent Growth[1]

Betsy M. Sutherland,[2,3] Janet S. Cimino, Neva Delihas, Alice G. Shih,[3] and Rowena P. Oliver[3]

Biology Department, Brookhaven National Laboratory, Upton, New York 11973

ABSTRACT

We have developed a system for ultraviolet light (UV) transformation of human embryonic cells to anchorage-independent growth. The procedure involves multiple UV irradiations, post irradiation growth, and plating in soft agar. Transformants are obtained at frequencies from 1 to 80 per 10^5 cells at UV exposures to 25 J/sq m. The resulting transformants can be subcultured on solid surfaces. The cells show crisscrossing and piling up; they reach 2- to 5-fold higher saturation densities than the parental cells. Some subcultures show increased plating efficiency in soft agar and increased life span. The susceptibility of the UV transformation process to apparent photoenzymatic reversal implies that pyrimidine dimers play a role in its induction.

INTRODUCTION

UV induces skin cancer in experimental animals and in humans (3). However, it has been difficult to develop an *in vitro* model for human UV oncogenesis, e.g., UV transformation of human cells, due to the apparent refractory nature of human cells to *in vitro* transformation with agents other than viruses. [We here use "*in vitro* transformation" in accord with the nomenclature of the terminology committee of the Tissue Culture Association (13).] Recently, however, transformation of human cells by chemicals and by ^{60}Co γ-irradiation has been reported by Kakunaga (6) and by Namba *et al.* (11), and preliminary reports of UV transformation of human cells have been presented (9, 18). Chan and Little (1) have also produced transformation of the mouse line C3H/10T½ by UV.

We thus sought to develop a system for UV transformation of human cells to anchorage-independent growth; our system involves multiple UV irradiations, cell growth before plating, and selection by growth in soft agar. The frequency of anchorage-independent growth increases as a function of increasing UV exposure. The frequency at a given UV exposure depends on the generation number of the cells.

An important tool in UV photobiology has been the "photoreactivation test" (19). Photoreactivating enzyme carries out the specific, light-dependent monomerization of cyclobutyl pyrimidine dimers in DNA (14, 15). This specificity of the enzyme for dimers allows its use as an analytical tool. If UV-induced biological damage is reversible in a true photoenzymatic reaction, then pyrimidine dimers are important in production of that

damage. Although UV transformation of rodent cells was reported in 1976 (1), the low level of photoreactivating enzyme in those cells, at best 10% that of normal human cells (23), and the dependence of photoreactivating enzyme levels on culture conditions (10, 21) have impeded its use in evaluating the role of dimers in UV transformation. We have used photoreactivation-competent cells grown under conditions favorable for photoreactivating enzyme production; in these cells, exposure to visible light immediately after UV greatly reduced the transformation frequency. These results imply that pyrimidine dimers are important in induction of UV transformation in human cells.

MATERIALS AND METHODS

Primary cultures of HESM's[4] were obtained from Flow Laboratories, Rockville, Md. Cells were grown in a Dulbecco's modified Eagle's medium (21) prepared in this laboratory and supplemented with 20% fetal calf serum (Irvine Scientific, Irvine, Calif.) Cells were checked for distribution of LDH isozymes and for karyotype to check for species and normal human chromosomal complement. Cells were checked weekly and before each experiment for the presence of *Mycoplasma* by the method of Chen (2). Cells were plated in 60-mm plastic culture dishes and allowed to grow overnight. The medium was removed, the cells were washed with two 2-ml rinses of phosphate-buffered saline (171 mM NaCl-3.36 mM KCl-1 mM Na_2HPO_4-3.68 mM KH_2PO_4), and then 1 ml of the solution was layered over the cells.

The cells were exposed to UV radiation from a low-pressure mercury arc with its principal emission at 254 nm. UV fluences were determined using a Jagger meter (5) calibrated against a Yellow Springs Instruments Company radiometer. For photoreactivation, cells were exposed to a 60-watt incandescent bulb at a distance of 21 cm. A 12-mm Plexiglas plate was placed between the cells and the bulb. Dark controls were placed in a light-proof box and placed beside the cells being illuminated to assure similarity of temperature between the 2 groups. Pilot experiments indicated that the temperatures of the 2 groups varied less than 1°. All exposures were carried out at room temperature.

The method of Macpherson (8) was used in all soft agar plating experiments. In brief, a 10-ml agar layer [80 ml 1.25% Bacto-agar, 20 ml of tryptone phosphate (Difco Laboratories, Inc.) broth, 20 ml of fetal calf serum, and 80 ml of a double strength Dulbecco's medium prepared according to Sutherland and Oliver (21)] was placed in a 60-mm plastic Petri dish and allowed to solidify. Medium was removed from the cells, and the cells were washed with two 2-ml portions of phosphate-

[1] This research was supported by Grants CA 23096 from the National Cancer Institute and NP-154B from the American Cancer Society and by the United States Department of Energy.

[2] Recipient of a Research Career Development Award (CA 00466) from the National Cancer Institute. To whom requests for reprints should be addressed.

[3] This investigation was begun while these authors were at the Department of Molecular Biology, University of California, Irvine, Calif.

Received July 30, 1979; accepted February 28, 1980.

[4] The abbreviations used are: HESM, human embryonic skin and muscle fibroblast; LDH, lactate dehydrogenase.

buffered saline. One ml of 0.05% trypsin and 0.51 mM EDTA in Hanks' balanced salt solution [containing, per liter, 0.4 g KCl, 0.006 g KH_2PO_4, 8 g NaCl, 0.35 g $NaHCO_3$, and 0.09 g $Na_2HPO_4 \cdot 7H_2O$ (pH 7.4 to 7.6)] were layered over the cells and removed as soon as the cells began to round up. A 2-ml portion of medium was added, and the cells were resuspended and counted in a hemocytometer for total cell number or for viable cells by their ability to exclude erythrosin B (6×10^{-5} g/ml). Cells were diluted so that 10^5 cells in 1 ml of medium were added to 2 ml of the agar mixture, and the resulting suspension (3 ml) was plated on a 60-mm dish containing the agar base. Dishes were scored after plating for the presence of cell clumps; each dish and lid was marked with a vertical orientation mark, and the location of each clump was marked on the dish lid with a green felt-tipped pen. Such clumps occurred rarely (0 to 2 per plate). Cells were allowed to grow for 14 days and then were scored for the presence of colonies which did not arise from cell clumps. (Clones were unambiguously distinct from the clumps, which were identified from the overlying green mark.)

Photoreactivating enzyme was determined by the rapid dimer assay of Sutherland and Chamberlin (20). In brief, cell extracts prepared by a 45-sec sonication of washed cells suspended in Buffer E (10 mM Tris, pH 7.0-0.1 mM EDTA, 0.1 mM dithiothreitol) were added to 0.2 ml 20 mM potassium phosphate buffer, pH 7.2, containing 0.1 mM dithiothreitol, 0.1 mM EDTA, and 30 to 100 pmol of UV-irradiated ^{32}P-labeled T-7 DNA. For each determination, one assay tube was placed in the dark, and another was exposed to photoreactivating light from a General Electric 150-watt spot lamp; both samples were maintained at 37° by immersion in a circulating water bath. The dimer content of the samples was then determined by nuclease digestion, adsorption of dimer-containing oligonucleotides to Norit, filtration, and counting. Photoreactivating activity is the difference in dimer content of the light and dark samples; units of photoreactivation are expressed as pmol/mg/hr. Protein concentrations were determined by the method of Lowry (7) using bovine serum albumin as the protein standard.

RESULTS

Production of Transformants. We have developed a protocol for UV induction of anchorage-independent growth by human cells. This procedure involved plating 1.0 to 3×10^5 cells/60-mm dish (Day 1) and allowing them to grow overnight at 37° in a 5% CO_2 atmosphere. The cells were washed, UV irradiated (one-third of the total exposure), and then supplied with fresh growth medium on Days 2, 3, and 4. The cells were allowed to grow (Days 5 to 9) and then they were plated in 0.33% soft agar and scored for cell clumps (Day 9). The cells were allowed to grow for an additional 14 days, and the resulting clones were counted on Day 23. Cells exposed to 254-nm radiation by this protocol grew into clones capable of anchorage-independent growth (Fig. 1b). Clones ranged from about 100 to several thousand cells. Fig. 1a shows a typical dish of unirradiated cells treated according to the procedure. These cells did not divide in the soft agar, although cellular metabolism continued, as judged by a slow drop in the pH of the agar overlay. Since we usually plated 10^5 cells/dish, we could have easily detected one clone in 10^5 cells; we estimate that these cells gave rise to clones at a frequency of approxi-

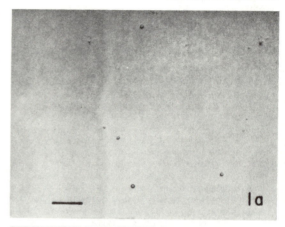

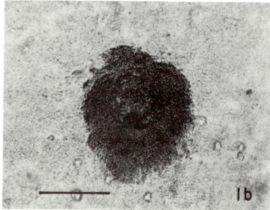

Fig. 1. *a*, unirradiated HESM cells 14 days after plating in soft agar. *b*, anchorage-independent clone rising from UV-irradiated HESM cells 14 days after plating in soft agar. *Bar*, 0.2 mm.

mately one clone per 20 to 50 plates of unirradiated cells, *i.e.*, one clone per 2×10^6 to 5×10^6 cells. Experiments in which 1, 2, or 5×10^6 cells were plated in one dish showed no clones; however, the presence of such high cell numbers might be unfavorable for growth.

We examined 2 features of this procedure to determine their necessity in the production or observation of transformants, namely, the multiple irradiation schedule and the growth period before plating in agar. Results of 2 independent experiments indicated that, although transformants were obtained when the UV was administered in a single dose, the transformation frequency was increased by a factor of about 4 when the same total UV dose was given in 2 (one-half the total dose for each exposure) or 3 (one-third the total dose for each exposure) exposures. Thus, all experiments in this series were carried out using 3 exposures. We also examined the requirement for growth of the irradiated cells before plating in agar. In an experiment in which the 5-day growth period yielded 20 transformed clones/10^5 cells (for 10 J/sq m total UV exposure), cells plated immediately after UV gave less than 2 clones/10^5 cells, and cells plated after growth for 3 days gave 6.5 clones/

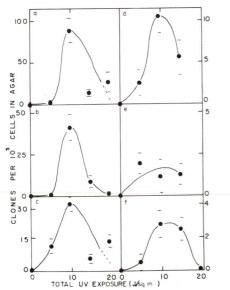

Chart 1. Number of anchorage-independent clones in HESM cells per 10^5 cells plated in soft agar as a function of UV exposure. Each *panel* represents an independent experiment; each *point* is the average of at least 5 replicate plates. *a*, passage 3; *b*, passages 10 and 11; *c*, passage 14; *d*, passage 12; *e*, passage 24; *f*, passage 25. *Bars*, S. D.

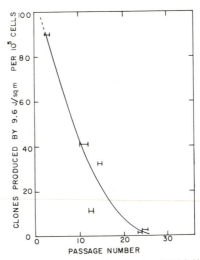

Chart 2. Production of anchorage-independent clones by 9.6 J/sq m as a function of cell passage number. The length of the *horizontal bar* represents the span of passages for cells used in an individual experiment. Data taken from Chart 1.

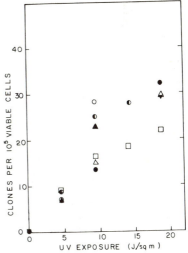

Chart 3. The number of anchorage-independent clones per 105 viable cells as a function of UV exposure. Each set of symbols represents an independent experiment; each *point* is the average of at least 5 replicate plates. Typical standard deviations include (for experiment shown as □): 0 J/sq m, ±<0.1; 5 J/sq m, ±3; 10 J/sq m, ±4; 15 J/sq m, ±4.9; 20 J/sq m, ±4.3. All values per 10^5 cells.

10^5 cells. We thus allowed the cells to grow in the plastic culture dishes for 5 days before plating in agar.

Frequency of Transformation. In a series of experiments designed to determine a dose-response curve for the frequency of UV transformants as a function of UV exposure, we found that the shapes of the dose-response curves were similar but that the frequency of transformation varied from experiment to experiment. Chart 1 shows 6 experimental dose-response curves for UV transformation of HESM's. The maximum frequency of transformation in each case occurred at about 10 J/sq m, but the value of the maximum varied from approximately 80 to 3 clones per 10^5 cells. We noted that the highest transformation frequencies seemed to occur in cells of the lowest passage number; Chart 2 shows that there is a strong dependence of transformation frequency on passage number, with almost no transformants evident after passage 25.

The data in Charts 1 and 2 represent frequency of transformants per 10^5 cells plated in agar. It was likely that the decreased transformation frequencies at higher UV exposures resulted from cell killing and that transformation frequencies based on a measure of viable cells might indicate if this were true. Since even unirradiated HESM cells form colonies very poorly on plastic or glass surfaces (10%), we regard such data as giving information only on a small fraction of the population. We have thus used vital dye exclusion (6×10^{-5} g of erythrosin B per ml); this method gives good correlation with colony-forming ability in V-79 cells, which form colonies efficiently on solid surfaces. Chart 3 shows that this is indeed the case and that the frequency of transformation increases up to at least 25 J/sq m. Under the multiple exposure protocol, virtually no loss of viable cells was observed at 5 J/sq m, with about 10 to 25% loss at 10 J/sq m.

Photoreversal of Induction of Transformation. Normal human cells grown on the Dulbecco's modified Eagle's medium used in these experiments contain photoreactivating enzyme and can photoreactivate dimers in cellular DNA. We tested HESM's for photoreactivating enzyme activity and found the specific activity of the enzyme (605 pmol/mg/hr) to be about the same as other normal human cells (600 to 700 pmol/mg/hr) (22). We then examined the photoreversibility of the induction of transformants by UV. For each of the 3 irradiation

1936

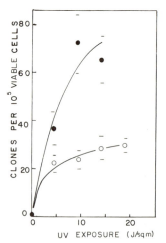

Chart 4. Reduction of frequency of anchorage-independent clones by photoreactivating light exposure of UV-irradiated HESM's. For each irradiation period, cells were exposed to 254 nm radiation, then kept in the dark (●) or exposed to photoreactivating light (○) for 15 min. Bars, S. D.

procedures, cells were exposed to UV and then kept in the dark or illuminated for 15 min with light from a 60-watt incandescent bulb. Chart 4 shows the results of such an experiment; post-UV exposure of the cells to photoreactivating illumination results in a marked decrease in the frequency of transformation produced by a given UV exposure. Photoreactivating illumination alone had no effect on the cells.

Characterization of the Transformants. Clones growing in the soft agar were picked individually, pipetted vigorously to remove the clone from the agar, and placed in 1 ml of medium in a T-25 plastic culture flask. Clones attached to the plastic surface and individual cells growing out from the clone were visible after overnight growth. The resulting cultures were first tested for contamination by bacteria, fungi, or *Mycoplasma*. We also determined the pattern of electrophoretic mobility of the isozymes of LDH to assure that the transformants were indeed human cells. Shannon and Macy (16) have pointed out that " . . . isoenzyme analysis has been useful in determining the species of presumably 'transformed' cells that have been submitted [to the American Type Culture Collection] for identification. In all cases thus far, determination of the G6PD [glucose-6-phosphate dehydrogenase] and LDH patterns have shown that such cultures contained cells predominantly (if not all) from another species." Table 1 shows the LDH isozyme patterns for HESM cells (parental stock) and for cells from transformed clone 1868 in comparison with the standards of human lung fibroblast. (WI-38), normal human skin fibroblast (Le San, American Type Culture Collection 1229), rat heart, rat muscle, Chinese hamster (V-79), and Chinese hamster ovary cells. It is clear that both the HESM stock and the 1868 transformant show LDH isozyme patterns characteristic of human cells.

We next examined the growth properties of cells resulting from the anchorage-independent clones. The transformants grew to 2- to 5-fold higher density than did the stock cells and showed both crisscrossing and piling up of cells. We also examined the serum dependence of the transformants relative

to the stock cells; no differences were found between the stock and the 5 transformants tested. Stock HESM's undergo approximately 50 to 55 doublings under our conditions. We examined 20 isolates from clones in agar for life span under normal growth conditions. One of 20 underwent no more doublings than did the stock HESM's. Three isolates underwent 70 doublings before the generation time increased greatly, and 2 had reached their 91st and 95th generation, respectively, when they were lost due to contamination resulting from sterile hood malfunction.

We assessed the ability of cultures derived from the clones in agar to plate in soft agar. Table 2 shows that the individual isolates varied greatly in replating efficiency in agar, ranging from a 1.32-fold increase, essentially the same as stock cells, to a 84-fold increase, almost 100 times that of the HESM stocks. In collaboration with S-I. Shin, Albert Einstein College of Medicine, we also tested one isolate for tumorigenicity in immune-deficient mice. Cells were injected into each of 3 nude

Table 1

LDH isozyme characterization of parental and UV-transformed human cells

Cells were suspended in phosphate-buffered saline and sonicated for 45 sec on Setting 5 of a Kontes sonicator. Samples were subjected to electrophoresis in a 5% polyacrylamide gel with a 2.5% polyacrylamide stacking gel. Electrophoresis was carried out in 25 mM Tris-0.19 M glycine at 3 ma/tube for 4 hr. LDH activity was visualized by the method of Shannon and Macy (16). All positions were determined to the nearest mm, and the fractional positions were determined by dividing the migration distance by that of the tracking dye.

Cell or tissue type	LDH isozyme position (fractional migration)					
	1	2	3	4	5	6
Standards						
Rat						
Heart	0.08	0.23	0.37	0.56	0.69	0.92
Muscle	0.08	0.24	0.40	0.58	0.72	0.92
Chinese hamster						
CHO[a]	0.03	0.92				
V-79	0.03	0.92				
Human						
WI-38	0.04	0.21	0.35	0.50	0.94	
CRL1229	0.05	0.21	0.33	0.48	0.92	
Parental						
HESM	0.06	0.21	0.34	0.46	0.89	
Transformed						
18-68	0.05	0.21	0.34	0.48	0.92	

[a] CHO, Chinese hamster ovary cells.

Table 2

Rates of growth in soft agar of 6 transformed subcultures

Three clones derived from UV-irradiated HESM stocks by soft agar plating were subcultured to give 6 cell subcultures. Without further irradiation, each subculture was treated with trypsin, counted, and plated in soft agar. After 2 weeks, the number of clones was determined; each value is the average of data obtained in at least 2 independent experiments with 6 replicate plates/subculture.

Clone	Growth in soft agar (clones/10⁵ cells)	Subculture	Growth in soft agar (clones/10⁵ cells)	Increase in cloning efficiency
1-2 fcspp4	1.9	B3A	162 ± 21[a]	84
		B3B	5.4 ± 1	3.7
107 csdpl	10.5	BG6Gla	13.9 ± 2.1	1.3
		B6G	111 ± 15	10.5
1-7 hspp3	10.5	B5B2a	362 ± 27	34.3
		B5ta22a	76 ± 7	7.2

[a] Mean ± S.D.

mice, but no tumors had appeared 3 to 4 months later, when the mice died (presumably of other causes).

DISCUSSION

We have developed a system for the UV induction of anchorage-independent growth in human cells. The system involves multiple UV exposures, a growth period before plating, and selection by growth in soft agar. It is based on 3 principles: (a) there is a high incidence of sunlight-induced skin cancer in individuals with histories of chronic sunlight exposure (12); (b) in many bacterial mutagenesis experiments, cells must be allowed a growth period for expression of a mutant gene before they are subjected to selection procedures requiring expression of that gene; and (c) the abilty to undergo anchorage-independent growth has been closely correlated with tumorigenicity (17). We examined the first 2 points to determine their necessity for production of anchorage-independent growth. We found that division of the UV into 2 or more exposures increased the transformation frequency about 4-fold. Namba et al. (11) also found that multiple ^{60}Co γ-irradiations were necessary for transformation of human (WI-38) cells. We also found that the frequency of transformation to anchorage-independent growth was increased by allowing the cells to grow for several days before plating in agar. The growth under conditions allowing anchorage may be analogous to growth under permissive conditions before selection under nonpermissive conditions. In addition to possible expression of genes permitting anchorage-independent growth, the preplating growth period may allow an increase in the number of such transformants relative to the anchorage-dependent population. For both ^{60}Co and chemical transformation of human cells, growth periods after treatment were required for appearance of transformed cells.

Are the anchorage-independent clones we observe due to induction of transforming event(s) by the UV or to selection by UV of the radiation-resistant fraction of cells which are coincidentally able to grow without anchorage? Because of the strong dependency of transformation frequency on passage number, it was not possible for us to test the transformation frequency in cultures derived from individually isolated cells, as Kakunaga (6) did for chemical transformation of human cells. However, 2 lines of evidence argue against selection as the source of the transformants: (a), we obtained transformants even at UV doses where there was little or no loss of cell viability; (b) we observed photoreactivation of induction of transformation even at these low UV exposures. This implies that the transformation event was separable from killing and thus that the anchorage-independent cells did not arise from selection of a radioresistant, anchorage-independent subpopulation.

The transformants to anchorage-independent growth show some characteristics of normal cells and some characteristics of human cells transformed by other agents. They are able to grow in soft agar; they do not maintain the monolayer configuration of the parental HESM's but rather show crisscrossing and piling up (the rapidly growing cultures of the transformants are frequently characterized by the presence of macroscopically visible 3-dimensional protrusions of cells from the cell surface); they reach higher saturation densities than do the parental cells; and they show an extended life span. However, their plating frequency in soft agar is lower than for many oncogenic cell lines; some, at least, of the cultures derived from the anchorage-independent clones have a finite life span; and the cell isolate tested apparently was nontumorigenic. It is possible that these cells represent an intermediate on the way to oncogenic transformation or that oncogenic cells were produced but were overgrown by nononcogenic cells during the post-UV proliferative period.

Photoreactivating enzyme carries out the specific light-dependent monomerization of cyclobutyl pyrimidine dimers in DNA. This specificity allows the use of photoreactivation as a test of the role of dimers in the production of biological damage. If the production of UV-induced biological damage can be reversed in a true photoenzymatic event, then pyrimidine dimers are important in the induction of that damage. Hart and Setlow (4) have presented evidence that dimers play a major role of induction of tumors by UV in the fish Poecilia formosa. However, the low level of photoreactivating enzyme in murine cells and the strong dependence of photoreactivating enzyme levels in human cells on culture conditions have deterred examination of photoreactivation of transformation in mammalian cells. We have used photoreactivation-competent cells and culture conditions favorable for production of the photoreactivating enzyme. The results of Chart 4 indicate that the induction of transformation is photoreversible. Photoreactivating light given alone or before UV irradiation did not affect the transformation frequency. These results imply that pyrimidine dimers are important in transformation of human cells UV.

REFERENCES

1. Chan, G. L., and Little, J. B. Induction of oncogenic transformation in vitro by ultraviolet light. Nature (Lond.), 264: 442–444, 1976.
2. Chen, T. R. Microscopic demonstration of mycoplasma contamination in cell cultures and cell culture media. TCA Man. 1: 229–232, 1976.
3. Epstein, J. H. Ultraviolet carcinogenesis. In: A. C. Giese (ed.), Photophysiology, Vol. 5, pp. 235–273. New York: Academic Press Inc., 1970.
4. Hart, R., Setlow, R. B., and Woodhead, A. Evidence that pyrimidine dimers in DNA can give rise to tumors. Proc. Natl. Acad. Sci. U. S. A., 74: 5574–5578, 1977.
5. Jagger, J. A small and inexpensive ultraviolet dose-rate meter useful in biological experiments. Radiat. Res., 14: 394–403, 1961.
6. Kakunaga, T. Neoplastic transformation of human diploid fibroblast cells by chemical carcinogens. Proc. Natl. Acad. Sci. U. S. A., 75: 1334–1338, 1978.
7. Lowry, O. H., Rosebrough, N. J., Farr, A. L., and Randall, R. J. Protein measurement with the Folin phenol reagent. J. Biol. Chem., 193: 265–275, 1951.
8. Macpherson, I. Soft agar techniques. In: P. F. Kruse, Jr., and M. K. Patterson, Jr. (eds.), Tissue Culture: Methods and Applications, pp. 276–280. New York: Academic Press Inc., 1973.
9. McCloskey, J. A., and Milo, G. In vitro transformation of normal diploid human cells by UV and X-rays. Abstracts, Fifth Annual Meeting of the American Society for Photobiology, San Juan, Puerto Rico, p. 110, 1977.
10. Mortelmans, K., Cleaver, J. E., Friedberg, E. C., Paterson, M. C., Smith, B. P., and Thomas, G. H. Photoreactivation of thymine dimers in uv-irradiated human cells: unique dependence on culture conditions. Mutat. Res., 44: 433–446, 1977.
11. Namba, M., Nishitani, K., Kimoto, T. Carcinogenesis in tissue culture 29: neoplastic transformation of a normal human diploid cell strain, W1-38, with Co-60 gamma rays. Jpn. J. Exp. Med., 48: 303–311, 1978.
12. Robbins, J. H., Kraemer, K. H., Lutzner, M. A., Festoff, B. W., and Coon, H. G. Xeroderma pigmentosum: an inherited disease with sun sensitivity, multiple cutaneous neoplasms, and abnormal DNA repair. Ann. Intern. Med., 80: 221–248, 1974.
13. Schaeffer, W. I. Proposed usage of animal tissue culture terms (revised 1978): usage of vertebrate cell, tissue and organ culture terminolgy. TCA Man. 4: 779–786, 1978.
14. Setlow, J. K. The molecular basis of biological effects of ultraviolet radiation and photoreactivation. Curr. Top. Radiat. Res., 2: 195–248, 1966.
15. Setlow, J. K., Boling, M. E., and Bollum, F. J. The chemical nature of photoreactivable lesions in DNA. Proc. Natl. Acad. Sci. U. S. A. 53: 1430–1436, 1965.
16. Shannon, J. E., and Macy, M. L. Enzymatic fingerprinting. In: P. F. Kruse,

Jr., and M. K. Patterson, Jr. (eds.), Tissue Culture: Methods and Applications, pp. 804–807. New York: Academic Press Inc., 1973.

17. Shin, S.-I, Freedman, V. H., Risser, R., and Pollack, R. Tumorigenicity of virus-transformed cells in *nude* mice is correlated specifically with anchorage independent growth *in vitro*. Proc. Natl. Acad. Sci. U. S. A., *72:* 4435–4439, 1975.

18. Sutherland, B. M. Photoreactivation: evaluation of pyrimidine dimers in ultraviolet radiation-induced cell transformation, Natl. Cancer Inst. Monogr., *50:* 129–132, 1978.

19. Sutherland, B. M. Photoreactivation in mammalian cells. Int. Rev. Cytol., Suppl. 8, 301–334, 1978.

20. Sutherland, B. M., and Chamberlin, M. J. A rapid and sensitive assay for pyrimidine dimers in DNA. Anal. Biochem., *53:* 168–176, 1973.

21. Sutherland, B. M., and Oliver, R. Culture conditions affect photoreactivating enzyme levels in human fibroblasts. Biochim. Biophys. Acta, *442:* 358–367, 1976.

22. Sutherland, B. M., Oliver, R., Fuselier, C. O., and Sutherland, J. C. Photoreactivation of pyrimidine dimers in the DNA of normal and xeroderma pigmentosum cells. Biochemistry, *15:* 402–406, 1976.

23. Sutherland, B. M., Runge, P., and Sutherland, J. C. DNA photoreactivating enzyme from placental mammals: origin and characteristics. Biochemistry, *13:* 4710–4715, 1974.

Proc. Natl. Acad. Sci. USA
Vol. 77, No. 12, pp. 7262–7266, December 1980
Cell Biology

Relationship between x-ray exposure and malignant transformation in C3H 10T½ cells

(carcinogenesis/epigenetic mechanism/transformation frequency)

ANN R. KENNEDY*, MAURICE FOX†, GARY MURPHY*, AND JOHN B. LITTLE*

*Laboratory of Radiobiology, Department of Physiology, Harvard University, School of Public Health, 665 Huntington Avenue, Boston, Massachusetts 02115;
and †Department of Biology, Massachusetts Institute of Technology, 77 Massachusetts Avenue, Cambridge, Massachusetts 02139

Communicated by S. E. Luria, August 12, 1980

ABSTRACT The appearance of transformed foci after x-irradiation of the C3H 10T½ line of murine cells requires extensive proliferation followed by prolonged incubation under conditions of confluence. When the progeny of irradiated cells are resuspended and plated to determine the number of potential transformed foci, the absolute yield is constant over a wide range of dilutions and is similar to that observed in cultures that have not been resuspended. In addition, for cells exposed to a given x-ray dose, the number of transformed foci per dish is independent of the number of irradiated cells. These observations suggest that few, if any, of the transformed clones occur as a direct consequence of the x-ray exposure and challenge the hypothesis that transformed foci are the clonal products of occasional cells that have experienced an x-ray-induced mutational change. Rather, it appears that at least two steps are involved. We suggest that exposure to x-rays results in a change, for example, the induction or expression of some cell function, in many or all of the cells and that this change is transmitted to the progeny of the surviving cells; a consequence of this change is an enhanced probability of the occurrence of a second step, transformation, when these cells are maintained under conditions of confluence.

The mouse-embryo-derived cell line C3H 10T½ is widely used to measure the effects of radiation and chemical carcinogens on the formation of foci of transformed cells. When cells that have been exposed to these damaging agents are allowed to grow to confluence and then further incubated for several weeks, clones of cells that have lost contact inhibition become apparent on the background of confluent cells. When cells cloned from these transformed foci are inoculated into syngeneic mice, they are found to be tumorigenic (1).

The yield of transformed foci increases as a function of x-ray exposure, up to a dose of about 400 rads (1 rad = 1 × 10⁻² gray) (1, 2). Further increases in x-ray exposure (to as much as 1400 rads) result in little or no increase in the yield of transformed foci (1, 2).

Extensive cellular proliferation is critical for the appearance of recognizable foci of transformed cells (1–3). In addition, the cell density at plating of the exposed cells has been reported to play a role in the expression of transformation (4–8).

In this paper, we describe experiments designed to examine the requirement for cellular proliferation and the effects of cell density on the development of malignantly transformed foci after an x-ray exposure of 400–600 rads. To investigate the mode of multiplication of cells having the potential to form transformed foci, x-ray-exposed cells were allowed to grow to confluence, about 13 generations, and resuspended; various

dilutions of these resuspended cells were seeded again to permit growth and assessment of the yield of transformed foci. In addition, we investigated the influence of varying the number of irradiated cells on the ultimate yield of transformed foci.

We found that, with the progeny of irradiated cells, the number of transformed foci is independent of the number of progeny cells seeded and that the number of transformed foci detected per confluent dish does not increase with the number of irradiated cells giving rise to the confluent population. We interpret these results in terms of a two-step process. The first step involves a change that is induced in all of the surviving cells that are exposed to x-rays in this dose range. This change is transmitted to the progeny of the irradiated cells and makes them more prone to undergo the second step, transformation, many generations later.

MATERIALS AND METHODS

We used the C3H mouse-embryo-derived cell line (10T½, clone 8) isolated and characterized by Reznikoff *et al.* (8, 9) and adapted in our laboratory for studies of radiation-induced transformation (1–3). Stock cultures were maintained in 60-mm petri dishes and passaged by subculturing at a 1:20 dilution every 7 days. The cells used were in passages 9 to 14. They were grown in a humidified 5% CO_2/95% air atmosphere at 37°C in Eagle's basal medium supplemented with 10% heat-inactivated fetal calf serum and antibiotics. Cells were seeded on replicate 100-mm petri dishes (1–400 viable cells each) and irradiated 24 hr later. Irradiation was carried out at room temperature with a 100-kV Philips MG-100 industrial x-ray generator operating at 9.6 mA and having a dose rate of 78 rads/min. The transformation frequencies were the same for a given radiation exposure, whether the cells remained in the dishes in which they were irradiated or were resuspended immediately after irradiation and seeded in fresh dishes (2), thus excluding any possible persisting contribution from a radiation effect on the plastic of the petri dish. Plating efficiencies in each group of routine experiments were determined from dishes seeded at a cell density one-fifth that used for the transformation assay and counted 10 days after irradiation. Types 2 and 3 foci were scored as transformants; type 3 cells are tumorigenic in 80–100% of inoculated mice, and type 2 cells are tumorigenic in 60–75% of inoculated mice (1).

RESULTS

The protocol for x-ray transformation experiments is described schematically in Fig. 1. Because the capacity for cell proliferation is critical for the phenotypic expression of x-ray transformation in 10T½ cells (1, 2), the irradiated cells were incubated to allow about 12–13 rounds of cell division before they

Cell Biology: Kennedy *et al.*

Proc. Natl. Acad. Sci. USA 77 (1980)   7263

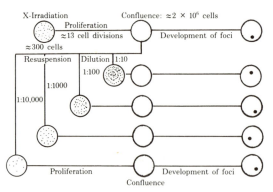

FIG. 1. Schematic description of development of malignant transformation *in vitro*. Specific time course shown in the top line is for 10T½ cells. The number of cells initially seeded was such that, taking into account the normal plating efficiency and the lethality of the x-ray treatment for 10T½ cells, about 300 viable cells per 100-mm petri dish (left) would result. The cells were irradiated the following day and then allowed to proliferate until confluence was reached (10–14 days later), when cell division ceased (center). Dense, piled-up, transformed foci appeared 4–5 weeks later (right) overlying the confluent monolayer (group A of Table 1). The lower lines describe the results of resuspension of individual dishes and reseeding at different dilutions. As each dish became nearly confluent, its contents were trypsinized, resuspended, diluted, and reseeded into new dishes (groups B–F of Table 1). The cells in each of the reseeded dishes then proliferated until confluence was reached a second time, and incubation was continued for 4 additional weeks until foci developed. In dilution experiments such as these each dish that is resuspended and diluted will give rise to one dish at each successive dilution (a total of as many as 4 dishes). Each of the 10 original irradiated dishes to be diluted was handled similarly.

reached confluence (about 2 × 10⁶ cells per 100-mm dish), and this was followed by 4–5 weeks of incubation under confluent conditions.

Several of the experiments described here involved reseeding of the progeny of irradiated cells once they had reached the confluent state of growth. Cells were seeded in replicate dishes (usually 25–40 per experiment) at a density that would result in about 300 viable cells after exposure to 400 rads [surviving fraction about 20% after 400 rads (1) and plating efficiency about 25% for 10T½ cells]. The irradiated cells were allowed to proliferate until they became nearly confluent, about 10⁶ cells per dish. At this point, some of the dishes were left undisturbed, while others were trypsinized and the cells in them were suspended and reseeded at various dilutions (Table 1). Cells harvested from each dish (irradiated) were treated separately to provide one reseeded dish at each successive dilution (see Fig. 1). The dishes containing various dilutions of reseeded cells were then returned to the incubator, and the cells were allowed to grow to confluence, followed by further incubation for the full period for the appearance of transformed foci (about 6 weeks). The frequency of appearance of transformants can be estimated from the number of foci per dish. Alternatively, if the appearance of foci is assumed to be Poisson distributed, the frequency of dishes on which no transformed foci appear can be used to calculate the mean yield. The latter procedure avoids the ambiguities that would derive from the occasional appearance of a transformed cell before a population of confluent cells is resuspended. Such a transformed cell may have proceeded through several divisions and thus be represented at a high frequency among the resuspended cells, like the "jackpot" clones of mutants described by Luria and Delbrück (10). Such presumed jackpot clones are evident in one or two of the sets

of dishes derived from the 1:10 and 1:30 dilutions of resuspended confluent cells (groups B and C, Table 1).

From the fraction of dishes in each group that did not have transformants [P(0)], the average number (λ) of transformed foci per dish was calculated according to the Poisson distribution: $P(0) = e^{-\lambda}$. The 95% confidence interval for λ was estimated on the basis of a table of exact confidence intervals for a true binomial distribution of P(0). (The confidence limits are not symmetrical around λ because the Poisson distribution is skewed.)

It was found that the total number of foci per dish was approximately constant even though the dilution range was more than three orders of magnitude. Furthermore, the number of transformed foci per dish was similar to that observed in the dishes containing the undisturbed progeny of irradiated cells. Thus, irradiated cells that had been allowed to grow through 13 generations did not appear to contain an increased number of cells capable of yielding transformed foci.

Next, we investigated the transformation yield when the initial cell densities were less than the usual number (300–400) of viable cells per dish. The effects of radiation doses of 600 rads on cultures having initial cell densities ranging from ≈1 to ≈200 cells per dish are shown in Table 2. The mean yield of transformed foci per dish was constant over a range of two orders of magnitude of initial cell densities. Thus, a plot of the relationship between the average number (λ) of transformed foci per dish and the number of viable cells initially seeded or reseeded per dish shows that there is no significant change in the number of transformed cells per dish as a function of the number of irradiated cells per dish (Fig. 2). Rather, the yield appears to be related to the numbers of cells present on the plate when confluence is reached. The implications of this observation are discussed below.

At x-ray doses of less than 400–600 rads, the yield of transformed foci depends on the dose. Thus, in an experiment in which various cell densities were exposed to doses of 100 rads, the mean number of foci per dish was about 0.1 (seven foci among 85 dishes at cell densities of 3–86 cells per dish), substantially less than was observed after a dose of 600 rads.

Unirradiated controls for these low cell density or subcultured cell populations included dishes seeded with 300 viable cells (10 dishes per experiment) and dishes seeded with 1–10 cells (two experiments involving 10 dishes each). In addition, dishes seeded with 300 cells allowed to grow to confluence were then resuspended and seeded at a 1:1000 dilution, 300 cells per dish (two experiments, 10 dishes each), and allowed the full expression period for detection of transformed foci (6 weeks). No transformation was observed in any of these control cultures.

DISCUSSION

We have previously reported that, when irradiated cells (about 300 viable cells per dish) are permitted to grow to confluence (about 2 × 10⁶ cells per dish) and are resuspended and reseeded at 300 cells per dish and incubated for 6 weeks to permit regrowth to confluence and the development of transformed foci, the yield of transformed foci per dish is the same as that on dishes containing undisturbed cells (11, 12). This observation does not contradict the notion that an occasional cell among the survivors of the x-ray exposure could have been altered in a heritable way and that the frequency of its progeny among all cells on the dish had remained unchanged during the growth to confluence. However, the observations reported in this paper show that the yield of transformed foci per dish is constant, even when the confluent cells are resuspended and reseeded over a range of ≈25 to 35,000 cells per dish. Similarly, in experiments in which the initial cell density at the time of x-ray exposure was

Table 1. Transformation as a function of dilution at confluence: 400 rads

Experiment	Dilution at confluence	Viable cells per dish, no.	Dishes, no.	Fraction of dishes containing transformants	Total foci observed, no. Type 3	Types 2 and 3	Ratio of foci to dishes	Fraction of dishes without transformants, P(0)	λ (95% confidence interval*)
Group A†									
1	—	420	14	8/14 (0.57)	5	10			
2	—	440	27	21/27 (0.78)	4	33			
3	—	247	16	4/16 (0.25)	4	6			
4	—	96	10	5/10 (0.50)	2	6			
5	—	300	17	7/17 (0.41)	8	11			
6	—	120	20	6/20 (0.30)	3	6	72/104 (0.7)	53/104 (0.51)	0.7 (0.92–0.49)
Group B‡									
1	1:10	35,000	10	8/10 (0.80)	0	83§			
2	1:10	27,000	10	4/10 (0.40)	0	11	94/20 (4.7)	8/20 (0.40)	0.9 (1.61–0.45)
Group C									
2	1:30	9,000	10	3/10 (0.30)	2	21	21/10 (2.1)	7/10 (0.70)	0.4 (1.05–0.07)
Group D									
1	1:100	3,500	10	5/10 (0.50)	4	7			
2	1:100	2,700	10	4/10 (0.40)	0	9	16/20 (0.8)	11/20 (0.55)	0.6 (1.17–0.26)
Group E									
1	1:1,000	350	10	3/10 (0.30)	3	5			
3	1:1,000	216	20	5/20 (0.25)	3	5			
4	1:1,000	60	10	4/10 (0.40)	4	4			
5	1:1,000	325	19	3/19 (0.16)	2	3			
6	1:1,000	240	15	5/15 (0.33)	3	5	22/74 (0.3)	54/74 (0.73)	0.3 (0.48–0.17)
Group F	1:10,000	25¶	10	2/10 (0.20)	2	4	4/10 (0.4)	8/10 (0.80)	0.2 (0.82–0.03)

* Estimated by using exact confidence intervals for a binomial distribution of P(0).
† Group A not reseeded; initial cell density shown.
‡ Cell density shown is for after reseeding.
§ More than half of the 83 colonies were on 2 dishes (distribution among 8 dishes: 1, 1, 4, 8, 8, 11, 19, 31).
¶ For this group, number of cells per dish was determined on the cell suspension used for the transformation assay itself.

varied, the yield of transformants per dish appeared to be constant and independent of the number of cells that were initially exposed to x-irradiation.

When irradiated cells have grown to confluence, their progeny have frequencies of transformation (number of transformants per dish) independent of dilution and equal to the frequencies observed on dishes in which the cells have remained undisturbed. This suggests that the cell alteration that results in the formation of a clone of transformed cells is not the immediate, direct consequence of the exposure to x-rays (e.g.,

Table 2. Transformation as a function of cell density: 600 rads

Cells per dish, no.	Viable cells per dish, no.	Dishes, no.	Fraction of dishes containing transformants	Total foci observed, no. Type 3	Types 2 and 3	Ratio of foci to dishes	Fraction of dishes without transformants, P(0)	λ (95% confidence interval*)
100–400	260	19	7/19 (0.37)	4	8			
	165	20	10/20 (0.50)	6	12			
	135	20	5/20 (0.25)	4	5			
	114	20	6/20 (0.30)	2	6			
	113	20	6/20 (0.30)	3	6	37/99 (0.37)	65/99 (0.66)	0.41 (0.58–0.29)
50–100†	100	17	9/17 (0.53)	6	10			
	90	14	4/14 (0.29)	2	4			
	82	13	3/13 (0.23)	1	3			
	52	20	4/20 (0.20)	3	4	21/64 (0.33)	44/64 (0.69)	0.37 (0.59–0.24)
10–30†	26	20	9/20 (0.45)	7	13			
	20	15	5/15 (0.33)	4	6	19/35 (0.54)	21/35 (0.60)	0.51 (0.87–0.27)
<5†‡	4	13	6/13 (0.46)	4	6			
	4	11	4/11 (0.36)	3	5			
	3	16	5/16 (0.31)	2	6			
	3	8	4/8 (0.50)	2	4			
	2	16	5/16 (0.31)§	2	5			
	1	5	2/5 = 0.40§	1	2			
	1	6	1/6 = 0.17§	0	1	29/75 (0.40)	48/75 (0.64)	0.45 (0.65–0.29)

* Estimated by using exact confidence intervals for a binomial distribution of P(0).
† Number of cells per dish was determined from the actual dishes used for the transformation assay and terminated at 10 days (i.e., not from a normal plating efficiency—a 1/5 dilution of the cell suspension used for the transformation assay).
‡ A few dishes that appeared to be confluent with transformed cells were counted as one focus. At this cell density, several of the foci were larger than normally observed, sometimes covering the entire dish.
§ Based on the number of dishes that reached confluence (i.e., had at least one cell per dish).

237

Cell Biology: Kennedy *et al.*

Proc. Natl. Acad. Sci. USA 77 (1980) 7265

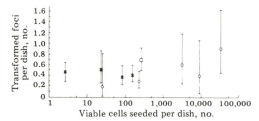

FIG. 2. Relationship between average number of transformed foci per dish and average number of viable cells seeded per dish. □, 400 rads, no dilution (group A of Table 1); ○, 400 rads, various dilution groups (groups B–F of Table 1); ✕, 600 rads, initial cell density (Table 2). Bars indicate 95% confidence interval.

a mutational event). If the occasional transformed colony were the result of a change induced by x-rays in an occasional cell, resuspension of the descendants of that cell would be expected to give rise to many transformed foci when the cells were reseeded at high cell densities. The number of transformed foci per dish would be expected to decrease when the cells were reseeded at progressively lower densities. Neither expectation was realized. The further observation that the yield of transformants (measured in terms of the number of transformants per dish) is insensitive to the initial cell density similarly contradicts the expectation that x-ray damage is the determining event in the formation of a transformed cell.

Both sets of observations can, however, be accounted for by assuming that a two-stage process is responsible for radiation-induced transformation. We assume that exposure to x-rays, in the dose range of 400–600 rads, is sufficient to produce (in all or nearly all of the surviving cells) a functional change that is inherited by their progeny. We further assume that a consequence of this functional change is an increased probability that the cells, when maintained under conditions of confluence, will sport a cell capable of forming a transformed focus. The reduced yield of transformants evident at lower x-ray doses could reflect a reduced fraction of cells manifesting this functional change.

Other investigators, calculating the transformation yield on the basis of the frequency per surviving cell, have reported an apparent increase in yield with decreasing cell densities for $10T\frac{1}{2}$ cells (4, 5, 8). This observation is similar to that reported here. It has also been suggested that a minimum colony size is necessary for potentially transformed cells to escape the suppressive effects of normal cells and express themselves as transformants (13). To account for our observations by such an effect would require a remarkable coincidence—i.e., that the suppressing effect be precisely balanced by the dilution effect over a wide range of dilutions and of growth experience (13 to 29 cellular generations).

Other investigators, examining the transformation yield from small numbers of treated cells have found large fractions of dishes having transformed foci. In transformation experiments using one cell at risk per dish and carcinogenic chemicals, Mondal and Heidelberger showed that, for C3H mouse prostate cells (14, 15), most or all of the dishes manifested transformed foci. A similar finding has recently been reported by Terzaghi and Nettesheim (16) for the development of tumorigenic potential in carcinogen-exposed cells of the rat tracheal mucosa. Although a large fraction of cultures containing tracheal epithelial cells exposed *in situ* to dimethylbenzanthracene had neoplastic potential on explantation and growth *in vitro*, only a small number of tumors developed from carcinogen-exposed cells left in the host animals (16).

We wish to suggest that the change that is detected as a transformed focus in these experiments is not a direct consequence of the exposure to radiation or chemical carcinogens. Rather, we suggest that, for the dose range that has been used, the exposure results in a cellular alteration such as a functional or metabolic change common to most or all of the surviving cells and that this change is transmitted to their progeny during subsequent growth. One consequence of this change is an enhanced probability of a second event, perhaps mutational, that is expressed as a transformed clone. This second event appears to occur for the most part during maintenance of the cultures under confluent conditions.

Evidence suggesting that the cellular alteration leading to malignancy may involve an epigenetic change has been discussed by Braun (17). The primary consequence of radiation exposure could be such an epigenetic change. The biochemical nature of this change remains, of course, as yet unspecified.

Although the suggested first step could also reflect a genetic change, a mutation common to most or all of the cells exposed to these radiation doses seems unlikely. On the other hand, physiological changes inherited through many cell generations are characteristic of differentiation processes in higher organisms. The mechanisms involved are obscure, but possible models in which an induced physiological change is inherited during clonal growth in the absence of any genetic change (18) have been described for bacterial systems.

In one such case, Novick and Weiner (19) showed that there is a concentration range of an inducer of the β-galactosidase operon of *Escherichia coli* insufficient to induce the operon genes but sufficient to maintain, for many generations, a maximum induced level of β-galactosidase in bacteria previously induced by exposure to a high concentration of inducer. A second product of the β-galactosidase operon, the β-galactoside permease, provides an explanation for the inheritance of this physiological change—i.e., that, when induced levels of the permease are present, the bacteria can concentrate the inducer even when its external concentration is too low to produce induction in uninduced bacteria. Another well-analyzed case of a persistent nongenetic change has been reported for the bacteriophage λ regulatory system in lysogens of *E. coli* (20).

The x-ray-induced functional change postulated as the initial alteration in irradiated $10T\frac{1}{2}$ cells may be a novel example of a persistent nongenetic change in a mammalian cell line.

This research was supported by Grant CA-22704 from the National Cancer Institute, Department of Health, Education and Welfare and by Grant ES-00002 from the National Institute of Environmental Health Sciences.

1. Terzaghi, M. & Little, J. B. (1976) *Cancer Res.* **36**, 1367–1374.
2. Little, J. B. (1977) in *Origins of Human Cancer*, eds. Hiatt, H. H., Watson, J. D. & Winston, J. A. (Cold Spring Harbor Laboratory, Cold Spring Harbor, NY), pp. 923–939.
3. Kennedy, A. R., Mondal, S., Heidelberger, C. & Little, J. B. (1978) *Cancer Res.* **38**, 439–443.
4. Haber, D. A., Fox, D. A., Dynan, W. S. & Thilly, W. G. (1977) *Cancer Res.* **37**, 1644–1648.
5. Haber, D. A. & Thilly, W. G. (1978) *Life Sci.* **22**, 1663–1674.
6. Bertram, J. S. (1977) *Cancer Res.* **37**, 514–523.
7. Han, A. & Elkind, M. M. (1979) *Cancer Res.* **39**, 123–130.
8. Reznikoff, C. A., Bertram, J. S., Brankow, D. W. & Heidelberger, C. (1973) *Cancer Res.* **33**, 3239–3249.
9. Reznikoff, C. A., Brankow, D. W. & Heidelberger, C. (1973) *Cancer Res.* **33**, 3231–3238.
10. Luria, S. E. & Delbrück, M. (1943) *Genetics* **28**, 491–511.

11. Kennedy, A. R., Murphy, G. & Little, J. B. (1980) *Cancer Res.* **40,** 1915–1920.

12. Kennedy, A. R. & Little, J. B. (1980) in *Radiation Biology in Cancer Research*, eds. Meyn, R. E. & Withers, H. R. (Raven, New York).

13. Bell, G. I. (1976) *Science* **192,** 569–572.

14. Mondal, S. & Heidelberger, C. (1970) *Proc. Natl. Acad. Sci. USA* **65,** 219–225.

15. Mondal, S. (1980) in *Mammalian Cell Transformation by Chemical Carcinogens,* eds. Mishra, N., Dunkel, V. & Mehlman, M. (Senate, Princeton Junction, NJ).

16. Terzaghi, M. & Nettesheim, P. (1970) *Cancer Res.* **39,** 4003–4010.

17. Braun, A. C. (1977) in *The Story of Cancer: on its Nature, Causes, and Control* (Wesley, Reading, MA).

18. Dulbecco, R. (1979) *Microbiol. Rev.* **43,** 443–452.

19. Novick, A. & Weiner, M. (1957) *Proc. Natl. Acad. Sci. USA* **43,** 553–566.

20. Calef, E., Avitabile, A., delGuidice, L., Marchelli, C., Menna, T., Neubauer, Z. & Soller, A. (1971) in *The Bacteriophage Lambda,* ed. Hershey, A. D. (Cold Spring Harbor Laboratory, Cold Spring Harbor, NY), pp. 609–620.

239

Supplementary Readings

1961–1970

Basilico, C., Y. Matsuya, and H. Green. 1970. The interaction of polyoma virus with mouse–hamster somatic hybrid cells. *Virology 41:* 295–305.

Benjamin, T. 1966. Virus-specific RNA in cells productively infected or transformed by polyoma virus. *J. Mol. Biol. 16:* 359–373.

Chen, T.T. and C. Heidelberger. 1969. Quantitative studies on the malignant transformation of mouse prostate cells by carcinogenic hydrocarbons in vitro. *Int. J. Cancer 4:* 166–178.

Dulbecco, R. 1970. Behavior of tissue culture cells infected with polyoma virus. *Proc. Natl. Acad. Sci. 67:* 1214–1220.

Freeman, A.E., C.H. Calisher, P.J. Price, H.C. Turner, and R.J. Huebner. 1966. Calcium sensitivity of cell cultures derived from adenovirus-induced tumors. *Proc. Soc. Exp. Biol. Med. 122:* 835–840.

Freeman, A.E., P.H. Black, E.A. Vanderpool, P.H. Henry, J.B. Austin, and R.J. Huebner. 1967. Transformation of primary rat embryo cells by adenovirus type 2. *Proc. Natl. Acad. Sci. 58:* 1205–1212.

Freeman, A.E., P.H. Black, R. Wolford, and R.H. Huebner. 1967. Adenovirus type 12–rat embryo transformation system. *J. Virol. 1:* 362–367.

Hanafusa, H. 1969. Rapid transformation of cells by Rous sarcoma virus. *Proc. Natl. Acad. Sci. 63:* 318–325.

Macintyre, E. and J. Ponten. 1967. Interaction between normal and transformed fibroblasts in culture. *J. Cell Sci. 2:* 309–322.

Montagnier, L. 1968. Cancerologie. Correlation entre la transformation des cellules BHK 21 et leur resistance aux polysaccharides acides en milieu gelifie. *C. R. Acad. Sci. 267:* 921–924.

Pollack, R., H. Green, and G.J. Todaro. 1968. Growth control in cultured cells. *Proc. Natl. Acad. Sci. 60:* 126–133.

Rein, A. and H. Rubin. 1968. Effects of local cell concentrations upon the growth of chick embryo cells in tissue culture. *Exp. Cell Res. 49:* 666–678.

Sachs, L. and D. Medina. 1961. In vitro transformation of normal cells by polyoma virus. *Nature 189:* 457–458.

Stoker, M. 1964. Regulation of growth and orientation in hamster cells transformed by polyoma virus. *Virology 24:* 165–174.

Stoker, M. and I. Macpherson. 1961. Studies on transformation of hamster cells by polyoma virus in vitro. *Virology 14:* 359–370.

Todaro, G.J., H. Green, and B.D. Goldberg. 1964. Transformation of properties of an established cell line by SV40 and polyoma virus. *Proc. Natl. Acad. Sci. 51:* 66–73.

1971–1980

Balk, S.D., P.I. Polimeni, B.S. Hoon, D.N. LeStourgeon, and R.S. Mitchell. 1979. Proliferation of Rous sarcoma virus-infected, but not of normal, chicken fibroblasts in a medium of reduced calcium and magnesium concentration. *Proc. Natl. Acad. Sci. 76:* 3913–3916.

Birg, F., R. Dulbecco, M. Fried, and R. Kamen. 1979. State and organization of polyoma virus DNA in transformed rat cell lines. *J. Virol. 29:* 633–648.

Brugge, J. and J. Butel. 1975. Role of SV40 gene A function in maintenance of transformation. *J. Virol. 15:* 619–635.

DiPaolo, J.A., P.J. Donovan, and R.L. Nelson. 1971. In vitro transformation of hamster cells by polycyclic hydrocarbons: Factors influencing the number of cells transformed. *Nat. New Biol. 230:* 240–242.

Freeman, A., P. Price, R. Bryans, R. Gordon, R. Gilden, G. Kelloff, and R. Huebner. 1971. Transformation of rat and hamster cells by extracts of city smog. *Proc. Natl. Acad. Sci. 68:* 445–449.

Gordon, R.J., R.J. Bryan, J.A. Rhim, C. Demoise, R.G. Wolford, A.E. Freeman, and R.J. Huebner. 1973. Transformation of rat and mouse embryo cells by a new class of carcinogenic compounds isolated from particles in city air. *Int. J. Cancer 12:* 223–232.

Green, H., R. Wang, C. Basilico, R. Pollack, T. Kusamo, and S. Salas. 1971. Mammalian somatic cell hybrids and their susceptibility to viral infection. *Fed. Proc. 30:* 930–934.

Hennings, H., D. Michael, C. Cheng, P. Steinert, K. Holbrook, and S.H. Yuspa. 1980. Calcium regulation of growth and differentiation of mouse epidermal cells in culture. *Cell 19:* 245–254.

Isom, H.C., M.J. Tevethia, and J.M. Taylor. 1980. Transformation of isolated rat hepatocytes with simian virus 40. *J. Cell Biol. 85:* 651–659.

Kelley, S., M.A.R. Bender, and W.W. Brockman. 1980. Transformation of BALB/c-3T3 cells by tsA mutants of simian virus 40: Effect of transformation technique on the transformed phenotype. *J. Virol. 33:* 550–552.

Kimura, G. and A. Itagaki. 1975. Initiation and maintenance of cell transformation by SV40: A viral genetic property. *Proc. Natl. Acad. Sci. 72:* 673–677.

Lofroth, G., E. Hefner, I. Alfheim, and M. Moller. 1980. Mutagenic activity in photocopies. *Science 209:* 1037–1039.

Martin, R. and J. Chou. 1975. SV40 functions required for the establishment and maintenance of malignant transformation. *J. Virol. 15:* 599–612.

Oey, J., A. Vogel, and R. Pollack. 1974. Intracellular cyclic AMP concentration responds specifically to growth regulation by serum. *Proc. Natl. Acad. Sci. 71:* 694–698.

Osborn, M. and K. Weber. 1975. The SV40 A gene function and the maintenance of transformation. *J. Virol. 15:* 636–644.

Price, P.J., W.A. Suk, P.C. Skeen, M.A. Chirigos, and R.J. Huebner. 1975. Transforming potential of the anticancer drug adriamycin. *Science 187:* 1200–1201.

Purchio, A.F., E. Erikson, J. Brugge, and R. Erikson. 1978. Identification of a polypeptide encoded by the avian sarcoma virus src gene. *Proc. Natl. Acad. Sci. 75:* 1567–1571.

Reddy, V.B., B. Thimmappaya, R. Dhar, K.N. Subramanian, B.S. Zain, J. Pan, P.K. Ghosh, M.L. Celma, and S.M. Weissman. 1978. The genome of simian virus 40. *Science 200:* 494–502.

Ringold, G.N., P.R. Shank, H.E. Varmus, J. Ring, and K.R. Yamamoto. 1979. Integration and transcription of mouse mammary tumor virus DNA in rat hepatoma cells. *Proc. Natl. Acad. Sci. 76:* 665–669.

Roehm, C. and A. Lipton. 1973. Depletion of serum growth factors by 3T3 mouse fibroblasts and viral transformants. *Nat. New Biol. 245:* 115–116.

Sambrook, J. 1972. Transformation by polyoma virus and simian virus 40. *Adv. Cancer Res. 16:* 141–180.

Sambrook, J. and R. Pollack. 1974. Isolation and culture of cells: Basic methodology for cell culture-cell transformation. *Methods Enzymol. 32:*583-592.

Smith, H.S., C.D. Scher, and G.J. Todaro. 1971. Induction of cell division in medium lacking serum growth factor by SV40. *Virology 44:* 359–370.

Steinberg, B. and R. Pollack. 1979. Anchorage independence: Analysis of factors affecting the growth and colony formation of wild type and dl.54/59 mutant SV40 transformed lines. *Virology 99:* 302–311.

Stoker, M.G.P. 1973. Role of diffusion boundary layer in contact inhibition of growth. *Nature 246:* 200–203.

Tegtmeyer, P. 1975. Function of SV40 gene A in transforming infection. *J. Virol. 15:* 613–618.

Weber, J. 1973. Clonal variability of adenovirus-transformed cells. *J. Cell Sci. 13:* 421–427.

Weiss, S.R., H.E. Varmus, and J.M. Bishop. 1977. The size and genetic composition of cells producing avian sarcoma-leukosis viruses. *Cell 12:* 983–992.

Wyke, J.A. and M. Linial. 1973. Temperature-sensitive avian sarcoma viruses: A physiological comparison of twenty mutants. *Virology 53:* 152–161.

Chapter 4 Mechanisms of Growth Control

Abercrombie, M. and J.E.M. Heaysman. 1954. Observations on the social behavior of cells in tissue culture. II. "Monolayering" of fibroblasts. *Exp. Cell Res.* 6: 293–306.

Holley, R.W. and J.A. Kiernan. 1968. "Contact inhibition" of cell division in 3T3 cells. *Proc. Natl. Acad. Sci.* 60: 300–304.

Balk, S.D., J.F. Whitfield, T. Youdale, and A.C. Braun. 1973. Roles of calcium, serum, plasma, and folic acid in the control of proliferation of normal and Rous sarcoma virus-infected chicken fibroblasts. *Proc. Natl. Acad. Sci.* 70: 675–679.

Rifkin, D.B., L.P. Beal, and E. Reich. 1975. Macromolecular determinants of plasminogen activator synthesis. *Cold Spring Harbor Conf. Cell Proliferation* 2: 841–847.

Collett, M.S. and R.L. Erikson. 1978. Protein kinase activity associated with the avian sarcoma virus *src* gene product. *Proc. Natl. Acad. Sci.* 75: 2021–2024.

Lee, L.S. and I.B. Weinstein. 1978. Epidermal growth factor, like phorbol esters, induces plasminogen activator in HeLa cells. *Nature* 274: 696–697.

Folkman, J. and A. Moscona. 1978. Role of cell shape in growth control. *Nature* 273: 345–349.

Farmer, S.R., A. Ben-Ze'ev, B.-J. Benecke, and S. Penman. 1978. Altered translatability of messenger RNA from suspended anchorage-dependent fibroblasts: Reversal upon cell attachment to a surface. *Cell* 15: 627–637.

Steinberg, B., R. Pollack, W. Topp, and M. Botchan. 1978. Isolation and characterization of T antigen-negative revertants from a line of transformed rat cells containing one copy of the SV40 genome. *Cell* 13: 19–32.

Lane, D.P. and L.V. Crawford. 1979. T antigen is bound to a host protein in SV40-transformed cells. *Nature* 278: 261–263.

Rohrschneider, L.R. 1980. Adhesion plaques of Rous sarcoma virus-transformed cells contain the *src* gene product. *Proc. Natl. Acad. Sci.* 77: 3514–3518.

The papers in this chapter provide data for some hypotheses that have been offered to explain and to link to each other the anchorage, serum, morphological change, and calcium dependence assays of transformation. In this way, they also provide a rational entry to the underlying molecular biology of growth control.

Normal fibroblastic cells must anchor and spread in order to grow. As they accumulate, they eventually cover their substrate. If normal cells were unable to spread on each other, then growth would cease and a monolayer would result. But do cells recognize each other's presence? **Abercrombie** and **Heaysman** show that they do, at least well enough to cease moving when they bump into each other. (For current studies of cytoplasmic organization during cell movements see Chapter 11.) Careful measurement by time-lapse photography of chick fibroblast cells migrating toward each other from two plasma clots showed that cells at the junction of the advancing edges of the two explants slow their movements. After touching, these cells did not overlap each other. Although cell division was not measured as carefully as was cell movement, Abercrombie and Heaysman hint that cell division at the junction region is reduced. In discussing the possible causes of the cessation of movement, and possibility of division, at the junction of two regions of outgrowth, they attempt to rule out local starvation by arguing that food, i.e., plasma, is equally available to cells elsewhere in each clot, and such cells continue to grow when the junctional cells stop growing and moving.

Holley and **Keirnan** directly confront this argument and neatly override it. 3T3 cells are very contact-inhibited, both for movement and for division. In Eagle's medium rather than a plasma clot, the final cell density attained by 3T3 cells was directly proportional to the concentration of serum fed to the cells. Furthermore, SV40-transformed 3T3 cells did not need as much serum as untransformed 3T3 cells to reach any given cell density. Holley and Kiernan conclude that the density assay of transformation (Chapter 3) is a reflection of this difference in serum requirement. Their report of an active factor from urine has not yet been fully followed up, but the factor's stability to boiling makes it possible that it is made up of trace metal ions or small polypeptides (Chapter 1) adherent to the high-molecular-weight protein they purified as reported in this paper.

Can all the other transformed phenotypes be the consequence of a diminished requirement for a more efficient use of serum? **Balk, Whitfield, Youdale,** and **Braun** show that normal chick fibroblasts in any cell density require calcium to initiate replication. The requirement is absolute, but the amount of calcium needed is much greater in plasma than in serum, suggesting that platelets contain compounds that enhance the capacity of normal cells to use exogenous calcium. Rous-transformed chick fibroblasts lose both their calcium requirement and any requirement for platelet factors.

Many transformed cells secrete increased amounts of proteases, in particular plasminogen activator. This protease generates plasmin from the serum proenzyme plasminogen. Plasmin, the proteolytic enzyme that breaks down fibrin clots, is therefore present in cultures of many transformed cells. Production of plasminogen activator by transformed cells raises the question of whether a differentiated property of normal cells has been amplified as a result of transformation. Plasminogen activator is produced by many normal differentiated cells as part of the body's capacity to dissolve clots. **Rifkin, Beal,** and **Reich** use avian sarcoma virus (ASV) (Chapter 3) to ask whether production of plasminogen activator is under host or viral regulation in the transformed cell. Chick cells infected with temperature-sensitive ASV appear fully transformed at low temperature and quite similar to normal cells at high temperature. Plasminogen activator production in temperature-sensitive transformed cells was extremely sensitive to temperature, with synthesis ceasing soon after shift up and reinitiating soon after shift down. Rifkin et al. also show that new plasminogen activator synthesis begins with new cellular transcription and that actinomycin blocks the expected loss of plasminogen activa-

tor after shift up. The last result is quite reminiscent of Tompkin's observation on tyrosine aminotransferase (TAT) in liver cells (Chapter 2) and argues that plasminogen activator synthesis is under both cellular and viral regulation.

The avian sarcoma virus first described by Rous (RSV) makes a 60,000-dalton protein called src. **Collett** and **Erikson** show that src is a kinase, that is, a phosphoprotein capable of transferring phosphate from adenosine triphosphate to other proteins, for example, one of the chains of immunoglobulin during immunoprecipitation. The normal cellular substrate for src is still unknown, although many candidates have been offered, most recently a set of other cellular protein kinases, including membrane-associated ion pumps. Remarkably, phosphorylations caused by many tumor virus proteins are of the amino acid tyrosine, rather than of the more common recipients serine and threonine. We know that phosphorylation often permits proteins to bind calcium and that calcium binding can in turn alter their function. Perhaps, therefore, the capacity of RSV-transformed cells to grow in low calcium will turn out to be the most direct sign of src activity, although the molecular links between kinase activity and cell growth per se remain for now obscure.

The intimate link of differentiation to growth control is most directly brought out by the simultaneous effects on both of a set of compounds called tumor promoters. Tumor promoters are a class of chemical compounds that cannot by themselves induce tumors but that dramatically increase the tumorigenic capacity of chemical carcinogens (Chapter 3). **Lee** and **Weinstein** show that EGF and the classic promoter TPA both induce plasminogen activator secretion in HeLa cells. Both inducers work at nanomolar concentrations, the low range at which hormones are active. Perhaps TPA acts in the cell membrane by mimicking the binding of EGF to EGF receptors (see Chapter 10).

Normal fibroblastic cells spread out on their substrate. Change in cell shape then occurs in the cell cycle as cells round for mitosis (see Chapter 6). To bring an entire population of cells at all points in the cycle to a single degree of spreading, **Folkman** and **Moscona** use the plastic of soft contact lenses. Cells are not able to spread at all on this plastic. Folkman and Moscona precisely varied the adhesiveness of the tissue culture dishes by applying different concentrations of this plastic. Cell spreading thus became a controlled variable while serum and medium were kept constant. They found cell shape to be tightly coupled to DNA synthesis and growth. The flatter the cells, the more they grew. The monolayer saturation density of 3T3 cells can now be explained as a consequence of packing. At a monolayer density, cells that cannot spread on each other no longer have any way to spread out enough to permit further division. Obviously, anchorage-independent transformed cells must lose this cell-shape-dependent growth control. However, the reason why excess serum at least partially relieves this cell-shape dependence in 3T3 colonies is not made any clearer by these studies.

So far, growth control has been measured as the absence of cell division. **Farmer, Ben-Ze'ev, Benecke,** and **Penman** look inside anchorage-dependent cells that are kept from division by suspension in a viscous fluid. They find that messenger RNAs (mRNA) for many proteins are in a stable but untranslatable state in suspended normal cells. Immediately upon reattachment to substrate, these mRNAs proceed to translate a set of proteins similar, but not identical, to the ones made at all times by anchored, growing normal cells. Interestingly, actin, a major component of the cell's cytoskeleton, was made in great excess upon reattachment, a time when the cytoskeleton undergoes the dramatic reorganization of cell spreading (see Chapter 11).

The remaining papers in this chapter deal with the question of the role of viral gene products in the disruption of normal growth control: Where are these proteins? Why must they be present to maintain a state of diminished serum requirement, contact inhibition, and anchorage requirement?

SV40 codes for two proteins in a transformed cell (Chapter 8). One is large T antigen, a DNA-binding phosphoprotein that may initiate DNA replication in the transformed cell. The other is small T antigen, whose role in transformation is currently unresolved. Both are translated from mRNAs derived from one RNA transcribed from integrated early SV40 DNA (see Chapter 7).

Steinberg, Pollack, Topp, and **Botchan** use revertant cell lines derived from an SV40-transformed rat cell line to study the requirement for viral gene products in the maintenance of the transformed state. If viral genes are required to maintain a particular transformed phenotype,

one class of growth-controlled revertant cells should include cell lines in which viral gene expression has been altered. The rat cell line they used was haploid for SV40; that is, only one copy of SV40 DNA was present per cell genome. Using negative selection (see Chapter 7), they selected a series of phenotypic revertant clones. All clones lacked both viral T antigens. Some also lacked viral DNA, but others showed no detectable change in viral DNA. Since the common response of these transformed cells to selection for increased growth control was loss of viral gene expression, Steinberg et al. argue that such expression is indeed necessary for most of the transformed phenotype.

It is likely that T antigen interacts with a cellular protein or proteins involved in cellular DNA synthesis in normal and in spontaneous tumor cells. Such cell proteins would have to be under strict cell-cycle control in normal cells (see Chapter 6). If their presence meant that a cell would make DNA, then they should be specifically enriched in transformed cells. **Lane** and **Crawford** describe a 54,000-dalton phosphoprotein that binds tightly to T antigen of SV40. Similar, but not identical, proteins are found in SV40-transformed cells of many species and also in tumors such as teratocarcinomas (see Chapter 5), which have never seen SV40. Perhaps in such tumors host proteins resembling T antigen will be found to interact with the 54,000-dalton protein.

Src, the protein encoded by RSV, is a kinase. **Rohrschneider** localizes this kinase to adhesion plaques, which lie under the adherent membrane of a spread cell. (Adhesion plaques are the regions of attachment of a spread cell to its substrate.) Only functional src is localized there, since temperature-sensitive src-transformed cells lose this localization at high temperature. Although the proteins phosphorylated by src in the transformed cell are still not well described, we can reasonably guess from the papers in this chapter that src in some way alters the cell's shape by direct interaction with cell surface and cytoskeletal molecules (see Chapters 10 and 11). Once again, a serious question remaining is how to link such shape changes to changes in serum requirement.

As we go to press, an important article on this subject has just been published by Spector, O'Neal, and Racker. See Supplementary Readings List for full citation.

Experimental Cell Research, 6, 293–306 (1954)

OBSERVATIONS ON THE SOCIAL BEHAVIOUR OF CELLS IN TISSUE CULTURE

II. "MONOLAYERING" OF FIBROBLASTS

M. ABERCROMBIE and JOAN E. M. HEAYSMAN

Department of Anatomy and Embryology, University College London, England

Received July 5, 1953

Many tissue culturists have noticed that in hanging drop cultures made with a liquid medium the migration of fibroblasts is restricted to the interface between cover slip and liquid. A sheet of cells which is mainly one cell thick grows out in this plane. It is generally agreed that the cells are confined to the interface because their mechanism of movement requires a solid substratum. There has not however, to our knowledge, been comment on the fact that a cell in a liquid medium culture is presented not only with the interface between the cover slip and the medium, but with another interface, between the exposed surfaces of neighbouring cells and the medium; and that this second interface seems little used by the migrating cells, at least in the early stages of a culture's life.

The absence of comment on this apparent restriction of cell movement is perhaps to be explained by an impression that the cells have little opportunity to move over each other's surfaces. They seem to be migrating at similar speeds and predominantly radially from the explant, and such a tendency to uniformity of movement must reduce the chances of two cells crossing one over the other. Measurements show, however, that the cells are not very highly correlated in their movements. We referred in our previous paper (1) to the wide variation in linear velocity between different cells at the same time in the same culture. Direction of movement also, as we shall show in a subsequent paper, is by no means uniform, although the predominant direction is away from the explant. We therefore suggested in our previous paper that in liquid medium there is some restriction on freedom of movement of fibroblasts over each other's surfaces, which is partly or wholly responsible for the tendency to form a single layer of cells at the interface, a mode of behaviour which we termed "monolayering". The purpose of the present paper is to investigate monolayering further, and to bring forward evidence that it is in fact due to a mutual restriction of movement, with perhaps far-reaching implications for the analysis of cell behaviour.

Such tendency as there is for neighbouring fibroblasts to move in the same direction at the same speed may itself, like the monolayering, be an outcome of the suggested mutual restriction of movement. At the outset it is however necessary to take account of the possibility that such correlated movement is an outcome of other influences. These other influences, if they exist, must then contribute to the occurrence of monolayering, and this would greatly complicate the investigation. The difficulty can be largely avoided if we ensure that all the fibroblasts observed do not move predominantly in the same direction by placing two explants close together in a culture. By this means we make two advancing sheets of cells move towards each other and collide in the space between the explants. This is the experimental situation we have used throughout the present work.

METHOD

Cultures were made by the hanging drop technique in equal amounts of (a) fowl plasma and (b) extract of 7-day chick embryos diluted to 25 per cent with Pannett and Compton's (7) Saline. Explants were fragments of the ventricle of 7 to 10 day old chick embryos. Two explants were placed 0.5 to 1 mm apart on each cover slip. The plasma clot between the explants was cut away before incubation. The space so caused filled immediately with exudate, and this inter-explant region was the site of most of our observations. Those on living cultures were made by phase contrast microscopy at 38°, in some instances with the help of time-lapse filming; those on killed cultures after fixation in 4 per cent formaldehyde-saline and staining in alum haematoxylin.

RESULTS

The following is a brief qualitative description, based mainly on the examination of films, of what happens between the two confronted explants in what we shall term the inter-explant area. The two sheets of cells, each a constantly changing irregular two-dimensional meshwork, looser in texture at the periphery than centrally, advance towards each other, narrowing down the empty space between. Isolated cells may cross over from one sheet to the other. Contact is then established here and there between the peripheral cells of the two sheets and steadily becomes more general. The loose texture of each periphery is gradually lost, the intercellular spaces becoming smaller. So far there is no noticeable diminution of the steady trend of the two sheets outward from their respective explants. But as empty spaces fill, concerted movement becomes more localised, except at the lateral margins of the inter-explant area, i.e. the sides not bordered by the explants. There some of the cells take part in a new trend towards the lateral unoccupied space. As the intercellular spaces become reduced to sizes considerably smaller than the area of a single cell, general trends of movement cease to be detectable, except laterally. The cells maintain only an

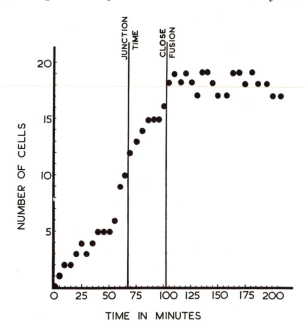

Fig. 1. The number of cells in an area of interface between two explants during the approach and junction of the two sheets of cells, recorded at five-minute intervals. "Junction time" is the time when opposing cells first make contact; "close fusion" indicates the time when the cells have become closely packed throughout the area observed.

uncoordinated oscillation. The continuous sheet of cells which now covers the inter-explant area is still largely a "monolayer", though with some overlapping of cell processes and here and there of entire cells.

For the validation and further analysis of these qualitative impressions we have undertaken several different measurements of cell behaviour in the inter-explant area.

Cell population. The number of cells observed within a square area of the interface, 0.12 mm² in extent, situated in the centre of the inter-explant area, has been counted at five-minute intervals during the approach and junction of the two sheets of cells in a number of cultures. A typical example is shown in Fig. 1. In this culture readings were started just before any cells had entered the area of observation. As cells invaded from both sides, the population of the area rapidly and steadily rose. The time at which the

Experimental Cell Research 6

first junction between cells of the two sides was established in the observed area is shown by the first vertical line ("junction time"). The population continued to rise for a period after the junction time, the cells becoming more tightly packed, but when no space bigger than a cell remained ("close fusion") it appeared to stabilise.

Eight culture areas have been followed in this way from the time of the first appearance of cells in them to long after the time of close fusion. The regression coefficients for population against time have been calculated for each culture before and after first junction. Of the readings before junction, which start as soon as cells enter the observed area, the first quarter have been discarded, because one or a few scattered peripheral cells frequently enter the area in advance of the main influx, so that the initial increase of population may be very slow. Since increase of population continues undiminished for a period after the junction time, while the larger free spaces are filling, the first hour after the junction time has been arbitrarily rejected from the calculations, and the post-junction regression based on the subsequent readings. The post-junction regression coefficient has been subtracted from the pre-junction regression coefficient for each culture. The mean of these differences was 0.21 ± 0.07 (the pre-junction coefficients being larger) which is significantly different from zero ($t = 2.88$, 7 degrees of freedom, $0.02 < P < 0.05$). Our sample of cultures therefore clearly shows a check in the rise of population in the inter-explant area after junction of the two sheets of cells has occurred (Table I).

The mean regression coefficient before junction (0.32 ± 0.07) is significantly different from zero ($t = 4.9$, 7 degrees of freedom, $P < 0.01$). After junction it is 0.11 ± 0.05 which is not significantly different from zero ($t = 2.3$, 7 degrees of freedom, $0.05 < P < 0.1$). Nevertheless we consider that more extensive data would show a significant increase in population of the post-junction group. This is partly because the necessarily arbitrary definition of the group means that some cultures will be included which are still filling up vacant spaces, partly because we have other evidence that a very slow immigration may continue even when packing has become very close, and partly because mitosis continues. The population therefore probably does not stabilise completely after junction, but it very nearly does.

A check in population increase would of course eventually occur if two monolayers of cells simply continued migration regardless of each other after they had met. In that case the population would stabilise at a minimum of approximately twice the level it had reached at first junction. Twelve cultures were available in which the level of apparent stabilisation (i.e. the mean level in each culture where at least the last 4 available readings showed no upward trend) could be compared with the population level at the time of first junction (Table I). Putting the latter at 100, the mean level of apparent stabilisation was 158 ± 7, which is significantly different from 200 ($t = 5.9$, 11 degrees of freedom, $P < 0.001$). The check in population increase appears

TABLE I

Rate of increase with time of cell population in an area between the explants, before junction of the outgrowing cells, and some time after junction. For the same cultures, and for four additional ones, the population within the area at the time of junction (mean of the readings between which junction occurred) and when approximate stabilisation has occurred (mean of several readings) are given in the last two columns.

Regression of population on time			Population	
Before junction	After junction	Difference	At junction	At stabilisation
0.373	0.186	0.187	12.0	17.8
0.530	− 0.083	0.613	16.5	18.9
0.259	0.057	0.202	11.0	16.7
0.249	0.333	− 0.084	10.5	18.0
0.349	0.078	0.271	11.0	16.8
0.411	0.018	0.393	11.0	17.5
0.201	0.048	0.153	11.0	15.5
2.208	0.230	− 0.022	7.5	14.0
–	–	–	8.5	18.1
–	–	–	10.5	14.7
–	–	–	9.5	13.5
–	–	–	11.0	17.9

therefore to be inconsistent with the superposition of two monolayers; nor was any suggestion of this detectable by watching the films.

Observation of the films makes it clear that the main check to population increase occurs after junction. It is important to consider whether any foreshadowing of the check can be detected before the first junction is made. Thirty-three cultures were followed up to their junction time for this purpose. For each culture the regression of population on time was calculated in the pre-junction period, discarding as before the first quarter of the readings, which eliminates any initially slow rise, and also omitting from the calculation the last quarter of the readings just before junction when mutual interference might be expected to make itself felt. By extrapolation from this

Experimental Cell Research 6

regression the expected population at the observed junction time was calculated and compared with the actual population at this time. The mean difference (expected minus observed population) was 0.07 ± 0.60, which is obviously not significant. We were therefore unable to detect any check to population growth before junction had occurred: the population at first junction can be effectively predicted from the slope of the steady population rise some time before. The mean observed population was 12.0 so that the 5 per cent confidence limit of the difference is about 10 per cent of the mean.

The observed check in the steady rise of population in an area between the two explants must occur either by cells starting to leave the observed area nearly as fast as they enter it, or by cells almost ceasing to enter it. One obvious possibility may be disposed of at once: that cells start to leave the area by dropping off the cover slip, and are replaced by continued migration from the explant. In fact, not one of the many hundred fibroblasts which have been followed for long periods after junction has ever been lost in this way; nor are cells found loose in the medium after culture.

Velocity. The velocity of cell movement was measured directly in living cultures by plotting the position of a given cell every five minutes from an eye-piece grid on to graph paper; a magnification in the microscope of 120 times was used, and the distance between every pair of successive points measured. Readings were made in 10 cultures some time after junction of the two sheets of cells, when the population in the inter-explant area had become closely packed. Each culture provided velocities of several cells, in two positions; (a) between the explants and (b) in the freely expanding sheet of cells, also in an area of liquid medium, at either side of the pair of explants. Cells in both situations were used only if they were in contact with 4 or 5 other cells, in order to standardise the known effects of contact on velocity (1). In situation (a), mean velocity (305 measurements) was $11.1 \pm 0.7 \ \mu$/hour; in situation (b) (241 measurements) it was $41.9 \pm 1.8 \ \mu$/hour. The difference is obviously significant. The mean speed in situation (b) compares satisfactorily with that obtained by a different method in our previous work (1) for cells in a freely moving sheet in the inter-explant area before junction; mean velocity for a contact number of 4, 4.5 or 5 was there $49.3 \ \mu$/hour.

The velocities in the inter-explant area after junction are so low that the question must arise as to whether the cells are in fact stationary, the recorded movement consisting merely of errors of observation and measurement. This is not so, since in films of cultures in which the two sheets of cells had joined closely together the cells can be seen to be moving independently, and very occasionally to change their immediate neighbours. The variations in the cell population within a given area which occur after close packing has been attained are also evidence of continued movement.

Direction of movement. The frequency with which a cell moves in a given direction has been estimated by plotting the course of the cell as for the velocity measurements (but a magnification of 320 times was used) and recording the direction of the line connecting its position at the beginning and end of each five-minute interval. The direction was noted as falling within one of four quadrants: away from the explant, towards it, to left or to right. The same cells (15 cells from 8 cultures) have been observed both before and after junction, so that it is known from which explant they have come. Of 310 observations made up to the time of junction (behaviour is unaltered right up to the time of junction), movement was away from the explant in 145, to the left in 87, to the right in 64 and towards the explant in 14. These frequencies differ of course significantly amongst themselves ($\chi^2 = 114$, 3 degrees of freedom, $P < 0.001$). Of 126 observations made more than one hour after junction of the two sheets of cells had first occurred, movement was away from the explant in 36, to the left in 33, to the right in 35 and towards the explant in 22. The distribution of movements after junction is significantly different from that obtained before ($\chi^2 = 28$, 3 degrees of freedom, $P < 0.01$). Furthermore, after junction the frequency of movement does not differ significantly as between the four directions ($\chi^2 = 4.0$, 3 degrees of freedom, $P > 0.2$). We believe, however, that more extensive data and more refined analysis would show a significant though slight excess of movement away from the explant at the expense of movement towards it, for the same reasons that led us to consider that complete stability of population after junction would not be demonstrable.

It might be argued that the observed decrease of velocity would itself produce, as an artefact, the equalisation of directional frequency, since random errors of measuring technique become more important the slower the speed. The plotted tracks of the cells give no indication, however, that after junction there is any trend away from the explant underlying the observed irregularity of direction. It may be added that after junction the cells do not circle or undergo any other detectable patterned movement which could produce the equal distribution of directions actually observed.

We may therefore conclude that after junction movement ceases to be mainly away from the explant, and comes to differ little more than randomly from equality in all directions. In combination with the slowing of velocity which occurs after junction, we have here the probable explanation of the check in population increase in the inter-explant area: cells substantially cease to enter or to leave the area.

We now turn to consider the causes of the change of behaviour at the time of junction. It is necessary first of all to dispose of the possibility that the change merely reflects the cessation of "growth" which marks the end of the life-span of a culture. (Our data on cell velocity have already provided some evidence against this.) Measurements of the radial extent of the sheet of cells which had migrated from each explant were taken shortly before junction in 6 cultures, when on the average four fifths of the inter-explant area had been covered by the migrating cells. Measurements taken at the open sides of the two explants showed no significant difference from those taken in the inter-explant area, averaging 310 μ and 324 μ respectively; the mean difference between these measurements on the 12 explants was 14 ± 27 μ. Measurements made on 8 other cultures earlier before fusion showed the same equality; growth appears to be substantially the same on all sides of an explant up to the time of junction. Three to 18 hours after junction, the radial extent at the sides of the explants averaged 640 μ and much exceeded half the distance between the two explants which averaged 407 μ; the mean difference of 12 pairs of measurements was 234 ± 95 μ. Lateral outgrowth after junction had therefore significantly surpassed inter-explant outgrowth in extent ($t = 2.5$, 11 degrees of freedom, $0.02 < P < 0.05$). But since the packing of the cells was no longer the same in the two situations, a more stringent test is given by comparing, in cultures 3 to 18 hours after junction, the number of cells in a strip of given width between the two explants with the number in a strip of similar width extending to the periphery of the outgrowth at the sides of each explant. Two such estimates were usually made at the side of each explant; the position of the strips was selected "blindly" without reference to the extent of growth. Data from 18 explants are available and only one showed fewer cells at the sides than in the inter-explant area. In the lateral strips the population averaged 1.5 times that of half the inter-explant strip, and the difference is significant ($t = 4.5$, 17 degrees of freedom, $P < 0.001$). The change of behaviour between the two explants which brings migration into this area almost to a standstill shortly after junction is not therefore merely a reflection of the ageing of the whole culture; migration clearly continues in the rest of the culture.

Can the behavioural change in question be due to a particular pattern of metabolite concentration established between the two explants? An effect of the absolute concentration of some stimulatory or inhibitory substance which the explants consume or produce is the first possibility. It must be rejected, because half-way between the two explants at the time of junction the concentration of a metabolic product cannot significantly exceed, nor the concentration of a consumable metabolite fall below, the concentration in the immediate neighbourhood of the explant. Yet concerted movement away from an explant is not inhibited in its immediate neighbourhood, either in the inter-explant area just before close fusion or at the sides after close fusion. A second possibility is that we have to deal with absolute metabolite concentrations engendered exclusively by the outgrowing cells. But these should operate as much in the densely populated part of a migrating sheet of cells as in the dense population after junction; yet one moves with an active trend and the other does not. Absolute concentrations can therefore hardly be invoked in any form. There remain concentration gradients. If the two explants either consume or produce some substance, the concentration gradient of the substance will be zero along a line between the two explants parallel to their confronted faces. An all-or-nothing directional response to such a gradient would pile the cells up at the line of zero gradient; this does not occur. We can also exclude the more likely kind of reaction to such a gradient, *viz.*, that the intensity of the directional response of the cell varies with the slope of the gradient. It this were so, the diminishing slope should slow the advance of the cells before they meet, so that junction should be delayed or incomplete for a time. This is of course precisely how monocytes behave in such a situation (2), though apparently the implications of this behaviour have not previously been pointed out. Fibroblasts behave quite differently. Their change in behaviour begins only after contact has been made, as shown for increase of cell population in an area, radial extent of outwandering, frequency of the different directions of movement and (1) velocity of displacement of individual cells.

Since the influence of diffusible substances in the medium seems incapable of providing a plausible explanation of our observations, and the same arguments would apply to any action at a distance, it seems that the actual junction of the two sheets of cells must be responsible for the change in behaviour which follows. There seems no other likely explanation than our original supposition that there is a restriction of the freedom of movement of cells over each other's surfaces. When they are faced with a situation such that in order to continue their previous trend they must move over each other, they cease their previous trend and remain as a monolayer. In order to support this hypothesis, we must now produce evidence that after junction of the two sheets of cells a tendency to form a monolayer really occurs. That is certainly the impression derived from qualitative observation, but

Experimental Cell Research 6

some overlapping of cells takes place. This makes it necessary to compare quantitatively the actual distribution of cells with the distribution which might reasonably be expected if there is no restriction on overlapping of cells.

Since after junction the cells continue moving, though slowly; and since they are not moving in an orderly way, the directions in which they are going at any moment being uncorrelated; it may be supposed that if they did not interfere with each other's movement they would be, or would soon become, randomly distributed in two dimensions, with a certain number of the cells overlapping when viewed in plan. If they are assumed to be randomly distributed in this way, the expected number of cell overlaps can be calculated for a given cell population in a given area. By a cell overlap we mean any instance where one cell is partly or wholly superimposed on another cell. For convenience we have used overlapping of nuclei rather than of whole cells. The nuclear area affects the chances of overlap occurring, and for the calculation we require therefore to know the *effective area* of a nucleus, that is, the area such that if the mid-point of another nucleus falls within it an overlap will occur. In each culture, fixed some time after junction had occurred and stained, the mean diameters of a large sample of nuclei were measured in an expanse of known area between the two explants. The mean nuclear radius, r, being used, the effective area of a nucleus is given by $\pi (2r)^2$ (assuming that nuclei are circular in plan, and disregarding the variance of their radii). The chances of overlap of any pair of nuclei in a given expanse of culture are then given by the effective area of one nucleus divided by the total area (A) of the expanse of culture observed. If the given expanse contains n cells (the number of cells is easily counted), the number of possible pairs of cells is $\dfrac{n(n-1)}{2}$. The expected number of overlaps for a given area of any culture can therefore be calculated from the expression $\dfrac{\pi(2r)^2}{A} \cdot \dfrac{n(n-1)}{2}$ and compared with the actual number of overlaps observed.

Observed and expected values were obtained for 24 areas in 9 separate cultures, fixed from 3 to 24 hours after first junction (Table II). The observed values average 0.33 of the expected.

We are indebted to Dr. N. W. Please for pointing out how this difference should be tested for significance. Since the expected value may be assumed to be Poisson distributed, its variance is equal to itself, and the difference may be tested by $\dfrac{\text{exp.} - \text{obs.}}{\sqrt{\text{exp.}}}$, the probability of which may be read in a

TABLE II

Overlapping of nuclei in the inter-explant area after junction of the outgrowths. The first column shows the total number of cells in the area observed. The effective area of one nucleus is given as a percentage of the total area observed. For calculation of expected number of overlaps, see text. The final column is the difference between observed and expected number of overlaps divided by the standard deviation of the expected number. In no case does the probability that the difference occurs by chance exceed 0.01.

No. of cells	Effective nuclear area	Expected no. of overlaps (E)	Observed no. of overlaps (O)	$\dfrac{E-O}{\sqrt{E}}$
148	0.64	69.2	20	5.91
177	0.57	88.4	22	7.06
152	0.51	58.1	10	6.31
163	0.54	71.8	24	5.64
131	0.82	70.0	17	6.34
133	0.57	50.4	12	5.41
123	0.60	44.9	9	5.36
183	0.71	118.1	35	7.65
149	0.53	58.2	22	4.75
138	0.54	51.0	16	4.90
85	0.59	21.0	6	3.27
122	0.72	53.1	20	4.54
126	0.54	42.4	8	5.28
200	0.46	90.9	56	3.66
232	0.51	136.3	70	5.68
260	0.47	158.6	75	6.64
147	0.72	77.7	25	5.98
168	0.65	91.3	26	6.83
126	0.63	49.3	12	5.31
244	0.55	161.8	50	8.79
286	0.49	200.2	85	8.14
110	0.94	56.2	32	3.23
154	0.53	62.0	38	3.05
130	1.10	92.9	39	5.59

table of normal deviates. Every observed value proves to be significantly less than its expected value at the 0.01 level of probability. Indeed, if the expected value is made half that calculated above, the population of observed values is still significantly less, i.e. the mean of the normal deviates used for testing differs significantly from zero (P<0.001). The distribution of cells in these cultures is then such that many fewer nuclear overlaps occur than is to be expected from a random distribution. The evidence therefore supports the hypothesis of a restriction of free movement of fibroblasts over each other's surfaces.

Experimental Cell Research **6**

DISCUSSION

Our results lead to the conclusion that there is some restriction of the movement of fibroblasts over each other's surfaces. It is a restriction which does not apparently operate at a distance to hinder the mutual *approach* of fibroblasts. Fibroblasts therefore freely make contact with each other, and in fact adhere together (5), forming the meshwork so characteristic of their growth. The restriction operates only after contact has been established. There is evidence (1) that the initial reaction to contact is a slight acceleration of the movement of the cells towards each other. After that there must usually occur a prohibition of further movement in this direction. The prohibition does not invariably occur in the conditions we have investigated, since there develops some overlapping of cells; and we have as yet no conclusive information as to how important the prohibition is in cultures grown for more than 36 hours, or grown in different media.

This directional prohibition of movement we shall refer to briefly as "contact-inhibition". Two hypotheses may be suggested as to how it works: (a) The surface activity by which a fibroblast moves may be inhibited in the neighbourhood of the place on its surface where an adhesion has formed with another fibroblast. (b) The exposed surface of a fibroblast parallel to the cover slip may be of such a kind that no adhesion or spreading of another cell surface can usually occur on it; thereby it would differ from the lateral surface by which the cells adhere to each other. A differentiation of some kind between exposed and mutually adherent cell surfaces, with which this phenomenon might be compared, occurs normally in epithelia. At present no decisive evidence can be offered for either of these hypotheses.

By discouraging movement in certain directions contact-inhibition may, as we previously suggested (1), be an explanation of the inverse relationship which we found between velocity and number of contacting cells. We suggested another explanation of this inverse relationship in the mutual mechanical interference caused by adhesion amongst discordantly moving cells. This is a mode of cell interaction which should be kept distinct from contact-inhibition. Distinct too are the repulsive effects discovered by Twitty (8) and his colleagues in neural crest cells. These latter effects are mediated by diffusing substances, and therefore act at a distance. We believe that a similar sort of mutual restriction of movement by diffusing substances may also be important in the behaviour of macrophages; but we have not been able to demonstrate it in the behaviour of fibroblasts.

We must now outline some of the consequences which contact-inhibition may have for the explanation of fibroblast behaviour in liquid media. In the first place, contact-inhibition must make cells move, if they move at all, predominantly away from the explant. Movement which involves overlapping will be discouraged, movement into spaces clear of cells will be preferentially selected, and hence there will be a general trend of migration into the open spaces around the explant. The commonly accepted postulation of diffusion gradients is not then required to explain why cells spend most of their time travelling in one main direction in relation to the explant—away from it. The direction of movement is controlled not by the position of the explant but by the position of free space.

A second inference now follows: whatever the original shape of a culture, it will, as it grows, tend to assume a circular form (when seen in plan). This must happen because cells tend to move towards free space; the direction of movement in which they interfere least with each other is at right angles to the culture periphery. If there is an indentation in the periphery of a culture, cells will converge into it from both sides, and proceed to fill it. If there is a projection from the periphery, cells will disperse from it to both sides, and reduce its prominence. In the first case the number of cells lying within the segment of the culture defined by the indentation will increase more rapidly than the number in neighbouring segments. Conversely, in the second case it will increase less rapidly. Only a circular form will maintain itself, because only then is each part of the periphery of the culture situated on the same radius as the free space it will invade. While contact inhibition can in this way account for much of the tendency to a circular form, further observation will probably show that the process has additional complications.

The long-standing puzzle as to why fibroblast cultures rapidly assume and regenerate to a circular form has been approached previously by experiments on cultures in fibrin clots (3, 4, 6). Without further evidence obtained in such different conditions we will not discuss this work, but we believe that here too the hypothesis of contact-inhibition may be of explanatory value. We also require additional information as to the conditions in which contact-inhibition can occur before we apply it to problems of the distribution of cells in the whole organism, such as the lamellar arrangement so often found in connective tissue, and to the starting, directing and stopping of migratory movements in embryo and adult. Some of the latter problems have been discussed by Weiss (9). Our "contact-inhibition" clearly comes within the general category of Weiss' "coaptation".

Experimental Cell Research 6

SUMMARY

1. Observations have been made on cell behaviour in the area between two embryonic chick heart explants, placed 0.5 to 1 mm apart in liquid medium tissue culture.

2. The cell population in a given area in this situation rises steadily from zero as cells from both explants invade it, but then almost ceases to rise soon after the opposing outgrowths have met. There is no detectable diminution of rate of rise of population while the cells are approaching, before the outgrowths actually join.

3. The velocity of individual cell movement after junction of the outgrowths is much slower between the explants than at either side of the explants.

4. The direction of movements of cells in the region between the explants is predominantly outward from their respective explants until the two outgrowths of cells join, when it becomes little more than randomly different from equal in all directions.

5. It is concluded that outgrowth into the area between the two explants almost ceases soon after junction has been established between the opposing sheets of cells. After this junction, the number of cells in a radial strip joining the two explants becomes exceeded by the sum of two radial strips one at the side of each explant. Culture outgrowth therefore continues laterally after it has substantially ceased between the two explants.

6. After junction the cells between the two explants tend to remain as a "monolayer". The number of instances where one nucleus overlaps another is much fewer than is to be expected if the cells were randomly distributed.

7. It is concluded that fibroblasts avoid moving over each other's surfaces. Such "contact-inhibition" of movement can explain why it is that fibroblasts normally migrate predominantly radially from an explant and that the whole culture tends rapidly to become circular in plan whatever its initial form.

This work was made possible by a grant from the Nuffield Foundation, for which we should like to express our gratitude. Our thanks are also due to Professor J. Z. Young for his criticism of the manuscript.

REFERENCES

1. ABERCROMBIE, M., and HEAYSMAN, J. E. M., *Exptl. Cell Research*, **5**, 111 (1953).
2. CARREL, A., and EBELING, A. H., *J. Exptl. Med.*, **36**, 365 (1922).
3. EPHRUSSI, B., *Arch. anat. microscop.*, **29**, 95 (1933).
4. FISCHER, A., *Arch. Pathol. Anat. Physiol.* (*Virchow's*), **279**, 94 (1930).
5. KREDEL, F., *Bull. Johns Hopkins Hosp.*, **40**, 216 (1927).
6. MAYER, E., *Wilhelm Roux' Arch. Entwicklungsmech. Organ.*, **130**, 382 (1933).
7. PANNETT, C. A., and COMPTON, A., *Lancet*, **1**, 381 (1924).
8. TWITTY, V. C., *Growth Symposium*, **9**, 133 (1949).
9. WEISS, P., *Quart. Rev. Biol.*, **25**, 177 (1950).

Reprinted from the Proceedings of the National Academy of Sciences
Vol. 60, No. 1, pp. 300–304. May, 1968.

"CONTACT INHIBITION" OF CELL DIVISION IN 3T3 CELLS*

By Robert W. Holley and Josephine A. Kiernan

BIOCHEMISTRY AND MOLECULAR BIOLOGY, CORNELL UNIVERSITY, ITHACA, NEW YORK

Communicated by Renato Dulbecco, February 26, 1968

The 3T3 cell, an established line of mouse fibroblast cell, has been considered to be extremely sensitive to "contact inhibition" of cell division. Under the usual culture conditions, 3T3 cells grow rapidly in sparse culture, but cell division stops after the cells become confluent, at approximately 10^6 cells per 6-cm dish. The cell monolayer has a typical "cobblestone" appearance. Todaro, Lazar, and Green[1] have described studies of the effect of serum on cell division in these "contact-inhibited" cells. The addition of serum to an inhibited culture leads to a rise in RNA synthesis, followed in a few hours by protein synthesis, and eventually by some DNA synthesis. Todaro, Lazar, and Green[1] have inferred that a factor in serum overcomes "contact inhibition" of cell division. In the experiments described below, we find that the characteristic "contact-inhibited" cell density observed for 3T3 cells is a fortuitous result of growing the cells in a medium that contains 10 per cent calf serum. The final cell density, after cell division stops, is directly proportional to the amount of serum added to the medium.[2] The experiments suggest that serum contributes a factor or factors required by 3T3 cells for cell division. Viral-transformed 3T3 cells have a greatly reduced requirement for the serum factor(s).

Whether serum factors offer an explanation of "contact inhibition" of cell division in other instances remains to be determined. It is pertinent that Temin[3] has found that an insulin-like factor in serum is required for cell division by cultured chick cells. Transformation of chick cells by Rous sarcoma virus lowers the requirement for this serum factor.

Materials and Methods.—The 3T3 cell line was obtained from Dr. Marguerite Vogt. The cell line had been obtained originally from Dr. Howard Green and had been cloned recently by Dr. Vogt to maintain the typical "cobblestone" appearance. During prolonged culture, 3T3 cells gradually lose their high requirement for serum and grow to higher cell densities. Therefore, the cell line was maintained at $-90°C$, and cells used in the experiments were not cultured over 8 weeks. The cells were grown in enriched Eagle's medium, as used in Dulbecco's laboratory.[4] To count the cells, the medium was removed from the dishes, the cell layer was trypsinized with half the concentration of trypsin used during transfer of the cells, and the cells were counted in the trypsin solution by means of a hemacytometer. Counts were on duplicates. Experiments were replicated at least three times. The standard error observed for replicate counts was approximately 10%.

Assay for growth factor: Approximately 10^5 3T3 cells were plated per 6-cm plastic dish in 5 ml of medium with 6% calf serum. Solutions to be assayed were added 24–48 hr later. (The solutions to be assayed were sterilized by 3-min UV irradiation with a germicidal lamp rather than by filtration, since the growth factor appears to be adsorbed on Millipore filters under some conditions.) Counts of the final number of cells per dish were made at 5 days, usually about a day after growth had stopped.

Partial purification of the growth factor from human urine: The urine was frozen immediately, to avoid microbial action, and was lyophilized. The residue was dissolved in water to give one tenth the original volume, and the solution, with suspended solids, was dialyzed overnight at 4°C against 0.05 N sodium chloride, in the presence of chloroform.

The dialyzed solution was centrifuged to remove insoluble material and was lyophilized. The residue from the second lyophilization was redissolved in water to give 1% of the original volume of urine. Solids were removed by centrifugation, and approximately 4.5 ml of the clear solution was chromatographed on a 2 × 30-cm column of Sephadex G 25 (Fine) packed in 0.1 N NaCl. The active material was excluded by the gel and came off the column in association with the initial ultraviolet-absorbing light yellow band. Fractions comprising the initial peak (total of 15–20 ml) were combined, and the active material was adsorbed on a 0.7 × 25-cm column of DEAE-cellulose (Whatman, micro-granular, DE32) packed in 0.1 M Tris–chloride buffer, pH 7.5. The column was eluted with a linear gradient prepared from 40 ml of 0.1 M Tris–chloride, pH 7.5, and 38 ml of 1 N NaCl in the Tris buffer. The activity for growth of 3T3 cells came off the column just after the main ultraviolet-absorbing band, at approximately 0.5 N NaCl. Solutions of the growth factor were stored at −20°C.

Results and Discussion.—Preliminary experiments, in which serum was added after the 3T3 cells had become confluent, were consistent with the results described by Todaro, Lazar, and Green.[1] However, interpretation of the results was changed by experiments in which extra serum was added early in the growth of the cultures. Figure 1 shows growth curves of 3T3 cells in media that contained 10, 20, and 30 per cent calf serum. The final cell count varies directly with the amount of serum added.[2] The growth curves in the presence of high serum show no evidence of a break in growth rate when the cells become confluent at approximately 6 × 10⁵ cells per dish.

If 3T3 cells are grown in medium that contains 30 per cent calf serum and the medium is changed frequently, the cells grow to a density of approximately 6 × 10⁶ cells per dish before the cell layer detaches. At this high cell density, which is approximately ten times the density of a confluent monolayer, the 3T3 cells are very tightly packed and are piled on top of each other.

It is clear, therefore, that 3T3 cells grow readily to cell densities far above the density of a confluent monolayer. Nevertheless, different interpretations of the

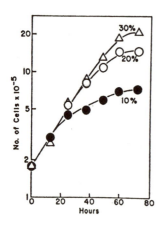

FIG. 1.—Growth curves of 3T3 cells in media containing 10, 20, and 30% calf serum.

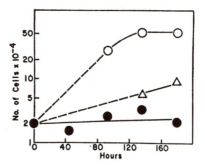

FIG. 2.—Growth curves of 3T3 cells in medium depleted by growth of 3T3 cells until growth stopped: *solid circles*, without fresh calf serum; *open circles*, with 10% fresh calf serum; *triangles*, depleted medium replaced at 20 hr and at 90 hr with medium depleted by a 3-day exposure to confluent 3T3 cells.

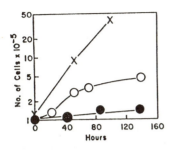

FIG. 3.—Growth curves in medium depleted by growth of 3T3 cells until growth stopped; *solid circles*, 3T3 cells without fresh calf serum; *open circles*, 3T3 cells with 10% fresh calf serum; *crosses*, SV3T3 cells without fresh calf serum.

results are possible. One interpretation is that serum contains a growth factor that is required by 3T3 cells. Alternatively, serum may contain a factor that overcomes contact inhibition. To distinguish between these two alternatives, studies were made of the growth of sparse 3T3 cells in depleted medium. As shown in Figures 2 and 3, sparse 3T3 cells, plated at approximately 10^4 or 10^5 cells per dish, grow very slowly in medium taken from confluent 3T3 cells, unless fresh calf serum is added. In contrast to the behavior of 3T3 cells, SV 40-viral-transformed 3T3 cells (SV3T3 cells) grow very well in the depleted medium (Fig. 3).

In most of these experiments, sparse cells were plated in medium that had been depleted by growth of 3T3 cells until growth had stopped. Growth of sparse cells is significantly better in medium that has been depleted by a three-day exposure to confluent cells (Fig. 2), though growth is still poor compared with that observed after the addition of serum. It is possible that the method of depletion of the medium is responsible for the variation between our results and those of Todaro, Lazar, and Green[1] that sparse cells grow to confluency in depleted medium.

If serum contains a growth factor that is required by 3T3 cells, it should be possible to limit growth at a cell density below the confluent cell density simply by limiting the amount of serum in the medium. This can be done easily. In medium that contains 1 per cent serum, growth of 3T3 cells stops at approximately 10^5 cells per dish.

The growth of 3T3 cells under the usual culture conditions thus seems to be limited by the exhaustion of one or more factors in serum. The active material in calf serum is nondialyzable, and appears to be of relatively high molecular weight (approximately 100,000) by gel filtration. Activity is lost on pronase treatment of the serum. Serum heated in a boiling water bath forms a semisolid product that is inactive in supporting the growth of 3T3 cells, but there is a possibility that active material is trapped within the gel.

Mouse serum has been found to be approximately ten times as active as calf serum, and rat serum has intermediate activity, suggesting that the active material may show some species specificity. Serum from adrenalectomized rats and from thyroidectomized rats has normal activity. Serum from hypophysectomized rats has approximately half the activity of normal rat serum one week after hypophysectomy. Assays of a number of commercial preparations of hormones, including ACTH, follicle-stimulating hormone, gonadotropin, growth hormone, insulin, luteinizing hormone, oxytocin, and prolactin, have been negative.

Commercial thyrotropic hormone (Sigma Chemical Co.) has been found to contain considerable activity. It is approximately 100 times as active as calf serum, on a milligram protein basis; however, the fact that the amount of the preparation required to give a significant growth response (approximately 0.1 unit of thyrotropic hormone activity) is at least 100 times the amount of thyrotropic hormone expected[5] in the volume of serum that has equivalent activity suggests that the active material is an impurity in the thyrotropic hormone preparation. The active material in the commercial thyrotropic hormone preparation has the same gel-filtration properties as the active material in calf serum. The activity is destroyed by pronase. In contrast with serum, a solution of the thyrotropic hormone preparation can be heated in a boiling water bath ten minutes without coagulation, and most of the activity for growth of 3T3 cells survives.

Commercial human chorionic gonadotropin (Sigma Chemical Co.) also contains activity. Again, the amount of activity is low enough to suggest that the active material is an impurity in the preparation.

The fact that human chorionic gonadotropin is isolated from pregnant urine suggested that the active material might be a normal constituent of urine. Urine turns out to be an excellent source. The activity of rat urine and human urine, on a volume basis, is approximately equal to that of serum, though urine is toxic at high levels. The active material in urine has been concentrated and partially purified by lyophilization, dialysis, lyophilization, gel filtration, and DEAE-cellulose chromatography as described in the *Materials and Methods* section. Table 1 summarizes the results. With 3T3 cells, the isolated material gives a growth response at the microgram level. The activity is destroyed by pronase, but is relatively stable to heating at 100°. The active material isolated from urine is excluded by Sephadex G25, but has a significantly lower molecular weight than the active material in serum.

The maximum growth of 3T3 cells obtained by the addition of the growth factor isolated from urine or from commercial thyroptropic hormone is less than the maximum growth that can be obtained by the addition of calf serum. This suggests that there are other factors in calf serum that are required by 3T3 cells. This possibility is being investigated.

Summary.—Evidence is presented that under the usual conditions for culture of 3T3 cells, exhaustion of essential growth factor(s) present in the serum in the medium is responsible for cessation of growth (contact inhibition of cell divi-

TABLE 1. *Purification of growth factor from urine.*

Additions to 5 ml medium	A_{280} units*	No. of cells per 6-cm dish ($\times 10^{-5}$)	Relative activity†
None (control)	—	6.6	—
0.5 ml calf serum	38	16.4	1
0.2 ml human urine	15	11.5	1.3
0.02 ml after second lyophilization	2.3	11.8	9
0.01 ml Sephadex-G25 column fraction	0.02	9.0	450
0.02 ml DEAE–cellulose column fraction	0.004	9.6	2900

* One unit of absorbance at 280 mμ is defined as the amount of material that gives an absorbance of 1.0 at 280 mμ when dissolved in 1.0 ml and with a light path of 1.0 cm.

† The increase in number of cells (above the control) per A_{280} unit of material, with the increase observed for calf serum set equal to one.

sion) as the cells become crowded. Viral-transformed 3T3 cells have a greatly reduced requirement for the serum factor(s) and thus do not show "contact inhibition" of cell division. Material that is active in stimulating the growth of 3T3 cells has been found in commercial preparations of thyrotropic hormone and human chorionic gonadotropin, and in urine. A procedure for partial purification of the active material from human urine gives material that shows growth-stimulating activity at the microgram level.

* The work was started in Dr. Renato Dulbecco's laboratories at the Salk Institute for Biological Studies. The advice of Dr. Dulbecco and Dr. Marguerite Vogt is gratefully acknowledged. The work was supported by the National Science Foundation.

[1] Todaro, G. J., G. K. Lazar, and H. Green, *J. Cell. Comp. Physiol.*, 66, 325 (1965).

[2] These results are in agreement with a recent report by Todaro, G., Y. Matsuya, S. Bloom, A. Robbins, and H. Green, in *Growth Regulating Substances for Animal Cells in Culture*, ed. V. Defendi and M. Stoker (Philadelphia: Wistar Institute Press, 1967), p. 87.

[3] Temin, H. M., *J. Cell Physiol.*, 69, 377 (1967); in *Growth Regulating Substances for Animal Cells in Culture*, ed. V. Defendi and M. Stoker (Philadelphia: Wistar Institute Press, 1967), p. 103.

[4] Vogt, M., and R. Dulbecco, these PROCEEDINGS, 49, 171 (1963).

[5] Robbins, J., J. E. Rall, and P. G. Condliffe, in *Hormones in Blood*, ed. C. H. Grace and A. L. Bacharach (New York: Academic Press, 1961).

Reprinted from

Proc. Nat. Acad. Sci. USA
Vol. 70, No. 3, pp. 675–679, March 1973

Roles of Calcium, Serum, Plasma, and Folic Acid in the Control of Proliferation of Normal and Rous Sarcoma Virus-Infected Chicken Fibroblasts

(heat-inactivated plasma/autonomy/wound hormone)

S. D. BALK, J. F. WHITFIELD, T. YOUDALE, AND ARMIN C. BRAUN

Division of Biological Sciences, National Research Council of Canada, Ottawa, Ontario K1A OR6; and The Rockefeller University, New York, N.Y. 10021

Contributed by Armin C. Braun, December 5, 1972

ABSTRACT   In a culture medium of pH 7.4 and a folic acid concentration of 100 μg/liter that contains 5% heat-inactivated chicken plasma rather than serum, the rate of proliferation of normal chicken fibroblasts is determined by the concentration of calcium. Proliferation, rapid when the calcium concentration is physiological, decreases when the calcium concentration is reduced. At a very low calcium concentration, in this culture medium, normal fibroblasts are maintained without proliferation, whereas those infected with Rous sarcoma virus proliferate rapidly. This proliferative inactivity of normal fibroblasts does not involve contact-inhibition, since the effect is observed at low, as well as higher, culture densities. When a physiological amount of calcium is added to cultures of normal fibroblasts that have been maintained in very low calcium–plasma medium for 3 days, labeled thymidine uptake and protein synthesis are strongly stimulated, and cell division follows. The use of heat-inactivated chicken serum, instead of plasma, in this medium appears to strongly sensitize normal fibroblasts to the mitogenic action of calcium.

In a plasma-containing culture medium of physiological calcium concentration and a folate concentration of 5 μg/liter, neither normal nor Rous sarcoma virus-infected fibroblasts proliferate to an appreciable extent. The use of serum, however, instead of plasma results in rapid proliferation of both normal and infected cells, as does increase in the folate concentration of the plasma-containing medium to 100 μg/liter.

The fact that while normal fibroblasts are maintained without proliferation in low calcium–plasma medium, Rous sarcoma virus-infected fibroblasts proliferate rapidly, indicates that the effect of calcium is regulatory rather than permissive. These results suggest that the proliferation of normal fibroblasts is initiated by a cellular function involving calcium, and that the autonomous proliferation of the neoplastic fibroblasts results either from increased calcium uptake or from an alteration or a bypass of that function. The results also suggest that serum contains a mitogenic factor(s) not present in plasma, possibly a "wound hormone" for fibroblasts.

A large amount of evidence has appeared that indicates that calcium and the principal hormones of the calcium homeostatic system are major physiological regulators of the proliferation of thymic lymphoblasts and bone-marrow erythroid cells, and perhaps of liver parenchymal cells and peripheral lymphocytes (1). It has also been shown, by S.D.B., that calcium ion concentration *in vitro* controls the proliferation of normal chicken fibroblasts in chicken plasma-containing medium, but does *not* control the proliferation of these cells after they have been infected by the Schmidt–Ruppin strain of Rous sarcoma virus (2, 3). The ability of calcium to regulate normal fibroblast proliferation is obscured in con-

ventional, but less physiological, serum-containing medium (2, 3). This effect is due to the ability of serum to strongly stimulate the proliferation of normal cells.

Since a considerable alteration of a calcium-dependent proliferative control system follows infection of cells by an oncogenic virus, elucidation of the nature of the action of calcium could yield a greater understanding of the loss of proliferative control that is the essence of the neoplastic state. Therefore, in the present study we have located the calcium-sensitive stage of the normal fibroblast's growth-division cycle. Moreover, we will show that the mitogenic properties possessed by serum, but not by plasma, stem from an apparent sensitization of normal cells to the mitogenic action of calcium, as well as from a permissive effect on the proliferation of both normal and infected cells.

MATERIALS AND METHODS

The basic materials and methods used in the experiments reported here were described (2, 3).

Incubation Conditions—Carbon Dioxide. Some of the proliferative effects described previously (2, 3) were of variable magnitude. It was determined that this variability was related to differences in culture medium pH, and that these differences involved, in turn, inadequate control of CO_2 tension in the incubation chambers. For the experiments reported here, the CO_2 content of the humidified air–CO_2 mixture that passed through the incubator chambers was monitored with a Fyrite CO_2 Indicator (Bacharach Instrument Co., Pittsburgh, Pa.) and maintained at 5%. With 5% CO_2 in the incubation atmosphere, the pH of the culture media used in these experiments was 7.4.

Synthetic Medium. The components of the synthetic medium used in these experiments, as modified from those described previously (2, 3), are listed in Table 1.

Calcium-free synthetic medium was prepared by mixture of the following components, in appropriate proportions: glass-distilled water; $10\times$ salt solution, including lactate and pyruvate; $50\times$ amino acids in 1 N HCl; $100\times$ glutamine, asparagine·H_2O and tryptophan; $1000\times$ vitamins; $1000\times$ Ampicillin; 100 mM $MgSO_4·7H_2O$; 1 N NaOH (NaOH was added to neutralize the HCl that was used to dissolve the amino acids. The amount of NaCl in the $10\times$ salt solution was reduced so that the NaCl content of the finished medium was that given in Table 1). After mixture, the completed,

NRCC-12924

TABLE 1. *Composition of synthetic medium, mg/liter*

NaCl 6800	Cystine 24	Tyrosine 36
NaHCO₃ 2200	Glutamic acid 10	Valine 46
KCl 445	Glycine 17	Thiamine·HCl 0.10
NaH₂PO₄·H₂O 101	Histidine 31	Pyridoxal·HCl 0.05
CaCl₂ 0–160	Hydroxyproline 10	Nicotinamide 0.10
MgSO₄·7H₂O 168	Isoleucine 52	Riboflavin 0.01
Glucose 4000	Leucine 52	d-Biotin 0.01
Lactic acid 100	Lysine 47	Calcium pantothe-
Sodium pyruvate 6	Methionine 15	nate 0.15
Glutamine 60	Phenylalanine 32	Choline chloride 1
Asparagine 10	Proline 30	Folic acid 0.005*
Arginine 70	Serine 12	Inositol 2
Alanine 30	Threonine 48	Ampicillin 100
Aspartic acid 10	Tryptophan 10	

* In medium for primary and secondary cultures, folic acid was used at 5 μg/liter. In experiments, folate was used at concentrations of 5–100 μg/liter. All experimental media also contained hypoxanthine and xanthine, each at 10 μM.

calcium-free synthetic medium was thoroughly bubbled with 5% CO_2 in air.

Calcium was added to culture media from a 100 mM CaCl₂ stock. Folic acid in excess of 5 μg/liter was added, by dilution, from a 100 mg/liter sodium folate stock. Hypoxanthine and xanthine were included in media used for experiments at 10 μM each, a physiological concentration. They were added from a 10 mM stock in 0.1 N NaOH.

Preparation, Growth, and Passage of Cultures; Infection with Schmidt-Ruppin Rous Sarcoma Virus. Primary and secondary cultures were grown, and secondary cultures were infected with Rous sarcoma virus, in synthetic medium containing 1.44 mM calcium, with 10% heat-inactivated commercial chicken serum. Primary cultures were prepared, from the pectoral muscles of 8-week-old COFAL-negative cockerels, by the method given previously (2, 3). No supplementary

folic acid was used for infection of secondary cultures with sarcoma virus.

Secondary cultures of normal and sarcoma virus-infected fibroblasts were passaged to yield the tertiary cultures of normal and infected cells used for experiments. Cells for tertiary cultures were seeded into 35-mm plastic tissue culture dishes. The seeding medium was 91% calcium-free synthetic medium–4% 2.5 mM EGTA in calcium-free synthetic medium–5% heat-inactivated plasma (total calcium, 2.5 mM). On the day after seeding, tertiary cultures were changed to test media. Test media were changed on day 2 and on each day thereafter.

Components and Construction of Experimental Media. Heat-inactivated plasma and heat-inactivated serum were prepared as described (2, 3).

Test media for the experiments represented in Figs. 1–4 were made with very low calcium–plasma medium ($[Ca^{2+}] \rightarrow 0$), or very low calcium–serum medium ($[Ca^{2+}] \rightarrow 0$) as a base, and addition of appropriate amounts of calcium or folate. Very low calcium–plasma medium was composed of 90% calcium-free synthetic medium (containing hypoxanthine and xanthine at 10 μM), 5% 2.5 mM EGTA in calcium-free synthetic medium, and 5% heat-inactivated plasma. Since the total calcium concentration of cockerel plasma was 2.5 mM (10 mg %), the concentration of available calcium in the very low calcium–plasma medium should approach zero. Although the serum and plasma used in the experiment represented in Fig. 4 were prepared from the same lot of pooled cockerel blood, the serum contained calcium at a total concentration of 9.5 mg %, rather than 10 mg % as did the plasma. Accordingly, very low calcium–serum medium ($[Ca^{2+}] \rightarrow 0$) for this experiment was composed of 90% calcium-free synthetic medium, 4.75% 2.5 mM EGTA in calcium-free synthetic medium, and 5% heat-inactivated serum.

Cell Counts. Test media were aspirated from dishes and 1 ml of calcium- and magnesium-free balanced salt solution, with 0.10% Difco trypsin and 0.01% Dow Corning Antifoam-F, was added. After 1 hr in the incubator, two drops of calf serum and two drops of 20-times physiological concentra-

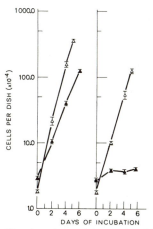

FIG. 1. Proliferation of normal (▲———▲) and Rous sarcoma virus-infected (△———△) chicken fibroblasts in plasma-containing medium at physiological (1.4 mM) (*left*) and very low ($[Ca^{2+}] \rightarrow$ 0) (*right*) calcium concentrations. Mediumc ontained 0.1 mg/liter of folic acid, 10 μM hypoxanthine, and 10 μM xanthine.

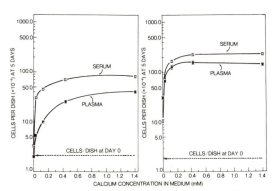

FIG. 2. Effect of calcium concentration on the proliferation of normal (*left*) and Rous sarcoma virus-infected (*right*) fibroblasts in plasma- and serum-containing media (see Fig. 1.). Heat-inactivated serum and heat-inactivated plasma were prepared from the same lot of pooled cockerel blood.

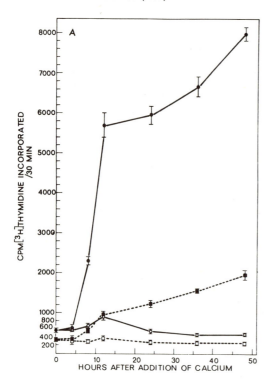

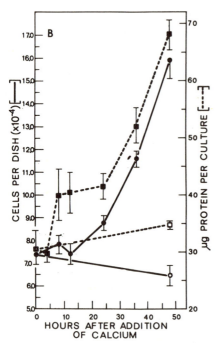

Cultures of normal fibroblasts were maintained for 3 days in
very low calcium–plasma medium ([Ca²⁺] → 0). At "zero"
time, enough sterile 100 mM CaCl₂ solution was added to some

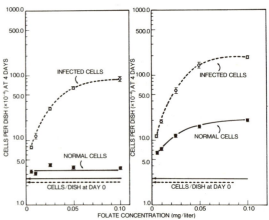

FIG. 4. Effect of folic acid concentration on the proliferation
of normal and Rous sarcoma virus-infected fibroblasts in plasma-
containing medium at physiological (1.2 mM) (*right*), and very
low ([Ca²⁺] → 0) (*left*) calcium concentrations. Medium contained 10 μM hypoxanthine and 10 μM xanthine.

tion calcium and magnesium solution were added. Contents of
the dishes were triturated 21 times with a disposable Pasteur
pipette, then transferred to a tube containing 8.9 ml of Isoton
diluting fluid (Coulter Electronics Inc., Hialeah, Fla.). The
diluted cell suspensions were counted with a Coulter model
B Electronic Cell Counter with a 100-μm aperture. All experimental points reported (cell counts, thymidine incorporation,
and protein per culture) represent the means ± SEM of
determinations on four culture dishes.

Thymidine Incorporation. Cells were exposed to 5.0 μCi
of [³H]thymidine (20 Ci/mmol) per ml of medium for 30
min. The monolayers were then washed twice with normal
saline. Cultures were trypsinized and triturated as above,
but with the addition of 10 mM unlabeled thymidine to the
trypsinizing solution. After trypsinization and trituration,
contents of the dishes (1.1 ml) were transferred to 15-ml
conical centrifuge tubes containing 1.1 ml of cold 14%
HClO₄ (perchloric acid), and the tubes were shaken to mix
their contents. The centrifuge tubes containing the extraction mixtures were held at 4°.

The tubes were centrifuged, and the supernatants were
transferred to scintillation vials containing 10 ml of Omnifluor (New England Nuclear)–dioxane cocktail and counted
in a Beckman LS 255 scintillation counter. These counts were
from cold HClO₄-soluble material, and were considered to
reflect thymidine pool sizes.

The precipitate was washed three times with cold 7%
HClO₄, then held at 80° for 30 min in 1 ml of 7% HClO₄.
The tubes were spun at 2000 rpm in an International PRJ
centrifuge and the supernatants were transferred to scintillation vials containing 10 ml of Omnifluor–dioxane cocktail
for counting. These hot HClO₄-soluble counts were considered
to represent incorporation of thymidine into DNA.

of the dishes to give a final calcium concentration of 1.5 mM
(*closed symbols*). A: *Solid lines;* hot acid-soluble cpm; *dashed lines,*
cold acid-soluble cpm.

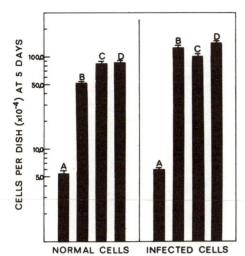

NORMAL CELLS   INFECTED CELLS

Fig. 5. Stimulation of fibroblast proliferation by chicken serum, but not chicken plasma, or by folic acid in a culture medium of physiological calcium concentration (1.4 mM). Media contained hypoxanthine and xanthine, each at 10 μM. Heat-inactivated serum and heat-inactivated plasma were prepared from the same lot of pooled cockerel blood. The zero-time values for culture dishes of normal and infected cells were, respectively, $2.8 \pm 0.1 \times 10^4$ and $3.0 \pm 0.1 \times 10^4$ cells per dish. *A*, plasma–5 μg/liter folate; *B*, plasma–100 μg/liter folate; *C*, serum, 5 μg/liter folate; *D*, serum, 100 μg/liter folate.

Determination of Protein per Culture. Test media were aspirated and monolayers were washed twice with normal saline. 1.0 ml of 1 N NaOH was added, and the dishes were incubated for 1 hr at 41°. Contents of the dishes were triturated and then transferred to 15-ml conical centrifuge tubes, which were held at room temperature. Protein was measured according to the method of Lowry *et al.* (4), with bovine-albumin fraction V as a standard.

DISCUSSION OF RESULTS

Normal chicken fibroblasts did not multiply in a medium containing chicken plasma and a very low calcium concentration ($[Ca^{2+}] \rightarrow O$), but cells infected with the Rous sarcoma virus proliferated rapidly under these conditions (Fig. 1). More detailed examination of the calcium dependence of normal cell proliferation showed that maximum proliferation occurred when the extracellular calcium concentration was about 0.5 mM (Fig. 2). However, the proliferative activity of infected cells, which was already high at very low calcium concentrations, reached its maximum when the extracellular calcium concentration was only about 0.1 mM (Fig. 2). Therefore, infection with Rous virus appears either to increase calcium uptake by the cells or to functionally alter or largely bypass a normal, calcium-dependent system for the control of proliferation.

The use of serum instead of plasma obscured the proliferative differences between normal and Rous virus-infected fibroblasts. The experimental curves shown in Fig. 2 suggest that this result was due to a capacity of serum to strongly sensitize normal cells to the action of calcium. Thus, normal cells

did not proliferate in very low calcium medium containing serum ($[Ca^{2+}] \rightarrow O$), but elevation of the calcium concentration, only slightly, to 25 μM increased their proliferative activity to its half-maximal value (Fig. 2). In contrast to its striking effect on normal cells, the substitution of serum for plasma did not substantially alter the response of infected cells to variation of the calcium concentration between 0.025 and 1.4 mM (Fig. 2).

When a physiological amount of calcium was added to cultures of normal fibroblasts that had been maintained, without proliferation, in very low calcium–plasma medium ($[Ca^{2+}] \rightarrow O$), labeled thymidine uptake and protein synthesis were strongly stimulated, and cell division followed (Fig. 3). Between 4 and 8 hr after restoration of a physiological extracellular calcium concentration (1.5 mM), there was a striking increase in the rate of incorporation of [³H]thymidine into hot HClO₄-soluble material (DNA), without an equivalent increment in the cellular content of cold acid-soluble thymidine (the thymidine pool) (Fig. 3*A*). Coincident with the initiation of DNA synthesis, protein synthesis began, and sometime between 8 and 20 hr later, the cells began to divide (Fig. 3*B*). These observations suggest that calcium controls the proliferation of normal fibroblasts by initiating some critical process in the phase of the cell cycle (G1) prior to DNA synthesis that determines the operations of the necessarily large complex of reactions leading to mitosis and cytokinesis.

Once calcium has activated the growth of normal fibroblasts, the subsequent level of proliferative activity is determined by the availability of folic acid (Fig. 4). Folate concentrations as high as 0.1 mg/liter did not stimulate proliferatively inactive normal fibroblasts in very low calcium–plasma medium (Fig. 4), but did stimulate the proliferation of cells that had been "activated" by culture in 1.2 mM calcium (Fig. 4). Rous sarcoma virus-infected cells, on the other hand, are proliferatively active in very low calcium–plasma medium, and folate was consequently able to stimulate their proliferation in the presence or absence of calcium (Fig. 4). Therefore, manipulation of the folate concentration can greatly magnify the calcium-dependent difference between the proliferative activities of normal and infected cells.

Serum has a second property that plasma lacks, in addition to its apparent ability to sensitize normal cells to calcium: In a plasma-containing culture medium of physiological calcium concentration and folate concentration 5 μg/liter, neither normal nor Rous virus-infected fibroblasts proliferate to an appreciable extent. The use of serum, however, instead of plasma, results in rapid proliferation of both normal and infected cells, as does an increase in the folate concentration of the plasma-containing medium to 0.1 mg/liter (Fig. 5).

In summary, the fact that while normal fibroblasts are maintained without any proliferation in low calcium–plasma medium, Rous virus-infected fibroblasts proliferate rapidly, indicates that the effect of calcium is regulatory rather than permissive. These results suggest that the proliferation of normal fibroblasts is initiated by a cellular function involving calcium. The autonomous proliferation of the neoplastic fibroblasts appears, on the other hand, to result either from an increased capacity for calcium uptake or an alteration or bypass of the normal calcium-dependent proliferation-control system.

The best demonstration of the difference between the proliferative properties of normal and infected cells is obtained

Proc. Nat. Acad. Sci. USA 70 (1973)

with a "very low calcium" medium containing chicken plasma and 0.1 mg of folic acid per liter. The universal practice of using serum, instead of plasma, obscures this difference, since serum contains a factor(s) that first appears to greatly sensitize normal cells to calcium and then, like folate, promotes the subsequent multiplication of the activated cells. In the animal, this serum factor(s) would appear only locally and transiently in a blood-clot in an injured tissue, where it might act as a "wound hormone" that could initiate, or at least promote, cell proliferation (2, 3).

S.D.B. was supported by Special Fellowship no. 1 FO3 CA-53849-01 from The National Cancer Institute of the U.S. Public Health Service. Issued as N.R.C.C. no. 12924.

1. Whitfield, J. F., Rixon, R. H., MacManus, J. P. & Balk, S. D. (1972) *In Vitro* 8, in press.

2. Balk, S. D. (1971) *Proc. Nat. Acad. Sci. USA* 68, 271–275.

3. Balk, S. D. (1971) *Proc. Nat. Acad. Sci. USA*, 68, 1689–1692.

4. Lowry, O. H., Rosebrough, N. J., Farr, A. L. & Randall, R. J. (1951) *J. Biol. Chem.* 193, 265–275.

Reprinted from Proteases and Biological Control
Cold Spring Harbor Conferences on Cell Proliferation
Vol. 2, pp. 841–847. 1975

Macromolecular Determinants of Plasminogen Activator Synthesis

Daniel B. Rifkin, Leslie P. Beal* and E. Reich

Department of Chemical Biology, The Rockefeller University
New York, New York 10021

It has been demonstrated that enhanced production of plasminogen activator (PA) accompanies neoplastic transformation of primary or early passaged fibroblasts. Thus transformation by MSV, RSV or SV40 is associated with large (usually at least 50-fold) increases of plasminogen activator content above the level found in normal cells. This increase in PA does not occur following infection with cytocidal viruses or leukemia viruses and is therefore related to transformation rather than to virus infection or cell lysis (Unkeless et al. 1973; Ossowski et al. 1973b). The activation of plasminogen that takes place in cultures of transformed cells has been shown to influence the cellular phenotype (Ossowski et al. 1973a,b, 1974); thus growth in semisolid media, cell migration and cell morphology are reduced significantly if the formation and/or activity of plasmin are inhibited. Certain of these parameters of plasminogen activator production and cell phenotype are described elsewhere in this volume (Reich; Ossowski, Quigley and Reich; Pollack et al.; Goldberg, Wolf and Lefebvre; Christman et al.; Danø and Reich; Roblin, Chou and Black).

In view of the association between plasminogen activator synthesis and transformation, we have initiated experiments to define some of the factors that may be involved in control of activator synthesis. We have employed chick embryo fibroblasts (CEF) infected with a temperature-sensitive mutant of Rous sarcoma virus, ts 68. Cells infected with ts 68 and grown at 41°C are normal both by morphological and biochemical criteria, whereas the same cells grown at 36°C are transformed by the same criteria (Kawai and Hanafusa 1971). Because these phenotypic properties are fully reversible within a few hours following appropriate shifts in temperature, this system is a very favorable one for exploring variables that might govern activator production.

---

* Present Address: New York University, School of Medicine, New York, New York 10016.

METHODS

Secondary chick embryo fibroblasts were infected as described previously (Rifkin and Reich 1971) with virus generously provided by S. Kawai. Eagle's minimal medium supplemented with 10% fetal bovine serum was used throughout. The infected cells were subcultured and plated into 60-mm petri dishes at a density of 8×10^5 cells per 60-mm dish. The cells were incubated at either 41 or 36°C for 24 hours before the initiation of the experiment. At the beginning of each experiment, the medium was removed and replaced with fresh medium containing the appropriate drug(s) warmed to the proper temperature. The culture dishes were then incubated at the indicated temperature for the desired periods. At the end of an experiment, the medium was removed, the cell monolayer first washed twice with isotonic buffer and then removed with a policeman, and the cells collected by centrifugation at 2000g for 3 minutes. After centrifugation, the supernatant solution was removed by aspiration and the cells were dissolved in 0.3 ml of 0.50% Triton X-100 in 0.1 M Tris-HCl pH 8.1. Fifty microliters of this cell extract was added to a 35-mm petri dish containing 2 ml of Tris buffer (0.1 M, pH 8.1) supplemented with chicken plasminogen (5 μg/ml); the petri dish was coated with ^{125}I-fibrin (10 μg/cm^2; 50,000 cpm) (Unkeless et al. 1973). The values for each time point represent an average of two determinations performed on the cells removed from each of two separate petri dishes. The dishes were incubated for 3 hours at 37°C, and the supernatant fluid was then assayed for solubilized ^{125}I in a gamma-counter. The drugs were obtained from the following sources: actinomycin, Merck Sharp and Dohme, Rahway, N.J.; 5-bromotubercidin, Drs. H. B. Wood and R. Engle, Drug Development Branch, NCI, Bethesda, Md.; cycloheximide, Sigma Chemical Company, St. Louis, Mo.

RESULTS

When CEF-ts 68 cells were shifted from 41 to 36°C, increases in plasminogen activator could first be detected after 2–3 hours (Fig. 1). The intracellular activity then increased linearly for approximately 5 hours, after which it remained constant. The kinetics of increase in plasminogen activator were similar to other biochemical changes that accompany similar temperature shifts of CEF-ts 68 cells (Kawai and Hanafusa 1971). The increase in intracellular activator occurred 2–4 hours before changes in extracellular fibrinolysis (Unkeless et al. 1973); this difference is likely due to several factors, such as extracellular dilution, inhibition by components of the medium and transit time out of the cell (D. B. Rifkin, unpubl.).

Also shown in Figure 1 are the levels of plasminogen activator in cells that were shifted from the permissive to the nonpermissive temperature. The loss of activity was rapid, with a half-time of 1–2 hours. This might have been produced in several ways, including the possibility that the activator was itself temperature sensitive. Previous work had indicated that the properties of plasminogen activator were determined by the cell rather than by the transforming agent (Ossowski et al. 1973b), and a temperature-sensitive enzyme would therefore be an unexpected finding; nevertheless, an experiment was

Figure 1

Intracellular levels of plasminogen activator in chick embryo fibroblasts infected with RSV ts 68. Chick embryo fibroblasts infected with RSV ts 68 were grown either at 41 or 36°C for at least one week, then trypsinized and plated at the appropriate temperature one day before initiation of the experiment. At zero time, the medium was removed from the cultures and replaced with fresh medium that had been warmed to the appropriate temperature; the cultures were then placed either at 36 or 41°C. At the indicated times, the medium was removed and the cells washed, scraped and centrifuged. The cell pellets were assayed for plasminogen activator as described in the text. (•——•) 36 → 41°C, RSV ts 68; (■——■) 41 → 36°C, RSV ts 68; (o) CEF, 41 or 36°C; (□) RSV-SR-A, 41 or 36°C.

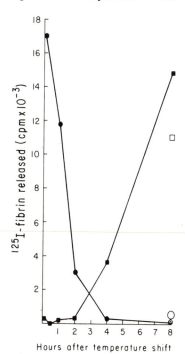

performed (by S. T. Rohrlich) to test for this (Fig. 2). Serum-free conditioned medium was prepared from cultures infected and transformed either by wild-type Schmidt-Ruppin virus or by ts 68 (at 36°C); separate aliquots of the media were then incubated at 36 and 41°C, respectively, and the plasminogen activator activity assayed after different time intervals. The plasminogen activator activity from both cultures does not decrease significantly at either incubation temperature (Fig. 2).

To obtain additional insight into the factors that might govern the loss of plasminogen activator, the effects of several inhibitors were examined; these were added to the cultures at the time of the shift from the permissive (36°C) to the nonpermissive temperature (41°C) (Fig. 3). When cycloheximide addition accompanied the shift to 41°C, the subsequent loss of enzyme was accelerated: over one-half of the activity had disappeared within 30 minutes. In contrast, cells treated with actinomycin retained a large proportion of intracellular activator for long periods. This retention of activity was not affected by actinomycin concentration, since cells treated with higher (10 μg/ml) or lower (0.4 μg/ml) concentrations responded in a similar way (D. B. Rifkin and L. P. Beal, unpubl.). These results suggest (1) that the maintenance of intracellular enzyme depends on continuing protein synthesis, and (2) that enzyme disappearance following the temperature shift requires RNA synthesis. The rapid early drop in cellular enzyme content in the presence of actinomycin varied in different experiments, and the basis for this variation so far remains undefined.

As a further test for the presumed RNA synthesis requirement, cells were treated with 5-bromotubercidin, another inhibitor of DNA-dependent RNA synthesis. This compound is known to inhibit the synthesis of both mRNA

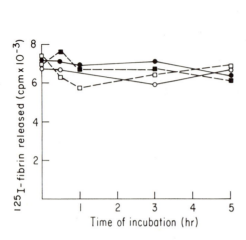

Figure 2

Temperature sensitivity of RSV-SR-A and RSV ts 68 induced plasminogen activators. Serum-free conditioned medium was prepared from chick embryo fibroblasts infected with either wild-type or temperature-sensitive virus at least one week before the initiation of the experiment and grown at 36°C. This serum-free conditioned medium was then incubated at either 36 or 41°C for the indicated times and assayed for plasminogen activator. (●——●) RSV-SR-A, PA incubated at 41°C; (○——○) RSV-SR-A, PA incubated at 36°C; (■---■) RSV ts 68, PA incubated at 41°C; (□---□) RSV ts 68, PA incubated at 36°C.

and rRNA, but in contrast to actinomycin, the effect of BrTu is fully reversible (Brdar, Rifkin and Reich 1973). The action of BrTu on plasminogen activator levels (Fig. 4) was very similar to that of actinomycin; enzyme levels remained high following the temperature shift up in the presence of either inhibitor. The reversibility of the BrTu effect is seen in Figure 5. In this experiment, the rapid early loss of enzyme accounted for approximately 50% of the total, and the remainder was stable in the presence of BrTu. However, removal of BrTu either at 2 or 5 hours after the temperature shift was followed by a further rapid drop in the level of activator. This is a further indication of the requirement for RNA synthesis in the disappearance of plasminogen activator that accompanies the loss of transformation.

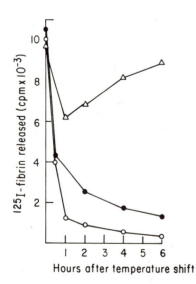

Figure 3

The effect of inhibitors of macromolecule synthesis on plasminogen activator levels in chick embryo fibroblasts infected with ts 68. Chick embryo fibroblasts infected with ts 68 were grown at 36°C. At zero time, the culture medium was removed and replaced with medium warmed to 41°C. One-third of the cultures received medium containing actinomycin (1 μl/ml); one-third, medium containing cycloheximide (20 μg/ml); and one-third, medium alone (controls). The cultures were then placed at 41°C, and at the appropriate times, the cells were scraped and assayed for plasminogen activator as described in the text. (●——●) Control; (△——△) actinomycin (1 μl/ml); (○——○) cycloheximide (20 μl/ml).

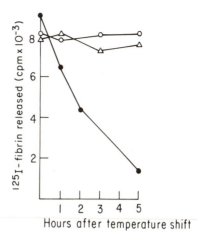

Figure 4
The effect of bromotubercidin on the intracellular level of plasminogen activator. Chick embryo fibroblasts infected with ts 68 and grown at 36°C were treated in a manner similar to those described in the legend to Figure 3, except that one set of cultures received medium containing bromotubercidin (10 μg/ml); one set, actinomycin (1 μg/ml); and the third served as control. (●——●) Control; (o——o) actinomycin; (△——△) bromotubercidin.

The maintenance of elevated levels of activator when RNA synthesis is blocked following the upward temperature shift depends on continuing protein synthesis. This is shown by cycloheximide additions to actinomycin-treated cultures in which high intracellular levels of activator normally persisted for many hours (Fig. 6). The effect of cycloheximide led to a rapid drop in cellular activator content, indicating that the enzyme level was being maintained by de novo protein synthesis. This conclusion is also implied by the observation that in addition to persisting within cells, plasminogen activator was continually secreted into the medium by actinomycin-treated cultures at 41°C (D. B. Rifkin, unpubl.).

DISCUSSION

Changes in incubation temperature of cultures infected with the temperature-sensitive virus ts 68 are known to condition the expression of the transformed

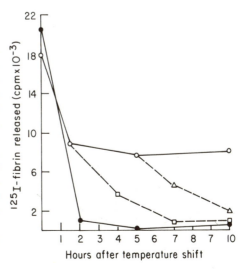

Figure 5
The reversible effect of bromotubercidin. Cultures of chick embryo fibroblasts infected with ts 68 were treated precisely as described in the experiments illustrated in Figures 3 and 4. However, at 2 and 5 hours after the temperature shift, the bromotubercidin was removed from some sets of cultures and replaced with normal medium. (●——●) Control; (o——o) bromotubercidin (10 μg/ml); (□----□) bromotubercidin washed out at 2 hours; (△---△) bromotubercidin washed out at 5 hours.

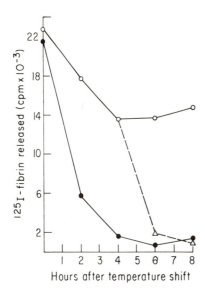

Figure 6
The effect of actinomycin and cycloheximide on the intracellular level of plasminogen activator. Cultures of chick embryo
fibroblasts infected with ts 68 and grown
at 36°C were treated as described in previous experiments. To one set of cultures
which had received actinomycin at zero
time, cycloheximide was added at 4 hours
after the temperature shift. (●——●) Control; (o——o) actinomycin (10 μg/ml);
(△---△) cycloheximide (20 μg/ml).

phenotype. As already reported elsewhere (Unkeless et al. 1973), these
changes are associated with marked fluctuations in plasminogen activator production. The results presented above show that these changes occur very
rapidly following temperature shifts, and the formation of plasminogen activator in such cultures is closely correlated with the expression of the viral transforming function.

Previous results (Unkeless et al. 1973a) had demonstrated the requirement for mRNA synthesis to precede plasminogen activator production following downward temperature shifts to conditions permissive for transformation. The present results show a comparable requirement for RNA synthesis
to precede termination of activator formation in the converse situation, i.e.,
when cultures are shifted upward to temperatures that are nonpermissive for
transformation, but we do not know the nature of the relevant RNA species.

The ability of actinomycin to limit the disappearance of inducible enzymes
following transfer to noninducing conditions has been reported for several
systems (Tomkins et al. 1972; Vilcek and Havell 1973; Whitlock and Gelboin 1973). Several interpretations have been proposed as explanations of
these phenomena, including possible decreases in catabolism of specific proteins (Reel, Lee and Kenney 1970) and inhibition of synthesis of an inferred
RNA species that turns over rapidly and might function as an inhibitor of
translation (Tomkins et al. 1972). Our data are as yet insufficient to discriminate between these hypotheses or to eliminate other plausible explanations.

Acknowledgments

Daniel B. Rifkin holds a Faculty Research Award PR-99 from the American
Cancer Society. This work was supported by grants from the National In-
stitutes of Health, CA-13138, CA-08290-10, and from the Council for
Tobacco Research.

REFERENCES

Brdar, B., D. B. Rifkin and E. Reich. 1973. Studies of Rous sarcoma virus: Effects of nucleoside analogues on virus synthesis. *J. Biol. Chem.* **248:**2397.

Kawai, S. and H. Hanafusa. 1971. The effects of reciprocal changes in temperature on the transformation state of cells infected with a Rous sarcoma virus mutant. *Virology* **46:**470.

Ossowski, L., J. P. Quigley and E. Reich. 1974. Fibrinolysis associated with oncogenic transformation. Morphological correlates. *J. Biol. Chem.* **249:**4312.

Ossowski, L., J. P. Quigley, G. M. Kellerman and E. Reich. 1973a. Fibrinolysis associated with oncogenic transformation. Requirement of plasminogen for correlated changes in cellular morphology, colony formation in agar, and cell migration. *J. Exp. Med.* **138:**1056.

Ossowski, L., J. C. Unkeless, A. Tobia, J. P. Quigley, D. B. Rifkin and E. Reich. 1973b. An enzymatic function associated with transformation of fibroblasts by oncogenic viruses. II. Mammalian fibroblast cultures transformed by DNA and RNA tumor viruses. *J. Exp. Med.* **137:**112.

Reel, J. R., K-L. Lee and F. T. Kenney. 1970. Regulation of tyrosine α-keto-glutarate transaminase in rat liver. *J. Biol. Chem.* **245:**5800.

Rifkin, D. B. and E. Reich. 1971. Selective lysis of cells transformed by Rous sarcoma virus. *Virology* **45:**172.

Tomkins, G. M., B. B. Levinson, J. D. Baxter and L. Dethlefsen. 1972. Further evidence for posttranscriptional control of inducible tyrosine aminotransferase synthesis in cultured hepatoma cells. *Nature New Biol.* **239:**9.

Unkeless, J. C., A. Tobia, L. Ossowski, J. P. Quigley, D. B. Rifkin and E. Reich. 1973. An enzymatic function associated with transformation of fibroblasts by oncogenic viruses. I. Chick embryo fibroblast cultures transformed by avian RNA tumor viruses. *J. Exp. Med.* **137:**85.

Vilcek, J. and E. A. Havell. 1973. Stabilization of interferon RNA activity by treatment of cells with metabolic inhibitors and lowering of the incubation temperatures. *Proc. Nat. Acad. Sci.* **70:**3909.

Whitlock, J. P. and H. U. Gelboin. 1973. Induction of aryl hydrocarbon (benzo [a] pyrene) hydroxylase in liver cell culture by temporary inhibition of protein synthesis. *J. Biol. Chem.* **248:**6114.

Reprinted from

Proc. Natl. Acad. Sci. USA
Vol, 75, No. 4, pp. 2021–2024, April 1978
Microbiology

Protein kinase activity associated with the avian sarcoma virus *src* gene product

(*src* gene product p60*^{src}*/immunoprecipitation/avian sarcoma virus temperature-sensitive mutant/avian sarcoma virus-transformed mammalian cells/[γ-^{32}P]ATP)

MARC S. COLLETT AND R. L. ERIKSON

Department of Pathology, University of Colorado Medical Center, Denver, Colorado 80262

Communicated by David M. Prescott, February 6, 1978

ABSTRACT Incorporation of phosphorus from [γ-^{32}P]ATP into protein was catalyzed by specific immunoprecipitates from avian sarcoma virus (ASV)-transformed avian and mammalian cells. This incorporation was observed only when antiserum from tumor-bearing rabbits able to specifically precipitate the ASV sarcoma gene product, p60 *^{src}*, was used to immunoprecipitate antigens from transformed cell lysates. Immunoprecipitates of extracts from normal cells or cells infected with a transformation-defective ASV mutant showed no activity in this assay, nor did any immune complexes formed with normal rabbit serum and any of the cell extracts tested. The expression of the protein kinase activity (ATP:protein phosphotransferase, EC 2.7.1.37) was growth temperature-dependent in cells infected with an ASV mutant temperature-sensitive for transformation. These results on an enzymatic activity associated with the ASV transforming protein are discussed in terms of protein phosphorylation as a mechanism for viral transformation.

Four genes have been identified and mapped on the avian sarcoma virus (ASV) genome (1, 2). Three of these genes—*gag*, *pol*, and *env*—code for the virion structural proteins, including the group-specific (gs) viral core proteins, the RNA-directed DNA polymerase, and the envelope glycoproteins, respectively. The fourth gene, designated *src*, has been identified through the isolation of temperature-sensitive (ts) and deletion mutants defective in transformation *in vitro* and sarcoma induction *in vivo* (3–7).

Recent work from this laboratory has resulted in the identification of a protein of molecular weight (M_r) 60,000 that appears to be the product of the ASV *src* gene (8–10). Determination that this protein is actually the product of the *src* gene is based on the following data: (*i*) It was detected as a nonstructural, transformation-specific antigen in ASV-transformed chicken cells, ASV-transformed mammalian cells, and ASV-induced mammalian tumor cells, by immunoprecipitation of radiolabeled cell extracts with serum from ASV-tumor-bearing rabbits. (*ii*) *In vitro* cell-free translation of the 3′ third of nondefective ASV viral RNA, the region of the genome that contains the *src* gene, resulted in the synthesis of a polypeptide of M_r 60,000. (*iii*) The polypeptide of M_r 60,000 made *in vitro* by cell-free translation and the transformation-specific antigen isolated by immunoprecipitation of all types of ASV-infected cell extracts tested are identical as determined by peptide analyses (ref. 10; J. S. Brugge, E. Erikson, M. S. Collett, and R. L. Erikson, unpublished results). We feel that it is consistent with these data to conclude that the protein of M_r 60,000 is the product of the ASV *src* gene and consequently give it the following designation: p60*^{src}*.

In order to further elucidate the mechanism of ASV-induced oncogenesis it is necessary to determine the function of the p60*^{src}* polypeptide. We report here an enzymatic activity ascribable to the ASV *src* gene product: protein phosphorylation. In this preliminary communication we describe the identification and association of protein kinase activity (ATP: protein phosphotransferase, EC 2.7.1.37) with p60*^{src}*.

MATERIALS AND METHODS

Cells and Virus. Chicken embryo fibroblasts were prepared from 11-day-old embryos (Spafas, Inc., Roanoke, IL). The Schmidt–Ruppin (SR) strain of ASV, subgroup D, was originally obtained from J. Wyke, and transformation-defective (td) SR-ASV and ASV-NY68 were obtained from H. Hanafusa.

Baby hamster kidney cells transformed by SR-ASV (BHK-SR3/1a) and a morphological revertant cell line (BHK-SR3/11R) (11) were provided by H. Varmus.

Normal European field vole (*Microtus agrestis*) cells and vole cells transformed with SR-ASV originally by P. Vogt (clone 1-T) (12) were provided by A. Faras.

Preparation of Antiserum and Immunoprecipitation. Antiserum was obtained from New Zealand rabbits in which tumors had been induced by injection of purified SR-ASV as previously described (8, 13). Several preparations of antiserum, referred to as TBR serum, were employed in the studies reported here. All were able to immunoprecipitate virion structural proteins as well as the transformation-specific antigen of M_r 60,000 (8, 13).

For the preparation of cell extracts, unlabeled cell cultures grown on 100-mm dishes were washed several times with STE buffer (0.15 M NaCl/0.05M Tris·HCl, pH 7.2/0.001 M EDTA), scraped from the dishes, and either lysed in 1.2 ml of RIPA buffer [1.0% Triton X-100/1.0% sodium deoxycholate/0.1% sodium dodecyl sulfate (NaDodSO₄)/0.15 M NaCl/0.05 M Tris·HCl, pH 7.2] or sonicated in 0.6 ml of 0.01 M Tris·HCl, pH 7.2. All of the above buffers contained 100 kallikrein inactivator units/ml of Trasylol (FBA Pharmaceuticals). The cell lysates were then clarified at 100,000 g for 30 min and the supernatants (adjusted to RIPA buffer) were incubated with 10–30 μl of serum. After 30 min at 4°, 100–200 μl of a 10% suspension of the protein A-containing bacterium, *Staphylococcus aureus*, strain Cowan I, were added to adsorb the immune complexes (14). The bacteria were washed four times with RIPA buffer,

Abbreviations: ASV, avian sarcoma virus; *src*, designation for an ASV gene responsible for transformation of fibroblasts; p60*^{src}*, designation for the protein product of the ASV *src* gene; ts, temperature-sensitive; td, transformation-defective; M_r, molecular weight; TBR, serum from rabbits bearing SR-ASV-induced tumors; NaDodSO₄, sodium dodecyl sulfate; SR, Schmidt–Ruppin strain of ASV, subgroup D; Pr, Prague strain of ASV, subgroup C; B77, Bratislava strain of ASV, subgroup C; BH-ASV(RAV-50), Bryan strain of ASV, pseudotype Rous-associated virus 50, subgroup D.

once with 0.15 M NaCl/0.05 M Tris·HCl, pH 7.2, and then prepared for the protein kinase activity assay.

Assay for Protein Kinase Activity. The detection of protein kinase activity in immunoprecipitates involved the resuspension of the bacteria-bound immune complexes directly in the reaction mixture. The reaction mixtures (25 μl) contained 20 mM Tris·HCl, pH 7.2, 5 mM MgCl$_2$, and 0.4–1.2 μM [γ-^{32}P]ATP (310–870 Ci/mmol). [γ-^{32}P]ATP, made according to the procedure of Glynn and Chappell (15), was initially and generously supplied by T. Walker, B. Pace, and N. Pace. Reactions were generally incubated at 30° for 10 min, at which time an equal volume of 2× sample buffer [0.14 M Tris·HCl, pH 6.8/22.4% (vol/vol) glycerol/6% NaDodSO$_4$/0.02% bromophenol blue/10% (vol/vol) 2-mercaptoethanol] was added. After heating at 95° for 1 min, the mixtures were centrifuged and the supernatants were recovered. Samples were then taken for determination of trichloroacetic acid-precipitable radioactivity or polyacrylamide gel analysis.

Polyacrylamide Gel Electrophoresis and Autoradiography. Samples were analyzed by electrophoresis through a discontinuous slab gel system with the buffer systems described by Laemmli (16). Gels were stained, destained, and dried onto Whatman 3MM paper (8). Radioactivity was detected by exposure to Kodak X-Omat R film with and without the use of Du Pont Cronex Lightning Plus intensifying screens.

RESULTS

Detection of a Protein Kinase Activity in Immunoprecipitates from ASV-Transformed Avian and Mammalian Cells. Antigens immunoprecipitable from ASV-infected cell extracts by TBR serum include precursors to the mature internal structural proteins, their intermediate and mature cleavage products, the viral DNA polymerase, small amounts of the virion membrane glycoprotein, and the viral nonstructural, transformation-specific antigen of M_r 60,000 (8–10, 13). The protein of M_r 60,000 is believed to be the product of the ASV *src* gene (10) and will hereafter be designated p60^{src}.

To determine if immunoprecipitated p60^{src} contained protein phosphorylating activity, bacteria-bound immune complexes were subjected to standard protein kinase assay conditions. It is not unreasonable to expect enzymatic activity in such complexes because of the observation that antisera from both tumor-bearing rabbits and tumor-bearing hamsters were able to precipitate the viral DNA polymerase but failed to inhibit its enzymatic activity (13). Thus, extracts from normal chick cells, avian leukosis virus (RAV-2)-infected chick cells, ASV-transformed chick cells, and chick cells infected with a td mutant of ASV were immunoprecipitated with either normal rabbit serum or TBR serum. The bacteria-bound immune complexes were incubated with [γ-^{32}P]ATP in the absence of any exogenous protein substrates. The reaction was terminated, the bacteria were separated from the IgG and precipitated antigens with electrophoresis sample buffer, and the latter were subjected to NaDodSO$_4$/polyacrylamide gel electrophoresis (Fig. 1). It can be clearly seen that under these assay conditions, protein phosphorylation can be detected only in extracts of ASV-transformed cells immunoprecipitated with TBR serum (Fig. 1, track 3b). A single protein band of approximate M_r 53,000 was highly phosphorylated. Because this radiolabeled material comigrates with the Coomassie blue-stained IgG heavy chain band, and also comigrates with the stained high molecular weight undissociated IgG when analyzed on gels in the absence of 2-mercaptoethanol (data not shown), we believe the phosphorylated protein to be an immunoglobulin. Further indica-

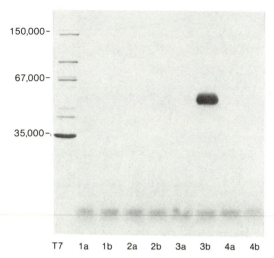

FIG. 1.   Detection of a protein kinase activity in immunoprecipitates of chick cell lysates. Cell extracts were prepared from chick embryo fibroblast cultures that were either uninfected (tracks 1), infected with the avian leukosis virus RAV-2 (tracks 2), infected with SR-ASV (tracks 3), or infected with a transformation-defective deletion mutant of SR-ASV (tracks 4). Each extract (1800–3600 μg of protein) was immunoprecipitated with either normal rabbit serum (a tracks) or TBR serum (b tracks), and a portion of the bacteria-bound immune complexes was incubated in the protein kinase reaction mixture. After termination of the reaction by heating to 95° for 1 min in sample buffer and pelleting of the bacteria, the supernatant was subjected to electrophoresis in a discontinuous NaDodSO$_4$/polyacrylamide slab gel. The stacking gel contained 4.2% acrylamide and 0.1% bisacrylamide, and the separation gel was 10% polyacrylamide (38:1 acrylamide/bisacrylamide), 11 cm long. The figure represents an autoradiogram of the dried gel. Phage T7 virion proteins are included as molecular weight markers.

tions that phosphorylation is occurring in the antigen–IgG complex are the observations that the kinetics of the reaction are extremely rapid (complete within 1 min at 23°) and that among the few tested, such as histone and casein, exogenously added substrates are not phosphorylated. That the radioactive band is indeed a phosphorylated protein was demonstrated by tryptic peptide and phosphoamino acid analyses. The phosphorylated protein was found to contain a single tryptic phosphopeptide containing exclusively phosphothreonine (data not shown).

To further demonstrate that the phosphorylating activity detected in ASV-infected chick cell extracts immunoprecipitated with TBR serum was due to the presence of the viral sarcoma gene product, several mammalian cell lines were subjected to a similar analysis. Baby hamster kidney cells transformed by SR-ASV (BHK-SR3/1a) also demonstrated the IgG phosphorylating activity when cell extracts were immunoprecipitated with TBR serum (Fig. 2, track 1b). However, when a morphological revertant subclone of the BHK-SR3/1a cells (17) was analyzed, no phosphorylating activity was detected in the immunoprecipitates (Fig. 2, track 2b). Analysis of radiolabeled antigens immunoprecipitated with TBR serum has revealed that the expression of p60^{src} in these revertant cells is undetectable (data not shown).

A second mammalian cell type investigated for the presence of the sarcoma-specific protein phosphorylating activity was the European field vole cell. Both normal and SR-ASV-transformed vole cells were analyzed and, as is seen in Fig. 2, (track

Microbiology: Collett and Erikson

Proc. Natl. Acad. Sci USA 75 (1978)   2023

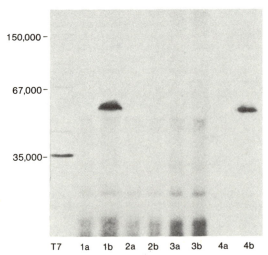

FIG. 2. Presence of a protein kinase activity in immunoprecipitates of SR-ASV-transformed mammalian cells. Cell extracts were prepared from cultures of SR-ASV-transformed baby hamster kidney cells (BHK-SR3/1a) (tracks 1), a normal phenotypic revertant clone of these cells (BHK-SR3/11R) (tracks 2), normal field vole cells (tracks 3), and SR-ASV-transformed field vole cells (tracks 4). Immunoprecipitation with either normal (a tracks) or TBR serum (b tracks) and assay of phosphorylating activity were as described in the legend to Fig. 1.

4b) only the transformed cells immunoprecipitated with TBR serum contained the protein kinase activity.

Correlation of the Protein Kinase Activity and the Immunoprecipitation of p60^{src}. The immunoprecipitation of p60^{src} by the preparations of TBR serum used in these experiments is strain specific. Antibody produced in rabbits bearing tumors induced by SR-ASV does not immunoprecipitate a transformation-specific protein from cells transformed by other strains of ASV (J. S. Brugge and R. L. Erikson, unpublished results). Therefore, to determine if a protein kinase activity was present in immunoprecipitates of chick cells transformed by the Prague strain (Pr-ASV), the Bratislava strain (B77-ASV), and the Bryan strain [BH-ASV(RAV-50)] of ASV, cell lysates were prepared, immunoprecipitated with TBR serum, and subjected to the protein kinase assay. The results, presented in Table 1 along with a summary of the data presented in Figs. 1 and 2, indicate that even though the virus-infected cells are transformed, the protein kinase activity is not found in cell extracts from which no p60^{src} can be immunoprecipitated. It should be pointed out here that because the *src* genes of the various ASV strains are very similar, as judged by molecular hybridization (18) and cell-free translation of subgenomic viral RNA (9, 10), similar gene products are probably present in cells transformed by other strains of ASV, but they are not detected because of a lack of crossreaction with the currently available antisera.

Temperature Sensitivity of the Protein Kinase Activity in Chick Cells Infected with a Temperature-Sensitive Transformation Mutant of ASV. To demonstrate more directly that the protein kinase activity is the result of expression of the ASV *src* gene, studies were undertaken to determine if the observed protein phosphorylation was thermosensitive in chick cells infected with a ts mutant in the *src* gene. Previous experiments have suggested that the *src* gene product encoded by such mutants is irreversibly denatured at 41°, the nonpermissive

Table 1. Correlation of the protein kinase activity and the immunoprecipitation of p60^{src}

Virus	Cell	Transformation	p60^{src}	Protein kinase activity
—	Chick	−	−	−
RAV-2	Chick	−	−	−
td-SR-ASV	Chick	−	−	−
SR-ASV	Chick	+	+	+
SR-ASV	Hamster	+	+	+
SR-ASV	Hamster revertant	−	−	−
—	Vole	−	−	−
SR-ASV	Vole	+	+	+
Pr-ASV	Chick	+	−	−
B77-ASV	Chick	+	−	−
BH-ASV (RAV-50)	Chick	+	−	−

Uninfected cell cultures or cultures transformed with the various strains of ASV were either radiolabeled with [^{35}S]methionine (for the detection of p60^{src}) or left unlabeled (for the detection of protein kinase activity). Cell extracts were prepared and immunoprecipitated with TBR serum as described in *Materials and Methods*. NaDodSO$_4$/polyacrylamide gel electrophoresis was used to determine the presence (+) or absence (−) of [^{35}S]methionine-labeled p60^{src} (8). Protein kinase activity was determined as described in the legend to Fig. 1.

temperature (6). Parallel cultures of chick cells infected with nondefective SR-ASV and the SR-ASV ts mutant NY68 (6) were maintained at both 35° and 41°. Cell extracts were prepared from each of the four cultures, immunoprecipitated with TBR serum, and then analyzed for protein kinase activity in the bacteria–immune complex assay. Cells infected with nondefective ASV showed a slight (2-fold) increase in phosphorylating activity in *src* protein immunoprecipitates when grown at 41° compared to those grown at 35° (Table 2). In contrast to these results, the NY68-infected cells, when grown at the nonpermissive temperature (41°), showed a dramatic decrease in phosphorylating activity (Table 2). These data indicate that the expression of the protein kinase activity detected in *src* protein immunoprecipitates is dependent on the temperature at which

Table 2. Growth temperature-dependent expression of *src* protein-immunoprecipitated phosphorylating activity in chick cells infected with a ts transformation mutant of ASV

Virus	Growth temperature, °C	Phosphorylating activity	
		^{32}P incorporated, fmol/mg protein	Normalized values
SR-ASV (nd)	35	16.6	1.00
SR-ASV (nd)	41	35.8	2.16
SR-NY68	35	19.8	1.00
SR-NY68	41	1.7	0.09

Parallel cultures of chick cells infected with nondefective (nd) SR-ASV and cells infected with the ts transformation mutant of SR-ASV, NY68, were maintained at either the permissive (35°) or nonpermissive (41°) temperature. Cell extracts were prepared from the four cultures, samples were taken for determination of protein content, and the remainder was immunoprecipitated with TBR serum as described in *Materials and Methods*. Phosphorylating activity by the bacteria-bound immunoprecipitated complexes was also determined. The resulting activity values, determined by quantitation of the phosphorylated IgG bands from a polyacrylamide gel, are normalized with respect to the amount of cell extract protein used for immunoprecipitation and the activity present in the respective cells grown at 35°.

chick cells infected with an ASV ts mutant in the *src* gene are grown.

DISCUSSION

The results presented in this communication demonstrate that a protein kinase is the product of the ASV *src* gene, or is closely associated with it. Biosynthetic radiolabeling permits the detection by immunoprecipitation of only one polypeptide in nd virus-infected cells which is not in td virus-infected cells, and that is the polypeptide with a M_r of 60,000 previously identified as the product of the ASV *src* gene (10). Using TBR serum to immunoprecipitate p60^{src}, we have shown here that a protein kinase activity is associated only with cells, both avian and mammalian, infected with transforming virus, that expression of this activity is growth temperature-dependent in cells infected with a *src* gene ts mutant virus, and that the protein kinase activity is dependent on the immunoprecipitation of p60^{src}. Additional experiments in this laboratory further support the association of the protein kinase activity with p60^{src}; the enzymatic activity and p60^{src} cosediment during glycerol gradient centrifugation and coelute from ion exchange columns run singly or sequentially. However, the possibility remains that a highly active kinase that cannot be detected by biosynthetic radiolabeling specifically associates with p60^{src} and, therefore, appears to be specifically immunoprecipitable. Even if this proves to be the case, such a result may permit a better understanding of ASV-induced oncogenesis. The purification of the relevant enzymes from cells infected by nondefective and ts virus will serve to better resolve these issues.

The role of protein phosphorylation in the functional regulation of a variety of cellular processes is well documented (see ref. 19 for a review and further references). Phosphotransferase reactions may be catalyzed by enzymes that are dependent on the presence of cyclic nucleotides for maximal activity (cyclic nucleotide-dependent protein kinases) or their activity may be unaffected by cyclic nucleotides (phosphoprotein kinases) (19). Both classes of protein kinases are generally capable of phosphorylating a number of substrates *in vitro* that cannot be considered the normal *in vivo* targets. Consequently, the phosphorylation of IgG observed here probably provides no clues as to the biologically significant phosphorylation in ASV-induced transformation.

In this regard, it has been previously suggested that the product of the ASV *src* gene may be directly or indirectly involved in a reversible chemical modification of components involved in the maintenance of cell structure (20). This idea is based on indirect experiments employing ts mutants of ASV and inhibitors of protein synthesis. These experiments indicated that all the components necessary for normal cellular architecture were present, although disaggregated, in transformed cells, and were reassembled upon inactivation of the ts *src* gene product, without the necessity of protein synthesis. Thus, one might speculate, for example, that a component(s) of the cytoskeleton complex when phosphorylated may fail to interact properly with other proteins, resulting in disaggregation of the protein networks essential for normal cellular structure and function. When the product of a ts *src* gene is inactivated at the nonpermissive temperature, phosphatases could restore the affected protein(s) to its normal composition, permitting reassembly. Indeed there is evidence that protein phosphorylation influences shape changes in erythrocyte membranes (21), and perhaps analogous events may occur in fibroblasts.

However, in order to begin to understand the circuits involved in oncogenic transformation, the direct identification of the target of the *src* protein kinase in infected cells will be necessary.

R.L.E. acknowledges helpful discussions on possible mechanisms of transformation with Renzo Rendi. We also wish to thank A. J. Faras, H. Hanafusa, and H. E. Varmus for providing cells and virus, E. Erikson for assistance with some of the experiments, and J. Brugge for providing the antisera used in this study. This investigation was supported by Grants CA-21117 and CA-15823 from the National Institutes of Health, and Grant VC-243 from the American Cancer Society. M.S.C. is a Fellow in Cancer Research supported by Grant DRG-181-F of the Damon Runyon–Walter Winchell Cancer Fund.

1. Wang, L.-H., Duesberg, P., Beemon, K. & Vogt, P. K. (1975) *J. Virol.* **16**, 1051–1070.
2. Joho, R. H., Stoll, E., Friis, R. R., Billeter, M. A. & Weissmann, C. (1976) in *Animal Virology*, eds. Baltimore, D., Huang, A. S. & Fox, C. F. (Academic, New York), pp. 127–145.
3. Vogt, P. K. (1971) *Virology* **46**, 939–946.
4. Biggs, P. M., Milne, B. S., Graf, T. & Bauer, H. (1973) *J. Gen. Virol.* **18**, 399–403.
5. Toyoshima, K. & Vogt, P. K. (1969) *Virology* **39**, 930–931.
6. Kawai, S. & Hanafusa, H. (1971) *Virology* **46**, 475–479.
7. Martin, G. S. (1970) *Nature* **227**, 1021–1023.
8. Brugge, J. S. & Erikson, R. L. (1977) *Nature* **269**, 346–348.
9. Purchio, A. F., Erikson, E. & Erikson, R. L. (1977) *Proc. Natl. Acad. Sci. USA* **74**, 4661–4665.
10. Purchio, A. F., Erikson, E., Brugge, J. S. & Erikson, R. L. (1978) *Proc. Natl. Acad. Sci. USA* **75**, 1567–1571.
11. Macpherson, I. A. (1965) *Science* **143**, 1731–1733.
12. Chen, Y. C., Hayman, M. J. & Vogt, P. K. (1977) *Cell* **11**, 513–521.
13. Brugge, J. S., Erikson, E. & Erikson, R. L. (1978) *Virology* **84**, 429–433.
14. Kessler, S. W. (1975) *J. Immunol.* **115**, 1617–1624.
15. Glynn, I. M. & Chappell, J. B. (1964) *Biochem. J.* **90**, 147–149.
16. Laemmli, U. K. (1970) *Nature* **227**, 680–685.
17. Deng, C.-T., Stehelin, D., Bishop, J. M. & Varmus, H. E. (1977) *Virology* **76**, 313–330.
18. Stehelin, D., Guntaka, R. V., Varmus, H. E. & Bishop, J. M. (1976) *J. Mol. Biol.* **101**, 349–365.
19. Rubin, C. S. & Rosen, O. M. (1975) *Annu. Rev. Biochem.* **44**, 831–887.
20. Ash, J. F., Vogt, P. K. & Singer, S. J. (1976) *Proc. Natl. Acad. Sci. USA* **73**, 3603–3607.
21. Birchmeier, W. & Singer, S. J. (1977) *J. Cell Biol.* **73**, 647–659.

Reprinted from Nature, Vol. 274, No. 5672, pp. 696–697. 1978.

presence of plasminogen, indicating that we were dealing with a plasminogen activator. EGF or TPA added directly to the assay rather than to cells did not cause fibrinolysis.

Figure 1 shows the effect of increasing concentrations of EGF on cell-associated plasminogen activator in HeLa cultures. Induction was detected at a concentration of EGF as low as 1.25 ng ml^{-1} (2×10^{-10} M) and was maximal at 10 ng ml^{-1} (1.7×10^{-9} M). The induced levels were 7–12 times those present in untreated cultures. The time courses of induction of plasminogen activator in cell lysates and in media collected from HeLa cultures are shown in Fig. 2a and b. An increase in cell-associated plasminogen activator was observed as early as 4 h after the addition of EGF and continued to rise during the next 24 h (Fig. 2a). This was paralleled by a marked increase in plasminogen activator in the media of the HeLa cell cultures, presumably reflecting increased secretion of the protease. Actinomycin D (10 μg ml^{-1}) and cycloheximide (100 μg ml^{-1}) completely blocked EGF induction of plasminogen activator (Table 1), indicating that the induction requires both RNA and protein synthesis. These results, and the fact that both cell-associated and extracellular levels of plasminogen activator are increased (Fig. 2), provide evidence that EGF causes an increase in *de novo* synthesis of plasminogen activator. When TPA and EGF were tested in combination, at concentrations (5 ng ml^{-1} of each) which were suboptimal for each agent, an additive effect on plasminogen activator production was observed. When TPA (10 ng ml^{-1}) was tested in combination with a dose (10 ng ml^{-1}) of EGF which produced maximum induction of plasminogen activator, there was no augmentation or inhibition of the induction (Table 1). Phorbol, which lacks tumour-promoting activity and does not induce plasminogen activator[2,7,8], did not enhance or inhibit EGF induction of plasminogen activator, even when tested at 100 ng ml^{-1} (Table 1).

At present, the implications of these findings in terms of the mechanism of action of EGF are unclear. Our results do not apply to all cell types, for thus far we have not detected significant induction of plasminogen activator by murine EGF in chick embryo fibroblasts, a Chinese hamster cell line (CHO), a rat hepatoma cell line (HTC) or 3T3 mouse fibroblasts, even though the latter cell line is responsive to the growth stimula-

Epidermal growth factor, like phorbol esters, induces plasminogen activator in HeLa cells

STUDIES on the action of the tumour-promoting compound 12-*O*-tetradecanoyl-phorbol-13-acetate (TPA), and related plant diterpenes, in cell culture systems have revealed several unusual biological properties[1–3]. These include the induction of plasminogen activator production[2] as well as several other phenotypic changes resembling those seen in cells transformed by chemical carcinogens or viruses[1]. Furthermore, these compounds are extremely potent inhibitors of several types of terminal differentiation[1,4,5], all these effects being exerted by very low concentrations of the active compounds, in the range of 10^{-9} M. In addition, the activities of a series of such compounds, with respect to their cell culture effects, in general parallel their activities as promoters in the mouse skin two-stage carcinogenesis system[1,6]. We have previously postulated that the highly pleiotropic effects of these compounds may be due to their ability to usurp the effector system of an endogenous growth regulatory hormone[1,3,7,8]. The polypeptide hormone epidermal growth factor (EGF) is a possible candidate for this putative substance as it shares several biological properties with the phorbol esters[9–14]. Consistent with this hypothesis are recent results from our laboratory indicating that TPA is a potent inhibitor of the binding of ^{125}I-EGF to its cellular receptors[3]. We report here that EGF and TPA resemble each other in their ability to serve as potent inducers of plasminogen activator production in HeLa cell cultures. Our results further suggest that the mechanism of EGF induction of this protease is also similar to that of induction produced by TPA and related tumour promoters.

Plasminogen activator was assayed by methods reported elsewhere[15,16], and specific details are given in the legends to Figs 1 and 2. A cell lysate was used to measure total cell-associated activity, and extracellular activity was assayed in serum-free medium collected from cell cultures. Lysis of ^{125}I-labelled fibrin by cell samples was completely dependent on the

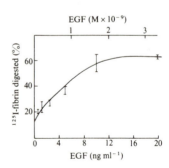

Fig. 1 Dose–response study of plasminogen activator induction by EGF in HeLa cells. Subconfluent cultures of HeLa, approximately 6×10^6 cells per 9-cm dish, were exposed to various concentrations of EGF in Eagle's minimum medium (MEM). After 18 h the cells were washed twice with phosphate-buffered saline (PBS) and collected into 2 ml of 10 mM Tris-Cl (pH 8), 10 mM NaCl and 0.5% Triton X-100. Aliquots containing 0.01 A_{280} units were added, in triplicate assays, to ^{125}I-fibrin-coated plates in the presence of 5 μg ml^{-1} human plasminogen in assay buffer containing 0.1 M Tris-Cl (pH 8) and 10 mM NaCl. After a 1.5-h incubation at 37 °C, the solubilised ^{125}I-fibrin was collected and radioactivity was determined in a Searle Analytic 92 liquid scintillation counter programmed for ^{125}I. Percentage of fibrin digested was determined, using a trypsin-treated fibrin plate as the 100% value. The background, which was determined by omitting the sample, was subtracted from all assays.

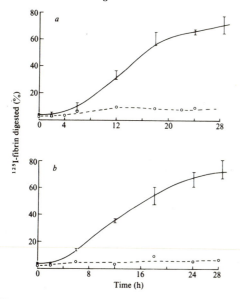

Fig. 2 Time course of induction of plasminogen activator by EGF. *a*, Cell lysate assay. EGF in MEM was added to culture plates of HeLa at zero time, and at the subsequent times indicated, cell lysates were prepared and assayed as in Fig. 1. *b*, Cell medium assay. Instead of cell lysates, the medium of the cells exposed to EGF was collected at the indicated times and centrifuged to remove any detached cells. The supernatant fluid (0.01 A_{280} units) was then assayed for plasminogen activator as described for cell lysates in Fig. 1. - - -, Untreated HeLa; ——, HeLa treated with 20 ng ml^{-1} EGF.

tory effects of EGF[17]. The analysis of 3T3 is complicated, however, by the fact that even untreated cultures contain an appreciable level of plasminogen activator. EGF has growth stimulating effects on HeLa and other human cell cultures[18]. There is evidence that this stimulation requires an initial binding of EGF to cells, following which there is proteolysis of the bound EGF and an apparent decrease in EGF receptors[18]. Our results raise the possibility that these effects may be related to EGF induction of plasminogen activator or possibly other cell-associated proteases.

Several aspects of the induction of plasminogen activator by EGF resemble those previously described with TPA and related compounds[2,6]. Both classes of agents act in the range 10^{-8}–10^{-10} M and lead to an increase in both cell-associated and extracellular plasminogen activator. They also display similar time courses of induction and with both classes of agents the induction process requires RNA and protein synthesis. As mentioned above, EGF and TPA also share several other biological effects[7-14]. These include growth stimulation, enhancement of sugar transport, induction of ornithine decarboxylase and prostaglandin synthesis, and promotion of mouse skin carcinogenesis. Further studies are required to determine the extent to which these similar effects reflect similar biochemical mechanisms of action and to elucidate the significance of such effects in terms of the processes of growth control and tumour promotion.

We thank Ms Deborah DeFeo for assistance in the plasminogen activator assays, and Drs G. Todaro and J. DeLarco for providing samples of murine EGF. This research was supported by Contract NO-1-CP-2-3234 from the NCI, D.H.E.W.

LIH-SYNG LEE
I. BERNARD WEINSTEIN

*Institute of Cancer Research and Division
of Environmental Sciences,
Columbia University College of Physicians and Surgeons,
701 West 168th Street,
New York, New York 10032*

Received 11 May; accepted 19 June 1978.

1. Weinstein, I. B. & Wigler, M. *Nature* **270**, 659–660 (1977).
2. Wigler, M. & Weinstein, I. B. *Nature* **259**, 232–233 (1976).
3. Lee, L. S. & Weinstein, I. B. *Science* (in the press).
4. Yamasaki, H. *et al. Proc. natn. Acad. Sci. U.S.A.* **74**, 3451–3453 (1977).
5. Diamond, L. *et al. Nature* **269**, 247–248 (1977).
6. Wigler, M., DeFeo, D. & Weinstein, I. B. *Cancer Res.* **38**, 1434–1437 (1978).
7. Weinstein, I. B., Wigler, M., Fisher, P. B., Sisskin, E. & Pietropaolo, C. in *Carcinogenesis*, Vol. 2 (eds Slaga, T. J., Sivak, A. & Boutwell, R. K.) 313–333 (Raven, New York, 1978).
8. Weinstein, I. B., Wigler, M. & Pietropaolo, C. in *Origins of Human Cancer, Cold Spring Harbor Conferences on Cell Proliferation*, IV (eds Hiatt, H. H., Watson, J. D. & Wingston, J. A.) 751–772 (Cold Spring Harbor Labs, 1977).
9. Levine, L. & Hassid, A. *Biochem. biophys. Res. Commun.* **79**, 477–484 (1977).
10. Hoober, J. K. & Cohen, S. *Biochim. biophys. Acta* **138**, 347–356 (1967).
11. Stastny, M. & Cohen, S. *Biochim. biophys. Acta* **204**, 578–589 (1970).
12. Rose, S. P., Stahn, R., Passovoy, D. S. & Hershman, H. *Experientia* **32**, 913–914 (1976).
13. Driedger, P. E. & Blumberg, P. M. *Cancer Res.* **37**, 3257–3265 (1977).
14. Yuspa, S. H. *et al. Nature* **262**, 402–404 (1976).
15. Unkeless, J. C., Dano, K., Kellerman, G. M. & Reich, E. *J. biol. Chem.* **249**, 4295–4305 (1974).
16. Ossowski, L., Unkeless, J. C., Tobia, A., Quigley, J. P. & Reich, E. *J. exp. Med.* **137**, 112–126 (1973).
17. Rose, S. P., Pruss, R. M. & Herschman, H. R. *J. cell. comp. Physiol.* **86**, 593–598 (1975).
18. Cohen, S., Carpenter, G. & Lembach, K. J. *Adv. metabolic Disorders* **8**, 265–284 (1975).

Table 1 Effects of various reagents on the production of plasminogen activator in HeLa cells

Reagents	Plasminogen activator (% fibrinolysis)
None	12
EGF (10 ng ml^{-1})	73
EGF (10 ng ml^{-1}) + actinomycin D (10 µg ml^{-1})	13
EGF (10 ng ml^{-1}) + cyclohexamide (100 µg ml^{-1})	2
EGF (5 ng ml^{-1})	37
TPA (5 ng ml^{-1})	42
EGF (5 ng ml^{-1}) + TPA (5 ng ml^{-1})	75
TPA (10 ng ml^{-1})	88
EGF (10 ng ml^{-1}) + TPA (10 ng ml^{-1})	90
Phorbol (100 ng ml^{-1})	8
Phorbol (100 ng ml^{-1}) + EGF (5 ng ml^{-1})	36

Fibrinolysis assays were carried out on cell lysates using freshly prepared ^{125}I-fibrin plates, as described in Fig. 1, except that 8 µg ml^{-1} of human plasminogen was used for each assay and the extent of fibrinolysis was measured after a 3-h incubation.

Reprinted from Nature, Vol. 273, No. 5661, pp. 345–349. 1978. © 1978 Macmillan Journals, Ltd.

Role of cell shape in growth control

Judah Folkman & Anne Moscona

Department of Surgery, Children's Hospital Medical Center and The Harvard Medical School, Boston, Massachusetts 02115

Tissue culture plastic adhesivity was precisely varied by applying different concentrations of poly(2-hydroxyethyl methacrylate). The extent of cell spreading was thus accurately controlled so that cells cultured on these substrata could be held at any one of a graded series of quantitated cell shapes. Cell shape was found to be tightly coupled to DNA synthesis and growth in nontransformed cells. These findings suggest a mechanism that is important in growth control of mammalian cells, and provide a more fundamental interpretation of such phenomena as density dependent inhibition of cell growth and anchorage dependence.

OUR hypothesis that appropriate cell shape is critical for DNA synthesis by normal cells[1–3] was proposed to explain 'anchorage dependence', a term that describes the inability of normal cells to grow unless attached to a substratum. Most primary fibroblasts proliferate if attached to glass or plastic, but do not grow in suspension culture[4]. Furthermore, we felt that a problem associated with 'anchorage dependence' might also be explained; although normal cells will not grow in suspension culture, they will attach to glass fibrils dispersed in the suspension culture, and proliferation occurs if the fibrils are long enough[5]. However, when the fibrils are less than 20 μm long, proliferation ceases despite the fact that the cells are still anchored. A similar problem is posed by the observation that although many cells do not adhere to bacteriologic plastic, those that do, proliferate very slowly or not at all.

We considered that cell shape was an important variable in all these situations. Cells in suspension culture are spherical; cells on tissue culture plastic are very flat or extremely extended, and cells on bacteriologic plastic, or those attached to glass fibrils less than 20 μm long, maintain a foreshortened configuration that lies somewhere between spherical and flat. Despite its explanatory value, there was little quantitative data to support this theory. There was no easy way to hold cells indefinitely, at various degrees of spreading. Also, comparative measurements of cell shape were cumbersome. We have now solved these difficulties, and describe here new experimental evidence in support of this hypothesis.

Control of cell shape by variation of substratum adhesiveness

We have found that cell shape can be precisely varied by a simple modification of substratum adhesiveness. The adhesiveness of plastic tissue culture dishes can be permanently reduced in a graded manner by applying increasing concentrations of poly(2-hydroxyethyl methacrylate) (poly(HEMA)). When an alcoholic solution of poly(HEMA) is pipetted into a plastic culture dish, a thin, hard, sterile film of optically clear polymer remains tightly bonded to the plastic surface after the alcohol evaporates. Serial dilutions of the alcohol–polymer solution, introduced into each dish at a constant volume, result in decreasing thicknesses of the polymer film. These films vary between 35 and 0.0035 μm in thickness, with a correspondingly increased adhesiveness. Cells plated on the thickest layer of

polymer (35 μm) are the least adherent, and most closely approach a spheroidal configuration. On thinner polymer layers, the cells become flatter (Fig. 1). At least 12 cell conformations from spherical to flat can be achieved by this method. Once a population of cells has reached shape equilibrium, the cells maintain an average shape, specific for each polymer thickness, throughout an experiment and for at least 7 d.

Cell shape was measured by two methods: a horizontal dimension (cell diameter) was measured with the micrometer

Dilution of poly (HEMA) solution	Mean cell diameter (μm)	Dilution of poly (HEMA) solution	Mean cell diameter (μm)
1.00	10 ± 2	6×10^{-3}	27 ± 8
1×10^{-1}	9 ± 2	4×10^{-3}	33 ± 10
8×10^{-2}	9 ± 2	1×10^{-3}	37 ± 12
4×10^{-2}	16 ± 6	5×10^{-4}	46 ± 11
1×10^{-2}	24 ± 7	1×10^{-4}	46 ± 13
8×10^{-3}	27 ± 11	Plastic	51 ± 7

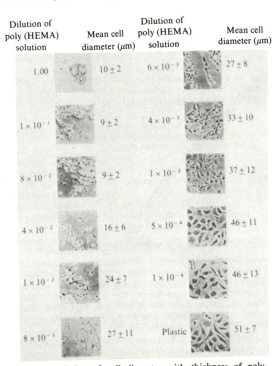

Fig. 1 Variation of cell diameter with thickness of poly-(HEMA) substratum. The 2.1-cm⁻² wells of Falcon 3008 culture plates were coated with dilutions of poly(HEMA). 6 g poly(HEMA) was dissolved in 50 ml 95% ethanol and the mixture turned slowly overnight at 37 °C to dissolve the polymer completely. The viscous solution was centrifuged for 30 min at 2,500 r.p.m. to remove undissolved particles. This stock was then diluted with 95% ethanol from 10⁻¹ (that is, 1.00 ml poly(HEMA + 9.00 ml ethanol) to 10⁻⁴. 200 μl was pipetted into the microwells so that each concentration was represented in quadruplicate. Sterile technique was used in a 37 °C warm room and plates were dried on a level bench free of vibrations. The wells were allowed to dry for 48 h with the lids in place. If there was a break in sterility, exposure to ultraviolet light before use did not harm the poly(HEMA) films. Bovine aortic endothelial cells[21] were plated at 15,000 per cm², and incubated for 24 h. 20 cells in each well were then measured with a Nikon profile projector at ×200. The longest axis of each cell body was measured; pseudopod extensions were not included in the measurement.

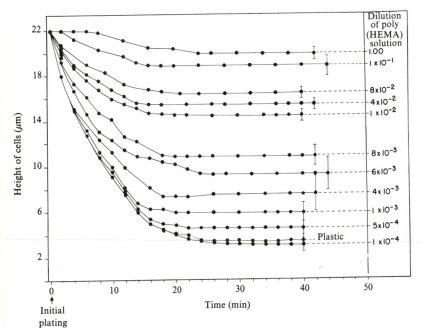

Fig. 2 Variation of cell height with thickness of poly(HEMA) substratum immediately after cells are plated. The micrometer of an inverted phase microscope was connected by a pulley to an angular transducer which was fed to a digital read-out device calibrated to read directly in μm. Poly(HEMA) films of graded thickness were prepared as in Fig. 1. Endothelial cells were plated at 10,000 per cm^2 per well. Just before adding the cells, a fine scratch was made on the poly(HEMA) surface. The lid was sealed with parafilm and the plate was placed on the microscope stage at 37 °C. The height of each of 5 cells was measured to $\pm 0.1 \mu m$ by focusing on the body of the cell. Each experiment was repeated in quadruplicate. The shape at equilibrium did not change over a 7-d period. Although poly(HEMA) films swell slightly after hydration, all thicknesses of poly(HEMA) reached swelling equilibrium less than 1 min after media were added.

stage of a Nikon profile projector (Fig. 1), and a vertical dimension (cell height) was obtained by using the fact that cells freshly plated on plastic continually go out of focus as they flatten and spread. Therefore, the fine-focus micrometer on a standard inverted phase microscope was attached to a rotary transducer so that measurements in the vertical dimension could be recorded by digital read-out to $\pm 0.1 \mu m$. This permitted rapid measurement of the height of a cell, its height decreasing as the cell flattened. The floor of the culture dish was calibrated to 'zero' by making fine scratches on it with a glass knife or a glass wool swab, and also by focusing on microspheres of 20 μm diameter resting on the surface. The time required for cells to reach shape equilibrium by measuring the vertical dimension is shown in Fig. 2.

Variation of DNA synthesis with cell shape

The number of cells capable of DNA synthesis was determined for the population of cells on each thickness of poly(HEMA) substratum by incubating the cells with ^{3}H-thymidine after they had reached shape equilibrium. For all three lines tested (bovine endothelial cells, WI-38 cells and A-31 cells), incorporation of ^{3}H-thymidine was found to be inversely proportional to the height of the cell (Fig. 3). Autoradiographs were also made of each cell population and confirmed the results from scintillation counting (Figs 4 and 5). These experiments, in which cells were not crowded and in fact were not in contact with each other, led to the following question. If DNA synthesis is inhibited as cells change toward a more spheroidal conformation, could this shape modulation account for the

decrease in DNA synthesis associated with crowding of cells? It is well known that DNA synthesis gradually stops as cells become confluent[6]. It has also been shown[7] that mitoses continue for a little while after contact. It was largely for this reason that Stoker and Rubin proposed the term 'density dependent inhibition of growth'[8].

Is density dependent inhibition of cell growth mediated by cell shape?

Endothelial cells were plated at various densities on tissue culture plastic. As the cells grew to confluence, they came out of focus in the reverse order from their plating as each cell was laterally compressed by its neighbours. However, this change occurred over several days. The most confluent cells attained the greatest mean height and also incorporated the least amount of ^{3}H-thymidine. The most striking finding, however, was that whenever a poly(HEMA) substratum was chosen with the appropriate adhesivity to hold sparsely plated WI-38 cells at the same height as their confluent counterparts on plastic, the incorporation of ^{3}H-thymidine for the two populations was similar (Table 1). In other words, sparse cells held at rounded shapes approached levels of DNA synthesis that closely matched that of confluent cells. In this experiment, cell shape was driven toward the spheroidal configuration by two different mechanisms: first, the cells on poly(HEMA) were not in contact with each other but they became rounded or hexagonal because of a reduction in substratum adhesiveness; second, by contrast, the cells on plastic were on a maximally adhesive substratum

Table 1 Effect of crowding on cell shape and DNA synthesis (WI-38 cells)

Dilution of poly (HEMA)	Poly (HEMA) Cell density (cells per cm²)	³H Incorporation (c.p.m.)	Cell height (μm)	Plastic Cell density (cells per cm²)	³H Incorporation (c.p.m.)
1.00	15,000	7±5	22	250,000 (confluent)	7±1
4×10^{-2}	15,000	50±5	15	100,000 (confluent)	55±9
4×10^{-3}	15,000	250±15	6	30,000 (subconfluent)	254±23

WI-38 cells were plated at the densities indicated on uncoated plastic (Falcon 3008) and also on a series of poly(HEMA) substrates of graded thickness. Cell height, number and ^{3}H-thymidine incorporation (c.p.m.) were measured as in Fig. 3. When poly(HEMA) plates of different adhesivity containing sparsely plated cells were selected for cells whose height most closely matched the height of cells on plastic at various densities, the incorporation of ^{3}H-thymidine was more equivalent for these pairs of plates than for any other combination of plates. Similar results were obtained with endothelial cells.

but could not flatten because they were being crowded by their neighbours. However, the resulting reduction in DNA synthesis was similar in both situations. This experiment was repeated with endothelial cells with similar findings.

The result was substantiated by a second experiment. Swiss 3T3 cells were sparsely plated on bacteriologic plastic, and their mean height and percentage ^{3}H-thymidine incorporation was measured. Swiss 3T3 cells were also seeded on tissue culture plastic at pre-confluent density and allowed to grow to a density at which their average height was equivalent to the height of the cells on bacteriologic plastic, then ^{3}H-thymidine incorporation was determined. It was evident that the crowded cell population on tissue culture plastic with a mean height most closely approximating to the height of sparse cells on bacteriologic plastic, contained a ratio of DNA synthesising cells that most closely matched the population on bacteriologic plastic (Table 2).

The idea that density dependent inhibition of growth is partly mediated by cell shape was then corroborated by a third experiment. When wounds are made in a confluent monolayer of cultured cells, the surviving cells at the wound edge, and those in a few rows behind the edge, show an increased incorporation of ^{3}H-thymidine[9]. After the cells have closed the gap, DNA synthesis returns to the low level of the other confluent cells. Therefore, wounds were made in a confluent monolayer of A-31 cells cultured on plastic. Focal depth measurements that began immediately after wounding showed that the first row of cells at the wound edge began to flatten within 30–60 min, as these cells gradually spread out onto the exposed substratum (Fig. 6). Within one hour the first row of cells had flattened by

Table 2 The effect of crowding on cell shape and DNA synthesis (Swiss 3T3 cells on bacteriologic plastic compared with tissue culture plastic)

Time after plating (h)	No. of cells	c.p.m. per 1,000 cells	Mean height (μm)	Range
		Bacteriological dishes		
60	1.26×10^5	76.9	12	(8–14)
72	1.51×10^5	60.0	12	(8–14)
94	1.93×10^5	45.0	12	(8–14)
		Tissue culture dishes		
60	2.63×10^5	132.0	5	(1–8)
72	3.00×10^5	51.2	12	(9–14)
94	3.85×10^5	17.8	17	(10–20)

27 bacteriological dishes (Falcon 1008) and 27 tissue culture dishes (Falcon 3005) were seeded with 5×10^4 cells. Three dishes in each group were scratched for cell height measurements. At each time point, the mean cell height is the average for 20 cells, with 10 measurements per cell. Cell number and ^{3}H-thymidine incorporation were also measured at each time point. ^{3}H-thymidine incorporation in Swiss 3T3 cells grown to confluence on tissue culture plastic until they were sufficiently crowded to resemble the shape of sparsely plated cells on bacteriologic plastic, was equivalent to the ^{3}H-thymidine incorporation of sparse cells on bacteriologic plastic.

54% of their original height, the next row by 42% and the third row by 30%. Beyond that, cells were still at the same height as the remainder of the confluent layer. Although ^{3}H-thymidine was not added in this experiment, the expected labelling index for A-31 cells of these respective shapes would predict the highest labelling index for the first row with a subsequent decrease in labelling of the cells further back.

Controls

It might be argued that the decreasing DNA synthesis in cells plated on thicker poly(HEMA) layers could be an artefact of sublethal cytotoxicity on the part of the polymer. This possibility was ruled out by five experiments: (1) glass coverslips coated with poly(HEMA) were placed on top of cells that had already been plated on plastic. Time lapse films showed that cells directly beneath the poly(HEMA) had the same mitotic rate as cells remote from it. (2) Glass cylinders were coated with thick poly(HEMA) and cemented to plastic dishes which were then seeded with cells. Auto-radiographs showed that cells growing on the plastic surface came directly in apposition to the wall of poly(HEMA) without any zone of decreased ^{3}H-thymidine labelling. (3) Deep scratches were made in thick poly(HEMA) layers. The cells would not spread or grow on top of the poly(HEMA), but they did stretch and proliferate deep in the scratched troughs, while surrounded by poly(HEMA). (4) Endothelial cells were plated in wells in which the substratum was either thick poly(HEMA) (35 μm) or 1.0% Agarose. Cells settled on the surface but could not attach, and continued to roll around. Every day cells were 'rescued' from a poly(HEMA) well and from an Agarose well by aspiration with a pipette, and plated on tissue culture plastic. For each day that cells had remained spherical, plating efficiency on plastic decreased. By the 8–9th day, plating efficiency was zero, and cells from both the poly(HEMA) wells and the Agarose wells were considered dead. This indicated that poly(HEMA) per se is as nontoxic as Agarose, and that the eventual loss of plating efficiency was the result expected for nontransformed cells maintained in the completely rounded state (like suspension culture). (5) Finally, transformed cells (SV 3T3) grew almost as rapidly on thickest poly(HEMA) as on 10^{-4} polymer or plastic despite the fact that their cell shape varied with substrate adhesiveness in the same proportion as for nontransformed cells. This indicates that in transformed cells, cell shape and cell growth may be uncoupled.

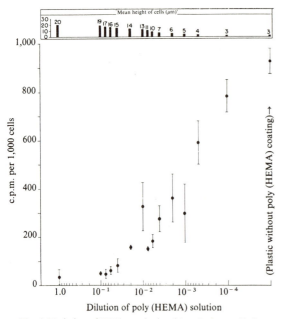

Fig. 3 Variation of DNA synthesis with cell shape : Endothelial cells. Endothelial cells were plated on poly(HEMA) substrates and plastic as in Fig. 2. Four wells were used for each poly(HEMA) dilution. Cell height was measured in the 1st well. Cells in the 2nd well were counted at the end of the experiment (48 h). Cells in the 3rd and 4th wells were exposed to ^{3}H-thymidine (1 μCi ml⁻¹) for 42 h, and c.p.m. per 1,000 cells were determined on precipitated DNA. Similar results were obtained with WI-38 and A-31 cells. The wide standard deviations in wells of poly(HEMA) 10^{-2} to 10^{-3} are partly due to microscopic ripples at the periphery of each well, mostly related to rapid drying of the polymer film.

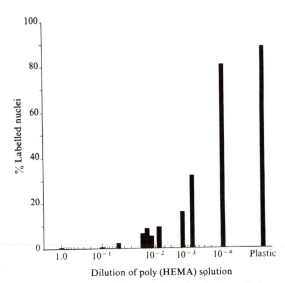

Fig. 4 Variation of DNA synthesis with cell shape : Endothelial cells (autoradiography). Cells were treated as in Fig. 3, except that at the end of ^{3}H-thymidine incubation, all the cells in each well were removed by brief trypsinisation to a corresponding microwell (Falcon 3040) of uncoated plastic. After 2 h, these attached cells were prepared for autoradiography, developed after 5 d, and 400 cells in each well counted, and expressed as % labelled nuclei. Similar experiments were carried out with WI-38 and A-31 cells.

Discussion

Our experiments show that for nontransformed mammalian cells, the shape of the cell is tightly coupled to DNA synthesis. As cells of various lines are brought from the extremely flat shape toward a spheroidal conformation, fewer cells incorporate ^{3}H-thymidine. Finally, when cells are almost completely spherical, but still barely attached to the substratum, they fail to enter the S phase, analogous to floating cells. Cell shape is controlled either by substratum adhesiveness or by the combination of substratum adhesiveness and crowding by neighbouring cells. It should be emphasised that time-lapse films show that for substrata of low adhesivity, equilibrium shape is actually the average shape maintained by each cell as it continually spreads out and retracts. However, even the extent of this transient spreading is less when the adhesivity of the substratum is diminished.

The mechanism by which poly(HEMA) reduces the adhesivity of polystyrene is not entirely known. Poly(HEMA) is a hydrophilic hydrogel of neutral charge. It may act by reducing the net negative electrostatic charge of the polystyrene, but other mechanisms are tenable. Although scanning electron micrographs of all thicknesses of poly(HEMA) showed a smooth homogeneous surface, it is possible that tiny spicules of plastic, below the resolution of the scan, protrude as multiple contact points to which cells stick. Increasing thicknesses of poly(HEMA) would then act to decrease the number of available contacts with plastic. The important conclusion is that whatever the mechanism by which poly(HEMA) can reduce adhesivity of polystyrene, cells respond by a change in shape.

These experiments suggest that several different phenomena of cell culture may be mediated by cell shape. For cells subject to anchorage dependence, shape is modulated by substrate adhesiveness, and only when cells are spread to the appropriate degree can DNA synthesis proceed. Density dependent inhibition of growth may result from the close packing of cells by their neighbours, result-

ing in a cell shape that does not allow DNA synthesis. As cells become confluent, mitoses may continue briefly after cell contact because of the time required to bring about close cell packing. The subsequent cell shape changes apparently occur as the pressure generated by each mitosis is attenuated throughout the monolayer. Zetterberg and Auer described the reduction in area occupied by each cell as cell density increased[6]. Furthermore, the stimulation of cell growth by wounding of a monolayer may also result from modulation of cell shape at the wound edge. As the leading edge of cells spreads out onto the open substratum, cells in the interior seem to become unpacked, and partially begin to stretch and flatten although to a much lesser degree than cells at the edge (Fig. 6). Dulbecco's finding[10] that cells at the wound edge have a higher labelling index than cells in the next interior row may be the result of this stepwise unpacking and flattening. Additional evidence is found in the demonstration that the spreading and movement of a whole row of cells into an open wound precedes DNA synthesis[11].

The relation of serum growth factors to cell shape

If cell shape is so tightly associated with the onset of DNA synthesis, how then is the influence of serum growth factors to be explained? The addition of fresh serum or of purified growth factors[12] or the increased velocity of medium flow[13] can sometimes overcome the restriction of growth imposed by high cell density. This has led to a humoral hypothesis which suggests that high cell density limits the availability of medium components such as growth factors present in the serum[14]. However, the humoral hypothesis cannot explain the inhibition of cell growth in sparse cultures on substrata of low adhesivity (that is, poly(HEMA) 10^{-2}).

This difficulty can be resolved by considering that growth control associated with cell shape and the sensitivity of cells to growth factors may be linked. Thus, higher concentrations of growth factors would be required for DNA synthesis as cell conformation was converted from flat toward hexagonal, and finally to spheroidal, where no concentration of serum would be sufficient for growth. There is

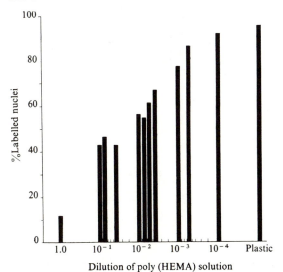

Fig. 5 Variation of DNA synthesis with cell shape : A-31 cells (autoradiography). BALB/3T3 (A-31) cells (late passage) were incubated with ^{3}H-thymidine as in Fig. 4. In contrast to endothelial cells, this A-31 population already has a few cells able to grow without anchorage. Further passage from high density might be expected to select transformed cells capable of growth in suspension culture.

One hour after wounding

Fig. 6 Change in cell shape after wounding of confluent monolayer. BALB/3T3 (A-31) cells (early passage) were allowed to grow to early confluence on tissue culture plastic (Falcon 3001), on which a fine scratch had previously been made for cell height calibration. The height of cells over the whole plate before wounding was $14.8 \pm 0.5\ \mu m$. One hour after the monolayer had been wounded with a glass knife, cell heights were measured again. Cells at the wound edge had spread out and flattened down to $6.8\ \mu m$. Cells in the next interior row had also begun to flatten, but to a lesser extent and without loss of contact.

scattered evidence for this idea in the literature. Such a relationship between cell shape and sensitivity to growth factors would explain why cells at the edge of a wound require less serum for DNA synthesis than cells a few rows behind them[10]. Also, this concept would explain why BHK21 cells grown on plastic were 60 times more sensitive to serum than the same cells grown in suspension culture[15] and why Paul et al. found serum requirements to be reduced in anchored cells[16]. This hypothesis could be tested by culturing cells on substrata of graded adhesivity with increasing serum concentrations.

Theoretical considerations

These data do not distinguish whether cell shape 'controls' cell proliferation or whether both parameters are dependent on a third factor related in some way to the substrate. Also unclear are the important variables associated with the rounded cell that prevent proliferation. The disappearance of bundles of microfilaments[17], the decreased attachment points to the substratum, or the reduction in the rate of transport of certain nutrients[18,19] are all possibilities. The method of holding large populations of cells at any given average shape, as described here, may provide a new approach to these problems. Because transformed cells can proliferate without anchorage[20] or while being held in any cell shape, this difference from nontransformed cells may also be amenable to study by poly(HEMA). It is possible that the various stages of transformation will be classifiable in a more quantitative way according to the ability of cells to grow on increasing thicknesses of poly(HEMA).

Finally, normal cell growth *in vivo* may possibly be regulated by a complex balance of adhesive forces between cells and between cell and substratum (that is, basement membrane) that combine to modulate cell shape, and permit or prevent DNA synthesis. In contrast, a hallmark of malignancy is continual proliferation despite cell crowding. We have reported the importance of angiogenesis capacity for tumour cells to surmount the diffusion problems of crowding[22]. The present study suggests an antecedent property; the capacity of a tumour cell to overcome the effects of close cell packing and thus continue DNA synthesis.

We thank Drs A. Kornberg and R. Tucker for discussions, Drs A. Pardee, R. Pollack and H. Green for reviewing this manuscript, Drs D. Ausprunk and C. Handenschild for the scanning electron micrographs, J. Wilson, Mrs C. Butterfield, B. Smith and C. Cobb for assistance, Mrs P. Breen for typing the manuscript and Dr S. Ronel (Hydron) for poly(2-hydroxyethyl methacrylate). This work was supported by grant CA14019-04 from the US NCI and a grant from Monsanto Co.

Received 19 December 1977; accepted 14 April 1978.

1. Folkman, J. & Greenspan, H. P. *Biochim. biophys. Acta* **417**, 211–236 (1975).
2. Folkman, J. in *Advances in Pathobiology, Cancer Biology II, Etiology and Therapy* Vol. 4, (eds Fenoglio, C. & King, D. W.) 12 (Stratten Intercontinental, 1976).
3. Folkman, J. in *Recent Advances in Cancer Research : Cell Biology, Molecular Biology, and Tumour Virology* Vol. 1 (ed. Gallo, R. C.) 119–130 (CRC, Cleveland, 1977).
4. Stoker, M. G. P., O'Neill, C., Berryman, S. & Waxman, V. *Int. J. Cancer* **3**, 683–693 (1968).
5. Maroudas, N. G., O'Neill, C. H. & Stanton, M. F. *Lancet* i, 807–809 (1973).
6. Zetterberg, A. & Auer, G. *Expl Cell Res.* **62**, 262–270 (1970).
7. Martz, E. & Steinberg, M. S. *J. Cell Physiol.* **79**, 189–210 (1972).
8. Stoker, M. G. P. & Rubin, H. *Nature* **215**, 171–172 (1967).
9. Dulbecco, R. & Stoker, M. G. P. *Proc. natn. Acad. Sci. U.S.A.* **66**, 204–210 (1970).
10. Dulbecco, R. *Nature* **227**, 802–806 (1970).
11. Sholley, M. M., Gimbrone, M. & Cotran, R. *Lab. Invest.* **36**, 18–25 (1977).
12. Mierzejewski, K. & Rozengurt, E. *Nature* **269**, 155–156 (1977).
13. Stoker, M. G. P. *Nature* **246**, 200–203 (1973).
14. Holley, R. W. *Nature* **258**, 487–490 (1975).
15. Clarke, G. D., Stoker, M. G. P., Ludlow, A. & Thornton, M. *Nature* **227**, 798–801 (1970).
16. Paul, D., Henahan, M. & Walter, S. *J. natn. Cancer Inst.* **53**, 1499–1503 (1974).
17. Vasiliev, Ju. M. & Gelfand, I. M. *Expl Cell Res.* **97**, 241–248 (1976).
18. Pardee, A. B. *Natn. Cancer Inst. Monogr.* **14**, 7–20 (1964).
19. Bissell, M. J., Farson, D. & Tung, A. S. C. *J. Supramolec. Struct.* **6**, 1–12 (1977).
20. Shin, S., Freedman, V., Risser, R. & Pollack, R. *Proc. natn. Acad. Sci. U.S.A.* **72**, 4435–4439 (1975)
21. Birdwell, C. R., Gospodarowicz, D. & Nicolson, G. L. *Nature* **268**, 528–531 (1977).
22. Folkman, J. & Cotran, R. in *Int. Rev. exp. Path.* **16** (eds Richter, G. W. & Epstein, M. A.) 207–248 (Academic, New York, 1976).

Cell, Vol. 15, 627–637, October 1978, Copyright © 1978 by MIT

Altered Translatability of Messenger RNA from Suspended Anchorage-Dependent Fibroblasts: Reversal upon Cell Attachment to a Surface

Stephen R. Farmer, Avri Ben-Ze'ev,
Bérnd-Joachim Benecke and Sheldon Penman
Biology Department
Massachusetts Institute of Technology
Cambridge, Massachusetts 02139

Summary

Anchorage-dependent cells, when forced into suspension culture, display a repertoire of dramatic, coordinated regulatory phenomena. Message production promptly decreases 5 fold but the cells maintain a constant amount of poly(A)$^+$ by means of a concomitant stabilization of mRNA against decay. Protein synthesis shuts down much later and the mRNA is stored in a nonfunctioning state. In this study, the inactive mRNA is extracted from suspended cells and shown to have aberrant translation properties. Well defined polypeptides are apparently no longer synthesized when this mRNA directs protein formation in either reticulocyte or wheat germ-derived heterologous translation systems. Rather, shortened peptides are formed by this mRNA and these become smaller as mRNA is used from cells suspended for longer periods of time. Very few focused spots are formed when the aberrant polypeptides are analyzed in two-dimensional electrophoresis.

The sedimentation properties of suspended cell mRNA and the size of poly(A) are unchanged from control monolayer cells. Cross-hybridization of cDNA transcribed from a control cell message population with suspended cell mRNA shows that all sequences are present in normal concentrations. While most identifiable spots disappear from the two-dimensional gel electropherograms of the protein products produced by suspended cell mRNA, a few polypeptides are still synthesized in relatively normal amounts. Conserved polypeptides are found in products of both the reticulocyte and wheat germ systems, but they are different products in each case. The lesion in the suspended cell mRNA does not seem to be at the 5' termini, since synthesis of the shortened peptides is fully sensitive to inhibition by pm^{7}G.

Cells that contain extensively modified message can resume protein synthesis when allowed to reattach to a solid substrate. There is an apparent remodification of mRNA to normal translatability within a few hours of cell reattachment, since mRNA from recovering cells quickly resumes directing relatively normal patterns of polypeptide synthesis in vitro. The restoration of normal message function occurs even when new message formation is blocked with actinomycin.

Cells recovering after reattachment synthesize supranormal amounts of a few major proteins involved with cell structure, as shown in these studies by an increased amount of translatable sequences which encode these proteins. The most apparent enhanced message is that coding for actin. mRNA from recovering cells produces in vitro several times more actin relative to other proteins than does control cell mRNA. The enhancement of actin mRNA is not seen in the message population of cells that reattach in the presence of actinomycin. The results suggest a morphologically related induction of gene expression.

Introduction

The requirement of some cells for a solid substrate upon which to grow has been known for many years and is sometimes termed "anchorage dependence" (MacPherson and Montagnier, 1964; Stoker et al., 1968). Its loss is frequently used as an assay for viral transformation. There has been surprisingly little study of the biochemical concomitants of the forced suspension of anchorage-dependent cells. The inhibition of cellular processes in suspended anchorage-dependent cells and their restoration upon attachment to a solid substrate affords insight into regulatory processes not easily studied by other means. In addition, the response of anchorage-dependent cells to a solid surface is an easily studied example of the control of cell metabolism by plasma membrane configuration.

Previous studies have shown that suspending anchorage-dependent cells has profound effects upon macromolecular metabolism during suspension. In particular, protein synthesis inhibition and its resumption after cell reattachment involve an unusual mechanism which may be of significance in other biological systems.

The cells used in these studies are 3T6 fibroblasts, an established anchorage-dependent line. When anchorage is prevented by suspension in methocel, DNA synthesis stops (Otsuka and Moskowitz, 1975) and mRNA production drops precipitously, although HnRNA labeling is unaffected. There is a gradual but eventually drastic decline in protein synthesis, so that after 72 hr in suspension culture, protein synthesis amounts to only 10–15% of the expected rate for cells growing in monolayer. mRNA is withdrawn from active polyribosomes and is stored in a quasi-stable form (Benecke, Ben-Ze'ev and Penman, 1978).

This report discusses the inactive stabilized mRNA of suspended anchorage-dependent cells. Protein synthesis inhibition in the suspended fibroblasts does not appear to be due to generalized cell damage, and synthesis recovers rapidly when cells are permitted to reattach to a solid substrate

(Benecke et al., 1978). Rather, the inhibition appears to result from a regulatory mechanism that withdraws messenger RNA from active polyribosomes and stores it in a quasi-stable, untranslated form. The properties of this stored messenger RNA are altered so that they no longer direct the production of normal products in heterologous in vitro protein synthesizing systems. This modification is apparently reversed, however, within a few hours after cell reattachment to a solid substrate.

Results

In Vitro Translation Products of Suspended Cell mRNA

The altered properties of messenger RNA from suspended cells first became apparent in examining the translation products produced by these molecules in a cell-free protein synthesizing system derived from rabbit reticulocytes. The mRNA from normally growing fibroblasts directs the synthesis of an extensive array of major cellular proteins. In contrast, the mRNA from suspended cells appears altered so that the production of normal polypeptides is reduced in a translation assay, and this alteration becomes progressively greater as cells are kept in suspension culture.

The different patterns of polypeptides produced by equal amounts of poly(A)$^+$ mRNA [determined by poly(A) content, see below] from normally growing (monolayer) and suspended fibroblasts and translated in an mRNA-dependent reticulocyte cell-free system are shown in one-dimensional electropherograms in Figure 1. The densitometry tracings of prefogged fluorograms show very different polypeptide distributions for mRNA preparations from cells suspended for different lengths of time. These distributions change progressively with the length of time that cells are kept in suspended culture. The mRNA from growing monolayer cells directs the synthesis of proteins whose electrophoretic pattern closely resembles that of the polypeptides synthesized by cells in vivo. In contrast, the fluorogram of polypeptides synthesized by mRNA from cells suspended for 30 hr shows a pronounced shift to products of lower molecular weight and a dramatic disappearance of high molecular weight monodisperse peaks, such as the 43K dalton protein tentatively identified as actin.

Messenger RNA of cells suspended for 48 hr directs even fewer discrete products, and the shift to smaller polypeptide size is even more pronounced. Total incorporation appears reduced, due in part to very small peptides migrating off the analyzed region of the acrylamide gel. After 65 hr of cell suspension, no discrete polypeptides can be detected in the mRNA-directed products of in vitro translation and total incorporation is much reduced.

Mixing experiments show that the altered translation properties of suspended cell mRNA are not due to the presence of a factor that affects the reticulocyte translation system. Suspended and monolayer mRNA are translated separately, then equal amounts of each are mixed and also translated. The products of translation are analyzed on SDS-polyacrylamide gels. The insert to Figure 1 shows densitometry scans of the 43K dalton (actin-like) region of the resultant gels, and shows that the presence of the suspended mRNA population in the mixed message assay in no way affects the expected translation of the actin mRNA from the monolayer population. Thus it seems improbable that there is a diffusable factor that co-purifies with suspended cell poly(A)$^+$ mRNA and affects the translation efficiency of the reticulocyte system.

More information about the altered translation properties of suspended cell mRNA is provided by two-dimensional gel electrophoresis. Figures 2A and 2B compare the electropherograms obtained from the translation of equal amounts of monolayer and suspended cell poly(A)$^+$ message in a rabbit reticulocyte system. Here again there is a strong reduction in the amount of material produced by the suspended cell mRNA that will focus to well defined polypeptide spots. In particular, the major protein product, which co-migrates with purified actin, has now become a faint spot. It should be noted, however, that not quite every polypeptide has been suppressed. A few relatively strong spots are still observed in the suspended cell electropherogram, and these correspond to polypeptides present in the products of monolayer RNA at about the same absolute intensity. Thus it appears that the reduction in polypeptide directing capability is not the same for every message sequence, and four spots (indicated by arrows) are consistently found to be of similar intensity whether produced by monolayer or suspended cell mRNA. These spots are not present as endogenous products of the reticulocyte system but rather appear to be products of messages that are still translated to produce full-sized products. These conserved polypeptides indicate that simple overall degradation is not the cause of the altered message properties.

The reduction in the intensity of focused spots in the two-dimensional electropherogram produced by the suspended cell mRNA (see Figures 2A and 2B) is much greater than the reduction in TCA-precipitable polypeptides (see figure legend). As suggested by the one-dimensional electropherogram of Figure 1, it appears that the reticulocyte system, when translating suspended cell mRNA, produced aberrant products which are not discrete and which no longer give distinct spots in the electropherogram.

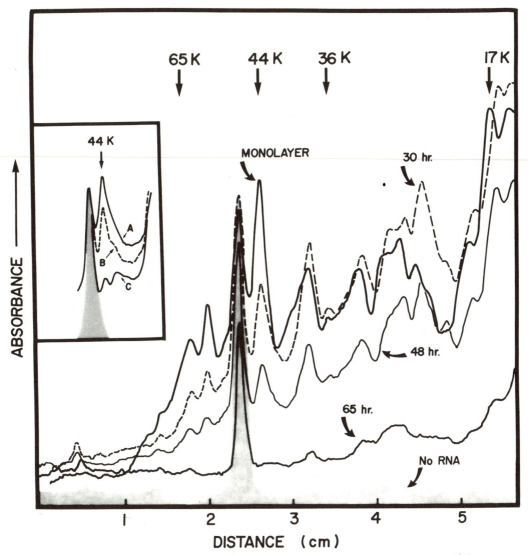

Figure 1. Electrophoretic Analysis of in Vitro Translation Products Directed by mRNA from Monolayer and Suspended Cells

Samples [equivalent to the same amount of input poly(A)⁺ RNA] were taken from the mRNA-dependent reticulocyte lysate and analyzed on SDS 7–17% gradient polyacrylamide slab gels, as described in Experimental Procedures.

 Amounts of acid-insoluble radioactivity layered were: (1) monolayer: 33,200 cpm; (2) 30 hr suspension: 22,500 cpm; (3) 48 hr suspension: 18,000 cpm; (4) 65 hr suspension: 8,500 cpm; (5) no RNA: 2,500 cpm. The dry gels were exposed to X-ray film for two days and the fluorograms were then scanned at 500 mµ.

 The insertion shows a comparison of the actin region of fluorograms from a mixing experiment, in which 200 ng of monolayer (A) and 72 hr suspended (C) mRNA and 100 ng of each mixed (B) were translated and analyzed as above. Note that there is half as much monolayer mRNA in (B) as there is in (A).

The normally inactive mRNA from suspended cells shows altered translation properties in the heterologous reticulocyte system. Another heterologous translation system might be affected differently. Translation of suspended cell message in a wheat germ-derived cell-free system gives similar

overall results, but the products differ in detail.

 The wheat germ translation products directed by normal monolayer and suspended cell mRNA are analyzed by two-dimensional electrophoresis and shown in Figures 2C and 2D. The overall pattern of polypeptides from normal mRNA is significantly

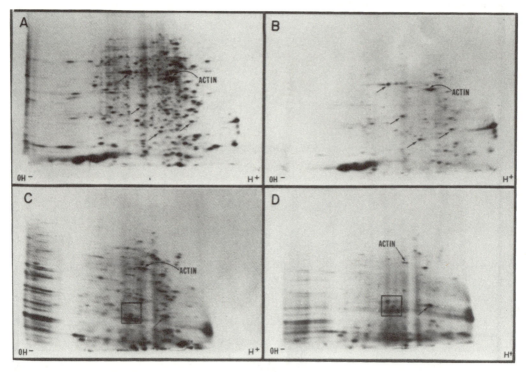

Figure 2. Two-Dimensional Gel Analysis of Reticulocyte (A, B) and Wheat Germ (C, D) Cell-Free Translation Products Primed with Monolayer and Suspended Cell mRNA

In vitro translations were as described in Experimental Procedures. Samples analyzed are equivalent to the same amount of input poly(A)⁺ RNA. Amounts of acid-insoluble radioactivity layered were: Monolayer — 55,000 cpm (A) and 50,000 cpm (C); Suspended — 25,000 cpm (B) and 38,000 cpm (D). The dry gels were exposed to X-ray films for five days.

different from that produced by the reticulocyte system. Many relative spot intensities are greatly changed, and in this system the actin spot appears to take two resolvable forms. We have drawn attention to the enclosed triplet of spots found at seven o'clock relative to actin, and also to the spot indicated by the small arrow. These spots are relatively undiminished in the suspended cell-directed products, although a previously undetected strong fourth spot appears in the enclosed group. The conservation of a limited number of message products is similar to the result found when using reticulocyte lysate. This observation suggests that the polypeptides are different, that even the conserved polypeptides are produced by altered messages and that a few messages are coincidentally read correctly in the two systems, although the possibility of different post-translational modification of the same peptide cannot be ruled out.

The Similarity of Sequences in Monolayer and Suspended Cell mRNA

Although the translation products are altered, the

sequences in suspended cell poly(A)⁺ mRNA are essentially the same as those in control monolayer message, at least with respect to complementary DNA (cDNA) cross-hybridization, which measures the 3′ ends of the message. The data in Figure 3 show that cDNA made from normal monolayer mRNA hybridizes to suspended cell message with practically the same kinetics and plateau value as to its own template. This finding indicates that the same sequences are present in the monolayer and suspended cell mRNA, although there could be a limited change in relative abundances.

The cross-hybridization data also establish the validity of using poly(A) content as a measure of the amount of mRNA. The relative mass of mRNA is measured in the hybridization experiments, as in the translation assays, by poly(U) binding. A change in poly(A) size in the suspended cells would lead to an altered apparent messenger complexity. This measurement indicates that poly(A) size remains constant in suspended cells, and this is confirmed by the direct measurement of poly(A) size shown in the insert to Figure 4. It should also

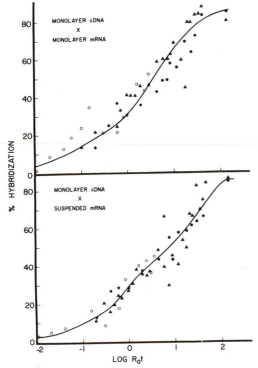

Figure 3. Hybridization of 3T6 Monolayer and Suspended Cell mRNA with cDNA to Monolayer mRNA

Hybridizations were performed as described in Experimental Procedures, using two different concentrations of mRNA: (○) 50 μg/ml; (●) preparation 1, (▲) preparation 2 – both at 500 μg/ml.

be noted that the total amount of poly(A)-containing RNA in the suspended cells is the same as in monolayer cells (Figure 4).

Intactness of Suspended Cell mRNA

Some other mechanisms, such as massive degradation, might be advanced to account for the altered coding properties of suspended cell mRNA. The following measurements indicate that no gross alterations of mRNA are detectable and that the 5′ ends of mRNA are probably present and functional. First, the size of poly(A)+ mRNA was measured by its sedimentation distribution in a sucrose density gradient. Gradient fractions were assayed by poly(U) hybridization. Little or no change was observed in the distribution of poly(A)-containing molecules from 72 hr suspended cells compared to normal monolayer RNA, as shown in Figure 4. Thus any degradation of these preparations is insufficient to alter gross physical size. These suspended cell message molecules, however, despite their essentially unchanged size, can only direct the synthesis of polypeptides of 10,000 daltons molec-

ular weight or less in the reticulocyte cell-free system, as compared to an average of 40K daltons for normal monolayer mRNA.

A more subtle possible alteration in message would involve a modification of the 5′ terminal "cap" structure. This would render the suspended message molecules incapable of associating with ribosomes in a functional linkage, except at greatly reduced rates (Zan-Kowalczewska et al., 1977). A direct measurement of the rate of association of suspended cell mRNA with the ribosomes of a reticulocyte lysate, however, indicates that this aspect of messenger function is probably normal, as shown by the data in Figure 5. Messenger RNA is labeled in 3T6 fibroblasts and the cultures are then split. Half of the cultured cells are under suspension conditions for 72 hr while the remainder are kept in subconfluent monolayers. Poly(A)+ RNA is purified from control and suspended cells and incubated in reticulocyte lysate for 2 and 6 min under protein synthesizing conditions, after which the entire lysate is layered on sucrose gradients and centrifuged. Nonspecific mRNA association with ribosomes is suppressed by making the ionic strength of the sucrose gradients equal to 0.5 M NaCl. This high salt concentration dissociates nonfunctioning ribosomes into subunits, and only ribosomes stabilized by a nascent polypeptide complex continue to sediment at 80S or greater (Zylber and Penman, 1970). The high ionic strength should inhibit nonspecific association of mRNA with ribosomes or other rapidly sedimenting structures.

The gradient profiles in Figure 5 show that the mRNA from suspended cells enters rapidly sedimenting structures (that is, those associated with multiple ribosomes) at about the same rate as the mRNA from monolayer cells. Thus the alteration in translation properties is not reflected in the early steps of protein synthesis, which involve the association of message with ribosomes and the formation of the first peptide linkages. Further evidence that the rapidly sedimenting mRNA is associated with ribosomes is provided by the use of the drug anisomycin. Addition of this inhibitor of peptide elongation (Lodish, 1971) to the reticulocyte lysate incubation mixture allows the association of both labeled mRNAs with 80S structures at the same rate, but blocks the formation of larger aggregates (data not shown).

Thus it appears that mRNA from suspended cells associates with ribosomes at normal rates and directs the synthesis of proteins that are apparently not completed in a normal fashion. As expected, incorporation into TCA-precipitable material by reticulocyte lysate primed with suspended cell message is inhibited much less than the incorporation into discrete bands or spots in one- or two-dimensional gel electrophoresis. This can be seen in

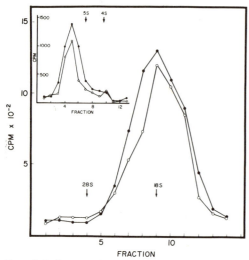

Figure 4. Sedimentation Distribution of Poly(A)⁺ RNA from Monolayer and Suspended Cells

Unlabeled cytoplasmic poly(A)⁺ RNA from equal amounts of monolayer and 72 hr suspended cells was centrifuged through SDS-sucrose gradients. Each fraction was analyzed for poly(A) content by poly(U) hybridization. The insert shows a size determination of the poly(A) in the mRNA from both groups on a 10% polyacrylamide gel. Precise details are given in Experimental Procedures. (●——●) Monolayer, (○—— ○) suspended.

Figure 6, which compares the total incorporation directed by 72 hr suspended cell mRNA with that directed by mRNA from monolayer cells. The suppression of total incorporation in these measurements amounts to between 25 and 40% of the control. This is considerably less suppression than is seen by observing specific polypeptides in the electropherograms. Aside from the reduced incorporation, the synthesis directed by suspended cell message appears to follow a normal time course. In addition, both the reticulocyte and wheat germ systems saturate at the same input mass of suspended mRNA as normal message, although the amino acid incorporation at saturation is greatly reduced.

The apparent change in the translation properties of suspended cell mRNA requires a reinvestigation of the ionic requirements of the cell-free system when translating this material. The activity of the system as a function of the K⁺ ion concentration is shown in Figure 7A. The salt optimum for the system primed by suspended cell message is essentially the same as that for normal monolayer mRNA, although the shapes of the salt dependence curves are not exactly identical. The significance of this small alteration is not known.

Most of the messenger RNA from suspended cells appears to continue to function, at least in the early steps of protein synthesis, as shown by the data in Figure 5. This mRNA behaves as though it has the usual 5′ terminal modification or cap structure, since incorporation in a wheat germ system primed with this material is greatly inhibited in the presence of pm⁷G (Hickey, Weber and Baglioni, 1976). This is shown in Figure 7B for different K⁺ ion concentrations (Kemper and Stolarsky, 1977; Weber et al., 1977). The apparent retention of a cap corresponds with the ribosome binding studies shown in Figure 5, although these data do not rule out partial removal or modification of 5′ ends from some message molecules.

Recovery of Normal Translation Properties after Cell Reattachment

Suspended cells need only to reattach to a solid substrate to regain their normal macromolecular metabolism (Benecke et al., 1978). Protein synthesis returns to normal rates within a few hours, long before mRNA production has risen from its low value in suspended cells. The in vivo studies strongly suggest that cellular protein synthesis resumed largely by reutilizing the message previously stored in an untranslated form in the suspended cells.

It would be expected that modifications of mRNA would be removed when these molecules resume functioning. The experiments shown in Figure 8 indicate that upon cell reattachment, normal translation properties are restored to the preexisting mRNA, which now produces full-sized products in reticulocyte lysate.

Figure 8 compares the proteins synthesized by suspended and reattached cell mRNA in a reticulocyte system. As before, there is a pronounced reduction of defined spots in the translation products of suspended cell message. An equal amount of mRNA from cells that have reattached to a solid substrate for 6 hr, however, even in the presence of actinomycin, yields a qualitatively different translation pattern (shown in Figure 8B). Although new message formation is completely blocked by the drug, there is an increased amount of protein in attached cell mRNA products which focuses to discrete spots. An even more dramatic restoration of normal products is seen in Figure 8C for a similar amount of mRNA from cells that have reattached without any drug present. It should be noted that new mRNA production is still very low during the 6 hr reattachment period (~ 20% of normal) and probably cannot account for the restoration of the qualitatively normal polypeptide pattern seen.

Most (80–85%) of the poly(A)⁻ RNA of suspended cells is found free of ribosomes in postpolysomal regions of sucrose gradients, while the remaining 15–20% is on polysomes. Message gradually reenters polysomes during the recovery period. When

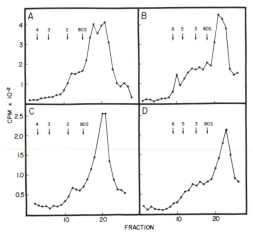

Figure 5. Association of Labeled Monolayer and Suspended 3T6 Cell mRNA with Reticulocyte Ribosomes

Subconfluent monolayer 3T6 cells were labeled with 10 μCi/ml ³H–uridine for 16 hr. Half were then transferred to and maintained in suspension culture for 72 hr. Poly(A)⁺ RNA was extracted from both the labeled monolayer and suspended cells essentially as described in Experimental Procedures, except that an additional step was included to remove labeled ribosomal RNA. The poly(A)⁺ fraction eluted from oligo(dT) was heated to 65°C for 5 min in elution buffer [10 mM Tris (pH 7.4), 0.05% SDS] and then rapidly cooled to 20°C. NaCl was added to give a concentration of 400 mM and SDS to 0.5%, and this fraction was then recycled over oligo(dT). This step reduced ribosomal RNA contamination to <1%.

Reticulocyte cell-free systems (100 μl, not nuclease-treated) contained either 10,000 cpm monolayer mRNA (A, B) or 5000 cpm suspended mRNA (C, D) (0.4 μg RNA in each case). Assays were supplemented with methionine to 200 μM and were incubated for 2 min (A, C) or 6 min (B, D) at 30°C.

Reaction tubes were then chilled on ice and 0.8 ml of HSB [0.5 M NaCl, 5 mM MgCl₂, 10 mM Tris (pH 7.4)] were added. The total volume was then layered onto 13 ml 15–30% (w/v) linear sucrose gradients, made up in HSB and centrifuged in a Beckman SW40 rotor at 4°C at 40,000 rpm. Gradients were fractionated into scintillation vials and bleached with hydrogen peroxide before counting. Arrows represent optical density peaks of the endogenous reticulocyte polysomes.

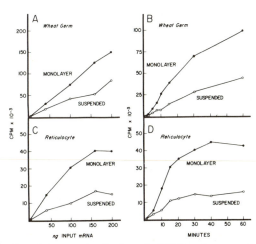

Figure 6. Incorporation Kinetics and mRNA Dose Responses of Wheat Germ Extract and mRNA-Dependent Reticulocyte Lysate Primed with Monolayer and Suspended Cell mRNA

In vitro translations were performed as described in Experimental Procedures. (A, C) Effect of input poly(A)⁺ RNA concentration on translation. Incubations at each concentration were for 60 min. (B, D) Kinetics of translation. Amount of input mRNA in each assay was 150 ng.

Incorporation due to endogenous protein synthesis (wheat germ <7000 cpm, reticulocyte lysate <2000 cpm) measured in parallel incubations without added mRNA has been subtracted from each determination. (●——●) Monolayer, (○——○) suspended.

Discussion

The mRNA obtained from suspended 3T6 fibroblasts behaves very differently from the RNA found in normally growing monolayer cells. A reversible modification appears to take place in the internal region of mRNA, so that the translation products produced by reticulocyte or wheat germ lysates are no longer full-sized or discrete. Both the reduction in incorporation and the shift to smaller peptides (see Figure 1), and the much greater reduction of focusable peptide spots as compared with TCA-precipitable material (shown in Figure 2) suggest that the heterologous lysate ribosomes begin polypeptides but terminate them early when translating the suspended cell message. The putative modification proceeds progressively as cells are incubated for longer times in suspension culture and approximately parallels the decrease in protein synthesis seen in vivo (Benecke et al., 1978).

The modification of mRNA appears to take place within the cells and does not seem to be due simply to increased mRNA degradation during extraction. This fact is supported by the normal size of the mRNA prepared from the suspended cells, and by its normal binding rate to reticulocyte ribosomes, suggesting intact 5' ends. Furthermore, a few polypeptides (noted in Figure 2) are synthesized in vitro

the mRNA from the polysomal and postpolysomal fractions of recovering cells is translated in vitro, the polysomal mRNA shows normal translatability while RNA from the postpolysomal fraction is still largely untranslatable (data not shown). Apparently the fraction of message that is mobilized onto polysomes in the cell is restored to normal translatability.

It is interesting that the products of the message from recovering cells in Figure 8C are not exactly the same quantitatively as those from normal cellular message. The synthesis of actin is noticeably increased relative to other proteins, and this increased actin message may be due to new gene transcription.

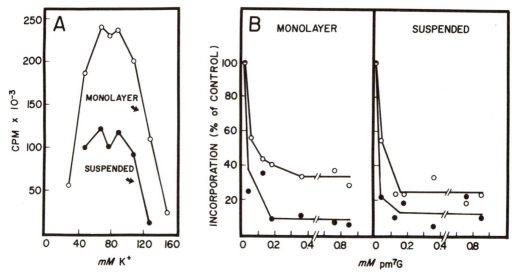

Figure 7. The Response of Wheat Germ Lysate to Altered Potassium Ion and to pm⁷G When Primed with Monolayer and Suspended Cell mRNA

(A) Potassium concentrations were varied by addition of potassium acetate; an initial concentration of 28 mM KCl was contributed by the wheat germ extract. (B) pm⁷G inhibitions were peformed at two different K⁺ concentrations in the wheat germ extract cell-free system. (O——O) At 68 mM K⁺, (●——●) at 108 mM K⁺.

by suspended cell mRNA at the same rate as by the mRNA from control monolayer cells. It seems improbable that a subclass of messages would be impervious to degradative attack, suggesting that this is not the principal source of message modification. It may be noted, however, that the surviving polypeptides are not the same in the wheat germ and reticulocyte systems, suggesting the possibility that these are also encoded by modified message (the modification happening to be read correctly by the heterologous systems), assuming that posttranslational peptide modification does not account for the observed differences.

It is significant that the apparent modification seen in heterologous lysate translation of suspended mRNA appears to be completely reversible. Simply allowing cells to attach to a solid substrate — that is, removing them from methocel and plating them on tissue culture plastic — leads to a return of normal message properties, as seen in the data in Figure 8. A considerable although not complete recovery occurs in the presence of actinomycin, showing that the recovery of normal message properties is not due simply to the production of new RNA to replace the previously modified message. This finding is in agreement with the results of previous studies performed in vivo, which show that the stored untranslated message present in suspended cells reenter polyribosomes upon cell reattachment. The recovery facilitated by actinomycin is not completely normal,

however, as shown by the incomplete recovery of protein synthesis in vivo (A. Ben-Ze'ev, S. Farmer, D. Lawler and S. Penman, manuscript in preparation), and the pattern in Figure 8B differs from the normal both quantitatively and qualitatively. This difference may be due to a need for new transcription and translation to complete the process of message recovery, as suggested by experiments to be reported elsewhere. Certainly the induction of specific new message sequences seems to occur, and actin synthesis, signified by the major spot in Figure 8C, appears to be greatly enhanced, possibly by extensive transcription of the actin genes during recovery.

An important question that cannot yet be answered is whether the mRNA modification found in the suspended cells is the cause or simply a correlate of protein synthesis inhibition. The level of modification, as assayed crudely by the fidelity of heterologous system translation, appears to follow the in vivo protein synthesis rate closely. The mRNA alteration increases during the decline of protein synthesis after cells are put in suspension, and decreases during the subsequent rise of protein synthesis upon reattachment. Nevertheless, it is difficult to see how an internal modification could so completely affect messenger function in the intact cell. It seems reasonable to assume that protein synthesis in the suspended cell does not proceed by mistranslating messages, and indeed those peptides still produced in the intact cell

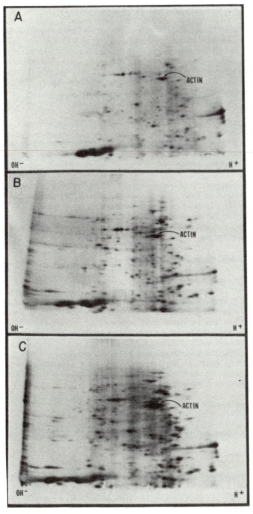

Figure 8. Translation Products Directed by Suspended and Recovery mRNA

3T6 cells were maintained in suspension culture for 72 hr. Half were then placed in normal medium and allowed to attach to plastic culture for 5 hr in the presence of actinomycin D (2 μg/ml) or media alone. The presence of the drug had no effect on the replating efficiency of these cells over the 5 hr period. Neither were there any apparent morphological effects. Poly(A)+ RNA was isolated as described in Experimental Procedures.

In vitro translations and gel analysis were performed as described in Figure 2. Samples analyzed are equivalent to the same amount of input poly(A)+ RNA. Amounts of acid-insoluble material layered were: (A) suspended − 25,000 cpm; (B) recovery and actinomycin D − 56,000 cpm; (C) recovery alone − 61,500 cpm. The dry gels were exposed to X-ray film for five days.

appear to be of normal size. Rather, the modified message appears to be totally withdrawn from translation and does not even form initiation complexes with ribosomes within the cell. Of course,

other mechanisms affecting protein synthesis may operate at the level of mRNA-associated proteins (Civelli et al., 1976; for review, see Shafritz, 1977) or by changes in the polypeptide initiation factors as, for example, has been described for the heme-deficient reticulocyte (Farrell et al., 1977). Alternatively, the modifications may occur sufficiently close to the polypeptide initiation site as to inhibit message association with the endogenous ribosomes; for this to be the case, it must be assumed that the reticulocyte system handles these messages differently than the ribosomes in intact cells. Certainly the two heterologous systems, reticulocyte and wheat germ, behave differently with respect to the modified message, and it is conceivable that the reticulocyte system does not completely duplicate the protein synthesis system of an intact somatic mammalian cell. A possible approach to solving this problem would involve the development of the nuclease-treated endogenous system from the fibroblast cells themselves.

One apparently simple messenger modification does not seem to occur, that being a modification or removal of the 5' cap structures from mRNA. Although such an alteration would account for some of the results reported here, it cannot explain the normal rate of message association with both reticulocyte and wheat germ ribosomes and the high rate of nondiscrete polypeptide formation. A partial removal of caps might also be expected to result in an altered level of input mRNA saturation in the in vitro systems, and this does not occur. Rather, it appears that the initial steps of peptide formation proceed at normal rates and that the lesions appear later in peptide chain elongation.

There are a number of biological phenomena involving the level of message activity which may be related to the putative modifications reported here. One particularly relevant report by Raj and Pitha (1977) indicates the existence of different forms of interferon mRNA. Interferon message from poly(rl)-poly(rc)-induced human fibroblast cells produced interferon in Xenopus oocytes but not in a wheat germ cell-free system. Superinduction of these cells with cyclohexamide, however, results in a change in the interferon message, which now produces functional interferon in the wheat germ system as well. This behavior in different heterologous systems resembles that reported here, and we believe that a similar mechanism may be operative in both cases.

The presence of masked messages in oocytes (Gross et al., 1973), the inhibition of host cell message translation in virus-infected cells (Liebowtz and Penman, 1971) and the selective inhibition of viral mRNA translation by interferon (Gupta et al., 1974; Yakobson et al., 1977) are examples of translational controls whose mechanisms have not been elucidated. Factors have been postulated to account for selective translation but none have been satisfactorily demonstrated. At present the relation of the many phenomena of altered translation selectivity to those reported here is purely speculative, but the experimental test is obvious, and the possible role of message modification in a wide variety of biological phenomena will doubtless be ascertained shortly.

Experimental Procedures

Detailed procedures for the maintenance of 3T6 cells in subconfluent monolayer and suspension cultures, and for the isolation, quantitation and size analysis of poly(A)+ RNA are all documented in Benecke et al. (1978).

Cell-Free Protein Synthesis Assays
Translation in Wheat Germ Extracts
Preincubated wheat germ S30 extracts were prepared according to Roberts and Paterson (1973), and translations were performed in 25 μl reaction volumes (unless stated otherwise) at 25°C, as described previously in Kaufmann et al. (1977).
Translation in mRNA-Dependent Reticulocyte Lysate
Reticulocyte lysates were prepared from New Zealand white rabbits made anemic by seven daily injections of neutral phenylhydrazine (10 mg/kg body weight). Rabbits were bled by cardiac puncture after a four day recovery period and reticulocyte lysates were prepared as described by Hunt, Vanderhoff and London (1972). These lysates were rendered mRNA-dependent by controlled treatment with micrococcal nuclease as outlined by Pelham and Jackson (1976). Translations were performed in 25 μl volumes (unless stated otherwise) at 37°C, as described previously in Kaufmann et al. (1977).

Polyacrylamide Gel Electrophoresis
Dodecylsulphate-Polyacrylamide Gradient Slab Gels
Samples (5–10 μl) of cell-free protein synthesis assays were analyzed on 7–17% gradient polyacrylamide slab gels containing sodium dodecylsulphate according to Laemmli (1970), as described previously by Kaufmann et al. (1977).
Two-Dimensional Gel Electrophoresis
Isoelectrofocusing in the first dimension was essentially as detailed by O'Farrell (1975), except that the pH gradient was extended from 3 to 10. Samples (10–25 μl) of the cell-free protein synthesis assays were freeze-dried before being dissolved in the isoelectrofocusing sample buffer. The second dimension was electrophoresed through a 10–15% polyacrylamide slab gel, as above.

After electrophoresis, gels were impregnated with PPO (Bonner and Laskey, 1974), dried and exposed to prefogged Kodak X-Omat film at −70°C (Laskey and Mills, 1975).

Synthesis of cDNA
This procedure was a slight modification of the one described previously (Kaufmann et al., 1977). 75 μCi of ^{3}H-dCTP (New England Nuclear, 24–25 Ci/mM) were dried down under vacuum in a siliconized tube. 20 μl of the reaction mix were then added, bringing the final concentrations of reagents to (in 25 μl) 50 mM Tris (pH 8.3), 20 mM DTT, 6 mM MgAc$_2$, 60 mM NaCl, 1 mM dATP, 1 mM dGTP, 1 mM dTTP, 122 μg/ml Act D and 5 μg/ml (dT)$_{12-18}$ (Collaborative Research). 1 μg (in 2 μl) of mRNA and 20 units of AMV reverse transcriptase were added. Reaction was at 37°C for 30 min.

The cDNA was then isolated from the final reaction mix by a new method developed by M. Rosbash (manuscript in preparation). Briefly, this procedure involves ethanol precipitation of the cDNA from the reaction mix in the presence of 1.5 M ammonium acetate, which prevents the co-precipitation of protein. Final reaction volume was 23 μl; to this were added 75 μl 10 mM Tris, 10 mM MgAc$_2$, 300 μl 2 M ammonium acetate and 10 μl tRNA (4.5 mg/ml). Precipitation was effected by the addition of 1 ml ethanol. Two further ethanol precipitations were needed to remove contaminating ^{3}H-dCTP completely. The cDNA was then subjected to alkaline hydrolysis in 0.3 N NaOH at 37°C for 3 hr. After neutralization with HCl, the purified cDNA was obtained by ethanol precipitation, using purified tRNA as carrier.

Hybridization of mRNA with cDNA
Hybridizations were performed essentially as described by Kaufmann et al. (1977) with slight modifications. A large excess (1000 fold) of mRNA was mixed with sufficient cDNA to give 500–1000 cpm per point, and in such a volume as to give 40–50 points. This mixture was dispensed into capillary tubes (siliconized), each containing 1 μl. Hybridizations were carried out at 70°C. Individual capillary tubes were removed and broken at the various time points, and their contents were expelled into S1 nuclease buffer at 4°C. Hybridization was carried out at mRNA concentrations of up to 500 μg/ml and for up to 24 hr. Determination of S1 nuclease-resistant material was as described previously (Williams and Penman, 1975).

Acknowledgments

This study was undertaken with support from grants provided by the NSF and the NIH. Part of this work was performed while S.R.F. and A.B.Z. held research training fellowships awarded by the International Agency for Research on Cancer. S.R.F. is now a fellow of the Muscular Dystrophy Association of America.

We gratefully acknowledge the expert technical assistance of Donna Lawler.

Received May 26, 1978; revised July 14, 1978

References

Benecke, B.-J., Ben-Ze'ev, A. and Penman, S. (1978). Cell *14*, 931–939.

Bonner, W. M. and Laskey, R. A. (1974). Eur. J. Biochem. *46*, 83–88.

Civelli, O., Vincent, A., Burik, J. F. and Scherrer, K. (1976). FEBS Letters *72*, 71–76.

Farrell, P. J., Balkow, K., Hunt, T., Jackson, R. J. and Trachsel, H. (1977). Cell *11*, 187–200.

Gross, K. W., Jacobs-Lorena, M., Baglioni, C. and Gross, P. R. (1973). Proc. Nat. Acad. Sci. USA *70*, 2614–2618.

Gupta, S. L., Grazidei, W. D., III, Weideli, H., Sopori, M. L. and Lengyel, P. (1974). Virology *57*, 49–63.

Hickey, E. D., Weber, L. A. and Baglioni, C. (1976). Proc. Nat. Acad. Sci. USA *73*, 19–23.

Hunt, T., Vanderhoff, G. and London, I. M. (1972). J. Mol. Biol. *66*, 471–481.

Kaufmann, Y., Milcarek, C., Berissi, H. and Penman, S. (1977). Proc. Nat. Acad.Sci. USA *74*, 4801–4805.

Kemper, B. and Stolarsky, L. (1977). Biochemistry *16*, 5675–5680.

Laemmli, U. K. (1970). Nature *227*, 680–685.

Laskey, R. A. and Mills, A. D. (1975). Eur. J. Biochem. *56*, 335–341.

Liebowitz, R. and Penman, S. (1971). J. Virol. *8*, 661–668.

Lodish, H. F. (1971). J. Biol. Chem. *246*, 7131–7138.

MacPherson, I. and Montagnier, L. (1964). Virology 23, 291–294.

O'Farrell, P. H. (1975). J. Biol. Chem. 250, 4007–4021.

Otsuka, H. and Moskowitz, M. (1975). J. Cell Physiol. 87, 213–220.

Pelham, H. R. B. and Jackson, R. J. (1976). Eur. J. Biochem. 67, 247–256.

Raj, N. B. K. and Pitha, P. M. (1977). Proc. Nat. Acad. Sci. USA 74, 1483–1487.

Roberts, B. E. and Paterson, B. M. (1973). Proc. Nat. Acad. Sci. USA 70, 2330–2334.

Shafritz, D. A. (1977). In Molecular Mechanism of Protein Synthesis, H. Weissbach and S. Petska, eds. (New York: Academic Press), pp. 555–601.

Stoker, M., O'Neill, C., Berryman, S. and Waxman, V. (1968). Int. J. Cancer 3, 683–693.

Weber, L. A., Hickey, E. D., Nuss, D. L. and Baglioni, C. (1977). Proc. Nat. Acad. Sci. USA 74, 3254–3258.

Williams, J. G. and Penman, S. (1975). Cell 6, 197–206.

Yakobson, E., Prives, C., Hartman, J. R., Winocour, E. and Revel, M. (1977). Cell 12, 73–81.

Zan-Kowalczewska, M., Bretner, M., Sierakowska, H., Szczesna, E., Filipowicz, W. and Shatkin, A. J. (1977). Nucl. Acids Res. 4, 3065–3081.

Zylber, E. A. and Penman, S. (1970). Biochim. Biophys. Acta 204, 221–229.

Cell, Vol. 13, 19–32, January 1978, Copyright © 1978 by MIT

Isolation and Characterization of T Antigen-Negative Revertants from a Line of Transformed Rat Cells Containing One Copy of the SV40 Genome

Bettie Steinberg and Robert Pollack
SUNY at Stonybrook
Stonybrook, New York 11794
William Topp and Michael Botchan
Cold Spring Harbor Laboratory
Cold Spring Harbor, New York 11724

Summary

Negative selection with FUdR produced revertants from the transformed rat line 14B, which contains one insertion of the SV40 viral genome (Botchan, Topp and Sambrook, 1976). 14B contains nuclear T antigen, grows to a high density, grows in low serum and is anchorage-independent. The revertants fall into three classes with regard to viral DNA sequences: the SV40 DNA is retained; the SV40 DNA is retained but has undergone a deletion; and the SV40 DNA is lost, generating a cured cell. This heterogeneity is not a result of long-term passage. The revertants arise with a frequency of one in 8.4×10^5 cells after as few as 12 passages. All three classes of revertants are T antigen-negative, density-sensitive, more serum sensitive than 14B and anchorage-dependent. These data argue for a direct role of the functioning viral genome in the maintenance of the transformed state, and that with 14B, the phenotypes of transformation are not virus gene dosage-dependent.

Introduction

Nonpermissive cells, infected with SV40, give rise to transformed cell lines. These cells lose some or all of the normal growth controls, they may grow in low serum concentrations, grow beyond confluence in high serum, grow when suspended in methocel and form tumors when injected into animals.

SV40-transformed cells also display a number of antigenic differences from the parent cell. Most notable of these is the presence of T antigen(s) in the nucleus. One of these proteins has been shown to be the product coded for by the A complementation group of SV40 (Tegtmeyer et al., 1975; Tenen, Baygell and Livingston, 1975). The SV40 T antigen (or A gene product) is a polypeptide of approximately 94,000 daltons and is expressed both before and after the onset of viral DNA replication in the viruses lytic cycle (Tooze, 1973). The presence of this T antigen is believed necessary for the establishment of transformation (Tegtmeyer, 1975; Martin and Chou, 1975) and may also be necessary for the continued expression of the transformed phenotype (Tegtmeyer, 1975; Brugge

and Butel, 1975; Martin and Chou, 1975; Osborn and Weber, 1975). Based in part on the observation that temperature-sensitive mutations in the A gene render the initiation of viral DNA replication temperature-sensitive in its lytic cycle (Tegtmeyer, 1972) and in part on changes observed in the initiation of cell DNA synthesis in SV40-transformed cells, many investigators have proposed that the SV40 T antigen acts directly as an initiator of host cell DNA synthesis (Tegtmeyer, 1972; Butel, Brugge and Noonan 1974; Osborn and Weber, 1975; Martin and Oppenheim, 1977).

At the present time, we do not know whether all the aspects of transformation require a direct viral role for maintenance, or whether the virus is needed only for the initiation of some of the transformed phenotypes. Moreover, we do not know which genes of the virus are required for the various transformation characteristics. Conditional mutations in the A gene, which is required for transformation, have been of some use in this respect. Revertants should allow us to probe these questions further. Furthermore, if viral genes are required to maintain a particular transformed phenotype, one class of revertants should include cell lines whose viral DNA sequences and viral gene expression have been altered. Our understanding of the mechanisms of reversion and of transformation itself has been limited, however, by the fact that most transformed cells contain multiple copies of the virus. Clearly then, if a transformed cell contains multiple copies of dominant viral genes, reversion from transformation must frequently occur by mutations in other cellular functions. It has therefore not been possible to determine with certainty whether reversion resulted from an alteration in a viral genome, the host genome or some interaction between the two. It has also not been possible to determine whether a reduction in the number of copies of active SV40 DNA in the revertants is a major factor in the phenomenon of reversion. Previous studies have not detected a significant loss of viral DNA sequences from revertant cells containing multiple copies of viral DNA (Ozanne, Sharp and Sambrook, 1973).

Revertants from transformed cells have been isolated by exposing the transformed cells to a variety of killing agents, including FUdR (Pollack, Green and Todaro, 1968; Nomura et al., 1973), concanavalin A (Ozanne and Sambrook, 1971; Culp and Black, 1972), colchicine (Vogel, Risser and Pollack, 1973) and BUdR (Vogel and Pollack, 1973), under conditions where only transformed cells could grow.

One such group of revertants from SV40-transformed cells was selected for density inhibition. These density revertants, isolated independently by Pollack, Green and Todaro (1968), Pollack and Burger (1969), Grimes and Black (1971), Culp and Black (1972), Wicker et al. (1972), Risser and Pollack (1973) following treatment with a variety of killing agents, have a number of characteristics in common. All the cells grow to a much lower density than the transformed parent. At confluence, the number of cells in mitosis is greatly reduced. They generally have an organized cytoarchitecture, as revealed by indirect immunofluorescence, similar to nontransformed cells (Pollack, Osborn and Weber, 1975). Most of these revertants have a marked reduction in their ability to grow in methocel and are less tumprigenic. Most will still grow in low serum. They still contain SV40 DNA, however, and infectious virus can be rescued from the cells. Moreover, the revertants generally exhibit a marked increase in their DNA content and increase in ploidy. Perhaps most important of all, they still express the SV40 T antigen.

At the time this work was initiated, all revertants not only contained SV40 DNA, but also synthesized T antigen. The persistence of T antigen(s) in revertants can be interpreted in several different ways. It can mean that T antigen(s) is necessary but not sufficient for the transformation phenotype, and that the cell can revert by a cellular alteration. Alternatively, it can mean that T antigen(s) is not involved in maintenance of the transformed cell. A third explanation is that the antigenicity is present but the product is not functional.

There are reports in the literature which support the concept that reversion can correlate with a loss of viral T antigen. Basilico and Zouzias (1976) have reported a temperature-sensitive revertant which ceases T antigen synthesis at confluence at high temperature, when a temperature-sensitive cellular mutation permits the cells to block in G_0. Furthermore, Kelly and Sambrook (1974) isolated cytochalasin B-resistant variants from a cell line containing multiple copies of SV40. These variants had lost one copy of the SV40 genome and were transiently T-negative. They did not, however, maintain the properties characteristic of revertants. Finally, Marin and MacPherson (1969) isolated revertants from a polyoma transformant which lacked viral antigens, and it was suggested by these investigators that the cell reversion was initiated by the loss of the viral genome.

An established rat embryo line fully transformed by SV40 DNA, line 14B, has been reported to contain only one copy of the SV40 genome per cell (Botchan et al., 1976). In this paper, we describe density-sensitive revertants, isolated from this cell line with FUdR, which completely lack T antigen. In some of these revertants, the region of the integrated viral DNA that codes for T antigen(s) has been mutated.

Results

Expression of T Antigen in Revertant Cells

Seven density-sensitive colonies were isolated by FUdR selection from 14B using the following protocol. The transformed cells were cloned 3 times — first by picking isolated colonies, then by isolating single cells from this colony in microwells — grown for 20 doublings, recloned once by picking an isolated colony and then grown again for approximately 20 doublings. The transformed cells were then exposed to FUdR, and seven surviving colonies with flat "revertant" morphologies were picked as described in Experimental Procedures. All the colonies that were selected, which were subsequently found to be revertants with respect to many of the properties of SV40 transformation (see below), have in common a lack of T antigen expression. This SV40-coded protein is absent from all seven revertant cell lines as measured by immunofluorescence, and is below the limits of detection when labeled cell proteins are immunoprecipitated with anti-T sera.

Figure 1A displays the parent transformed cell line 14B, stained for T antigen by indirect immunofluorescence. The nuclei show the characteristic bright fluorescence of SV40-transformed cells. In contrast, the revertant varients of 14B, line FL[1] 1–3, FL[1] 3–8 and FL[1] 3–5, are shown in Figures 1B, 1C and 1D, respectively. These cell lines and the other four are negative for T antigen expression under a variety of conditions. Basilico and Zouzias (1976) had shown that a cellular mutation could render T antigen expression temperature-sensitive when the cells were confluent. We therefore tried varying incubation temperature and cell density testing 14B, Rat-1 (the parent of 14B) and three of the revertants for T antigen immunofluorescence. The results of this experiment are shown in Table 1. The transformed cell line, 14B, remained T antigen-positive under all conditions, while Rat-1 and the three revertant lines never expressed T antigen.

These observations were extended by immunoprecipitating the T antigen with anti-T sera (provided by Dr. R. Tjian) after labeling total proteins with either ^{32}P– or ^{35}S–methionine. Figure 2A is an autoradiogram of such ^{32}P–labeled immunoprecipitates after gel electrophoresis. From line 14B, a major band at 95,000 daltons is seen. This protein has a similar mobility to the phosphorylated protein immunoprecipitated from lytically infected monkey cells. T antigen is known to be a phosphorylated protein (Tegtmeyer, Rundell and Collins, 1977). No phosphorylated protein could be detected in any of the revertant cell lines. That this immunoprecipitated protein with an apparent molecular weight 95,000 daltons is T antigen is substantiated by

301

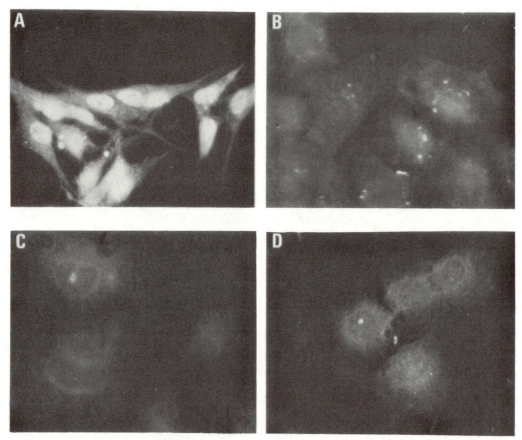

Figure 1. Immunofluoroescent Assay for SV40 T Antigen

The SV40 T antigen was detected by indirect immunofluorescence as described in Experimental Procedures. (A) is a photograph of the fluorescence seen from the transformed parent 14B. (B, C and D) represent fluorescent images of revertant cell lines FL¹ 1–3, FL¹ 3–8 and FL¹ 3–5, respectively.

the finding that it is not immunoprecipitated by preimmune sera. When total cell proteins were labeled with ^{35}S-methionine, other protein bands were seen after immunoprecipitation with anti-T sera and subsequent gel electrophoresis and autoradiography (Figure 2B). No T antigen-specific bands could be detected in the revertant cell lines. The pattern of labeling shown for the revertant FL¹ 3–5 is identical to that seen in its untransformed grandparent Rat-1 (data not shown). Several other protein bands could be detected in these ^{35}S-methionine-labeled immunoprecipitates from cells of line 14B. Presumably some of these represent proteolytic degradation products of the 95,000 dalton protein and/or other independent SV40 T antigen-containing polypeptides (see Experimental Procedures).

SV40 DNA in Revertants

The loss of T antigen expression in the revertant cell lines could be the result of an epigenetic change in the cells' regulatory machinery, or a genetic change in either the structural gene(s) coding for T antigen or its production. Analysis of the integrated viral DNA in the revertants and in the parent transformed cell line 14B was therefore performed by hybridization techniques previously described (Botchan et al., 1976). A continuous stretch of DNA sequences extending from position 20 clockwise to position 98 on the SV40 chromosome has been detected in line 14B by these techniques. Furthermore, there has been no indication of deletion or duplication of these viral sequences in this cell DNA. It has been inferred from these data that the SV40 DNA is joined to cell DNA by sequences mapping somewhere between position 98 clockwise to position 20 on the viral chromosome.

The restriction enzyme Bal I, an endonuclease isolated from Brevibacterium albidum, is of special

Table 1. Effect of Culture Conditions on T Antigen Expression

| | Fraction of Cells Expressing T Antigen | | | |
| | Confluent Cells[a] | | Sparse Cells[b] | |
Line	37°C	32.5°C	37°C	32.5°C
Rat 1	<0.001	<0.001	<0.001	<0.001
14B	0.96	0.96	0.96	0.96
FL¹ 1-4	<0.001	<0.001	<0.001	<0.001
FL¹ 3-3	<0.001	<0.001	<0.001	<0.001
FL¹ 3-8	<0.001	<0.001	<0.001	<0.001

[a] Cells plated at 2×10^4 cells per cm² and grown at desired temperature until confluence.
[b] Cells plated at 10^3 cells per cm² and grown at desired temperature for 3 days.

use in this type of analysis because its recognition sequence is not present in SV40 DNA. Thus when high molecular weight chromosomal DNA is extracted from cell line 14B and this DNA is hydrolyzed by Bal I, only one SV40 DNA fragment is created. This fragment contains the entire SV40 DNA insertion and the chromosomal DNA that is proximal to the virus on both sides of the insertion. The resulting fragment can be detected by the Southern blotting technique (Southern, 1975; Botchan et al., 1976; see also Ketner and Kelly, 1976). Analysis of the SV40 fragment created by Bal I restriction enzyme hydrolysis of revertant cell DNA is shown in Figure 3. The results of two separate experiments are shown in this figure. The bottom panel shows one band in the slot marked "14B" and a band at the same position for cell lines FL¹ 1-3 and FL¹ 1-4. In one lane, the equivalent of one copy per cell of SV40 DNA (2×10^{-6} μg per 2 μg of cell DNA) was mixed with FL¹ 1-4 DNA. Bands corresponding to forms II and III SV40 DNA are visible, as well as a very faint band at the position of form I DNA. In two other lanes, 2 μg of 14B DNA were mixed with 2 μg of revertant cell DNAs from lines FL¹ 1-3 and FL¹ 1-4. Only one intense band was seen in these lanes. In contrast to these results, no detectable SV40 DNA could be found in cell DNA from lines FL¹ 3-8 and FL¹ 3-2. That FL¹ 3-8 and FL¹ 3-2 had lost their copy of the SV40 DNA has been verified by independent experiments with several other enzymes including Eco RI, Hpa I and Pvu II (data not shown). We have not been able to detect any mutation in the SV40 DNA of cell lines FL¹ 1-3 and FL¹ 1-4 with the same set of enzymes. This, of course, does not rule out the possibility that small deletions and/or point mutations in the viral DNAs of these cells had occurred (see Figure 4). The results of a similar experiment are illustrated in the upper right panel of Figure 3. The panel in the upper left of this figure is a

picture of the total cell DNA in each slot stained with ethidium bromide before the transfer of these DNAs to nitrocellulose was executed. The slot labeled "R" contained 0.2 μg of Ad-2 DNA which serve as markers and 5×10^{-6} μg of SV40 DNA. Forms I, II and III SV40 are visible in the autoradiogram. Several other faint bands are also visible in the autoradiogram. These represent marker fragments of phage Mu DNA that were added to the DNA samples to ensure that transfer and detection efficiency were satisfactory and that the migration of DNA in the different lanes was equivalent. The slot labeled "3-5" also contained one tenth equivalent of free SV40 DNA. These various controls make it quite certain that cell lines FL¹ 3-8 and FL¹ 3-2 can contain no more than a fragment of SV40 DNA of length equivalent to 10% of the viral genome. In cell line FL¹ 3-5, a new DNA fragment is detected which has a faster mobility than that fragment seen in cell line 14B. The difference in mobilities between these bands indicates a difference of molecular weight between these two fragments of approximately 300,000 daltons. This result can mean that the SV40 sequence has translocated its chromosomal location or there has been a deletion of DNA. To learn more about the difference between the SV40 genomes of these two cells, the restriction enzyme Pvu II (from Proteus vulgaris) which cleaves SV40 DNA at positions 98.5, 32 and 72 was used. The results of this experiment are shown in Figure 4. Digestion of 14B DNA with endonuclease Pvu II yields a fragment of molecular weight 2.3×10^6 daltons, which is probably the fragment which contains the fusion sequences of cell DNA with SV40 DNA. Two other fragments with mobilities identical to the A and B fragments of SV40 cleaved with Pvu II are seen. The C fragment is missing in this digestion, presumably because it contains within it the host cell DNA-viral DNA recombination joint. This result, taken together with our previous analysis of this cell line (Botchan et al. 1976), must mean that the recombination joints map very close to position 98 on the viral chromosome, on both sides of the viral insertion. The DNA samples from revertant cell lines FL¹ 1-3 and FL¹ 1-4 show a pattern of fragments indistinguishable from their parent 14B. In the case of revertants FL¹ 3-5 and FL¹ 3-3, the A fragment is missing and a new fragment appears. In view of the results shown in Figure 3, the appearance of a new fragment in FL¹ 3-5 DNA with a mobility indicative of a size approximately 230,000 daltons smaller than the Pvu II A fragment of SV40, we must conclude that there has been a deletion of SV40 DNA within this cell. The new fragment in cell line FL¹ 3-3 is consistent with a deletion in the SV40 DNA sequences of this cell of

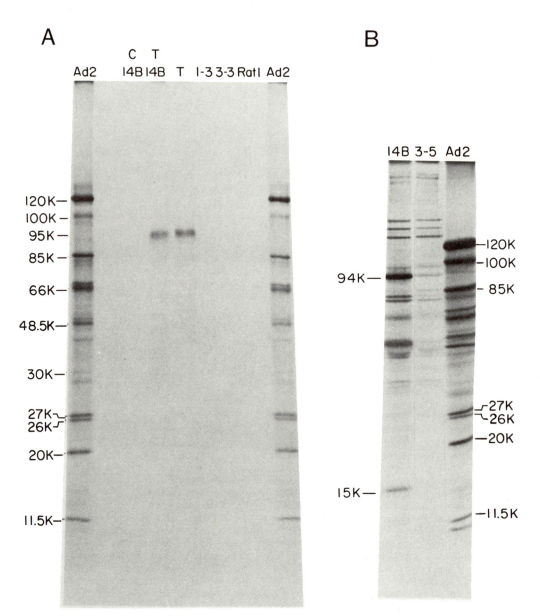

Figure 2. Autoradiographic Detection of SV40 T Antigen

Total cell proteins were pulse-labeled with either ^{32}P-phosphate (A) or with ^{35}S-methionine (B) as described in Experimental Procedures. Immunoprecipitated T antigens were prepared and fractionated by electrophoresis on 7–15% (A) or 5–20% (B) gradient polyacrylamide gels (Tjian and Pero, 1976).

(A) The slot labeled C/14B was a control immunoprecipitate prepared with preimmune anti-T sera and labeled 14B proteins. T/14B shows the protein band detected in transformed cells of 14B; this is to be compared to that band detected in SV40 lytically infected CV-1 cells which is shown in the slot labeled T. No phosphorylated protein could be detected in cell lines 1–3, 3–3 or Rat-1 after immunoprecipitation. Marker proteins are total adenovirus 2 proteins (Ad2) labeled (with ^{35}S-methionine) late in the lytic infection of Hela cells.

(B) Two SV40-specific protein bands could be detected in ^{35}S-methionine proteins of transformed cell line 14B. They are labeled 94K and 15K. These bands could not be detected in the proteins from cell line FL^1 3–5.

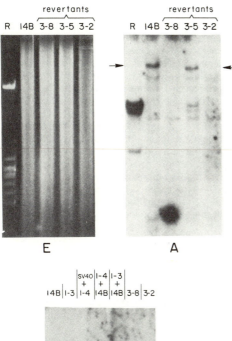

E

A

SV40
+
I-4 | I-3
+ | +
14B | I-3 | I-4 | 14B | 14B | 3-8 | 3-2

A

Figure 3. Detection of Fragments That Contain SV40 Sequences after Hydrolysis of Cell DNA with Endonuclease Bal I

(E) is a photograph of the total DNA in each slot stained with ethidium bromide after fractionation of the fragments on 0.7% agarose gels. Slot R is a control experiment which was a digestion of 0.2 of Ad2 DNA and 5×10^{-6} µg of SV40 DNA. Each cell DNA lane contained 2 µg of the various cell DNAs as well as ^{32}P-labeled marker Mu-phage fragments. Slot 3-5 also contained 2×10^{-7} µg of SV40 DNA.

The upper right panel (A) is an autoradiogram of the fragments detected on the nitrocellulose sheet after transfer of the fragments and hybridization with "nick-translated" SV40 DNA. The arrows point to the integrated DNA fragments seen in cell DNAs from line 14B and FL[1] 3-5.

The bottom panel (A) is also an autoradiogram of a similar experiment. Various mixtures of SV40 DNA and cell DNAs were made as indicated in the figure and text.

690,000 daltons. This has been confirmed by Bal I digestion of this DNA (data not shown). The deletions in FL[1] 3-5 and FL[1] 3-3 are 6 and 18%, respectively, of the SV40 genome and map within the coordinates of positions 32 and 72 on the viral chromosome. Although further work will be necessary to refine the map coordinates of these deletions, it is clear that they fall within that part of the inserted SV40 DNA which must code for the messenger RNA of T antigen (Khoury et al., 1975; Prives et al., 1977).

In summary, three different types of revertants have been isolated: those which seem to have lost their SV40 DNA; those which have suffered deletions within that region of the SV40 DNA which codes for T antigen mRNA; and those which show no gross rearrangements at all. Because only seven independent revertant clones were analyzed, we do not know the actual frequencies with which these different classes of revertants arise. We have chosen to examine the biological properties of one representative of each class in what follows.

Fluctuation Analysis

FUdR treatment may either induce the conversion of 14B to the revertant phenotype, or it may simply select for revertants which had appeared spontaneously in the population, or both. To distinguish among these possibilities, we did a fluctuation analysis, as originally described by Luria and Delbrück (1943).

Line 14B was originally transformed by a DNA infection of Rat-1, cloned through microwells, replated and repicked and grown for 20 generations. It was then recloned, grown for approximately 200 doublings (12 passages) and used for the fluctuation analysis, described below.

Two confluent plates of 14B were prepared by plating 10^5 cells per plate and growing the cells to a final concentration of approximately 7×10^6 cells per plate. These plates were treated with FUdR, and the surviving cells from each plate were diluted 1:40 and replated (Table 2A). The number of survivors in this part of the experiment reflects the average number of flat colony-forming cells in a large population of 14B. The standard deviation reflects the statistical variability in plating aliquots of this population. The reversion frequency, calculated as that fraction of cells in the original large population capable of generating flat colonies, is 1.5×10^{-4}. In a parallel experiment, the FUdR survivors formed colonies on glass coverslips and were stained for T antigen. 56% of the flat colonies were T antigen-negative. Thus the frequency of T-negative revertants is approximately 8×10^{-5}.

In the second part of the experiment, 20 separate plates of 14B were grown to confluence from very small numbers of cells (< 100 cells per plate).

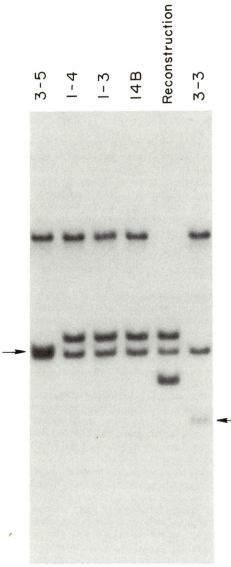

Figure 4. Detection of Fragments That Contain SV40 Sequences after Hydrolysis of Various Cell DNAs with Endonuclease Pvu II 50 μg of cell DNA were cleaved with endonuclease Pvu II, and the products were fractionated on alkaline-1.4% agarose gels (McDonnell et al., 1977). The slot labeled Reconstruction contained 50 × 10⁻⁶ μg of SV40 DNA and 1 μg of Ad2 DNA also cleaved with endonuclease Pvu II. The arrows point to bands seen in revertant cell lines FL¹ 3-5 and FL¹ 3-3 which are not present in the parent cell line 14B.

Each plate was then treated with FUdR. The survivors were diluted 1:25, and one aliquot from each plate was replated (Table 2B). If the appearance of cells capable of forming flat colonies is a random

event, it will occur in some of the populations at an earlier time than in others. Those populations in which a cell capable of forming a flat colony arose early or was present in the original small population will accumulate more flat colony-forming cells than those populations where the event occurred shortly before drug treatment. Such random events will lead to a very large variance in the distribution of flat colonies per dish. Table 2B shows such a variance. The deviation we observe in Table 2B is much higher than could be accounted for by plating variability in Table 2A. FUdR treatment therefore selects for preexisting revertants in the 14B population and does not induce them. The high degree of heterogeneity observed in Table 2B after 16 doublings of the population reflects the high frequency of reversion shown in Table 2A.

Cellular Morphology

The cellular morphology of 14B, Rat-1 and the three revertants, growing in colonies, is shown in Figure 5. Cells in Figures 5a, 5c, 5e, 5g and 5i were located at the edge of the colony, while those in Figures 5b, 5d, 5f, 5h and 5j were in the center of the colony.

It is apparent that 14B (Figures 5a and 5b) is not contact-inhibited, and that the cells pile up in the center of the colony. At the bottom of Figure 5a, the cells are only one cell thick, and are oriented in such a way that the longitudinal axis of each cell tends to radiate from the center of the colony (seen at the top of Figure 5a). In Figure 5b, it is possible to see the dense multilayer morphology at the center of the colony.

Rat-1 (Figures 5c and 5d) and FL¹ 1-4 (Figures 5e and 5f) look very similar. The cells grow as a monolayer, are approximately the same size and change shape from an elongated form to a more cuboidal shape when they are at confluence for a long time (that is, in the center of a colony). Lines FL¹ 3-3 (Figures 5g and 5h) and FL¹ 3-8 Figures 5i and 5j) also form flat monolayers. They contain larger cells than Rat-1 or FL¹ 1-4. Note that the center of the FL¹ 3-3 colony (Figure 5h) contains a few very large cells with two nuclei. This is the only revertant that commonly contains such cells.

Growth Properties

Table 3 summarizes a number of growth characteristics of the transformed cell line and the revertants. These properties are described individually.

Effects of Serum Concentration on Growth Rate and Saturation Density

Cell increase was followed for 2 weeks for each line with concentrations of fetal calf serum from

Table 2. Fluctuation Analysis of 14B Treated with FUdR

Plate Number	Number of Flat Colonies		Number of Flat Colonies (B)
	(A1)	(A2)	
1	24	20	25
2	21	34	77
3	17	30	72
4	39	22	20
5	22	35	700
6	33	39	14
7	23	29	300
8	31	27	64
9	30	31	65
10	25	35	118
11	18	36	28
12	28	40	89
13	22	28	40
14	25	28	700
15	29	40	22
16	28	41	23
17	30	40	12
18	32	36	30
19	36	30	75
20	16	32	100
Average	26.45	32.65	124.9
Variance	37.05	35.87	40079
Standard Deviation	6.08	5.99	200.2
Frequency of Flat Colonies	1.5×10^{-4}		
Frequency of Tag-Negative Colonies	8×10^{-5}		

(A) 14B was plated on two 60 mm plates at 10^5 cells per plate, grown for approximately six doublings, treated, diluted 1:40 and plated on 40 plates.
(B) 14B was plated on twenty 35 mm plates at 200 cells per plate, grown for 8 days, trypsinized and transferred to 60 mm plates, grown to confluence, treated, diluted 1:25 and replated.

0.1–10%. Rat-1 and the revertants did not increase significantly in cell number at serum concentrations below 1.0% (data not shown). The doubling time for each cell population at 10 and 1% was calculated from the linear portion of the growth curves (Table 3).

In 10% FCS, the transformed cells have a doubling time of 14 hr, while the normal cell line has a doubling time of 22 hr. All three revertants have doubling times similar to, but longer than, Rat-1. The differences among the lines are greater in 1% FCS. All the lines will grow in 1% serum, but the revertants have doubling times almost twice that of Rat-1, while the doubling time of the transformed cells is only two thirds that of Rat-1. The revertants are more like secondary rat embryo fibroblasts than like Rat-1 in their inability to grow in low serum.

Saturation densities were calculated from the number of cells per cm² when there was no further increase in growth curves. Line 14B grows very densely in 10% FCS, with the cells growing on top of each other to form a tangled mass. The morphology of the culture is similar in 1% serum, but the cells grow to a lower density. Rat-1 and the revertants cease to grow at confluence in both 10 and 1% FCS. The cells spread to a larger diameter in 1% FCS, which could contribute to the lower saturation density in low serum. With all three revertants, the saturation density is lower than for Rat-1. As with doubling times, the revertants are "more normal" than Rat-1. Experiments are currently in progress to determine whether Rat-1 and the revertants enter a stage of G_0 at confluence, in a manner similar to normal cells.

Plating Efficiency on Plastic
Plating efficiencies of the various lines are also shown in Table 3. They vary from 1–100% on plastic dishes. The cured cells, FL[1] 3–8, have the highest plating efficiency on plastic (100%), although Rat-1 and FL[1] 1–4 plate with reasonably high efficiency (80 and 62%, respectively). A growing culture of FL[1] 3–3 contains many cells with two nuclei (Figure 5h). If these are cells which are no longer capable of division, their presence would contribute to both the slow doubling time of the population and the low plating efficiency of this revertant line.

Growth without Anchorage
Line 14B grows in methocel (Table 3). Within 2 weeks after inoculation, most of the 14B colonies were 0.15–0.3 mm in diameter. After 3 weeks, each of the other cell lines had only a few colonies larger than 0.15 mm on plates containing 10^5 cells. In addition, Rat-1, FL[1] 3–3 and FL[1] 3–8 had numerous tiny colonies, approximately 0.05 mm in diameter, while FL[1] 1–4 had a few of these small colonies. These colonies presumably arose from cells that were only able to divide a few times in methocel.

The plating efficiency on methocel relative to that on plastic was calculated (Table 3). 14B had an RPE of 15%, more than three orders of magnitude higher than Rat-1, FL[1] 1–4 and FL[1] 3–8. FL[1] 3–3 had an RPE 10 times higher than Rat-1 and the other revertants, because it makes colonies on plastic with such low efficiency. Methocel growth was the only culture condition where FL[1] 3–3 grew as well as the other revertants. It is apparent that the growth characteristics of the revertants

307

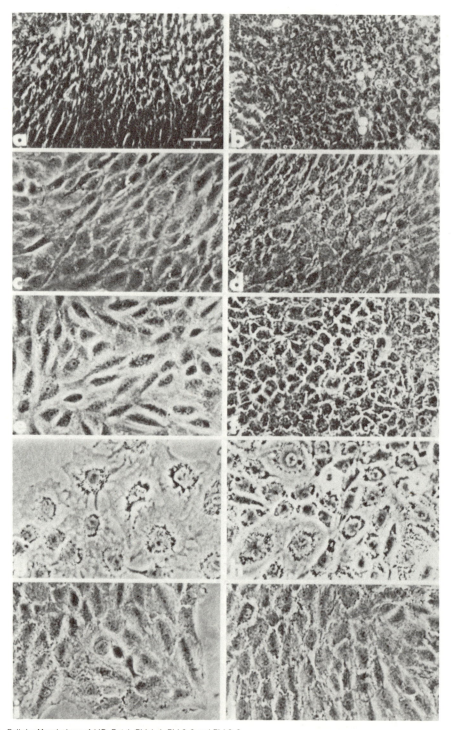

Figure 5. Cellular Morphology of 14B, Rat-1, FL¹ 1-4, FL¹ 3-3 and FL¹ 3-8

(a and b) 14B; (c and d) Rat-1; (e and f) FL¹ 1-4; (g and h) FL¹ 3-3; (i and j) FL¹ 3-8. (a, c, e, g and i) are from the edges of colonies, while (b, d, f, h and j) are from the centers of colonies. Bar=13 μ.

Table 3. Growth Properties of Cell Lines Derived From Rat-1 Cells

Cell Line	Doubling Time (Hr)		Saturation Density[a]		Plating Efficiency on Plastic[b] (10% FCS)	Plating Efficiency Methocel[b] (10% FCS)	Relative Plating Efficiency Methocel/Plastic
	10% FCS	1% FCS	10% FCS	1% FCS			
Rat-1	22	32	23	6.5	0.80	6×10^{-5}	7×10^{-5}
14B	14	22	83	28	0.35	0.05	0.15
FL^1 1-4	26	56	16	4.9	0.62	3×10^{-5}	5×10^{-5}
FL^1 3-3	35	68	3.4	0.4	0.11	5×10^{-5}	4.5×10^{-4}
FL^1 3-8	27	56	8.4	6.3	1.0	4×10^{-5}	4×10^{-5}

[a] Cells per $cm^2 \times 10^{-4}$.
[b] Number of colonies formed per number of cells inoculated.

are more like Rat-1 than like the transformed line 14B.

Cytoarchitectural Proteins

There appear to be differences in the distribution, localization and packing of actin cables between SV40-transformed 3T3 cells and normal 3T3 cells (Goldman, Yerna and Schloss, 1976; R. D. Goldman and R. Pollack, manuscript in preparation). This results in a reduction in the number of transformed cells containing cables which can be visualized by immunofluorescent staining. 9% of the cells in line 14B contained visible cables. Rat 1 (55% cables), FL^1 1-4 (60% cables), FL^1 3-3 (57% cables) and FL^1 3-8 (81% cables) approach the value for REF (84% cables).

Chen, Gallimore and McDougall (1976) have reported that adenovirus-transformed cells lack LETS (Hynes, 1974). In contrast, all our cell lines, including 14B, had large amounts of LETS over the surface of the colonies.

FUdR Sensitivity of Revertants

The possibility existed that the FUdR treatment selected for variants which were resistant to FUdR, rather than ones which were growth-controlled. To test this, sparse cultures of three revertant lines plus 14B and Rat-1 were exposed to concentrations of FUdR ranging from 3×10^{-3} μg/ml to 30 μg/ml. With all the cell lines, increasing the concentration of drug decreased the number of cells capable of forming colonies (Figure 6). Cells were treated for 4 days to ensure that all cells might undergo DNA synthesis since FL^1 3-3 grows with such a slow doubling time.

Lines 14B, Rat-1 and FL^1 1-4 all show a similar response to increasing drug concentration. Lines FL^1 3-3 and FL^1 3-8 are slightly more sensitive. Clearly, revertants are not FUdR-resistant cells.

Discussion

We have isolated stable nonconditional revertants

from an SV40-transformed cell line, 14B, which contains one copy of the viral genome in a pseudodiploid cell. These revertants are T antigen-negative, density-sensitive, anchorage-sensitive and nontumorigenic, and arise with a relatively high frequency. One group of revertants contains genome length SV40 DNA, one group has deletions in the SV40 DNA, while another group contains no detectable SV40 DNA at all. These results argue for a direct role for a functioning viral genome in the maintenance of the transformed state. The evidence for reversion through loss of an early SV40 gene function is, of course, circumstantial based upon the correlation that in all seven revertants selected, all had lost the ability to express the T antigen gene. Furthermore, in four out of seven examples, either the SV40 genome was lost or had suffered a deletion in the early region of the viral DNA. These results, coupled with the high frequency of reversion, make it seem rather probable that any *one* of a series of different events can result in the initiation of reversion.

The fact that some of these revertants can be retransformed at normal frequencies reinforces this notion (see below). On the other hand, it should be clear that we do not know what function of the early SV40 gene product(s) has been selected against. It seems probable that the T antigen may have functions other than that for which temperature-sensitive mutations are available. In addition, there exists the possibility that another SV40 early protein is made whose expression may be coordinate with T antigen expression; baroque possibilities which involve splicing of two separate genes or overlapping coding sequences which would fit into a scheme involving coordinate expression of the genes are well beyond the scope of this discussion. Suffice it to say that there is a considerable amount of data which shows that there are two early complementation groups, both of which are involved in transformation for the sister virus polyoma of SV40 (for example Eckhart, 1977; Fluck, Staneloni and Benjamin, 1977).

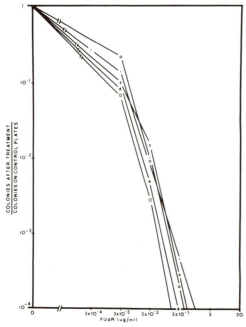

Figure 6. FUdR Sensitivity of 14B, Rat-1, FL¹ 1–4, FL¹ 3–3 and FL¹ 3–8.

(●——●) 14B; (×——×) Rat-1; (○——○) FL¹ 1–4; (△——△) FL¹ 3–3; (□——□) FL¹ 3–8.

14B contains only one copy of SV40 DNA, grows in low serum, grows to a high density, forms colonies in methocel, and forms rapidly growing tumors in nude mice and invasive malignant tumors in newborn Fisher rats (unpublished observations). Thus the maximal transformed phenotype need not require multiple copies of the SV40 genome, and the different phenotypes of transformation need not be a function of SV40 gene dosage.

The reverants isolated from 14B are unlike the density-inhibited revertants which have been reported previously (Pollack et al., 1968; Pollack and Burger, 1969; Culp and Black, 1972; Wicker et al., 1972). The class of revertants which have completely lost the SV40 genome are totally unable to express any SV40-directed modifications of growth control, and might be expected to resemble most closely the nontransformed grandparental line Rat-1. Instead, they are more like primary cells than the established line Rat-1. If the progenitors of 14B in the sojourn from a normal Rat-1 cell to a fully transformed cell became dependent upon viral functions, and in so doing dispensed with their normal regulatory pathways (in part), one might predict that loss of this viral gene would leave the cell unable to grow in the same manner as its unperturbed ancestors.

All the revertants we have described are T antigen-negative, while all earlier revertants were T antigen-positive. The T antigen-positive revertants were less growth-controlled than their normal grandparents (Culp and Black, 1972; Pollack and Vogel, 1973; Vogel et al., 1973), while these T antigen-negative revertants have regained all the normal growth regulation properties we have examined. Furthermore, a large percentage of the 14B colonies surviving even one FUdR cycle are flat, and of these a majority are T antigen-negative. This suggests that loss of early SV40 gene expression is a very efficient pathway to reversion, and that early SV40 genes therefore have a major role in the maintenance of the transformed state. While we do not know the mechanisms by which this loss of expression arose, our data suggest the following possibilities:

—The most obvious way for 14B to have lost T antigen is through the loss of the viral genome, generating the class of cured cells. This loss could be the result of excision of the viral genome or the loss of all or part of the chromosome containing SV40. The cured cell line FL¹ 3–8 has a modal chromosome number slightly lower than Rat-1, suggesting that these cells might have lost one chromosome. Banding studies are now in progress to answer this question.

—Alternately, there could have been a mutation in the SV40 genome which prevents the subsequent transcription or translation of a stable viral gene product.

—The cellular genome can be altered either genetically or epigenetically in such a way as to block viral gene expression.

Any of these possibilities would result in revertants if viral gene expression is required for the maintenance of transformation. At the present time, we cannot prove by direct arguments that any of the above mechanisms were the cause of reversion, but it seems probable the FL¹ 3–2, FL¹ 3–7 and FL¹ 3–8 were generated by the first category, FL¹ 3–3 and FL¹ 3–5 by the second category, and FL¹ 1–3 and FL¹ 1–4 by the third category. Superinfection of the revertants with SV40 DNA produces two very different sets of results (our unpublished data). Those revertants which had lost the SV40 genome (FL¹ 3–7 and FL¹ 3–8) and one of those with a deletion in the SV40 genome (FL¹ 3–5) were retransformed, as measured by focus formation and growth in methocel, as efficiently as Rat-1 (for example, 0.1% for the focus formation), presumably by the replacement of the lost or altered genome. The other revertants, which contain the SV40 genome, were not transformed at measurable frequencies. This would suggest that these cells may be unable to express the SV40 early region or are refractory to the SV40 functions.

Preliminary evidence shows, however, that when a large number of molecules of purified SV40 DNA are injected directly into the revertant nucleus, T antigen is synthesized (A. Graessman, personal communication). Additional experiments are currently being performed to define the relationship between T antigen expression, reversion and retransformation in these revertants.

A major difference between these lines and others (Pollack, Wolman and Vogel, 1970; Culp et al., 1971; Vogel, Ozanne and Pollack, 1975; Bradley and Culp, 1977) is their pseudodiploid chromosome number. Nearly all the revertants in the literature have grossly elevated chromosome numbers. In contrast, most cells of Rat-1, 14B, FL^1 1–4 and FL^1 3–8 have modal chromosome numbers very close to the normal rat diploid number of 42 (see Table 4). Chasin (1974) has shown that the frequency of mutant expression is a function of the number of positive alleles present within the cell genome which are involved with the phenotype in question. For example, the spontaneous mutation frequency from A$^+$/A$^+$ to A$^-$/A$^-$ for APRT is the square of the frequency of mutation from A$^+$/A$^-$ to A$^-$/A$^-$. He measured the spontaneous frequency of APRT loss in the heterozygote at 1.5×10^{-5}. The frequency of all T antigen-negative revertants from 14B is not too different from the heterozygote mutation frequency measured by Chasin. It may well be that the pseudodiploid character of Rat-1 and 14B, as well as the presence of a single copy of SV40 DNA per cell, allowed for mutation to a T-negative revertant state to occur in 14B at frequencies such that the revertants could be easily isolated. Thus pseudodiploid cell lines may more frequently spontaneously alter expression of cellular genes which are involved in viral gene expression than do cell lines which have multiple copies of these genes. Lines FL^1 1–3 and FL^1 1–4, revertants which show no detectable changes in their integrated viral DNA and are not readily retransformed, are putative examples of this sort of variant. In any case, it seems probable that the reversion to a T-negative state that we have measured at 8×10^{-5} is the sum of a few different processes.

This set of cell lines has made it possible for us to ask a number of questions about transformation which were difficult to ask with other transformed lines. We can now question the absolute requirements for viral products to maintain the transformed state. The reverants from this line should allow us to determine the cellular contributions to the maintenance of transformation, and the ways in which such cellular contributions can be modified to produce lines which appear even more normal than the original Rat-1 line. This type of system may allow us to develop genetic models for the regulation of viral DNA integrated into the eucaryotic genome.

Experimental Procedures

Culture Conditions
All experiments were performed in an atmosphere of 10% CO_2, 90% air, with a relative humidity of 100%. Unless specified differently, the temperature was maintained at 37°C. Cultures were grown in 60 mm plastic dishes (Falcon) in Dulbecco's modified Eagle's medium (Gibco H 21) supplemented with 100 units per ml of penicillin and streptomycin (Gibco) and 10% fetal calf serum [FCS (Reheis Laboratories)], unless a lower concentration is specified.

Cells
Rat-1, an established rat embryo line previously called F2408, has been described (Mishra and Ryan, 1973; Botchan et al., 1976). The SV40 transformant, 14B, was isolated as a dense focus following transformation of Rat-1 by purified SV40 DNA (Botchan et al., 1976).

Reversion
Revertants were isolated by plating 14B at 10^5 cells per 60 mm dish and growing for 5 days to yield dense plates containing approximately 5×10^6 cells. The plates were exposed to either 0.3 μg FUdR or 30 μg FUdR plus a 10 fold excess concentration of uridine (Pollack et al., 1968) for 2 days. The medium was then changed, and surviving cells were suspended in 0.05% trypsin with 5×10^{-4} EDTA (Gibco) and replated without further dilution. After culturing for 9 days, isolated colonies were picked using sterile cloning rings. Strains FL^1 1–3 and FL^1 1–4 were derived from plates exposed to 0.3 μg/ml FUdR plus 3 μg/ml uridine. Strains FL^1 3–3, 3–5, 3–5 and 3–8 were exposed to 30 μg/ml FUdR plus 300 μg/ml uridine.

Growth Curves
Cells were seeded at 1.2×10^4 cells per cm^2 in 10% FCS. After 6 hr, the medium was changed to the desired serum concentration. The number of cells per plate was counted daily in a Coulter counter after trypsinization. The medium was changed every 2 days.

Plating Efficiency
Cells were trypsinized, counted and diluted into 60 mm dishes at 10^2 and 10^3 cells per dish. The medium was changed twice weekly. After 14 days, the cells were fixed in 10% formalin in PBS and stained with hematoxylin, and the colonies were counted.

Growth in Methocel
The ability of cells to grow in the absence of anchorage was determined by plating 10^5, 10^4 and 10^3 cells per 60 mm dish in culture medium containing 1.3% methyl cellulose (Methocel, 4000 cps; Dow Chemical Co.) over a layer of 1% agar (Difco Bacto Agar) in the same medium (Risser and Pollack, 1974). Cells were fed twice weekly with an additional 2 ml of methocel

Table 4. Chromosome Number

Cell Line	% Pseudodiploid	% Pseudotetraploid
Rat 1	86	14
14B	80	20
FL^1 1–4	92	8
FL^1 3–3	<1	100
FL^1 3–8	90	10
REF	100	0

preparation and cultured for 3 weeks. Colonies > 0.15 mm in diameter (visible by eye) were scored using a dissecting microscope.

Immunofluorescent Assays

The cells were seeded on coverslips in 35 mm dishes at either 10^3 cells per cm^2 or 10^4 cells per cm^2 and grown for 2-3 days. At the end of this period, the dishes seeded at 10^3 cells per cm^2 were still sparse, while those seeded at 10^4 cells per cm^2 were confluent. The cells were then fixed in 100% methanol and stained with hamster anti-SV40 T antigen (NIH), followed by fluorescein-conjugated rabbit anti-hamster immunoglobulin G (Flow). Actin was determined by fixing sparse cells in 10% formaldehyde, post-fixing in acetone and staining with rabbit anti-actin, followed by goat anti-rabbit immunoglobulin G (Flow), as described by Pollack and Rifkin (1975). LETS was determined using rabbit anti-human cold-insoluble globulin, followed by fluorescein-conjugated goat anti-rabbit immunoglobulin G, as described by Chen et al. (1976). The rabbit anti-actin was a gift from Dr. K. Burridge, and the rabbit anti-human CIG was a gift from Dr. L. B. Chen, both of Cold Spring Harbor Laboratory.

FUdR Toxicity

The cells were plated at 2×10^3 cells per cm^2, incubated over night and treated with FUdR as previously described (Pollack et al., 1968).

Fluctuation Analysis

The frequency of appearance of "flat" colonies in clutures of 14B, following FUdR treatment, was determined by a fluctuation analysis as described by Pollack (1970), modeled on the procedure of Luria and Delbrück (1943). Confluent cultures of 14B (7 × 10^6 cells per 60 mm dish) were grown from either 10^2 or 10^5 cells. The cultures were treated for 2 days with 3 μg/ml FUdR + 30 μg/ml uridine, and then suspended with trypsin, diluted and plated. After 2 weeks, the colonies were fixed and stained, and the number of dense and "flat" colonies was counted.

In a parallel experiment, a confluent 60 mm plate was treated with FUdR ± uridine for 2 days, suspended with trypsin and diluted to a final volume of 16 ml. 2 ml aliquots were transferred to eight 35 mm dishes containing one 18 mm glass coverslip per dish. After 2 weeks, the coverslips were fixed and stained for T antigen, and mounted cell side up on slides. The number of T antigen-negative colonies per coverslip was counted. The slides were then stained with hemotoxylin, and the number of "flat" colonies was counted.

Immunoprecipitation of SV40 T Antigens

For labeling cell proteins with ^{32}phosphate (Ortho phosphoric acid carrier-free from New England Nuclear), transformed cells were grown to semiconfluence and starved for phosphate by incubation in phosphate-depleted media for 3-5 hr. The cells were then given 1 mCi of ^{32}P-orthophosphoric acid per ml in complete media and allowed to incorporate the label for 2 hr. Nuclei from these cells were then isolated, and the immunoprecipitates were prepared from extracts of these nuclei as described by Tjian (1978). The extraction buffer contained 0.01 M Tris-HCl, (pH 8.0) 1 mM EDTA, 0.4 M LiCl$_2$, 200 μg/ml PMSF, 0.5 mM dTT. Samples were analyzed on gradient 7-15% polyacrylamide gradient gels (Tjian and Pero, 1976). The ^{35}S-methionine protocol was as described by Prives et al. (1977), except that the cells were prestarved in methionine-free media and labeled for 2 hr in the presence of 250 μCi/ml of methionine in methionine-free media. The use of NP-40 in the extraction buffers increases the probability of proteolytic degradation of the 95,000 dalton protein (R. Tjian, personal communication). A predominant band, however, seen both in lytic infections and in transformed cells at an apparent molecular weight of 15,000 daltons which may be a second early SV40 protein is not seen without this detergent extraction (R. Hanich and M. Sleigh, personal communication).

Detection of Integrated Viral DNA in Chromosomal DNA of Cells

The blotting experiments described were performed according to procedures previously detailed by Botchan et al. (1976). For those results shown in Figure 4, the DNA samples were fractionated by electrophoresis in 1.4% alkaline-agarose gels. The conditions for this electrophoresis system are as described by McDonnell, Simon and Studier (1977). The capacity of this gel system is high. Each lane contained 50 μg of Pvu II-cleaved total cell DNA except for the slot labeled R, which contained the digestion products of 50×10^{-6} μg of SV40 DNA and 1 μg of Ad2 DNA. After transfer to nitrocellulose and hybridization with labeled SV40 DNA probe, the fragments were detected by autoradiography. The image was enhanced by the use of Dupont "Lightning-Plus" intensifier screens (P. Shank and S. Hughes, personal communication).

Chromosome Number

Determination of chromosome number was carried out by the method of Vogel et al. (1975).

Acknowledgments

We thank Dr. R. Tjian and Mr. P. Hanich for their generous help with the immunoprecipitation and gel electrophoresis assay of the antigen, and Mrs. K. Nyman for the chromosome analysis. The skillful technical help provided for by Mrs. S. Weirich and Mr. S. Zucker is also appreciated, as is the support for this work provided for by grants from the NSF and the NIH. B. S. is supported by a postdoctoral training grant from the National Cancer Institute.

The costs of publication of this article were defrayed in part by the payment of page charges. This article must therefore be hereby marked "*advertisement*" in accordance with 18 U.S.C. Section 1734 solely to indicate this fact.

Received September 26, 1977; revised October 28, 1977

References

Basilico, C. and Zouzias, D. (1976). Proc. Nat. Acad. Sci. USA 73, 1931-1935.

Botchan, M., Topp, W. and Sambrook, J. (1976). Cell 9, 269-287.

Bradley, W. and Culp, L. A. (1977). J. Virol. 21, 1228-1231.

Brugge, J. and Butel, J. (1975). J. Virol. 15, 617-635.

Butel, J. S., Brugge, J. S. and Noonan, C. A. (1974). Cold Spring Harbor Symp. Quant. Biol. 39, 25-36.

Chasin, L. (1974). Cell 2, 37-41.

Chen, L. B., Gallimore, P. and McDougall, J. (1976). Proc. Nat. Acad. Sci. USA 73, 3570-3574.

Culp, L. and Black, P. (1972). J. Virol. 9, 611-620.

Culp, L., Grimes, W. and Black, P. (1971). J. Cell. Biol. 50, 682-690.

Eckhart, W. (1977). Virology 77, 589-597.

Fluck, M., Staneloni, M. and Benjamin, T. L. (1977). Virology 77, 610-624.

Goldman, R. D., Yerna, M. J. and Schloss, J. A. (1976). J. Supramol. Structure 5, 155-183.

Hynes, R. O. (1974). Cell 1, 147-156.

Kelly, F. and Sambrook, J. (1974). Cold Spring Habor Symp. Quant. Biol. 39, 345-355.

Ketner, G. and Kelly, T. (1976). Proc. Nat. Acad. Sci. USA 73, 1102-1106.

Khoury, G., Martin, M. A., Lee, T. N. H. and Nathans, D. (1975). Virology 63, 263-272.

Luria, S. and Delbrück, M. (1943). Genetics 28, 491-511.

Marin, G. and MacPherson, I. (1969). J. Virol. 3, 146-149.

Martin, R. G. and Chou, J. Y. (1975). J. Virol. *15*, 599–612.

Martin, R. G. and Oppenheim, A. (1977). Cell *11*, 859–869.

McDonnell, M. W., Simon, M. N. and Studier, F. W. (1977). J. Mol. Biol. *110*, 119–146.

Mishra, N. and Ryan, W. (1973). Int. J. Cancer *11*, 123–130.

Nomura, S., Fishinger, P., Mattern, C., Gerwin, B. and Dunn, K. (1973). Virology *56*, 152–163.

Osborn, M. and Weber, K. (1975). J. Virol. *15*, 636–644.

Ozanne, B. and Sambrook, J. (1971). In The Biology of Oncogenic Viruses, L. Sivestre, ed. (New York: Elsevier).

Ozanne, B., Sharp, P. and Sambrook, J. (1973). J. Virol. *12*, 90–98.

Pollack, R. (1970). In Vitro *6*, 58–65.

Pollack, R. and Burger, M. M. (1969). Proc. Nat. Acad. Sci. USA *62*, 1074–1076.

Pollack, R. and Vogel, A. (1973). J. Cell. Physiol. *82*, 93–100.

Pollack, R. and Rifkin. D. (1975). Cell *6*, 495–506.

Pollack, R., Green, H. and Todaro, G. J. (1968). Proc. Nat. Acad. Sci. USA *60*, 126–133.

Pollack, R., Wolman, S. and Vogel, A. (1970). Nature *228*, 967–970.

Pollack, R., Osborn, M. and Weber, K. (1975) Proc. Nat. Acad. Sci. USA *72*, 994–998.

Prives, C., Gilboa, E. and Revel, M. and Winocour, E. (1977). Proc. Nat. Acad. Sci. USA *74*, 457–461.

Risser, R. and Pollack, R. (1974). Virology *59*, 477–489.

Southern, E. (1975). J. Mol. Biol. *98*, 503–518.

Tegtmeyer, P. (1972). J. Virol. *10*, 591–598.

Tegtmeyer, P. (1975). J. Virol. *15*, 613–618.

Tegtmeyer, P., Schwartz, M., Collins, J. and Rundell, K. (1975). J. Virol. *16*, 168–178.

Tegtmeyer, P., Rundell, K. and Collins, J. K. (1977). J. Virol. *21*, 647–657.

Tenen, D. G., Baygell, P. and Livingston, D. M. (1975). Proc. Nat. Acad. Sci. USA *72*, 4351–4355.

Tjian, R. (1978). Cell *13*, 165–179.

Tjian, R. and Pero, J. (1976). Nature *262*, 753–757.

Tooze, J. (1973). The Molecular Biology of Tumor Viruses (New York: Cold Spring Harbor Laboratory)

Vogel, A. and Pollack. R. (1973). J. Cell. Physiol. *82*, 189–198.

Vogel, A., Risser, R. and Pollack, R. (1973). J. Cell. Physiol. *82*, 181–188.

Vogel, A., Ozanne, B. and Pollack, R. (1975). In Mammalian Cells: Probes and Problems, C. R. Richard et al., eds., ERDA Symposium Series.

Wicker, R., Bourali, M. T. , Suarez, H. and Cassingena, R. (1972). Int. J. Cancer *105*, 632–640.

Reprinted from Nature, Vol. 278, No. 5701, pp. 261–263. 1979. © 1979 Macmillan Journals, Ltd.

T antigen is bound to a host protein in SV40-transformed cells

THE early region of the small DNA tumour virus, simian virus 40 (SV40), is known to code for at least two polypeptides, the t and T antigens ('small t' and 'large T'). Both these polypeptides are expressed in cells transformed by the virus[1–4], and the T antigen has been shown to be essential for both the initiation and maintenance of the transformed state[5–9]. We therefore need to know how this T protein interacts with components of the host cell in order to understand the mechanism of SV40-induced transformation. We report here that the T antigen in a line of SV40-transformed mouse cells forms an oligomeric complex with a specific cell coded protein.

When an extract of the SV40-transformed mouse cell line SVA31E7 is immunoprecipitated by as little as 1 μl of a rabbit antiserum raised against purified T (ref. 10), two polypeptides are clearly specifically immunoprecipitated (Fig. 1, tracks 0, 1). The most slowly migrating species has a molecular weight of 94,000 and is identical to the T antigen precipitated from extracts of SV40 lytically infected monkey cells[10,11]. The faster migrating species has an approximate MW of 53,000 (53 K) and has not been detected in SV40 lytically infected monkey cells. These same two polypeptides were also specifically immunoprecipitated by other anti-T sera: hamster anti-T sera (raised by inoculating the animals with the SV40-transformed hamster cell line, H65), mouse anti-T sera (raised by repeatedly immunising BALB/c mice with the SV40-transformed mouse cell line, SVT2 (ref. 12)), and anti-U sera, which were raised in monkeys and are specific for the C-terminal end of T (refs 13,14).

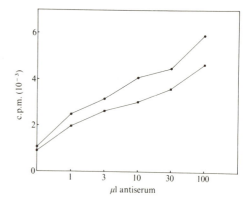

Fig. 2 The strips of dried polyacrylamide gel containing the T antigen (●) and 53K protein (■) from each track were excised, using the autoradiograph as a template, from the gel shown in Fig. 1. Each strip was solubilised[21] and counted. The c.p.m. present in each band in each track are plotted against the input of rabbit anti-T serum used to immunoprecipitate the cell extract.

The immunoprecipitation of the 53K protein by the rabbit anti-T serum has two possible explanations, either that the 53K protein shares common antigenic determinants with the gel-purified T antigen used to immunise the rabbit or that the protein is bound in a complex to another species that has such antigenic determinants. The 53K protein could then be specifically immunoprecipitated by the anti-T serum in a 'piggy-back' fashion, in the same way, for example, as antisera to human β2-microglobulin immunoprecipitate HLA antigen[15].

To distinguish between these two possibilities we first quantitated the ability of the rabbit anti-T serum to immunoprecipitate the 53K protein, and the T antigen from an extract of SVA31E7 cells. The precipitation ratio for the two proteins (c.p.m. in T/c.p.m. in 53K) remained essentially constant over a 100-fold range of anti-T serum input (Figs 1, 2). Therefore, either the serum has a very similar titre against both species, that is, the 53K protein and T share the majority of the antigenic determinants recognised on T by this antiserum, or the 53K protein is being immunoprecipitated because it is bound to T. Rabbit anti-T serum does not precipitate a detectable 53K band from untransformed mouse 3T3 A31 cells. This may be due to one or both of two reasons. Either the cells do not contain the 53K protein or, as there is no T with which it can complex, the 53K protein present is not precipitated by the rabbit antiserum specific for T. A similar experiment to that shown in Figs 1 and 2 with a lower input of cell extract showed that the ratio of T to 53K protein remained the same in serum excess.

It is known that the rabbit anti-T serum, which was raised against T antigen purified by SDS-polyacrylamide gel electrophoresis, recognises a set of determinants on T that are very stable, being resistant to heat, SDS and disulphide bond reduction (ref. 10 and D.P.L. and L.V.C., unpublished). Therefore, we used a direct binding radioimmunoassay to determine whether the isolated 53K protein possessed these antigenic determinants. The results of this assay (Fig. 3) indicate that 53K protein separated from T by SDS-gel electrophoresis is no longer precipitated by the rabbit anti-T serum, whereas the T antigen isolated in the same way is still fully reactive. This shows that the determinants recognised by the rabbit anti-T serum on T are absent from the 53K polypeptide. The immunoprecipitation of the 53K protein from cell extracts by this serum must therefore be due to the fact that it exists in a complex with T. Preliminary gel chromatography results are consistent with this idea. T and the 53K protein eluted together from Sephacryl S-200 (exclusion limit 0.25×10^6) in the excluded volume (Fig. 4). Sucrose gradient analysis of extracts of transformed cells has

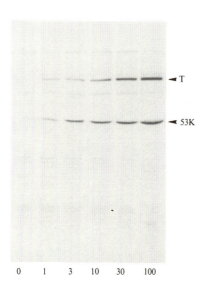

0   1   3   10   30   100

Fig. 1 Quantitation of T and 53K immunoprecipitation by rabbit anti-T serum. An equivalent aliquot of an NP40 cell extract of ^{35}S-methionine-labelled SVA31E7 cells was added to each of 6 tubes followed by normal rabbit serum and rabbit anti-T serum in the following respective amounts: track 0, 30 μl, 0 μl; track 1, 29 μl, 1 μl; track 3, 27 μl, 3 μl; track 10, 20 μl, 10 μl; track 30, 0 μl, 30 μl; track 100, 0 μl, 100 μl. After 3 h incubation at 4 °C, 100 μl of a 10% suspension of fixed *Staphylococcus aureus*[20] was added to each tube and following a further 10 min incubation the bacteria were washed 3 times in NET buffer[20] and collected by centrifugation. Bound proteins were eluted in 55 μl of sample buffer and 10 μl of each eluate loaded on a 7–20% linear gradient acrylamide gel. The dried gel was autoradiographed for 24 h.

shown that the 53K protein and a fraction of the T antigen sediment together at about 14–16S (F. McCormick, personal communication). Co-elution or co-sedimentation clearly do not prove that T and the 53K protein occur in the same complex but are consistent with the complex idea.

We then tested all the other sera which had immunoprecipitated T antigen and the 53K protein from the cell extract to see whether they could bind to the isolated 53K protein and T antigen. The anti-U serum and several different batches of hamster anti-T sera behaved like the rabbit anti-T polypeptide serum in that they efficiently re-immunoprecipitated the gel-purified T antigen but failed to bind the isolated 53K polypeptide. In contrast, the hyperimmune mouse anti-T serum specifically bound both the isolated 53K polypeptide and the T antigen (Fig. 3a, c). The binding of this mouse serum to the gel-purified T antigen from the SVA31E7 cells could be completely inhibited by an unlabelled extract of SV40-infected monkey cells but this treatment did not reduce its titre against

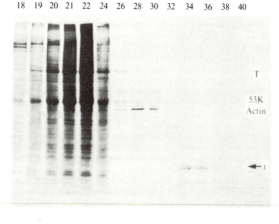

Fig. 4 Gel chromatography of transformed cell extract. Cell extract from ^{35}S-methionine-labelled SVA31E7 cells (10^9 c.p.m.) was applied to a Sephacryl S-200 column (1.5 cm × 34 cm) and eluted with buffer containing 0.1 M NaCl, 0.01 M Tris, pH 8.0, 1 mM dithiothreitol and 5% glycerol. One-ml fractions were collected and 0.1 ml of selected fractions immunoprecipitated with hamster anti-T serum. The immunoprecipitates were analysed by gel electrophoresis as described for Fig. 1. Each track is labelled with the fraction number, indicating the number of ml from addition of the sample to the column. The void volume of the column was estimated to be 20 ml. Much of the T antigen and 53K protein eluted around this position, whereas t was found around 35 ml and the 42,000 band, thought to be actin, at around 28 ml.

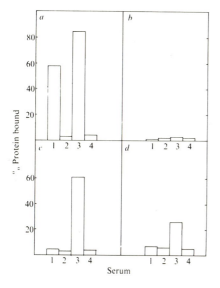

Fig. 3 Serum binding to purified polypeptides. Polypeptides for binding assays were purified from ^{35}S-methionine-labelled cell extracts by immunoprecipitation and SDS-polyacrylamide gel electrophoresis. Cells (PY6 or SVA31E7) were grown in 90-mm dishes in E4 medium with 10% calf serum, and labelled for 3 h with 500 μCi ^{35}S-methionine, then extracted in 0.5% NP40 Tris, pH 8.0, for 30 min on ice[10]. To the supernatant of the centrifuged extract, 100 μl of our hamster anti-T serum (for A31E7 cells) or 100 μl of mouse anti-T serum (for PY6 cells) was added. Three hours later 200 μl of *S. aureus* was added and the bacteria washed, collected and eluted as described for Fig. 1. The eluted proteins were separated on 15% polyacrylamide slab gels. The gel fragments containing the polypeptides of interest were located by autoradiography of the dried gel, excised with a scalpel and rehydrated in 2 ml of 0.1 M pH 9 bicarbonate buffer containing 0.1% SDS. Binding assays were carried out as described previously[10] except that 50 μl of *S. aureus* was used in the final step. a, Binding to T polypeptide from A31E7 cells by: (1) 10 μl rabbit anti-T serum, (2) 10 μl normal rabbit serum, (3) 10 μl mouse anti-T serum, (4) 10 μl normal mouse serum. b, Binding assay control. Binding to a 42K protein from A31E7 cells which is nonspecifically precipitated from cell extracts. Sera as in a. c, Binding to 53K protein specifically precipitated from A31E7 cells by sera as in a. d, Binding to 52K protein, specifically precipitated from PY6 cells with mouse anti-T serum, by sera as in a. Ordinate is the % of input ^{35}S-labelled protein bound to *S. aureus*. Correction for counter background only.

the 53K protein. Its binding of the 53K protein was, however, inhibited by unlabelled extracts of the SV40-transformed mouse cell lines, SVT2 (the cell line used to immunise the mouse) and SVA31E7 (the cell line from which the purified 53K polypeptide was prepared). Therefore, both the 53K protein and T antigen have unique non-cross-reactive antigenic determinants and the mouse serum contains some antibodies specific for the 53K protein and some specific for T antigen.

This conclusion has been supported by the results of in vitro synthesis experiments carried out in collaboration with E. Paucha. When polyadenylated mRNA from SVA31E7 cells was translated in the mouse L-cell protein synthesising system, our hamster anti-T serum immunoprecipitated the T and t polypeptides but not the 53K protein, whereas this same serum could immunoprecipitate both T and the 53K protein from in vivo labelled cell extracts. This is not due to a failure to translate the mRNA coding for the 53K protein, as when the same in vitro product was immunoprecipitated by the mouse anti-T serum (which contains antibodies to the isolated 53K protein), the 53K protein could then be detected in addition to the T and t polypeptides. This indicates that the T antigen and 53K protein do not assemble to form a complex in the in vitro conditions used in the translation reaction.

We have only encountered one other batch of serum able to bind the isolated 53K protein from SVA31E7 cells. This serum was raised in hamsters by inoculation with TSV5 Cl2 hamster transformed cells. Such reactivity explains why P. May et al. (in preparation), in contrast to the previous results of E. Paucha (personal communication), have been able to detect the 53K protein in their cell-free translation of polyadenylated mRNA from SV40-transformed mouse cells.

We postulate that the 53K protein is not virally coded because (1) we could not detect it in lytically infected monkey cells, (2) it shares no detectable antigenic determinants with T, but does have its own set of antigenic determinants (immunogenic in syngeneic mice), and (3) the DNA sequence and mRNA mapping data indicate that it would be difficult for the early region of

SV40 to encode a third polypeptide of 53,000 MW[16-18]. If the protein is not coded for by the virus, then it might be detectable in cells transformed by other viruses, by radiation, by chemical carcinogens or even in untransformed cells. In these cases it would not be complexed to T antigen and would therefore be precipitated only by antibodies specific for the 53K protein itself but not by antibodies to T antigen. An initial study supports this idea. When extracts of the polyoma-transformed mouse cell line PY6 were immunoprecipitated with the rabbit anti-T serum, no specific band was detected; immunoprecipitation of the same extract with the mouse anti-T serum (which contains antibodies to the 53K protein) did specifically bring down a band of ~52,000 MW. This cross-reactive protein, when isolated from the gel, was reprecipitated specifically by the mouse antiserum (Fig. 3d). Significantly, it was also specifically bound by the anti-TSV5 Cl2 serum but not by anti-U serum, rabbit anti-T polypeptide serum (Fig. 3d) or our hamster anti-T serum. The 52K protein from PY6 cells was always precipitated to a lesser extent than the 53K protein. For this reason we believe that the two proteins are related but not identical.

It will be necessary to establish further the nature of the 53K protein, but its specific binding to T antigen, and the presence of an antigenically related protein in polyoma-transformed cells imply that it may have a crucial role in the modulation of the transformed state. Thus, it is possible that it acts to neutralise some of the specific functions of T antigen; for example, it could render the cell non-permissive by preventing the T-dependent initiation of viral replication[18] or it may normally act as a regulator of certain cellular functions related to growth control and itself be neutralised by binding to T antigen. It is of prime importance to determine the level of this T antigen binding protein in untransformed cells and normal tissues and to see if it is induced by other carcinogenic agents.

We thank Dr A. M. Lewis, Jr for the monkey anti-U serum, Dr P. May for the hamster anti-TSV5Cl2 cell serum, and Drs P. May, E. Paucha and F. McCormick for permission to quote their unpublished results. PY6 and SVA31E7 cells were from Dr Y. Ito.

Note added in proof: The presence of a middle-T antigen species in SV40 transformed rat cells and SV40 transformed mouse cells has recently been reported[22-24] and a preliminary characterisation of the protein carried out[25].

D. P. LANE*

Department of Zoology,
Imperial College London SW7, UK

L. V. CRAWFORD

Department of Molecular Virology,
Imperial Cancer Research Fund,
London WC2, UK

Received 3 November 1978; accepted 30 January 1979.

* Present address: Cold Spring Harbor Laboratory, Cold Spring Harbor, New York.

1. Prives, C., Gilboa, A. & Revel, M. *Proc. natn. Acad. Sci. U.S.A.* **74**, 457–461 (1977).
2. Crawford, L. V. *et al. Proc. natn. Acad. Sci. U.S.A.* **75**, 117–121 (1978).
3. Sleigh, M. J., Topp, W. C., Hanich, R. & Sambrook, J. *Cell* **14**, 79–88 (1978).
4. Paucha E. *et al. Proc. natn. Acad. Sci. U.S.A.* **75**, 2165–2169 (1978).
5. Martin, R. C. & Chou, J. Y. *J. Virol.* **15**, 599–612 (1975).
6. Tegtmeyer, P., Schwartz, M., Collins, J. K. & Rundell, K. *J. Virol.* **16**, 168–178 (1975).
7. Brugge, J. S. & Butel, J. S. *J. Virol.* **15**, 619–635 (1975).
8. Kimura, G. & Itagaki, A. *Proc. natn. Acad. Sci. U.S.A.* **72**, 673–677 (1975).
9. Brockman, W. W. *J. Virol.* **25**, 860–870 (1978).
10. Lane, D. P. & Robbins, A. K. *Virology* **87**, 182–193 (1978).
11. Paucha, E., Harvey, R., Smith, R. & Smith, A. E. *INSERM Colloq.* **69**, 189–198 (1977).
12. Cicurel, L. & Croce, C. M. *J. Immun.* **119**, 850–854 (1977).
13. Lewis, A. M. & Rowe, W. P. *J. Virol.* **7**, 189–197 (1971).
14. Robb, J. A. *Proc. natn. Acad. Sci. U.S.A.* **74**, 447–451 (1977).
15. Grey, H. M. *et al. J. exp. Med.* **138**, 1608–1612 (1973).
16. Fiers, W. *et al. Nature* **273**, 113–120 (1978).
17. Reddy, V. B. *et al. Science* **200**, 494–502 (1978).
18. Berk, A. J. & Sharp, P. A. *Proc. natn. Acad. Sci. U.S.A.* **75**, 1274–1278 (1978).
19. Tegtmeyer, P. *J. Virol.* **10**, 591–598 (1972).
20. Kessler, S. W. *J. Immun.* **115**, 1617–1624 (1975).
21. Ward, S., Wilson, D. L. & Gilliam, J. J. *Analyt. Biochem.* **38**, 90–97 (1970).
22. Carroll, R. B., Goldfine, F. M. & Melero, J. A. *Virology* **87**, 194–198 (1978).
23. Gaudry, P., Rassoulzadegan, M. & Cuzin, F. *Proc. natn. Acad. Sci. U.S.A.* **75**, 4987–4991 (1978).
24. Edwards, C. A. F., Khoury, G. & Martin, R. G. *J. Virol.* **29**, 753–762 (1979).
25. Chang, C., Simmons, D. T., Martin, M. A. & Mora, P. T. *J. Virol.* (submitted).

Proc. Natl. Acad. Sci. USA
Vol. 77, No. 6, pp. 3514–3518, June 1980
Cell Biology

Adhesion plaques of Rous sarcoma virus-transformed cells contain the *src* gene product

(interference-reflection microscopy/pp60^{src} phosphoprotein/immunofluorescence microscopy/cytoskeleton/cellular "feet")

LARRY R. ROHRSCHNEIDER

Fred Hutchinson Cancer Research Center, 1124 Columbia Street, Seattle, Washington 98104

Communicated by Edwin G. Krebs, March 28, 1980

ABSTRACT Another intracellular location of the Rous sarcoma virus (RSU) *src* gene product (pp60^{src}) has been detected within RSV-transformed cells by indirect immunofluorescence. By using rabbit anti-tumor serum specific for pp60^{src}, a speckled pattern of fluorescence was found on the ventral surface of RSV (Schmidt–Ruppin strain)-transformed normal rat kidney cells. Several tests indicated that this pattern was specific for pp60^{src}. In addition, interference-reflection microscopy was used to visualize cellular adhesion plaques, which are the points at which cells attach to the substratum. Simultaneous immunofluorescence and interference-reflection microscopy indicated that the speckles of pp60^{src} fluorescence corresponded exactly to the adhesion plaque structures. The presence of pp60^{src} within the adhesion plaques was further demonstrated by indirect immunofluorescence on isolated adhesion plaques that remained bound to glass after removal of the cells. pp60^{src} also was observed in adhesion plaques of RSV-transformed chicken embryo fibroblasts (CEF) and mouse fibroblasts, as well as CEF infected with the temperature-sensitive RSV mutant tsNY68 and grown at permissive temperature. At nonpermissive temperature, pp60^{src} was not detectable in adhesion plaques of the tsNY68-infected CEF. Adhesion plaques serve as focal points of microfilament bundle attachment, and these results suggest that pp60^{src} interacts directly with cellular cytoskeletal components.

Rous sarcoma virus (RSV) has been a valuable tool for examining the mechanism of neoplastic transformation because it transforms a broad spectrum of avian and mammalian cells. The transforming function of the virus is encoded in a single gene called *src* (1) and, so far, only a single transforming protein has been identified as the *src* gene product (2). This protein is a 60,000 M_r phosphoprotein (termed pp60^{src}) (3–5) containing two major phosphorylation sites per molecule (6). One site is probably regulated by a cyclic AMP-dependent protein kinase, whereas the second site may involve autophosphorylation through a unique kinase activity associated with pp60^{src}. This kinase activity phosphorylates substrate proteins on tyrosine residues (7). That the kinase activity is indeed encoded in the *src* gene has been suggested by its invariable association with pp60^{src} (8, 9), its temperature dependence in T-class mutants (4–6, 8, 9), and its copurification with partially purified pp60^{src} (10, 11). These results all suggest that transformation in this system may result from mechanisms involving phosphorylation by pp60^{src}.

It is not known whether pp60^{src} interacts with many cellular target sites or whether a monofunctional pp60^{src} acts at a single target to effect transformation and the numerous alterations associated with transformed cells (12). Transformation by RSV, however, is known to disrupt microfilament bundles, and therefore a cytoplasmic target for pp60^{src} may directly or indirectly control the organization of microfilament bundles within the cytoplasm (13, 14). This possibility is compatible with

studies that have localized pp60^{src} to various cytoplasmic compartments (15–17); however, the functional significance of these locations in maintaining the transformed phenotype is not understood.

This paper presents results of further efforts to analyze the intracellular location of pp60^{src} by indirect immunofluorescence. In addition, a technique called interference-reflection microscopy has enabled the direct visualization of cellular plaques that are in contact with the substrate on which the cells grow (18, 19). These contact points are termed "adhesion plaques" or, more figuratively, "feet" and represent focal points of stress fiber organization (19–21). The results in this paper indicate that a functional pp60^{src} is associated with the cellular adhesion plaques in RSV-transformed cells.

METHODS AND MATERIALS

Cells and Viruses. Chicken embryo fibroblasts (CEF) and SR-NRK cells (RSV⁻ transformed normal rat kidney cells, clone A4B5G4) were as described (16, 22). The Schmidt–Ruppin strain of Rous sarcoma virus subgroup D (SR-RSV-D) was from a recently cloned laboratory stock. A mutant of SR-RSV-A, designated tsNY68, was obtained from Mike Weber (Univ. of Illinois). A line of SR-RSV-D-transformed BALB/c cells (SR-BALB) was obtained from S. Aaronson (National Institutes of Health) and recloned by end point dilution.

Immunofluorescence. Cells were grown on ethanol-sterilized glass coverslips, fixed 20 min in 4% (wt/vol) paraformaldehyde in phosphate-buffered saline (P$_i$/NaCl), rinsed in P$_i$/NaCl, and treated 3 min in 0.2% Triton X-100 in P$_i$/NaCl to make cells permeable. The sera and remainder of the technique were as described (16). All sera were absorbed just prior to use with 250 μg of unlabeled SR-RSV-D, which was then removed by centrifugation. For some experiments the cells were removed from the coverslips by an unpublished detergent treatment method that leaves the adhesion plaques attached to the glass. Briefly, cells were extracted with 0.2% Triton X-100 on ice, fixed in paraformaldehyde, then gingerly blown from the coverslip under a stream of P$_i$/NaCl.

Interference-Reflection Microscopy. The method of Curtis (18) was followed for setting the Zeiss Universal Photomicroscope for recording interference-reflection images. Illumination was from a HBO 200 W/4 high-pressure mercury lamp. The lamp field stop was closed and the green (BP 546) rhodamine filter was inserted into the optic path of the epifluorescence condenser III RS. A polarizing filter was used in the barrier filter slide to reduce the intensity of illumination, and the image was observed through a ×40 oil immersion objective.

Abbreviations: RSV, Rous sarcoma virus; SR-RSV-D, Schmidt–Ruppin strain of RSV subgroup D; CEF, chicken embryo fibroblasts; SR-NRK, normal rat kidney cells transformed with SR-RSV-D; pp60^{src}, the 60,000 M_r phosphoprotein product of the RSV transforming gene; P$_i$/NaCl, phosphate-buffered saline.

Cell Biology: Rohrschneider

Proc. Natl. Acad. Sci. USA 77 (1980) 3515

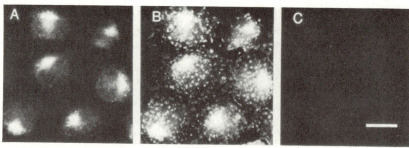

FIG. 1. Patterns of pp60^{src} localization within SR-NRK cells detected by indirect immunofluorescence. SR-NRK cells were fixed and stained with rabbit antibodies to pp60^{src} and rhodamine-conjugated goat antiserum to rabbit IgG. (*A*) Fluorescence pattern seen with focal plane set at mid-nuclear level or slightly higher, weak cytoplasmic and prominent perinuclear spot fluorescence evident. (*B*) Fluorescence pattern seen in same field of cells with focal plane set near ventral surface of cells; note speckled pattern of fluorescence. (*C*) SR-NRK cells allowed to react with normal rabbit serum in place of anti-tumor serum, focus near ventral cell surface. Bar in *C* = 20 μm.

Photographic Techniques. Rhodamine fluorescence was recorded on Kodak Tri-X pan film rated at ASA 1600 and developed in Diafine (Acufine, Chicago, IL). Interference-reflection images were recorded on Kodak Panatomic X film rated at ASA 128 and developed in Kodak Microdol X. All photographs were printed on high-contrast paper. Interference-reflection images are often presented as composites due to the limited field of view resulting from the necessity of closing the lamp field stop to increase image contrast.

RESULTS

Patterns of pp60^{src}-Specific Fluorescence in SR-NRK Cells. The rabbit anti-RSV-tumor serum used in these studies has been characterized by immune precipitation and immunofluorescence, and it appears to be specific for pp60^{src} when preabsorbed with unlabeled virus (16, 22, 23). The results in Fig. 1 demonstrate some of the patterns of pp60^{src}-specific fluorescence detectable within SR-NRK cells by using this preabsorbed antitumor serum. The fluorescence patterns that one sees are greatly dependent on where the plane of focus is set within the cells. When the focal plane rests about midnuclear level or

higher, a characteristic perinuclear spot of fluorescence is detectable. This spot is cytoplasmic but sometimes, depending upon the geometry of observation, appears flattened against the outer nuclear membrane (Fig. 1*A*). A weak diffuse cytoplasmic fluorescence also can be seen within the SR-NRK cells. In addition, pp60^{src}-specific fluorescence is usually found at gap-junctions between SR-NRK cells (16, 17); however, junction formation had not yet occurred in the relatively sparse culture shown in Fig. 1. Most of these fluorescence patterns for pp60^{src} have been observed previously (15–17, 22); however, the perinuclear distribution of pp60^{src} was not detected by immunoferritin electron microscopy (17).

In contrast to the results shown in Fig. 1*A*, a speckled pattern of fluorescence was seen in the same SR-NRK cells shown in Fig. 1*A* when the focal plane was shifted to the region of cell-substrate attachment (Fig. 1*B*). This speckled pattern was superimposed on the perinuclear spot pattern, and was not detectable in Fig. 1*A* due to the shallow depth of field of the objective lens and the relatively weak intensity of the speckles. This pattern was not peculiar to the rabbit anti-tumor serum, because antiserum from a marmoset bearing an RSV-induced

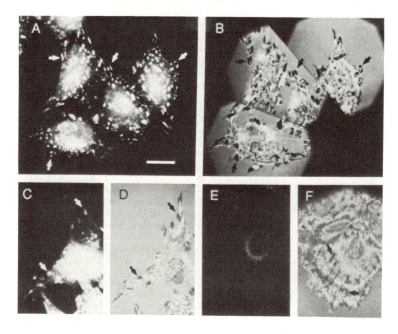

FIG. 2. Comparison of speckled pattern of pp60^{src} fluorescence with cellular adhesion plaques. Cells were fixed and pp60^{src} was stained by indirect immunofluorescence. The pp60^{src} immunofluorescence at the ventral cell surface was photographed and the adhesion plaques (dark spots) on the same field of cells were then recorded by interference reflection microscopy. (*A* and *C*) pp60^{src} fluorescence on SR-NRK cells; (*B* and *D*) interference reflection image of cells shown in *A* and *C*, respectively. Note that *B* is a composite of three interference reflection fields. (*E*) Fluorescence on NRK cells allowed to react with antiserum to pp60^{src}; (*F*) interference reflection image of NRK cells shown in *E*. Arrows emphasize prominent fluorescent speckles and their corresponding adhesion plaques. Bar in *A* = 20 μm.

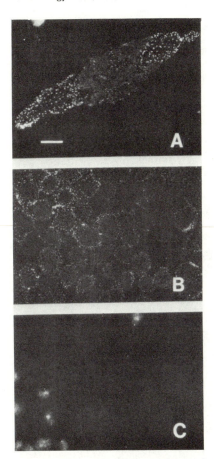

FIG. 3.   Detection of pp60^{src} fluorescence in coverslip-bound adhesion plaques after removal of SR-NRK cells. SR-NRK cells, grown on glass coverslips, were removed by a detergent treatment method that left the adhesion plaques bound to the coverslip. The adhesion plaques were fixed in 4% paraformaldehyde, rinsed, and stained with anti-tumor serum to pp60^{src} by indirect immunofluorescence. (A) The adhesion plaques from a single giant SR-NRK cell stained with anti-tumor serum. (B) A field of SR-NRK adhesion plaques stained with anti-tumor serum. (C) A field of SR-NRK adhesion plaques allowed to react with normal rabbit serum. Note that a few nuclei are present in C but they are totally absent from A and B. Bar in A = 20 μm.

tumor has given similar results (not shown). Normal rabbit serum did not stain the SR-NRK cells (Fig. 1C).

Comparison of the Speckled Pattern of pp60^{src} Fluorescence with Cellular Adhesion Plaques. The speckled pattern of pp60^{src} specific fluorescence was situated within a plane on or near the point of attachment of these cells to the glass substrate. This suggested that perhaps some component of cell-substrate attachment at the ventral surface of these cells could be associated with pp60^{src}. This possibility was explored further by utilizing an optical technique that permits the visualization of the points at which the cell attaches to the glass substrate (18–20). This optical technique is called interference-reflection microscopy, and the cellular contact points detected by this method have been called adhesion plaques or "feet."

The simultaneous use of interference-reflection microscopy and immunofluorescence microscopy enabled the direct comparison of pp60^{src} fluorescent speckles at the ventral surface of SR-NRK cells with adhesion plaques on the same cells. The

results in Fig. 2 A and C show the speckled pattern of pp60^{src} fluorescence on SR-NRK cells. The interference-reflection images of the identical fields are shown in Fig. 2 B and D, respectively. White arrows point to characteristic speckles of pp60^{src}-specific fluorescence, and black arrows point to respective regions of cells recorded by interference-reflection microscopy. Adhesion plaques can be seen as dark spots. Comparison of the respective fluorescence and interference-reflection images, using the arrows to aid alignment, reveals that each speckle of fluorescence corresponds to an adhesion plaque. This is most obvious at the cell periphery but is also seen directly under cells, where some adhesion plaques appear very small. The intensity of fluorescence within the adhesion plaques did not always correlate with the size of the adhesion plaque. This was again most evident directly under the cells and may be due to the fact that some plaques were in the process of formation while others were being dissolved at the time of fixation. The speckled pattern of pp60^{src} fluorescence was specific for SR-NRK cells and not detectable on NRK cells (Fig. 2E) even though prominent adhesion plaques could be seen (Fig. 2F).

To further confirm that the pp60^{src} was situated in the cellular adhesion plaques, cells were removed from the glass coverslips by using a detergent treatment method that leaves the "feet" behind, still attached to the glass coverslip. The SR-NRK "feet" still stained brightly when tested for pp60^{src} by immunofluorescence, using the rabbit anti-tumor serum (Fig. 3 A and B). The intensity, however, was less than that seen on whole cells, due perhaps to partial extraction of pp60^{src} during the isolation procedure. Normal serum did not stain these "feet" well (Fig. 3C), and likewise anti-tumor serum did not stain NRK "feet" (see Fig. 2 E and F).

Adhesion Plaques and pp60^{src} in Other RSV-Transformed Cells. The association of pp60^{src} fluorescence with cellular adhesion plaques was not limited to the epithelioid SR-NRK cells. Fluorescent adhesion plaques also were observed in RSV-transformed fibroblasts, but sometimes with more difficulty due to more intense cytoplasmic pp60^{src} fluorescence. The results in Fig. 4 A and B show pp60^{src} fluorescence and the exact interference-reflection image of SR-RSV-D-transformed CEF. The fluorescent spots, illuminating the location of pp60^{src}, correspond exactly to adhesion plaques as recorded by the interference-reflection technique. Furthermore, on focusing at different planes throughout these cells it was apparent that ruffled or blebbed structures containing pp60^{src} fluorescence were often situated directly above many adhesion plaques (not shown). The results in Fig. 4 C and D also demonstrate that, like SR-NRK and SR-CEF, adhesion plaques of SR-RSV-D-transformed mouse cells also contain pp60^{src}.

Association of pp60^{src} with Adhesion Plaques in tsNY68-Infected Chicken Cells. The association of an active pp60^{src} with cellular adhesion plaques was examined in CEF infected with a mutant of RSV (tsNY68) encoding a temperature-sensitive pp60^{src} (24). At the permissive temperature (35°C), tsNY68-infected CEF appear transformed and synthesize a pp60^{src} that contains kinase activity (4–6, 9, 15). At the nonpermissive temperature (41.5°C), these same cells appear morphologically normal and the pp60^{src}, even though synthesized in amounts comparable to those made under permissive temperature conditions, exhibits a greatly reduced kinase activity (5, 6, 15). The results in Fig. 4 E and G demonstrate the distribution of pp60^{src} fluorescence in tsNY68-infected CEF maintained at permissive temperature. Comparison of the fluorescence with the respective interference-reflection images (Fig. 4 F and H) shows that the tsNY68 pp60^{src} was associated with adhesion plaques at the permissive temperature. However, after the tsNY68-infected CEF had grown at nonpermissive

Cell Biology: Rohrschneider

Proc. Natl. Acad. Sci. USA 77 (1980)   3517

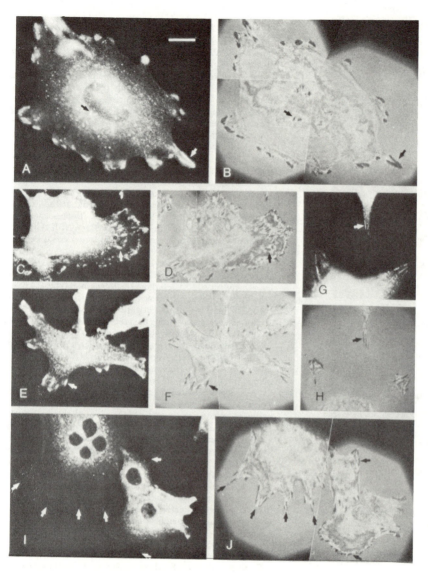

FIG. 4. Comparison of pp60^{src} fluorescence and adhesion plaque location in various RSV-transformed cells. Cells were fixed and stained for pp60^{src} by indirect immunofluorescence. The pp60^{src} fluorescence was photographed and then the interference reflection image of the same field of cells was recorded to show the adhesion plaques or "feet." (*A* and *B*) SR-CEF, fluorescence and "feet," respectively; (*C* and *D*) SR-BALB, fluorescence and "feet," respectively; (*E* and *F*) tsNY68-infected CEF grown at permissive temperature (35°C), fluorescence and "feet," respectively; (*G* and *H*) same as previous pair; (*I* and *J*) tsNY68-infected CEF grown at nonpermissive temperature (41.5°C), fluorescence and "feet," respectively. Arrows emphasize a few "feet" and corresponding location in the fluorescence photographs. Bar in *A* = 20 μm.

temperature for at least 24 hr, the localization of pp60^{src} within adhesion plaques was greatly reduced or totally eliminated (compare Fig. 4 *I* with *J*). In agreement with the results of others (5, 15), pp60^{src} was detectable at the nonpermissive temperature but situated only in what appeared to be the soluble cytoplasm even though conspicuous adhesion plaques were visible in the mutant-infected cells maintained at nonpermissive temperature.

DISCUSSION

The results in this paper bring together two observations pertinent to the mechanism of RSV-induced transformation. These observations were made through the combined use of indirect immunofluorescence microscopy and interference-reflection microscopy. First, by using rabbit anti-tumor serum and immunofluorescence an intracellular location of pp60^{src} in addition to previously observed cytoplasmic locations of pp60^{src} has been found. This location was visualized as a speckled pattern

of fluorescence on the ventral subsurface of RSV-transformed cells. This speckled pattern appeared specific for pp60^{src} and was not detected in uninfected cells, and previous results suggested that it was not on the external cell surface (16, 17). Therefore, pp60^{src} was probably contained on the cytoplasmic side of specific structures located on the ventral cell surface. Second, the speckled pattern of pp60^{src} had been observed previously (see figure 2g in ref. 16); however, the significance of this localization was not understood until comparison was made with cellular adhesion plaques visualized by interference-reflection microscopy. This analysis demonstrated that the speckles of pp60^{src} detected by immunofluorescence corresponded exactly with the organization of adhesion plaques on the ventral cell surface. This association was confirmed by demonstrating that pp60^{src} still could be detected in substrate-bound adhesion plaques after removal of the main cell body. Furthermore, the temperature-sensitive pp60^{src} synthesized in tsNY68-infected CEF was found within adhesion plaques only when the cells were grown at permissive temperature. The concordance between the activity of pp60[src] at permissive temperature (4–6, 9, 15) and the localization within adhesion plaques suggests that these sites somehow may be involved in maintaining the transformed phenotype.

Studies using interference-reflection microscopy have indicated that cells in culture exhibit different regions of separation from the substratum on which they grow (19, 21). The adhesion plaques, sometimes called focal contact areas, represent specialized cellular membrane regions of nearest approach to the substratum (10–15 nm) and are discrete sites of cell–substratum adhesion. Other broader and more uniform areas of separation (approximately 30 nm), called close contact areas, may also play a role in adhesion. The remainder of the ventral cell membrane is separated by a greater distance (100–140 nm) and probably does not participate in adhesion. Of the two regions that mediate cell–substratum adhesion, the adhesion plaques are perhaps more significant (19, 21) and were the only sites on the ventral surface where pp60[src] was detectable. This suggests that pp60[src] at these sites could influence cell–substratum adhesion. Such an effect has been reported for transformed cells (25, 26); however, it is not entirely clear whether the adhesion assays measured the actual cell–substratum adhesion through adhesion plaque sites or whether these assays simply measured the ability to separate the cell from the substratum-bound adhesion plaque. The effect of pp60[src] on this aspect of adhesion plaque function will require further study.

The adhesion plaques, while serving as static contact points to the substratum, also function in more dynamic aspects of cellular activity. In locomotion, the adhesion plaques have been observed to remain stationary with respect to the substratum. As the cell moves forward new adhesion plaques form close to the leading edge of the lamellapodium, while more posterior adhesion plaques at the trailing edge of the cell dissolve (20). On the cytoplasmic side of the ventral cell membrane, adhesion plaques have been found associated with the distal ends of microfilament bundles or "stress fibers" (19–21, 27). The adhesion plaques therefore serve as components of the cytoskeleton that anchor the fibrillar network to the substratum. In the advancing lamellae, adhesion plaque placement occurs synchronously with microfilament bundle formation (21). The microfilament bundles generally extend from the adhesion plaques under the nucleus toward the posterior of the cell (28, 29). Fibronectin has been postulated to participate in the formation of attachment plaques through transmembrane connections with microfilaments (30). This organization of microfilaments into bundles and their transmembrane anchorage at the adhesion plaque sites are believed necessary for cellular locomotion. The influence of pp60[src] on cell motility is not adequately understood; however, it is difficult to imagine that the presence of pp60[src] within adhesion plaques would not influence at least some aspects of cell locomotion.

In the tsNY68-infected cells, adhesion plaque formation occurred at both permissive and nonpermissive temperature, but only the adhesion plaques of cells maintained at permissive temperature contained pp60[src]. Some morphological differences were apparent between adhesion plaques of normal and RSV-tumor cells; however, the above results suggest that pp60[src] does not influence the formation of these structures but may affect their subsequent function. What that exact function may

be is currently unknown, but, perhaps, it could be related to the disorganization of microfilament bundles.

Finally, these results provide evidence that pp60[src] can be found associated with the adhesion plaque components of the cytoskeleton. These adhesion plaques serve as important bridges connecting the internal cytoskeleton with the external tactile environment. Therefore the presence of pp60[src] at these sites may have a profound and coordinate influence not only on internal and external functions but perhaps also on the ability of the cell to respond to the external environment.

I thank Charlotte Leitch for technical assistance, Kathy Shriver, Paul Neiman, and Bob Eisenman for helpful discussions and critical evaluation of the manuscript, and Bob Ramberg for use of the fluorescence microscope. This work was supported by Grant CA 20551 from the National Cancer Institute.

1. Wang, L. H. (1978) *Annu. Rev. Microbiol.* **32**, 561–592.
2. Brugge, J. S. & Erikson, R. L. (1977) *Nature (London)* **269**, 346–348.
3. Purchio, A. F., Erikson, E., Brugge, J. S. & Erikson, R. L. (1978) *Proc. Natl. Acad. Sci. USA* **75**, 1567–1571.
4. Collett, M. S. & Erikson, R. L. (1978) *Proc. Natl. Acad. Sci. USA* **75**, 2021–2024.
5. Levinson, A. D., Oppermann, H., Levintow, L., Varmus, H. E. & Bishop, J. M. (1978) *Cell* **15**, 561–572.
6. Collett, M. S., Erikson, E. & Erikson, R. L. (1979) *J. Virol.* **29**, 770–781.
7. Hunter, T. & Sefton, B. M. (1980) *Proc. Natl. Acad. Sci. USA* **77**, 1311–1315.
8. Erikson, E., Collett, M. S. & Erikson, R. L. (1978) *Nature (London)* **274**, 919–921.
9. Sefton, B. M., Hunter, T. & Beemon, K. (1979) *J. Virol.* **30**, 311–318.
10. Maness, P. F., Engeser, H., Greenberg, M. E., O'Farrell, M., Gall, W. E. & Edelman, G. M. (1979) *Proc. Natl. Acad. Sci. USA* **76**, 5028–5032.
11. Erikson, R. L., Collett, M. S., Erikson, E., & Purchio, A. F. (1979) *Proc. Natl. Acad. Sci. USA* **76**, 6260–6264.
12. Hanafusa, H. (1977) in *Comprehensive Virology*, eds. Fraenkel-Conrat, H. & Wagner, R. (Plenum, New York), Vol. 10, pp. 401–483.
13. McClain, D. A., Maness, P. F. & Edelman, G. M. (1978) *Proc. Natl. Acad. Sci. USA* **75**, 2750–2754.
14. Ash, J. F., Vogt, P. K. & Singer, S. J. (1976) *Proc. Natl. Acad. Sci. USA* **73**, 3603–3607.
15. Brugge, J. S., Steinbaugh, P. J. & Erikson, R. L. (1978) *Virology* **91**, 130–140.
16. Rohrschneider, L. R. (1979) *Cell* **16**, 11–24.
17. Willingham, M. C., Jay, G. & Pastan, I. (1979) *Cell* **18**, 125–134.
18. Curtis, A. S. G. (1964) *J. Cell Biol.* **20**, 199–215.
19. Izzard, G. S. & Lochner, L. R. (1976) *J. Cell Sci.* **21**, 129–159.
20. Abercrombie, M. & Dunn, G. A. (1975) *Exp. Cell Res.* **92**, 57–62.
21. Heath, V. P. & Dunn, G. A. (1978) *J. Cell Sci.* **29**, 197–212.
22. Rohrschneider, L. R. (1979) *Cold Spring Harbor Symp. Quant. Biol.* **44**, in press.
23. Rohrschneider, L. R., Eisenman, R. N. & Leitch, C. R. (1979) *Proc. Natl. Acad. Sci. USA* **76**, 4479–4483.
24. Kawai, S. & Hanafusa, H. (1971) *Virology* **46**, 470–479.
25. Weber, M. J., Hale A. H. & Losasso, L. (1977) *Cell* **10**, 45–51.
26. Willingham, M. C., Yamada, K. M., Yamada, S. S., Pouyssegur, J. & Pastan, I. (1977) *Cell* **10**, 375–380.
27. Geiger, B. (1979) *Cell* **18**, 193–205.
28. Henderson, D. & Weber, K. (1979) *Exp. Cell Res.* **124**, 301–316.
29. Revel, J. P. & Wolken, K. (1973) *Exp. Cell Res.* **78**, 1–14.
30. Hynes, R. O. & Destree, A. T. (1978) *Cell* **15**, 875–886.

Supplementary Readings

Before 1961

Furth, J. and H. Sobel. 1947. Neoplastic transformation of granulosa cells in grafts of normal ovaries into spleens of gonadectomized mice. *J. Natl. Cancer Inst.* 8: 7–16.

Levan, A. 1956. The significance of polyploidy for the evolution of mouse tumors. *Exp. Cell Res.* 11: 613–629.

1961–1970

Burger, M.M. 1970. Proteolytic enzymes initiating cell division and escape from contact inhibition of growth. *Nature* 227: 170–171.

Defendi, V., J. Lehman, and P. Kraemer. 1963. "Morphologically normal" hamster cells with malignant properties. *Virology* 19: 592–598.

Dulbecco, R. 1970. Topoinhibition and serum requirement of transformed and untransformed cells. *Nature* 227: 802–806.

Fisher, H. and J. Yeh. 1967. Contact inhibition in colony formation. *Science* 155: 582–583.

Jainchill, J.L. and G.J. Todaro. 1970. Stimulation of cell growth in vitro by serum with and without growth factor; relation to contact inhibition and viral transformation. *Exp. Cell Res.* 59: 137–146.

Nilausen, K. and H. Green. 1965. Reversible arrest of growth in G1 of an established fibroblast line (3T3). *Exp. Cell Res.* 40: 166–168.

Pollack, R., S. Wolman, and A. Vogel. 1970. Reversion of virus-transformed cell lines: Hyperploidy accompanies retention of viral genes. *Nature* 228: 967–970.

Schutz, L. and P. Mora. 1968. The need for direct cell contact in contact inhibition of cell division in culture. *J. Cell. Physiol.* 71: 1–6.

Stoker, M. 1967. Contact and short-range interactions affecting growth of animal cells in culture. *Curr. Top. Dev. Biol.* 2: 108–128.

Stoker, M.G.P., M. Shearer, and C. O'Neill. 1966. Growth inhibition of polyoma-transformed cells by contact with static normal fibroblasts. *J. Cell Sci.* 1: 297–310.

Todaro, G., G. Lazar, and H. Green. 1965. The initiation of cell division in a contact-inhibited mammalian cell line. *J. Cell. Comp. Physiol.* 66: 325–334.

Vasiliev, J.M., I.M. Gelfand, V.I. Guelstein, and E.K. Fetisova. 1970. Stimulation of DNA synthesis in cultures of mouse embryo fibroblast-like cells. *J. Cell. Physiol.* 75: 305–314.

1971–1980

Baker, J.B. and T. Humphreys. 1971. Serum-stimulated release of cell contacts and the initiation of growth in contact-inhibited chick fibroblasts. *Proc. Natl. Acad. Sci.* 68: 2161–2164.

Baker, J.B., G.S. Barsh, D.H. Carney, and D.D. Cunningham. 1978. Dexamethasone modulates binding and action of epidermal growth factor in serum-free cell culture. *Proc. Natl. Acad. Sci.* 75: 1882–1886.

Carpenter, G. and S. Cohen. 1976. Human epidermal growth factor and the proliferation of human fibroblasts. *J. Cell. Physiol.* 88: 227–238.

Carroll, R.B., L. Hager, and R. Dulbecco. 1974. Simian virus 40 T antigen binds to DNA. *Proc. Natl. Acad. Sci.* 71: 3754–3757.

Catt, K.J., J.P. Harwood, G. Aguilera, and M.L. Dufau. 1979. Hormonal regulation of peptide receptors and target cell responses. *Nature* 280: 109–116.

Ceccarini, C. and H. Eagle. 1971. Induction and reversal of contact inhibition of growth by pH modification. *Nat. New Biol.* 233: 271–273.

Culp, L.A. and P.H. Black. 1972. Contact-inhibited revertant cell lines isolated from SV40-transformed cells. III. Concanavalin A-selected revertant cells. *J. Virol.* 9: 611–620.

Culp, L.A., W.J. Grimes, and P.H. Black. 1971. Contact-inhibited revertant cell lines isolated from SV40-transformed cells. *J. Cell Biol.* 50: 691–708.

Davidson, R.L. and D. Horn. 1974. Reversible "transformation" of bromodeoxyuridine-dependent cells by bromodeoxyuridine. *Proc. Natl. Acad. Sci.* 71: 3338–3342.

deAsua, L.J. and E. Rozengurt. 1974. Multiple control mechanisms underlie initiation of growth in animal cells. *Nature* 251: 624–626.

Faulk, W.P., B.L. Hsi, and P.J. Stevens. 1980. Transferrin and transferrin receptors in carcinoma of the breast. *Lancet*, August 23, ii: 390–392.

Fischinger, P., S. Nomura, P. Peebles, D. Haapala, and R. Bassin. 1972. Reversion of MSV-transformed mouse cells: Variants without a rescuable sarcoma virus. *Science* 176: 1033–1035.

Friedmann, T., R. Doolittle, and G. Walter. 1978. Amino acid sequence homology between polyoma and SV40 tumor antigens deduced from nucleotide sequences. *Nature* 274: 291–293.

Gazdar, A.F., H.B. Stull, H.C. Chopra, and Y. Ikawa. 1974. Properties of flat variants of murine sarcoma virus transformed non-producer cells isolated after high-temperature passage. *Int. J. Cancer* 13: 19–26.

Gidwitz, S., W.A. Toscano, D.G. Toscano, W.J. Weber, and D.R. Storm. 1980. A comparison between adenylate cyclase solubilized from normal and Rous sarcoma virus-transformed chicken embryo fibroblasts. *Biochim. Biophys. Acta* 627: 1–6.

Harris, A. 1973. Behavior of cultured cells on substrata of variable adhesiveness. *Exp. Cell Res.* 77: 285–297.

Heldin, C.H., B. Westermark, and A. Wasteson. 1979.

Platelet-drived growth factor: Purification and partial characterization. *Proc. Natl. Acad. Sci.* 76: 3772–3776.

Henderson, E.E. and R. Ribecky. 1980. Transformation of human leukocytes with Epstein-Barr virus after cellular exposure to chemical or physical mutagens. *J. Natl. Cancer Inst.* 64: 33–40.

Hibbs, J.B., Jr. 1973. Macrophage nonimmunologic recognition: Target cell factors related to contact inhibition. *Science 180:* 868–870.

Hitotsumachi, S., Z. Rabinowitz, and L. Sachs. 1971. Chromosomal control of reversion in transformed cells. *Nature 231:* 411–514.

Holley, R.W. 1972. A unifying hypothesis concerning the nature of malignant growth. *Proc. Natl. Acad. Sci.* 69: 2840–2841.

Holley, R.W. 1975. Control of growth of mammalian cells in cell culture. *Nature 258:* 487–490.

Hopkins, C.R. 1980. Epidermal growth factor and mitogenesis. *Nature 286:* 205–206.

Jha, K.K., E.F. Gurney, L.A. Feldman, and H.L. Ozer. 1980. Expression of transformation in cell hybrids. III. Analysis of a revertant of SV-T2. *Cold Spring Harbor Symp. Quant. Biol.* 44: 689–694.

Jones, K.L. and D. Gospodarowicz. 1974. Biological activity of a growth factor for ovarian cells. *Proc. Natl. Acad. Sci.* 71: 3372–3376.

Kakunaga, T. and J.D. Crow. 1980. Cell variants showing differential susceptibility to ultraviolet light-induced transformation. *Science 209:* 505–507.

Klagsburn, M. 1978. Human milk stimulates DNA synthesis and cellular proliferation in cultured fibroblasts. *Proc. Natl. Acad. Sci.* 75: 5057–5061.

Lee, W.-H., K. Bister, A. Pawson, T. Robins, C. Moscovich, and P.H. Duesberg. 1980. Fujinami sarcoma virus: An avian RNA tumor virus with a unique transforming gene. *Proc. Natl. Acad. Sci.* 77: 2018–2022.

Levisohn, S.R. and E.B. Thompson. 1973. Contact inhibition and gene expression in HTC/L cell hybrid lines. *J. Cell. Physiol.* 81: 225–232.

McNutt, N.S., L.A. Culp, and P.H. Black. 1971. Contact-inhibited revertant cell lines isolated from SV40-transformed cells. II. Ultrastructural study. *J. Cell Biol.* 50: 691–708.

Mierzejewski, K. and E. Rozengurt. 1977. Density-dependent inhibition of fibroblast growth is overcome by pure mitogenic factors. *Nature* 269: 155–156.

Nigam, V., R. Lallier, and C. Brailovsky. 1973. Ganglioside patterns and phenotypic characteristics in a normal variant and a transformed back variant of a SV40-induced hamster tumor cell line. *J. Cell Biol.* 58: 307–316.

Nomura, S., P.J. Fischinger, C.F.T. Mattern, B.I. Gerwin, and K.J. Dunn. 1973. Revertants of mouse cells transformed by murine sarcoma virus. II. Flat variants induced by fluorodeoxyuridine and colcemid. *Virology 56:* 152–163.

O'Neill, C.H., P.N. Riddle, and P.W. Jordan. 1979. The relation between surface area and anchorage dependence of growth in hamster and mouse fibroblasts. *Cell 16:* 909–918.

Oey, J., A. Vogel, and R. Pollack. 1974. Intracellular cyclic AMP concentration responds specifically to growth regulation by serum. *Proc. Natl. Acad. Sci.* 71: 694–698.

Ossowski, L., J.C. Unkeless, A. Tobia, J. Quigley, D. Rifkin, and E. Reich. 1973. An enzymatic function associated with transformation of fibroblasts by oncogenic viruses. II. Mammalian fibroblast cultures transformed by DNA and RNA tumor viruses. *J. Exp. Med.* 137: 112–126.

Ozanne, B. and A. Vogel. 1974. Selection of revertants of Kirsten sarcoma virus transformed nonproducer BALB/3T3 cells. *J. Virol.* 14: 239–248.

Perbal, B. 1980. Transformation phenotype of polyoma virus-transformed rat fibroblasts: Plasminogen activator production is modulated by the growth state of the cells and regulated by the expression of an early viral gene function. *J. Virol.* 35: 420–427.

Pledger, W.J., C.D. Stiles, H.N. Antoniades, and C.D. Scher. 1977. Induction of DNA synthesis in BALB/c 3T3 cells by serum components. *Proc. Natl. Acad. Sci.* 74: 4481–4485.

Pollack, R., R. Risser, S. Conlon, and D. Rifkin. 1974. Plasminogen activator production accompanies loss of anchorage regulation in transformation of primary rat embryo cells by SV40 virus. *Proc. Natl. Acad. Sci.* 71: 4792–4796.

Prives, C., Y. Beck, and H. Shure. 1980. DNA binding properties of simian virus 40 T-antigens synthesized in vivo and in vitro. *J. Virol.* 33: 689–696.

Radke, K. and G.S. Martin. 1980. Transformation by Rous sarcoma virus: Effects of src gene expression on the synthesis and phosphorylation of cellular polypeptides. *Cold Spring Harbor Symp. Quant. Biol.* 44: 975–982.

Renger, H.C. and C. Basilico. 1972. Mutation causing temperature sensitive expression of cell transformation by a tumor virus. *Nature 69:* 109–114.

Revel, J.P., P. Hoch, and D. Ho. 1974. Adhesion of culture cells to their substratum. *Exp. Cell Res.* 84: 207–218.

Rifkin, D.B. and R. Pollack. 1977. Production of plasminogen activator by established cell lines of mouse origin. *J. Cell Biol.* 73: 47–55.

Roth, S. 1973. A molecular model for cell interactions. *Q. Rev. Biol.* 48: 541–563.

Rozengurt, E., A. Legg, and P. Pettican. 1979. Vasopressin stimulation of mouse 3T3 cell growth. *Proc. Natl. Acad. Sci.* 76: 1284–1287.

Rubin, A.H., M. Terasaki, and H. Sanui. 1978. Magnesium reverses inhibitory effects of calcium deprivation on coordination response of 3T3 cells to serum. *Proc. Natl. Acad. Sci.* 75: 4379–4383.

Rubin, A.H., M. Terasaki, and H. Sanui. 1979. Major

intracellular cations and growth control. *Proc. Natl. Acad. Sci.* 76: 3917–3921.

Rudland, P.S. 1978. Hormones and cell culture. *Nature* 276: 113–114.

Rudland, P.S., W. Seifert, and D. Gospodarowicz. 1974. Growth control in cultured mouse fibroblasts. *Proc. Natl. Acad. Sci.* 71: 2600–2604.

Sato, G. and L. Reid. 1978. Replacement of serum in cell culture by hormones. *Int. Rev. Biochem.* 20: 219–251.

Scher, C.D., M.E. Stone, and C.D. Stiles. 1979. Platelet-derived growth factor prevents G_0 growth arrest. *Nature* 281: 390–392.

Scher, C.D., W.J. Pledger, P. Martin, H. Antoniades, and C.D. Stiles. 1978. Transforming viruses directly reduce the cellular growth requirement for a platelet derived growth factor. *J. Cell. Physiol.* 97: 371–380.

Sefton, B.M., T. Hunter, and K. Beemon. 1980. Temperature-sensitive transformation by Rous sarcoma virus and temperature-sensitive protein kinase activity. *J. Virol.* 33: 220–229.

Shier, W.T. 1980. Serum stimulation of phospholipase A_2 and prostaglandin release in 3T3 cells is associated with platelet-derived growth-promoting activity. *Proc. Natl. Acad. Sci.* 77: 137–141.

Spector, D.H., K. Smith, T. Padgett, P. McCombe, D. Roulland-Dussoic, C. Moscovici, H.E. Varmus, and J.M. Bishop. 1978. Uninfected avian cells contain RNA related to the transforming gene of avian sarcoma viruses. *Cell* 13: 371–379.

Stephenson, J.R., R.K. Reynolds, and S.A. Aaronson. 1973. Characterization of morphologic revertants of murine and avian sarcoma virus-transformed cells. *J. Virol.* 11: 218–222.

Stiles, C.D., G.T. Capone, C.D. Scher, H.N. Antoniades, J.J. Van Wyk, and W.J. Pledger. 1979. Dual control of cell growth by somatomedins and platelet-derived growth factor. *Proc. Natl. Acad. Sci.* 76: 1279–1283.

Unkeless, J.C., A. Tobia, L. Ossowski, J.P. Quigley, D.B. Rifkin, and E. Reich. 1973. An enzymatic function associated with transformation of fibroblasts by oncogenic viruses. I. Chick embryo fibroblast cultures transformed by avian RNA tumor viruses. *J. Exp. Med.* 137: 85–111.

Vogel, A. and R. Pollack. 1973. Isolation and characterization of revertant cell lines. IV. Direct selection of serum-revertant sublines of SV40-transformed 3T3 mouse cells. *J. Cell. Physiol.* 82: 189–198.

Vogel, A. and R. Pollack. 1974. Isolation and characterization of revertant cell lines. VI. Susceptibility of revertants to retransformation by simian virus 40 and murine sarcoma virus. *J. Virol.* 14: 1404–1410.

Watt, T.S. and L.R. Gooding. 1980. Formation of SV40 specific and H-2 restricted target cell antigen by somatic cell fusion. *Nature* 283: 74–76.

Wiblin, C.N. and I. Macpherson. 1973. Reversion in hybrids between SV40-transformed hamster and mouse cells. *Int. J. Cancer.* 12: 148–161.

Wyke, J. 1971. A method of isolating cells incapable of multiplication in suspension culture. *Exp. Cell Res.* 66: 203–208.

Yang, N.S., K.T. Jorgensen, and P. Furmanski. 1980. Absence of fibronectin and presence of plasminogen activator in both normal and malignant human mammary epithelial cells in culture. *J. Cell Biol.* 84: 120–130.

1981–

Anderson, D.D., R.P. Beckmann, E.H. Harms, K. Nakamura, and M.J. Weber. 1981. Biological properties of "partial" transformation mutants of Rous sarcoma virus and characterization of their pp60[src] kinase. *J. Virol.* 37: 445–448.

Spector, M., S. O'Neal, and E. Racker. 1981. Regulation of phosphorylation of the β subunit of the Ehrlich ascites tumor Na^+K^+-ATPase by a protein kinase cascade. *J. Biol. Chem.* 256: 4219–4227.

Chapter 5 Tumorigenicity and Metastasis

Aaronson, S.A. and G.J. Todaro. 1968. Basis for the acquisition of malignant potential by mouse cells cultivated in vitro. *Science 162:* 1024–1026.

Troll, W., A. Klassen, and A. Janoff. 1970. Tumorigenesis in mouse skin: Inhibition by synthetic inhibitors of proteases. *Science 169:* 1211–1213.

Freedman, V.H. and S. Shin. 1974. Cellular tumorigenicity in nude mice: Correlation with cell growth in semi-solid medium. *Cell 3:* 355–359.

Boone, C.W. 1975. Malignant hemangioendotheliomas produced by subcutaneous inoculation of Balb/3T3 cells attached to glass beads. *Science 188:* 68–70.

Illmensee, K. and B. Mintz. 1976. Totipotency and normal differentiation of single teratocarcinoma cells cloned by injection into blastocysts. *Proc. Natl. Acad. Sci. 73:* 549–553.

Fidler, I.J. and M.L. Kripke. 1977. Metastasis results from preexisting variant cells within a malignant tumor. *Science 197:* 893–895.

Poste, G. and M.K. Flood. 1979. Cells transformed by temperature-sensitive mutants of avian sarcoma virus cause tumors in vivo at both permissive and nonpermissive temperatures. *Cell 17:* 789–800.

Israel, M.A., H.W. Chan, M.A. Martin, and W.P. Rowe. 1979. Molecular cloning of polyoma virus DNA in *Escherichia coli:* Oncogenicity testing in hamsters. *Science 205:* 1140–1142.

In the body, single cells of a growing tumor often have the capacity to initiate a new tumor elsewhere. When this occurs, the tumor is said to have metastasized. Can cultured cells initiate a new tumor in an animal? How closely can such a process mimic the appearance of metastases in the body? Final answers are not yet in, but some good starts are provided in the papers in this chapter.

The first problem faced in any sensible test of the tumorigenic capacity of cultured cell lines is the barrier the body throws up against any transplant: immune rejection of foreign cells. Thus, although the series of cell lines, 3T3, 312, etc., isolated by Todaro and Green from Swiss mouse embryos (Chapter 1) would seem to be a good set for testing the relationship of contact inhibition to tumorigenicity, **Aaronson** and **Todaro** open their paper by remarking that those cells, derived from an outbred mouse colony, could not be usefully reinjected into animals of the colony. Aaronson and Todaro therefore constructed an equivalent set from inbred embryos of the BALB/c mouse line. By injection of cell suspensions into the shoulders of these mice, they found a simple correlation between the saturation density of a cell line and its capacity to initiate tumors at the site of injection. SV40-transformed BALB3T3 lines were also tumorigenic, but rather weakly so. Apparently, cells transformed by SV40 develop a surface transplantation antigen that is virus-specific, and this antigen leads to their rejection by the immune system of the BALB mouse. The relation of this antigen to the large and small T antigens of SV40 is not well worked out, but presumably the virus-coded proteins somehow are present at the cell surface, as well as in the nucleus.

As reported in the previous chapter, Lee and Weinstein and Rifkin et al. showed that transformed cells produce plasminogen activator and that this production is stimulated in normal cells by tumor promoters. The classic tissue used to demonstrate promoter activity is the skin of a rabbit or mouse. Promoters alone do not cause tumors. A single low dose of a chemical carcinogen painted on the skin is also insufficient to make tumors. Promoters such as phorbol ester painted on every few days after a single low dose of chemical carcinogen cause tumors to appear. **Troll, Klassen,** and **Janoff** find that protease inhibitors block this promotion, a result that links the in vitro studies of Rifkin et al. and Lee and Weinstein to in vivo tumor promotion and suggests that such in vitro studies may have relevance to the disease.

Loss of contact inhibition is only one of many measures of the transformed state (Chapters 3 and 4). Whereas Aaronson and Todaro measured only the loss of saturation density, **Freedman** and **Shin** inject a large and varied enough set of cell lines to correlate tumorigenicity with many of the transformation assays described in Chapter 3. Freedman and Shin largely, but not entirely, bypass the problem of immune rejection by using mutant nude mice. The recessive mutation *nude* renders a mouse devoid of hair and of mature T cells. Lacking the latter, nude mice cannot mount a complete cell-mediated immune response to foreign cells. However, their B-cell response is excellent, and they also can kill tumor cells with their natural killer (NK) cells. Negative results as well as positive ones are informative in Freedman and Shin's study. Not all cell suspensions grew in the nude mouse, so the ones that did may be used to define cellular tumorigenicity. In particular, lines that lost contact inhibition or serum requirements but retained an anchorage requirement were unable to grow well as tumors. Freedman and Shin conclude that anchorage independence is the closest in vitro correlate of tumorigenicity in the nude mouse.

The short paper by **Boone** shows that anchorage dependence alone may keep BALB 3T3 cells from growing as tumors. Injection of BALB3T3 cells anchored on pea-size glass beads led to large, bloody tumors in BALB/c mice. Because these are tumors of blood vessel cells, Boone suggests that the BALB3T3 line originated from embryonic endothelial blood vessel cells and that the line has an intrinsic capacity to grow as a tumor, provided it is given an anchoring substrate such as the beads. Boone does not show whether or not any precrisis BALB/c cells would also make tumors if injected on beads. This is, of course, unlikely, given the observed failure of beads alone to induce tumors as they lie in a bed of dermal fibroblasts. Thus, we may reasonably think of loss of anchorage dependence as a necessary step in the acquisition of neoplastic potential, but not as the only growth control operating on normal cells to keep them from becoming tumors.

Teratomas are not like other tumors. They are a jumble of stemlike and differentiated cells growing as a disorganized mass. The differentiated cells are not tumorigenic. Each teratoma is descended from a tumorigenic stem cell, the embryonal carcinoma cell, which itself is a deranged descendent of a gonadal totipotent germ-line cell. An undifferentiated stem cell must be present to make the tumor, and only its undifferentiated descendents can be transplanted to make more tumors. **Illmensee** and **Mintz** show that the remarkable totipotency of teratocarcinoma stem cells extends even to embryogenesis. Single stem cells will grow into a tumor upon injection into a susceptible adult mouse. However, when injected into early embryos of the same mouse strain, the teratocarcinoma cell and its descendents are entrained into the normal tissues of the resulting mouse. The normal mouse that results from such an injected embryo is a mosaic of tumor-derived normal tissues and tissues derived from normal embryo cells. In this sense, the mouse may be said to have had four parents. Teratocarcinoma stem cells grow well in culture. Because of their capacity to enter normal embryogenesis, they can be used for genetic manipulation in culture (see Chapter 8) and then injected into embryos to construct mosaic mice that have cultured mutant tumor cells in all or most tissue cells. A serious question is raised here: Can a common mechanism of tumorigenicity account for the growth controls lost by a cell such as 3T3 and those acquired by the teratocarcinoma cell?

If a suspension of cultured cells is injected into a tail vein of a mouse instead of under the skin, the cells are rapidly carried to the heart and then pumped out into the lung, where they lodge in capillaries. In this assay, some, but not all, tumorigenic cells establish multiple small tumors in the lung. Because the properties of certain metastatic populations also include survival in the blood stream, escape from it in the lungs, and initiation of tumors there, this tail vein assay serves as one test for metastatic capacity among tumor cell lines. **Fidler** and **Kripke** use Luria-Delbrück fluctuation analysis to show that the metastatic capacity of the melanoma cell line B16 is heterogeneous, due to preexisting variation among the M16 cells. High metastatic capacity breeds true upon recloning of certain M16 cells. Thus, the ability to grow into a metastatic nodule is a cellular, genetically stable phenomenon. Like tumorigenicity, metastatic capacity is likely to have molecular underpinnings available for study in cultured cell clones.

This chapter closes with two recent reexaminations of the simple assay for tumorigenicity. These last two papers both suggest that a direct assay for tumorigenicity by viruses requires a smaller viral gene contribution than do assays for the establishment and maintenance of the different transformed states in culture. With Rous sarcoma virus (RSV), tumorigenicity was first seen as the appearance of tiny growths after viral infection of the chorioallantoic membrane (CAM), which lies next to the embryo of a fertilized egg. In the CAM, two, thin epithelial cell layers cover a dermal center. **Poste** and **Flood** overlaid the CAM with a small volume of medium containing cells to be tested for tumorigenicity. Some cell lines invaded the epidermis of the CAM and began to grow as tumors in the dermis. Within a few days they generated visible tumors. Is the CAM a fair model of an adult tissue? The answer to this is not yet clear, but Poste and Flood show that tumorigenicity on the CAM and tumorigenicity in animals are in most cases coordinate. The major exception seems to be with temperature-sensitive RSV-transformed cells. These cells grew on the CAM at temperatures well above the threshold for restriction of their in vitro phenotypes. Perhaps the chick embryo's plasma or CAM cell structure complements the temperature-sensitive lesion in *src* function in a way cell culture cannot.

Polyoma virus makes tumors in susceptible animals at the site of injection. Using recombinant DNA technology, **Israel, Chan, Martin,** and **Rowe** show that a portion of the polyoma DNA is fully tumorigenic in baby hamsters. They grew polyoma DNA in an *E. coli* plasmid and then cut this DNA with restriction enzymes, injected fragments of DNA into baby hamsters, and waited for tumors. They also assayed the various vectors for tumorigenicity. Bacteria carrying the polyoma DNA recombinant plasmid were not at all tumorigenic when fed to the hamsters. Polyoma DNA recovered from the plasmid was about as tumorigenic as polyoma DNA recovered from mouse cells. We may at least tentatively conclude from these results that no additional viral tumorigenicity resides in the manner in which the polyoma DNA is grown. Perhaps this is obvious in retrospect, but this paper served to alleviate the apprehensions of many scientists about the potential danger of moving sequences of DNA from tumor viruses into prokaryotes. The surprising result in this paper is that polyoma DNA cut by the restriction endonuclease *Eco*RI is tumorigenic. *Eco*RI breaks the polyoma gene for nuclear large T antigen, so this protein cannot be necessary for tumorigenicity.

Reprinted from Science, Vol. 162, pp. 1024–1026. 1968. © 1968 AAAS

Basis for the Acquisition of Malignant Potential by Mouse Cells Cultivated in vitro

Abstract. Balb/c mouse embryo lines maintained in culture for over 200 generations under conditions that minimize cell-cell contact do not become tumorigenic. Lines cultivated under conditions where there is extensive cell contact become tumor-producing within 30 generations. The tissue-culture property that correlates best with tumorigenicity is the loss of contact inhibition of cell division.

The mouse cell line 3T3 is one that has been used extensively for studies of cell growth and viral transformation. It was originally derived from random-bred Swiss mouse embryo cultures (1) and therefore cannot be used to correlate directly the properties observed in culture with the tumor-producing capacity of the cells. Recently lines with properties that are virtually identical to those of 3T3 have been developed from inbred Balb/c mouse embryo cultures (2). Like the original 3T3, they are continuous, aneuploid cells that are extremely sensitive to contact inhibition of cell division (3, 4) and grow to a very low saturation density (5). Lines that have acquired the ability to grow well at high cell density and are no longer sensitive to contact inhibition of cell division have also been derived from the same original pool of mouse embryo cells.

The purpose of the present experiments was to determine which in vitro properties are associated with tumorigenicity in the animal. The results show that lines that grow efficiently in the presence of extensive cell-cell con-tact readily produce tumors; those lines where cell division is arrested by neighboring cells in tissue culture are also unable to grow in the animal.

The cell lines used were all derived from a single pool of 14- to 17-day-old Balb/c mouse embryos. Briefly, by maintaining a schedule of cell transfer that minimizes cell-cell contact, lines have been developed that remain very sensitive to contact inhibition of cell division (Balb/3T3). From the same embryo cultures, Balb/3T12 lines have been established by using a transfer schedule in which cell-cell contact was extensive, thus selecting for cells better able to grow under crowded culture conditions.

A method which selects more rapidly for contact-insensitive cells in a heterogeneous population takes advantage of the ability of such cells to grow when inoculated onto monolayers of contact-inhibited cells such as 3T3 and Balb/3T3 (2). Balb/3T12 cells obtained by this method grew to very high saturation densities. Balb/3T3 and Balb/3T12 lines transformed by simian virus 40 (SV40) were obtained by infecting log phase cultures with a small plaque mutant of SV40 (6) and isolating clones by using the appropriate selective system (2). Dulbecco's modification of Eagle's medium supplemented with 10 percent calf serum (Colorado Serum Co.) was used in all experiments.

Both newborn and weanling Balb/c mice, the latter having been irradiated with 300 rads, were inoculated subcutaneously in the interscapular region. Rapidly dividing, subconfluent cultures of the lines to be tested were trypsinized, centrifuged at low speed, and resuspended in complete medium. Cell counts were performed in duplicate, and aliquots containing 10^5, 10^6, and 10^7 cells were injected. Animals were observed at weekly intervals for the appearance of progressively growing subcutaneous tumors at the site of inoculation. Though nodules of 1 to 2 mm could be detected, tumors were scored as positive only when they reached a size of 5 mm. The great majority of tumors, once having reached this size, continued to grow and eventually killed the animals.

The in vitro growth characteristics of the various cell lines tested included doubling time, cloning efficiency, saturation density, and colony-forming ability on confluent monolayers of contact-inhibited cells. Doubling times for all cultures used in the present experiments were 18 to 22 hours. By maintenance of cultures in such a way as to limit cell contact, the saturation density of Balb/3T3 cells has remained at 5×10^4 cells/cm² over a period of 10 months and through more than 200 cell generations since it was established. Balb/3T12 mass cultures after 30 cell generations in culture had saturation densities of 1.5 to 2.0×10^5 cells/cm². When less contact-inhibited clones were selected by growth on 3T3 monolayers, saturation densities of from 5 to 12 × 10^5 cells/cm² were obtained. Balb/3T12 sublines selected for their colony-forming ability on 3T3 monolayers were found to be much more efficient at forming colonies when reinoculated onto 3T3 monolayers.

The contact-sensitive Balb/3T3 line and the contact-insensitive Balb/3T12 lines were inoculated into animals after 30, 100, and 200 generations in culture. Table 1 shows that the inoculation of up to 10^7 Balb/3T3 cells per animal does not produce tumors. After 6 months and 100 cell generations in culture, the Balb/3T3 cells were found

Table 1. Tumor production by Balb/3T3 and Balb/3T12 cells after varying times in culture. NT, not tested.

Cell line	Time in culture (mo.)	Cell generations in culture*	Colony-forming ability (%)†		Saturation density ‡ (cells/cm², × 10⁻⁵)	No. of tumors/ No. of animals §	
			Alone	On 3T3 monolayer		10⁶ cells	10⁷ cells
Experiment 1							
Balb/3T3	3	30	30–50	0.0	0.5	NT	0/3
Balb/3T12	3	30	0.02–0.04	0.5–2	1.5–2.0	NT	4/6
Experiment 2							
Balb/3T3	6	100	30–50	0.0	0.5	0/18	0/8
Balb/3T12	6	100	0.1–0.2	40–80	5.0–7.5	17/40	25/28
Experiment 3							
Balb/3T3	9	200	30–50	0.0	0.5	0/8	0/8
Balb/3T12	9	200	5–10	80–100	10–12.5	8/10	10/10

* Log₂ of cumulative increment in cell number (see 1). † The range of values obtained from two or more experiments where 10^2, 10^3, and 10^4 cells were inoculated both onto bare plastic and onto 3T3 monolayers. The medium was changed twice weekly and the colonies were counted at 14 days.   ‡ The maximum cell number attained when 5×10^3 cells/cm² were inoculated onto 20-cm² petri dishes under conditions where the medium containing 10 percent calf serum was changed every 3 days. The saturation density was taken as that value where three successive cell counts at 2-day intervals showed no increase in cell number.   § Observation period: experiment 1, 11 months; experiment 2, 8 months; experiment 3, 5 months.

to have a high colony-forming ability when inoculated onto the bare surface of a petri dish but no ability to form colonies when inoculated onto a confluent monolayer of 3T3 cells. The Balb/3T12 line had a poor colony-forming ability by itself but formed progressively growing colonies with high efficiency on monolayers of contact-inhibited cells. While no animals inoculated with Balb/3T3 cells produced tumors, 17 of 40 animals developed tumors with 10^6 Balb/3T12 cells. That the number of generations in culture need not be related to oncogenicity is shown by the fact that Balb/3T12 after only 30 cell generations was tumorigenic, while Balb/3T3 has not produced tumors even after 200 generations in culture.

The dose response in animals for a number of lines, including SV40-transformed Balb/3T3 and Balb/3T12 lines, is shown in Table 2. Transformation of Balb/3T3 by SV40 conferred on the cell the ability to produce tumors in one-half of the animals when 10^7 cells were used. In all of the animals there was progressive growth of nodules during the first 2 weeks. In four of the six animals these eventually regressed. In one, a tumor had grown even larger than 5 mm in diameter before regressing. At a higher cell inoculation (2×10^7 cells per animal) five out of nine animals developed progressively growing tumors which eventually killed them.

Balb/3T12 sublines, selected for their colony-forming ability on 3T3 monolayers, produced tumors in eight out of ten animals within 3 months after inoculation of 10^6 cells. An SV40-transformed Balb/3T12 line that had been through a comparable number of generations in culture and had a similar saturation density and cloning efficiency on 3T3 monolayers when compared to Balb/3T12 was found to be roughly two logs *less* efficient in tumor induction than the non-SV40-transformed tumor cells. As was seen with SV-Balb/3T3, the animals inoculated with 10^6 and 10^7 SV-Balb/3T12 cells all developed small nodules, most of which eventually regressed.

On the basis of the data presented, it appears that the tumor-producing potential of a given cell line correlates most closely with its saturation density (that is, its ability to continue growing in the presence of extensive cell-cell contact). This can be seen in Fig. 1,

Table 2. Tumor production by mouse cell lines in irradiated weanling Balb/c mice.

Cell line*	Tumor incidence † with various cell doses ‡		
	5	6	7
Balb/3T3	0/10	0/8	0/8
Balb/3T12	2/10	8/10	10/10
SV-Balb/3T3	0/10	0/9	3/6
SV-Balb/3T12	0/10	0/9	3/5
T-Balb/3T12	10/10	4/4	

* Except for T-Balb/3T12, which is a tumor line derived from mass culture 3T12, all lines had been through approximately an equal number of cell generations in culture (200) at the time of inoculation into animals.   † Number of animals with tumors of 0.5 mm or greater per number of animals inoculated. Observation period, 3 months.   ‡ The values (namely, 5, 6, 7) are $\log_{10}$ of the cell dose.

where the latent period for 50 percent tumor incidence is plotted against the saturation density for a series of Balb/3T12 cell lines. The latent period decreased from 6 months to 1 month as the saturation density rose from 1.5×10^5 cells/cm² to 12×10^5 cells/cm². The line with the highest saturation density, an SV40-transformed Balb/3T3 line that had been selected for its high plating efficiency on 3T3 monolayers, produced tumors in 50 percent of the animals within 2 weeks. The dashed line in Fig. 1 represents the saturation density of Balb/3T3. Though over 50 animals have been inoculated, no tumors have developed.

While more tumorigenic variants can be selected from a mixed population, as for example by growth on a contact-inhibited monolayer, the acquisition of the property of transplantability can be

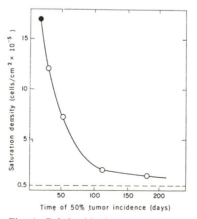

Fig. 1. Relationship between saturation density and tumor-forming ability. 10^7 cells were injected. ○, Balb/3T12 cells. ●, SV-Balb/3T3 cells. Dashed line represents saturation density of Balb/3T3 cells.

brought about directly by infection with a tumor virus such as SV40. Balb/3T3, a nontumorigenic mouse cell line, when transformed by SV40, loses contact inhibition of cell division, attains a high saturation density, and becomes malignant. The SV40-transformed cells initially grow well in vivo but are frequently found to regress. This failure of progressive tumor growth as well as the decrease in malignancy of SV-Balb/3T12 cells is most likely due to the increased antigenicity of the SV40-transformed cells. This may explain the apparent paradox between the inability of SV40 to produce tumors when injected directly in mice (7) and the high efficiency of SV40 transformation of mouse cells in tissue culture (8).

Although cells with increased saturation density and increased tumorigenicity are generally selected for by the standard procedures in tissue culture, the 3T3 transfer schedule avoids or at least minimizes this selective pressure. This regimen leads to continuous, aneuploid mouse cell lines that have not, even after hundreds of generations in culture, acquired detectable tumorigenicity. Selective conditions can also be employed to decrease the degree of malignancy of an already transplantable cell line. Pollock et al. (9) have shown that variants with reduced tumorigenicity can be obtained by using 5-fluorodeoxyuridine to select against cells that continue to divide in the presence of cell-cell contact. The less malignant variants are also found to have both a lower saturation density and a decreased plating efficiency on 3T3 monolayers (9).

Many long-established lines of spontaneously transformed mouse cells have been shown to contain viruses that are members of the mouse leukemia-sarcoma complex (10), while, on the other hand, primary cultures of Balb/c mouse embryo cells do not appear to contain these viruses (11). Kindig and Kirsten (12) have reported that 3T3 is unusual among permanent mouse lines in that it is apparently free of virus-like particles. Various Balb/3T3 and SV40-transformed Balb/3T3 lines and sublines are all, like 3T3, negative for mouse leukemia virus by complement-fixation (11). One subline of Balb/3T12, while negative at 30 generations, has been found to release mouse leukemia virus when tested after 100 cell generations. The relationship, if any, between the appearance of virus, the selection of non-contact-inhibited cells, and the acquisition of transplantability in this line needs further study.

That there exists a strong correlation between the ability of cells to continue dividing in the presence of extensive cell-cell contact in vitro and the ability of the same cells to produce tumors in the animal has been demonstrated above. The nature of the cellular changes involved are as yet not understood. They may include, for example, surface alterations that decrease the requirement for adhesion to the plastic substrate (*13*), the loss of ability to produce and/or respond to short-range inhibitory molecules (*14*), or a decreased dependence under conditions of cell crowding for a factor present in normal serum (*3, 15*). Whatever the basis for the relationship between the ability of the cells to grow in crowded cell cultures and their ability to produce tumors in the animal, the availability of simple means to select both for and against these properties by using cloned cell lines should permit a more systematic approach to the study of carcinogenesis.

Stuart A. Aaronson
George J. Todaro
Viral Carcinogenesis Branch,
National Cancer Institute,
Bethesda, Maryland 20014

References and Notes

1. G. J. Todaro and H. Green, *J. Cell Biol.* **17**, 299 (1963).
2. S. A. Aaronson and G. J. Todaro, *J. Cell Physiol.* **72**, 141 (1968).
3. G. J. Todaro, Y. Matsuya, S. Bloom, A. Robbins, H. Green, *Wistar Inst. Symp. Monogr.* **7**, 89 (1967).
4. The term "density dependent inhibition" has been suggested as an alternative to "contact inhibition of cell division" [M. G. Stoker and H. Rubin, *Nature* **215**, 171 (1967)].
5. The saturation density or maximum cell number attained per unit of surface area depends both on the genetic properties of the cell being tested and on the amount of serum provided to the culture. The higher the serum concentration and/or the more frequently the medium is changed, the greater will be the saturation density. For a given set of culture conditions, however, the saturation density is a highly reproducible cellular property.
6. K. K. Takemoto, R. L. Kirschstein, K. Habel, *J. Bacteriol.* **92**, 990 (1966).
7. K. K. Takemoto, R. C. Ting, H. L. Ozer, P. Fabisch, *J. Nat. Cancer Inst.*, in press.
8. G. J. Todaro and H. Green, *Virology* **28**, 756 (1966); P. H. Black, *ibid.*, p. 758.
9. R. E. Pollack, H. Green, G. J. Todaro, *Proc. Nat. Acad. Sci. U.S.* **60**, 126 (1968).
10. W. T. Hall, W. F. Anderson, K. K. Sanford, V. J. Evans, J. W. Hartley, *Science* **156**, 85 (1967).
11. J. W. Hartley, personal communication.
12. D. A. Kindig and W. H. Kirsten, *Science* **155**, 1543 (1967).
13. K. K. Sanford, B. E. Barker, M. W. Woods, R. Parshad, L. W. Law, *J. Nat. Cancer Inst.* **39**, 271 (1967).
14. R. R. Burk, *Nature* **212**, 1261 (1966); *Wistar Inst. Symp. Monogr.* **7**, 39 (1967).
15. H. M. Temin, *J. Cell Physiol.* **69**, 377 (1967); *Wistar Inst. Symp. Monogr.* **7**, 103 (1967); R. W. Holley and J. A. Kiernan, *Proc. Nat. Acad. Sci. U.S.* **60**, 300 (1968).
16. This work is supported by National Cancer Institute contract No. PH 43-65-641. We thank Clinton Thompson, Lillian Killos, Elaine Rands, and Lillian Murphy for their valuable technical assistance.

27 September 1968   ∎

Reprinted from Science, Vol. 169, pp. 1211–1213. 1970. © 1970 AAAS

Tumorigenesis in Mouse Skin:

Inhibition by Synthetic Inhibitors of Proteases

Abstract. Three synthetic inhibitors of proteases (tosyl lysine chloromethyl ketone, tosyl phenylalanine chloromethyl ketone, and tosyl arginine methyl ester) inhibit the tumorigenesis initiated in mouse skin by 7,12-dimethylbenz(a)anthracene and promoted by croton oil or its active principle, phorbol ester. These protease inhibitors, when applied directly to mouse skin, inhibit some of the irritant effects of the tumor promoter and are not toxic.

Two distinct biological processes are involved in chemical tumorigenesis in mouse skin—initiation by primary carcinogens and promotion by cocarcinogens (1, 2). The mechanisms of both processes are poorly understood. Initiators (primary carcinogens) require only a single application and may produce a somatic mutation or activate an oncogenic virus. These processes cannot be readily repaired or reversed once fixed in the genome. These proposed mechanisms of initiation are at least consonant with the observation that single initiating effects last virtually for the lifetime of the animal (2). The mechanism by which promoters (cocarcinogens) act is even less well understood. These agents require repeated application, and the response to them is modified by a number of exogenous factors such as caloric restriction (3), treatment with cortisone (4), or polyinosinic polycytidylic acid (5). We now show that specific inhibitors of proteases suppress promotion of tumor by croton oil or its purified active principle, phorbol myristate acetate (hereafter phorbol ester) (2). These results suggest that an endogenous protease may play a role in the mechanism of cocarcinogenesis.

We used the chloromethyl ketones of tosyl lysine (TLCK) and phenylalanine (TPCK), which inhibit trypsin and chymotrypsin, respectively, by forming a covalent adduct to histidine in the active site (6). In addition, both compounds reportedly inhibit the sulfhydryl protease, papain, by forming a covalent compound with the essential sulfhydryl group (7). We also used a competitive inhibitor for trypsin and papain, tosyl arginine methyl ester (TAME). These materials were applied to mouse skin along with the promoting agents, or separately; they reduced the immediate inflammatory effects of croton oil or phorbol ester, as well as tumor promotion by these compounds.

Tumorigenesis was initiated with 10 μg of 7,12-dimethylbenz(a)anthracene (DMBA) and promoted with 1 μg or 0.1 μg of phorbol ester (Schuchardt-

U.S.A., Katonah, N.Y.) or 5 μg of croton oil in 10 μl of acetone or dimethyl sulfoxide (DMSO) applied three times weekly to the ears of mice. Each experiment and control group contained 21 Carworth CF-1 strain animals. The time of first appearance of grossly visible tumor was noted, but only tumors larger than 1 mm and persisting over 30 days were scored. In the first experiment, 1 μg of phorbol ester was used as the promoter, and the effect of 10 μg of TLCK applied with the promoter three times a week was studied (Fig. 1A). In the control group (initiated with DMBA and then treated with phorbol ester alone) 50 percent of the animals bore tumors 50 days after promotion was begun, 100 percent had tumors at 90 days. In the group treated with TLCK, 50 percent of the animals bore tumors at 100 days; there was only 20 percent increase over this value at 200 days. In the second experiment, 1 μg of TLCK was used as the inhibitor in conjunction with 0.1 μg of phorbol ester; the onset

Table 1. Inhibition of tumorigenesis by protease inhibitors. All animals were given 10 μg of DMBA as initiator and then 5.0 μg of croton oil in acetone, applied three times weekly as promoter. The protease inhibitors TLCK, TPCK, and TAME were applied in DMSO three times weekly in 1.0-μg doses, 1 to 2 hours after applications of croton oil. All treatments were applied to ear skin of mice (the controls received DMSO alone). The average times of appearance of tumors in all three experimental groups are significantly different from the controls at P <.005; T indicates the number of tumor-bearing mice; S indicates the number of survivors.

Weeks on promotion	Inhibitor treatment							
	Control		TPCK		TLCK		TAME	
	T	S	T	S	T	S	T	S
10	8	19	0	21	0	21	0	21
12	10	19	0	21	0	21	0	21
14	11	19	0	21	1	21	3	21
16	11	19	0	21	4	21	5	21
18	11	19	0	21	4	21	5	21
20	11	19	0	21	4	21	5	21
22	11	19	0	21	5	21	5	21
24	11	19	1	21	5	21	5	21
30	11	19	1	21	5	21	5	21

of tumors was delayed by 100 days (Fig. 1B). The tumor response of control mice in experiment 2 (treated with 0.1 μg phorbol ester three times a week) was virtually identical to that of test mice in the first experiment (treated with 1 μg of phorbol ester three times a week, with 10 μg of TLCK as inhibitor) (Fig. 1, A and B). The incidence of tumors and the time of occurrence of first tumors were virtually the same. Thus, it can be suggested that 10 μg of TLCK inhibited 90 percent of the promoting activity of 1 μg of phorbol ester in the first experiment.

In a third experiment, 5 μg of croton oil was used as the promoter three times weekly, and the effects of 1 μg of TLCK, TPCK, or TAME were tested. In this experiment, TPCK virtually suppressed all tumor formation for 200 days, whereas TLCK and TAME delayed the appearance of visible tumors and reduced the numbers of tumor-bearing animals to less than 50 percent of the controls over the same period (Table 1).

Irritation was scored for intensity of erythema 24 hours after application of the promoting agent by a method similar to that used by Hecker and associates in their bioassay of promoting agents in croton oil (8). All three inhibitors of protease reduced erythema when applied for a week or longer. Inhibition of irritation was also observed upon histological examination of leukocyte infiltration in the early period after one application of phorbol ester (1 μg). Resultant leukocyte invasion was significantly reduced by a single application of 10 μg of TPCK or TLCK applied in DMSO 5 minutes after phorbol ester was applied (Fig. 2). Inhibition of erythema and leukocyte invasion was most consistently observed when the dose of antiprotease was ten times greater (by weight) than that of phorbol ester.

The inhibitors used may interfere with promotion by blocking proteolytic activity arising from interaction of the promoting agent with some tissue constituent. We were able to show that 2 hours after application of phorbol ester (1 μg) to the ears of STS mice (a tumor-sensitive strain) (3), the specific tissue activity against TAME, the synthetic substrate of trypsin, at neutral pH increases three- to fivefold over control values. The increased activity of the tissue homogenates could be completely inhibited by addition of 0.1M TLCK and partially by 0.1M TPCK (9). With respect to the source of protease, it has been noted that phorbol ester, croton oil, and Tween can release enzymes from rabbit liver lysosomes in vitro and that the effectiveness of these agents in this regard closely parallels their effectiveness as promoters of tumors (10).

On the other hand, Sivak et al. have failed to confirm this finding in mouse skin lysosomes (11), but isolation of undamaged lysosomes is notoriously diffi-

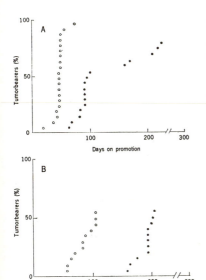

Fig. 1. Time patterns of appearance of skin tumors showing TLCK inhibition. Each point represents a single, tumor-bearing animal and is plotted according to the time of appearance of a grossly visible tumor. (A) Experiment 1. All animals were given an initiating treatment of 10 μg of DMBA in acetone. Three days after initiation, three applications per week of 1.0 μg of phorbol ester in acetone was begun. (○) Control animals; (●) animals that received 10.0 μg of TLCK (tosyl lysine chloromethyl ketone) together with phorbol ester. Average time of tumor appearance in controls was 51.6 days; in animals receiving TLCK it was increased to 122.4 days, a difference significant at $P < .1$. The average number of tumors per animal was 2.57 for controls and 1.90 for animals receiving TLCK. This difference was significant at $P < .05$ by a Student's t-test. (B) Experiment 2. All animals were treated as in experiment 1, except that doses of phorbol ester and TLCK were reduced tenfold. (○) Controls; (●) animals receiving TLCK. Average time of tumor appearance in controls was 83.9 days; in animals receiving TLCK the average time was 189.6 days. Average numbers of tumors per animal were 0.71 and 0.52 for control and treated groups, respectively, not a significant difference. The difference in average time of appearance is significant at $P < .005$.

cult in this tissue. As a determination of protease we measured the esterase activity with the synthetic substrate TAME. This substrate is hydrolyzed by many proteases, for example, trypsin, plasmin, thrombin, and papain, and its hydrolysis to tosylarginine and methanol serves as a sensitive method for assay of protease (12). Mouse ears were frozen in liquid nitrogen, pulverized, extracted with neutral buffer, and sonically disrupted for 15 seconds. Extracts of the ears of mice treated with phorbol ester contained increased TAME esterase compared to untreated controls. It seems unlikely that the mere liberation of enzyme previously bound to lysosome was responsible for the change induced by phorbol ester. Rather it may be that protease activity is related to the inflammatory changes induced by phorbol ester in mouse skin. As in other tissues increases in hydrolytic activity are ascribed to the influx of hydrolase-rich phagocytic cells. In addition to the acute influx of leukocytes into the target tissue (Fig. 2B) an early and sustained increase in vascular permeability occurs after treatment with phorbol ester (9).

The two-stage carcinogenesis in mouse skin can be considered a useful model for describing processes involved in all tumor formation. By analogy inhibitors of protease may be effective in delaying the propagation of cancer in other systems. Indeed, we have noted significant increase in the survival of tumor-sensitive mice with spontaneous breast cancer when treated with 1 mg of TPCK per week.

Possible mechanisms by which proteolytic enzymes, appearing in response to promoters of tumors, may act to enhance expression of carcinogenic transformation could include gene activation by removal of repressor substances (histones?) (10). A similar mechanism has been proposed for a tissue-bound TAME esterase during fertilization in sea urchin ova (13). Minute concentrations of TPCK and TLCK inhibit lymphocyte mitosis induced by phytohemagglutinin, perhaps by a similar block of gene derepression (14). The efficacy of inhibition of promotion by these protease inhibitors appears to be greater than that reported for injections of polyinosinic polycytidylic acid (5). In addition, the protease inhibitors did not exhibit toxicity over 45 weeks of treatment. Rubin's (15) identification of a factor causing overgrowth of Rous sarcoma in tissue

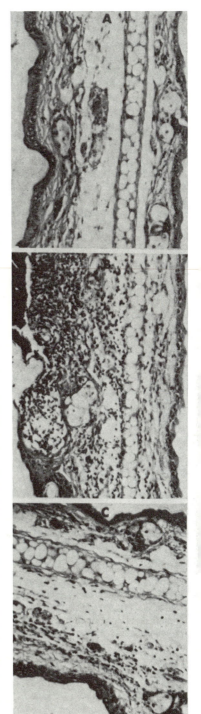

Fig. 2. Twenty-four-hour leukocytic infiltration in CF-1 strain mouse ears; hematoxylin and eosin (× 182). Animals were treated with (A) acetone followed by DMSO (control); (B) 1 μg of phorbol ester in acetone followed by DMSO; (C) 1 μg of phorbol ester in acetone followed by 10 μg of TPCK in DMSO.

1212

culture which could be replaced by the proteolytic enzyme trypsin points to the important role of proteases in growth.

WALTER TROLL
ARNOLD KLASSEN
AARON JANOFF

Departments of Environmental Medicine and *Pathology,*
New York University Medical Center,
New York 10016

References and Notes

1. I. Berenblum and P. A. Shubik, *Brit. J. Cancer* **8**, 109 (1954).
2. B. L. Van Duuren, *Progr. Exp. Tumor Res.* **11**, 31 (1969).
3. R. K. Boutwell, *ibid.* **4**, 207 (1964).
4. F. N. Ghadialli and H. N. Green, *Brit. J. Cancer* **8**, 291 (1954).
5. H. V. Gelboin and H. B. Levy, *Science* **167**, 205 (1970).
6. C. Shaw, M. Mares-Guia, N. Cohen, *Biochemistry* **4**, 2219 (1965).
7. J. R. Whitaker and Perez-Villasenor, *Arch. Biochem. Biophys.* **124**, 70 (1968).
8. E. Hecker and H. Bresch, *Z. Naturforsch.* **20b**, 216 (1965).
9. A. Klassen, A. Janoff, W. Troll, *Proc. Amer. Ass. Cancer Res.* **11**, 44 (1970).
10. G. Weissmann, W. Troll, B. L. Van Duuren, G. Sessa, *Biochem. Pharmacol.* **17**, 2421 (1968).
11. A. Sivak, F. Ray, B. L. Van Duuren, *Cancer Res.* **29**, 624 (1969).
12. S. Roffmann, U. Sunocca, W. Troll, *Anal. Biochem.,* in press.
13. W. Troll, A. Grossman, S. Chasis, *Biol. Bull.* **135**, 440 (1968).
14. R. Hirschhorn, J. Grossman, G. Weissman, W. Troll, in preparation.
15. H. Rubin, *Science* **167**, 1271 (1970).
16. Supported by PHS research grants CA-09568, CA-008491, and HE-08192; Core grant BSS-ES 00014 NCI-CA06989; Allied Chemical Corp.; and ACS grant IN-14K. A.J. is a career development award fellow of the PHS (K3-GM-6461).

13 April 1970; revised 8 June 1970   ∎

Cell, Vol. 3, 355–359, December 1974, Copyright © 1974 by MIT

Cellular Tumorigenicity in *nude* Mice: Correlation with Cell Growth in Semi-Solid Medium

Victoria H. Freedman and Seung-il Shin
Department of Genetics
Albert Einstein College of Medicine
1300 Morris Park Avenue
Bronx, New York 10461

Summary

Cultured cells derived from either normal or malignant tissues of several species have been tested by injection into the immune-deficient *nude* mouse in order to determine the cellular properties which are associated with tumorigenicity in vivo. Results show that one in vitro property consistently correlated with neoplastic growth in *nude* mice is the ability of the cell to form spherical colonies in a semi-solid growth medium such as methyl cellulose suspension. Cellular tumorigenicity is not determined solely by the malignancy of the tissue of origin, since cells derived from nonmalignant tissues become tumorigenic when they are no longer anchorage dependent for growth. In addition, acquisition of infinite growth potential in heteroploid cell lines is not in itself sufficient to confer tumorigenic capacity on the cells. These results suggest that the degree of cell growth in methyl cellulose is a useful parameter in vitro for predicting tumorigenicity in the animal, and also demonstrate the potential usefulness of the *nude* mouse for analysis of cellular malignancy irrespective of the tissue or species of origin.

Introduction

Normal cells maintained in culture can undergo "malignant transformation" either spontaneously (Sanford, Likely, and Earle, 1954), after infection by an oncogenic virus (Temin and Rubin, 1958), or following treatment with a chemical carcinogen (Chen and Heidelberger, 1969). The transformed cells acquire altered growth characteristics, and can be selected as a denser colony in a layer of contact sensitive cells (Todaro and Green, 1963) or in agar suspension culture (MacPherson and Montagnier, 1964).

The nature of the transformation process, however, remains poorly understood. In fact, "malignancy" on the cellular level has often been defined purely operationally in terms of greater cellular growth autonomy in vitro without regard to neoplastic growth potential in vivo (Todaro and Green, 1963). Since many cell culture lines are derived from noninbred animals, it has been necessary to assay cellular tumorigenicity in immune suppressed animals. Even if a cell line is descended from an inbred strain, it may express a new cellular antigen and then elicit immunologic rejection when injected into the animal of origin. In order to overcome this difficulty, a generally applicable in vivo assay for cellular malignancy would require a test system which is immunologically indifferent to the test cell. To facilitate further genetic analysis, the system should permit the recovery of the cells after growth as tumors in the animal. Through such a system it would be possible to assess cellular malignancy in terms of tumorigenicity, and eventually identify the specific cellular properties which determine this tumorigenicity.

The *nude* mouse is congenitally thymusless (Pantelouris, 1968), and deficient in thymus-dependent immunologic functions (Rygaard, 1969). Thus *nude* mice can support the growth of certain human tumor transplants (Rygaard and Povlsen, 1969; Povlsen et al., 1973), as well as a wide range of xenogeneic skin grafts (Manning, Reed, and Shaffer, 1973). Recently, Giovanella, Stehlin, and Williams (1974) reported that injection of cell cultures derived from human malignant tumors can give rise to metastasizing tumors in *nude* mice.

We have tested a large number of permanent cell lines and primary diploid cell strains from several species for tumor-forming capacity in *nude* mice. A preliminary report of these results has been presented elsewhere (Freedman and Shin, 1974).

Results and Discussion

Tumorigenicity in *nude* Mice

The results of tumorigenicity tests with a group of established heteroploid cell lines are summarized in Table 1. These cell lines represent 5 mammalian species, and were derived originally from both normal and malignant tissues. All established cell lines produced tumors in *nude* mice with the exception of the mouse 3T3 cell lines. However, none of the primary diploid fibroblasts from embryonic or adult tissues were tumorigenic, as demonstrated in Table 2. Giovanella et al. (1974) have suggested that *nude* mice can discriminate between "normal" and "malignant" cells by permitting the growth of injected cells derived from malignant tumors, while not permitting the growth of cells derived from normal tissues. Our results indicate, however, that tumorigenic cell lines need not be descended from malignant tissues, since cells derived from originally normal tissues can clearly become tumorigenic in culture (for example, L929 and BHK–21).

Second, these results show that the process of spontaneous establishment in culture (Hayflick and Moorhead, 1961), by which a diploid culture with a limited life span becomes a permanent heteroploid cell line with an infinite growth potential, is not in itself sufficient to confer tumorigenicity. This can be concluded from the fact that mouse 3T3 cell lines, which are aneuploid derivatives of normal mouse embryo cells (Todaro and Green, 1963; Aaronson and Todaro, 1968), failed to produce tumors in *nude* mice.

Third, tumorigenicity can be induced in nontumorigenic cells, whether diploid (for example, RK–1) or heteroploid (for example, 3T3), by transformation with an oncogenic virus such as SV40 (for example, RKT–TG1 and SV101). This is consistent with earlier observations by Aaronson and Todaro (1968).

Association of Tumorigenicity with Loss of Anchorage Dependence

When anchorage-dependent cells such as primary diploid fibroblasts or 3T3 are plated in a semi-solid growth medium containing soft agar or methyl cellulose, and thus prevented from attaching to the surface of the tissue culture dish, they fail to form growing colonies (Stoker et al., 1968; Risser and Pollack, 1974). On the other hand, loss of anchorage dependence and acquisition of the ability to grow in methyl cellulose have been shown to be associated with spontaneous or viral transformation (MacPherson and Montagnier, 1964; Stoker, 1968).

We therefore have tested representative cells, which had been screened for tumorigenicity in *nude* mice, for their ability to form colonies in methyl cellulose suspension. As Table 3 demonstrates, all of the tumorigenic cell lines grew in methyl cellulose with the same plating efficiency as on plastic substrate in liquid medium. In contrast, the nontumorigenic cells showed no detectable growth in methyl

Table 1. Tumorigenicity of Permanent Cell Lines in *nude* Mice

| Cell Line | Tissue of Origin of Cell Line | | | Tumorigenicity in *nude* Mice | |
	Species	Tissue of Origin	Malignancy of Tissue of Origin[a]	Incidence of Tumor Growth[b]	Tumorigenicity[c]
D98 (Berman and Stulberg, 1956)	Human	Adult bone marrow	N	4/4	+
HeLa (Gey, Coffman, and Kubicek, 1952)	Human	Adult epithelioid carcinoma	M	2/2	+
HS0643 (Hackett, 1973)[d]	Human	Adult lung carcinoma	M	4/4	+
L929 (Sanford, Earle, and Likely, 1948)	Mouse	Adult connective tissue	N	3/3	+
A9 (Littlefield, 1964)	Mouse	8–Azaguanine resistant mutant of L929	N	8/8	+
RAG (Klebe, Chen, and Ruddle, 1970)	Mouse	Adult renal adenocarcinoma	M	2/2	+
3T3 Swiss (Todaro and Green, 1963)	Mouse	Embryonic fibroblast	N	0/2	−
BALB/3T3 (Aaronson and Todaro, 1968)	Mouse	Embryonic fibroblast	N	0/2	−
SV101 (Todaro, Green, and Goldberg, 1964)	Mouse	3T3, transformed by SV40	N	3/4	+
Ehrlich (Lettré strain)	Mouse	Ascitic tumor	M	5/5	+
RKT–TG1 (Shin, 1970)[e]	Rabbit	Adult kidney, transformed by SV40	N	2/2	+
WOR–6 (Shin, 1972)[f]	Chinese hamster	Adult lung	N	2/2	+
BHK–21 (McPherson and Stoker, 1962)	Syrian hamster	Newborn Kidney	N	3/3	+
T6a (Marin and Littlefield, 1968)	Syrian hamster	8–Azaguanine resistant mutant of BHK, polyoma transformed	N	1/1	+

(a) malignant (M); nonmalignant (N).
(b) Number of *nude* mice that developed tumors/number of *nude* mice injected.
(c) tumorigenic (+); nontumorigenic (–).
(d) Established from a lung carcinoma of unknown origin, by Dr. Adeline Hackett of Naval Biochemical Research Laboratory, Oakland, California.
(e) SV40-transformed clonal derivative of a primary diploid kidney epithelial culture from a 3 month old female rabbit (RK–1 in Table 2), (Shin, unpublished).
(f) A ouabain-resistant mutant isolated from wg3h (Westerveld et al., 1971), a derivative of the DON line resistant to 8–azaguanine (Shin, unpublished).

cellulose. In further experiments involving a large number of cell lines, we have so far encountered no exception to the observed association between the tumorigenicity in *nude* mice and the cellular ability to proliferate in semi-solid growth medium. This is in agreement with earlier observations of MacPherson and Montagnier (1964), who found similar associations between the transplantability of a BHK–21 subline and its efficiency of plating in soft agar.

Spontaneous or induced transformation of cells in culture can cause permanent alterations in the physiologic and social behavior of cells, so that they are no longer subject to growth inhibition by high cell density, absence of anchorage (Stoker, 1968), or low serum concentration (Dulbecco, 1970). However, cellular sensitivity to each of these constraints is not coordinately controlled, and it is possible to measure the effects of each on cell growth separately (Risser and Pollack, 1974). The specific correlation of cellular tumorigenicity with anchorage dependence has now been confirmed in a separate series of experiments, in which nonselectively isolated SV40 transformants of mouse 3T3 and rat em-

bryonic fibroblasts were individually tested for tumorigenicity in *nude* mice (Freedman, Shin, Risser, and Pollack, manuscript in preparation). In this last series, cellular tumorigenicity was again correlated with colony formation in methyl cellulose, but not with serum requirement, saturation density, or the degree of expression of viral T antigen.

Whether the close association between cellular tumorigenicity and the loss of anchorage dependence applies also to established lymphoblastoid cell lines in culture has not yet been determined. Since the lymphoblastoid cells normally proliferate in vivo without anchorage, and are maintained in suspension cultures in vitro, the concepts of anchorage dependence or contact-mediated growth control may have no useful meaning. However, it has been shown by Povlsen et al. (1973) that implanted tissue sections of Burkitt's lymphoma can give rise to localized solid tumors in *nude* mice, and by Rygaard (personal communication) that at least some human lymphocyte culture lines established from normal peripheral blood cells are tumorigenic in this animal.

Absence of Host Cell Recruitment

To determine if the tumorigenic cells injected into *nude* mice could induce neoplastic transformation of the host cells or their migration into the growing tumor mass, we have analyzed the tumors which grew out from representative cell lines carrying well established cellular markers. The mouse cell A9 (Littlefield, 1964) and the Syrian hamster cell line T6a, a derivative of BHK–21 (Marin and Littlefield, 1968), are both deficient in hypoxanthine phosphoribosyltransferase (HPRT; EC 2.4.2.8.), and resistant to high levels of 8–azaguanine. In addition, both cell lines have characteristic karyotypes easily distinguishable from the normal diploid mouse karyotype. Tumors which developed from A9 and T6a were excised and processed to produce fresh cultures, and analyzed immediately. Both tumor-

Table 2. Tumorigenicity of Primary Diploid Cultures in *nude* Mice

	Tissue of Origin of Cell Strain		Incidence of Tumor Growth*
Cell Strain and Source	Species	Tissue of Origin	
S1434 (H. Klinger)	Human	Normal embryonic fibroblast	0/8
REF (R. Risser)	Rat	Normal embryonic fibroblast	0/2
RK–1 (S. Shin)	Rabbit	Normal adult kidney	0/2
PCSF (E. Robbins)	Chicken	Normal embryonic fibroblast	0/2

*Number of *nude* mice that developed tumors/number of *nude* mice injected.

Table 3. Association of Cellular Tumorigenicity in *nude* Mice with Efficiency of Plating in Methyl Cellulose

			Efficiency of Plating (EOP)	
Cell Line	Species	Tumorigenicity[a]	On Plastic Substrate (%)	In Methyl cellulose (%)
S1434	Human	–	<10	<0.001
D98	Human	+	15.5	20.9
HeLa	Human	+	90.1	88.8
3T3 (Swiss)	Mouse	–	39.0[b]	<0.001[b]
SV101	Mouse	+	38.0[b]	27.0[b]
A9	Mouse	+	45.5	50.1

[a]From Tables 1 and 2.
[b]Data from Risser and Pollack (1974).

Table 4. Retention of Cellular Phenotype after Passage of A9 Cells in *nude* Mice

Phenotype	Before Injection	Tumor-Derived Culture
Karyotype		
Chromosome number[a]	53.3 ± 9.4 (13)	52.9 ± 13.3 (33)
Range	48–78	42–102
Cells with marker chromosomes	62%	51%
HPRT Activity	Negative	Negative
Efficiency of Plating in:		
Normal growth medium	45.4%	50.5%
Growth medium + 8–azaguanine	50.5%	49.8%
Growth medium + 6–thioguanine	44.0%	45.5%
Growth medium + HAT[b]	$<10^{-5}$%	$<10^{-5}$%

(a) Mean ± S.D. (Number of metaphases counted).
(b) HAT: hypoxanthine (10^{-4}M), aminopterin (4×10^{-7}M), and thymidine (1.6×10^{-5}M).

derived cultures were indistinguishable from the original cell lines with regard to HPRT activity, resistance to 8–azaguanine, and karyotype, indicating that the genetic markers of the injected cells were preserved, and that the tumor growth in *nude* mice did not involve recruitment of the host cells through neoplastic transformation. Comparative data for the mouse cell mutant A9 are presented in Table 4. Crude tissue extracts were also prepared from tumors derived from BHK–21 (MacPherson and Stoker, 1962), and tested for the species-specific isozyme patterns of lactic dehydrogenase, glucose–6–phosphate dehydrogenase, and HPRT by electrophoresis on cellulose acetate gel according to the methods described by Meera Khan (1971) and van Diggelen and Shin (1974). Only the Syrian hamster isozymes of BHK–21 could be detected, suggesting that even in the tumor mass, few host-derived cells were present.

Characteristics of Tumors in *nude* Mice

It is noteworthy that the tumors which develop in *nude* mice as a result of the injection of cultured cells always grow as single, well-defined, encapsulated masses. Contrary to the reported observations of Giovanella et al. (1974), we have never observed macroscopic metasteses or any invasion of the surrounding host tissue upon dissection. This was seen most clearly when Ehrlich ascites tumor cells (Lettré strain) were injected intraperitoneally. Though this cell line always produces ascitic tumors in normal mice, in all 5 *nude* mice injected only solid tumors appeared in the inguinal area; no ascitic tumors could be detected.

Conclusion

These results show that the *nude* mouse is indeed an appropriate animal model for studies of cellular

tumorigenicity, and further suggest that the tumorigenicity in vivo of cultured cells may be determined by the same mechanism which enables the cell to overcome the requirement for anchorage in vitro.

Experimental Procedures

Maintenance of *nude* Mice
The colony of *nude* mice was initiated with a nucleus of heterozygous (*nu/+*) breeding pairs on BALB/c background from Dr. C. W. Friis of G1. Bomholtgard, Ry, Denmark. Cages, bedding, and other material that come into direct contact with the mice were autoclaved before use. The entire colony was maintained in sterile-air laminar flow cage racks (Carworth Co., New City, New York), isolated from the main animal facilities. Under these conditions, *nude* mice had an average life span of about 10 months.

Cell Culture
All cells were maintained in McCoy's 5a medium (GIBCO) plus 15% fetal bovine serum, except mouse 3T3 cell lines and SV101, which were grown in Dulbecco–Vogt modified Eagle's medium (GIBCO) plus 10% calf serum.

Test for Tumorigenicity in *nude* Mice
Monolayer cells were harvested by trypsinization, and 2×10^6 cells suspended in 0.2 ml of phosphate-buffered saline were injected subcutaneously into each *nude* mouse at 4–6 weeks of age. Ehrlich ascites tumor cells (Lettré strain) were collected by centrifugation and injected intraperitoneally in doses of 10^3, 10^4, and 10^5 cells per mouse. All mice were regularly examined for tumor development for at least 4 months following the injection.

Measurement of Cellular Plating Efficiency in Methyl Cellulose
Efficiency of plating in methyl cellulose suspension was determined by plating cells in 5a medium containing 1.2% Methocel (Dow Chemical Co., 4000 cps) plus 10% fetal calf serum, over a layer of 0.9% agar (Difco Bacto Agar) in the same culture medium, as described by Risser and Pollack (1974).

Isozyme Determination by Electrophoresis
The general procedures for enzyme electrophoresis on cellulose acetate gel (Cellogel, Reeve-Angel Co., 0.25 mm nominal thickness) were followed for lactic dehydrogenase and glucose-6–phosphate dehydrogenase (Meera Khan, 1971). Isozyme analysis of hypoxanthine phosphoribosyltransferase was carried out according to the method of van Diggelen and Shin (1974).

Acknowledgments

We are indebted to Drs. R. Pollack, R. Risser, and B. Bloom for helpful discussions. We wish to thank Dr. C. W. Friis for his generous donation of the *nu/* + breeding pairs, Dr. J. Rygaard for advice on maintenance of the *nude* mice, and Dr. O. van Diggelen for performing some of the electrophoresis. We also thank André Brown for excellent technical assistance. This work was supported by research grants from the National Institutes of Health and the National Science Foundation. V. H. F. is a pre-doctoral trainee in the Sue Golding Graduate Division of the Albert Einstein College of Medicine, and is supported by an N.I.H. Training Grant in Genetics.

Received September 9, 1974

References

Aaronson, S. A., and Todaro, G. J. (1968). Science *162*, 1024.

Berman, L., and Stulberg, C. S. (1956). Proc. Soc. Exp. Biol. Med. *92*, 730.

Chen, T. T., and Heidelberger, C. (1969). Int. J. Cancer *4*, 166.

Dulbecco, R. (1970). Nature *277*, 802.

Freedman, V. H., and Shin, S. (1974). Genetics *77s*, 23.

Gey, G. O., Coffman, W. D., and Kubicek, M. T. (1952). Cancer Res. *12*, 264.

Giovanella, B. C., Stehlin, J. S., and Williams, L. J. Jr. (1974). J. Nat. Cancer Inst. *52*, 921.

Hayflick, L., and Moorhead, P. S. (1961). Exp. Cell Res. *25*, 585.

Klebe, R. J., Chen, T. R., and Ruddle, F. H. (1970). J. Cell Biol. *45*, 74.

Littlefield, J. W. (1964). Science *145*, 709.

MacPherson, I., and Montagnier, L. (1964). Virology *23*, 291.

MacPherson, I., and Stoker, M. G. (1962). Virology *16*, 147.

Manning, D. D., Reed, N., and Shaffer, C. F. (1973). J. Exp. Med. *138*, 488.

Marin, G., and Littlefield, J. W. (1968). J. Virol. *2*, 69.

Meera Khan, P. (1971). Arch. Biochem. Biophys. *145*, 470.

Pantelouris, E. M. (1968). Nature *217*, 370.

Povlsen, C. O., Fialkow, P. J., Klein, E., Klein, G., Rygaard, J., and Wiener, F. (1973). Int. J. Cancer *11*, 30.

Risser, R., and Pollack, R. (1974). Virol. *59*, 477.

Rygaard, J. (1969). Acta Path. Microbiol. Scand. *77*, 761.

Rygaard, J., and Povlsen, C. O. (1969). Acta Path. Microbiol. Scand. *77*, 758.

Sanford, K. K., Earle, W. R., and Likely, G. D. (1948). J. Nat. Cancer Inst. *9*, 229.

Sanford, K. D., Likely, G., and Earle, W. (1954). J. Nat. Cancer Inst. *15*, 215.

Stoker, M. (1968). Nature *218*, 235.

Stoker, M., O'Neill, C., Berryman, S., and Waxman, V. (1968). Int. J. Cancer *3*, 683.

Temin, H., and Rubin, H. (1958). Virology *6*, 669.

Todaro, G. J., and Green, H. (1963). J. Cell Biol. *17*, 299.

Todaro, G. J., Green, H., and Goldberg, B. D. (1964). Proc. Nat. Acad. Sci. USA *51*, 66.

van Diggelen, O. P., and Shin, S. (1974). Biochem. Gen. *11*, in press.

Westerveld, A., Visser, R. P. L. S., Meera Khan, P., and Bootsma, D. (1971). Nature New Biol. *234*, 20.

Reprinted from Science, Vol. 188, pp. 68–70. 1975. © 1975 AAAS

Malignant Hemangioendotheliomas Produced by Subcutaneous Inoculation of Balb/3T3 Cells Attached to Glass Beads

Abstract. *The Balb/3T3 mouse embryo cell line has been frequently used in cancer research as representative of nontumorigenic cells with the characteristic in vitro properties of postconfluence inhibition of cell division, low saturation density, and anchorage dependence. On the reasoning that anchorage dependence might also apply in vivo, each of nine mice were subcutaneously inoculated with an average of 15,400 Balb/3T3 cells attached to two glass beads 3 millimeters in diameter. After 8 weeks, all the mice had developed large bloody tumors that microscopically proved to be hemangioendotheliomas. The inoculation of Balb/3T3 cells alone or beads alone produced no tumors. Transplants of each tumor into normal mice grew to kill the animal within 6 weeks. Tumor cells from collagenase-disaggregated tumor tissue had a plating efficiency of 21.2 percent compared to that of normal adult subcutaneous fibroblasts of less than 0.1 percent. The tumor cells in vitro closely resembled Balb/3T3 cells in appearance and were tumorigenic at a dose of 10^4 cells. A second, repeat experiment produced the same type of tumors grossly and microscopically in 17 of 25 mice between 99 and 211 days after inoculation of the Balb/3T3 cells attached to glass beads. These findings require a reassessment of the postulate that low saturation density, postconfluence of cell division, and anchorage dependence are characteristic in vitro properties only of nonneoplastic cells.*

The Balb/3T3 mouse embryo cell line [1] exhibits the properties of low saturation density (4×10^4 to 5×10^4 cells per square centimeter) and postconfluence inhibition of cell division [2, 3] as the result of being carried in continuous exponential growth at low cell density with minimal cell-cell contact. These properties have been correlated with the fact that the line is nontumorigenic when inoculated subcutaneously into syngeneic mice [4]. An earlier 3T3 line produced from random-bred Swiss mouse embryo [5] possesses the same in vitro properties as the Balb/3T3 line but cannot be tested for tumorigenicity because of the histocompatibility barrier. If Balb/3T3 cells are not carried in continuous exponential growth, they can become spontaneously neoplastically transformed [6]. For this reason, authentic clone A31 Balb/3T3 cells were obtained from the American Type Culture Collection, Rockville, Maryland, for use in these studies.

Balb/3T3 cells have been quite frequently used in cancer research as representative of nontumorigenic cells and have been compared with neoplastically transformed derivative lines, especially those transformed by SV40 virus, with regard to morphology [7], postconfluence inhibition of cell division [2, 3] and of locomotion [8], response to factors affecting growth [9] and locomotion [10], agglutinability by lectins [11], cyclic adenosine monophosphate metabolism [12], membrane transport of glucose, amino acids, and nucleo-

sides [13], ganglioside and glycosyl transferase content [14], and alterations in surface concentration of H2 antigens [15]. Because its low saturation density makes foci of piled up, morphologically transformed cells easy to see, the Balb/3T3 line has also been used in the study of neoplastic transformation in vitro by viruses [16, 17] and chemicals [18].

Balb/3T3 cells also have the property of anchorage dependence, or the inability to divide in vitro unless attached to a solid substrate [19]. I reasoned that anchorage dependence might also apply in vivo, and that Balb/3T3 cells might grow and produce tumors if they were inoculated subcutaneously attached to a solid substrate. The experiments to be described showed that this was indeed the case. Glass beads, 3 mm in diameter (Kimax, No. 5663-F28, Arthur H. Thomas Company, Philadelphia), were placed in a 60-mm plastic petri dish, and 5 ml of tissue culture medium (Dulbecco-Vogt modified Eagle's minimal essential medium with 10 percent fetal bovine serum) containing approximately 3×10^6 singly suspended, trypsin-detached Balb/3T3 cells (clone A31, passage 83, No. CCL 163, obtained from the American Type Culture Collection) were added. The next day, two beads with attached cells were inoculated subcutaneously under ether anesthesia into each of nine Balb/c mice by making a 1-cm incision through the skin with scissors, opening the subcutaneous space over one side of the back by

blunt dissection, dropping the beads into the space, and closing the incision with skin clips. The number of 3T3 cells attached per glass bead under these conditions averaged 7700, determined by counting in a hemocytometer the number of cells detached with trypsin from 500 beads. The calculated maximum possible number of cells per bead (surface area × saturation density) was 14,000 cells. Each of another group of 20 mice was inoculated subcutaneously with two glass beads that had been incubated in tissue culture medium without attached Balb/3T3 cells. A third group of animals consisted of four subgroups of ten mice each that were inoculated subcutaneously with Balb/3T3 cells suspended in tissue culture medium at doses of 10^3, 10^4, 10^5, and 10^6 cells per mouse. By the end of 8 weeks after inoculation, all nine mice inoculated with Balb/3T3 cells attached to glass beads had developed tumors approximately 2 cm in diameter, whereas the groups inoculated with glass beads alone or Balb/3T3 cells alone had no tumors (nor did they develop tumors during 4 months of subsequent observation). When the skin was reflected, all of the tumors were similar in gross and microscopic appearance. They were of a dark reddish-brown color, covered with a thin fibrous capsule, and were fluctuant to touch. They could be separated from the fascia overlying the muscles of the back by blunt dissection but were firmly attached to the skin. Cut sections revealed a large central area of friable reddish-brown tissue mixed with similarly colored fluid indicative of old hemorrhage. The two glass beads were found loosely enmeshed in this tissue. The peripheral area of the cut section was 1 to 4 mm thick and consisted of glistening, pale pink translucent tissue. Microscopic sections showed that this peripheral area was composed of tightly packed, highly undifferentiated, large, polygonal tumor cells that in some areas were interspersed between anastomosing small blood channels that were frequently seen to be lined by tumor cells (Fig. 1). It appeared that these blood channels had ruptured to produce the degenerating tumor tissue and old hemorrhage seen in the central portion of the tumor. Groups of tumor cells at the border of the tumor were invading the surrounding muscular and fibroareolar tissue. The microscopic diagnosis was malignant hemangioendothelioma [20].

Small portions of each of the nine tumors were trocar-transplanted subcutaneously into ten mice. The transplants all grew to form large tumors that killed the animals within 6 weeks. The appearance of the transplanted tumors closely resembled the original tumors in gross and microscopic appearance, indicating that the central areas of hemorrhage in the original tumors were not secondary to the presence of the glass beads.

Taking advantage of the known high plating efficiency of Balb/3T3 cells [30 to 50 percent (1)], I proceeded to recover the tumor cells in vitro free of normal stromal cells by planting the cells from collagenase-disaggregated tumor tissue at 100 cells per petri dish. The mean number of colonies appearing in 12 dishes 2 weeks later was 21.2, standard error 0.87 (plating efficiency 21.2 percent). For comparison, I determined that the mean number of colonies in ten petri dishes 2 weeks after planting 10^4 cells from collagenase-disaggregated connective tissue, scraped from the dorsal skin of adult Balb/c mice, was 5.0, standard error 1.33 (plating efficiency 0.05 percent). I also determined the plating efficiency of collagenase-disaggregated fibroblasts that had proliferated and grown into an approximately 1-cm wad of nylon wool implanted subcutaneously 3 weeks previously. In this case the mean number of colonies in nine petri dishes planted at 10^4 cells per dish was 8.7, standard error 0.7 (plating efficiency 0.087 percent). Thus, the very low plating efficiency of normal connective tissue cells (less than 0.1 percent) assured me that the colonies derived from plating tumor tissue cells at 100 cells per dish were practically all tumor cells. The tumor cell colonies were all similar in appearance, making it unlikely that the tumor tissue was made up of a mixture of host-derived and Balb/3T3-derived tumor cells.

In addition to their high plating efficiency, the tumor cells also resembled Balb/3T3 cells in their morphology, being polygonal and tending to form a monolayer with a tile mosaic-like pattern. The appearance of clonal colonies of the tumor cells was distinctive. The peripheral half of their radial extent consisted of a monolayer of cells closely resembling Balb/3T3 cells, while in their central portion a layer of rounded, refractile cells rested on top of the monolayer. In mass culture the tumor cells formed a monolayer of

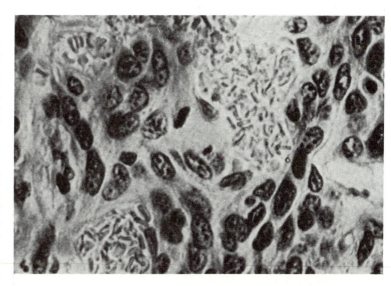

Fig. 1. Histologic section of tumor developing from Balb/3T3 cells attached to glass beads. The tumor cells have formed vascular channels containing red blood cells. One of the tumor cells bordering the lumen of a vascular channel is in the anaphase stage of mitosis. The microscopic diagnosis is malignant hemangioendothelioma.

polygonal cells that continued to divide slowly to form rounded, refractile cells that adhered to the monolayer; this behavior was very different from that of adult connective tissue cells, which formed overlapping strata of parallel-oriented spindle-shaped cells, or that of a line of spontaneously transformed adult connective tissue cells, which formed a multilayered feltwork of crisscrossing spindle cells.

The tumorigenicity of the cultured tumor cells, determined by inoculating graded doses of cells subcutaneously into groups of 10 mice at each dose, was 2/10 at 10^4 cells per dose, 7/10 at 10^5 cells, 7/10 at 10^6 cells, and 10/10 at 10^7 cells per dose. The tumors all reached a size of 1.5 cm in average diameter within 3 weeks. They were very similar in both gross and microscopic appearance to the original tumors produced by inoculating the Balb/3T3 cells attached to glass beads.

A complete repeat experiment was performed in which two glass beads with attached Balb/3T3 cells were inoculated subcutaneously as before into each of 25 mice. Two humors appeared by the 99th day after inoculating, 11 more between the 152nd and 167th days, and four more by the 211th day, when the experiment was terminated. All of these tumors were similar in gross and microscopic appearance to those described in the first experiment.

In view of the fact that malignant hemangioendotheliomas appeared at the site of subcutaneous inoculation of Balb/3T3 cells attached to 3-mm glass beads, whereas inoculation of cells or glass beads alone produced no tumors, and adhering to the conventional definition of a neoplastic cell population as one that produces tumors when inoculated subcutaneously, Balb/3T3 cells must be considered neoplastic. This will require a reassessment of the current concept that postconfluence inhibtion of cell division, low saturation density, and anchorage dependence are characteristic in vitro properties only of nonneoplastic cells.

Since the saturation density of Balb/3T3 cells increases with increase in the serum concentration of the culture medium (21), the existence of a many-fold higher serum concentration in vivo seems a plausible, although of course unproved, explanation for some of the continued proliferation of the Balb/3T3 cells on the glass beads to the point where they finally no longer required attachment to a solid substrate in order to divide. This loss of anchorage dependence by the Balb/3T3 cells can be identified as an example of tumor progression (22).

The possibility that the tumors were derived from host cells rather than the glass-attached Balb/3T3 cells appears unlikely for the reasons that the Balb/3T3 cells alone or beads alone produced no tumors, the high plating efficiency of the tumor cells was like that of the Balb/3T3 cells, and the morphology of the tumor cells in culture was similar to that of Balb/3T3 cells.

There are now two pieces of evidence that the Balb/3T3 cell is a vascular endothelial cell: It looks like an endothelial cell by scanning electron microscopy (7), and it forms tumors that resemble those derived from vascular endothelial cells. The evidence is further supported by the resemblance in morphology and behavior between endothelial cells in vivo and Balb/3T3 cells in vitro. Endothelial cells forming the walls of capillaries appear as a tightly adherent monolayer of nondividing polygonal cells that are activated to grow by wounding and that do not grow as a multilayer, since this would tend to occlude the lumen of the capillary.

It appears that the in vitro properties of the Balb/3T3 cloned line of probable endothelial cells should not be used as the standard for nontumorigenic mouse cells, not only because the cells represent just one of dozens of embryonic cell types, but also because they are neoplastic when inoculated subcutaneously attached to a solid substrate.

CHARLES W. BOONE

Cell Biology Section, Viral Biology Branch, National Cancer Institute, Bethesda, Maryland 20014

References and Notes

1. S. A. Aaronson and G. J. Todaro, *J. Cell. Physiol.* **72**, 141 (1968).
2. E. Martz and M. S. Steinberg, *ibid.* **79**, 189 (1972).
3. Other terms for this phenomenon have been contact inhibition of cell division (*1*), density dependent inhibition of growth [M. G. Stoker and H. Rubin, *Nature (Lond.)* **215**, 171 (1967)], and topoinhibition [R. Dulbecco, *ibid.* **227**, 802 (1970)].
4. S. A. Aaronson and G. J. Todaro, *Science* **162**, 1024 (1968).
5. G. J. Todaro and H. Green, *J. Cell Biol.* **17**, 299 (1963).
6. C. W. Boone, E. L. Lundberg, T. Orme, R. Gillette, *J. Natl. Cancer Inst.* **51**, 1731 (1973).
7. K. R. Porter, G. J. Todaro, V. Fonte, *J. Cell Biol.* **59**, 633 (1973).
8. M. H. Gail and C. W. Boone, *Exp. Cell Res.* **64**, 156 (1971).
9. A. Lipton, D. Paul, M. Henehan, I. Klinger, R. W. Holley, *ibid.* **74**, 466 (1972); P. S. Rudand, W. Eckhart, D. Gospodarowicz, W. Seifert, *Nature (Lond.)* **250**, 337 (1974).
10. A. Lipton, I. Klinger, D. Paul, R. W. Holley, *Proc. Natl. Acad. Sci. U.S.A.* **68**, 2799 (1971); R. R. Burk, *ibid.* **70**, 369 (1973).
11. C. H. O'Neill and M. E. Burnett, *Exp. Cell Res.* **83**, 247 (1974); W. Eckhart, R. Dulbecco, M. M. Burger, *Proc. Natl. Acad. Sci. U.S.A.* **68**, 283 (1971); G. L. Nicolson, *Nat. New Biol.* **233**, 244 (1971).
12. S. Bannai and J R. Sheppard, *Nature (Lond.)* **250**, 62 (1974); L. J. deAsua, E. Rozengurt, R. Dulbecco, *Proc. Natl. Acad. Sci. U.S.A.* **71**, 96 (1974).
13. W. E. C. Bradley and L. A. Culp, *Exp. Cell Res.* **84**, 335 (1974); K. J. Isselbacher, *Proc. Natl. Acad. Sci. U.S.A.* **69**, 585 (1972); R. Kram, P. Mamcnt, G. M. Tomkins, *ibid.* **70**, 1432 (1973).
14. R. O. Brady and P. T. Mora, *Biochim. Biophys. Acta* **218**, 308 (1970); L. M. Patt and W. J. Grimes, *J. Biol. Chem.* **249**, 4157 (1974); S. Roth and D. White, *Proc. Natl. Acad. Sci. U.S.A.* **69**, 485 (1972).
15. E. Tsakraklides, C. Smith, J. H. Kersey, R. A. Good, *J. Natl. Cancer Inst.* **52**, 1499 (1974).
16. J. L. Jainchill and G. J. Todaro, *Exp. Cell Res.* **59**, 137 (1970).
17. S. A. Aaronson, J. L. Jainchill, G. J. Todaro, *Proc. Natl. Acad. Sci. U.S.A.* **66**, 1236 (1970).
18. J. A. DiPaolo, K. Takano, N. C. Popescu, *Cancer Res.* **32**, 2686 (1972); M. A. Basombrio and R. T. Prehn, *Int. J. Cancer* **10**, 1 (1972).
19. M. Stoker, C. O'Neill, S. Berryman, V. Waxman, *Cancer* **3**, 683 (1968).
20. The consultative advice and opinion of Dr. Harold L. Stewart, Division of Cancer Cause and Prevention, National Cancer Institute, in making the diagnosis of malignant hemangioendothelioma is acknowledged.
21. R. W. Holley and J. A. Kiernan, *Proc. Natl. Acad. Sci. U.S.A.* **60**, 300 (1968).
22. L. Foulds, *Neoplastic Development* (Academic Press, New York, 1969).

9 December 1974 ∎

70

Reprinted from
Proc. Nat. Acad. Sci. USA
Vol. 73, No. 2, pp. 549–553, February 1976
Cell Biology

Totipotency and normal differentiation of single teratocarcinoma cells cloned by injection into blastocysts

(mouse teratoma/embryonal carcinoma/embryoid body core cells/allophenic mice/genetic mosaicism)

KARL ILLMENSEE AND BEATRICE MINTZ*

Institute for Cancer Research, Fox Chase, Philadelphia, Pennsylvania 19111

Contributed by Beatrice Mintz, November 19, 1975

ABSTRACT   A definitive test for developmental totipotency of mouse malignant teratocarcinoma cells was conducted by cloning singly injected cells in genetically marked blastocysts. Totipotency was conclusively shown in an adult mosaic female whose tumor-strain cells had made substantial contributions to all of the wide range of its somatic tissues analyzed; the clonally propagated cell lineage had therefore differentiated in numerous normal directions. The test cells were from "cores" of embryoid bodies of a euploid, chromosomally male (X/Y), ascites tumor grown only *in vivo* by transplantation for 8 years. The capacity of cells from the same source to differentiate, in a phenotypic male, into reproductively functional sperms, has been shown in our previous experiments [(1975) *Proc. Nat. Acad. Sci. USA* 72, 3585–3589]. Cells from this transplant line therefore provide material suitable for projected somatic and germ-line genetic analyses of mammalian differentiation based on "cycling" of mutation-carrying tumor cells through developing embryos. In some animals obtained from single-cell injections, tumor-derived cells were sporadically distributed in developmentally unrelated tissues. These cases can be accounted for by delayed and haphazard cellular integration, and by a marked degree of sustained cellular developmental flexibility in early mammalian development, irrespective of certain classical "germ-layer" designations. All mosaic mice obtained have thus far been free of teratomas. In one case, the injected stem cell contributed only to the pancreas and gave rise to a malignancy resembling pancreatic adenocarcinoma. The high modal frequency of euploidy in these individually tested cells thus tends to indicate that a near-normal chromosome complement is sufficient for total restoration of orderly gene expression in a normal embryonic environment; it may also be necessary for teratoma stem-cell proliferation to be terminated there.

Teratomas, unlike other tumors, often comprise a variety of tissues, derived from teratocarcinoma, or embryonal carcinoma, stem cells (1, 2). Pluripotency of the stem cells was demonstrated when solid tumors of varied tissue composition were obtained from transfers of single cells to graft hosts (3). However, the consistent absence of certain tissues (4), and the immaturity and aberrations of others, in teratomas has until recently left open the question whether embryonal carcinoma cells are in fact developmentally totipotent. Full realization of normal developmental potentialities would be expected to occur, for embryonal carcinoma as for normal embryonal cells, only in the normally organized environment of an early embryo. We therefore undertook experiments in which mouse teratocarcinoma cells were injected into the cavity of genetically marked blastocysts (5, 6). The donor cells were from the "cores" of teratoma embryoid bodies grown only *in vivo* as a transplantable ascites since the tumor (OTT 6050) was isolated in 1967 (7). This source seemed particularly promising because the cells had retained pluripotency, as seen in the tissues formed in solid growths, and because chromosomal examination disclosed that they were still very largely euploid (5, 6), as in the tumor of origin (8).

In our first series of experiments (5, 6), small groups of five embryoid body core cells were injected into blastocysts. There they developed in a completely orderly fashion for the first time in their 8-year transplant history, and contributed substantially to production of tumor-free healthy mice with a wide array of tumor-derived tissues, including some never seen in the solid tumors themselves. Adult tissue-specific products (e.g., immunoglobulins, hemoglobin, liver proteins, melanin, etc.), of the genetic variant types characteristic of the inbred strain in which the tumor had originated, but not of the blastocyst strain, attested to normal function. Cells derived from the teratoma, which in this case is of X/Y sex chromosome type, also gave rise to sperms from which many normal progeny were obtained. These results strongly indicated that individual embryonal carcinoma cells probably had the full range of developmental potentialities; and that teratocarcinogenesis entails changes in gene function rather than gene structure.

In the present series of experiments, a final test of teratoma-stem-cell totipotency was sought by introduction of *single* embryoid body core cells into blastocysts, to allow clonal propagation of the donor cell. Single-cell injections have the further advantage that they test the developmental consequences of any chromosomal or genetic variations among the source cells. We report here definitive evidence for developmental totipotency of single teratocarcinoma cells. Production of teratoma-free animals with many tumor-derived normal tissues indicates that retention of euploidy in the tumor cells is a sufficient, and possibly a necessary, condition for restoration of completely normal gene expression in an appropriate environment.

MATERIALS AND METHODS

Teratocarcinoma Cells. The OTT 6050 teratoma was experimentally produced by Stevens in 1967 (7), by grafting a 6-day chromosomally male (X/Y) embryo (8) of the 129/Sv *Sl^J C P* inbred strain (to be referred to as 129) under the testis capsule. The embryo became disorganized and formed a teratoma which, after intraperitoneal injection into another recipient, was converted to a modified ascites form. "Embryoid bodies" in the ascites consist, when small-size, of a "core" of embryonal carcinoma cells surrounded by a "rind" of yolk sac epithelium. The *core cells* were used for blastocyst injections. They were isolated, as before (6), from embryoid bodies of less than 100 μm size, or from a modified source (to be described elsewhere) which yielded similar results.

Injections into Blastocysts. The microinjection proce-

Abbreviations: GPI, glucosephosphate isomerase; IDH, isocitrate dehydrogenase.
* Address reprint requests to this author.

549

Table 1. Tissue contributions[a] clonally derived from single teratocarcinoma cells (129 strain) after injection into blastocysts (WH or C57BL/6-*b/b* strain)

Case, sex	Autopsy age	Coat	Blood cells	Brain	Spleen	Heart	Skeletal muscle	Kidneys	Reproductive tract	Liver	Stomach, intestines, pancreas	Thymus	Lungs	Other
A♀	4 wk	2	10	20	40	40	40	40	40	20	40	40	33	
B♀	3 wk	0	10	40	40	40	0	33	0	33	0	33	50	b
				33	*35*	*25*		*5*		*30*		*25*	*35*	c
C♂	3 wk	0	0	80	0	0	0	0	80	10	0	40	10	d
D♀	1 d		0	0	0	0	0	50	0	60	33	0	25	
E♀	3 wk	0	0	0	0	5	0	0	0	0	5	10	0	e
F♀	1 d		10	← Whole-body homogenate 33 →										f
G♂	Alive	2	10											
H♂	1 wk	0	0	0	0	0	60	0	0	0	0	0	40	
I♀	1 d		0		40	0	0			0	0		50	g
J♂	Alive	0	15											
K♀	Alive	0	10											
L♂	Alive	5	0											
M♀	Alive	2	0											
N♂	1 wk	0	0	0	0	0	0	0	0	0	0	0	75	
O♂	2 wk	0	0	0	0	0	0	0	0	0	5	0	0	h

a Coat strain-specific markers were at the *agouti* locus (in injections into WH blastocysts), expressed in hair follicle dermis, or at both the *agouti* locus and the *black* locus, expressed in melanocytes (in C57BL/6-*b/b* blastocyst injections). All other tested tissues were biochemically analyzed for GPI electrophoretic allelic strain variants. In addition, some tests (*italics*, second line of case B) were conducted for IDH allelic variants, because absence of IDH expression in blood cells permits blood contamination to be ruled out as a source of any 129-strain component in IDH-tested tissues. All data are given in percent 129 type.
b Salivary glands, 40.
c Salivary glands, 5.
d (Hepatitis).
e Placenta, 0.
f (*W/W* anemia).
g Carcass, 0 (*W/W* anemia).
h Pancreatic adenocarcinoma, 100. The pancreatic tumor found in this animal had some histologically normal elements, and the homogenate contained 50% of each GPI strain (Fig. 5, slot b). Therefore, the actual cancerous component was apparently all of the histocompatible 129 strain-type (i.e., teratocarcinoma-derived) as shown in the *corrected* entry (100%) in the table, and the normal component was blastocyst-derived. A small amount of ostensibly normal pancreas was included with stomach and intestines in the GPI test of a homogenate of those tissues; the 5% 129-strain GPI in that test was probably due only to some accompanying pancreatic adenocarcinoma.

dure (9) was again used to entrap a teratocarcinoma cell in the blastocyst cavity near the inner cell mass. Most recipient blastocysts were of the WH (ICR subline) inbred strain, in which only the *W* (*dominant white spotting*) locus is segregating; matings of +/*W* parents were used. Both +/+ and +/*W* segregants are normal; *W/W* individuals ordinarily die of a severe anemia within a few days after birth. A few blastocysts were of the C57BL/6-*b/b* strain (to be referred to as C57-*b/b*), which differs from C57BL/6 only by mutation of the *black* (*B*) coat-color allele to *brown* (*b*). Injected blastocysts, after a few hours' incubation (6), were surgically transferred to uteri of pseudopregnant ICR albino females.

Tissue Genotypic Analyses. The 129 strain of tumor origin is *black* (*B/B*) for a melanocyte marker and *white-bellied agouti* (A^w/A^w) for a dermal hair-follicle marker. The WH blastocyst strain is *black* and *non-agouti* (*a/a*); the C57-*b/b* blastocyst strain is *brown* (*b/b*) and *non-agouti*. With the glucosephosphate isomerase (GPI; D-glucose-6-phosphate ketol-isomerase, EC 5.3.1.9) marker, tumor- and blastocyst-derived cells can be distinguished in blood cell lysates and in homogenates of soft tissues, by starch gel electrophoresis (10) of allelic forms (*Gpi-1* locus). The 129 strain

is homozygous for the slow-migrating $Gpi\text{-}1^a$ type, WH and C57-*b/b* strains for the fast $Gpi\text{-}1^b$ type. Another marker, NADP-dependent isocitrate dehydrogenase [IDH; *threo*-D_s-isocitrate:NADP$^+$ oxidoreductase (decarboxylating), EC 1.1.1.42], which is not expressed in blood cells, was also used for independent tests of some tissue homogenates, by starch gel electrophoresis (11). The 129 strain is homozygous for the slow-migrating $Id\text{-}1^a$ variant, the WH and C57-*b/b* strains for the fast-migrating $Id\text{-}1^b$ type.

RESULTS

Of a total of 161 injected and surgically transferred blastocysts, 71 (44%) survived. Three were from C57-*b/b*- and 68 from WH-strain blastocysts. The survivors (33 females, 34 males, four of unrecorded sex) included five living fetuses sacrificed on day 15 of gestation; 16 live animals delivered by caesarean section at term, for genotypic comparisons with their placentas, and sacrificed within a day's time; and 50 older postnatal animals. Of the 50 postnatal animals, 24 were autopsied at 1–7 weeks of age; the remaining 26 are still alive and now range from 9 to 16 weeks of age. Analyses

Cell Biology: Illmensee and Mintz

Proc. Nat. Acad. Sci. USA 73 (1976)   551

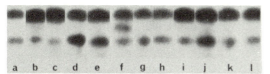

FIG. 1. GPI allelic strain variants in starch gel electrophoresis of tissue homogenates. Slot (a), a 1:1 mixture of controls: 129-strain-type, slow-migrating; WH type, fast-migrating. Tissues from experimental animal A: blood (b), brain (c), spleen (d), heart (e), skeletal muscle, showing a hybrid enzyme band due to heterokaryon formation (f), kidneys (g), reproductive tract (h), liver (i), pooled stomach, intestines, and pancreas (j), thymus (k), and lungs (l). The 129-strain component of all these normal tissues was derived from a single teratocarcinoma cell after injection into a WH-strain blastocyst. (Origin and anode are lowermost.)

have been completed on all except the 26 living animals, which have had only coat and blood cells genotypically classified; five have coat and/or blood mosaicism and some of the remaining 21 may prove later to have some internal tissue mosaicism. Among all 71 animals produced, whether partially or completely analyzed, there were 21 (30%) with 129-strain cells in one or more tissues (Tables 1 and 2). In addition, six resorption sites were found during caesarean sections; with the GPI marker, they had 0, 40%, 75%, 80%, 80%, and 85%, respectively, of the 129 type. Histological samples were indistinguishable from those in spontaneous resorptions among controls.

The results summarized in Table 1 clearly demonstrate that the clonal lineage of a single embryonal carcinoma cell from an embryoid body core can differentiate in widely diverse directions, contributing to all tissues thus far tested, as in case A (Fig. 1). Inasmuch as GPI is expressed in blood cells as well as other tissues which themselves contain blood, it should be emphasized that this marker alone does not prove mosaicism in tissues other than blood, if the animal also has blood mosaicism. Evidence for tissue-specific mosaicism was therefore obtained from the following sources: (*i*) the 129 type of GPI was absent from the bloods of several animals but present, unequivocally, in their other tissues (cases C, D, E, H, I, and N). (*ii*) Quantitative estimates of GPI isozymes indicated that a number of tissues possessed the 129-strain GPI type at appreciably higher levels than accountable for by that animal's blood (e.g., 50% 129 type in the lungs, and only 10% in the circulating blood, of animal B) (Fig. 2). (*iii*) IDH, which is not expressed in blood cells, was used as an independent marker in tissue homogenates from cases A and B (Table 1 and Fig. 3), where it confirmed the presence of tissue-specific 129-strain contributions. (IDH-isozyme estimates for case A are omitted from Table 1

Table 2. Placenta compared with total-body contributions* (estimated from percent 129-strain-type GPI) clonally derived from single 129-strain teratocarcinoma cells after injection into WH-strain blastocysts

Case	Sex	Placenta†	Body
P	♂	50	10‡
Q	♂	40	40
R	♀	40	25
S	♀	10	75
T	♂	10	33

* From cases delivered by caesarean section at term (P-S) or on day 15 of gestation (T). An additional placenta (U) had 50% 129-type GPI; the animal itself was cannibalized.
† Each surrogate mother had only the WH-strain GPI type and, therefore, did not contribute any of the 129-strain placental cells. The bloods of animals Q and S had only WH-type GPI and also did not account for any of their placental 129-type GPI.
‡ W/W anemia was present in this animal.

because the gels were technically inadequate for quantitation.) (*iv*) In skeletal muscle, the occurrence of an intermediate or hybrid enzyme band in both the GPI and IDH tests proved that cells of both strains had specifically differentiated as myoblasts and had fused and formed heterokaryons with heterodimeric enzyme, as previously shown in muscle of allophenic mice (11). (*v*) Coat mosaicism in four animals (all from WH-strain blastocysts) provided further evidence for a tissue-specific, as well as a normally functioning, 129-strain contribution, by means of allelic differences at the *agouti* locus, expressed in hair follicles. The 129-strain element in the coats was notably small, comprising only a few abbreviated *agouti* stripes or patches on the face (cases A and G) or dorsum (cases M and L) on a *non-agouti* ground. (*vi*) Differentiation into placental tissues also occurred (Table 2). The 129-type GPI placental contribution was shown in all cases (P-U) not to be of maternal origin (Fig. 4) and, in at least two cases (Q and S), not to be due to blood from the fetus.

No animal has thus far shown any indication of a teratoma. This confirms the results of our previous study in which 93 survivors, from blastocysts injected with groups of five embryoid body core cells, were also free of teratomas (ref. 6, and unpublished results). Among the few ailments was a case (animal C) of severe diffuse hepatitis resembling the not uncommon disease caused by mouse hepatitis virus. There is no reason to ascribe the condition to the 10% of the liver derived from the teratoma cell. There were five cases of genetically caused W/W anemia (from injected blastocysts of +/W parents). Three of these animals (Tables 1 and

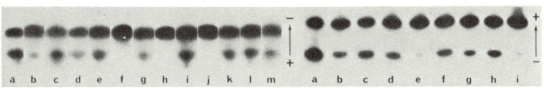

FIG. 2 (*left*). GPI strain-specific allelic variants in starch gel electrophoresis of tissue homogenates. Slot (a), a 1:1 mixture of 129 (slow-migrating-type) and WH (fast-type) controls. Tissues from normal experimental animal B: blood (b), brain (c), spleen (d), heart (e), skeletal muscle (f), kidneys (g), reproductive tract (h), liver (i), pooled stomach, intestines, and pancreas (j), thymus (k), lungs (l), and salivary glands (m). All 129-type contributions arose from one teratocarcinoma cell placed in a WH blastocyst.

FIG. 3 (*right*). Verification of specific-tissue mosaicism, irrespective of blood content, of the same animal as in Fig. 2, by means of an independent marker not expressed in blood. IDH allelic strain variants in starch gel electrophoresis of tissue homogenates. Slot (a), a 1:1 mixture of 129 (slow-migrating-type) and WH (fast-type) controls. The experimental animal's tissues are brain (b), spleen (c), heart (d), kidneys (e), liver (f), thymus (g), lungs (h), and salivary glands (i).

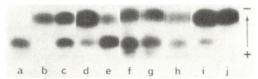

FIG. 4. Comparisons of placenta and whole-body strain composition (including blood cells in each) after injection of single teratocarcinoma cells (129 strain) into blastocysts (WH strain) followed by caesarean delivery at term. GPI isozymic strain variants in starch gel electrophoresis of tissue homogenates from 129-strain (slot a) and WH-strain (b) controls and a 1:1 control mixture (c), and from the placenta (d) and body (e) of case S; placenta (f) and body (g) of case Q; and placenta (h) and body (i) of case P. The surrogate mother of the first case (j) and each of the other cases was of the fast GPI type and therefore did not contribute to the 129-strain GPI in the placenta.

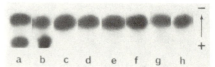

FIG. 5. GPI strain-specific variants in starch gel electrophoresis of tissue homogenates from a 1:1 mixture of 129 (slow-migrating-type) and C57BL/6-*b/b* (fast-migrating) controls (slot a) and of a tumor (b) found in the pancreas of animal O, obtained after injection of one 129-strain teratocarcinoma cell into a C57BL/6-*b/b* blastocyst. Histologically, the tumor mass contained both malignant and normal cells; transplant tests showed the malignant component to be all of the 129 strain. All other tissues tested were solely blastocyst-derived; they included blood (c), brain (d), skeletal muscle (e), kidneys (f), liver (g), and lungs (h).

2) proved to have 129-type cells which, in at least one case (F), were identified in the blood. (A fourth W/W had only blastocyst-derived cells, and the fifth was cannibalized.)

The sole instance of an abnormality directly attributable to cells from the teratoma lineage was striking and without precedent. Animal O, obviously ill and retarded since birth, was killed at 2 weeks of age and found to have a conspicuous pancreatic tumor. In histological examination, kindly performed by Dr. R. P. Custer of this Institute, it resembled a pancreatic adenocarcinoma, predominantly of the acinar type but also with ductal epithelial dysplasia resembling carcinoma *in situ*. During the autopsy, pieces were grafted subcutaneously into recipients of the C57BL/6-*b/b* strain, from which the blastocyst had been obtained, and the 129 teratoma-strain; these have some histocompatibility allelic differences at the major locus (*H-2ᵇ* and *H-2ᵇᶜ*, respectively) (12), and at minor loci. Three of four grafts grew in 129-strain hosts; none of five grafts survived in C57-*b/b* hosts. Therefore, although a homogenate of the tumor contained 50% of each GPI strain-type (Fig. 5, slot b), the actual malignant component was apparently all of the 129 strain (Table 1). This conclusion is consistent with the fact that the tumor mass included some histologically normal elements, probably largely of C57-*b/b* strain composition, as judged from the GPI results. The animal had no other tumors and had only C57-*b/b* cells in its other tissues. (A very slight amount of 129-type GPI in pooled stomach, intestines, and an ostensibly normal part of the pancreas was, in all likelihood, due to contamination with some 129-strain pancreatic tumor cells.)

DISCUSSION

A conclusive demonstration of teratocarcinoma-cell developmental totipotency would require that a *single* cell be shown to form all fully differentiated somatic tissues. The single embryoid body core cell injected into a blastocyst in case A (Table 1 and Fig. 1) has now provided such evidence, based on examination of virtually all major tissues in the resultant animal. The data thus support and extend our earlier conclusion of probable totipotency, from experiments in which groups of five embryoid body core cells were introduced into blastocysts (5, 6). Cells derived from this X/Y tumor can also form fully functional sperms in a phenotypic male (6). From evidence in X/X ↔ X/Y allophenic mice (13), X/Y primordial germ cells would not be expected to progress to the gamete stage in a phenotypic female such as animal A. In the strict sense, a functional germ-line contribution may not be essential for proof of totipotency: totipotent embryo

cells of certain sterile genotypes (e.g., W/W) form mice with all somatic tissues but a deficiency of germ cells. Nevertheless, possession by teratoma cells of the capacities for full germ-line (6) as well as somatic differentiation is fortunate, as both are indispensable for realizing one of the long-range aims underlying this work.

The aim alluded to, and outlined earlier (5, 6), is to utilize mutation-carrying teratocarcinoma cells as a new experimental tool for the analysis of mammalian differentiation *in vivo*, by "cycling" them through mice via blastocyst injections. Presumptive or known mutations could then be assessed for full developmental and biochemical expression in the soma; the heritable status of the variant could be shown by its transmission to progeny; and its location in relation to other genes could be mapped through recombination in the germ-line.

Inclusion of some blastocysts of the lethally anemic W/W genotype in the present work was intended as an exploration of one facet of these projected genetic studies: the use of lethal genes. Most lethals are probably deleterious because only one or a few, rather than all, cell types are defective. Replacement or interaction of the defective cells with genetically normal ones could permit the animal to be "rescued" and the analyses to go forward. An example of rescue by specific-cell replacement is the fully viable W/W ↔ $+/+$ allophenic mouse, in which sufficient normal ($+/+$) embryo cells have occupied the hematopoietic system (14). The same principle is being tested here, albeit in some future experiments the lethal genes may be introduced in the teratoma cells. Three mice from injected W/W blastocysts were in fact found to have tumor-derived cells (Tables 1 and 2), identified in at least one case (F) as having made some contribution to the blood. Since the animal was killed for study, its chances for survival are unknown, but the results are encouraging for the further use of lethal-normal combinations.

In both our 5-cell and our 1-cell injection series, the initial number of teratoma cells is close to the 3-cell number estimated (14) to comprise the precursor cells of the embryo proper, as distinct from future extraembryonic components. Yet the 5-cell injections yielded 16% (15/93) of the survivors with teratoma-derived cells (ref. 6, and unpublished data), the 1-cell injections at least 30% (21/71, with 21 living mice almost wholly unanalyzed). While some of the increase may be attributable to improved technique (reflected in the 33% compared to 44% survival rates), it is probably due chiefly to the observed relatively greater tendency for a single injected cell to remain near the inner cell mass when deposited there, hence to be integrated into the embryo-forming region.

Cases of sporadic distribution of clonally derived 129-strain cells, often in developmentally unrelated tissues

(Table 1), are strikingly more frequent than previously observed in conventional allophenic mice from aggregates of two blastomere strains (14–16). This disparity could conceivably result from a tendency toward delayed teratoma-cell integration into the embryo, perhaps due to initial differences in donor and host cellular adhesiveness (5). There may by then be a small donor clone whose cells are scattered to relatively few places, where development proceeds according to local cues. This could also account for the low incidence and small-area coverage of coat mosaicism (Table 1), again in contrast to allophenic mice. Melanocyte determination in the mouse [from 17 pairs of neural crest cells (17)] and hair-follicle-dermis determination [from some 85 pairs of somite cells (14)] are believed to occur on about late day 6 of embryonic life, in precursor cells flanking the mid-dorsum and strung out along the full length of the embryo. If in fact a teratoma cell injected into a day 3 blastocyst tends often to undergo late assimilation and limited distribution, its daughter cells would be least often represented in those tissues, such as the coat, that originate from a relatively far-flung system of specific precursor cells. Coat markers may therefore be the least favorable ones to reveal teratoma-cell participation. This is consistent with the results of Brinster (18), who, in the only previous attempt to inject embryonal carcinoma cells into blastocysts, used coat color as the only marker (thereby precluding tests for totipotency). He obtained only one animal, out of 60 survivors, with a slight tumor-strain contribution in its coat.

The frequent sporadic distribution of tumor-derived tissues (Table 1) also serves as a *caveat* for future experiments with mutagenized teratocarcinoma cells: in order to demonstrate any *specific* "restriction" in differentiation of these cells, that could be attributed to their mutant status, it would first be necessary to survey a large sample of mosaic animals.

Prolonged capacity of tumor-derived cells to undergo assimilation and subsequent differentiation also implies that the donor cells had continued to remain totipotent for some time. This would not be surprising, in view of the experimental origin of the teratoma from a 6-day postimplantation embryo (7). Moreover, the tumor stem cells have been found to resemble certain undifferentiated cells of the early postimplantation stage, rather than preimplantation-stage cells, in their ultrastructure (19) and alkaline phosphatase content (20); their soluble protein profiles also differ from those of morula cells (5). The evidence thus strongly supports the probability that some totipotent cells are normally still present as late as day 6 of embryonic life. The candidate totipotent cells have been tentatively identified as "ecto-meso-derm" (19, 20), according to conventional "germ-layer" terminology. While germ layers normally arise in characteristic locations (for still unknown reasons) and normally undergo orderly movements ending in specific differentiations, experiments with many species (21), including mammalian embryos (22), have long shown that layers are substantially intermovable and capable of giving rise to structures usually formed from the others. Totipotency of individual "ecto-mesoderm" cells is consistent with such flexibility.

The only abnormal derivative obtained here was the pancreatic tumor in case O. Localization of tumor-strain cells and tumor specifically to the pancreas may possibly signify that the single donor cell was already restricted to the status

of a pancreatic stem cell which either had an intrinsic defect, or else underwent *de novo* neoplastic conversion. This case demonstrates that the requirement that single cells be injected into blastocysts, for proof of developmental totipotency, is realistic, as some specialized types of stem cells or some irreversibly malignant cells may be included if large numbers of cells are introduced.

Note Added in Proof. In a recent report from another laboratory [Papaioannou, V., McBurney, M., Gardner, R. & Evans, M. (1975) *Nature* **258**, 70–73], blastocysts were injected with 20 to 40 teratocarcinoma cells each, from three *in vitro* cell lines from other teratomas. Of a total of 11 mosaic mice with several somatic tissue contributions from the injected cells, as judged from GPI and pigment markers, most of the animals, obtained from two of the cell lines, also developed tumors. The remaining line, characterized by 40 chromosomes and X/O sex chromosome constitution, yielded one mouse without a tumor.

This work was supported by USPHS Grants HD-01646, CA-06927, and RR-05539, and by an appropriation from the Commonwealth of Pennsylvania. We thank Mrs. Claire Cronmiller for excellent technical assistance.

1. Stevens, L. C. (1967) *Adv. Morphog.* **6**, 1–31.
2. Pierce, G. B. (1967) in *Current Topics in Developmental Biology*, eds. Moscona, A. A. & Monroy, A. (Academic Press, New York), Vol. 2, pp. 223–246.
3. Kleinsmith, L. J. & Pierce, G. B., Jr. (1964) *Cancer Res.* **24**, 1544–1551.
4. Stevens, L. C. & Hummel, K. P. (1957) *J. Nat. Cancer Inst.* **18**, 719–747.
5. Mintz, B., Illmensee, K. & Gearhart, J. D. (1975) in *Symp. on Teratomas and Differentiation*, eds. Sherman, M. I. & Solter, D. (Academic Press, New York), 59–82.
6. Mintz, B. & Illmensee, K. (1975) *Proc. Nat. Acad. Sci. USA* **72**, 3585–3589.
7. Stevens, L. C. (1970) *Dev. Biol.* **21**, 264–382.
8. Dunn, G. R. & Stevens, L. C. (1970) *J. Nat. Cancer Inst.* **44**, 99–105.
9. Lin, T. P. (1966) *Science* **151**, 333–337.
10. Gearhart, J. D. & Mintz, B. (1972) *Dev. Biol.* **29**, 27–37.
11. Mintz, B. & Baker, W. W. (1967) *Proc. Nat. Acad. Sci. USA* **58**, 592–598.
12. Snell, G. D., Graff, R. J. & Cherry, M. (1971) *Transplantation.* **11**, 525–530.
13. Mintz, B. (1969) in *Birth Defects: Original Article Series 5* (National Foundation, New York), pp. 11–22.
14. Mintz, B. (1970) in *Symp. Int. Soc. Cell Biol.*, ed. Padykula, H. (Academic Press, New York), Vol. 9, pp. 15–42.
15. Mintz, B. (1971) in *Symp. Soc. Exp. Biol.*, eds. Davies, D. D. & Balls, M. (Cambridge University Press, New York), Vol. 25, pp. 345–370.
16. Mintz, B. (1974) *Annu. Rev. Genet.* **8**, 411–470.
17. Mintz, B. (1967) *Proc. Nat. Acad. Sci. USA* **58**, 344–351.
18. Brinster, R. L. (1974) *J. Exp. Med.* **140**, 1049–1056.
19. Damjanov, I., Solter, D., Belicza, M. & Skreb, N. (1971) *J. Nat. Cancer Inst.* **46**, 471–480.
20. Damjanov, I., Solter, D. & Skreb, N. (1971) *Z. Krebsforsch.* **76**, 249–256.
21. Oppenheimer, J. M. (1940) *Quart. Rev. Biol.* **15**, 1–17.
22. Levak-Svajger, B. & Svajger, A. (1974) *J. Embryol. Exp. Morphol.* **32**, 445–459.

Reprinted from Science, Vol. 197, pp. 893–895. 1977. © 1977 AAAS

Metastasis Results from Preexisting Variant Cells Within a Malignant Tumor

Abstract. Clones derived in vitro from a parent culture of murine malignant melanoma cells varied greatly in their ability to produce metastatic colonies in the lungs upon intravenous inoculation into syngeneic mice. This suggests that the parent tumor is heterogeneous and that highly metastatic tumor cell variants preexist in the parental population.

The question of why cancer cells metastasize is one of the most important issues in tumor biology. In human cancer it is the process of metastasis, the formation of secondary tumor foci at distant sites, that eventually defeats the efforts of both surgeon and clinical oncologist. In spite of the importance of this phenomenon, little is known about the pathogenesis of metastatic foci or their relationship to the primary tumor. Studies with transplantable tumors in rodents have shown that both host factors and properties of the tumor cells can contribute to the success or failure of the metastatic process (*1*).

Earlier studies with the B16 melanoma in syngeneic C57BL mice showed that the majority of tumor cells injected intravenously die very rapidly in the circulation, and only about 0.1 percent survive and yield metastases (*2*). Further experiments suggested that the survival of these few tumor cells was not a random occurrence, but was due to certain unique properties of the surviving cells (*3*). In this study, we wished to determine whether these unique metastatic cells preexisted in the tumor cell population, or whether they arose during metastasis by a process of adaptation to local environmental conditions. If highly metastatic variant cells could be shown to preexist in the parent population, this would support the suggestion by Nowell (*4*) that tumor cell variants arise within developing tumors, are subjected to host selection pressures, and are responsible for the emergence of new sublines with increased malignant potential.

To distinguish between these possibilities, we performed an experiment similar in design to the classical fluctuation test devised by Luria and Delbrück to distinguish between selection and adaptation in the origin of bacterial mutants (*5*). In our study a cell suspension of the B16 melanoma parent line was divided into two parts. One portion was used to inject syngeneic C57BL/6 mice intravenously. The other portion was used to produce clones, which were then also injected intravenously into groups of C57BL/6 mice (Fig. 1). Eighteen days af-

ter the tumor cells were injected, the number of lung metastases in each recipient was counted. If the number of metastatic foci in the lungs of the mice receiving the cloned sublines was similar to the number of foci seen in mice receiving the parent line, this would indicate that the parent population was homogeneous and that the metastatic foci probably resulted from adaptation during the process of metastasis. Alternatively, if the cloned sublines gave rise to widely different numbers of lung colonies, this would suggest that the parent tumor was heterogeneous and that cells of both high and low metastatic potential preexisted in the parent population.

This experiment was performed with the transplantable B16 melanoma, which originated spontaneously in a C57BL/6 mouse in 1954. We obtained it from Jackson Laboratory, Bar Harbor, Maine, in 1970 and passaged it several times in syngeneic recipients prior to establishing it in cell culture (*6*). After four to five passages in vitro, this parent line was frozen and stored in liquid nitrogen until used in these studies. The cells were thawed and cultured for three passages in vitro to obtain the starting material for this experiment. Seventeen single cell clones were derived by a combination of the soft agarose and microculture techniques (*7*). All cell suspensions for intravenous injection were prepared in an identical manner by light trypsinization from cultures in exponential growth

Table 1. Metastases resulting from intravenous injection of cells from parent B16 tumor line and its in vitro cloned sublines.

Source of B16	Number of mice per group	Number of pulmonary tumor colonies per mouse*	Number of pulmonary tumor colonies		Number of animals with extrapulmonary metastases
			Median	Range	
Parent line	60	8, 20, 24, 26, 27, 27, 27, 28, 28, 30, 30, 30, 30, 31, 31, 32, 33, 33, 33, 34, 34, 35, 36, 36, 37, 37, 39, 40, 40, 40, 41, 43, 44, 44, 44, 44, 44, 46, 46, 46, 49, 56, 56, 59, 60, 62, 64, 66, 66, 67, 68, 69, 72, 72, 72, 73, 78, 84, 98, 131	40.5	8–131	8/60 ovary, 6/60 liver, 3/60 gut, 11/60 lymph nodes, 2/60 adrenal, 1/60 heart, 4/60 kidney, 1/60 nasal sinuses
Clone 16	10	2, 2, 2, 3, 3, 4, 6, 9, 11, 15	3.5	0–15	0/10
Clone 15	11	2, 3, 3, 4, 5, 5, 6, 8, 9, 15, 20	5	2–20	1/11 lymph node
Clone 12	9	0, 0, 1, 4, 6, 10, 16, 27, 34	6	0–34	0/9
Clone 24	9	5, 6, 7, 9, 10, 11, 13, 23, 29	10	5–29	1/9 ovary, 1/9 liver, 1/9 lymph node
Clone 19	10	0, 3, 4, 8, 9, 17, 20, 22, 33, 42	13	0–42	0/10
Clone 7	10	0, 5, 9, 10, 16, 18, 19, 27, 29, 43	17	0–43	0/10
Clone 21	8	1, 5, 8, 15, 21, 22, 23, 48	18	1–48	1/8 lymph node
Clone 18	11	0, 1, 1, 2, 8, 36, 41, 42, 70, 75, 91	36	0–91	0/11
Clone 5	10	2, 25, 31, 44, 45, 46, 67, 74, 78, 171	45.5	2–171	0/10
Clone 6	9	5, 41, 42, 89, 99, 115, 115, 149, 232	99	5–232	0/9
Clone 17	9	104, 110, 132, 144, 150, 151, 173, 206, 210	150	104–210	0/9
Clone 3	9	160, 166, 196, 208, 214, 229, 241, 261, 450	214	160–450	1/9 lymph node
Clone 1	9	73, 114, 153, 165, 237, 272, 273, 290, 321	237	73–321	0/9
Clone 2	10	7, 28, 206, 218, 241, 268, 336, 353, 378, 450	254.5	7–450	0/10
Clone 13	9	50, 62, 79, 91, 260, 306, 320, 338, 350	260	50–350	2/9 ovary, 1/9 liver
Clone 14	9	All nine >500	>500		2/9 ovary
Clone 9	10	All ten >500	>500		2/10 adrenal, 1/10 kidney, 1/10 brain, 6/10 lymph node, 2/10 liver

*Mice were injected intravenously with 50,000 viable tumor cells and killed 18 days later.

phase (3). The cells were washed in culture medium, resuspended in physiologic saline, and pipetted gently to dissociate any cell clumps. Only cell suspensions of more than 95 percent viability, as measured by trypan-blue exclusion, and that were free of cell aggregates were used for injection. C57BL/6 mice were injected in the lateral tail vein with 5×10^4 viable tumor cells in a volume of 0.2 ml. Eighteen days after injection, the mice were killed. The number of pulmonary tumor colonies in each animal was counted in double-blind fashion under a dissecting microscope by two independent observers. The B16 melanoma, which appears as superficial black nodules, grows preferentially in the lungs after intravenous injection (3). Complete autopsies were performed, and all suspected extrapulmonary metastases were confirmed by microscopic examination of fixed histological sections.

Table 1 lists the number of pulmonary nodules per mouse obtained from intravenous injection of the parent B16 line and each of the 17 clones. The metastatic

Table 2. Pulmonary metastases resulting from intravenous injection of cells from B16 clones 21 and 24 and their subclones.

B16 source	Number of mice injected	Number of pulmonary tumor colonies	
		Median	Range
Parent clone 21	8	18	1–48*
Subclone 21-d	9	6	0–45
21-b	10	24	0–135
21-c	10	38.5	0–118
21-a	10	42.5	0–138
Parent clone 24	9	10	5–29*
Subclone 24-b	10	3.5	0–30
24-a	10	8	0–48
24-c	9	9	0–96

*No statistically significant differences were detected between parent clones and their subclones or among the related subclones using a two-tailed Mann-Whitney U test (8).

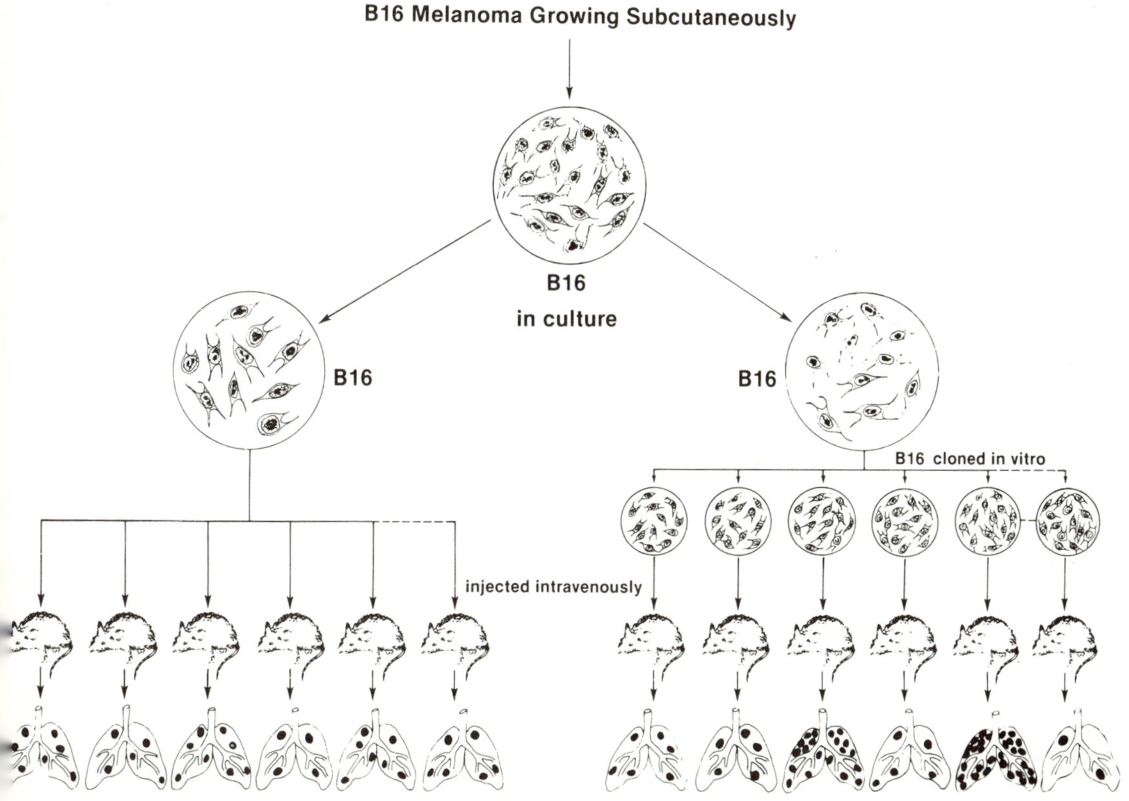

potential of the clones, as seen from the median number of pulmonary colonies, differs dramatically from that of the parent B16 line. In fact, only two clones, 18 and 5, are indistinguishable from the parent line, based on the Mann-Whitney U test (8). There was also considerable

variation among the clones in the number and sites of extrapulmonary metastases. We conclude, therefore, that the cells with a high metastatic potential are present within the parent B16 line prior to their injection into animals.

However, the variability among the clones could have resulted from the process of cloning rather than from heterogeneity of the parent tumor. To test whether the cloning procedure could be responsible for generating these variants, clones 21 and 24 were recloned to produce several subclones. If the cloning procedure were responsible for introducing the variation, then the subclones should also exhibit wide variation in metastatic potential. The distributions of the number of lung colonies produced by the groups of subclones do not differ statistically from each other or from their respective parent clones (Table 2). This suggests that the process of cloning is not the major factor responsible for the variability of the clones seen in Table 1. We conclude that the parent B16 tumor is extremely heterogeneous with respect

B16 Melanoma Growing Subcutaneously

B16 in culture

B16

B16

B16 cloned in vitro

injected intravenously

pulmonary tumor colonies per mouse (60 mice)

pulmonary tumor colonies (10 mice per clone)

Fig. 1. Scheme for demonstrating that metastatic variants preexist within a malignant tumor. B16 melanoma grown in vitro was divided into two parts. One part was injected intravenously into syngeneic C57BL/6 mice and the other was used to produce several clones. Once established, the clones were also injected intravenously into syngeneic mice. All tumor cell suspensions were prepared in an identical manner. Mice were injected intravenously with 5×10^4 viable tumor cells and were killed 18 days later. The number of pulmonary and extrapulmonary metastases in each mouse was determined with the use of a dissecting microscope.

894

349

to the metastatic potential of its individual cells. This high degree of heterogeneity is probably attributable to the fact that the B16 melanoma has existed as a transplanted tumor for more than 20 years. It is quite likely that many variants would arise during this period by the process of mutation and selection and by epigenetic mechanisms (4). In addition, the process of metastasis is a complex one with many sequential steps. It begins with the invasion of tissues and vessels by cells originating in the primary cancer. After their entry into the circulation, most cells are arrested in the first capillary bed encountered, but some continue and are trapped in other organs. After this arrest, the tumor cells must invade the parenchyma, proliferate, establish a vascular supply, and escape host defense mechanisms in order to develop into secondary foci. A cell that acquires an increased ability to survive any one of these steps would be viewed as having an increased metastatic potential. Thus, there are probably many different pathways by which a cell could acquire an increased or decreased capacity to form a new colony at a distant site.

The possible existence of highly metastatic variant cells within a primary tumor may have important consequences for cancer therapy. Efforts to design effective therapeutic agents and procedures should be directed toward the few, albeit fatal, metastatic subpopulations. Continuing efforts to eradicate the bulk of neoplastic cells, without regard to their biological behavior, are likely to be unproductive. Perhaps the highly metastatic clones described in our study would be useful tools for testing new therapeutic approaches to cancer.

ISAIAH J. FIDLER
MARGARET L. KRIPKE
Basic Research Program, National
Cancer Institute Frederick Cancer
Research Center,
Frederick, Maryland 21701

References and Notes

1. I. Zeidman, Cancer Res. 17, 157 (1957); E. R. Fisher and B. Fisher, Arch. Pathol. 83, 321 (1967); I. J. Fidler, in Cancer, F. F. Becker, Ed. (Plenum, New York, 1976), vol. 4, pp. 101–131.
2. I. J. Fidler, J. Natl. Cancer Inst. 45, 775 (1970).
3. _____, Nature (London) New Biol. 242, 148 (1973); Cancer Res. 35, 218 (1975).
4. P. C. Nowell, Science 194, 23 (1976).
5. S. E. Luria and M. Delbrück, Genetics 28, 491 (1943).
6. Cells were grown on plastic in Eagle's minimum essential medium supplemented with 10 percent fetal calf serum, vitamin solution, sodium pyruvate, nonessential amino acids, penicillin-streptomycin, and L-glutamine, designated complete minimum essential medium (CMEM) (Flow Laboratories, Rockville, Md.).
7. A 1 percent mixture of agarose was prepared in distilled, deionized sterile H_2O. The mixture was boiled and agitated, then cooled to 37°C and mixed with equal volume of double strength CMEM. A layer of CMEM and agarose was placed in a petri dish (120 by 20 mm) and the remaining mixture was kept warm at 37°C. A single cell suspension of the B16 melanoma in CMEM was adjusted to contain 2000 viable cells per milliliter. The cell suspension was diluted with an equal amount of the agarose and CMEM mixture. One milliliter of the final suspension (1000 cell/ml) was added to each petri dish, and the mixture was allowed to harden. The dishes were incubated at 37°C (5 percent CO_2) overnight. After that time, the dishes were examined under an inverted microscope and the positions of isolated single cells were noted and marked. Single cells were removed with a Pasteur pipette and placed in a Microtest II well (Falcon Plastics). Twelve hours later the wells were examined, and those with an attached single cell were identified. Tumor colonies resulting from these single cells were propagated and serially transferred to vessels of increasing size.
8. S. Siegel, Nonparametric Statistics for the Behavioral Sciences (McGraw-Hill, New York 1956), pp. 116–126.
9. We thank Z. Barnes and J. Connor for technical assistance. Supported by National Cancer Institute contract N01-C0-25423 with Litton Bionetics, Inc.

16 May 1977

Cell, Vol. 17, 789–800, August 1979, Copyright © 1979 by MIT

Cells Transformed by Temperature-Sensitive Mutants of Avian Sarcoma Virus Cause Tumors in Vivo at Permissive and Nonpermissive Temperatures

George Poste and Marian K. Flood
Department of Experimental Pathology
Roswell Park Memorial Institute
Buffalo, New York 14263

Summary

Chick embryo (CE) fibroblasts and normal rat kidney (NRK) cells transformed by temperature-sensitive (ts) mutants of avian sarcoma virus (NY68, LA23, LA24, LA25, LA29, LA31, Gl201, Gl202, Gl251, Gl253) induce tumors on the chorioallantoic membrane (CAM) of chick eggs at temperatures that correspond to the permissive and nonpermissive temperatures used to induce conditional expression of the "transformed" phenotype in these cells when cultured in vitro. Chick embryo cells infected with transformation-defective mutants of ASV (td101, td108) or RAV-50 were nontumorigenic under the same conditions, as were nontransformed CE and NRK cells. This indicates that the CAM is not an unusually susceptible substrate for cell growth and that the ability of tsASV-transformed cells to form tumors at nonpermissive temperatures reflects their true tumorigenicity. In contrast, a ts mutant chemically transformed rat liver cell line, ts-223, only formed tumors on the CAM under permissive conditions. The wild-type parent cells (W-8) of this mutant produced tumors at both permissive and nonpermissive temperatures. Direct implantation of microprobe thermometers into tumors caused by tsASV-transformed cells at nonpermissive temperatures confirmed that tumor formation occurred in a stable temperature environment and was not due to temperature fluctuations which might have created semi-permissive or permissive conditions for tumor growth. Cells isolated from tumors formed at nonpermissive temperatures and recultured in vitro displayed temperature-dependent hexose transport and colony formation in agar similar to the original parent cell inoculum. Similarly, virus recovered from tumors at nonpermissive temperatures retained the ts mutation.

Introduction

Avian sarcoma viruses (ASV) have been used extensively to induce neoplastic transformation of cells in vitro (Vogt, 1977). The ability of these viruses to transform both avian and mammalian cells, the availability of viral mutants with altered abilities to transform cells, and detailed knowledge of the structure of the virus genome make ASV useful for studying the events that accompany cellular transformation. Several conditional, temperature-sensitive (ts) mutants of ASV (tsASV) have been isolated that fail to induce or to maintain cellular transformation under nonpermis-

sive conditions (Wyke, 1975; Friis, 1978). Mutants of this type are referred to as belonging to class T (for transformation). When grown at permissive temperatures, cells transformed by these mutants show many of the phenotypic changes seen in cells transformed by wild-type (wt) ASV, but when cultured under nonpermissive conditions, some or all of the "transformed" properties are lost and the cells more closely resemble untransformed cells. Class T mutants can replicate normally at both temperatures, suggesting that a virus-encoded, heat-labile, nonstructural protein is responsible for expression of the transformed phenotype under permissive conditions. The lesion in class T mutants involves the src region of the virus genome (Vogt, 1977; Friis, 1978), and the phenotypic alterations which are expressed conditionally in cells transformed by these mutants are thus presumed to result from the action (direct or indirect) of a src gene product which is functional only at permissive temperatures. Since the ability of ASV to transform cells in vitro and to induce sarcomas in vivo is determined by the src region of the ASV genome (Vogt, 1977), the alterations in cell function expressed at permissive temperatures by cells transformed by class T mutants have been interpreted as being intimately related to the neoplastic process and can thus serve as in vitro markers of the neoplastic phenotype. If this interpretation is correct, then tumor formation in vivo by cells transformed by class T mutants should occur only at permissive temperatures. This important question, however, remains to be answered. The lack of information on this point stems from the practical problem posed by the need to assay the tumorigenicity of tsASV-transformed cells over a temperature range (33–41°C) that falls outside the span of normal body temperatures in mammals (36–38°C) and birds (40–42°C) conventionally used in tumorigenicity assays.

In this paper, we report that the tumorigenicity of cells transformed by ts mutants of ASV can be assayed in vivo at any temperature between 33°C and 41°C by studying their capacity to invade and proliferate in the chorioallantoic membrane (CAM) of embryonated chick eggs. This method enables the permissive and nonpermissive temperatures used in studies with these cells in vitro to be duplicated in vivo. We report here that avian and mammalian cells transformed by ts mutants of ASV cause tumors on the CAM at both nonpermissive and permissive temperatures. In contrast, a ts mutant chemically transformed rat liver cell line (Yamaguchi and Weinstein, 1975) produced tumors only under permissive conditions.

Results

Tumorigenicity of Cells Transformed by ts Mutants of ASV

Chick embryo (CE) cells transformed by ts mutants of ASV induce tumor formation when inoculated onto the

Table 1. Tumorigenicity of Chick Embryo Fibroblasts Infected with Avian Sarcoma Virus Strains

Virus Strain[a]	Frequency of CAM Tumors[b, c]		
	35°C	38°C	41°C
None, unin-fected con-trol cells	0/43 (0%)	0/36 (0%)	0/80 (0%)
tsNY68-SR-A	12/15 (80%)		14/18 (78%)
tsLA23-PR-A	11/12 (92%)		11/14 (79%)
tsLA24-PR-A	8/10 (80%)		9/12 (75%)
tsLA25-PR-A	10/13 (77%)		11/15 (73%)
tsLA29-PR-A	8/8 (100%)		6/8 (75%)
tsLA31-PR-A	7/10 (70%)		7/11 (64%)
tsGI201-PR-A	10/12 (83%)		10/12 (83%)
tsGI202-PR-A	8/9 (89%)		7/10 (70%)
tsGI251-PR-A	11/12 (92%)		8/12 (67%)
tsGI253-PR-A	9/12 (75%)		9/10 (90%)
td101-SR-A	0/10 (0%)	0/10 (0%)	0/10 (0%)
td108-SR-A	0/10 (0%)	0/8 (0%)	0/10 (0%)
wtSR-A		10/12 (83%)	
wtPR-A		12/12 (100%)	
wtPR-C		7/10 (70%)	
wtB77-C		8/9 (89%)	
RAV-50	0/10 (0%)	0/10 (0%)	0/8 (0%)

[a] Abbreviated as described in Experimental Procedures.
[b] 10 day old eggs were inoculated with 1×10^6 cells in 50 μl of serum-free medium, incubated at the indicated temperature for 7 days and then examined for macroscopic tumor nodules (>1 mm diameter) on the CAM. All nodules were confirmed as being tumors by histologic examination.
[c] Number of eggs developing tumors/total number of eggs inoculated with cells. The figures in parentheses are the percentage of inoculated eggs developing tumors.

CAM of embryonated eggs incubated at 35°C or 41°C (Table 1; Figure 1). These temperatures correspond to the permissive (35°C) and nonpermissive (41°C) conditions used to induce conditional expression of so-called ''transformed'' properties in these cells when cultured in vitro. The possibility that tumor formation results from transformation of CAM cells by infectious virus released from the inoculated cells rather than from proliferation of the inoculated cells is considered improbable for two reasons. First, tumor formation at permissive (33°C) and nonpermissive temperatures (39.5°C) also occurs in eggs inoculated with nonvirogenic NRK cells transformed by ts mutants of ASV (Table 2). Although fusion between transformed NRK cells and permissive CAM cells could conceivably rescue infectious virus from NRK cells, infectious virus was not detected in chorioallantoic fluids harvested from eggs inoculated with tsASV-transformed NRK cells incubated at either 33° or 39.5°C (results not shown). Tumor formation at per-

missive (33°C) and nonpermissive (39°C) temperatures was also seen in eggs inoculated with nonvirogenic F2408 rat cells transformed by LA334 (Table 2). Second, eggs infected with ts mutants of ASV (2 $\times 10^2$–10^3 ffu) incubated at 41°C for up to 7 days fail to develop macroscopic tumor nodules. Although histologic alterations occur in the CAM of eggs infected with free virus (see below), these differ from the lesions produced by transformed cells.

Measurements at 8 hr intervals using a microprobe thermometer established that the temperature on the surface of the CAM in eggs inoculated with CE and NRK cells transformed by ts mutants was stable during the period of tumor growth and did not fluctuate more than ±0.5°C from the temperatures indicated in Tables 1 and 2. In addition, in eggs inoculated with a large number (5 $\times 10^6$) of tsASV-transformed CE or NRK cells, the tumor nodules were of sufficient size (>5mm diameter 5 days after inoculation) to allow implantation of a microprobe thermometer (MT-3) directly into the tumor. These experiments revealed that the intratumor temperature did not vary by more than ±0.4°C from the temperature recorded simultaneously from surrounding normal regions of the CAM. These results demonstrate that tumor formation by tsASV-transformed cells at nonpermissive temperatures occurred in a stable temperature environment and was not due to major temperature fluctuations which might have created permissive or semi-permissive conditions favoring cell proliferation.

Nontransformed CE cells or CE cells infected with RAV-50 or transformation-defective strains of ASV (td101, 108) failed to form tumors on the CAM at any temperature between 35° and 41°C (Table 1). The inability of these cells to cause tumors indicates that the CAM is not an unusually susceptible tissue substrate for cell growth, and that tumor formation by tsASV-transformed cells at both permissive and nonpermissive temperatures accurately reflects ther tumorigenic properties. This conclusion is also supported by the finding that a chemically transformed rat liver cell line, ts-223, which displays temperature-dependent expression of the transformed phenotype (Yamaguchi and Weinstein, 1975), produced tumors on the CAM at the permissive temperature (36°C) but was nontumorigenic at the nonpermissive temperature (40°C) (Table 2). In contrast, the wild-type parent cells (W-8) of this mutant produced tumors at both temperatures.

Titration of the number of ASV-transformed cells required to induce tumor formation revealed that the tumorigenicity of chick cells transformed by ts mutants LA23, 24, 25, GI251 and GI253 was comparable to that of cells transformed by the appropriate wild-type virus strains, but cells transformed by NY68, GI201 and GI202 were less oncogenic than cells transformed by wild-type virus (Table 3). However, for each individual mutant, the tumorigenicity of transformed cells

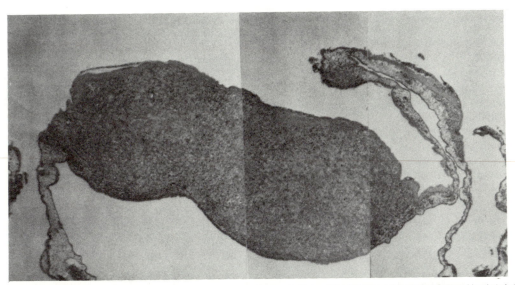

Figure 1. Low Power Photomicrographic Montage of CAM from Egg Inoculated with 3×10^6 *ts*LA31-Transformed NRK Cells and Incubated at 39.5°C for 5 Days

Figure shows extensive accumulation of tumor cells within the mesodermal layer of the CAM and marked enlargement of the CAM at the site of tumor formation. Stain: hematoxylin and eosin. Magnification 82×.

Table 2. Tumorigenicity of Virus- and Chemically Transformed Rat Cell Lines

Cell Type	Frequency of CAM Tumors[a]				
	33°C	34°C	36°C	37°C	39.5°C
Untransformed NRK cells	0/24 (0%)			0/18 (0%)	0/24 (0%)
LA7-NRK	11/12 (92%)				9/11 (82%)
LA23-NRK	10/10 (100%)				7/10 (70%)
LA24-NRK	10/12 (83%)				8/12 (67%)
LA25-NRK	9/12 (75%)				7/10 (70%)
LA31-NRK	7/10 (70%)				7/10 (70%)
*wt*B77-NRK	8/11 (73%)			11/12 (92%)	10/12 (83%)
LA334-F2408 rat cells		11/12 (92%)			9/11 (82%)
W-8 rat liver cells			7/9 (78%)	10/12 (83%)	9/12 (75%)
ts-223 rat liver cells			10/12 (83%)		0/11 (0%)

[a] Determined as in Table 1.

was identical at permissive and nonpermissive temperatures (Table 3).

Histological Characterization of CAM Tumors

The CAM is composed of three layers (Figure 2). The outer epithelium, which is situated beneath the egg shell membrane, is of ectodermal origin and derived from the chorion. The inner epithelium, which lines the allantoic cavity, is derived from the allantois and is of endodermal origin. A mesodermal layer is interposed between the two epithelial layers and is composed of a loose matrix of fibroblasts and a network of blood vessels. The overall thickness of the CAM differs widely (150 μ to 1 mm) due largely to variation in the size of the mesodermal vascular bed in different regions.

Histologic examination of sections of tumor nodules formed at permissive and nonpermissive temperatures by *ts*ASV-transformed chick and NRK cells revealed extensive accumulation of tumor cells within the mesodermal layer of the CAM (Figures 1, 3, 4, and 5). The ectodermal epithelium overlying these lesions was structurally intact, suggesting that invasion of tumor cells into the mesoderm did not involve overt tissue destruction and probably occurred via infiltration of tumor cells between adjacent epithelial cells (see Trin-

Table 3. Titration of the Minimum Number of Avian Sarcoma Virus-Transformed Cells Required for Formation of CAM Tumors

Virus Strain	Cell Type	Minimum Number of Cells Causing Tumors in 50% of Inoculated Eggs[a]				
		33°C	35°C	37°C	39.5°C	41°C
tsNY68-SR-A	CEF		10^4 (8/12)[b]			10^4 (8/12)
tsLA23-PR-A	CEF		10^2 (9/10)			10^2 (7/10)
tsLA24-PR-A	CEF		10^1 (11/14)			10^1 (6/11)
tsLA25-PR-A	CEF		10^2 (8/12)			10^2 (9/12)
tsGl201-PR-A	CEF		10^5 (9/12)			10^5 (7/11)
tsGl202-PR-A	CEF		10^4 (9/12)			10^4 (7/12)
tsGl251-PR-A	CEF		10^2 (10/14)			10^2 (8/12)
tsGl253-PR-A	CEF		10^2 (10/12)			10^2 (7/11)
wtPR-A	CEF			10^2 (7/11)		
wtSR-A	CEF			10^1 (7/11)		
tsLA7-B77-C	NRK	10^2 (8/12)			10^2 (8/12)	
tsLA23-PR-A	NRK	10^2 (8/11)			10^2 (7/11)	
tsLA31-PR-A	NRK	10^2 (7/12)			10^2 (8/12)	
wtB77-C	NRK			10^2 (10/12)		

[a] 10 day old eggs were inoculated with 10 fold serial dilutions of cells in a standard inoculum volume of 50 μl of serum-free medium to achieve inocula ranging from 1 to 1×10^6 cells per egg. The lowest number of cells producing a tumor nodule (2 mm or greater in diameter) after incubation for 7 days at the indicated temperature was measured. Nodules were confirmed as being tumors by histologic examination. A minimum of three eggs were inoculated with each dilution of cells, and the results are derived from three separate experiments.
[b] Number of eggs developing tumors after inoculation with the indicated number of cells/total number of eggs inoculated.

kaus, 1976). Squamous metaplasia of the ectodermal epithelium and formation of keratin "pearls" were prominent in tumors induced by NRK cells transformed by both wild-type ASV and ts mutants (not shown), but these changes were rarely seen in neoplastic lesions induced by CE cells transformed by the same virus strains. Similar changes and invagination of ectodermal epithelium around invading tumor cells and keratinization of the internalized epithelium to form "pearls" in the mesoderm have also been reported in other studies with rodent tumor cell lines inoculated onto the CAM (for references see Leighton, 1967).

The histologic appearance of tumors induced by cells transformed by ts mutants of ASV was indistinguishable from that caused by cells of the same type transformed by wt virus growing on the CAM at 38°C. In addition, no major histologic differences were detected between lesions produced by cells of the same type transformed by different ts mutants, or between lesions caused by the same cell type grown at permissive or nonpermissive temperatures. Tumors formed by transformed cells differed, however, from the CAM lesions (pocks) induced by free virus. As described by other investigators (Keogh, 1938; Prince, 1958; Biquard and Vigier, 1972), pock formation was seen in eggs infected with wtASV at 37–38°C or ts mutants of ASV at 35–37°C, but did not occur in eggs infected with ts mutants of ASV at 41°C. Reduced pock formation (approximately 40% of con-

trols infected at 35°C) was seen, however, in eggs infected with ts mutants at 35°C and incubated at this temperature for 2 days before switching to 41°C for 5–7 days. Histologically, the pocks produced by wtASV and ts mutants differ from the lesions induced by cells transformed by these agents. Pocks induced by free virus were characterized by initial proliferation and vacuolization of the ectodermal epithelium of the CAM followed by the development of small foci of tumor cells 2–3 days after infection. These foci always developed in close association with the ectoderm, and invasion into the deeper regions of the mesoderm was not seen until at least 4–5 days after infection (Figure 5). This contrasts with the lesions produced by transformed cells which are characterized by rapid invasion of tumor cells into the mesoderm to form colonies as early as 24 hr after inoculation. Subsequent proliferation of these colonies results in extensive accumulation of tumor cells which occupy the entire thickness of the mesoderm (Figure 1). The ectodermal epithelium does not, however, show the characteristic proliferation and vacuolization seen in CAMs infected with free virus (Figure 5).

The mitotic index in neoplastic lesions produced in the mesoderm of the CAM by ts-transformed CE and NRK cells at nonpermissive temperatures was identical to that in lesions produced by ts-transformed cells at permissive temperatures and wt-transformed cells at 37°C (Table 4). This indicates that tumor formation by ts-transformed cells at nonpermissive temperatures

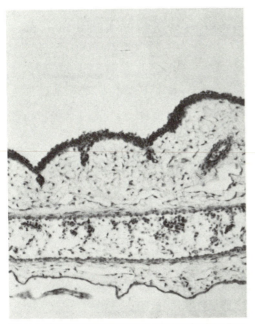

Figure 2. Photomicrograph of the CAM of 15 Day Old Chick Eggs

Figure shows the ectodermal epithelium (top), a well vascularized mesodermal layer and the inner endodermal epithelium. Stain: hematoxylin and eosin. Magnification 218×.

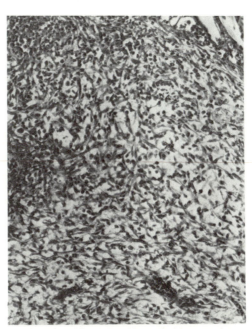

Figure 4. Photomicrograph of the Mesodermal Layer of CAM Inoculated with 1 × 10⁵ *ts*LA31-Transformed Chick Cells and Incubated at 41°C for 4 Days

Figure shows extensive accumulation of transformed fibroblasts. Stain: hematoxylin and eosin. Magnification 336×.

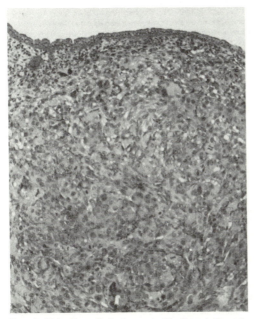

Figure 3. Higher Power Photomicrograph of the Tumor in Figure 1

Figure shows occupation of the mesoderm by tumor cells and lack of morphologic change in the overlying ectodermal epithelium. Stain: hematoxylin and eosin. Magnification 441×.

does not result simply from passive invasion of cells into the mesoderm but also involves active replication of invading cells within the CAM.

Characterization of Cells and Virus Strains Isolated from Tumors Formed at Nonpermissive Temperatures

To exclude the possibility that tumor formation at nonpermissive temperatures induced by *ts*ASV-transformed cells might be due to cell variants with a reduced *ts*-phenotype present in the original cell inoculum, cells were isolated from tumor nodules, recultured in vitro, and tested to determine whether they displayed temperature-dependent expression of "transformed" properties in a fashion similar to the original parent cells. These experiments revealed that cells isolated from tumors induced at nonpermissive temperatures exhibit significant temperature-dependent changes in hexose transport (Table 5) and colony formation in agar (Table 6) when cultured in vitro at permissive and nonpermissive temperatures. These experiments demonstrate that the cells responsible for tumor formation at nonpermissive temperatures retain a *ts*-phenotype and indicate that tumor formation is not due to selective proliferation in ovo of cell variants with a reduced *ts* phenotype.

Experiments were also performed to examine the properties of virus recovered from *ts*ASV-transformed

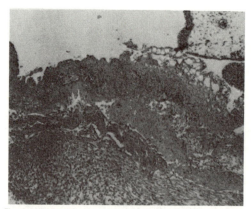

Figure 5. Photomicrograph of Virus-Induced "Pock" on CAM Infected with the tsLA31 Mutant of Avian Sarcoma Virus (1 × 10³ ffu) and Incubated at 35°C for 6 Days

Figure shows vacuolation and hyperplasia of ectodermal epithelium and proliferation of mesodermal fibroblasts. Stain: hematoxylin and eosin. Magnification 211×.

Table 4. Mitotic Index of Virus-Transformed Cell Populations in the Mesodermal Layer of the Chick Chorioallantoic Membrane

Virus Strain[a]	Cell Type[b]	Mitotic Index[c] 33°C	35°C	37°C	38°C	39°C	41°C
wtSR-A	CEF				1.39		
wtPR-A	CEF				1.18		
tsNY68-SR-A	CEF		1.07				0.96
tsLA31-PR-A	CEF		1.14				1.22
tsGI251-PR-A	CEF		1.47				1.29
wtB77-C	NRK			1.43			
tsLA7-B77-C	NRK	1.08				1.19	
tsLA25-PR-A	NRK	1.37				1.07	
tsLA31-PR-A	NRK	1.28				1.33	

[a] Abbreviated as described in Experimental Procedures.
[b] (CEF) chick embryo fibroblasts; (NRK) normal rat kidney.
[c] Number of cells in mitosis per 1000 transformed cells measured as described in Experimental Procedures.

Table 5. Hexose Uptake by ASV-Transformed Cells and Cells Isolated from CAM Tumors Produced by ASV-Transformed Cells

Cell Line	2–Deoxyglucose Uptake (cpm/mg Protein)[a] Permissive Temperature[b]	Nonpermissive Temperature[c]	P/NP Ratio[d]
Uninfected CEF	1,900	2,100	0.9
LA23-CEF	33,800	5,100	6.6
GI251-CEF	29,500	3,900	7.6
wtPR-A-CEF	28,400	26,100	1.1
CAM-LA23-CEF[e]	23,200	4,200	5.5
CAM-GI251-CEF	18,000	3,800	4.7
CAM-wtPR-A-CEF	26,200	28,800	0.9
Uninfected NRK	1,300	1,400	0.9
LA23-NRK	9,200	2,700	3.4
LA29-NRK	8,400	2,400	3.5
wtB77-NRK	7,500	7,300	1.0
CAM-LA23-NRK[e]	6,500	2,400	2.7
CAM-LA29-NRK	6,200	2,100	2.9
CAM-wtB77-NRK	5,700	6,800	0.8

[a] Mean values derived from three separate experiments.
[b] 35°C for CEF and 33°C for NRK cells.
[c] 41°C for CEF and 39.5°C for NRK cells.
[d] Ratio of hexose uptake, permissive (P) over nonpermissive (NP) temperature.
[e] Cells isolated from chorioallantoic tumor nodules formed at nonpermissive temperatures following inoculation of 10 day old eggs with 5 × 10⁶ CEF or NRK cells transformed by the indicated ASV strain. Tumor-derived cells were cultured in vitro for 5 days at 37°C followed by incubation at the indicated temperature for 2 days before assay of hexose uptake.

cells following tumor formation at nonpermissive temperatures. Virus recovered by homogenization of CAM tumors from eggs inoculated with tsASV-transformed CE cells and incubated at 41°C for 6 days displayed a ts phenotype similar to parent virus (Table 7). Similarly, virus rescued by cell fusion from cells isolated from CAM tumors formed by tsASV-transformed NRK cells at nonpermissive temperatures also exhibited a ts phenotype (Table 7).

These results eliminate the possibility that tumor formation under nonpermissive conditions might have resulted from transformation of CAM cells by new progeny virus with a reduced temperature sensitivity produced by genetic interaction between the original ts mutant virus and endogenous (host) sarcoma-specific genes.

Discussion

The present experiments indicate that the CAM of the embryonated chick egg provides a convenient and simple system for assaying the tumorigenicity of cells transformed by ts mutants of ASV. Embryonated eggs can be maintained without adverse effect at any temperature between 31° and 42°C. This allows the tumorigenicity of tsASV-transformed cells to be assayed at temperatures which correspond to the permissive (33–35°C) and nonpermissive (39–41°C) temperatures used to induce conditional expression of "transformed" properties in these cells in vitro.

Our results indicate that avian and mammalian cells transformed by ts mutants of ASV are tumorigenic in ovo at both permissive and nonpermissive temperatures. By direct implantation of microprobe thermometers into CAM tumors produced by tsASV-transformed cells, we have been able to monitor the temperature at which tumor formation occurs and, to

Table 6. Colony Formation in Agar by ASV-Transformed NRK Cells and Cells Isolated from CAM Tumors Produced by ASV-Transformed NRK Cells

Cell Line	Log$_{10}$ Colonies in Agar per 10^6 Cells[a]	
	Permissive Temperature (33°C)	Nonpermissive Temperature (39.5°C)
Uninfected NRK	0.0	0.0
LA23-NRK	5.20	0.0
LA31-NRK	4.45	1.20
*wt*B77-NRK	4.68	5.04
CAM-LA23-NRK[b]	2.78	0.0
CAM-LA31-NRK	4.15	1.08
CAM-*wt*B77-NRK	3.95	3.46

[a] Colonies counted 14 days after seeding. The results represent mean values derived from three separate experiments, except for CAM-LA31-NRK which are derived from two experiments.
[b] Cells isolated from CAM tumors formed at nonpermissive temperatures as described in Table 4 and incubated at the indicated temperature for 2 days before assaying colony formation at the same temperature.

Table 7. Temperature-Dependent Focus-Forming Activity of Virus Strains Recovered from Cells Isolated from CAM Tumors Formed at Nonpermissive Temperatures by Cells Transformed by ASV Strains

Cell Line	Titer (ffu/ml) of Recovered Virus[a]	
	35°C	41°C
CAM-LA23-CEF	3 × 10^4	<10^1
CAM-*wt*PR-A-CEF	6 × 10^4	4 × 10^4
CAM-LA23-NRK	5 × 10^2	<10^1
CAM-*wt*B77-NRK	3 × 10^2	3 × 10^2

[a] 10 day old eggs were inoculated with 5 × 10^6 cells transformed by the indicated virus strains and incubated for 5 days at 39.5°C (NRK) or 41°C (CEF). CAM tumors were then excised and dispersed to single cells. 2 × 10^7 cells isolated from tumors induced by transformed CEF cells were homogenized, and cell-free filtrates were assayed for focus-forming activity on cultures of CEF at 35° or 41°C. Cells isolated from tumors induced by transformed NRK cells were plated into multiwell dishes and grown to confluency, and aliquots of 4 × 10^6 cells were fused with 1 × 10^7 CEF using inactivated Sendai virus (see Experimental Procedures). Cell-free supernatants were harvested from fused cultures after 7 days and assayed for focus-forming activity on CEF monolayers at 35° or 41°C.

confirm that stable nonpermissive conditions were established. By re-isolating cells from tumors formed at nonpermissive temperatures and showing that they still express the *ts* phenotype when cultured in vitro, we have excluded the possibility that tumor formation was caused by the emergence of cells with a wild-type-transformed phenotype. In addition, virus recovered from cells isolated from tumors formed at nonpermissive temperatures retained the *ts* mutation. This eliminates the possibility that tumor formation resulted from transformation of CAM cells by new progeny virus having a reduced temperature sensitivity formed by genetic interaction between the original *ts* mutant and endogenous (host cell) sarcoma-specific genes (see Hanafusa et al., 1977).

Studies in this (Poste et al., 1979) and other laboratories (Leighton, 1967; Easty and Easty, 1974; Scher, Haudenschild, and Klagsbrun, 1976) using a wide range of non-neoplastic and neoplastic cells have shown that the ability of cells to invade and proliferate within the mesodermal layer of the CAM correlates with their tumorigenicity in laboratory rodents. The present data showing that nontransformed CE and NRK cells and cells transformed by *td* mutants of ASV fail to form CAM tumors are consistent with these findings. Collectively, these data indicate that the CAM is not an unusually susceptible substrate for cell growth and that the ability of *ts*ASV-transformed cells to form tumors at nonpermissive temperatures reflects their true tumorigenic potential. This conclusion is further strengthened by the present finding that a chemically transformed cell line (ts223) which displays temperature-dependent expression of trans-

formed properties when grown in vitro (Yamaguchi and Weinstein, 1975) will only produce CAM tumors at permissive temperatures. In contrast, the wild-type parent cells (W-8) from which this mutant line was derived are tumorigenic at both permissive and nonpermissive temperatures.

Previous studies have shown that *ts* mutants of ASV will cause tumors in chickens when inoculated into the wing web (Toyoshima, Owada, and Kozai, 1973; Becker et al., 1977) or intra-abdominally (Purchase et al., 1977). The chicken has a normal body temperature of 42°C and should thus represent a nonpermissive environment. Becker et al. (1977) reported that *ts* mutants of ASV were less oncogenic than *wt* virus strains, but Purchase et al. (1977) found that the oncogenicity of *ts* mutants was identical to *wt* strains. However, neither of these studies investigated whether the virus recovered from tumors in chickens infected with *ts* mutants retained the *ts* phenotype. Consequently, the possibility cannot be excluded that tumor formation was caused by new progeny virus with a *wt* phenotype that arose by genetic interaction between *ts* mutants and endogenous sarcoma-specific sequences present in chicken cells in analogous fashion to tumor formation induced by *td* mutants of ASV (Hanafusa et al., 1977). An additional interpretational problem concerns the use of the wing web for assaying the tumorigenicity of *ts* mutants of ASV. The temperature of the wing web, and similar superficial body sites, can fluctuate substantially and may be several degrees lower than the "core" body temperature of 42°C. The possibility therefore exists that tumor formation in the wing web might be occurring in a semipermissive environment. The use of the CAM avoids this problem, and the present data indicate that

Table 8. Temperature-Dependent Expression of "Transformed" Properties in Chick Embryo and NRK Cells Transformed by *ts* Mutants of Avian Sarcoma Virus

Property[a]	References (Numbers = CEF; Letters = NRK)[b]									
	NY68	LA23	LA24	LA25	LA29	LA31	GI201	GI202	GI251	GI253
Altered morphology	1	2	2		2	2				
Increased saturation density	1	2,3 A,B	2,3 A,B	A,B	2,3 A,B	2,3 A,B	4	4		
Colony formation in agar	1	2,3,A	2,3,A	A	2,3,A	2,A	4	4		
Increased susceptibility to lectin agglutination		B	B	B	B		4	4	4	4
Increased hexose transport	1,5	3,A	3,6,7	3,5	3,A	A	4	4	4	4
Altered surface morphology/ultrastructure	8–10		A							
Changes in surface glycopeptides	11,12	A		A	A	13,A				
Reduction or loss of fibronectin	5	6	6	5			4	4	4	4
Decreased plasma membrane hematoside		14	14	14	14					
Decreased GM3 glycolipid				C						
Decreased fluorescamine binding			7							
Increased release of plasminogen activator	11						4	4	4	4
Reduced adenyl cyclase activity		15			15					
Reduced pp60[src] protein kinase activity	16									
Disorganization of cytoskeletal elements	9,17	D								

[a] Expressed only at permissive temperatures.

[b] (1) Kawai and Hanafusa (1971); (2) Wyke and Linial (1973); (3) Kurth et al. (1975); (4) Becker et al. (1977); (5) Vaheri and Ruoslahti (1974); (6) Hynes and Wyke (1975); (7) Parry and Hawkes (1978); (8) Ambros, Chen and Buchanan (1975); (9) Wang and Goldberg (1976); (10) Gilula, Eger and Rifkin (1975); (11) Weinstein et al. (1979); (12) Stone, Smith and Joklik (1974); (13) Isaka et al. (1975); (14) Hakomori, Wyke and Vogt (1977); (15) Yoshida, Owada and Toyoshima (1975); (16) Collett and Erickson (1978); (17) Edelman and Yahara (1976); (A) Chen et al. (1977); (B) Kurth (1975); (C) Lingwood, Ng and Hakomori (1978); (D) Ash, Vogt and Singer (1976).

*ts*ASV-transformed cells are tumorigenic under conditions where the nonpermissive temperature is rigorously maintained and that virus recovered from these lesions retains the *ts* phenotype. Our findings therefore confirm and extend previous suspicions that *ts* mutants of ASV retain their oncogenic potential at nonpermissive temperatures.

If, as current evidence suggests (Vogt, 1977), tumor formation in vivo by ASV requires expression of the *src* gene, then the present results indicate that some form of functional *src* gene product is being produced in *ts*ASV-transformed cells at both permissive and nonpermissive temperatures.

Erikson and his colleagues (Collett and Erikson, 1978; Brugge, Steinbaugh and Vogt, 1978a; Brugge et al., 1978b; Collett, Erikson and Erikson, 1979) have recently identified the product of the *src* gene as a 60,000 dalton phosphoprotein, designated pp60[src], which acts as a protein kinase. Partial proteolysis mapping (Brugge et al., 1978a; Collett et al., 1979) indicates that pp60[src] has two major sites of phosphorylation, one involving a phosphoserine residue located on the aminoterminus of the molecule and the other a phosphothreonine located on the carboxyterminal portion. Of particular interest in view of our results that cells transformed by *ts*NY68 are tumorigenic at both permissive and nonpermissive temperatures is the observation that similar amounts of pp60[src] are synthesized at both temperatures in cells transformed by this mutant (Brugge et al., 1978a; Levinson et al., 1978). However, protein kinase activity and phosphorylation of the threonine residue are both reduced in pp60[src] synthesized at nonpermissive temperatures. Assuming that these events are duplicated in transformed cells growing in vivo, it would appear that neither of these two properties of pp60[src] is involved in tumor formation. This lack of correlation between tumorigenicity and pp60[src] kinase activity argues against the proposal (Levinson et al., 1978) that ASV-induced oncogenesis might be caused by aberrant phosphorylation of cellular proteins by pp60[src]. Phosphorylation of pp60[src] may instead be important in determining its ability to interact with cellular components. If this is the case, phosphorylation of the serine residue would appear to be of more importance in allowing pp60[src] to induce the cellular alterations required for tumor formation since this residue is phosphorylated at both nonpermissive and permissive temperatures and thus correlates with tumorigenicity.

It remains to be seen, however, whether pp60[src] synthesized by other *ts* mutants behaves similarly.

The possibility that there may be several forms of ts mutation affecting the src gene is suggested by the fact that different ts mutants cause differing patterns of phenotypic alteration in transformed cells (see Friis, 1978; Weber and Friis, 1979). Recently, Rübsamen, Friis and Bauer (1979) have identified significant differences in the thermolability of p60src produced by different ts mutants, but their relationship to the function of p60src still remains to be defined.

When grown at permissive temperatures, CE and NRK cells transformed by the ts mutants used in this study display "transformed" properties similar to those seen in cells transformed by wtASV, but when cultured at nonpermissive temperatures they lose some or all of these properties (Table 8). If tsASV-transformed cells growing in vivo behave in the same fashion, then the present results showing that they cause tumors at permissive and nonpermissive temperatures would mean that none of the properties listed in Table 8 are essential for tumor formation. If all of the properties listed in Table 8 result from the action (direct or indirect) of the ASV src gene product, then the functional dissociation of tumorigenicity from these properties at nonpermissive temperatures would indicate that under these conditions the src gene product retains its capacity to induce tumorigenic conversion but is crippled with respect to the functions needed to induce the cellular alterations listed in Table 8.

An alternative interpretation, however, is that the CAM might somehow act as a fully permissive environment in which any defect in the src gene product synthesized under nonpermissive conditions can be overcome to allow full expression of the transformed phenotype, that is, the properties in Table 8 and tumor formation. This situation would contrast with events in cultured ts-transformed cell populations where the absence of the hypothetical permissive element(s) provided by the CAM would impose strict conditional expression of transformed properties. An analogy might be drawn between this possibility and the behavior of certain hormone-dependent breast tumor cell lines (Minesita and Yamaguchi, 1965; Desmond, Wolbers and Sato, 1976). Expression of transformed properties in these cells in vitro can be modulated by androgens, and tumorigenicity in vivo is expressed only in male mice and not in females or castrated males.

If tumor formation at nonpermissive temperatures by tsASV-transformed cells results from the special permissive status of the CAM, this would presumably mean that the src gene product produced in this environment is no longer crippled. No information is available as to whether the src gene product synthesized at nonpermissive temperatures is irreversibly crippled or whether its activity can be fully or partially restored by conditions that would have to operate independently of temperature. Recent observations showing that tsASV-transformed cells grown in vitro will express transformed properties at nonpermissive temperatures when exposed to phorbol esters raise the possibility that some aspects of crippled src gene function might be salvageable, although other interpretations of this phenomenon are also possible (see Bissell, Hatie and Calvin, 1979). If host-derived factors enable tsASV-transformed cells to form tumors in the CAM at nonpermissive temperatures, then a similar mechanism might also contribute to the oncogenicity of tsASV mutants in chickens at 42°C (see above).

We consider that CAM tumors offer a convenient alternative to tumor formation in chickens for studying questions relating to src gene function in ASV-induced tumors in vivo. The rapid formation of tumors on the CAM, their easy accessibility and the lack of infiltrating host cells in these lesions dictate that CAM tumors provide an inexpensive source of tumor tissue for characterization of src gene product synthesized by cells in vivo. As reported here, tumor formation on the CAM can be studied over a wider temperature range than in chickens, thus facilitating direct comparison of src gene products in tumors caused by ts mutants and by wt virus. The ability to isolate large numbers of tumor cells from CAM lesions after only a few days is also helpful for experimental comparison of the properties of ASV-transformed cells grown in vivo and in vitro. This short period of growth in vivo dictates that tumor cells can be compared with in vitro cell populations that will have had little time to undergo phenotypic change as a result of further cultivation in vitro. This contrasts with tumor formation in chickens in which the longer time (weeks to months) required for tumor formation provides greater opportunities for the development of phenotypic differences between cells growing in vivo and in vitro which may have no relationship to tumorigenicity and result merely from prolonged growth of the cells in different environments.

Experimental Procedures

Viruses

The following viruses were used to transform chick embryo fibroblasts and/or NRK cells: Schmidt-Ruppin strain Rous sarcoma virus, subgroup A (SR-A); tsNY68, a ts mutant derived from SR-A (Kawai and Hanafusa, 1971); td101 and td108, which are transformation-defective (td) mutants of SR-A (Kawai, Duesberg and Hanafusa, 1977); Prague strain Rous sarcoma virus, subgroup A (PR-A); tsLA23, tsLA24, tsLA25, tsLA29 and tsLA31, which are ts mutants derived from 5–azacytidine-mutagenized PR-A (Wyke, 1973); tsGI201, tsGI202, tsGI253 and tsGI253, which are ts mutants derived from PR-A mutagenized with BrdU (Becker et al., 1977); B77-C-Bratislava strain of avian sarcoma virus, subgroup C (B77-C); tsLA7 and tsLA334, which are ts mutants derived from B77-C (Toyoshima and Vogt, 1969); Prague strain Rous sarcoma virus, subgroup C (PR-C); and Rous-associated virus 50 (RAV-50), avian sarcoma virus, subgroup D (Hanafusa and Hanafusa, 1966). In this paper ts mutants are identified simply by a letter prefix and their number (for example, NY68 = tsNY68 SR-A).

Cell Culture and Transformation

Chick embryo fibroblasts (CEF) were prepared from 10–12 day old gs⁻ embryos (Spafas Inc., Norwich, Connecticut) and cultured in Ham's F10 medium supplemented with 10% tryptose phosphate broth, 4% calf serum and 1% chick serum (Vogt, 1969). Following viral transformation, 1% dimethylsulfoxide (DMSO) was added to the culture medium. Normal rat kidney (NRK) cells were cultured in the same growth medium as chick cells, except that chick serum and DMSO were omitted (Chen, Hayman and Vogt, 1977).

Secondary CEF were infected with virus strains at a multiplicity of 0.5–2 focus-forming units (ffu) per cell and incubated at 37°C (wild-type virus) or the permissive temperature of 35°C (ts mutants) until evidence of cell transformation was detected 3–4 days post-infection. To enhance virus absorption, all infections were carried out in the presence of 2 μg/ml polybrene. Cultures were then subcultured and maintained for 4 days at either 37°C (wild-type virus strains) or 41°C (ts mutants). Examination of duplicate cultures infected with ts mutants maintained at permissive temperatures (35°C) revealed that >80% of the cells were infected. The transformed cells were then transferred to 60 mm plastic petri dishes and incubated for a further 48 hr at 37°C (wild-type virus strains) or 35° or 41°C (ts mutants) before assaying their tumorigenicity. Cells incubated at 37° or 35°C were seeded at 8×10^5 cells per dish and those incubated at 41°C were seeded at 1.5×10^6 cells per dish.

NRK cells transformed by wild-type virus and ts mutants were provided by P. Vogt. The methods used to transform these cells and the properties of transformed cell populations have been described by Chen et al. (1977). These cells were cultured at 37°C (wild-type virus) or at 33° or 39.5°C (permissive and nonpermissive temperatures, respectively, for ts mutants).

A line of rat cells (F2408) transformed by the tsLA334 mutant of B77-C, transformed originally by J. Wyke, were provided by R. O. Hynes and cultured in Dulbecco's modified Eagle's medium plus 10% fetal calf serum.

W-8 rat liver cells transformed by N-acetoxy-2-acetylaminofluorene and a ts mutant line, ts-223, isolated from mutagenized W-8 cells and showing temperature-dependent expression of transformed properties in vitro (Yamaguchi and Weinstein, 1975) were provided by I. B. Weinstein and cultivated in Ham's F12 medium with 10% fetal calf serum.

All culture media and sera were obtained from GIBCO (Grand Island, New York).

Assay of "Transformed" Properties in Cultured Cells

Colony formation in agar was determined as described by Wyke and Linial (1973).

Hexose transport was measured using the method of Martin et al. (1971) as modified by Chen et al. (1977). Cells in 100 mm dishes were washed 3 times with prewarmed glucose-free Hank's saline and then incubated for 10 min in 5 ml glucose-free Hank's saline containing 20 μCi/ml 2-deoxy-³H-glucose (spec. act. 50 Ci/mM; Amersham-Searle, Arlington Heights, Illinois). The dishes were then washed 4 times with ice-cold Hank's saline and dried, and the cells were suspended in 5 ml Folin C reagent. An aliquot was taken for determination of protein content. Additional 1.0 ml aliquots were mixed with 0.15 ml of 5% trichloracetic acid followed by the addition of 10 ml of scintillation fluid (Aquasol; New England Nuclear), and the radioactivity was measured in a Packard Tri-Carb scintillation counter and sugar uptake expressed as cpm ³H-deoxyglucose per mg cell protein.

Tumorigenicity Assay

Newly fertilized eggs were incubated at 38°C in a humidified atmosphere (relative humidity approximately 75%) in an egg incubator (Favorite Incubator; Leahy Mfg. Co., Higginsville, Missouri) until day 8 after fertilization, when the chorioallantoic membrane (CAM) was dropped using the false air sac method (Leighton, 1967). The window created in the shell overlying the dropped CAM was sealed with Scotch tape, and the eggs were incubated for a further 48 hr at the temperature to be used in the tumorigenicity assay. Differing numbers

of cells (see Results) or virus (5×10^2–10^3 ffu) were then layered onto the surface of the dropped CAM in 20–50 μl of serum-free medium. Cells were preincubated for 48 hr at the temperature of the tumorigenicity assay before layering onto the CAM. The window in the shell was then resealed with tape and the eggs were incubated at the appropriate temperature for 1–7 days. The surface of the dropped CAM was examined daily for the formation of neoplastic lesions using a stereoscopic dissecting microscope (magnification 8–40×). Samples of CAM were also removed for histopathology at daily intervals after inoculation of cells. Tissue blocks were fixed in 10% formaldehyde in 0.15 M phosphate-buffered saline (pH 7.4), followed by preparation of 5 μ sections stained with hematoxylin and eosin.

The mitotic index of transformed cell populations within the mesodermal layer of the CAM was determined by microscopic counts of the number of mitotic cells identified in forty microscope fields of 5 μ thick sections of the CAM stained with hematoxylin and eosin. Necrotic areas and nontumor structures such as blood vessels or cords of invaginated ectodermal epithelium were avoided and not included in the counting fields. The total area of the 40 fields examined was 1.5 mm² and contained a mean of 9300 cells for transformed NRK cells and 11,400 for transformed CEF cells. The mitotic index is expressed as the number of mitoses per 1000 cells. For counting purposes, a photomicrograph (430×) of each field was made and overlaid with clear plastic film; nuclei were marked off as they were counted.

Temperature Measurements in Ovo

The temperature on the surface of the CAM was measured using a microprobe thermometer constructed within a blunt tip 29 gauge needle (0.013 in diameter) (MT-4 microprobe; Bailey Instruments, Saddlebrook, New Jersey) placed directly onto the CAM through the window in the egg shell and coupled to a BAT-C (Bailey Instruments) digital readout portable multichannel thermometer with a 0.01°C resolution. The thermocouple wire from the microprobe to the thermometer (core:copper and constantan; insulation:kapton) was taped to the egg shell and the egg, together with the thermometer unit, and placed in the incubator; temperature measurements were made at intervals over several days. Calibration experiments were performed in which the actual temperatures recorded on the CAM at each temperature setting on the incubator were compared in order to define the exact incubator control settings needed to achieve specific temperatures in ovo.

In eggs inoculated with transformed cells, temperature measurements were made simultaneously from the surface of the dropped CAM as described above; from within tumor nodules using a 23 gauge needle (sharp tip) microprobe thermometer (MT-3; Bailey Instruments) implanted directly into the tumor, and from the general atmosphere within the incubator using a MT-1 microprobe thermometer (Bailey Instruments). All three measurements could be made simultaneously and displayed via three separate input channels on the BAT-C recording thermometer unit.

Isolation of Cells from Tumors

10 day old eggs inoculated with 5×10^6 CE or NRK cells transformed by wild-type ASV or ts mutants were incubated for 7 days at either 38°C (wild-type transformants) or the appropriate permissive and nonpermissive temperatures (ts transformants). This size of cell inoculum facilitated formation of very large tumor nodules on the CAM (>0.5 cm diameter 5 days after inoculation). Tumors were excised, washed twice with cold F10 medium containing 2% bovine serum albumin and gentamycin (50 μg/ml), and cut into 250 μ tissue slices with a Smith-Farquhar tissue chopper. Slices were then incubated in Ca²⁺-Mg²⁺-free HBSS containing 1 mg/ml 2× crystallized trypsin (Sigma Chemical, St. Louis, Missouri) for 30 min to yield single cell suspensions. After enzymatic dissociation, cell suspensions were washed twice in F10 medium containing soybean trypsin inhibitor, and resuspended in complete growth medium and plated into individual wells of a Costar (#3524) 24-well tissue culture cluster dish. In the case of cells derived from tumors induced by transformed NRK cells, the trypsin-dispersed cell suspensions were incubated with

mouse anti-chicken IgG and guinea pig complement (Miles Labs, Kankakee, Illinois) for 30 min at 37°C to lyse contaminating CAM cells before being plated into the dishes.

Following growth to confluency in the multiwell dishes, the isolated cells were subcultured and grown in 25 cm² plastic flasks (Falcon) for experiments on the temperature-dependent expression of various phenotypic properties.

Rescue of Virus from NRK Cells

Tumors formed on the CAM at nonpermissive temperatures by NRK cells transformed by *ts* mutants of RSV were excised; the NRK cells were separated from chick cells as described above and the surviving NRK cell fraction was cultured in vitro at either 33° or 39.5°C. After a period of growth and between 1–3 subcultivations, the recovered NRK cells were fused with chick embryo fibroblasts (4×10^6 NRK cells; 1×10^7 chick embryo fibroblasts) by incubation with 1500 HAU of ultraviolet-inactivated Sendai virus for 2 hr at 37°C (Poste, 1973). The fused cell mixture was then plated into 60 mm plastic petri dishes and incubated at 35°C. Cell-free supernatants were harvested from these cultures at 2 or 7 days after plating and assayed for their ability to induce focus formation on chick embryo fibroblasts (Vogt, 1969).

Acknowledgments

This work was supported by grants from the National Cancer Institute. We thank D. Graham, W. Chichester and C. Somerville for their excellent technical assistance; Drs. P. Vogt, H. Hanafusa, I. B. Weinstein, R. O. Hynes and R. R. Friis for provision of virus strains and/or transformed cell populations; and Drs. R. L. Erikson and R. R. Friis for providing preprints of their recent work.

The costs of publication of this article were defrayed in part by the payment of page charges. This article must therefore be hereby marked "*advertisement*" in accordance with 18 U.S.C. Section 1734 solely to indicate this fact.

Received February 12, 1979; revised May 21, 1979

References

Ambros, V. R., Chen, L. B. and Buchanan, J. M. (1975). Surface ruffles as markers for studies of cell transformation by Rous sarcoma virus. Proc. Nat. Acad. Sci. USA *72*, 3144–3148.

Ash, J. E., Vogt, P. K. and Singer, S. J. (1976). Reversion from transformed to normal phenotype by inhibition of protein synthesis in rat kidney cells infected with a temperature-sensitive mutant of Rous sarcoma virus. Proc. Nat. Acad. Sci. USA *73*, 3603–3606.

Becker, D., Kurth, R., Critchley, D., Friis, R. R. and Bauer, H. (1977). Distinguishable transformation-defective phenotypes among temperature-sensitive mutants of Rous sarcoma virus. J. Virol. *21*, 1042–1055.

Biquard, J. M. and Vigier, P. (1972). Characteristics of a conditional mutant of Rous sarcoma virus defective in ability to transform cells at high temperatures. Virology *47*, 444–465.

Bissell, M. J., Hatie, C. and Calvin, M. (1979). Is the product of the src gene a promoter? Proc. Nat. Acad. Sci. USA *76*, 348–352.

Brugge, J. S., Steinbaugh, P. J. and Erikson, R. L. (1978a). Characterization of the avian sarcoma virus protein p60^{src}. Virology *91*, 130–140.

Brugge, J. S., Erikson, E., Collett, M. S. and Erikson, R. L. (1978b). Peptide analyses of the transformation-specific antigen from avian sarcoma virus transformed cells. J. Virol. *26*, 773–782.

Chen, Y. C., Hayman, M. J. and Vogt, P. K. (1977). Properties of mammalian cells transformed by temperature-sensitive mutants of avian sarcoma virus. Cell *11*, 513–521.

Collett, M. S. and Erikson, R. L. (1978). Protein kinase activity associated with the avian sarcoma virus src gene product. Proc. Nat. Acad. Sci. USA *75*, 2021–2024.

Collett, M. S., Erikson, E. and Erikson, R. L. (1979). Structural

analysis of the avian sarcoma virus transforming protein: sites of phosphorylation. J. Virol., in press.

Desmond, W. J., Jr., Wolbers, S. J. and Sato, G. (1976). Cloned mouse mammary cell lines requiring androgens for growth in culture. Cell *8*, 79–86.

Easty, D. M. and Easty, G. C. (1974). Measurement of the ability of cells to infiltrate normal tissues in vitro. Br. J. Cancer *29*, 36–49.

Edelman, G. M. and Yahara, J. (1976). Temperature sensitive changes in surface modulating assemblies of fibroblasts transformed by mutants of Rous sarcoma virus. Proc. Nat. Acad. Sci. USA *73*, 2047–2051.

Friis, R. R. (1978). Temperature sensitive mutants of avian RNA tumor viruses; a Review. Curr. Topics Microbiol. Immunol. *79*, 259–291.

Gilula, N. B., Eger, R. R. and Rifkin, D. B. (1975). Plasma membrane alteration associated with malignant transformation in culture. Proc. Nat. Acad. Sci. USA *72*, 3594–3598.

Hakomori, S.-I., Wyke, J. A. and Vogt, P. K. (1977). Glycolipids of chick embryo fibroblasts infected with temperature-sensitive mutants of avian sarcoma virus. Virology *76*, 485–493.

Hanafusa, H. and Hanafusa, T. (1966). Determining factor in the capacity of Rous sarcoma virus to induce tumors in mammals. Proc. Nat. Acad. Sci. USA *55*, 532–538.

Hanafusa, H., Halpern, C. C., Buchhagen, D. L. and Kawai, S. (1977). Recovery of avian sarcoma virus from tumors induced by transformation-defective mutants. J. Exp. Med. *146*, 1735–1747.

Hynes, R. O. and Wyke, J. A. (1975). Alterations in surface proteins in chicken cells transformed by temperature-sensitive mutants of Rous sarcoma virus. Virology *64*, 492–504.

Isaka, T., Yoshida, M., Owada, M. and Toyoshima, K. (1975). Alterations in membrane polypeptides of chick embryo fibroblasts induced by transformation with avian sarcoma viruses. Virology *65*, 226–237.

Kawai, S. and Hanafusa, H. (1971). The effects of reciprocal changes in temperature on the transformed state of cells infected with a Rous sarcoma virus mutant. Virology *46*, 470–479.

Kawai, S., Duesberg, P. H. and Hanafusa, H. (1977). Transformation-defective mutants of Rous sarcoma virus with src gene deletions of varying length. J. Virol. *24*, 910–914.

Keogh, E. V. (1938). Ectodermal lesions produced by the virus of Rous sarcoma. Br. J. Exp. Pathol. *19*, 1–9.

Kurth, R. (1975). Differential induction of tumour antigens by transformation-defective virus mutants. J. Gen. Virol. *28*, 167–177.

Kurth, R., Friis, R. R., Wyke, J. A. and Bauer, H. (1975). Expression of tumor-specific surface antigens on cells infected with temperature-sensitive mutants of avian sarcoma virus. Virology *64*, 400–408.

Leighton, J. (1967). The Spread of Cancer: Pathogenesis, Experimental Methods, Interpretations (New York: Academic Press).

Levinson, A. D., Oppermann, H., Levintow, L., Varmus, H. E. and Bishop, J. M. (1978). Evidence that the transforming gene of avian sarcoma virus encodes a protein kinase associated with a phosphoprotein. Cell *15*, 561–572.

Lingwood, C. A., Ng, A. and Hakomori, S. (1978). Monovalent antibodies directed to transformation-sensitive membrane components inhibit the process of viral transformation. Proc. Nat. Acad. Sci. USA *75*, 6049–6053.

Martin, G. S., Venuta, S., Weber, M. J. and Rubin, H. (1971). Temperature-dependent alterations in sugar transport in cells infected by a temperature sensitive mutant of Rous sarcoma virus. Proc. Nat. Acad. Sci. USA *68*, 2739–2741.

Minesita, T. and Yamaguchi, K. (1965). An androgen-dependent mouse mammary tumor. Cancer Res. *25*, 1168–1175.

Parry, G. and Hawkes, S. P. (1978). Detection of an early surface change during oncogenic transformation. Proc. Nat. Acad. Sci. USA *75*, 3703–3707.

Poste, G. (1973). Anucleate mammalian cells: applications in cell biology and virology. In Methods in Cell Biology, 7, D. Prescott, ed. (New York: Academic Press), pp. 211–249.

Poste, G., Flood, M. K., Kirsh, R. and Chichester, W. (1979). Selection and isolation of tumor cell variants with enhanced invasive and metastatic properties. Cancer Res., in press.

Prince, A. M. (1958). Quantitative studies on Rous sarcoma virus. III. Virus multiplication and cellular response following infection of the chorioallantoic membrane of the chick embryo. Virology 5, 435–457.

Purchase, H. G., Okazaki, W., Vogt, P. K., Hanafusa, H., Burmester, B. P. and Crittenden, L. B. (1977). Oncogenicity of avian leukosis viruses of different subgroups and mutants of sarcoma viruses. Infect. Immun. 15, 423–428.

Rübsamen, H., Friis, R. R. and Bauer, H. (1979). src gene product from different strains of avian sarcoma virus: kinetics and possible mechanism of heat inactivation of protein kinase activity from cells infected by transformation-defective, temperature-sensitive mutant and wild-type virus. Proc. Nat. Acad. Sci. USA 76, 1–5.

Scher, C. D., Haudenschild, C. and Klagsbrun, M. (1976). The chick chorioallantoic membrane as a model system for the study of tissue invasion by viral transformed cells. Cell 8, 373–382.

Stone, K. R., Smith, R. E. and Joklik, W. K. (1974). Changes in membrane polypeptides that occur when chick embryo fibroblasts and NRK cells are transformed with avian sarcoma viruses. Virology 58, 86–100.

Toyoshima, K. and Vogt, P. K. (1969). Temperature sensitive mutants of avian sarcoma virus. Virology 39, 930–931.

Toyoshima, K., Owada, M. and Kozai, Y. (1973). Tumor producing capacity of temperature sensitive mutants of avian sarcoma viruses in chicks. Biken J. 16, 103–110.

Trinkaus, J. P. (1976). On the mechanism of metazoan cell movements. In The Cell Surface in Animal Embryogenesis and Development, G. Poste and G. L. Nicolson, eds. (Amsterdam: North-Holland), pp. 225–329.

Vaheri, A. and Ruoslahti, E. (1974). Disappearance of a major cell-type specific surface glycoprotein antigen (SF) after transformation of fibroblasts by Rous sarcoma virus. Int. J. Cancer 13, 579–586.

Vogt, P. K. (1969). Focus assay of Rous sarcoma virus. In Fundamental Techniques in Virology, K. Habel and N. P. Salzman, eds. (New York: Academic Press), pp. 198–211.

Vogt, P. K. (1977). The genetics of RNA tumor viruses. In Comprehensive Virology, 9, R. R. Wagner and H. Fraenkel-Contrat, eds. (New York: Plenum Press), pp. 341–455.

Wang, E. and Goldberg, A. R. (1976). Changes in microfilament organization and surface topography upon transformation of chick embryo fibroblasts with Rous sarcoma virus. Proc. Nat. Acad. Sci. USA 73, 4065–4069.

Weber, M. J. and Friis, R. R. (1979). Dissociation of transformation parameters using temperature-conditional mutants of Rous sarcoma virus. Cell 16, 25–32.

Weinstein, I. B., Wigler, M., Fisher, P. B., Sisskin, E. and Pietropaolo, C. (1979). Cell culture studies on the biologic effects of tumor promotors. In Carcinogenesis, 2, T. J. Slaga, A. Sivak and R. K. Boutwell, eds. (New York: Raven Press), in press.

Wyke, J. A. (1973). The selective isolation of temperature sensitive mutants of Rous sarcoma virus. Virology 52, 587–590.

Wyke, J. A. (1975). Temperature sensitive mutants of avian sarcoma viruses. Biochim. Biophys. Acta 417, 91–121.

Wyke, J. A. and Linial, M. (1973). Temperature sensitive avian sarcoma viruses: a physiological comparison of twenty mutants. Virology 53, 152–161.

Yamaguchi, N. and Weinstein, I. B. (1975). Temperature sensitive mutants of chemically transformed epithelial cells. Proc. Nat. Acad. Sci. USA 72, 214–218.

Yoshida, M., Owada, M. and Toyoshima, K. (1975). Strain specificity of changes in adenylate cyclase activity in cells transformed by avian sarcoma viruses. Virology 63, 68–76.

Reprinted from Science, Vol. 205, pp. 1140–1142. 1979. © 1979 AAAS

Molecular Cloning of Polyoma Virus DNA in

Escherichia coli: Oncogenicity Testing in Hamsters

Abstract. *Inoculation of suckling hamsters with 2 × 10⁸ live cells of* Escherichia coli *K12 strain χ1776, carrying the complete genome of polyoma virus in a recombinant plasmid, failed to induce tumors in any of 32 recipients. Also, lambda phage DNA and particles with a monomeric insert of polyoma DNA did not induce tumors. Purified recombinant plasmid DNA, as well as phage particles and DNA containing a head-to-tail dimer of polyoma DNA, showed a low degree of oncogenicity, comparable to that of polyoma DNA prepared from mouse cells. These findings support the previous conclusions, based on infectivity assays in mice, that propagation of polyoma virus DNA as a component of recombinant DNA molecules in* E. coli *K12 reduces its biologic activity many orders of magnitude relative to the virus itself.*

We have recently reported the results of a series of risk assessment experiments involving derivatives of *Escherichia coli* K12 bearing recombinant DNA that contains the complete genome of polyoma virus (*1, 2*). In those experiments, the circular polyoma viral DNA was converted to the linear form by cleavage with single-cut restriction enzymes, ligated to plasmid or lambda phage vectors, and propagated in *E. coli* K12. The *E. coli* that contained recombinant DNA, as well as the purified recombinant DNA, were tested for their ability to produce polyoma infection. Although the recombinant molecules contained complete polyoma genomes which were infectious when enzymatically excised from the recombinant molecules, *E. coli* carrying these molecules consistently failed to induce polyoma infection, even when massive numbers were fed or injected into mice, a highly sensitive indicator system for productive polyoma infection. Similar results were recently reported by Fried *et al.* (*3*).

As a further step in evaluating the biologic activity of the polyoma-plasmid and polyoma-lambda recombinant DNA host-vector systems, we have tested their ability to induce tumors in suckling hamsters. These animals are highly sensitive to tumor induction by polyoma virus (4); and further, intact virions are not required for tumorigenesis, since we have recently shown that naked DNA (5) and even subgenomic fragments (6) of viral DNA can induce tumors in them. Thus, inoculation of hamsters provides a bioassay system that can detect biologic activity of a subgenomic portion of the viral DNA, thereby complementing and extending the mouse infectivity assays.

The recombinant DNA materials we tested (1, 2) include plasmid pBR322 containing the complete polyoma genome inserted at the Eco RI or Bam HI site, and λgtWES · λB phage with monomeric or head-to-tail dimeric inserts of the complete polyoma genome at the Eco RI site. The LP strain of polyoma (7) was used as the source of polyoma DNA; this strain is quite tumorigenic in hamsters, the mean tumor-producing dose (TPD$_{50}$) being 4×10^3 plaque-forming units of virus (5). The recombinant plasmid systems were studied by inoculation of hamsters with live E. coli carrying the recombinant molecules or with purified recombinant DNA before and after treatment with restriction enzymes. The phage system was studied by injection of hamsters with recombinant-containing particles or DNA. Experimental materials were inoculated subcutaneously into the backs of 1-day-old hamsters. The animals were observed for tumor formation for 5 months (8). Table 1 gives the results of the inoculations with live χ1776 carrying the recombinant plasmids. None of the 32 animals developed tumors after injection of 2×10^8 live bacteria; assuming a copy number of ten plasmids per cell, the inoculum contained an amount of polyoma DNA equivalent to that contained in 1700 TPD$_{50}$ of polyoma virus (9). Since ten copies per cell is a minimum estimate, and since the χ1776 may have multiplied after injection, the actual dose of polyoma DNA per hamster may have been considerably larger. The dose of χ1776 administered to the baby hamsters is close to the maximum that could be tolerated, since the hamsters were ill for 2 days after the injection; furthermore, this inoculum corresponds to the mean lethal dose of χ1776 for suckling mice (10).

The tumorigenicity of recombinant DNA's, recombinant phage particles and nonrecombinant virus–derived polyoma DNA is shown in Table 2. Uncleaved, supercoiled plasmid DNA's induced tumors in 2 of 27 recipients. When the recombinant plasmids were cleaved with the restriction enzyme used for

Table 1. Inoculation of suckling hamsters with χ1776 carrying polyoma DNA in recombinant plasmids. Plasmids pPB5 and pPB6 are recombinant plasmids consisting of pBR322 with the complete polyoma genome inserted at the Bam HI site in the two possible orientations; likewise, pPR18 and pPR21 are pBR322 with the polyoma genome inserted in both orientations at the Eco RI site (1). Equal numbers of organisms in the mid-log phase of growth from each of the cultures were suspended in saline; portions of the suspension containing 2×10^8 live E. coli K12 were injected subcutaneously in the backs of 1-day-old hamsters. Animals were examined for tumors twice weekly over a 5-month period.

Inoculum	Tumors*
χ1776 (pPB5) + χ1776 (pPB6)	0/23
χ1776 (pPR18) + χ1776 (pPR21)	0/9

*Ratio of the number of hamsters with tumors to the number of hamsters treated.

their construction, the preparations induced tumors in 9 of 17 animals. This activity is equivalent to that of linear forms of polyoma DNA prepared from infected mouse cells (5, 6) (Table 2, last line). No tumors developed in the hamsters receiving lambda phage DNA containing the monomeric inserts, but 3 of 16 hamsters receiving recombinant molecules containing the dimeric insert developed tumors. Similarly, injection of phage particles resulted in tumors in 2 of 12 hamsters injected with dimer-containing particles, and no tumors in the case of the monomeric insert. In general, the onco-

genic activity of the intact recombinant DNA and phage materials was comparable to, or less than, that of an equivalent amount of virus-derived polyoma DNA. This, in turn, is four to five orders of magnitude less than the activity of polyoma virus itself.

With regard to the relevance of these studies to risk assessment, the tests with live E. coli are clearly the most pertinent in that they are the only tests that in any way mimic a possible naturally occurring event. The tests with purified DNA and phage particles are important in allowing much larger amounts of DNA to be tested than can be administered in live E. coli, and, in our view, they are of value chiefly for indicating the extreme limits of what might be obtained after injection of intact organisms. The fact that the recombinant DNA molecules were biologically active when injected as cell-free materials, but not when contained in the live E. coli host, suggests that the bacterial cell constitutes an effective barrier against the transfer of its DNA to eukaryotic cells.

Since polyoma virus is so highly oncogenic in hamsters, and since noninfectious subgenomic segments of polyoma DNA are oncogenic in this host, these experiments constitute a "worst case" analysis; that is, the experimental design maximizes the chances for obtaining positive results. Thus, it is all the more striking that neither the live χ1776 har-

Table 2. Evaluation of the tumorigenicity of recombinant and nonrecombinant polyoma DNA's in suckling hamsters. DNA's were diluted in saline to give the equivalent of 0.45 to 0.5 μg of polyoma DNA per 0.03 ml of inoculum. Day-old Syrian hamsters were inoculated subcutaneously and checked twice weekly for the development of tumors over a 5-month period. The DNA's of polyoma plasmids pPR18, pPR21, pPB5, and pPB6 (see Table 1) were tested either in the supercoiled configuration (uncleaved) or after cleavage with the restriction enzyme used for the insertion (1, 6). Lambda-polyoma monomer recombinants λ-PY3 and λ-PY63 contain the polyoma genome in opposite orientations; the two phages or their DNA's were combined in equal proportions for the injections. The dimer-containing phage λ-PY30-5B (2) was separated from the monomer-containing phages by density-gradient centrifugation prior to injection or extraction of DNA. Phage particle inocula contained 10^{10} or 7×10^9 plaque-forming units, respectively, per 0.03 ml for the monomer- and dimer-containing phages. Nonrecombinant polyoma DNA was isolated from mouse cells infected with large-plaque polyoma virus (5). Tumors in animals inoculated with recombinant DNA were invariably subcutaneous tumors at the site of inoculation; the great majority of those examined histologically were fibrosarcomas.

Inoculum	Number of hamsters with tumors/number tested			
	DNA			Phage particles
	Un-cleaved	Cleaved		
		Eco RI	Bam HI	
Recombinant DNA				
Plasmid system				
pPR18 + pPR21	1/11	6/9		
pPB5 + pPB6	1/16		3/8	
Phage system				
Monomer insert (λ-PY3 + λ-PY63)	0/20			0/8
Dimer insert (λ-PY30-5B)	3/16			2/12
Nonrecombinant DNA				
Polyoma DNA	4/73 (19 percent)	29/64 (45 percent)	11/35 (31 percent)	

boring recombinant plasmids nor the phage preparations containing the monomeric polyoma DNA insert induced any tumors.

As in the studies of mouse infectivity, the dimer-containing phage materials showed the most biologic activity (18 percent tumor response) of the various recombinant materials tested, probably reflecting the ability of these molecules to generate intact polyoma genomes by recombinational mechanisms. The mouse studies showed that the potentially infectious dimer-containing phage DNA did not lead to infection when administered as infected *E. coli* in the latent period (2). While this mode of infection was not studied in the hamster system, the extremely low efficiency of lambda phage production in the absence of aerobic conditions, as would be the case in the tissues of the animal, makes it quite likely that the same reduction in activity would have been seen.

Whereas the experiments presented are in need of extension to other host-vector combinations, they do add to the reassuring conclusions of the earlier mouse infectivity studies (1, 2). The findings that no tumors were induced with the viable plasmid-containing bacteria not only provides further evidence for the safety of cloning viral genomes in *E. coli* but also provides for the safety of cloning other postulated oncogenic gene segments.

MARK A. ISRAEL, HARDY W. CHAN
MALCOLM A. MARTIN
Recombinant DNA Research Unit,
National Institute of Allergy and
Infectious Diseases, Bethesda,
Maryland 20205

WALLACE P. ROWE
Laboratory of Viral Diseases,
National Institute of Allergy and
Infectious Diseases

References and Notes

1. M. A. Israel, H. W. Chan, W. P. Rowe, M. A. Martin, *Science* **203**, 883 (1979).
2. H. W. Chan, M. A. Israel, C. Garon, W. P. Rowe, M. A. Martin, *ibid.*, p. 887.
3. M. Fried, B. Klein, K. Murry, J. Tooze, W. Boll, C. Weissmann, *Nature (London)* **279**, 811 (1979).
4. B. E. Eddy, S. E. Stewart, R. Young, G. B. Mider, *J. Natl. Cancer Inst.* **20**, 747 (1958); M. Stoker, *Br. J. Cancer* **14**, 679 (1960).
5. M. A. Israel, H. W. Chan, S. L. Hourihan, W. P. Rowe, M. A. Martin, *J. Virol.* **29**, 990 (1979).
6. M. A. Israel, D. Simmons, S. L. Hourihan, W. P. Rowe, M. A. Martin, *Proc. Natl. Acad. Sci. U.S.A.*, in press.
7. A large plaque variant (LP) of polyoma virus [M. Vogt and R. Dulbecco, *Virology* **16**, 41 (1962)] was provided by Dr. T. Benjamin.
8. Hamsters inoculated with live *E. coli* or with phage particles were held under P4 physical containment conditions, while hamsters receiving DNA were held under P3. All host-vector systems were EK2.
9. This calculation is based on the assumption that the ratio of particles to infectivity is 300.
10. S. B. Levy, N. Sullivan, S. Gorbach, *Nature (London)* **274**, 395 (1978).

12 April 1979

Supplementary Readings

Before 1961

Biskind, M.S. and G.S. Biskind. 1944. Development of tumors in the rat ovary after transplantation into the spleen. *Proc. Soc. Exp. Biol. Med. 58:* 176–179.

Gannon, W. 1929. Organization for physiological homeostatis. *Physiol. Rev. 3:* 345–352.

1961–1970

Ambrose, E.J., J.A. Dudgeon, D.M. Easty, and G.C. Easty. 1961. The inhibition of tumour growth by enzymes in tissue culture. *Exp. Cell Res. 24:* 220–227.

Finch, B. and B. Ephrussi. 1967. Retention of multiple development of potentialities by cells of a mouse testicular teratocarcinoma during prolonged culture. *Proc. Natl. Acad. Sci. 57:* 615–621.

Flanagan, S.P. 1966. "Nude," a hairless gene with pleiotropic effects in the mouse. *Genet. Res. 8:* 295–309.

Furth, J. 1967. Pituitary cybernetics and neoplasia. *Harvey Lect. 63:* 47–71.

Huebner, R. and G. Todaro. 1969. Oncogenes of RNA tumor viruses as determinants of cancer. *Proc. Natl. Acad. Sci. 64:* 1087–1094.

Pantelouris, E. 1968. Absence of thymus in a mouse mutant. *Nature 217:* 370–371.

1971–1980

Alberts, D.S., H.S.G. Chen, B. Soehlen, S.E. Salmon, E.A. Surwit, L. Young, and T.E. Moon. 1980. In vitro clonogenic assay for predicting response of ovarian cancer to chemotherapy. *Lancet,* August 16, *ii:* 340–342.

Artzt, K. 1972. Breeding and husbandry of "nude" mice. *Transplantation 13:* 547–549.

Bernstine, E.G., M.L. Hooper, S. Grandchamp, and B. Ephrussi. 1973. Alkaline phosphatase activity in mouse teratoma. *Proc. Natl. Acad. Sci. 70:* 3899–3903.

Bregula, U., G. Klein, and H. Harris. 1971. The analysis of malignancy by cell fusion. II. Hybrids between Ehrlich cells and normal diploid cells. *J. Cell Sci. 8:* 673–680.

Brown, R.A., J.B. Weiss, I.W. Tomlinson, P. Phillips, and S. Kumar. 1980. Angiogenic factor from synovial fluid resembling that from tumors. *Lancet,* March 29, *i:* 682–685.

Carl, P.L., P.K. Charkravarty, J.A. Katzenellenbogen, and M.J. Weber. 1980. Protease-activated "prodrugs" for cancer chemotherapy. *Proc. Natl. Acad. Sci. 77:* 2224–2228.

Christman, J.K., S. Silagi, E.W. Newcomb, S.C. Silverstein, and G. Acs. 1975. Correlated suppression by 5-bromodeoxyuridine of tumorigenicity

and plasminogen activator in mouse melanoma cells. *Proc. Natl. Acad. Sci. 72:* 47–50.

Fialkow, P.J., G.M. Martin, G. Klein, P. Clifford, and S. Singh. 1972. Evidence for a clonal origin of head and neck tumors. *Int. J. Cancer 9:* 133–142.

Folkman, J., E. Merler, C. Abernathy, and G. Williams. 1971. Isolation of a tumor factor responsible for angiogenesis. *J. Exp. Med. 133:* 275–288.

Gearhart, J.D. and B. Mintz. 1974. Contact-mediated myogenesis and increased acetylcholinesterase activity in primary cultures of mouse teratocarcinoma cells. *Proc. Natl. Acad. Sci. 71:* 1734–1738.

Gimbrone, M., S. Leapman, R. Cotran, and J. Folkman. 1972. Tumor dormancy in vitro by prevention of neovascularization. *J. Exp. Med. 136:* 261–276.

Giovanella, B.C., S.O. Yim, J.S. Stehlin, and L.J. Williams. 1972. Development of invasive tumors in the "nude" mouse after injection of cultured human melanoma cells. *J. Natl. Cancer Inst. 48:* 1531–1533.

Kisch, A.L., R.O. Kelley, H. Crissman, and L. Paxton. 1973. Dimethyl sulfoxide-induced reversion of several features of polyoma transformed baby hamster kidney cells (BHK-21). *J. Cell Biol. 57:* 38–53.

Klein, G., U. Bregula, F. Wiener, and H. Harris. 1971. The analysis of malignancy by cell fusion. I. Hybrids between tumor cells and L cell derivatives. *J. Cell Sci. 8:* 659–672.

Klein, G., B.C. Giovanella, T. Lindahl, P.J. Fialkow, S. Singh, and J.S. Stehlin. 1974. Direct evidence for the presence of Epstein-Barr virus DNA and nuclear antigen. *Proc. Natl. Acad. Sci. 71:* 4737–4741.

Langman, R.E. 1980. Transformation and tumorigenesis. *Nature 283:* 246–248.

Lehman, J.M., W.C. Speers, D.E. Swartzendruber, and G.B. Pierce. 1974. Neoplastic differentiation: Characteristics of cell lines derived from a murine teratocarcinoma. *J. Cell. Physiol. 84:* 13–28.

Levine, A., M. Torosian, A. Sarokhan, and A. Teresky. 1974. Biochemical critera for the in vitro differentiation of embryoid bodies produced by a transplantable teratoma of mice. *J. Cell. Physiol. 84:* 311–318.

Levy, S.B., B. Marshall, D. Rowse-Eagle, and A. Onderdonk. 1980. Survival of *Escherichia coli* host-vector systems in the mammalian intestine. *Science 209:* 391–394.

Liotta, L.A., K. Tryggvason, S. Garbisa, I. Hart, C.M. Foltz, and S. Shafie. 1980. Metastatic potential correlates with enzymatic degradation of basement membrane collagen. *Nature 284:* 67–68.

Manning, D.D., N.D. Reed, and C. Shaffer. 1973.

Maintenance of skin xenografts of widely divergent phylogenetic origin on congenitally athymic (nude) mice. *J. Exp. Med. 138:* 488–494.

Manton, K.G. and E. Stallard. 1980. A two-disease model of female breast cancer: Mortality in 1969 among white females in the United States. *J. Natl. Cancer Inst. 64:* 9–16.

Pantelouris, E.M. 1973. Athymic development in the mouse. *Differentiation 1:* 437–450.

Pellicer, A., E.F. Wagner, A. El Kareh, M.J. Dewey, A.J. Reuser, S. Silverstein, R. Axel, and B. Mintz. 1980. Introduction of a viral thymidine kinase gene and the human β-globin gene into developmentally multipotential mouse tetracarcinoma cells. *Proc. Natl. Acad. Sci. 77:* 2098–2102.

Povlsen, C.O. and J. Rygaard. 1971. Heterotransplantation of human adenocarcinomas of the colon and rectum to the mouse mutant nude. *Acta Pathol. Microbiol. Scand. 79:* 159–169.

Povlsen, C.O. and J. Rygaard. 1972. Heterotransplantation of human epidermal carcinomas to the mouse mutant "nude." *Acta Pathol. Microbiol. Scand. 80:* 713–717.

Raff, M.C. 1973. τ-Bearing lymphocytes in nude mice. *Nature 246:* 350–351.

Stern, P., M. Gidlund, A. Orn, and H. Wigzell. 1980. Natural killer cells mediate lysis of embryonal carcinoma cells lacking MHC. *Nature 285:* 341–342.

Stutman, O. 1974. Tumor development after 3-methylcholanthrene in immunologically deficient athymic nude mice. *Science 183:* 534–536.

Swartzendruber, D. and J. Lehman. 1975. Neoplastic differentiation. *J. Cell. Physiol. 85:* 173–178.

Teresky, A.K., M. Marsden, E.L. Kuff, and A.J. Levine. 1974. Morphological criteria for the in vitro differentiation of embryoid bodies produced by a transplantable teratoma of mice. *J. Cell. Physiol. 84:* 319–332.

Wiener, F., G. Klein, and H. Harris. 1971. The analysis of malignancy by cell fusion. *J. Cell Sci. 8:* 681–692.

Chapter 6 The Cell Cycle and DNA Replication

Stanners, C.P. and J.E. Till. 1960. DNA synthesis in individual L-strain mouse cells. *Biochim. Biophys. Acta 37:* 406–419.

Huberman, J.A. and A.D. Riggs. 1968. On the mechanism of DNA replication in mammalian chromosomes. *J. Mol. Biol. 32:* 327–341.

Rao, P.N. and R.T. Johnson. 1970. Mammalian cell fusion: I. Studies on the regulation of DNA synthesis and mitosis. *Nature 225:* 159–164.

Smith, J.A. and L. Martin. 1973. Do cells cycle? *Proc. Natl. Acad. Sci. 70:* 1263–1267.

Rossow, P.W., V.G.H. Riddle, and A.B. Pardee. 1979. Synthesis of labile, serum-dependent protein in early G_1 controls animal cell growth. *Proc. Natl. Acad. Sci. 76:* 4446–4450.

Pardoll, D.M., B. Vogelstein, and D.S. Coffey. 1980. A fixed site of DNA replication in eucaryotic cells. *Cell 19:* 527–536.

A growing population of cultured cells doubles in about a day. The physical separation of two daughter cells at mitosis takes only an hour or so. During the rest of the day a cell is creating a copy of itself. Although to the eye this process is not at all as striking as the process of dividing, the biochemical events of cell doubling and their timing relative to one another are of major importance. These events are analyzed by a variety of techniques in the papers in this chapter.

Stanners and **Till** use autoradiography to see which cells on a dish are synthesizing DNA. Cells making DNA incorporate tritiated thymidine into their nuclei, where it remains after fixation. When fixed cells are overlaid with film emulsion, the radioactive nuclei expose the emulsion and so become covered by silver grains when the emulsion is developed. With this marker for DNA synthesis, and with the morphological signs of mitosis, Stanners and Till were able to break the cell cycle into four parts. In an exponentially increasing population of mouse tumor cells, a gap of time (G1) separates mitosis (M) from DNA synthesis (S) and another gap (G2) separates S from the next M. In a tour de force, they were also able to exclude gaps in S and to measure the times spent by a cell in each of the four parts of the cycle, without the use of any other techniques.

Huberman and **Riggs** examine replication of cell DNA. Autoradiography is again the tool of choice, but here the labeled cells are gently dissolved in detergent before being exposed to film. From the disrupted nuclei, the DNA spreads out. As a result, long lines of silver grains are found upon development of the emulsion. Pulse labeling shows that DNA replication initiates from many points, or "origins," and that it then proceeds from each origin in both directions along the double helix, generating "eyes" or bubbles of newly made DNA. The sequence and secondary structure of origins of replication in mammalian cells are currently a subject of intense study. Growth-controlled fibroblasts are blocked somewhere before S, indicating that deprivation of serum or anchorage in some way blocks initiation of DNA synthesis. Perhaps viral transforming gene products such as SV40 large T antigen (Chapter 4) initiate cellular DNA replication directly to override such blocks.

HeLa cells can be blocked at the beginning of S by thymidine or at the end of M by nitrous oxide treatment. Such cells can be fused to one another by coalescence of their cell membranes with an inactivated virus. The products of fusion are multinucleated, and their cytoplasms and nuclei will be from different parts of the cell cycle. From analysis of S and M in the nuclei of heterokaryons, as these fusion products are called, **Rao** and **Johnson** were able to make three interesting points about the regulation of the HeLa cell cycle. First, cells in S are able to entrain the nuclei of cells in G1 into a premature S. Second, mitoses are self-synchronizing. No matter from which two parts of the cycle cells were chosen, mitoses always occurred at once in both nuclei of a heterokaryon. Third, comparison of binucleate with trinucleate fusion products showed a dosage effect. The more S nuclei, the more rapidly G1 nuclei are recruited into DNA synthesis. This suggests that in HeLa cells, diffusible molecules appearing in S may act positively to control DNA synthesis.

If an exponentially growing population of cells is synchronized after G1, then it remains synchronous through the next mitosis. However, passage through G1 always destabilizes synchrony and leads to a broadened distribution of cycle times by the next S. Growth controls, such as serum starvation, cause populations to accumulate in G1, and a similar heterogeneity in S occurs when such populations are released by provision of serum or anchorage. The two papers that follow offer somewhat different approaches to explain this intrinsic heterogeneity in the G1 part of the cycle.

Imagine that cells may spend any amount of time in a special part of G1, during which they are not directed toward further steps in the cell cycle. Furthermore, imagine that all cells of a clone kept in the same culture conditions have a constant probability of leaving this state and entering the rest of the cycle. **Smith** and **Martin** argue that the heterogeneity of intermitotic times is due to such a probabilistic period in G1. The major prediction of their hypothesis is that in growth-controlled or self-synchronized populations, all cells are equivalent; no separate compartments of cycling cells and blocked cells would exist. Transformation would be an increase in transition probability for the entire cell population rather than a shift in state of most cells from quiescence to cycling.

But what makes the rare cell divide in a quiescent population if all cells are equivalent? **Rossow, Riddle,** and **Pardee** postulate that a critical concentration of a signaling molecule must accumulate before a cell can progress to S. They poisoned growing cells with very low levels of cyclohexamide, an inhibitor of protein synthesis. The population grew more slowly, and extensions in doubling times by up to 70% were accounted for by elongation in the G1 period. The rest of the cycle was unaffected by this treatment. Rossow et al. show that such a population of cells is intrinsically heterogeneous. They argue that cells remain blocked or restricted before S until they synthesize enough of an unstable regulatory compound to trigger transition through the cycle to the next restriction. If this is correct, then growth control may work by down-regulation of the synthesis of such a regulatory compound. In this regard, the observation by Farmer et al. (Chapter 4) that translation of mRNA is suspended in anchorage-deprived cells may be significant. It is tempting to speculate that virus-coded molecules such as T and src mimic or bypass such endogenous cellular regulatory molecules.

The last paper may resolve one paradox of the cell cycle. In S, the DNA of the cell is replicated at hundreds or thousands of initiation points and replication forks. Therefore, it must be a skein of many loops. In mitosis, the DNA of the cell is folded and packed into chromosomes, which split apart to send equivalent copies of DNA to each daughter cell (see Chapter 8). How can S occur in such a way as to prepare for mitotic separation of the two daughter copies of the cell's DNA? **Pardoll, Vogelstein,** and **Coffey** show that the nucleus of an S cell has an insoluble protein matrix onto which are attached the growing points of newly replicated DNA. Apparently, the enzymes of DNA replication are attached to this matrix, and the DNA shuttles through the matrix in loops as it is made. One can easily imagine the matrix becoming part of the mitotic chromosome and the two daughter loops of newly synthesized DNA being separated from one another at mitosis.

DNA SYNTHESIS IN INDIVIDUAL L-STRAIN MOUSE CELLS

C. P. STANNERS* AND J. E. TILL

*Department of Medical Biophysics, University of Toronto, and Physics Division,
The Ontario Cancer Institute, Toronto, Ontario (Canada)*

(Received April 21st, 1959)

SUMMARY

The time relationship between DNA synthesis and mitosis has been determined for L-strain mouse cells cultivated *in vitro*. DNA synthesis was detected autoradiographically by following the uptake of [³H]thymidine into the cell nucleus. DNA synthesis in this cell system was found to take place in an approximately linear fashion over a single period of six to seven hours, ending three to four hours before mitosis, for a total generation time of twenty hours.

A mathematical treatment for exponentially multiplying cultures is presented.

INTRODUCTION

It is obvious that a cell must double its content of deoxyribose nucleic acid (DNA) before it divides. For many years it was believed that the cell synthesized this DNA during prophase and metaphase, so that when the two sets of chromosomes separated at anaphase, each had a full complement of DNA. Recently, many workers have shown, by the use of photometric and autoradiographic techniques, that DNA synthesis actually occurs during interphase over a period which ends at a certain time before mitosis, and which has a definite length[1-7]. This was found to be true for many different types of cells, both animal and plant, but the exact timing of the DNA synthesis period appeared to vary with the cell type. We have investigated, by the method of autoradiography, DNA synthesis in exponentially multiplying mouse cells, cultivated *in vitro*. This investigation is preliminary to studies of the perturbing influence of various chemical and physical agents upon DNA metabolism in the cell.

MATERIALS AND METHODS

Detection of DNA synthesis

Since thymidine has been shown to be a specific precursor of DNA[8], the uptake of [³H]thymidine (³HTDN) into the cell nucleus was used as a measure of DNA synthesis. The β-decay of tritium was detected autoradiographically. The low energy of tritium β-particles affords excellent resolution on the autoradiographs[9] (see Fig. 1).

* Graduate student, Department of Medical Biophysics, University of Toronto, and Fellow in Radiation Physics of the National Cancer Institute of Canada.

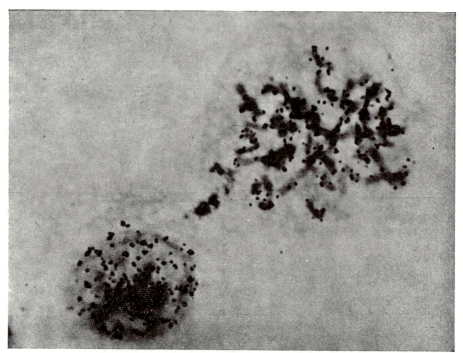

Fig. 1. Autoradiograph of two L cells after 16-h exposure to [³H]thymidine. One cell is in meta-phase, the other in interphase. The metaphase cell has synthesized one complement of DNA in the presence of label. Film exposure time was 15 days.

Cells

EARLE's L-strain mouse cells (NCTC Clone 929) were used[10]. Cells were cultivated continuously in the logarithmic growth phase, using suspension cultures[11]. It has been shown that under these conditions, almost 100 % of the cells are viable[12]. Growth medium for stock cultures consisted of a 20 % solution of horse serum in CMRL 1066[13]. For labelling of cellular DNA, ³HTDN at a specific activity of 360 mC/mmole* was added to the growth medium to a concentration of 0.25 µC/ml. In one experiment a concentration of 0.5 µC/ml was used. The effective specific activity of the ³HTDN was reduced 40-fold due to the presence of 10 µg/ml of unlabelled thymidine in the growth medium. The concentration of ³HTDN used was safely below the radiation toxic level reported by PAINTER, DREW AND HUGHES[14] for HeLa cells, and no effect of addition of ³HTDN on the growth rate of the cultures was observed.

Preparation of slides for autoradiography

After addition of ³HTDN, the culture was sampled at hourly intervals. The samples taken were small, so that a single suspension culture could be used for an entire experiment. The cells of each sample were immediately dilated in hypotonic saline and fixed in acetic–alcohol, whereupon aliquots were smeared on duplicate slides and allowed to dry. The slides were then washed in acetic–alcohol, stained with 2 % acetic orcein, washed again in alcohol to remove excess stain, and then washed in running water for 3 to 4 h to prepare them for autoradiography.

* Schwarz Laboratories, Mt. Vernon, N.Y.

Biochim. Biophys. Acta, 37 (1960) 406–419

Autoradiography

After washing, the slides were ringed with egg albumin to prevent lifting of the photographic emulsion, and the emulsion was applied, using the method of MESSIER AND LEBLOND[15], in which the slides are dipped directly into liquid bulk emulsion*. The emulsion was placed in direct contact with the cells, without an interposed film of celloidin, in order to ensure maximum efficiency of detection of the low energy tritium β-particles[16]. Direct contact between emulsion and cells did not introduce spurious grains, as unlabelled cells showed a normal background grain count over their nuclei. In one experiment, the stripping film technique of PELC[17] was used.

For exposure, the slides were stored in light tight boxes at 4°. Exposure time was usually two weeks.

Counting methods

Fifty metaphases were scored to determine the percentage of metaphases which were labelled in each sample. The percentage of labelled cells was determined by scoring 500 interphase nuclei. An interphase nucleus or a metaphase was considered to be labelled if it was covered with 5 or more grains, the average background being about 2 grains/nucleus. The generation time of the cells was determined by measuring the cell concentration in the growing culture at various times with a hemocytometer. Mitotic indices were measured by counting the number of cells in mitotis in 3000 cells. In our material, the great majority of cells in mitosis were found to be in metaphase; one slide showed, for example, 73 metaphases, 7 prophases, and 2 anaphases. Our mitotic indices are therefore, to a close approximation, the percentage of cells seen in metaphase. Any cell showing a single group of clearly discernible chromosomes in the absence of a nuclear membrane was considered to be in metaphase.

RESULTS

From the recent reports of interphase DNA synthesis in various types of cells referred to in the introduction, it was assumed at the outset that every cell in the exponentially growing culture passes repetitiously through a mitotic cycle of the type illustrated in Fig. 2. If cells are considered to move clockwise around the cycle, they pass successively through a postmitotic non-synthetic period (G_1), a period of active synthesis of DNA (S), and a premitotic non-synthetic period (G_2)[1, 3, 4, 7]. It remained to verify this pattern of DNA synthesis in our cells and to measure the actual values of G_2, S, and G_1. We also investigated how DNA synthesis proceeds with time over the S period.

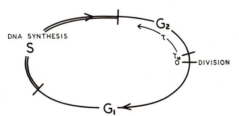

Fig. 2. The mitotic cycle of actively multiplying cells, showing the DNA synthetic period and its time relation to division (cells move clockwise around the cycle).

---

* Ilford Nuclear Research Emulsion, type G5, in gel form, Ilford, Inc., Ilford, Essex.

Biochim. Biophys. Acta, 37 (1960) 406–419

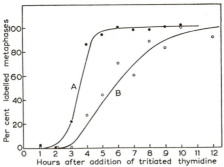

Fig. 3. The percentage of labelled metaphases for two cultures (denoted A and B) at various times after addition of [³H]thymidine.

Determination of the G_2 period

The time required for labelled DNA to show up in metaphase chromosomes should give a measure of the length of the G_2 period. The ratio of labelled metaphases to total metaphases, labelled or unlabelled ("percent labelled metaphases") was measured for each hour after addition of ³HTDN. Fig. 3 gives the results of two experiments. It is to be seen that in both cases virtually no labelled metaphases are seen for 3 h after addition of label. That this is not a lag in the uptake of label is shown by the observation that many *interphase* nuclei have taken up label within 1 h after addition of ³HTDN (see Figs. 8 and 10). Indeed, the lag in uptake of label is probably negligible[7]. The 3-h delay in the appearance of label in cells in metaphase means that the period between completion of DNA synthesis and metaphase, *i.e.*, the G_2 period (see Fig. 2), is at least 3 h long. This, however, is a minimum period only. If all the cells had the same 3-h G_2 period, all metaphases would be labelled, though only slightly so, when this period had elapsed; that is, all metaphases seen at time G_2 after the addition of label would have just been completing DNA synthesis when ³HTDN was added, and would thereby all be labelled. The percent labelled metaphases curve would, after a 3-h interval, rise abruptly to 100 %. The fact that it does not indicates that there is a variation among cells in the length of the G_2 period. Since, as will be shown, the duration of metaphase is short, about 0.5 h, the variation seen in Fig. 3 is only in a small part due to the time required for cells to pass through metaphase. A larger variation was observed in experiment B than in experiment A.

By taking the derivative of the percent labelled metaphase curves, the frequency distributions relating the numbers of cells with a given G_2 period are obtained. These are presented in Fig. 4. In experiment A the result is a sharp Gaussian curve with a peak at 3.5 h; in experiment B it is a more flattened curve skewed to the right, with a peak near 4 h. The reason for the difference between the two experiments has not been ascertained, but is probably due to differing culture conditions.

The generation time was found to be 20 h for both experiments. Thus, for a generation time of 20 h, the G_2 period has a most probable value of from 3 to 4 h.

Determination of the S period

The length of the G_2 period has now been found. The length of the S period can be determined in two ways.

Biochim. Biophys. Acta, 37 (1960) 406–419

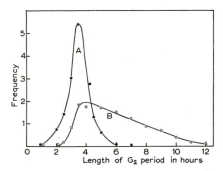

Fig. 4. Frequency distributions giving the relative number of cells with a given G_2 period for cultures A and B.

Grain counts over metaphases: The grain count over metaphases seen at increasing times after addition of label, should be zero (neglecting background) until the G_2 period has elapsed, then should increase for a time equal to the S period, after which it should increase no further since all the cells in metaphase after this time would have synthesized a full complement of DNA in the presence of label. The time between the first increase in the grain count over background and the time when the grain count saturates gives the length of the S period, and the intermediate grain counts give the manner in which DNA synthesis proceeds with time over the S period. Unfortunately, all the metaphases do not have the same number of grains over them at any time after addition of label, because there is variation in the length of the G_2 period, there is autoradiographic variation, and there is possible variation in DNA synthesis rates amongst cells. If, however, a distribution of the number of metaphases which have a grain count within a given interval is plotted for each time, the peak grain count values of these distributions should show the time dependence which would have been observed had there been no variation in the metaphase grain counts.

The slides of experiment A, which showed the smaller variation in the G_2 period, were selected for this experiment. Duplicate slides were used, one set being exposed for 15 days, the other for 8 days. Grains were counted over 50 metaphases generally, for each hour after addition of ^{3}HTDN, and a histogram of the number of metaphases within a given grain count interval plotted. Fig. 5 shows two such histograms for

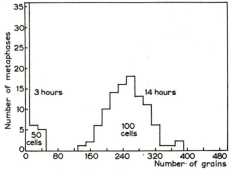

Fig. 5. Grain counts over individual metaphases; histograms of the number of metaphases within a given grain count interval for 3 h and for 14 h after addition of [^{3}H]thymidine.

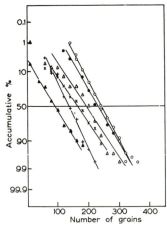

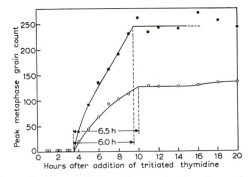

Fig. 7. Peak metaphase grain count values for various times after addition of [³H]thymidine. Film exposure time for the upper curve was 15 days, lower curve, 8 days.

Fig. 6. The percentage of metaphases with a grain count up to and including the abscissa grain count values, plotted on probability paper. A Gaussian distribution appears as a straight line with its peak value at the 50% intercept. Six lines are given, corresponding to six times after addition of [³H]thymidine: ▲, 5 h after addition of ³HTDN; +, 6 h; ×, 7 h; △, 8 h; ●, 9 h; ○, 14 h.

3 and 14 h after addition of label. The 14-h distribution is a Gaussian, as were the distributions for all other times where all the metaphases were labelled.

The experimentally determined grain count distributions for each successive hourly sample were plotted on probability paper. Several of these are depicted in Fig. 6. When plotted on probability paper, a Gaussian distribution appears as a straight line, the peak value being given by the 50% intercept of the line. The peak grain counts determined in this way for each hour after addition of ³HTDN is plotted in Fig. 7. This method of determination of the peak grain count was not applied to the distributions for hours 1 to 4, since a Gaussian distribution would not be expected until all metaphases were labelled.

In Fig. 7, the upper curve represents peak grain counts as a function of time for a 15-day exposure, and the lower curve for an 8-day exposure of duplicate slides from the same experiment (A). Since grains are most conveniently counted when they number about 100 over a metaphase, the upper curve is more reliable for $t < 8$ h and the lower curve is more reliable for $t > 8$ h. It is seen that the metaphase grain count saturates in both cases 9.5 to 10 h after addition of ³HTDN. The DNA synthetic period is therefore 6 to 6.5 h long.

From Fig. 7, it is seen that DNA synthesis is close to being linear with time over the S period. The curves are slightly concave which may indicate the DNA synthesis occurs at a slightly greater rate near the end of the S period than at the beginning, though the significance of the shapes of the curves is doubtful. From the lower curve, which gives a more reliable saturation plateau, it is seen that the peak metaphase grain count is virtually constant from 10 to 20 h, which is the generation time. The slight rise occurring from 16 to 20 h, which represents an increase of about 6% in the total grain count, could be due either to a second period of DNA synthesis at a very reduced rate, or to the presence of a few cells in the population which have short enough generation times to have entered the S period a second time. In point of fact, since the standard deviation of the peak grain count values for the lower curve is

Biochim. Biophys. Acta, 37 (1960) 406–419

about 4 grains, the rise is not statistically significant. Determinations, to be described below, of the percentage of interphase cells which take up label, yield results that are in agreement with the single period hypothesis. Thus, the grain counts over meta-phases as a function of time indicate that DNA synthesis in L-cells occurs at a near linear rate over a single 6.0 to 6.5-h period which ends 3 to 4 h before mitosis.

Percent labelled cells: The percent of labelled interphase nuclei measured each hour after addition of ^{3}HTDN should give an independent measure of the S period, and should be consistent with the values of G_1 and G_2 measured by the metaphase grain count method. The disadvantage of the percent labelled cells method is that it is very sensitive to any synchronization within the culture. In order to evaluate S, the degree of synchronization must be determined.

Asynchronous culture

We shall define an asynchronous culture as one in which the number of cells increases in a perfectly smooth exponential fashion with time, that is, one in which there is a constant fraction of cells undergoing division. For such a culture the frequency distribution of cells around the mitotic cycle (Fig. 2), or the number of cells in each portion of the cycle, is an exponential function[18] (see APPENDIX) with twice as many cells leaving mitosis as entering it, and the mitotic index is constant. It will be seen that a determination of the mitotic index as a function of time provides a sensitive indicator of the degree of synchronization present in a culture.

In experiment B, the mitotic index was found to be nearly constant with a value of about 1 % over a generation time of 20 h. It can thus be treated as an asynchronous culture. The length of time τ_m which the cells spend in mitosis for such a culture can be evaluated by applying eqn. (7) of the APPENDIX:

$$\tau_m = \frac{MT}{0.693}$$

where $M \times 100$ is the mitotic index and T is the generation time. This equation yields a value of τ_m of about 0.3 h. This is probably a minimum value for τ_m since slides that have not undergone preparation for autoradiography show mitotic indices of from 2 to 3 % *i.e.* a τ_m of about 0.7 h. However, since in this experiment the mitotic index obtained under constant conditions of slide preparation was found not to vary, it is valid to treat the culture as asynchronous.

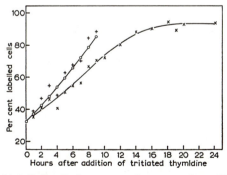

Fig. 8. The percentage of labelled cells for an asynchronous culture (B) at various times after addition of [^{3}H]thymidine. ×, experimental curve; ○, the theoretical curve calculated from eqn. (14) using $S = 7.1$ h. For comparison, the corrected experimental points, +, are shown.

Biochim. Biophys. Acta, 37 (1960) 406–419

Determinations were made of the percentage of cells showing more than 5 grains above their nuclei ("percent labelled cells") for each hour after addition of ^{3}HTDN. Changes in the percentage of labelled cells could be due to flow of unlabelled cells into the S period and division of already labelled cells, yielding an increase, or to division of unlabelled cells, yielding a decrease. Since we wish to relate the percentage of labelled cells to the S period only, the experimental values obtained were corrected to what would have been observed had no multiplication taken place (see APPENDIX). A further correction was made for the fact that 5 % of the cells in the culture were unable to synthesize DNA. This is evident from the observation that the percent labelled cells curve levels off at 95 % after 16 h, as shown in Fig. 8. This correction is straightforward and the details will not be given. As discussed in the APPENDIX, further corrections may be avoided by using only those values for times less than G_1.

From each determination, a value of S was calculated by applying eqn. (15) derived in the APPENDIX:

$$S = \frac{1}{a} \ln \left[L(t)_{\text{corr}} + e^{aG_2} \right] - (G_2 + t), \qquad t < G_1$$

where $a = \ln 2/T$, T is the generation time, and $L(t)_{\text{corr}} \times 100$ is the corrected percentage of labelled cells. ^{3}HTDN was added at $t = 0$.

The results are tabulated in Table I. The constancy of S serves as a test of the formulation.

TABLE I

EVALUATION OF THE S PERIOD

Time h	Asynchronous culture (B)		
	Experimental percentage labelled cells	Corrected percentage labelled cells	Value of S (from eqn. (15)) h
1	36	39	7.2
2	42	47	7.6
3	47	55	7.9
4	41	49	6.0
5	51	63	7.3
6	55	68	7.1
7	57	71	6.5
8	67	85	7.6
			Mean: 7.1 h

From Table I, the average value found for S was 7.1 h. Using this average value for S in eqn. (14) of the APPENDIX the "theoretical" curve shown in Fig. 8 was constructed. Also shown are the corrected points and the uncorrected experimental curve.

Comparison of the curves shows that the results are consistent with the hypothesis that a single period of DNA synthesis exists for all cells, the period being about 7 h in duration and ending approx. 4 h before division.

Partially synchronized culture

If the mitotic index of a culture is not constant with time, the frequency distribution relating the number of cells within each portion of the cell cycle is not exponential. This can have a very pronounced effect upon quantities which depend upon the distribution, such as the percent labelled cells.

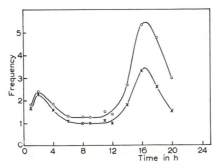

Fig. 9. Lower curve: the mitotic index at times after addition of [³H]thymidine for a partially synchronized culture (A). Upper curve: the derived frequency distribution $n(t, 0)$, giving the relative number of cells in each portion of the mitotic cycle for the same culture.

The mitotic index of experiment A was found to vary with time as shown in Fig. 9. The slides of this experiment were used to determine the S period by the metaphase grain count method, though it should be emphasized that the mitotic variation would have no effect upon this determination, since the treatment is not dependent on the distribution of cell numbers in the various portions of the mitotic cycle. Nevertheless, as a check on the grain count method, an independent estimate of the duration of the S period may be obtained from a count of the percent labelled cells, and this does depend strongly on the distribution of cell numbers throughout the mitotic cycle. An analysis of this dependence is outlined in the APPENDIX.

Assuming that all the cells move around the mitotic cycle at the same rate, the relative number of cells in each portion of the cycle at $t = 0$ is given by $n(t,0)$, where t represents the time separating a particular cell from division. $n(t,0)$ is related to the measured mitotic index $M(t)$ according to eqn. (9) of the APPENDIX:

$$ n(t,0) = \frac{1}{\tau_m} N_0 \, M(t) \exp \frac{1}{\tau_m} \int_0^t M(t) \mathrm{d}t $$

where N_0 is the total number of cells anywhere in the cycle at time $t = 0$, and

$$ \tau_m = \frac{1}{\ln 2} \int_0^T M(t) \mathrm{d}t $$

where T is the generation time.

The measured values for M(t) obtained from the slides of experiment A, presented in Fig. 9, yield a value of 0.5 h for τ_m. The values derived for n(t,0) using eqn. (9) are also plotted in Fig. 9.

The percentage of labelled cells were corrected to what would have been observed had no cell multiplication taken place. The procedure for this correction is given in the APPENDIX. Since the experimental $L(t)$ curve reached virtually 100 % after about 15 h, no correction was required to allow for cells unable to synthesize DNA. The corrected percent labelled cell values, as a function of time, $L(t)_{corr} \times 100$, can be predicted from $n(t,0)$ using eqn. (13) of the APPENDIX, by assuming a value for S:

$$ L(t)_{corr} = \frac{\int_{G_2}^{G_2 + S + t} n(t,0) \mathrm{d}t}{N_0} , \qquad t < G_1 $$

The value of S which gave best agreement between corrected experimental and

Biochim. Biophys. Acta, 37 (1960) 406–419

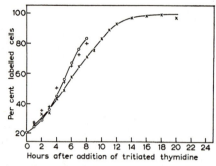

Fig. 10. The percentage of labelled cells for a partly synchronized culture (A) at various times after addition of [³H]thymidine. ×, experimental curve; O, the theoretical curve calculated from eqn. (13) using $S = 8.0$ h. For comparison, the corrected experimental points, +, are shown.

TABLE II

EVALUATION OF THE S PERIOD

Time h	Partly synchronized culture (A)		
	Experimental percentage labelled cells	Corrected percentage labelled cells	Value of S (from eqn. (13)) (h)
1	26	27	8.3
2	33	36	8.8
3	34	38	8.2
4	44	51	8.4
5	49	55	7.8
6	58	65	7.8
7	65	73	7.6
8	71	80	7.4
			Mean: 8.0 h

theoretical $L(t)$ values, was determined for each hour, the integration being done graphically. These values are tabulated in Table II. The calculated values of $L(t)$ obtained using the mean value of S from Table II are plotted in Fig. 10, along with the corrected points and uncorrected experimental curve.

The mean value obtained for S was 8 h. This is somewhat larger than the value of 6.0 to 6.5 h obtained for metaphase grain counts performed on the same set of slides. This discrepancy may be due either to the assumptions inherent in the analysis, or to errors in the mitotic index values. It is quite difficult to obtain a reliable measure of the mitotic index, due to statistical errors and preparational inconsistencies.

DISCUSSION

The mitotic cycle found for L-strain mouse cells is similar to those reported by other workers for mammalian cells cultivated *in vitro*[4,7] in that they are characterized by a long post mitotic non-synthetic phase (G_1) involving approximately half of the total generation time, followed by a period of DNA synthesis (S) with a duration of about one-third of the generation time, followed in turn by a short premitotic non-synthetic period (G_2) which lasts for about one-fifth of the generation time. The division process itself is rapid, requiring less than an hour for completion.

Biochim. Biophys. Acta, 37 (1960) 406–419

The evidence also favours a linear rate of synthesis during the S period. The slight decrease in slope apparent in the curves of Fig. 7 would imply a decreased rate of DNA synthesis during the earlier stages of the period, since moving towards the right in Fig. 7 represents moving to earlier times in the mitotic cycle, *i.e.* away from mitosis.

The quantitative interpretation of much data relating to exponential cultures is only possible through an analysis similar to that given in the APPENDIX for the rate of accumulation of labelled cells. The procedure is complicated by the fact that the frequency distribution of cells around the cycle is exponential rather than constant. For example, suppose the percentage of labelled cells after a brief exposure to label for an asynchronous culture with a generation time of 20 h, was found to be 30 %. The length of the S period is not given by taking $0.30 \times 20 = 6$ h which assumes a constant distribution, but by eqn. (15), which yields $S = 6.7$ h, a difference of over 10 %. An exponential distribution, also, can be expected only when the mitotic index is constant with time. If it is not, the distribution must be evaluated (see APPENDIX), since cyclic quantities are very dependent upon it. For example, in one partially synchronized culture, the initial percent of labelled cells was found to be 20 %, which, using eqn. (15), gives $S = 4.7$ h. When the proper distribution was evaluated, S was calculated to be 7.3 h.

From the measured values of the rate, duration and location within the mitotic cycle of the S period, the average DNA content of a cell in relation to its premitotic and postmitotic DNA content can be calculated. This calculation is presented in the APPENDIX, and yields the result that the average content is 65 % of the premitotic content.

ACKNOWLEDGEMENT

This work was aided in part by a grant from the National Cancer Institute of Canada.

APPENDIX

Distribution of cell numbers at different portions of the mitotic cycle

Asynchronous case: In an asynchronous, exponentially multiplying culture, the cell concentration $N(t)$ as a function of time t is given by:

$$N(t) = N_0 e^{at} \tag{1}$$

Where N_0 is the initial concentration and $a = 0.693/T$, T being the doubling time.

To locate the position of any given cell within the mitotic cycle of Fig. 2, a variable τ, where $0 < \tau < T$ is assigned, representing the time separating the cell from division. At $\tau = G_2 + S$, the cell begins DNA synthesis which lasts a time S; at $\tau = G_2$ the cell ceases DNA synthesis and enters the premitotic non-synthetic period of duration G_2; at $\tau = \tau m$, the cell enters mitosis and at $\tau = 0$ the cell divides. Thus, the G_2 period is defined as the period from completion of synthesis to division.

Let $n(t, \tau)$ be the number of cells per unit time at time t flowing through a point on the cycle which precedes division by a time τ. At $\tau = 0$, the number of cells dividing in a time interval t to $t + dt$ is $dN(t)$, the number of new cells appearing:

$$n(t, 0)\, dt = dN(t) \tag{2}$$

From eqn. (1), $$n(t, 0) = aN_0 e^{at} = aN(t) \tag{3}$$

Assuming a uniform rate of flow of cells around the cycle, a cell which is at position τ at a time t will have moved to $\tau = 0$ at time $t + \tau$, so that:

$$n\,(t + \tau,\, 0) = n(t,\, \tau) \qquad (4)$$

which yields, using eqns. (1) and (3),

$$n\,(t,\, \tau) = aN(t)\,e^{a\tau} \qquad (5)$$

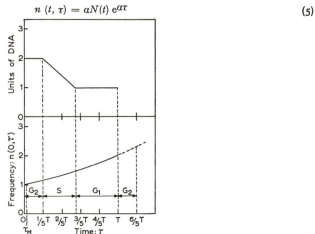

Fig. 11. Lower graph: the frequency distribution of an asynchronous culture giving the relative number of cells in each part of the mitotic cycle and showing the approximate position of the G_2, S and G_1 periods. Upper graph: The DNA content, in arbitrary units, of cells in the G_2, S and G_1 periods.

In Fig. 11, $n(t, \tau)$ is plotted for constant t. It is apparent that

$$N(t) = \int_{\tau = 0}^{T} n(t,\, \tau)\,\mathrm{d}\tau$$

which is the area between $\tau = 0$ and $\tau = T$ in Fig. 11.

Defining the mitotic index, $M(t) \times 100$, as the percent of cells in the interval $\tau = 0$ to $\tau = \tau_m$, we have:

$$M(t) = \frac{\displaystyle\int_0^{\tau_m} n(t,\, \tau)\,\mathrm{d}\tau}{\displaystyle\int_0^{T} n(t,\, \tau)\,\mathrm{d}\tau} \qquad (6)$$

$$= e^{a\,\tau_m} - 1$$

$$\simeq a\,\tau_m,\quad \tau_m \ll T \qquad (7)$$

The mitotic index is constant with time.

Partially synchronous case: To evaluate $n(t, 0)$ for the case where the growth curve is not smoothly exponential, one makes use of eqn. (6):

$$M(t) = \frac{\displaystyle\int_0^{\tau_m} n\,(t,\, \tau)\,\mathrm{d}\tau}{N(t)} \simeq \frac{n(t,\, 0)\,\tau_m}{N(t)}, \qquad \tau_m \ll T$$

so that $\quad n(t,0) \simeq \dfrac{1}{\tau_m}\,N(t)\,M(t) \qquad (8)$

Using eqn. (2), which still applies, and integrating, the result is obtained:

$$n(t,\, 0) = \frac{1}{\tau_m}\,N_0\,M(t)\,\exp \frac{1}{\tau_m}\int_0^t M(t)\mathrm{d}t \qquad (9)$$

Biochim. Biophys. Acta, 37 (1960) 406–419

From measured values of $M(t)$, $n(t, 0)$ may be calculated. To evaluate τ_m, the following boundary conditions which arise from the assumption in eqn. (4) are applied:

$$M(t + T) = M(t)$$

$$n(t + T, 0) = 2n(t, 0)$$

From eqn. (9):

$$\tau_m = \frac{1}{\ln 2} \int_0^T M(t)dt \qquad (10)$$

The percentage of cells taking up label

Asynchronous case: Referring to Fig. 11 and using eqn. (4), the percent of labelled cells $L(t) \times 100$ as a function of time t is given by:

$$L(t) = \frac{\int_{G_2}^{G_2 + S + t} n(t,0)dt}{N(t)} \quad , \text{ for } t < G_2$$

$$= \frac{\int_t^{G_2 + S + t} n(t,0)dt + \int_{T+G_2}^{T + t} n(t,0)dt}{N(t)} \quad , \text{ for } t > G_2 \qquad (11)$$

where label is added at time $t = 0$.

The second integral in eqn. (11) for $t > G_2$ arises from the fact that after time G_2 has elapsed, cells that have picked up label begin to divide, thus increasing the percentage of labelled cells. It is assumed in eqn. (11) that all cells have the same G_2 period. The theoretical treatment is improved by correcting for the division of labelled and unlabelled cells, taking into account the variation in G_2 times. The first labelled cells to arrive in division are spread out in time in the way depicted in Fig. 4 in the text. From the experimental data of Fig. 3 and the observed $L(t)$ values, the percentage of labelled cells which would have been observed had no cell division taken place, can be obtained. It has the form

$$L(t)_{corr} = \frac{1}{N_0} \left[L(t)_{exper} N(t) - \int_0^t P(t - \tau_m) aN(t)dt \right] \qquad (12)$$

Where $aN(t)dt$ is the increase in cell number over a time interval dt, and $P(t - \tau_m)$ is the fraction of labelled cells entering division, taken from Fig. 3. Allowance is made for the fact that metaphase precedes division by time τ_m. The first term in the bracket represents the total number of labelled cells that are seen, and the second term represents the number of these that are due to division of already labelled cells.

After a time G_1 has elapsed, a further correction would be required to allow for cells which have divided after addition of label and which have not been in the S period in the presence of label. There is little point in extending the treatment for times larger than G_1, however, since the spread in S and G_1 times would have to be accurately known or the uncertainty introduced in making such a correction would be large.

Thus for $t < G_1$, eqn. (11) becomes:

$$L(t)_{corr} = \frac{\int_{G_2}^{G_2 + S + t} n(t,0)dt}{N_0} \qquad (13)$$

Biochim. Biophys. Acta, 37 (1960) 406–419

Using eqn. (3), this gives:

$$L(t)_{\text{corr}} = e^{aG_2}\left[e^{a(s\,+\,t)} - 1\right], \qquad t < G_1 \tag{14}$$

From which one obtains:

$$S = \frac{1}{a}\ln\left(L(t)_{\text{corr}} + e^{aG_2}\right) - (G_2 + t) \tag{15}$$

For each corrected experimental $L(t)$ value, a value for S can be computed.

Partially synchronous case: Using the $n(t, 0)$ values obtained from eqn. (9), the experimental $L(t)$ values can be corrected for division by applying eqn. (12) substituting $n(t,0)$ for $aN(t)$. Eqn. (13) still applies, and using $n(t,0)$ again, values of $L(t)_{\text{corr}}$ may be calculated for various values of S, and the S chosen which gives best agreement between calculated and experimental values. The integration is done graphically.

Calculation of average DNA content per cell for an asynchronous exponential culture

Assuming a linear synthesis of DNA over the S period and referring to Fig. 11, the total DNA in the culture at time t is given by the sum of three terms: first, a term for the cells in the G_2 period which all have a DNA content of 2 units (2D), second, a term for the cells in the G_1 period which have a DNA content of 1 unit (1D) and third, a term for those in the S period which have a DNA content ranging linearly from 1 unit to 2 units. The average DNA/cell is thus given by:

$$\frac{\text{Total DNA}}{N(t)} = \frac{1}{N(t)}\left\{\int_0^{G_2} 2\,Dn(t,\tau)d\tau + \int_{G_2+S}^T Dn(t,\tau)d\tau\right.$$
$$\left. + \int_{G_2}^{G_2+S} D\left[2 + \frac{G_2}{S} - \frac{\tau}{S}\right]n(t,\tau)\,d\tau\right\} \tag{16}$$

The time (t) dependence naturally drops out of eqn. (16) so that the average DNA/cell is constant with time. If S is 7 h, G_2 4 h and T is 20 h, eqn. (16) gives the average DNA per cell to be 1.30 D or 0.65 the premitotic content.

REFERENCES

[1] H. H. SWIFT, *Physiol. Zoöl.*, 23 (1950) 169.
[2] P. M. B. WALKER AND H. B. YATES, *Proc. Roy. Soc. (London) B*, 140 (1952) 274.
[3] A. HOWARD AND S. R. PELC, *Heredity, Suppl.* 6, (1952) 261.
[4] L. G. LAJTHA, R. OLIVER AND F. ELLIS, *Brit. J. Cancer*, 8 (1954) 367.
[5] J. H. TAYLOR AND R. D. MCMASTER, *Chromosoma*, 6 (1954) 489.
[6] S. HORNSEY AND A. HOWARD, *Ann. N.Y. Acad. Sci.*, 63 (1956) 915.
[7] R. B. PAINTER AND R. M. DREW, *Lab. Invest.*, 8 (1959) 278.
[8] M. FRIEDKIN, D. TILSON AND D. ROBERTS, *J. Biol. Chem.*, 220 (1956) 627.
[9] J. H. TAYLOR, P. S. WOODS AND W. L. HUGHES, *Proc. Natl. Acad. Sci. U.S.*, 43 (1957) 122.
[10] K. K. SANFORD, W. R. EARLE AND G. D. LIKELY, *J. Natl. Cancer Inst.*, 9 (1948) 229.
[11] A. F. GRAHAM AND L. SIMINOVITCH, *Proc. Soc. Exptl. Biol. Med.*, 89 (1955) 326.
[12] R. B. L. GWATKIN, J. E. TILL, G. F. WHITMORE, L. SIMINOVITCH AND A. F. GRAHAM, *Proc. Natl. Acad. Sci. U.S.*, 43 (1957) 451.
[13] R. C. PARKER, L. N. CASTOR AND E. A. MCCULLOCH, in *Cellular Biology, Nucleic Acids and Viruses*, New York Academy of Sciences, New York, 1957, p. 303.
[14] R. B. PAINTER, R. M. DREW AND W. L. HUGHES, *Science*, 127 (1958) 1244.
[15] B. MESSIER AND C. P. LEBLOND, *Proc. Soc. Exptl. Biol. Med.*, 96 (1957) 7.
[16] E. A. TONNA AND E. P. CRONKITE, *Stain Technol.*, 33 (1958) 255.
[17] I. DONIACH AND S. R. PELC, *Brit. J. Radiol.*, 23 (1950) 184.
[18] E. O. POWELL, *J. Gen. Microbiol.*, 15 (1956) 492.

Biochim. Biophys. Acta, 37 (1960) 406–419

J. Mol. Biol. (1968) **32**, 327–341

On the Mechanism of DNA Replication in Mammalian Chromosomes

JOEL A. HUBERMAN AND ARTHUR D. RIGGS†

*Division of Biology, California Institute of Technology
Pasadena, California, U.S.A.*

(*Received 7 August 1967, and in revised form 2 November 1967*)

We have combined the techniques of pulse-labeling and DNA autoradiography to investigate the mechanism of DNA replication in the chromosomes of Chinese hamster and HeLa cells. Our results prove that the long fibers of which chromosomal DNA is composed are made up of many tandemly joined sections in each of which DNA is replicated at a fork-like growing point. In Chinese hamster cells most of these sections are probably less than 30 μ long, and the rate of DNA replication per growing point is 2·5 μ per minute or less.

In addition, we have taken advantage of the apparent slowness of equilibration with external thymidine of the internal thymidine triphosphate pool in Chinese hamster cells to determine the direction of DNA synthesis at the conclusion of the pulse in pulse-chase experiments. We have found, unexpectedly, that replication seems to proceed in opposite directions at adjacent growing points. Furthermore, adjacent diverging growing points appear to initiate replication at the same time.

1. Introduction

These studies were undertaken in the hope of increasing our understanding of the mechanism by which the large amount of DNA in the chromosomes of higher organisms is replicated. Previous autoradiographic experiments (Cairns, 1966; Huberman & Riggs, 1966; Sasaki & Norman, 1966) have shown that the DNA of mammalian chromosomes is arranged in the form of long fibers. Maximum fiber lengths of 500 μ (from HeLa cells) and 1800 μ (from Chinese hamster cells) have been reported by Cairns (1966) and Huberman & Riggs (1966), respectively, for DNA from cells lysed with detergent, while Sasaki & Norman (1966) have found DNA fibers more than 2 cm long from nuclei of human lymphocytes lysed without detergent.

In addition, Cairns (1966) has reported the results of pulse experiments which suggest that the long DNA fibers are composed of many separately replicated, tandemly joined sections. One of the aims of our studies was to verify this finding of Cairns. Another aim was to determine whether or not the DNA in each of the separate sections is replicated at a fork-like growing point similar to the growing point for *Escherichia coli* DNA (Cairns, 1963*a*).

Our results do, indeed, show that chromosomal DNA fibers are subdivided into separate sections which are replicated at fork-like growing points. In addition, our results also suggest the unexpected conclusion that DNA is replicated in opposite

† Present address: Salk Institute, La Jolla, Calif., U.S.A.

327

directions at adjacent growing points. That is, the direction of replication apparently alternates along the length of the long DNA fibers.

In the remainder of this paper we shall use the term replication section to refer to the stretch of DNA replicated by a single growing point. We shall also use the word autoradiogram to refer to any apparently continuous line of grains presumably caused by decay of tritium incorporated into DNA and the word fiber to refer to a single DNA molecule or a single series of DNA molecules joined end-to-end.

2. Materials and Methods

In general the procedures employed were similar to those we have described previously (Huberman & Riggs, 1966). Modifications are given in Figure and Plate legends.

Incubations with pronase were carried out as described previously (Huberman & Riggs, 1966), except that the DNA was dialyzed against a higher concentration of pronase (1 mg/ ml.), and that, after the Millipore filters (Millipore Corp.) were glued to glass slides, they were exposed to a formaldehyde-saturated atmosphere for 36 hr and then covered with a thin Parlodion film.

3. Results

(a) Cold pulse-labeling experiments

(i) Significance of tandem arrays

When Chinese hamster cells were briefly exposed to [³H]thymidine, then immediately lysed and the released DNA subjected to autoradiography, tandem arrays of DNA autoradiograms, similar to those reported by Cairns (1966), were sometimes found (Plate I). In order to prove that these tandem arrays were the result of separate replication sections in single DNA fibers and not the result of side-by-side aggregation of several separate DNA fibers, each containing a single replication section, we performed the following experiments.

We reasoned that if all the DNA fibers were completely labeled by growing the cells in high specific activity [³H]thymidine for 24 hours, then side-by-side aggregates could be distinguished by their higher grain density; we also reasoned that if, during the 24-hour period, we exposed the cells to [³H]thymidine of a lower specific activity for a short interval, we could distinguish the autoradiograms of DNA replicated during that short interval by their lower grain density.

If each DNA fiber contained just one replication section, then after a cold pulse experiment of the type described here, one would expect to find no more than one low grain density (cold) region in any long DNA autoradiogram. On the other hand, if each DNA fiber could contain several replication sections, then after a cold pulse experiment one would expect to find some single long DNA autoradiograms containing several cold regions.

The actual labeling sequences used in these experiments are summarized in Fig. 1. When the stripping films were exposed for sufficient time to bring out the low specific activity regions (three months or more), long DNA autoradiograms containing several regions of low grain density were frequently found for cold-pulsed DNA but not for control DNA (Plate II). Thus Cairns' (1966) hypothesis that the long DNA fibers are composed of many tandemly joined replication sections is proved.

(ii) Rate of DNA replication during DNA synthesis

Further information can be obtained from these experiments. The average generation time of the Chinese hamster cells under our growth conditions is about 17 hours.

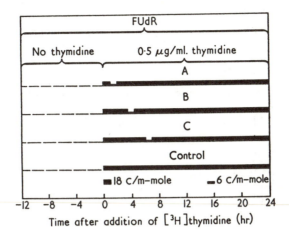

FIG. 1. Labeling schedule for cold pulse experiments.

Chinese hamster cells were grown as described in the legend to Plate I. After a 12-hr pretreatment with FUdR (0·1 μg/ml.), [³H]thymidine (18 c/m-mole; 0·5 μg/ml.) was added to 4 separate Petri plates (A, B, C and control). At various times after the initial addition of [³H]thymidine, as indicated in the Figure, the medium containing [³H]thymidine at 18 c/m-mole was removed from the plates and replaced, for 1 hr, by medium containing [³H]thymidine (0·5 μg/ml.) at 6 c/m-mole. After 24 hr the cells were harvested by trypsinization and diluted to 400 cells/ml. in isotonic saline. Lysis and autoradiography were performed as in the legend to Plate I.

DNA synthesis requires, on the average, about six hours (Taylor, 1960; Hsu, Dewey & Humphrey, 1962). Thus we can calculate that the 12-hour pretreatment with FUdR† (which inhibits thymidine monophosphate biosynthesis) must have blocked nearly two-thirds of the cells at the beginning of DNA synthesis and the remaining cells somewhere in the DNA synthesis period (see Hsu, 1964, for the use of FUdR in synchronizing Chinses hamster cell cultures). Consequently during the cold pulse of Experiment A (Fig. 1), which occurred between one and two hours after the relief of the FUdR block by the addition of [³H]thymidine, the majority of cells were near the beginning of their DNA synthesis period. Similarly, during the cold pulse of experiment B, most cells were in the middle of DNA synthesis, while during the cold pulse of experiment C most cells were either at the end of DNA synthesis or had completed it. If the assumption is made that the size of the regions of low grain density is a measure of the rate of DNA replication per growing point, then our results can provide an answer to the question: Does the rate of DNA replication change during the period of DNA synthesis?

The length distributions of low grain density regions for experiments A, B and C are shown in Fig. 2. Low grain density regions less than 20 μ long were not easily resolved. However, for the low grain density regions of resolvable size there is no large difference among the three distributions. Thus there is apparently no large change in the rate of DNA replication per growing point during DNA synthesis.

(iii) *Distribution of replicating DNA*

In cold pulse experiments A and B, approximately 12% of the total length of autoradiograms was of low grain density; in experiment C, regions of low grain density

† Abbreviations used: FUdR, 5-fluorouracil deoxyriboside; BUdR, 5-bromouracil deoxyriboside.

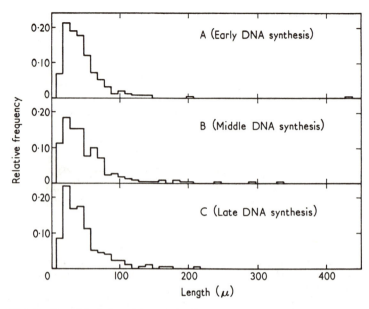

FIG. 2. Distribution of lengths of low grain density regions in cold pulse autoradiograms. Chinese hamster cells were labeled as described in Fig. 1 (A, B and C) and their DNA subjected to autoradiography. The autoradiograms were examined by dark-field microscopy at a magnification of 100. The contours of low grain density regions were traced with a camera lucida, and the lengths of the tracings were then measured and corrected for magnification.

contributed about 7% of the total length. However, in experiments A and B about two-thirds of the autoradiograms contained no regions of low grain density at all; in experiment C four-fifths of the autoradiograms contained no regions of low grain density.

Thus, during any one-hour period, regions of DNA synthesis are not distributed equidistantly along the DNA fibers in chromosomes. The non-uniform distribution of regions of DNA synthesis is apparent in Plate II.

(b) *Effect of pronase on autoradiogram lengths*

We have previously found (Huberman & Riggs, 1966) that low concentrations of the proteolytic enzyme, pronase, acting on the long DNA fibers in solution prior to autoradiography have no effect on autoradiogram lengths. In addition, Macgregor & Callan (1962) have shown that neither of the proteolytic enzymes trypsin nor pepsin is able to break amphibian lampbrush chromosomes when employed at a concentration of 250 μg/ml. for up to two hours at room temperature.

Nevertheless, we felt that the finding that each long DNA fiber may contain many separate replication sections made imperative an investigation with higher concentrations of pronase. In our earlier experiments we were unable to use higher pronase concentrations because some residual pronase, adsorbed to the Millipore filters, digested the stripping film. Therefore, in order to prevent this problem, we inactivated the residual pronase with formaldehyde vapors and further protected the stripping film with a thin Parlodion film placed between it and the Millipore filter. We now report that incubation of the long DNA fibers (from cells exposed to [³H]thymidine for 35 hours) with about 100 μg/ml. of pronase for up to six hours at 34°C has no effect on either the maximum length or general length distribution of the autoradiograms. Thus any linkers connecting the separate replication sections, or connecting other points in the long DNA fibers, must be resistant to pronase. Large protein linkers are therefore extremely unlikely. In this connection it is interesting that, according to Davern (1966), the *E. coli* chromosome also contains no pronase-sensitive linkers.

(c) *Thirty-minute pulse-labeling experiments*

(i) *Does chromosomal DNA replicate at a fork?*

For complete autoradiographic visualization of a replication fork like that of *E. coli* (Cairns, 1963a), it is necessary that all three branches of the fork be labeled. This requires that cells complete more than one generation while incorporating [³H]-thymidine. We have found, however, that extensive incorporation of [³H]thymidine at specific activities adequate for DNA autoradiography completely prevents division of Chinese hamster cells. Consequently we have been forced to use a less direct method in our attempt to demonstrate the presence or absence of replication forks in Chinese hamster chromosomal DNA.

An appropriate method was suggested by an experiment of Cairns (1963a). He compared the DNA autoradiograms from *E. coli* cells exposed to [³H]thymidine for a short time and then immediately lysed (simple pulse) with the DNA autoradiograms from *E. coli* cells exposed first to [³H]thymidine and then to non-radioactive thymidine and then lysed (pulse–chase). In the simple pulse experiment, the autoradiograms seemed to be the result of *two* DNA fibers lying side-by-side; in contrast, the autoradiograms in the pulse–chase experiment appeared to have been produced by *single* DNA fibers. Cairns suggested that the different appearance of the two types of autoradiograms was the consequence of replication at a fork-like growing point; the labeled regions of DNA in the simple pulse experiment would be held together at the replication fork and might well appear side-by-side in autoradiograms, while the labeled regions of DNA in the pulse–chase experiment would be separated from the replication fork by a considerable length of unlabeled DNA and would be much more likely to appear separated in autoradiograms. These possibilities are shown diagrammatically in Fig. 3.

We reasoned that fork-type replication in Chinese hamster cells, if present, should produce the same phenomenon. To test this possibility we performed an analogous experiment with Chinese hamster cells. We used a pulse time of 30 minutes and a chase time of 45 minutes. In the case of the simple pulse, all the media to which the cells were exposed after removal from the pulse medium contained FUdR to prevent cellular synthesis of unlabeled thymidine monophosphate.

In one respect our results were different from those Cairns (1963a) obtained with *E. coli*. Very few of the autoradiograms obtained after simple pulse labeling had an *appearance* suggesting that they might be the result of two DNA fibers lying side-by-side. That is, most such autoradiograms were just single lines of grains with no evidence of separation into two lines of grains. In those few cases where some separation occurred, the separation appeared to be in the middle of the labeled regions rather than at their ends (see below). It is true, however, that many of the autoradiograms obtained after simple pulse labeling had a *grain density* corresponding to that expected for two DNA double helices lying side-by-side if each double helix were labeled in a

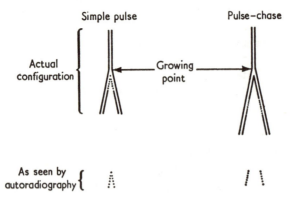

FIG. 3. The effect of replication at a fork-like growing point on the autoradiograms seen after simple pulse and pulse–chase experiments.

single polynucleotide chain. In Fig. 4 are shown histograms of grain-density frequency for both the simple pulse autoradiograms and the pulse–chase autoradiograms. Since a grain density of about 2·5 grains per μ is just slightly less (after correction for specific activity, exposure time, and base composition) than the grain density obtained by Cairns (1963b) for a double helix of *E. coli* DNA labeled in a single polynucleotide chain, each autoradiogram having approximately this grain density was probably produced by a double helix containing a single labeled chain. Likewise, autoradiograms having a grain density of about five grains per μ were probably produced by two double helices lying side-by-side and each containing a single labeled chain. The broadness of the distributions can be attributed partly to the statistical error involved in measuring grain density in the shorter autoradiograms and partly to actual errors in grain counting. The higher grain densities may have been underestimated due to the difficulty in distinguishing grains in crowded areas. It is clear, however, that half or more of the simple pulse autoradiograms were probably produced by two labeled chains (the other simple pulse autoradiograms could be explained as the result of breakage). This fact suggests that the two daughter chains are synthesized at about the same time and are held together for a short time after they are synthesized. It is also clear that most of the pulse–chase autoradiograms were produced by single labeled DNA chains (a few pulse–chase autoradiograms produced by two labeled chains might be expected as a result of incomplete chain separation). Thus, although the two daughter chains are held together for a short time after they are synthesized, they later become free to separate from each other. All these properties are consistent with replication at fork-like growing points. Further evidence for the fork-like nature of the growing points will be discussed below.

(ii) *Grain density gradients*

Comparison of typical simple pulse autoradiograms (Plate I) with typical pulse–chase autoradiograms (Plate III) shows another difference besides those mentioned above. Whereas the ends of the simple pulse autoradiograms are always distinct, the ends of the pulse–chase autoradiograms are frequently indistinct because of a gradual decline in grain density from the full grain density expected for one or two labeled chains of DNA in the interior of the autoradiograms to undetectable grain density at the ends.

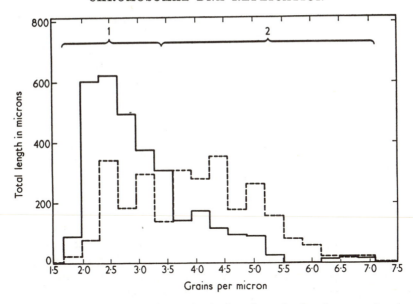

FIG. 4. Frequencies of grain densities in simple pulse and pulse–chase experiments.

Chinese hamster cells were labeled as described in the legend to Plate I (simple pulse) or Plate III (pulse–chase) and their DNA subjected to autoradiography. Exposure time was 4 months. Grain densities were measured with bright-field microscopy at a magnification of 1000 in all the auto- radiograms visible over large areas of single Millipore filters. Measurements were made by counting all the grains of each autoradiogram and dividing by the length of the autoradiogram (determined with an eyepiece micrometer). The total length of all autoradiograms in the sample (more than 100 autoradiograms for both distributions shown) having a given grain density is plotted in the Figure as a function of grain density. ————, Pulse–chase; ––––––, simple pulse. The brackets in the Figure indicate the grain densities presumed to correspond to 1 or 2 labeled polynucleotide chains.

We interpret this reduction in grain density to be the result of gradual change in the specific activity of the intracellular thymidine triphosphate pool upon replacement of the [³H]thymidine in the external medium with non-radioactive thymidine. Accord- ingly, for the grain-counting experiments above, we counted grains only in internal regions of the pulse–chase autoradiograms where no decline in grain density was evident.

If our interpretation of the grain density gradients is correct, then the direction of decline in grain density is the direction in which DNA was being synthesized at the beginning of the cold chase. Therefore we were surprised to find that many of the autoradiograms had grain densities declining at *both* ends, in *opposite* directions (Plate III). In fact, when tandem arrays of pulse–chase autoradiograms were exam- ined, about 90% of the internal (and therefore unbroken) autoradiograms proved to have declining grain densities at both ends (Plate IV). Thus all these autoradiograms must be the result of at least two growing points proceeding in opposite directions, and most of the pulse–chase autoradiograms which lack gradients at one or both ends (Plate III) must have been produced by broken DNA fibers.

The apparent exceptions to the rule of double gradients in tandem arrays are of two types. The first type is rare (about 3% of the internal autoradiograms) and of uncertain significance. Examples of this type are shown in Plate V(a), (b) and (c). The central structure in Plate V(a) may be an internal autoradiogram without grain density gradients at either end. In Plate V(b) and (c) are examples of internal autoradiograms apparently lacking gradients at *one* end.

The second kind of apparent exception is more frequent (about 7%). Examples are shown in Plates VI(c) and VII(a). Here the distinct autoradiogram ends are paired opposite each other in adjacent autoradiograms, and there are no grains between the paired ends. This configuration suggests that the unlabeled areas between the paired ends are the result of DNA replication completed *before* treatment with FUdR. Since the intracellular thymidine triphosphate pool would be depleted in the presence of FUdR, no time would be required for pool equilibration after introduction of [³H]thymidine and no grain density gradients would be seen in the resulting auto-radiograms at the points of resumption of DNA synthesis.

(iii) *Sister double helices*

Notice that all the structures in Plate VI are examples of partial separation of sister double helices. They can be identified as such by the correspondence of grain density patterns in the parallel autoradiograms. This correspondence suggests that corresponding regions of the two labeled polynucleotide chains were synthesized at identical times. Certainly it means that within stretches of DNA longer than the autoradiographic limit of resolution (about $5\,\mu$) DNA was not synthesized first on one parental chain as template and then on the other; rather both parental chains must have acted as template at once. Fork-like growing points provide by far the simplest explanation of this behavior. In fact, the fork-like points of attachment of the sister double helices in Plate VI may be growing points.

Another example of sister double helix separation is shown in Plate V(d). The grain density in the heavily-labeled regions on the top is much less than that expected for a single chain labeled during the pulse (shown in the heavily labeled regions on the bottom). Thus the heavily-labeled regions on the top may represent points where DNA synthesis was initiated after the pulse, during the period of pool equilibration. If this is true then Plate V(d) offers an example of neighboring replication sections beginning replication at different times.

(iv) *The size of replication sections*

We have used the tandem arrays of autoradiograms resulting from the 30-minute pulse and pulse–chase treatments to obtain an estimate of the size of the replication sections in Chinese hamster chromosomal DNA. Separate estimates were made for the simple pulse autoradiograms and for the pulse–chase autoradiograms. To avoid errors resulting from possible breaks in DNA fibers at the ends of tandem arrays, only *internal* autoradiograms of the tandem arrays were used for measurement unless, in the case of pulse–chase autoradiograms, a grain density gradient was present at the end of the array. Due to the fact that most internal autoradiograms in these experiments are the result of at least two growing points (see above), direct measurement of the size of individual replication sections was not possible. Instead, the center-to-center distances between internal or unbroken external autoradiograms of tandem arrays were measured (see Fig. 5). These distances are assumed to correspond to the sum of two adjacent replication sections. Further rationale for this procedure is given in the Discussion section.

Histograms showing the frequency distributions of center-to-center distances are shown in Fig. 6. Note that measurements of simple pulse autoradiograms and of

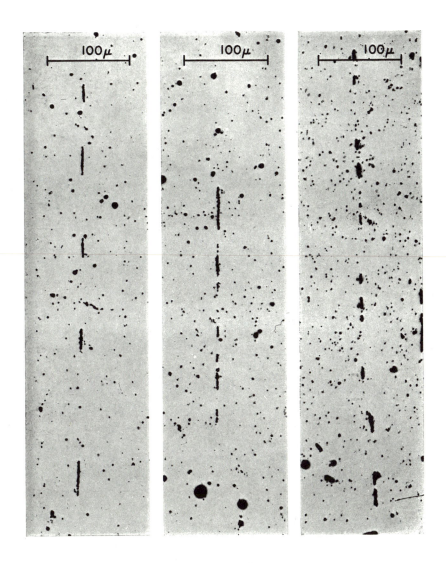

PLATE I. Tandem arrays of autoradiograms.

Cells of Chinese hamster fibroblast strain B14FAF28 (a gift from Dr T. C. Hsu) were grown as monolayer cultures on plastic Petri dishes in Eagle's medium supplemented with 10% calf serum. After a 12-hr pretreatment with FUdR (0·1 μg/ml.), [³H]thymidine (18 c/m-mole, Nuclear Chicago) was added to 0·5μg/ml. 30 min later the cells were harvested by trypsinization and diluted to 1 × 10⁴ cells/ml. in isotonic saline. The solutions used for trypsinization and dilution contained FUdR (0·1 μg/ml.). The cells were diluted tenfold into lysis medium (1·0 M-sucrose–0·05 M-NaCl–0·01 M-EDTA, pH 8·0), lysed by dialysis against 1% sodium dodecyl sulfate in lysis medium, then dialyzed further against dialysis medium (0·05 M-NaCl–0·005 M-EDTA, pH 8·0). The released DNA was trapped on Millipore VM filters which had served as dialysis membranes. It was then subjected to autoradiography.

Exposure time with Kodak AR10 autoradiographic stripping film (Eastman Kodak Co.) was 4 months. Picture taken by dark-field microscopy.

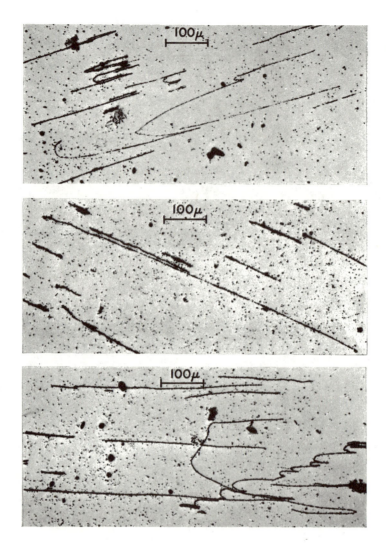

PLATE II. Cold pulse autoradiograms.

Chinese hamster cells were labeled as described in Fig. 1 (A, B and control) and their DNA was subjected to autoradiography. Exposure time was 6 months. Dark field.

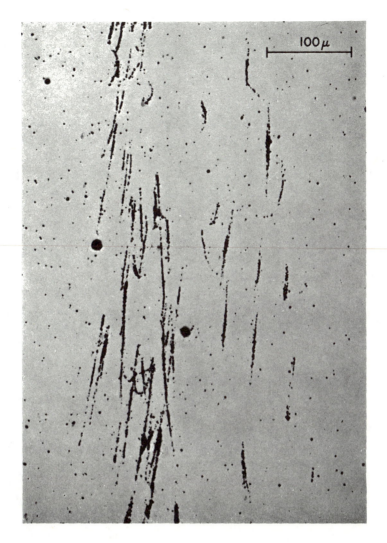

PLATE III. Typical autoradiograms produced in a pulse–chase experiment.

Chinese hamster cells were labeled as described in the legend to Plate I, except that after 30 min of incubation with [³H]thymidine and FUdR, the medium containing [³H]thymidine and FUdR was removed and replaced with medium containing non-radioactive thymidine (5 µg/ml.) and no FUdR. Incubation was continued for 45 min. The cells were then harvested by trypsinization and diluted to 1×10^4 cells/ml. in isotonic saline. Lysis and autoradiography were performed as in the legend to Plate I. Exposure time was 4 months. Dark field.

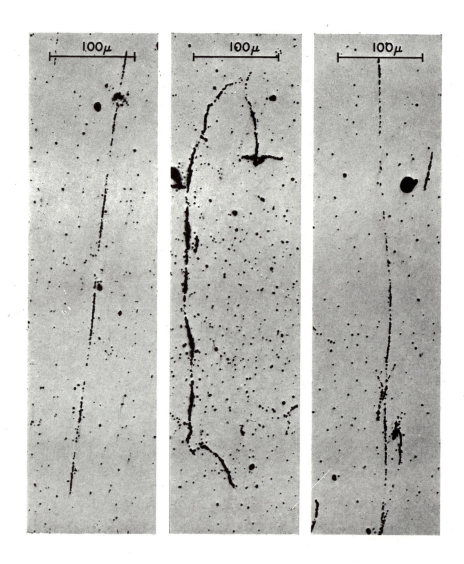

PLATE IV. Tandem arrays of autoradiograms produced in a pulse–chase experiment. Labeling and autoradiography were performed as in the legend to Plate III. Dark field.

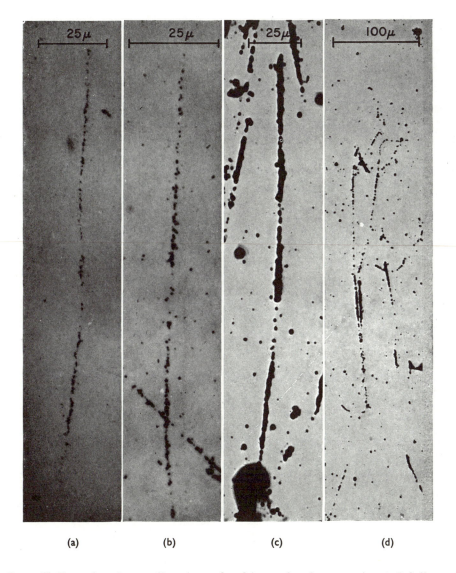

PLATE V. Examples of autoradiograms produced in a pulse–chase experiment. Labeling and autoradiography were performed as in the legend to Plate III.

(a) Example of autoradiogram apparently without grain density gradients at either end. Picture taken by bright-field microscopy.

(b) Example of autoradiogram apparently lacking a grain density gradient at one end. Bright field.

(c) Example of autoradiogram apparently lacking a grain density gradient at one end. Dark field.

(d) Example of apparent asynchronous initiation of DNA replication in neighboring replication sections. Note that this is also an example of sister double helices. Dark field.

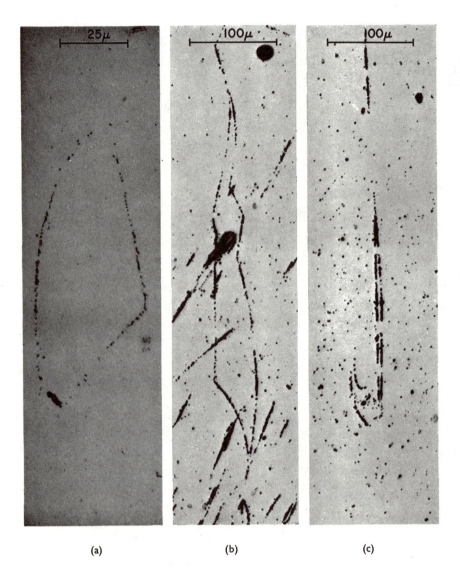

(a) (b) (c)

PLATE VI. Examples of sister double helices. Labeling and autoradiography were performed as in the legend to Plate III.

(a) Bright field.

(b) and (c) Dark field.

400

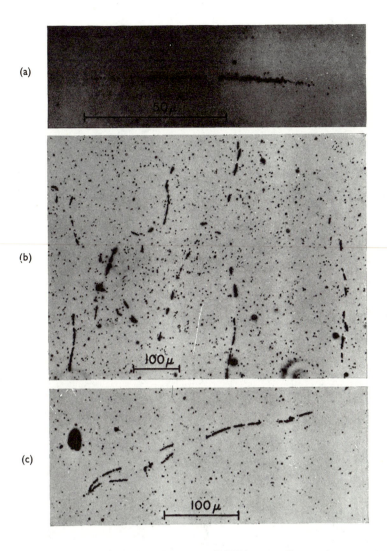

PLATE VII. (a) Example of Chinese hamster cell DNA autoradiogram interrupted by a region where replication presumably took place before addition of FUdR. Conditions were identical to those of Plate III. Bright field.

(b) Example of small, closely spaced autoradiograms of HeLa cell DNA. HeLa S3 cells were grown under conditions identical to those for Chinese hamster cells (legend to Plate I). After a 15-hr pretreatment with FUdR (0·1 μg/ml.), [³H]thymidine (18 c/m-mole) was added to 2 μg/ml. 1 hr later the medium containing [³H]thymidine and FUdR was removed and replaced with medium containing non-radioactive thymidine (5 μg/ml.) and no FUdR. Incubation was continued for 45 min. The cells were then harvested by trypsinization and diluted to 1×10^4 cells/ml. in isotonic saline. Lysis and autoradiography were performed as in the legend to Plate I. Exposure time was 3 months. Dark field.

(c) Example of sister double helices in HeLa cell DNA. Conditions were identical to those of Plate VII(b) except that the cells were exposed to [³H]thymidine for 2 hr. Dark field.

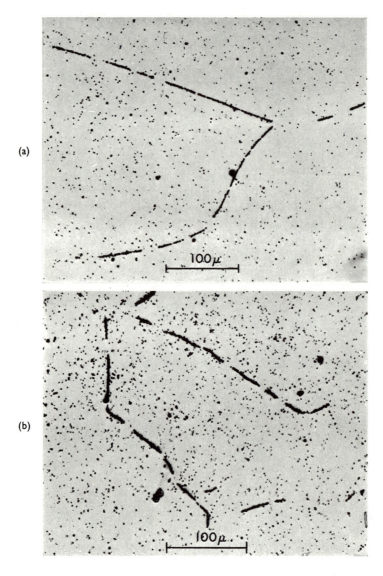

PLATE VIII. Examples of sister double helices in HeLa cell DNA.

(a) Conditions identical to those of Plate VII(c). Dark field.

(b) Conditions identical to those of Plate VII(b) except that there was no pretreatment with FUdR, and FUdR was not present during the pulse. Exposure time was 7·5 months. Dark field.

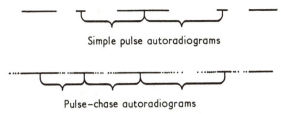

Simple pulse autoradiograms

Pulse–chase autoradiograms

FIG. 5. Method of measurement of the size of pairs of replication sections from tandem arrays of autoradiograms.

The horizontal lines indicate representative autoradiograms. Grain density gradients at the ends of pulse–chase autoradiograms are indicated by dots. The brackets show the lengths which were measured (center-to-center distances between internal or unbroken external autoradiograms).

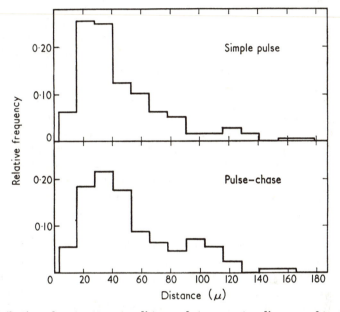

FIG. 6. Distribution of center-to-center distances between autoradiograms of tandem arrays.

Chinese hamster cells were labeled either by the simple pulse method (described in the legend to Plate I) or the pulse–chase method (described in the legend to Plate III), and their DNA subjected to autoradiography. Center-to-center distances between autoradiograms of tandem arrays were measured according to the criteria of Fig. 5. Measurements were made using an eye-piece micrometer and bright-field microscopy at a magnification of 1000.

pulse–chase autoradiograms gave similar results. The most frequently observed distances are between 15 and 60 μ, which would correspond to replication section sizes of about 7 to 30 μ.

The accuracy of these measurements is not certain. The frequency of distances less than 15 μ may have been underestimated due to difficulties in resolution. In addition, the measurements were based on the assumption that each pair of adjacent diverging replication sections in the DNA fibers forming the tandem arrays would be represented by a separate autoradiogram. The possibility that adjacent pairs might have completed replication during the pulse, so that their autoradiograms would be fused, and the possibility that some pairs might not have begun to replicate during the pulse would thus result in an exaggerated estimate of their size.

Since all these uncertainties would cause an over-estimate of pair size, one can conclude that most pairs of adjacent replication sections are at least as small as indicated by the histograms of Fig. 6.

(d) *Estimation of replication rate*

(i) *Estimates from the 30-minute pulse–chase experiment*

We shall use the term replication rate to refer to the rate at which DNA is synthesized at an individual growing point. This rate will be measured in units of μ of movement of the growing point along the parental DNA double helix per minute (μ/min).

The possibility of autoradiogram fusion discussed above and uncertainties in the time of stopping and starting of replication in individual replication units make estimates of replication rate based on the lengths of autoradiograms produced after a 30-minute pulse extremely uncertain. However, the pulse–chase autoradiograms have provided two alternative methods for estimation of replication rate. First, in most cases the grain densities at the ends of the pulse–chase autoradiograms appear to decrease monotonically, suggesting that, within the gradient region, replication was proceeding at a single growing point throughout the equilibration period. Monotonic gradients 20 to 40μ long were frequently found (Plate III), while monotonic gradients as much as $100\,\mu$ long were occasionally found (Plate IV(c)). Since the *maximum* possible equilibration time is 45 minutes (chase time), a *minimum* estimate of replication rate is 0·4 to 0·9 μ per minute for the frequently observed gradients and up to 2 μ per minute for the occasionally observed gradients. The variations found in lengths of monotonic gradients can be explained as the result of heterogeneity between cells in pool equilibration time or as the result of heterogeneity of replication rate or both.

In addition, if our interpretation of the sharp interruptions in the autoradiograms shown in Plates VI(c) and VII(a) is correct, then the ends without grain density gradients represent points where replication started at the beginning of the 30-minute pulse. Likewise the positions where the grain density begins to decline in such autoradiograms represent the points to which replication had proceeded by the end of the pulse. We conclude that each branch of such autoradiograms represents a region synthesized during the entire pulse. We have found samples of such autoradiograms with branches from 15 to 37 μ long. On the assumption that each branch is the result of a single growing point, these lengths correspond to replication rates of 0·5 to 1·2 μ per minute.

(ii) *The maximum possible replication rate*

To measure the maximum possible rate of DNA replication, Chinese hamster cells were exposed to [³H]thymidine for pulse times of 5, 10, 20, 60 and 120 minutes and their DNA subjected to autoradiography. The maximum autoradiogram lengths found after these pulse times increased more or less linearly with pulse time at a rate of 5 to 10 μ per minute.

These measurements set an upper limit to the replication rate of Chinese hamster DNA. Assuming that each autoradiogram is the product of only two growing points, then the maximum possible rate of replication is 2·5 to 5 μ per minute. Although the maximum autoradiogram lengths found after the longer pulse times are probably the result of autoradiogram fusion, the fact that autoradiograms 25 μ long are found even after the shortest pulse time of five minutes suggests that Chinese hamster cell DNA may, indeed, occasionally replicate as rapidly as 2·5 μ per minute.

(e) *Experiments with HeLa cells*

In order to compare DNA replication in HeLa cells with that in Chinese hamster cells, we conducted a series of experiments in which we exposed HeLa cells and Chinese hamster cells to [^{3}H]thymidine for times of 1, 2, 4, 8 and 18 hours. For all these experiments the pulse with [^{3}H]thymidine was followed by a 45-minute chase with non-radioactive thymidine. Because we also wanted to investigate the possible effect of FUdR on DNA replication in these cells, some of the experiments were carried out in the absence of FUdR.

In general, the HeLa cell autoradiograms and the Chinese hamster cell autoradiograms were similar in size and appearance. This was especially true for pulse times of four hours or longer. For pulse times of one or two hours, however, some differences were evident. First, although the HeLa cells were subjected to the same kind of pulse–chase treatment as the Chinese hamster cells, grain density gradients were difficult to observe in the HeLa cell autoradiograms. What might have been short grain density gradients were found only occasionally. We should point out that, even for the Chinese hamster cells, grain density gradients were not nearly so frequent in these longer experiments as they were in the 30-minute pulse–chase experiments. This low frequency of gradients has not been investigated further.

Also, although the absence or presence of FUdR had no detectable effect on the size or type of autoradiograms produced by Chinese hamster DNA, the DNA from HeLa cells exposed to FUdR for 15 hours and then pulse-labeled for one hour produced a much greater proportion of very short autoradiograms, usually close together in tandem arrays, than did DNA from Chinese hamster cells or from HeLa cells not exposed to FUdR. An example of these short HeLa cell autoradiograms is shown in Plate VII(b).

The fact that FUdR did not alter the appearance of Chinese hamster cell autoradiograms suggests that it has no effect on DNA autoradiography in Chinese hamster cells other than the intended ones of preventing thymidine monophosphate biosynthesis and blocking most cells at the beginning of DNA synthesis. If FUdR has only these effects in HeLa cells, then our finding that DNA from HeLa cells treated with FUdR for 15 hours and then pulse-labeled for one hour produces a high proportion of small, close autoradiograms suggests that the DNA which replicates earliest in HeLa cells may be unusually rich in small replication sections. However, the possibilities that the thymidine starvation produced by FUdR may induce replication at an abnormally large number of growing points (Pritchard & Lark, 1964) in HeLa cells, or may allow accumulation of damage which would result in repair replication (Taylor, Haut & Tung, 1962) cannot be excluded.

Like the Chinese hamster DNA, the HeLa DNA produced autoradiograms suggesting the partial separation of sister double helices. Three examples are shown in Plates VII(c) and VIII.

4. Discussion

(a) *Summary of results*

The autoradiographic evidence we have presented above suggests that the process of DNA replication in Chinese hamster chromosomes has the following attributes.

(1) The long fibers of chromosomal DNA are made up of many tandemly joined replication sections, as proposed by Cairns (1966). (See Plates II, V(d) and VI.)

(2) If any linkers connect the replication sections or other points in the long DNA fibers, they must be resistant to pronase.

(3) DNA is replicated at fork-like growing points (Plates V(d) and VI; Fig. 4).

(4) Usually, perhaps always, DNA is synthesized in opposite directions at adjacent growing points with the result that, in our experiments, most unbroken auto-radiograms were the result of two or more growing points (Plate IV).

(5) Most replication sections are less than $30\,\mu$ long. Many are so short that they are difficult to resolve by autoradiography.

(6) Neighboring replication sections can begin replication at different times (Plate V(d)).

(7) At any one time, regions of DNA synthesis are not distributed equidistantly along the chromosomal DNA fibers (Plate II).

(8) The rate of DNA replication per growing point is $2\cdot5\,\mu$ per minute or less. This rate does not vary greatly during the period of DNA synthesis (Fig. 2).

(b) Bidirectional replication

(i) Arrangement of replication sections

The fact that DNA is usually synthesized in opposite directions at adjacent growing points implies that all or nearly all replication sections must be arranged as shown in Fig. 7(a). In this diagram, each replication section is bounded by an origin (marked O) and a terminus (marked T). Within each section, DNA replication would proceed from the origin to the terminus. Each replication section would share its origin with one adjacent section and share its terminus with the other adjacent section.

(ii) The unit of control

The fact that adjacent replication sections share origins or termini suggests the possibility that initiation of replication may be controlled at the level of adjacent *pairs* of sections rather than at the level of individual sections. In fact, at least three models for control of the initiation of DNA replication can be imagined.

(1) Control at the level of the replication section. In this case, each section might initiate replication independently of all its neighbors.

(2) Control at the level of converging pairs of replication sections. In this case adjacent sections sharing a terminus would initiate replication together.

(3) Control at the level of diverging pairs of replication sections. In this case, adjacent sections sharing an origin would initiate replication together.

Each of these models makes certain predictions for the types of internal autoradiograms which might be found in tandem arrays after a pulse–chase experiment, but only the third model predicts that all internal autoradiograms either should have grain density gradients at both ends (Plate IV) or should be arranged in pairs with adjacent distinct ends (Plates VI(c) and VII(a)). With rare exceptions (Plate V(a), (b) and (c)), these are the only kinds of internal autoradiograms found. Even the exceptions can be accounted for by the third model if certain assumptions are made about the nature of termini (see below). Furthermore, structures like those in Plate V(d) where neighboring *diverging* pairs of replication sections appear to have started replication at different times are definitely contrary to the predictions of the second model.

The available evidence thus favors the third model above. We propose the term replication unit to mean the basic unit of control in the initiation of replication— presumably an adjacent pair of diverging replication sections. Note that our measurements of center-to-center distances between internal autoradiograms (Fig. 6) are of the order of size of such replication units.

(iii) *The nature of origins and termini*

It is now well established that, at the level of resolution of whole chromosome autoradiography, regions of initiation and termination of DNA replication are reproducibly and heritably controlled (see Huberman, 1967, for review). However, neither whole chromosome autoradiographic studies nor the studies presented in this paper have been able to determine whether sites of initiation (origins) are reproducible at the atomic level. The possibility remains that sites of initiation may occur anywhere within relatively long stretches of the DNA fibers.

Likewise, there are at least two ways in which DNA replication could be terminated at the ends of replication sections. One possibility is that there are specific sites along the DNA fibers where replication, from either direction, is stopped. The other possibility is that replication continues until adjacent converging growing points meet. According to the first possibility one might expect to find, as a result of termination of replication during the pulse in a pulse–chase experiment, occasional internal autoradiograms lacking grain density gradients at one or both ends. Autoradiograms of the type shown in Plate V(a), (b) and (c) may be examples of such termination. However, these structures are so rare (about 3% of the internal autoradiograms) that the existence of specific termini must remain in doubt.

(iv) *The bidirectional model*

Despite the uncertainty noted above about the reproducibility of origins and termini, the fact remains that, in any given replication, each growing point must have both a starting site and a stopping site. Even if one attributes no more to origins and termini than this, one can, by using the results presented in this paper, obtain a better model of the way in which chromosomal DNA fibers are replicated than was previously possible. This model, which we shall call the bidirectional model, is shown in the diagram in Fig. 7; its basic feature is that the replication unit consists of two diverging replication sections. Note that we have assumed in presenting the model that termini are sites where replication stops rather than sites where converging growing points meet. This assumption must be considered tentative. Otherwise the various features of the model seem well supported by the experimental evidence.

(c) *The rate of DNA replication*

(i) *Comparison of results*

The rate of DNA replication in the chromosomes of mammalian cells has previously been measured in two ways. Cairns (1966 and personal communication) measured the lengths of the autoradiograms produced after exposure of HeLa cells to [^{3}H]-thymidine for 45 minutes and 180 minutes, and concluded that the average length increased at a rate of $0\cdot5\,\mu$ per minute or less. Taylor (personal communication) measured the BUdR pulse time required before Chinese hamster DNA molecules were fully converted to hybrid density, and estimated from his results that the BUdR-labeled region increased in length at a rate of one to two μ per minute. According to

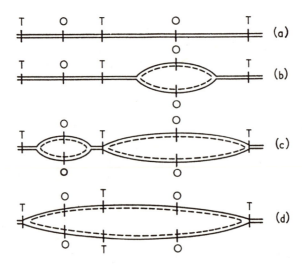

FIG. 7. Summary of the bidirectional model for DNA replication.

Each pair of horizontal lines represents a section of a double helical DNA molecule containing two polynucleotide chains (————, parental chain; ------, newly synthesized chain). The short vertical lines represent positions of origins (O) and termini (T). The diagrams represent different stages in the replication of two adjacent replication units:

(a) Prior to replication; (b) replication started in right-hand unit; (c) replication started in left-hand unit and completed at termini of right-hand unit; (d) replication completed in both units; sister double helices separated at the common terminus.

the bidirectional model, these measurements should be considered over-estimates of the actual rate of DNA replication per growing point, because some of the measured lengths, if not most, probably were the result of *two* growing points.

These measurements are in fairly good agreement with the results we have obtained by a variety of methods. Measurements of the maximum autoradiogram lengths obtained after short pulses suggest that the rate of DNA replication must be less than about 2·5 μ per minute. Measurements of the length of DNA synthesized during equilibration of the thymidine triphosphate pool suggest rates of replication from 0·4 to 2 μ per minute.

Probably the most reliable estimates are provided by measurements of the lengths of the branches in autoradiograms such as those in Plates VI(c) and VII(a) where it is likely that DNA synthesis was going on throughout the 30-minutes pulse. Such measurements suggest replication rates from 0·5 to 1·2 μ per minute.

The uncertainties involved in all these measurements mean that a precise estimate of the replication rate in Chinese hamster cells is still impossible. However, one can draw some general conclusions from the measurements made so far.

All the measurements suggest that there is probably no one replication rate but rather a range of replication rates. Whether or not the replication rate in a single replication unit is always constant is not known.

Also, it is interesting that, despite the fact that mammalian replication sections are much smaller than bacterial chromosomes, the rate of bacterial DNA replication is much larger (30 μ per minute as measured by Cairns, 1963a) than the rate of mammalian DNA replication. Perhaps the necessity for synthesizing and organizing the many chromosomal proteins of mammalian cells requires a lower rate of replication.

We gratefully acknowledge the advice and assistance of Dr Giuseppe Attardi in whose laboratory this work was done. This investigation was supported by U.S. Public Health Service grants GM–11726, 5–F1–GM–21,622 and F2–HD–22,991 and by the Arthur McCallum Fund.

REFERENCES

Cairns, J. (1963a). *J. Mol. Biol.* **6**, 208.

Cairns, J. (1963b). *Cold Spr. Harb. Symp. Quant. Biol.* **28**, 43.

Cairns, J. (1966). *J. Mol. Biol.* **15**, 372.

Davern, C. I. (1966). *Proc. Nat. Acad. Sci., Wash.* **55**, 792.

Huberman, J. A. (1967). In *Some Recent Developments in Biochemistry*. Taipei: Academia Sinica, in the press.

Huberman, J. A. & Riggs, A. D. (1966). *Proc. Nat. Acad. Sci., Wash.* **55**, 599.

Hsu, T. C. (1964). *J. Cell Biol.* **23**, 53.

Hsu, T. C., Dewey, W. C. & Humphrey, R. M. (1962). *Exp. Cell Res.* **27**, 441.

Macgregor, H. C. & Callan, H. G. (1962). *Quart. J. Micro. Sci.* **103**, 173.

Pritchard, R. H. & Lark, K. G. (1964). *J. Mol. Biol.* **9**, 288.

Sasaki, M. S. & Norman, A. (1966). *Exp. Cell Res* **44**, 642.

Taylor, J. H. (1960). *J. Biophys. Biochem. Cytol.* **7**, 455.

Taylor, J. H., Haut, W. F. & Tung, J. (1962). *Proc. Nat. Acad. Sci., Wash.* **48**, 190.

Reprinted from Nature, Vol. 225, No. 5228. pp. 159–164. 1970. © 1970 Macmillan Journals, Ltd.

Mammalian Cell Fusion : I. Studies on the Regulation of DNA Synthesis and Mitosis

by

POTU N. RAO
ROBERT T. JOHNSON

Eleanor Roosevelt Institute for Cancer Research
and Department of Biophysics,
University of Colorado Medical Center,
Denver, Colorado 80220

DNA synthesis and mitosis are inducible in multinucleate HeLa cells formed by fusion between cells in different phases of the life cycle.

FUSION of two cells in different phases of the mitotic cycle is a means by which the regulatory mechanisms for the initiation of DNA synthesis and mitosis can be examined and related to specific stages of the cell cycle. One can then ask the question: to what extent can a cell's progress in the life cycle be altered by fusion with other cells in the same or different stages ?

Synchronization and Cell Fusion

HeLa cells were grown as suspension cultures at 37° C in Eagle's minimal essential medium supplemented with sodium pyruvate, glutamine and 5 per cent foetal calf serum. The spinner flasks were gassed with a mixture of 5 per cent CO_2 in air. Cells were maintained in exponential growth by diluting the suspension to a concentration of 2×10^5 cells/ml. every day with fresh, pre-warmed medium. At 37° C, these cells have a generation time of 21·8 h with a DNA synthetic (S) period of 7·0 h; a pre-DNA-synthetic ($G1$) period of 10·4 h; a post-DNA-synthetic ($G2$) period of 3·5 h and a mitotic duration of 0·9 h (ref. 1). For the purpose of fusion, cells were synchronized by the excess thymidine double block technique as described earlier[1]. A synchronized population of cells in S period was obtained by collecting cells 1 h after the reversal of the second thymidine block, whereas the $G2$ cells were collected 6 h

after the reversal of a similar block. An early $G1$ population was obtained by collecting the cells 2 h after the release of an N_2O block following the reversal of a single excess thymidine block[2].

We carried out three types of experiments involving fusion of $G1$ with S cells, $G1$ with $G2$ cells, and S with $G2$ cells (Table 1). One member of each cell pair was labelled with ^{3}H-thymidine (0·05 μCi/ml.; 6·7 Ci/mmole) previous to fusion, to permit its identification in the resulting hybrid. About 5×10^6 cells of each of the two synchronized populations were placed in a total volume of 1 ml. of Hanks basal salt solution, without glucose, containing 1,000 haemagglutinating units of ultraviolet inactivated Sendai virus. The virus–cell mixture was kept at 4° C for 15 min and at 37° C for 20 min according to the procedures described by Harris et al.[3]. After the completion of cell fusion, the suspension was diluted 1 : 50 with fresh medium, dispensed into forty-eight sterile plastic culture dishes (30 mm in diameter) and incubated at 37° C. For each experiment, the dishes were divided into three sets of sixteen. To one set ^{3}H-thymidine (0·2 μCi/ml.; 6·7 Ci/mmole) and colcemide ($6·7 \times 10^{-7}$ M) were added to study the pattern of DNA synthesis in the fused cells while mitosis was blocked. To the second set of dishes colcemide alone was added to study the rate of mitotic accumulation.

Table 1. PROTOCOL FOR OBTAINING SYNCHRONIZED POPULATIONS OF HeLa CELLS FOR THE THREE TYPES OF FUSION EXPERIMENTS

Time (h)	Spinner-A for prelabelled S and $G2$ populations	Spinner-B for $G2$ population	Spinner-C for $G1$ population	Spinner-D for $G1$ population
0	Start spinner-A; add ^{3}H-thymidine (0·05 μCi/ml.)	—	—	—
24	Place excess thymidine block	Start spinner-B; place excess thymidine block	—	—
30	—	—	Start spinner-C; place excess thymidine block	—
41	Release first thymidine block; add ^{3}H-thymidine (0·1 μCi/ml.)	Release first thymidine block	—	Start spinner-D; place excess thymidine
50·5	Place second thymidine block	Place second thymidine block	—	—
51	—	—	Release thymidine block; place cells in plastic culture dishes; incubate at 37° C	—
55	—	—	Transfer these dishes into N_2O chamber	—
62	—	—	—	Release thymidine block; place cells in plastic culture dishes; incubate at 37° C
64	—	—	Release N_2O block; place cells in a spinner	—
65	Divide the cell suspension into three spinners-A1, A2 and A3; release second thymidine block in A1	Release second thymidine block	—	—
66	—	—	Fuse cells from spinner-A1 with those from spinner-C to give $G1/S^*$ fusion (Expt. II)	Transfer these dishes into N_2O chamber
70	Release second thymidine block in spinners A2 and A3	—	—	—
71	—	Fuse cells from spinner-A2 with those from spinner-B to give a $S^*/G2$ fusion (Expt. I)	—	—
75	—	—	—	Release N_2O block; place cells in a spinner.
77	—	—	—	Fuse cells from spinner-A3 with those from spinner-D to give $G1/G2^*$ fusion (Expt. III)

* Prelabelled population

Nothing was added to the third set, which served as the control and permitted study of mitotic synchrony among the nuclei in the multinucleate cells.

At regular intervals the dishes were trypsinized. From each sample 0.75×10^5 to 1×10^5 cells were taken in a volume of about 0·5 ml. and centrifuged at 600 r.p.m. for 10 min directly on a microscope slide by means of a cytocentrifuge (Shandon–Elliot, England). The cells were fixed for 5 min in absolute ethanol : glacial acetic acid (3:1) and were given three 10 min extractions with cold 5 per cent trichloroacetic acid. The slides were then processed for autoradiography according to the procedures previously described[4].

Heterophasic and Homophasic Multinucleate Cells

We use the term heterophasic multinucleate cell to describe the result of fusion between similar cells in different phases of the life cycle (Fig. 1A, B, for example). Homophasic multinucleate cells would be formed by the fusion of cells in the same phase of the cell cycle (Fig. 1C). The following effects of fusion were studied. (a) Initiation of DNA synthesis in G1 and G2 nuclei after fusion with cells in G1, S or G2. (b) Suppression of DNA synthesis in S and G1 nuclei after fusion with cells in G1, S or G2. (c)

Acceleration in the time of entry into mitosis of G1 and S nuclei by the G2 and, conversely, the delay of entry into mitosis of G2 after fusion with cells in G1, S or G2.

In every case, a lightly prelabelled phased cell population was fused with an unlabelled one. Then a large dose (0·2 µCi/ml.) of ³H-thymidine and colcemide was added immediately to one set of dishes and samples were taken at various periods of incubation to study the induction of DNA synthesis in the fused cells. The density of labelling after fusion was sufficiently greater than that of the prelabelled cells before fusion to leave little or no ambiguity about whether a given labelling had occurred before or after fusion.

The following procedure was used in scoring the effect of fusion on DNA synthesis. Before fusion the cells of each population were mononucleate and either labelled or unlabelled (L or U). The frequency of multinucleate cells in the HeLa cell suspension before fusion ranged from 3–5 per cent. After fusion about 30 per cent of the mixed population consisted of multinucleate cells of various classes (Table 2). As expected, among the fused cells binucleates were most numerous followed by tri, tetra and higher multinucleate cells in that order of decreasing frequency. For example, the binucleate cells can be either

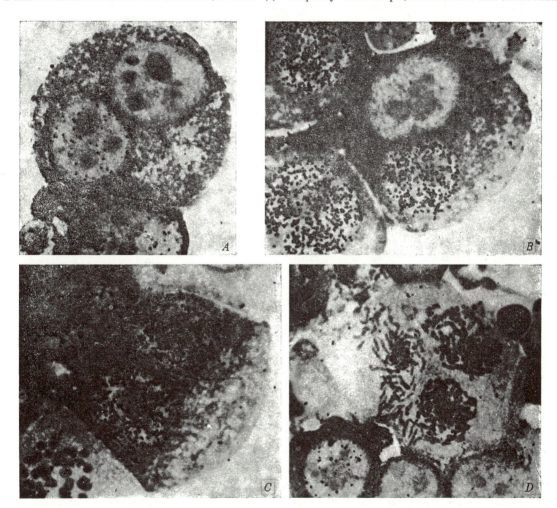

Fig. 1. A, Heterophasic S/G2 binucleate cell at $t = 0$ after fusion. The S nucleus was prelabelled with ³H-thymidine. B, Heterophasic S/G2 binucleate cell at $t = 6$ h after fusion and incubation with ³H-thymidine. The increased intensity of labelling of the S nucleus as compared with that in A arises from continued DNA synthesis after fusion. There was no uptake of ³H-thymidine by the G2 nucleus. C, Homophasic S/S binucleate cell at $t = 6$ h after fusion and incubation with ³H-thymidine. The intensity of labelling in each of the nuclei is comparable with that in the S nucleus in B. D, Heterophasic G1/2G2 trinucleate cell in synchronous mitosis (no colcemide treatment was given). G2 nuclei were prelabelled. Note a slightly less condensed state of the chromosomes of the unlabelled (G1) nucleus.

2

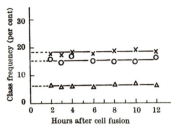

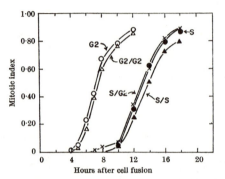

Fig. 2. The frequency of homo and heterophasic binucleate cells in the $S/G2$ fusion as a function of time after continuous labelling with ^{3}H-thymidine. The S nucleus was prelabelled. The mononucleate $G2$ class (O) is presented for comparison. ×, $S/G2$; △, $G2/G2$.

Fig. 3. Mitotic accumulation functions for the mono and binucleate S and $G2$ classes are compared with the heterophasic $S/G2$ binucleate class. Note that the pattern of mitotic accumulation in $S/G2$ is similar to that of the S parent.

U/U, L/L or L/U if two unlabelled, two labelled or one labelled and one unlabelled cells, respectively, were fused together. When the fused cells are incubated with ^{3}H-thymidine, if the unlabelled nuclei do not incorporate ^{3}H-thymidine the frequencies of the classes U/U, L/L and L/U remain constant. Otherwise their frequencies would undergo a change. The class frequencies plotted in Figs. 2, 4A and 6A were obtained in the following manner.

$$\text{Class frequency} = \frac{N_c \times 100}{N_{sp}}$$

where N_c is the number of cells in a given class at a given time and N_{sp} is the total number of cells in that sub-population at that time (Table 2).

Table 2. VARIOUS CLASSES OF CELLS THAT ARE FORMED BY A RANDOM CELL FUSION BETWEEN LABELLED AND UNLABELLED POPULATIONS

Sub-population	Homophasic	Fusion classes Heterophasic	
1. Mononucleate cells (parental types only)	L; U		
2. Binucleate cells	LL; UU	LU	
3. Trinucleate cells	3L; 3U	2L/U; L/2U	
4. Tetranucleate cells	4L; 4U	3L/U; 2L/2U; L/3U	
5. Pentanucleate*			

L = Prelabelled nucleus. U = Unlabelled nucleus.
* The cells with five or more nuclei were too few to be included in this study.

A change in the frequency of any class was brought about by the initiation of DNA synthesis as monitored by the incorporation of ^{3}H-thymidine in one or more nuclei of a multinucleate cell. For example, the frequency of the class L/U would decrease if the unlabelled nucleus of the heterophasic binucleate cells incorporated ^{3}H-thymidine. The decrease in the frequency of the L/U would result in the proportional increase of the class L/L. In this case there is a possibility that uptake of labelled thymidine in one of the nuclei of a binucleate cell of the homophasic U/U class would result in a cell of the L/U class and this cannot be distinguished from the original L/U class formed by the

fusion of a labelled cell with an unlabelled one. By studying the changes in the frequency of each class, however, one could detect DNA synthesis in one of the nuclei of the U/U class. Our findings indicate that there is a high degree of synchrony in DNA synthesis among the nuclei of the homophasic binucleate cells. Either both or none of the nuclei would synthesize DNA in a cell of the U/U class. The most important feature of this type of analysis is the built-in controls in the system. The kinetics of the heterophasic classes can readily be compared with the unfused mononucleate parents and fused homophasic classes.

From the class frequency curves the relative rate of induction of DNA synthesis was calculated and expressed as the percentage of increase in labelling as shown here.

$$\text{Increase in labelling} = \frac{Nt_0 - Nt_n}{Nt_0} \times 100$$

where Nt_0 is the number of cells in a given class at $t = 0$ h (soon after cell fusion) and Nt_n is the number of cells in that class at $t = n$ h after fusion. For the determination of mitotic index each class was treated as a separate entity.

$S/G2$ Fusion (Experiment I)

A prelabelled S population was fused with an unlabelled $G2$ population and the patterns of DNA synthesis and mitotic accumulation were studied in the fused and unfused cells. Because the $G2$ population had completed DNA synthesis before cell fusion, no change in the frequency of $G2$ or $G2/G2$ classes was expected nor observed (Fig. 2). The absence of any decrease in the frequency of the class $S/G2$ was an indication that there was no induction of DNA synthesis in the $G2$ nucleus. On the other hand, the $G2$ component of this heterophasic cell did not inhibit the DNA synthesis in the S nucleus as shown by the increased intensity of labelling (compare Fig. 1A with B and C).

The mitotic accumulation function[1,5] was studied in the set of dishes to which colcemide alone was added. The mononucleate $G2$ population was the first to enter mitosis, closely followed by the homophasic binucleate ($G2/G2$) cells (Fig. 3). Similarly, the S and S/S classes arrived at mitosis about 6 h later in that sequence. The heterophasic

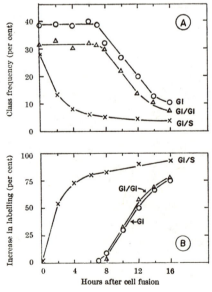

Fig. 4. Induction of DNA synthesis in the $G1$ nucleus of the $G1/S$ fusion. A, Class frequencies of $G1$, $G1/G1$ and $G1/S$ as a function of time after fusion and continuous incubation with ^{3}H-thymidine. The S nuclei were prelabelled. B, Rate of induction of DNA synthesis in the $G1$ nucleus of the $G1/S$ fused cells.

3

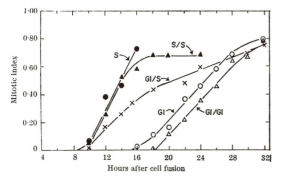

Fig. 5. Mitotic accumulations from the *G1/S* fusion, showing the time interval between the parental *S* and *G1* classes and the intermediate nature of the heterophasic *G1/S* binucleate cells.

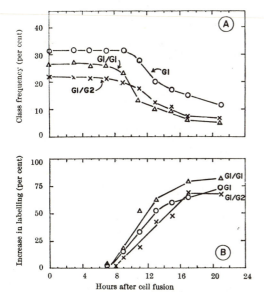

Fig. 6. *A*, The frequencies of homo and heterophasic binucleate cells in the *G1/G2* fusion as a function of time after continuous labelling with ³H-thymidine. The *G2* nuclei were prelabelled. The mononucleate *G1* class is shown for comparison. *B*, Rate of induction of DNA synthesis in the heterophasic *G1/G2* binucleate cells is compared with those of *G1* parent and *G1/G1* homophasic binucleate cells.

cells (*S/G2*) arrived at mitosis about the same time as mononucleate *S* cells. The *S/G2* cells, however, were definitely ahead of the *S/S* class in their entry into mitosis, indicating a slight pull of *S* nucleus by the *G2* component towards mitosis.

G1/S Fusion (Experiment II)

In this cross the *S* population was prelabelled. Results on the induction of DNA synthesis in the *G1* nucleus of the *G1/S* fused cell are presented in Fig. 4. The frequency of *G1* and *G1/G1* classes did not change until about 8 h after fusion, when they started entering into the *S* period as indicated by the decrease in the frequency. By contrast there was a sharp decrease in the frequency of the *G1/S* class beginning at *t* = 0 h, indicating the induction of DNA synthesis in the *G1* nucleus (Fig. 4*A*). DNA synthesis was induced in more than 50 per cent of the *G1* nuclei of the heterophasic *G1/S* cells within 2 h after cell fusion (Fig. 4*B*). The 50 per cent labelling point was reached in *G1* and *G1/G1* classes about 12 h after fusion. Clearly there was a rapid induction of DNA synthesis in the nucleus of a *G1*

cell soon after fusion with a cell that was synthesizing DNA. This was further confirmed by the mitotic accumulation curve for the *G1/S* class of cells which lies between those for the *S/S* and *G1/G1* classes (Fig. 5). The earlier induction of DNA synthesis in the *G1* nucleus of the *G1/S* cells resulted in an earlier initiation of mitosis in the heterophasic cells than in the homophasic *G1/G1* cells.

G1/G2 Fusion (Experiment III)

A prelabelled *G2* population was fused with a *G1* population. The frequencies of the various classes (Fig. 6*A*) studied in order to measure the rate of initiation of DNA synthesis (Fig. 6*B*) indicated that there was no significant difference between the heterophasic *G1/G2* and homophasic *G1/G1* in the pattern of labelling. In other words, the *G2* component of the heterophasic binucleate (*G1/G2*) cell had no effect on the normal course of DNA synthesis in the *G1* nucleus. The mitotic accumulation function for *G1/G2* class was quite similar to that of *G1/G1* (Fig. 7). About 20–35 per cent of the *G1*, *G1/G1* and *G1/G2* cells entered mitosis much earlier than expected because of contamination of the *G1* population of this experiment with some *G2* cells as a result of technical difficulties (compare curves *G1* and *G1/G1* of Fig. 7 with those in Fig. 5). This asynchrony in the *G1* population was responsible for the initial mitotic accumulation as shown in Fig. 7. This did not, however, prevent us from observing that in general the *G2* component did not affect the *G1* nucleus in its progression through the mitotic cycle. It was the *G2* nucleus of *G1/G2* cell that was delayed in entering into mitosis by its *G1* component.

Dosage Effect

The metabolic activity of a multinucleate cell formed by fusion between cells in different phases of the cell cycle was

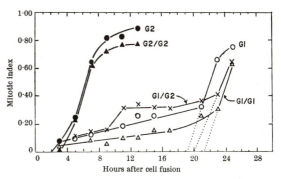

Fig. 7. Mitotic accumulations for the *G1/G2* fusion, showing the interval between the parental *G1* and *G2* types. The mitotic accumulation function for the heterophasic *G1/G2* is similar to the *G1* parent.

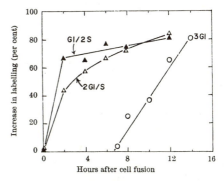

Fig. 8. Dosage effect on the induction of DNA synthesis in the *G1* nuclei of the heterophasic trinucleate cells of the *G1/S* fusion. The *S* nuclei were prelabelled.

4

greatly influenced by the ratio of cells in advanced to early states. The dose response of the induction of DNA synthesis and the initiation of mitosis was studied in tri and tetranucleate heterophasic cells. In the $G1/S$ fusion the rate of induction of DNA synthesis in $G1$ nuclei was faster in the $G1/2S$ class than in $2G1/S$ (Fig. 8). The dosage effect of the S component on the $G1$ is illustrated by the time taken by a given class of cells to reach a labelling index of 50 per cent. This time was 1·5 h for $G1/2S$, 1·75 h for $G1/S$ and 3·0 h for the $2G1/S$ class. About 75 per cent of the $G1$ nuclei in the heterophasic cells were synthesizing DNA by the time the homophasic ($3G1$) cells started entering S period.

In the $S/G2$ and $G1/G2$ fusions, no induction of DNA synthesis in the $G2$ nucleus was observed even when the ratio of S or $G1$ to $G2$ was high. Similarly, an increase in the proportion of $G2$ to S or $G1$ component did not result in the inhibition of DNA synthesis in the S nucleus nor in the progression of the $G1$ nucleus into the S period.

The initiation of mitosis in the multinucleate HeLa cells was also dose dependent as observed in $S/G2$ and $G1/G2$ fusions. The data on the mitotic accumulation of the two mononucleate parents and the homophasic multinucleate cells of the $S/G2$ fusion are presented in Fig. 9A; for the corresponding heterophasic multinucleate cells see Fig. 9B. The greater the ratio of $G2 : S$ nuclei in a fused cell, the earlier was the initiation of mitosis in all the nuclei of that cell. For example, a mitotic index of 0·50 was reached by the 4S class at 14·2 h after cell fusion, whereas it was 13·3 h for $G2/S$, 11·6 h for $2G2/S$, 10·7 h for $3G2/S$ and 8·5 h

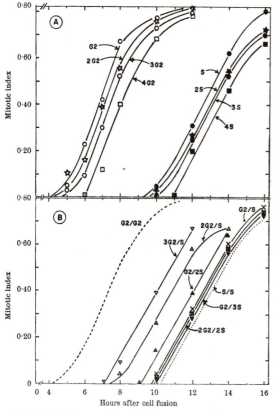

Fig. 9. A, Mitotic accumulation functions of mono, bi, tri and tetranucleate homophasic cells of the $S/G2$ fusion. The S nuclei were prelabelled. The greater the number of nuclei in a cell the slower was the onset of mitosis. B, Dosage effect of the $G2$ or S component on the rate of mitotic accumulation in the heterophasic multinucleate cells of the $S/G2$ fusion. The homophasic binucleate $G2/G2$ and S/S are represented by dashed and dotted lines respectively.

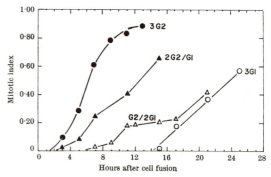

Fig. 10. Dosage effect of the $G2$ component on the rate of mitotic accumulation of the heterophasic trinucleate cells of the $G2/G1$ fusion. The $G2$ nuclei were prelabelled.

for the 4$G2$ class. Even though the pulling effect—with respect to mitosis—of the $G2$ over the $G1$ component in cells with a 1 : 1 ratio was negligible, an increase in the ratio of $G2 : G1$ did bring about earlier onset of mitosis in heterophasic trinucleate cells (Fig. 10). The $2G2/G1$ class of cells entered mitosis quite early, however, partly because of the contamination of the $G1$ population with $G2$ cells.

Mitotic Synchrony

A very high degree of mitotic synchrony was observed among the nuclei in fused cells in the control set of dishes to which no colcemide or ^{3}H-thymidine was added. In the $G1/S$ and $G1/G2$ fusions the mitotic synchrony was almost perfect (Fig. 1D). Asynchronous mitoses were observed in only five out of 1,000 fused cells in mitosis. In these two fusions there was no significant difference between homophasic and heterophasic cells with regard to mitotic synchrony, but in the $S/G2$ fusion some degree of mitotic asynchrony (about 11 per cent) was observed in heterophasic cells but not in the homophasic cells. Further analysis of this asynchrony revealed that in the fused cells the $G2$ nuclei were always in a more advanced phase of mitosis than the S nuclei.

The phenomenon of anomalous chromosome condensation, or the "pulverization" of chromosomes as it was first termed by Sandberg et al.[6], was predominant in the $S/G2$ fusion but rarely seen in $G1/S$ or $G1/G2$ fusions.

Induction of DNA Synthesis

There was obviously a rapid induction of DNA synthesis in the $G1$ nuclei of the $G1/S$ fused cells (Fig. 4B). The rate of induction was dependent on the ratio of S to $G1$ nuclei (Fig. 8). The greater the number of S components in the fused cell the faster was the induction of DNA synthesis in the $G1$ nuclei. The $G1$ component of the heterophasic $G1/S$ cell did not inhibit DNA synthesis that was in progress in the S nucleus. These findings are generally in accord with those in other systems[3,7–12]. The rapid induction of DNA synthesis in the $G1$ nuclei in a dose-dependent manner suggests that certain substances which are present in the S component probably migrate into $G1$ nucleus and cause the initiation of DNA synthesis. There is some evidence of such factors in the studies of Thompson and McCarthy[13]

In the virus-fused HeLa cells we have found that the $G2$ nucleus and cytoplasm had no effect on the normal course of DNA synthesis of an S nucleus (Fig. 1A and B). These heterophasic cells then complete the cycle and reach mitosis. No matter what the ratio of $G2 : S$ nuclei in a fused cell, the S nuclei continued DNA synthesis. In these cells there was no re-induction of DNA synthesis in the $G2$ nucleus. Studies on *Physarum polycephalum*[14], *Amoeba proteus*[15] and *Stentor coeruleus*[9] resulted in similar findings except for those studies by Prescott and Gold-

5

stein[16,17]. We show that the presence of a $G2$ component did not inhibit a $G1$ nucleus from entering the S phase (Fig. 6A and B). Even cells with high ratios of $G2 : G1$ nuclei show normal progression of the $G1$ nuclei into the S period. The nuclei of these $G1/G2$ multinucleate cells subsequently entered mitosis with almost perfect synchrony (Fig. 1D).

With respect to DNA synthesis, therefore, we show that $G1$ nuclei in the presence of an S component can rapidly move into S phase, thus eliminating most of the $G1$ period in a dose-dependent manner. The inducers for DNA synthesis are present in cells synthesizing DNA and are probably transmitted through the cytoplasm to the $G1$ nucleus following $G1/S$ cell fusion. These inducers cannot initiate a second round of DNA synthesis in $G2$ nuclei of the HeLa cells. The fact that DNA synthesis continued in S nuclei in the presence of a $G1$ or $G2$ component suggests the absence of any inhibitors of DNA synthesis during these phases. We conclude that there is a positive control of DNA synthesis in HeLa cells with the appearance of inducer substances at the start of the S phase. These substances are not produced in either $G1$ or $G2$ phase cells, and their absence during these periods—rather than the presence of specific repressor materials—is reflected in the absence of DNA replication during these phases of the cell cycle.

Initiation of Mitosis

Mitotic accumulation functions revealed that a $G2$ cell, when fused with a $G1$ or S cell in a 1 : 1 ratio, was always delayed until the latter was ready to enter into mitosis (Figs. 3 and 7). In the fused cells the $G2$ component imparted some advantage with regard to an early onset of mitosis to the $G1$ or S nucleus. In the sub-population of binucleate cells, the heterophasic cells ($G2/S$ and $G2/G1$) always reached a mid point (0·50) in the mitotic accumulation a little earlier than the homophasic cells of the late parent-$G1/G1$ or S/S (Figs. 3 and 7). The mitotic inducing effect of the $G2$ cells was even more obvious with an increase in the dosage of the $G2$ component in the fused cells (Figs. 9B and 10). In these cases the $G2$ phase of the S or $G1$ nucleus was essentially shortened in much the same way as the $G1$ phase was shortened by the S component in the $G1/S$ fusion. In the $G1/S$ fusion, unlike the $S/G2$ or $G1/G2$ fusions, the mode of mitotic accumulation for the heterophasic $G1/S$ cells was intermediate between the two parents because of the rapid induction of DNA synthesis in the $G1$ nucleus (Fig. 5). In other words, the pre-DNA synthetic period of the $G1$ nucleus was greatly reduced under the influence of the S component.

In the heterophasic $S/G2$ cells the $G2$ nuclei were delayed in entering mitosis by the presence of the S component and the greater the initial dose of the S component the greater was the delay (Fig. 9B). This may be interpreted as an equilibration of mitotic inducer substances between all the nuclei of a fused cell, so that the time of entry into mitosis depends on the critical concentration of the inducer substances and hence the ratio of $G2$ to S components present initially. These conclusions are also applicable to the $G1/G2$ cells (Fig. 10).

The equilibration of materials in multinucleate cells leading to dosage effects on DNA synthesis and mitosis has been described by many workers using many different systems, for example, Amoeba[17]; Stentor[18–20]; Pelomyxa[21]; Physarum[22–25]; HeLa–Ehrlich ascites heterokaryons[26]. In a series of experiments with Physarum, Rusch et al.[24] found that substances which induce mitosis were synthesized during the interphase, and reached critical levels just before mitosis. Coalescence of early and late plasmodia led to a speeding up of the lagging nuclei and a retardation of the advanced nuclei, so that complete synchrony of nuclear division was always produced. When the sizes of the late and early plasmodia were varied, the speeding up of the lagging nuclei was proportional to the size of the advanced plasmodium used. In addition, mitotic gradients were observed between partially overlapping late and early plasmodia, indicating that a gradient of mitotic inducers was indeed present. Such dosage effects in Physarum strongly resemble our findings, and suggest that the mechanisms responsible for the initiation of mitosis in Physarum have much in common with those in mammalian cells.

In our fusion experiments, we have found extremely low levels of asynchronous mitoses among the multinucleate cells, in agreement with the findings of others[26–29]. Giménez-Martin et al.[30] examined the asynchrony of nuclear entry into mitosis in caffeine induced multinucleate onion root tip cells: as the number of nuclei per cell increased the extent of asynchronous passage into prophase among the nuclei also increased. The subsequent passage into metaphase was, however, always highly synchronous.

The nuclei in most of the naturally occurring and artificially created multinucleate cells and syncytia show a high order of synchrony of DNA synthesis and of entry into mitosis[31–35]. We have shown that even though the cells were widely separated in their metabolic states, when fused together they achieve a rapid coordination and reach mitosis synchronously during the first mitotic cycle itself. The fact that most of the multinucleate systems achieve synchrony rather quickly suggests that there are some factors operating in this direction. The induction of DNA synthesis and the initiation of mitosis are probably the two main forces operating in order to bring about synchrony in multinucleate systems.

We thank Dr Theodore T. Puck for critical reading of the manuscript; Professor Henry Harris, of the University of Oxford, for kindly supplying us with Sendai virus; and George Barella and David Peakman for technical assistance. R. T. J. is a recipient of a Damon Runyon postdoctoral fellowship. The investigation was aided by a US Public Health Service grant from the National Institute of Child Health and Human Development.

Received September 22, 1969.

[1] Rao, P. N., and Engelberg, J., in Cell Synchrony Studies in Biosynthesis Regulation (edit. by Cameron, I. L., and Padilla, G. M.), 332 (Academic Press, New York and London, 1966).
[2] Rao, P. N., Science, 160, 774 (1968).
[3] Harris, H., Watkins, J. F., Ford, C. E., and Schoefl, G. I., J. Cell Sci., 1, 1 (1966).
[4] Rao, P. N., and Engelberg, J., Science, 148, 1092 (1965).
[5] Puck, T. T., and Steffen, J., Biophys. J., 3, 379 (1963).
[6] Sandberg, A. A., Sofuni, T., Takagi, N., and Moore, G. E., Proc. US Nat. Acad. Sci., 56, 105 (1966).
[7] Graham, C. F., Arms, K., and Gurdon, J. B., Devel. Biol., 14, 349 (1966).
[8] Gurdon, J. B., Proc. US Nat. Acad. Sci., 58, 545 (1967).
[9] De Terra, N., Proc. US Nat. Acad. Sci., 57, 607 (1967).
[10] Graham, C. F., J. Cell Sci., 1, 363 (1966).
[11] Jacobson, C. E., Exp. Cell Res., 53, 316 (1969).
[11] Johnson, R. T., and Harris, H., J. Cell Sci. (in the press).
[13] Thompson, L. R., and McCarthy, B. J., Biochem. Biophys. Res. Commun., 30, 166 (1968).
[14] Guttes, S., and Guttes, E., J. Cell Biol., 37, 761 (1968).
[15] Ord, M. J., Nature, 221, 964 (1969).
[16] Prescott, D. M., and Goldstein, L., Science, 155, 469 (1967).
[17] Goldstein, L., and Prescott, D. M., in The Control of Nuclear Activity (edit. by Goldstein, L.), 3 (Prentice-Hall, New Jersey, 1967).
[18] Weisz, P. B., J. Exp. Zool., 131, 137 (1956).
[19] De Terra, N., Exp. Cell Res., 21, 41 (1960).
[20] Tartar, V., J. Exp. Zool., 163, 297 (1966).
[21] Daniels, E. W., J. Exp. Zool., 117, 189 (1951).
[22] Guttes, E., Guttes, S., and Rusch, H. P., Fed. Proc., 18, 479 (1959).
[23] Guttes, E., and Guttes, S., Experientia, 19, 13 (1963).
[24] Rusch, H. P., Sachsenmaier, W., Behrens, K., and Gruter, V., J. Cell Biol., 31, 204 (1966).
[25] Guttes, E., Devi, V. R., and Guttes, S., Experientia, 25, 615 (1969).
[26] Johnson, R. T., and Harris, H., J. Cell Sci. (in the press).
[27] Johnson, R. T., and Harris, H., J. Cell Sci. (in the press).
[28] Oftebro, R., and Wolf, I., Exp. Cell Res., 48, 39 (1967).
[29] Oftebro, R., Scand. J. Clin. Lab. Invest., 22, 79 (1968).
[30] Giménez-Martin, G., Lopez-Saéz, J. F., Moreno, P., and González-Fernández, A., Chromosoma, 25, 282 (1968).
[31] Mazia, D., in The Cell (edit. by Brachet, J., and Mirsky, A. E.), 77 (Academic Press, New York and London, 1961).
[32] González, M. A., Genet. Ibérica, 19, 1 (1967).
[33] Cone, C. D., J. Theoret. Biol., 22, 365 (1969).
[34] Braun, R., Mittermayer, C., and Rusch, H. P., Proc. US Nat. Acad. Sci., 53, 924 (1965).
[35] Fawcett, D. W., Ito, S., and Slautterback, D., Biophys. Biochem. Cytol., 5, 453 (1959).

Proc. Nat. Acad. Sci. USA
Vol. 70, No. 4, pp. 1263–1267, April 1973

Do Cells Cycle?

(cell kinetics/control of growth/DNA replication/cell culture)

J. A. SMITH AND L. MARTIN

Departments of Hormone Biochemistry and Hormone Physiology (Endocrine Group), Imperial Cancer Research Fund,
Lincoln's Inn Fields, London WC2A 3PX England

Communicated by Arthur B. Pardee, February 6, 1973

ABSTRACT We propose that a cell's life is divided into
two fundamentally different parts. Some time after
mitosis all cells enter a state (A) in which their activity is
not directed towards replication. A cell may remain in the
A-state for any length of time, throughout which its
probability of leaving A-state remains constant. On leaving A-state, cells enter B-phase in which their activities
are deterministic, and directed towards replication.
Initiation of cell replication processes is thus random, in
the sense that radioactive decay is random. Cell population
growth rates are determined by the probability with
which cells leave the A-state, the duration of the B-phase,
and the rate of cell death. Knowledge of these parameters
permits precise calculation of the distribution of intermitotic times within populations, the behavior of synchronized cell cultures, and the shape of labeled mitosis
curves.

Subdivision of the intermitotic period into G_1, S, and G_2 (1)
has stimulated attempts to analyze the processes controlling
the progression of cells from one mitosis to the next. Our aim
is to show that much of the information gathered may
require reinterpretation because of a fundamental mis-
conception about the nature of G_1.

Analyses of the fraction of labeled mitoses (FLM) at
various times after exposure to a pulse of [³H]thymidine
([³H]dT) demonstrate that the durations of S and G_2 are
characteristic of particular cell types, and usually do not show
much intrapopulation variation (2). Both phases can be regarded as deterministic and specifically related to division.
However, the duration of G_1 is extremely variable, both
between different cell types and within homogeneous populations (3), and this variability accounts for most of the
variation of the intermitotic period. Changes in generation
time are also usually attributed to changes in G_1 duration.
G_1 has therefore attracted particular interest as the period in
which proliferation is regulated. Implicit in the cell-cycle concept is the idea of continuous progression through a chain of
events leading to division, for "cycle" implies "an interval
during which one sequence of a regularly recurring succession
of events is completed" (Webster's Dictionary). There are
some rapidly proliferating cells, for which the concept seems
appropriate, but even here the distribution of intermitotic
times has a curiously wide spread. Applied to slowly proliferating cells, the concept is confused, for although G_1 is
regarded as an extensible progression of events leading

towards division, the experimental data force the conclusion
that this progress can be arrested at some stage or stages (4).
Reexamination of published data suggests an alternative
interpretation of the nature of G_1. Most of the relevant data
are about the *frequency* of division. Now a frequency may
reflect a regularly recurring process or it may reflect events
occurring randomly with a certain probability (e.g., radioactive decay). Pursuing this thought, we arrived at the model
illustrated in Fig. 1.

We propose that the intermitotic period is composed of an
A-state and a B-phase. The B-phase includes the conventional
S, G_2, and M phases. Whether it also includes part of G_1 is discussed below. Some time after mitosis the cell enters the
A-state, in which it is not progressing towards division. It
may remain in this state for *any* length of time, throughout
which its probability of entering B-phase is constant. This
"transition probability" (P) may be supposed to be a characteristic of the cell type but capable of modification by environmental factors.

Distribution of Generation Times. The proposed hypothesis
predicts wide variation in the duration of G_1 within populations. This variability has had to be considered when
trying to estimate mean cycle times, etc., and considerable
effort has gone into determining its statistical distribution.
The data have not led to any precise formulation, and the
most that can be claimed is that the distribution is usually
skewed to the right and the variance is large (2). Nevertheless,
information about individual generation times can be used to
decide whether a "probability" model is reasonable. If the
intermitotic period were of uniform duration in a population,
the cells would all divide at the same age. If the proportion
(α) of the initial population (N) remaining in interphase were
plotted against age, Fig. 2A would result. If intermitotic times
were normally distributed, Fig. 2B would result. This curve,
plotted with a logarithmic ordinate (Fig. 2C) shows that the
probability of division increases continuously with age. This
result is implicit in the cell-cycle concept. However, if the
population were characterized by a transition probability
(P) and a B-phase of duration T_B, α would decline expo-

Abbreviations: G_1, the interval between mitosis (M) and DNA
synthesis; S, the period of DNA synthesis; G_2, the interval between S and M; T_x, the duration of a period x; FLM, the fraction
of labeled mitoses; LI, labeling index.

FIG. 1.

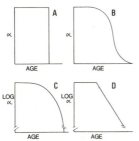

FIG. 2. The proportion of cells remaining in interphase as a function of age. Explanation in text.

$$\alpha = [N - (\text{no. cells already divided})]/N \qquad [1]$$

nentially beginning at time T_B. On a logarithmic ordinate this gives a straight line (Fig. 2D), and the probability of division remains constant.

Fig. 3 shows the frequency distributions of intermitotic times for various cells obtained by time-lapse cinematography of exponentially growing cultures 6–10. We have expressed these data as log α against age. All show exponential decay after a lag, demonstrating that initiation of cell replication processes occurs at random and not at the end of a "regularly recurring succession of events." The term "cell cycle" is therefore inappropriate.

Experimental Determination of T_B and P. The above examples were rapidly proliferating cells. No similar studies seem to have been attempted with slowly growing populations. However, if one assumes that the transition probability can take any value and is environmentally modifiable, the model can be used to describe many types of proliferative behavior. P clearly *is* variable in different populations, as is T_B (see above). That both are also subject to environmental modification can be found from data (11) for *Euglena* grown in different culture conditions. Clearly T_B and P are of great significance for the kinetic analysis of growth. Both can be obtained by the method described above where

$$P = [(\alpha_t - \alpha_{t+\Delta t})/\Delta t]/\alpha_t \qquad [2]$$

The data given below in Fig. 3 do not fit the predicted curve (Fig. 2D) exactly, since there is an initial downward curvature before linearity is reached. This curvature is due to

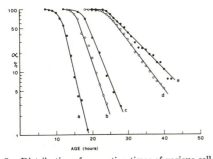

FIG. 3. Distribution of generation times of various cell types in culture, (a) rat sarcoma (6); $P(\text{hr}^{-1}) = 0.45$; $T_B(\text{hr}) = 9.5 \pm 1.0$, (b) HeLa S3 (7); $P = 0.32$; $T_B = 14 \pm 0.8$, (c) mouse fibroblasts (8); $P = 0.30$; $T_B = 15 \pm 1.2$, (d) L5 cells (9); $P = 0.18$; $T_B = 22.5 \pm 1.4$, (e) HeLa (10); $P = 0.14$; $T_B = 23 \pm 0.8$.

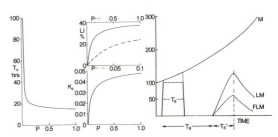

FIG. 4. (*left*) Curves for a hypothetical cell type relating transition probability (P), population doubling time (T_d), labeling index (LI), and cell production rate constant (K_p), calculated with $T_B = 15$ hr and $T_S = 8$ hr, from Eqs. 3, 4, 10, and 12.

FIG. 5. (*right*) Theoretical basis for the form of FLM curves. After a pulse of [^{3}H]dT, no mitoses will be labeled before time T_G; then, in a period $= T_M$ all mitoses become labeled. Throughout the period T_S, only labeled cells enter mitosis. After this the number of labeled mitoses (LM) falls to zero. By this time the number of labeled *cells* will have doubled, by division. No cells re-enter mitosis before a period $= T_B$. If LM at time t_i is n, then the number of these cells in mitosis during a later time interval is $2N_{t_i}\{\exp(-K_{\text{trans}}[t + 1]) - \exp(-K_{\text{trans}}t)\}$ where $t = 0$ when $t - t_i = T_B$. This equation is used to calculate the contribution of successive groups of labeled mitoses in the first peak to the second. It is convenient to take groups at hourly intervals (note that one uses the *area* under the curve). FLM = LM/M. M increases exponentially as $M_0 \exp(K_p t)$.

variation in T_B. The minimum T_B is the time the first cells enter division, and the maximum the time at which the curve becomes linear. If we assume a normal distribution between these limits, the coefficients of variation are about 10%.

T_B and P can also be obtained from cells synchronized by the procedure of Terasima and Tolmach (12), by counting the cells at intervals after seeding. This procedure gives a cumulative curve of generation times that can be treated in the same way as the time-lapse data. These methods can be used only for cells in culture. However T_B can be obtained from FLM curves, provided the second peak is sufficiently well defined (see Fig. 5). T_B is then the time between the beginning of the first and second peaks of labeled mitoses. Then P may be computed from the cell production rate constant (K_p) (13) and T_B.

$$K_p = \ln 2/T_d \qquad [3] \qquad\qquad K_p = LI/T_s \qquad [4]$$

We shall describe the derivation of P in detail. It is simplest if B-phase is all premitotic and cell loss is negligible. Growth is then exponential, and the increase in cell number with time is:

$$N_{t_1} - N_{t_0} = N_{t_0} \exp\{K_p(t_1 - t_0)\} - N_{t_0} \qquad [5]$$

where N_{t_1} and N_{t_0} are the number of cells at times t_1 and t_0. At any given time a fraction of the cells, N_A/N, is in the A-state. With constant K_p, N_A/N remains constant, and the increase in N_A is:

$$N_{A_{t_1}} - N_{A_{t_0}} = N_{A_{t_0}} \exp\{K_p(t_1 - t_0)\} - N_{A_{t_0}} \qquad [6]$$

For derivation of P it is necessary to know N_A/N as a function of K_p and T_B. Consider an interval $t_1 - t_0 = T_B$. During this time all cells that divide must have been at some stage of

B-phase at time t_0; therefore

$$N_{t_1} - N_{t_0} = N_{Bt_0} = N_{t_0} \exp(K_p T_B) - N_{t_0} \quad [7]$$

By definition $N_B + N_A = N$, so, from [7]

$$\{N \exp(K_p T_B) - N\} + N_A = N$$

Dividing through by N,

$$\{\exp(K_p T_B) - 1\} + N_A/N = 1$$

Rearranged, this becomes

$$N_A/N = 2 - \exp(K_p T_B) \quad [8]$$

(This equation shows that N_A/N is a function of both K_p and T_B, and also, as $K_p \to 0$, $N_A/N \to 1$, i.e., when growth rate is zero, all cells are in A-state. Conversely, when $\exp(K_p T_B) \to 2$, $N_A/N \to 0$, i.e., virtually all cells will be in B-phase when growth is maximal.)

The rate of cell production at time t is $K_p N_t$. Clearly, all the cells dividing at time t must have undergone the A–B transition at time $t - T_B$. If we introduce a rate constant for transition, K_{trans}:

$$K_{trans} N_{A_{t-T_B}} = K_p N_t \quad [9]$$

From [8]

$$N_{A_{t-T_B}} = 2N_{t-T_B} - N_{t-T_B} \exp(K_p T_B) \quad [a]$$

$$N_t = N_{t-T_B} \exp(K_p T_B) \quad [b]$$

Substituting **a** and **b** in [9] and dispensing with subscripts to N

$$K_{trans} = K_p \exp(K_p T_B)/2 - \exp(K_p T_B) \quad [10]*$$

The cohort of cells in the A-state at time t_0 will decay exponentially according to:

$$N_{A_t} = N_{A_{t_0}} \exp(-K_{trans} t) \quad [11]$$

P equals the proportion of cells "lost" from A-state per unit time:

$$P = 1 - \exp(-K_{trans}) \quad [12]$$

If cell loss occurs randomly, its only effect is to reduce the rates of increase of N and N_A equally. Thus N_A/N remains the same, and the derivation of P is unchanged. In practice, cell loss must be taken into account when K_p is estimated (13, 14).

Fig. 4 shows that when P is high, variations in P have relatively little effect on growth rate. Fine control of growth by varying P would therefore require fairly low Ps. It is interesting that its values, even in rapidly growing cells in culture (Fig. 3), were all below 0.5 hr^{-1}.

Constituents of B-Phase and "Position" of the A-State. The above data say nothing about the position of the A-state, but it is accepted that variation in generation times occurs mainly in G_1. B-phase obviously includes S, G_2, M, and probably some part of G_1. This is so for human amnion cells in which T_B was 16 hr and $S + G_2 + M$ only 9 hr, and in which no cells less than 6 hr old incorporated [³H]dT (5). In

---

* If B phase is divided into post- and premitotic phases, a different expression for K_{trans} is obtained, but numerical substitution gives the same answer as Eq. **10**.

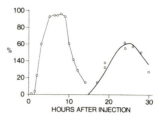

Fig. 6. FLM curve for BICR/M1 tumor (18). Cell loss was negligible. Volume doubling time was 23 hr. Assuming this to equal T_d, $K_p = 0.03$ hr^{-1} (Eq. **3**). From the experimental points, $T_B = 14$ hr, $K_{trans} = 0.1$ hr^{-1} (Eq. **10**). The second peak was calculated from successive hourly groups of labeled mitoses in the first peak, as described in Fig. 5.

synchronous cultures, also, some minimum time is required between mitosis and the beginning of S. This part of G_1 could be either a postmitotic period or one immediately preceding DNA synthesis, or both. The need for postmitotic reorganization seems intuitively probable, and there is evidence currently taken to indicate that a significant set of events immediately precedes DNA synthesis (15). This has been regarded as an important phase of the cycle in the regulation of growth. Our thesis, of course, is that the initiation of DNA synthesis is random and not regulated at all in the usual sense. It is nevertheless important to know when the A–B transition occurs, because the sequence of events in B-phase is interesting in its own right and because one wishes to know what occurs at transition. Unfortunately, we cannot find any data enabling us to decide whether B-phase begins before initiation of DNA synthesis or not.

Reinterpretation of FLM Curves. This technique consists in labeling cells in S with a "pulse" of [³H]dT and discovering the time course with which they subsequently pass through mitosis (16). At a time equal to G_2, cells that were at the end of S when labeled enter mitosis; thereafter, for a time equal to T_s, only labeled cells enter M, so the fraction of labeled mitoses (FLM) = 100%. Once cells labeled early in S have passed through mitosis, the FLM falls to zero and remains so until cells labeled late in S again enter mitosis. If the cell cycle is invariant, there follows a peak of labeled mitoses identical with the first, and so on *ad infinitum*. In practice, no such thing happens (16). The first peak of labeled mitoses usually fits reasonably well, but the second seldom if ever reaches 100% and is more widely spread. Such "damping" is always observed and may be of any degree. In extreme cases, no second peak occurs at the expected time (17–19). This result is attributed to variability of cell-cycle time. It is nevertheless maintained that such data yield values for the mean cycle time, measured as the time between the maxima of the first and second peaks. Analysis of this kind of data has reached a high level of sophistication (2, 20–23).

In the hypothesis proposed here the first peak of labeled mitoses is interpreted as before, but thereafter the analysis is different. The cells labeled at the end of S appear first in the FLM curve. Since T_B is the minimum intermitotic time, it is only after this that these cells will again enter mitosis. These cells will contribute to the pool of labeled mitoses at a rate determined by P. Their contribution will rise abruptly after T_B and then decline exponentially. Successive groups will behave in the same way; thus the total of labeled mitoses in

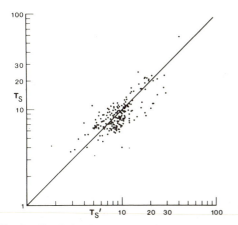

FIG. 7. Correlation between T_S and $T_{S'}$ in FLM curves. Of 348 published, 199 contained sufficient data to estimate both T_S and $T_{S'}$. The *line* shows the theoretical regression if T_B were invariate. Variation in T_B spreads the second peak, thus leading to higher estimates of $T_{S'}$ and lowering the slope of the regression of T_S on $T_{S'}$. Calculated slopes were 0.73 and 0.79 for normal and neoplastic tissues, respectively.

the second peak at any time can be calculated as the sum of a staggered set of curves of this type. The total rises for a period equal to T_S, whatever value P may take; the maximum height of the second peak is, however, a function of P.

Since the distribution of generation times is defined by P and T_B, it follows that FLM curves can be calculated from data obtained by other methods (Fig. 5). In practice it is easier to calculate the second peak from points on the first (Fig. 6). This method also takes some account of variability in B-phase. Published FLM curves confirm the predicted relationship between T_S and the time ($T_{S'}$) taken for the second peak to reach a maximum (Fig. 7).

"Proliferative Pools." Since variation in P alone simply alters the height of the second FLM peak without changing its position, the distance between peaks is not equal to nor a function of "mean generation time." Clearly, we must expect discrepancies between estimates of generation time based on FLM curves and those on, e.g., knowledge of T_S and labeling index. Where growth rate is high the discrepancy may be slight, since as P increases, the FLM curve approaches the "classical." With slowly proliferating tissues, large discrepancies have frequently been observed (13) and interpreted in terms of "resting phase," "proliferative pool," and "growth fraction." Such concepts propose that within morphologically homogeneous populations only some cells are cycling. The remainder are conceived as fertile, "out of cycle," but capable of re-entry after appropriate stimulation. They are sometimes considered to be in a special "G_0" phase (13).

If the existence of the A-state is admitted, it is unnecessary to postulate two distinct modes of behavior for cells in a homogeneous population, and the method commonly used for their detection is invalid. This is not to say that proliferative pools do not exist, and it is worth considering other methods of detection. One is to label cells in S and wait for the labeled group to lose its synchrony. The fraction of labeled mitoses

should then approach the labeling index, unless there is a subpopulation of nonproliferative cells (24). A second is to label continuously with [³H]dT. The labeling index will eventually reach 100% unless there is a subpopulation of nondividing cells. There are well-recognized methodological difficulties in both methods (13). In practice the "growth fraction" so measured tends towards unity as the interval between labeling and estimation increases. Although this result is expected, because the "nonproliferating" fraction is supposed to arise from the "decycling" of proliferating cells, it makes the choice of times for estimation of growth fraction arbitrary. In any case, unless "nonproliferating cells" can be shown to be fertile, they may be regarded as moribund.

Our proposal could be regarded as postulating proliferative pools as an inherent property of all cell populations, in that there will always be some cells engaged in processes leading to division and some that are not. Clearly this is not what is usually meant; namely that a distinct subpopulation with zero transition probability coexists with the proliferating population. In attempting to overcome certain difficulties in the "G_0" concept Burns and Tannock (17) proposed a model formally identical to ours. Unfortunately they failed to generalize their model, applying it only to slowly proliferating cells. Neither did they adduce firm evidence in its favor, and their paper has not received the attention it deserves, even in the restricted field to which it was applied.

Synchronous Cell Culture. Our hypothesis makes the kinetics of synchronized cultures predictable and increases their usefulness. Methods for synchronizing cells in culture have been extensively used in the study of biochemical changes through the cycle (25). However, it has not been

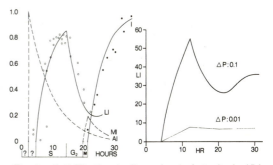

FIG. 8. (*left*) Kinetics of cells synchronized at mitosis. AI is the proportion of cells in A-state. Cells enter A-state some unknown time after mitosis. AI declines according to Eq. 11. Labeling index (LI) rises for a period = T_S as $1 - \exp(-K_{trans} \cdot t)$, where t is the time after the *minimum* G_1 period. Thereafter LI is given by $\{1 - \exp(-K_{trans} \cdot t)\} - \{1 - \exp(-K_{trans} \cdot [t - T_S])\}$. Similar curves can be calculated for the mitotic or G_2 indices. The proportional increase in cell number (I) beginning at time T_B is $1 - \exp(-K_{trans}[t - T_B])$. Curves I and LI were calculated from data for L5 cells (9). The value of $P = 0.18$ hr^{-1} was obtained from the distribution of intermitotic times (see Fig. 3). T_S was taken as 10 hr, by examination. MI = mitotic index.

FIG. 9. (*right*) Changes in LI when P, initially 0, is increased. Calculation of the first peak is as described in Fig. 8 and of the secondary rise as in Fig. 5, with $T_S = 8$ hr and $T_B = 14$ hr. The lag between stimulation and increased LI could be the time required to increase P, a fixed pre-S period, or both.

possible to maintain good synchrony, and attempts to improve synchrony by cloning (26) or fractionation of mitotic cells according to size (27) have failed. The existence of an indeterminate A-state accounts for this failure; the degree of synchrony is a function of P, and complete synchrony is possible only if $P = 1$. After synchronizing of cells in mitosis, the proportions of cells in S, G_2, or M rise to maxima in times equal to the durations of the phases (Fig. 8) because the number of cells entering decreases continuously, while the number leaving at a given time is the number that entered at a previous time equal to the duration of the phase. In Fig. 8, the cells behave as predicted.

Changes in Transition Probability. The growth rate of a population of cells is determined by P, T_B, and the rate of cell loss. It seems likely that variation in P is the major means of regulation. Fig. 9 shows the effects of small and large abrupt changes in P on the [^{3}H]dT labeling index. The larger change produces a quasi-synchronous burst of DNA synthesis. The response to the smaller change is so heavily damped that the labeling index simply rises to a new level. Both kinds of response have been observed (28–31).

Although it is premature to inquire closely into the mechanisms underlying the transition probability, it is pertinent to ask whether changes in P arise from generalized changes in cellular economy or from specific processes that "set" its value. In the simplest case, transition from A-state to B-phase could depend on a critical amount of a single substance, regulated by a number of linked feedback loops. The instantaneous amount of the "initiator" would vary cyclically (32). If the "initiator" were rare, the variation would be subject to considerable biochemical noise, and the threshold would be exceeded at irregular intervals. The mean amount of "initiator" and the amplitude of its fluctuations, and hence P, could be modified in many ways.

1. Howard, A. & Pelc, S. R. (1951) *Exp. Cell Res.* 2, 178–187.
2. Nachtwey, D. S. & Cameron, I. L. (1968) in *Methods in Cell Physiology*, ed. Prescott, D. M. (Academic Press, New York), Vol. III, pp. 213–257.
3. Prescott, D. M. (1968) *Cancer Res.* 28, 1815–1820.
4. Epifanova, O. I. & Tershikh, V. V. (1969) *Cell Tissue Kinet.* 2, 75–93.
5. Sisken, J. E. & Morasca, L. (1965) *J. Cell Biol.* 25, 179–189.
6. Dawson, K. B., Madoc-Jones, H. & Field, E. O. (1965) *Exp. Cell Res.* 38, 75–84.
7. Marin, G. & Bender, M. A. (1966) *Exp. Cell Res.* 43, 413–423.
8. Killander, D. & Zetterberg, A. (1965) *Exp. Cell Res.* 38, 272–284.
9. Terasima, T., Fujiwara, Y., Tanaka, S. & Yasukawa, M. (1968) in *Cancer Cells in Culture*, ed. Katsuta, H. (University of Tokyo Press, Tokyo), pp. 73–83.
10. Hsu, T. C. (1961) *Tex. Rep. Biol. Med.* 18, 31–33.
11. Cook, J. R. & Cook, B. (1962) *Exp. Cell Res.* 28, 524–530.
12. Terasima, T. & Tolmach, L. J. (1963) *Exp. Cell Res.* 30, 344–362.
13. Lala, P. K. (1971) in *Methods in Cancer Research*, ed. Busch, H. (Academic Press, New York), Vol. 6, pp. 3–95.
14. Steel, G. G. (1968) *Cell Tissue Kinet.* 1, 193–207.
15. Baserga, R. (1968) *Cell Tissue Kinet.* 1, 167–191.
16. Quastler, H. & Sherman, F. G. (1959) *Exp. Cell Res.* 17, 420–438.
17. Burns, F. J. & Tannock, I. F. (1970) *Cell Tissue Kinet.* 3, 321–334.
18. Gibbs, S. J. & Casarett, G. W. (1969) *Radiat. Res.* 40, 588–600.
19. Epifanova, O. I. (1966) *Exp. Cell Res.* 42, 562–577.
20. Barrett, J. C. (1966) *J. Nat. Cancer Inst.* 37, 443–450.
21. Steel, G. G. (1972) *Cell Tissue Kinet.* 5, 87–100.
22. Steel, G. G. & Hanes, S. (1971) *Cell Tissue Kinet.* 4, 93–105.
23. Steel, G. G., Adams, K. & Barrett, J. C. (1966) *Brit. J. Cancer* 20, 784–799.
24. Mendelsohn, M. L., Dohan, F. C. & Moore, H. A. (1960) *J. Nat. Cancer Inst.* 25, 477–484.
25. Mitchison, J. M. (1971) *The Biology of the Cell Cycle* (Cambridge University Press, Cambridge).
26. Peterson, D. F. & Anderson, E. C. (1964) *Nature* 203, 642–643.
27. Fox, T. O. & Pardee, A. B. (1970) *Science* 167, 80–82.
28. Smith, J. A. & King, R. J. B. (1972) *Exp. Cell Res.* 73, 351–359.
29. Martin, L., Finn, C. A. & Trinder, G. (1973) *J. Endocrinol.* 56, 133–144.
30. Barka, T. (1965) *Exp. Cell Res.* 37, 662–679.
31. Temin, H. M. (1970) *J. Cell. Physiol.* 75, 107–120.
32. Goodwin, B. C. (1963) *Temporal Organisation in Cells: A Dynamic Theory of Cellular Control Processes* (Academic Press, London).

Reprinted from

Proc. Natl. Acad. Sci. USA
Vol. 76, No. 9, pp. 4446–4450, September 1979
Cell Biology

Synthesis of labile, serum-dependent protein in early G_1 controls animal cell growth

(3T3 cells/restriction point/cycloheximide/cell cycle/protein synthesis)

PETER W. ROSSOW, VERONICA G. H. RIDDLE, AND ARTHUR B. PARDEE*

Department of Pharmacology, Harvard Medical School; and Laboratory of Tumor Biology, Sidney Farber Cancer Institute, 44 Binney Street, Boston, Massachusetts 02115

Contributed by Arthur B. Pardee, May 21, 1979

ABSTRACT   We present a model to account for several major observations on growth control of animal cells in culture. This model is tested by means of kinetic experiments which show that exponentially growing animal cells whose ability to synthesize total protein has been inhibited with cycloheximide (by up to 70%) grow at rates approximately proportional to their rates of protein synthesis. However, virtually the entire elongation of the cell cycle occurs in the part of the G_1 phase that depends on a high concentration of serum in the medium. This part of the cycle has earlier been suggested to lie prior to the restriction point—i.e., the point beyond the main regulatory processes of G_1. The remainder of the cycle, from restriction point to mitosis, is markedly insensitive to these concentrations of cycloheximide as well as to growth regulation. We quantitatively account for the specific lengthening of that part of the cycle involved in growth regulation by assuming that cells must accumulate a specific protein in a critical amount before they can proceed beyond the restriction point. The lability of this protein (half-life about 2 hr) makes its accumulation unusually sensitive to inhibition of total protein synthesis by cycloheximide. Its production appears to depend on growth factors provided by serum. The model can also account for greater variations of G_1 durations as the growth of cell populations is made slower. It also predicts two sorts of quiescence: one of cells slowly traversing G_1, in slightly suboptimal conditions; the other of cells that enter G_0 under inadequate conditions. Transformation of different sorts could create cells with altered variables for initiation, synthesis, or inactivation of the regulatory protein or could altogether eliminate the need for the protein.

The proliferation rate of animal cells is largely determined by the relative times that they spend in the cell cycle as opposed to quiescence. In each cycle an initiation event is required to start another round of the cell cycle; in its absence cells enter a quiescent state in which they remain until initiation occurs (1). Growth appears to be determined by the cyclic occurrence of this initiation event; completion of the cycle is relatively unaffected by conditions of growth (2). Similarly, for bacterial growth (3) and for synthesis of nucleic acids and proteins, control is by frequency of initiation; subsequent events proceed at constant rates.

We have defined completion of the growth-controlling event in the cell cycle as the restriction point R (1). When a growing culture (of 3T3 mouse cells) is shifted to a condition (medium containing 0.5% serum) that does not support continued growth (4, 5), R can be calculated from the fraction of cells that cannot leave the G_1 phase to lie about 2 hr before the beginning of DNA synthesis (6). Work with chicken cells earlier indicated that R lies somewhere in mid-G_1 (7).

The role of R point control is accentuated by studies of cells transformed to tumorigenicity by DNA tumor viruses (8–10). Whereas nontumorigenic cells become quiescent within one generation after the serum concentration is reduced, these transformed cells have lost their growth control and grow in media containing very low concentrations of serum. More remarkably, they (unlike the untransformed cells) do not accumulate with a G_1 DNA content; rather, they continue to traverse the entire cycle slowly. They have lost the requirement for serum factors that normal cells possess for passing the R point.

The complex nature of cell growth control is indicated by the many conditions that make cells become quiescent with a G_1 DNA content. Aside from insufficient serum, cells are arrested at high density or when they are put into suspension culture. The underlying biochemistry is affected by various drugs that arrest normal, but not transformed, cells (11). Factors that modulate cyclic nucleotide concentrations, as well as inhibitors of RNA and protein synthesis, prevent passage through G_1, and so does an insufficient supply of an essential nutrient, including an essential amino acid (12–14). Slight inhibition of protein synthesis by streptovitacin A (11) or the related compound cycloheximide (15) specifically acts in the G_1 phase to prevent or markedly delay DNA initiation.

Considering these many observations, we propose that the control of growth at the R point depends on the ability of the cell to accumulate a particular "initiator" (R) protein in a critical amount. Following the remarkable experiments of Schneiderman et al. (12) on the delays of initiation of DNA synthesis after pulses of cycloheximide given at high concentrations, we propose that the R protein is quite labile and, hence, unusually sensitive to conditions that limit protein synthesis. Serum growth factors are proposed to induce this protein. After the protein has accumulated in sufficient quantity, DNA synthesis and the remainder of the cycle continue relatively unaffected by serum or moderate inhibition of protein synthesis.

In the present work, we have tested this hypothesis by experiments in which we have combined partial inhibition of protein synthesis with lowering of the concentration of serum. We arrive at a quantitative model consistent with data on cell cycle durations and variations.

MATERIALS AND METHODS

Cell Culture. Swiss 3T3 cells were obtained from Howard Green (Massachusetts Institute of Technology) and BALB/c 3T3 clone A31 cells were obtained from Charles D. Scher (Sidney Farber Cancer Institute). Cells were routinely grown at 37°C in water-saturated 10% CO_2/90% air atmosphere in Dulbecco's modification of Eagle's medium (GIBCO H21, high glucose) supplemented with 10% calf serum, 100 units of pen-

Abbreviation: R, restriction point.
* To whom reprint requests should be addressed at Sidney Farber Cancer Institute.

Cell Biology: Rossow *et al.*

Proc. Natl. Acad. Sci. USA 76 (1979)   4447

icillin per ml, and 100 μg of streptomycin per ml. Both cell lines were determined to be free of pleuropneumonia-like organism (PPLO) by the ratio of [³H]uridine to [³H]uracil incorporation into RNA (16).

Cytofluorimetry. Cells were prepared for cytofluorimetry essentially by the method of Fried *et al.* (17). Analysis was performed with a model 4800A Cytofluorograf equipped with a model 2102 multichannel analyzer with a distribution integration capability (Biophysics Systems, Ortho Instruments, Westwood, MA).

Cell Counting. Cells were removed from duplicate plates or flasks by use of trypsin and were resuspended in 20 ml of Hanks' balanced salt solution containing 0.5% formalin to fix them. Cell numbers were determined with a Coulter Counter model B.

Cell Labeling and Autoradiography. Cells were labeled in 35-mm petri dishes by the addition of 2 μCi of [*methyl-*³H]-thymidine per ml incubated in the medium, and they were incubated for the indicated times. Trichloroacetic acid-insoluble radioactivity was determined after twice rinsing the plates with 5% ice-cold trichloroacetic acid and incubating at 37°C for 30 min with 0.5 ml of 0.1 M NaOH. The radioactivity in 0.2 ml was measured in either an Intertechnique model SL30 or a Beckman model 9000 liquid scintillation spectrophotometer. Alternatively, the fraction of labeled cells was determined by fixing the cells onto the petri dish with MeOH/HOAc, 2:1 (vol/vol). The dried plates were then coated directly with Kodak NTB 2 emulsion, exposed for 3 days, and then developed. A minimum of 400 total cells was used for each determination.

Materials. Calf serum was obtained from Flow Laboratories (McLean, VA) and powdered media from GIBCO. Radio-chemicals were obtained from New England Nuclear. Cyclo-heximide was obtained from Sigma and was prepared as a 100× concentrated stock in distilled H₂O and sterilized by filtration.

RESULTS

Inhibition of Cell Growth. Inhibition of growth was determined by exposing cultures of exponentially growing 3T3 cells to cycloheximide at various concentrations and determining the number of cells in the culture at intervals. Exponential growth was observed with low doses of cycloheximide over a 4-day period (Fig. 1). Growth was arrested at doses of the drug above 0.5 μg/ml, and substantial death occurred at about 1.0 μg/ml. Results from growth-rate experiments with Swiss 3T3 cells were in excellent agreement with these findings for A31 (data not shown). The relative rates of growth were obtained from the slopes of these curves and are shown in Fig. 2.

The inhibition of protein synthesis was determined by exposing cultures of exponentially growing cells to cycloheximide at various concentrations and to [³H]leucine at a final specific activity of 5 μCi/ml. After 24 hr of exposure, the radioactivity incorporated into trichloroacetic acid-insoluble material was determined and is expressed as percent of the control incorporation in the absence of drug (Fig. 2). In other experiments, the magnitude of the inhibitory effect at a given dose of cy-cloheximide was shown to be the same for times ranging from 0.5 to 24 hr (data not shown). The relative rates of growth and of protein synthesis are similar at cycloheximide concentrations up to 0.5 μg/ml.

Specific Lengthening of G₁ by Cycloheximide. The distribution of cells within the cycle can tell us if the cells growing exponentially in the presence of cycloheximide were perturbed at some specific phase or, rather, were slowed down throughout the cycle. Fig. 3 presents DNA histograms determined cy-

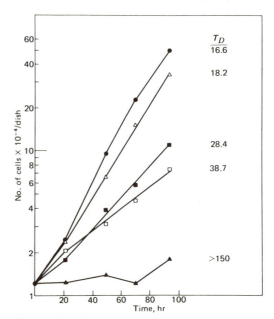

FIG. 1. Growth curves of BALB/c 3T3 cells grown in several concentrations of cycloheximide. T_D, population doubling time (in hr). Cycloheximide concentration (in μg/ml): ●, 0; △, 0.01; ■, 0.05; □, 0.1; ▲, 0.5.

tofluorographically for a set of exponentially growing Swiss 3T3 cells. The proportion of cells with a G₁ DNA content increased as the cycloheximide concentration increased (18). Quantitatively similar results were obtained for A31 cells (data not shown)

The accumulation of cells in the G₁ phase can be interpreted in two ways. All of the cells in the cultures containing cyclo-heximide could be traversing the G₁ phase more slowly than the controls. Alternatively, the elongated average doubling time could be due to a subpopulation of cells that are out of the cycle completely (e.g., in a G₀ state that has a G₁ DNA content), while the remaining cells are cycling with nearly the normal gener-

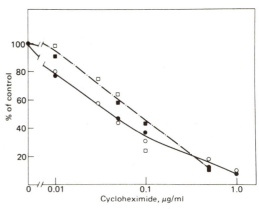

FIG. 2. Relative rates of protein synthesis and growth as a function of cycloheximide concentration. Closed symbols, Swiss 3T3; open symbols, BALB/c A31; circles, protein synthesis; squares, growth rate.

Proc. Natl. Acad. Sci. USA 76 (1979)

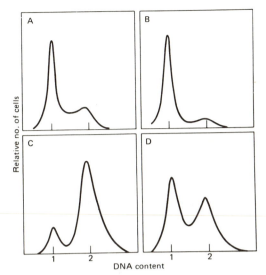

FIG. 3.   Representative histograms of exponentially growing Swiss 3T3 cells and of cells after shiftdown into 0.5% serum and 0.05 μg of Colcemid per ml. (*A*) No cycloheximide, exponential growth; (*B*) 0.05 μg of cycloheximide per ml, exponential growth; (*C*) no cycloheximide, 20 hr after shiftdown; (*D*) 0.05 μg of cycloheximide per ml, 20 hr after shiftdown.

ation time. If the latter is correct, the G_0 cells should exit from the G_1 peak of the histogram after cycloheximide removal according to G_0-to-S kinetics rather than M-to-S kinetics. For 3T3 cells, the minimum duration of G_0 to S is about 14 hr (19). The average duration of M to S is about 5 hr (6). When exponential cultures of either Swiss 3T3 or A31 cells in media containing up to 0.05 μg of cycloheximide per ml were shifted to media lacking the drug and containing 10% serum, all cells moved out of G_1 and into S phase within 12 hr. Even after growth with 0.1 μg of cycloheximide per ml, most cells had moved out of G_1 within 12 hr, and all had moved out by 20 hr (in the presence of Colcemid to prevent re-entry into G_1). Furthermore, G_0 cells in the presence of cycloheximide should very infrequently exit from the "G_1" peak of the flow microfluorimeter pattern. Colcemid was added to an exponential population of Swiss 3T3 cells growing in the presence of 0.03 μg of cycloheximide per ml (doubling time 21 hr) in order to catch cells that entered G_2 phase. Within 24 hr thereafter, all of the cells had left G_1 and had arrived at G_2 phase. This result, as well as the rapid movement of G_1 cells into G_2 after cycloheximide is removed, argues against a sizable population of the cells in cultures containing up to 0.1 μg of cycloheximide per ml being initially in a G_0 state.

The average durations of G_1 and of the remainder of the cycle can be calculated from histograms of the sort shown in Fig. 3 *A* and *B*. One assumes the distribution of cells in the population to follow the von Foerster equation, which takes into account the decreasing number of cells with increasing age in the cycle. The calculation shows that the S + G_2 + M duration is increased only slightly by cycloheximide and that almost all of the increase in cycle duration occurs in the G_1 period (18). These observations are extended in the following section of this article.

Elongation Is Prior to the Restriction Point. We were especially interested to determine whether the elongation of the G_1 phase by cycloheximide reflects inhibition of cycle ini-

tiation events during the serum-sensitive portion of G_1. To this end, exponentially growing cells were separated into pre- and post-restriction subpopulations, according to method I of Yen and Pardee (6). A primary block imposed by 0.5% serum arrests (in G_1) cells that have not yet traversed the serum-sensitive R point, and a secondary Colcemid block prevents those cells that are beyond the R point step from re-entering G_1, by arresting them at mitosis. After 20–24 hr of this treatment, the initial exponential population is separated into these two fractions, with G_1 and G_2 DNA contents, respectively. Elongation of the cycle prior to R should be reflected by an increase in the final G_1 peak relative to the G_2 peak.

This method requires that the cells must quickly be serum-starved by the shiftdown (18). We used a high cell density of 10^4 cells per cm^2 so that the small quantities of serum factors present in 0.5% serum were rapidly used by the cells (6). That the main process sensitive to cycloheximide is temporally located no later than the last serum-sensitive step in G_1 is shown by results of the kind in Fig. 3 *C* and *D*. The histograms obtained after 20 hr of shiftdown show that the fraction of 3T3 cells with a final G_1 DNA content was greater the higher the cycloheximide concentration. A31 cells gave qualitatively the same result.

The location of the R point was calculated (6) from the relative areas under the peaks by using the von Foerster equation to take into account the age distribution of the exponential population (Fig. 4). These calculations showed that for both cell lines the value of $T_{M \to R}$ (time from mitosis to R), the serum-sensitive portion of the cycle, increased dramatically with increasing cycloheximide concentration, in parallel with the increase of total generation time. In contrast, the value for the remainder of the cycle ($T_{R \to M}$) increased only slightly. For the Swiss cells a 4-fold increase in T_D is largely accounted for by a 30-fold increase in $T_{M \to R}$, with only a 1.3-fold increase in $T_{R \to M}$. For A31 cells the values were 2.7-, 6.5-, and 1.2-fold, respectively. The reciprocals of these times (the rate) are plotted against rates of protein synthesis (see Fig. 5).

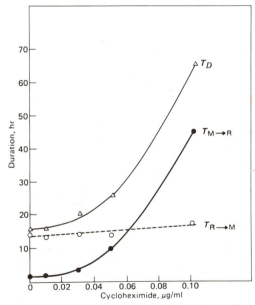

FIG. 4.   Elongation of pre- and post-restriction point portions of the cell cycle by cycloheximide. T_D, population doubling time; $T_{M \to R}$, time from mitosis to R; $T_{R \to M}$, time from R to mitosis.

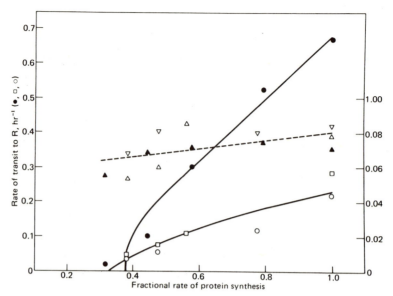

FIG. 5.   Rate of cell cycle transit as a function of fractional rate of protein synthesis. Swiss 3T3: ●, rate from M to R; ▲, rate from R to M. BALB/c A31: ○, rate from M to R, method I; □, rate from M to R, method II; ▽, rate from R to M, method I; △, rate from R to M, method II.

The above experiments were repeated by an alternative method (6) that avoids Colcemid. The re-entry of G_2 cells into G_1 is corrected for by including the increase in cell number, as each dividing cell becomes two cells. These results corroborate the results obtained with method I (Fig. 5).

DISCUSSION

Events required for cell cycle initiation were preferentially inhibited by low concentrations of cycloheximide. The G_1 phase, and particularly its initial, prerestriction point portion, were shown to be specifically lengthened manyfold. In striking contrast, the remainder of the cycle was only slightly lengthened; the effects of cycloheximide on the duration of late G_1, S, G_2, and M were even less affected than was total protein synthesis. The serum-independent portion of G_1, after the restriction point, remained at about 2 hr in the presence of cycloheximide; this interval might be required for protein synthesis and serum-independent events preparatory to DNA synthesis. The observations by Highfield and Dewey (13) and by Brooks (15), that DNA synthesis was inhibited only after about 1–2 hr after addition of cycloheximide, permitted the same explanation. These preparatory events and the subsequent parts of the cycle appear to depend upon proteins that are synthesized in sufficient amounts so that their function is not limiting if total protein synthetic capacity is inhibited by 50% for several generations. The serum-dependent, cycloheximide-sensitive, and serum-independent, cycloheximide-insensitive parts of the cycle are related to the A and B states, respectively, of Smith and Martin (2). The A state is defined as the one in which the cell probabilistically performs an initiation event that determines the rate of growth and the variation of cells within the population (2).

Berlin and Schimke (20) suggested that enzymes whose levels are rate determining will have a rapid rate of turnover, thereby making them respond rapidly to environmental alterations. Schneiderman *et al.* (12) proposed, from their data on delays of entry of synchronized CHO cells into S phase after exposures to high concentrations of cycloheximide, that the transit through G_1 requires synthesis of a labile protein (half-life 2 hr) in a critical amount. Brooks (15) also concluded from his studies of kinetics of entry of 3T3 cells into S, starting at quiescence, that a labile protein was required. He stressed the importance of this protein in setting the transition probability of the culture.

Our results fit a simple quantitative model [consistent with the concept of Schneiderman *et al.* (12)]. Assume the quantity of this protein (r) changes with time (t) according to

$$dr/dt = a - br,$$

in which a is the rate constant of total protein synthesis and b is the constant for rate of degradation of r. Integration and solution for the time (T) required to make the critical amount of protein (R), assuming that $r = 0$ when $t = 0$, gives

$$T = (-1/b) \ln (1 - bR/a).$$

In the presence of cycloheximide the rate of protein synthesis becomes fA, where f is the fraction of the uninhibited rate A. The quantity $(1 - bR/fA) = 0$ when f reaches a value $(F) = bR/A$. That is, F is the fractional rate of protein synthesis that makes T infinite. We can thus determine the constant (bR/A) that is equal to F, and thereby

$$T = (-1/b) \ln (1 - F/f).$$

We have plotted (Fig. 5) the rate of transit from M to R (the reciprocal of T) against the fractional rate of protein synthesis (f). This plot would be a straight line through the origin if protein R were stable ($b = 0$). Instead, the rate of transit was severely diminished as the rate of protein synthesis decreased; it became 0 (e.g., $T = \infty$) when $f = 0.38$. Using $F = 0.38$, we fitted the data with the line shown, which gives the half-life of R [equal to $(1/b) \ln 2$] as 2.2 hr. The curve is very sensitive to the choice of the half-life; values of 2.0 or 2.5 hr produced curves far from the data points. This half-life is very close to the value of 1.9 hr we obtained (21) by the method of Schneiderman *et al.* for 3T3 cells. It is also close to the value of Schneiderman *et al.* of 2.0 hr for CHO cells (12). Thus, the model based on accumulation of a labile protein and our data are quantitatively consistent. In accord with this model, serum growth factors could modulate the cellular content of protein R by affecting the rate constants a or b.

Variation of cycle duration and of G_1 phase within a culture is another property related to the sensitive growth control mechanism (2). Variation has been proposed to depend on a probabilistically occurring event that acts within otherwise similar cells at different times after division. The kinetic data do not clearly favor this mechanism over another one based on biochemical differences between the cells (22). Variation must have a biochemical basis because it is increased at lower serum concentrations (23) or in the presence of low concentrations of cycloheximide (15). Our model can account for these observations if we assume that cells differ by a factor of about 2-fold in their rates of protein R induction. If, for example, we give the fastest cells a rate 1.3 A and the slowest a rate 0.6 A, where A is the average rate, we calculate values of T as 1.0 and 3.0 hr, respectively. If the rates are inhibited by 1/3 with cycloheximide, the range of T becomes 2.0–7.0 hr. Inhibition thus moderately increases T for the fastest cells (by 1 hr), but considerably increases T for the slowest cells (by 4 hr). These results are reasonable when compared with data for variation of the fastest and slowest 10% of 3T3 cells (23) and the effects of cycloheximide on this variation (15).

From this model we can conceive quiescent cells as residing either in the G_1 state of the cell cycle or, alternatively, in G_0, depending on conditions. Moderate inhibition of protein synthesis, as in the present experiments, could hold the cells in a balance of synthesis and degradation that allows only very slow progress towards S phase. The cells would still be in G_1 phase. More severe restriction that nearly permanently prevents cells from reaching R would at the same time cause other proteins to become inactive and put the cells into a G_0 state, which is biochemically different from G_1 (24) and from which the cells could recover only after a long lag. Even more severe inhibition could lead to cell death. These considerations also apply to the consequences of amino acid deprivation, which can put some cells into G_0 (14), like serum deprivation (1), or other cells into a state from which they can more rapidly emerge (25–27).

This model predicts the existence of a serum-sensitive, rapidly turning over protein unique to early G_1. We can ask whether decontrolled growth of cancer cells results from modifications of the R system, such as from reduced requirements of serum factors for induction (28), or increased stability of the labile regulatory protein, R. An evident candidate for the R protein is the kind of protein produced by uninfected cells that is similar to proteins coded by viral genes such as *src* (29), *onc* (30), or gene A of simian virus 40 (31).

We thank M. Addonizio and D. Lehtomaki for providing excellent technical assistance and D. Schneider for preparing the manuscript for publication. This work was supported by U.S. Public Health Service Grant GM 24571 (to A.B.P.) and National Cancer Institute Fellowship 5-F32-CA 05595-02 (to P.W.R.). V.G.H.R. is an Aid for Cancer Research Fellow.

1. Pardee, A. B. (1974) *Proc. Natl. Acad. Sci. USA* **71**, 1286–1290.
2. Smith, J. A. & Martin, L. (1973) *Proc. Natl. Acad. Sci. USA* **70**, 1263–1267.
3. Helmstetter, C. E., Cooper, S., Pierucci, O. & Revelas, E. (1968) *Cold Spring Harbor Symp. Quant. Biol.* **33**, 809–822.
4. Holley, R. W. & Kiernan, J. A. (1968) *Proc. Natl. Acad. Sci. USA* **60**, 300–304.
5. Hayashi, I. & Sato, G. H. (1976) *Nature (London)* **259**, 132–134.
6. Yen, A. & Pardee, A. B. (1978) *Exp. Cell Res.* **116**, 103–113.
7. Temin, H. M. (1971) *J. Cell. Physiol.* **78**, 161–170.
8. Martin, R. G. & Stein, S. (1976) *Proc. Natl. Acad. Sci. USA* **73**, 1655–1659.
9. Bartholomew, J. C., Yokota, H. & Ross, P. (1976) *J. Cell. Physiol.* **88**, 277–286.
10. Paul, D. (1973) *Biochem. Biophys. Res. Commun.* **53**, 745–753.
11. Pardee, A. B. & James, L. J. (1975) *Proc. Natl. Acad. Sci. USA* **72**, 4994–4998.
12. Schneiderman, M. H., Dewey, W. C. & Highfield, D. P. (1971) *Exp. Cell Res.* **67**, 147–155.
13. Highfield, D. P. & Dewey, W. C. (1972) *Exp. Cell Res.* **75**, 314–320.
14. Tobey, R. A. & Ley, K. D. (1970) *J. Cell Biol.* **46**, 151–157.
15. Brooks, R. F. (1977) *Cell* **12**, 311–317.
16. Schneider, E. L., Stanbridge, E. J. & Epstein, C. J. (1974) *Exp. Cell Res.* **84**, 311–318.
17. Fried, J., Perez, A. G. & Clarkson, B. D. (1976) *J. Cell Biol.* **71**, 172–181.
18. Riddle, V. G. H., Pardee, A. B. & Rossow, P. W. (1979) *J. Supramol. Struct.*, in press.
19. Riddle, V. G. H., Dubrow, R. & Pardee, A. B. (1979) *Proc. Natl. Acad. Sci. USA* **76**, 1298–1302.
20. Berlin, C. M. & Schimke, R. T. (1965) *Mol. Pharmacol.* **1**, 146–156.
21. Riddle, V. G. H., Rossow, P. W., Boorstein, R. J., Addonizio, M. L., & Pardee, A. B. (1979) in *Molecular and Cellular Basis of Lymphocyte Activation* (Elsevier/North-Holland, Amsterdam), in press.
22. Pardee, A. B., Shilo, B. Z. & Koch, A. L. (1979) in *Conference on Hormones and Cell Growth* (Cold Spring Harbor Laboratory, Cold Spring Harbor, NY), in press.
23. Shields, R. & Smith, J. A. (1977) *J. Cell. Physiol.* **91**, 345–356.
24. Baserga, R. (1976) *Multiplication and Division in Mammalian Cells* (Dekker, New York).
25. Burstin, S. J., Meiss, H. K. & Basilico, C. (1974) *J. Cell. Physiol.* **84**, 397–408.
26. Yen, A. & Pardee, A. B. (1978) *Exp. Cell Res.* **114**, 389–395.
27. Pledger, W. J., Stiles, C. D., Antoniades, H. N. & Scher, C. D. (1978) *Proc. Natl. Acad. Sci. USA* **75**, 2839–2843.
28. Cherington, P. V., Smith, B. L. & Pardee, A. B. (1979) *Proc. Natl. Acad. Sci. USA* **76**, 3937–3941.
29. Oppermann, H., Levinson, A. D., Varmus, H. E., Levinlow, L. & Bishop, J. M. (1979) *Proc. Natl. Acad. Sci. USA* **76**, 1804–1808.
30. Duesberg, P. H. & Vogt, P. K. (1979) *Proc. Natl. Acad. Sci. USA* **76**, 1633–1637.
31. Tenen, D. G., Garewal, H., Haines, L., Hudson, J., Woodard, V., Light, S. & Livingston, D. M. (1977) *Proc. Natl. Acad. Sci. USA* **74**, 3745–3749.

Cell, Vol. 19, 527–536, February 1980, Copyright © 1980 by MIT

A Fixed Site of DNA Replication in Eucaryotic Cells

Drew M. Pardoll, Bert Vogelstein and
Donald S. Coffey
Cell Structure and Function Laboratory
The Oncology Center
Johns Hopkins University
School of Medicine
Baltimore, Maryland 21205

Summary

We studied the role of the nuclear matrix (the skeletal framework of the nucleus) in DNA replication both in vivo and in a cell culture system. When regenerating rat liver or exponentially growing 3T3 fibroblasts are pulse-labeled with ³H–thymidine and nuclear matrix is subsequently isolated, the fraction of DNA remaining tightly attached to the matrix is highly enriched in newly synthesized DNA. After a 30 sec pulse labeling period and limited DNAase I digestion, the matrix DNA of 3T3 fibroblasts, which constitutes 15% of the total DNA, contains approximately 90% of the labeled newly synthesized DNA. Over 80% of this label can be chased out of the matrix DNA if the pulse is followed by a 45 min incubation with excess unlabeled thymidine. These and other kinetic studies suggest that the growing point of DNA replication is attached to the nuclear matrix. Studies measuring the size distribution of the matrix DNA also support this conclusion. Reconstitution controls and autoradiographic studies indicate that these results are not due to preferential, nonspecific binding of nascent DNA to the matrix during the extraction procedures. Electron microscopic autoradiography shows that, as with intact nuclei, sites of DNA replication are distributed throughout the nuclear matrix. A fixed site of DNA synthesis is proposed in which DNA replication complexes are anchored to the nuclear matrix and the DNA is reeled through these complexes as it is replicated.

Introduction

Almost twenty years ago, Jacob, Brenner and Cuzin (1963) recognized that a structural framework for replication of DNA would be quite valuable to a cell. In their original hypothesis, they postulated that DNA sequences involved in the initiation of DNA synthesis were bound to a supporting element within the bacterial cell. This physical attachment of the daughter DNA strands to the cell wall would facilitate the subsequent partitioning of the chromosomes to the daughter cells. This hypothesis has been substantiated for several bacterial species, and a number of genetic and biochemical techniques have been used to show that the procaryotic replication complex is attached to the cell membrane (Ganesan and Leder-

berg, 1965; Smith and Hanawalt, 1967; Sueoka and Quinn, 1968; Tremblay, Daniels and Schaechter, 1969; Worcel and Burgi, 1972).

A similar mechanism would have equal utility in eucaryotes, where an enormous amount of DNA must be ordered spatially during replication such that the daughter strands remain untangled yet coupled in a precise fashion for later entry into mitosis. Hence many investigators have carried out studies analogous to those in bacteria to determine whether replicating DNA in eucaryotes is associated with the nuclear membrane (Comings and Kakefuda, 1968; Williams and Ockey, 1970; Hanaoka and Yamada, 1971; Pearson and Hanawalt, 1971; Huberman, Tsai and Deich, 1973). Although this area is somewhat controversial, most investigators have found no preferential association of replicating DNA with the nuclear membrane (for review see Lewin, 1974).

Recent work has shown, however, that other structural elements exist in the nucleus in addition to the nuclear membrane. In particular, the nucleus of a wide variety of eucaryotic cells contains an underlying skeleton, often called the nuclear matrix (Berezney and Coffey, 1974; Shelton, 1976; Comings and Okada, 1976; Herlan and Wunderlich, 1976; Keller and Riley, 1976; Hodge et al., 1977; Herman, Weymouth and Penman, 1978; Miller, Huang and Pogo, 1978). This matrix is revealed by sequentially extracting purified nuclei with detergents, hypotonic low magnesium buffers and high salt, and digesting with nucleases. The resultant structure retains the same size and shape of the original nucleus and can be seen to contain a residual peripheral lamina with nuclear pore complexes, residual nucleoli and an internal fibrogranular network (Figure 1). It consists largely of protein components representing a small subset of acidic polypeptides (5–10% of total nuclear protein), RNA and DNA (Berezney and Coffey, 1977). As shown in this study, the fraction of total DNA remaining attached to the matrix can be made to vary between 0.5 and 90% by varying the extent of DNAase I digestion.

We have investigated whether the nuclear matrix in eucaryotes replaces the bacterial membrane as a supporting structure for DNA synthesis. Using biochemical and autoradiographic methods, we have studied DNA synthesis in vivo with the regenerating rat liver and in cell culture with 3T3 fibroblasts. Our results suggest that the matrix provides fixed sites for the attachment of replication complexes and that the DNA is reeled through these sites as it is replicated.

Results

Association of Newly Replicated DNA with the Nuclear Matrix of Regenerating Rat Liver

Using the regenerating rat liver system, we first sought to determine whether newly made DNA was bound to the nuclear matrix and, if so, the spatial distribution of

Figure 1. Electron Micrograph of 3T3 Fibroblast Nuclear Matrix

Nuclear matrix was obtained from 3T3 fibroblasts as described in Experimental Procedures. Samples for electron microscopy were fixed in glutaraldehyde, postfixed in OsO_4 and stained with uranyl acetate and lead citrate. The nuclear matrix is seen to consist of a residual lamina, a residual nucleolus and an internal fibrogranular network. Bar = 1 μ.

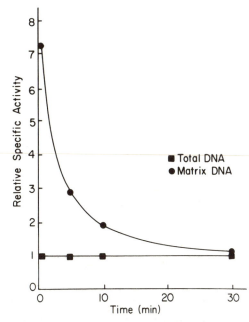

Figure 2. Kinetics of Association of Newly Replicated DNA with the Nuclear Matrix of Regenerating Rat Liver

Regenerating rat livers were labeled for various times after injection of 200 μCi ^{3}H–thymidine into the portal vein of partially hepatecto-mized rats. Nuclei were isolated and incubated for 12 hr in TM buffer. Matrix was then isolated by extracting nuclei with low Mg buffer, high salt buffer and Triton X-100. The DNA content of each fraction was quantitated by diphenylamine assay. Specific activities of the matrix DNA relative to total DNA are shown for various pulse times.

matrix-associated nascent DNA.

[3]H–thymidine was injected into the portal veins of 24 hr partially hepatectomized rats. At various times after injection, the livers were removed and minced, and nuclei were isolated according to the method of Berezney and Coffey (1977). Incubation of the purified liver nuclei for 12 h in TM [10 mM Tris (pH 7.4), 5 mM $MgCl_2$] buffer at 4°C allows an endogenous DNAase to nick the nuclear DNA. During the subsequent matrix extraction procedure, 95–99% of the DNA is removed; the residual DNA is tightly bound to the nuclear matrix. The specific activity of the matrix DNA (relative to that of total DNA) as a function of pulse time is shown in Figure 2. At short pulse times the matrix DNA has a very high specific activity (relative to total DNA), indicating that it is highly enriched in newly synthesized DNA. After longer pulse times, the relative specific activity of the matrix DNA decreases toward a value of 1.

In a nonsynchronous system such as the regenerating rat liver, DNA is in all phases of replication; that is, some replicons have just initiated while others are at various stages of elongation (Figure 3). The data in Figure 2 are compatible with the attachment of the growing points of replicating DNA to the nuclear matrix. The possibility that only the initiation points in each replicon are bound to the nuclear matrix is eliminated by data presented below.

It was of particular interest to examine where the

newly made DNA was located within the matrix structure. Hence we performed electron microscopic autoradiography on matrix isolated from pulse-labeled regenerating rat liver. In 80% of the labeled matrices examined, most of the silver grains were located in the interior of the matrix (Figure 4). In the other 20%, there was increased labeling of the periphery or the nucleolar region. Similar distributions were found in autoradiograms of whole nuclei. Those cases showing increased labeling at the periphery or residual nucleolar region probably represent nuclei in the late stages of S phase, during which time peripheral heterochromatin and nucleolar DNA are replicated (Huberman et al., 1973). Hence the location of newly made DNA on the matrix parallels that of the intact nucleus.

These data are compatible with the results of previous investigations which have shown labeling in the interior as well as at the periphery of nuclei (Huberman et al., 1973). Other investigators (Fakan and Hancock, 1974) have shown that there is preferential localization of newly labeled DNA at the border between condensed chromatin and interchromatinic areas in exponentially dividing mouse cells (P815). Since this border is not well defined in the rat liver nuclear

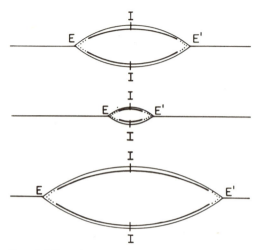

Figure 3. Different Stages of Replicon Replication during Pulse Labeling

When asynchronously growing cells are pulse-labeled with ³H-thymidine, the label is incorporated at the replicating fork of replicons that are in various stages of elongation. Replicons in three different stages of elongation are depicted schematically. Sites of replicon initiation (I) and chain elongation (E, E') are marked. Newly synthesized DNA that has incorporated label is shown by a dotted line.

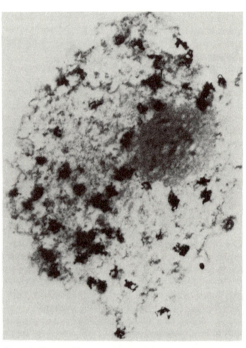

Figure 4. Electron Microscopic Autoradiogram of Nuclear Matrix from ³H-Thymidine Pulse-Labeled Regenerating Rat Liver

Regenerating rat livers were pulse-labeled in vivo for 5 min with ³H-thymidine, and nuclear matrices were isolated as in Figure 2. Matrix was prepared for electron microscopy by fixation in 4% glutaraldehyde, postfixation in 2% OsO₄, sectioning and staining with lead citrate. Grids were coated with autoradiographic emulsion, exposed for 60 days and then developed. Magnification 19,000X.

matrix, our results are not inconsistent with these findings.

Study of Matrix-Associated DNA Replication in 3T3 Fibroblasts

To study matrix-associated DNA replication further, we have focused on a cell culture system in which pulse labeling and pulse chase experiments can be performed more readily. Since isolated 3T3 fibroblast nuclei have virtually no endogenous nuclease activity, exogenous DNAase must be added to nick the DNA and thereby allow its removal during the matrix extraction procedures. Between 0.5 and 90% of the DNA remains associated with the nuclear matrix, depending on the concentration and duration of DNAase I treatment. Using DNAase I in the preparation of nuclear matrix from 3T3 fibroblast nuclei, we have obtained a structure similar to rat liver nuclear matrix consisting of a peripheral lamina, residual nucleoli and an internal fibrogranular network (Figure 1). The SDS-PAGE profile of 3T3 nuclear matrix is similar to that of rat liver nuclear matrix, showing five major protein bands with molecular weights between 42,000 and 78,000 daltons and 10–20 minor bands; no histones remain (D. M. Pardoll and B. Vogelstein, unpublished data).

The DNA of rapidly growing unsynchronized 3T3 fibroblasts was uniformly labeled by allowing at least two generations of growth in media containing ¹⁴C-thymidine. The cells were then pulse-labeled with ³H-thymidine and the nuclei were isolated (see Experi-

mental Procedures). If DNAase I was added to nuclei before matrix isolation, the resultant matrix DNA showed a slight (2 fold) enrichment in newly synthesized DNA. The DNA released from the nuclei during the DNAase I incubation was highly (10 fold) enriched in ³H label, indicating that nascent DNA in the intact nucleus is extremely sensitive to DNAase I. This result has been reported for HeLa cells (Seale, 1975) and is probably due to the relative absence of nucleosomes on nascent DNA (Weintraub, 1973; Riley and Weintraub, 1979). To show an attachment of newly made DNA to a subnuclear structure, we attempted to circumvent the differential susceptibility of nascent DNA to DNAase I in intact nuclei by first removing the histones from the DNA (with the 2 M NaCl extraction step), thereby rendering the newly synthesized and bulk DNA equally susceptible to nuclease. When matrices were prepared in this way (adding DNAase I after 2 M NaCl extraction), the residual matrix DNA had a very high relative specific activity (that is, a high ³H/¹⁴C ratio relative to total DNA). In all further experiments, DNAase I incubations were carried out after the high salt extraction. Using various times of

DNAase I digestion to control the amount of DNA released from the matrix, we found that the relative specific activity of the matrix DNA from a 1 min pulse labeling period increased as the amount of DNA on the matrix decreased (Figure 5). This is what would be expected if the growing points of replicating DNA were attached to the matrix. One would also expect that with increased labeling times, some of the ^{3}H-labeled nascent DNA would be far enough from matrix attachment points that it would be severed from the matrix upon DNAase I treatment. This prediction was confirmed in the 10 and 30 min pulse experiments shown in Figure 5. With the 30 min pulse, there is very little enrichment in the matrix DNA. If the relative specific activity is plotted against pulse time, as in Figure 6, the result appears very similar to that in the regenerating rat liver system (Figure 2).

Size Distribution of Matrix DNA

To characterize further the nascent DNA bound to the matrix, we utilized gel electrophoresis to analyze its size distribution. Matrix from 1 min pulse-labeled fibroblasts was prepared using enough DNAase I treatment so that 6% of the total DNA remained matrix-bound. This DNA, which contained approximately 40% of the ^{3}H-labeled nascent DNA, was purified by SDS-proteinase K treatment followed by phenol extraction. The matrix DNA was then sized on a 1% agarose gel. Figure 7 shows the distribution of ^{3}H

counts as well as the ^{3}H/^{14}C ratio from slices along the gel.

Several conclusions can be drawn from Figure 7. First, the newly made DNA covers a very broad size distribution, suggesting that the DNAase I action on the histone-depleted DNA was essentially random. The fact that much of the DNA is in the form of short pieces (500–1500 bp) indicates that the DNAase I often cuts very close to the matrix. Second, the ^{3}H/^{14}C ratio increases with decreasing DNA size until it reaches a value of approximately 3000 bp, at which point it plateaus. This value is roughly equal to the number of base pairs replicated within 1 min (Edenberg and Huberman, 1975). If DNA replication were occurring on the matrix, one would predict that all matrix-bound DNA fragments shorter than the number of bases replicated during a 1 min labeling period should be uniformly labeled and thus have a constant ^{3}H/^{14}C ratio. Hence the above result is in concordance with a matrix-bound DNA replication site.

30 Sec Pulse Labeling with 3T3 Fibroblasts

One might expect all of the newly replicated DNA to be associated with the matrix if the matrix is the site of DNA replication. A question that arises from the

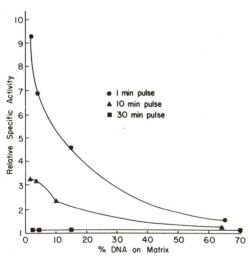

Figure 5. Association of Newly Replicated DNA with 3T3 Fibroblast Matrix after DNAase I Digestion

Rapidly growing 3T3 fibroblasts were pulse-labeled for 1, 10 and 30 min with ^{3}H-thymidine and nuclei and then matrix was isolated. Varying amounts of DNA were released from matrix by incubation with DNAase I for different periods of time. Matrix was then pelleted, dissolved in 1% SDS and counted. The specific activities of the matrix DNA relative to total DNA specific activity are shown as a function of the percentage of total nuclear DNA remaining attached to the matrix.

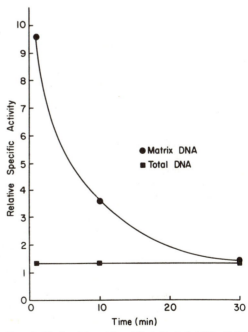

Figure 6. Kinetics of Association of Newly Replicated DNA with the Nuclear Matrix of 3T3 Fibroblasts

3T3 fibroblasts were pulse-labeled for various times with ^{3}H-thymidine and matrix was isolated as in Figure 5. Matrix was then incubated for 25 min with DNAase I (400 U/ml), pelleted, resuspended in 1% SDS and counted. Specific activity of matrix DNA relative to total DNA specific activity is shown for the various pulse times.

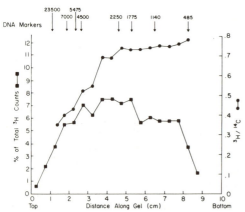

Figure 7. Size Distribution of 1 Min Pulse-Labeled Matrix DNA from 3T3 Fibroblasts

3T3 fibroblasts were pulse-labeled for 1 min with ^{3}H–thymidine and matrix was isolated as in Figure 5. Matrix was then incubated for 12 min with DNAase I (400 U/ml) and pelleted. Matrix DNA, which represented 7% of total DNA, was purified using SDS-proteinase K incubation followed by phenol extraction. The DNA was electrophoresed on a 1% agarose gel. The gel was cut into 5 mm slices which were dissolved in boiling 1 M HCl and counted in Aquasol™. The percentage of total ^{3}H counts and ^{3}H/^{14}C ratio are shown for each gel slice. DNA size markers are indicated at the top.

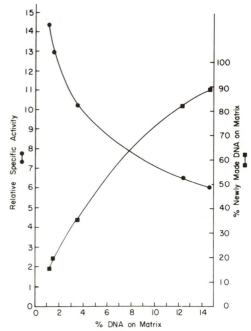

Figure 8. Association of Newly Replicated DNA with 3T3 Fibroblast Matrix after 30 Sec Pulse Labeling Period

Rapidly growing 3T3 fibroblasts were pulse-labeled for 30 sec with ^{3}H–thymidine and matrix was isolated as in Figure 5. Matrix was incubated for various times with DNAase I (400 U/ml), pelleted, resuspended in 1% SDS and counted. The percentage of total ^{3}H label on the matrix and specific activities of the matrix DNA relative to total DNA specific activity are shown as a function of the percentage of total nuclear DNA remaining attached to the matrix.

data of Figure 5 is why only 18–63% of the newly replicated DNA is associated with the matrix at DNAase I digestion conditions that leave 2–14% of the total DNA on the matrix. The fact that much of the matrix DNA is shorter than the length replicated during a 1 min labeling period (Figure 7), however, indicates that the DNAase I often cuts within the labeled DNA and thus releases label from the matrix. Hence it should be possible to retain a higher percentage of label on the matrix with a shorter labeling period.

Figure 8 shows the results of a 30 sec pulse testing this possibility. The result is that when matrix is incubated with DNAase I such that 85% of DNA is released from it, the remaining matrix DNA contains 90% of the newly synthesized DNA. In addition, using the same argument as presented above, after a 30 sec labeling period the ratio of ^{3}H/^{14}C in matrix DNA fragments should plateau at a fragment length corresponding to the number of base pairs replicated in 30 sec. Indeed, when the experiment in Figure 7 was carried out with a 30 sec pulse labeling, this plateau was reached at approximately 1600 bp (data not shown), which is close to the estimated number of base pairs replicated in 30 sec (Edenberg and Huberman, 1975) and is about half the number of base pairs at which the plateau occurs for the 1 min pulse labeling (Figure 6).

Pulse-Chase Labeling of 3T3 Fibroblasts

If the DNA replication site were bound to the matrix, it should be possible to chase most of the ^{3}H pulse label off the matrix by incubating cells in media containing an excess of unlabeled thymidine following the pulse. Pulse-chase experiments, however, are difficult to perform in intact mammalian cells. When pulse-labeled fibroblasts were washed and incubated with medium containing 40 µM unlabeled thymidine (100 times the concentration of ^{3}H–thymidine used for the pulse), label incorporation continued at one third the normal rate for 5–10 min before leveling off. The rate of replication, as assayed by simultaneous ^{14}C–deoxyadenosine incorporation, was not altered during the chase period (our unpublished results). Pulse-chase experiments were carried out despite the difficulties in performing an ideal chase. Table 1 shows that a 30 sec pulse followed by a 45 min chase results in the removal of most (83%) of the ^{3}H cpm from the matrix DNA. This is a necessary criterion in establishing that the growing forks are always associated with the nuclear matrix, and that a given DNA sequence is transiently associated with the matrix when it is part of the growing fork.

Reconstitution Controls

We have shown with the above data that newly replicated DNA seems to be associated with the nuclear

Table 1. Results of Pulse-Chase Experiments with 3T3 Fibroblasts

		Total DNA (^{14}C cpm)	Total DNA (%)[a]	Newly Synthesized DNA (^{3}H cpm)	Newly Synthesized DNA (%)[b]	Specific Activity Relative to Total DNA
30 Sec pulse[c]	matrix DNA	510	1.9	2,480	24.7	13.0
	non-matrix DNA[e]	26,300	98.1	7,570	75.3	0.8
	total DNA	26,800	100	10,000	100	1.0
45 Min pulse[c]	matrix DNA	630	2.6	32,100	2.8	1.1
	non-matrix DNA[e]	23,700	97.4	1,130,000	97.2	1.0
	total DNA	24,400	100	1,160,000	100	1.0
30 Sec pulse- 45 min chase[d]	matrix DNA	550	2.4	430	1.4	0.6
	non-matrix DNA[e]	22,400	97.6	30,100	98.6	1.0
	total DNA	22,900	100	30,500	100	1.0

[a] Based on ^{14}C cpm.
[b] Based on ^{3}H cpm.
[c] Cells grown in ^{14}C–thymidine (0.005 μCi/ml) for 48 hr were pulse-labeled with 20 μCi/ml ^{3}H–thymidine.
[d] Chase was performed by incubation of pulse-labeled cells in medium containing 40 μM unlabeled thymidine (100 × concentration of ^{3}H–thymidine used for the pulse).
[e] DNA released during isolation of matrix.

matrix, presumably by specific attachments of DNA (or nonhistone chromatin proteins) to matrix proteins at the growing fork. The possibility remained, however, that this association was the result of nonspecific binding of newly replicated DNA to matrix components. To investigate this possibility, two types of experiments were carried out. In reconstitution experiments, we added purified matrix DNA or DNA released during matrix preparation ("non-matrix DNA") to suspensions of nuclei during various stages of matrix extraction. When DNA was added before or during detergent washing, during low magnesium buffer extraction, during high salt extraction or before or after DNAase I digestion, there was no indication of preferential binding of newly made DNA (^{3}H-labeled) or matrix DNA with nuclear matrices (Table 2). None of the DNA preparations bound significantly to nuclei or matrix.

Although this result shows that purified newly replicated DNA has no great affinity for matrix, it is possible that the incompletely deproteinated DNA released during actual matrix preparation might be binding to matrix nonspecifically. If this were true, then one might expect that all matrices in a given preparation would bind some of the released newly synthesized DNA. Autoradiography of nuclei and matrix isolated from 1 min pulse-labeled 3T3 fibroblasts was carried out to determine the distribution of label among matrices. Only 23% of the matrix spheres were labeled, a value the same as the percentage of labeled nuclei (24%). Hence rebinding of released newly replicated chromatin in a nonspecific manner cannot account for our data.

Table 2. Binding of Matrix and Non-matrix DNA to Nuclei and Nuclear Matrix

	cpm Bound to Nuclei at Various Stages of Matrix Extraction (%)[a]			
	Matrix DNA		Non-matrix DNA	
Steps at Which Labeled DNA Was Added[b]	[3]H	[14]C	[3]H	[14]C
Intact nuclei	2.0	2.3	1.6	1.7
During extraction with low magnesium buffer	<1.0	<1.0	<1.0	<1.0
During extraction with high salt buffer	<1.0	<1.0	<1.0	<1.0
During DNAase I digestion	<1.0	<1.0	<1.0	<1.0
Purified matrix[c]	1.2	1.4	1.4	1.6

[a] 3T3 cells were labeled with ^{14}C-thymidine for 48 hr and pulse-labeled with ^{3}H-thymidine for 1 min. Nuclear matrix was prepared from these cells as described in Experimental Procedures. DNA was then purified from the matrix and from the pooled nuclear extracts ("non-matrix DNA") using SDS-proteinase K and phenol-chloroform extraction.
[b] Unlabeled nuclei (1 × 10[6] per ml) at various stages of matrix preparation were incubated with the indicated labeled DNA (5 μg/ml) for 1 hr at each step. After centrifuging at 800 × g for 15 min the percentage of cpm pelleted was determined by scintillation counting.
[c] After washing out DNAase I with high salt buffer.

Discussion

We have presented evidence that the nuclear matrix of regenerating rat liver and 3T3 fibroblasts is associated with newly synthesized DNA. This has been demonstrated by nicking the nuclear DNA with either endogenous nuclease in the case of rat liver or exogenous DNase I in the case of 3T3 fibroblasts, and showing that the DNA which remained bound to the nuclear matrix was enriched in newly made DNA. Furthermore, we have shown that the smaller the amount of DNA on the matrix, the greater the enrichment (Figure 5). With suitable pulse labeling times and nuclease digestions, 90% of the newly synthesized DNA is associated with the matrix when only 15% of the total DNA remains with the matrix (Figure 8). As the growing fork progresses (during a chase period), the previously labeled DNA loses its matrix association (Table 1).

The fact that the matrix DNA is enriched in newly replicated DNA for a short period after its synthesis (Figures 2 and 6; Table 1) indicates that replicons in some phase of replication (initiation, elongation or termination) are anchored to the matrix. Several lines of evidence, however, rule out the possibility that only initiation or termination account for this enrichment. Attachment of DNA to the matrix at the initiation (or termination) site alone does not easily explain the result that as more DNA is digested from the matrix, the specific activity of the matrix DNA rises (Figures 5 and 8). The data in Figure 8 also show that the great majority of newly synthesized DNA is matrix-associated when less than 15% of the total DNA bound to the matrix. In a nonsynchronous system such as we have used, DNA will be in all stages of replication, and therefore only a small number of replicons will initiate (or terminate) during a 30 sec pulse period. Hence, if the initiation or termination sites alone were matrix-bound, the matrix could only contain the bulk of newly synthesized DNA if the labeled DNA strands attached to the matrix were long enough to include the growing points. Numerous sizing experiments (Figure 7 and our unpublished data), however, show that the matrix DNA fragments are 500–10,000 bp; that is, far shorter than the length of a replicon (50,000–150,000 bp) (Edenberg and Huberman, 1975). Our data thus imply that virtually all replication forks are attached to the nuclear matrix. This conclusion is further supported by the fact that the specific activity of matrix DNA plateaus at a fragment size approximately equal to the number of base pairs replicated during the pulse labeling period (Figure 7).

An alternative explanation for our findings and others (Berezney and Coffey, 1975; Dijkwel, Mullenders and Wanka, 1979) is that newly replicated DNA binds to the matrix nonspecifically at some time during the extraction procedure. We carried out several experiments to address this possibility. Reconstitution experiments gave no indication that newly replicated DNA would bind to nuclei or nuclear matrix, even when added at all stages of the preparative procedure. It was important that these experiments were carried out with newly replicated DNA since it is known that

such DNA is somewhat different than total DNA, presumably due to the presence of single-stranded regions (Painter and Schaeffer, 1969; Berger and Irvin, 1970; Habener, Bynum and Shack, 1970). Autoradiographic experiments show that the proportion of labeled nuclear matrices is similar to that of labeled nuclei in the same experiment. This diminishes the possibility that nascent chromatin released during the extraction procedures binds in a nonspecific manner to the nuclear matrices in suspension. Nevertheless, we cannot completely rule out the possibility that the nascent DNA becomes trapped within or bound to only the S phase nuclei during some phase of the extraction procedure and that this association did not really exist in situ.

The concept of a matrix-bound replication complex has some interesting implications with regard to the topology of DNA organization in the nucleus. It is known that the eucaryotic genome is divided into thousands of independently replicating replicons and that DNA synthesis in each replicon is bidirectional with respect to an initiation point (Huberman and Riggs, 1968). To reconcile a fixed site of replication with bidirectional synthesis, there must be two adjacent fixed replication complexes for each replicon. In other words, E and E' of Figure 3 would be adjacent. In this model, the DNA would be reeled in from both sides as the newly replicated strands loop out between the two complexes (Figure 9B). Hence loops of DNA would be created automatically and connected to the matrix as part of the replication process. This is in contrast to the more common view that two replication complexes are moving in opposite directions to create the observed bidirectionality. Indeed, if the large number of enzymes discovered to function in procaryotic DNA replication (Wickner and Hurwitz, 1974; Schekman et al., 1975) are also involved in eucaryotic DNA replication, it might be more energetically economical that a replication complex containing all these enzymes be fixed, with the DNA being reeled through them as it is replicated, rather than the reverse.

Many investigators have shown that adjacent replicons replicate simultaneously (reviewed by Hand, 1978). This fact can be accommodated suitably with the model if pairs of fixed replication sites from adjacent, simultaneously replicating replicons are in close proximity, thereby yielding clusters of replication complexes anchoring alternating replicated and nonreplicated loops (Figure 9C). This proposed arrangement of DNA attached to the matrix in loops is appealing in that it suggests a manner in which the replicated daughter strands are spatially ordered for segregation in preparation for mitosis.

A number of reports within the past three years have suggested that interphase chromatin is arranged in such a loop organization (reviewed by Comings, 1978; Georgiev, Nedospasov and Bakayev, 1978).

A. FIXED SITE OF REPLICATION

Figure 9. Proposed Model for Fixed Site of DNA Replication on the Nuclear Matrix

(A) Single fixed replication complex with the DNA reeled through as it is replicated. (B) Pair of adjacent fixed replication complexes allowing bidirectional DNA synthesis. The DNA is reeled in from both sides and newly replicated DNA loops out in between. (C) Cluster of fixed replication complexes. DNA is bound via the replication complexes in alternating replicated and nonreplicated loops. Arrows represent direction of DNA movement.

B. BIDIRECTIONAL REPLICATION COMPLEX

DNA LOOPS BEFORE REPLICATION

REPLICATED DNA LOOPS

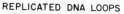

C. CLUSTER OF FIXED REPLICATION COMPLEXES (REPLISOME)

DNA LOOPS BEFORE REPLICATION

Benyajati and Worcel (1976), studying interphase nuclei from D. melanogaster, reported a ''folded chromosome'' with supercoiled DNA loops held by a supporting structure and concluded that both RNA and nonhistone proteins were required to maintain the DNA in the folded loop conformation. Cook, Brazell and Jost (1976) have obtained similar results using nuclei from HeLa cells. Ide et al. (1975), working with tissue culture cells derived from mouse mammary carcinoma, centrifuged nuclei through a neutral sucrose gradient containing SDS and actinomycin D. The resulting structures had a high sedimentation rate due to the folded nature of the supercoiled DNA loops. The residual structure to which the presumed DNA loops were bound had an SDS-polyacrylamide gel pattern qualitatively similar to that of nuclear matrix, with a major band at about 70K daltons and a smaller band at about 55K daltons.

Paulson and Laemmli (1977) have recently found that HeLa metaphase chromosomes contain a residual protein scaffolding. Extraction of chromosomes with dextran sulfate and heparin removed all histones and yielded a residual structure to which the DNA was attached in loops. The loop sizes were estimated to be 10–50 μ, a range which coincides with the distribution of S phase replicon lengths (Gautschi, Kern and Painter, 1973). The attachments of DNA to the scaffold were resistant to 2 M NaCl, just as are the matrix-DNA attachments. It is intriguing to speculate on the relation between the DNA loops on the chromosomes observed by Laemmli and the proposed loop structure for interphase nuclei.

Experimental Procedures

Animals and Cells

Male Sprague-Dawley (Madison, Wisconsin) rats weighed 200–250 g. 3T3 cells (obtained from T. Kelly, Johns Hopkins School of Medicine) were grown in Eagle's minimal essential medium supplemented with 10% fetal bovine serum (MEM-10). Partial hepatectomies were performed as previously described (Higgins and Anderson, 1931).

Pulse Labeling

Pulses in the rat were performed 24 hr after partial hepatectomy, at which time DNA synthesis is maximal (Fabrikant, 1968). 200 μCi ^{3}H–thymidine (50–60 Ci/mmole; New England Nuclear, Boston, Massachusetts) were injected into the hepatic portal vein. Thymidine incorporation was halted by rapid excision and mincing of the liver in TM buffer [10 mM Tris (pH 7.4), 5 mM MgCl$_2$] at 4°C.

Pulse labeling of 3T3 fibroblasts was performed with exponentially growing cells (20–40% confluence) that had been prelabeled for 48 hr with ^{14}C–thymidine (50–60 mCi/mmole; New England Nuclear) at an input concentration of 0.005 μCi per ml of medium. Old medium was removed and fresh 37°C MEM-10 containing 20 μCi/ml ^{3}H–thymidine (50–60 Ci/mmole) was added for the pulse. Thymidine incorporation was halted by removing the medium containing label and covering the cells with 4°C PBS [0.15 M NaCl, 0.01 M Na$_3$PO$_4$ (pH 6.8)] containing 25 mM EDTA. Cells were removed from the flask with a rubber policeman. Pulse-chase experiments were performed by removing the medium containing ^{3}H–thymidine and replacing it with medium containing 40 μM unlabeled thymidine.

Isolation of Nuclei

Rat liver nuclei were isolated according to the method of Berezney and Coffey (1977). 3T3 fibroblast nuclei were isolated by suspending cells in HEPES buffer [1 mM MgCl$_2$, 0.5 mM CaCl$_2$, 0.022 M sucrose, 0.05 M HEPES (pH 7.8)], adding NP40 to a final concentration of 1% and vortexing for 5 min followed by ten gentle strokes with a Dounce homogenizer (loose fitting pestle). Nuclei were separated from cytoplasmic debris by centrifuging for 5 min at 500 × g.

Matrix Isolation

Rat liver nuclear matrix was isolated according to the procedure of Berezney and Coffey (1974) using sequential extraction with low magnesium buffer [10 mM Tris (pH 7.4), 0.2 mM $MgCl_2$], high salt buffer [2 M NaCl, 0.2 mM $MgCl_2$, 10 mM Tris (pH 7.4)] and 1% Triton X-100. 3T3 nuclear matrix was prepared by incubating purified nuclei in low salt buffer followed by slow addition of NaCl to a final concentration of 2 M. DNAase I (Worthington, Freehold, New Jersey) was added to this solution (at a final concentration of 400 U/ml) for 5–30 min (depending on the amount of digestion desired), and the resultant matrices were centrifuged for 15 min at 2000 × g. α–phenylmethyl-sulfonyl fluoride (1 mM) was present in all extraction buffers to minimize protease degradation of the matrix protein.

DNA Isolation and Gel Electrophoresis

DNA was purified from matrix and high salt supernatant using the proteinase K procedure of Gross-Bellard, Oudet and Chambon (1973).

Electrophoresis of DNA was carried out in slab gels of 1% agarose (Sigma, St. Louis, Missouri) using a Hoeffer (San Francisco, California) electrophoresis apparatus. Samples were loaded in 10 mM sodium phosphate buffer (pH 6.8) containing 5% glycerol and 0.02% bromophenol blue. Samples were run for 12 hr at 25 V. Bacteriophage λ DNA digested with Eco RI and SV40 DNA digested with Hind III and Hae III were used as size markers. Marker DNAs and restriction enzymes were purchased from Bethesda Research Laboratories (Bethesda, Maryland).

Autoradiography

For electron microscopic autoradiography, ^{3}H-labeled rat liver nuclear matrix was fixed in 4% glutaraldehyde, post-fixed with 2% OsO_4, embedded, sectioned to 0.1 micron and stained with lead citrate. Sections were coated with Ilford L-4 emulsion (Ilford, England) and exposed for 60 days before development.

For light microscopic autoradiography, ^{3}H-labeled 3T3 fibroblast nuclei and matrix were centrifuged onto glass slides, fixed with 4% glutaraldehyde and coated with NTB2 emulsion (Kodak). Slides were developed after 5 days of exposure. Nuclei or matrix were scored as labeled if they contained more than five grains. The average grain count of labeled nuclei and matrix were 160 ± 20 and 60 ± 12, respectively.

Acknowledgments

We thank William Lennarz, Jr. for performing the electron microscopic autoradiography, Adam Leaderman for assistance with the rat liver experiments and Joel Shaper for valuable discussions. This work was supported in part by NIH grants and a gift from Bristol Myers Company. D.P. is a Medical Scientist Trainee and is supported by an NIH grant. The electron micrographs were prepared by Ms. Carol J. Tillman through the Core Electron Microscopy Laboratory of the Johns Hopkins Population Center, with the support of a grant from the USPHS.

The costs of publication of this article were defrayed in part by the payment of page charges. This article must therefore be hereby marked "*advertisement*" in accordance with 18 U.S.C. Section 1734 solely to indicate this fact.

Received August 27, 1979; revised November 16, 1979

References

Benyajati, C. and Worcel, A. (1976). Isolation, characterization and structure of the folded interphase genome of Drosophila melanogaster. Cell 9, 393–407.

Berezney, R. and Coffey, D. S. (1974). Identification of a nuclear protein matrix. Biochem. Biophys. Res. Commun. 60, 1410–1417.

Berezney, R. and Coffey, D. S. (1975). Nuclear protein matrix: association with newly synthesized DNA. Science 189, 291–293.

Berezney, R. and Coffey, D. S. (1977). Nuclear matrix: isolation and characterization of a framework structure from rat liver nuclei. J. Cell Biol. 73, 616–637.

Berger, H. and Irvin, J. L. (1970). Changes in the physical state of DNA during replication in regenerating liver of the rat. Proc. Nat. Acad. Sci. USA 65, 152–159.

Comings, D. E. (1978). Compartmentalization of nuclear and chromatin protein. In The Cell Nucleus, 4, H. Busch, ed. (New York: Academic Press), pp. 345–371.

Comings, D. E. and Kakefuda, T. (1968). Initiation of deoxyribonucleic acid replication at the nuclear membrane in human cells. J. Mol. Biol. 33, 225–229.

Comings, D. E. and Okada, T. A. (1976). Nuclear Proteins. III. The fibrillar nature of the nuclear matrix. Exp. Cell Res. 103, 341–360.

Cook, P. R., Brazell, I. A. and Jost, E. (1976). Characterization of nuclear structures containing superhelical DNA. J. Cell Sci. 22, 303–324.

Dijkwel, P. A., Mullenders, L. and Wanka, F. (1979). Analysis of the attachment of replicating DNA to a nuclear matrix in mammalian interphase nuclei. Nuc. Acids Res. 6, 219–230.

Edenberg, H. J. and Huberman, J. A. (1975). Eucaryotic chromosome replication. Annual Review of Genetics, 9, H. Roman, ed. (Palo Alto: Annual Reviews, Inc.), pp. 245–284.

Fabrikant, J. I. (1968). The kinetics of cellular proliferation in regenerating liver. J. Cell Biol. 36, 551–565.

Fakan, S. and Hancock, R. (1974). Localization of newly synthesized DNA in a mammalian cell as visualized by high resolution autoradiography. Exp. Cell Res. 83, 95–102.

Ganesan, A. T. and Lederberg, J. (1965). A cell membrane bound fraction of bacterial DNA. Biochem. Biophys. Res. Commun. 18, 824–835.

Gautschi, J. R., Kern, R. M. and Painter, R. B. (1973). Modification of replicon operation in HeLa cells by 2,4-dinitrophenol. J. Mol. Biol. 80, 393–403.

Georgiev, G. P., Nedospasov, S. A. and Bakayev, V. U. (1978). Supranucleosomal levels of chromatin organization. In The Cell Nucleus, 6, H. Busch, ed. (New York: Academic Press), pp. 3–34.

Gross-Bellard, N., Oudet, P. and Chambon, P. (1973). Isolation of high-molecular weight DNA from mammalian cells. Eur. J. Biochem. 36, 32–38.

Habener, J. F., Bynum, B. S. and Shack, J. (1970). Destabilized secondary structure of newly replicated HeLa DNA. J. Mol. Biol. 49, 157–170.

Hanaoka, F. and Yamada, M. (1971). Localization of the replication point of mammalian cell DNA at the membrane. Biochem. Biophys. Res. Commun. 42, 647–653.

Hand, R. (1978). Eucaryotic DNA: organization of the genome for replication. Cell 15, 317–325.

Herlan, G. and Wunderlich, F. (1976). Isolation of a nuclear matrix from Tetrahymena macronuclei. Cytobiologie 13, 291–296.

Herman, R., Weymouth, L. and Penman, S. (1978). Heterogeneous nuclear RNA-protein fibers in chromatin-depleted nuclei. J. Cell Biol. 78, 663–674.

Higgins, G. M. and Anderson, R. M. (1931). Experimental pathology of the liver. I. Restoration of the liver of the white rat following partial surgical removal. Arch. Pathol. 12, 186.

Hodge, L. D., Mancini, P., Davis, F. M. and Heywood, P. (1977). Nuclear matrix of HeLa S_3 cells. J. Cell Biol. 72, 194–208.

Huberman, J. A. and Riggs, A. D. (1968). On the mechanism of DNA replication in mammalian chromosomes. J. Mol. Biol. 32, 327–341.

Huberman, J. A., Tsai, A. and Deich, R. A. (1973). DNA replication sites within nuclei of mammalian cells. Nature 241, 32–26.

Ide, T., Nakane, M., Anzai, K. and Andoh, T. (1975). Supercoiled DNA folded by non-histone proteins in cultured mammalian cells. Nature 258, 445–447.

Jacob, F., Brenner, S. and Cuzin, F. (1963). On the regulation of DNA replication in bacteria. Cold Spring Harbor Symp. Quant. Biol. *28*, 329–348.

Keller, J. M. and Riley, D. E. (1976). Nuclear ghosts: a nonmembranous structural component of mammalian cell nuclei. Science *193*, 399.

Lewin, B. M. (1973). *Gene Expression*, 2 (London: John Wiley and Sons), pp. 60–64.

Miller, T. E., Huang, C. and Pogo, A. O. (1978). Rat liver nuclear skeleton and small molecular weight RNA species. J. Cell Biol. *76*, 692–704.

Painter, R. B. and Schaeffer, A. (1969). State of newly synthesized HeLa DNA. Nature *221*, 1215–1217.

Paulson, J. R. and Laemmli, U. K. (1977). The structure of histone-depleted metaphase chromosomes. Cell *12*, 817–828.

Pearson, G. D. and Hanawalt, P. C. (1971). Isolation of DNA replication complexes from uninfected and adenovirus-infected HeLa cells. J. Mol. Biol. *62*, 65–80.

Riley, D. and Weintraub, H. (1979). Conservative segregation of parental histones during replication in the presence of cycloheximide. Proc. Nat. Acad. Sci. USA *76*, 328–332.

Schekman, R., Weiner, J., Weiner, A. and Kornberg, A. (1975). Ten proteins required for conversion of φX174 single-stranded DNA to duplex form *in vitro*. J. Biol. Chem. *250*, 5859–5865.

Seale, R. L. (1975). Assembly of DNA and protein during replication in HeLa cells. Nature *255*, 247–249.

Shelton, K. R. (1976). Selective effects of nonionic detergent and salt solutions in dissolving nuclear envelope protein. Biochim. Biophys. Acta *455*, 973–982.

Smith, D. W. and Hanawalt, P. C. (1967). Properties of the growing point region in the bacterial chromosome. Biochim. Biophys. Acta *149*, 519–531.

Sueoka, N. and Quinn, W. G. (1968). Membrane attachment of the chromosome replication origin in *B. subtilis*. Cold Spring Harbor Symp. Quant. Biol. *33*, 695–706.

Tremblay, G. Y., Daniels, M. J. and Schaechter, M. (1969). Isolation of a cell membrane-DNA-nascent RNA complex from bacteria. J. Mol. Biol. *40*, 65–76.

Weintraub, H. (1973). The assembly of newly replicated DNA into chromatin. Cold Spring Harbor Symp. Quant. Biol. *38*, 247–256.

Wickner, S. and Hurwitz, J. (1974). Conversion of φX174 viral DNA to double-stranded form by purified *Escherichia coli* proteins. Proc. Nat. Acad. Sci. USA *71*, 4120–4124.

Williams, C. A. and Ockey, C. H. (1970). Distribution of DNA replicator sites in mammalian nuclei after different methods of cell synchronization. Exp. Cell Res. *63*, 365–372.

Worcel, A. and Burgi, E. (1972). On the structure of the folded chromosome of *E. coli*. J. Mol. Biol. *71*, 127–148.

Supplementary Readings

Before 1961

Swift, H.H. 1960. The desoxyribose nucleic acid content of animal nuclei. *Biochim. Biophys. Acta.* *37:* 406–419.

1961–1970

Bootsma, D., L. Budke, and O. Vos. 1964. Studies on synchronous division of tissue culture cells initiated by excess thymidine. *Exp. Cell Res. 33:* 301–309.

Miller, R. and R. Phillips. 1969. Separation of cells by velocity sedimentation. *J. Cell. Physiol. 73:* 191–202.

Petersen, D.F. and E.C. Anderson. 1964. Quantity production of synchronized mammalian cells in suspension culture. *Nature 203:* 642–643.

Pfeiffer, S.E. and L.J. Tolmach. 1967. Selecting synchronous populations of mammalian cells. *Nature 212:* 139–142.

Puck, T.T. 1964. Studies of the life cycle of mammalian cells. *Cold Spring Harbor Symp. Quant. Biol. 29:* 167–176.

Puck, T.T. and J. Steffen. 1963. Life cycle analysis of mammalian cells. *Biophys. J. 3:* 379–397.

Robbins, E. and N.K. Gonatas. 1964. Electron micrographs of mitotic HeLa cells. *J. Cell Biol. 21:* 429–463.

Schindler, R., L. Ramseier, J. Schaerf, and A. Grieder. 1970. Studies on the division of mammalian cells. III. Preparation of synchronously dividing cell populations by isotonic sucrose gradient centrifugation. *Exp. Cell Res. 59:* 90–96.

Schroeder, T.E. 1968. Cytokinesis: Filaments in the cleavage furrow. *Exp. Cell Res. 53:* 272–318.

Sisken, J.E. and R. Konosita. 1961. Timing of DNA synthesis in the mitotic cycle in vitro. *J. Biophys. Biochem. Cytol. 9:* 509–518.

Tobey, R.A., E.C. Anderson, and D.F. Peterson. 1967. Properties of mitotic cells prepared by mechanically shaking monolayer cultures of Chinese hamster cells. *J. Cell. Physiol. 70:* 63–68.

1971–1980

Brooks, R.F., D.C. Bennett, and J.A. Smith. 1980. Mammalian cell cycles need two random transitions. *Cell 19:* 493–504.

Burstin, S.J., H. Meiss, and C. Basilico. 1974. A temperature-sensitive cell cycle mutant of the BHK cell line. *J. Cell. Physiol. 84:* 397–407.

Gerace, L. and G. Biobel. 1980. The nuclear envelope lamina is reversibly depolymerized during mitosis. *Cell 19:* 277–287.

Grimes, W.J. and J.L. Schroder. 1973. Dibutyryl cyclic adenosine 3′,5′ monophosphate, sugar transport, and regulatory control of cell division in normal and transformed cells. *J. Cell Biol. 56:* 487–491.

Johnson, R. and P. Rao. 1971. Nucleo-cytoplasmic interactions in the achievement of nuclear synchrony in DNA synthesis and mitosis in multinucleate cells. *Biol. Rev. Camb. Philos. Soc. 46:* 97–156.

Roscoe, D.H., H. Robinson, and A.W. Carbonell. 1973. DNA synthesis and mitosis in a temperature sensitive Chinese hamster cell line. *J. Cell. Physiol. 82:* 333–338.

Shall, S. and A.J. McClelland. 1971. Synchronization of mouse fibroblast LS cells grown in suspension culture. *Nat. New Biol. 229:* 59–61.

Smith, B.J. and N.M. Wigglesworth. 1972. A cell line which is temperature-sensitive for cytokinesis. *J. Cell. Physiol. 80:* 253–260.

Smith, B.J. and N.M. Wigglesworth. 1975. Studies on a Chinese hamster line that is temperature sensitive for the commitment to DNA synthesis. *J. Cell. Physiol. 84:* 127–134.

Todo, A., A. Strife, J. Fried, and B.D. Clarkson. 1971. Proliferative kinetics of human hematopoietic cells during different growth phases in vitro. *Cancer Res. 31:* 1330–1340.

Zigman, S. and P. Gilman, Jr. 1980. Inhibition of cell division and growth by a redox series of cyanine dyes. *Science 208:* 188–191.

Chapter 7 Chromosomes, Chromatin, and Gene Expression

Tjio, J.H. and A. Levan. 1956. The chromosome number of man. *Hereditas 42:* 1–6.

Caspersson, T., L. Zech, and C. Johansson. 1970. Analysis of human metaphase chromosome set by aid of DNA-binding fluorescent agents. *Exp. Cell Res. 62:* 490–492.

Pardue, M.L. and J.G. Gall. 1970. Chromosomal localization of mouse satellite DNA. *Science 168:* 1356–1358.

Olins, A. and D. Olins. 1974. Spheroid chromatin units. *Science 183:* 330–331.

Weintraub, H. and M. Groudine. 1976. Chromosomal subunits in active genes have an altered conformation. *Science 193:* 848–856.

Paulson, J.R. and U.K. Laemmli. 1977. The structure of histone-depleted metaphase chromosomes. *Cell 12:* 817–828.

Prives, C., E. Gilboa, M. Revel, and E. Winocour. 1977. Cell-free translation of simian virus 40 early messenger RNA coding for viral T-antigen. *Proc. Natl. Acad. Sci. 74:* 457–461.

Berk, A.J. and P.A. Sharp. 1978. Spliced early mRNAs of simian virus 40. *Proc. Natl. Acad. Sci. 75:* 1274–1278.

Seidman, M.M., A.J. Levine, and H. Weintraub. 1979. The asymmetric segregation of parental nucleosomes during chromosome replication. *Cell 18:* 439–449.

Early, P., J. Rogers, M. Davis, K. Calame, M. Bond, R. Wall, and L. Hood. 1980. Two mRNAs can be produced from a single immunoglobulin μ gene by alternative RNA processing pathways. *Cell 20:* 313–319.

In mitosis, the DNA of a cell is partitioned, folded, and condensed with a set of binding proteins into chromosomes. At the end of mitosis, the chromosomes unravel into skeins of DNA still complexed with many proteins. This complex, called chromatin, fills the reassembled nuclei. During S, and possibly elsewhere in the interphase of the cell cycle, DNA is also bound to an insoluble nuclear matrix. As the cell cycle progresses, some, but not all, regions of DNA are transcribed. From these transcripts, the proteins of the cell are made. In this chapter we examine the very different structures containing DNA in mitotic chromosomes, and in interphase nuclei, where transcription and the cellular processing of RNA occur.

How many chromosomes are there in a mitotic human cell? This simple question is definitively answered in the modest but clear paper by **Tjio** and **Levan**. At the time this paper was written, the human chromosome number had for some decades been believed to be 48. Improving the preparation of mitotic cells for staining, Tjio and Levan eliminated the problem of overlap and spillover of chromosomes from one nucleus into the microscope field of another. They then counted 46 chromosomes per human cell and bucked the tide to say so. Their last paragraph is a model of cautious pride.

In the paper by Tjio and Levan, chromosomes are homogeneous, sausagelike bodies that differ from each other only in size and in the location of the centromeric constriction by which they are attached to the spindle (see Chapter 11). The stunningly high density of information in their DNA is obscured by their condensed state and by the limits of resolution of the microscope. In the next two papers, regions of chromosomes are resolved in cultured cells. **Caspersson, Zech,** and **Johansson** soaked the chromosomes in a fluorescent dye. The dye binds to

DNA, and so the chromosomes fluoresce. Fluorescence is not even, however: Bands traverse most chromosomes, and, by their bands, chromosomes of similar shape can be distinguished from one another. Thus, by banding, each of the chromosomes seen by Tjio and Levan can be given a unique name. The bands suggest that the DNA of a chromosome must be packaged in a nonrandom way.

In the paper by Caspersson et al., fluorescence seems to be brightest in regions of heterochromatin. **Pardue** and **Gall** show that heterochromatic DNA in the centromere region of mouse chromosomes contains many copies of a highly repeated sequence, satellite DNA. They isolated the satellite band from radioactive cellular DNA by virtue of its unique buoyant density. Labeled satellite DNA was then denatured into single strands by heat and applied to chromosomes that had been spread on glass plates. The repeated DNA sequence of the satellite is about 400 base pairs, long enough to rehybridize to the many equivalent copies of this sequence in the chromosomal preparations. Because the number of copies of the repeat is so great, rehybridization is relatively efficient, and autoradiography localized the labeled DNA to regions around the chromosomal centromeres. At least in the case of this form of heterochromatin, we may reasonably suppose that a quinacrine dye localized many copies of a short, repeated sequence of DNA. As the next papers show, proteins also play a major role in chromosome structure.

When the chromosomes loosen and unravel at the end of mitosis, cellular DNA does not lose all its DNA-binding proteins or become a simple double helix. Rather, the DNA remains complexed to a set of proteins, including the histones. The packaging of eukaryotic chromatin is a formidable problem even in nonmitotic cells, since interphase nuclear DNA concentrations are in the range of 1–10 mg/ml. Clearly, coiling or folding of the DNA is necessary to explain this packing. **Olins** and **Olins** describe the first order of chromatin packing, the histone-DNA complex called a nucleosome. They show that cellular DNA wraps around histones to form a series of beads, separated by stretches of DNA. Note the compound used to lyse the interphase nuclei for electron microscopy: "Joy" household detergent. In the past few years, nucleosome packaging has been resolved into the current picture of a repeating unit of 140–200 base pairs of DNA associated with two self-complementary protein tetramers containing one each of the four major histones.

Are the nucleosomes homogeneous? Clearly, the DNA sequences must vary from one to the next, and as **Weintraub** and **Groudine** show, gene expression also modifies nucleosome structure. They find that the nucleosomes of transcriptionally active DNA sequences are diagnostically sensitive to certain deoxyribonucleases. Using the DNA sequences of the chicken globin gene as a specific probe, Weintraub and Groudine show that nucleosomes of this sequence are degraded by deoxyribonuclease I if the chromatin is extracted from red cell nuclei but not if it is extracted from fibroblast nuclei or from a population of red cell precursors.

When chromatin condenses into chromosomes, nucleosomes must themselves be packaged in higher orders of organization for mitosis. **Paulson** and **Laemmli** go directly to the problem of high-order structure by removing the histones from mitotic chromosomes. The DNA-protein complexes that remain are, by electron microscopy, a set of quite beautiful objects. The DNA of histone-depleted chromosomes is attached in long loops to a matrix of proteins that retains the overall form of the chromosomes. It is interesting to speculate that these might be related to the scaffold-DNA loops in replicating nuclei (Chapter 6). In any event, Paulson and Laemmli's pictures show that some nonhistone proteins are organized into a scaffold responsible for the higher-order structure of each chromosome.

Although chromosome structure clearly plays a role in gene expression, the process itself must begin with transcription of RNA from specific sequences of DNA. The next two papers study the expression of the transforming genes of SV40 in lytically infected and transformed cells. Together, they raise and resolve a paradox of eukaryotic gene expression: More protein is coded for in a stretch of SV40 DNA than that sequence can account for by the simple ratio of three bases to one amino acid.

SV40 DNA is an entirely known closed circular sequence of 5243 base pairs which includes a single origin of replication. Transcription initiates in both directions from this origin and terminates about half way around the circle. Thus, each strand of the circular DNA is the source of a transcript that runs halfway from the origin. One strand, called early, is transcribed in the lytic cycle before viral DNA synthesis begins. **Prives, Gilboa, Revel,** and **Winocour** recovered early mRNAs from infected cells by their capacity to hybridize to SV40 DNA. They then examined the coding capacity of these natural mRNAs by comparing the products synthesized by them in a cell-free protein translation system to the ones synthesized in the same translation system primed with a genome-length direct transcript made from the appropriate strands of SV40 DNA by *E. coli* RNA polymerase (cRNA). They found that mRNAs from infected and transformed cells directed the synthesis of two specific proteins, of molecular weights 17,000 and 90,000. The larger has the same molecular weight as the nuclear T antigen of SV40 (Chapters 3 and 4). The smaller is an unexpected protein, which at that time was not reported to be present in transformed cells. RNA transcribed by *E. coli* RNA polymerase did not direct the synthesis of the large T antigen.

For the early region of SV40, cellular mRNA coded for two different proteins. This result requires that transcript and mRNA from cells have different sequences, but how is that possible?

Berk and **Sharp** answer the question in an elegant analysis of RNA-DNA hybrid molecules. They isolated early mRNAs by hybridization, as Prives et al. had done, but then they hybridized them to radioactive SV40 DNA strands. They then attacked the hybrid molecules with nucleases that degrade free ends of RNA or DNA, and finally with S1, an endo-deoxyribonuclease that degrades single-stranded DNA. Loops of single-stranded DNA can exist in DNA-RNA hybrids only if the DNA has sequences that are missing from mRNA. S1 cut into the DNA of their DNA-RNA hybrids, showing that SV40 early mRNA lacked sequences that were present in SV40 early DNA. Finally, Berk and Sharp showed that the two viral proteins are translation products of two alternatively spliced mRNAs derived from one early transcript and that both proteins indeed exist in both transformed and lytically infected cells. The role of these two proteins in transformation is currently a subject of heated debate (Chapter 3), but there is no doubt that the SV40 early region has served to illuminate a major regulatory event in eukaryotic gene expression: posttranscriptional intragenic RNA splicing.

In SV40, replication and transcription occur in close proximity on the same stretch of DNA.

Seidman, Levine, and **Weintraub** reveal a surprising and intimate linkage between the two events for SV40, and perhaps for cells. As normal chromatin replicates, new histones are synthesized and added to DNA to make new nucleosomes. The DNA replicates semi-conservatively: One strand of parental DNA is the template for and becomes part of each of the two daughter DNA molecules. However, nucleosome replication is conservative: All parental nucleosomes go to one daughter DNA molecule and all new ones go to the other. The daughter DNA duplex receiving new histones is the one whose new DNA strand is replicated discontinuously in the 3′ to 5′ direction.

As we have seen, the SV40 origin of replication is also the site from which RNA transcription initiates. In both directions out from the origin, mRNAs are transcribed from opposing strands of the SV40 DNA. Thus, in each direction, only one strand serves as template for transcription and only one strand receives old nucleosomes during replication. Seidman et al. showed that the strand that codes for mRNA is in fact the one that receives parental nucleosomes. Is this the case for cellular DNA as well? Apparently yes, for the strand that was identical to cDNA made from pooled, stable, cellular cytoplasmic mRNA was also the one retaining old nucleosomes during cellular DNA replication.

This startling result, if confirmed, could provide a molecular basis for the unequal mitosis characteristic of stem cell/differentiating cell systems (Chapter 2). It is reasonable to assume that at least some regulatory DNA-binding proteins interact specifically with naked DNA rather than with chromatin. Such proteins will be able to bind only to one of the two daughter DNA duplexes at replication: the one with the 3′ to 5′ discontinuously replicated new strand and the newly assembled nucleosomes. The other new DNA duplex will bind old nucleosomes and need never be naked. Clearly, subsequent gene expression in the daughter cell carrying the former DNA duplex can differ from that in the parent cell, whereas the latter daughter cell is likely to remain a regulatory copy of the parental cell, i.e., a stem cell.

Are new facts about gene transcription and RNA processing that are learned from SV40 relevant to cellular DNA? One might argue, for instance, that the SV40 genome, constrained to be packaged in its capsid, has evolved clever linkages of replication and transcription or even ways to get two proteins from one transcript. But is such wit a part of a cell's repertoire as well? **Early, Rogers, Davis, Calame, Bond, Wall,** and **Hood** show how a lymphoid B cell uses precisely such tactics to make either one or both of two forms of an immunoglobulin. One form of the IgM molecule is secreted, the other membrane-bound. They both originate from one gene, which is transcribed into a single transcript. Splices occur, and intervening sequences are removed as it becomes an mRNA. When all splices come out, the resulting mRNA codes first for a leader sequence which pushes the protein through the membrane (see Chapter 10), then for the variable and constant regions of the heavy chain of IgM, and finally for a hydrophobic sequence which prevents the whole molecule from slipping through the membrane. This is how surface IgM is made. Sometimes the last splice is prevented. When this occurs, the same transcript now finishes with a new protein sequence absent from the surface molecule. A protein made this way fails to stick in the membrane and is then secreted as the heavy-chain serum IgM. Despite an apparently great excess of DNA in the cultured cell, one transcript can be used to make more than one protein. Thus, regulation of gene expression in cultured cells sometimes occurs by posttranscriptional double entendres much as it does in SV40.

Separat ur Hereditas 42 (1956)

THE CHROMOSOME NUMBER OF MAN

By *JOE HIN TJIO* and *ALBERT LEVAN*

ESTACION EXPERIMENTAL DE AULA DEI, ZARAGOZA, SPAIN, AND CANCER CHROMOSOME
LABORATORY, INSTITUTE OF GENETICS, LUND, SWEDEN

WHILE staying last summer at the Sloan-Kettering Institute, New York, one of us tried out some modifications of Hsu's technique (1952) on various human tissue cultures carried in serial *in vitro* cultivation at that institute. The results were promising inasmuch as some fairly satisfactory chromosome analyses were obtained in cultures both of tissues of normal origin and of tumours (LEVAN, 1956).

Later on both authors, working in cooperation at Lund, have tried still further to improve the technique. We had access to tissue cultures of human embryonic lung fibroblasts, grown in bovine amniotic fluid; these were very kindly supplied to us by Dr. RUNE GRUBB of the Virus Laboratory, Institute of Bacteriology, Lund. All cultures were primary explants taken from human embryos obtained after legal abortions. The embryos were 10—25 cm in length. The chromosomes were studied a few days after the *in vitro* explantation had been made.

In our opinion the hypotonic pre-treatment introduced by Hsu, although a very significant improvement especially for spreading the chromosomes, has a tendency to make the chromosome outlines somewhat blurred and vague. We consequently tried to abbreviate the hypotonic treatment to a minimum, hoping to induce the scattering of the chromosomes without unfavourable effects on the chromosome surface. Pre-treatment with hypotonic solution for only one or two minutes gave good results. In addition, we gave a colchicine dose to the culture medium 12—20 hours before fixation, making the medium 50×10^{-9} mol/l for the drug. The colchicine effected a considerable accumulation of mitoses and a varying degree of chromosome contraction. Fixation followed in 60 % acetic acid, twice exchanged in order to wash out the salts left from the culture medium and from the hypotonic solution that would otherwise have caused precipitation with the orcein. Ordinary squash preparations were made in 1 % acetic orcein. For chromosome counts the squashing was made very mild in order to keep the chromosomes in the metaphase groups. For idiogram studies a more thorough squashing was preferable. In many cases single cells were squashed

1 — *Hereditas 42*

under the microscope by a slight pressure of a needle. In such cases it was directly observed that no chromosomes escaped.

THE CHROMOSOME NUMBER

With the technique used exact counts could be made in a great number of cells. Figs. 1 *a* and *b* represent typical samples of the appearance of the chromosomes at early metaphase (*a*) and full metaphase (*b*), showing the ease with which the counting could be made. In Table 1 the numbers of counts made from the four embryos studied are recorded.

TABLE 1. *Number of exact chromosome counts made.*

Embryo No.	Number of cultures	Number of counts
1	5	15
2	10	98
3	3	119
4	4	29
Total	22	261

We were surprised to find that the chromosome number 46 predominated in the tissue cultures from all four embryos, only single cases deviating from this number. Lower numbers were frequent, of course, but always in cells that seemed damaged. These were consequently disregarded just as the solitary chromosomes and the groups with but a few chromosomes, which were frequent. In some doubtful cases the numbers 47 and 48 were counted (in four cases not included in the table). This may be due to one or two solitary chromosomes having been pressed into a 46-chromosome plate at the squashing. It is also possible that deviating numbers may originate through non-disjunction, thus representing a real chromosome number variation in the living tissue. This kind of variation will probably increase as a consequence of the change in environment for the tissue involved in the *in vitro* explantation. Hsu (1952) reports a certain degree of such variation in his primary cultures. LEVAN (1956), studying long-carried serial subcultures, found hypotriploid stemline numbers in two of them, and a near-diploid number in a third culture. In this culture one cell with 48 chromosomes was analysed. Naturally, at that time, this was thought to represent the normal diploid number.

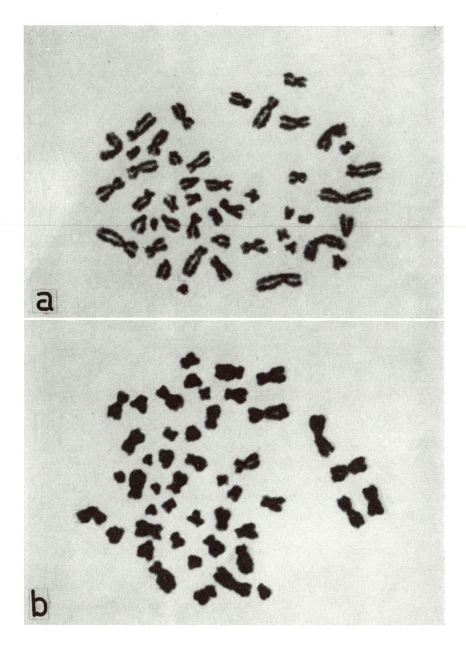

Fig. 1. Colchicine-metaphases of human embryonic lung fibroblasts grown *in vitro*. *a*: early metaphase, *b*: full metaphase. The two cells are from embryos 2 and 3 (Table 1), respectively. — ×2300.

CHROMOSOME MORPHOLOGY

Some data on the chromosome morphology of the 46 human chromosomes will be communicated here. The detailed idiogram analysis will be postponed, however, until we are able to study individuals of known sex, the sex of the present embryos being unknown. The comparative study of germline chromosomes in spermatogonial mitoses constitutes an urgent supplement to the present work.

In Fig. 2 four cells are analysed ranging from late prophase (*a*) to late c-metaphase (*d*). The chromosomes of metaphases with moderate colchicine contraction vary in length between 1 and 8 μ (Fig. 2 *b*), but the entire range of variation of Fig. 2 is from 1 to 11 μ. The chromosome morphology is roughly concordant with the observations of earlier workers, as, for instance, the idiogram of Hsu (1952). The chromosomes may be divided into three groups: M chromosomes (median-submedian centromere; index long arm : short arm 1—1,9), S chromosomes (subterminal centromere; arm index 2—4,9), and T chromosomes (nearly terminal centromere; arm index 5 or more).

The M and S chromosomes are present in about equal numbers (twenty of each), while six T chromosomes are found. The classification of the three groups is arbitrary, of course, since gradual transitions of arm indices occur between the three groups. Certain submedian M chromosomes are hard to distinguish from some of the S chromosomes, and the most asymmetric S chromosomes approach the T group.

The chromosomes are easily arranged in pairs, but only certain of these pairs are individually distinguishable. Thus, the M chromosomes include the three longest pairs, which can always be identified. The two longest pairs are different: the second having a decidedly more asymmetric location of its centromere. The two or three smallest M pairs are also recognizable. Between the three longest and the three shortest pairs there are four intermediate pairs that cannot be individually recognized.

The S chromosomes are hardly identifiable, since they form a series of gradually decreasing length. The largest pair, however, is characteristic. Certain chromosomes were seen to have a small satellite on their short arms. Secondary constrictions, too, have been observed now and then, so that it may be hoped that the detailed morphologic study will lead to the identification of more chromosome pairs. The T chromosomes are recognizable; they constitute three pairs of middle-sized chromosomes. Unlike the mouse chromosomes, the human T chromosomes evidently have a small shorter arm.

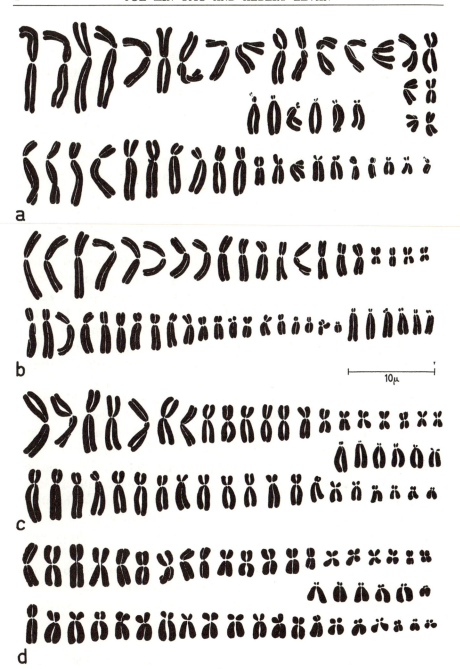

Fig. 2. Four idiogram analyses of human embryonic lung fibroblasts grown *in vitro*. The chromosomes have been grouped in three classes: M (top row), S (bottom row), and T (in between, except in *b*, where T is at the end of the S row). Within each class the chromosomes have been roughly arranged in diminishing order of size. — ×2400.

CONCLUSION

The almost exclusive occurrence of the chromosome number 46 in one somatic tissue derived from four individual human embryos is a very unexpected finding. To assume a regular mechanism for the exclusion of two chromosomes from the idiogram at the formation of a certain tissue is unlikely, even if this assumption cannot be entirely dismissed at this stage of inquiry. Our experience from one somatic tissue in mice and rats, *viz.*, regenerating liver, speaks against this assumption. The exact diploid chromosome set was always found in regenerating liver.

After the conclusion had been drawn that the tissue studied by us had 46 as chromosome number, Dr. EVA HANSEN-MELANDER kindly informed us that during last spring she had studied, in cooperation with Drs. YNGVE MELANDER and STIG KULLANDER, the chromosomes of liver mitoses in aborted human embryos. This study, however, was temporarily discontinued because the workers were unable to find all the 48 human chromosomes in their material; as a matter of fact, the number 46 was repeatedly counted in their slides. We have seen photomicrographs of liver prophases from this study, clearly showing 46 chromosomes. These findings suggest that 46 may be the correct chromosome number for human liver tissue, too.

With previously used technique it has been extremely difficult to make counts in human material. Even with the great progress involved in Hsu's method exact counts seem difficult, judging from the photomicrographs published (Hsu, 1952 and elsewhere). For instance, we think that the excellent photomicrograph of Hsu published in DARLINGTON's book (1953, facing p. 288) is more in agreement with the chromosome number 46 than 48, and the same is true of many of the photomicrographs of human chromosomes previously published.

Before a renewed, careful control has been made of the chromosome number in spermatogonial mitoses of man we do not wish to generalize our present findings into a statement that the chromosome number of man is $2n=46$, but it is hard to avoid the conclusion that this would be the most natural explanation of our observations.

Acknowledgements. — We wish to express our sincere thanks to the Swedish Cancer Society for financial support of this investigation, and to Dr. RUNE GRUBB for supplying us with tissue cultures.

SUMMARY

The chromosomes were studied in primary tissue cultures of human lung fibroblasts explanted from four individual embryos. In all of them the chromosome number 46 was encountered, instead of the expected number 48. Since among 265 mitoses counted all except 4 showed the number 46, this number is characteristic of the tissue studied. The possible bearing of this result on the chromosome number of man is discussed.

Institute of Genetics, Lund, January 26, 1956.

Literature cited

DARLINGTON, C. D. 1953. The facts of life. — London, 467 pp.

HSU, T. C. 1952. Mammalian chromosomes *in vitro*. — The karyotype of man. — J. Hered. 43: 167—172.

LEVAN, A. 1956. Chromosome studies on some human tumors and tissues of normal origin, grown *in vivo* and *in vitro* at the Sloan-Kettering Institute. — Cancer (in the press).

Reprinted from Experimental Cell Research, Vol. 62, pp. 490–492. 1970

Analysis of human metaphase chromosome set by aid of DNA-binding fluorescent agents

T. CASPERSSON, L. ZECH and C. JOHANSSON, *Institute for Medical Cell Research and Genetics, Medical Nobel Institute, Karolinska Institutet, 104 01 Stockholm, Sweden*

Characteristic fluorescence distribution patterns for human chromosomes treated by certain DNA-reacting fluorescent agents have been described [1]. By far the best results were obtained with Quinacrine Mustard (QM) which binds by intercalation as well as by way of its alkylating groups. The QM-fluorescence method is useful also for the analysis of interphase nuclei, as certain chromosome parts remain strongly QM-reacting, notably part of the Y-chromosome [2, 4]. The method can thus be of help in determination of cellular sex.

The fluorescence patterns of the metaphase chromosomes were determined photoelectrically and grouped in "types". We have now been able to correlate these types with the conventional numbering. The main QM-independent criteria used were chromosome length, centromere index and labelling time with tritiated thymidine. At the same time the observational material has been widened to comprise measurements of about 5 000 chromosomes from 14 healthy subjects. The patterns, which are quite reproducible in chromosomes of similar degrees of contraction, are presented in fig. 1. The range of variability etc. will be presented in a later publication. The typical pattern of chromosomes 4, 5 and X has already been published [3].

The proper technique for the recognition of

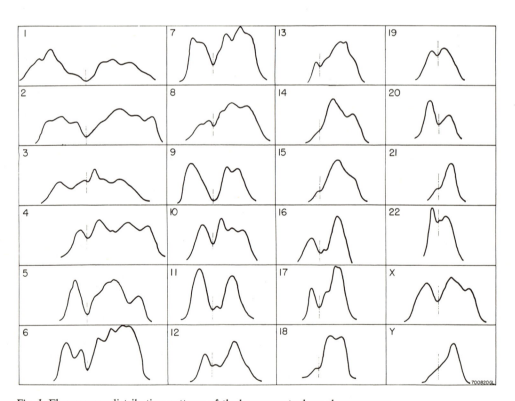

Fig. 1. Fluorescence distribution patterns of the human metaphase chromosomes.

Exptl Cell Res 62

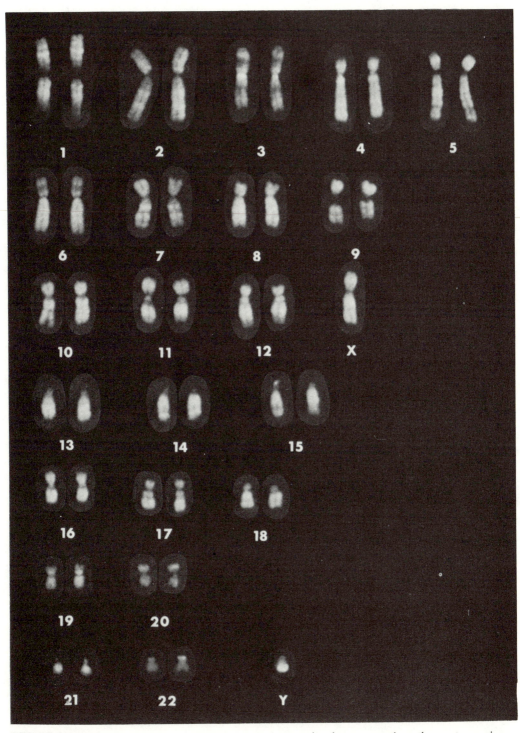

Fig. 2. Fluorescence micrographs of human metaphase, printed so as to show the most conspicuous QM-bands. ×2 500.

Exptl Cell Res 62

Table 1.

No.	GMC	CI	AR	SC
1	+	(long arm identified by AR)		
2	+			
3	+			
4	+		+	
5	+		+	
6	+	+		
7	+	+		
8	+	+		
9	+	+		+
10	cannot morphologically be distinguished from 12			
11	+	+		(+)
12	see 10			
X	+	+	+	
13	+		+	
14	+		+	
15	+		+	
16	+	+		(+)
17	+	+	+	
18	+	+	+	
19	cannot morphologically be distinguished from 20			
20	see 19			
21	cannot morphologically be distinguished from 22			
22	see 21			
Y	+			

the fluorescence patterns is photoelectric evaluation [1] and this technique is applicable to practically any QM-preparation. Recent alterations in QM-staining and photographic techniques have so improved contrast that in good preparations the majority of the chromosomes can be identified visually in a fluorescence micrograph by means of their strongly fluorescent bands. For complete analysis, measurements are needed. Fig. 2 shows a complete set of chromosomes with many intrachromosomal details visible. The visually observable bands can be directly compared with the curves in fig. 1.

Table 1 lists the independent criteria for the determination of the chromosome numbers which were used for the correlation of fluorescence patterns with chromosome numbers. GMC means gross morphological characters including length. CI denotes centromere index; AR, autoradiography, and SC, secondary constriction.

In fig. 1 one of the two very typical QM-patterns for 10 and 12 has been taken as representing 10 and the other 12, for these chromosomes cannot be distinguished by conventional methods. The same holds true for 19/20 and 21/22. Attention should be drawn to the fact that the fluorescence of the *short* arms of 13, 15 and sometimes 14 and 22 may vary considerably (cf [2]).

Fourier analysis of the fluorescence patterns—as yet the C group—has shown that the individual curves can be described accurately enough with only 10–14 coefficients which makes automated analysis feasible and reasonable.

REFERENCES

1. Caspersson, T, Zech, L, Johansson, C & Modest, E J, Chromosoma 30 (1970) 215.
2. Caspersson, T, Zech, L & Johansson, C, Exptl cell res 60 (1970) 315.
3. Caspersson, T, Zech, L & Johansson, C, Exptl cell res 61 (1970) 474.
4. Zech, L, Exptl cell res 58 (1969) 463.

Received June 12, 1970

Reprinted from Science, Vol. 168, pp. 1356–1358. 1970. © 1970 AAAS

Chromosomal Localization of Mouse Satellite DNA

Abstract. *Hybridization of radioactive nucleic acids with the DNA of cytological preparations shows that the sequences of mouse satellite DNA are located in the centromeric heterochromatin of the mouse chromosomes. Other types of heterochromatin in the cytological preparations do not contain satellite DNA.*

One characteristic which distinguishes the DNA of higher organisms from that of bacteria is the presence of families of repeated nucleotide sequences. These repeated sequences are found in multiplicities ranging from 10^2 to 10^6 per genome (*1*), but very little is known about their function or their organization within the chromosomal complement. Recently a technique which makes possible the cytological localization of specific nucleotide sequences has been developed (*2*). This localization is accomplished by hybridizing the DNA of cytological preparations with radioactive nucleic acid. The regions in the preparation to which the radioactive nucleic acid has bound are then detected by autoradiography. Such a technique permits a direct investigation of the distribution of families of repeated sequences within the genome.

Perhaps the most thoroughly studied fraction of repetitive DNA, with the exception of the sequences coding for ribosomal RNA, is the mouse satellite DNA. Therefore we chose mouse satellite DNA for our first investigations of the cytological localization of multiply repeated DNA sequences. We show that this fraction is located in the centromeric heterochromatin of the mouse chromosomes, a fact which we have briefly reported (*3*).

Mouse satellite DNA forms a band slightly separated from the main peak when mouse DNA is spun to equilibrium in a CsCl density gradient (*4*). It makes up about 10 percent of the total DNA, regardless of the tissue from which the DNA has been prepared, and is found in about the same proportion in tissue culture lines (*5*). From renaturation kinetics it has been estimated that mouse satellite DNA consists of approximately 10^6 copies per genome of a sequence some 400 nucleotide pairs in length (*6*). It is possible that the copies are not all identical. However, the rapid reassociation seen after denaturation indicates a high degree of homogeneity

(*6*). Although the sequences of the mouse satellite make up 10 percent of the mouse DNA they do not seem to code for a corresponding fraction of the RNA in the tissues which have been studied. Flamm, Walker, and McCallum (*7*) were unable to detect any hybridization of satellite DNA to RNA from mouse liver, spleen, or kidney. Recently Harel *et al.* (*8*) have reported that rapidly labeled RNA from some of the same tissues did bind to satellite DNA. The coding properties of this fraction require further study.

Our hybridization experiments on the localization of mouse satellite DNA have been done in two ways. First, we have applied fractions of radioactive mouse DNA to cytological preparations of mouse tissue culture cells (*9*).

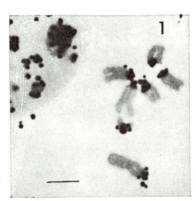

Fig. 1. Autoradiograph of a mouse tissue culture preparation after cytological hybridization with radioactive RNA copied in vitro from mouse satellite DNA. The RNA has bound to the centromeric heterochromatin of the chromosomes and to the chromocenters of the interphase nucleus on the left. The DNA of this preparation was denatured in situ by treatment with 0.07N NaOH. The slide was then incubated with radioactive RNA for 10 hours at 66°C. The preparation was treated with ribonuclease to remove RNA that was not specifically bound and then coated with autoradiographic emulsion. The RNA had a calculated specific activity of 7×10^7 disintegrations min^{-1} μg^{-1}. Slide stained with Giemsa. Exposure, 5 days; $\times 2000$; scale length, 5 μm.

The radioactive DNA was extracted from tissue cultures of the mouse A9 line grown in medium containing [³H]thymidine (*3*). Satellite DNA was separated from the rest of the mouse DNA by silver-ion–cesium sulfate (*10*) density gradient centrifugation. The DNA was denatured with heat before use in the hybridization reaction. This DNA had a specific activity of 200,000 cpm μg^{-1} as determined by spotting known amounts of the DNA on a nitrocellulose filter and counting the filter in toluene fluor in a scintillation counter. In the second type of experiment we applied radioactive RNA, transcribed in vitro from mouse DNA, to cytological preparations of both mouse testis and mouse tissue culture (*2*). Mouse liver DNA was fractionated by silver-ion–cesium sulfate centrifugation. The satellite DNA and the main peak DNA were transcribed separately with *Escherichia coli* RNA polymerase (*11*) and tritiated ribonucleotide triphosphates. This complementary RNA had a calculated specific activity of 7×10^7 disintegration min$^{-1}\mu$g^{-1}. In all experiments the DNA of the cytological preparations was denatured by treatment with NaOH before hybridization. Autoradiographs were exposed for several days when hybridization was done with complementary RNA and for several months when hybridization was done with radioactive DNA.

The normal mouse chromosomal complement consists of 20 pairs of telocentric or acrocentric chromosomes (*12*). Each chromosome has a region next to the centromere which can be identified as heterochromatin by its staining properties. In some mouse tissue culture lines a few of the chromosomes are metacentric and have presumably arisen by fusion of two of the chromosomes of the normal com-

The Sertoli cells of the testis contain what has been described as a compound nucleolus, consisting of an acidophilic body and one or more basophilic bodies. The basophilic bodies bound satellite DNA, indicating that in these cells the centromeres are closely associated into small groups near the nucleolus.

In the mouse both the X and the Y chromosomes show heterochromatization at various times, yet the binding of mouse satellite DNA indicates that

plement. There is heterochromatin adjacent to the centromere on both arms of these chromosomes. All of our preparations are stained with Giemsa after the development of the autoradiograph. We find that with this stain the centromeric heterochromatin on chromosomes that have been treated with NaOH stains more densely than the rest of the chromosome.

We obtained similar results from both the DNA-DNA and the DNA-RNA hybridization experiments. Satellite DNA and its complementary RNA bound only to the centromeric heterochromatin of the chromosomes (13). Only one chromosome in our preparations appeared consistently unlabeled. Because the unlabeled chromosome was small and more heterochromatic than the other chromosomes in the testis preparations, we believe that it is the Y chromosome. In parallel experiments radioactive DNA from the main band and complementary RNA hybridized with many regions distributed over the entire chromosomal complement, an indication that the euchromatic regions of the chromosomes had been denatured and were capable of binding complementary nucleic acid sequences.

Localized binding of satellite DNA was evident throughout the cell cycle in both mitosis and meiosis. The binding was in the centromeric heterochromatin in all of the stages in which the chromosomes are condensed. In other stages the positions of the centromeres could be followed because of the localization of the satellite binding. In interphase nuclei, satellite DNA bound preferentially to the chromocenters, indicating that the majority of the centromeric regions are associated in these deeply staining chromatin blocks. In spermatids the centromeric regions were concentrated in a single mass in the central region. This mass also differs from the rest of the nuclear DNA in its binding of Giemsa stain.

satellite sequences are present only in the centromeric regions of the X. Figure 4 shows a pachytene spermatocyte. The heterochromatic X and Y make up the "sex vesicle" (14). Ribonucleic acid copied from the mouse satellite DNA is bound to the centromeric heterochromatin of the autosomes and to the tip of the sex vesicle (arrow in Fig. 4).

Our studies confirm and extend previous work on the localization of mouse satellite DNA. Maio and Schildkraut

(15) separated isolated metaphase chromosomes from mouse L cells into several size classes by sedimenting the chromosomes through sucrose density gradients. They found that each size class contained approximately the same proportion of satellite DNA as did the entire genome, which suggests that satellite DNA makes up a constant proportion of the DNA of the individual chromosomes. Furthermore, although 70 percent of the DNA could be extracted from the chromosomes with 2M NaCl, the satellite DNA re-

mained bound to the insoluble protein matrix of the chromosomes, an indication that the protein associations of satellite DNA differ from those of main-band DNA. Yasmineh and Yunis (16) found that most of the DNA extracted from isolated heterochromatin of the mouse liver and brain cells was satellite DNA. The euchromatin prepared in these same experiments contained only DNA of main-band density.

Our results clearly demonstrate that heterochromatin is not a single

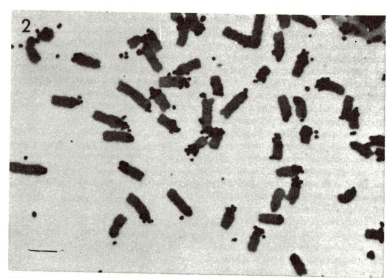

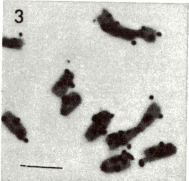
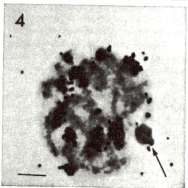

Fig. 2. Autoradiograph of a tissue culture preparation similar to that in Fig. 1. In these heteroploid tissue cultures the number of grains per chromosome is variable but no chromosome is consistently unlabeled. In testis squashes (not shown here) the Y chromosome appears to be unlabeled. Exposure, 5 days; × 1750; scale length, 5 μm.
Fig. 3. Autoradiograph of a mouse tissue culture preparation after cytological hybridization with radioactive RNA copied in vitro from mouse main peak DNA. This RNA has bound to regions of the chromosome other than the centromeric heterochromatin. The conditions of denaturation and hybridization are the same as for Fig. 1. Exposure, 23 days; × 2700; scale length, 5 μm.   Fig. 4. Autoradiograph of a mouse pachytene spermatocyte hybridized with radioactive RNA copied in vitro from mouse satellite DNA. The densely stained heterochromatic regions of the autosomes are heavily labeled but the sex vesicle, containing the heterochromatic X and Y chromosomes, shows only four grains and these are localized over the tip (arrow). The conditions of denaturation and hybridization are the same as for Fig. 1. Exposure, 10 days; × 1700; scale length, 5 μm.

category. It has been recognized for some time that certain chromosome regions maintain their heterochromatic character at all times. This type of heterochromatin has been termed constitutive and is usually considered genetically inert. Other chromosome regions or entire chromosomes, such as the mammalian X (17), become heterochromatic only in some particular stages or tissues. The second type of heterochromatin has been termed facultative or functional and is often associated with a turning-off of the genes on the chromosomes involved. The regions adjacent to the centromere are usually thought of as constitutive heterochromatin. This heterochromatin contains a specific type of DNA which is not present in the facultative heterochromatin of the sex chromosomes. It has also been reported that mouse chromosomes have constitutive heterochromatin at the telomeres (18). If these regions do contain satellite DNA sequences, these sequences are too few to be detected on our slides.

Although we have localized mouse satellite DNA in the centromeric heterochromatin, this localization does not establish a function for either satellite DNA or heterochromatin. It seems that this function is one which is necessary to the chromosome since the proportion of satellite DNA is maintained in established mouse cell lines even though the chromosomes have undergone other morphological change. We have found the same pattern of satellite hybridization in all of the mouse cell lines which we have studied. In this respect it becomes important to establish whether the apparent lack of hybridization at the Y centromere indicates a quantitative or qualitative difference from the other chromosomes. It is possible that centromeric heterochromatin might bind the proteins of the spindle microtubules during mitosis. We have investigated the binding of purified microtubular subunits to mouse satellite DNA in vitro (19) but our preliminary experiments showed no binding under the conditions used in our assay (20).

Centromeric heterochromatin has been described in the chromosomes of many animals and plants. It seems possible, therefore, that the centromeres are regularly adjacent to regions of repetitive DNA. Experiments in this laboratory have shown that the fly *Rhynchosciara hollaenderi* has a satellite of repetitive DNA. This satellite, like that of the mouse, has been localized in the centromeric heterochromatin by cytological hybridization (21). Cytological hybridization has also shown the presence of repetitive DNA at the centromeres in *Triturus viridescens* (22) and *Drosophila melanogaster* (23). It now seems important to determine whether noncentromeric heterochromatin consists of repetitive DNA with nucleotide sequences different from those of centromeric heterochromatin (24).

MARY LOU PARDUE
JOSEPH G. GALL

Department of Biology,
Yale University,
New Haven, Connecticut 06520

References and Notes

1. R. J. Britten and D. E. Kohne, *Science* **161**, 529 (1968).
2. J. G. Gall and M. L. Pardue, *Methods Enzymol.*, in press; *Proc. Nat. Acad. Sci. U.S.* **63**, 378 (1969); H. John, M. Birnstiel, K. Jones, *Nature* **223**, 582 (1969); M. Buongiorno-Nardelli and F. Amaldi, in press.
3. M. L. Pardue and J. G. Gall, *Proc. Nat. Acad. Sci. U.S.* **64**, 600 (1969).
4. S. Kit, *J. Mol. Biol.* **3**, 711 (1961).
5. J. J. Maio and C. L. Schildkraut, *ibid.* **24**, 29 (1967).
6. M. Waring and R. J. Britten, *Science* **154**, 791 (1966).
7. W. G. Flamm, P. M. B. Walker, M. McCallum, *J. Mol. Biol.* **40**, 423 (1969).
8. J. Harel, N. Hanania, H. Tapiero, L. Harel, *Biochem. Biophys. Res. Commun.* **33**, 696 (1968).
9. All of the mouse cells were kindly provided by Dr. F. Ruddle, Yale University.
10. G. Corneo, E. Ginelli, C. Soave, G. Bernardi, *Biochemistry* **7**, 4373 (1968).
11. R. R. Burgess, *J. Biol. Chem.* **244**, 6160 (1969).
12. T. C. Hsu and K. Benirschke, *An Atlas of Mammalian Chromosomes* (Springer-Verlag, New York, 1967).
13. If we assume that all of the satellite sequences are evenly distributed among the centromeric regions of 39 mouse chromosomes we are detecting the presence of 10^{10} daltons of satellite DNA on a single chromosome.
14. L. Sachs, *Ann. Eugen.* **18**, 255 (1954).
15. J. J. Maio and C. L. Schildkraut, *J. Mol. Biol.* **40**, 203 (1969).
16. W. G. Yasmineh and J. J. Yunis, *Biochem. Biophys. Res. Commun.* **35**, 779 (1969); *Exp. Cell Res.* **59**, 69 (1970).
17. S. Ohno, W. D. Kaplan, R. Kinosita, *Exp. Cell Res.* **18**, 415 (1959); M. F. Lyon, *Nature* **190**, 372 (1961).
18. S. Ohno, W. D. Kaplan, R. Kinosita, *Exp. Cell Res.* **15**, 426 (1958).
19. These experiments were done in collaboration with J. Olmsted and Dr. J. Rosenbaum.
20. W. Gilbert and B. Müller-Hill, *Proc. Nat. Acad. Sci. U.S.* **58**, 2415 (1967).
21. R. Eckhardt, unpublished results.
22. G. Barsacchi, unpublished results.
23. J. Shaklee, unpublished results.
24. Since submission of this report, K. W. Jones has published observations similar to ours [*Nature* **225**, 912 (1970)].
25. Supported by PHS grants GM 12427 and GM 397. We thank Mrs. C. Barney for technical assistance. This work will be submitted by M.L.P. in partial fulfillment of requirements for the doctoral degree at Yale University.

16 February 1970

Reprinted from Science, Vol. 183, pp. 330–331. 1974. © 1974 AAAS

Spheroid Chromatin Units (ν Bodies)

Abstract. *Linear arrays of spherical chromatin particles (ν bodies) about 70 angstroms in diameter have been observed in preparations of isolated eukaryotic nuclei swollen in water, centrifuged onto carbon films, and positively or negatively stained. These bodies have been found in isolated rat thymus, rat liver, and chicken erythrocyte nuclei. Favorable views also reveal connecting strands about 15 angstroms wide between adjacent particles.*

The packaging of DNA within eukaryotic chromosomes continues to be a formidable structural problem. Packing ratios greater than 100/1 (DNA length/chromatid length) are not uncommon for metaphase chromosomes (*1*). The DNA concentrations within localized regions of interphase nuclei may approximate 200 mg/ml or more (*2*). Acutely aware of this problem, investigators have postulated multiple orders of coiling or folding of a fundamental nucleohistone molecule (*1, 3*). Several models have been derived from low-angle x-ray diffraction studies, including: four DNA molecules packed into a single nucleohistone fibril (*4*); a single DNA double helix and associated proteins folded into an irregular superhelix 80 to 120 Å in diameter and 45 Å in pitch (*5*); and a single DNA-protein fiber constrained into a superhelix 100 Å in diameter and 120 Å in pitch (*6*). Ultrastructural studies have also yielded a profusion of models. Spreading of chromosomes on a Langmuir trough frequently yields fibrils about 250 Å in diameter, although differences due to tissue type, presence of chelating agents, and method of dehydration and drying have been reported (*7*). Direct adsorption of sheared chromatin onto microscope grids has revealed a network of fibers approximately 100 Å wide with numerous side branches 80 to 200 Å in length (*5*). Spraying of chromatin onto a grid yields a network of fibers (*8*) and separated filaments (20 to 30 Å in diameter) containing numerous "nodular" elements about 150 Å in diameter (*9*). Thin sections of nuclei and chromosomes reveal fragments of threads frequently 100 to 200 Å wide (*3, 10, 11*). Bram and Ris (*5*) regard the 250-Å fiber as a folding (or doubling) of a superhelix, due to divalent metal ions, and interpret the thin-section data as artifacts of chelation by buffer ions. Lampert (*12*) views the 250-Å filament as a folding of the superhelix of Pardon and Wilkins (*6*), and explains the thin-section data in terms of shrinkage due to fixation. Despite this divergence of views, there is a consensus that multiple levels of coiling or folding are required to explain the observed variation in chromatin fiber widths.

We have attempted to visualize chromatin structure by methods different from those cited above. Interphase nuclei were isolated from fresh rat thymus (*13*), rat liver (*2*), and chicken erythrocytes (*2*), washed and centrifuged twice in CKM buffer (*14*) and once in 0.2*M* KCl, suspended in 0.2*M* KCl at a concentration of approximately 10^8 nuclei per milliliter, and diluted 200-fold into distilled H_2O. Nuclei were allowed to swell for 10 to 15 minutes, then made 1 percent in formalin (*p*H 6.8 to 7.0). Fixation proceeded for at least 30 minutes. All operations, up to this point, were at 0°

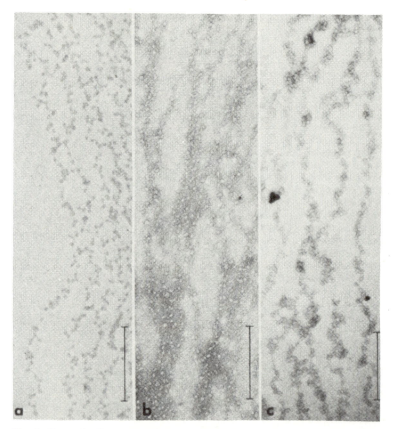

Fig. 1. Chromatin fibers spilling out of ruptured nuclei. The degree of fiber swelling and the proximity of individual ν bodies to each other varies within different regions of a single nucleus. Scale bars, 0.2 μm. (a) Rat thymus chromatin, positively stained with a mixture of 4 percent aqueous phosphotungstic acid and 95 percent ethanol (3 : 7), rinsed in 95 percent ethanol, and dried in air. (b) Rat thymus chromatin, negatively stained with 0.5 percent ammonium molybdate, adjusted to *p*H 7.4 to 8.0 with ammonium hydroxide. (c) Chicken erythrocyte chromatin, negatively stained as in (b). Clustering of ν bodies is most evident in (c), where groups of three or more are readily visualized. Connecting strands are most easily seen in (b).

to 4°C. Aliquots of the swollen and fixed nuclei were centrifuged through 10 percent formalin (*p*H 6.8 to 7.0) onto carbon-covered grids, rinsed in dilute Kodak Photo-Flo, and dried in air, a technique developed by Miller and co-workers (*15*). When examined after positive staining, chromatin fibers could be readily visualized streaming out of ruptured nuclei. Fibers were often very long (about 8 μm in length), unbranched, and in parallel arrays, and revealed irregularly distributed thick and thin regions. Frequently, views of chromatin fibers (Fig. 1a) show spherical particles, ν bodies (*16*), 60 to 80 Å in diameter, connected by thin filaments (about 15 Å wide). Less stretched regions of chromatin revealed apparent packing of ν bodies. Analysis of fiber widths from positively stained preparations showed peaks at 75 to 100, 125 to 150, and possibly 225 to 250 Å (Fig. 2), consistent with the ranges of fiber widths described earlier. Better visualization of the ν bodies and connecting strands has been obtained by the use of negative stains (Fig. 1, b and c). The thickened fiber regions were seen to represent clusters of ν bodies. Measurements of diameters of ν bodies for the different tissues employed yielded the following average diameters and standard deviations: rat thymus, 83 ± 23 Å; rat liver, 60 ± 16 Å; and chicken erythrocyte, 63 ± 19 Å. Connecting strands, for rat thymus chromatin, exhibited average widths of 15 ± 4 Å. Figure 2 demonstrates that the distribution of diameters of ν bodies, measured from negatively stained preparations, superimposes on the lowest peak of fiber diameters calculated from positively stained materials.

For a number of reasons we believe that this appearance of chromatin fibers as "particles on a string" is related to the native configuration and is not an artifact of the preparative procedures. Washing of isolated nuclei in CKM buffer (*14*) and in 0.2*M* KCl appears to remove some nonhistone but no histone protein (*17–19*) although histone migration along the DNA cannot be eliminated. However, nuclei so treated reveal the same spectrum of chromatin fiber widths after fixation and thin sectioning (*2*) as those observed for fixed and sectioned whole tissue (*10*). Swelling of nuclei in water leads to stretching and thinning of chromatin fibers, as revealed after fixation in water and thin sectioning (*20*), and the disappearance of several of the low-angle x-ray

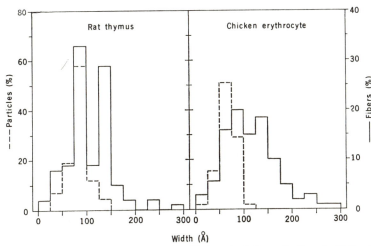

Fig. 2. Histograms comparing widths of positively stained chromatin fibers (solid lines) with diameters of ν bodies (broken lines) from negatively stained preparations. Random sampling of fiber widths was obtained by superimposing a lattice of lines 1 inch apart on the photographic print and measuring the width of any fiber intersecting the grid lines. The positively stained preparations were rat thymus ($N = 100$ samples) and chicken erythrocyte ($N = 360$). The diameters of ν bodies were measured only when their edges were clearly defined and not overlapping another ν body. The preparations for measurements of ν bodies were rat thymus ($N = 114$) and chicken erythrocyte ($N = 200$). A calibration grid (54,864 lines per inch) was photographed with each set of micrographs, printed, and measured simultaneously with the sample photographs.

reflections (*21*). Since addition of divalent metal ions to water-swollen nuclei and chromatin does produce essentially normal low-angle x-ray reflections (*2, 21*) and a partial return of ultrastructural morphology (*20*), the structural changes exhibit some reversibility. Fixation with glutaraldehyde does not markedly perturb low-angle x-ray reflections (*2*), although similar data for formaldehyde are not available. In order to demonstrate that nuclear isolation and washing are not essential for visualization of the chromatin units, fresh whole chicken erythrocytes were disrupted in cold 0.3 percent Joy detergent, followed by swelling, fixation, and centrifugation onto carbon films, a method developed by Miller and Bakken (*22*). Although the spreading of chromatin fibers was not as good as that shown in Fig. 1, particles resembling ν bodies were observed throughout the chromatin.

Assuming that the ν bodies described here represent a real packaging of nucleohistone (*23*), we calculate some of their expected physical properties. For an average particle diameter of 70 Å, a spherical shape, and a partial specific volume of 0.68 cm³/g (*24*), we estimate an approximate molecular weight of 160,000 per ν body. Further, we assume that every ν body has at least one histone of each of the five classes, and that the sum of their mo-

lecular weights is 84,000 (*25*). Therefore, the DNA would have a calculated molecular weight of about 80,000 and a total length of about 400 Å, packed into the spheroid particle 70 Å in diameter. Thus, a packing ratio (DNA length/particle diameter) of about 6/1 might be expected. If there is significant dehydration and shrinkage of the ν bodies, the calculated particle molecular weight would have to be increased. The dimensions of ribosomal particles and subunits, measured by electron microscopy, are roughly half of the calculated hydrated volumes from hydrodynamic measurements (*26*). It would be conceivable, therefore, for each ν body to contain two of each type of histone molecule complexed with a double-stranded DNA with a molecular weight of about 160,000. Further packaging of the DNA might then represent a folded or helical close packing of the spherical ν bodies under the influence of metal cations and noncovalent interaction. Studies should be directed toward fragmentation of chromatin and isolation of particles with properties complementary to the ν bodies.

ADA L. OLINS, DONALD E. OLINS
University of Tennessee–Oak Ridge Graduate School of Biomedical Sciences, and Biology Division, *Oak Ridge National Laboratory, Oak Ridge, Tennessee 37830*

331

455

References and Notes

1. E. J. DePraw, *DNA and Chromosomes* (Holt, Rinehart & Winston, New York, 1970).
2. D. E. Olins and A. L. Olins, *J. Cell Biol.* **57**, 715 (1972).
3. H. Ris and D. F. Kubai, *Annu. Rev. Genet.* **4**, 263 (1970).
4. V. Luzzati and A. Nicolaieff, *J. Mol. Biol.* **1**, 127 (1959); *ibid.* **7**, 142 (1963).
5. S. Bram and H. Ris, *ibid.* **55**, 325 (1971).
6. J. F. Pardon and M. H. F. Wilkins, *ibid.* **68**, 115 (1972).
7. J. G. Gall, *Chromosoma* **20**, 221 (1966); S. L. Wolfe, *J. Cell Biol.* **37**, 610 (1968); H. Ris, in *Handbook of Molecular Cytology*, A. Lima-de-Faria, Ed. (North-Holland, Amsterdam, 1969), p. 221; B. R. Zirkin, *J. Ultrastruct. Res.* **36**, 237 (1971); A. J. Solari, *Exp. Cell Res.* **67**, 161 (1971).
8. R. F. Itzaki and A. J. Rowe, *Biochim. Biophys. Acta* **186**, 158 (1969).
9. H. S. Slayter, T. Y. Shih, A. J. Adler, G. D. Fasman, *Biochemistry* **11**, 3044 (1972).
10. J. S. Kaye and R. McMaster-Kaye, *J. Cell Biol.* **31**, 159 (1969); H. G. Davies, *J. Cell Sci.* **3**, 129 (1968); A. C. Everid, J. V. Small, H. G. Davies, *ibid.* **7**, 35 (1970).
11. B. R. Zirkin and S.-K. Kim, *Exp. Cell Res.* **75**, 490 (1972).
12. F. Lampert, *Nature (Lond.)* **234**, 187 (1971).
13. V. G. Allfrey, V. C. Littau, A. E. Mirsky, *J. Cell Biol.* **21**, 213 (1964).
14. The buffer consisted of 0.05*M* sodium cacodylate, *p*H 7.5; 0.025*M* KCl; 0.005*M* MgCl$_2$; and 0.25*M* sucrose.
15. O. L. Miller, Jr., and B. R. Beatty, *Science* **164**, 955 (1969); *J. Cell Physiol.* **74** (Suppl. 1), 255 (1969); O. L. Miller, Jr., B. A. Hamkalo, C. A. Thomas, Jr., *Science* **169**, 392 (1970).
16. The Greek letter ν is suggested to denote these spheroid chromatin bodies. This notation was considered to be appropriate since they are both new and nucleohistone. Alternatively, we suggest that they be referred to as "deoxyribosomes" in order to emphasize their particular particulate analogy with ribosomes.
17. Solvents similar to these have been shown to extract ribonucleoprotein particles and nuclear sap proteins [H. Busch, *Histones and Other Nuclear Proteins* (Academic Press, New York, 1965)] including the RNA-containing interchromatin granules (*18, 19*). Furthermore, preliminary observations (D. E. Olins and E. B. Wright, unpublished) indicate that proteins extractable from chicken erythrocyte nuclei by washing in CKM buffer and 0.2*M* KCl contain no detectable histones, as judged by gel electrophoresis in buffers containing sodium dodecylsulfate [K. Weber and M. J. Osborn, *J. Biol. Chem.* **244**, 4406 (1969)].
18. A. Monneron and Y. Moulé, *Exp. Cell Res.* **51**, 531 (1968).
19. A. Monneron and W. Bernhard, *J. Ultrastruct. Res.* **27**, 266 (1969).
20. K. Brasch, V. L. Seligy, G. Setterfield, *Exp. Cell Res.* **65**, 61 (1971); A. L. Olins and D. E. Olins, unpublished observations.
21. R. A. Garrett, *Biochim. Biophys. Acta* **246**, 553 (1971).
22. O. L. Miller, Jr., and A. H. Bakken, *Acta Endocrinol. Suppl. 168* (1972), p. 155.
23. It is considered unlikely that the ν bodies and connecting strands described in this report might correspond to nuclear ribonucleoprotein particles. Interchromatin granules (200 to 250 Å in diameter) and perichromatin granules (400 to 450 Å in diameter) are considerably larger than the ν bodies ($\sim$70 Å in diameter) (*19*). Interchromatin granules would be expected to be extracted during the washes in CKM buffer (*18*). Furthermore, ν bodies are observed in considerably greater numbers than would be expected for the known ribonucleoprotein particles.
24. G. Zubay and P. Doty, *J. Mol. Biol.* **1**, 1 (1959).
25. J. H. Diggle and A. R. Peacocke, *FEBS (Fed. Eur. Biochem. Soc.) Lett.* **18**, 138 (1971). From their data we take the following molecular weights: F2C, 20,800; F2B, 14,400; F2A2, 16,800; F2A1, 19,000; and F3, 13,000. The substitution of F1 histone for F2C in the rat tissue does not appreciably change the sum of molecular weights. Adding one molecule of each histone yields a total molecular weight of 84,000.
26. W. E. Hill, G. P. Rossetti, K. E. Van Holde, *J. Mol. Biol.* **44**, 263 (1969).
27. A preliminary report of this work was presented at the meeting of the American Society of Cell Biology, Miami, 15 November 1973. At the same meeting, C. L. F. Woodcock reported the observation of similar spherical particles in chromatin fibers. The authors thank O. L. Miller, Jr., and B. A. Hamkalo for advice and criticism and M. Hsie for excellent technical assistance. This work was sponsored, in part, by the Atomic Energy Commission under contract with Union Carbide Corporation, and, in part, by NIGMS grant 1 R01 GM 19334-01 to D.E.O. One of us (D.E.O.) is a recipient of NIGMS research career development award 5 K04 GM 40441-03.

19 July 1973   ∎

332

Reprinted from Science, Vol. 193, pp. 848–856. 1976. © 1976 AAAS

Chromosomal Subunits in Active Genes Have an Altered Conformation

Globin genes are digested by deoxyribonuclease I in red blood cell nuclei but not in fibroblast nuclei.

Harold Weintraub and Mark Groudine

Knowledge of the structure of DNA has provided many insights into its biological function (*1*). In higher cells, a detailed understanding of the structure of chromatin will probably provide analogous insights into how genes are regulated. Already, there are a number of important observations demonstrating a relation between the structure of chromatin and its biological activity (*2, 3*).

The packaging of most of the nuclear DNA is now thought to be based on repeating units of about 180 to 200 base pairs of DNA associated with specific complexes of histones (*4, 5*), possibly two self-complementary tetramers each containing one of the four major histones (*6*). These two tetramers could define the twofold axis of symmetry within the nucleosome. These complexes interact through 70 to 90 amino acid residues at their carboxyl terminal end to produce a tight, trypsin-resistant core (*7*). The positively charged histone amino terminal residues extend outward from this core and define what may prove to be a "kinked" or "coiled" pathway for the DNA (*5, 8*) about the histone complexes. These so-called "particles-on-a-string" or "nu" bodies constitute the primary level of folding for the bulk of the chromosome. Through their mutual interactions higher levels of DNA packaging can be achieved, although details of this organization are not known. At present there is no proof that nu bodies are homo-

Dr. Weintraub is an assistant professor in the Department of Biochemical Sciences, Frick Laboratories, Princeton University, Princeton, New Jersey 08540. Dr. Groudine was a visiting fellow in the same department and is now at the Department of Radiation Oncology, University of Washington Hospital, Seattle 98105.

SCIENCE, VOL. 193

geneous (9) either in composition or in conformation. Also lacking is concrete evidence that nu bodies are associated with active genes (10) and, if so, whether they are in the same conformation as those nu bodies associated with inactive genes. The important finding that *Escherichia coli* RNA polymerase preferentially transcribes globin genes from reticulocyte chromatin but not from liver or brain chromatin (11) reflects the fact that some structural aspect of an activated gene is different in different tissues. However, whether this difference occurs only at the 5' end of the gene, perhaps at a promotor region, or throughout the entire length of the gene is not known. Similarly, it is not clear whether the tissue-specific differences in accessibility revealed by RNA polymerase arise from differences in the basic nu body configuration or in the way some or all of the nu bodies in a transcription unit are packaged into higher levels of organization.

In this article, we show that active genes are likely to be packaged by histones but that these histones are in an altered conformation, one that renders the associated DNA extremely sensitive to digestion by pancreatic deoxyribonuclease I.

Digestion of Chicken Erythrocyte Nuclei by Deoxyribonuclease I

The kinetics of digestion of chicken erythrocyte nuclei by pancreatic deoxyribonuclease I (12) are shown in Fig. 1. The bottom insets show the corresponding patterns of resistant single-stranded

DNA fragments displayed on a denaturing acrylamide gel. The kinetics reveal a rapid initial digestion of about 15 percent of the DNA followed by a slower digestion that levels off at about 50 percent, and an even slower process that leads to digestion beyond 50 percent. The characteristic pattern of resistant products (between 20 and 160 bases in our standard gels) appears very early in digestion and persists through 50 percent digestion with no apparent increase or decrease in the intensity of any one particular band. Beyond 50 percent digestion, the larger bands disappear before the smaller bands do. This disappearance is accompanied by an accumulation of DNA between the usually well defined DNA peaks. We interpret this to mean that the enzyme is sequentially digesting a base or two at a time when digestion proceeds beyond 50 percent, and that this phase of the digestion process probably represents "nibbling" (digestion of a base at a time) that occurs in DNA regions that are intimately protected by proteins, presumably histones. In contrast, the repeating pattern of denatured DNA fragments between 20 and 160 bases probably represents the cutting at very accessible regions between and within individual nu bodies. It is not clear to us why the larger fragments fail to break down into the smaller ones before digestion reaches 50 percent and nibbling ensues. The conversion of larger fragments into smaller ones would have been predicted if the fragments arose from the statistical cleavage of a number of accessible sites within a homogeneous population of nu bodies.

Tissue-Specific, Preferential Digestion of Active Genes

Staphylococcal nuclease shows no preferential digestion of specific nuclear DNA sequences, in particular sequences coding for active genes (13). Since it is clear that pancreatic deoxyribonuclease has a much higher affinity for sites within nu bodies, and hence might be expected to differentiate between similar nu bodies, we decided to investigate whether pancreatic deoxyribonuclease preferentially digested the DNA coding for active genes. There are a number of indications that this might be occurring during the digestion of nuclei with this enzyme. We have shown that ribosomal DNA in nuclei is especially sensitive to deoxyribonuclease (14); in addition, Billing and Bonner (15) have shown that RNA labeled for a short period is rapidly released on mild digestion of nuclei with either deoxyribonuclease I or II. Finally, Berkowitz and Doty (16) have shown that a putative active fraction of sheared chromatin (isolated as a slow-sedimenting fraction on sucrose gradients) is much more sensitive to deoxyribonuclease I than is bulk chromatin.

To investigate this question further, we have prepared a complementary DNA (cDNA) probe to globin messenger RNA (mRNA) isolated from the reticulocytes of the adult chicken. The details of the purification and analysis and the characteristics of the cDNA have previously been described (17); in particular it has been demonstrated that the cDNA made to adult globin mRNA can be used to detect globin mRNA coding for embryon-

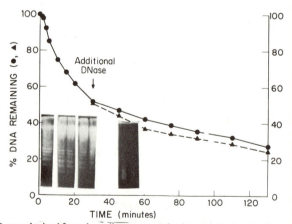

Fig. 1. Digestion of chick erythrocyte nuclei with pancreatic deoxyribonuclease I. (Curves) Avian red blood cells were isolated as described (17). The cells were washed twice in phosphate-buffered saline (PBS) (Grand Island), and the nuclei were isolated by suspension in reticulocyte standard buffer (RSB) (0.01M tris-HCl, pH 7.4; 0.01M NaCl; 3 mM MgCl₂) containing 0.5 percent NP-40 (British Drug House). The nuclei were washed several times in RSB and then digested at a DNA concentration of 1 mg/ml at 37°C with pancreatic deoxyribonuclease I (Sigma) (20 μg/ml) for increasing periods of time. The percentage of DNA remaining (●——●) was determined either by precipitation with cold 7 percent perchloric acid and subsequent measurement of the absorbancy at 260 nm of the acid soluble and insoluble fraction or by sedimenting the nuclei at low speed and measuring the release of material absorbing at 260 nm. Either method gave essentially the same results. Addition of fresh deoxyribonuclease (20 μg/ml) when the digestion begins to level off (at about 50 percent) does not affect the course of subsequent digestion. (▲——▲) The time course of digestion and the pattern of resistant fragments (see below) is the same in chromatin isolated either by sonication or by mild treatment with nuclease. Substitution of CaCl₂ for MgCl₂ also did not affect the course of digestion or pattern of resistant fragments. (Electrophoretic patterns) The deoxyribonuclease-resistant DNA was obtained from the sedimented nuclei at intervals during the digestion. The following were added to the nuclear pellet (final concentration) EDTA (10 mM), sodium dodecyl sulfate (SDS) (0.1 percent), and protease (Aldrich) (500 μg/ml). The mixture was incubated for 60 minutes at 37°C, boiled for 5 minutes, added directly to sample buffer (50 percent glycerol, 0.01 percent bromphenol blue), and loaded onto a 6 percent acrylamide slab gel containing 98 percent formamide and 20 mM sodium phosphate at pH 7.0. Electrophoresis was conducted for 5 hours at 160 volts in a running buffer containing 20 mM sodium phosphate, pH 7.0. The gels were stained with ethidium bromide (2 μg/ml) and the denatured DNA bands (bottom insets) were photographed through a red filter after illumination with ultraviolet light. *DNase*, deoxyribonuclease.

ic globin polypeptide chains (*17*). Thus, embryonic red cell RNA saturates about 70 percent of the cDNA probe prepared from adult red cell globin mRNA, while adult red cell RNA saturates the labeled probe at about 95 percent (*17*). The cross reaction of the adult probe with embryonic RNA is qualitatively attributable to the fact that the adult line of red cells synthesize three main types of globin chains, two of which are identical to chains synthesized by the embryonic line of red cells and the third bears about a 50 percent homology by tryptic peptide analysis (*18*) to two other embryonic globin chains.

When the DNA in nuclei from adult chick erythrocytes—this line of red blood cells contains the adult globin polypeptide chains—is digested with pancreatic deoxyribonuclease I so that 10 percent is acid soluble, approximately 75 percent of the globin cDNA probe fails to hybridize with the remaining 90 percent of the purified erythrocyte DNA (Fig. 2). Similarly, after the same amount of digestion, DNA isolated from the embryonic line of erythroblasts protects only about 50 percent of the cDNA at saturation. In contrast, undigested DNA isolated from either embryonic or adult red cells saturates the probe at more than 94 percent. These experiments show that, in the different red cell lines, specific globin sequences are particularly sensitive to digestion. We present evidence below that embryonic but not adult-specific globin sequences and digested in embryonic red cells and adult but not embryo-specific sequences are digested in adult red cells. The same preparations of digested nuclei from either the adult or embryonic line of red cells contains the DNA sequences coding for ovalbumin mRNA (Fig. 2). Most importantly, nuclei from cultured chick fibroblasts, or freshly isolated chick brain, after 10 to 20 percent digestion (more extensive amounts of digestion were not tested), retain the DNA sequences coding for both adult and embryonic globin (Fig. 3). In addition, the fibroblast nuclei retain the DNA sequences coding for ovalbumin (Fig. 2). In contrast, digestion of red cell nuclei to DNA fragments of approximately the same size by a different nuclease, staphylococcal nuclease, results in no preferential digestion of any globin sequences (Fig. 3). This was first shown by Axel *et al.* (*13*) and recently extended by Lacy and Axel (*19*).

In summary, digestion of nuclei with pancreatic deoxyribonuclease reveals that specific globin sequences are preferentially degraded in erythroid cells, but not in nonerythroid cells. Similarly, the ovalbumin gene is not preferentially digested in cells that do not produce ovalbumin.

Identification of Globin Genes Digested in Adult and Embryonic Nuclei

The failure of DNA isolated from red cell nuclei treated with deoxyribonuclease I to saturate the cDNA probe (Fig. 2) could be due (i) to the specific degradation of a unique subset of globin genes, as we have suggested, or (ii) to an overall reduction in DNA sequences complementary to all sequences present in the cDNA, resulting in a situation where the cDNA is in excess. To exclude the second possibility, we have performed several of the reannealing experiments shown in Fig. 2 with one-tenth the amount of driver DNA (that is, the DNA which determines the rate of the reaction) and the same amount of cDNA. Under these conditions, no change is observed in the level of saturation of the cDNA. This demonstrates that specific globin DNA sequences are digested by pancreatic deoxyribonuclease and also that specific sequences are resistant.

As was mentioned previously, our cDNA probe contains sequences complementary to three adult globin mRNA

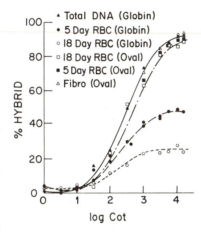

Fig. 2. Preferential digestion of active genes by pancreatic deoxyribonuclease I. Red blood cells (RBC) were obtained by vein puncture from 18-day (containing adult-type globin chains) and 5-day (containing embryo-type globin chains) chick embryos. Fibroblasts were dissected from the region of the developing breast muscle of 11-day chick embryos and grown in culture (*17*). Nuclei from 18-day embryo RBC's, 5-day embryo RBC's, and 11-day cultured chick embryo fibroblasts were isolated in RSB containing 0.5 percent NP-40 and digested with pancreatic deoxyribonuclease I until 10 to 20 percent of the DNA was soluble in acid (legend to Fig. 1). DNA was prepared as follows. Nuclei were centrifuged and suspended in 0.1 percent SDS, 100 μg of pronase per milliliter, and 5 mM EDTA overnight at 37°C. The sample was extracted several times with equal volumes of a mixture of phenol and chloroform (1 : 1), and several times with a mixture of chloroform and isoamyl alcohol (24 : 1). The resultant aqueous phase was made 0.1M with respect to NaCl, and the nucleic acid was precipitated overnight at −20°C with two volumes of 95 percent ethanol. The nucleic acid was recovered by centrifugation for 30 minutes at 10,000 rev/min (HB-4 head of a Sorvall RC-5 centrifuge), suspended in 10 mM NaCl, 10 mM tris-HCl (pH 7.4), and incubated for 30 minutes at 37°C with ribonuclease A (20 μg/ml) (Worthington) that had been boiled for 30 minutes. The preparation was again extracted with the phenol-chloroform and then chloroform–isoamyl alcohol mixtures and the extract was precipitated with ethanol. The DNA concentration was determined by absorbancy at 260 nm in a Zeiss spectrophotometer (DNA at 1 mg/ml gave 20 A_{260}). Total DNA was prepared directly from 18-day red cell nuclei and sonicated so that the average length corresponded to 500 nucleotides. The cDNA complementary to globin mRNA from adult reticulocytes was prepared as described (*17*). Hybridizations were conducted with an excess of DNA to cDNA (1 × 10⁷ to 2 × 10⁷ : 1) and analyzed (*17*). The DNA samples suspended in a mixture of 0.3M NaCl, 50 mM tris-HCl (pH 7.4), and 0.1 percent SDS and ranging in concentration from 1 to 20 mg/ml were denatured by heat and annealed at 65°C (at 1000 count/min per 5 μl of reaction mixture) to [³H]deoxycytidine- and [³H]thymidine-labeled cDNA (5 × 10⁷ to 8 × 10⁷ count/min per microgram). Under these conditions the calculated ratio of globin DNA to globin cDNA was 10 : 1 to 15 : 1. Polypropylene tubes overlaid with paraffin oil were used for the hybridizations. At intervals (from 5.7 minutes to 96 hours), 10-μl samples of reaction mixtures were pipetted into 400 μl of a mixture of 30 mM sodium acetate (pH 4.5), 0.15M NaCl, 1 mM ZnSO₄, and 10 μg of denatured DNA from salmon sperm. Half (200 μl) of the above mixture was immediately precipitated with trichloroacetic acid (TCA) and the other half was incubated with partially purified S1 nuclease at 45°C for 40 minutes. The percentage of hybridization was determined by comparison of the TCA-precipitable radioactivity remaining after S1 digestion to that precipitable in the undigested samples. The S1 background radioactivity ranged from 2 to 6 percent. The percentage of hybridized cDNA is plotted as a function of the C_0t, which represents the concentration of deoxyribonucleotide (moles) times the time of digestion (seconds) per liter (*28*). The same plateaus were obtained at concentrations of driver DNA from 2 to 20 mg/ml in a reaction volume of 50 μl, to which cDNA (5000 count/min) was added. (▲——▲) Total DNA hybridized to globin cDNA; (●——●) 5-day red cell DNA treated with deoxyribonuclease I and hybridized to globin cDNA; (○——○) 18-day red cell DNA treated with deoxyribonuclease I and hybridized to globin cDNA; (□——□) 18-day red cell DNA treated with deoxyribonuclease I and hybridized to ovalbumin cDNA; (■——■) 5-day red cell DNA treated with deoxyribonuclease I and hybridized to ovalbumin cDNA; (△——△) fibroblast DNA treated with deoxyribonuclease I and hybridized to ovalbumin cDNA.

850

molecules. The resistant DNA obtained after embryonic erythroblast nuclei are mildly treated with deoxyribonuclease I saturates the cDNA probe at about 50 percent. Our hypothesis is that only actively transcribing globin genes are digested. Since two of three adult globin genes are also active in the embryonic red cell line, we predict that two adult globin genes are digested in embryonic cells and the third is resistant. To test this, polysomal RNA from embryonic red cells was added to the hybridization mixture which also contained ³H-labeled cDNA from adult globin mRNA and partially digested DNA from embryonic red cell nuclei. Under these conditions, the cDNA is fully saturated at about 98 percent (Fig. 4). Thus, embryonic red cell RNA fully complements the deficiency in globin DNA sequences resulting from the digestion. This implies that embryonic sequences are absent and the adult-specific sequences are present in the DNA from the embryonic red cell nuclei treated with deoxyribonuclease I.

A similar type of analysis suggests that adult-specific sequences are preferentially digested in the adult line of red cell nuclei. Figure 2 shows that the cDNA is saturated at about 25 percent by DNA isolated from adult red cell nuclei treated with pancreatic deoxyribonuclease. Our hypothesis in this case is that only adult-specific globin genes are digested. In principle, if this is true, then none of the probe should be protected from S1 nuclease. We believe that the partial protection of the probe by DNA obtained from adult nuclei treated with deoxyribonuclease I can be explained by two embryonic globin genes that bear a marked homology to the β-globin gene in the adult (18). These embryo-specific globin genes (named ε and ρ) would be present in the deoxyribonuclease I–resistant fraction from adult red cell nuclei and, by virtue of their homology to the adult β-globin, would be expected to partially protect some of the cDNA. Since the adult β-gene codes for about 50 percent of the adult globin chains, it is not unreasonable that it constitutes about 50 percent of our cDNA population. In addition, peptide analysis has shown about a 50 percent identity in tryptic peptides between the β-globin in adults and the ε- and ρ-globin chains in the embryo (18). Thus, if our logic is correct, we predict that the digested adult nuclei should saturate about 25 percent of the cDNA probe (50 percent × 50 percent = 25 percent). To test this, embryonic red cell RNA was added to the hybridization mixture containing the cDNA and DNA from treated adult nuclei. Under these condi-

tions, the probe is saturated at about 70 percent (Fig. 4). This is precisely the saturation achieved by pure embryonic red cell RNA alone (17), an indication that the deoxyribonuclease I–resistant DNA from adult red cells contains no adult-specific globin sequences (Table 1).

The sensitivity of the globin genes to deoxyribonuclease I could be related to the fact that they are extremely active in transcription and are therefore not typical of most of the genes actively being transcribed. The chicken genome contains endogenous DNA sequences complementary to those of avian tumor viruses (ATV). A cDNA probe made against a specific type of avian tumor virus, avian myeloblastosis virus RNA (AMV), hybridizes with chicken DNA with a sequence homology of about 60 percent and a log $C_0t_{1/2}$ of about 1.75. This corresponds to about eight to ten copies of endogenous ATV DNA per cell. In separate experiments the total RNA from embryonic chick erythroblasts was hybridized to the ³H-labeled AMV cDNA. The kinetics of hybridization indicated that from one-tenth to two copies of endoge-

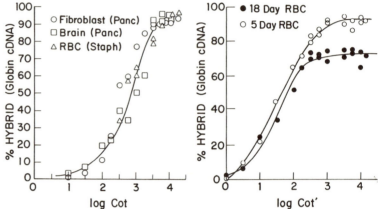

Fig. 3 (left). Retention of inactive genes after treatment of nuclei with pancreatic deoxyribonuclease. As was described (legend to Fig. 2), 18-day RBC nuclei and 11-day fibroblast nuclei were isolated from chick embryos. Brains were dissected from 11-day chick embryos, incubated for 30 minutes in PBS (Gibco) with 0.1 percent trypsin, washed repeatedly with PBS to eliminate contaminating RBC's, and lysed by homogenization in RSB with 0.5 percent NP-40. Fibroblast and brain nuclei were digested so that 20 percent of the DNA was acid soluble (legend to Fig. 1). The 18-day RBC's were digested until 50 percent of the DNA was acid soluble with staphylococcal nuclease (50 μg/ml) for 15 minutes at 37°C in RSB supplemented with $10^{-4}M$ CaCl₂. More than 80 percent of the acid insoluble DNA from this preparation consisted of 20 to 145 base pairs, as analyzed on 6 percent acrylamide gels. The isolation of resistant DNA from the digested nuclear preparations and the conditions and analysis of hybridization, including DNA concentrations and times of incubation, were as described (Fig. 2 legend). (O——O) Fibroblast DNA digested with deoxyribonuclease I and hybridized to globin cDNA; (□——□) brain DNA digested with deoxyribonuclease I and hybridized to globin cDNA; (△——△) 18-day red blood cell DNA digested with staphylococcal nuclease and hybridized to globin cDNA.   Fig. 4 (right). Hybridization kinetics of globin cDNA and mixtures of 5-day erythroid RNA and DNA from 18-day or 5-day RBC nuclei digested with pancreatic deoxyribonuclease. Embryonic erythroid RNA was prepared from the cytoplasm of RBC's from 5-day chick embryos. The 5-day RBC's were obtained by vein puncture, washed repeatedly with autoclaved PBS, and lysed in autoclaved RSB supplemented with 0.5 percent NP-40 and mouse liver ribonuclease inhibitor (17). Nuclei were sedimented in a table-top centrifuge, and the supernatant was extracted several times with equal volumes of a solvent mixture containing phenol-chloroform (1 : 1) and then several times with a mixture of chloroform and isoamyl alcohol (24 : 1). The aqueous phase was made 0.1M with respect to NaCl, and the nucleic acid was precipitated with two volumes of 95 percent ethanol (overnight at −20°C) and recovered by centrifugation for 60 minutes at 11,000 rev/min (HB-4 head of a Sorvall RC-5 centrifuge); it was suspended in a solution of 10 mM NaCl, 10 mM tris-HCl (pH 7.4), and 5 mM MnCl₂, incubated with ribonuclease-free deoxyribonuclease (10 μg/ml) (Worthington) for 30 minutes at 37°C, and again extracted with the above solvent mixtures. The RNA was precipitated with ethanol and suspended in a solution of 10 mM NaCl, 10 mM tris-HCl (pH 7.4), and the amount of RNA was determined at A_{260} in a Zeiss spectrophotometer (RNA at 1 mg/ml = 24 A_{260}). The nuclei were digested with pancreatic deoxyribonuclease (10 μg/ml) (Worthington) for 30 minutes at 37°C, and again extracted with the conditions, and analysis of hybrid formation are described in Fig. 2. The ratio of DNA to RNA in the hybridization mixture was 10 : 1. The nucleic acid concentration ranged from (per milliliter) 19 mg of DNA and 1.9 mg of RNA to 1.9 mg of DNA and 0.19 mg of RNA. Because of the minor contribution of RNA to the total nucleic acid concentration, the RNA concentration was not considered in the calculation of C_0t'. (●——●) DNA from pancreatic deoxyribonuclease–treated 18-day red blood cells plus RNA from 5-day embryonic red blood cells hybridized to globin cDNA. (O——O) DNA from pancreatic deoxyribonuclease–treated 5-day red blood cells plus RNA from 5-day embryonic red blood cells hybridized to globin cDNA.

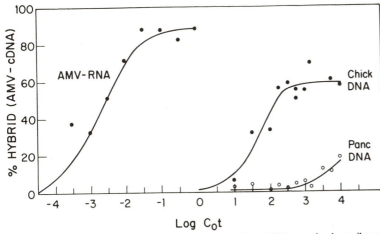

Fig. 5. Preferential digestion of endogenous avian tumor virus (ATV) genes by deoxyribonuclease I, in the embryonic line of chick erythroblasts. Preparations of DNA and procedures for digestion and hybridization were as described (Figs. 1 and 2). The AMV cDNA was prepared by the method described (*17*) for globin cDNA. The template for the reaction was the 35S RNA prepared from purified virus (a gift from J. Beard). The specific activity of the probe was about 4×10^7 to 6×10^7 count/min per microgram, and the probe hybridized to about 90 percent with its template, with a log $C_0t_{1/2}$ of about −2.75.

nous RNA sequences were present per cell. Similar very low but detectable levels of accumulation of RNA sequences from endogenous chicken viruses have been reported by Hanafusa *et al.* (*20*). Despite the fact that this gene has a low activity in erythroblasts, it is nevertheless sensitive to digestion by pancreatic deoxyribonuclease in isolated nuclei (Fig. 5). From the $C_0t_{1/2}$ of the reactions, it is possible to calculate that these endogenous viral sequences are about 100 times more sensitive to deoxyribonuclease I than the bulk of the nuclear DNA. Thus, two specific classes of active genes are preferentially digested by deoxyribonuclease I—the very active globin genes and the less active endogenous RNA tumor virus genes.

Digestion of a Specific Class of Nuclear DNA Sequences

The next question is whether after 20 percent digestion of nuclei by deoxyribonuclease I a specific 20 percent of the DNA sequences is absent from the remaining DNA. Total [³H]thymidine-labeled chick DNA was hybridized with a 10,000-fold excess of driver DNA obtained after mild deoxyribonuclease

treatment of embryonic red cell nuclei. Figure 6 shows that 78 percent of the total ³H-labeled DNA is protected at saturation when DNA from nuclei treated with deoxyribonuclease I is used to drive the reaction. In contrast, when the driver DNA is DNA obtained from the 11S monomers produced by staphylococcal nuclease treatment of nuclei, 94 percent of the labeled DNA is protected from S1 nuclease. Addition of total DNA (but not deoxyribonuclease I–treated DNA) to the reaction after 78 percent saturation is achieved increases the extent of hybridization to 95 percent (Fig. 6 and Table 2A). This suggests that the smallest deoxyribonuclease-digested fragments do not inhibit the hybridization of longer fragments and that the failure to reach full saturation is due to the absence of a particular subset of sequences in the deoxyribonuclease-treated DNA. The saturation value decreases to 65 percent after 52 percent digestion (Table 2).

Whether the specific nuclear DNA sequences that are preferentially digested by deoxyribonuclease I are related to sequences that are actively being transcribed was tested as follows. The DNA from nuclei digested to 20 percent acid solubility was hybridized to total [³H]thymidine-labeled tracer DNA. After saturation had been reached (78 percent), nuclear RNA was added to the hybridization reaction. Under these conditions, the saturation value increases to 89 percent (Fig. 6 inset). The difference in saturation between the reactions that occur in the presence and absence of added nuclear RNA (78 versus 89 percent) is very reproducible and is probably due to RNA-DNA hybrid formation

Table 1. Observed and predicted saturation of globin cDNA by various nucleic acid preparations. Globin gene terminology is based on the tryptic peptide analysis of Brown and Ingram (*18*), as is the representation of globin chains in the adult (18-day) and embryonic (5-day) red blood cell (RBC) populations; (+) indicates either the presence of the DNA sequences coding for particular globin chains or the presence of the globin polypeptide chains in the respective embryonic or adult populations; (−) indicates the absence of the globin genes or the absence of the globin polypeptide chains. In the nucleic acid mixtures containing both DNA and RNA, the contribution of each nucleic acid to the protection of cDNA from S1 nuclease digestion is represented by the subscripts D and R, respectively. Saturation values are taken from Figs. 2 and 3.

Item	Globin genes						Saturation of adult cDNA (%)	
	π (α-like)	α_A	α_D	ε (β-like)	ρ (β-like)	β	Observed	Predicted
cDNA from adult RBC*		35%	15%			50%		
Active genes in adult RBC	−	+	+	−	−	+		
Active genes in embryonic RBC	+	+	+	+	+	−		
Postulated sequences remaining after treatment of:								
(A) Embryonic RBC (Fig. 2)	−	−	−	−	−	+	50	50
(B) Adult RBC (Fig. 2)	+	−	−	+	+	−	28	25†
(C) Embryonic RBC plus embryonic RBC RNA (Fig. 3)	+$_R$	+$_R$	+$_R$	+$_R$	+$_R$	+$_D$	94	100
(D) Adult RBC plus embryonic RBC RNA	+$_{D,R}$	+$_R$	+$_R$	+$_{D,R}$	+$_{D,R}$	+$_{D,R}$	72	75

*Complementary DNA was prepared from adult globin mRNA (*17*). The percentage representation of specific globin genes in the cDNA population is based on the analysis of Brown and Ingram (*18*), and on the assumption that a stoichiometric relation exists between the template mRNA and cDNA product.   †The predicted 25 percent saturation of cDNA by DNA of deoxyribonuclease I–digested adult RBC nuclei is based on the shared tryptic peptides of the adult β-chain and the embryonic ε-and ρ-chains.

SCIENCE, VOL. 193

461

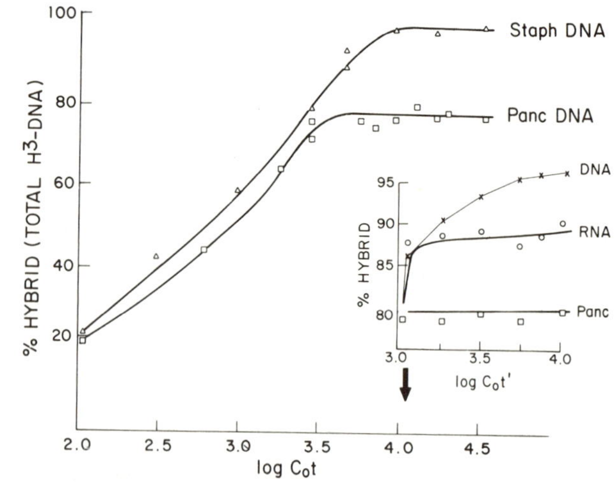

Fig. 6. Kinetics of reassociation of chick total ^{3}H-labeled DNA and DNA from nuclease-treated nuclei of 18-day RBC's. Total ^{3}H-labeled DNA was prepared by incubation of RBC's from 4-day chick embryos with [^{3}H]deoxythymidine. Cells were obtained by vein puncture, washed several times in medium F-12 (Gibco), and incubated for 5 hours in F-12 with [^{3}H]deoxythymidine (50 μc/ml; 16 c/mmole; New England Nuclear). The preparations of total DNA and of DNA from nuclei of 18-day RBC's digested with pancreatic deoxyribonuclease I were as described (Fig. 2); and the conditions of staphylococcal nuclease digestion were as described in Fig. 3 except that digestion was for 10 minutes to produce a population of fragments with a weight average molecular weight of 150 bases. Nuclear RNA was prepared from 5-day RBC's by lysis of these cells in the presence of mouse liver ribonuclease inhibitor (18), with an autoclaved solution of 0.5 percent NP-40 in RSB. The nuclei were washed several times in autoclaved RSB and lysed by gentle homogenization in 20 volumes of 0.15M NaCl, 0.05M sodium acetate (pH 5.1), and 0.3 percent SDS. The nuclear lysate was extracted three times with equal volumes of a mixture of phenol and chloroform (1 : 1) and numerous times with a mixture of chloroform and isoamyl alcohol (24 : 1). Ethanol precipitation, elimination of DNA, and determination of RNA concentration were as described (Fig. 4). Hybridization and analysis of hybrid formation were as described (Fig. 2). The concentration of labeled DNA was 10^{-4} that of driver DNA. Total ^{3}H-labeled DNA (5000 count/min per 5 μl of hybridization mixture) was annealed with DNA (20 mg/ml) from deoxyribonuclease I digestion of 18-day RBC nuclei (□——□) or DNA (20 mg/ml) from 11S monomers obtained from staphylococcal nuclease digestion of nuclei from 18-day RBC's (△——△). All points are plotted as DNA $C_0 t$. (Inset) After the reaction with the digested DNA from 18-day RBC's reached saturation (log $C_0 t = 4.0$), an equal volume of the digested DNA was added to the reaction at 20 mg/ml (□——□), or an equal volume of total DNA at 20 mg/ml (x——x), or an equal volume of nuclear RNA at 10 mg/ml (○——○). The reaction was allowed to proceed until a second saturation was achieved. The arrow shows when the additional nucleic acids were added to the original hybridization reaction.

between nuclear RNA and its ^{3}H-labeled template DNA since treatment of the hybrids with ribonuclease H (which specifically degrades RNA in DNA-RNA hybrids) makes an additional 15 percent of the hybrids sensitive to S1 nuclease (Table 2A).

The experiments described in Fig. 6 suggest that pancreatic deoxyribonuclease preferentially digests nuclear DNA sequences active in transcription of nuclear RNA. To verify this, [^{3}H]thymidine-labeled nuclei were partially digested with staphylococcal nuclease to produce a random population of small, resistant DNA fragments of predominantly 180 and 360 base pairs. This labeled DNA was hybridized to saturation with DNA obtained from red cells treated with pancreatic deoxyribonuclease, and the unhybridized labeled DNA (about 20 percent of the total) was isolated by passing the mixture over hydroxyapatite (HAP). According to the data in Fig. 6, the single-stranded flow-through DNA from HAP should be enriched in those sequences coding for active genes. To test this, excess total nuclear RNA was hybridized with the HAP flow-through DNA. About 48 percent of the labeled DNA is protected from S1 nuclease at saturation, and in a control experiment, nuclear RNA saturated only 9.8 percent of the total ^{3}H-labeled DNA (Table 2B). Thus the HAP flow-through DNA is enriched in DNA sequences complementary to nuclear RNA.

These conclusions can be tested in another way. Nuclear RNA, which has a kinetic complexity five to ten times greater than cytoplasmic RNA (21), was labeled for a short period and then hybridized to an excess of total red cell DNA or to an excess of embryo red cell DNA that had been treated with deoxyribonuclease I (Table 2C). Whereas 91 percent of the labeled RNA was protected from ribonuclease A digestion after being saturated by total DNA and 86 percent was protected by the staphylococcal nuclease DNA fragments, only 16 percent of the labeled RNA was protected at saturation by the pancreatic deoxyribonuclease I–treated DNA. Since 70 to 90 percent of the nuclear RNA is predominantly one major kinetic class of sequences (21), these experiments also suggest that pancreatic deoxyribonuclease preferentially digests much of the DNA coding for nuclear transcripts.

Altered Histone Conformation

Associated with the Globin Genes

It is possible that the preferential digestion of the globin genes by deoxyribonuclease I is related to the way the 11S monomers are packaged into higher order structures within the cell nucleus. In order to test this possibility, the purified 11S monomers were mildly digested with pancreatic deoxyribonuclease in the presence of 3 mM MgCl$_2$ until 15 percent of the DNA was acid soluble. When the

isolated DNA was hybridized to the globin cDNA, the adult-specific globin genes were absent (Table 2D). Thus, although the 11S monomers contain all the globin sequences, in the appropriate ionic conditions the proteins protecting the globin genes adopt a configuration that renders the associated globin DNA sensitive to digestion by pancreatic deoxyribonuclease. If these proteins are histones, then perhaps they are modified, either by direct chemical modification or by association with nonhistone proteins.

The sensitivity of monomers to pancreatic deoxyribonuclease apparently rules out a class of explanations for the digestion of active genes by deoxyribonuclease I based on the higher order packaging of monomers within the nucleus. The following experiment also supports the view that monomer packaging is not the basis for the digestion of active genes. The 11S monomers isolated from 18-day red cell nuclei were treated with trypsin (22) to remove 20 to 30 residues from the NH$_2$-terminus of each of the histones (7). The trypsin was then inactivated by addition of soybean trypsin inhibitor, and the particles were redigested with staphylococcal nuclease. The resistant DNA (about 20 percent of the total) was isolated and hybridized to tracer amounts of total ^{3}H-labeled DNA and to globin ^{3}H-labeled cDNA. Whereas more than 80 percent of the total ^{3}H-labeled DNA hybridizes to the resistant DNA, only 25 percent of the globin cDNA forms stable hybrids. (We have made no attempt to show that the 25 percent hybridization represents cross reaction with inactive embryo genes as we describe in Fig. 4). Thus, active globin 11S monomers become sensitive to staphylococcal nuclease after but not before treatment with trypsin, suggesting again that the conformation of transcriptionally "active" monomers is different from that of the inactive ones. The increased accessibility of actively transcribing genes to staphylococcal nuclease after treatment with trypsin is in good agreement with the in vivo experiments of Roberts and Kroeger (23), who showed

Table 2. Observed saturation of trace amounts of ³H-labeled probes by vast excesses of driver nucleic acids. (A) The treatment of nuclei, isolation of DNA, and reannealing were as described (Fig. 6). Percentage digestion was determined by the absorbancy at 260 nm of the perchloric acid soluble fraction. The 11S monomers were prepared as described (D) below. The weight average molecular weight of the DNA fragments decreased from 200 bases at 10 percent digestion to about 80 bases at 52 percent digestion. The data for the 19 percent digestion are taken from Fig. 6. In the remaining experiments, DNA was at a concentration of 15 mg/ml and hybridization was assayed as described (Fig. 6). For ribonuclease H digestions, the hybridized mixture of total ³H-labeled DNA, the DNA from nuclei treated with deoxyribonuclease I, and the nuclear RNA was desalted by passage over Sephadex G25 equilibrated with a solution of 40 mM tris-HCl (pH 7.7), 4 mM MgCl₂, and 1 mM dithiothreitol. The sample was treated with ribonuclease H (Miles; 7 units) in the presence of bovine serum albumin (30 µg/ml) and glycerol (4 percent) for 30 minutes at 37°C. This procedure was followed by digestion with S1 nuclease, as described. (B) The 4-day RBC's were incubated with [³H]thymidine (Fig. 6). Monomers were prepared from these labeled cells (D, below). The DNA from ³H-labeled monomers were annealed with excess DNA from RBC nuclei treated with deoxyribonuclease I (Fig. 1) as described (Fig. 2). At log C_0t of 4.25, approximately 80 percent of the ³H-labeled DNA was hybridized as assayed by S1 digestion. Single-stranded DNA was isolated by HAP chromatography. Portions of the reaction mixture (100,000 count/min) were pipetted into 0.15M sodium phosphate (PB) (pH 7.0) and placed on a water-jacketed column containing 10 g of HAP (Bio-Rad-DNA grade) at 60°C. Repeated washings with 0.15M PB resulted in the elution of single-stranded DNA (approximately 20,000 count/min). This fraction is referred to as "HAP flow-through DNA." It is 98 percent sensitive to S1 nuclease and contains more than 90 percent single-copy sequences. The remaining double-stranded DNA (80,000 count/min) was eluted with 0.48M PB. The HAP flow-through DNA was desalted by passage over Sephadex G25 (equilibrated in 0.1M NaCl, 0.01M tris-HCl, pH 7.4) and precipitated with 100 µg of (carrier) yeast transfer RNA (tRNA) (Sigma) in two volumes of 95 percent ethanol. The HAP flow-through DNA was recovered by centrifugation for 30 minutes (11,000 rev/min, HB-4 head of a Sorvall RC5 centrifuge) and suspended in a solution of 10 mM NaCl, 10 mM tris-HCl (pH 7.4). Nuclear RNA (12 mg/ml) was prepared from 10-day-old chick embryo reticulocytes (Fig. 6 legend) and used to drive the reaction against total ³H-labeled DNA or the HAP flow-through fraction. The hybridization conditions and analysis of hybrid formation were as described (Fig. 2). Seven percent of the total DNA and 1 percent of the HAP flow-through DNA behaved as foldback DNA, as assayed by S1 nuclease, and was subtracted from the observed saturation values to yield the figures of 9.8 percent and 48 percent, respectively. Saturation was achieved by log Cr_0t = 3, although points were taken to log Cr_0t = 4 (Cr_0t, moles of ribonucleotide per liter × seconds). (C) ³H-Labeled nuclear RNA was prepared by incubation of 4-day chick embryo RBC's for 30 minutes in medium F-12 (Gibco) with [³H]uridine (100 µc/ml) (New England Nuclear; 16 c/mmole). The isolation of nuclear RNA and the preparation of the staphylococcal nuclease limit digest were as described. Total DNA and DNA from nuclei treated with deoxyribonuclease I (isolated as described above) were annealed with tracer ³H-labeled nuclear RNA (5000 count/min per 5 µl of the hybridization mixture). At intervals, a sample (10 µl) of the hybridization mixture was pipetted into 400 µl of double-strength SSC (SSC consists of 0.14M NaCl, 0.014M sodium citrate); half of this sample was immediately precipitated with trichloroacetic acid, and the other half was incubated for 30 minutes at 37°C with previously boiled ribonuclease A (20 µg/ml) (Worthington). The percentage hybridization was then determined (Fig. 2 legend). Of the labeled RNA 6 percent was resistant to ribonuclease in the absence of DNA and this was subtracted from the observed saturation values obtained at log Cr_0t = 4.25. (D) Conditions and analysis of hybridization for reactions with trace amounts of ³H-labeled globin cDNA and the isolation of total DNA and the DNA from digested RBC and fibroblast were as described (Fig. 2 legend). The posterior one-third of de-embryonated blastoderms (area vasculosa) were dissected from approximately 500 24-hour chick embryos, washed several times in PBS (Gibco), and gently lysed by homogenization in RSB with 0.5 percent NP-40. The resultant nuclei were then treated with pancreatic deoxyribonuclease to 20 percent acid solubility and the DNA was isolated as described (Fig. 1 legend). Monomers were prepared by incubation of 18-day RBC nuclei with staphylococcal nuclease (50 µg/ml) (Worthington) for 5 minutes at 37°C in RSB supplemented with 10⁻⁴M CaCl₂. The nuclei were then centrifuged and washed in 0.075M NaCl, 0.02M EDTA, and 0.01M tris-HCl (pH 7.4). The resultant pellet was suspended in 5 mM sodium phosphate (pH 6.8), and insoluble material was removed by low speed centrifugation. Of the soluble material, 90 percent sedimented at 11S and 10 percent at 15S. Removal of contaminating 15S material had no effect on our results. Monomers were treated with pancreatic deoxyribonuclease in either 5 mM sodium phosphate (pH 6.8) or in RSB. The isolation of DNA was as described (Fig. 2 legend). (E) The 18-day RBC monomers were prepared as described above. Trypsinized monomers were prepared by the incubation of RBC monomers with trypsin (100 µg/ml) in RSB for 20 minutes at 37°C. After the addition of soybean trypsin inhibitor (200 µg/ml) (Worthington), the suspension was made 10⁻⁴M with respect to CaCl₂ and digested with staphylococcal nuclease (50 µg/ml) for 30 minutes at 37°C. The isolation of the remaining DNA (50 to 80 bases in length) and the conditions and analysis of hybridization were as described (Fig. 2 legend). About 20 percent of the DNA was resistant after these procedures; yet, this resistant DNA saturates more than 80 percent of the labeled total DNA, an indication that the inactive regions of the chromosome are randomly covered by the trypsin-resistant histone cores.

Item	Saturation* (%)
(A) Tracer ³H-labeled DNA (total) annealed with driver DNA from	
11S monomers	94 ± 3†
18-day RBC DNA—10% digestion with deoxyribonuclease I	85 ± 4
18-day RBC DNA—19% digestion with deoxyribonuclease I	78 ± 3
18-day RBC DNA—52% digestion with deoxyribonuclease I	65 ± 5
18-day RBC DNA—19% digestion with deoxyribonuclease I	78 ± 3
plus DNA—19% digestion with deoxyribonuclease I	79 ± 4
plus total DNA	95 ± 4
plus nuclear RNA	89 ± 3
plus nuclear RNA and then ribonuclease H	74 ± 4
(B) Driver nuclear RNA annealed with tracer	
³H-DNA (HAP flow-through)	48 ± 8
³H-DNA (total)	9.8 ± 1
(C) Tracer ³H-nuclear RNA annealed with driver	
DNA (total)	91 ± 6
18-day RBC DNA (staphylococcal nuclease, limit digest)	85 ± 5
18-day RBC DNA digestion with deoxyribonuclease I	16 ± 5
(D) Tracer ³H-labeled globin cDNA annealed with driver	
DNA (total)	93 ± 3
18-day RBC DNA digestion with deoxyribonuclease I	25 ± 3
Fibroblast DNA digestion with deoxyribonuclease I	94 ± 3
24-hour chick blastoderm DNA digestion with deoxyribonuclease I	94 ± 4
18-day RBC monomers (deoxyribonuclease I–5 mM sodium phosphate; 0.5 mM MgCl₂)	91 ± 4
18-day RBC monomers (deoxyribonuclease I–3 mM MgC₂; 10 mM NaCl)	25 ± 4
(E) Driver DNA from 18-day RBC monomers treated with	
trypsin and then staphylococcal nuclease and annealed with tracer	
³H-globin-labeled cDNA	25 ± 5
³H-DNA (total)	83 ± 4

*Saturation refers to the plateau in percent hybridization of tracer ³H-labeled probe by driver DNA or RNA. All experiments were taken out to log C_0t or log Cr_0t = 4.25. The plateau was defined in all cases by 4 to 10 points.   †Mean ± standard deviation of the mean.

854

bar

SCIENCE, VOL. 193

y

463

that, after injection of trypsin into salivary glands, the puffed regions were morphologically more sensitive than the unpuffed regions and were consequently even more accessible to transcription by the endogenous RNA polymerase.

Implication for Gene Regulation

Experiments with staphylococcal nuclease and deoxyribonuclease II have led to the conclusion that active genes are probably associated with histones (10, 13, 19), while our experiments are best understood if the conformation of these histones is different from that of the bulk of the histones—a conformation that renders the associated DNA particularly sensitive to digestion by pancreatic deoxyribonuclease I in the proper ionic environment. While our experiments with the globin cDNA show that the nontranscribed strand of the DNA double helix is digested by deoxyribonuclease I, the experiments with nuclear RNA (Table 2) and those previously described with ribosomal RNA (14) show that the transcribed strand of the double helix is also digested by deoxyribonuclease I.

Much of our data involves hybridization with very small DNA fragments 20 to 250 bases in length (the weight average molecular weight is about 150 to 200 bases). Almost certainly, the smallest fragments do not participate in the hybridization reaction; consequently, our estimate of the effective DNA concentration during renaturation is likely to be slightly high. The hybrids formed between self-annealed fragments from .the pancreatic deoxyribonuclease digestion of nuclei have a melting temperature (T_m) (measured by hyperchromicity and by thermal elution from HAP) of 77°C in $0.15M$ sodium phosphate buffer at pH 7.0 (data not shown). This is significantly lower than the T_m of 84°C that we observe for self-annealed fragments that are much longer (500 base pairs). Although we cannot offer a conclusive explanation for these results, it is likely that the difference in T_m is a consequence of the small size of the DNA fragments from digestion. Nevertheless, the difference in T_m suggests the interesting possibility that active genes may yield a subclass of deoxyribonuclease I–resistant fragments that are protected by histones, but smaller in size than that needed for stable hybrid formation at 65°C. This is probably not the case since the preferential digestion of the globin genes can still be demonstrated even after the hybridization reaction is performed under less

stringent conditions at 50°C (data not shown). Given these reservations, we think our interpretation of the deoxyribonuclease digestions is probably correct since the biological controls are so striking. Thus, in adult erythrocytes the staphylococcal nuclease fragments retain globin sequences, while the DNA fragments from pancreatic deoxyribonuclease digestion, which are about the same size, do not contain hybridizable adult globin DNA. Similarly, the same-sized deoxyribonuclease I fragments from fibroblasts or brain retain the globin and ovalbumin sequences, whereas those from embryonic and adult red cells retain only the ovalbumin sequences and not their respective activated globin sequences.

It is unlikely that the described preferential digestion of active genes is a consequence of the transcription process per se since the adult globin genes are sensitive to the nuclease in mature adult erythrocytes that have stopped synthesizing RNA. This observation also demonstrates that the inactivation of the avian red cell during erythroid development is not necessarily a consequence of an altered chromosome structure imposed at this primary level of DNA folding.

Since our cDNA is complementary to the globin structural gene and since this region of the transcription unit is clearly in an altered conformation (as revealed by deoxyribonuclease I digestion) in red blood cells, but not in brain or fibroblasts, a simple model of gene activation involving only the activation of some promotor sequence at the 5' end of the transcription unit can probably be excluded. We therefore propose that gene activation requires the assembly of an altered subunit structure throughout the entire transcription unit. The mechanism by which an altered subunit structure is propagated across the entire length of a transcription unit represents the primary conceptual problem raised by these results [see also (24)].

The deoxyribonuclease-resistant DNA protects less of the tracer ³H-labeled total DNA as digestion of nuclei is increased (Table 2A). Similar effects are not observed with DNA fragments obtained after digestion of nuclei with staphylococcal nuclease. Interpretation of these data is complicated by the fact that the DNA fragments become progressively smaller as the digestion proceeds; nevertheless, $C_0t_{1/2}$ for the reaction of those sequences that do hybridize is not significantly different from control values, and the saturation values increase by no more than 4 to 8 percent when the

reaction mixture is shifted to less stringent conditions (50°C) after reaching saturation at 65°C. These observations suggest the possibility that there may actually be a spectrum of transcribing (or potentially transcribable) chromosomal structures which can be defined by their sensitivity to deoxyribonuclease I.

The findings from these experiments with pancreatic deoxyribonuclease raise the possibility that gene activation during the development of the red cell lineage may involve the sequential assembly of a different type of chromosome structure (one that is deoxyribonuclease I sensitive) about the specific genes to be activated. To study this question, we have isolated a population of precursor erythroid cells from the developing yolk sac of 25-hour chick embryos. This population has been reported to contain more than 80 percent precursor red blood cells (25). When nuclei from these cells were digested with deoxyribonuclease I, the globin genes were not preferentially digested (Table 2D), and the kinetics of hybridization were essentially the same as those obtained from total DNA (Fig. 2). Thus, between 25 and 35 hours of development (when hemoglobin first appears in the chick embryo) there appears to be a new type of chromosome structure imposed on the globin genes in cells within the erythroid lineage. Since the chromosome is assembled at the time of DNA replication (26), it is not unreasonable that this new type of structure is actually dictated as the globin genes are replicating (14, 27). These observations are consistent with previous findings (17) that globin RNA sequences are not detectable in embryos at 25 hours of development, but begin to appear at 35 hours in coordination with the appearance of detectable hemoglobin. Thus, even though posttranscriptional controls are important in gene regulation (21), a major component of regulation is a transcriptional one mediated through chromosome structure.

Summary

Ten percent digestion of isolated nuclei by pancreatic deoxyribonuclease I preferentially removes globin DNA sequences from nuclei obtained from chick red blood cells but not from nuclei obtained from fibroblasts, from brain, or from a population of red blood cell precursors. Moreover, the nontranscribed ovalbumin sequences in nuclei isolated from red blood cells and fibroblasts are retained after mild deoxyribonuclease I digestion. This suggests that active genes are preferentially digested by deoxyribonuclease I. In contrast, treatment of red cell nuclei with staphylococcal nuclease results in no preferential digestion of active globin genes. When the 11S monomers obtained after staphylococcal nuclease digestion of nuclei are then digested with deoxyribonuclease I, the active globin genes are again preferentially digested. The results indicate that active genes are probably associated with histones in a subunit conformation in which the associated DNA is particularly sensitive to digestion by deoxyribonuclease I.

References and Notes

1. J. D. Watson and F. H. C. Crick, *Nature (London)* 171, 964 (1953).
2. M. F. Lyon, *ibid.* 190, 372 (1961); E. B. Lewis, *Adv. Genet.* 3, 73 (1950); W. K. Baker, *ibid.* 14, 133 (1968); H. D. Berendes, *Int. Rev. Cytol.* 35, 61 (1973); M. Ashburner, *Cold Spring Harbor Symp. Quant. Biol.* 37, 655 (1973); J. H. Frenster, *Nature (London)* 206, 680 (1965); B. J. McCarthy, J. T. Nishiura, D. Doenecke, D. S. Nasser, C. B. Johnson, *Cold Spring Harbor Symp. Quant. Biol.* 38, 763 (1973); R. T. Simpson, *Proc. Natl. Acad. Sci. U.S.A.* 71, 2740 (1974); K. Marushige and J. Bonner, *ibid.* 68, 2941 (1971).
3. J. Gottesfeld, R. F. Murphy, J. Bonner, *Proc. Natl. Acad. Sci. U.S.A.* 72, 4404 (1975).
4. A. L. Olins and D. E. Olins, *Science* 183, 330 (1974); C. L. F. Woodcock, *J. Cell Biol.* 59, 3689 (1973); J. P. Baldwin, P. G. Boseley, M. Bradbury, K. Ibel, *Nature (London)* 253, 245 (1975); R. D. Kornberg and J. O. Thomas, *Science* 184, 865 (1974); C. G. Sahasrabuddhe and K. E. Van Holde, *J. Biol. Chem.* 249, 152 (1974); J. A. D'Anna and I. Isenberg, *Biochemistry* 13, 2098 (1974); S. C. R. Elgin and H. Weintraub, *Annu. Rev. Biochem.* 44, 725 (1975); D. Hewish and L. Burgoyne, *Biochem. Biophys. Res. Commun.* 52, 504 (1973); M. Noll, *Nature (London)* 251, 249 (1974); R. Axel, W. Melchior, B. Sollner-Webb, G. Felsenfeld, *Proc. Natl. Acad. Sci. U.S.A.* 71, 4101 (1974); B. M. Honda, D. L. Baillie, E. P. M. Candido, *FEBS Lett.* 48, 156 (1974); D. R. Oosterhof, J. C. Hozier, R. L. Rill, *Proc. Natl. Acad. Sci. U.S.A.* 72, 633 (1975); P. Oudet, M. Gross-Bellard, P. Chambon, *Cell* 4, 281 (1975); H. J. Li, *Nucleic Acid Res.* 2, 1275 (1975); V. V. Bakayev, A. A. Melnickov, V. D. Osicka, A. J. Varshavsky, *ibid.*, p. 1401; H. G. Martinson and B. J. McCarthy, *Biochemistry* 14, 1073 (1975); J. D. Griffith, *Science* 187, 1202 (1975).
5. J. L. Germond, B. Hirt, P. Oudet, M. Gross-Bellard, P. Chambon, *Proc. Natl. Acad. Sci. U.S.A.* 72, 1843 (1975); R. Clark and G. Felsenfeld, *Nature (London)* 229, 101 (1971); J. P. Langmore and J. C. Wooley, *Proc. Natl. Acad. Sci. U.S.A.* 72, 2691 (1975).
6. H. Weintraub, K. Palter, F. Van Lente, *Cell* 6, 85 (1975).
7. H. Weintraub and F. Van Lente, *Proc. Natl. Acad. Sci. U.S.A.* 71, 4249 (1974).
8. F. H. C. Crick and A. Klug, *Nature (London)* 255, 530 (1975).
9. In the strictest sense all nu bodies cannot be homogeneous since the histones themselves are not homogeneous. Thus, histones are extensively modified [A. Ruiz-Carrillo, L. Wangh, V. Allfrey, *Science* 190, 117 (1975)] and are also genetically polymorphic (L. H. Cohen, K. M. Newrock, A. Zweidler, *ibid.*, p. 994).
10. Recent papers by Gottesfeld *et al.* (*3*) and Lacy and Axel (*19*) as well as older experiments of Axel *et al.* (*13*) make it very likely that histones are associated with actively transcribed regions of DNA.
11. R. Axel, H. Cedar, G. Felsenfeld, *Proc. Natl. Acad. Sci. U.S.A.* 70, 2029 (1973); R. S. Gilmour and J. Paul, *ibid.*, p. 3440; A. W. Steggles, G. N. Wilson, J. A. Kantor, *ibid.* 71, 1219 (1974); T. Barrett, P. Maryanka, P. Hamlyn, H. Gould, *ibid.*, p. 5057.
12. R. F. Itzhaki, *Biochem. J.* 125, 221 (1971); A. E. Mirsky, *Proc. Natl. Acad. Sci. U.S.A.* 68, 2945 (1971); M. Noll, *Nucleic Acid Res.* 1, 1573 (1974); D. Oliver and R. Chalkley, *Biochemistry* 13, 5093 (1974); T. Pederson, *Proc. Natl. Acad. Sci. U.S.A.* 69, 2224 (1972).
13. R. Axel, H. Cedar, G. Felsenfeld, *Cold Spring Harbor Symp. Quant. Biol.* 38, 773 (1973).
14. H. Weintraub, in *Results and Problems in Cell Differentiation*, J. Reinert and H. Holtzer, Eds. (Springer-Verlag, Berlin, 1975), vol. 7, p. 27.
15. R. J. Billing and J. Bonner, *Biochim. Biophys. Acta* 281, 453 (1972).
16. C. Berkowitz and P. Doty, *Proc. Natl. Acad. Sci. U.S.A.* 72, 3328 (1975).
17. M. Groudine, H. Holtzer, K. Scherrer, A. Therwath, *Cell* 3, 243 (1974); M. Groudine and H. Weintraub, *Proc. Natl. Acad. Sci. U.S.A.* 72, 4464 (1975).
18. J. Brown and V. Ingram, *J. Biol. Chem.* 249, 3960 (1974).
19. E. Lacy and R. Axel, *Proc. Natl. Acad. Sci. U.S.A.* 72, 3978 (1975).
20. H. Hanafusa, W. S. Hayward, J. H. Chen, T. Hanafusa, *Cold Spring Harbor Symp. Quant. Biol.* 39, 1139 (1974).
21. K. Scherrer and L. Marcaud, *J. Cell. Physiol.* 72 (Suppl. 1), 181 (1968); L. Grouse, M. D. Chilton, B. J. McCarthy, *Biochemistry* 11, 798 (1972); I. R. Brown and R. B. Church, *Dev. Biol.* 29, 73 (1972); E. H. Davidson and R. J. Britten, *Q. Rev. Biol.* 48, 565 (1973); M. J. Getz, G. D. Birnie, B. D. Young, E. MacPhail, J. Paul, *Cell* 4, 121 (1975); B. R. Hough, M. J. Smith, R. J. Britten, E. H. Davidson, *ibid.* 5, 291 (1975).
22. After treatment with trypsin, the repeating pattern of monomer, dimer, trimer (and so on), DNA fragments generated by partial digestion of nuclei with staphylococcal nuclease is largely preserved. In addition, more than 90 percent of the 11S monomers, after treatment with trypsin, retain a trypsin-resistant core composed of interacting histone COOH-terminal cleavage fragments (H. Weintraub, in preparation).
23. M. Roberts and H. Kroeger, *Experientia* 20, 326 (1957).
24. K. Yammamoto and B. Alberts, *Ann. Rev. Biochem.*, in press.
25. M. Wenk, *Anat. Rec.* 169, 453 (1971).
26. H. Weintraub, *Cold Spring Harbor Symp. Quant. Biol.* 38, 247 (1973); R. L. Searle and R. T. Simpson, *J. Mol. Biol.* 94, 479 (1975).
27. H. Holtzer, H. Weintraub, R. Mayne, B. Mochan, *Curr. Top. Dev. Biol.* 9, 299 (1973).
28. R. J. Britten and D. Kohne, *Science* 161, 529 (1968).
29. We thank N. Powe and R. Blumental for technical assistance, the National Science Foundation and American Cancer Society for support, R. Axel for the ovalbumin cDNA, V. Vogt for the S1 nuclease, and A. J. Levine for critically reading the manuscript. M.G. thanks the Medical Scientific Training Program of the University of Pennsylvania and the National Institutes of Health.

Cell, Vol. 12, 817–828, November 1977, Copyright © 1977 by MIT

The Structure of Histone-Depleted Metaphase Chromosomes

James R. Paulson and U. K. Laemmli
Department of Biochemical Sciences
Princeton University
Princeton, New Jersey 08540

Summary

We have previously shown that histone-depleted metaphase chromosomes can be isolated by treating purified HeLa chromosomes with dextran sulfate and heparin (Adolph, Cheng and Laemmli, 1977a). The chromosomes form fast-sedimenting complexes which are held together by a few nonhistone proteins.

In this paper, we have studied the histone-depleted chromosomes in the electron microscope. Our results show that: the histone-depleted chromosomes consist of a scaffold or core, which has the shape characteristic of a metaphase chromosome, surrounded by a halo of DNA; the halo consists of many loops of DNA, each anchored in the scaffold at its base; most of the DNA exists in loops at least 10–30 μm long (30–90 kilobases).

We also show that the same results can be obtained when the histones are removed from the chromosomes with 2 M NaCl instead of dextran sulfate. Moreover, the histone-depleted chromosomes are extraordinarily stable in 2 M NaCl, providing further evidence that they are held together by nonhistone proteins.

These results suggest a scaffolding model for metaphase chromosome structure in which a backbone of nonhistone proteins is responsible for the basic shape of metaphase chromosomes, and the scaffold organizes the DNA into loops along its length.

Introduction

Our research efforts are currently directed toward understanding the higher order structure of eucaryotic chromosomes. Considerable progress recently has been made in understanding how histones fold the chromosomal DNA into the basic nucleohistone fiber (see, for example, Oudet, Gross-Bellard and Chambon, 1975; Finch and Klug, 1976). Our work has been motivated, however, by the idea that other components, probably nonhistone proteins, must be responsible for the higher order folding of the basic chromatin fiber. Our approach to the problem of higher order structure, therefore, is to remove the histones from chromosomes and to study the structure and biochemistry of the DNA-protein complexes which remain.

In the accompanying paper (Adolph et al., 1977a), we have shown that histone-depleted chromo-

somes can be isolated on sucrose gradients following treatment with dextran sulfate and heparin. These complexes are highly sensitive to SDS, urea and proteases, but they are insensitive to high salt or RNAase. SDS-polyacrylamide gel electrophoresis reveals that they contain essentially no histones, and up to 30 major nonhistone proteins. These results suggest that the DNA is maintained in a highly folded, fast-sedimenting structure by nonhistone proteins.

In order to get information on the higher order structure of metaphase chromosomes, we have studied these histone-depleted chromosomes by electron microscopy. The removal of histones should aid in understanding how the nonhistones fold the chromosomal DNA by allowing the DNA to be spread out. When histones are present, the adherence of nucleohistone fibers to one another, and the electron density of the chromosomes, do not permit a view of the underlying structures.

In this paper, we report what we have learned from electron microscope studies of histone-depleted chromosomes spread with cytochrome c.

Results

Preparation of Histone-Depleted Chromosomes for Electron Microscopy

In the preceding paper (Adolph et al., 1977a), we have described the preparation of histone-depleted metaphase chromosomes. When chromosomes are treated with 2 mg/ml dextran sulfate and 0.2 mg/ml heparin, more than 99% of the histones and many of the nonhistone proteins are removed. These histone-depleted chromosomes form compact, fast-sedimenting structures which can be isolated on a sucrose gradient.

When we attempted to prepare this material for electron microscopy by spreading with cytochrome c, two problems arose. The first problem was to obtain histone-depleted chromosomes in sufficient concentration to make routine studies convenient. Since all the DNA exists in only a few very large structures, a very high DNA concentration is needed to see a reasonable number of particles on one electron microscope grid. We solved this problem by sedimenting the chromosomes to a cushion of Metrizamide (see Experimental Procedures). By this method, we can easily obtain a fraction containing 10–50 μg/ml DNA, whereas we can regularly obtain only 1 μg/ml DNA from the peak fraction of a sucrose gradient.

The second problem was that when we used 2 mg/ml dextran sulfate to remove the histones, enough dextran sulfate contaminated the sample to disturb the spreading procedure. Presumably the dextran sulfate binds nearly all the cytochrome c so that a monolayer cannot form. The problem is more serious for the ammonium acetate spreading method than for the carbonate method (see Experimental Procedures), since much more cytochrome c binds to polyanions at pH 6.5 than at pH 10. We solved this problem, however, by using a lower concentration of dextran sulfate to remove the histones.

Thus we routinely prepared histone-depleted chromosomes for electron microscopy by treating chromosomes with 0.2 mg/ml dextran sulfate and 0.02 mg/ml heparin and sedimenting them to a cushion of 0.6 M Metrizamide in gradient buffer. We emphasize that these modifications of the basic procedure of Adolph et al. (1977a) are only a convenience and do not affect the results. We obtained the same results by spreading samples from continuous sucrose gradients. We ran these gradients as described by Adolph et al. (1977a), except that we omitted detergents from the body of the gradient. We also obtained the same results when we used 2 mg/ml dextran sulfate, provided we diluted samples 10–15 times in gradient buffer to reduce the concentration of contaminating dextran sulfate before spreading. As will be shown below, we can obtain the same results without using dextran sulfate at all, if we remove histones by treating chromosomes with 2 M NaCl.

The histone-depleted chromosomes have the same sedimentation properties whether we treat them with 0.2 mg/ml or 2 mg/ml dextran sulfate (Adolph et al., 1977a). Moreover, we have verified, by using the Metrizamide cushion procedure (J. R. Paulson, unpublished data), and also by using continuous sucrose gradients (Adolph et al., 1977a), that 0.2 mg/ml dextran sulfate is sufficient to remove 99% of the histones.

We used two methods of spreading with cytochrome c in these studies (see Experimental Procedures). Both methods gave essentially the same results, but each had special advantages. We preferred the ammonium acetate method because it gave the highest contrast to the DNA strand. The carbonate method was useful in early work, however, because it is rarely affected by contaminating dextran sulfate and because it enables one to spread many different samples in a short period of time. In addition, this method sometimes gives a very informative view of the DNA loops in the histone-depleted chromosomes.

Histone-Depleted Chromosomes Consist of a Protein Scaffold Surrounded by a Halo of DNA

Figure 1 shows a representative micrograph of a histone-depleted chromosome. The chromosome consists of a central darkly staining scaffold or core which is surrounded by a halo of DNA. A

magnified view of the central scaffold and part of the DNA halo is shown in Figure 2.

The shape of the scaffold is strikingly similar to the characteristic morphology of metaphase chromosomes, and this fact makes it immediately clear that each chromosome is sedimenting as an intact structure. We find very few free DNA ends around the edges of the DNA halo, and very little free DNA in the background on these grids. This shows that the chromosomes have not been subjected to excessive shear. Complexes such as the one shown in Figure 1 are essentially the only DNA-containing particles we find on these grids. We occasionally find particles which do not contain DNA (presumably cell debris) in these preparations, but they probably do not contribute to the protein pattern of histone-depleted chromosomes purified on sucrose gradients (Adolph et al., 1977a).

It should be noted that in these studies we used only chromosome preparations in which the chromosome morphology was well preserved (as judged by phase-contrast microscopy) for preparation of histone-depleted chromosomes. When chromosomes have been severely stretched or sheared (compare Stubblefield and Wray, 1971), the scaffold is generally found to be pulled apart into fragments and its organization is not discernable.

Two additional examples of chromosome scaffolds are shown in Figure 3. The core is sometimes dense, as in Figure 3a, and sometimes opened up into a fibrous network, as in Figure 3b and Figure 2. In nearly all of the histone-depleted chromosomes, however, the scaffold has the shape of a metaphase chromosome with paired chromatids connected at a centromere. On a typical grid, from which the micrograph in Figure 1 was taken, we examined 194 particles. Of these, 74% (144) had the sister chromatid scaffolds tightly associated, as in Figure 2 and Figure 3b. Another 15% (30) had the sister chromatid scaffolds separated and only held together by a few thin fibers, as in Figure 3a. The remaining 10% (20) appeared to be individual chromatids which had been separated from their sisters. Breaks in a chromatid scaffold are quite rare, occurring in only about 4% (7) of the particles.

When 2 M urea is included in the hypophase, breaks in the scaffold become much more frequent as the scaffold disintegrates. This is in keeping with our finding that the histone-depleted chromosomes are partially unfolded by 2 M urea and completely unfolded by 4 M urea (Adolph et al., 1977a). In the example shown in Figure 4, the scaffold has already begun to dissociate after 3 min in contact with 2 M urea.

Occasionally, we see thick bundles of fibers, similar to those observed by Kavenoff and Bowen

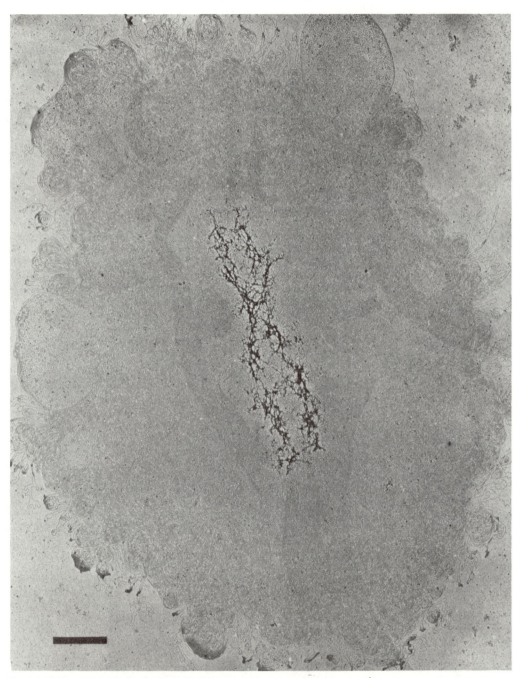

Figure 1. Electron Micrograph of a Histone-Depleted Metaphase Chromosome from Hela

Chromosomes were treated with dextran sulfate and heparin, purified by sedimentation to a cushion of 0.6 M Metrizamide and spread on 0.125 M ammonium acetate (see Experimental Procedures). The cytochrome c monolayer was sampled 1 min after spreading.

The chromosome consists of a central, densely staining scaffold or core surrounded by a halo of DNA extending 6–9 μm outward from the scaffold. The low magnification makes it difficult to see the individual DNA strands except along the edge of the DNA halo. The spider-like flecks in the background (for example, in the upper right corner) are contaminating dextran sulfate.

The bar in this and subsequent micrographs represents 2 μ.

Figure 2. Higher Magnification of the Chromosome Scaffold and Part of the Surrounding DNA of the Chromosome Shown in Figure 1
Note that the scaffold appears to consist of a dense network of fibers. Although individual DNA strands can be easily resolved at this magnification, the large amount of DNA makes it impossible to trace any given strand. Bar = 2 μ.

469

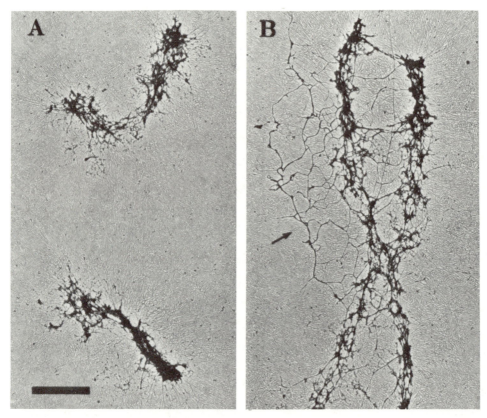

Figure 3. Central Scaffolds of Two Histone-Depleted Chromosomes

Chromosomes were prepared as described in the legend to Figure 1. (A) illustrates a situation seen in about 15% of the chromosomes in which the chromatids are separated and only connected by a few thin fibers. In (B) note the open fibrous appearance of the core. The thick fibrous extensions on the left of the core (arrow) are seen occasionally, and are believed to arise artifactually by collapse and aggregation of DNA strands during dehydration in ethanol (compare Lang, 1969, 1973). Bar = 2 μ.

(1976). An example is seen in Figure 3b (arrow). We believe that these thick fibers are not important structurally, but that they are artifactually produced by ethanol dehydration. Since so much DNA is present in these chromosomes, it is probable that sometimes some of the DNA is sterically prevented from adsorbing to the cytochrome monolayer. As has been shown by Lang (1969, 1973), this DNA could collapse and aggregate into thick fibers when placed in 90% ethanol.

DNA Is Attached to the Scaffold in Loops
When we spread histone-depleted chromosomes on 0.125 M ammonium acetate, the halo of DNA extends outward from the scaffold an average of 10–12 μm radially, although values of 6–20 μm have been observed. In this respect, the chromosome shown in Figure 1 is atypical, since the DNA

extends outward only 6–9 μm. In fact, we selected this chromosome for this reason, since larger chromosomes are more difficult to reproduce photographically. In other respects, however, it is typical.

The fairly uniform radius of the DNA halo suggests that the majority of the DNA in these chromosomes exists in loops of approximately the same length, and we would like to determine the lengths of these loops. The radius of the halo tells us that many loops must be at least 20–24 μm long, but there are too many DNA strands piled on top of one another to trace any individual loop, as can be seen from Figure 2.

In some preparations spread by the carbonate method, the DNA loops are well separated (Figure 5). We cannot account for the DNA in these chromosomes in the loops, and indeed the scaffold is surrounded by dense material (Figure 5, bottom)

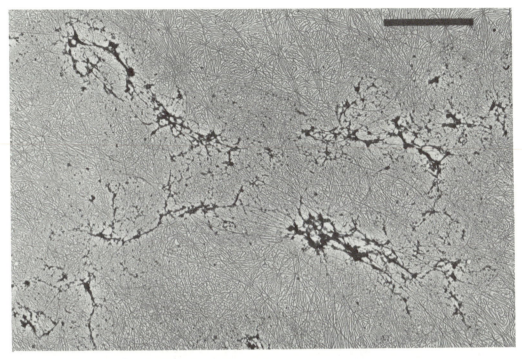

Figure 4. Central Scaffold of a Histone-Depleted Chromosome Spread on a Hypophase Containing Urea

Purified histone-depleted chromosomes were spread by the ammonium acetate method (see Experimental Procedures), except that the hypophase contained 0.125 M ammonium acetate and 2 M urea. The cytochrome c monolayer was sampled after 3 min.

Note that after only brief contact with urea, the scaffold is beginning to disintegrate. This is most easily seen by comparing this scaffold with the one shown in Figure 2. Many other scaffolds on this same grid had dissociated even further. Bar = 2 μ.

which probably consists of incompletely spread DNA. It is possible that in these chromosomes, we have not completely removed the histones, and hence the DNA does not completely unfold.

We took advantage of these unusual preparations to obtain a minimum estimate of the DNA loop size. We measured all the DNA loops along one side of a small chromatid and plotted the results in Figure 6. The number average length of a loop is 14 μm or 42 kilobases [assuming that 1 μm of DNA equals about 3000 base pairs (Chow, Scott and Broker, 1975)]. When the average is weighted according to the amount of DNA in each loop, however, the result is 23 μm (70 kb). This represents the length of the loop in which an average DNA nucleotide finds itself. It is clear from Figure 6b that the majority of the DNA is in loops between 10 and 30 μm long (30–90 kb).

We would like to emphasize that these methods give only a minimum estimate for the DNA loop size. Other micrographs of histone-depleted chromosomes spread by the ammonium acetate methods make clear that some loops in excess of 60

μm do occur. Loops which are very evident, such as those in Figure 5, occur in only a few preparations in which the DNA does not appear to be completely unfolded. Thus we believe 10–30 μm may be an underestimate of the actual loop size, since the DNA loops may not be completely unraveled. In particular, the large number of loops <5 μm long in Figure 6 may be misleading.

An especially important observation is that a loop generally returns to the scaffold adjacent to its point of origin (Figure 5). For the 76 loops measured in Figure 6, 57% (43) have both ends emanating from the core immediately adjacent to each other. Another 28% (21) have six or fewer DNA strands in between.

These results clearly show that the DNA is attached to the scaffold in loops, and that both ends of a DNA loop appear to be anchored at the same place in the scaffold. The majority of the DNA exists in loops that are at least 10–30 μm long (30–90 kb).

We have not directly observed supercoiling in the DNA loops. It is probable, however, that under

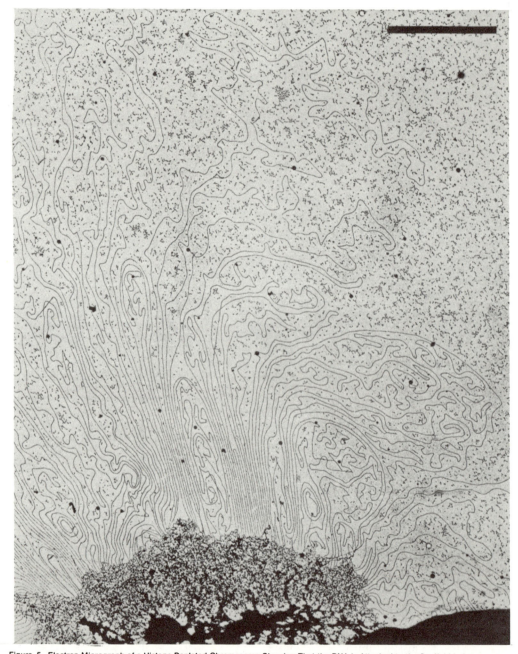

Figure 5. Electron Micrograph of a Histone-Depleted Chromosome Showing That the DNA Is Attached to the Scaffold in Loops

Chromosomes were treated with dextran sulfate and heparin, purified and spread with cytochrome c by the carbonate method (see Experimental Procedures). The contrast of this micrograph was enhanced by copying onto high-contrast plates as described by Chow et al. (1975).

Note that both ends of a DNA loop appear to emanate from adjacent points in the scaffold. A histogram of the loop lengths in this chromosome is shown in Figure. 6. Bar = 2 μ.

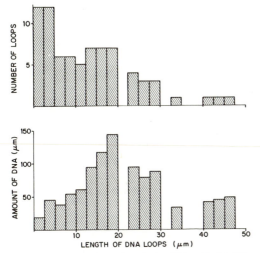

Figure 6. Histogram of Loop Lengths in a Histone-Depleted Metaphase Chromosome

All 76 loops along one side of the chromatid shown in Figure 5 were measured. In the upper panel, the number of loops falling into each size class is plotted. In the lower panel, the total length of DNA occurring in loops of each size class is plotted. This plot makes clear that most of the DNA exists in loops at least 10–30 μm long, although loops up to 47 μm long were observed.

the conditions used to isolate the metaphase chromosomes (pH 6.5, 0.5 mM CaCl$_2$), many nicks are introduced into each loop (Wray, Stubblefield and Humphrey, 1972).

Removing Histones from Chromosomes with 2 M NaCl

It is well known that histones can be removed from DNA with high salt (Spelsberg and Hnilica, 1971). Although we have mainly used the dextran sulfate-heparin procedure to prepare histone-depleted chromosomes, similar results can be obtained by treating chromosomes with 2 M NaCl.

The method described in Experimental Procedures is modified as follows: all step gradient solutions are as before except that they contain 2 M NaCl; chromosomes are added to a solution of 2 M NaCl, 0.1% NP-40, 10 mM EDTA, 10 mM Tris–HCl (pH 9.0) and 1 mM PMSF, gently mixed and layered on top of the step gradient. Centrifugation is begun after a 30 min incubation and the rest of the method is as before. No dextran sulfate or heparin is used.

Figure 7 shows an example of a 2 M NaCl-treated chromosome purified by sedimentation to a Metrizamide cushion. The scaffold is fairly dense and has a somewhat different texture than that of the dextran sulfate-treated chromosomes. Only part of the surrounding DNA is shown. We have not yet characterized the 2 M NaCl-treated chromosomes

biochemically, but they give the same profile as dextran sulfate-treated chromosomes when sedimented through sucrose gradients (K. W. Adolph, unpublished observations).

One could argue that metaphase chromosomes are stabilized mainly by side-to-side interactions between the chromatin fibers, and that our scaffolds consist merely of a few histones at the base of the loops which have not been extracted. The following experiment, however, strongly suggests that this is not the case.

Histone-depleted chromosomes prepared by treatment with 2 M NaCl were spread within 2 hr after removal from step gradients, and again 24 hr later. All the chromosomes were intact in both samples, and there were no significant differences between them. In other words, the histone-depleted chromosomes are completely stable for more than 24 hr in 2 M NaCl, as judged by electron microscopy. This strongly suggests that nonhistone proteins, rather than a few remaining histones, stabilize the chromosome scaffold.

Discussion

In this paper, we have examined histone-depleted metaphase chromosomes in the electron microscope by spreading with cytochrome c. Our results show that:

—The histone-depleted chromosomes consist of a central scaffold, which has the basic shape characteristic of metaphase chromosomes, surrounded by a halo of DNA. As we have shown elsewhere, the scaffold consists mainly of a few nonhistone proteins, and few or no proteins are present in the halo (Adolph et al., 1977a, 1977b). The scaffold is extraordinarily stable in 2 M NaCl but disintegrates rapidly when spread on a hypophase containing urea. Since Adolph et al. (1977a) have shown by sedimentation studies that histone-depleted chromosomes fall apart in 2 M or 4 M urea, the rapid dissociation of the scaffold following contact with urea is consistent with the notion that the scaffold is holding the DNA in a compact conformation.

—The chromosomal DNA is organized into many loops which are anchored in the scaffolding. Both ends of a loop appear to be anchored at the same place, since the DNA strand emanates radially from the scaffold and returns to an adjacent point. Most of the DNA exists in loops at least 10–30 μm long, although loops >60 μm in length have been observed. It is interesting to note that similar loop sizes have been observed in E. coli (Kavenoff and Ryder, 1976), and the same range of loop sizes has been inferred from sedimentation studies of eucaryotic interphase nuclei (Cook and Brazell, 1975; Benyajati and Worcel, 1976).

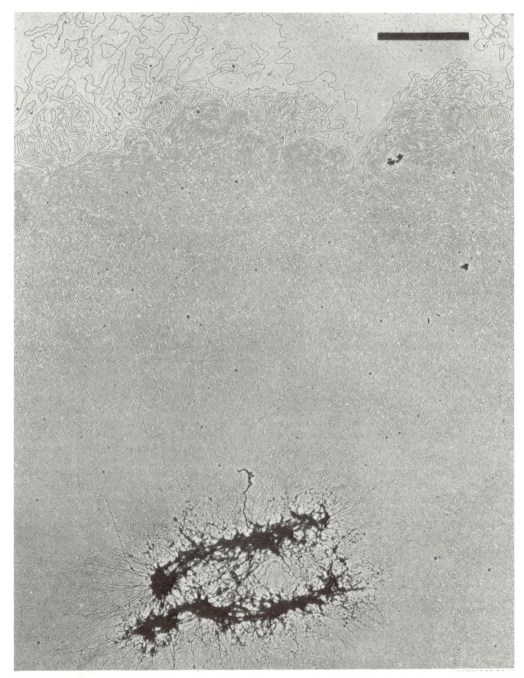

Figure 7. Histone-Depleted Chromosome Prepared by Treating Chromosomes with 2 M NaCl, Showing the Chromosome Scaffold and Part of the Surrounding DNA

Chromosomes were prepared as described in the text and spread on 0.125 M ammonium acetate (see Experimental Procedures). No dextran sulfate or heparin was used.

These chromosomes are virtually indistinguishable from the dextran sulfate-treated chromosomes, either by sedimentation studies (K. W. Adolph, unpublished observations) or in the electron microscope.

—Histone-depleted chromosomes have a similar structure whether the histones are removed with dextran sulfate and heparin or with 2 M NaCl.

These results suggest a scaffolding model for metaphase chromosome structure. In this model, chromosomes receive their basic shape from a scaffolding or backbone of nonhistone proteins which maintains the chromosome shape even when histones are removed, and which organizes the DNA into loops along its length. In the simplest (unineme) model, a single DNA molecule is folded many times, and the folds are maintained by periodic attachments to the scaffold. In this way the DNA molecule could be laid down from one end of the chromatid to the other, without need for any longitudinal fibers.

The DNA loops, of course, do not exist as free DNA but are tightly compacted by histones into 0.5–2 μm loops of 300 Å nucleohistone fiber (see, for example, Rattner, Branch and Hamkalo, 1975; Finch and Klug, 1976; Hozier, Renz and Nells, 1977). It is clear that many such loops attached to a scaffold would give the appearance of a dense bundle of folded chromatin fibers, as is seen in the micrographs of DuPraw (1970). Alternatively, these loops of 300 Å fibers could be further twisted to form microconvules, 500–600 Å in diameter (Daskal et al., 1977). One of the most important features of the scaffolding model, however, is that the compaction of the DNA by the histones into the nucleohistone fiber, and the higher order folding of that fiber by the nonhistone scaffolding proteins, are independent levels of structure. The implications of our data for possible models of chromosome structure have been discussed more fully by Laemmli et al. (1977).

The similarity in electron density between the scaffold and chromatin may explain why the scaffold has not been detected in previous electron microscope studies of intact chromosomes, either in whole mount preparations or in thin sections. There are, of course, suggestions of a chromosome scaffold or core in the literature, particularly in the work of Abuelo and Moore (1969), Stubblefield and Wray (1971), and Sorsa (1973). One of the difficulties in interpreting these electron micrographs, however, is that structures resembling a chromosome core could be an artifact induced by stretching.

Several investigators have proposed chromosome models involving a chromosome core. For example, Stubblefield and Wray (1971) suggested that a few central core chromatin fibers organize the rest of the chromatin into loops of "epichromatin." Sorsa (1974) has reviewed evidence for chromosome models involving an axial DNA filament.

In the scaffolding model, however, the core is not simply composed of a chromatin fiber or fibers, but is an independent structure, made up of a fibrous network of nonhistone proteins. In this respect, the scaffolding model is more like older models such as that of Taylor (1967).

From the electron micrographs presented here, we cannot determine how much DNA is present in the scaffold. We have, however, demonstrated elsewhere that the scaffold may be isolated as a structurally independent entity by treating the chromosomes with micrococcal nuclease before removing the histones (Adolph et al., 1977b). These scaffolds isolated by sucrose gradient sedimentation contain <0.1% of the chromosomal DNA, and they have a similar appearance to the central structure (scaffold) seen in the electron micrographs presented in this paper. The isolated scaffolds also have the same stability properties and the same protein composition as the histone-depleted chromosomes. These experiments show that the scaffold seen in the center of the histone-depleted chromosomes is a structurally independent entity and biochemically distinct from the rest of the chromosomes.

Our results say nothing about possible DNA links between different chromosomes, since the chromosome isolation procedure (Wray and Stubblefield, 1970) would undoubtedly shear such links if they exist. We also have no information on the supposed double nature of the chromosome core (compare Stubblefield and Wray, 1971). The scaffolding model can be easily elaborated to explain uninemy, binemy, chromosome replication and so forth, but we have no experimental data to support such speculations.

In conclusion, the results reported here, together with the results of Adolph et al. (1977a, 1977b), strongly suggest that nonhistone proteins are responsible for the higher order structure of eucaryotic chromosomes, and that these proteins are organized into a structurally independent entity which we call the chromosomal scaffold. Our future work will be directed toward determining which nonhistone proteins are structural elements of the chromosomal scaffold, how they are arranged and what role they play during the rest of the cell cycle.

Experimental Procedures

Cell Culture and Isolation of Metaphase Chromosomes

HeLa S_3 cells were grown in suspension, labeled with ^{3}H–thymidine and blocked in metaphase with colchicine as described by Adolph et al. (1977a). Metaphase chromosomes were then prepared in chromosome isolation buffer [0.1 mM PIPES (piperazine-N,N'-bis-2-ethane sulfonic acid) (pH 6.5), 0.5 mM CaCl$_2$, 1 M hexylene glycol] by the method of Wray and Stubblefield (1970). Chromosomes were used within 1 hr for preparation of histone-depleted chromosomes. Only preparations judged by phase-contrast microscopy to have well preserved chromosome morphologies were used.

Isolation of Histone-Depleted Metaphase Chromosomes

Histone-depleted chromosomes were routinely prepared for electron microscopy as follows. Between 5 and 25 μl of chromosomes containing 2–10 μg of DNA were mixed by gentle rolling with 1 ml of a solution containing 0.2 mg/ml sodium dextran sulfate 500 (Pharmacia, lot number 263), 0.02 mg/ml heparin (Sigma, number H-3125), 10 mM EDTA, 10 mM Tris–HCl (pH 9.0), 0.1% Nonidet P-40 and 1 mM PMSF (phenylmethylsulfonyl fluoride). Immediately after mixing, this solution was carefully layered onto the top of a step gradient using a plastic pipette from which the tip had been cut to have an internal diameter of 2 mm.

The step gradients consisted of three layers beneath the dextran sulfate layer: 1 ml of 2.5% sucrose, 10 mM EDTA, 10 mM Tris–HCl (pH 9.0), 0.1% NP-40 and 1 mM PMSF; 3 ml of 5% sucrose in gradient buffer [10 mM EDTA, 10 mM Tris–HCl (pH 9.0), 0.1 M NaCl], and 0.3 ml of 0.6 M Metrizamide (Nyegaard and Company, Oslo) in gradient buffer.

Following incubation in the dextran sulfate solution for 30 min at 4°C, chromosomes were sedimented to the Metrizamide cushion by centrifugation for 5 min at 500 × g and then 90 min at 3000 X g. Fractions were collected from the top of the tube using a wide bore plastic pipette, and part of each fraction was counted to verify the position of the ^{3}H-thymidine-labeled DNA. Generally, an 0.2 ml sample was obtained which had a DNA concentration of 10–50 μg/ml.

Preparation of Grids for Electron Microscopy

The following method of making Parlodion-coated grids gives films which are free of holes and strong even on 100 or 200 mesh grids. Parlodion (3.5% in amyl acetate) was stored over Linde molecular sieve (Type 4A) to eliminate moisture. Grids were placed on a stainless steel screen covered by 1 cm of water in a Buchner funnel, and a film was created by touching a drop of Parlodion solution to the water surface. The film was allowed to dry until wrinkles formed at the edges and until the center of the film turned gold and then clear. The film was then lowered onto the grids and screen by draining the water from the funnel. The screen was removed from the funnel, blotted from underneath to remove adhering droplets of water, and immediately placed in a dry oven at 60°C for 3 hr or overnight. Grids were stored in a dessicator and used within 2–3 days. Cytochrome c monolayers were picked up using the side of the Parlodion film which had faced the air.

Spreading Chromosomes with Cytochrome c

Two methods were used, which will be referred to as the ammonium acetate method and the carbonate method. Special care was taken in handling the histone-depleted chromosomes. Mixing operations were performed by gently rolling in a volume of at least 100 μl, and pipetting operations were always done using plastic pipettes from which the tips had been cut to give an internal diameter of 1–2 mm. Considerable care was also taken in the cleanliness of solutions used for spreading. For all purposes relating to electron microscopy, we used only high-purity water (Hydro, Inc., Durham, North Carolina). All containers were thoroughly rinsed with this water before use, and all stock solutions were filtered through 0.2 μm Nalgene filters. The teflon block was cleaned with soap and water and thoroughly rinsed with ethanol, acetone and water before use. Glass slides for ramps were cleaned with acid, soap and water, ethanol and acetone, rinsed with water and stored in water until just before use when they were rinsed with the hypophase solution. In the ammonium acetate method, sterile plastic 35 X 10 mm petri dishes were used for hypophases. We found that cytochrome c solutions could be stored for several months at 4°C, but care had to be taken to avoid contaminating them with any dust. For this reason,

pipette tips were thoroughly rinsed with water before inserting into the stock cytochrome solution.

The ammonium acetate method was a modification of the procedure of Kavenoff and Ryder (1976). Chromosomes were diluted as desired with gradient buffer [10 mM EDTA, 10 mM Tris–HCl (pH 9.0), 0.1 M NaCl], and to a 200 μl sample we added 20 μl of cytochrome c solution [1 mg/ml in 5 M ammonium acetate (pH 6.5)]. After gentle mixing, about 50 μl of this sample was spread on a hypophase of 0.125–0.5 M ammonium acetate (pH 6.5) using a clean microscope slide as a ramp. The cytochrome monolayer was sampled beginning 1 min after spreading. Spreading was performed at room temperature.

The carbonate method was a modification of the method of Inman and Schnös (1970) worked out by M. L. Wong in this laboratory. To a 50 μl sample of histone-depleted chromosomes (diluted as desired in gradient buffer), we added consecutively: 50 μl freshly prepared carbonate buffer (0.4 ml H$_2$O, 40 μl 1 M Na$_2$CO$_3$, 32 μl 0.126 M Na$_2$ EDTA), 100 μl formamide (Matheson, Coleman and Bell), and 30 μl cytochrome c (1 mg/ml in H$_2$O). After each addition, the sample was thoroughly mixed by gentle rolling. About 10 μl were spread by the droplet method on 0.4 ml water in the teflon tray described by Inman and Schnös (1970). In this method, a 10 μl droplet was formed using a segment of thin-walled teflon tubing attached to a Hamilton syringe, and the droplet was lightly touched to the surface of the hypophase. No ramp was used. Spreading was carried out at 4°C and samples were picked up 1–10 min after spreading.

In both methods, samples were spread within 2 hr after removal from the gradients, unless otherwise noted.

Contrast Enhancement

Immediately after samples had been picked up on Parlodion-coated grids, they were rinsed for 10 sec in 90% ethanol, stained for 20 sec in a fresh 250 fold dilution into 90% ethanol of the stock stain solution, rinsed for 10 sec in 2-methyl butane and air-dried. Finally, samples were rotary-shadowed with platinum from an angle of 8°.

The stock solution of stain was 0.05 F uranyl acetate in 0.05 N HCl (Davis and Davidson, 1968) which can be stored in the dark at 4°C for up to a month. An aliquot of this stock solution was filtered through an 0.3 μ Millipore filter shortly before use.

Electron Microscopy

Samples were viewed in a Phillips EM300, photographed at original magnifications from 2000–10,000X on Kodak Electron Microscope Film (no. 4489) developed in D19.

Acknowledgments

We are grateful to M. L. Wong and R. Kavenoff for technical advice, and to W. R. Baumbach and S. M. Cheng for preparing metaphase chromosomes. This research was supported by grants from the NSF and from the USPHS. J.R.P. is a Fellow in Cancer Research of the Damon Runyon-Walter Winchell Fund.

The costs of publication of this article were defrayed in part by the payment of page charges. This article must therefore be hereby marked "advertisement" in accordance with 18 U.S.C. Section 1734 solely to indicate this fact.

Received July 22, 1977; revised August 31, 1977

References

Abuelo, J. G. and Moore, D. E. (1969). J. Cell Biol. 41, 73–90.

Adolph, K. W., Cheng, S. M. and Laemmli, U. K. (1977a). Cell 12, 805–816.

Adolph, K. W., Cheng, S. M., Paulson, J. R. and Laemmli, U. K. (1977b). Proc. Nat. Acad. Sci. USA, in press.

Benyajati, C. and Worcel, A. (1976). Cell 9, 393–407.

Chow, L. T., Scott, J. M. and Broker, T. R. (1975). Electron Microscopy of Nucleic Acids (Cold Spring Harbor, New York: Cold Spring Harbor Laboratory).

Cook, I. and Brazell, P. (1975). J. Cell Sci. *19*, 261–279.

Daskal, Y., Mace, M. L., Wray, W. and Busch, H. (1976). Exp. Cell Res. *100*, 204–212.

Davis, R. W. and Davidson, N. (1968). Proc. Nat. Acad. Sci. USA *60*, 243–250.

DuPraw, E. J. (1970). DNA and Chromosomes (New York: Holt, Rinehart and Winston).

Finch, J. T. and Klug, A. (1976). Proc. Nat. Acad. Sci. USA *73*, 1897–1901.

Hozier, J., Renz, M. and Nells, P. (1977). Cold Spring Harbor Symp. Quant. Biol. *42*, in press.

Inman, R. B. and Schnös, M. (1970). J. Mol. Biol. *49*, 93–98.

Kavenoff, R. and Bowen, B. C. (1976). Chromosoma *59*, 89–101.

Kavenoff, R. and Ryder, O. (1976). Chromosoma *55*, 13–25.

Laemmli, U. K., Cheng, S. M., Adolph, K. W., Paulson, J. R., Brown, J. R. and Baumbach, W. R. (1977). Cold Spring Harbor Symp. Quant. Biol., in press.

Lang, D. (1969). J. Mol. Biol. *46*, 209.

Lang, D. (1973). J. Mol. Biol. *78*, 247–254.

Oudet, P., Gross-Bellard, M. and Chambon, P. (1975). Cell *4*, 281–300.

Rattner, J. B., Branch, A. and Hamkalo, B. A. (1975). Chromosoma *52*, 329–338.

Sorsa, V. (1973). Hereditas *75*, 101–108.

Sorsa, V. (1974). Hereditas *79*, 109–116.

Spelsberg, T. C. and Hnilica, L. S. (1971). Biochim. Biophys. Acta *228*, 202–211.

Stubblefield, E. and Wray, W. (1971). Chromosoma *32*, 262–294.

Taylor, J. H. (1967). In Molecular Genetics, Part II, J. H. Taylor, ed. (New York, London: Academic Press) pp. 95–135.

Wray, W. and Stubblefield, E. (1970). Exp. Cell Res. *59*, 469–478.

Wray, W., Stubblefield, E. and Humphrey, R. (1972). Nature New Biol. *238*, 237–239.

Proc. Natl. Acad. Sci. USA
Vol. 74, No. 2, pp. 457–461, February 1977
Biochemistry

Cell-free translation of simian virus 40 early messenger RNA coding for viral T-antigen

(radioimmunoprecipitations/RNA·DNA-Sepharose hybridization/wheat-germ protein-synthesizing system/ sodium dodecyl sulfate polyacrylamide gels)

CAROL PRIVES, ELI GILBOA, MICHEL REVEL, AND ERNEST WINOCOUR

Department of Virology, The Weizmann Institute of Science, Rehovot, Israel

Communicated by André Lwoff, November 2, 1976

ABSTRACT   Simian virus 40 (SV40) mRNA was isolated by hybridization of cytoplasmic RNA, from SV40-infected BS-C-1 monkey cells early in lytic infection, to SV40 DNA immobilized on Sepharose. The early viral mRNA, when added to a wheat-germ translation system, directed the synthesis of a unique class of products including a 90,000 molecular weight (M_r) polypeptide. It was found that this 90,000 M_r product as well as a prominent 17,000 M_r polypeptide could be specifically immunoprecipitated with hamster antiserum to SV40 T-antigen, but not with hamster control serum. Similar immunoprecipitation of extracts of SV40-infected cells with hamster anti-T serum yielded 90,000 M_r and 17,000 M_r polypeptides; these polypeptides were not found in immunoprecipitates of uninfected cell extracts. SV40 cRNA, prepared by asymmetric transcription of plaque-purified SV40 DNA, directed the cell-free synthesis of several products, including a 70,000 M_r polypeptide that could be specifically immunoprecipitated with anti-T serum. However, no T-antigen-related polypeptide was found in infected cells that corresponded in size to the major immunoprecipitated cRNA product.

Complementation studies with temperature-sensitive mutants of the tumor-causing simian virus 40 (SV40) indicate that it may have as few as three primary gene products (1). Only one of these, the A gene product, has been directly linked to the process of virus-induced neoplastic transformation (2–5). Cells in the early phase of productive lytic infection and virus-transformed cells exhibit some common features. In each case, a species of virus-specific RNA is induced, which hybridizes to the same strand of SV40 DNA (6, 7) and to the same fragments of SV40 DNA generated by restriction endonucleases (8, 9). At least three virus-associated antigens, T, U, and TSTA, have been detected both early in lytic infection and in transformed cells (10). The T-antigen has so far been characterized primarily immunologically, employing sera from animals bearing SV40-induced tumors (11–13). T-antigen has been partially purified (14–16) and is now considered to be a DNA-binding protein (17–20) with the molecular weight of the denatured polypeptide estimated to be from 70,000 (21) to 100,000 (22). Comparison of the properties of the T-antigen after infection with either wild-type or A-mutant virus at the restrictive temperature has supplied evidence that this antigen (or a part thereof) is encoded by viral DNA (22–24). To demonstrate directly that the information for the SV40 T-antigen is encoded in the viral genome, and to study its expression further, we have investigated the cell-free translation of early SV40 mRNA.

We report that SV40 early mRNA from infected cells, and SV40 cRNA transcripts made in vitro, direct the synthesis in the wheat-germ cell-free system of polypeptides that can be specifically immunoprecipitated by antiserum to SV40 T-antigen. The products directed by "in vivo" mRNA and synthetic cRNA differ, however, in their electrophoretic mobility on sodium dodecyl sulfate/polyacrylamide gels. The former products correspond to the T-antigen immunoprecipitated from extracts of infected cells and the largest of these has an estimated molecular weight of 90,000. A preliminary description of these findings has been given previously (25).

MATERIALS AND METHODS

Wheat germ was a gift from Z. Schildhaus of the Bar Rav Mill, Tel-Aviv, Israel. [^{35}S]Methionine (300 Ci/mmol) and [5,6-^{3}H]uridine (40 Ci/mmol) were purchased from Amersham. Antiserum to SV40 T-antigen was obtained from Flow Laboratories. Goat anti-hamster serum was a gift from M. Fogel.

Cells and Viruses. Plaque-purified SV40 (strain 777) was grown in the BS-C-1 line of African green monkey cells as previously described (26).

Cell Extracts. Cultures (10^7 cells) of uninfected or SV40-infected cells (multiplicity of infection = 2 × 10^9 plaque-forming units/ml) were labeled from 47 to 48 hr post infection with [^{35}S]methionine, (20 μCi/ml of methionine-free medium). After the cells were washed 3 times with cold phosphate-buffered saline, 2 ml of IP extraction buffer containing 25 mM N-2-hydroxyethylpiperazine-N'-2-ethanesulfonic acid (Hepes) at pH 8.0, 0.14 M NaCl, 0.5 mM magnesium acetate, and 0.5% Nonidet P-40 was added. The cell extracts were collected and centrifuged at 5000 × g for 2 min to separate nuclear and cytoplasmic portions. Nuclei were suspended in 1 ml of electrophoresis sample buffer [0.1 M Tris·HCl at pH 6.8, 1% sodium dodecyl sulfate, 0.7 M 2-mercaptoethanol, and 10% (vol/vol) glycerol] and sonicated 1 min at half maximum power in a Raytheon sonicator. The cytoplasmic portion was either adjusted to electrophoresis sample buffer conditions or centrifuged for 40 min at 40,000 rpm in a Spinco 40 rotor to obtain a 40,000 post-ribosomal supernatant fraction.

RNA Preparations. To prepare early SV40 RNA, 10^9 BS-C-1 cells were infected with SV40 (multiplicity of infection = 2 × 10^9 plaque-forming units/ml) in the presence of 20 μg/ml of cytosine arabinonucleoside as well as 5 μCi of [5,6-^{3}H]uridine. Cytoplasmic RNA was extracted 16–18 hr after infection and hybridized to 100 μg of SV40 DNA linked (27) to Sepharose 4B for 24–36 hr. The RNA was eluted by successive treatments with buffered 98% formamide and 1.0 M NaCl as described (27). The two eluate fractions were pooled and precipitated in the presence of 10 μg of rabbit liver tRNA, and collected in a small volume suitable for addition to the translation system (25–50 μl). SV40 cRNA was synthesized as previously described (27) and found to be at least 98% asymmetrically transcribed from SV40 form I (covalently closed, circular, superhelical) DNA.

The Wheat-Germ System. Wheat-germ extract was prepared and cell-free protein synthesis was assayed essentially as

Abbreviations: SV40, simian virus 40; Hepes, N-2-hydroxyethylpiperazine-N'-2-ethanesulfonic acid; M_r, molecular weight.

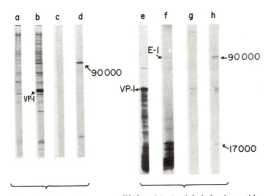

FIG. 1.   Autoradiograms of [^{35}S]methionine labeled polypeptides synthesized *in vivo* and *in vitro*. Left-hand group: polypeptides were extracted from (a) uninfected cells, or (b) SV40-infected cells 48 hr post infection and then subjected to immunoprecipitation with hamster anti-SV40-serum, yielding (c) from uninfected and (d) from infected cells. Right-hand group: polypeptides synthesized in the wheat-germ system (25 μl assays) directed by (e) 0.5 μg of late SV40 mRNA, 5 μl of reaction mix applied to the gel; (f) early mRNA selected from 2×10^8 SV40-infected cells, 25 μl of reaction mix applied to the gel. The cell-free products were coelectrophoresed with hamster anti-SV40-T immunoprecipitates of cell extracts from (g) uninfected, and (h) SV40-infected cells.

described by Roberts and Paterson (28) with the following modifications: The wheat germ was ground in 15 ml of grinding buffer, the preincubation step was eliminated, the pH of Hepes buffer used in the assays was 8.0, and spermine·HCl was added to a final concentration of 80 μM.

Polyacrylamide Gel Electrophoresis. Cell extracts, cell-free products, or immunoprecipitates which were suspended in electrophoresis sample buffer were subjected to electrophoresis through 10–20% gradient slab polyacrylamide gels as described (29). The gels were dried and exposed to Kodak RP54 x-ray film for various time intervals.

Immunoprecipitations. To 0.3 ml of 40,000 supernatant portions of either cytoplasmic extracts or wheat-germ system reaction mixtures diluted to 0.3 ml with IP extraction buffer, 5 μl anti-SV40-T serum or hamster serum of equivalent gamma globulin content was added. After incubation for 60 min at 30°, 50 μl of goat anti-hamster serum was added and the incubation was continued for 60 min at 30° followed by 16 hr at 4°. The samples were centrifuged at $5000 \times g$ for 2 min and the pellets washed twice in phosphate-buffered saline, suspended in 50 μl of electrophoresis sample buffer, and heated for 10 min at 90°.

RESULTS

Immunoprecipitation of T-antigen from SV40-Infected Cells. The characterization of the cell-free products of early SV40 mRNA relies mainly on immunological procedures, since the early SV40-associated proteins are defined at present by their antigenic specificity. When the proteins of [^{35}S]methionine-labeled SV40-infected monkey cells were reacted with anti-SV40-T-serum and fractionated by gel electrophoresis, a large polypeptide, whose molecular weight (M_r) was estimated at 90,000, as well as another 17,000 M_r polypeptide, was immunoprecipitated from the rest of the cellular and viral proteins (Fig. 1b and d). Other polypeptides close in electrophoretic mobility to the major 90,000 M_r protein are observed in various proportions depending on the cell lines used after immunoprecipitation. Neither immunoprecipitation of a similar

quantity of [^{35}S]methionine-labeled cytoplasm of uninfected cells (Fig. 1c and g) with anti-T serum nor immunoprecipitation of the labeled cytoplasm of infected cells with equivalent amounts of control hamster serum yielded the 90,000 M_r polypeptide. Tegtmeyer *et al.* (22) have described a 100,000 M_r polypeptide which is specifically immunoprecipitated from the cytoplasm of infected cells by hamster anti-SV40-T serum, which is most likely the same as or related to the 90,000 M_r polypeptide that we isolate. We consider the 90,000 M_r polypeptide to be the SV40 T-antigen; from its estimated size it appears to represent most of the coding capacity of the early region of the SV40 genome.

Cell-Free Product Directed by Early SV40 mRNA That Had Been Selected by Hybridization. We next analyzed the cell-free products whose synthesis was directed by early SV40 mRNA. BS-C-1 cells were infected with SV40 in the presence of cytosine arabinonucleoside, which inhibits expression of late RNA and associated late functions (30). The RNA was then hybridized to 100 μg of SV40 DNA linked to Sepharose by a previously described procedure (27). In a typical experiment, 5×10^8 cells, of which one-fifth were labeled with 5 mCi of [^{3}H]uridine, yielded 1.8×10^8 cpm total RNA, from which 3.4 $\times 10^4$ cpm were selected by hybridization (i.e., 0.019% of the total), a value only slightly above the background. However, when a sample of the RNA selected by SV40 DNA-Sepharose was tested by hybridization to SV40 DNA immobilized on filters, 3–5% of the input hybridized, compared to only 0.1% of similarly processed RNA from mock-infected cells.

The hybridized early SV40 RNA was added to the wheat-germ system and the cell-free products were analyzed by sodium dodecyl sulfate/polyacrylamide gel electrophoresis. The products synthesized (Fig. 1f) were different in size and distribution from those directed by late SV40 mRNA (Fig. 1e), in which the major capsid protein VP-1 is predominant (29). The early SV40 mRNA directs a number of products, of which the largest is a prominent polypeptide whose molecular weight was estimated as 90,000, and which we call the E-1 polypeptide. The E-1 polypeptide is not seen among the products directed by uninfected BS-C-1 mRNA (data not shown). The E-1 product was found to migrate with the anti-T-immunoprecipitated 90,000 M_r polypeptide from SV40-infected cells (Fig. 1f and h).

Further identification of the E-1 polypeptide as T-antigen was obtained after immunoprecipitation of the cell-free products of early SV40 mRNA with hamster anti-SV40-T serum or hamster control serum. Fig. 2 shows that the major 90,000 M_r cell-free product is specifically immunoprecipitated with hamster anti-T serum but not with hamster control serum. Another polypeptide product directed by early mRNA, and which is also found in infected cells, is a 17,000 M_r polypeptide. This polypeptide is immunoprecipitated by anti-T serum, although a slight reaction with hamster control serum has been observed. Neither products of the endogenous wheat-germ protein-synthesizing system, nor globin mRNA, tobacco mosaic virus mRNA, nor mRNA from uninfected cells yielded any specific immunoprecipitable material when treated and analyzed as described above.

The efficiency of the hybridization selection of early mRNA was demonstrated in an experiment in which the translation products of SV40-selected and nonselected RNA (that RNA which does not hybridize to SV40 DNA-Sepharose) were compared (Fig. 3). Early mRNA selected by hybridization to SV40 DNA-Sepharose directed the synthesis of a series of products including the prominent E-1 polypeptide. The nonselected RNA gave rise to a different spectrum of products and

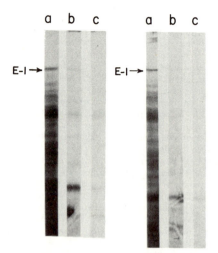

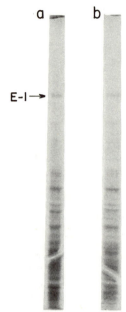

FIG. 2. Immunoprecipitation of [³⁵S]methionine-labeled polypeptides synthesized *in vitro*. The two groups represent autoradiograms of two separate experiments. (a) Polypeptides directed by SV40 early mRNA as described in Fig. 1f; (b) immunoprecipitation of (a) with hamster anti-SV40-T serum; (c) immunoprecipitation of (a) with normal hamster serum.

the E-1 polypeptide was totally absent. Thus, the hybridization technique provides an effective means for efficiently selecting and partially purifying early SV40 T-antigen mRNA.

Translation of SV40-Selected mRNA from Transformed Cells. Early mRNA from lytically infected monkey cells was

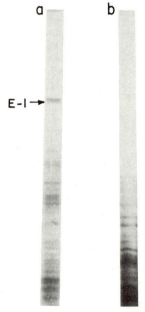

FIG. 3. Autoradiograms of [³⁵S]methionine-labeled polypeptide products of early-SV40-selected and nonselected mRNA (quantities as in Fig. 1f). (a) Polypeptides directed by early mRNA prepared by selective hybridization to SV40 DNA; (b) polypeptides directed by 0.2 μg of poly(A)-containing nonannealed RNA, i.e., that RNA which did not hybridize to SV40 DNA and which had been subjected to oligo(dT)-cellulose chromatography (31).

FIG. 4. Cell-free products of SV40 RNA from BS-C-1 cells early in lytic infection and from SV80 transformed cells. Autoradiograms of [³⁵S]methionine-labeled polypeptides directed by (a) SV40 early mRNA from infected BS-C-1 cells (quantities as in Fig. 1f); (b) SV40-specific RNA from SV80 cells. Cytoplasmic RNA from 10⁹ SV80 cells was hybridized to 100 μg of SV40 DNA-Sepharose; hybridization-selected RNA from 2 × 10⁸ cells was added per 25 μl of wheat-germ system reaction mixture, all of which was subjected to polyacrylamide gel electrophoresis.

compared to virus-specific RNA from SV40-transformed cells. The transformed line used was SV80, which is reported by Henderson *et al.* (15) to contain a higher proportion of T-antigen than other lines tested, as measured by complement fixation. RNA was extracted from approximately equal numbers of SV80 cells and infected BS-C-1 cells and hybridized to SV40 DNA-Sepharose as described above. Comparison of the cell-free products synthesized showed that both contained similar if not identical products, including the major early E-1 product (Fig. 4). Thus, SV40-selected mRNA from a line of virus-transformed cells directs similar products to that directed by early virus-specific mRNA extracted from permissive cells.

Immunoprecipitation of Cell-Free Products Whose Synthesis Is Directed by SV40 cRNA. An analysis of the cell-free products directed by synthetic SV40 cRNA was carried out in order to compare this alternate source of SV40 early RNA to infected cells. cRNA was prepared as described previously (27), and was found to be 98% asymmetrically transcribed. Analysis on polyacrylamide gels in comparison with known prokaryotic RNA standards showed the cRNA to be heterogeneous in size, the major component being approximately 19 S, with a small proportion of unit length RNA (28 S) (Fig. 5A). The activity of the cRNA in the wheat-germ system was found to be approximately equal to that of other biological mRNAs such as globin mRNA, or mRNA from BS-C-1 cells. The *in vitro* products of cRNA were rarely greater than 70,000 daltons in mass. The cell-free products of SV40 cRNA were subjected to immunoprecipitation with SV40 anti-T serum as described above. It was found (Fig. 5B) that a 70,000 M_r polypeptide, among the

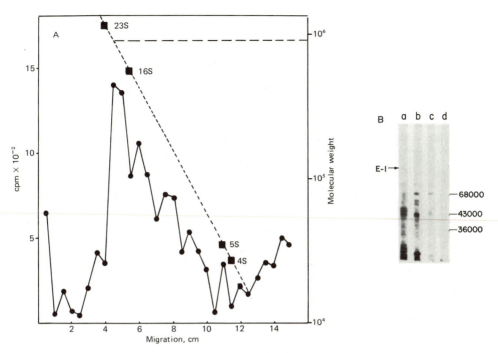

FIG. 5. (A) Size estimation of SV40 cRNA. [5,6-³H]Uridine-labeled SV40 cRNA was prepared as described (27). 15,000 cpm was applied to polyacrylamide agarose slab gels (27) and run concurrently with indicated *Escherichia coli* RNA markers (■ - - - ■). The gels were subjected to 350 V for 3 hr, after which 2 mm slices were dissolved in NCS and radioactivity was measured by liquid scintillation counting (●——●).
(B) Cell-free products of SV40 cRNA. Autoradiograms of [³⁵S]methionine-labeled polypeptides: products directed by (a) SV40 early mRNA as in Fig. 1f; (b) SV40 cRNA whose size distribution is shown in (A); (10 μg was added per 50 μl of wheat-germ system reaction mixture, 5 μl of which was subjected to polyacrylamide gel electrophoresis). Immunoprecipitation of 15 μl of SV40 cRNA cell-free products in (b) with (c) hamster anti-SV40-T serum; (d) hamster control serum.

largest and most prominent of the cRNA products, could be specifically immunoprecipitated with hamster anti-T serum but not with hamster control serum. No counterpart to the 70,000 M_r immunoprecipitated SV40 cRNA products could be detected among the immunoprecipitated polypeptides of SV40-infected cells.

DISCUSSION

We have shown that a preparation of infected cell RNA, 30- to 50-fold enriched for early SV40 mRNA, directs the synthesis of a 90,000 M_r product (the E-1 polypeptide) which can be specifically immunoprecipitated by hamster anti-SV40-T serum and which comigrates electrophoretically with a similarly immunoprecipitated polypeptide isolated from virus-infected cells. The specificity of the immunoprecipitation technique is shown by experiments in which the E-1 product directed by early SV40 mRNA is immunoprecipitated by hamster anti-SV40-T serum and not by hamster control serum, and by the fact that a comigrating cellular 90,000 M_r polypeptide can be immunoprecipitated from infected, but not from uninfected cells. Since both the cellular and cell-free products have the same electrophoretic mobility and are specifically immunoprecipitated by anti T serum, it is likely that they are the same polypeptide and that they represent the primary translation product of the early mRNA. The 17,000 M_r polypeptide, which was similarly immunoprecipitated from the cell-free products of SV40 early mRNA as well as from SV40-infected cells, may be related to the 90,000 M_r product or may represent a nonspecific immunoprecipitation reaction. Evidence in favor

of the first alternative has recently been obtained: the tryptic digests of the 17,000 M_r and 90,000 M_r polypeptides, isolated from cells, have some common peptides (C. Prives and L. Lipot, unpublished experiments). The 17,000 M_r product may be part of the native T-antigen complex, a proteolysis product of T-antigen, or an additional virus-specific antigen such as U-antigen (32). It is not known if it is the product of a cistron separate from that coding for the 90,000 M_r polypeptide.

Polypeptides directed by SV40 cRNA in a cell-free translation system must be virus specific because the RNA is transcribed *in vitro* from plaque-purified SV40 DNA. We have observed that different preparations of SV40 cRNA in the wheat-germ system directed the synthesis of products whose electrophoretic mobilities in the gel were somewhat variable. Generally, polypeptides of 60,000–70,000 M_r were specifically immunoprecipitated by SV40 anti-T serum; this class had no size counterpart among similarly immunoprecipitated polypeptides from infected cells. Nevertheless, the fact that SV40 cRNA can direct the cell-free synthesis of a product which can be specifically immunoprecipitated with SV40 anti-T serum provides proof that SV40 T-antigen is at least partly virus coded. Similar findings have been reported for SV40 cRNA in a linked transcription–translation system by Roberts *et al.* (33) and for SV40 cRNA of genome length by Smith *et al.* (34). The discrepancy in size between the cRNA-directed product and that of early SV40 mRNA is not presently understood. Since early SV40 mRNA directs the synthesis of a larger product that corresponds in size and immunospecificity to a polypeptide that can be isolated from infected cells, the translation of viral RNA isolated from infected cells, rather than synthetic cRNA, may be the preferred approach for studying the metabolism of T-antigen mRNA.

It is noteworthy that it is possible, by hybridization selection, to partially purify and translate an mRNA class that comprises an extremely small proportion of the cells' total RNA. Early mRNA is only 1–2% of late virus mRNA (35), which itself is not greater than 10% of cellular mRNA even late in infection (36). Thus, early SV40 mRNA does not represent more than 0.1% of the total mRNA of the infected cell—a proportion too small to provide detectable products in the cell-free translation system without the prior selection process. Hybridization to immobilized SV40 DNA provides a minimal 20-fold enrichment of SV40-specific sequences in the RNA added to the translation system. Because it is unlikely that all of the nonspecific RNA in this enriched class is mRNA (the source of input RNA being total cytoplasmic RNA from infected cells), the 3–5% SV40 RNA in this fraction (see *Results*) is likely to be an underestimate of the proportion of messenger RNA or other translatable RNA added to the wheat-germ system.

The selection of an early SV40 mRNA class coding for T-antigen, as well as the isolation of T-antigen from cells by immunoprecipitation, opens up several new approaches. In the one example presented herein, the SV40-specific RNA from virus-transformed human cells directs the synthesis of products, including the E-1 90,000 M_r T-antigen polypeptide, similar to those directed by SV40 early mRNA extracted from lytically infected monkey cells. However, T-antigens isolated from different host species have been shown to vary in size (37, 38). Moreover, variations in the virus cell-transformation system have been reported (39) as well as the isolation of revertant offspring of transformed cells, all of which can now be analyzed as to their virus-specific mRNA and related cell-free translation products. The overproduction of the 100,000 M_r A protein in cells infected by the Ts A SV40 mutant at restrictive temperature (22) can be investigated by the approach described in this paper. Thus, the isolation and translation of SV40 early mRNA coding for SV40 T-antigen can provide an additional approach towards understanding virus-induced cell transformation.

Drs. Haim Aviv and David Givol are thanked for many useful discussions during the course of these studies. The technical assistance of L. Lipot, T. Koch, and B. Danovitch is gratefully acknowledged. The work was supported by National Cancer Institute Contract no. N01 CP 33220.

1. Chou, J. Y. & Martin, R. G. (1974) *J. Virol.* **13**, 1101–1109.
2. Tegtmeyer, P. (1975) *J. Virol.* **15**, 613–618.
3. Chou, J. Y. & Martin, R. G. (1975) *J. Virol.* **15**, 145–150.
4. Brugge, J. S. & Butel, J. (1975) *J. Virol.* **15**, 619–635.
5. Osborn, M. & Weber, K. (1975) *J. Virol.* **15**, 636–644.
6. Sambrook, J., Sharp, P. A. & Keller, W. (1972) *J. Mol. Biol.* **70**, 57–71.
7. Khoury, G., Byrne, J. C., Takemoto, K. K. & Martin, M. A. (1973) *J. Virol.* **11**, 54–60.
8. Khoury, G., Martin, M. A., Lee, T. N. H. & Nathans, D. (1975) *Virology* **63**, 263–272.
9. Khoury, G., Howley, P., Nathans, D. & Martin, M. (1975) *J. Virol.* **15**, 433–437.
10. Kelly, T. J., Jr., & Lewis A. M., Jr., (1973) *J. Virol.* **12**, 643–652.
11. Black, P. H., Rowe, W. P., Turner, H. C. & Huebner, R. J. (1963) *Proc. Natl. Acad. Sci. USA* **50**, 1148–1156.
12. Pope, J. H. & Rowe, W. P. (1964) *J. Exp. Med.* **120**, 121–128.
13. Rapp, F., Kitihara, T., Butel, J. S. & Melnick, J. L. (1964) *Proc. Natl. Acad. Sci. USA* **52**, 1138–1142.
14. Livingston, D. M., Henderson, I. C. & Hudson, J. (1974) *Cold Spring Harbor Symp. Quant. Biol.* **39**, 283–289.
15. Henderson, I. C. & Livingston, D. M. (1974) *Cell* **3**, 65–70.
16. Osborn, M. & Weber, K. (1974) *Cold Spring Harbor Symp. Quant. Biol.* **34**, 267–276.
17. Carroll, R. B., Hager, L. & Dulbecco, R. (1974) *Proc. Natl. Acad. Sci. USA* **71**, 3754–3757.
18. Reed, S. I., Ferguson, J., Davis, R. W. & Stark, G. R. (1975) *Proc. Natl. Acad. Sci. USA* **72**, 1605–1609.
19. Jessel, D., Hudson, J., Landau, T., Tenen, D. & Livingston, D. M. (1975) *Proc. Natl. Acad. Sci. USA* **72**, 1960–1964.
20. Spillman, T., Spomer, B. & Hager, L., (1975) *Fed. Proc.* **34**, 527.
21. Del Villano, B. C. & Defendi, V. (1973) *Virology* **51**, 34–46.
22. Tegtmeyer, P., Schwartz, M., Collins, J. K. & Rundell, K. (1975) *J. Virol.* **16**, 168–178.
23. Tenen, D. G., Baygell, P. & Livingston, D. M. (1975) *Proc. Natl. Acad. Sci. USA* **72**, 4351–4355.
24. Alwine, J. C., Reed, S. I., Ferguson, J. & Stark, G. R. (1975) *Cell* **6**, 529–533.
25. Prives, C., Aviv, H., Gilboa, E., Winocour, E. & Revel, M. (1975) *Methodol. Exp. Physiol. Physiopathol. Thyroidiennes, Colloq.* **47**, 305–312.
26. Lavi, S. & Winocour, E. (1972) *J. Virol.* **9**, 309–316.
27. Gilboa, E., Prives, C. L. & Aviv, H. (1975) *Biochemistry* **14**, 4215–4220.
28. Roberts, B. E. & Paterson, B. M. (1973) *Proc. Natl. Acad. Sci. USA* **70**, 2330–2334.
29. Prives, C. L., Aviv, H., Paterson, B. M., Roberts, B. E., Rozenblatt, S., Revel, M. & Winocour, E. (1974) *Proc. Natl. Acad. Sci. USA* **71**, 302–306.
30. Butel, J. S. & Rapp, F. (1965) *Virology* **27**, 490–495.
31. Aviv, H. & Leder, P. (1972) *Proc. Natl. Acad. Sci. USA* **69**, 1408–1412.
32. Lewis, A. M., Jr., Levin, M. J., Wiese, W. H., Crumpacker, C. S. & Henry, P. H. (1969) *Proc. Natl. Acad. Sci. USA* **63**, 1128–1135.
33. Roberts, B. E., Gorecki, M., Mulligan, R. C., Danna, K. J., Rozenblatt, S. & Rich, A. (1975) *Proc. Natl. Acad. Sci. USA* **72**, 1922–1926.
34. Smith, A. E., Bayley, S. T., Wheeler, T. & Mangel, W. F. (1975) *Methodol. Exp. Physiol. Physiopathol. Thyroidiennes, Colloq.* **47**, 331–338.
35. Aloni, Y., Winocour, E. & Sachs, L. (1968) *J. Mol. Biol.* **31**, 415–429.
36. Prives, C. (1973) in *Tumor Virus-Host Cell Interaction*, ed. Kolber, A. (Plenum Press, New York), pp. 127–139.
37. Carroll, R. B. & Smith, A. E. (1976) *Proc. Natl. Acad. Sci. USA* **73**, 2254–2258.
38. Ahmad-Zadeh, C., Allet, B., Greenblatt, J. & Weil, R. (1976) *Proc. Natl. Acad. Sci. USA* **73**, 1097–1101.
39. Risser, R. & Pollack, R. (1974) *Virology* **59**, 477–489.

Reprinted from

Proc. Natl. Acad. Sci. USA
Vol. 75, No. 3, pp. 1274–1278, March 1978
Biochemistry

Spliced early mRNAs of simian virus 40

(S₁ endonuclease/exonuclease VII/transcription mapping/viable deletion mutants)

ARNOLD J. BERK AND PHILLIP A. SHARP

Department of Biology and Center for Cancer Research, Massachusetts Institute of Technology, Cambridge, Massachusetts 02139

Communicated by J. D. Watson, January 9, 1978

ABSTRACT   Biochemical methods are presented for determining the structure of spliced RNAs present in cells at low concentrations. Two cytoplasmic spliced viral RNAs were detected in CV-1 cells during the early phase of simian virus 40 (SV40) infection. One is 2200 nucleotides in length and is composed of two parts, 330 and 1900 nucleotides, mapping from ~0.67 to ~0.60 and from ~0.54 to ~0.14, respectively, on the standard viral map. The other is 2500 nucleotides long and also is composed of two parts, 630 and 1900 nucleotides mapping from ~0.67 to ~0.54 and from ~0.54 to ~0.14, respectively. Correlation of the structure of these mRNAs with the structure of the early SV40 proteins, small T antigen (17,000 daltons) and large T antigen (90,000 daltons), determined by others suggests that: (i) translation of the 2500-nucleotide mRNA yields small T antigen; (ii) translation of the 2200-nucleotide mRNA proceeds through the splice point in the RNA to produce large T antigen (and thus large T antigen is encoded in two separate regions of the viral genome); and (iii) the DNA sequences between ~0.67 and ~0.60 present in both mRNAs are translated in the same reading frame in both mRNAs to yield two separate gene products that have the same NH₂-terminal sequence. Therefore, expression of the early SV40 genes is partially controlled at the level of splicing of RNAs.

The genome of simian virus 40 (SV40) is a closed circular duplex DNA molecule approximately 5000 base pairs in length. The genetic information encoded in approximately half of this DNA is expressed during the early phase of infection in the form of two proteins (1–3). The larger of these, large T antigen, has a molecular weight estimated by sodium dodecyl sulfate/polyacrylamide gel electrophoresis to be 90,000–100,000 (4), while the other, small T antigen, has a molecular weight of approximately 17,000 (1–3). The small T antigen shares a considerable fraction of its amino acid sequence with the large T antigen, suggesting that portions of these two proteins are encoded in the same DNA sequence (2, 3, 5).

The sequences present in the early mRNAs that encode these two proteins are transcribed in the counterclockwise direction from the early region of the genome mapping between roughly 0.67 and 0.17 unit lengths from the single EcoRI site (6, 7) (Fig. 1). Both early proteins can be translated in vitro from mRNAs sedimenting at 19 S, the small T antigen mRNA sedimenting slightly faster than the large T antigen mRNA (3). These two mRNAs therefore must contain extensively overlapping sequences because each of them is approximately the size of the entire early region.

Recently, several paradoxical observations have been made concerning the organization of DNA sequences that encode the two early gene products. The basis of this paradox is that nearly the entire early region (2500–2600 nucleotides in length) would be required to encode a protein with the estimated molecular weight of large T antigen, 90,000–100,000. Yet, large deletions

within the early region mapping between 0.59 and 0.54 (8) do not result in the production of an altered large T antigen (ref. 1; M. J. Sleigh, W. C. Topp, R. Hanich, and J. Sambrook, personal communication). Different examples of these deletion mutants (1) or of an independently isolated series of similar mutants (Sleigh et al., personal communication) either fail to induce a detectable small T antigen or induce an altered protein that has a decrease in molecular weight roughly corresponding to the size of the deletion. Therefore, although the two early proteins are apparently encoded in partially overlapping DNA sequences, sequences mapping between 0.54 and 0.59 must code for small T antigen exclusively. In addition, DNA sequence studies (9) have shown that a transcript of the early strand would contain two termination codons in each of all three reading frames near position 0.54. In an attempt to clarify this puzzling situation, we have determined the structure of the early SV40 mRNAs by using new methods developed in our laboratory (10).

Recently it has been shown that the sequences comprising a single covalently continuous viral mRNA molecule may arise from well-separated regions on a viral genome (11–16). Such mRNA molecules are referred to as "spliced mRNAs." In this work, we find two early SV40 mRNAs that are also spliced. The map positions of the splice points and of the sequences present in these early mRNAs can account for the shared amino acid sequence of the large and small T antigens, the occurence of termination codons in all three reading frames at 0.54, the phenotype of the deletion mutants, and the relative sizes of the large and small T antigen mRNAS.

MATERIALS AND METHODS

Preparation of RNA. Subconfluent plates of CV-1 cells were infected at 37° with 5–10 plaque forming units per cell of SV40 strain 777. After 60 min, Dulbecco's modified medium containing 10% calf serum and cytosine arabinoside (20 μg/ml) was added. After 18 more hr of incubation at 37°, the cells were harvested by scraping, and cytoplasmic RNA was isolated as described (10).

Preparation of ³²P-Labeled DNA. Closed circular SV40 DNA labeled in vivo with ³²P had a specific activity of 1–3 × 10⁶ cpm/μg and was prepared as described (17). Alternatively, SV40 DNA was labeled by nick-translation (19) except for the following modifications: all four of the α-³²P labeled dNTPs were present at a specific activity of 20–40 Ci/mmol, and DNase I, 8 pg/ml, was added to the reaction which proceeded for 45 min at 15°. Approximately 30% of the dNTPs were polymerized into DNA which was >50% full-length SV40 strands. Final specific activities were 1–3 × 10⁷ cpm/μg. Labeled SV40 DNA was digested to completion with the restriction endonucleases indicated in the text and figure legends. EcoRI, Bam I, and Bgl I were purified in our labora-

Abbreviations: SV40, simian virus 40; t_m, melting temperature.

Biochemistry: Berk and Sharp

Proc. Natl. Acad. Sci. USA 75 (1978)   1275

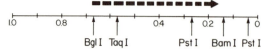

FIG. 1. Restriction map of SV40 DNA. The circular genome is represented as a line broken at the single *Eco*RI site defined as 0/1.0 and is divided into fractional unit lengths. The broken arrow represents the approximate region of the genome expressed during the early phase of infection and the direction of transcription.

tory. *Taq* I was a gift from C. Cole, and *Pst* I was purchased from New England Biolabs.

Hybridization. Restriction endonuclease-cleaved ^{32}P-labeled SV40 DNA or restriction fragments, purified as described (10), were ethanol-precipitated along with 10–20 μg of carrier yeast RNA. The pellets were briefly dried and dissolved in 20–100 μl of 80% formamide/0.4 M NaCl/0.04 M 1,4-piperazinediethanesulfonic acid (Pipes), pH 6.4/1 mM EDTA (20). Aliquots of the ethanol-precipitated early SV40 cytoplasmic RNA were pelleted and briefly air-dried. The RNA pellets were then dissolved in the hybridization buffer in which the ^{32}P-labeled SV40 DNA previously had been dissolved. Final concentrations of SV40 DNA and cytoplasmic RNA were 5 or 10 μg/ml and 2.5 to 5 mg/ml, respectively. The hybridization mixture was placed in capped 1.5-ml polypropylene tubes and incubated at 60° for 10 min and then at 49° for 3 hr, except where indicated.

Endonuclease S₁ Digestion. An aliquot of the hybridization mixture was removed and diluted into 10 volumes of 0.25 M NaCl/0.03 M NaOAc/1 mM ZnSO₄/5% glycerol/thermally denatured salmon sperm DNA, (20 μg/ml) chilled to 0°. S₁ purified by the method of Vogt (21) was added, 1 unit per ml of S₁ reaction mix, and the solution was incubated at 45° for 30 min. S₁-resistant material was ethanol-precipitated.

Exonuclease VII Digestion. An aliquot of the hybridization mixture was removed and diluted into 10 volumes of 0.03 M KCl/0.01 M Tris, pH 7.4/0.01 M Na₃EDTA, chilled to 0°. An amount of *Escherichia coli* exonuclease VII, purified by the method of Chase and Richardson (22), sufficient to digest 2 μg of thermally denatured linear SV40 DNA to 95% acid solubility in 60 min at 45° in this buffer was added per ml of exonuclease VII reaction mix and the solution was then incubated at 45° for 60 min. Exonuclease VII-resistant material was ethanol-precipitated. Exonuclease VII was a generous gift from S. Goff.

Gel Electrophoresis. Neutral agarose gels were 1.4% agarose in 0.04 M NaOAc/0.05 M Tris/2 mM EDTA, pH 8.3. Alkaline agarose gels were 1.4% agarose in 0.03 M NaOH/2 mM EDTA (23). Acrylamide gels were gradients of 5% acrylamide/0.17% bisacrylamide to 10% acrylamide/0.3% bisacrylamide with a 2.4% acrylamide/0.008% bisacrylamide sample well. The gel contained 8 M urea/0.04 M NaOAc/0.05 M Tris/2 mM EDTA, pH 8.3. The length of single-stranded DNA fragments or RNA·DNA duplex segments was determined by their mobility relative to standards of restriction fragments of Ad2 DNA. The assumed lengths of these *Hin*dIII and *Sma* I Ad2 fragments are reported in the legend to Fig. 3. The relative mobility of intact SV40 DNA in either the alkaline or neutral gel yields a length of 5000 base pairs. Recent sequence data suggest a length of approximately 5225 base pairs. The errors for the length of fragments cited in the text represent the experimental reproducibility of the relative mobilities of bands. Because the mobility of a fragment is probably affected by base sequence composition as well as length, the lengths and errors reported here cannot necessarily be directly converted into absolute numbers of nucleotides in a specific sequence.

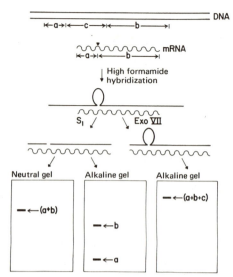

FIG. 2. Strategy for the analysis of spliced mRNA structure by gel electrophoresis of endonuclease S₁- and exonuclease VII-digested RNA·DNA hybrids.

RESULTS

The strategy for the analysis of viral mRNA structure is depicted in Fig. 2 and is as follows. Total cytoplasmic RNA is hybridized to ^{32}P-labeled viral DNA under conditions above the melting temperature (t_m) of the viral DNA but below the t_m of the RNA·DNA hybrid (20). If viral mRNAs are spliced, hybrid structures will result that are similar to those observed in the electron microscope by Berget *et al.* (11). RNA·DNA duplex is flanked by single-stranded DNA, and loops of nonhybridized single-stranded DNA result at splice points in the mRNA. When these structures are treated with the single-strand specific endonuclease S₁ (21), the single-stranded DNA is hydrolyzed, resulting in a fully duplex structure with discontinuities in the DNA at the splice points. When the S₁-treated hybrids are resolved by electrophoresis through neutral agarose gels, a band is observed migrating as expected for a duplex DNA molecule equal in length to the total mRNA (Fig. 2, lengths a + b). When these bands are excised and analyzed further by electrophoresis after denaturation, the resulting single-stranded DNA fragments observed are equal in length to the sequences spliced together to form the mRNA molecule. We refer to RNA segments transcribed from a contiguous set of DNA sequences as colinear transcripts (Fig. 2, lengths a and b). If the initial hybridization is performed using DNA digested with restriction endonucleases, patterns are produced that define the map positions of the colinear transcripts (10).

The mRNA·genome DNA hybrids are also analyzed by digestion with single-strand specific exonuclease VII of *E. coli* (ref. 22; S. Goff and P. Berg, personal communication). This exonuclease, which digests processively in both the 5′ and 3′ directions (24), removes the DNA single strands extending beyond the 5′ and 3′ ends of the mRNA but does not remove the single-stranded DNA loops that result at splice points. By analyzing the length of the exonuclease VII-resistant DNA in alkaline agarose gels, the length along the genome between the 5′ sequence and the 3′ sequence present in the mRNA is determined (Fig. 2, lengths a + b + c). Because lengths a and b

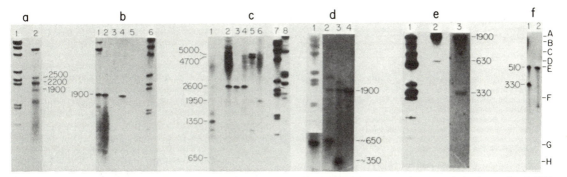

FIG. 3.   Autoradiograms of gels of endonuclease S₁- and exonuclease VII-digested hybrids. (*a*) Neutral agarose gel of early cytoplasmic RNA hybridized to ^{32}P-labeled SV40 DNA cut at 0 with *Eco*RI and digested with endonuclease S₁ (track 2). Track 1, marker *Sma* I digest of Ad2 DNA. The band closest to the top of track 2 is full-length linear SV40 DNA resulting from renaturation of intact strands during quenching of the hybridization before treatment with S₁. (*b*) Alkaline agarose gel of S₁-digestion products of early RNA hybridized to: SV40 DNA cut at 0 with *Eco*RI (track 1); SV40 DNA cut at 0.14, 0.57, and 0.67 with *Bam* I, *Taq* I, and *Bgl* I (track 2); the DNA fragment from 0.67 to 0.14 (track 3); the DNA fragment from 0.14 to 0.57 (track 4); the DNA fragment from 0.57 to 0.67 (track 5) (all in the clockwise direction). Track 6, marker *Sma* I digest of Ad2 DNA. (*c*) Alkaline agarose gel of: S₁-digestion products of early RNA hybridized to SV40 DNA cut at 0.04 and 0.27 with *Pst* I (track 1). Exonuclease VII digestion products of early RNA hybridized to SV40 DNA cut at: 0 with *Eco*RI (track 2); 0 and 0.14 with *Eco*RI and *Bam* I (track 3); 0.67 with *Bgl* I (track 4); 0.57 with *Taq* I (track 5); 0.27 and 0.04 with *Pst* I (track 6). Also shown: marker digests of Ad2 DNA digested with *Sma* I (track 7) and *Hind*III (track 8). The lengths (kilobases) of the Ad2 *Sma* I restriction fragments are: A, 7.04; B, 6.27 (A and B run as a doublet); C, 5.22; D, 4.24; E, 2.84; F, 2.31; G, 2.21; H, 1.51; I, 1.33; J, 1.05; K, 0.63. The lengths of the Ad2 *Hind*III restriction fragments are: A, 8.23; B, 5.08; C, 3.33; D, 3.19; E, 3.12 (C, D, and E run as a triplet); F, 2.73; G, 2.63 (F and G run as a doublet); H, 2.21; I, 2.03; J, 1.30; K, 0.95. The bands at the top of the gel in *b* track 1 and *c* track 2 are due to a small fraction of SV40 DNA that was not linearized by *Eco*RI digestion. From the top of the gel downward they represent single-strand circles, single-strand linears, and form I DNA. (*d*) The 2500 (track 2), 2200 (track 3), and 1900 (track 4) nucleotide RNA·DNA hybrid bands cut from a preparative neutral agarose gel, similar to that shown in *a*, after hybridization to nick-translated DNA. The gel slices were melted, diluted with an equal volume of water, and layered on an alkaline agarose gel. Track 1, *Sma* I Ad2 markers. (*e*) The 2500 (track 2) and 2200 (track 3) nucleotide RNA·DNA hybrids prepared as in *d* after hybridization to *in vivo* labeled DNA denatured with alkali and layered on a 5–10% 8 M urea gradient acrylamide gel. Track 1, marker *Hae* III restriction fragments of SV40 DNA: A, 1465; B, 820; C, 550; D, 370; E, 335; F₁ₐ, 315; F₁ᵦ, 310; F₂, 300; G, 220; H, 165 nucleotides. (*f*) S₁ products of early RNA hybridized to the nick-translated *Taq* I/*Bgl* I B fragment alkali-denatured and electrophoresed on a 5–10% 8 M urea gradient acrylamide gel (track 1). Untreated denatured *Taq* I/*Bgl* I B fragment (track 2). The positions of marker *Hinf* SV40 restriction fragments are shown: A, 1790; B, 1120; C, 760; D, 570; E, 490; F, 235; G, 110; H, 75 nucleotides.

have been determined, the genome sequence, c, between the colinear transcripts can be calculated.

As a first step in this analysis, we defined the map coordinates of the long (>1000 nucleotides) colinear transcripts present in early cytoplasmic RNA. Early SV40 cytoplasmic RNA was hybridized to ^{32}P-labeled SV40 DNA cut once in the late region at 0 with *Eco*RI. After S₁ treatment and electrophoresis on alkaline agarose gels, a major band of 1900 nucleotides was reproducibly observed (Fig. 3*b*, track 1). Often, there was a smear of DNA present on the gel below the major 1900-nucleotide band, as shown in this example. A major 1900-nucleotide band was also reproducibly observed when early RNA was hybridized to DNA cut at 0.67 (*Bgl* I), 0.14 (*Bam* I), or 0.57 (*Taq* I), or at all three of these positions (Fig. 3*b*, track 2). The same band was observed when the DNA fragment mapping from 0.57 (*Taq* I) to 0.14 (*Bam* I) was used in the hybridization (Fig. 3*b*, track 4). It should be noted that, when purified fragments were used in the hybridization, the smear of DNA below the 1900-nucleotide band was not observed. The 1900-nucleotide early colinear transcript must map from 0.54 ± 0.01 to 0.14 ± 0.01 because hybridization to DNA cut at 0.27 and 0.04 (*Pst* I) results in bands migrating at lengths of 1350 and 650 nucleotides on alkaline agarose gels (Fig. 3*c*, track 1). Hybridization to SV40 DNA cut with *Hpa* I (at positions 0.175, 0.375, and 0.73) resulted in bands migrating at the position of the 0.375–0.175 fragment and at 800 nucleotides, in agreement with this mapping position (data not shown).

When an excess of *Eco*RI-cut DNA was hybridized to early cytoplasmic RNA and the products were digested with exonuclease VII, a fragment of DNA 2600 nucleotides long was

generated (Fig. 3*c*, track 2). A 2600-nucleotide band also was observed after exonuclease VII treatment of early RNA hybridized to DNA cut with *Eco*RI plus *Bam* I and to DNA cut with *Bgl* I (Fig. 3*c*, tracks 3 and 4). This exonuclease VII-resistant fragment maps from 0.67 ± 0.02 to 0.14 ± 0.01 because bands of length 1950 and 650 nucleotides are generated by this procedure when the hybridized DNA is initially cut with *Pst* I (at sites 0.27 and 0.04) (Fig. 3*c*, track 6). From these results we conclude that the early SV40 mRNAs detected by these methods must contain sequences at their 5′ ends mapping at 0.67 ± 0.02 and sequences at their 3′ ends mapping at 0.14 ± 0.01. Because the 1900-nucleotide colinear transcript maps from 0.54 ± 0.01 to 0.14 ± 0.01, we conclude that the early SV40 mRNAs must have additional sequences spliced to the 5′ end of the 1900-nucleotide colinear transcript.

To ensure that the processive digestion of single-stranded SV40 DNA by exonuclease VII had not been blocked by an inverted repeat in the DNA sequence, which occurs near position 0.67 (25), the following control experiment was performed. ^{32}P-Labeled SV40 DNA cut at 0.67 with *Bgl* I was hybridized to a 10-fold molar excess of the unlabeled SV40 DNA *Hpa* I fragment mapping from 0.175 to 0.375 and the hybridization was allowed to proceed until 80% of the fragment had renatured. The products of the hybridization were digested with exonuclease VII under the same conditions used above and the resistant fragments were analyzed by alkaline gel electrophoresis. Autoradiography revealed a faint band comigrating with full-length SV40 DNA and a dense band comigrating with the *Hpa* I fragment (data not shown). These results indicate that, under the conditions of digestion used, there are no blocks

Biochemistry: Berk and Sharp

Proc. Natl. Acad. Sci. USA 75 (1978) 1277

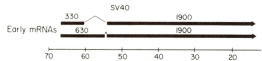

FIG. 4. Structure of the early SV40 mRNAs as determined by the data presented in Fig. 3. Heavy lines represent genome sequences included in the mRNAs and arrowheads indicate the 3' direction. The caret-shaped symbol indicates that sequences are covalently joined in the mRNA molecule by a 3'-5' phosphodiester bond. Numbers above the heavy lines represent lengths in nucleotides. Numbers below the narrow line indicate SV40 genome map coordinates in hundreths of a unit.

to the processive digestion of single-stranded SV40 DNA by exonuclease VII in sequences mapping counterclockwise from 0.67 to 0.375 and clockwise from 0.67 to 0.175.

To determine the nature of the sequences spliced to the 5' end of the 1900-nucleotide colinear transcript, the following experiments were performed. SV40 DNA cut at 0 (*Eco*RI) was hybridized to early RNA and treated with S_1, and the products were run on a neutral agarose gel. Three bands were reproducibly observed at 2500 nucleotides, 2200 nucleotides, and 1900 nucleotides (Fig. 3a, track 2). In addition, a broad band was often observed at 1550–1650 nucleotides. To determine the sizes of the colinear transcripts from which the RNA molecules in these RNA·DNA hybrids are composed, these bands were cut from a preparative gel and analyzed by electrophoresis through an alkaline agarose gel. The 2500-nucleotide RNA·DNA hybrid gave rise to two single-stranded DNA fragments migrating at 1900 and ~650 nucleotides (Fig. 3d, track 2). The 2200-nucleotide band gave rise to two bands migrating at 1900 and ~350 nucleotides, and the 1900-nucleotide band yielded a single band migrating at 1900 nucleotides (Fig. 3d, tracks 3 and 4). Lengths of the smaller colinear transcripts were estimated more precisely on acrylamide gels (Fig. 3e) to be 630 and 330 nucleotides. These same lengths were observed when the initial hybridization was to DNA cut at 0.67 with *Bgl* I. Faint bands are also observed on the alkaline agarose gel migrating at 2500 and 2200 nucleotides (Fig. 3d, tracks 2 and 3). These are due to background transferred from the preparative neutral agarose gel, because no colinear transcripts longer than 1900 nucleotides were identified earlier (Fig. 3b). When similar experiments were performed on the broad 1550- to 1650-nucleotide band cut from a neutral agarose gel of S_1-treated hybrid (Fig. 3a), only the broad 1550–1650 band was again observed (data not shown).

Most simply interpreted, these results demonstrate that there are two early mRNAs, each of which is spliced (Fig. 4). The larger of these is a 2500-nucleotide mRNA composed of two parts. At the 5' end of this mRNA is a colinear transcript of 630 nucleotides with its 5' end at ~0.67. This is spliced to a 1900-nucleotide colinear transcript with its 5' end at 0.54 ± 0.01 and its 3' end at 0.14 ± 0.01. The smaller of the two abundant early mRNAs, a 2200-nucleotide mRNA, is similarly composed of two parts. At the 5' end is a colinear transcript of 330 nucleotides also having its 5' terminus at ~0.67. It is also spliced to the 1900-nucleotide colinear transcript mapping from 0.54 ± 0.01 to 0.14 ± 0.01. We interpret the 1900-nucleotide band on the neutral S_1 gel to be due to partial S_1 cutting at the splice point in these mRNAs. We do not know the origin of the broad band observed in some neutral S_1 gels migrating with a mobility of roughly 1550–1650 nucleotides or the origin of the minor bands observed in some tracks on the alkaline S_1 gels.

To test these deduced structures we analyzed the products of early cytoplasmic RNA hybridized to SV40 DNA cut at 0.57

(*Taq* I) by digestion with exonuclease VII. Hybridization of the 2500-nucleotide mRNA to *Taq* I-cut DNA should result in a circular hybrid molecule in which both ends of the hybridized early DNA strand are in duplex regions. Thus, the entire strand should be resistant to exonuclease VII digestion. Hybridization of the 2200-nucleotide mRNA to the early strand of *Taq* I-cut DNA should also result in a circular hybrid molecule. However, in this case the ends of the hybridized single-stranded DNA should not be base-paired because DNA sequences from 0.54 to 0.60 are not present in the RNA. Consequently, a total of 300 nucleotides should be removed from the 5' and 3' ends of the early DNA strand before hybrid region blocks further exonuclease digestion. These expected products 5000 and 4700 nucleotides long were observed as the most prominent bands on an alkaline agarose gel of exonuclease VII-digested hybrids between early RNA and *Taq* I-cut SV40 DNA (Fig. 3c, track 5).

The deduced structure of the early SV40 RNAs includes sequences mapping from 0.67 (*Bgl* I) to 0.57 (*Taq* I). Yet, when this isolated restriction fragment was hybridized to early RNA at 49°, no S_1-protected DNA was observed (Fig. 3b, track 5). The explanation for this observation is that the t_m of this restriction fragment in the hybridization buffer is 36°, ~10° lower than the calculated melting temperature (t_m) of the total genome (data not shown). When hybridization of early RNA to this restriction fragment is performed at 40° rather than at 49° and the products are treated with S_1, protection of a 330-nucleotide fragment is observed (Fig. 3f) as predicted by the proposed structure of the 2200-nucleotide mRNA. Protection of the full length *Taq/Bgl* fragment is also observed, as expected from the proposed structure of the 2500-nucleotide mRNA.

The SV40 DNA sequence near position 0.54 contains a stretch of 18 A·T base pairs (9). We were concerned that the S_1-sensitive site observed in the DNA at position 0.54 in RNA·DNA hybrid of the 2500-nucleotide mRNA might be due to cutting at this A·T(U)-rich region in a perfectly base-paired duplex, rather than the result of a splice in the 2500-nucleotide mRNA. Therefore, a control experiment was performed in which cRNA transcribed from SV40 DNA by *E. coli* RNA polymerase was hybridized to the *Hind* II-III A restriction fragment (0.43–0.66) and the products were digested with S_1 under the conditions used in this study. The full-length restriction fragment was protected from S_1 digestion—i.e., there was no cutting at position 0.54. Therefore, the S_1-sensitive site in the DNA of the 2500-nucleotide mRNA·SV40 DNA hybrid is due to a splice in this mRNA at map position 0.54.

DISCUSSION

Fig. 4 represents the deduced structure of the early SV40 mRNAs detected in this work. The structure of these mRNAs may provide an explanation for the seemingly paradoxical observations discussed in the introduction. The small T antigen may be translated from the 2500-nucleotide mRNA. The termination codons present in all three reading frames near position 0.54 (9) in this RNA would not be present in the 2200-nucleotide early mRNA which does not contain RNA sequences from 0.60 to 0.54. Therefore, if translation is initiated at the same AUG near the 5' termini of the 2500-nucleotide and 2200-nucleotide mRNAs, the translation products of both mRNAs would contain the same NH_2-terminal sequences. Translation of the 2500-nucleotide mRNA would be terminated near 0.54, giving rise to small T antigen. Translation of the 2200-nucleotide mRNA would continue beyond the deleted terminators to yield the large T antigen. If this were the case, then deletions in the genome between 0.60 and 0.54 would affect the translation product of the 2500-nucleotide mRNA, the small T antigen, but would have no effect on the translation product of the 2200-nucleotide mRNA, the large T antigen. As mentioned above, mutants with large deletions in the interval between 0.54 and 0.59 induce a wild-type large T antigen but either fail to produce a detectable small T antigen or produce an altered polypeptide with a molecular weight less than 17,000 (ref. 1; Sleigh *et al.*, personal communication). This model for the translation of the two observed spliced mRNAs is also consistent with the observation that the small T antigen mRNA sediments slightly faster than the large T antigen mRNA (3).

A similar model for the structure of the early SV40 mRNAs and their translation was proposed to explain the pattern of early protein synthesis induced by SV40 deletion mutants (1). Final confirmation of this model will require determination of the NH_2-terminal sequences of large and small T antigens and a comparison of these sequences with the DNA sequence in the region of 0.67 (25).

A number of observations suggest that the organization of genetic information in the early region of the mouse papova virus (polyoma virus) may be analogous to that found in SV40. Benjamin (26) selected host-range mutants of polyoma virus that are not able to grow on 3T3 cells but are able to grow on a line of 3T3 cells transformed by polyoma virus. The prototype of this group has a deletion in a region of the genome analogous to the SV40 deletions mentioned above (27). Like the SV40 mutants, they fail to induce detectable levels of small T antigens that are observed in cells infected with wild-type polyoma virus, but produce a wild-type large T antigen (28).

If translation occurs through a splice point in the 2200-nucleotide mRNA as suggested by the model presented above, then the process by which the two colinear transcripts are joined must be precise. Furthermore, the chemistry of joining at the splice point must be a 5′-3′ phosphodiester bond. It is noteworthy that deletions in the genome throughout the interval between 0.59 and 0.54 do not affect the production of large T antigen or, therefore, of its message. This suggests that these intervening sequences do not play a role in the process by which this spliced mRNA is produced. Therefore, the process of splicing may be specified by relatively short sequences in the genome in the immediate vicinity of the sequences joined at a splice point. Evidence has been obtained that strongly suggests that the spliced late Ad2 mRNAs are generated from long initial transcripts by removal of RNA sequence between splice points (29, 30). If this is also the case for the early SV40 mRNAs, then clearly the relative abundance of these two early mRNAs would be controlled by post-transcriptional splicing. This would allow coordinate control of expression of the two early mRNAs by regulation of transcription initiation at a single promoter and yet allow another mechanism to control the relative abundance of these two messages and, hence, the relative cellular concentrations of their translation products.

We thank C. Cole for helpful discussion and for the gift of *Taq* I, S. Goff for the gift of exonuclease VII of *E. coli*, and S. Goff and S. Weissman for communication of results prior to publication. P.A.S. gratefully acknowledges a grant (VC-151A) and a Faculty Research Award from the American Cancer Society and a Cancer Center Core Grant (CA-14051). A.J.B. gratefully acknowledges a Helen Hay Whitney Foundation Postdoctoral Fellowship (F-340).

1. Crawford, L. V., Cole, C. N., Smith, A. E., Paucha, E., Tegtmeyer, P., Rundell, K. & Berg, P. (1978) *Proc. Natl. Acad. Sci. USA* **75**, 117–121.
2. Prives, C., Gilboa, E., Revel, M. & Winocour, E. (1977) *Proc. Natl. Acad. Sci. USA* **74**, 457–461.
3. Paucha, E., Harvey, R., Smith, R. & Smith, A. E. (1977) *INSERM, Collog.* **49**, in press.
4. Rundell, K., Collins, J. K., Tegtmeyer, P., Ozer, H. L., Lai, C. J. & Nathans, D. (1977) *J. Virol.* **21**, 636–646.
5. Simons, D. T. & Martin, M. A. (1978) *Proc. Natl. Acad. Sci. USA* **75**, 1131–1135.
6. Khoury, G., Martin, M. A., Lee, T. N. H., Danna, K. J. & Nathans, D. (1973) *J. Virol.* **78**, 377–389.
7. Sambrook, J., Sugden, B., Keller, W & Sharp, P. A. (1973) *Proc. Natl. Acad. Sci. USA* **70**, 3711–3715.
8. Shenk, T. E., Carbon, J. & Berg, P. (1976) *J. Virol.* **18**, 664–671.
9. Thimmapaya, B. & Weissman, S. M. (1977) *Cell* **11**, 837–843.
10. Berk, A. J. & Sharp, P. A. (1977) *Cell* **12**, 721–732.
11. Berget, S. M., Moore, C. & Sharp, P. A. (1977) *Proc. Natl. Acad. Sci. USA* **74**, 3171–3175.
12. Chow, L. T., Gelinas, R. E., Broker, T. R. & Roberts, R. J. (1977) *Cell* **12**, 1–8.
13. Klessig, D. F. (1977) *Cell* **12**, 9–21.
14. Aloni, Y., Dhar, R., Laub, O., Horowitz, M. & Khoury, G. (1977) *Proc. Natl. Acad. Sci. USA* **74**, 3686–3690.
15. Hsu, M.-T. & Ford, J. (1977) *Proc. Natl. Acad. Sci. USA* **74**, 4982–4985.
16. Celma, M. L., Dhar, R. & Weissman, S. M. (1977) *Nucleic Acids Res.* **4**, 2549–2559.
17. Sambrook, J., Sharp, P. A. & Keller, W. (1972) *J. Mol. Biol.* **70**, 57–71.
18. Rigby, P. W., Dieckmann, M., Rhodes, C. & Berg, P. (1977) *J. Mol. Biol.* **113**, 237–251.
19. Maniatis, T., Jeffrey, A. & Kleid, D. G. (1975) *Proc. Natl. Acad. Sci. USA* **72**, 1184–1188.
20. Casey, J. & Davidson, N. (1977) *Nucleic Acids Res.* **4**, 1539–1552.
21. Vogt, V. M. (1973) *Eur. J. Biochem.* **33**, 192–200.
22. Chase, J. W. & Richardson, C. C. (1974) *J. Biol. Chem.* **249**, 4545–4552.
23. McDonell, M. W., Simon, M. N. & Studier, F. W. (1977) *J. Mol. Biol.* **110**, 119–146.
24. Chase, J. W. & Richardson, C. C. (1974) *J. Biol. Chem.* **249**, 4553–4561.
25. Subramanian, K. N., Reddy, B. V. & Weissman, S. M. (1977) *Cell* **10**, 497–507.
26. Benjamin, T. L. (1970) *Proc. Natl. Acad. Sci. USA* **67**, 394–399.
27. Feunteun, J., Sompayrac, L., Fluck, M. & Benjamin, T. (1976) *Proc. Natl. Acad. Sci. USA* **73**, 4169–4173.
28. Schaffhausen, B. S., Silver, J. E. & Benjamin, T. L. (1978) *Proc. Natl. Acad. Sci. USA*, **75**, 79–83.
29. Goldberg, S., Weber, J. & Darnell, J. E. (1977) *Cell* **10**, 617–621.
30. Berget, S. M., Berk, A. J., Harrison, T. & Sharp, P. A. (1977) *Cold Spring Harbor Symp. Quant. Biol.* **42**, in press.

Cell, Vol. 18, 439–449, October 1979, Copyright © 1979 by MIT

The Asymmetric Segregation of Parental Nucleosomes during Chromosome Replication

Michael M. Seidman* and Arnold J. Levine†
Princeton University
Department of Biochemical Sciences
Princeton, New Jersey 08544
Harold Weintraub
The Hutchinson Cancer Research Center
Seattle, Washington 98104

Summary

SV40 DNA replicated in the presence of cyclohexi-mide was more sensitive to staphylococcal nu-clease digestion and had a lower superhelical den-sity than viral DNA replicated in the absence of this drug. These data indicate that fewer nucleosomes are associated with progeny SV40 DNA molecules after DNA replication in the absence of protein synthesis and that these nucleosomes are derived from the parental histones. We designed an exper-iment to determine whether these parental SV40 nucleosomes segregate to the leading side of the replication fork where DNA synthesis is continuous, the lagging side of the fork where DNA synthesis is discontinuous or randomly to both sides of the fork. The results indicate that the parental histones dis-tributed themselves asymmetrically, preferentially (80–90%) segregating with the leading side of both SV40 DNA replication forks during bidirectional rep-lication in the absence of protein synthesis. In the case of SV40, the same parental DNA strands are the templates for the leading side of DNA replication at both forks as well as the templates for the infor-mational or coding strand of early and late viral mRNA synthesis. Based on this correspondence, we designed an experiment to test whether chicken cells growing in culture and replicating their DNA in the absence of protein synthesis segregated their parental histones asymmetrically to the progeny DNA strand that also coded for stable nuclear RNA transcripts. The results of these experiments indi-cate that, like SV40, parental cellular histones seg-regate asymmetrically and are preferentially asso-ciated with those DNA template strands that code for stable nuclear RNA species detected by hybrid-ization to single-copy DNA.

Introduction

The replication fork of a DNA molecule is asymmetric, consisting of a leading side where DNA synthesis is continuous and a lagging side where discontinuous replication occurs (Alberts and Sternglanz, 1977). The site of assembly of nucleosomes onto this newly

* Present address: Laboratory of Molecular Carcinogenesis, Bldg. 37, NIH, Bethesda, Maryland 20205.
† Present address: Department of Microbiology, State University of New York at Stony Brook, Stony Brook, New York 11794.

replicated DNA appears to occur at or near the repli-cation fork (Cremisi, Chestier and Yaniv, 1977; Wor-cel, Han and Wong, 1978). Two distinct types of experiments indicate that the assembly of nucleo-somes and the segregation of parental nucleosomes (derived from the regions ahead of the replication fork) onto the daughter DNA molecules occur by a nonran-dom process. Using density-labeled amino acids, Lef-fak, Grainger and Weintraub (1977) showed that the parental histone octamer (and multimers thereof) seg-regated from the unreplicated DNA to the progeny DNA molecules as an intact unit, free of newly synthe-sized histones. Similarly, the newly assembled histone octamers on the progeny DNA were essentially free of parental histone contributions (Leffak et al., 1977). Second, when DNA was replicated in the absence of protein synthesis, the parental nucleosomes segre-gated to only one of the two daughter molecules, leaving the other daughter strand free of nucleosomes (Weintraub, 1973; Seale, 1976; Weintraub, 1976; Riley and Weintraub, 1979).

The basic asymmetry of the replication fork and the nonrandom segregation of the parental nucleosomes onto the daughter DNA strands led Riley and Wein-traub (1979) to propose that the leading side of the replication fork is covered by the parental histones while the lagging side of the fork receives the newly synthesized nucleosomes. In this paper, we present direct biochemical evidence consistent with the hy-pothesis that the parental nucleosomes do indeed segregate to the leading side of the replication fork when protein synthesis is inhibited by cycloheximide. We chose the SV40 chromosome as a model system to study this question. It is known that SV40 DNA is associated with cellular histones in a nucleosome conformation (Griffith, 1975) and replicates bidirec-tionally from a unique origin (Fareed, Garon and Salz-man, 1972). With the exception of the viral A gene product T antigen, the viral replicative machinery is derived from the host cell, so it is probable that SV40 chromosome replication is very similar to the replica-tion of cellular chromatin. In the presence of cyclo-heximide, both cellular and viral DNA chain elongation continues, although viral and cellular initiation events appear to be blocked (Kang et al., 1971). Chromatin synthesized in the absence of protein synthesis is deficient in its histone content, as shown for viral DNA by a reduction in the superhelical density of SV40 DNA (Bourgeaux and Bourgeaux-Ramoisy, 1972; Jaenisch and Levine, 1973) and for cellular chromatin by an increased sensitivity to staphylococcal nuclease digestion (Seale, 1976; Weintraub, 1976). Since the free histone pool is thought to be very small, the DNA synthesized in the presence of cycloheximide appears to be associated only with the parental nucleosomes. Thus the use of this drug offers an opportunity to study the segregation of parental nucleosomes in the absence of newly made histones.

In this paper, we demonstrate that the nuclease-resistant SV40 DNA daughter strands replicated in the absence of protein synthesis hybridize predominantly to the template strand on the leading side of the replication fork. This interpretation of our results is consistent with the idea that the parental histones preferentially segregate to and protect the leading and not the lagging DNA strand from the nuclease digestion. In a more general sense, this asymmetric segregation of parental histones to one of the daughter chromosomes provides a vehicle for transmitting and segregating chromosomal information to daughter cells during development (Tsanev and Sendov, 1971; Alberts, Worcel and Weintraub, 1976; Weintraub et al., 1978; Newrock et al., 1979). A question of some significance, then, is the relationship between the template strands for the leading side of the replication fork and the information-coding strand for mRNA. In the case of SV40, it is known that the same parental DNA strand is the template for both viral mRNA synthesis and the leading side (continuous synthesis) of the replication fork during DNA replication. This observation has been extended from the specific case of SV40 to the more general case of eucaryotic cellular chromatin. The results of experiments to test this idea suggest that in cellular chromatin the parental nucleosomes also segregate to the template DNA strand that is transcribed into the stable nuclear RNA species detected by hybridization analysis.

Results

Experimental Protocol: the Fate of Parental Nucleosomes during DNA Replication

A DNA replication fork is composed of three arms: an unreplicated duplex which lies ahead of the fork, and two daughter DNA helices, one, termed the leading side, where synthesis is continuous, and the other, the lagging side, where synthesis is discontinuous. An analysis of replicating SV40 and other eucaryotic chromosomes (McKnight and Miller, 1977; Seidman, Garon and Salzman, 1978) indicated that nucleosomes were present on the unreplicated region and both of the daughter helices surrounding the replication fork. For the purposes of this discussion, it is assumed that the parental nucleosomes (those on the unreplicated DNA) are transferred to the daughter DNA as the fork passes through a nucleosome. Our basic experiment (Figure 1) was designed to answer the following question: after DNA replication has occurred in the presence of cycloheximide, are the parental SV40 nucleosomes associated with the daughter double helices on the leading DNA strand, the lagging DNA strand or both sides of the fork? This experiment was performed by labeling the daughter DNA strands of replicating SV40 chromosomes with ^{3}H–thymidine under conditions in which new histone synthesis was blocked by inhibiting protein synthesis

with cycloheximide. The ^{3}H–thymidine-labeled DNA of the daughter strands was then digested with staphylococcal nuclease to yield a population of DNA fragments which were resistant to nuclease by virtue of their association with recycled parental histones. These nuclease-resistant DNA fragments consisted of a population of ^{3}H–thymidine-labeled daughter strands and unlabeled parental strands. The experimental problem was then to identify the nature of the labeled daughter strands; that is, were they complementary to the template strands for the leading or the lagging side of the replication fork?

To accomplish this, the nuclease-resistant, ^{3}H–thymidine-labeled daughter DNA strands were characterized by hybridization to the separated single strands of SV40 DNA derived from restriction enzyme fragments located on either side of the origin of SV40 DNA replication. The hybridization protocol (Figure 1) detected the nuclease-resistant progeny DNA fragments derived from both replication forks of the SV40 chromosome (replication is bidirectional, proceeding from a single origin). Measuring the relative amounts of the nuclease-resistant daughter DNA strands derived from the leading and lagging side templates at each fork permitted the distinction between random and asymmetric modes of nucleosome segregation to the newly synthesized DNA (see Figure 1).

The specific restriction enzyme fragments used in this experiment were obtained by digesting purified SV40 DNA with Bam HI (which cleaves at 0.14 map units) and Hpa II (which cleaves at 0.735 map units). This double digest produced a large A fragment containing 59% of the genome and encoding the sequences for the viral early genes. A smaller B fragment was also produced, containing 41% of the genome and encoding the information for the late genes of this virus. The individual single strands of these DNA fragments were separated by alkaline pH denaturation followed by electrophoresis on agarose gels, then transferred to nitrocellulose paper using the procedure of Southern (1975). By using labeled viral mRNA as a hybridization probe, it has been possible to identify the individual DNA single strands which are templates for mRNA (Birkenmeier, May and Salzman, 1977; see Figure 1). Since the direction of mRNA synthesis is the same as the direction of DNA synthesis (5' to 3' and away from the origin site on both sides of the genome), the mRNA coding strands are also the template DNA strands for the leading side of each replication fork.

Staphylococcal Nuclease Digestion of DNA

To carry out this experiment, BSC-1 cells were infected with SV40, and at 36–40 hr after infection these cell cultures were pulse-labeled with ^{3}H–thymidine in the presence or absence of cycloheximide, as described in Experimental Procedures. The nuclei from these cells were prepared and digested with

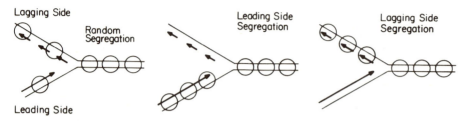

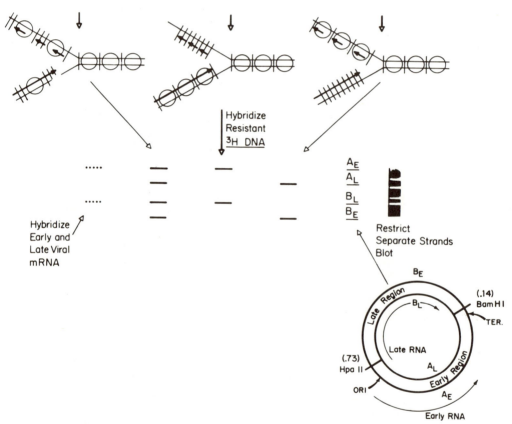

Figure 1. Fate of Parental Nucleosomes during DNA Replication: Design of the Experiment

The structures of the three possible modes of segregation of nucleosomes after replication in cycloheximide and the consequences of digestion by staphylococcal nuclease on these three modes of nucleosome segregation are presented. The leading side of the replication fork is indicated by the unbroken line and arrow. The lagging or discontinuous side is represented by the dashed line and arrows. No special relationship is implied between the small fragments of DNA on the lagging side and the nucleosomes. The positions of the origin and termination of replication and the HpA II and Bam HI sites are given, as well as the position and orientation of early and late viral messages. The pattern of the separated single strands of the fragments from the double digest of SV40 DNA is shown schematically and also in the ethidium-stained gel to the right of the schematic. The A_E strand is the template for the leading side of the replication fork on the early side of the origin. A_L is the template for the lagging side of the fork on the early side. B_L is the template for the leading side of the replication fork on the late side of the origin, while B_E is the template for the lagging side of the fork on the late side. A_E is also the coding strand for viral early mRNA, and B_L is the coding strand for late mRNA. The predicted hybridization results for each of the possible fork configurations and the hybridization of early and late mRNA are also presented.

staphylococcal nuclease at 37°C. The kinetics of digestion of the labeled DNA synthesized in the presence or absence of cycloheximide are presented in

Figure 2A. Approximately 50% of the labeled DNA in nuclei derived from cells not treated with cycloheximide was digested to acid-soluble products. The la-

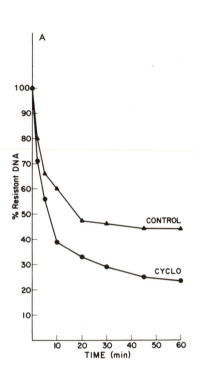

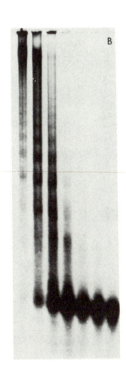

Figure 2. Staphylococcal Nuclease Digestion of Nuclei from Infected Cells Labeled with ³H–Thymidine in the Presence of Cycloheximide

(A) Kinetics of digestion. At various times during the digestion of nuclei from SV40-infected cells, aliquots from the control or cycloheximide-treated cell cultures were removed and the percentage of TCA-resistant ³H–labeled DNA was determined.

(B) Agarose gel pattern of nuclease-resistant DNA fragments. After 2, 5, 10, 20, 30, 45 and 60 min of nuclease digestion of infected cell nuclei, aliquots were removed and the DNA fragments were resolved by electrophoresis on a 2.5% agarose gel. This pattern is from a cell culture treated with cycloheximide. The pattern of digestion of the control sample was similar, although the production of monomer fragments was not as rapid. The gels were fluorographed.

beled DNA from the drug-treated cells was more sensitive to this nuclease and reached a resistant plateau of digestion at 25%. Figure 2B presents the agarose gel electrophoresis pattern of the labeled DNA, resistant to staphylococcal nuclease and derived from the cycloheximide-treated cells. At early times in the digestion, the familiar "ladder" pattern was observed, while at completion of the digestion, monomer nucleosome-sized fragments were produced. Although not shown, the agarose gel electrophoresis pattern of DNA from the digestion of nuclei derived from cells not treated with cycloheximide was similar to that observed in Figure 2B. It is estimated that about 80% of the labeled DNA in the nucleus and applied to this gel was viral DNA (found in a 1M NaCl-SDS supernatant fraction; Hirt, 1967).

The results presented in Figure 2 are similar to those reported by Seale (1976) and Weintraub (1976) for cellular chromatin. The increased nuclease sensitivity of the DNA labeled in the presence of cycloheximide was consistent with the idea that this DNA was not protected from nuclease digestion by histones to the same extent as was DNA in chromosomes synthesized in the absence of this drug. The appearance of the staphylococcal-resistant DNA in monomer nucleosome-sized fragments suggested that this DNA existed in a nucleosomal configuration.

Superhelix Density Analysis

Another quantitative measure of histone DNA interaction is provided by the superhelix density of covalently closed circular DNA (Germond et al., 1975). In this experiment, parallel infected cultures of BSC-1 cells were labeled for 15 min with ³H–thymidine in the presence or absence of cycloheximide. Then both sets of cultures were extracted by the method of Hirt (1967) to selectively obtain the viral DNA. After deproteinization, the DNA was analyzed by electrophoresis on a 1.5% agarose gel. The results of this experiment are presented in Figure 3. (A) Contains a preparation of SV40 DNA isolated from virions which consist of fully superhelical Form I DNA and a smaller amount of the relaxed Form II DNA. (B) Contains the sample from the control cells not treated with the drug. Because of the relatively slow replication time of SV40 DNA (Levine, et al., 1970), a substantial amount of the viral DNA labeled with ³H–thymidine for 15 min was still in the replicative form. In this gel system the replicative intermediates migrated slightly more quickly than Form II DNA. The molecules that had completed replication during the labeling period migrated in the gel with the mobility of the fully superhelical Form I DNA marker. There is also a small amount of Form II and Form III DNA (the linear form which migrates between Form I and Form II) observed in these preparations.

491

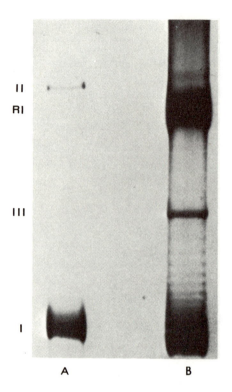

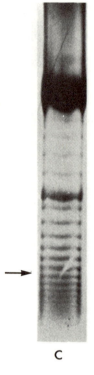

Figure 3. Superhelix Density Analysis of Viral DNA

Cells treated or untreated with cycloheximide were incubated with ^{3}H–thymidine for 15 min. DNA extracts were prepared by the method of Hirt (1967) and the DNA was deproteinized, phenol-extracted and ethanol-precipitated. The samples were analyzed on a 1.5% agarose gel. (A) Marker DNA isolated from virions. Fully superhelical Form I DNA and relaxed Form II DNA are in this sample. (B) Viral DNA from infected cells. (C) Viral DNA from the cells treated with cycloheximide. The positions of Form I, Form II, Form III (linears) and replicating molecules (RI) are shown. The arrow indicates the center of the mass distribution of the bands determined by densitometer analysis.

Lane C contains the analysis of the viral DNA labeled in the presence of cycloheximide. Cycloheximide treatment slowed the conversion of replicating molecules to Form I DNA (Jaenisch and Levine, 1973), and the majority of the labeled molecules were still in the replicating form. Those that were converted to the closed circular form during the labeling period, however, showed a unimodal distribution of supercoiled molecules, with a reduction in superhelix density of about 50% relative to the control sample. The decrease in superhelix density results from the failure to synthesize new histones in the presence of cycloheximide. The retention, on average, of half the number of supercoils presumably reflects the recycling of parental histones onto newly replicated viral DNA. These data are in agreement with previous reports (Yu and Cheevers, 1976) and are consistent with the conclusions of the staphylococcal nuclease digestion (Figure 2).

Hybridization Analysis of SV40 DNA Resistant to Nuclease Digestion

The above experiments demonstrate that SV40 DNA replicated in the absence of protein synthesis contains a lower histone content and that a smaller proportion of the DNA is protected from nuclease attack. To analyze the origin of the viral DNA synthesized in the presence of cycloheximide and resistant to nuclease digestion, we performed the following experiment. The staphylococcal nuclease-resistant DNA was obtained from nuclei of SV40-infected cells labeled with ^{3}H–thymidine in the presence of cycloheximide (see Figure 2 and Experimental Procedures). This DNA was purified, denatured and then allowed to reanneal in the presence of a nitrocellulose strip containing the separated strands of the A and B fragments (from a Bam HI, Hpa II digest of SV40 DNA, Figure 1). The fluorograph of this nitrocellulose strip is presented in figure 4A. It is clear from these results that the hybridization of the ^{3}H–thymidine-labeled, nuclease-resistant fragments to the separated DNA strands of both the A and B fragments was distinctly asymmetric. Densitometer tracings of this autoradiogram (Figure 4) showed that 82% of the labeled viral DNA that was resistant to nuclease treatment hybridized to the A fragment single strand that was a template for the leading side of the replication fork. Approximately 95% of the hybridization to the B fragment single strand (from the other replication fork) was also to the template for the leading side synthesis of DNA. This experiment has been carried out several times and in all cases 80–90% of the labeled viral DNA which was synthesized in the presence of cycloheximide and resistant to nuclease digestion hybridized to the tem-

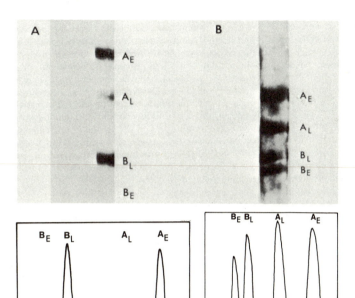

Figure 4. Hybridization of Nuclease-Resistant DNA (A) and DNA Not Treated with Nuclease (B)

(A) The staphylococcal nuclease-resistant DNA, labeled in cycloheximide, was deproteinized, denatured and reannealed in the presence of a nitrocellulose strip onto which separated strands of the A and B fragments from the BAM HI-HpA II digest had been transferred. The A_E strand is the template for the leading side of the replication fork in the early coding region of the genome; it is also the coding strand for early message. The B_L strand is the template for the leading side in the late coding region and is also the coding strand for late message. After hybridization the strip was fluorographed and exposed to X-ray film.

(B) Infected cells were labeled with ^{3}H–thymidine in the presence of cycloheximide exactly as in (A). Nuclei were prepared and the DNA was purified directly, sheared and hybridized as in (A). A different preparation of separated strands was used in this experiment.

plate strand for the leading side of the replication fork.

Several control experiments were performed. SV40-infected cells were treated with cycloheximide and labeled with ^{3}H–thymidine as before (see Experimental Procedures). Analysis of the labeled viral DNA on alkaline sucrose gradients showed that the daughter DNA strands were of nearly genome length in molecules that had not finished replication. This suggested that normal chain elongation occurs during the cycloheximide treatment. Nuclei were prepared from such cells and the DNA was purified, sonicated to a size of about 0.5 kb (there was no nuclease digestion) and hybridized to the separated single strands of the A and B fragments on the nitrocellulose strips. In this case (Figure 4B), the hybridization of the labeled viral DNA to the separated single strands of the A or B fragments was equal, indicating that the nuclease digestion was essential for the asymmetric hybridization result. Nuclei from SV40-infected cells labeled with ^{3}H–thymidine in the absence of cycloheximide were obtained and digested with staphylococcal nuclease. The nuclease-resistant DNA was purified and hybridized to the separated DNA strands. Hybridization of the labeled DNA to the separated strands of each fragment was again equivalent, indicating that only DNA replication in the absence of protein synthesis results in a nuclease protection of the leading side of the replication fork. These experiments also demonstrate that the asymmetry in the hybridization re-

sults obtained in Figure 4a is not the result of the cycloheximide treatment or the staphylococcal nuclease digestion per se. It is clear that DNA is synthesized on both the leading and lagging sides of the replication forks in the presence of cycloheximide, but that the DNA on the leading side of the fork acquires a preferential protection from nuclease digestion when it is made in the absence of protein synthesis. These data, taken together with the results presented in Figures 2 and 3, strongly suggest that the protection of the leading-side, newly synthesized DNA is derived from histones—necessarily parental histones.

Relationship between Parental Histone Segregation and the Template DNA Strands for Transcription of Cellular mRNA

In the case of SV40, the template strands for the leading side of DNA synthesis at both replication forks also serve as the template strands for early and late viral mRNA. Consequently, the daughter DNA strands, labeled in the presence of cycloheximide and resistant to nuclease digestion, have the same polarity or sense as the viral mRNA. One would therefore expect no hybridization between the nuclease-protected, leading-strand progeny DNA at the fork and viral mRNAs. This reasoning suggested an experimental approach for studying the relationship between templates for leading strand synthesis and the templates for mRNA or stable nuclear RNA in cellular chromatin. To test the possibility that cellular DNA sequences were organized with the parental histones segregating to the DNA template strand that codes for stable nuclear RNA, we performed the following experiment. Growing chick MSB cells were incubated with ^{3}H–thymidine and cycloheximide for 30 min. Under these conditions the parental histone octamers segregate to only one of the two daughter DNA double helices, leaving the other DNA strand relatively free of histones (Riley and Weintraub, 1979). The nucleosome-free DNA side of the replication fork was then preferentially digested with staphylococcal nuclease and the labeled resistant DNA was purified. This DNA had a single-stranded length of 140–160 bases and hybridized normally to total chicken cell DNA (Figure 5b). In contrast, the nuclease-resistant DNA fraction (purified to obtain single-copy DNA sequences; see Experimental Procedures) did not hybridize with nuclear RNA.

The failure of this ^{3}H–thymidine-labeled progeny DNA to hybridize with nuclear RNA was not due to a failure of these cells to replicate complementary-strand DNA in the presence of cycloheximide. Single-copy DNA obtained from cells labeled with ^{3}H–thymidine in the presence of cycloheximide, but not digested with staphylococcal nuclease, hybridized to nuclear RNA to the same extent (11–12%) as total single-copy DNA from these cells (Figure 5a). In another control experiment, cells were labeled with ^{3}H–thymidine for two generations and then treated with cycloheximide for 30 min in the absence of labeled thymidine. The nuclei were isolated from these cells and the chromatin was digested to a limit (monomers) with staphylococcal nuclease. Purified single-copy DNA from this preparation hybridized normally to both DNA and nuclear RNA (data not shown). Thus treatment with cycloheximide does not appear to sensitize DNA transcribing stable nuclear RNA to the staphylococcal nuclease treatment. In a third control experiment, cells were treated with cycloheximide and labeled with ^{3}H–thymidine for a 30 min period. At that time, the cycloheximide block and labeling of the DNA

was reversed by removing the cells from the inhibitor and ^{3}H–thymidine. Within 15 min after the removal of the protein synthesis inhibitor, the labeled progeny strand DNA regained its normal sensitivity to nuclease (50% digestion limit), and when single-copy DNA was purified from this preparation it hybridized normally to nuclear RNA. As a final control for the hybridization reaction, the nature of the nucleic acid hybrids were verified to be true DNA-RNA hybrids. Prior treatment of the nuclear RNA preparation with 0.3 N NaOH (overnight at 37°C) or RNAase A (50 μg/ml; 30 min at 37°C in 2 × SSC) completely destroyed its ability to form hybrids with the labeled DNA. In addition, treatment of the hybrids formed between labeled DNA and unlabeled RNA, with RNAase A at low ionic strength (0.01 × SSC), destroyed the hybrids, as assayed by subsequent S1 nuclease treatment. These experiments suggest that the nuclease-resistant daughter DNA strand made in the presence of cycloheximide is of the same polarity as hnRNA and therefore does not hybridize with the hnRNA.

Preferential Release of DNA by Hae III Digestion of Nuclei

Half of the cellular DNA synthesized in the presence of cycloheximide appears to be protected from staphylococcal nuclease digestion, presumably because parental nucleosomes asymmetrically protect one of the progeny strands at the replication fork (in the case of SV40, the leading side of the fork). Furthermore, the daughter strand protected from nuclease digestion by parental nucleosomes is synthesized in the same polarity as is stable nuclear RNA in chick MSB cells or mRNA of SV40. As a further test of these ideas, an experiment was designed to preferentially isolate the newly synthesized DNA from the side of the replication fork not protected by the parental histones. Growing MSB cells were labeled with ^{3}H–thymidine for 30 min in the presence of cycloheximide. The nuclei from these cells were then treated with the restriction endonuclease Hae III, which should preferentially digest

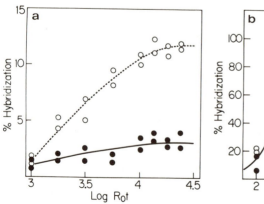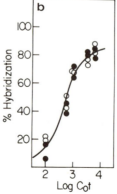

Figure 5. Preferential Digestion of Labeled DNA Sequences Complementary to hnRNA after Replication in the Presence of Cycloheximide

MSB cells were labeled with ^{3}H–TdR (100 μCi/ml) for 30 min in the presence of cycloheximide. Nuclei were isolated and digested to 75% acid solubility with staphylococcal nuclease. The resistant single-copy DNA was hybridized to nuclear RNA (a) or total chicken DNA (b). (●) DNA from cycloheximide-treated, nuclease-treated cells; (○) DNA from cycloheximide-treated cells (not nuclease-treated) sheared to about 200 bp. The specific activity of the labeled DNA was estimated at about 2 × 10[5] cpm/μg. Reactions were in a total volume of 10 μl with 10[5] cpm of DNA and 18 mg/ml of RNA.

and release from the nuclei labeled DNA not protected by parental histones. When this was done, 36% (an average of five determinations) of the labeled DNA synthesized in the presence of cycloheximide was released from the nuclei. In contrast, less than 1% of the labeled DNA synthesized in the absence of cycloheximide was released from nuclei by Hae III digestion. The cellular DNA released from nuclei by Hae III digestion and the remaining DNA in the nuclear pellet were then separately purified and redigested with the Hae III restriction endonuclease. These two DNA preparations were then hybridized to the stable (unlabeled) nuclear RNA from MSB cells. The results of this experiment, presented in Figure 6, demonstrate that the Hae III-sensitive (released from nuclei by Hae III digestion), single-copy DNA was enriched in DNA sequences complementary to nuclear RNA. The labeled DNA that remains in the nuclear pellet after Hae III digestion is depleted in sequences complementary to nuclear RNA. As a control, the small amount of labeled DNA released from control cells showed only a slight increase (to 12%) in hybridization to nuclear RNA.

Two types of control experiments strongly indicate that the labeled cellular DNA replicated in the absence of protein synthesis and released by the Hae III restriction endonuclease from the nuclei was derived from the side of the replication fork that did not receive parental proteins. First, the native chromosomal DNA released by Hae III digestion of nuclei derived from cells treated with cycloheximide migrated on polyacrylamide gels with an average mobility identical to that of the same preparation treated with SDS and pronase (naked DNA). Thus the Hae III digestion products released from nuclei obtained from cells synthesizing DNA in the absence of protein synthesis had an electrophoretic mobility characteristic of DNA and not of nucleoprotein (which migrates much more slowly in these gels). Second, while the total labeled DNA from nuclei prepared from cells incubated in the presence of cycloheximide is digested to a 25% limit by staphylococcal nuclease, the chromatin released by Hae III digestion of these nuclei is digested to more than 97% by the staphylococcal nuclease. These results strongly suggest that the Hae III endonuclease acts preferentially on the DNA that does not receive the parental proteins (histones).

Discussion

The results of these experiments with the SV40 chromosomes indicate that after DNA replication in the absence of protein synthesis, the viral DNA on the leading side of the replication fork is preferentially resistant to staphylococcal nuclease digestion, while the DNA on the lagging side of the fork is more sensitive. Both the products of the staphylococcal nuclease digest (Figure 2) and the lowered superhelical density of the progeny SV40 Form I DNA (Figure 3) argue that the protection of SV40 DNA replicated in the absence of protein synthesis is provided by the parental histones. It appears that during DNA replication, preexisting histones preferentially segregate and form nucleosomes on the leading but not on the lagging side of the replication fork. Unless this asymmetric segregation of histones is induced by the treatment with cycloheximide, one would expect that newly synthesized histones would usually associate with the daughter DNA on the lagging side of the replication fork. The lagging side of the SV40 replication fork is synthesized discontinuously by the addition of 4S fragments which average 200–230 bases in size (Kaufman, Anderson and DePamphlis, 1977). It may not be fortuitous that this is approximately the size of a nucleosomal domain.

The conclusion that parental nucleosomes segregate to the leading side of a replication fork rests upon the results of the hybridization experiment presented in Figure 4. At the time of the pulse labeling with ^{3}H–thymidine, about 1% or less of the viral DNA was replicating, so the vast majority of the viral DNA was not labeled during the incubation with ^{3}H–thymidine and cycloheximide. The nuclease digestion then yields a population of unlabeled DNA fragments hav-

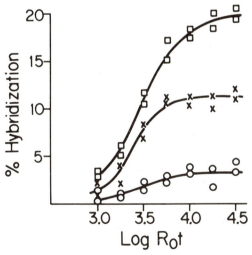

Figure 6. Hae III Nuclease Yields a Preferential Release from Nuclei of Actively Transcribed DNA Sequences Synthesized in the Presence of Cycloheximide

MSB cells were labeled with ^{3}H–TdR (100 μCi/ml) for 30 min in the presence of cycloheximide. Nuclei (at 1 mg/ml of DNA) were isolated and digested at 37°C with Hae III (1000 U/ml) in a volume of 50 μl for 30 min. Nuclei were pelleted and the DNA in the supernatant (□) and the DNA remaining in the pellet (○) was purified and hybridized to an excess of hnRNA (18 mg/ml), along with a sample of total labeled chicken DNA (×). The specific activity of the released DNA was estimated as 10^6 cpm/μg. The specific activity of the other two preparations was about 2×10^5 cpm/μg. Reactions were in a total volume of 10 μl with 10^5 cpm of input DNA. Single-copy DNA was used for each reaction.

ing equal amounts of leading and lagging side DNA strands from all regions of the viral genome in addition to the labeled DNA fragments. Since the nitrocellulose strips contain less than 0.5 μg of viral DNA and the hybridization solution contains about 10 μg of viral DNA (at 1–10 μg/ml), it is clear that the hybridization reaction is being driven by the unlabeled DNA in solution. The $Cot_{1/2}$ for SV40 DNA is 2×10^{-3} moles per liter per sec. By making the appropriate corrections for the ionic strength and fragment sizes of the DNA, it can be calculated that the $t_{1/2}$ for the reannealing of the viral DNA in this reaction was between 1–10 min, which is far less than the 8–12 hr used in these experiments. It is clear, therefore, that the hybridization reaction has gone to completion and that the asymmetry in the hybridization of labeled DNA in solution to the separated DNA strands on the nitrocellulose paper (which occurs at a slower rate than in solution) truly reflects the unequal levels of labeled leading- and lagging-side daughter strands after nuclease digestion.

Although there is substantial bias in the hybridization results in all these experiments, there is always some hybridization to the lagging-side template strand. Control experiments ruled out the possibility that this was due to hybridization of labeled cellular DNA or SV40 cross-strand hybridization due to mismatched hybrids. Some of the hybridization to the lagging strand of the A fragment is probably due to the overlap of this fragment into the "late" side of the genome, at both the origin and the termination regions. It seems probable, however, that there were also some stable nucleosomes formed on the lagging side of the fork. While it will be of considerable interest to determine whether this protection on the lagging side occurs at specific DNA sequences, there are also several "trivial" explanations for the finding. Thus hybridization to the lagging side could have been the result of residual histone synthesis (the protein synthesis inhibition was not better than 95%), a small histone pool or the variability inherent in a complicated biological system. Another possibility is that histones migrate from the cellular chromatin or slide along the viral chromosome, as suggested by Cremisi et al. (1977) and Beard (1978). The pronounced asymmetry of the hybridization indicates that if there is inter- or intrastrand nucleosome migration in vitro or in vivo, it is not extensive. This conclusion was supported by an experiment in which the 15 min ^{3}H–thymidine pulse labeling in the presence of cycloheximide was followed by a 30 min chase in unlabeled medium containing cycloheximide. The nuclease digestion pattern (Figure 2) and the DNA hybridization results (Figure 4) were identical to those obtained for the pulse label alone in cycloheximide.

While these results suggest that parental histones are associated with the DNA on the leading side of the replication fork, they offer no explanation for this phenomenon. It is conceivable that the histones leave the DNA as the fork passes through and then simply associate with the side of the replication fork that is the first to become double-stranded. In previous work it was found that histones from parental nucleosomes did not mix with the newly synthesized histones and that neighboring octamers remained adjacent to each other throughtout DNA replication (Leffak et al., 1977). Since histone octamers are not stable at physiological ionic strength, it seems probable that parental histones do not leave the DNA during replication.

The conservative segregation of nucleosomes during replication permits the introduction of asymmetry into the replicative process at the level of the chromosome. One daughter gene sequence will retain the nucleosome "pattern" of the parent while the other, associated with newly synthesized (and possibly modified) chromosomal proteins (Newrock et al., 1977), may be responsive to different regulatory signals (Tsanev and Sendov, 1971; Alberts et al., 1976; Weintraub, et al., 1978). Thus an important feature of gene organization may be the relationship between nuclear RNA coding strands and leading side template strands in eucaryotic replication. If the arrangement of genes and origins in cellular chromosomes follows the SV40 example, then the labeled cellular DNA strands synthesized in the presence of cycloheximide and resistant to staphylococcal nuclease should not hybridize with the stable cellular nuclear transcripts in hnRNA (steady-state nuclear RNA).

This experimental approach has been applied to the more general case of eucaryotic cells growing in cell culture. When cellular DNA is replicated in the absence of protein synthesis, the nucleosomes (parental) are associated with only one side of the replication fork (Riley and Weintraub, 1979). The side of the replication fork that is not protected from staphylococcal nuclease digestion contains the newly synthesized DNA complementary to the stable (steady-state) nuclear RNA detected by the DNA-RNA hybridization experiments. The parental proteins (nucleosomes) associated with cellular DNA sequences transcribing stable nuclear RNA therefore fail to segregate (protect from staphylococcal nuclease) with the nascent DNA strand complementary to this hnRNA. It is important to point out that the nuclear RNA used here is a probe for only one of the two nascent DNA strands synthesized in the presence of cycloheximide, and that these experiments therefore do not directly assay (by hybridization) the newly synthesized DNA strand that has the same polarity as nuclear RNA. Because of this fact, it remains a formal possiblity that for DNA sequences being actively transcribed into stable nuclear RNA, both sides of the replication fork were devoid of nucleosomes—that is, were not protected from staphylococcal nuclease—although this was not detected by previous electron microscopic studies (Riley and Weintraub, 1979).

If the preferential segregation of parental proteins (nucleosomes) to the leading side of a replication fork observed with SV40 can be generalized to the cellular chromatin results, then an interesting chromosome organization emerges. Since replication in both SV40 and cellular DNA is bidirectional, both the origin of DNA replication and the active promoter for transcription of stable nuclear transcripts must be located at the 5' side of any transcriptional unit that produces stable nuclear RNA. Consequently, replication origins associated with transcriptional units would not be chosen at random and transcription units would be wholly contained between replication origins. Clearly, additional experiments using defined genes as probes are required to confirm this hypothesis.

The proposed relationship between DNA coding sequences and the origins of DNA replication may have an important role in the regulation and expression of genes. Any rearrangements of gene coding sequences relative to the position of an origin of DNA replication by inversion or transposition or the utilization of new origins of DNA replication (Levine, 1972) would alter the sense of a coding strand of DNA with respect to its function as a leading or lagging side of a template DNA during replication. The possible significance of these types of events has been discussed above and by other investigators (Tsanev and Sendov, 1971; Levine, 1972; Alberts et al., 1976; Newrock et al., 1977; Weintraub et al., 1978).

Experimental Procedures

Cells and Virus

A continuous cell line of African Green Monkey cells, BSC-1, which is permissive for SV40 virus infection, was used in these experiments. These cells were infected with the SV40 temperature-sensitive mutant virus tsB11 (Tegtmeyer et al., 1974) and then maintained at the restrictive temperature (39.5°C). It has recently been shown that encapsidation of SV40 virus is very efficient in vivo (Garber, Seidman and Levine, 1978). To avoid problems with complete or partial encapsidation of viral DNA labeled during cycloheximide treatment (M. M. Seidman, unpublished observations), the ts B11 mutant was used. At the restrictive temperature this mutant produces a thermolabile VP-1 protein, the major capsid protein, and encapsidation does not occur. Viral DNA synthesis, the formation of progeny Form I DNA and viral chromatin were unaffected by this mutation (Tegtmeyer, 1974; M. M. Seidman, E. A. Garber and A. V. Levine, manuscript in preparation).

Cycloheximide Treatment and Pulse Labeling

At 36–40 hr post-infection, the infected cells were incubated at 0°C in medium containing 50 μg/ml cycloheximide and 100 μCi/ml ^{3}H–thymidine (40–60 Ci/mmole; NEN) for 5 min. This procedure had the advantage of permitting both cycloheximide and ^{3}H-thymidine to equilibrate into the cells. After the 0°C incubation, the medium was replaced by medium of the same composition at 39.5°C and the cells were incubated at 39.5°C for 15 min. Protein synthesis in cells treated in this fashion was reduced to a few percent of control values, while viral DNA synthesis was normal on both sides of the replication fork (Perlman and Huberman, 1977), although the conversion of late-replicating molecules to Form I DNA was reduced, as expected (Jaenisch and Levine, 1973). The 15 min labeling period was chosen so as to minimize the possibility of reinitiating DNA synthesis in the molecules labeled during the cycloheximide treatment. Control cells

(no cycloheximide) were labeled at 39.5°C with ^{3}H–thymidine for 1–4 hr to insure complete maturation of viral chromatin.

Staphylococcal Nuclease Digestion

After labeling, the cells were washed with phosphate-buffered saline and then with 100 mM KCl in 10 mM PIPES buffer (pH 6.8) and then scraped from the plates and homogenized in the buffer. Nuclei were collected by centrifugation, washed once in PIPES-KCl and then suspended at a final concentration of 20 OD260 u/ml. The nuclear suspension was adjusted to 0.5 mM CaCl$_2$ and staphylococcal nuclease was added to a final concentration of 20 μg/ml. The nuclei were incubated at 37°C and aliquots were withdrawn at various times for TCA precipitation or gel analysis.

Agarose Gel Electrophoresis

At appropriate times during digestion, aliquots of the nuclear suspension were removed and adjusted to 0.1% SDS, 5 mM EDTA and 100 μg/ml Proteinase K (Boehringer-Mannheim). After incubation for several hours, the samples were adjusted to 10% sucrose, 0.1% bromphenol blue and electrophoresed in a 2.5% agarose gel in 40 mM Tris-HCl, 20 mM sodium acetate, 1 mM EDTA (pH 7.5). The gel was fluorographed as described by Laskey and Mills (1975), with ethanol used as the solvent. The superhelix density analysis of samples from the Hirt extracts (Hirt, 1967) was performed by electrophoresis in 1.5% agarose gels.

Preparation of Filters and Hybridization

Purified SV40 Form I DNA was digested with BAM HI and HpA II. The strands were separated as described by Birkenmeier, May and Salzman, 1977). The separated strands were transferred to nitrocellulose by the method of Southern (1975) and used for hybridization. The samples of digested nuclei were incubated with 200 μg/ml Proteinase K, 0.1% SDS and 5 mM EDTA for 12–16 hr and then phenol-extracted and ethanol-precipitated. The precipitated DNA was dissolved in 6 × SSC and used for hybridization as described by Botchan, Topp and Sambrook (1976). After the washing steps, the nitrocellulose strips were dried and soaked briefly in 2% PPO in methylnaphthylene. The strips were dried and autoradiographed with X-ray film.

Hybridization of DNA to hnRNA

hnRNA was isolated from MSB cells (Riley and Weintraub, 1979) as described previously (Weintraub and Groudine, 1976). Cells were incubated in suspension in RPMI medium (Grand Island Biological) supplemented with 10% fetal calf serum and 1% Pennstrep (Grand Island Biological) and labeled with 100 μCi/ml of ^{3}H-thymidine (50 Ci/mM; ICN), usually in the presence of 100 μM cycloheximide (Sigma). Nuclei were isolated and digested in RSB [0.01 M NaCl, 0.01 M Tris-HCl (pH 7.4), 5 mM MgCl$_2$] with either staphlococcal nuclease (5 μg/ml) or DNAase I (10 μg/ml), and the DNA was purified as described by Weintraub and Groudine (1976). Hae III (BRL) digestions of nuclei were in 20 mM Tris (pH 7.4), 60 mM NaCl, 7 mM MgCl$_2$, 100 μg/ml BSA and 1 mM β–mercaptoethanol. Hybridizations were performed in 0.4 M NaCl, 50 mM Tris-HCl, (pH 7.4) and 0.1% SDS at 68°C. Reactions were monitored by resistance to nuclease S1 (BRL) as described by Weintraub and Groudine (1976). Single-copy DNA was purified by allowing the labeled DNA to reassociate to a Cot of 100 and purifying the unhybridized DNA (usually about 80–85% of the input) by passage over hydroxyapatite columns (BioRad) at 68°C in 0.12 M Na phosphate (pH 6.8) and 0.1% SDS. The unhybridized DNA was dialyzed extensively against 0.01 M Tris-HCl (pH 7.4), 0.1 M NaCl and then ethanol-precipitated and hybridized to hnRNA. DNA size determination was assayed by acrylamide gel (8%) electrophoresis in 8 M urea, 0.04 M Tris-HCl (pH 7.8), 2 mM EDTA, 0.02 mM NaAcetate. Samples were denatured at 100°C for 10 min before loading.

Acknowledgments

We thank Dr. I. M. Leffak, who was involved in the early phase of this work, and Dr. J. Stalder for advice and helpful discussions. The technical assistance of A. K. Teresky, C. McIver and K. Godfrey is acknowledged. This research was supported by two grants from the NIH. M.M.S. was a postdoctoral fellow on an NIH training grant.

The costs of publication of this article were defrayed in part by the payment of page charges. This article must therefore be hereby marked *"advertisement"* in accordance with 18 U.S.C. Section 1734 solely to indicate this fact.

Received June 25, 1979

References

Alberts, B. and Sternglanz, R. (1977). Recent excitement in the DNA replication problem. Nature *269*, 655–662.

Alberts, B., Worcel, A. and Weintraub, H. (1976). On the biological implications of chromatin structure. In The Organization and Expression of the Eukaryotic Genome, E. M. Bradbury and K. Javaherian, eds. (New York: Academic Press), pp. 165–19.

Beard, P. (1978). Mobility of histones on the chromosome of Simian Virus 40. Cell *15*, 955–967.

Birkenmeier, E. H., May, E. and Salzman, N. P. (1977). Characterization of SV40 tsA58 transcriptional intermediates at restrictive temperatures: relationship between DNA replication and transcription. J. Virol. *22*, 702–710.

Botchan, M., Topp, W. and Sambrook, J. (1976). The arrangement of Simian Virus 40 sequences in the DNA of transformed cells. Cell *9*, 269–287.

Bourgeaux, P. and Bourgeaux-Ramoisy, D. (1972). Is a specific protein responsible for the supercoiling of polyoma DNA? Nature *235*, 105–107.

Cremisi, C., Chestier, A. and Yaniv, M. (1977). Assembly of SV40 and polyoma mini-chromosomes during replication. Cold Spring Harbor Symp. Quant. Biol. *42*, 409–416.

Fareed, G. C., Garon, C. F. and Salzman, N. P. (1972). Origin and direction of Simian Virus 40 deoxyribonucleic acid replication. J. Virol. *10*, 484–491.

Garber, E. A., Seidman, M. M. and Levine, A. J. (1978). The detection and characterization of multiple forms of SV40 nucleoprotein complexes. Virology *90*, 305–316.

Germond, J. E., Hirt, B., Oudet, P., Gross-Bellard, M. and Chambon, P. (1975). Folding of the DNA double helix in chromatin like structures from Simian Virus 40. Proc. Nat. Acad. Sci. USA *72*, 1843–1847.

Griffith, J. D. (1975). Selective extraction of polyoma DNA from infected mouse cell cultures. Science *187*, 1202–1203.

Hirt, B. (1967). Selective extraction of polyoma DNA from infected mouse cell cultures. J. Mol. Biol. *26*, 365–369.

Jaenisch, R. and Levine, A. J. (1973). DNA replication in SV40 infected cells VII. Formation of SV40 catenated and circular dimers. J. Mol. Biol. *73*, 199–212.

Kang, H. S., Eshbach, T. B., White, D. A. and Levine, A. J. (1971). Deoxyribonucleic acid replication in SV40 infected cells. J. Virol. *7*, 112–120.

Kaufman, G., Anderson, S. and DePamphlis, M. L. (1977). RNA primers in Simian Virus 40 replication. J. Mol. Biol. *116*, 549–567.

Laskey, R. A. and Mills, A. D. (1975). Quantitative film detection of ^{3}H and ^{14}C in polyacrylamide gels by fluorography. Eur. J. Biochem. *56*, 335–341.

Leffak, I. M., Grainger, R. and Weintraub, H. (1977). Conservative assembly and segregation of nucleosomal histones. Cell *12*, 837–845.

Levine, A. J. (1972). A model for the maintenance of the transformed state by polyoma and SV40. Prospectives in Virology, *VIII* (New York: Academic Press), p. 61.

Levine, A. J., Kang, H. S. and Billheimer, F. E. (1970). DNA replication in SV40 infected cells. J. Mol. Biol. *50*, 549–568.

McKnight, S. L. and Miller, O. L., Jr. (1977). Electron microscopic analysis of chromatin replication in the cellular blastoderm Drosophila melanogaster embryo. Cell *12*, 795–804.

Newrock, K. M., Alfageme, C. R., Nardi, R. V. and Cohen, L. H. (1977). Histone changes during chromatin remodeling in embryogenesis. Cold Spring Harbor Symp. Quant. Biol. *42*, 421–431.

Perlman, D. and Huberman, J. A. (1977). Asymmetric Okazaki piece synthesis during replication of Simian Virus 40 DNA in vivo. Cell *12*, 1029–1043.

Riley, D. and Weintraub, H. (1979). Conservative segregation of parental histones during replication in the presence of cycloheximide. Proc. Nat. Acad. Sci. USA *76*, 328–332.

Seale, R. L. (1976). Studies on the mode of segregation of histone Nu bodies during replication in HeLa cells. Cell *9*, 423–429.

Seale, R. L. and Simpson, R. T. (1975). Effects of cycloheximide on chromatin biosynthesis. J. Mol. Biol. *94*, 479–501.

Seidman, M. M., Garon, C. L. and Salzman, N. P. (1978). The relationship of SV40 replicating chromosomes to two forms of the non-replicating SV40 chromsome. Nucl. Acids Res. *5*, 2877–2893.

Southern, E. (1975). Detection of specific sequences among DNA fragments separated by gel electrophoresis. J. Mol. Biol. *98*, 503–517.

Tegtmeyer, P., Robb, J. A., Widmer, C. and Ozer, H. L. (1974). Altered protein metabolism in infection by the late tsB11 mutant of Simian Virus 40. J. Virol. *14*, 997–1007.

Tsanev, R. and Sendov, B. (1971). Possible molecular mechanism for cell differentiation in multicellular organisms. J. Theoret. Biol. *30*, 337.

Weintraub, H. (1973). The assembly of newly replicated DNA into chromatin. Cold Spring Harbor Symp. Quant. Biol. *38*, 247–258.

Weintraub, H. (1976). Cooperative segregation of Nu bodies during chromosome replication in the presence of cycloheximide. Cell *9*, 419–422.

Weintraub, H. and Groudine, M. (1976). Chromosomal subunits in active genes have an altered conformation. Science *193*, 848.

Weintraub, H., Flint, S. J., Leffak, I. M., Groudine, M. and Grainger, R. M. (1978). The generation and propagation of variegated chromosome structures. Cold Spring Harbor Symp. Quant. Biol. *42*, 401–407.

Worcel, A., Han, S. and Wong, M. L. (1978). Assembly of newly replicated chromatin. Cell *15*, 969–977.

Yu, K. and Cheevers, W. (1976). DNA synthesis in polyoma virus infection. J. Virol. *17*, 402–414.

Cell, Vol. 20, 313–319, June 1980, Copyright © 1980 by MIT

Two mRNAs Can Be Produced from a Single Immunoglobulin μ Gene by Alternative RNA Processing Pathways

P. Early, J. Rogers,* M. Davis, K. Calame,
M. Bond, R. Wall* and L. Hood
Division of Biology
California Institute of Technology
Pasadena, California 91125
*Molecular Biology Institute
and Department of Microbiology and Immunology
UCLA School of Medicine
Los Angeles, California 90024

Summary

As shown in the accompanying paper, μ chains of the membrane-bound (μ_m) and secreted (μ_s) forms of IgM are encoded by two species of mRNA. Cloned cDNAs produced from the two μ mRNAs of M104E mouse myeloma tumors differ only at their 3' ends, which encode either the μ_m or μ_s C terminus. In this paper, we show that both μ_m and μ_s mRNAs are produced from transcripts of a single μ gene. The last 187 nucleotides of μ_s mRNA are derived from DNA contiguous with the 3' end of the sequence encoding the $C_\mu 4$ domain. The μ_m cDNA clone does not include these 187 nucleotides, but instead contains 392 nucleotides derived from two exons located 1850 bp 3' to the $C_\mu 4$ sequence. Comparison of genomic and cDNA sequences show that in μ_m mRNA, an RNA splice of 1850 nucleotides joins a site in the coding sequence at the end of $C_\mu 4$ with a site at the beginning of the first membrane-specific exon. A second RNA splice of 118 nucleotides joins sequences transcribed from the first and second membrane-specific exons. The differences observed between μ_m and μ_s cDNAs suggest that developmental control of the site at which poly(A) is added to transcripts of the μ gene determines the relative levels of μ_m or μ_s chain synthesis. We discuss possible models for the control of μ gene transcripts and the significance of this form of developmentally regulated RNA processing for the evolution of eucaryotic "split genes."

Introduction

IgM antibody molecules exist in two separate forms, as monomeric membrane-bound receptors on lymphocytes and as secreted pentameric effectors of humoral immunity in the bloodstream. The μ heavy chains of IgM molecules contain four constant region domains, plus a short C terminal segment (Kehry et al., 1979). In secreted IgM molecules, the C terminal segment of μ chains forms disulfide bonds to a J chain linking each pentamer together. Monomeric IgM molecules on the surfaces of lymphocytes are attached to the cell membrane near their C terminal ends. Comparative analyses of membrane-bound (μ_m) and

secreted (μ_s) μ chain polypeptides suggest that they differ only in their respective C terminal segments (M. Kehry et al., manuscript submitted; reviewed by Rogers et al., 1980).

The preceding paper presents sequences from two cDNA clones ($\mu 6$ and $\mu 12$) which separately encode the μ_m and μ_s chains from M104E mouse myeloma tumors (Rogers et al., 1980). Both cDNA clones contain identical sequences encoding the $C_\mu 3$ and $C_\mu 4$ domains. However, the μ_m and μ_s cDNA clones encode distinct C terminal segments of 20 amino acids in the case of μ_s chains and 41 amino acids in μ_m chains (Rogers et al., 1980). The C terminal segment encoded by the μ_s cDNA clone is identical to that determined from the complete amino acid sequence of M104E μ chains (Kehry et al., 1979). While the four C_μ domains are encoded by separate exons (Calame et al., 1980; Gough et al., 1980), previous evidence has suggested that the coding sequence for the μ_s C terminal segment is contiguous with the 3' end of the $C_\mu 4$ exon (Calame et al., 1980). R loop mapping experiments suggest that the μ_m C terminal segment is encoded by a separate exon in the C_μ gene (Rogers et al., 1980).

In this paper we examine the genomic origins of μ_m and μ_s mRNAs. We find that both mRNAs are transcribed from the same μ gene and that two additional RNA splices generate μ_m mRNA. DNA rearrangement does not play a direct role in controlling μ_m or μ_s synthesis. A form of developmentally regulated RNA processing is responsible for determining the relative levels of μ_m and μ_s mRNAs produced in different cell types.

Results

Both μ_m and μ_s mRNAs Must Be Derived from a Single Germline C_μ Gene

In determining the genomic origins of μ_m and μ_s mRNAs, we asked whether the two mRNAs are derived from one or two C_μ genes. As stated elsewhere (Calame et al., 1980), the haploid BALB/c mouse germline genome contains a single C_μ gene. Ten independently isolated germline C_μ genomic clones all contained the same C_μ gene and flanking sequences (M. Davis, K. Calame and P. Early, unpublished results). Hybridization of labeled μ_s ($\mu 12$) plasmid DNA to Southern (1975) blots of germline or embryo DNAs digested with one of four different restriction enzymes showed only the pattern of bands expected from the cloned germline C_μ gene (Davis et al., 1980; M. Davis, unpublished results). These observations, plus the fact that restriction maps and the DNA sequences of large portions of the μ_s and μ_m cDNA clones are identical (Rogers et al., 1980), lead to the conclusion that both the μ_m and μ_s constant regions must be encoded by the exons of the single germline C_μ gene.

3' Terminal Sequences of μ_m and μ_s cDNAs Hybridize to Separate Exons in the C_μ Gene

To determine the germline locations of the sequences present in the μ_m and μ_s cDNA clones, we hybridized the labeled cDNA clones to a Southern blot of restriction fragments from ChSpμ7, a germline C_μ genomic clone (Davis et al., 1980). A restriction map of this C_μ clone (Calame et al., 1980) is shown in Figure 1. The 4.3 kb Xba I fragment of ChSpμ7 indicated in Figure 1 was isolated and digested with various restriction enzymes, and the fragments were separated on 8% polyacrylamide gels. Southern blots of these gels were hybridized with either μ_m (μ6) or μ_s (μ12) plasmid DNAs which had been labeled with ^{32}P by nick translation (Maniatis, Jeffrey and Kleid, 1975).

As shown in Figure 2, both the μ_s and μ_m cDNA clones hybridized with restriction fragments derived from the C_μ3 and C_μ4 exons. The μ_s cDNA clone also hybridized with restriction fragments containing the C_μ2 exon and secreted C terminal segment sequences (Calame et al., 1980; Rogers et al., 1980). The μ_s cDNA clone (μ12) does not contain V_H or C_μ1 sequences (Calame et al., 1980) and the μ_m cDNA clone (μ6) does not contain V_H, C_μ1 or C_μ2 sequences (Rogers et al., 1980).

The μ_m cDNA clone did not hybridize with restriction fragments containing the secreted C terminal sequences 3' to C_μ4 (Figure 2). However, the μ_m cDNA clone did hybridize with restriction fragments derived from a region at the 3' end of the 4.3 kb Xba I fragment, about 2 kb 3' to the C_μ4 exon, as can be seen by comparing the fragment sizes in Figure 2 with the restriction map in Figure 1. The μ_m cDNA clone also hybridized with the 1.4 kb Xba I fragment (Figure 1) located immediately 3' to the 4.3 kb Xba I fragment

in the germline C_μ clone, ChSpμ7 (data not shown). The μ_s cDNA clone did not hybridize with this 1.4 kb Xba I fragment. These results, in conjunction with R loop mapping (Rogers et al., 1980), suggest that the different C terminal and 3' untranslated sequences of the μ_m and μ_s cDNA clones are encoded by separate exons about 2 kb apart within the germline C_μ gene.

Specific μ_m mRNA Sequences Are Derived from Two Exons Located 1850 bp 3' to the C_μ4 Exon

The region of the ChSpμ7 germline C_μ clone surrounding the 3' end of the 4.3 kb Xba I fragment (Figure 1) was sequenced by the technique of Maxam and Gilbert (1977) to define precisely the relationship of this portion of the genome to the μ_m cDNA clone. As shown in Figure 3, this region includes two exons which encode the complete M segment and contain the 3' untranslated sequences of the μ_m cDNA clone. These will be referred to as the M exons. The first M exon encodes amino acids 557–595 of the M104E μ_m chain, while the second M exon encodes amino acids 596–597 and also contains the termination codon (UGA) and 3' untranslated region of the μ_m cDNA clone. An intron of 118 nucleotides separates the two M exons. The RNA splice site at the 5' end of the first M exon and the splice sites between the two M exons obey the GT . . . AG rule (Breathnach et al., 1978; Catterall et al., 1978).

μ_s mRNA Is Derived from Sequences Contiguous with the C_μ4 Exon

Figure 4 shows the nucleotide sequence of a region in the ChSpμ7 C_μ clone contiguous with the 3' end of the C_μ4 exon. As suggested elsewhere by restriction mapping (Calame et al., 1980), this region includes

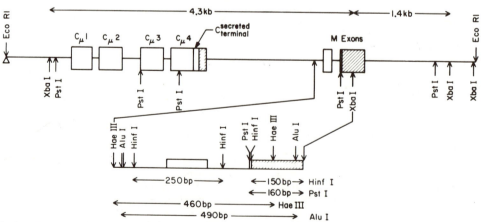

Figure 1. Restriction Map of the C_μ Gene in the Germline ChSpμ7 Clone

The Eco RI fragment shown is 6.7 kb long, with the direction of transcription from left to right. The triangle at the left end signifies a synthetic Eco RI site created by the Eco RI linker library technique (Maniatis et al., 1978). Raised boxes are exons and the horizontal lines are introns or nontranscribed sequences. Shading denotes 3' untranslated sequences present in either μ_m or μ_s mRNAs. The expanded diagram in the lower portion of the figure shows the M exon restriction fragments detected by hybridization to the μ_m cDNA clone μ6 (Figure 2).

cleotides 5′ to the terminal poly(A) sequence (Proudfoot and Brownlee, 1976). Poly(A) addition to both μ_m and μ_s mRNAs occurs within the sequence TCACT located at the 3′ termini of both cDNA sequences in the genomic clone. The presence of poly(A) in both cDNA clones and the similarity of the 3′ terminal sequences to other poly(A) addition sites (Proudfoot and Brownlee, 1976) indicate that the $\mu6$ and $\mu12$ cDNA clones end at the normal 3′ termini of μ_m and μ_s mRNAs, respectively.

μ_m and μ_s mRNAs Are Derived from Transcripts of One μ Gene

Having defined the relationship of the μ_m and μ_s cDNA clones to the exons of the germline C_μ gene, it is necessary to consider whether rearrangement of the distinct μ_m and μ_s DNA sequences occurs in the genomes of cells producing secreted or membrane-bound IgM molecules. In an IgM-producing cell, DNA rearrangements will have created a V_H gene associated with a C_μ gene (Davis et al., 1980; Early et al., 1980). This type of DNA rearrangement presumably occurs only on one chromosome, since a single V_H region is expressed by each cell (reviewed by Goding, 1978). Additional DNA rearrangements within the expressed C_μ gene might lead to μ_m or μ_s synthesis. If this were the case, differentiated B cells which produce both μ_m and μ_s mRNAs, such as those in the M104E myeloma tumor (Rogers et al., 1980), could be expected to contain two distinct copies of the expressed C_μ gene.

The existence of two distinguishable copies of the C_μ gene in M104E DNA was tested in the Southern blot shown in Figure 5. For comparison, both embryo and M104E DNAs were digested with Eco RI, electrophoresed on an agarose gel, blotted to a nitrocellulose filter and hybridized with labeled $\mu12$ plasmid DNA. Only one C_μ band is visible in either embryo or M104E DNA. This observation eliminates the possibility that the μ_m and μ_s mRNAs of the M104E myeloma tumor are produced from two copies of the expressed C_μ gene with distinguishable Eco RI cleavage patterns. Note that this result excludes any deletions within one of a putative two copies of the expressed C_μ gene, perhaps to remove the M exons, since any such deletions would alter the size of the Eco RI fragment containing the resultant C_μ gene (Figure 1).

In Figure 5, the single C_μ Eco RI fragment in M104E DNA is about 9.9 kb in length, compared to 12.2 kb for the germline fragment. The lack of a band of germline length suggests that the unexpressed C_μ gene has been lost in M104E, a phenomenon frequently observed with myeloma tumors (Davis et al., 1979; Cory and Adams, 1980; Gough et al., 1980). The reduction in length of the C_μ Eco RI fragment in the expressed M104E C_μ gene is not the direct result of V_H joining, since the J_H gene segment (J_{H107}) used in M104E is located 5′ to the germline C_μ Eco RI

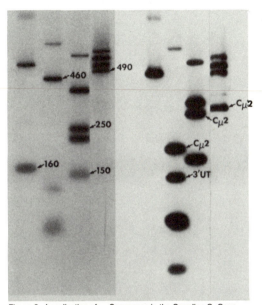

Figure 2. Localization of μ_m Sequences in the Germline C_μ Gene

The 4.3 kb Xba I fragment of ChSpμ7 (Figure 1) was digested with the indicated restriction enzymes and electrophoresed on 8% polyacrylamide gels with pBR322 markers. Southern blots from separate gels were hybridized with either μ_m (μ6) or μ_s (μ12) cDNA plasmids labeled with ³²P by nick translation. Restriction fragments derived from $C_\mu3$ and $C_\mu4$ exons hybridized to both plasmids (Calame et al., 1980; Rogers et al., 1980; J. Rogers, unpublished results) and are not labeled. Note that the gel on the right (μ12) has been electrophoresed longer than the gel on the left (μ6). As indicated, restriction fragments which hybridized only to μ12 originate either from the $C_\mu2$ exon or the specific μ_s 3′ sequence (Calame et al., 1980; Rogers et al., 1980). The fragments whose lengths are shown hybridized only with μ6, and all of these can be derived from the restriction map of the M exons shown in Figure 1.

the C terminal sequence of the μ_s cDNA clone (μ12) directly adjoining the $C_\mu4$ exon. The entire C terminal and 3′ untranslated regions of the μ_s cDNA clone μ12 (Calame et al., 1980; Rogers et al., 1980) are present as a continuous nucleotide sequence which terminates 187 nucleotides 3′ to the $C_\mu4$ exon.

The sequence of the ChSpμ7 genomic clone which includes the sequence at the 3′ end of μ12 shows no extensive homology with the two M exons or the sequence following them. The genomic sequence following the 3′ end of the M exons is also notably more T-rich than the sequence following the 3′ end of μ12 (Figure 3). However, both the μ_m (μ6) and μ_s (μ12) cDNA clones (Rogers et al., 1980) contain the apparent poly(A) addition signal AATAAA located 19 nu-

5' GGCCTGTTCTGTGCCTCCGTCTAGCTTGAGCTATTAGGGGACCAGTCAATACTCGCTAAGATTCTCCAGAACCAT

CAGGGCACCCCAACCCTTATGCAAATGCTCAGTCACCCCAAGACTTGGCTTGACCCTCCCTCTCTGTGTCCCTTC

557
GluGlyGluValAsnAlaGluGluGluGlyPheGluAsnLeuTrpThrThrAlaSerThrPheIleValLeu
ATAGAGGGGGAGGTGAATGCTGAGGAGGAAGGCTTTGAGAACCTGTGGACCACTGCCTCCACCTTCATCGTCCTC

595
PheLeuLeuSerLeuPheTyrSerThrThrValThrLeuPheLys
TTCCTCCTGAGCCTCTTCTACAGCACCACCGTCACCCTGTTCAAGGTAGTATGGTTGTGGGGCTGAGGACACAGG

GCTGGGACAGGGAGTCACCAGTCCTCACTGCCTCTACCTCTACTCCCTACAAGTGGACAGCAATTCACACTGTCT

596597
ValLys
CTGTCACCTGCAGGTGAAATGACTCTCAGCATGGAAGGACAGCAGAGACCAAGAGATCCTCCCACAGGGACACTA

CCTCTGGGCCTGGGATACCTGACTGTATGACTAGTAAACTTATTCTTACGTCTTTCCTGTGTTGCCCTCCAGCTT

TTATCTCTGAGATGGTCTTCTTTCTAGACTGACCAAAGACTTTTTGTCAACTTGTACAATCTGAAGCAATGTCTG

poly(A)
GCCCACAGACAGCTGAGCTGTAAACAAATGTCACATGGAAATAAATACTTTATCTTGTGAACTCACTTTATTGTG

AAGGAATTTGTTTTGTTTTTCAAACCTTTCCTGCGGTGTTGACAG   3'

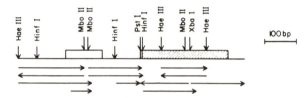

Figure 3. Sequence of the M Exons

The μ_m amino acids encoded by these exons are numbered by homology with human μ chains (Kehry et al., 1979). M104E μ_m mRNA actually encodes 593 amino acids, excluding the N terminal signal peptide. The UGA termination codon is boxed. GT ... AG sequences at RNA splice sites are underlined. The 3' terminus of μ_m cDNA (μ6) is indicated by the poly(A) addition point. A hexanucleotide sequence associated with polyadenylation is underlined (Proudfoot and Brownlee, 1976). The lower portion of the figure shows the strategy used in determining this sequence.

fragment (Early et al., 1980). Similar alterations in a C_μ Eco RI fragment, evidently not the direct result of V_H joining, have been seen in HPC76, in IgM-producing myeloma tumor (Gough et al., 1980). In that case, analysis of a C_μ genomic clone from the myeloma tumor shows that the alterations have not affected C_μ gene organization 3' to C_μ4, including the M exons (compare the restriction map of our germline C_μ clone in Figure 1 with that in Figure 4 of Gough et al., 1980).

Figure 5 also shows that the 9.9 kb C_μ Eco RI fragment in M104E DNA is nearly identical in size to a cloned C_μ Eco RI fragment in Ch603μ35. The cloned Ch603μ35 C_μ gene (Calame et al., 1980) was derived from the M603 IgA-producing myeloma tumor, which contains low levels of the germline 12.2 kb C_μ Eco RI fragment (Figure 5). The Ch603μ35 clone has deleted DNA 5' to the C_μ gene during cloning, a phenomenon we have observed consistently when recombinant Charon 4A phage containing the C_μ gene are grown in DP50supF (Davis et al., 1980; M. Davis, unpublished results). The size of the deletions 5' to the C_μ gene which occurred in cloning Ch603μ35 are similar to the apparent deletions in the M104E and HPC76 myeloma tumors. In the myeloma tumors, as in the bacterial hosts of Ch603μ35, deletions 5' to the C_μ gene may not be specifically involved in the regulation of μ gene expression, but may only reflect a propensity for certain sequences (perhaps repeats) to undergo

deletion during their replication over many cell generations. However, we cannot rule out the possibility that DNA rearrangements 5' to the C_μ gene, while not altering the organization of the M exons or the μ_s C terminal coding sequence, could indirectly affect the relative levels of synthesis of the two forms of μ mRNA.

Normal mouse spleen lymphocytes synthesize both μ_m and μ_s polypeptide chains (Vassalli et al., 1979). Spleen lymphocytes selected for IgM synthesis by the fluorescence-activated cell sorter contain expressed μ genes with the same arrangement of μ_m and μ_s sequences seen in the C_μ germline clone depicted in Figure 1 (P. Early and C. Nottenburg, unpublished observations). These observations and those discussed for the IgM-producing myeloma tumors indicate that DNA rearrangements do not directly determine μ_m or μ_s chain synthesis. Accordingly, both μ_m and μ_s mRNAs must be produced from transcripts of a single μ gene.

Discussion

RNA Splicing from the 3' C_μ4 Boundary Occurs only in μ_m mRNA

The sequences of the C_μ genomic clone and the μ_s cDNA clone both contain CTGGTAAAC at the boundary of C_μ4 and the μ_s C terminal segment (Figure 4). In contrast, the μ_m cDNA sequence at the same posi-

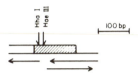

```
      550
     ArgThrValAspLysSerThrGlyLysProThrLeuTyrAsnValSerLeuIleMetSerAspThrGlyGlyThr
5'   AGGACCGTGGACAAGTCCACTGGTAAACCCACACTGTACAATGTCTCCCTGATCATGTCTGACACAGGCGGCACC
```

```
      576
     CysTyr
     TGCTATTGACCATGCTAGCGCTCAACCAGGCAGGCCCTGGGTGTCTAGTTGCTCTGTGTATGCAAACTAACCATG
```

poly(A)

```
     TCAGAGTGAGATGTTGCATTTTATAAAAATTAGAAATAAAAAAATCCATTCAAACGTCACTGGTTTTGATTATA
```

```
     CAATGCTCATGCCTGCTGAGACAGTTGTGTTTTGCTTGCTCTGCACACACCCTGC   3'
```

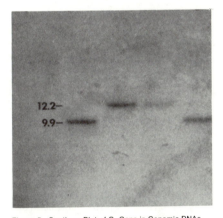

tion, CTGAGGGGG (Rogers et al., 1980), is evidently the result of an RNA splice between the C_μ4 sequence and the sequence TCCCTTCATAG/AGGGGG at the 5' boundary of the first M exon (Figure 3). A slash is used to denote the splice junction, in accordance with the GT ... AG rule (Breathnach et al., 1978; Catterall et al., 1978). The μ_s C terminal coding sequence which follows the C_μ4 splice site (CTG/GTAAAC) and the intron sequence at the 5' boundary of the first M exon (TCCCTTCATAG/AG) both exhibit complementarity to the U1 sequence, as has been noted for other splicing junctions (Reddy et al., 1974; Rogers and Wall, 1980; Lerner et al., 1980). This suggests that U1 or a similar small nuclear RNA may be involved in the splicing of μ_m mRNA. The 1850 nucleotide splice to generate μ_m mRNA is the first example in eucaryotic cells of an RNA splice to remove sequences which would otherwise encode a polypeptide; that is, the C terminal segment of the μ_s chain.

Figure 6 shows a diagram of the splicing patterns deduced for μ_m and μ_s mRNAs. Since the μ12 and μ6 cDNA clones lack the variable region and some constant region sequences, it cannot be stated conclusively that the μ_m and μ_s mRNAs from the M104E myeloma tumor are identical 5' to C_μ3. However, available information on the sizes of the two forms of μ mRNA and the nature of the polypeptides they produce suggests that the μ_m and μ_s mRNAs are identical from their 5' ends to codon 556 (Figure 4) (Rogers et al., 1980; M. Kehry et al., manuscript submitted).

The μ_m mRNA is produced by two more RNA splices than is μ_s mRNA (Figure 6). The first of these splices occurs at the boundary of C_μ4 with the μ_s C terminal sequence, so that the coding region for the μ_s C terminal segment is eliminated from μ_m mRNA. Instead, a longer hydrophobic membrane-bound C terminal segment, the M segment, is derived from the two M exons located 1850 bp 3' to the C_μ4 exon. The precursor to μ_s mRNA does not undergo RNA splicing

at the 3' end of the C_μ4 sequence (Figure 6). The μ_s mRNA terminates 187 nucleotides 3' to the C_μ4 exon, eliminating the possibility of an RNA splice to the M exon sequence.

Developmental Regulation of RNA Processing Directs the Synthesis of μ_m or μ_s mRNAs

Although production of μ_m as opposed to μ_s mRNA involves two additional RNA splices, it is unlikely that any developmental regulation of the splicing process itself takes place in the μ transcripts. The addition of poly(A) to the 3' ends of nuclear RNAs is a rapid process which generally precedes RNA splicing, as shown in adenovirus 2 (Nevins and Darnell, 1978) and

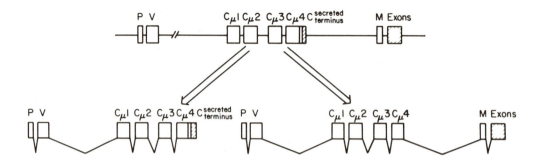

secreted mu mRNA   **membrane mu mRNA**

Figure 6. Splicing Patterns Deduced for μ_m and μ_s mRNAs

The μ_m and μ_s mRNAs are assumed to be identical 5' to $C_\mu 4$, although the $\mu 6$ and $\mu 12$ cDNA clones do not contain this complete sequence. Raised boxes indicate exons. 3' untranslated sequences are shaded. P refers to the signal peptide exon and V to the rearranged V_H exon. Bent lines indicate RNA splicing between exons.

SV40 (Lai, Dhar and Khoury, 1978) as well as in immunoglobulins (Schibler, Marcu and Perry, 1978; Gilmore-Hebert and Wall, 1979). This observation suggests that the control of μ_s mRNA synthesis depends on the addition of poly(A) to RNA transcripts 187 nucleotides 3' to the $C_\mu 4$ exon. When polyadenylation at this point occurs before splicing to the M exons can take place, μ_s mRNA is produced. Otherwise, elongation of the transcript continues and RNA splicing produces μ_m mRNA.

Nuclear RNA 3' ends for poly(A) addition may be created either by cleavage of nascent transcripts or by termination of transcription. The 3' ends of some adenovirus 2 late RNAs in vitro appear to be produced by termination (J. Manley, personal communication). On the other hand, RNA cleavage generates the 3' ends of a majority of late transcripts in adenovirus 2 (Nevins and Darnell, 1978; Fraser et al., 1979) and SV40 (Ford and Hsu, 1978; Lai et al., 1978). In these cases, transcription continues for more than 1000 nucleotides 3' to some cleavage sites. Accordingly, either RNA cleavage or termination of transcription could be the mechanism producing the 3' ends of μ mRNAs.

The control of polyadenylation in μ transcripts could be exercised by either positive or negative regulators of RNA cleavage or termination. These would presumably interact differentially with nucleotides in the C_μ gene near either the first (μ_s) or second (μ_m) sites of polyadenylation. Fusion of the B lymphoma W279, which produces both μ_m and μ_s mRNAs, with the myeloma cell line MPC11 results in a hybridoma line which only synthesizes μ_s mRNA (Raschke, Mather and Koshland, 1979; Rogers et al., 1980). This observation suggests that a positive regulator contributed by the myeloma cell is responsible for inducing polyadenylation at the first site to produce μ_s mRNA. In plasma cells secreting IgM, the change from predominantly μ_m to predominantly μ_s mRNA synthesis

might occur by the production or release of a positive regulator molecule which either enhances cleavage of transcripts 187 nucleotides 3' to the $C_\mu 4$ sequence or interacts with RNA polymerase II to cause termination of transcription at this point. Various ratios of μ_m and μ_s mRNAs could be synthesized, depending on the concentration of the regulator.

Perhaps the simultaneous expression of μ and δ chains with identical V_H regions in a single B lymphocyte at an early stage of development (reviewed by Goding, 1978) can be explained by an extension of some μ transcripts to include the C_δ sequences. The C_δ sequences might be spliced directly to V_H, eliminating the intervening C_μ sequences. This idea is particularly attractive because the C_δ gene segment is located about 6 kb 3' to the M exons in germline DNA (K. Moore and T. Hunkapiller, personal communication), whereas the spacing between the C_γ genes is more than 20 kb (M. Davis and S. Kim, personal communication).

RNA Splicing and the Evolution of Developmentally Regulated Polyadenylation Sites

The presence of alternative membrane and secreted C terminal sequences adds flexibility to the C_μ gene. The products of a single expressed μ gene can exist either as membrane-bound or serum antibodies. The evolution of this dual role for a single gene apparently depended on the existence of RNA splicing between exons encoding functionally distinct portions of the μ polypeptide. An early ancestor of the C_μ gene probably included an M exon (Rogers et al., 1980). Messenger RNA produced from this gene encoded an exclusively membrane-bound cell surface receptor protein. This hypothesis is supported by phylogenetic studies on certain invertebrates which indicate that immune-like reactions are mediated by cell surface recognition molecules and not by secreted proteins (Marchalonis and Cone, 1973).

The evolution of a second role for the ancestral μ polypeptide, that of a secreted protein, probably began with a mutation in the intron preceding the M exon. This mutation caused some RNA transcripts of the ancestral μ gene to be polyadenylated prior to the M exon. Messenger RNA produced from these transcripts would no longer have undergone splicing to the M exon, but instead would have contained a new 3' terminus derived from part of the former intron. When translated into protein, the new type of mRNA would have produced the N terminal portion of the ancestral μ polypeptide, but with the M segment replaced by a series of amino acids derived from the former intron sequence. This novel form of the ancestral μ polypeptide could be secreted, since it retained an N terminal "signal peptide" but lacked the membrane-binding M segment. Subsequent evolution presumably optimized the sequence of the μ_s C terminal segment for interactions with an aqueous environment and J chain (Koshland, 1975). In time, developmental control regulating polyadenylation at either the first or second site in the C_μ gene was also acquired.

Developmental regulation of polyadenylation sites may also have evolved in other eucaryotic "split genes." The potential flexibility this provides for gene expression may be an evolutionary advantage for the eucaryotic form of gene organization. Such flexibility would be in addition to the role that separated exons may have played as independent genetic elements, able to recombine to form new "split genes" (Gilbert, 1978; Davis et al., 1979). Perhaps as the result of such processes, the M exons or similar sequences may be found to occur in a variety of genes encoding membrane-bound proteins.

Experimental Procedures

The germline C_μ clone ChSpμ7 was isolated from a Charon 4A phage library derived from BALB/c mouse sperm DNA partially digested with Hae III and Alu I and ligated to synthetic Eco RI linkers (Maniatis et al., 1978; Davis et al., 1980). The cells used to prepare DNA were assayed by light microscopy to be greater than 99% sperm (Joho et al., 1980). In addition to the 6.7 kb Eco RI fragment shown in Figure 1, ChSpμ7 contains an 8 kb Eco RI fragment 3' to the C_μ gene (Calame et al., 1980). Ch603μ35 was isolated from a Charon 4A phage library containing M603 myeloma DNA partially digested with Eco RI (Early et al., 1979; Calame et al., 1980). Recombinant plasmids μ12 and μ6 contain M104E μ_s and μ_m cDNA sequences cloned in the Pst I site of pBR322 by dG:dC tailing (Calame et al., 1980; Rogers et al., 1980).

Restriction endonucleases and polynucleotide kinase were obtained from New England Biolabs and Boehringer-Mannheim. BA85 Nitrocellulose filter sheets were from Schleicher and Schuell, and γ-^{32}P-ATP (~9000 Ci/mmole) was from ICN. Manipulations of organisms containing recombinant DNA were performed under P2/EK2 or P2/EKI containment conditions in compliance with the NIH Guidelines.

Acknowledgments

This work was supported by NIH grants to L. H. and R. W. P. E. and M. D. are suported by an NIH training grant. K. C. is supported by an NIH fellowship. We thank Marilyn Kehry and Richard Douglas for helpful discussions.

The costs of publication of this article were defrayed in part by the payment of page charges. This article must therefore be hereby marked "*advertisement*" in accordance with 18 U.S.C. Section 1734 solely to indicate this fact.

Received March 17, 1980

References

Breathnach, R., Benoist, C., O'Hare, K., Gannon, F. and Chambon, P. (1978). Proc. Nat. Acad. Sci. USA 75, 4853–4857.

Calame, K., Rogers, J., Early, P., Davis, M., Livant, D., Wall, R. and Hood, L. (1980). Nature 284, 452–455.

Catterall, J. F., O'Malley, B. W., Robertson, M. A., Staden, R., Tanaka, Y. and Brownlee, G. G. (1978). Nature 275, 510–513.

Cory, S. and Adams, J. M. (1980). Cell 19, 37–51.

Davis, M., Early, P., Calame, K., Livant, D. and Hood, L. (1979). In Eukaryotic Gene Regulation, R. Axel, T. Maniatis and C. F. Fox, eds. ICN-UCLA Symposium (New York: Academic Press), pp. 393–406.

Davis, M. M., Calame, K., Early, P. W., Livant, D. L., Joho, R., Weissman, I. L. and Hood, L. (1980). Nature 283, 733–739.

Early, P. W., Davis, M. M., Kaback, D. B., Davidson, N. and Hood, L. (1979). Proc. Nat. Acad. Sci. USA 76, 857–861.

Early, P., Huang, H., Davis, M., Calame, K. and Hood, L. (1980). Cell 19, 981–992.

Ford, J. P. and Hsu, M.-T. (1978). J. Virol. 28, 795–801.

Fraser, N. W., Nevins, J. R., Ziff, E. and Darnell, J. E. (1979). J. Mol. Biol. 129, 643–656.

Gilbert, W. (1978). Nature 271, 501.

Gilmore-Hebert, M. and Wall, R. (1979). J. Mol. Biol. 135, 879–891.

Goding, J. W. (1978). In Contemporary Topics in Immunobiology, 8, N. L. Warner and M. D. Cooper, eds. (New York: Plenum Press), pp. 203–243.

Gough, N. M., Kemp, D. J., Tyler, B. M., Adams, J. M. and Cory, S. (1980). Proc. Nat. Acad. Sci. USA 77, 554–558.

Joho, R., Weissman, I. L., Early, P., Cole, J. and Hood, L. (1980). Proc. Nat. Acad. Sci. USA 77, 1106–1110.

Kehry, M., Sibley, C., Fuhrman, J., Schilling, J. and Hood, L. (1979). Proc. Nat. Acad. Sci. USA 76, 2932–2936.

Koshland, M. E. (1975). Adv. Immunol. 20, 41–69.

Lai, C.-J., Dhar, R. and Khoury, G. (1978). Cell 14, 971–982.

Lerner, M. R., Boyle, J. A., Mount, S. M., Wolin, S. L. and Steitz, J. A. (1980). Nature 283, 220–224.

Maniatis, T., Jeffrey, A. and Kleid, D. G. (1975). Proc. Nat. Acad. Sci. USA 72, 1184–1188.

Maniatis, T., Hardison, R. C., Lacy, E., Lauer, J., O'Connell, C., Quon, D., Sim, G. K. and Efstratiadis, A. (1978). Cell 15, 687–701.

Marchalonis, J. J. and Cone, R. E. (1973). Aust. J. Exp. Biol. Med. Sci. 51, 461–488.

Maxam, A. M. and Gilbert, W. (1977). Proc. Nat. Acad. Sci. USA 74, 560–564.

Nevins, J. R. and Darnell, J. E., Jr. (1978). Cell 15, 1477–1493.

Proudfoot, N. J. and Brownlee, G. G. (1976). Nature 263, 211–214.

Raschke, W. C., Mather, E. L. and Koshland, M. E. (1979). Proc. Nat. Acad. Sci. USA 76, 3469–3473.

Reddy, R., Ro-Choi, T. S., Henning, D. and Busch, H. (1974). J. Biol. Chem. 249, 6486–6494.

Rogers, J. and Wall, R. (1980). Proc. Nat. Acad. Sci. USA 77, 1877–1879.

Rogers, J., Early, P., Carter, C., Calame, K., Bond, M., Hood, L. and Wall, R. (1980). Cell 20, 303–312.

Schibler, U., Marcu, K. B. and Perry, R. P. (1978). Cell 15, 1495–1509.

Southern, E. M. (1975). J. Mol. Biol. 98, 503–517.

Vassalli, P., Tedghi, R., Lisowska-Bernstein, B., Tartakoff, A. and Jaton, J.-C. (1979). Proc. Nat. Acad. Sci. USA 76, 5515–5519.

Supplementary Readings

Before 1961

Darlington, C. and L. LaCour. 1940. Nucleic acid starvation of chromosomes in *Trillium*. *J. Genet.* 40: 185–214.

Hsu, T. 1954. Cytological studies on HeLa, a strain of cervical carcinoma. *Texas Rep. Biol. Med.* 12: 833–846.

Vogt, M. 1959. A study of the relationship between karyotype and phenotype in cloned lines of strain HeLa. *Genetics* 44: 1257–1270.

1961–1970

Britten, R.J. and D.E. Kohne. 1968. Repeated sequences in DNA. *Science* 161: 529–540.

Huberman, J. and G. Attardi. 1966. Isolation of metaphase chromosomes from HeLa cells. *J. Cell Biol.* 31: 95–105.

Laskey, R. and J. Gurdon. 1970. Genetic content of adult somatic cells tested by nuclear transplantation from cultured cells. *Nature* 228: 1332–1334.

Maio, J. and C. Schildkraut. 1969. Isolated mammalian metaphase chromosomes. II. Fractionated chromosomes of mouse and Chinese hamster cells. *J. Mol. Biol.* 40: 203–216.

O'Riordan, M., J. Robinson, K. Buckton, and H. Evans. 1970. Distinguishing between the chromosomes involved in Down's syndrome (trisomy 21) and chronic myeloid leukemia (PH1) by fluorescence. *Nature* 230: 167–168.

Shiraishi, Y. 1970. The differential reactivity of human peripheral leukocyte chromosomes induced by low temperature. *Jpn. J. Genet.* 45: 429–442.

1971–1980

Arrishi, F. and T. Hsu. 1971. Localization of heterochromatin in human chromosomes. *Cytogenetics* 10: 81–86.

Bell, G.I., R.L. Pictet, and W.J. Rutter. 1980. Sequence of the human insulin gene. *Nature* 284: 26–32.

Caspersson, T., M. Hulten, J. Lindsten, A.J. Therkelsen, and L. Zech. 1971. Identification of different Roberstonian translocations in man by quinacrine mustard fluorescence analysis. *Hereditas* 67: 213–220.

Craik, C.S., S.R. Buchman, and S. Beychok. 1980. Characterization of globin domains: Heme binding to the central exon product. *Proc. Natl. Acad. Sci.* 77: 1384–1388.

Cozzarelli, N.R. 1980. DNA gyrase and the supercoiling of DNA. *Science* 207: 953–960.

Davidson, R.L., E.R. Kaufman, C.P. Dougherty, A.M. Ouellete, C.M. DiFolco, and S.A. Latt. 1980. Induction of sister chromatid exchanges by BUdR is largely independent of the BUdR content of DNA. *Nature* 284: 74–76.

Davis, M.M., K. Calame, R.W. Early, D.L. Livant, R. Joho, I.L. Weissman, and L. Hood. 1980. An immunoglobulin heavy chain gene is formed by at least two recombinational events. *Nature* 283: 733–739.

Fyrberg, E.A., K.L. Kindle, N. Davidson, and A. Sodja. 1980. The actin genes of *Drosophila*: A dispersed multigene family. *Cell* 19: 365–378.

Goudie, R.B., D.R. Goudie, H. Dick, and M.A. Ferguson-Smith. 1980. Unstable mutations in vitiligo, organ-specific autoimmune diseases, and multiple endocrine adenoma/peptic-ulcer syndrome. *Lancet*, August 9, ii: 285–287.

Griffith, J.D. 1975. Chromatin structure, deduced from a minichromosome. *Science* 187: 1202–1203.

Gruss, P. and G. Khoury. 1980. Rescue of a splicing defective mutant by insertion of an heterologous intron. *Nature* 286: 573–578.

Itoh, N. and H. Okamoto. 1980. Translational control of proinsulin synthesis by glucose. *Nature* 283: 100–102.

Jakobivits, E.B., S. Bratosin, and Y. Aloni. 1980. A nucleosome-free region in SV40 mitochromosomes. *Nature* 285: 263–265.

Jalal, S.M., A. Markvong, and T.C. Hsu. 1975. Differential chromosomal fluorescence with 33258 Hoechst. *Exp. Cell Res.* 90: 443–444.

Kollar, E.J. and C. Fisher. 1980. Tooth induction in chick epithelium: Expression of quiescent genes for enamel synthesis. *Science* 207: 993–995.

Kraemer, P.M., D.F. Petersen, and M.A. Van Dilla. 1971. DNA constancy in heteroploidy and the stem line theory of tumors. *Science* 174: 714–717.

Latt, S.A. 1974. Sister chromatid exchanges, indices of human chromosome damage and repair. *Proc. Natl. Acad. Sci.* 71: 3162–3166.

Lomholt, B. and J. Mohr. 1971. Human karyotyping by heat-Giemsa staining and comparison with fluorochrome techniques. *Nat. New Biol.* 232: 109–110.

Pontecorvo, G. 1971. Induction of directional chromosome elimination in somatic cell hybrids. *Nature* 230: 367–369.

Radke, K.L., C. Colby, J. Kates, H.M. Krider, and D.M. Prescott. 1974. Establishment and maintenance of the interferon-induced antiviral state. *J. Virol.* 13: 623–630.

Rogers, J. and R. Wall. 1980. A mechanism for RNA splicing. *Proc. Natl. Acad. Sci.* 77: 1877–1879.

Shortle, D. and D. Nathans. 1978. Local mutagenesis: A method for generating viral mutants with base substitutions in preselected regions of the viral genome. *Proc. Natl. Acad. Sci.* 75: 2170–2174.

Sumner, A., H. Evans, and R. Buckland. 1971. New

technique for distinguishing between human chromosomes. *Nat. New Biol. 232:* 31–32.

Ullrich, A., T.J. Dull, A. Gray, and I. Sures. 1980. Genetic variation in the human insulin gene. *Science 209:* 612–615.

Wang, H.C. and S. Fedoroff. 1972. Banding in human chromosomes treated with trypsin. *Nat. New Biol. 235:* 52–53.

Chapter 8 Somatic Cell Genetics

Szybalski, W. and M.J. Smith. 1959. Genetics of human cell lines. I. 8-Azaguanine resistance, a selective "single-step" marker. *Proc. Soc. Exp. Biol. Med.* 101: 662–666.

Sorieul, S. and B. Ephrussi. 1961. Karyological demonstration of hybridization of mammalian cells *in vitro*. *Nature* 190: 653–654.

Littlefield, J.W. 1964. Selection of hybrids from matings of fibroblasts in vitro and their presumed recombinants. *Science* 145: 709–710.

Weiss, M.C. and H. Green. 1967. Human-mouse hybrid cell lines containing partial complements of human chromosomes and functioning human genes. *Proc. Natl. Acad. Sci.* 58: 1104–1111.

Kao, F.-T. and T.T. Puck. 1968. Genetics of somatic mammalian cells. VII. Induction and isolation of nutritional mutants in Chinese hamster cells. *Proc. Natl. Acad. Sci.* 60: 1275–1281.

Chasin, L.A. and G. Urlaub. 1975. Chromosome-wide event accompanies the expression of recessive mutations in tetraploid cells. *Science* 187: 1091–1093.

Alt, F.W., R.E. Kellems, J.R. Bertino, and R.T. Schimke. 1978. Selective multiplication of dihydrofolate reductase genes in methotrexate-resistant variants of cultured murine cells. *J. Biol. Chem.* 253: 1357–1370.

The minimal requirements for systematic mapping of the genes of cultured cells include a way to mutagenically induce changes in cellular DNA, a set of heritably stable markers that are easy to select for and against, and a means of direct mating that entirely bypasses the animal's need for meiotic recombination. The papers in this chapter show how these three requirements are met in culture.

In normal mammalian cells, synthesis of deoxyribonucleotides for DNA proceeds via two alternative pathways. If free bases from cell breakdown are present, then these can be recycled into nucleotides by the enzymes of a salvage pathway. Provided that cofactors for methylation, such as tetrahydrofolic acid, are available, de novo synthesis of the four bases goes on independently of this salvage pathway. Even when one pathway is lost by mutation, the other pathway permits survival. This allows easy direct selection of cells that have lost one or the other pathway. For example, the purine salvage pathway is rendered toxic to the cells by purine analogs such as 8-azaguanine because this analog is metabolized by the salvage enzyme hypoxanthine-guanine phosphoribosyl transferase (HGPRT) into toxic compounds. Cells characteristically become resistant to 8-azaguanine toxicity by loss of HGPRT. Such cells survive the drug and are still able to make DNA via the de novo pathway. Compounds like 8-azaguanine are, in other words, selective: Descendents of survivors of a dose of the drug will sometimes be resistant thereafter.

By fluctuation analysis, **Szybalski** and **Smith** find that about one in 10^4 cells of a cloned mouse tumor line are able to grow in toxic concentrations of 8-azaguanine and that the drug selects, but does not induce, these variants. Although Szybalski and Smith did not measure the enzyme directly, resistant colonies bred true, suggesting that the loss of enzyme was genetically stable.

Mutations and markers are enough for molecular mapping, but a full genetic analysis requires a cross of some sort. Sendai virus fuses cells, and the multinucleate products live long enough to engage in DNA synthesis and even mitosis (Chapter 6). In the next two papers a further step in this asexual form of mating is described, the recovery of clones of uninucleate cells derived from a fusion and carrying at least some of the chromosomes from each of two different "parental" cells. **Sorieul** and **Ephrussi** mix and culture together, for three months, equal numbers of two different mouse tumor lines. Each tumor line had a characteristic marker chromosome not present in the other. Simple cocultivation apparently was sufficient to produce hybrid cells, since after a few months the population contained many cells with a mixture of chromosomes that included both markers.

Easy isolation of clones of hybrid cells was not possible by simple cocultivation. With a selection against both parental cell types, only hybrid cells can survive and multiply to form colonies, and their isolation is easy. The next papers introduce the most commonly used double selection.

Azaguanine-resistant cells live solely by de novo synthesis of purines. For methylation, this pathway requires the cofactor tetrahydrofolic acid and therefore the enzyme dihydrofolate reductase (DHFR) to synthesize it. Normal cells can live without DHFR activity or, as Szybalski and Smith showed, they can live without HGPRT, but they cannot live without both. DHFR is inhibited by the drug aminopterin. So long as the bases hypoxanthine and thymidine are also present so that HGPRT and related pyrimidine enzymes can function, aminopterin has no effect on normal cells. Azaguanine-resistant cells, however, die in aminopterin, since they lack one of the enzymes necessary for incorporation of exogenous purines. This sort of alternate pathway selection tactic works with a similar pair of selective killers along a parallel set of two pyrimidine synthetic pathways. Bromodeoxyuridine (BrdU) kills normal cells after it is metabolized by the thymidine salvage enzyme thymidine kinase (TK). BrdU-resistant cells lack TK. Aminopterin in the presence of thymidine is therefore toxic to TK⁻ cells, because added thymidine is not useable without TK. Thus, we have two entirely different mutants, both of which die in the same medium. This doubly selective medium is called HAT (hypoxanthine, aminopterin, thymidine).

Littlefield shows that HAT can indeed yield hybrids as the sole survivors of a cocultivated population. He mixed TK⁻ cells and HGPRT⁻ cells and let them grow together for four days in regular medium. Uninucleate hybrids formed in those four days. When the medium was supplemented with HAT, the hybrids grew into surviving colonies. The cells in each colony had a chromosome number about equal to that of the sum of the chromosome numbers of the parental cells, and each colony was stable thereafter in its ability to grow in HAT.

At least one active gene from each parental cell line must have contributed to the viability of Littlefield's hybrids. **Weiss** and **Green** carry out a variation on his protocol and are able to isolate human-mouse hybrid cell lines. As before, HAT medium and a TK⁻ mouse cell line are used. Instead of an HGPRT⁻ line for the other parent, they chose the normal human precrisis fibroblast WI38 (Chapter 1). After coincubation and addition of HAT, the TK⁻ mouse cells died, and the WI38 monolayer was overgrown in a few places by dense colonies of cells with a morphology reminiscent of the TK⁻ mouse cells. Reconstruction experiments proved that no TK⁻ cell could have survived the HAT treatment. Indeed, surviving colonies initially contained chromosomes from both mouse and human parents. So far, this was neat, but not surprising. However, with time in culture, human chromosomes were lost from the hybrid lines.

This instability opened a path to gene mapping by synteny: Hybrids kept in HAT medium always retain one specific chromosome from the human parent. It is reasonable to guess that this chromosome contains the gene for TK. Indeed, when hybrids with this single human chromosome are given a dose of BrdU, all survivors lose it. In most interspecific hybrids, chromosomes from one parent are preferentially segregated. With a set of such hybrids, any gene whose product can be distinguished as to parental species can be mapped to a chromosome by correlation of the presence of product with the presence of chromosome from that parent.

So far, we have seen the thymidine analog BrdU used as a selective drug. **Kao** and **Puck** use it in a general selection system for isolation of auxotrophic mutant cells. DNA that has incorporated BrdU becomes fragmented when illuminated with light, so TK$^+$ cells that are growing in low amounts of BrdU can be killed when blue light hits them. Nongrowing TK$^+$ cells are not killed under these conditions, since BrdU is not taken into DNA unless cells are cycling.

To get mutants requiring glycine, for example, Kao and Puck expose Chinese hamster ovary (CHO) cells first to the mutagen ethylmethylsulfonate and then to a few days in glycineless medium. Should any glycine-requiring mutant cells arise, they will be unable to grow, and so they will survive when the culture is next given BrdU and irradiated with blue light.

The high frequency of the glycine-requiring mutants isolated by Kao and Puck implies that the cells they worked with have only one copy of a chromosome carrying a gene whose loss will lead to glycine requirement. In general, high mutation frequencies in such experiments are troublesome. Since cultured cells are expected to be at least diploid, the frequency of recessive mutation should reflect the need to knock out two copies of some gene, and therefore it ought to be very low. **Chasin** and **Urlaub** focus on this gene-dosage problem in a study of HGPRT$^-$ mutation in CHO cells. Since HGPRT is sex-linked, there is only one active gene in diploid CHO cells. To be sure of starting with a cell with two active HGPRT genes, they constructed tetraploid derivatives of CHO cells. Surprisingly, the mutation frequency for HGPRT$^+$ to HGPRT$^-$ was not a great deal lower for tetraploid than for diploid CHO cells. By linkage studies with another nonselected X-linked marker enzyme, they were able to show that each time a tetraploid CHO cell mutated to HGPRT$^-$, one of the two active X chromosomes was lost from the cell. Thus, when cultured cells show high frequencies of mutation for presumably diploid genes, one copy is mutated, but the other may simply be lost along with the rest of its chromosome by some mechanism that is as yet not well understood.

Chromosome loss after hybridization or mutation is a powerful tool for gene mapping, but it is a bit disconcerting that a piece of chromatin as large as a whole chromosome may be so easily lost by cultured cells. **Alt, Kellems, Bertino,** and **Schimke** show an even more unexpected sort of genomic instability in cultured cells: drug resistance by the amplification of a gene. In their paper, Alt et al. examine the behavior of the gene for DHFR. Recall that this enzyme is inhibited by aminopterin (here called by its clinical name, methotrexate) and that methotrexate is toxic at high concentrations for normal cells unless free bases are provided.

Methotrexate by itself is an antitumor drug, but methotrexate-resistant tumor cells often arise, and these cells limit the drug's effectiveness. The same resistance arises in cultures maintained in gradually increasing concentrations of methotrexate. In both kinds of resistant cells, the specific activity of DHFR can become hundreds of times higher than normal. Alt et al. find that the gene copy number for DHFR is elevated in such resistant cells to an amount equivalent to the increase in specific activity of the enzyme and the expected parallel increase in mRNA sequences for DHFR. This phenomenon suggests that gene duplications may be amenable to direct manipulation and opens the possibility of constructing in vitro systems for the study of sequence evolution in multigene families. The mechanism of amplification here remains a mystery.

Reprinted from Proceedings of the Society for Experimental Biology and Medicine
Vol. 101, pp. 662–666. 1959

Genetics of Human Cell Lines I. 8-Azaguanine Resistance, a Selective "Single-Step" Marker.* (25053)

Waclaw Szybalski and M. Joan Smith   (Introduced by Vernon Bryson)
Inst. of Microbiology, Rutgers State University, New Brunswick, N. J.

From the methodological point of view, the mammalian cell grown *in vitro* can be regarded as a unicellular microorganism. Thus, methods developed for quantitative work with microbes can be adapted to the genetic study of human cells, provided suitable selective markers are available. Plating and colony-counting technic, so useful for assay of viable microbial cells, was introduced into tissue culture methodology by Puck *et al.*(1). Our purpose was to find a suitable mutational system for mammalian cells in which, under selective conditions, mutant cells survive and form well developed colonies while parental population is completely inhibited or eliminated. Mutation from 8-azaguanine (AG) sensitivity to resistance satisfied the foregoing criteria(2). Moreover, the property of 8-azaguanine resistance appears to be an excellent genetic marker, since it is not associated with modifications in morphological or cultural characteristics of cells either in presence or absence of the selective agent.

Materials and methods. Strain *Detroit-98* (D98), derived from human sternal marrow by Berman and Stulberg(3), was kindly supplied by Dr. H. Moser of Cold Spring Harbor Labs. A single-cell-derived clone D98S was used throughout these studies. *Media.* The basic medium was essentially that described by Eagle(4), containing 10% com-

* This investigation was supported by Research Grant No. 3492 of U. S. Public Health Service, Nat. Inst. of Health.

plete horse serum. Sterilization was effected by filtration through Selas No. 2 filter. *Cultivation and plating procedures.* The methods were based on those developed by Puck *et al.*(1) and adapted for strain D98 by Moser and Tomizawa(5). Cells were grown on glass surface of 60-mm Petri dishes at 37°C in 5-10% CO_2 atmosphere. The inoculum was prepared from a 3- to 6-day-old culture of vigorously growing cells, washed serum-free with balanced salt solution before detachment and dispersion in the same solution containing 0.25% pancreatin (Nutritional Biochemical Corp., Cleveland) (5-minute incubation at 37°C followed by gentle manual shaking). The action of pancreatin was terminated by addition of serum-containing medium, after which the cell suspension was filtered aseptically through fine cheesecloth to remove any cell clumps. The cell density of this essentially monodispersed suspension was assayed in a hemacytometer. For colony counts, cells appropriately diluted in 5 ml of serum-containing medium were plated and incubated 6 to 12 days, with a complete medium change every 2 to 3 days. Growth was assayed either by counting and measuring colonies or by protein determination. *Protein determination* is based in part on protein staining method of Durrum(6) and its modifications for electrophoretic(7) and cytochemical(8) purposes. Adaptation of these methods for quantitative protein determination in mammalian cell cultures, suggested by Dr. A. W. Kozinski of this laboratory, permits determination of total protein without disturbing colonies on the glass. Thus a colony count may be performed on the same plate after completion of protein determination. The assay is based on spectrophotometric determination of bromphenol blue selectively bound by the protein component of $HgCl_2$-fixed cells and subsequently eluted with alkaline acetone. The procedure consists of: (1) thorough washing of glass-attached cells with balanced salt solution to remove all serum proteins; (2) 15-minute staining at 37°C with aqueous solution containing 0.1% bromphenol blue and 5% $HgCl_2$; (3) three consecutive washes with 0.5% acetic acid each lasting 6 minutes; (4) quantita-

tive elution with 3-ml portions of aqueous acetone (20 ml acetone 0.2 ml 10 N NaOH, 5 ml water, freshly prepared); (5) adjusting the collected extracts to 10-ml volume and measuring optical density at 595 mμ wave length against a blank prepared similarly from cell-free plate preincubated with serum-containing medium. The actual amount of cell protein/plate is ascertained from predetermined calibration curves. The same procedure was also used for cell suspensions, with centrifugation as an added step. *Colony count.* After fixation in Bouin's fluid or after protein determination, colonies were stained with Giemsa solution. The dry dishes were then inserted into a projector (Bausch & Lomb, "Tri-Simplex" micro-projector), and well-focused images of colonies were scored by an electronic counter (built by Philip D. Mintz of this laboratory).

Results. When above procedures are followed, strain D98S exhibits a consistent plating efficiency of greater than 0.6, *i.e.*, over 60% of microscopically assayed cells give rise to well defined colonies (Fig. 1), which can easily be counted, providing their number does not exceed 5000/plate. Above this figure, colonies coalesce, and total protein synthesis after 14 days' incubation becomes less than proportional to the number of cells in the inoculum (Fig. 2).

A series of plates was prepared, each seeded with approximately 5000 cells, to which AG was added in graded concentrations. Fig. 3 represents the survival curve 1 of the wild-

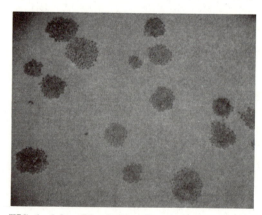

FIG. 1. 6-day-old colonies of strain D98S grown on glass, Bouin fixed, Giemsa stained.

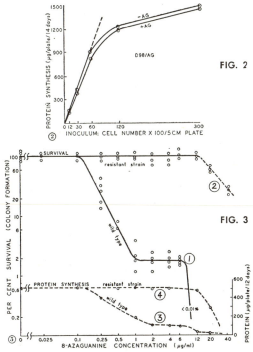

FIG. 2

FIG. 3

FIG. 2. Relationship between inoculum size (colony-forming cells) and total growth determined as amount of protein synthesized (14 days) by AG-resistant strain D98/AG grown in presence (+AG) or absence (–AG) of 8 μg AG/ml medium.

FIG. 3. 12-day growth determined as amount of protein synthesized and survival (colony formation) of AG-sensitive (wild type—1, 3) and AG-resistant (2, 4) isolates derived from strain D98S grown at increasing concentrations of AG. Inoculum was 5000 cells/plate.

type strain D98S. It is apparent that in presence up to 0.1 μg AG/ml survival and colony-forming ability of each cell is virtually unaffected. There is a sharp drop in plating efficiency between 0.1 and 1 μg AG/ml, followed by a plateau extending to approximately 10 μg AG/ml, at which level 1 to 2% of cells survive and form colonies. These colonies were isolated and found to be stable, AG-resistant clones (D98/AG) as reflected by survival curve 2. Their growth, as measured by protein synthesis, was not affected by AG in concentrations up to 12 μg/ml (curve 4), while protein synthesis of the wild-type strain was increasingly suppressed above 0.1 μg/ml.

Thus the wild-type population contains approximately 1 to 2% mutant cells, exhibiting to 100-fold increase in AG resistance without impairment of plating efficiency or growth rate either in the presence or in the absence of AG (Fig. 2 and 3). Fig. 4 shows appearance of colonies on 2 plates seeded with equal inoculum of AG-sensitive strain D98S, incubated in absence (A) and in presence (B) of AG (6 μg/ml).

Resistance to AG appears to be a highly stable property. Subculture for over 150 generations, either in absence (curve 3) or in presence (curve 2) of AG (6 μg/ml) did not alter level of resistance (Fig. 5). By subculturing resistant lines at high concentrations of AG, strains with slightly higher resistance could be isolated (Fig. 5, curve 4).

Determination of mutation rates was based on Luria-Delbrück variance test(9), modified for mammalian cells grown on glass. Each colony grown in absence of AG and firmly attached to glass may be considered as analogous to one "test tube culture"(9). On addition of the drug (AG), only colonies containing at least one resistant cell would give rise to secondary resistant colonies, while others would slough off the glass and be lost during medium changes. Mutation rate based on preliminary experiments, employing the P_o method(9) corresponds to 5 x 10^{-4}/cell/generation.

Properties of AG-resistant cells. AG-sensitive cells were found to exhibit the same inhibitory threshold for AG and 8-azaguanosine (10^{-7} M). Neither guanine (1 to 50 μg/ml) nor adenine nor thymine antagonize the inhibitory action of 1 to 50 μg/AG/ml. AG-resistant lines are approximately 100 times more resistant to AG, but only 2-3 times more resistant to 8-azaguanosine.[†] They were also found to be more resistant to 5-fluorouracil and to ultraviolet (UV) light (Fig. 6 and Fig. 7, curve 3). It was therefore of interest to

[†] Strain D98/AG contains approximately 0.01% mutants (D98/AGR) resistant in a single (genetic) step to over a 100-fold higher concentration of 8-azaguanosine (AGR), with concomitant further increase in AG resistance (over 10-fold). Direct mutations from AG sensitivity (D98S) to AGR resistance (D98/AGR) were not detected in populations as large as 10^7 cells. Thus AG and AGR resistances behave as two sequential mutational steps.

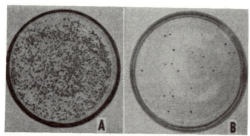

FIG. 4. Colony formation on 2 plates seeded with identical inocula (4000 cells) of AG-sensitive cells (strain D98S) in absence (A) or presence (B) of 6 μg AG/ml medium.

determine whether UV would increase mutation frequency from AG sensitivity to resistance or would select for the more UV-resistant, AG-resistant cells(10). The wild-type population was irradiated with increasing doses of UV and plated in absence (curve 1) or in the presence (curves 2 and 2a) of 6 μg AG/ml. It is apparent that AG-resistant cells not exposed to AG prior to irradiation, exhibit a "single-hit" survival curve (curves 2 and 2a), while UV survivals of parental strain D98S (curve 1) and of the established resistant line D98/AG (curve 3) are characterized by "multi-hit" curves. Curve 4 represents UV survival of

the AG-resistant line at slightly inhibitory concentrations of AG (12 μg/ml). The shape of survival curve appears unchanged. It may therefore be concluded that the newly emerging AG-resistant cells, pre-existing in the wild-type population, contain only one AG-resistant "complement" (chromosome?), while the number of these "complements" increases during establishment of the resistant sublines. Further genetic and cytological studies may explain this phenomenon and indicate whether processes analogous to nondisjunction or mitotic crossing-over could be involved.

Summary. A cloned wild-type population (strain D98S) derived from human bone marrow cells, which is inhibited by 8-azaguanine (AG) in concentrations in excess of 0.1 μg/ml, contains approximately 1 to 2% cells resistant to AG in 100-fold higher concentrations. A single-step mutational process (in broad meaning of the word) appears to be involved. The AG-resistant lines are stable upon prolonged subculture either in presence or absence of drug. Mutation rate from sensitivity to resistance is of the order of 5×10^{-4}/cell/generation. AG resistance is a use-

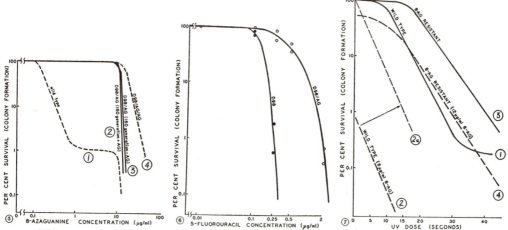

FIG. 5. Survival of AG-sensitive (1) and of AG-resistant mutant lines at increasing concentrations of AG. Curves 2 and 3 represent survival of AG-resistant strains after subculture for 160 generations in presence (2) or absence (3) of 5 μg AG/medium. Survival of more resistant line D98/AG/AG, obtained by serial subculture of D98/AG at 20 to 40 μg AG/ml, is depicted by curve 4.

FIG. 6. Survival of AG-sensitive (D98) and AG-resistant (D98/AG) strains at increasing concentrations of 5-fluorouracil.

FIG. 7. Ultraviolet survival of AG-sensitive (1, 2) and AG-resistant (3, 4) cell lines grown in absence (1, 3) or presence (2, 4) of AG. Curve 2a is parallel to curve 2. UV intensity 120 μW/cm².

ful genetic marker, since at selective AG concentrations (a wide plateau at 1-10 μg/ml) colony formation by resistant cells is not impaired and sensitive cells are destroyed. A method for protein determination is described, based on measurement of eluted bromphenol blue from HgCl$_2$-fixed, glass-attached cells.

1. Puck, T. T., Marcus, P. I., Cieciura, S. J., *J. Exp. Med.,* 1956, v103, 273.

2. Szybalski, W., *Microb. Genetics Bull.,* 1958, No. 16, 30.

3. Berman, L., Stulberg, C. S., Proc. Soc. Exp. Biol. and Med., 1956, v93, 730.

4. Eagle, H., *J. Exp. Med.,* 1955, v102, 37.

5. Moser, H., Tomizawa, K., *Ann. Rep. Biol. Lab. Cold Spring Harbor,* 1957-1958, v68, 33.

6. Durrum, E. L., *J. Am. Chem. Soc.,* 1950, v72, 2943.

7. Kunkel, H. G., Tiselius, A., *J. Gen. Physiol.,* 1951, v35, 89.

8. Mazia, D., Brewer, P. A., Alfert, M., *Biol. Bull.,* 1953, v104, 57.

9. Luria, S. E., Delbrück, M., *Genetics,* 1943, v28, 491.

10. Szybalski, W., Smith, M. J., *Fed. Proc.,* 1959, v18, 336.

Received May 22, 1959.   P.S.E.B.M., 1959, v101.

Reprinted from Nature, Vol. 190, No. 4776, pp. 653–654. 1961. © 1961 Macmillan Journals, Ltd.

Karyological Demonstration of Hybridization of Mammalian Cells *in vitro*

IN a recent publication, Barski, Sorieul and Cornefert[1] described the formation of cells with a hybrid karyotype in mixed cultures of two mouse cell strains *in vitro*. The experiments to be described show that this phenomenon is easily reproducible and that it is not limited to a pair of strains of strictly defined karyotypes.

The two cell lines used in our experiments are strains *NCTC* 2555 and *NCTC* 2472 of Sanford, Likely and Earle[2], ultimately derived from a single cell of mouse subcutaneous adipose tissue, as received from Dr. G. Barski by kind permission of Drs. Wilton R. Earle and Katherine K. Sanford. Both strains are populations composed predominantly of hypotetraploid cells but containing varying proportions of cells of higher ploidy[3]. The karyotypes of the hypotetraploid cells of the two strains, as established in the course of our experiments, are given in Table 1 and illustrated by Figs. 1,*A* and *C*.

It can be seen that the two cell types can be unmistakably identified by the presence of (*a*) a high number of bi-armed chromosomes, some of which represent good 'markers', in the cells of line *NCTC* 2555 ; (*b*) at least one extra long chromosome and a somewhat shorter one (not shown in Table 1, but visible in Fig. 1,*C*) with a sub-median secondary construction in the cells of line *NCTC* 2472.

Mixed cultures were initiated on January 3, 1961, by inoculating equal numbers of cells of the two lines into two media (a modified 199 (ref. 4) and *NCTC* 109 (ref. 5)) containing 10 per cent of heat-inactivated horse serum. Dispersion of cells for transfer was accomplished either mechanically or by trypsinization. The cultures were incubated at 37° C. in an atmosphere of 5 per cent carbon dioxide in air. Control cultures of the two strains separately were treated

Table 1

	NCTC 2555	*NCTC* 2472
Total number of chromosomes, modal	57	55
Total number of chromosomes, variation extremes	51–64	51–59
Number of biarmed chromosomes, modal	15	1
Number of biarmed chromosomes, variation extremes	11–16	1–3
Number of extra long telocentrics, modal	0	1
Number of extra long telocentrics, variation extremes	0	1–3
Minutes	?	0

518

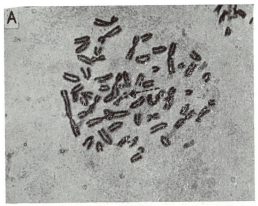

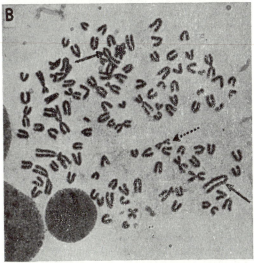

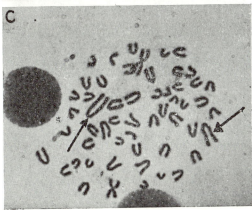

Fig. 1. *A*. Metaphase with 58 chromosomes (of which 15 are bi-armed) in a cell of strain *NCTC* 2555. Note characteristic metacentric (arrow) present in almost all cells of the strain. *B*. Representative example of hybrid cell with 112 or 113 chromosomes, of which 16 or 17 are bi-armed. Note the presence of the three 'markers' (arrows) shown in *A* and *B*
C. Metaphase with 54 chromosomes (2 bi-armed) in a cell of strain *NCTC* 2472. Note the two 'marker' telocentrics: the extra long one and the somewhat shorter, submedially constricted one (arrows)

similarly.   Karyological analyses of the mixed populations and of cells of each line grown separately were carried out at intervals on aceto-orcein squashes of colchicine-treated cultures.

At the time of writing (March 28), the occurrence of hybrid cells was observed in all four of the karyo-logically analysed mixed populations (two in medium 109 and two in medium 199).   The hybrid cells are recognized by the total number of chromosomes and the number of bi-armed chromosomes expected from the fusion of hypotetraploid cells of the two strains, as well as by the presence of 'marker' chromosomes (Fig. 1,*B*).   The frequencies of hybrid cells in the four mixed populations are at present approximately 1, 1, 5 and 10 per cent, the highest value having been observed in medium 109.

These results confirm the observations of Barski *et al.*, and furthermore show that the formation of hybrid cells is easily reproducible and is not restricted to two rigorously defined karyotypes : the latter is clear from the fact that strain *NCTC* 2472 used by us and strain *N* 1 *T* (derived from *NCTC* 2472) used by Barski *et al.* have somewhat different karyotypes. Thus the long constricted chromosome of line *NCTC* 2472 used by us has never been observed in Barski's line *N* 1 *T* ; the cells of the latter strain very seldom contain bi-armed chromosomes, whereas those of line *NCTC* 2472 used by us all carry at least one such chromosome.

It may be pointed out also that, of the two strains involved in the mixed cultures of Barski *et al.*, at least one was re-isolated from a tumour induced by the inoculation of the original *in vitro* line.   On the other hand, the cells used in our experiments never under-went animal passage since their explantation by Sanford *et al.* It is clear, therefore, that the tendency to cell fusion is not a result of animal passage.

These observations suggest that hybridization *in vitro* of other types of somatic cells should be equally possible.   If this hope is justified, hybridization may become a useful tool for the investigation of a number of problems of somatic cell genetics, of oncology and of virology[6].   Examples of problems which are common to these fields and which hybridization *in vitro* could help resolving are : the genetic mechanism of varia-tions in resistance to drugs and other chemothera-peutic agents observed frequently in tumour cells[7]; and the possible occurrence of zygotic induction[8] in virogenic cell lines.

The practicability of the method will depend to a large extent on the establishment of selective tech-niques for the detection of hybrid cells.   Its applic-ability to the study of the mechanisms of cellular differentiation will require further the establishment of techniques for overcoming barriers of tissue specificity.

The technical assistance of Mlle. M. T. Thomas is gratefully acknowledged. This work was supported by a grant from the Rockefeller Foundation.

SERGE SORIEUL
BORIS EPHRUSSI

Laboratoire de Génétique Physiologique du Centre
National de Recherche Scientifique,
Gif sur Yvette (Seine et Oise), France.

[1] Barski, G., Sorieul, S., and Cornefert, F., *C.R. Acad Sci., Paris*, **251**, 1825 (1960).

[2] Sanford, K. K., Likely, G. D., and Earle, W. R., *J. Nat. Cancer Inst.*, **15**, 215 (1954).

[3] Chu, E. H. Y., Sanford, K. K., and Earle, W. R., *J. Nat. Cancer Inst.*, **21**, 729 (1958).

[4] Morgan, J. F., Morton, H. J., and Parker, R. C., *Proc. Soc. Exp. Biol. Med.*, **73**, 1 (1950).

[5] Evans, V. J., Bryant, J. C., McQuilkin, W. T., Fioramonti, M. C., Sanford, K. K., Westfall, B. B., and Earle, W. R., *Cancer Res.*, **16**, 87 (1956).

[6] Lederberg, J., *J. Cell. and Comp. Physiol.*, **52**, Supp. 1, 383 (1958). Schultz, J., *Ann. N.Y. Acad. Sci.*, **71**, (6), 994 (1958).

[7] Hauschka, T. S., *J. Cell. and Comp. Physiol.*, **52**, Supp. 1, 197 (1958).

[8] Jacob, F., and Wollman, E. L., *Ann. Inst. Pasteur*, **91**, 486 (1956).

Reprinted from Science, Vol. 145, pp. 709–710. 1964. © 1964 AAAS

Selection of Hybrids from Matings of Fibroblasts in vitro and Their Presumed Recombinants

Abstract. When two clonal lines of mouse fibroblasts, each containing a drug-resistant marker, are grown together for 4 days, hybrid cells can be detected by selective conditions. These hybrid cells are presumed to be the result of mating. By the same method evidence can be obtained which suggests that mating may be followed by segregation.

Barski, Sorieul, and Cornefert (*1*) first demonstrated that fusion can occur between cultured mouse cells; Ephrussi and Sorieul (*2*) have since studied this phenomenon extensively by karyotype analyses, having in mind its possible utility for the genetic analysis of mammalian cells. Gershon and Sachs (*3*) recently confirmed the occurrence of fusion, using mouse cells with different histocompatibility antigens. In the study reported here, the detection of far fewer hybrid cells than were found previously has been possible through the use of two clonal lines of mouse fibroblasts (L cells), each containing a drug-resistant marker.

Cell culture and analytical methods were the same as those used previously (*4*); the assay for thymidine kinase activity in cell extracts was a modification of that for guanylic acid–inosinic acid pyrophosphorylase (*4*). Cells of the A3-1 line lack pyrophosphorylase and are resistant to 3 μg of 8-azaguanine per milliliter (*4*), and those of the B34 line lack thymidine kinase and are resistant to 30 μg of 5-bromodeoxyuridine (BUDR) per milliliter (*5, 6*). These markers do not appear to be linked (*7*). Cultures of the B34 line contain a small number of revertant cells (*6*) and therefore are maintained in the presence of BUDR. For the mating experiments the cells were washed to remove BUDR and resuspended in medium containing 10 μg of thymidine per milliliter. Then an equal number of these cells and A3-1 cells were grown together for 4 days in a suspension culture. Four or more replicate petri dishes containing 5×10^6 cells from this culture, and dishes containing the same number of B34 cells (similarly washed and grown with thymidine) or A3-1 cells, were incubated with medium containing $4 \times 10^{-7}M$ aminopterin, $3 \times 10^{-6}M$ glycine,

$1.6 \times 10^{-5}M$ thymidine, and $1 \times 10^{-4}M$ hypoxanthine, in which only cells with both enzymes can survive to form colonies (*4, 7, 8*). The medium was replenished after 1, 3, and 6 days, and the petri dishes were examined for colonies after about 14 days.

In three experiments no colonies were present in the dishes containing A3-1 cells, while in the dishes containing B34 cells the frequency of cells (presumably revertant) which survived to form colonies averaged 8×10^{-7} (3×10^{-7}, 7×10^{-7}, and 14×10^{-7}). In the mixed cultures the frequency of viable cells averaged 5.6×10^{-6} (2.6×10^{-6}, 7.0×10^{-6}, and 7.2×10^{-6}), or 14 times the number of revertant B34 cells which would have been present in the mixed cultures.

Clonal lines were established from six such colonies from the mixed cultures. One (M3) was similar morphologically to B34, and in all other respects appeared to be a revertant of B34. Another (M11) contained two types of cells, and has not been studied extensively. The other four (M1, M4, M12, and M13) were composed of large cells similar to A3-1 morphologically, and mostly mononuclear; they contained approximately double the usual amount of DNA, RNA, and protein per cell. Chromosome counts on these lines are given in Fig. 1, and the activities of thymidine kinase and guanylic acid–inosinic acid pyrophosphorylase per cell are shown in Table 1. These data indicate that M1, M4, M12, and M13 contained about twice the usual number of chromosomes and both enzymes, presumably through the fusion of an A3-1 cell and a B34 cell.

If cell fusion is to be useful for the genetic analysis of mammalian cells, as is a comparable process in fungi (*9*), the hybrid cells must segregate to form cells of lower chromosome number. Although perhaps due to artifacts, the occasional low chromosome counts in the fused cell lines (Fig. 1) suggested segregation, as suspected also by Ephrussi and Sorieul (*2*). Furthermore, by selection with 8-azaguanine or with 8-azaguanine and BUDR it has been possible to detect in the M4 culture two cell types which might have arisen by segregation. (For technical reasons BUDR alone could not be used as a selective agent in petri dishes.) In two experiments the frequencies of 8-azaguanine–resistant cells in the M4 culture were 5.9 and $7.8 \times$

Table 1. Enzymatic activities in cell extracts. Results expressed as arbitrary units per cell.

Line	Thymidine kinase	Guanylic acid–inosinic acid pyrophosphorylase
A3-1	100	1
B34	3	100
M1	98	172
M4	62	96
M12	81	109
M13	117	167

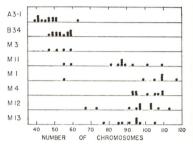

Fig. 1. Chromosome counts on the parental (A3-1 and B34) and the "mated" cell lines. No count greater than 64 was found in the parental lines.

10^{-4}, and the frequencies of doubly resistant cells, which grew quite slowly, were 2.6 and 3.8×10^{-6}. Some of the 8-azaguanine–resistant cells might have arisen by mutation of M4 cells (*4*), and studies of the chromosome numbers of such cells will help to answer this question. The frequency of the doubly resistant cells is considerably more than might have been expected (about 10^{-10}) from coincidental mutations in the M4 cells (*10*). Thus it seems probable that the doubly resistant cells are due to segregation with recombination. If this process can be validated, perhaps methods will be found to increase its frequency, as well as the frequency of fusion, and it may become possible to extend these techniques to diploid cultures.

JOHN W. LITTLEFIELD
*John Collins Warren Laboratories
of the Huntington Memorial Hospital
of Harvard University, Massachusetts
General Hospital, Boston*

References and Notes

1. G. Barski, S. Sorieul, F. Cornefert, *Compt. Rend.* **251**, 1825 (1960).
2. B. Ephrussi and S. Sorieul, in *Approaches to the Genetic Analysis of Mammalian Cells*, D. J. Merchant and J. V. Neel, Eds. (Univ. of Michigan Press, Ann Arbor, 1962), pp. 81–97.
3. D. Gershon and L. Sachs, *Nature* **198**, 912 (1963).
4. J. W. Littlefield, *Proc. Natl. Acad. Sci. U.S.* **50**, 568 (1963).

5. S. Kit, D. R. Dubbs, L. J. Piekarski, T. C. Hsu, *Exptl. Cell Res.* **31**, 297 (1963).
6. J. W. Littlefield and P. F. Sarkar, *Federation Proc.* **23**, 169 (1964).
7. J. W. Littlefield, unpublished material
8. W. Szybalski, E. H. Szybalska, G. Ragni, *Natl. Cancer Inst. Monograph* **7**, 75 (1962).
9. G. Pontecorvo and E. Käfer, *Advan. Genet.* **9**, 71 (1958).
10. The frequency of 10^{-10} was estimated from previous work (*4*) and unpublished studies.
11. Supported by USPHS grant CA-04670-05. This is publication No. 1173 of the Cancer Commission of Harvard University. We acknowledge the assistance of Susan J. Whitmore.

4 May 1964

Reprinted from the Proceedings of the National Academy of Sciences
Vol. 58, No. 3, pp. 1104–1111. September, 1967.

HUMAN-MOUSE HYBRID CELL LINES CONTAINING PARTIAL COMPLEMENTS OF HUMAN CHROMOSOMES AND FUNCTIONING HUMAN GENES*

By Mary C. Weiss† and Howard Green

DEPARTMENT OF PATHOLOGY, NEW YORK UNIVERSITY SCHOOL OF MEDICINE

Communicated by Boris Ephrussi, June 26, 1967

This paper will describe the isolation and properties of a group of new somatic hybrid cell lines obtained by crossing human diploid fibroblasts with an established mouse fibroblast line. These hybrids represent a combination between species more remote than those previously described (see, however, discussion of virus-induced heterokaryons). They are also the first reported hybrid cell lines containing human components and possess properties which may be useful for certain types of genetic investigations.

Interspecific hybridizations involving rat-mouse,[1] hamster-mouse,[2, 3] and Armenian hamster–Syrian hamster[4] combinations have been shown to yield populations of hybrid cells capable of indefinite serial propagation. Investigations of the karyotype and phenotype of such hybrids have shown that both parental genomes are present[2, 5] and functional.[6] In every case, some loss of chromosomes has been observed; this occurred primarily during the first few months of propagation, usually amounted to approximately 10–20 per cent of the complement present in newly formed hybrid cells, and involved chromosomes of both parents. Recent studies have provided evidence of preferential loss of chromosomes of one parental species in interspecific hybrids.[2, 5]

A more extreme example of this preferential loss has been encountered in the human-mouse hybrid lines to be described, in which at least 75 per cent, and in some cases more than 95 per cent, of the human complement has been lost. It has been possible to relate the human characteristics of the hybrid phenotype to the number of human chromosomes retained. This has been shown for the colonial morphology of the hybrid and for the human antigens in the hybrid cell membrane, which were detected by mixed cell agglutination. Furthermore, information has been obtained with respect to the chromosomal localization of the thymidine kinase gene.

Materials and Methods.—Cell lines: LM (TK⁻) cl 1-D (hereafter referred to as cl 1-D), a subline of mouse L cells, was isolated by Dubbs and Kit[7] and provided by Dr. B. Ephrussi. This clone, deficient in thymidine kinase, is resistant to 30 μg/ml of 5-bromodeoxyuridine (BUDR).

WI-38, a diploid strain of human embryonic lung fibroblasts,[8] was provided by the American Type Culture Collection and was propagated in monolayer culture in this laboratory for 10 to 20 cell generations before hybridization.

Culture method and selection of hybrids: All cultures were maintained in standard growth medium (Dulbecco and Vogt's modification of Eagle's minimal medium containing 10% calf serum), in some cases supplemented with 30 μg/ml of BUDR, or hypoxanthine ($1 \times 10^{-4} M$), aminopterin ($4 \times 10^{-7} M$), thymidine ($1.6 \times 10^{-5} M$) (HAT). The selective system used was originally described by Littlefield[9] and has been modified by Davidson and Ephrussi[10] for use with a biochemically marked cell line combined with a contact-inhibited strain of diploid fibroblasts. The mouse cell line cl 1-D, deficient in thymidine kinase, is unable to grow in medium containing HAT; no spontaneous reversion to HAT resistance has been observed in this cell line, and it seems most likely that a deletion involving this gene has occurred.[11] The human diploid strain, containing

1104

thymidine kinase activity, is able to grow in this medium but it forms only a relatively thin cell layer, while hybrid cells are able to pile up against this background and form discrete colonies.

Karyotype of parental cells: Cl 1-D is characterized by the presence of 51 (50–55) chromosomes, 9 (8–10) of which are large metacentrics. Among the latter is the D chromosome[12] which, owing to the presence of a secondary constriction, appears to be dicentric and has no equivalent in the human complement. As can be seen in Figure 2a, most of the chromosomes of cl 1-D are telocentric.

WI-38 is a diploid human strain;[8] the cells contain 46 chromosomes, the pairs of which may be grouped into seven classes (Denver Classification). Of these, only two (A and G) contain chromosomes which are likely to be confused with the mouse chromosomes (Fig. 2b).

Demonstration of cell surface antigens by mixed hemagglutination: These experiments were carried out using a modification of the method described by Kelus, Gurner, and Coombs.[13] Immune sera were obtained from rabbits following injection of WI-38 cells in Freund's adjuvant. Complement was inactivated at 56° and the serum absorbed twice with a total of 6×10^6 cl 1-D cells per ml of serum. In order to test for the presence of human antigens, 2×10^5 trypsinized parental or hybrid cells were washed in buffered salt solution containing 0.1% bovine serum albumin, incubated with 0.1 ml antiserum for 1 hr at room temperature, again washed three times and mixed with 0.1 ml of a 2% suspension of washed human red blood cells. The mixture was centrifuged gently, resuspended, and a drop pipetted onto a glass slide for examination. A cover slip placed over the drop was allowed to settle for 5 min, so that erythrocytes attached to the surface of a test cell were forced to its perimeter and appeared as an encircling ring. In controls for specificity of the agglutination, mouse erythrocytes were employed in place of human. Mouse species-specific antigens were identified by the use of rabbit antimouse ascites tumor antiserum and mouse erythrocytes.

Results.—Production and identification of hybrid cells: Cultures were initiated with mixtures of 2×10^6 cl 1-D cells and 1×10^4 WI-38 cells. After four days of growth in standard medium, the cultures were placed in selective medium (HAT). The cl 1-D cells degenerated within seven days, leaving a single layer of human cells; after 14 to 21 days, hybrid colonies (Fig. 1) could be detected growing on the human cell monolayer. A number of these were isolated and grown to mass culture. In other cases the entire culture was transferred; within a few weeks the hybrid cells over-grew the remaining human fibroblasts and in the course of serial cultivation all human cells disappeared from the population.

Of three independent experiments performed, all yielded hybrid colonies, with a frequency of approximately one per 2×10^4 WI-38 cells. In all hybrids examined 20 generations after their formation, the same karyotypic pattern was found; all (or nearly all) of the expected mouse chromosomes were present, but of the human chromosomes only a minority remained, varying in number from 2 to 15 in the

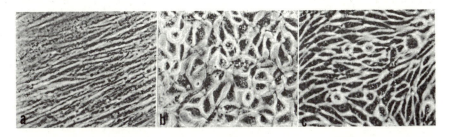

Fig. 1.—Phase contrast photomicrographs of living cells of (a) WI-38; (b) cl 1-D; (c) hybrid clone HM-2 P2. The parallel orientation of the highly elongated WI-38 cells is in contrast to the random orientation of the highly refractile and less fusiform cells of cl 1-D. The cells of the hybrid clone, while highly refractile, have a degree of orientation intermediate between cl 1-D and WI-38.

various populations. The mean numbers of human chromosomes found in six independent hybrid clones were as follows: 6.5, 2.0, 11.6, 10.9, 3.0, and 3.1. An occasional hybrid mitosis contained more than one copy of the same human chromosome, but most contained only one member of a given pair. The karyotype of a hybrid cell of clone HM-2 is shown in Figure 2c.

Loss of human thymidine kinase gene(s) from hybrid cells: Since survival of cells in HAT medium requires the presence of thymidine kinase, all hybrid cells selected in these experiments presumably contain the human gene(s) for this enzyme. This conclusion is supported by the fact that the cells can be killed by growth in the presence of BUDR. During continued propagation in HAT, any variants which may have lost this gene are eliminated. However, such variants occur with high frequency and were obtained selectively in a single step by transfer of the population to a medium containing BUDR, which eliminated most of the population and permitted the growth only of cells without thymidine kinase activity. As expected, these BUDR-resistant variant hybrid cells were not able to grow in HAT medium. They did retain many, but not all, of the human chromosomes which were present before the selection (see below). Cells resistant to BUDR also appeared in cultures propagated in standard medium without HAT and seemed to enjoy some selective advantage, for after several months of serial propagation they amounted to about 50 per cent of the population, as measured by HAT sensitivity. As no revertants to HAT resistance were obtained, BUDR resistance probably occurred by deletion.

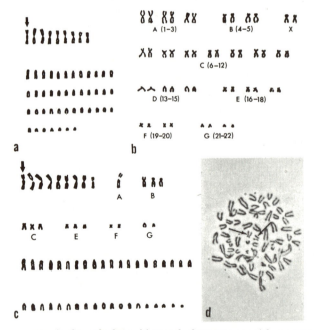

Fig. 2.—Chromosomes of Cl 1-D, WI-38, and hybrids. (a) Karyotype of the mouse parental line, cl 1-D. Nine long metacentric chromosomes are present, one of which (arrow, the D chromosome) is identifiable in all chromosome preparations. The karyotype also shows 43 telocentric chromosomes, none of which contains distinguishable satellites. (b) Karyotype of a cell of the WI-38 strain (see also ref. 8). Comparison with (a) shows that only chromosomes of groups A (long metacentrics) and G (small acrocentrics) are likely to be confused with chromosomes of cl 1-D. (c) Karyotype of hybrid clone HM-2. All chromosomes have been classified as to their origin, 14 of them being identified as human and 48 as mouse. In this karyotype, representatives from each of the major human chromosome groups except D are present. However, in other mitoses of HM-2, group D chromosomes were present. The A group chromosome is the only large bi-armed chromosome with a nonmedian centromere and is therefore probably a member of the no. 2 pair. (d) Mitosis of a hybrid cell of HM-1 after 60 generations of growth in HAT medium. The human chromosomes indicated by arrows are of group 6-12-X and are thought to carry the thymidine kinase gene. Two other human chromosomes are present, one of group D and one of group G.

Since most of the hybrid lines carried few human chromosomes, almost all of which could be distinguished from those of mouse origin, we attempted to correlate thymidine kinase activity with the presence of a specific human chromosome. Table 1 shows that nearly every cell in populations grown in HAT medium contains at least one chromosome of the group 6–12 and X, whereas chromosomes of this group are relatively rare in hybrid cells propagated in BUDR. In some of the hybrid metaphases, a submetacentric chromosome of this group has a peculiar configuration (Fig. 2d), in that there is slight separation of the chromatids in the region of the centromeres.[14] As shown in Table 1, this chromosome was identified in 15 of the 40 mitoses examined from HAT-grown populations and was not detected in any of the 34 BUDR-grown cells. This evidence would suggest that loss of the thymidine kinase gene occurs by loss of the relevant chromosome, and that this chromosome belongs to the 6–12-X group.

Although very soon after their formation human-mouse hybrids already showed extensive loss of human chromosomes, hybrid clones isolated later were found to be relatively stable karyotypically. However, it was expected that their human chromosome content would depend to some extent upon the selective conditions employed; therefore after 30 generations in HAT medium, hybrid cells were propagated in (1) HAT, (2) standard medium, and (3) BUDR. Two clonal hybrid populations were chosen for this study, HM-1 and HM-2, characterized by the presence of 6 and 11 human chromosomes, respectively.

Figure 3 shows the chromosomal changes which occurred during four to six months of serial culture. In both lines there was extensive loss of human chromosomes correlated with number of generations of propagation. The populations grown in HAT medium, checked after 20–60 generations, showed very little change, but after 85 generations there was a decrease by one half in the number of human chromosomes. This change appeared more marked in the absence of HAT or in BUDR medium. Under these conditions, a proportion of the cells (10–25%) of a number of clones lost all identifiable human chromosomes.

Figure 3b also shows the chromosomal composition of three subclones isolated from the HM-2 HAT-grown population after 85 generations. Two of these clones (HM-2 P5 and HM-1 P2), which were selected on the basis of humanlike colonial morphology, contained a larger number (9–12) of human chromosomes than the average of the parent population, while the third clone, which strongly resembled cl 1-D, nevertheless contained approximately six human chromosomes. It is clear that in all hybrid populations very little mouse genetic material was lost;

TABLE 1

CHROMOSOMES OF HAT-RESISTANT AND BUDR-RESISTANT HYBRID CLONES

Hybrid line	Total no. of cells counted	No. of cells containing member of 6-12-X group	Average no. of human chromosomes per cell
HM-1 HAT	20	17* (7)	6.0
HM-1 BUDR	20	5 (0)	3.4
HM-2 HAT	20	18* (8)	4.9
HM-2 BUDR	14	2 (0)	1.8

In parentheses are the number of cells containing the chromosome of distinctive appearance described in text.

* Failure to identify 6-12-X chromosome(s) in all cells of these HAT-grown populations could be due to chromosomal rearrangements, such as translocations. A cell which has lost the thymidine kinase-bearing chromosome might still be able to complete one or two cell divisions in HAT medium.

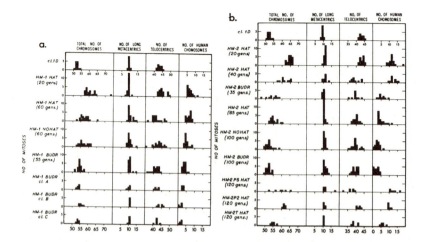

FIG. 3.—Histograms showing evolution of the karyotype of (*a*) HM-1 and (*b*) HM-2. The chromosomes of the hybrids are grouped into the following classes: (1) total number of chromosomes; (2) number of long metacentrics (this class is composed almost entirely of chromosomes of mouse origin but might also include one or two group A human chromosomes); (3) mouse telocentrics (in some cases, members of the group G human chromosomes would be placed in this class if the chromatin distal to the centromere were not visible as two distinct masses); and (4) number of human chromosomes. It is likely that these values are low by perhaps one or two chromosomes, due to failure to distinguish group A, and perhaps group G, chromosomes. Numbers on abscissa refer to bars immediately to their right.

on the average there was a decrease in number of mouse telocentrics by three or four and no decrease in number of long metacentrics. Some chromosomal rearrangements may have occurred among the mouse telocentrics in a minority of the population of HM-1, since those cells which contain fewer than expected telocentric chromosomes contain more than the expected number of long metacentrics and it appears likely that these changes were the result of centric fusion of telocentrics. (For a more detailed discussion of this process see ref 2.) However, in most hybrid populations no evidence of chromosomal rearrangement was found.

Cell membrane antigens and hybrid karyotype: The presence of human species-specific antigens in the cell membrane of hybrid cells was determined by mixed cell agglutination. A cell was considered positive if two thirds of its perimeter was covered with erythrocytes. By this criterion 60 per cent of the cells of WI-38 were positive under the standard conditions. Values given by hybrid cells are shown in Figure 4 on an arbitrary scale denoted as relative agglutination index (R.A.I.; WI-38 = 1.0) plotted against the mean number of human chromosomes per cell. Cl 1-D gave a value of 0.005, which can be considered as background due to nonspecific adherence of erythrocytes. All hybrids gave agglutination values in excess of this, and the values increased with increasing number of human chromosomes up to 12, at which point the R.A.I. became equal to that of the WI-38 parent. The R.A.I. rose very slowly with chromosome number up to about five chromosomes per cell, and then more sharply until the maximum was reached at 12 chromosomes. Similar results were obtained when less stringent criteria were used for scoring agglutinations; under these conditions a more nearly linear relation was observed

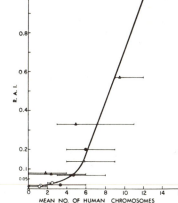

Fig. 4.—Human cell membrane antigens in hybrid lines. Ordinate gives relative agglutination index calculated from the fraction of cells showing positive hemagglutination, taking the values given by WI-38 as equal to 1.0. Abscissa gives mean human chromosome number per cell and the range for each population. Since the subclones have the narrowest range of human chromosome number, their agglutination values are probably the most significant.

- ● HM-1
- ○ HM-1 subclones
- ▲ HM-2
- △ HM-2 subclones
- ✕ cl 1-D control

between human chromosome number and R.A.I., but the background values given by cl 1-D were much higher.

The agglutination values obtained from the hybrids using antimouse cell serum and mouse erythrocytes were practically identical with those given by cl 1-D under the same conditions. This is consistent with the fact (Fig. 3) that virtually the entire cl 1-D genome is present in all the hybrids.

In control experiments no mixed agglutination was observed when hybrid cells were tested with antimouse serum and human erythrocytes, or with antihuman serum and mouse erythrocytes.

Poliovirus infection of hybrid cells: Somatic cell hybrids may be useful for the study of the process of viral infection. For example, it has been shown in heterokaryons made between cells of two different species that virus unable to multiply in one parental cell type is able to do so in the common cytoplasm.[15] However, in our hybrids containing up to 12 human chromosomes poliovirus infection at high multiplicity did not lead to any detectable cytocidal effect, under conditions in which the parental strain WI-38 was completely destroyed within 24–36 hours. While poliovirus itself does not successfully infect mouse cells, L-cells, from which cl 1-D is derived, are known to support viral multiplication very well following infection with poliovirus RNA.[16] Therefore, it seems most likely that failure of the virus to kill the hybrid cells is due to inability of the virus to penetrate the cell membrane, due either to lack of essential human elements or to interference by mouse elements.

Discussion.—The production and properties of a number of different interspecies hybrid lines, including some made between diploid strains and established cell lines, have been summarized by Ephrussi.[17] It has also been shown by Harris and Watkins[18] that virus-induced heterokaryons between HeLa (human) and Erhlich ascites (mouse) cells underwent at least one or two divisions as some hybrid mitoses could be identified. Though no established lines of hybrid cells developed from these heterokaryons, Yerganian and Nell[4] were able by this method to produce established lines of rodent cell hybrids. It might have seemed doubtful whether a permanent hybrid line could be made between a human fibroblast strain and an established mouse line since, in addition to the remoteness of the two species, human fibroblasts do not spontaneously develop into established lines as do those of rodents, but uniformly die out after 50–100 cell generations.[8, 19]   Though it turned out that these factors did not prevent the formation of hybrid lines, they may have been the cause of the very extensive loss of human chromosomes from the hybrid cells.[20]   This loss began before 20 generations after fusion and most likely during the first few divisions, as the colonial morphology did not change appreciably from the time colonies were first identified to the time pure cultures were obtained.

After the development of hybrid populations, continued subculture results in continuing though slow elimination of human chromosomes.   Since nearly the entire cl 1-D complement is retained, the human genes are probably not essential for viability, with the exception of that providing thymidine kinase activity, which is necessary as long as the cells are cultivated in HAT.   Omission of HAT from the medium permits variants lacking thymidine kinase to survive, as indicated by the appearance of HAT-sensitive cells.   When a hybrid population is placed in BUDR, cells bearing thymidine kinase are eliminated and the entire population becomes HAT-sensitive.

These losses of the thymidine kinase gene appear to occur by loss of the relevant chromosome(s) since: (1) there is a small decline in mean human chromosome number associated with continued growth in the absence of HAT, and in the BUDR-resistant populations;   (2) there is a rather strong correlation between the absence of a definite chromosome of the 6-12-X group, BUDR resistance, and HAT sensitivity;   and (3) there are no obvious chromosomal rearrangements by which genetic material of that chromosome might have been retained.

It has been shown in a number of cases that antigens characteristic of the parent cells are expressed in the hybrids.   Among the antigens which follow this rule are H-2 antigens of the mouse,[21] and polyoma virus-induced T-antigens and surface antigens.[22]   It is therefore not surprising that human species-specific cell membrane antigens could be detected in all of the hybrids described here.   Even the presence of a very small number of human chromosomes (<4) produced an R.A.I. of 2–15 times background.   The genes for surface antigens appear to be widely distributed among the different human chromosomes, since agglutinability of the hybrid cells increased progressively with number of human chromosomes.

Clones of hybrid cells selected on the basis of humanlike morphology (HM-2P2 and HM-2P5) contained the largest number of human chromosomes (9.5 and 12) and gave the highest agglutination values.   On the other hand, a clone selected on the basis of absence of humanlike morphology (HM-2T) had six human chromosomes and a corresponding R.A.I.   This suggests that some human chromosomes may have little to do in determining colonial morphology, though they do determine cell membrane composition.   Similar hybrids may permit the chromosomal localization of other human genes expressed at the cellular level, such as those for blood group antigens and enzymes whose physical properties are different from those of mouse determination.   It should also be possible to isolate clones of hybrid cells which have lost all human chromosomes and which might be useful for the investigation of nonchromosomal genes.

Summary.—Cocultivation of a human diploid cell strain with a thymidine kinase deficient mouse cell line has led to the formation of hybrid cell lines, which were isolated in selective medium.   The hybrid lines contained substantially the entire mouse genome and a greatly reduced complement of human chromosomes. The functioning of the human genes was shown by the presence of human antigens on the surface of the hybrid cells. The agglutinability of the cells in mixed hemagglutination tests depended on the number of human chromosomes contained (up to 12), indicating that the genes for surface membrane antigens are widely distributed among the human chromosomes. The human gene for thymidine kinase permits the hybrid cells to grow in medium containing aminopterin, but variants lacking this function may be selected in medium containing 5-bromodeoxyuridine. These variants appear to have lost a human chromosome of the 6-12-X group and it is suggested that this chromosome contains the thymidine kinase gene. Continued growth of hybrid lines results in slow elimination of human chromosomes. Study of clones containing a small number of human chromosomes should permit the localization of other human genes.

It is a pleasure to acknowledge the kind assistance of Mr. C. deSzalay and Dr. L. J. Scaletta and the valuable advice of Drs. G. J. Todaro and Z. Ovary.

* Aided by grant CA 06793, postdoctoral fellowship 1-F2-GM-34,679 (M.W.), and award 4-K6-CA-1181 (H.G.), all from the U.S. Public Health Service.

† Present address: Carnegie Institute of Washington, Department of Embryology, 115 W. University Parkway, Baltimore, Maryland.

[1] Ephrussi, B., and M. C. Weiss, these Proceedings, 53, 1040 (1965).

[2] Scaletta, L. J., N. Rushforth, and B. Ephrussi, *Genetics*, in press.

[3] Davidson, R. L., B. Ephrussi, and K. Yamamoto, these Proceedings, 56, 1437 (1966).

[4] Yerganian, G., and M. B. Nell, these Proceedings, 55, 1066 (1966).

[5] Weiss, M. C., and B. Ephrussi, *Genetics*, 54, 1095 (1966).

[6] *Ibid.*, p. 1111.

[7] Dubbs, D. R., and S. Kit, *Exptl. Cell Res.*, 33, 19 (1964).

[8] Hayflick, L., and P. S. Moorhead, *Exptl. Cell Res.*, 25, 585 (1961).

[9] Littlefield, J., *Science*, 145, 709 (1964).

[10] Davidson, R., and B. Ephrussi, *Nature*, 205, 1170 (1965).

[11] Kit, S., D. R. Dubbs, L. J. Piekarski, and T. C. Hsu, *Exptl. Cell Res.*, 31, 297 (1963).

[12] Hsu, T. C., *J. Natl. Cancer Inst.*, 25, 1339 (1960).

[13] Kelus, A., B. W. Gurner, and R. R. A. Coombs, *Immunology*, 2, 262 (1959).

[14] This configuration was not seen in the human parent cell and may have occured as a consequence of the new environment of the chromosome in the hybrid cell. It is also possible that this chromosome arises through translocation and is only part human, but this seems unlikely as it was seen in three hybrid populations of independent origin.

[15] Koprowski, H., F. C. Jensen, and Z. Steplewski, these Proceedings, 58, 127 (1967).

[16] Holland, J. J., B. H. Hayer, L. C. McLaren, and J. T. Syverton, *J. Exptl. Med.*, 112, 821 (1960).

[17] Ephrussi, B., "Phenotypic Expression," *In Vitro* (Baltimore: Williams and Wilkins, 1967), vol. 2, p. 40.

[18] Harris, H., and J. F. Watkins, *Nature*, 205, 640 (1965).

[19] Todaro, G. J., and H. Green, *Proc. Soc. Exptl. Biol. Med.*, 116, 688 (1964).

[20] A more general discussion of some of the mechanisms of loss of chromosomes in interspecific somatic hybrids has been given in: Ephrussi, B., and M. C. Weiss, in *Control Mechanisms in Developmental Processes*, ed. M. Locke (New York: Academic Press, in press).

[21] Spencer, R. A., T. S. Hauschka, D. B. Amos, and B. Ephrussi, *J. Natl. Cancer Inst.*, 33, 893 (1964).

[22] Defendi, V., B. Ephrussi, H. Koprowski, and M. C. Yoshida, these Proceedings, 57, 299 (1967).

Reprinted from the PROCEEDINGS OF THE NATIONAL ACADEMY OF SCIENCES
Vol. 60, No. 4, pp. 1275–1281. August, 1968.

GENETICS OF SOMATIC MAMMALIAN CELLS, VII. INDUCTION AND ISOLATION OF NUTRITIONAL MUTANTS IN CHINESE HAMSTER CELLS*

BY FA-TEN KAO AND THEODORE T. PUCK†

DEPARTMENT OF BIOPHYSICS, UNIVERSITY OF COLORADO MEDICAL CENTER, DENVER

Communicated May 27, 1968

In a previous paper[1] a method was suggested for isolation of nutritionally deficient mutants in mammalian cells (Fig. 1). A large cell population is exposed to 5-bromodeoxyuridine (BUdR) in a medium lacking certain nutrilites. Those cells competent to grow in the given medium incorporate BUdR into their DNA and are subsequently killed by an exposure to near-visible light. The deficient mutants do not incorporate the brominated analogue and are unaffected by the illumination. These are then grown up into colonies by replacement of the

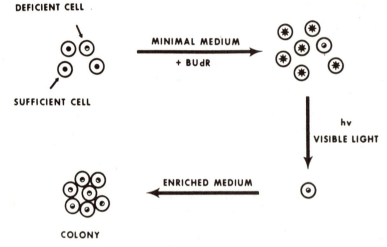

FIG. 1.—Schematic representation of the BUdR-visible light technique for isolation of nutritionally deficient mutant clones. The mixed cell population is exposed to BUdR in a deficient medium in which only the prototrophs can grow. These alone incorporate BUdR into their DNA and are killed on subsequent exposure to a standard fluorescent lamp. The medium is then changed to a composition lacking BUdR and enriched with various nutrilites, and the deficient cells grow up into colonies. About 10^{-4} of the slowly growing prototrophs may also escape killing by this method, so each colony is tested for its ability to grow in enriched but not in deficient medium.

nutritionally deficient medium with one enriched with various metabolites. The resulting colonies can be isolated and their nutritional requirements determined. The present report describes application of this method to the induction and isolation of mutants of Chinese hamster cells.

Materials and Methods.—The parental cell from which all the cultures described here were derived is the proline-requiring, Chinese hamster ovary cell, CHO/Pro⁻, which has only 21 chromosomes.[2] In some experiments, the subclone, CHO/Pro⁻ Kl, which possesses a stemline of only 20 chromosomes, was utilized. F12 medium[3] was used for routine cell cultivation and single-cell plating,[4] except where a deficient form of this medium was employed, as described. All plates were supplemented either with 10% fetal calf serum.

1275

or, when the introduction of small-molecular-weight nutrilites was to be avoided, with an equivalent amount of the macromolecular fraction alone, as described in earlier reports.[1, 2] The treatment with BUdR and visible light was carried out as described previously. After treatment in a minimal medium with BUdR, cells were illuminated with a standard fluorescent lamp, and then reincubated in an enriched medium consisting of complete F12 plus 10% whole fetal calf serum. This procedure results in colony formation from only 10^{-4} of the nutritionally sufficient cells, and from approximately half of the cells that have a nutritional deficiency like that of the proline-requiring mutant.[1] Consequently the practice was adopted of exposing about 10^6 cells to the BUdR-visible light procedure, and of testing approximately 100 of the colonies developing in the enriched medium for their ability to grow in a deficient medium.

Experimental Results.—(1) *Preparation of a minimal growth medium:* The standard F12 medium was originally developed to produce maximal growth rates for Chinese hamster cells in the absence of protein and consequently contains nutrilites that do not materially affect the growth rate or plating efficiency under the conditions employed here. Experiments were carried out to determine the minimal nutrilites required to produce maximal growth rate and plating efficiency for the standard CHO cell when supplemented with the macromolecular fraction of fetal calf serum. It was found that the following components could be omitted from F12 without affecting growth significantly: glycine, alanine, aspartic acid, glutamic acid, thymidine, hypoxanthine, inositol, vitamin B_{12}, and lipoic acid. The minimal medium obtained by omission of the nine indicated metabolites is called F12D.

(2) *Spontaneous mutation to auxotrophy:* Tests were carried out to determine whether spontaneous mutation in an appreciable frequency occurs at any of the loci responsible for nutritional independence with respect to the nine metabolites omitted from F12. No such spontaneous mutations were found when 10^6 cells of each clone were examined by the BUdR-visible light technique (Table 1). If the efficiency of detection of proline-deficient mutants in the BUdR-visible light procedure also holds for the nine metabolites investigated here, the individual frequencies of spontaneous mutation to deficiency for any one of these metabolites would appear to be $\leq 2 \times 10^{-6}$.

TABLE 1. *Yield of deficient mutants obtained in separate experiments with the two mutagens and the two standard clones employed.*

Mutagen	Cell employed	No. of clones surviving treatment with BUdR + near-visible light	No. of clones tested	No. of deficient mutant clones found
None	CHO/Pro⁻	129	100	0
None	Kl subclone	125	98	0
MNNG	CHO/Pro⁻	160	100	1
MNNG	Kl subclone	172	84	17
EMS	Kl subclone	175	91	8
EMS	Kl subclone	198	84	21

The MNNG was employed at a concentration of 0.50 $\mu g/ml$ for 16 hr, which leaves about 26% of the original cell population viable; while the EMS was used in a concentration of 200 $\mu g/ml$ for 16 hr, which leaves 78% of the cells viable. After exposure to the mutagen, 10^6 surviving cells were subjected to the BUdR-visible light technique, the colony survivors were counted, and approximately 100 of these tested for ability to grow in F12 but not in F12D.

(3) *Mutagenesis to auxotrophy by ethyl methanesulfonate (EMS) and N-methyl-N'-nitro-N-nitrosoguanidine (MNNG)*: Single cell survival curves were first constructed for each of these compounds, whose effectiveness in producing gene mutations in bacterial systems is well recognized.[5] Two hundred cells were inoculated into each of a series of Petri dishes in F12 supplemented with 10 per cent fetal calf serum. After six hours of incubation, EMS (Eastman Organic Chemicals, Rochester, N.Y.) or MNNG (Aldrich Chemical Co., Milwaukee, Wis.) was added in an appropriate concentration, and incubation was continued for an additional 16 hours, which is slightly more than one generation period. Thereafter, the medium was removed, the cells were washed three times, fresh growth medium was added, and incubation was continued for another five to seven days, after which the surviving colonies were counted. The results shown in Figure 2 demonstrate that, whereas the survival curve for EMS treatment has a large initial shoulder, the survival curve for MNNG approximates a simple exponential relationship with a much higher cell toxicity than that of the other mutagen.

Mutagenesis experiments were performed with CHO/Pro⁻ and its K1 subclone. In each experiment approximately $2-4 \times 10^6$ cells were treated for 16 hours with 200 μg/ml of EMS or with 0.50 μg/ml of MNNG, conditions providing the largest concentration of each drug at which a reasonably high cell survival is still achieved (78 and 26% for the EMS and MNNG, respectively). After this period, the medium was removed, the cells washed, and fresh growth medium re-

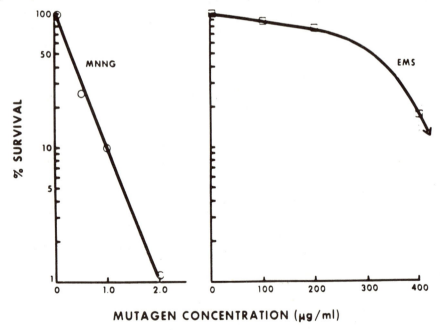

Fig. 2.—Survival curves for Chinese hamster cells exposed for 16 hr at 37°C to each of the mutagens of this study. Both cell clones utilized responded in similar fashion. The arrow in the EMS curve represents the fact that at 600 μg/ml of EMS, the plating efficiency was zero.

placed.   The cells were grown for an additional six days in order to allow full development of any end-point mutations.   The cells were trypsinized and 10^6 survivors were distributed among 40 plates, each finally containing 2.5×10^4 viable cells, and were then treated with BUdR and near-visible light in standard fashion for isolation of deficient mutants.   The results are summarized in Table 1, and indicate effective mutagenesis from both agents.

(4)   *Analysis of the induced nutritional requirements:*   The new nutritional requirements for 44 mutant clones here isolated are shown in Table 2.   Two

TABLE 2.   *Nutritional supplements required by the various newly isolated deficient mutants.*

Clone from which mutant originated	Mutagen employed	No. of Mutants Found with the Indicated Nutritional Deficiencies			
		Glycine	Glycine + thymidine + hypoxanthine	Inositol	Other requirements
CHO/Pro⁻	MNNG	1	0	0	0
Kl subclone	MNNG	4	10	1	2
Kl subclone	EMS	27	2	0	0

All mutants grow with plating efficiencies approximately 100% and zero in F12 and F12D, respectively.

clones have not yet been completely analyzed but appear to have different requirements from the others.   All clones display a plating efficiency of zero in F12D medium and virtually 100 per cent when the medium is supplemented with the agents shown.   All clones possess, in addition, the proline requirement described earlier that existed in the original cell culture from which these clones were derived.

Further measurements on these mutants are in progress.   The response to glycine by one of the mutants here isolated is shown in Figure 3 and a curve demonstrating the effect of varying glycine concentrations is shown in Figure 4. Spontaneous reversion to glycine independence for this mutant was sought, but no revertants have yet been found in a test population of 2×10^7 cells.   In another glycine-requiring mutant, one revertant was found in a test population of 5×10^7 cells.   Hence the revertant frequencies for these particular mutants appear to lie in the neighborhood of 3×10^{-8}, a value which permits good resolution in genetic experiments.

Mutagenesis with these agents appears to be unaccompanied by the extensive chromosomal alterations characteristic of the action of mutagens like X rays on mammalian cells.   Thus karyotypic analysis of five of the deficient mutants here described have revealed no detectable changes in number or structure of the chromosomes, as compared with the parental cell.

(5)   *Mutagenesis to prototophy at the proline locus:*   Both mutagens were also tested for their ability to produce revertants to proline-independence in the CHO/Pro⁻ cell.   In this case it was necessary only to plate the cells in a proline-deficient medium after treatment with the mutagenic agent, and to count the colonies which appear.   The results of five experiments have been pooled in Table 3, comparing the effectiveness of each mutagen.   EMS appears to be more effective than MNNG as a mutagen for this particular locus.

Discussion.—These experiments demonstrate that mutagenesis by agents

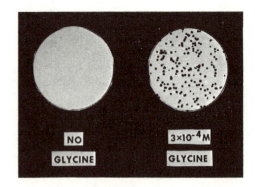

FIG. 3.—Demonstration that the glycine-requiring mutants induced by the chemical agents here employed yield no colonies in the absence of glycine but 100% plating efficiency in its presence. The prototroph has 100% plating efficiency in either case.

known to produce single gene mutations in bacteria and isolation of the resulting nutritionally deficient or sufficient forms can be carried out in simple fashion in mammalian cells. The techniques appear to permit production and quantitation of mutations to auxotrophy as well as to prototrophy.

The frequent appearance of the requirement for glycine in auxotrophic cells isolated after treatment with EMS and MNNG is noteworthy. In previous papers we have discussed the need for cells with specific chromosomal monosomies to overcome the difficulties of genetic analysis with diploid cells.[2, 6] In the present experiments one of the glycine mutants originated in the 21-chromosome CHO/Pro⁻ cell, while all of the remainder (including the mutant with the hypoxanthine and the thymidine requirements) arose from the 20-chromosome, K1 subclone. However, analysis revealed that even the glycine-deficient mutant from CHO/Pro⁻ possesses only 20 chromosomes.[6a]

As a working hypothesis, we presume that genes necessary for glycine biosynthesis are contained in one of the missing chromosomes of the cells under study, so that mutation of such loci on the remaining homologous chromosome resulted in expression of the glycine deficiency. One possible explanation for the high

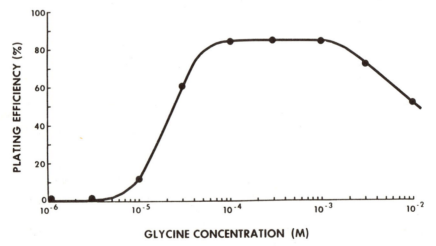

FIG. 4.—Dose-response curve to glycine by one of the glycine-requiring mutants of this study.

TABLE 3. *Test of EMS and MNNG to produce reversion to proline independence of the CHO/Pro⁻ cell.*

Mutagen	Total no. viable cells tested ($\times 10^6$)	No. of revertants	Reversion frequency ($\times 10^{-6}$)
None	15.2	18	1.2
EMS (200 μg/ml for 16 hr)	9.2	49	5.3
MNNG (0.5 μg/ml for 16 hr)	5.6	16	2.9

frequency of recurrence of the glycine deficiency may be that glycine is the only member of the nine metabolites searched for whose biosynthetic genes lay on the missing chromosomal members. This hypothesis will be tested by repeating these studies on Chinese hamster cells with no detectable chromosomal deficiencies and no heterozygosity. These conditions should be ascertainable by careful chromosomal analysis and by use of cells from the Yerganian Chinese hamsters which have been inbred for long periods so as to achieve a homozygous condition in the overwhelming majority to their genes.[7]

A systematic method which might provide large numbers of chromosomal monosomies in Chinese hamster cells and so permit extension of these studies will be described elsewhere.[8] An alternative possible approach to biochemical genetic studies on such cell cultures, utilizing specific antimetabolites that should greatly increase the number of usable mutations, has also been presented.[6]

The biochemistry underlying the glycine-deficiency mutation remains to be determined. Glycine is usually not required by mammalian cells, although such a requirement has been described in a primary monkey kidney culture.[9] Serine is the major source of glycine in mammalian cells, and the conversion is accomplished by the enzyme serine hydroxymethylase.[10] The one glycine-requiring mutant so far tested failed to utilize increased amounts of serine (10^{-4} to 10^{-1} M) to substitute for glycine. It is possible then that the new deficiency may reflect a serine hydroxymethylase defect.

The conversion of serine to glycine in mammalian cells requires reduced folic acid as a coenzyme, and the one-carbon moiety removed from serine is utilized in pyrimidine synthesis according to the following scheme:[11]

$$\text{Folic acid} \xrightarrow{\text{Folic acid reductase}} \text{Tetrahydrofolate (THF)},$$

$$\text{THF} + \text{serine} \xrightarrow{\text{Serine hydroxymethylase}} \text{Glycine} + \text{N}^5, \text{N}^{10}\text{-methylene THF},$$

$$\text{N}^5, \text{N}^{10}\text{-methylene THF} + \text{dUMP} \xrightarrow{\text{Thymidylate synthetase}}$$

$$\text{Thymidine monophosphate} + \text{dihydrofolate}.$$

The THF is also needed for the eventual production of inosinic acid, which, however, can be formed from hypoxanthine if it is supplied in the medium. Therefore, the mutants which require glycine, thymidine, and hypoxanthine for growth may represent deficiencies in folic acid reductase, so that all three metabolites become necessary. This provisional interpretation is supported by the fact that amethopterin, which inhibits folic acid reductase in mammalian cells, has also been found to introduce a requirement for these same three metabolites.[12]

Test for identity of the various glycine deficiencies may be possible by means of complementation studies with the use of the heterokaryon-production and cell-fusion techniques.[13, 14]   The low frequency of spontaneous reversion of the glycine-requiring mutants here described also makes these cells admirably suited for study of viral transduction and DNA transformation, and for quantitation of mutagenesis by drugs, radiation, and other agents.

The difference in the lethal activity displayed by the two mutagens here studied is noteworthy, and indicates important differences in their mode of action that may be susceptible to further biochemical analysis in this cell system. Studies are continuing, directed at the collection of further mutants, quantitation of mutagenic action by various agents, analysis of the underlying biochemistry of the genetic markers produced, and exploration of the relationships between a variety of mutagenic and carcinogenic agents on mammalian cells.

Summary.—Mutagenesis in Chinese hamster cells by ethyl methanesulfonate and N-methyl-N'-nitro-N-nitrosoguanidine, agents that produce single gene mutations in microorganisms, is described.   Auxotrophic mutants deficient in glycine, glycine + thymidine + hypoxanthine, and inositol were isolated by means of the technique that destroys prototrophs by exposure to BUdR followed by illumination with near-visible light.   Mutants so obtained are stable and exhibit low reversion rates so that they are useful for many kinds of genetic experiments.   Mutation to prototrophy at the proline locus was also achieved with these mutagens.   Application of these findings to various problems in mammalian cellular genetics has been indicated.

Grateful acknowledgement is made of the competent technical assistance of Miss Judy Hartz.

* From the Eleanor Roosevelt Institute for Cancer Research and the Department of Biophysics (contribution no. 335), University of Colorado Medical Center, Denver, Colorado. This investigation was aided by a grant no. 1 PO1 HD02080-02 from the National Institutes of Health, U.S. Public Health Service and by a grant no. DRG-337L from the Damon Runyon Memorial Fund for Cancer Research, Inc.

† American Cancer Society Research Professor.

[1] Puck, T. T., and F. T. Kao, these PROCEEDINGS, 58, 1227 (1967).

[2] Kao, F. T., and T. T. Puck, *Genetics*, 55, 513 (1967).

[3] Ham, R. G., these PROCEEDINGS, 53, 288 (1965).

[4] Ham, R. G., and T. T. Puck, in *Methods in Enzymology*, ed. S. P. Colowick and N. O. Kaplan (New York: Academic Press, 1962), vol. 5, p. 90.

[5] Loveless, A., and S. Howarth, *Nature*, 184, 1780 (1959); Zamenhof, S., L. H. Heldenmuth, and P. J. Zamenhof, these PROCEEDINGS, 55, 50 (1966).

[6] Puck, T. T., and F. T. Kao, these PROCEEDINGS, 60, 561 (1968).

[6a] Analysis of the karyotypes of these clones will be reported elsewhere.   However, such analysis is less illuminating than one could hope because the original progenitor CHO/Pro⁻ has undergone several aberrations which make difficult the identification of the missing parts of the chromosomal complement in it and its subclones.

[7] Yerganian, G., *J. Natl. Cancer Inst.*, 20, 705 (1958).

[8] Cox, D., and T. T. Puck, in preparation.

[9] Eagle, H., A. E. Freeman, and M. Levy, *J. Exptl. Med.*, 107, 643 (1958).

[10] Huennekens, F. M., and M. J. Osborn, *Advan. Enzymol.*, 21, 369 (1959).

[11] Lieberman, I., and P. Ove, *J. Biol. Chem.*, 235, 1119 (1960).

[12] Hakala, M. T., and E. Taylor, *J. Biol. Chem.*, 234, 126 (1959).

[13] Harris, H., and J. F. Watkins, *Nature*, 205, 640 (1965).

[14] Ephrussi, B., and M. C. Weiss, these PROCEEDINGS, 53, 1040 (1965).

[15] Kao, F. T., and T. T. Puck, *Abstracts*, 12th Annual Meeting of the Biophysical Society, Pittsburgh (1968).

Reprinted from Science, Vol. 187, pp. 1091–1093. 1975. © 1975 AAAS

Chromosome-Wide Event Accompanies the Expression of Recessive Mutations in Tetraploid Cells

Abstract. *Mutants resistant to 6-thioguanine were induced in pseudotetraploid hybrid Chinese hamster cells homozygous wild-type at the locus for hypoxanthine phophoribosyl transferase but heterozygous for the linked marker glucose-6-phosphate dehydrogenase. About half of these mutants had concomitantly lost the wild-type allele for glucose-6-phosphate dehydrogenase, as expected if mutation plus chromosome segregation had occurred.*

The interpretation that variant clones of cultured animal cells arise from mutations is complicated by the fact that the cells of higher organisms can also give rise to somatically heritable phenotypic changes by epigenetic processes. The latter are presumed to underlie cellular differentiation. If mutation in cultured cells is to be studied as a phenomenon per se, and if it is to be used as a tool to illuminate cell function, it is important to be able to distinguish mutation from epigenetic processes in any particular system.

There are now several systems in which the demonstration of an altered gene product in variant cells provides strong evidence for a mutational event (*1*). In most of these systems, a constitutive and ubiquitous enzyme has been studied, which makes it less likely that variants would arise epigenetically. However, in several of the same or similar systems, predictions of variant frequencies as a function of gene dosage, made on the assumption that purely mutational events were occurring, have not always been borne out. Thus Harris (*2*) found no difference in the spontaneous mutation rate to 8-azaguanine resistance in diploid, tetraploid, and octaploid Chinese hamster cells, and Mezger-Freed (*3*) found no difference between haploid and diploid frog cells in the frequency of 5-bromodeoxyuridine–resistant variants. Resistance to each of

these drugs is usually associated with a recessive loss in enzyme activity (*4–6*), although this point was not tested in the experiments cited. In the case of recessive mutations, it should be necessary to mutate each copy of the gene in question before a mutant phenotype can be expressed and recognized. If each mutation is an independent event, the mutant frequency should be an exponential function of the gene dosage; for example, the frequency with two genes should be approximately the square of the frequency in the single gene case. The failure of this prediction has led to the idea that mutation is not operating in these systems (*2, 3*).

We have previously examined this question, comparing mutation frequencies induced by ethyl methanesulfonate (EMS) in (pseudo) diploid and tetraploid Chinese hamster ovary (CHO) cells (*7*). Mutants resistant to 6-thioguanine (TG) were induced at a frequency of 2.5×10^{-4} in diploid cells, while in tetraploids the figure was 0.9×10^{-5}. Drug resistance was shown to be recessive and associated with the loss of hypoxanthine phosphoribosyltransferase (HPRT) activity. Since the gene specifying this enzyme is X-linked in humans (*8*) and probably also in Chinese hamster cells (*9, 10*) there is presumably only one active gene in diploid cells and two active genes in tetraploids (*11*). While the 25-fold difference was highly

Table 1. Distribution of G6PD phenotypes among TG-resistant mutant colonies derived from a tetraploid hybrid clone. The hybrid clone Y143 was induced to mutate with EMS and TG-resistant cells selected as previously described (7). Colonies that grew in TG were stained directly for G6PD activity (10).

Experiment	Frequency of TG-resistant colonies	Colonies (No.)	
		G6PD⁺	G6PD⁻
1	5×10^{-6}	40	59
2	1.1×10^{-6}	19	20
3	4.2×10^{-6}	53	71

significant, the mutant frequency in the tetraploids was still about 50 times higher than that expected on the basis of two independently occurring gene mutations. Rather than invoking an epigenetic explanation, we proposed a mechanism combining the spontaneous loss of one chromosome bearing an HPRT gene with mutation of the single remaining gene (7). The numerical results were consistent with such an interpretation based on segregation rates of the HPRT⁺ (TG-sensitive) phenotype from heterozygous tetraploid hybrid cells (12).

This explanation leads to at least three predictions: (i) TG-resistant mutants derived from tetraploid cells should all have lost one particular chromosome; (ii) since spontaneous chromosome loss is constantly occurring, segregants should accumulate with time, and the induced mutation rate should increase with clonal age; and (iii) markers linked to the HPRT locus should also have been segregated in mutants derived from tetraploids. The first prediction has been difficult to test, since most of these mutants have lost one to five chromosomes, and even within a clone the distribution of chromosome numbers in tetraploids is not as narrow as that found in diploid CHO cells (7).

The second point was examined by cloning a tetraploid subline of CHO cells that had been produced by treatment of the diploid with low doses of Colcemid (7, 13). The clonal population was grown in nonselective medium, and the EMS-induced mutation rate was measured at intervals over a period of 3 months. As can be seen in Fig. 1, the general trend (with the exception of one point) indicates increased mutability with clonal age. Although this result is in agreement with the segregation plus mutation hypothesis it does not represent strong evidence in its favor, as almost any proposal involving a spon-

taneous change in a cell as a prerequisite for mutability would predict such a time-dependent increase.

The third prediction is a much more stringent one, and can be tested by exploiting a recently isolated CHO mutant lacking glucose-6-phosphate dehydrogenase (G6PD) activity (10). The wild-type alleles of G6PD and HPRT are linked in CHO cells, since they cosegregate from heterozygous hybrid cells more than 92 percent of the time (10). If the segregation of one chromosome bearing the HPRT gene is almost always involved in the generation of TG-resistant (HPRT⁻) mutants from tetraploid cells, then segregation of the linked G6PD locus should also occur. This prediction can be tested by using as a tetraploid cell a hybrid heterozygous for G6PD and homozygous wild type for HPRT. Such a hybrid was therefore constructed, the parental lines being two mutant subclones of CHO cells, strain 43-64 (glyB⁻) and strain Y113 (glyA⁻ G6PD⁻) (10). Hybrids were selected on the basis of the complementing glycine auxotrophies carried by the two parental lines (14). One hybrid clone (strain Y143) was recloned in selective (glycine-free) medium and used for subsequent experiments. Its genetic constitution can be represented as

$$\frac{glyA^+ \ glyB^- \ HPRT^+ \ G6PD^+}{glyA^- \ glyB^+ \ HPRT^+ \ G6PD^-}$$

The hybrid has a modal chromosome number of 40, close to the sum of those of the parents (42 chromosomes). Its phenotype is G6PD⁺, the G6PD⁻ phenotype being a recessive trait (10). The G6PD phenotype of cells and colonies can be clearly demonstrated by using a histochemical stain for enzyme activity (10).

If random segregation of either of the chromosomes bearing the HPRT and G6PD loci were occurring before mutation, then 50 percent of the HPRT⁻ mutants derived from this hybrid should be G6PD⁺ and 50 percent G6PD⁻. The results of three separate experiments are shown in Table 1. There is good agreement with the predicted 50-50 distribution of G6PD⁺ and G6PD⁻ colonies among the mutants. It can also be seen that the average induced mutation frequency in these near-tetraploid hybrid cells is about 1/70 of that usually found in diploid cells (2.5×10^{-4}) (7) and is similar to the frequency found in tetraploid cells of similar clonal age that had

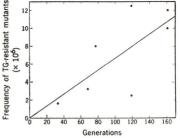

Fig. 1. Frequency of EMS-induced mutation as a function of clonal age. The pseudotetraploid line 5T11 of CHO cells was recloned and grown in monolayer culture in F-12 medium with 10 percent fetal calf serum. Cells were passaged when they reached confluence: at least 2×10^6 cells were transferred at each passage. Subpopulations were periodically induced to mutate and selected for TG-resistant cells as described previously (7). The total number of mutant colonies scored was 114, with a range of 5 to 25 in individual experiments.

been formed by low-Colcemid treatment (Fig. 1).

Although these results do implicate one chromosome-wide event in the generation of TG-resistant mutants from tetraploid cells, they do not clearly discriminate between chromosome loss, homozygosis through mitotic recombination, and X chromosome inactivation as a mechanism. A careful study of the karyotypes of these mutants should be able to demonstrate the first possibility. The results are not consistent with the occurrence of two point mutations as the basis for the TG-resistant phenotype, and they argue against a mechanism that specifically turns off HPRT activity in these cells by an epigenetic process.

There is little doubt that stable, heritable phenotypic changes that represent epigenetic alterations in regulatory programs can be demonstrated in cultured animal cells. The most convincing examples are the extinction (15) or induction (16) of differentiated cell products in hybrids formed between different types of specialized cells. It is equally clear that mutational changes can also be induced in cultured animal cells. If mutation is to be useful as a means to study gene regulation in culture, it will be necessary to distinguish these two types of change within one system.

Lawrence A. Chasin
Gail Urlaub
Department of Biological Sciences,
Columbia University,
New York 10027

References and Notes

1. A. Albrecht, J. Biedler, D. Hutchinson, *Cancer Res.* **32**, 1539 (1972); V. Chan, G. Whitmore, L. Siminovitch, *Proc. Natl. Acad. Sci. U.S.A.* **69**, 3119 (1972); A. Beaudet, D. Roufa, C. Caskey, *ibid.* **70**, 320 (1973); L. H. Thompson, J. L. Harkins, C. P. Stanners, *ibid.*, p. 3094; J. Sharp, N. Capecchi, M. Capecchi, *ibid.*, p. 3145; L. Chasin, A. Feldman, M. Konstam, G. Urlaub, *ibid.* **71**, 718 (1974); D. S. Secher, R. G. Cotton, C. Milstein, *FEBS (Fed. Eur. Biochem. Soc.) Lett.* **37**, 311 (1973); B. Birshtein, J.-L. Preud'homme, M. Scharff, in *The Immune System: Genes, Receptors, and Signals*, E. Sircarz and C. F. Fox, Eds. (Academic Press, New York, 1974), pp. 339–351.
2. M. Harris, *J. Cell Physiol.* **78**, 177 (1971).
3. L. Mezger-Freed, *Nat. New Biol.* **235**, 245 (1972).
4. E. Chu, P. Brimer, K. Jacobson, E. Merriam, *Genetics* **62**, 359 (1969).
5. J. W. Littlefield, *Science* **145**, 709 (1964).
6. Y. Matsuya, H. Green, C. Basilico, *Nature (Lond.)* **220**, 1199 (1968).
7. L. Chasin, *J. Cell Physiol.* **82**, 299 (1973).
8. V. McKusick and G. Chase, *Annu. Rev. Genet.* **7**, 435 (1973).
9. A. Westervald, P. Visser, M. Freeke, D. Bootsma, *Biochem. Genet.* **7**, 33 (1972).
10. M. Rosenstraus and L. Chasin, *Proc. Natl. Acad. Sci. U.S.A.*, in press.
11. M. Lyon, *Biol. Rev.* **47**, 1 (1972).
12. L. Chasin, *Nat. New Biol.* **240**, 50 (1972); G. Marin, *Exp. Cell Res.* **57**, 29 (1969).
13. M. Harris, *Exp. Cell Res.* **66**, 329 (1971).
14. F. Kao, L. Chasin, T. Puck, *Proc. Natl. Acad. Sci. U.S.A.* **64**, 1284 (1969).
15. R. Davidson, B. Ephrussi, K. Yamamoto, *J. Cell Physiol.* **72**, 115 (1969); J. Schneider and M. Weiss, *Proc. Natl. Acad. Sci. U.S.A.* **68**, 127 (1971); S. Silagi, *Cancer Res.* **27**, 1953 (1967).
16. G. Darlington, H. Bernhard, F. Ruddle, *Science* **185**, 859 (1974); J. Peterson and M. Weiss, *Proc. Natl. Acad. Sci. U.S.A.* **69**, 571 (1972).
17. We thank Maurice Rosenstraus for valuable assistance. Supported by grants V117A and V117B from the American Cancer Society.

11 November 1974

∎

THE JOURNAL OF BIOLOGICAL CHEMISTRY
Vol. 253, No. 5, Issue of March 10, pp. 1357–1370, 1978
Printed in U.S.A.

Selective Multiplication of Dihydrofolate Reductase Genes in Methotrexate-resistant Variants of Cultured Murine Cells*

(Received for publication, November 4, 1977)

FREDERICK W. ALT,‡ RODNEY E. KELLEMS, JOSEPH R. BERTINO,§ AND ROBERT T. SCHIMKE

From the Department of Biological Sciences, Stanford University, Stanford, California 94305

The rate of dihydrofolate reductase synthesis in the AT-3000 line of methotrexate-resistant murine Sarcoma 180 cells is approximately 200- to 250-fold greater than that of the sensitive, parental line. We have purified cDNA sequences complementary to dihydrofolate reductase mRNA and subsequently used this probe to quantitate dihydrofolate reductase mRNA and gene copies in each of these lines. Analysis of the association kinetics of the purified cDNA with DNA from sensitive and resistant cells indicated that the dihydrofolate reductase gene is selectively multiplied approximately 200-fold in the resistant line. A similar analysis of a partially revertant line of resistant cells indicated that the loss of resistance observed when the AT-3000 line is grown in the absence of methotrexate is associated with a corresponding decrease in the dihydrofolate reductase gene copy number. In each of these lines the relative number of dihydrofolate reductase gene copies is proportional to the cellular level of dihydrofolate reductase and dihydrofolate reductase mRNA sequences.

We have also studied parental and methotrexate-resistant lines of L1210 murine lymphoma cells. Both resistance and an associated 35-fold increase in the level of dihydrofolate reductase appear to be stable properties of the resistant L1210 line since we find no decrease in either parameter in over 100 generations of growth in the absence of methotrexate. Once again, we find that the increased levels of dihydrofolate reductase in the methotrexate-resistant L1210 line are associated with a proportional increase in the number of dihydrofolate reductase gene copies. In this case the dihydrofolate reductase gene copy number appears to be relatively stable in the resistant line. Therefore, we conclude that selective multiplication of the dihydrofolate reductase gene can account for the overproduction of dihydrofolate reductase in both stable and unstable lines of methotrexate-resistant cells.

The resistance of both human neoplasms (1) and various lines of cultured cells (2–11) to the 4-amino analogs of folic acid is often associated with an increase in the cellular content of dihydrofolate reductase. We have been studying the overproduction of this enzyme in variant lines of murine Sarcoma 180 cells that were selected by a step-wise procedure for growth in the presence of high concentrations of methotrexate (a folic acid analogue). Dihydrofolate reductase comprises as much as 6% of the soluble protein in the methotrexate-resistant AT-3000 line, representing an increase of more than 200-fold over the level in the sensitive, parental cells (12). Purified dihydrofolate reductase from resistant cells appears to be identical to that from sensitive cells; and, in addition, the relative half life of the enzyme is similar to these lines (12). We have demonstrated that the increased level of dihydrofolate reductase in resistant cells is due to an increased rate of enzyme synthesis (12), and that, in turn, this increase is correlated with increased cellular levels of translatable dihydrofolate reductase mRNA (13).

One of the most interesting characteristics of the AT-3000 line is that high levels of resistance are lost when these cells are grown in the absence of methotrexate (3). Loss of resistance is associated with a decrease in the level of dihydrofolate reductase (3, 12), and a corresponding decrease in both the rate of dihydrofolate reductase synthesis (12) and the level of the specific mRNA activity (13). Several lines of evidence suggest that these decreases are due to the instability of the variation (mutation?) which leads to increased enzyme synthesis (12). Instability is also a characteristic of methotrexate resistance in a number of other cell lines (9, 14). In contrast, in certain lines of methotrexate-resistant baby hamster kidney (BHK) cells (15) as well as in other resistant lines (2), resistance and increased levels of dihydrofolate reductase appear to be stable characteristics and do not decline when the cells are grown in the absence of the drug. However, other properties of methotrexate resistance in BHK cells appear to be similar to those of Sarcoma 180 cells, including high rates of dihydrofolate reductase synthesis (16) and high levels of translatable dihydrofolate reductase mRNA (17). Various considerations of the possible mechanisms that could lead to stable or unstable changes in the phenotypic expression

* This research was supported by Grant GM 14931 from the National Institute of General Medical Sciences, National Institutes of Health and Grant CA 16318 from the National Cancer Institute, National Institutes of Health. The costs of publication of this article were defrayed in part by the payment of page charges. This article must therefore be hereby marked "*advertisement*" in accordance with 18 U.S.C. Section 1734 solely to indicate this fact.

‡ Present address, Center for Cancer Research, Massachusetts Institute of Technology, 77 Massachusetts Ave., Cambridge, Mass. 02139.

§ Permanent address, Department of Pharmacology, Sterling Hall of Medicine, Yale University, 333 Cedar Street, New Haven, Conn. 06510.

of cultured cells have been discussed (18, 19).

We report here the purification of cDNA sequences complementary to dihydrofolate reductase mRNA of murine origin, and subsequent use of this probe to quantitate dihydrofolate reductase mRNA and gene copies in a number of different cell lines. We have examined both sensitive and methotrexate-resistant lines of Sarcoma 180 cells, as well as a partially revertant line that was derived by growing resistant cells in the absence of methotrexate for 400 cell doublings (12). In addition, we have also studied parental and methotrexate-resistant lines of L1210 murine lymphoma cells. Both resistance and associated high levels of dihydrofolate reductase appear to be stable properties of the L1210 lines since we find no decrease in either parameter over several hundred generations of growth in the absence of methotrexate. We find that in both the stable (L1210) and unstable (S-180) lines of resistant cells, increased levels of dihydrofolate reductase and dihydrofolate reductase mRNA are associated with a proportional increase in the number of dihydrofolate reductase gene copies. When unstable lines are grown in the absence of selection, loss of resistance is associated with a decrease in the dihydrofolate reductase gene copy number.

EXPERIMENTAL PROCEDURES

Materials – Sources of most of the reagents have been given previously (12, 13). Oligo(dT)-cellulose and oligo(dT)$_{12-18}$ were purchased from Collaborative Research; micrococcal nuclease, salmon sperm DNA, and calf thymus DNA from Sigma; S1 nuclease from Miles; [^{3}H]leucine (5 Ci/mmol) from New England Nuclear; [^{3}H]deoxycytidine triphosphate (20 Ci/mmol) from Amersham/Searle; Chelex 100 and Bio-Gel hydroxylapatite from Bio-Rad. Purified reverse transcriptase (Lot no. G-1176, 39,216 units/mg) was supplied by Dr. J. W. Beard (Life Sciences Inc., St. Petersburg, Florida) and methotrexate by Dr. Paul Davignon, Pharmaceutical Resources Branch, National Cancer Institute.

Generously provided as gifts were purified ovalbumin mRNA and *E. coli* tRNA from Dr. Gray Crouse (Stanford University), and purified chicken oviduct DNA from Dr. Henry Burr (Stanford University).

Cell Culture – The Sarcoma 180 cell line and the 3000-fold methotrexate-resistant AT-3000 subline were grown as described previously (12) except that thymidine and glycine were omitted from the medium of the resistant cells. A partially phenotypic revertant line, Rev-400, was obtained by growing the AT-3000 line for 400 cell doublings in methotrexate-free medium. Some characteristics of the "revision" phenomenon have been described previously (12), and further details are described under "Results."

Suspensions of L1210 murine lymphoma cells (L1210S) were grown in Fischer's Medium for Leukemic Cells of Mice (GIBCO) containing 10% horse serum. A 5000-fold methotrexate-resistant subline (L1210RR) and a 25,000-fold resistant subline (L1210 RR500) were grown in the same medium supplemented with 100 μM and 500 μM methotrexate, respectively. For some of the experiments described the L1210RR line was grown for approximately 100 cell doublings (over 10 months) in methotrexate-free medium. Further characteristics of these lines will be described under "Results" and elsewhere.[1]

Determination of the Relative Rate of Dihydrofolate Reductase Synthesis – The relative rate of dihydrofolate reductase synthesis was determined as described previously by direct immunoprecipitation of the enzyme from extracts of pulse-labeled cells (12).

RNA Preparation – Total cytoplasmic RNA was prepared from each of the cell lines as described previously (13). These preparations were used immediately or stored in liquid nitrogen.

Poly(A)-containing RNA was prepared by oligo(dT)-cellulose chromatography of total cytoplasmic RNA. RNA was dissolved in 10 mM Tris/Cl (pH 7.4) and 0.5% sodium dodecyl sulfate, heated at 68° for 5 min and rapidly cooled in an ethanol-ice bath. This solution was then adjusted to 0.4 M NaCl and oligo(dT)-cellulose chromatography

---

[1] C. Lindquist and J. Bertino, manuscript in preparation.

was carried out essentially as described by Aviv and Leder (20). The bound RNA fraction was eluted with 10 mM Tris/Cl (pH 7.4) and 0.5% sodium dodecyl sulfate, adjusted to 400 mM NaCl, and precipitated overnight at -20° by the addition of 2 volumes of ethanol. The precipitates were dissolved in a minimal volume of H_2O and stored in liquid N_2.

Polysome Preparation – The various lines were grown in roller bottles and were fed with fresh medium 4 h prior to harvest. Cells were rinsed once with ice cold Hanks' balanced salts solution plus 50 μg/ml cycloheximide, scraped from the bottles with rubber policemen, and washed three times by centrifugation through the same salt solution. Homogenization (13) and preparation of polysomes by the "cushion" method was as described previously by Palacios *et al.* (21), except that the homogenization buffer contained 10 mM MgCl$_2$. Polysomes were dialyzed for 12 h against 25 mM Tris/Cl (pH 7.1), 25 mM NaCl, 5 mM MgCl$_2$, and 1 mg/ml sodium heparin (Buffer A) and then stored in liquid nitrogen for subsequent use.

Antibody Purification – Rabbit anti-dihydrofolate reductase γ-globulin, prepared against purified dihydrofolate reductase protein as described previously (12), was purified by affinity chromatography on dihydrofolate reductase-Sepharose. Conditions for preparation of the resin and affinity chromatography were essentially as described by Shapiro *et al.* (22). Bound γ-globulin, eluted with 4.5 M MgCl$_2$, was enriched approximately 100-fold for anti-dihydrofolate reductase activity. The purified antibody preparation was made ribonuclease-free by passage through a column of DEAE-cellulose overlaid with CM-cellulose (21).

Iodination of Anti-dihydrofolate Reductase Globulin – Anti-dihydrofolate reductase globulin was iodinated by the lactoperoxidase method essentially as described by Taylor and Schimke (23). Iodinated antibody was made ribonuclease-free as described above.

Binding of Iodinated Anti-dihydrofolate Reductase Antibody to Polysomes – Prior to incubation, polysomes prepared as described above were thawed at 4° and centrifuged for 10 min at 5000 $\times$ g to remove particulate material. Reaction mixtures containing 30 A_{260} units of polysomes and 1.3 μg of iodinated anti-dihydrofolate reductase (specific radioactivity 77,000 cpm/μg) in 2 ml of Buffer A were incubated for 50 min at 0°. Polysomes were then reisolated from the reaction mixture by the "cushion" method as described above and sedimented through a linear sucrose gradient (0.5 M to 1.5 M in 11 ml of Buffer A) for 1.8 h at 4°. Gradients were fractionated and monitored for A_{260} with an Isco model 640 density gradient fractionator equipped with an ultraviolet flow monitor. For scintillation counting, 0.5-ml fractions were dissolved in 10 ml of Instagel (Packard).

Isolation of Dihydrofolate Reductase-synthesizing Polysomes – Indirect immunoprecipitation of polysomes was carried out essentially as described by Shapiro *et al.* (22). Resistant cell (AT-3000) polysomes at a final concentration of 10 to 15 A_{260} units/ml in 25 mM Tris/Cl (pH 7.1), 4 mM MgCl$_2$, 150 mM NaCl, 750 μg/ml sodium heparin, and 0.5% w/v Triton X-100 and sodium deoxycholate were incubated with optimal concentrations of purified rabbit anti-dihydrofolate reductase γ-globulin (20 μg/A_{260} unit of polysomes) for 60 min at 0°. The antibody-nascent chain complex was then precipitated by incubation with goat anti-rabbit γ-globulin (80 μg/μg of rabbit γ-globulin) for an additional 90 min at 0°. The precipitated complex was pelleted and washed as described by Shapiro *et al.* (22). Pellets were resuspended in 25 mM Tris/Cl (pH 7.1), 5 mM EDTA, 6 mM MgCl$_2$, 25 mM NaCl, 1 mg/ml sodium heparin, and 1% sodium dodecyl sulfate, and RNA was extracted by the phenol/chloroform procedure described previously (13).

RNA-dependent Rabbit· Reticulocyte Lysates – Micrococcal nuclease-treated rabbit reticulocyte lysates were prepared by a modification of the procedure described by Pelham and Jackson (24) as follows: standard rabbit reticulocyte lysate reaction mixtures were prepared as described previously (13) except that [^{3}H]leucine and RNA were omitted. Aliquots (325 μl) of this mixture were combined with 3.3 μl of 100 mM CaCl$_2$ (final concentration, 1 mM) and 3.3 μl of a 1 mg/ml solution of micrococcal nuclease (final concentration 10 μg/ml), and incubated for 15 min at 25°, at which time nuclease action was inhibited by the addition of 7 μl of 100 mM ethylene glycol bis(β-aminoethyl ether)*N,N'*-tetraacetic acid (final concentration, 2 mM). The nuclease-treated reticulocyte lysate mix prepared in this fashion was either used immediately or stored for up to 2 weeks in liquid nitrogen with no significant loss of activity.

Typical *in vitro* protein synthesis assays consisted of 60 μl of nuclease-treated lysate reaction mix, 4.6 μl of 200 μM [^{3}H]leucine (specific radioactivity, 5 Ci/mmol), and 25.4 μl of an aqueous solution of RNA. Following incubation for 1 h at 25°, the reaction was terminated by the addition of 36 μl of 0.1 M leucine and 14 μl of a mixture of 10% (w/v) sodium deoxycholate and 10% (w/v) Triton X-100. Stimulation of total protein synthesis was determined as the difference between trichloroacetic acid-precipitable radioactivity appearing in reactions that had received RNA and that in reactions to which no RNA had been added. Incorporation into dihydrofolate reductase was measured by specific immunoprecipitation as described previously (13) and expressed as a percentage of the total stimulated trichloroacetic acid-precipitable radioactivity in the lysate reaction. Under these conditions, incorporation into total trichloroacetic acid-precipitable radioactivity and dihydrofolate reductase was linear with time for up to 90 min with added poly(A)-containing RNA to 15 μg/ml, and with added total RNA to at least 50 μg/ml. Typical stimulation for a standard translation assay was approximately 100,000 cpm/μg of poly(A)-containing RNA, a level 20- to 30-fold greater than background.

Sodium Dodecyl Sulfate-Polyacrylamide Gel Electrophoresis of Lysate Products – After termination of lysate reactions, aliquots were removed and mixed with an equal volume of dissolving buffer, boiled for 3 min, and subjected to sodium dodecyl sulfate-polyacrylamide gel electrophoresis as described previously (12). Subsequent to electrophoresis, gels were soaked for 16 h in a liter of 7.5% acetic acid and 5% methanol (with one change of solution) in order to remove soluble radioactivity. Gels were then sliced and prepared for scintillation counting as described previously (12).

cDNA Preparation – cDNA was prepared essentially as described by Buell *et al.* (25). The reactions were carried out in 20-μl volumes and contained: 50 mM Tris/Cl (pH 8.3), 140 mM KCl, 30 mM β-mercaptoethanol, 10 mM MgCl$_2$, 100 μg/ml oligo(dT), 0.5 mM dGTP, dATP, and dTTP, 0.5 mM [³H]dCTP (20 Ci/mmol), 16 units of avian myeloblastosis virus reverse transcriptase, and 3 μg of poly(A)-containing RNA prepared from either total cytoplasmic-resistant cell RNA, or RNA extracted from immunoprecipitated dihydrofolate reductase synthesizing resistant cell polysomes.

Reactions were incubated at 42° for 1 h and stopped by the addition of 120 μl of 0.3 M NaOH. After a further incubation at 37° for 20 h, samples were neutralized with 1 N HCl and sodium dodecyl sulfate was added to a final concentration of 0.1%. The reaction mixtures were then extracted with 2 volumes of CHCl₃, and the aqueous phase passed over a small (8-ml) column of G-100 Sephadex which was previously equilibrated with H₂O. The void volume was pooled and concentrated by ethanol precipitation.

In all cases the yield was approximately 10⁶ cpm of cDNA per μg of added RNA. Based on the specific radioactivity of the [³H]dCTP and assuming equal representation of all four bases, this corresponds to approximately 0.1 μg of cDNA synthesized per μg of added RNA. Approximately 7 to 8% of the trichloroacetic acid-precipitable radioactivity in the cDNA preparation was resistant to treatment with S1 nuclease.

RNA/cDNA Hybridizations – All analytical RNA/cDNA hybridizations were done in 20 mM Tris/Cl (pH 7.7), 600 mM NaCl, 2 mM EDTA, and 0.2% sodium dodecyl sulfate except where noted otherwise. Reaction mixtures of 2 to 40 μl were overlaid with mineral oil in plastic tubes and incubated at 68°. The quantities of [³H]cDNA and RNA used in these reactions are described in appropriate figure legends. In all cases, final R_0t values were corrected to standard salt conditions (26).

At the end of the incubation, reaction mixtures were diluted into 1 ml of buffer containing 30 mM Na (C₂H₃O₂) (pH 4.5), 3 mM ZnSO₄, 300 mM NaCl, and 10 μg/ml denatured salmon sperm DNA. Each sample was divided into two aliquots: one was digested for 30 min at 45° with 8 μg/ml of S1 nuclease, and the other incubated identically, but without S1 nuclease. After digestion, 100 μg/ml of carrier calf thymus DNA was added to both S1-treated and control samples and nucleic acids precipitated with an equal volume of 10% trichloroacetic acid containing 1% sodium pyrophosphate at 4° for 15 min. Precipitates were collected on Millipore filters, washed three times with 5% trichloroacetic acid, dried, and counted in 10 ml of ScintiLene (Fisher).

Hybrid formation was scored as the amount of trichloroacetic acid-precipitable radioactivity remaining after S1 treatment and expressed as a percentage of the untreated control value. Depending on the cDNA preparation from 1.5 to 8% of the trichloroacetic acid-precipitable counts were resistant to S1 treatment in the absence of added RNA. In all experiments, the appropriate percentage of endogenous S1 resistance was subtracted from treated and control

values before calculation of the per cent hybridization. In calculating R_0t values, we assumed an average value of 346 g of RNA nucleotides per mol.

DNA Preparation – The 27,000 × g pellets (containing nuclei) resulting from standard RNA preparations (13) were stored at $-20°$. Approximately 5 ml of frozen nuclear pellet was thawed and gently homogenized in 50 ml of 0.15 M NaCl, 0.1 M EDTA (pH 8.0), 0.6 M sodium perchlorate, and 1.0% sodium dodecyl sulfate by five strokes in a dounce homogenizer (loose pestle). The homogenate was slowly stirred at 25° for 30 min, extracted with 2 volumes of chloroform, and DNA was spooled from the aqueous phase after the addition of 2 volumes of ice cold ethanol.

Spooled DNA was dissolved in 10 mM Tris/Cl (pH 7.4) and then treated with 60 μg/ml pancreatic ribonuclease (boiled for 10 min in 20 mM NaCl prior to use) for 2 h at 37°. Sodium dodecyl sulfate and proteinase K were then added to a final concentration of 0.2% and 60 μg/ml, respectively, and the incubation continued for another 5 h at 37°. The solution was then extracted with 2 volumes of CHCl₃ and the aqueous phase precipitated overnight at $-20°$ with 2 volumes of ethanol. Precipitated DNA was pelleted by centrifugation for 5 min at 2000 × g, lyophilized, and dissolved in 100 mM sodium acetate (pH 7.8). DNA was then sheared by passage through the needle valve of a French pressure cell at a pressure of 20,000 p.s.i. Divalent cations were removed by passing the sheared DNA preparations over a small (10 ml) volume of Chelex (equilibrated with 100 mM sodium acetate, pH 7.8), and the DNA was subsequently ethanol-precipitated as described above and redissolved in 20 mM Tris/Cl (pH 7.4) and 1 mM EDTA. 1 M NaOH was then added to a final concentration of 0.3 M and the solution was incubated for 22 h at 37°, at which time the base was neutralized by the addition of an equivalent amount of 1 N HCl.

These preparations were then stored at 4° until subsequent use as described below. All of the DNA samples prepared in this fashion sedimented as symmetrical peaks on isokinetic alkaline sucrose gradients (see below) with a calculated size of approximately 450 base pairs.

Sedimentation Analysis of DNA – DNA was analyzed by sedimentation through isokinetic alkaline sucrose gradients prepared as described by McCarty *et al.* (27) using 5% and 29.4% sucrose containing 0.1 N NaOH and 0.9 M NaCl. The molecular size of the DNA was calculated from S value as described by Studier (28).

cDNA/DNA Association Reactions – DNA/DNA associations were done in reaction mixtures containing 25 mM Tris/Cl (pH 7.4), 1 mM EDTA, 300 mM NaCl, 50 pg of [³H]cDNA (500 cpm), and 500 μg of cellular DNA (prepared as described above) in a final volume of from 0.05 ml to 1.1 ml. Reaction mixtures were overlaid with mineral oil in plastic tubes, heated to 102° for 10 min in an H₂O/ethylene glycol bath, cooled, and incubated at 68° for various times in order to achieve the desired C_0t values.

Single- and double-stranded DNA were then fractionated by chromatography on hydroxylapatite. Reaction mixtures were diluted into 5 ml of 0.12 M NaPO₄ (pH 6.8) and passed over a column containing 1 g of hydroxylapatite (boiled for 5 min in 5 ml of 0.12 M NaPO₄ prior to use and equilibrated in the same buffer) which was maintained at 60° with a recirculating water bath. Single-stranded DNA was eluted with 0.12 M NaPO₄ (pH 6.8) and double-stranded material subsequently eluted with 0.5 M NaPO₄ (pH 6.8). The single- and double-stranded fractions were monitored for A_{260}, and the DNA then was precipitated by the addition of carrier calf thymus DNA to 25 μg/ml and 0.1 vol of 100% trichloroacetic acid. Trichloroacetic acid-precipitable material was collected and counted as described above. In order to calculate DNA concentration, an A_{260} absorbance of 1 was assumed to correspond to DNA concentrations of 43 μg/ml and 50 μg/ml, respectively, for single- and double-stranded DNA fractions. The per cent double-stranded in each sample was determined by dividing the amount of DNA or [³H]cDNA recovered in the double-stranded fraction by the total amount recovered in the double- and single-stranded fractions. In calculating C_0t values we assumed an average value of 332 g of DNA nucleotides per mol.

RESULTS

Purification of Dihydrofolate Reductase-specific cDNA

Immunoprecipitation of Dihydrofolate Reductase-synthesizing Polysomes – In order to further study the factors responsible for the accumulation of high levels of translatable dihydrofolate reductase mRNA in methotrexate-resistant cells, we needed a cDNA probe complementary to dihydrofolate reductase mRNA. The usual method for the preparation of such a reagent has involved purification of a specific mRNA and subsequent synthesis of a complementary cDNA. Dihydrofolate reductase mRNA contains poly(A), allowing easy separation from rRNA, but its sedimentation rate on sodium dodecyl sulfate or denaturing sucrose gradients is not sufficiently distinct from that of total poly(A)-containing RNA to permit a significant additional purification by size fractionation (13). Therefore, we have employed the specific polysome immunoprecipitation procedure described by Shapiro *et al.* (22) to enrich for dihydrofolate reductase-synthesizing polysomes. The initial step in this procedure involved incubation of purified (100-fold) anti-dihydrofolate reductase antibody with resistant cell polysomes. The data in Fig. 1*a* demonstrate that this procedure results in the specific binding of the antibody to a size class of resistant cell polysomes (5 to 7 ribosomes) expected for those engaged in the synthesis of dihydrofolate reductase ($M_r = 20,000$). However, only a low level of apparently nonspecific binding is observed with polysomes from sensitive cells (Fig. 1*b*), where the rate of dihydrofolate reductase synthesis is below the resolution level of this technique. These results suggest that the incubation procedure results in the binding of purified antibody specifically to dihydrofolate reductase nascent chains. Subsequent to the initial binding reaction, the resulting antibody·nascent chain·polysome complexes were precipitated with a second antibody directed against the first antibody (see "Experimental Procedures" for details). We estimated the purification achieved by this procedure by translating the poly(A)-containing RNA extracted from the immunoprecipitated polysomes in the mRNA-dependent rabbit reticulocyte lysate (25). At the end of the incubation, samples of the total lysate reaction mix were analyzed by electrophoresis on sodium dodecyl sulfate-polyacrylamide gels. Fig. 2*a* indicates the very low background observed in this system in the absence of added RNA. In contrast, addition of purified ovalbumin mRNA resulted in the stimulation of a single peak of incorporated radioactivity with a mobility characteristic of authentic ovalbumin (Fig. 2*b*). This result indicates that the generation of incomplete or fragmented polypeptide chains is not a problem with this system. Furthermore, the specificity of the assay is evidenced by the fact that in this experiment more than 95% of the stimulated incorporation was precipitable with anti-ovalbumin antibody (data not shown). The addition of polysomal poly(A)-containing RNA from resistant cells resulted in the synthesis of a broad size distribution of proteins of which approximately 1.9% were precipitable by anti-dihydrofolate reductase antibody (data not shown). This value corresponds well to our estimate of the relative rate of dihydrofolate reductase synthesis as a per cent of total protein synthesis in this line.[2] Poly(A)-containing RNA extracted from the immunoprecipitated polysomes stimulated incorporation into a single major peak of radioactivity which co-migrated with added dihydrofolate reductase marker (Fig. 2*d*). In this experiment, 25% of the stimulated incorporation was precipitable by anti-dihydrofolate reductase antibody (data not shown). Comparison of the relative incorporation into dihydrofolate reductase

[2] We have previously described dihydrofolate reductase synthesis as a per cent of soluble protein synthesis. In these lines, soluble protein accounts for approximately 20 to 30% of the total protein synthesis (data not shown).

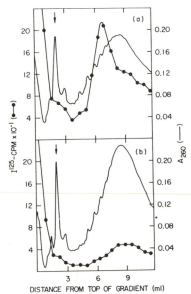

Fig. 1. Binding of anti-dihydrofolate reductase antibody to polysomes. The binding of ^{125}I-labeled anti-dihydrofolate reductase antibody to polysomes from AT-3000 (*a*) and S-3 (*b*) cells was examined as described under "Experimental Procedures." ^{125}I-radioactivity, ●——●; A_{260}, ——.

in the experiments presented in Fig. 2, *c* and *d* indicates an approximately 10-fold purification of dihydrofolate reductase mRNA by the polysome precipitation procedure.

cDNA Synthesis from the Partially Purified Dihydrofolate Reductase mRNA — cDNA was prepared from the partially purified dihydrofolate reductase mRNA resulting from the immunoprecipitation procedure and then analyzed by hybridization to excess poly(A)-containing RNA from sensitive or resistant cells (Fig. 3*a*). Comparison of the kinetics of these reactions indicates that approximately 15 to 20% of the cDNA sequences hybridize to mRNA sequences that are considerably more abundant in resistant cells than in sensitive cells. This percentage roughly corresponds with our estimate of the proportion of dihydrofolate reductase mRNA sequences in the partially purified RNA preparation from which the cDNA was synthesized (see above). However, these data indicate that this cDNA preparation is not pure enough for use as an analytical reagent. In order to further enrich for cDNA sequences complementary to the dihydrofolate reductase mRNA, we devised the purification scheme described below.

Purification of cDNA Sequences Complementary to Dihydrofolate Reductase mRNA — As a further means of purification of dihydrofolate reductase-specific sequences in the cDNA preparation, we exploited the large and apparently specific increase in the abundance of dihydrofolate reductase mRNA sequences in the RNA population of resistant as compared to sensitive cells. Analysis of the soluble proteins produced by sensitive and resistant cells suggested that the only major difference between the two is the overproduction of dihydrofolate reductase (12). Furthermore, we have used a reticulocyte lysate *in vitro* translation assay to demonstrate that most, if not all, of the several hundred-fold increase in the level of dihydrofolate reductase synthesis in resistant cells can be

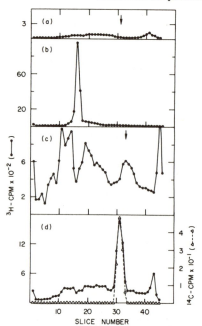

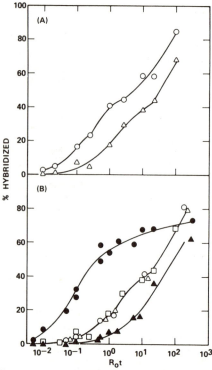

Fig. 2. Electrophoretic analysis of mRNA-dependent rabbit retic-ulocyte reaction products. Aliquots from mRNA-dependent lysate reactions stimulated with (*a*) no added RNA, (*b*) 1.5 μg of purified ovalbumin mRNA, (*c*) 1.25 μg of poly(A)-containing RNA prepared from AT-3000 cell polysomes, (*d*) 1 μg of poly(A)-containing RNA prepared from immunoprecipitated dihydrofolate reductase-synthe-sizing polysomes (see "Experimental Procedures" for details) were examined by sodium dodecyl sulfate-polyacrylamide gel electropho-resis as described under "Experimental Procedures." Other aliquots of the stimulated lysate reactions were used for the specific immu-noprecipitation of ovalbumin (*reaction b*) as described by Rhoads *et al.* (29) and dihydrofolate reductase (*reactions c* and *d*) as described previously (12). The incorporation into each of these proteins relative to total stimulation was measured as described under "Experimental Procedures." ³H-labeled lysate reaction products, ●——●; authentic ¹⁴C-labeled dihydrofolate reductase, △- - -△. *Arrows* mark the migration of added ¹⁴C-labeled dihydrofolate reductase in *panels a* and *c*.

Fig. 3. *A*, hybridization of cDNA prepared from partially purified dihydrofolate reductase mRNA to poly(A)-containing RNA from resistant and sensitive S-180 cells. Poly(A)-containing RNA ex-tracted from the S-3 (△——△) and AT-3000 (○——○) cell lines was reacted with 60 pg (600 cpm) of [³H]cDNA prepared from partially purified dihydrofolate reductase mRNA (see text). Similar quanti-ties of RNA from sensitive and resistant lines, ranging from 0.1 μg to 1 μg per sample were used to drive hybridization reactions that were stopped at corresponding R_0t values. Other reaction conditions and measurement of the extent of hybridization by S1 nuclease hydrolysis are described under "Experimental Procedures." Endog-enous S1 resistance of this cDNA preparation was approximately 8%. *B*, hybridization of partially purified cDNA from sensitive and resistant cells. The [³H]cDNA recovered in the single- and double-stranded fractions resulting from *Step A* of the purifica-tion procedure outlined in Fig. 4 was hybridized to excess poly(A)-containing RNA from S-3 and AT-3000 cells. Reaction conditions were essentially as described in *A*. Hybridization to AT-3000 RNA: [³H]cDNA from single- (●——●) and double-stranded (○——○) fractions; hybridization to S-3 RNA: [³H]cDNA from single- (▲——▲) and double-stranded (△——△) fractions. Hybridization of unfractionated [³H]cDNA to S-3 RNA is reproduced from *A* (□——□).

attributed to a similar increase in the level of dihydrofolate reductase mRNA activity (13). Therefore, we estimate that dihydrofolate reductase mRNA sequences are as much as 200 to 300 times more abundant in resistant cells than in sensitive cells, whereas other mRNA sequences are probably present in similar abundance in the two cell types. Furthermore, since the cDNA was prepared from resistant cell RNA that was enriched an additional 10-fold for dihydrofolate reductase mRNA sequences, the dihydrofolate reductase-specific se-quences in the cDNA preparation could be as much as 2000- to 3000-fold more abundant than the complementary se-quences in sensitive cell poly(A)-containing RNA. These esti-mates provide the rationale for Step A of the dihydrofolate reductase-specific cDNA purification procedure that is out-lined in Fig. 4. The cDNA preparation was incubated with a 30-fold mass excess of sensitive cell poly(A)-containing RNA to a R_0t value sufficiently high to ensure completion of the reaction (see legend to Fig. 4). Under these conditions, cDNA complementary to mRNA sequences that are present in simi-lar abundance in resistant and sensitive cells (or in greater

abundance in sensitive cells) should be driven into hybrids by the excess sensitive cell RNA. However, the majority of the cDNA sequences that are complementary to mRNA sequences present in far greater abundance in resistant cells than in sensitive cells (relative to the 30-fold mass excess of sensitive cell RNA) will remain single-stranded at the end of the reaction. Therefore, based on the estimates described above, these unhybridized cDNA sequences should be greatly en-riched for sequences complementary to dihydrofolate reduc-tase mRNA.

Subsequent to the hybridization reaction, single- and double-stranded material was separated by chromatography on hydroxylapatite, RNA removed by alkaline hydrolysis, and the cDNA from both fractions analyzed by hybridization to excess RNA from sensitive and resistant cells. The cDNA

Step A

1. Poly(A)-containing RNA from resistant cells (enriched for dihydrofolate reductase sequences)

$\downarrow$

[^{3}H]cDNA

$\downarrow$

2. Hybridized to 30-fold mass excess of poly(A)-containing RNA from sensitive cells (final R_0t = 1600 mol-s/liter)

$\downarrow$

3. Nonhybridized [^{3}H]cDNA recovered by hydroxylapatite chromatography

Step B

1. [^{3}H]cDNA selected by Step A hybridized to large excess of resistant cell poly(A)-containing RNA (final R_0t = 0.8 mol-s/liter)

$\downarrow$

2. Hybridized [^{3}H]cDNA recovered by hydroxylapatite chromatography

Fig. 4. Purification of cDNA sequences complementary to dihydrofolate reductase mRNA. *Step A*, hybridization to a limited excess of sensitive cell poly(A)-containing RNA. Approximately 200 ng of [^{3}H]cDNA that was prepared from resistant cell (AT-3000) poly(A)-containing RNA extracted from partially purified dihydrofolate reductase-synthesizing polysomes (see "Experimental Procedures" and text for details) was hybridized to 6 μg of sensitive cell (S-3) poly(A)-containing RNA (30-fold mass excess of RNA). The final reaction volume was 20 microliters, and the other conditions were as described under "Experimental Procedures." The extent to which the preparative reaction approached maximum hybridization was estimated at various times (R_0t values) by measuring the S1 nuclease resistance of control samples identical to the preparative reaction just described except that only 400 pg of [^{3}H]cDNA was used (15,000-fold mass excess of RNA). At an R_0t of 1600 mol-s/liter, a value where the cDNA in the control samples was essentially 100% S1 nuclease-resistant, the preparative sample was diluted with 68 μl of H_2O containing 15 μg each of native and denatured salmon sperm DNA (sheared to 400 base pairs) and 12 μl of 1 M $NaPO_4$ (final concentration, 0.12 M). Single- and double-stranded material was then fractionated by chromatography on hydroxylapatite essentially as described under "Experimental Procedures." Approximately 23% of the radioactivity failed to bind to the column in 0.12 M $NaPO_4$. This represented single-stranded material since greater than 93% was sensitive to S1 nuclease digestion. The remainder of the cDNA was eluted in the double-stranded fraction with 0.4 M $NaPO_4$ and was essentially 100% resistant to S1 nuclease. RNA was removed from the double-stranded fraction by base hydrolysis (see "Experimental Procedures" for details) and the cDNA from each of these fractions was tested by hybridization to excess poly(A)-containing RNA from resistant and sensitive cells. (See text and legend to Fig. 3B for details.) *Step B*, low R_0t fractionation of cDNA-resistant cell poly(A)-containing RNA hybrids. Approximately 30 ng of [^{3}H]cDNA, prepared as described in *Step A*, were hybridized to 165 μg of poly(A)-containing RNA from resistant (AT-3000) cells (approximately 100-fold excess of dihydrofolate reductase-specific RNA sequences) in a 600-μl reaction mixture containing 0.12 M $NaPO_4$, 1 mM EDTA, and 0.1% sodium dodecyl sulfate. After incubation at 68° to a R_0t of 0.8, the reaction mix was diluted to 3.7 ml with 0.12 M $NaPO_4$ plus 25 μg each of native and denatured salmon sperm DNA. At this point, 43% of the cDNA was resistant to S1 nuclease. Single- and double-stranded material was fractionated by hydroxylapatite chromatography as described above, and approximately 35% of the radioactivity was recovered in the 0.5 M $NaPO_4$ (double-stranded) fraction. RNA was removed from this fraction by alkaline hydrolysis as described under "Experimental Procedures." Following neutralization, 25 μg of *E. coli* tRNA carrier was added and $NaPO_4$ was removed by chromatography on Sephadex G-100. The void volume was pooled, concentrated by ethanol precipitation, and dissolved in H_2O.

recovered in the double-stranded fraction should contain sequences present at a similar abundance in both cell types. As expected, this cDNA fraction hybridized to RNA from sensitive (Fig. 3B, $\triangle$——$\triangle$) and resistant (Fig. 3B, $\bigcirc$——$\bigcirc$) cells with kinetics that were essentially identical to each other and to those with which the unfractionated cDNA preparation hybridized to RNA from sensitive cells (Fig. 3B, $\square$——$\square$). However, most of the cDNA recovered in the single-stranded fraction hybridized to excess RNA from resistant cells (Fig. 3B, $\bullet$——$\bullet$) at a rate approximately 200-fold greater than to that of sensitive cells (Fig. 3B, $\blacktriangle$——$\blacktriangle$). This difference is consistent with our estimate of the relative level of dihydrofolate reductase mRNA sequences in these cell types. We recovered approximately 23% of the unfractionated cDNA in the single-stranded fraction, and of this about 65 to 70% had highly accelerated kinetics when hybridized to RNA from resistant cells as opposed to that of sensitive cells. Assuming that the relative abundance of sequences in the cDNA preparation is representative of the mRNA population from which it was derived, this recovery roughly corresponds to that expected for dihydrofolate reductase-specific sequences. The maximum hybridization observed with this cDNA fraction was never above 80%. Presumably, the explanation for this result is that in selecting for single-stranded material after the hybridization described above, we also enrich for any nonhybridizable material present in the unfractionated cDNA preparation.

As a final purification step (*Step B*, Fig. 4), the cDNA selected in Step A was hybridized to a 140-fold mass excess of resistant cell poly(A)-containing RNA, to a final R_0t of 0.8 mol-s/liter. Hybridized sequences were then isolated by chromatography on hydroxylapatite. As can be seen in Fig. 3B, cDNA complementary to RNA sequences that are highly abundant in resistant cells (putative dihydrofolate reductase-specific sequences) are hybridized at this R_0t and are therefore selected. However, both the low level of cDNA sequences that appear to hybridize to less abundant RNA sequences and the nonhybridizable material selected by the previous step are excluded. Approximately 35 to 40% of the cDNA was recovered in the double-stranded fraction in this step. When analyzed by alkaline, isokinetic sucrose gradient centrifugation (see "Experimental Procedures" for details) this material sedimented as a symmetrical peak at 5.4 S, with a calculated size of approximately 350 bases. The specificity of this purified cDNA fraction was then analyzed as described below.

Specificity of the Purified cDNA – The cDNA selected by the final step of the purification procedure (Step B) should represent the portion of the cDNA resulting from the previous step that had highly accelerated kinetics when hybridized to RNA from resistant cells as opposed to that of sensitive cells. As expected, this material still hybridizes to excess poly(A)-containing RNA from resistant cells at a rate approximately 200-fold greater than to that of sensitive cells (Fig. 5). However, these hybridization reactions now approach 100% with kinetics suggestive of a single, pseudo-first order reaction. This result suggests, but does not prove, that the purified cDNA preparation consists mainly of sequences complementary to a single species of mRNA. (See below for further discussion of this point.) Since the purification procedure would enrich for any cDNA sequence complementary to mRNA present at high abundance in the resistant but not the sensitive cells employed in the procedure, we further defined the specificity of the purified cDNA by analyzing the hybridization of this material to poly(A)-containing RNA

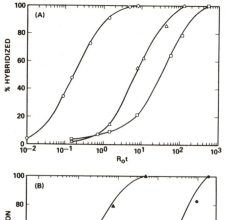

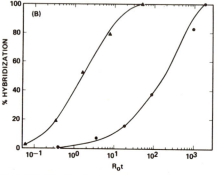

FIG. 5. *A*, hybridization of purified cDNA to RNA from sensitive, resistant, and partially revertant lines of S-180 cells. Poly(A)-containing RNA isolated from S-3 (0.17 μg to 12 μg/sample, □——□), AT-3000 (0.17 μg to 2.2 μg/sample, ○——○), and Rev-400 (0.17 μg to 5 μg/sample, △——△) were reacted with 30 pg (300 cpm) of the purified [³H]cDNA (selected as described in the legend to Fig. 4) and the extent of hybridization at the indicated $R_o t$ values measured by hydrolysis with S1 nuclease. (See "Experimental Procedures" for details.) Endogenous resistance of the purified cDNA to S1 nuclease hydrolysis was approximately 1.5%. *B*, hybridization of purified cDNA to RNA from sensitive and methotrexate-resistant lines of mouse L1210 lymphoma cells. Poly(A)-containing RNA isolated from the L1210S (0.055 μg to 5.5 μg/sample, ●——●), and L1210 RR500 (0.004 μg ot 4 μg/sample, ▲——▲) cell lines were reacted with 40 pg (400 cpm) of purified [³H]cDNA as described in the legend to Fig. 4.

extracted from several other cell types in which dihydrofolate reductase levels vary widely as a function of methotrexate resistance.

By growing resistant sarcoma 180 cells in the absence of methotrexate for 400 cell doublings (12), we have established a partially revertant line (Rev-400) in which the level of dihydrofolate reductase has declined to an apparently stable[3] value approximately 10-fold greater than that of sensitive cells (Table I). This decrease is also accompanied by a decrease in the relative synthesis of dihydrofolate reductase (12) and the level of translatable dihydrofolate reductase mRNA (13). Comparison of the $R_o t_{1/2}$ value for the reaction of the purified cDNA with RNA from partially revertant cells (Fig. 5*a*) to those observed in the reactions with sensitive and resistant cell RNA indicates that in each of these lines the abundance of mRNA sequences complementary to the purified cDNA is proportional to the relative level of dihydrofolate reductase (Table I). In addition, we have also examined hybridization of

[3] R. Kaufman, unpublished observation.

TABLE I
*Relative level of dihydrofolate reductase activity, mRNA and gene
copies in S-180 and L1210 lines*

The origin of each of the lines noted above is described under "Experimental Procedures" and "Results." In each column, values are normalized to those of the sensitive line which was taken as 1. Dihydrofolate reductase-specific activity was assayed as described previously (12). The relative abundance of dihydrofolate reductase mRNA sequences in the S-180 lines was determined from the inverse of $R_o t_{1/2}$ values for the reactions shown in Fig. 5*A*. The relative number of dihydrofolate reductase gene copies in the S-180 and L1210 lines was determined from the inverse of the $C_o t_{1/2}$ values of the reactions shown in Figs. 8 and 9, respectively. In order to estimate the $C_o t_{1/2}$ of the reaction with L1210S DNA (Fig. 9), we assumed this reaction would proceed to the same extent as the others.

Line	Relative dihydrofolate reductase		
	Specific activity	mRNA sequences	Gene copies
S-180			
S-3	1	1	1
AT-3000	250	220	180
Rev-400	10	7	10
L1210			
S	1		1
RR(+mtx)	35		45
RR(−mtx)	35		35

the purified cDNA to excess RNA from both murine L1210 lymphoma cells, as well as a 25,000-fold methotrexate-resistant subline, the RR500 (Fig. 5*b*). Relative to the parental line, the RR500 subline has an approximately 80-fold greater level of dihydrofolate reductase activity that is associated with an increase in both the level of dihydrofolate reductase synthesis and translatable dihydrofolate reductase mRNA (data not shown). The data in Fig. 5*b* indicate that the purified cDNA hybridizes to excess RNA from the L1210 RR500 line at a rate approximately 100-fold greater than that observed with RNA from the sensitive parental line. Therefore, sequences complementary to the purified cDNA are again present at a level proportional to the relative dihydrofolate reductase content of these two cell types. Thus these results, which link the abundance of RNA sequences complementary to the purified cDNA to dihydrofolate reductase levels in a variety of different cell lines, strongly suggest that the purified cDNA preparation consists specifically of sequences complementary to dihydrofolate reductase mRNA. This conclusion is substantiated by two independent lines of evidence which are described below.

We did not size-fractionate either the RNA or the cDNA in the purification procedure. Therefore, another criterion of the specificity of the purified cDNA would be to show that it is specifically complementary to RNA the size of dihydrofolate reductase mRNA. Thus, total RNA from resistant cells was fractionated on isokinetic sucrose gradients, and an equal portion of each fraction was hybridized to purified cDNA under conditions where the per cent hybridization is roughly proportional to the concentration of complementary RNA sequences (30). Other aliquots were used to assay both total and dihydrofolate reductase mRNA activity. As shown by the data in Fig. 6, the purified cDNA hybridizes specifically to a size class of RNA that is distinct from that of total mRNA activity and identical to that of translationally active dihydrofolate reductase mRNA.

Finally, we analyzed the hybridization of the purified cDNA

to both total polysomal RNA from resistant cells, as well as RNA from the same preparation in which dihydrofolate reductase sequences were either enriched or depleted by immunoprecipitation of dihydrofolate reductase-synthesizing polysomes. As shown in Fig. 7, the immunoprecipitation procedure specifically enriched for RNA sequences complementary to the purified cDNA. This result links the specificity of the purified cDNA to the previously demonstrated specificity of the anti-dihydrofolate reductase antibody (12). More importantly, however, the abundance of RNA sequences complementary to the purified probe was directly proportional to the level of dihydrofolate reductase mRNA in each of these fractions (Table II). Therefore, RNA sequences complementary to the purified cDNA are enriched identically to dihydrofolate reductase mRNA by immunoprecipitation of dihydrofolate reductase-synthesizing polysomes.

In summary, in all of the kinetic analyses described above the purified cDNA hybridized with excess RNA to essentially 100% with kinetics suggestive of a single, pseudo-first order reaction. Although this observation suggests that the cDNA is complementary to a single species of mRNA, indistinguishable reaction kinetics would be observed if the cDNA preparation consisted of several different sequences, all of which had complements present at identical abundance in the driver RNA. However, in all RNA preparations that we have examined, the rate at which the purified cDNA hybridized was proportional to the level of dihydrofolate reductase mRNA activity. This was true both in experiments where the abundance of dihydrofolate reductase mRNA varied due to biological factors (*i.e.* in resistant, sensitive, and revertant cells) as well as in experiments where the abundance of these sequences was experimentally manipulated (*i.e.* gradient fractionation or immunoprecipitation). We feel that it is extremely unlikely that any other mRNA would respond identically to translationally active dihydrofolate reductase mRNA with respect to all of these criteria. Therefore, we conclude that our purified cDNA preparation is comprised specifically of sequences complementary to dihydrofolate reductase mRNA.

We have also used the procedure described in Fig. 4 to purify cDNA sequences that were prepared from total poly(A)-containing RNA of resistant (AT-3000) cells that was not further enriched for dihydrofolate reductase mRNA. The cDNA purified in this way was again dihydrofolate reductase-

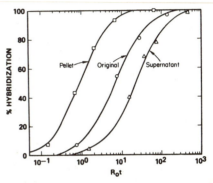

Fig. 7. Hybridization of purified cDNA to RNA extracted from immunoprecipitated dihydrofolate reductase-synthesizing polysomes. Dihydrofolate reductase-synthesizing polysomes were immunoprecipitated from 200 A_{260} units of AT-3000 cell polysomes as described under "Experimental Procedures." Total RNA was extracted from the supernatant and pellet fractions resulting from this procedure, as well as from a reserved sample of the original unfractionated polysomes. RNA from each of these fractions was then reacted with 30 pg (300 cpm) of purified cDNA and the extent of hybridization at the indicated R_0t values determined by hydrolysis with S1 nuclease (see "Experimental Procedures" for details). Hybridization of cDNA to RNA extracted from: pellet (0.115 to 11.5 μg/sample, □——□); supernatant (1.6 to 16 μg/sample, △——△); original polysomes (2.6 μg to 26 μg/sample, ○——○).

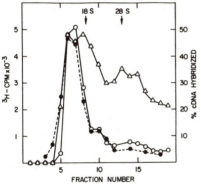

Fig. 6. Hybridization of purified cDNA to size-fractionated resistant cell RNA. Total cytoplasmic RNA from AT-3000 cells (200 μg) was fractionated on isokinetic sucrose gradients as previously described (13). Each fraction was adjusted to contain 0.3 M NaCl and 15 μg of *E. coli* tRNA carrier, and nucleic acids were subsequently precipitated overnight at −20° by the addition of 2 volumes of ethanol. Precipitates were washed twice with 70% ethanol plus 0.1 M NaCl, lyophilized, and dissolved in 100 μl of H₂O. Equal (25 μl) aliquots from each fraction were assayed in the mRNA-dependent reticulocyte lysate system for stimulation of incorporation into total trichloroacetic acid-precipitable material (△——△) and dihydrofolate reductase (○——○) as described under "Experimental Procedures." Other equal aliquots (3 μl) of each fraction were reacted with 25 pg (250 cpm) of purified [³H]cDNA for 45 min in a final reaction volume of 30 μl. Other reaction conditions and measurement of the extent hybridization in each sample (●- - -●) by S1 nuclease hydrolysis are described under "Experimental Procedures." These conditions were devised so that the maximum extent of hybridization in any sample was less than 50%. Therefore, the per cent hybridization of the [³H]DNA is roughly proportional to the concentration of complementary RNA sequences in the corresponding gradient fraction (30).

TABLE II

Hybridization to partially purified dihydrofolate reductase mRNA

Samples from each of the fractions described in Fig. 7 were also assayed for stimulation of incorporation into dihydrofolate reductase in the mRNA-dependent rabbit reticulocyte lysate system which was then expressed as a per cent of the total stimulated trichloroacetic acid-precipitable radioactivity as described under "Experimental Procedures." Also shown is the inverse of the $R_0t_{1/2}$ values of each of the corresponding hybridization reactions which is proportional to the abundance of complementary sequences in the RNA sample used to drive the reaction. In order to facilitate comparison of the inverse of $R_0t_{1/2}$ and the per cent dihydrofolate reductase synthesis, the values in each column were normalized to the value for the original fraction which was set equal to 1. Normalized values are shown in parentheses.

Sample	$1/R_0t_{1/2}$	% dihydrofolate reductase synthesis
Original	0.14 (1)	3.8 (1)
Supernatant	0.045 (0.32)	1.2 (0.32)
Pellet	1.1 (7.7)	22.5 (5.9)

specific as judged by the criteria described above. This result confirms our assumption that only the level of dihydrofolate reductase mRNA sequences are greatly increased in resistant cells. Furthermore, this result also indicates that the approximately 200-fold increase in the abundance of dihydrofolate reductase mRNA sequences in resistant cells (Fig. 5a) is sufficient to allow purification of dihydrofolate reductase-specific cDNA by this method.

This general approach to specific cDNA purification has been used to purify cDNA sequences complementary to RNA sequences that were absent in mutant cells (31) or viruses (32). Our results indicate that this approach can be extended to situations where it is possible to obtain RNA preparations that have been enriched for specific sequences (for example, by induction or partial purification).

Selective Multiplication of Dihydrofolate Reductase Genes in Resistant Lines

Selective Gene Multiplication in Unstable Lines of Methotrexate-resistant Cells — One possible mechanism consistent with the marked instability of the overproduction of dihydrofolate reductase in methotrexate-resistant lines of Sarcoma 180 cells (12) is selective multiplication of the dihydrofolate reductase structural gene (33). In order to test this possibility, DNA prepared from the nuclei of sensitive, resistant, and revertant cells was denatured and allowed to reanneal in the presence of a trace amount of dihydrofolate reductase specific cDNA. The per cent association at various C_0t values was then determined by fractionation of the double- and single-stranded material on hydroxylapatite. We detected no significant differences in the renaturation of the driver DNA from each of these cell types (Fig. 8, - - -) and, in all of these reactions, association of the dihydrofolate reductase-specific cDNA went to approximately 80 to 85% completion with kinetics characteristic of a unique, second order reaction. Association of the purified cDNA with sensitive cell DNA (Fig. 8, O———O) occurred over roughly the same C_0t range as observed for renaturation of the unique sequence fraction of the genomic DNA suggesting that dihydrofolate reductase genes are present, on the average, at no more than a few copies per cell in this line. However, the dihydrofolate reductase-specific cDNA associated with DNA from resistant cells (Fig. 8, △———△) at a rate approximately 200-fold greater than that with which it associated with sensitive cell DNA. In addition, similar relative rates were obtained when these reactions were assayed by S1 nuclease hydrolysis (data not shown). Thus, the dihydrofolate reductase structural gene is selectively multiplied approximately 200-fold in resistant cells, a level roughly in proportion to the relative increase in the content of dihydrofolate reductase and dihydrofolate reductase mRNA in this variant line (Table I).

Analysis of the association kinetics of the specific cDNA to DNA from partially revertant cells (Fig. 8, □———□) indicates that the dihydrofolate reductase gene copy number is unstable in resistant cells. Comparison of the kinetics of this reaction to those observed for the reaction of the cDNA to DNA from resistant and sensitive cells (Fig. 8, Table I) demonstrates that the number of dihydrofolate reductase gene copies in the partially revertant line has declined to a value approximately 10-fold greater than that of the sensitive cells. Once again, this value is proportional to the level of dihydrofolate reductase in revertant cells relative to sensitive and resistant cells (Table I).

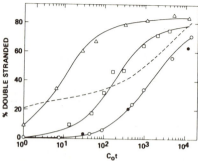

Fig. 8. Association kinetics of purified cDNA with DNA from sensitive, resistant, and partially revertant lines of S-180 cells. DNA was prepared from S-3, AT-3000, and Rev-400 as described under "Experimental Procedures." An aliquot was removed from the S-3 preparation at a point in the preparation immediately preceding NaOH treatment and adjusted to contain 20 μg of poly(A)-containing RNA from resistant cells per mg of DNA. This sample was then processed through the final DNA preparation steps identically to the others. Based on the yield of RNA and DNA from these lines, this ratio of added RNA to DNA approximately represents the relative level of RNA to DNA in these cells. DNA from each of these preparations was melted and allowed to reanneal in the presence of a trace amount of purified [³H]cDNA and the extent of association at various C_0t values measured by chromatography on hydroxylapatite (see "Experimental Procedures" for details). When incubated in the absence of driver DNA, approximately 2% of the [³H]cDNA was retained by hydroxylapatite. Total and double-stranded recoveries of [³H]cDNA from each sample were corrected for this value before calculation of the present double-stranded. The reassociation of the driver DNA from each of these preparations occurred with essentially identical kinetics which are summarized by the *dashed line*. Association of purified [³H]cDNA with DNA from S-3, O———O; S-3 processed with added poly(A)-containing RNA from AT-3000, ●———●; AT-3000, △———△; and Rev-400, □———□.

In order to show that contamination of the DNA preparations by cellular RNA could not have artifactually led to these results, we demonstrated that the rate with which the dihydrofolate reductase-specific probe associates with sensitive cell DNA was not affected by the addition of a large excess of resistant cell poly(A)-containing RNA to the sensitive cell DNA at a point in the DNA preparation procedure immediately preceding base treatment (Fig. 8, ●———●). Since the amount of added RNA represented considerably more than the maximum possible level of RNA contamination at this point (see legend to Fig. 8), this experiment demonstrates that the base hydrolysis step is sufficient to remove any contaminating RNA.

Selective Gene Multiplication in Stable Lines of Methotrexate-resistant Cells — The 5000-fold methotrexate-resistant L1210RR line of murine L1210 lymphoma cells contains an approximately 35-fold increase in the level of dihydrofolate reductase relative to the sensitive, parental line (Table I). We have also shown that this increase is associated with an increase in the rate of dihydrofolate reductase synthesis and the level of dihydrofolate reductase mRNA activity (data not shown). Increased dihydrofolate reductase levels appear to be a stable property of this resistant line, since we find no significant decrease in this parameter after more than 100 cell doublings in the absence of methotrexate (Table I). The stability of increased dihydrofolate reductase levels observed when these and other lines of methotrexate-resistant cells were grown in the absence of methotrexate (15) suggested that the alteration leading to increased enzyme synthesis might be a regulatory mutation. In order to test the generality of the selective gene multiplication phenomenon, we quantitated the relative number of dihydrofolate reductase genes in

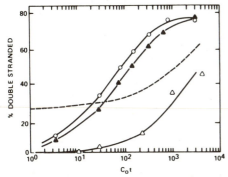

FIG. 9. Association kinetics of purified cDNA with DNA prepared from sensitive and methotrexate-resistant lines of L1210 cells. DNA prepared from various lines of L1210 cells was denatured and allowed to reanneal in the presence of a trace amount of purified [^{3}H]cDNA, and the extent of association at indicated C_0t values determined by chromatography on hydroxylapatite (see "Experimental Procedures" for details). Reassociation of the driver DNA summarized for all three reactions, – – –; association of the purified [^{3}H]cDNA with DNA from L1210S, △——△; L1210RR, ○——○; and L1210RR grown for approximately 100 cell doublings in the absence of methotrexate, ▲——▲.

the L1210 lines just described. The data in Fig. 9 indicate that the dihydrofolate reductase-specific cDNA associates with DNA from the L1210RR line grown in the presence of methotrexate (○——○) at a rate approximately 45-fold more rapid than that observed with DNA from the sensitive parental line (△——△), indicating that the relative number of dihydrofolate reductase genes is approximately 45-fold greater in the resistant line. Again, the relative number of dihydrofolate reductase gene copies is roughly proportional to the relative level of dihydrofolate reductase in these two lines (Table I). We observed only a slight, and probably not significant, decrease (20 to 25%) in the rate with which the probe associated to DNA from the L1210RR line that had been grown in the absence of methotrexate. Therefore, the dihydrofolate reductase gene copy number appears to be relatively stable in this line of methotrexate-resistant cells (Table I).

Thermal Stability of Duplexes between Dihydrofolate Reductase-specific cDNA and DNA from Different Cell Types – The thermal denaturation characteristics of duplexes formed between dihydrofolate reductase-specific cDNA and DNA from either sensitive cells, resistant cells, or mouse liver are essentially indistinguishable (Fig. 10a). The T_m values for these reactions range between 81.5° and 82.5°, and in each, the melting profile occurs as a single transition over a relatively narrow temperature range. The T_m of the driver DNA was similar for DNA from S-180 cells ($T_m = 84$) and human placenta ($T_m = 82$) (Fig. 10b). These reactions also proceeded to the same extent (see legend to Fig. 10). However, although the human DNA contains sequences which can anneal with dihydrofolate reductase-specific cDNA prepared from a murine cell line, the extent of duplex formation (see legend to

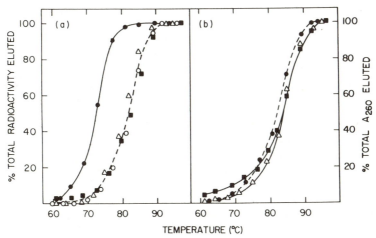

TEMPERATURE (°C)

FIG. 10. Thermal denaturation of duplexes formed between dihydrofolate reductase-specific DNA and DNA from various sources. 100 pg of dihydrofolate reductase-specific [^{3}H]cDNA (1000 cpm) was annealed with approximately 1 mg of DNA from AT-3000 (resistant) cells (final $C_0t = 1000$), S-3 (sensitive) cells (final $C_0t = 20,000$), mouse liver (final $C_0t = 14,000$), aborted human placenta (final $C_0t = 10,000$), and chicken oviduct (final $C_0t = 10,000$). The final reaction volume was 200 μl, the temperature 68°, and other conditions were as described under "Experimental Procedures." At the indicated C_0t values, reactions were diluted into 5 ml of 0.12 M NaPO$_4$ and adsorbed to 1-g columns of hydroxylapatite which were maintained at 60° with a recirculating water bath. The column was washed with 5 ml of 0.12 M NaPO$_4$ at 60° and subsequently the

temperature of the column and wash buffer was raised to 97° in increments of approximately 3°. At each step, the washing procedure was repeated. The resulting fractions were monitored for A_{260} and trichloroacetic acid-precipitable radioactivity as described under "Experimental Procedures." The final percentage of the driver DNA and [^{3}H]cDNA, respectively, that were recovered as double-stranded in each reaction are listed in parentheses below following the appropriate reaction symbols. *Panel a* shows elution of [^{3}H]cDNA and panel *b* show elution of driver DNA from reactions driven with DNA from: AT-3000, ○——○ (50, 82); S-3, △——△ (77, 60); mouse liver, ■——■ (77, 56); human placenta, ●——● (80, 26), and chicken oviduct (85, 0). Data are presented as the cumulative elution of DNA with increasing temperature.

Fig. 10) and the stability of the duplexes ($T_m = 72°$) was considerably less than those formed with DNA from murine sources (Fig. 10a). Under the relatively stringent conditions used for these reactions, we observed no duplex formation between dihydrofolate reductase-specific cDNA and chicken oviduct DNA (see legend to Fig. 10).

These results suggest that there is little difference in the nucleotide sequence of individual dihydrofolate reductase genes in resistant cells and, furthermore, that these sequences have diverged little from homologous sequences in sensitive cells or mouse liver (confirming the murine origin of the multiplied dihydrofolate reductase genes in resistant cells). Although there is significant divergence between the nucleotide sequences of the human and murine genes, there appears to be sufficient homology to allow use of the murine probe to analyze methotrexate resistance in human tumors.

DISCUSSION

We have shown that in methotrexate-resistant lines of Sarcoma 180 and L1210 murine lymphoma cells increased synthesis of dihydrofolate reductase is associated with increased copies of the dihydrofolate reductase structural gene. More recently we have also found that increased dihydrofolate reductase synthesis in methotrexate-resistant 3T6 cells is also accompanied by a corresponding increase in the number of dihydrofolate reductase gene copies.[4] These methotrexate-resistant lines represent the first reported examples of mammalian cells in which a structural gene that codes for a protein is selectively multiplied.

In order to understand the processes involved with the selective multiplication of dihydrofolate reductase genes in the resistant lines, we should first examine the role of methotrexate in this phenomenon. This folic acid analogue strongly and specifically inhibits dihydrofolate reductase in a competitive manner (34), and therefore indirectly inhibits the *de novo* synthesis of purines, thymidylate, and glycine (35). Hence, exposure to sufficiently high concentrations of methotrexate kills dividing cells. Resistance to this analogue has been found to result from any of several mechanisms (36), but the most frequently reported for mammalian cells is an increased cellular content of dihydrofolate reductase (2–11). In this case resistant cells accumulate sufficiently high dihydrofolate reductase levels to maintain some free enzyme activity in the presence of the drug (37). Highly resistant lines with greatly increased levels of dihydrofolate reductase (such as those described in this report) have never been selected in a single step. The common method for obtaining such lines involves either gradually increasing the concentration of methotrexate in the medium (3) or progressing in several steps (6), each step using a 10- to 20-fold greater concentration of the drug than is required to inhibit growth by 50%. In the latter case it is possible to estimate the frequency of resistant variants in the cell population. It has been our observation, as well as those of other laboratories (6, 38), that in the initial step this frequency is low (less than 1 in 10^6).

Several lines of evidence suggest that methotrexate does not act directly to induce or maintain the increased synthesis of dihydrofolate reductase in resistant lines. 1. Simple exposure of cells to methotrexate (without selection) has no effect

[4] R. E. Kellems, F. W. Alt, and V. Morhen, unpublished observation.

on dihydrofolate reductase synthesis (12). 2. In the methotrexate-resistant lines of L1210 cells that we have studied, as well as a number of other resistant lines (2, 15), increased dihydrofolate reductase synthesis is a stable property and does not decline when cells are grown in the absence of the drug. 3. The kinetics of the decrease in dihydrofolate reductase synthesis observed when unstable lines of resistant cells are grown in the absence of methotrexate do not correspond to dilution of the drug from the cells (12). 4. The decrease in dihydrofolate reductase synthesis observed when unstable lines are grown in the absence of methotrexate is also observed when cells are grown in the continued presence of the drug, but supplemented with a purine source, thymidine, and glycine.[3] Biedler *et al.* (39) have suggested that methotrexate may have mutagenic properties, possibly due to the inhibition of the synthesis of nucleic acid precursors. Such properties could influence the rate (or mechanism) with which resistant variants arise. However, fluctuation analyses done with L1210 cells indicated that in this line methotrexate-resistant variants are generated spontaneously during growth in the absence of the drug (40). In addition, this drug was found to have no mutagenic properties as judged by the *Salmonella* microsome test (41). Therefore, in summary, all available evidence indicates that methotrexate acts only as a selective agent and has no direct role in the resistance (gene multiplication) process.

We propose that exposure of sensitive cells to methotrexate selects for those cells in the population harboring spontaneous multiplications (duplications) of the dihydrofolate reductase structural gene and as a result, increased levels of dihydrofolate reductase. Of course, there are many other conceivable genetic alterations that could lead to increased dihydrofolate reductase levels, including those generating an absolute increase in the transcription rate of the gene or an increased stability of the specific mRNA. However, in all of the lines that we have studied (including 3T6 lines), we observe a proportionality between the relative level of dihydrofolate reductase activity and the relative number of dihydrofolate reductase gene copies (Table I). This result suggests that there is little difference between the activities of individual dihydrofolate reductase genes in sensitive and resistant cells, and that selective gene multiplication is the most important, if not the only mechanism leading to increased dihydrofolate reductase accumulation by these highly resistant lines. A possible explanation for the predominance of this mechanism is that the dihydrofolate reductase gene is expressed at or near the maximum possible activity in sensitive cells, and therefore no type of genetic alteration could greatly increase this activity. Alternatively, the events which lead to duplication or multiplication of dihydrofolate reductase genes may occur at a higher frequency than other types of genetic alterations that would lead to increased expression of a limited number of gene copies.

Clearly, in order to understand the mechanism by which these genes are multiplied, as well as why their number is relatively stable in some lines and not in others, it will be necessary to determine the location and molecular arrangement of the multiple gene copies in the various cell lines. Are they chromosomal or extrachromosomal, and do they exist in tandem arrays or at many locations in the genome? An interesting observation that may reflect on these questions was made by Biedler and her colleagues who consistently detected the appearance of a large homogeneously staining region associated with specific chromosomes of highly metho-trexate-resistant lines of Chinese hamster lung cells (42).[5] In addition, resistance and corresponding high dihydrofolate reductase levels were unstable when these lines were grown in the absence of methotrexate; and, significantly, the size of the chromosomal alteration decreased in parallel to the decrease in enzyme activity (14). It is tempting to speculate that such a chromosomal alteration might correspond to a tandem array of dihydrofolate reductase genes. In some of their resistant lines the specifically altered region represented as much as 6% of the chromosomal complement (14), considerably more than would be necessary to account for an increase in dihydrofolate reductase gene copy number corresponding to the increased enzyme content of the line (approximately 200-fold). However, in bacteria, selected duplication of a specific gene can extend far beyond the vicinity of that gene and involve as much as 20% of the bacterial chromosome (43). If genes other than those coding for dihydrofolate reductase are multiplied in the resistant Sarcoma 180 lines that we have studied, they apparently are not expressed. As judged by both comparison of proteins synthesized by sensitive and resistant cells (12), as well as by the specificity of the cDNA purification procedure (see above), the large increase in dihydrofolate reductase synthesis appears to be unique.

De novo duplication of specific genes in bacteria and phages occurs with relatively high frequencies (43–47); and in these cases duplications appear to be in tandem. A well known example of tandem duplications in eukaryotic cells occurs as a result of unequal crossing over at the bar locus in *Drosophila* (48, 49). If the initial event selected in methotrexate resistance were a tandem duplication of the dihydrofolate reductase gene or alternatively if the genes already existed in multiple, tandem copies in sensitive cells, expansion of the tandem array might occur by homologous but unequal crossover events between dihydrofolate reductase genes on homologous chromosomes (50). However, a more likely mechanism would involve unequal exchanges between sister chromatids. Sister chromatid exchange has been demonstrated to occur in a variety of organisms (51–53) and in *Drosophila* unequal sister chromatid exchanges presumably lead to changes in the number of tandem repeats at the bar locus (54) as well as the number of ribosomal genes at the bobbed locus (55). This process has been discussed in detail as a mechanism for the evolution of repeated DNA sequences (56), the coincidental evolution of members of multi-gene families (57, 58), and the magnification-reduction of the ribosomal gene copy number in *Drosophila* (58, 59). One attractive feature of such a mechanism for the selective multiplication of dihydrofolate reductase genes in methotrexate-resistant cells is that it would be consistent with the multi-step selection procedure necessary to generate these lines.

Alternatively, selective multiplication of dihydrofolate reductase genes may occur by a mechanism which at least initially generates extrachromosomal copies of the multiplied genes. The classic example of such a process in eukaryotic cells is the amplification of ribosomal genes in amphibian oocytes (60). In this case, amplification is specifically regulated as part of a developmental sequence and occurs extrachromosomally, apparently by a rolling circle replication

mechanism (61). Other possible amplification mechanisms include reverse transcription of the specific mRNA (62, 63) or disproportionate replication of specific genes (64). The former mechanism may be involved in the production of extrachromosomal copies of mouse mammary tumor virus genes (65) while the latter has recently been implicated in the production of large numbers of extrachromosomal copies of SV40 DNA from the integrated viral genome.[6] One common feature of these mechanisms is that large increases in the number of specific genes might be obtained in a single selective step. In this regard, it will be interesting to measure the absolute number of dihydrofolate reductase gene copies in sensitive cells, and the maximum increase in that number obtainable in a single step.

A selective increase in the number of dihydrofolate reductase genes in resistant cells might also be achieved by the retention of specific chromosomal fragments. Although there are apparently no specific differences between the karyotypes of sensitive and resistant lines (38),[7] chromosome transfer experiments indicate that chromosomal fragments retained by host cells are frequently so small that they may be cytologically undetectable (66, 67). Similarly to dihydrofolate reductase genes in unstable lines, such transferred genetic elements are usually lost rapidly from host cells (1 to 10% loss per generation) (68–71), but can be maintained indefinitely by growth under appropriate selective conditions (70). In addition, prolonged growth of host cells under selective conditions leads to the emergence of lines which stably express the transferred characteristic (66, 70).

Finally, by analogy to bacterial systems, duplication or subsequent multiplication of specific genes (by many of the mechanisms considered above) may also be promoted by flanking sequences (*e.g.* translocatable elements or viral sequences) (72, 73). Such sequences (insertion elements) may be involved in the accumulation of R-factors containing multiple r-determinant segments in chloramphenicol-resistant lines of *Proteus mirabilis*. This phenomenon also shares many features with methotrexate-resistance in Sarcoma 180 cells. High levels of chloramphenicol resistance are unstable in the absence of selection and result from increased production of chloramphenicol transacetylase in association with the selective multiplication of the r-determinant carrying the gene for this enzyme (74, 75).

An intriguing question is why the multiple gene copies are stable in some lines and unstable in others. One possibility is that the multiplication process occurs by a different mechanism in these lines, but recent results indicate that this need not be the case. After growth in the presence of methotrexate for an additional 2 years, the highly unstable lines of methotrexate-resistant S-180 cells described previously (12) appear to have become much more stable.[3] This phenomenon can be explained as follows: whatever the mechanism of gene loss (see discussion below), unstable lines of resistant cells are presumably constantly generating cells with decreased numbers of dihydrofolate reductase genes. Growth of such lines in the presence of methotrexate would maintain a certain average level of dihydrofolate reductase gene copies per cell by eliminating those cells in which the gene copy number (and correspondingly dihydrofolate reductase levels) had decreased below that necessary for survival. Under these conditions,

---

[5] A similar chromosomal alteration was recently observed in methotrexate-resistant Chinese hamster ovary cells. L. Chasin, personal communication.

[6] M. Botchan, personal communication.
[7] J. Nunberg and R. Kaufman, unpublished observation.

cells in which the gene copy number had become more stable would have an obvious selective advantage (more of their progeny would survive) and eventually outgrow the population.

Loss of chromosomal genes might be associated with specific chromosomal deletions or fragmentations (70). In addition, if the multiple gene copies exist in clusters of tandem repeats in resistant lines, instability in their numbers could be due to the same general types of processes which were considered above for their multiplication. Thus, unequal crossover events would generate as reciprocal products both a cell with increased numbers of dihydrofolate reductase gene copies and one in which the number was reduced; a decrease in the average number of dihydrofolate reductase gene copies per cell would result if in the absence of methotrexate, cells which devoted less of their energy to the production of unnecessarily high levels of dihydrofolate reductase had a selective growth advantage. Tandem duplications in bacterial cells are usually quite unstable (43, 44, 46), presumably due to crossover events between repeats on the same chromosome. Loss of dihydrofolate reductase genes might occur by a similar process, and by analogy to bacterial systems in which such repeats are much more stable in Rec A⁻ lines (43, 44), stabilization could result from the loss of an enzymatic function that was involved in their excision or exchange. Stabilization might also occur by the inactivation of flanking sequences (by excision or mutation) involved in the multiplication process or by translocation of clustered genes to multiple sites in the genome. Extrachromosomal genes might be lost by a number of different mechanisms. Unstable genetic elements resulting from chromosome transfer experiments (see above) are thought to be extrachromosomal, and recent evidence suggests that stability results from integration of the transferred fragment into the genome of the host cell (67). Stability of extrachromosomal genes, whatever their origin, might be achieved through such a mechanism.

All of our studies were done with murine cell lines; therefore in order to assess the generality of the selective gene multiplication phenomenon it will be necessary to know the mechanism of increased dihydrofolate reductase accumulation in cell lines derived from other organisms (5, 15). The selection of cell lines resistant to highly specific inhibitors of other key enzymes should allow extension of this approach to many different genes. For example, Kempe *et al.* have shown that in certain hamster cell lines, resistance to a specific inhibitor of aspartate transcarbamylase is associated with increased cellular content of that enzyme (76). More recently, this group has found that resistant lines synthesize the enzyme at a greater rate and contain increased levels of the specific mRNA.[8] It will by interesting to know if these lines also contain increased numbers of aspartate transcarbamylase gene copies.

We do not know if the processes leading to selective multiplication of dihydrofolate reductase genes in the permanent cell lines that we have studied have a role in normal cells. Certainly, a mechanism for generating spontaneous and random duplications of genetic material might be important for evolutionary flexibility (77), as well as for the generation of multigene families (33). The unstable lines of resistant Sarcoma 180 cells may, in fact, provide a good model system for studying the evolution and maintenance of multi-gene families; since under appropriate growth conditions it is possible

[8] R. Padgett, G. Wahl, and G. Stark, personal communication.

to select lines in which the dihydrofolate reductase gene copy number is increased, decreased, or fixed. In systems where it has been studied, selective gene multiplication as a mechanism for the synthesis of large amounts of differentiated cell proteins has not been observed (78–80). However, other lines of evidence suggest that various types of genomic alterations including duplications, deletions, and translocations may underlie a number of controls of differentiation (see Ref. 81 for further discussion of this point). Our results add further evidence to support the concept that the genome of higher organisms is not constant, but can undergo a variety of changes.

Acknowledgments – We are indebted to Henry Burr and Marvin Wickens for their advice and assistance, to Andi Justice for growing the cells used for many of these experiments, and to Keiko Nakanishi Alt for assistance in the preparation of this manuscript. We are especially grateful to Gray Crouse and Randy Kaufman for advice and thoughtful criticism throughout the course of this work.

REFERENCES

1. Bertino, J. R., Donohue, D. M., Simmons, B., Gabrio, B. W., Silber, R., and Huennekens, F. M. (1963) *J. Clin. Invest.* 42, 466–475
2. Fischer, G. A. (1961) *Biochem. Pharmacol.* 7, 75–80
3. Hakala, M. T., Zakrzewski, S. F., and Nichol, C. A. (1961) *J. Biol. Chem.* 236, 952–958
4. Kashet, E. R., Crawford, E. J., Friedkin, M., Humphreys, S. R., and Golding, A. (1964) *Biochemistry* 3, 1928–1931
5. Littlefield, J. W. (1969) *Proc. Natl. Acad. Sci. U. S. A.* 62, 88–95
6. Friedkin, M., Crawford, E. S., Humphreys, S. R., and Golding, H. (1962) *Cancer Res.* 22, 600–606
7. Perkins, J. P., Hillcoat, B. L., and Bertino, J. R. (1967) *J. Biol. Chem.* 242, 4771–4776
8. Sartorelli, A. C., Both, B. A., and Bertino, J. R. (1964) *Arch. Biochem. Biophys.* 108, 53–59
9. Jackson, R. C., and Huennekens, F. M. (1973) *Arch. Biochem. Biophys.* 154, 192–198
10. Courtenay, V. D., and Robins, A. B. (1972) *J. Natl. Cancer Inst.* 49, 45–53
11. Biedler, J. L., Albrecht, A. M., Hutchison, D. J., and Spengler, B. A. (1972) *Cancer Res.* 32, 153–161
12. Alt, F. W., Kellems, R. E., and Schimke, R. T. (1976) *J. Biol. Chem.* 251 3063–3074
13. Kellems, R. E., Alt, F. W., and Schimke, R. T. (1976) *J. Biol. Chem.* 251, 6987–6993
14. Biedler, J. L., and Spengler, B. A. (1976) *J. Cell Biol.* 70, 117a
15. Nakamura, H., and Littlefield, J. W. (1972) *J. Biol. Chem.* 247, 179–187
16. Hanggi, U. J., and Littlefield, J. W. (1976) *J. Biol. Chem.* 251, 3075–3080
17. Chang, S. E., and Littlefield, J. W. (1976) *Cell* 7, 391–396
18. Thompson, L. H., and Baker, R. M. (1973) *Methods Cell Physiol.* 6, 209–281
19. Demars, R. (1974) *Mutat. Res.* 24, 335–364
20. Aviv, H., and Leder, P. (1972) *Proc. Natl. Acad. Sci. U. S. A.* 69, 1408–1412
21. Palacios, R., Palmiter, R. D., and Schimke, R. T. (1972) *J. Biol. Chem.* 247, 2316–2321
22. Shapiro, D. J., Taylor, J. M., McKnight, G. S., Palacios, R., Gonzalez, C., Kiely, M. L., and Schimke, R. T. (1974) *J. Biol. Chem.* 249, 3665–3671
23. Taylor, J. M., and Schimke, R. T. (1974) *J. Biol. Chem.* 249, 3597–3601
24. Pelham, H. R. B., and Jackson, R. J. (1976) *Eur. J. Biochem.* 67, 247–256
25. Buell, G., Wickens, M., Payvar, F., and Schimke, R. T. (1978) *J. Biol. Chem.*, in press
26. Britten, R. J., Graham, D. E., and Neufeld, B. R. (1974) *Methods Enzymol.* 29, 363–418
27. McCarty, K. S., Jr., Volmer, R. T., and McCarty, K. S. (1974)

Anal. Biochem. **61**, 165–183

28. Studier, W. F. (1965) *J. Mol. Biol.* **11**, 373–390
29. Rhoads, R. E., McKnight, G. S., and Schimke, R. T. (1973) *J. Biol. Chem.* **248**, 2031–2039
30. Fan, H., and Baltimore, D. (1973) *J. Mol. Biol.* **80**, 93–117
31. Ramirez, F., Nutter, C., O'Donnel, J. V., Canale, V., Bailey, G., Sangvensermsvi, T., Maniatis, G., Marks, P., and Bank, A. (1975) *Proc. Natl. Acad. Sci. U. S. A.* **72**, 1550–1554
32. Stehelin, D., Guntaka, R., Varmus, H., and Bishop, J. M. (1976) *J. Mol. Biol.* **101**, 349–365
33. Hood, L., Campbell, J. H., and Elgin, S. C. R. (1975) *Ann. Rev. Genet.* **9**, 306–353
34. Werkeiser, W. C. (1961) *J. Biol. Chem.* **236**, 888–893
35. Huennekens, F. M. (1963) *Biochemistry* **2**, 151–159
36. Blakeley, R. L. (1969) *The Biochemistry of Folic Acid and Related Pteridines,* pp. 139–187, North Holland, Amsterdam
37. Hakala, M. T. (1965) *Biochim. Biophys. Acta* **102**, 198–209
38. Hakala, M. T., and Ishihara, T. (1962) *Cancer Res.* **22**, 987–996
39. Biedler, J. L., Albrecht, A. M., and Hutchinson, D. J. (1965) *Cancer Res.* **25**, 246–257
40. Law, L. W. (1952) *Nature* **169**, 628–629
41. Benedict, W. F., Baker, M. S., Haroun, L., Choi, E., and Ames, B. N. (1977) *Cancer Res.* **37**, 2209–2213
42. Biedler, J. L., and Spengler, B. A. (1976) *Science* **191**, 186–187
43. Anderson, R. P., Miller, C. G., and Roth, J. R. (1976) *J. Mol. Biol.* **105**, 201–218
44. Folk, W. R., and Berg, P. (1971) *J. Mol. Biol.* **58**, 595–610
45. Hill, C. W., and Combriato, G. (1973) *Mol. & Gen. Genet.* **127**, 197–214
46. Hariuchi, T., Hariuchi, S., and Novick, A. (1963) *Genetics* **48**, 157–169
47. Emmons, S. W., MacCosham, U., and Baldwin, R. L. (1975) *J. Mol. Biol.* **91**, 133–146
48. Sturtevant, A. H. (1925) *Genetics* **10**, 117–147
49. Bridges, C. B. (1936) *Science* **83**, 210–211
50. Stern, C. (1936) *Genetics* **21**, 625–730
51. Taylor, J. H., Woods, P. S., and Hughs, W. L. (1957) *Proc. Natl. Acad. Sci. U. S. A.* **43**, 122–128
52. Marin, G., and Prescott, D. M. (1964) *J. Cell Biol.* **21**, 159–167
53. McClintock, B. (1941) *Cold Spring Harbor Symp. Quant. Biol.* **9**, 72–80
54. Peterson, H. M., and Laughnan, J. R. (1963) *Proc. Natl. Acad. Sci. U. S. A.* **50**, 126–133
55. Schalet, A. (1969) *Genetics* **63**, 133–153
56. Smith, G. P. (1976) *Science,* **191**, 528–535
57. Smith, G. P. (1973) *Cold Spring Harbor Symp. Quant. Biol.* **38**, 507–513
58. Tartof, K. D. (1973) *Cold Spring Harbor Symp. Quant. Biol.* **38**, 491–500
59. Tartof, K. D. (1974) *Proc. Natl. Acad. Sci. U. S. A.* **71**, 1272–1276
60. Brown, D. D., and Dawid, I. B. (1968) *Science* **160**, 272–280
61. Hourcade, D., Dressler, D., and Wolfson, J. (1973) *Cold Spring Harbor Symp. Quant. Biol.* **38**, 537–550
62. Baltimore, D. (1970) *Nature* **226**, 1209–1211
63. Temin, H., and Mizutani, S. (1970) *Nature* **226**, 1211–1213
64. Tartof, K. D. (1975) *Ann. Rev. Genet.* **9**, 370
65. Ringold, G. M., Yamamoto, K. R., Shaulo, P. R., and Varmus, H. E. (1977) *Cell* **10**, 19–26
66. Willecke, K., Lange, R., Kruger, A., and Reber, T. (1976) *Proc. Natl. Acad. Sci. U. S. A.* **73**, 1274–1278
67. Fournier, R. E. K., and Ruddle, F. H. (1977) *Proc. Natl. Acad. Sci. U. S. A.* **74**, 3937–3941
68. McBride, O. W., and Ozer, H. L. (1973) *Proc. Natl. Acad. Sci. U. S. A.* **70**, 1258–1262
69. Willecke, K., and Ruddle, F. H. (1975) *Proc. Natl. Acad. Sci. U. S. A.* **72**, 1792–1796
70. Degnen, G. E., Miller, I. L., Eisenstadt, J. M., and Adelberg, E. A. (1976) *Proc. Natl. Acad. Sci. U. S. A.* **73**, 2838–2842
71. Spandidos, D. A., and Siminovitch, L. (1977) *Proc. Natl. Acad. Sci. U. S. A.* **74**, 3480–3484
72. Cohen, S. N. (1976) *Nature* **263**, 731–738
73. Kleckner, N. (1977) *Cell* **11**, 11–23
74. Rownd, R., Kasamatu, H., and Michel, S. (1971) *Ann. N. Y. Acad. Sci.* **182**, 188–206
75. Perlman, D., and Rownd, R. H. (1975) *J. Bacteriol.* **123**, 1013–1034
76. Kempe, T. D., Swyrd, E. A., Bruist, M., and Stark, G. R. (1976) *Cell* **9**, 541–550
77. Ohno, S. (1970) *Evolution of Gene Duplication,* Springer-Verlag, New York
78. Packman, S., Aviv, H., Ross, J., and Leder, P. (1972) *Biochem. Biophys. Res. Commun.* **49**, 813–819
79. Suzuki, Y., Gage, L. P., and Brown, D. D. (1972) *J. Mol. Biol.* **70**, 637–649
80. Sullivan, D., Palacios, R., Stavezer, J., Taylor, J. M., Faras, A. J., Kiely, M. L., Summers, N. M., Bishop, J. M., and Schimke, R. T. (1973) *J. Biol. Chem.* **248**, 7530–7539
81. Schimke, R. T., Alt, F. W., Kellems, R. E., Kaufman, R., and Bertino, J. R. (1977) *Cold Spring Harbor Symp. Quant. Biol.* **42**, in press

Supplementary Readings

Before 1961

Barski, G., S. Sorieul, and F. Cornefert. 1960. Production dans des cultures in vitro de deux souches cellulaires in association, de cellules de caratere "hybrids." *C.R. Acad. Sci. 251:* 1825–1830.

Brockman, R.W. and P. Stutts. 1960. A mechanism of resistance to 6-thioguanine. *Fed. Proc. 19:* 313.

Chu, E.H.Y. and N.H. Giles. 1958. Comparative chromosomal studies on mammalian cells in culture. I. The HeLa strain and its mutant clonal derivatives. *J. Natl. Cancer Inst. 20:* 383–401.

Djordjevic, B. and W. Szybalski. 1960. Genetics of human cell lines. III. Incorporation of 5-bromo- and 5-iododeoxyuridine into the deoxyribonucleic acid of human cells and its effect on radiation sensitivity. *J. Exp. Med. 112:* 509–531.

Lederberg, J. and N. Zinder. 1948. Concentration of biochemical mutants of bacteria with penicillin. *J. Amer. Cancer Soc. 70:* 4267–4268.

Puck, T.T. and H.W. Fisher. 1956. Genetics of somatic mammalian cells. I.Demonstration of the existence of mutants with different growth requirements in a human cancer cell strain (HeLa). *J. Exp. Med. 104:* 427–434.

Vogt, M. 1958. A genetic change in a tissue culture line of neoplastic cells. *J. Cell. Comp. Physiol. 52:* 271–285.

1961–1970

Boyle, J. and K. Raivivolo. 1970. Lesch-Nyhan syndrome: Preventive control by prenatal diagnosis. *Science 169:* 688–689.

Coon, H.G. and M.C. Weiss. 1969. Sendai-produced somatic cell hybrids between L cell strains and between liver and L cells. *Wistar Symp. 9:* 83–96.

Davidson, R. and B. Ephrussi. 1970. Factors influencing the "effective mating rate" of mammalian cells. *Exp. Cell Res. 61:* 222–226.

Enders, J. and J. Neff. 1968. Cytopathogenicity in monolayer hamster cell cultures fused with beta propiolactone-inactivated Sendai virus. *Proc. Natl. Acad. Sci. 127:* 260–267.

Freed, J.J. and L. Mezger-Freed. 1970. Stable haploid cultured cell lines from frog embryos. *Proc. Natl. Acad. Sci. 65:* 337–344.

Harris, H. 1965. Behavior of differentiated nuclei in heterokaryons of animal cells from different species. *Nature 206:* 583–588.

Harris, H. and J.F. Watkins. 1965. Hybrid cells derived from mouse and man: Artificial heterokaryons of mammalian cells from different species. *Nature 205:* 640–646.

Kao, F., L. Chasin, and T. Puck. 1969. Genetics of somatic mammalian cells. X. Complementation analysis of glycine-requiring mutants. *Proc. Natl. Acad. Sci. 64:* 1284–1291.

Kao, F., L. Chasin, and T. Puck. 1972. Genetics of somatic mammalian cells: Demonstration of a human esterase activator gene linked to the AdeB gene. *Proc. Natl. Acad. Sci. 69:* 3273–3277.

Kao, F.-T., R.T. Johnson, and T.T. Puck. 1969. Complementation analysis on virus-fused Chinese hamster cells with nutritional markers. *Science 164:* 312–314.

Klebe, R.J., T. Chen, and F.H. Ruddle. 1970. Mapping of human genetic regulator element by somatic cell genetic analysis. *Proc. Natl. Acad. Sci. 66:* 1220–1227.

Koprowski, H., F.C. Jensen, and Z. Steplewski. 1967. Activation of production of infectious tumor virus SV40 in heterokaryon cultures. *Proc. Natl. Acad. Sci. 58:* 127–133.

Krooth, R. and E. Sell. 1970. The action of Mendelian genes in human diploid cell strains. *J. Cell. Physiol. 76:* 311–330.

Ladda, R.L. and R.D. Estensen. 1970. Introduction of a heterologous nucleus into enucleated cytoplasms of cultured mouse L-cells. *Proc. Natl. Acad. Sci. 67:* 1528–1533.

Marin, G. and J. Littlefield. 1968. Selection of morphologically normal cell lines from Py-BHK hybrids. *J. Virol. 2:* 69–77.

Okada, Y. 1962. Analysis of giant polynuclear cell formation caused by HVJ virus from Ehrlich's ascites tumor cells. *Exp. Cell Res. 26:* 98–107.

Puck, T. and F. Kao. 1967. Genetics of somatic mammalian cells. V. Treatment with 5-bromodeoxyuridine and visible light for isolation of nutritionally deficient mutants. *Proc. Natl. Acad. Sci. 58:* 1227–1234.

Seegmiller, J.E., F.M. Rosenbloom, and W.N. Kelley. 1967. Enzyme defect associated with a sex-linked human neurological disorder and excessive purine synthesis. *Science 155:* 1682–1684.

Thompson, L.H., R. Mankovitz, R.M. Baker, J.E. Till, L. Siminovitch, and G.F. Whitmore. 1970. Isolation of temperature-sensitive mutants of L-cells. *Proc. Natl. Acad. Sci. 66:* 377–384.

Watkins, J. and R. Dulbecco. 1967. Production of SV40 virus in heterokaryons of transformed and susceptible cells. *Proc. Natl. Acad. Sci. 58:* 1369–1403.

Yerganaian, G. and M.B. Nell. 1966. Hybridization of dwarf hamster cells by UV-inactivated Sendai virus. *Proc. Natl. Acad. Sci. 55:* 1066–1073.

1971–1980

Boone, T., T. Chen, and F. Ruddle. 1972. Assignment of three human genes to chromosomes and evidence for translocation between human and mouse chromosomes in somatic cell hybrids. *Proc. Natl. Acad. Sci. 69:* 510–514.

Chasin, L. 1973. The effect of ploidy on chemical mutagenesis in cultured Chinese hamster cells. *J. Cell. Physiol.* 82: 299–308.

Chasin, L., A. Geldman, M. Konstam, and G. Urlaub. 1974. Reversion of a Chinese hamster cell auxotrophic mutant. *Proc. Natl. Acad. Sci.* 71: 718–722.

Chu, E., N. Sun, and C. Chang. 1972. Induction of auxotrophic mutants by treatment of Chinese hamster cells with BrdU and black light. *Proc. Natl. Acad. Sci.* 69: 3459–3463.

Croce, C.M., K. Huebner, A.J. Girardi, and H. Koprowski. 1974. Rescue of defective SV40 from mouse-human hybrid cells containing human chromosome 7. *Virology* 60: 276–281.

Elsevier, S., R. Kucherlapati, E.A. Nicholas, R.P. Creagan, R.E. Giles, F.H. Ruddle, K. Willecke, and J.K. McDougall. 1974. Assignment of the gene for galactokinase to human chromosome 17 and its regional localization to band q21-22. *Nature* 251: 633–635.

Follett, E., C. Pringle, W. Wunner, and J. Skehel. 1974. Virus replication in enucleate cells: Vesicular stomatitis virus and influenza virus. *J. Virol.* 13: 394–399.

Grzeschik, K. 1973. Utilization of somatic cell hybrids for genetic studies in man. *Humangenetik* 19: 1–40.

Harris, M. 1971. Mutation rates in cells at different ploidy levels. *J. Cell. Physiol.* 78: 177–184.

McKinnel, R.G., L.M. Steven, Jr., and E. Ellgaard. 1973. Serum electrophoresis of genetic replicate leopard frogs produced by nuclear transplantation. *Differentiation* 1: 1753–1755.

Miller, O., P. Allderdice, D. Miller, W. Brey, and B. Migeon. 1971. Human thymidine kinase gene locus. Assignment to chromosome 17 in a hybrid of man and mouse cells. *Science* 173: 244–245.

O'Brien, J., S. Okada, D. Fillerup, M. Veath, B. Adornato, P. Brennen, and J. Leroy. 1971. Tay-Sachs disease: Prenatal diagnosis. *Science* 172: 61–64.

Pennington, T.H. and E.A. Follett. 1974. Vaccinia virus replication in enucleate BSC-1 cells: Particle production and synthesis of viral DNA and proteins. *J. Virol.* 13: 488–493.

Pollack, R. and R. Goldman. 1973. Synthesis of infective poliovirus in BSC-1 monkey cells enucleated with cytochalasin B. *Science* 179: 915–916.

Poste, G. 1972. Enucelation of mammalian cells by cytochalasin B. Characterization of anucleate cells. *Exp. Cell Res.* 73: 273–286.

Poste, G. and P. Reeve. 1971. Formation of hybrid cells and heterokaryons by fusion of enucleated and nucleated cells. *Nat. New Biol.* 229: 123–125.

Poste, G., B. Schaeffer, P. Reeve, and D.J. Alexander. 1974. Rescue of SV40 from SV40-transformed cells by fusion with anucleate monkey cells and variation in the yield. *Virology* 50: 85–95.

Rosenstraus, M. and L. Chasin. 1975. Isolation of mammalian cell mutants deficient in glucose-6-phosphate dehydrogenase activity. *Proc. Natl. Acad. Sci.* 72: 493–497.

Velazquez, A., F. Payne, and R. Krooth. 1971. Viral induced fusion of human cells. I. Quantitative studies on the fusion of human diploid fibroblasts induced by Sendai virus. *J. Cell. Physiol.* 78: 93–110.

Veomett, G., D. Prescott, J. Shay, and K. Porter. 1974. Reconstruction of mammalian cells from nuclear and cytoplasmic components separated by treatment with cytochalasin B. *Proc. Natl. Acad. Sci.* 71: 1999–2002.

Wright, W.E. and H. Hayflick. 1972. Formation of anucleate and multinucleate cells in normal and SV40 transformed WI-38 by cytochalasin B. *Exp. Cell Res.* 74: 187–194.

Yamaguchi, N. and I.B. Weinstein. 1975. Temperature-sensitive mutants of chemically transformed epithelial cells. *Proc. Natl. Acad. Sci.* 72: 214–218.

Chapter 9 Transfer of Chromosomes and Genes

McBride, O.W. and H.L. Ozer. 1973. Transfer of genetic information by purified metaphase chromosomes. *Proc. Natl. Acad. Sci. 70:* 1258–1262.

Graham, F.L., A.J. van der Eb, and H.L. Heijneker. 1974. Size and location of the transforming region in human adenovirus type 5 DNA. *Nature 251:* 687–691.

Graessman, M., A. Graessman, and C. Mueller. 1977. The biological activity of different early simian virus 40 DNA fragments. *INSERM Colloq. 69:* 233–240.

Wigler, M., A. Pellicer, S. Silverstein, and R. Axel. 1978. Biochemical transfer of single-copy eucaryotic genes using total cellular DNA as donor. *Cell 14:* 725–731.

Klobutcher, L.A. and F.H. Ruddle. 1979. Phenotype stabilisation and integration of transferred material in chromosome-mediated gene transfer. *Nature 280:* 657–660.

Shih, C., B.-Z. Shilo, M.P. Goldfarb, A. Dannenberg, and R.A. Weinberg. 1979. Passage of phenotypes of chemically transformed cells via transfection of DNA and chromatin. *Proc. Natl. Acad. Sci. 76:* 5714–5718.

Cooper, G.M., S. Okenquist, and L. Silverman. 1980. Transforming activity of DNA of chemically transformed and normal cells. *Nature 284:* 418–421.

To study the regulation of gene expression or the structures of genes themselves, it is useful to have the ability to move DNA itself around from cell to cell. To do so, the DNA must survive extraction and purification intact and be presented to cells in such a way that expression occurs. In this chapter we see the rapid development of the methods of transferring genes into cultured cells. Coupled with powerful new methods for isolating, cloning, and sequencing specific genes, gene transfer has, in less than a decade, gone from a dubious pursuit to the preferred way to ask many sorts of genetic and biochemical questions of eukaryotic cells.

Mouse A9 cells lack hypoxanthine-guanine phosphoribosyl transferase (HGPRT) activity, and they revert to the expression of this activity only with the greatest reluctance. No more than 1 cell in 10^9 generates a colony in HAT medium. **McBride** and **Ozer** use this low background to demonstrate stable transfer of an active gene, which is carried into a cell as part of a whole chromosome. Normal Chinese hamster cells contain their own HGPRT. Since it is encoded by a hamster gene, it is not the same protein as the mouse HGPRT in A9 cells. Chromosomes isolated from mitotic hamster cells were fed to A9 cells. The A9 cells were permitted three days to take these up and then they were shifted into HAT medium. About 1 in 10^7 cells gave rise to colonies. The HGPRT in each colony clearly was of hamster origin. The gene was stably transferred into some clones, but other clones very rapidly lost the hamster enzyme.

The utility of McBride and Ozer's procedure is limited by its low efficiency. The next two papers present two very different improvements in efficiency: direct transfer and microinjection of DNA. Adenoviral DNA is large enough to code for dozens of genes (Chapter 4). **Graham, van der Eb**, and **Heijneker** ask whether fragments of adenoviral DNA can transform cells. To get adenoviral DNA into cells, they coprecipitate it with calcium chloride. Mixed with calcium chloride, a solution containing DNA yields a powdery precipitate that falls onto a monolayer of cells in a dish and is then taken up by the cells. An excess of carrier DNA seems to be necessary, suggesting that the uptake step may in some way degrade the incoming DNA. For mixtures of adenoviral DNA fragments, transformation frequency of primary rat cells is proportional to DNA dose. Exodeoxyribonucleases, which destroy the native ends of the adenoviral genome, destroy transformation activity. Graham et al. use these observations to show that a mere 5% of the adenoviral DNA is capable of transforming cells as efficiently as a randomly fragmented whole adenoviral genome, provided it includes the native left end of the genome. Thus, once again, transformation capacity can reside in a fraction of the entire genome of a virus (Chapter 5).

In size, in topological linkage of replication to transcription, and in splicing, the transforming region of adenovirus resembles the early region of SV40 (Chapters 4 and 7). The SV40 early region contains the origin of SV40 DNA replication and encodes, as well, many functions, including transformation and induction of cellular and viral DNA synthesis. Initiation of viral DNA synthesis is temperature-sensitive in group-A (tsA) mutants, which lie in this region (Chapters 3 and 4). To test which parts of the early region encode which functions, **Graessman, Graessman,** and **Mueller** inject specific fragments of SV40 into monkey and mouse cells and assay for immediate cellular responses. Using restriction endonucleases, they made a small number of different specific fragments from SV40 DNA. Then, molecules of each fragment were microinjected directly into the nuclei of cultured cells, with an apparatus that delivered 10^{-11} ml/nucleus. This rather small amount of additional volume does not seem to bother the cell. Viral early genes were active in every cell receiving either uncut viral DNA or a fragment encoding the entire early region, and the fragment coding for the entire early region complemented tsA mutants. A fragment coding for two thirds of the early region did not complement tsA mutants but did suffice both for synthesis of a detectable nuclear T antigen and for induction of cellular DNA synthesis.

The genomes of very big viruses code for many enzymes. The genes for these enzymes make valuable markers for gene transfer. For example, herpes virus makes a thymidine kinase (TK) that is active in mouse cells, and if sheared herpes DNA is added in calcium phosphate to TK⁻ mouse cells in HAT medium, it will rescue a few of the cells. Is the DNA of the herpes-encoded enzyme integrated in a host chromosome of a TK⁺ cell recovered from such a gene transfer experiment?

Because viral DNA may be easily made into a very radioactive probe, the herpes tk gene can be located in the DNA from the genome of a recipient cell by blot-transfer hybridization. **Wigler, Pellicer, Silverstein,** and **Axel** show first that one such recipient clone has only one copy of herpes tk, in one place in the cell genome. They then use DNA from that transferent cell clone as a donor for a second tk gene transfer experiment, again using calcium phosphate precipitation and HAT selection. Colonies surviving HAT in this experiment again had virus-encoded tk and integrated viral DNA by blot hybridization analysis. Although the exact location and stability of the integrated tk gene are not yet completely determined, it is clear that a gene can be transferred and tracked from one cell to another while remaining transcriptionally active.

Can a cell take up a whole active chromosome? If so, synteny (Chapter 8) should link sets of markers on the same chromosome. **Klobutcher** and **Ruddle** transfer genomic pieces large enough to express many markers of a single chromosome. Some of these must be almost as large as whole chromosomes. Human chromosomes were transferred into mouse TK⁻ cells and recipients selected for TK. Coexpression of human galactokinase and type-1 procollagen genes, both of which are linked to the tk gene on chromosome 17, showed that these markers remain linked in the recipient cell. When chromosomes were prepared, clones with stable donor-gene expression had an additional chromosomal fragment. Klobutcher and Ruddle hypothesize that transferred chromosomes lose their centromeres in the process of transfer and that stabilization of gene expression reflects the acquisition by a host chromosome of a donor chromosomal fragment whose size can be up to 1% of the size of the host's genome. Each stable clone contains this inserted chromosomal fragment in a different chromosome, suggesting the possibility that integration of donor sequences can occur in more than one place, as we know it may in certain viral transformations (Chapters 3 and 4).

The last two papers in this chapter use DNA to transfer the phenotype of oncogenic transformation. Although sequences from unknown viruses cannot be rigorously excluded, both studies find that no viral sequences need be intentionally present for a successful transfer to occur. However, the two studies profoundly differ as to the nature of the host sequences responsible for the event.

We have seen in Chapter 3 that chemicals can transform cultured fibroblasts and that the process of transformation resembles mutagenesis. **Shih, Shilo, Goldfarb, Dannenberg,** and **Weinberg** use a variety of chemically transformed mouse cell lines as donors in a DNA transfer and the NIH3T3 line as a recipient in a density transformation assay (Chapter 3). Dense foci arise. They presumably are the result of successful transfer of DNA responsible for the transformed phenotype from the chemically transformed cell to an NIH3T3 cell. The background of spontaneous dense foci in mock-infected cultures was high, about 1 in 10^6 inoculated cells. DNA from BALB3T3, NIH3T3, mouse liver, and 10 of the 15 chemically transformed lines gave no foci above this background. The DNA from one methylcholanthrene-transformed BALB/3T3 line, however, generated about tenfold more foci than background frequency.

Foci cloned from this successful experiment yielded cellular DNA that was again able to transform NIH3T3 at above the background frequency, suggesting that a DNA sequence from the chemically transformed cell had become stably associated with the DNA of an NIH3T3 cell after transfection. Apparently, then,

a transformed phenotype was transferred via high-molecular-weight DNA, as if it were the result of a discrete dominant allele. Either a specific new sequence was expressed in these mutagenized transformed cells, or some preexisting sequence was divorced by mutagenesis from cellular regulation.

If the latter hypothesis is correct, then normal and chemically transformed cells both ought to be able to donate transforming sequences, provided these are separated from host regulatory sequences. **Cooper, Okenquist,** and **Silverman** test this hypothesis in the same donor-recipient cell set as Shih et al. used. However, the anchorage-independence assay was chosen here. With NIH3T3 there is a low background of one spontaneous colony from about 10^9 cells. Unsheared DNA from normal cells had no capacity to transfer anchorage independence to NIH3T3. Cooper et al. unlinked donor genes from each other by randomly shearing the donor DNA to an average size of 2×10^6 molecular weight, approximately 1% of the size of the DNA used by Shih et al. Sheared DNAs from three chemically transformed lines and from three untransformed lines were all equally able to transfer anchorage independence, albeit very inefficiently. Sheared DNA from E. coli was unable to transfer anchorage independence, and sheared salmon sperm DNA was even more inefficient than mammalian DNA.

Taken together, these last two papers suggest that changes in position of DNA sequences may alone be sufficient to generate oncogenic transformed cells. DNA movement in cultured cells is currently of great interest to those studying differentiation of the immune system (Chapters 2 and 7) as well as to those studying disease. That it occurs at all is surprising, but given that it occurs, it would be equally surprising were it to serve no useful purpose in the body.

Proc. Nat. Acad. Sci. USA
Vol. 70, No. 4, pp. 1258–1262, April 1973

Transfer of Genetic Information by Purified Metaphase Chromosomes

(Chinese hamster and mouse fibroblasts/HeLa cells/hypoxanthine phosphoribosyl transferase)

O. WESLEY McBRIDE AND HARVEY L. OZER*

Laboratory of Biochemistry, National Cancer Institute, National Institutes of Health, Bethesda, Maryland 20014

Communicated by Alton Meister, February 15, 1973

ABSTRACT   Transfer of genetic information from isolated mammalian chromosomes to recipient cells has been demonstrated. Metaphase chromosomes isolated from Chinese hamster fibroblasts were incubated with mouse A₉ cells containing a mutation at the hypoxanthine–guanine phosphoribosyl transferase (*hprt*) locus. Cells were plated in a selective medium, resulting in death of all unaltered parental A₉ cells. However, colonies of cells containing hypoxanthine phosphoribosyl transferase (EC 2.4.2.8) appeared with a variable frequency of about 10^{-6} to 10^{-7}. The enzyme from these cells was indistinguishable from that from Chinese hamster cells, as shown by DEAE-cellulose chromatography and gel electrophoresis, and differed clearly from the mouse enzyme. The colonies, thus, did not result from reversion of A₉ parental cells to wild type, but appeared to represent progeny of individual cells that had ingested chromosomes, replicated, and expressed the *hprt* gene. These colonies differed from each other in stability of expression of the transferred gene.

A means for genetic mapping of mammalian chromosomes is provided by a combination of the technique of cell fusion and karyotypic analysis of resultant hybrid clones by the recently developed quinacrine (3) and Giemsa banding procedures (4). Techniques for the direct transfer of genetic information from subcellular particles to cells could provide a complementary method for genetic mapping. This would eliminate the necessity of awaiting segregation of chromosomes, thereby reducing the possibility of chromosomal rearrangements. Mammalian metaphase chromosomes appear eminently suitable for this purpose since a meaningful biological fractionation of genes is present in chromosomes, and numerous methods (5–7) have been described for isolation of these particles. Chromosomal DNA might be somewhat better protected from degradation during cellular uptake than free DNA due to its compact structure and its association with proteins and RNA. The introduction of intact chromosomes into cells could circumvent problems of integration of DNA into the host genome; subsequent replication and expression of chromosomal genes should be analogous to the steps following cell fusion.

Evidence exists that isolated metaphase chromosomes can penetrate into mammalian cells *in vitro* (8–16), but most of the chromosomal DNA is subsequently degraded (8–12). Previous information suggesting that mammalian chromosomes can be replicated after uptake is extremely sparse (14–16), and no evidence has been provided for expression of this new genetic information by the host cell.

This paper presents the first evidence that both replication and expression of chromosomal genes can occur after the uptake of mammalian metaphase chromosomes. Moreover, permanent transfer of this new genetic information results, although the frequency of this gene transfer is low.

MATERIALS AND METHODS

Cell Cultures. Cells used were: (*1*) wild-type Chinese hamster fibroblasts (V-79), recently cloned; (*2*) mouse fibroblasts (L₉₂₉); (*3*) HeLa cells; and (*4*) mouse L-cell lines A₉ and B₈₂, deficient in hypoxanthine phosphoribosyl transferase (HPRT; EC 2.4.2.8) and thymidine kinase (EC 2.7.1.21), respectively (17). Cells were maintained in monolayer cultures at 37° in a gas-flow (7% CO₂–air), humidified incubator, in Eagle's minimal essential medium (MEM) containing twice the usual concentration of amino acids and vitamins. Cells were also grown in suspension culture, in Eagle's medium without calcium, or in Ham's F-10 medium (18). All media were supplemented with 10% fetal-calf serum, 4 mM glutamine, penicillin (50 μg/ml), and streptomycin (50 μg/ml).

Isolation and Purification of Metaphase Chromosomes. Chromosomes were isolated under sterile conditions from [³H]-dT-labeled cells (0.2 mCi/liter) as described (19), by slight modifications of either the procedure of Mendelsohn *et al.* (5) at pH 3 or the method of Maio and Schildkraut (6) at pH 7. Chromosomes were subsequently separated from intact cells and most debris by ultracentrifugation through a layer of 80% sucrose (w/v). Nuclei were then removed by unit-gravity sedimentation at pH 7, essentially as reported for fractionation of nuclei (20). Final chromosome preparations contained about one nucleus per thousand cell equivalents of chromosomes (i.e., one nucleus per 25,000 chromatid pairs). For each preparation, the molecular weight of dissociated, single-stranded, chromosomal DNA was determined by velocity sedimentation in alkaline sucrose density gradients (21), with ¹⁴C-labeled simian virus 40 (SV40) DNA I and II (22) as markers. The molecular weight of chromosomal DNA decreases progressively on storage of the chromosomes, even at 5°, although chromosome morphology remains good. Thus, chromosomes were used in gene-transfer experiments immediately after isolation.

Incubation of A₉ Cells with Chromosomes. Purified metaphase chromosomes (about 1 cell equivalent per recipient cell) from Chinese hamster cells were dispersed with A₉ mouse fibroblasts (6 × 10⁶/ml) in complete Eagle's MEM spinner medium, containing 12 μg/ml of poly-L-ornithine (molecular weight 70,000; Mann Research) in a sterile, siliconized, glass culture tube. The tube was equilibrated with 5% CO₂–air and incubated for 2 hr at 37° while rolling in a nearly horizontal position at 10 rpm. Aliquots of 5 × 10⁵ cells were transferred to 100-mm plastic dishes (Falcon) containing 10 ml

Abbreviations: HPRT, hypoxanthine phosphoribosyl transferase (EC 2.4.2.8); *hprt*, gene directing synthesis of HPRT; HAT, hypoxanthine–amethopterin–thymidine, selective growth medium of Littlefield (1); MEM, Eagle's minimal essential medium (2).

*Present address: Worcester Foundation for Experimental Biology, Shrewsbury, Mass. 01545.

1258

of complete MEM. After 3 days of incubation, the medium was replaced with HAT medium, and the plates were refed with this selective medium at 3- to 4-day intervals for 6 weeks. Colonies that appeared during this interval were cloned in metal cylinders, removed by treatment with trypsin, and recultured in HAT medium.

Enzyme Assays. Cells were washed with 0.15 M NaCl, suspended in 0.01 M Tris·HCl (pH 7.4) (6×10^7 cells per ml), and lysed by freezing and thawing. The assay for HPRT activity was basically that described by Harris and Cook (23), involving conversion of [8-^{14}C]hypoxanthine substrate to [^{14}C]IMP product, which was collected on DEAE-cellulose disks (Whatman DE-81). The reaction product of the hamster and wild-type mouse HPRT, as well as that of extracts of the experimental clones, was confirmed to be [^{14}C]IMP by thin-layer chromatography on cellulose; both solvent systems B and C of Ciardi and Anderson (24) were used. Purity (>97–99%) of the [^{14}C]hypoxanthine substrate was ascertained in the same manner.

DEAE-Cellulose Chromatography. Micro-granular DEAE-cellulose (Whatman DE-52) was washed (25), equilibrated with starting buffer [0.01 M Tris·HCl (pH 8.7)], and packed in a 5×140 mm column. Enzyme extract (about 5 mg of protein) was applied to the column at 5° and sequentially eluted at a constant flow-rate (8 ml/hr) with 4 ml of starting buffer, a 60-ml linear (0–225 mM) NaCl gradient followed by 3 ml of 0.4 M NaCl (both containing starting buffer), and finally with 6 ml of 2 M NaCl-0.1 M Tris·HCl (pH 8.8). 1-ml Fractions were assayed immediately for HPRT activity and later for protein concentration, conductivity, and pH.

Gel Electrophoresis. Gels were prepared by a modification of the procedure of Bakay and Nyhan (26). An 8% poly-

TABLE 1. *HPRT-positive colonies after incubation of A_9 cells with chromosomes*

Experiment	Total no. of cells[a] ($\times 10^{-6}$)	Positive plates/ total plates[b]	DNA[c] mol. wt. ($\times 10^{-6}$)
1	6	5/12	30
2	25	2/50[d]	3
3	25	2/49[e]	30
4	50	0/100[f]	20
5	6	1/12	25
6	10	0/20[f]	1.5–30[g]
7	9	1/18	30
8A	10	1/20[h,i]	30–130
8B	10	3/20[h]	30–130

Chromosomes used in experiments 2, 4, and 8 were isolated at pH 7; otherwise chromosomes were isolated at pH 3. [a] Total number of A_9 cells incubated and subsequently plated. [b] Number of plates with one or more colonies per total number of plates inoculated. [c] Molecular weight of single-stranded DNA in the chromosome preparations (see *Methods*). [d] Colonies not confirmed by cloning and growth in HAT medium. [e] HeLa chromosomes incubated with A_9 recipient cells. [f] Incubation medium contained 2 mM CaCl₂ (monolayer medium). [g] Very heterogeneous molecular weight. [h] Ratio of cell equivalents of chromosomes to recipient cells was 10:1 in experiment 8B and 1:1 in experiment 8A; the ratio was about 1:1 in the other experiments. [i] This colony was a revertant.

acrylamide gel ($70 \times 100 \times 2$ mm) was formed between two glass plates, and a 5% stacking gel was added. A cellulose acetate strip containing sample slots was pushed into the upper gel. Protein extracts (about 15 μg of protein) were mixed with bovine-serum albumin (25 μg per application) and sucrose (10% w/v) and layered under upper-tray buffer in the slots. Bromphenol blue (0.05 ml saturated solution per liter) was included in the upper tray as a tracking dye. Electrophoresis was conducted at 300 V until albumin migrated to the bottom of the running gel. The gel was reacted with substrate (30 min at 37°), and the [^{14}C]IMP product was precipitated with 0.1 M LaCl₃-0.1 M Tris·HCl (pH 7.0) (26). Autoradiography was performed after repeated washing and dehydration of the gel (27).

Other Assays. Protein concentration was determined by the procedure of Lowry *et al.* (28) with bovine-serum albumin as the standard, and most assays were kindly performed by Mr. Miles Otey with a Technicon Autoanalyzer. Conductivity and pH measurements were performed at room temperature (24°). Particle concentrations were determined with an electronic counter (Celloscope).

Karyotypes. Cells were exposed to colcemid (0.2 μg/ml) for 3 hr, swollen in 1% Na citrate (20 min at 37°) or 75 mM KCl (20 min at 25°), and fixed with methanol–acetic acid (3:1). The fixed cells were applied to a cold, moist slide and spread by flaming before staining with crystal violet.

RESULTS

Isolation of HPRT-Positive Colonies after Incubation of A_9 Cells with Chromosomes. Isolated Chinese hamster chromosomes were incubated in suspension (see *Methods*) with mouse A_9 cells before plating the cells and the subsequent addition of selective medium (Table 1). Colonies appeared at a relatively low frequency of 10^{-6} in experiment 1 and about 10^{-7} in the combination of all experiments. A positive result was also obtained (experiment 3) with chromosomes isolated from HeLa cells. Since migration of cells may occur, resulting in satellite colonies, only one colony was scored for any plate, irrespective of the actual number of colonies observed, and each colony that was further analyzed was cloned from a separate plate. Local overgrowth of unaltered cells may also occur early before the addition of HAT, and slowly regress or persist for long intervals. Therefore colonies were cloned and cultured in the same selective medium and colonies lost in the cloning process were considered false positives. In two experiments designed to detect gene transfer from hamster chromosomes to A_9 cells, we used inactivated Sendai virus to mediate the transfer, but were unsuccessful.

Reversion of A_9 Cells. A control incubation performed in experiment 1 (Table 1), by use of identical procedures without chromosomes, resulted in no colonies. Larger numbers of cells (1.37×10^9) have also been plated in selective HAT medium, and only two revertant colonies were found. Thus, the A_9 cells have a very low rate of reversion that appears to be lower than the gene-transfer frequency, even considering that under the experimental conditions, cells could have doubled about three times before the HAT selective medium was added.

Detection of the Product of Gene Transfer In Vitro. The clones obtained in experiment 1 of Table 1 were propagated

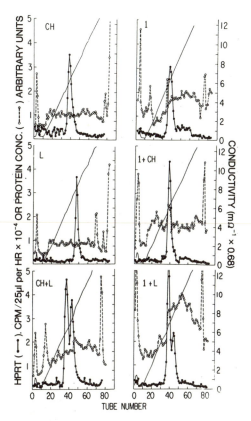

FIG. 1. Chromatography of crude enzyme extracts isolated from Chinese hamster (*CH*), mouse (*L*), Clone 1 cells, and artificial mixtures of these solutions. Extracts (3–15 mg of protein) were applied to columns of DEAE-cellulose and eluted with a gradient of NaCl in 0.01 M Tris·HCl (pH 8.7). Fractions of 1 ml were collected and assayed for HPRT activity (●——●), protein (O– – –O), and conductivity (——). Enzyme activity is plotted at 0.5 the normal scale for L, 1 +CH, and 1 + L. Each unit of protein concentration (*left ordinate*) represents 100 μg/ml (CH, L, 1 +CH) or 70 μg/ml (CH + L, 1, 1 + L).

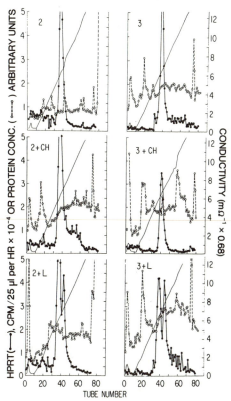

FIG. 2. DEAE-cellulose chromatography at 5° of HPRT extracts isolated from clones 2 and 3, and artificial mixtures of these solutions with Chinese hamster (*CH*) or mouse (*L*) extracts. Enzyme activity is plotted at 0.5 the usual scale for 3 +CH. Each unit of protein concentration (*left ordinate*) represents 100 μg/ml (2, 2 + L, 3 +CH) or 70 μg/ml (2 +CH, 3, 3 + L). See Fig. 1 and *Methods* for further details.

in suspension culture in HAT medium, and the high-speed (100,000 × g) supernatant fluids of freeze–thaw lysates were examined directly for HPRT activity (Table 2). The specific activities were similar to those obtained from extracts of Chinese hamster fibroblasts or wild-type mouse L₉₂₉ cells, or the closely related, B₈₂ thymidine-kinase-mutant L cells.

DEAE-Cellulose Chromatography of HPRT Extracts. Chromatography demonstrated a single peak of HPRT activity for both the mouse and hamster parental species (Fig. 1). However, the hamster HPRT is adequately resolved from mouse (L) enzyme when compared directly by elution position or conductivity at the point of emergence, or by mixture of the extracts before chromatography (*lower left*, Fig. 1). Clones from experiment 1 of Table 1 were similarly analyzed. Chromatography of an extract of clone 1 revealed a single peak of HPRT activity occurring in the position appropriate for hamster enzyme. The mixture of clone 1 extract with hamster HPRT again resulted in the single peak of activity,

whereas the mixture with mouse enzyme disclosed two peaks of HPRT at the appropriate positions for each species.

Similar results are presented in Fig. 2 with two other clones (from experiment 1) alone or as artificial mixtures with hamster or mouse HPRT. Approximately equal quantities (enzyme activity) of each of the two components were used in all mixtures. The elution positions and half-widths as measured by elution volumes or conductivities of eluate for all of these clones, individually or mixed with hamster extract, are virtually identical with that obtained with the hamster enzyme alone. It therefore appears unlikely that the patterns result from three revertant clones that all fortuitously show chromatographic behavior very similar to that of the hamster enzyme. Furthermore, HPRT in extracts of thymidine-kinase-mutant L-cells (B₈₂) and the single A₉ revertant exhibit chromatographic behavior (not shown) identical with that of wild-type mouse HPRT. A fourth clone from experiment 1 (Table 1) was lost before chromatography, but it appeared to contain hamster enzyme, as determined by electrophoresis on cylindrical acrylamide gels. The fifth clone was lost before further study.

Chromatographic analyses (not illustrated) of the clone

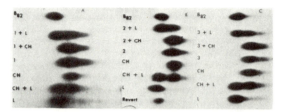

Fig. 3. Gel electrophoresis of HPRT extracts on vertical slabs of polyacrylamide at 5°. The individual extracts and artificial mixtures are identified by the same symbols as Figs. 1 and 2, and the A₉ revertant is also shown (B).

from experiment 5 of Table 1 and the three clones from experiment 8B of Table 1 also demonstrated the hamster-enzyme profile, while the clone from experiment 8A exhibited a profile that is identical with that of mouse HPRT, and therefore represents a revertant.

Acrylamide Gel Electrophoresis of HPRT. Mouse HPRT has a greater electrophoretic mobility than the hamster enzyme, from which it is adequately separated (Fig. 3). HPRT in extracts of clones 1, 2, and 3 from experiment 1 of Table 1 were all identical in mobility with hamster enzyme, when run alone or when mixed with hamster extract, whereas two bands resulted when artificial mixtures of these extracts with mouse HPRT were subjected to electrophoresis. Furthermore, the HPRT produced by the A₉ revertant (Fig. 3B) is not electrophoretically distinguishable from the wild-type mouse enzyme. Evidence (not shown) that the radioactive spots reflect the location of HPRT activity is provided by the fact that the spots were markedly attenuated when the gel was reacted with substrate at 5° rather than 37°, as well as by the fact that no radioactivity could be detected when 5-phosphoribosylpyrophosphate was omitted from the reaction mixture. No radioactivity was observed under any condition when an extract of A₉ cells was subjected to electrophoresis. Gel electrophoresis (not shown) also demonstrated that the HPRT products of the clone from experiment 5 and the three clones from experiment 8B of Table 1 were indistinguishable from the Chinese hamster enzyme, whereas the product of the clone from experiment 8A had the same electrophoretic mobility as the mouse HPRT.

Karyotypes. Histograms of the numbers of total chromosomes (Fig. 4) and biarmed chromosomes (not shown) in the clones from experiment 1 of Table 1 were closely similar to that of the parental A₉ cells. Karyotypes of all experimental cell lines clearly differ from the Chinese hamster karyotype, which exhibits a narrow mode of 23 chromosomes,

TABLE 2. *HPRT activity*

Cell type	Specific activity*	Cell type	Specific activity*
CH-V-79	196, 226, 354, 322	Clone 2	274, 254
L₉₂₉	134, 143, 151, 193	Clone 3	149
B₈₂	43, 142, 105	Clone 4	155
A₉	<0.01, <0.02	A₉ revertant	141
Clone 1	65, 160		

* nmol of IMP/hr per mg of protein.

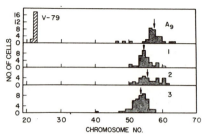

Fig. 4. Histogram of chromosomes in parental cell lines (V-79 and A₉) and in clonal lines (1, 2, and 3) from experiment 1 of Table 1. *Arrows* indicate the median number of chromosomes in each line.

indicating that none of these lines could have arisen by contamination of the cultures with hamster cells.

Stability of Genotype after Chromosome Transfer. The clones of experiment 1 of Table 1 were grown in selective HAT medium for several generations. After a shift to nonselective MEM spinner medium, the growth of each line was continued in suspension cultures for 2 months. Aliquots were removed at intervals for determination of plating efficiencies in MEM, HAT, and 20 μM 8-azaguanine, and the plating efficiencies in HAT relative to those in MEM are shown in Fig. 5. Clones 1 and 2 exhibited no detectible change in plating efficiency in the selective HAT medium during this entire interval, whereas there was a very rapid accumulation of HPRT-deficient cells when clone 3 was cultured in nonselective medium. The curve for clone 3 suggests that about 10–20% of the cells lose the *hprt* gene at each division. Similar reversion behavior has been reported by Schwarz *et al.* (29) for cells containing *hprt* on a chromosome fragment. The instability exhibited by clone 3 would be highly unlikely if it had arisen by reversion (back-mutation) of A₉ cells rather than by gene transfer, unless the parental A₉ cells had very marked selective growth advantage in MEM relative to the revertant.

DISCUSSION

The evidence for transfer of genetic information from ingested metaphase chromosomes to recipient cells and expression of this information by recipient mammalian cells can be summarized as: (1) A relatively high frequency of appearance

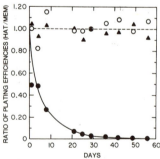

Fig. 5. Stability of the *hprt⁺* genotype in colonies after removing selective pressure. The plating efficiency in selective HAT medium compared to that in nonselective medium is plotted as a function of the time interval after the cells were removed from HAT medium. The cell lines depicted are clones 1 (○), 2 (△), and 3 (●——●) from experiment 1 of Table 1.

of colonies in selective medium when chromosomes are present, compared to the very low reversion frequency under similar conditions. (2) The rapid loss of the *hprt* gene by one of the clones (clone 3 of experiment 1), which is unexpected if the colony arose by reversion of parental cells. (3) The physical characterization of the enzyme (HPRT) product as indistinguishable from the chromosomal species and clearly different from the parental species, as shown by DEAE-cellulose column chromatography and acrylamide gel electrophoresis. This last point is the most convincing one.

Other possible explanations that have been considered, but appear extremely unlikely or completely inconsistent with the results, include the following:

(*1*) Reversion is inconsistent with any of the three points mentioned above and is especially refuted by the physical characterization of the gene product. Some revertants could occur involving mutation at a locus for a charged residue, resulting in a gene product that differed from the parental type. However, the possibility that the product of all revertants could be completely indistinguishable from that of the donor chromosome species by two methods of characterization seems remote. Furthermore, the single authentic revertant that was analyzed produced an HPRT that was not distinguishable from that of the parental (mouse) species. Schwarz *et al.* (29) also reported that an A₉ revertant produced HPRT that was electrophoretically identical to wild-type mouse enzyme.

(*2*) The possibility that the cultures were contaminated with a few wild-type (L₉₂₉) mouse cells is excluded by the characterization of the enzyme product and by the absence of similar colonies in control cultures.

(*3*) Contamination of the cultures with a few Chinese hamster cells or incomplete removal of these cells during the process of chromosome isolation is excluded by the fact that the karyotypes of the resultant clones were similar to that of the mouse species and totally different from that of the hamster species. Furthermore, no viable intact cells were detected under the conditions of chromosome isolation.

(*4*) It is unlikely that spontaneous fusion of any intact cells (surviving chromosome isolation and purification procedures) or nuclei with A₉ mouse cells is responsible for the observed results in view of the low number of intact cells and nuclei in the preparations. Furthermore, experiments performed under conditions more favorable for cell fusion, involving the use of inactivated Sendai virus, failed to result in colony formation. Also, no evidence for persistence of large numbers of hamster chromosomes was found in karyograms.

(*5*) Transformation of the cells by naked DNA or nucleoprotein cannot be excluded but it is considered unlikely. Most DNA or nucleoprotein would have been removed during the chromosome isolation by several centrifugations at 1000 × *g* for 30 min.

(*6*) Some nonspecific effect of added chromosomes is excluded by the physical characterization of the enzyme product as hamster type and the absence of gene transfer when chromosomes containing low molecular weight DNA were used (19). Any stimulation by degraded chromosomal products is unlikely since they would be rapidly diluted out during growth.

(*7*) Contamination of cultures with viruses or mycoplasma coding for an active HPRT is excluded by inability to culture mycoplasma from the clones, failure to observe HPRT-positive colonies in control cultures, and physical characterization of the enzyme product. However, the possibility that a transducing virus was present cannot be excluded.

There are several possible explanations for the low frequency of gene transfer observed (see ref. 19). The ability to demonstrate any gene transfer in the present experiments results from the use of a selective system using recipient cells with an extremely low reversion frequency and chromosomes isolated from a different species, thereby permitting positive identification of the species of origin of the gene product.

Mammalian chromosome uptake *in vitro*, particularly combined with the use of fractionated chromosomes, could provide a powerful tool for genetic mapping. However, the utility of this procedure would be increased by the development of methods for a greater efficiency of transfer and expression of the genetic information. The possible application of this technique to "gene modification" is open to considerably greater skepticism (30).

We thank Mrs. Susan Bridges for expert technical assistance.

1. Littlefield, J. W. (1964) *Science* **145,** 709–710.
2. Eagle, H. (1959) *Science* **130,** 432–437.
3. Caspersson, T., Zech, L. & Johansson, C. (1970) *Exp. Cell Res.* **60,** 315–319.
4. Drets, M. E. & Shaw, M. W. (1971) *Proc. Nat. Acad. Sci. USA* **68,** 2073–2077.
5. Mendelsohn, J., Moore, D. E. & Salzman, N. P. (1968) *J. Mol. Biol.* **32,** 101–112.
6. Maio, J. J. & Schildkraut, C. L. (1969) *J. Mol. Biol.* **40,** 203–216.
7. Hearst, J. E. & Botchan, M. (1970) *Annu. Rev. Biochem.* **39,** 151–182.
8. Chorazy, M., Bendich, A., Borenfreund, E., Ittensohn, O. L. & Hutchison, D. J. (1963) *J. Cell Biol.* **19,** 71–77.
9. Whang-Peng, J., Tjio, J. H. & Cason, J. C. (1967) *Proc. Soc. Exp. Biol. Med.* **125,** 260–263.
10. Ittensohn, O. L. & Hutchison, D. J. (1969) *Exp. Cell Res.* **55,** 149–154.
11. Kato, H., Sekiya, K. & Yosida, T. H. (1971) *Exp. Cell Res.* **65,** 454–462.
12. Burkholder, G. D. & Mukherjee, B. B. (1970) *Exp. Cell Res.* **61,** 413–422.
13. Ebina, T., Kamo, I., Takahashi, K., Homma, M. & Ishida, N. (1970) *Exp. Cell Res.* **62,** 384–388.
14. Yosida, T. H. & Sekiguchi, T. (1968) *Mol. Gen. Genet.* **103,** 253.
15. Sekiguchi, T., Sekiguchi, F., Satake, S. & Yosida, T. H. (1969) *Symp. Cell Biol.* **20,** 223.
16. Sekiguchi, T., Sekiguchi, F., Satake, S. & Yosida, T. H. (1969) *Jap. J. Med. Sci. Biol.* **22,** 72–73.
17. Littlefield, J. W. (1966) *Exp. Cell Res.* **41,** 190–196.
18. Ham, R. G. (1963) *Exp. Cell Res.* **29,** 515–526.
19. McBride, O. W. & Ozer, H. L. (1973) in *Possible Episomes in Eukaryotes Le Petit Colloquia on Biology and Medicine* (North-Holland, Amsterdam), Vol. 4, in press.
20. McBride, O. W. & Peterson, E. A. (1970) *J. Cell Biol.* **47,** 132–139.
21. Abelson, J. & Thomas, C. A., Jr. (1966) *J. Mol. Biol.* **18,** 262–291.
22. Sebring, E. D., Kelly, T. J., Jr., Thoren, M. M. & Salzman, N. P. (1971) *J. Virol.* **8,** 478–490.
23. Harris, H. & Cook, P. R. (1969) *J. Cell Sci.* **5,** 121–133.
24. Ciardi, J. E. & Anderson, E. P. (1968) *Anal. Biochem.* **22,** 398–408.
25. Himmelhoch, S. R. (1971) in *Methods in Enzymology*, ed. Jakoby, W. B. (Academic Press, New York), Vol. 22, pp. 273–286.
26. Bakay, B. & Nyhan, W. L. (1971) *Biochem. Genet.* **5,** 81–90.
27. Fairbanks, G., Jr., Levinthal, C. & Reeder, R. H. (1965) *Biochem. Biophys. Res. Commun.* **20,** 393–399.
28. Lowry, O. H., Rosebrough, N. J., Farr, A. L. & Randall, R. J. (1951) *J. Biol. Chem.* **193,** 265–275.
29. Schwartz, A. G., Cook, P. R. & Harris, H. (1971) *Nature New Biol.* **230,** 5–8.
30. Fox, M. S. & Littlefield, J. W. (1971) *Science* **173,** 195.

Reprinted from Nature, Vol. 251, No. 5477, pp. 687–691. 1974. © 1974 Macmillan Journals, Ltd.

Size and location of the transforming region in human adenovirus type 5 DNA

F. L. Graham* & A. J. van der Eb

Laboratory for Physiological Chemistry,

H. L. Heijneker

Laboratory of Molecular Genetics, State University of Leiden, Leiden, The Netherlands

Fragments of human adenovirus type 5 DNA as small as one million daltons can transform cells in vitro. The DNA segment which induces transformation is located between 1 and 6% from the left end of the Ad 5 DNA molecule.

HUMAN adenoviruses contain linear double stranded DNA of molecular weight 20×10^6 to 25×10^6 (refs 1, 2). They can be classified into three subgroups: A, B, and C which vary in their degree of oncogenicity[3]. Viruses of subgroup C (to which adeno 5 belongs) are nononcogenic in that they do not give rise to tumours after injection into newborn hamsters[4]; they can, however, transform cells *in vitro*[5,6] and in certain conditions the transformed cells can induce tumours in animals[7,8].

We have recently shown that Ad 5 DNA is infectious for human KB cells[9] and that it can transform rat cells *in vitro*[10]. Although DNA from simian adenovirus 7 had previously been reported to be oncogenic in newborn hamsters[11] our observations provided the first demonstration of transformation *in vitro* by adenovirus DNA. The development of an *in vitro* assay[10] afforded the opportunity for quantitative studies of transformation by adenovirus DNA.

While examining the effects of various treatments on the biological activity of Ad 5 DNA we observed that transforming activity was much more resistant to shearing than was infectivity: DNA preparations which were rendered noninfectious by shearing still retained undiminished transforming activity. We concluded that we might be observing transformation by DNA fragments. The studies presented here were undertaken to prove that Ad 5 DNA fragments of less than genome size have transforming activity, to determine the minimum size of fragments which retain the ability to transform cells, and to map the location of the transforming (T) segment within the viral DNA molecule.

The assays for biological activity were carried out using a modification of what we now call the calcium technique[9]. This method is based on the formation of a coprecipitate of DNA and calcium phosphate (formed by adding $CaCl_2$ to solutions of DNA in phosphate-containing buffer) which on addition to cell cultures becomes associated with the cells. DNA uptake then takes place at 37° C by a process which again requires calcium ions. The most important modification which we have introduced in the technique is the addition of a carrier DNA to all assays. This was shown to improve the efficiency of the assay[9] and has the important advantage of producing a linear dose response (ref. 12 and Fig. 1), thus simplifying quantitative calculations of specific activities.

Transformation by sheared Ad 5 DNA

Ad 5 DNA which was sheared to fragments having a sedimentation coefficient ($S_{20,w}$) of 22S (molecular weight 8×10^6) had the same transforming activity as unsheared DNA (Fig. 1) yet infectivity of the sheared DNA could not be detected when

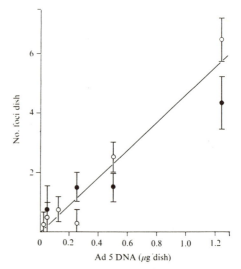

Fig. 1 Dose response for transformation of primary rat kidney cells by intact (○) and sheared (●) Ad 5 DNA. The DNA was extracted from purified Ad 5 as described previously[1]. Shearing was carried out by passing 200 µg ml⁻¹ solutions of Ad 5 DNA in 0.1×SSC (15 mM NaCl, 1.5 mM sodium citrate) 10 times through a needle (0.72 mm inside diameter × 30 mm long) using maximum hand pressure on a syringe with a 15 mm diameter plunger. All sedimentation coefficients were determined by band sedimentation in 1 M NaCl at 20° C at an initial DNA concentration of approximately 25 µg ml⁻¹ in an analytical ultracentrifuge equipped with ultraviolet optics and a photoelectric scanner. Coefficients have been corrected for viscosity but not for concentration dependence. Intact Ad 5 DNA sedimented at 30S and DNA sheared as described above sedimented at 22S. The assays for transforming activity were based on previous studies on infectivity[9] and transforming activity[10] of Ad 5 DNA and were carried out on subconfluent cultures of primary rat kidney cells prepared by trypsin dispersion of kidneys from week old Wistar rats. Viral DNA was diluted into HEPES buffered saline (HeBS: 8.0 g l⁻¹ NaCl, 0.37 l⁻¹ KCl, 0.125 g l⁻¹ Na₂HPO₄.2H₂O, 1.0 g l⁻¹ dextrose, 5.0 g l⁻¹ N-2-hydroxyethylpiperazine-N'-2-ethanesulphonic acid, final pH 7.05) containing salmon sperm DNA at 10 µg ml⁻¹. $CaCl_2$ was added to 125 mM and the mixture was incubated at room temperature for 10–20 min during which time a precipitate formed. Aliquots (0.5 ml per dish) of the resulting suspension (consisting of a co-precipitate of DNA and calcium phosphate) were then added to rat kidney cells growing in 60 mm plastic Petri dishes containing 5 ml Eagle's minimum essential medium supplemented with 10% calf serum (MEM+10% CS) and the dishes were placed in a 37° C, 5% CO_2 incubator. After 3–4 h at 37° C the medium was removed and replaced with fresh MEM +10% CS and the incubation at 37° C continued. After 3–4 d the medium was replaced with low calcium medium[5] (MEM for suspension cultures+0.1 mM $CaCl_2$+5% CS) and this was refreshed at 3–4 d intervals. At 3 weeks, by which time few normal cells remained attached to the dishes, the cultures were fixed, stained with Giemsa, and transformed colonies were counted. Each point represents the average of 4 dishes and the error bars represent ±1 s.d.

* Present address: Departments of Biology and Pathology, McMaster University, Hamilton, Ontario L8S 4L8, Canada.

assayed in conditions which resulted in 20–30 PFU μg⁻¹ for intact DNA. Furthermore, permissive cells (hamster and human) which could not be transformed by intact DNA (due to the resulting productive infection which spread to all the cells in the culture) could be transformed by noninfectious sheared DNA. These observations showed that transformation did not require infectious DNA molecules and suggested that DNA fragments smaller than a viral genome had the ability to transform. They did not, however, eliminate the possibility that transforming activity was associated with a shear-resistant (for example, circular) molecular form of Ad 5 DNA. To rule out this possibility intact and sheared DNA preparations were fractionated through sucrose gradients and fractions were assayed for transforming activity.

Activity should sediment at the same rate before and after shearing if it were associated with a shear resistant form of Ad 5 DNA. The results shown in Fig. 2 clearly demonstrate that this was not the case. Transforming activity of unsheared preparations sedimented with the same velocity as intact DNA and

shearing reduced the sedimentation velocity and transforming activity of the DNA to the same extent (Fig. 2a and b). The results shown in Fig. 2c indicate that even DNA fragments smaller than 15S (3×10^6 daltons) could transform with an efficiency as high as that of intact DNA. Thus transforming activity appeared to be retained even after breaking the DNA to fragments as small as 1/8 of the viral genome. It was now of considerable interest to determine more precisely the degree to which Ad 5 DNA could be fragmented before all transforming activity was lost.

Minimum size of transforming fragments

If transformation by Ad 5 DNA required the integrity of some segment (the T segment) of length L then transforming activity of randomly fragmented DNA should decrease as the fragment size approached L and should be zero for fragments less than or equal to L in molecular weight. To obtain an estimate of L, Ad 5 DNA was extensively fragmented by shearing or sonication (resulting in modal sedimentation coefficients of 11.5S and 12S, respectively), fractionated by velocity sedimentation, and assayed for transforming activity.

The results, plotted in Fig. 3a and b, showed that for both sheared and sonicated preparations, transforming activity was found only in fractions containing DNA fragments sedimenting faster than 10S. To ensure that the molecular weight of the DNA fragments had not changed between the time they were prepared and the time they were assayed for transforming activity, and to determine as accurately as possible the molecular weight of DNA fragments in the region where transforming activity approached zero, peak fractions from the gradients in Fig. 3a and b were concentrated after transforming activity had been measured and sedimentation coefficients were redetermined in the Model E ultracentrifuge. The resulting values (shown in Fig. 3a and b, and differing only slightly from the modal values obtained before fractionation) were then used to calibrate the gradients to determine molecular weights from the equations in the legend to Fig. 3.

The graph of specific transforming activity plotted versus molecular weight in Fig. 3c shows that fragmentation of Ad 5 DNA, whether by shearing or sonication, began to reduce transforming activity when the fragment size fell below about 2×10^6 daltons and activity was completely abolished at slightly below 1×10^6 daltons. If loss of activity at this point was due to the introduction of breaks into the T segment then the size of the T segment must be approximately 10^6 daltons. But since there may be other explanations for absence of transforming activity below 10^6 daltons (for example, DNA fragments smaller than 10^6 daltons may be inherently lacking in any biological activity in our assay system) we can only conclude that the T segment is not greater than about 10^6 daltons in length.

The possibility that transformation by small fragments required the interaction of two or more DNA segments can be ruled out since such a mechanism would result in a dose response of 2nd or higher order. Yet we observed no significant departure from a first order dose response with decreasing fragment size. It would appear, therefore, that only one small segment of the Ad 5 DNA molecule is required for transformation and that the molecular weight of this segment is not more than about 10^6 daltons.

The properties of cells transformed by small DNA fragments are now under investigation. We can only say that such transformed cells do not appear to differ morphologically from those transformed by intact DNA. One line of rat kidney cells transformed by fragments of molecular weight 1.5×10^6 is now at the 27th passage in culture; several transformed hamster lines (transformed by DNA fragments of molecular weight about 8×10^6) have been maintained in culture for 5–23 passages and one transformed human cell line is now at passage 51. Furthermore, of five transformed hamster cell lines tested for tumorigenicity, all induced tumours when injected into newborn hamsters. Thus DNA fragments seem to be capable of inducing

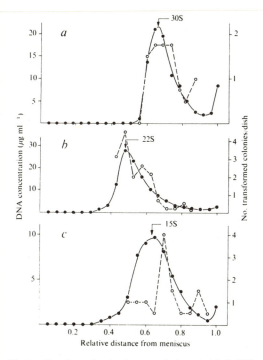

Fig. 2 Transformation by intact and sheared Ad 5 DNA fractionated by velocity sedimentation. Aliquots of ³H-Ad 5 DNA (2,000 c.p.m. μg⁻¹) were layered on to 5–22% exponential sucrose gradients containing 1 M NaCl, 0.01 M Tris, pH 8.1, and 0.5 mM disodium ethylenediaminetetraacetate which had been prepared by pumping sucrose solutions from a 70 ml mixing chamber connected to a reservoir containing 50% sucrose. Gradients were centrifuged at 4 C for 22 h (a and b) or 34 h (c) at 25,000 r.p.m. in a SW 27 rotor and fractionated from the top. DNA concentrations were determined by counting 25 μl and transforming activity was measured as described in the legend to Fig. 1 (50 μl DNA was assayed per dish). Recovery of radioactivity was 60–70%. The sedimentation coefficients given in the figure were determined by band sedimentation in the analytical ultracentrifuge before the DNA was fractionated through sucrose gradients. a, Intact DNA. Transforming activity was measured in two separate experiments, two dishes/fraction. The points therefore represent the mean of four dishes. b, DNA sheared to 22S according to the legend to Fig. 1. Each point represents the mean of two separate assays on two dishes each. c, DNA sheared by passage 20 times through a 0.42 mm × 22 mm long needle using maximum hand pressure on a syringe with a 9 mm plunger. Fractions were assayed once for transformation, 2 dishes per assay. ●, DNA concentration; ○, Transforming activity.

569

not only stable, but also oncogenic transformation. By indirect immunofluorescence using serum from tumour-bearing hamsters, we have demonstrated T antigen in all of several DNA-transformed cell lines tested, including one transformed by fragments of molecular weight 1.5×10^6. (The serum was obtained from hamsters bearing tumours induced by Ad 5 transformed cells and was pretested against transformed rat cells which had previously been shown positive for T antigen[10].)

Location of the T segment

The finding that fragments of Ad 5 DNA had the ability to transform cells suggested the feasibility of mapping the T segment

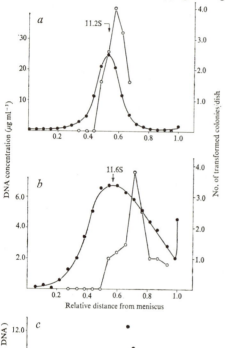

Fig. 3 Transformation by extensively fragmented Ad 5 DNA. a, DNA at 200 µg ml⁻¹ was sheared for 3 min in a Sorvall omnimixer micro attachment by a stainless steel blade rotating at 40,000 r.p.m. A sedimentation coefficient of 11.5S was determined for this DNA before it was fractionated on sucrose gradients. DNA was sedimented by centrifugation at 23,000 r.p.m. for 44 h on gradients prepared according to the legend to Fig. 2. After fractionating and assaying aliquots for transforming activity (50 µl per dish) the peak fraction was concentrated and the sedimentation coefficient again measured to give the value shown in the figure. b, DNA at 100 µg ml⁻¹ was sonicated in a MSE sonicator for 15 s, at 20,000 c.p.s., amplitude 2 µm. This resulted in a DNA preparation which sedimented in the analytical ultracentrifuge as a rather broad band having a modal sedimentation coefficient of approximately 12S. (The coefficient shown in the figure was determined for the combined fractions 10 and 11 after assaying for transformation.) ●, DNA concentration; ○, transforming activity, mean of 5 dishes per point. c, Specific transforming activity of Ad 5 DNA fragments calculated from panels A (○) and B (●) as a function of molecular weight. Molecular weights were calculated from the equations $D = kS$ and $S = 0.0882 \, M^{0.346}$ (ref. 13) where D is distance sedimented and k was determined from the sedimentation coefficients shown in panels A and B.

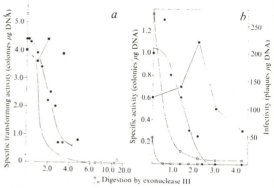

Fig. 4 a, Transforming activity of Ad 5 DNA after treatment by exo III or exo III+S1. Enzymatic digestions were carried out in the following way: To 1 ml ³H-Ad 5 DNA (10⁴ c.p.m. µg⁻¹, 300–500 µg ml⁻¹ in 0.1 × SSC) was added 0.1 ml of a buffer containing 33 mM each of Tris, pH 8.0, MgCl₂, and 2-mercaptoethanol. Digestion was initiated by the addition of 0.1 ml exonuclease III (2000 U ml⁻¹, 1.5 × 10⁵ U mg⁻¹ in 40 mM potassium phosphate buffer, pH 6.5, containing 1 mM 2-mercaptoethanol and 0.4 M KCl) and incubation at 20° C. Exonuclease III was prepared as described[23] with additional purification by phosphocellulose chromatography and Sephadex G100 gel filtration[24]. The extent of digestion was monitored by removing 10 µl aliquots and determining the fraction of acid soluble radioactivity. At various times 80 µl aliquots were removed and exonucleolytic digestion was stopped by adding 20 µl 0.15 M sodium acetate buffer, pH 4.2, containing 5 mM ZnSO₄ and 1.5 M NaCl. Single-stranded ends were subsequently degraded by adding 20 µl single-strand specific nuclease S1 from *Aspergillus oryzae* (300 U ml⁻¹, 1.5 × 10⁵ U mg⁻¹, in 20 mM sodium acetate buffer, pH 4.6, containing 0.1 mM ZnSO₄, 0.3 M NaCl, and 5% glycerol). S1 purification was essentially as described by Vogt[25]. The reaction was carried out at 37° C for 30 min and was terminated by the addition of 10 µl 0.1 M EDTA, pH 8.0. (The % digestion by S1 equalled the amount of DNA made acid soluble by exo III.) Finally, HeBS containing 0.5 mM EDTA was added to give a final DNA concentration of 50 µg ml⁻¹ and the samples were dialysed against HeBS + 0.5 mM EDTA. Assays for transforming activity were carried out according to the legend to Fig. 1, 2 µg DNA per dish, 5 dishes per point. (△) Untreated DNA; (■) exo III treated DNA; (○, ●), exo III-S1 treated DNA. b, Infectivity and transforming activity of Ad 5 DNA treated with exo III or exo III+S1. Enzyme digestions were carried out as in a except that unlabelled Ad 5 DNA was used and the extent of digestion by exo III was determined by measuring the amount of deoxynucleotides incorporated into the treated DNA by DNA polymerase I (purified from an *E. coli* pol A mutant deficient in 5′–3′ exonucleolytic activity[23]). Transforming activity (■, ●) and infectivity (□, ○) were assayed as described in the legend to Fig. 1 except that infectivity assays were carried out on primary human embryonic kidney (HEK) cells or on a line of HEK cells transformed by Ad 5 DNA. ■, □, exo III treated DNA; ●, ○, exo III+S1 treated DNA.

within the viral DNA molecule and of purifying this segment. The first step which we took in this direction was to determine which half of the molecule contained transforming activity. Since the DNA of Ad 5 (like that of Ad 2 (refs 14, 15)) has an asymmetric distribution of GC, it was possible to separate the two halves of Ad 5 DNA by equilibrium buoyant density centrifugation of half molecules in a Hg(II)-Cs₂SO₄ gradient according to the method of Nandi, Wang and Davidson[16]. DNA was sheared, fractionated on sucrose gradients to purify the halves which were then separated in a Cs₂SO₄ gradient in the presence of Hg(II). Fractions corresponding to AT or GC-rich halves were pooled, dialysed against HeBS, and aliquots were analysed by buoyant density centrifugation in CsCl in a Model E analytical ultracentrifuge. The results indicated that a single fractionation on a Hg(II)-Cs₂SO₄ gradient was sufficient to provide relatively pure preparations of AT-rich and GC-rich halves (the cross contamination was less than about 10%). When these were assayed for transforming activity it was found that the AT-rich halves had less than 5% of the transforming activity observed for the GC-rich fractions. (0.7 colonies per µg AT-rich

Table 1 Calculation of the location of the T segment

Experiment	Fraction of intact single strands	Average no. nicks/strand	Effective no. nicks/strand introduced by exo III*	No. 3′ ends per molecule†	% exo III digestion resulting in 50% reduction in transforming activity	Total % digestion by exo III followed by S1 at 50% reduction	Total % digestion per 3′end (distance from left end of Ad 5 genome to T segment)‡
1	70%	0.36	0.1	2.92	1.3	2.6	0.89
2	32%	1.14	0.1	4.48	2.2	4.4	0.98
3	100%	0	0.1	2.2	1.4	2.8	1.27

The fraction of intact single strands was determined by band sedimentation in alkali in the analytical ultracentrifuge. The average number of nicks per strand was calculated from this fraction assuming a Poisson distribution. The % digestion by exo III followed by S1 was double that by exo III alone.
* The exonuclease preparation had a low level of contaminating endonuclease activity introducing nicks which could then serve as substrates for exonuclease digestion.
† The number of 3′ends per molecule is 2 + the number of nicks per molecule.
‡ Calculated from the total % digestion divided by the number of 3′ ends per molecule.

DNA compared to 18 colonies μg⁻¹ for the GC-rich halves.) Thus the T segment appeared to be located in the GC-rich or left half of the Ad 5 DNA molecule, a finding which has been recently confirmed by studies on the transforming activity of DNA fragments obtained by treatment with the R1 restriction endonuclease from *E.coli*.

A more precise localisation of the T segment was obtained by assaying transforming activity of Ad 5 DNA treated with *E.coli* exonuclease III (exo III) (which removes mononucleotides from the 3′ end of polynucleotides in duplex molecules[17]) followed by single strand specific nuclease S1 from *Aspergillus oryzae*. The combination of these two enzymatic treatments has the effect of digesting the Ad 5 DNA molecule inward from the two ends and should result in a decrease in transforming activity when the digestion reaches the region required for transformation.

The results of two experiments are illustrated in Fig. 4a. Treatment of Ad 5 DNA with either exo III or S1 separately did not significantly affect transforming activity. The combination of exo III and S1 treatment initially had no effect on transforming activity but after an exo III digestion of about 1% (that is, a total digestion of 2% by the two enzymes), transforming activity dropped rapidly. In contrast, infectivity of Ad 5 DNA was immediately eliminated by exo III–S1 digestion (Fig. 4b); digestion of as little as 0.2% (data not shown) resulted in a complete loss of infectivity (> 200-fold decrease) indicating that all molecules were digested by the enzymatic treatment. Figure 4b shows also that infectivity was reduced by exo III alone, an effect which is presently being investigated in more detail to determine if, for example, it is related to the inverted terminal repetition contained in adenovirus DNAs[18,19]. The fact that transformation was not immediately affected by exo III–S1 treatment clearly shows that the extreme ends of the Ad 5 DNA molecule are not required for transformation.

We interpret the eventual loss of transforming activity after treatment of Ad 5 DNA with exo III and S1 as being the result of digestion into the T segment and in Table 1 have used the data of Fig. 4 to calculate the distance from the end of the Ad 5 DNA molecule to the beginning of the T segment. When account has been taken for single strand nicks (which are starting points for exonucleolytic digestion) the results from three experiments are in good agreement and give a mean value of 1.1%. From the fact that transforming activity was located on the GC-rich half of the Ad 5 DNA molecule, and from the results of Fig. 3 we conclude that the T segment begins at 1% from the left end of the Ad 5 genome and terminates at about 5–6%.

Implications

Transformation by DNA tumour viruses is often considered to be analogous to lysogenisation of bacteria by bacteriophage, a process thought to involve the integration of the phage genome into the host chromosome by a reciprocal recombination between the host DNA and a circular form of the phage DNA[20]. The results presented in Figs 2 and 3 provide clear evidence that

Ad 5 DNA fragments of considerably less than genome size have the ability to transform cells. As it is rather unlikely that such randomly generated fragments can circularise, these results suggest that transformation does not require a circular viral DNA molecule. That fragments as small as 10⁶ daltons can transform cells also implies that the proportion of the viral genome which is required to initiate and maintain transformation is less than about 5%. This is in reasonable agreement with results of attempts to determine the size of the transforming region by radiation studies[21], and is consistent with the fact that only about 4–10% of the viral genome is transcribed in rat cells transformed by Ad 2 (ref. 22). Thus, the T segment is large enough to code only for 1–2 average-sized proteins so that presumably not more than about two viral genes are involved in transformation by Ad 5 and these must be contiguous.

We have localised the T segment between 1 and 6% from the left end of the Ad 5 genome. Several lines of evidence are in agreement with this finding. First, Gallimore (P. H. Gallimore, personal communication) has shown that the left end of the Ad 2 genome from 0–14% was always present in several lines of Ad 2 transformed cells. Second, preliminary results indicate that treatment of Ad 5 DNA with two restriction enzymes which cleave in the region between 1 and 6% from the left end results in loss of transforming activity while treatment with a third enzyme which cleaves just to the right of this region leaves transforming activity undiminished and activity appears to be associated with the fragment originating from the left end of the Ad 5 DNA molecule (A. J. van der E., C. Mulder and F. L. G., unpublished).

As in the case of transformation by intact viruses, the mechanism by which the T segment induces cell transformation remains to be elucidated. However, the purification of the T segment, which now appears to be a possibility, should provide a simpler system with which to study the genes and gene products involved in oncogenic transformation.

We thank Miss R. H. Tjeerde and Mrs K. B. Postel for technical assistance and Drs S. O. Warnaar, S. Bacchetti, and J. Sussenbach for reading the manuscript. This work was supported in part by funds from Euratom and the Foundation for Medical Research of the Netherlands Organization for the Advancement of Pure Research (ZWO).

Received April 9; revised August 30, 1974.

[1] van der Eb, A. J., van Kesteren, L. W., and van Bruggen, E. F. J., *Biochem. biophys. Acta*, **182**, 530–541 (1969).
[2] Green, M., Pina, M., Kimes, R., Wensink, P. C., MacHattie, L. A., and Thomas, C. A., jun., *Proc. natn. Acad. Sci., U.S.A.*, **57**, 1302–1309 (1967).
[3] Schlesinger, R. W., in *Advances in Virus Research* (edit. by Smith, K. W., and Lauffer, M. A.) **14**, 1–61 (Academic Press, New York and London, 1969).
[4] Trentin, J. J., van Hoosier, G. L., and Samper, L., *Proc. Soc. exp. Biol. Med.*, **127**, 683–689 (1968).
[5] Freeman, A. E., Black, P. H., Vanderpool, E. A., Henry, P. H., Austin, J. B., and Heubner, R. J., *Proc. natn. Acad. Sci., U.S.A.*, **58**, 1205–1212 (1967).

6 Williams, J. F., and Ustacelebi, S., in *Ciba Foundation Symposium on Strategy of the Viral Genome* (edit. by Wolstenholme, G. E. W., and O'Connor, M.) 275–290 (Churchill and Livingstone, London, 1971).

7 Gallimore, P. H., *J. gen. Virol.*, **16**, 99–102 (1972).

8 Williams, J. F., *Nature*, **243**, 162–163 (1973).

9 Graham, F. L., and van der Eb, A. J., *Virology*, **52**, 456–467 (1973).

10 Graham, F. L., and van der Eb, A. J., *Virology*, **54**, 536–539 (1973).

11 Burnett, J. P., and Harrington, J. A., *Proc. natn. Acad. Sci., U.S.A.*, **60**, 1023–1029 (1968).

12 Graham, F. L., Veldhuisen, G., and Wilkie, N. M., *Nature*, **245**, 265–266 (1973).

13 Studier, F. W., *J. molec. Biol.*, **11**, 373–390 (1965).

14 Kimes, R., and Green, M., *J. molec. Biol.*, **50**, 203–206 (1970).

15 Doerfler, W., and Kleinschmidt, A. K., *J. molec. Biol.*, **50**, 579–593 (1970).

16 Nandi, U. S., Wang, J. C., and Davidson, N., *Biochemistry*, **4**, 1687–1696 (1965).

17 Richardson, C. C., Lehman, I. R., and Kornberg, A., *J. biol. Chem.*, **239**, 251–258 (1964).

18 Garon, F. C., Berry, K. W., and Rose, J. A., *Proc. natn. Acad. Sci., U.S.A.*, **69**, 2391–2395 (1972).

19 Wolfson, J., and Dressler, D., *Proc. natn. Acad. Sci., U.S.A.*, **69**, 3054–3057 (1972).

20 Signer, E. R., *Ann. Rev. Microbiol.*, **22**, 451–488 (1968).

21 Yamamoto, H., *Jap. J. Microbiol.*, **14**, 487–493 (1970).

22 Green, M., *Ann. Rev. Biochem.*, **39**, 701–756 (1970).

23 Heijneker, H. L., Ellens, D. J., Tjeerde, R. H., Glickman, B. W., van Dorp, B., and Pouwels, P. H., *Molec. gen. Genet.*, **124**, 83–96 (1973).

24 Jovin, T. M., Englund, P. T., and Bertsch, L. L., *J. biol. Chem.*, **244**, 2996–3008 (1969).

25 Vogt, V. M., *Eur. J. Biochem.*, **33**, 192–200 (1973).

Les Colloques de l'Institut National de la Santé et de la Recherche Médicale

Early proteins of oncogenic DNA viruses
Protéines précoces des virus oncogènes à DNA

INSERM, 5-7 juillet 1977, vol. 69, pp.233-240

THE BIOLOGICAL ACTIVITY OF DIFFERENT EARLY SIMIAN VIRUS 40 DNA FRAGMENTS

Monika Graessmann, Adolf Graessmann and Christian Mueller

Institut fuer Molekularbiologie und Biochemie der freien Universitaet
Berlin, Arnimallee 22, Berlin 33.

The Simian Virus 40 (SV40) tumor (T)-antigen is a virus coded
protein (1, 2), synthesized in productive and abortive infected cells (3).
This protein stimulates cellular DNA synthesis as shown by microinjection
of either early SV40 specific RNA or by microinjection of the purified
T-antigen into epitheloid cells of confluent primary mouse kidney cultures
or monkey kidney cells (TC 7) (4, 5).

Moreover efficient synthesis in terms of quality and quantity of
the T-antigen is required for viral DNA replication (6, 7, 8). The viral
DNA replication in turn is an essential step for late SV40 gene-expression
(8, 9).

The molecular weight determinations for the T-antigen range between
80 and 100 kilodaltons (10, 11), a size corresponding to the coding capacity
of the early viral genome region (0.175 - 0.655, figure 1) (12).

However, not the total early genome region has to be properly
expressed for the different early virus specific functions. Viable deletion
mutants, mapping between 0.54 and 0.57 have been constructed (13). Cells
infected with tsA mutants are stimulated for DNA synthesis but viral DNA
replication and capsid protein synthesis (V-antigen) cannot be demonstrated
at the non permissive temperature of 41.5°C (14, 15). Cells microinjected
with a linear SV40 DNA fragment, generated by cleavage of SV40 DNA form I
(DNA I) with the restriction endonucleases from Haemophilus parainfluenzae I
(Hpa I) and parainfluenzae II (Hpa II), representing about 60 % of the
early genome region, synthesized a protein(s) with T-antigen immunoreactivi-
ty (16).

To test whether early SV40 specific functions can be further
mapped, viral DNA fragments containing between 100 % and 58 % of the early
genome region were prepared by digestion of SV40 DNA I with appropriated
restriction endonucleases (figure 1) and microinjected into cells of confluent primary mouse kidney cultures and monkey kidney cells (TC7). The
recipient cells were tested for T- and V-antigen synthesis, ^{3}H-thymidine
incorporation and complementation of tsA7 at the non permissive temperature
of 41.5°C.

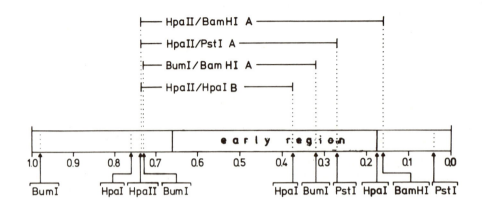

<u>Figure 1</u>. Assignment of microinjected DNA fragments to the physical map
of the SV40 viral genome.

        Linear SV40 DNA III as produced by cleavage with restriction
enzyme Eco RI (cleavage site 0.00 by convention) serves as reference. The
early genome region is indicated. Cleavage sites for the restriction enzymes Hpa I (17), Hpa II (17), Bum I (18), Pst I (19) and Bam HI (20) and
the location of the isolated fragments on the viral genome are shown.

        For preparation of viral DNA fragments SV40 DNA I, isolated from
SV40 (777) infected TC7 cells by the Hirt method (21), was digested under
standard conditions. In short, 50 μg DNA I were incubated with 50 units of
each of the appropriate restriction enzymes (Hpa I, Hpa II, Pst I and
Bam HI were purchased from New England Biolabs Ma., Bum I was a gift from
Cold Spring Harbor Laboratory ; 1 unit cleavages 1 μg lambda DNA completely
within 1 hour under the standard conditions) in 800 μl digestion buffer
(0.01 M Tris-HCl pH 7.4, 0.01 M $MgCl_2$, 0.006 M KCl, 0.001 M dithiothreitol,
0.1 % BSA) for 2 hours at 37°C. The digested DNA was deproteinized by
extraction with 1 vol chloroform/isoamylalcohol (24/1) and precipitated with
2 vol ethanol and 0.1 vol 1 M NaCl. Fragments were separated by electrophoresis on cylindrical 1.4 % agarose gels (17), stained with 2 μg/ml ethidium
bromide for 10 min. and bands, visualized with long wave UV light were cut
out. To remove the DNA gel pieces were extracted through a sterile 2.5 ml
plastic syringe into 5-10 vol of 0.01 M Tris-HCl pH 7.5, 0.01 M NaCl and
0.001 M Na_2 EDTA. After 24 hours at 4°C gel particles were removed by centrifugation at 30.000 g for 30 min. Supernatants were filtered through a
0.22 μm sterile Millipore membrane filter, freeze dried to 200 μl, freed of
ethidium bromide and equilibrated to 0.01 M Tris-HCl pH 7.4 on a 0.1 ml
Dowex 50 WX8/2 ml Sephadex-G25 column and further concentrated to 0.3-0.5 mg

234

DNA per ml injection buffer (22, 23).

Results

        The Hpa I/Hpa II SV40 DNA-B-fragment (0.375-0.735) is transcribed and translated into a polypeptide which contains the determinant group of the T-antigen. About 60 % of primary mouse kidney cells or TC7 cells micro-injected with the B-fragment, exhibited an intranuclear T-antigen specific fluorescence, not distinguishable from that in SV40 DNA I microinjected or virus infected cells as tested by the direct and indirect immunofluorescence technique (table 1).

        To test whether the Hpa I / Hpa II DNA-B-fragment stimulates also cellular DNA synthesis, cells of confluent primary mouse kidney cultures and TC7 cells were microinjected with this fragment. Thereafter the cells were further incubated in serum free Dulbecco's medium with ^{3}H-thymidine (0.1 µCi/ml) at 37°C. 24 hours after microinjection, cells were fixed and stained for T-antigen and processed for autoradiography. As shown in table 1, this fragment did not stimulate ^{3}H-thymidine incorporation over the background. However almost all T-antigen positive cells were stimulated for DNA synthesis after microinjection of the Hpa II/Pst I DNA-A-fragment (table 1, figure 1).

Biological response in TC7 cells upon microinjection of SV40 DNA I and specific SV40 DNA fragments

cells microin-jected with (0.3-0.5 mg/ml injection buffer)	% of viral genome	% of early region	number of injected cells (out of 100 microinjected cells) positive for			complementation of tsA7 virus at 41.5°C detected by V-antigen synthesis
			T-antigen*	V-antigen	T-antigen and 3-H thymidine incorporation*	
DNA I	100	100	100	99	100	+
Hpa II/Bam HI fragment A	58	100	64	0	58	+
Hpa II/Pst I fragment A	47	80	65	0	56	-
Bum I/Bam HI fragment A	41	69	63	0	24	-
Hpa I/Hpa II fragment B	36	58	61	0	2	-
mock injected	0	0	0	0	5	-

Table 1. Confluent cultures of primary mouse kidney and TC7 cells grown on

235

glass slides (10 x 50 mm, subdivided in squares of 1 mm^2), were microinjected under a phase contrast microscope (x 400) by help of microglass capillaries, having a diameter of 0.5 μm at the tip (4). T-antigen was detected by the direct and indirect immunofluorescence technique (T-anti sera : Flow Laboratory, Dr. G. Fey, Cold Spring Harbor Laboratory) and V-antigen with rhodamin B conjugated anti V-serum. For complementation studies, TC7 cells were microinjected with SV40 DNA I or the DNA-fragments and infected with tsA7 (1-2 PFU/cell) at 41.5°C and kept at this temperature until fixation. *) Results identical to those in monkey cells were obtained upon microinjection of SV40 DNA I and DNA-fragments into primary mouse kidney cells (T-antigen synthesis and stimulation of DNA synthesis).

None of these fragments induced V-antigen synthesis, confirming that the isolated DNA fragments were free of full length viral DNA, since every monkey cell (TC7) microinjected with 1-2 SV40 DNA I molecules synthesizes V-antigen (6).

Complementation of the tsA7 virus (V-antigen synthesis) at the non permissive temperature of 41.5°C was obtained by microinjection of the Hpa II/Bam H I DNA-A-fragment. This fragment contains the complete early genome region (table 1).

Summary

To map some of the early Simian Virus 40 functions further, different early viral DNA fragments were prepared by digestion of SV40 DNA I with appropriate restriction endonucleases (figure 1) and microinjected into primary mouse kidney cells and TC7 cells. The following functions could be mapped :
a.) The determinant group of the T-antigen is located between the map position 0.375 and 0.655. Microinjection of the Hpa I/ Hpa II SV40 DNA-B-fragment (0.375-0.735) into mouse and monkey cells induced the synthesis of a polypeptide, reacting with anti T-sera in an undistinguishable manner from that in virus infected cells. However further functions could not be attributed to this viral DNA fragment.
b.) Stimulation of cell DNA synthesis does not require the entire early SV40 genome region. About 90 % of the T-antigen positive cells incorporated ^{3}H-thymidine during 24 hours after microinjection of the Hpa II/Pst I DNA-A-fragment. This fragment contains 80 % of the early region (figure 1, table 1). Complementation of tsA7 virus at the non permissive temperature (41.5°C) was obtained by microinjection of the Hpa II / Bam H I DNA-A-fragment. This fragment contains the entire early genome region.
c.) Since stimulation of viral DNA replication is a prerequisite for late viral gene-expression (V-antigen synthesis), we may assume, that this function requires expression of the total early viral information. It is still unknown how cellular and viral DNA synthesis are stimulated by the SV40 early protein (T-antigen, A-protein). For stimulation of viral DNA replication, T-antigen has to be synthesized in large quantities (about 5×10^5 - 10^6 T-antigen molecules per cell) (7), which suggests that its function is stoichiometric and not catalytic.
d.) The early SV40 map region, required for the helper function for Adeno 2 in the non permissive monkey cells is discussed elsewhere (5).

236

Acknowledgments.

 We are grateful to Miss R. Bobrik, E. Guhl and H. Koch for skillful technical assistance. This work was supported by the Deutsche Forschungsgemeinschaft (gr 599/2, Gr 384/5).

Literature

1. Graessmann, A., Graessmann, M., Hoffmann, H., Niebel, J., Brandner, G. and Mueller, N. ; Inhibition by interferon of SV40 tumor antigen in cells microinjected with SV40 cRNA transcribed in vitro. FEBS Letters, 1974, 39, 249-251.

2. Roberts, B.E., Gorecki, M., Mudigan, R.C., Danna, J., Rozenblatt, S. and Rich, A. : Simian virus 40 DNA directs synthesis of authentic viral polypeptides in linked transcription translation cell free system. Proc. Nat. Acad. Sci. USA, 1975, 72, 1922-1926.

3. Tooze, J. : The molecular biology of tumor viruses. 1973, Cold Spring Harbor Laboratory, New York.

4. Graessmann, M. and Graessmann A. : Early simian virus 40 specific RNA contains information for tumor antigen formation and chromatin replication. Proc. Nat. Acad. Sci. USA, 1976, 73, 366-370.

5. Tjian, R.T.N., Fey, G.H. and Graessmann, A. : The biological activity of purified SV40 T-antigen microinjected into monkey cells. INSERM, 1977.

6. Graessmann, A., Graessmann, M. and Mueller, C. : Regulatory mechanism of simian virus 40 gene expression in permissive and non permissive cells. J. Virol., 1976, 17, 854-858.

7. Graessmann, A., Graessmann, M., Guhl, E. and Mueller, C. : The quantitative correlation between T-antigen synthesis and late SV40 gene expression in permissive and non permissive cells. In preparation.

8. Tegtmeyer, P. : Simian virus 40 deoxyribonucleic acid synthesis : The viral replicon. J. Virol., 1972, 10, 591-599.

9. Manteuil, S. and Girard, M. : Inhibitors of DNA synthesis : Their influence of simian virus 40 DNA. Virology, 1974, 60, 438-454.

10. Tegtmeyer, P., Schwartz, M., Collins, J.K. and Rundell, K. : Regulation of tumor antigen synthesis by simian virus 40 gene A. J. Virol., 1975, 16, 168-178.

11. Del Villano, B.C. and Defendi, V. : Characterization of SV40 T-antigen. Virology, 1973, 51, 34-46.

12. Khoury, G., Howley, P., Nathans, D. and Martin, M. : Post-transcriptional selection of simian virus 40 specific RNA. J. Virol., 1975, 433-437.

13. Shenk, T.E., Carbon, J. and Berg, P. : Construction and analysis of viable delection mutants of simian virus 40. J. Virol., 1976, 18, 664-671.

14. Chou, J.Y., Avila, J. and Martin, R.G. : Viral DNA synthesis in cells infected by thermosensitive mutants of simian virus 40. J. Virol., 1974, 14, 116-124.

15. Graessmann, A., Graessmann, M. and Mueller, C. : Regulatory function of simian virus 40 DNA replication for late viral gene expression. Submitted for publication.

16. Graessmann, A., Graessmann, M., Bobrik, R., Hoffmann, E., Lauppe, F. and Mueller, C. : Gene mapping of SV40 : The biological activity of specific viral DNA fragments produced by cleavage with Haemophilus parainfluenzae restriction endonuclease. FEBS letters, 1976, 61, 81-84.

17. Sharp, P.A. and Sambrook, J. : Detection of two restriction endonucleases activities in Haemophilus parainfluenzae using analytical agarose-ethidium bromide electrophoresis. Biochemistry, 1973, 12, 3055-3063.

18. Zain, B.S. and Roberts, R. : personal communication.

19. Smith, D.I., Blattner, F.R. and Davies, J. : The isolation and partial characterization of a new restriction endonuclease from Providencia stuartii. Nucleic Acid Res., 1976, 3, 343-353.

20. Ketner, G. and Kelly, T.J. : Integrated simian virus 40 sequences in transformed cell DNA : Analysis using restriction endonucleases. Proc. Nat. Acad. Sci., 1976, 73, 1102-1106.

21. Hirt, B. : Selective extraction of polyoma DNA from infected mouse cell culture. J. Mol. Biol., 1967, 26, 365-369.

22. Tanaka, T. and Weisblum, B. : Construction of a Colicin E 1-R factor composite plasmid in vitro : Means for amplification of deoxyribonucleic acid. J. Bacteriol., 1975, 121, 354-362.

23. Ganem, D., Nussbaum, A.L., DAVOLI, D. and Fareed, G.C. : Isolation, propagation and characterization of replication requirements of reiteration mutants of simian virus 40. J. Mol. Biol., 1976, 101, 57-83.

RESUME

L'ACTIVITE BIOLOGIQUE DE DIFFERENTS FRAGMENTS PRECOCES DU DNA DU VIRUS SIMIEN 40.

Pour préciser la cartographie des fonctions précoces de SV40 différents fragments précoces du DNA viral ont été préparés par digestion du DNA I de SV40 avec des endonucléases de restriction (Fig. 1) et microinjectés dans des cellules rénales de souris en culture

238

primaire et dans des cellules TC7. Les fonctions suivantes peuvent être cartographiées :

a. Le groupe déterminant de l'antigène T est localisé entre 0,375 et 0,655 unités cartographiques. Des cellules rénales de souris et des cellules de singe microinjectées avec le fragment B-Hpa I/Hpa II du DNA de SV40 (0,375-0,735 unités cartographiques) montrent une fluorescence intranucléaire spécifique de l'antigène T qui est pareille à celle observée dans des cellules infectées par le virus. D'autre part, d'autres fonctions ne pourraient être attribuées à ce fragment de DNA viral.

b. La stimulation de la synthèse de DNA cellulaire ne nécessite pas la totalité de la région précoce du génome de SV40. Environ 90 % de cellules antigène-T-positives incorporent ^{3}H-thymidine pendant 24 h après la microinjection du fragment A-Hpa II/Pft I du DNA. Ce fragment contient 80 % de la région précoce (Fig. 1 - Tableau I). La complémentation de tsA 7 à la température non permissive (41°5) a été obtenue par microinjection du fragment A-Hpa II/Bam HI du DNA. Ce fragment contient la région précoce entière du génome.

c. Puisque la stimulation de la réplication du DNA viral est une condition préalable de l'expression du gène viral tardif (synthèse de l'antigène V) nous pouvons supposer que cette fonction nécessite l'expression de la totalité de l'information virale précoce. On ne sait pas encore comment la synthèse du DNA cellulaire et viral est stimulée par la protéine précoce (antigène T, protéine A). Pour la stimulation de la réplication du DNA viral, l'antigène T doit être synthétisé en large quantité (environ $5 \times 10^5 - 10^6$ molécules d'antigène T par cellule) (réf. 7) ce qui suggère que sa fonction est stœchiométrique et non catalytique.

d. La région précoce de SV40 nécessaire pour la fonction auxiliaire de l'adéno 2 dans les cellules non permissives de singe est discutée ailleurs.

239

Cell, Vol. 14, 725–731, July 1978, Copyright © 1978 by MIT

Biochemical Transfer of Single-Copy Eucaryotic Genes Using Total Cellular DNA as Donor

Michael Wigler, Angel Pellicer, Saul Silverstein*
and Richard Axel
Institute of Cancer Research
and Department of Pathology
*Department of Microbiology
Columbia University
College of Physicians and Surgeons
701 West 168th Street
New York, New York 10032

Summary

Previous studies from our laboratories have demonstrated the feasibility of transferring the thymidine kinase (tk) gene from restriction endonuclease-generated fragments of herpes simplex virus (HSV) DNA to cultured mammalian cells. In this study, high molecular weight DNA from cells containing only one copy of the HSV gene coding for tk was successfully used to transform Ltk⁻ cells to the tk⁺ phenotype. The acquired phenotype was demonstrated to be donor-derived by analysis of the electrophoretic mobility of the tk activity, and the presence of HSV DNA sequences in the recipient cells was demonstrated. In companion experiments, we used high molecular weight DNA derived from tissues and cultured cells of a variety of species to transfer tk activity. The tk⁺ mouse cells transformed with human DNA were shown to express human type tk activity as determined by isoelectric focusing.

Introduction

The transfer of specific genes, free of chromosomal protein, may facilitate the analysis of the control of gene expression in complex eucaryotes. The availability of sensitive assay systems for transformation may ultimately allow the isolation of any gene for which selective growth conditions exist. To explore this possibility, we previously developed a transformation system for the thymidine kinase (tk) gene of herpes simplex virus, HSV-1. This system was chosen initially because the viral genome is orders of magnitude less complex than the cellular genome. Through a series of electrophoretic fractionations in concert with transformation assays, we isolated a unique 3.4 kb fragment of viral DNA which is capable of efficiently transferring tk activity to mutant Ltk⁻ cells (Wigler et al., 1977). Analysis of the transformed cell DNA in molecular hybridization experiments demonstrated that a single copy of the tk gene was covalently integrated into the DNA of all transformants (Pellicer et al., 1978).

The development of a system for the transfer of the HSV tk gene to mutant mouse cells has permitted us to extend these studies to unique cellular genes. In addition, the availability of cell lines bearing a single copy of the HSV tk gene has allowed us to trace the fate of this gene when DNA from these cells is used as donor in transformation experiments. We have found that high molecular weight DNA obtained from tk⁺ tissues and cultured cells from a variety of organisms can be used to transfer tk activity to tk⁻ mutant mouse cells. The resulting tk activity expressed in recipient cells is donor-derived.

Results

Transformation with Viral tk Integrated in Cellular DNA

Treatment of mutant mouse cells (Ltk⁻) deficient in thymidine kinase with the 3.4 kb Bam I restriction endonuclease fragment of HSV-1 DNA results in the appearance of numerous surviving colonies which stably express the tk phenotype (Wigler et al., 1977). By incorporating various improvements into the transformation protocol (see Experimental Procedures), we now routinely obtain efficiencies of approximately 1 colony per 10^6 cells per 40 pg of purified HSV tk gene. In the mammalian genome, a single-copy gene is present at less than one part per million. If we extrapolate from the transformation efficiency which we observe for the transfer of the viral tk gene and estimate the molecular weight of the haploid mouse genome to be 2×10^{12} daltons, we can expect to observe the transfer of a specific gene once per 10^6 cells per 30 μg of genomic DNA. Under our present transformation conditions, we can therefore expect to observe transfer of single-copy genes when total genomic DNA is used as donor.

Initial experiments designed to transfer the tk gene from cellular DNA to mutant tk⁻ cells were performed with donor DNA purified from HSV tk⁺-transformed Ltk⁻ mouse cells. The choice of this donor for initial studies was dictated by several considerations. First, we have previously shown that HSV tk⁺ cells contain only a single copy of the viral tk gene per cellular genome (Pellicer et al., 1978). Second, the properties of the viral enzyme are sufficiently different from those of the murine enzyme to allow characterization of the acquired tk activity by gel electrophoresis. Finally, the availability of purified restriction fragments containing the viral tk gene allows us to detect and analyze the physical state of the transferred gene in the DNA of the transformant.

The recipient cell chosen for these experiments

was Ltk⁻, clone D, a clone resistant to bromodeoxyuridine and deficient in cytoplasmic thymidine kinase (Kit et al., 1963). Ltk⁻ cells are unable to grow in medium containing HAT (hypoxanthine, aminopterin and thymidine), in which survival depends upon the presence of both salvage pathway enzymes, thymidine kinase and hypoxanthine-guanosine phosphoribosyl transferase (Littlefield, 1963). These cells have an exceedingly low rate of spontaneous reversion to the tk⁺ phenotype, which greatly facilitates the scoring of transformants.

High molecular weight DNA (>40 kb) was purified from a number of independently derived tk⁺ clones. This DNA was co-precipitated with calcium phosphate, and 20 μg were added to each culture dish containing 10⁶ cells. After 4 hr of exposure to DNA, cells were refed growth medium, and 20 hr later, cultures were refed growth medium containing HAT. Cultures were fed HAT medium every 2–3 days, and after 2 weeks, the surviving colonies were counted. In each experiment, DNA from Ltk⁻ was used as a control. Data from a series of transformation experiments are summarized in Tables 2 and 3.

Transformation was attempted using DNA purified from four independently derived clones of Ltk⁻ which contain the viral tk gene (Table 1). Transformation assays with DNA purified from the four HSV tk⁺ transformants gave rise to numerous colonies (Table 2). As expected, DNA obtained from Ltk⁻ was unable to transfer tk activity to Ltk⁻ cells. For clarity, we define primary transformants as the original HSV tk⁺ mouse cells which were derived following transfer of purified viral DNA. We define secondary transformants as tk⁺ cells obtained following transfer of cellular DNA extracted from primary transformants. It is apparent from Table 2 that the frequency of transformation varies for DNA derived from different sources. DNA derived from clones LH2b, LH7 and LHHB resulted in transformation frequencies 3–16 times greater than predicted.

dicted. DNA from clone LHH5-1 generated colonies at a frequency less than that predicted above.

Origin of the tk Activity in Secondary Transformants

The transformation frequencies which we observe (Table 2) range from one colony per 1×10^5 cells

Table 1. Derivation of Cell Lines

Cell Line	Transforming DNA[a]
LH2b	Bam I-generated, 3.4 kb doublet from HSV DNA
LH7	Bam I-generated, 3.4 kb doublet from HSV DNA
LHHB-1	Homogeneous Bam I-generated, 3.4 kb singlet from HSV DNA
LHH5-1	Hpa 8.3 kb fragment from HSV DNA
L(LHH5-1)1	Total cellular DNA from LHH5-1
L(LHH5-1)2	Total cellular DNA from LHH5-1

[a] See Pellicer et al. (1978).

Table 2. Transformation Data: HSV tk Gene

DNA Source	Total Colonies/Total Dishes	Relative Transformation Efficiency[a]
Ltk⁻	0/20	0.0
LH7	95/10 (9.5)	16.0
LH2b	16/9 (1.8)	3.0
LHHB	78/10 (7.8)	13.0
LHH5-1	4/20 (0.2)	0.3

[a] The calculation of transformation (colony per genome equivalents) efficiency was performed assuming that the tk gene is present only once per haploid genome. The total number of genome equivalents (in 20 μg of DNA) added per plate was directly determined from the genome size of donor. Transformation efficiency is then normalized to the transformation efficiency of the 3.4 kb Bam I HSV tk gene (one colony per 10⁶ cells per 40 pg of purified DNA).

Table 3. Transformation Data: Indigenous tk Gene

DNA Source	Total Colonies/Total Dishes	Relative Transformation Efficiency[a]
Ltk⁻ (Mouse Cells)	0/30	0.0
Drosophila Embryo Cells	0/10	0.0
Slime Mold	0/10	0.0
Salmon Sperm	0/10	0.0
LM (Mouse Cells)	63/10 (6.3)	10.5
Mouse Liver	28/4 (7.0)	12.0
CHO (Hamster Cells)	72/10 (7.2)	12.0
Chicken RBC	31/10 (3.1)	2.5
Calf Thymus	62/8 (7.8)	13.0
HeLa (Human Cells)	9/9 (1.0)	3.3

[a] See Table 2 for explanation.

to one colony per 5×10^6 cells. In our studies with the recipient cell Ltk⁻ over the past years, we have never observed a single spontaneous revertant. Our estimate of the rate of spontaneous reversion of Ltk⁻ to tk⁺ is <10⁻⁹. The appearance of even a single colony in cellular transformation experiments is therefore significant, and strongly suggests that expression of the tk⁺ phenotype results from the introduction and expression of foreign DNA. Nevertheless, the expression of tk activity in these transformed cells conceivably could result from either reversion or reactivation of wild-type enzyme rather than the introduction and expression of a new tk gene from donor DNA. Analysis of the electrophoretic properties of the tk activities of the transformed cells allows us to distinguish among these possibilities. The size and charge of the murine and viral tk activities are sufficiently different to permit separation by nondenaturating polyacrylamide gel electrophoresis. In Figure 1A, we observe that the relative mobility (R_f) of the wild-type murine tk activity is 0.18; the R_f of HSV-1 tk, however, is 0.4. Electrophoresis of the cytosol of two secondary transformants (see Table 1) demonstrates a single peak of tk activity with an R_f of 0.40, identical to that of the donor cell. Transformation with cellular DNA therefore results in the introduction and expression of the viral tk gene from donor DNA.

Physical Presence of the tk Gene in Secondary Transformants

The use of donor DNA derived from cells originally transformed with viral tk DNA allows a direct analysis of the physical state of the tk gene in recipient cells. It is possible to determine the size, number and arrangement of the HSV tk gene in transformed cell DNA by eluting restriction endonuclease-treated DNA from agarose gels onto nitrocellulose filters. Highly radioactive tk DNA is then annealed with these filters. The distribution of tk sequences within transformed cell DNA is determined by autoradiography. This experimental design derives from the powerful hybridization technology originally introduced by Southern (1975), which was previously used to demonstrate the presence of a single integrated copy of the HSV tk gene in Ltk⁻ cells transformed with HSV-1 DNA (Pellicer et al., 1978).

DNA fragments that contain the viral tk gene have been purified to homogeneity. One such fragment, 8.3 kb in length, is obtained following Hpa I digestion of HSV DNA. A clone of tk⁻ cells transformed with this fragment (LHH5-1) was the source of cellular DNA in one of the transformation experiments described above. In initial experiments, it was necessary to examine the organization of the tk gene in donor DNA.

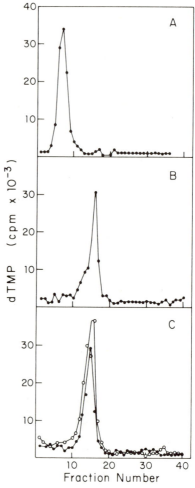

Figure 1. Electrophoretic Pattern of Thymidine Kinase Activities from Cytoplasmic Fractions of Various Cell Lines

The 30,000 × g supernatants of homogenates from four cell lines were applied to 5% polyacrylamide gels. The gels were electrophoresed and sliced into 2 mm slices. Each slice was assayed for thymidine kinase activity as described in Experimental Procedures. (A) LM (wild-type mouse extract), (B) LHH5-1 (primary transformant), (C) L(LHH5-1)1 and L(LHH5-1)2 (secondary transformants).

A restriction map of the Hpa I-generated 8.3 kb fragment of HSV-1 DNA is shown in Figure 2. This DNA fragment contains four sites of cleavage for the endonuclease Bam I. The structural gene for tk is entirely contained within the 3.4 kb Bam I fragment. If transformation resulted from the introduction of the intact 8.3 kb fragment, cleavage of cellular DNA with Bam I should generate five fragments homologous to the Hpa I fragment. Bam I-treated transformant DNA, however, contains only

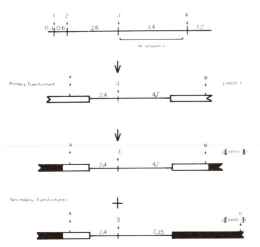

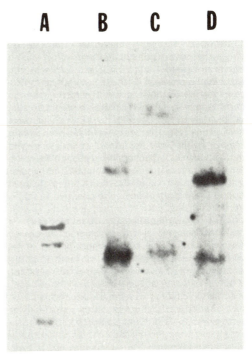

Figure 2. Restriction Map of the HSV tk Gene in Viral DNA and Cellular DNA from Primary and Secondary Transformants

The upper portion of this figure indicates the Bam I cleavage sites in the 8.3 kb fragment of viral DNA containing the tk gene. The tk structural gene sequence is noted at the leftward end of the 3.4 kb fragment. Transformation with the 8.3 kb fragment resulted in the loss of three Bam sites, with the retention of the tk structural gene sequence and BAM site number 3 generating a primary transformant. The organization of the Bam sites in DNA from the primary transformant was determined from Figure 3, slot B. Transformation with high molecular weight DNA from the primary transformant generated two secondary transformants. The organization of the tk gene and the Bam sites in these DNAs were determined from Figure 3, slots C and D. Bam sites labeled with numerals result from cleavage within viral DNA. Sites labeled with uppercase letters reflect sites in primary transformant DNA, and sites labeled with lowercase letters reflect sites in secondary transformant DNA. The size of the fragments is indicated in kb. This model is one of four logically equivalent models that fit the data from Figure 3.

Figure 3. Identification of HSV tk-Specific Sequences in Cells Transformed by Cellular DNA Containing Only One Copy of the HSV tk Gene

High molecular weight DNAs obtained from LHH5-1, L(LHH5-1)1 and L(LHH5-1)2 were cleaved with the enzyme Bam I and electrophoresed on 0.9% agarose gels. The DNA was denatured in situ, transferred to nitrocellulose filters and then annealed with ^{32}P–tk DNA. The 8.3 kb tk gene probe was derived by cleavage of HSV-1 DNA with Hpa I. Bam I-digested DNA from LHH5-1 [a primary transformant (slot B)], L(LHH5-1)1 [a secondary transformant (slot C)] and L(LHH5-1)2 [a secondary transformant (slot D)] are shown. As a reference, 0.5 ng of HSV-1 DNA were digested with Hpa I and Bam I and run in slot A.

two annealing fragments (Figure 3). These data suggest that nucleolytic attack of the 8.3 kb fragment occurred during the transformation process, resulting in the loss of three Bam I sites, and that the remaining fragment has integrated into host DNA (Figure 2).

High molecular weight cellular DNA from this primary transformant was used to transform fresh Ltk⁻ cells, generating two secondary transformants L(LHH5-1)1 and L(LHH5-1)2. The DNA from these clones was then cleaved with Bam I, and the organization of tk gene sequences was analyzed. The annealing profile observed with L(LHH5-1)2 is identical to that of donor DNA. In this instance, transformation resulted from the acquisition of DNA sequences retaining the original distribution of host sequences about the tk gene (Figure 2). The pattern observed with transformant L(LHH5-1)1 is more difficult to interpret. The low molecular weight band observed in donor DNA is preserved, but the second annealing fragment is increased in size. One possible interpretation for the change in

size of the tk fragment is described in Figure 2. These results demonstrate that secondary transformants contain at least one copy of the HSV tk gene.

Transformation with Indigenous Cellular Genes

These experiments have demonstrated the feasibility of transferring a unique gene without prior fractionation of the donor genome. We therefore attempted the transfer of indigenous cellular genes. High molecular weight DNA was isolated from LM, a line of mouse cells which expresses tk activity, and also from mouse liver. Transformation was carried out as described earlier, and after 2 weeks, colonies surviving in HAT medium were scored. With LM DNA, 65 colonies were observed in 10 culture dishes, and 28 colonies were observed in 4 culture dishes with mouse liver DNA (Table 3). In contrast, Ltk⁻ DNA failed to produce a single colony.

These experiments demonstrated the feasibility of intraspecific gene transfer. We next asked whether transformation could also be effected with DNA from distantly related eucaryotic organisms. High molecular weight DNA was purified from Dictyostelium, Drosophila embryo cultures, salmon sperm, chick erythrocytes, cultured hamster cells, calf thymus and HeLa cells. Chick, calf, hamster and human DNA generated numerous surviving colonies, while no transformation was observed with Dictyostelium, Drosophila or salmon DNA. We conclude that both intra- and interspecific transfer of the tk gene can be effected with high efficiency under our transformation conditions.

tk Activity of Transformants Is Donor-Derived

The appearance of surviving colonies following transformation assays with cellular DNA could result from reactivation of the murine tk, or from the introduction of a new wild-type tk gene coded for by donor DNA. As discussed earlier, the exceedingly low frequency of spontaneous reversion of the recipient cells, coupled with the inability to generate tk⁺ transformants using Ltk⁻ DNA as donor, argues strongly that the tk⁺ phenotype which we observe following transformation results from the introduction of a new structural tk gene into tk⁻ cells. The human tk enzyme displays biochemical properties distinct from those of the mouse, enabling us to determine the source of the tk expressed in transformants. The pI of human tk is 9.7, whereas the murine tk activity has a pI of 9.0 (Kit et al., 1974). Extracts of LM cells, HeLa cells and transformants generated with purified HeLa DNA were analyzed by isoelectric focusing in polyacrylamide slabs. The tk activity was localized by assaying the conversion of TdR to TMP in situ. The product of this reaction, ³H–TMP, was blotted out of the gel onto PEI plates which were then analyzed by fluorography. Figure 4 demonstrates that the pI of transformed cell tk is identical to that of human tk and differs from the more acidic murine tk. Transformation must therefore result from the expression of the donor tk gene.

Discussion

This study demonstrates the transfer of thymidine kinase activity to mouse Ltk⁻ cells using high molecular weight cellular DNA as donor. These experiments represent a logical extension of previous work in our laboratories, in which we demonstrated the stable transformation of tk⁻ cells to the tk⁺ phenotype using purified restriction endo-

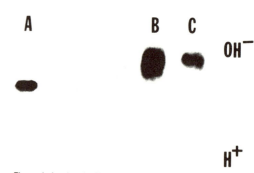

Figure 4. Isoelectric Focusing of Thymidine Kinase Activity in Gels

The 30,000 × g supernatants from homogenates of LM [a wild-type mouse cell (slot A)], HeLa [a human cell line (slot B)] and L(HeLa)-I [a tk⁻ mouse cell transformed using DNA from tk⁺ HeLa cells (slot C)] were focused on 4.5% acrylamide gels. Thymidine kinase activity was assayed in situ, and the product was blotted out onto PEI-cellulose and localized by fluorography as described in Experimental Procedures.

nuclease fragments of HSV-1 DNA as donor (Wigler et al., 1977).

A single copy of the viral tk gene has been found in all independently derived HSV tk⁺ clones examined (Pellicer et al., 1978). Our initial experiments demonstrating transfer of single-copy genes with cellular DNA were therefore performed with DNA purified from HSV tk⁺ cells. The addition of these DNA preparations to Ltk⁻ cells, followed by selection in HAT medium, resulted in the appearance of numerous surviving colonies. The maximum frequency of transformation observed was 10 colonies per 10⁶ cells per 20 μg of DNA, a frequency 40 fold higher than predicted from studies of transfection with the purified HSV tk gene. The tk activity expressed in these secondary transformants was demonstrated to be viral in origin by its electrophoretic mobility. We exploited the elegant technology of Southern (1975) to identify the number and location of the tk sequences that are liberated upon restriction endonuclease cleavage of transformed cell DNA. The data strongly suggest that the structural gene for tk is present in both primary and secondary transformants. The identification of viral tk activity and the detection of HSV tk gene sequences in the DNA of transformed Ltk⁻ cells demonstrates that the transformation which we observe using total cellular DNA as donor results from the introduction and expression of DNA sequences coding for the viral tk.

Successful transfer of the integrated HSV tk gene immediately suggested the possibility of transfer of the indigenous cellular tk gene. We have demonstrated that DNA from various species of mammals and birds can be used to transfer tk activity to murine Ltk⁻ cells. The maximum frequency of successful transformation observed is again 7 colonies per 10^6 cells per 20 μg of DNA. In control experiments, treatment of Ltk⁻ cells with Ltk⁻ DNA generated no colonies capable of survival in HAT ($<2 \times 10^{-8}$). We have been unable to detect gene transfer from DNA derived from Dictyostelium, Drosophila or salmon. There may be barriers to gene transfer between phyla or even between distantly related classes within the same phylum.

Clones capable of survival in HAT all displayed thymidine kinase activity. The tk activity of cells transformed with human DNA was characterized by isoelectric focusing, and shown to migrate with a pl identical to human tk. Thus the conclusion that cells take up and express single-copy genes from complex eucaryotic DNA appears to be firm.

The method which we have used to transfer the thymidine kinase gene can, in principle, be applied to any gene for which conditional selection criteria are available. In preliminary experiments, we have successfully transferred the phenotypic marker, ouabain resistance, from rodent cells to primate cells. In practice, the efficiency of gene transfer can be expected to be a function of the recipient cell, the source of the gene being transferred and the stringency of the selection criteria. In order for gene transfer to be readily detectable, it must occur at a frequency higher than the spontaneous rate of mutation of the recipient to the phenotype selected. The frequencies which we observe for the transfer of the tk gene to Ltk⁻ range from 2×10^{-7} to 1×10^{-5}. This is also the frequency range observed for spontaneous mutation at many interesting loci in cultured somatic cells. Improvements in transformation efficiency or prefractionation of donor DNA can be expected to extend the usefulness of this technique.

Transfer of single-copy genes in eucaryotes has also been achieved using metaphase chromosomes as donor (McBride and Ozer, 1973, Willecke and Ruddle, 1975). The transfer of single-copy genes using genomic DNA as donor has clear advantages: DNA can be obtained from interphase cells; genomic DNA can be cleaved with restriction enzymes and subsequently fractionated; distances between linked genes can, in theory, be precisely determined; and, most important, DNA-mediated gene transfer can be used as a bioassay allowing the purification and subsequent amplification of specific genes. Transformation with restriction endonuclease-cleaved, size-fractionated viral DNA fragments has allowed the purification of viral genes responsible for morphologic transformation (Graham et al., 1974) and the herpes simplex genes coding for thymidine kinase (Maitland and McDougall, 1977; Wigler et al., 1977). This approach, while successful with viral genomes, cannot be used to purify the single-copy genes of the vastly more complex mammalian genomes. An alternate approach involves the construction of phage libraries containing an entire eucaryote genome. Using transformation as a bioassay, clones bearing specific genes can then be identified. This approach may ultimately allow the isolation of genes for which specific hybridization probes are difficult to obtain. The availability of cloned genes and the ability to select mutants at these loci will facilitate the analysis of the nature of mutational events in higher eucaryotes.

Experimental Procedures

Cell Culture and Virus Production

Murine Ltk⁻ cells (clone D) (Kit et al., 1963) were obtained from Dr. Patricia Spear, and maintained in DME supplemented with 10% calf serum and 30 μg/ml of BUdR without antibiotics. Prior to transformation, cells were passed in DME with serum containing antibiotics without BUdR.

The F strain of HSV-1 was grown and titrated in monolayers of Vero cells as previously described (Nishioka and Silverstein, 1977).

Isolation and Fractionation of Viral DNA and Preparation of the tk Probe

Intact HSV DNA was isolated and digested with Hpa I. The restricted DNA was fractionated by electrophoresis through 0.7% agarose gels (40 × 20 × 0.3 cm) for 40 hr at 80 V. The 8.3 kb fragment containing the tk gene was extracted from the gel as previously described (Pellicer et al., 1978). The purified fragment containing the tk gene was labeled to high specific activity by nick translation as described by Maniatis, Jeffrey and Kleid (1975). The reaction mixture contained 50 mM Tris–HCl (pH 7.8), 5 mM MgCl₂, 10 mM 2-mercaptoethanol, 50 μg/ml BSA, 1 ng/ml DNAase I, 4.0 μM ³²P-deoxyribonucleotide triphosphates (200–300 Ci/mmole), 1–2 μg/ml of DNA fragment and 1 unit of DNA polymerase per 0.1 μg of DNA. The reaction was incubated at 15°C for 1 hr, and the product was isolated by phenol extraction and column chromatography on Sephadex G-50. The final product had a specific activity of 2–4 × 10⁸ cpm/μg.

Extraction of DNA

DNA was isolated from either frozen tissue or cultured cells. If tissue was the source of DNA, it was extracted as described by Axel, Cedar and Felsenfeld (1973), using the buffers of Marshall and Burgoyne (1976). DNA was extracted from cultured cells as previously described (Pellicer et al., 1978).

Transformation

The transformation protocol used was that described by Bacchetti and Graham (1977), with modifications. Sterile, ethanol-precipitated viral or high molecular weight eucaryotic DNA was gently resuspended in 1 mM Tris (pH 7.9), 0.1 mM EDTA. DNA (at 40 μg/ml) was adjusted to 250 mM CaCl₂ and added slowly to an equal volume of sterile 2× HBS [280 mM NaCl, 50 mM HEPES, 1.5 mM Na₂PHO₄ (pH 7.12)] with constant agitation. For transformation assays with the purified HSV DNA, high molecular weight salmon sperm DNA was used as carrier. The calcium phosphate-DNA precipitate was allowed to form for 30 min, and 1 ml of precipitate was added to the 10 ml of medium that covered the recipient cells. For each transformation experiment, recipient cells were plated from the same cell pool at 5 × 10⁵ per 10 cm plate 24 hr prior to the experiment, and refed growth medium 4 hr prior to the addition of DNA. After 4 hr of exposure to DNA, the medium was replaced with fresh medium, and the cells were allowed to incubate for an additional 20 hr, at which time the growth medium was changed to selective HAT medium (DME containing 10% calf serum, 15 μg/ml hypoxanthine, 0.2 μg/ml aminopterin, 5.0 μg/ml thymidine). The medium was changed the next day and then every third day for 2–3 weeks until tk+ clones developed. At this time, clones were picked from individual plates. Colonies were scored following formaldehyde fixation and staining with Giemsa. As a control, in each experiment 10 plates of recipient cells were treated with Ltk⁻ DNA under the transformation conditions described above.

Filter Hybridization
The DNA fragments from agarose slab gels (21 × 18 × 0.6 cm) were transferred to nitrocellulose filter sheets as described by Ketner and Kelly (1976). Annealing with labeled probe was carried out as previously described (Pellicer et al., 1978).

Derivation of Viral Transformed tk+ Cell Lines
Four independently derived biochemical transformants were used in these experiments. Two of these lines, LH2b and LH7, were derived by transformation with the Bam I-generated, 3.4 kb doublet of HSV-1 DNA. They differ only in the amount of viral DNA used to effect transformation. Each cell line contains only a single copy of the gene coding for the HSV-1-specified thymidine kinase as assayed by solution and filter hybridization (Pellicer et al., 1978). LHHB was derived by transformation with the unique 3.4 kb fragment obtained after digestion of the 8.3 kb Hpa I fragment with Bam. It too contains only a single copy of the HSV tk gene. LHH5-1 was derived following transformation with the purified 8.3 kb Hpa I fragment of HSV-1 DNA.

Electrophoresis of Thymidine Kinase Activity
Polyacrylamide gel electrophoretic analyses of the S100 fraction from tk+ cells were performed in 5% acrylamide gels as described by Lee and Cheng (1976). The running buffer was composed of 24 mM Tris, 191 mM glycine (pH 8.6) containing 5 mM 2-mercaptoethanol and 50 μM thymidine. Gels were electrophoresed at 2 ma per gel for 30 min, and the current was then raised to 3 ma per gel for the duration of the electrophoresis. At the termination of electrophoresis, gels were cut into 2 mm slices and assayed for tk activity by immersion into 100 μl of reaction mix, which was composed of 0.1 M Tris-maleate (pH 6.5), 25 mM KCl, 20 mM MgCl₂, 7 mM 2-mercaptoethanol, 10 mM ATP and 10 μl of ³H-TdR (spec. act. = 50 Ci/mmole). The slices were incubated at 37°C for 2 hr, and the amount of TMP produced was measured by spotting 50 μl of reaction mix onto DE-81 paper. The paper was washed 3 times with 4 mM ammonium formate and then with methanol to remove any residual TdR. Discs were counted in 5 ml of Econofluor (NEN) in a scintillation spectrometer.

Isoelectric Focusing of Thymidine Kinase Activity
Thymidine kinase activity was localized by isoelectric focusing essentially as described by Kit et al. (1974). The activity was assayed using a modification of the technique used by Chasin and Urlaub (1976) to assay hypoxanthine-guanosine phosphoribosyl transferase. Isoelectric focusing was performed in polyacrylamide slab gels 1.5 mm thick. The slabs were composed of 4.5% acrylamide, 0.15% N,N′-methylene-bisacrylamide, 7.6% sorbitol, 6% ampholine (pH 9–11) solution (LKB), 0.34% ampholine (pH 7–9) solution (LKB), 2 mM ATP, 2 mM 2-mercaptoethanol and 0.5% ammonium persulfate. The gels were cast and focused on an LKB multiphor apparatus across the short dimension. The anode solution was 0.1% ampholine (pH 7–9) and the cathode solution was 1% ampholine (pH 9–11). A constant voltage power supply was used to deliver 400 V to the gel, whose temperature was maintained by circulating cooled (2°C) water through a glass cooling plate. Small squares of 3MM paper saturated with extract were applied to the surface of the gel toward the cathode and focused for 4 hr. At the termination of the focusing, twice-concentrated tk reaction mix was applied to the gel surface and

allowed to soak into the gel at 37°C for 30 min. A sheet of plastic-backed polyethylemeimine-cellulose (PEI), cut to the size of the gel, was wetted with water, blotted to remove excess moisture and applied to the gel surface. Incubation was continued for an additional 90 min, at which time the gel and PEI sheet were inverted, and the reaction product (TMP) was blotted out onto the PEI sheet. After 30 min, the sheet was washed in running water for 20 min and dried. Tritium was visualized by fluorography. A solution of 2,5 diphenyl-oxazole in ether (100 mg/ml) was poured over the sheet and allowed to dry. The sheet was placed in contact with Cronex 2 DC X-ray film at −85°C for 4 days and then developed.

Acknowledgments

We wish to thank Mary Chen for excellent technical assistance. We thank Dr. Richard Firtel for providing us with slime mold DNA, and Drs. I. B. Weinstein and S. Spiegelman for their helpful criticism and support. This work was supported by a grant from the National Cancer Institute.

The costs of publication of this article were defrayed in part by the payment of page charges. This article must therefore be hereby marked "advertisement" in accordance with 18 U.S.C. Section 1734 solely to indicate this fact.

Received April 12, 1978

References

Axel, R., Cedar, H. and Felsenfeld, G. (1973). Proc. Nat. Acad. Sci. USA 70, 2029–2032.

Bacchetti, S. and Graham, F. L. (1977). Proc. Nat. Acad. Sci. USA 74, 1590–1594.

Chasin, L. A. and Urlaub, G. (1976). Somatic Cell Genet. 2, 453–467.

Graham, F. L., Abrahams, P. J., Mulder, C., Heijneker, H. L., Warnaar, S. O., de Vries, F. A. J., Fiers, W. and van der Eb, A. J. (1974). Cold Spring Harbor Symp. Quant. Biol. 39, 637–650.

Ketner, G. and Kelly, T. J. (1976). Proc. Nat. Acad. Sci. USA 73, 1102–1106.

Kit, S., Dubbs, D., Piekarski, L. and Hsu, T. (1963). Exp. Cell Res. 31, 291–312.

Kit, S., Leung, W. C., Trkula, D. and Jorgensen, G. (1974). Int. J. Cancer 13, 203–218.

Lee, L.-S. and Cheng, Y.-C. (1976). J. Biol. Chem. 251, 2600–2604.

Littlefield, J. (1963). Proc. Nat. Acad. Sci. USA 50, 568–573.

Maitland, N. J. and McDougall, J. K. (1977). Cell 11, 233–241.

Maniatis, T., Jeffrey, A. and Kleid, D. G. (1975). Proc. Nat. Acad. Sci. USA 72, 1184–1188.

Marshall, A. J. and Burgoyne, L. A. (1976). Nucl. Acids Res. 3, 1101–1110.

McBride, O. W. and Ozer, H. L. (1973). Proc. Nat. Acad. Sci. USA 70, 1258–1262.

Nishioka, Y. and Silverstein, S. (1977). Proc. Nat. Acad. Sci. USA 74, 2370–2374.

Pellicer, A., Wigler, M., Axel, R. and Silverstein, S. (1978). Cell 14, 133–141.

Southern, E. M. (1975). J. Mol. Biol. 98, 503–517.

Wigler, M., Silverstein, S., Lee, L.-S., Pellicer, A., Cheng, Y.-C. and Axel, R. (1977). Cell 11, 223–232.

Willecke, K. and Ruddle, F. H. (1975). Proc. Nat. Acad. Sci. USA 72, 1792–1796.

Reprinted from Nature, Vol. 280, No. 23, pp. 657–660. 1979. © 1979 Macmillan Journals, Ltd.

Phenotype stabilisation and integration of transferred material in chromosome-mediated gene transfer

L. A. Klobutcher
Department of Human Genetics

F. H. Ruddle
Department of Biology, Yale University, New Haven, Connecticut 06520

The human genes for thymidine kinase, galactokinase and procollagen type I, located on chromosome 17, have been transferred to cultured mouse cells. In unstable cell lines the transferred genetic material is shown to exist independently, whereas in stable cell lines the transferred material has become associated with chromosomes of the recipient cell. The size of the transferred genetic material exceeds 1% of the human haploid genome in some of the transformed cell lines. In addition, the data indicate a gene order of: centromere–galactokinase–(thymidine kinase, procollagen type I), on the long arm of human chromosome 17.

PURIFIED metaphase chromosomes have been used as vectors for the transfer of genetic information between cultured mammalian cells[1-7]. This process, referred to as chromosome-mediated gene transfer (CMGT), typically results in the transfer of subchromosomal amounts of genetic material[2-4]. Expression of the transferred genetic material (transgenome) can be either stable or unstable following return of gene transfer cell lines (transformants) to non-selective culture conditions[1,3,5-7] (for detailed reviews of the CMGT process see refs 8, 9).

To examine the nature of the CMGT process, we have transferred the human chromosome 17 gene thymidine kinase (TK, EC 2.7.1.75) to cultured mouse cells. In some instances co-transfer of the syntenic genes galactokinase[10] (GK, EC 2.7.1.6) and type I procollagen (PCI)[11] was detected. This three-gene segment has provided a means of defining the length of a transgenome and facilitated examination of the fate of the transgenome. In addition, we present data concerning the order of these three genes and assess the usefulness of CMGT as an intergenic mapping procedure. We have also observed donor chromosome fragments and co-transfer of asyntenic genes in some transformants.

Generation and characterisation of primary transformants

Human chromosomes isolated from HeLa S3 (TK$^+$) cells were applied to the mouse cell line LM(TK$^-$) which contains a non-reverting (<10^{-9}) mutant allele of TK. The CMGT procedure used[3] produced transformed colonies at an overall frequency of 2.3×10^{-7} per recipient cell (Table 1). In control experiments without donor chromosomes no colonies were observed.

Seven independent cell lines, designated TGT, were isolated from two transfer experiments. Two of the transformants,

2TGT2 and 2TGT4, were found to express the human form of GK (Table 2). Human PC I production was assayed by Ouchterlony double immunodiffusion using an antibody to the purified human protein[11]. Of the seven cell lines examined, only 2TGT4 expressed human PC I.

The stability of the transferred phenotype was assayed by cultivating the transformants in non-selective medium and periodically determining the proportion of cells able to form colonies in hypoxanthine, aminopterin and thymidine (HAT) selective medium (see Fig. 2 legend). Transformants TGT1, 2TGT1, 2TGT2 and 2TGT3 expressed the TK$^+$ phenotype stably, that is, >90% of the cells remained HAT resistant after 30 d in non-selective medium. Transformants 2TGT4 and 2TGT5 were unstable, losing the HAT-resistant phenotype with first order kinetics at a rate of 2.5% per cell generation. The remaining transformant, TGT2, seemed to be a mixture of stable and unstable cells.

Fig. 1 Alkaline Giemsa[12] stained metaphase spread of transformant 2TGT4 containing an independent human chromosome fragment (arrow). Human chromosomal material appears lightly stained, whereas mouse chromosomes stain darkly, with the exception of their centromeric regions (lightly stained). Twenty metaphase spreads were observed and photographed for each cell line analysed.

Karyotypic analyses of the transformants were carried out using the alkaline Giemsa staining technique[12]. This method produces a differential staining of rodent and human chromosomes and is effective in detecting small amounts of human chromosomal material in a predominantly mouse background. Of the seven transformed cell lines, two possessed detectable human chromosome fragments. Transformant TGT2 contained a large human fragment that was translocated to a

Table 1 Summary of TK transfer experiments

Expt	No. of LM(TK⁻) recipient cells	Cell equiv. of HeLa chromosomes	No. of flasks with colonies	Transfer frequency
1	2×10^7	3×10^7	2/20	1×10^{-7}
2	1×10^7	1×10^7	5/10	5×10^{-7}
3	2×10^7	0×10^7	0/20	0

Mitotic HeLa cells were obtained by treating cultures with 5 atm. nitrous oxide for 18 h. Chromosomes were isolated and applied to LM(TK⁻) cells according to the method of Willecke and Ruddle[3]. Following incubation, recipient cells were diluted with minimal essential medium (αMEM) plus 10% fetal calf serum, and 1×10^6 cells were plated per 75-cm² Corning culture flask. HAT selection was begun 24 h later[24]. Transformed colonies appeared after 2–3 weeks and were expanded for further analysis. All cell lines were negative for myco-plasma contamination by the culture method of Hayflick as modified by Barile et al.[25]. For the calculation of transfer frequency, each flask containing colonies was considered as one event, irrespective of the number of colonies present.

mouse chromosome. Transformant 2TGT4 contained an independent human fragment (Fig. 1). This fragment was similar in size to the long arm of human chromosome 17 and possessed a terminally constricted centromere-like region. G-banded chromosome preparations[13] revealed that this chromosome fragment possessed the banding pattern characteristic of the long arm of human chromosome 17. These observations of cytologically detectable chromosome fragments confirm previous work from our laboratory[14]. It now seems that there is a range of transgenome sizes, the largest exceeding 1% of the human haploid genome.

Co-transfer of asyntenic genes

The transformed cell lines were analysed for the expression of 20 additional human isozymes assigned to 16 different human chromosomes[15,16]. Five of the seven transformants were negative for all human isozymes examined. One cell line, TGT2, expressed the human form of adenylate kinase (AK1, EC 2.7.4.3), which has been assigned to human chromosome 9, and another transformant, 2TGT1, expressed human mannose phosphate isomerase (MPI, EC 5.3.1.8), a human chromosome 15 marker enzyme (Table 2).

Two types of experiments indicate that the asyntenic genes have become associated with the selected TK gene. First, nine independent subclones of TGT2 were generated under selective pressure for the TK gene. All these subclones continued to express the human form of AK1. Second, transformed cell lines which had co-transferred both syntenic and asyntenic genes were back-selected against TK by growth in medium with 5-bromodeoxyuridine (Table 2). Mass populations of surviving cells were analysed to reduce the effect of spurious dual segregation events. Back-selectants of transformants 2TGT2 and 2TGT4 lost the linked human chromosome 17 genes concomitant with TK. Similarly, back-selectants of 2TGT1 and TGT2 lost the natively asyntenic genes MPI and AK1. Karyotypic analysis of back-selectants derived from lines containing human chromosome fragments demonstrated loss of the fragments coincident with loss of the transferred markers.

There are at least two mechanisms which could result in the transfer and association of asyntenic genes with selected genes. The HeLa cell line used as the chromosome donor is typical of continuous tissue culture lines in possessing some chromosome rearrangements. Rearrangements involving chromosomes 17, 9 and 15 could have created artificial synteny groups in the donor cell line which were subsequently picked up in the transfer experiments. Alternatively, the CMGT process could cause co-transfer of genuinely asyntenic genes. Transfer of unlinked markers has been observed in bacterial transformation experi-

ments (congression)[17]. Such results are due to the cell's ability to ingest and express more than one piece of DNA. Mammalian cells are capable of ingesting more than one chromosome in CMGT conditions[18]. Uptake of more than one chromosome followed by partial degradation and reconstruction could explain co-transfer of asyntenic genes.

In other CMGT experiments we have observed five additional transformants expressing normally asyntenic genes (our unpublished results with Miller and Church). The first mechanism proposed would require extensive rearrangement of the HeLa karyotype to explain these results. As the HeLa cell line is basically triploid with a minimal number of rearrangements[19], we favour the second hypothesis. A rigorous test of the hypothesis will require transfer experiments using a primary cell line with a normal diploid karyotype as the chromosome donor. Such experiments are in progress.

The unstable transgenome and its conversion to stability

Various experiments have indicated that the transgenome in stable transformants has become associated with chromosomes

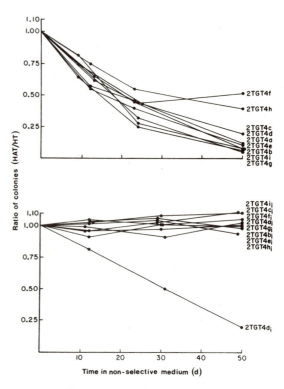

Fig. 2 Stability analyses of subclones of the 2TGT4 transformed cell line. *a*, Stability analyses of nine primary subclones, 2TGT4a–2TGT4i, of transformant 2TGT4. *b*, Second stability analyses of the nine subclones following their growth in non-selective medium for 75 d, and an additional period of growth with HAT-selective medium for 14 d. Eight of the nine subclones now stably express the transferred phenotype. Stability analyses were carried out by culturing transformed cell lines in non-selective medium for at least 30 d. At various times, 200 cells were plated in duplicate flasks with either HAT-selective medium or non-selective HT medium (medium supplemented with 0.1 mM hypoxanthine and 0.016 mM thymidine). After 9–10 d the resulting colonies were stained with 0.25% Wright's stain in methanol and counted. The number of colonies formed in HAT as against HT medium is taken as a measure of the fraction of cells retaining the transferred (TK⁺) phenotype.

588

of the recipient cell[20-23]. The precise nature of this association remains unclear. Even less is known about the state of the transgenome in unstable transformants. Prolonged culture in selective medium, however, often results in conversion from unstable to stable expression[5,7]. It has been postulated that stabilisation occurs infrequently as a unique event in a single cell. The resulting stable cell would thus acquire a growth advantage while in selective medium due to its ability to distribute the transgenome efficiently to daughter cells. Stable cells would then gradually overgrow the unstable cells, giving rise to a completely stable population.

Transformant 2TGT4, which expresses all three human chromosome 17 genes and contains a detectable human chromosome fragment, was chosen for investigating the nature of the unstable transgenome and its conversion to stability. Nine independent subclones (2TGT4a to 2TGT4i) of transformant 2TGT4 were derived and analysed for stability (Fig. 2). Six of the subclones were completely unstable, as more than 90% of the cells lost the HAT-resistance phenotype after 50 d of growth on non-selective medium. Subclones 2TGT4c, 2TGT4f and 2TGT4h did not meet this criterion for unstable expression and will be discussed separately. Each of the six unstable subclones continued to express human GK and PC I. Karyotypic analyses of each of these cell lines revealed a single human chromosome fragment indistinguishable from the fragment present in the 2TGT4 cell line. Thus, the unstable transgenome is an independent and constant entity, losing no genetic information and undergoing no gross structural changes.

Table 2 Summary of TK transformed cell lines

Cell line	TK	GK	PC I	AK1	MPI	Stability	Detectable human chromosomal material
Primary transformants:							
TGT1	+	−	−	−	−	S	−
TGT2	+	−	−	+	−	Mixed	+
2TGT1	+	−	−	−	+	S	−
2TGT2	+	+	−	−	−	S	−
2TGT3	+	−	−	−	−	S	−
2TGT4	+	+	+	−	−	U	+
2TGT5	+	−	−	−	−	U	−
Back-selected transformants:							
TGT2BS	−	ND	ND	−	−	ND	−
2TGT1BS	−	ND	ND	−	−	ND	ND
2TGT2BS	−	−	ND	ND	ND	ND	ND
2TGT4BS	−	−	−	ND	ND	ND	−

Human PC I expression was assayed by Ouchterlony double immunodiffusion using an antibody raised against purified human type I procollagen[11]. Human GK activity was detected by the starch gel electrophoretic method of Nichols et al.[26]. In addition to human AK1 and MPI, all transformants were assayed for the expression of 18 other human isozymes[15]. Back selection of primary transformed cell lines was carried out by growing cells in 30 µg ml^{-1} of 5-bromodeoxyuridine for 24 h. Culture medium was then removed and replaced with phosphate-buffered saline and the cultures exposed to long-wave UV light for 30 min. The cells were returned to growth in culture medium with 5-bromodeoxyuridine, and mass populations of surviving cells were analysed as back-selectants. +, Expression of human phenotype; −, expression of human phenotype not detected; U, unstable expression of transferred phenotype; S, stable expression of transferred phenotype; ND, not determined.

To study the conversion to stability, we isolated stable derivatives of the 2TGT4 subclones. Each of the nine subclones was grown on non-selective medium for 75 d to remove most unstable HAT-resistant cells from the population, and then returned to HAT-selective medium for 14 d to select and fix any pre-existing, small subpopulation of stable cells. The cell populations carried through this regime (designated 2TGT4a$_i$ to 2TGT4i$_i$) were again analysed for stability (Fig. 2). All the cell

Cell line	Chromosome	TK	GK	PC I
Parental line :				
2TGT4		+	+	+
Stable subclones :				
2TGT4a$_i$		+	+	+
2TGT4b$_i$		+	−	+
2TGT4c$_i$		+	−	+
2TGT4e$_i$		+	−	+
2TGT4f$_i$		+	+	+
2TGT4g$_i$		+	+	+
2TGT4h$_i$		+	+	+
2TGT4i$_i$		+	+	+

Fig. 3 Expression of human chromosome 17 genes and fate of the human chromosome fragment in the stabilised subclones of transformant 2TGT4. In each case the lightly stained human fragment, indicated by the bracket, has become associated with a darkly stained mouse chromosome. The human chromosome fragment has apparently been translocated to either a mouse chromosome's centromeric region (2TGT4a$_i$, 2TGT4b$_i$, 2TGT4c$_i$, 2TGT4e$_i$, 2TGT4f$_i$ and 2TGT4h$_i$) or to the distal portion of a mouse chromosome (2TGT4g$_i$ and 2TGT4i$_i$). Centromeric regions are identified by the arrows.

lines, with the exception of 2TGT4d$_i$, then expressed the transferred phenotype stably. Although all eight stable subclones continued to express human PC I, three of the cell lines no longer expressed human GK (Fig. 3). Karyotypic examination revealed that the human chromosome fragment had become associated with a mouse chromosome in all the stable subclones (Fig. 3). In each subclone the fragment was associated with a single mouse chromosome, but the mouse chromosome involved differed between cell lines. The rearrangements appeared either as centric fusions or terminal additions. When metaphase spreads were stained with the fluorescent dye, Hoechst 33258, each of the centric fusion type chromosomes had the brightly fluorescent centromeric region characteristic of mouse centromeric heterochromatin[13]. Neither the independent human fragment in transformant 2TGT4 nor any of its unstable subclones displayed centromeric fluorescence. These observations indicate that the mouse centromeric region has been retained during stabilisation.

Finally, if stabilisation originates in a single cell which then gradually overgrows the population, it should be possible to find cell populations which transiently contain both stable and unstable cells. During a stability test a mixed population would be expected to display a plateau at the frequency of stable cells in the population. The cell lines 2TGT4c, 2TGT4f and 2TGT4h seem to be mixed populations, as they displayed plateaus at 16%, 40% and 50% HAT-resistant cells, respectively (Fig. 2). Karyotypic analysis of these subclones supported this hypothesis. Two types of human chromosome fragment were observed in each of these cell lines. One subset of cells contained the independent human fragment, and a second subset contained the human fragment translocated to a mouse chromosome. The frequency of cells containing the translocated chromosome was similar to the stability test plateau frequency for each cell line. Translocation chromosomes were found in

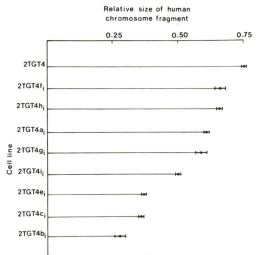

Relative size of human
chromosome fragment

Fig. 4 Histogram displaying the sizes of the human fragments present in the transformed cell line 2TGT4 and its eight stabilised subclones. Relative size values were determined by measuring both the human chromosome fragment and the long arm of a mouse marker chromosome in each metaphase spread. By taking the ratio of the size of the human fragment to the size of the marker chromosome a relative size value was obtained which corrects for variable condensation between metaphase spreads. An average size value was obtained from measurement of at least five metaphase spreads for each cell line. The bars represent ±1 s.e.m. The regions containing the genes for TK, GK and PC I are indicated at the base of the histogram (see text).

10%, 25% and 35% of the cells of transformants 2TGT4c, 2TGT4f and 2TGT4h, respectively.

These experiments provide a visual record of the stabilisation process. The association of the transgenome with recipient cell chromosomes may be viewed as the acquisition of host centromeric activity, ensuring the orderly distribution of transferred genes at mitosis. Stabilisation does not involve the homologous genetic site in the recipient, as the transgenome is shown to associate with several recipient cell chromosomes. These results agree with those of previous studies[22,23]. The earlier studies, however, did not rule out the possibility of multiple sites of association within a single cell. We have shown here that each stable cell contains only one copy of the transgenome. Moreover, the fact that each stabilised subclone has only one type of chromosome associated with the transgenome, as well as the observation of mixed populations with regard to stability, argues that stabilisation originates in a single cell which then overgrows the population.

Genetic mapping by CMGT

Following stabilisation, three of the 2TGT4 subclones did not express human GK (Fig. 3). Moreover, the length of the human fragment decreased following translocation and was correlated with the number of human genes retained (Fig. 4): the largest fragments were in cell lines retaining all three human chromosome 17 genes, and the smallest were in cell lines retaining only TK and PC I. To examine the nature of the breakage during stabilisation, chromosome preparations were examined by G-banding[13]. All stable transformants, except 2TGT4g$_i$ and 2TGT4i, which could not be examined by G-banding due to the similarity of the translocation chromosome to other mouse chromosomes, retained the darkly staining terminal band (17q22–17q25) and various amounts of the remainder of the long arm of human chromosome 17. These results suggest that breaks have occurred near the centromere-like end of the fragment during stabilisation. If we assume that stabilisation involves a single break and fusion of the fragment, and not a complex rearrangement, a deletion map is generated. The map indicates a gene order of centromere–GK–(TK, PC I) (Fig. 4). The data do not allow unambiguous determination of the order of TK and PC I, but failure to segregate PC I suggests that it is distal to TK.

Deletion mapping following stabilisation of visible chromosome fragments provides a means of deriving qualitative intergenic mapping data. Alternatively, the co-transfer frequency of non-selected genes in primary transformants may be a source of quantitative intergenic mapping data[5,6,9,14]. Although the number of cell lines analysed here is too small to assess directly the usefulness of CMGT in this regard, two observations indicate that only primary unstable transformants should be used to determine a reproducible co-transfer frequency. First, stabilisation coincides with a decreased transgenome size. The co-transfer frequency of a given gene in a population of stable transformants will therefore be lower than in a population of unstable transformants. Second, the consistent loss of the centromere-like end of the chromosome fragment in the stable subclones of transformant 2TGT4 suggests transgenomes may undergo stabilisation in a polar fashion. Polar effects would differentially influence the co-transfer frequency of genes proximal and distal to the selected gene.

Additional transfer lines will have to be examined to determine how generally applicable these findings are. The ability to generate large numbers of transformed cell lines through recent modifications of the transfer technique[14] will aid in both confirming these results and providing quantitative mapping data. The further development of CMGT as a mapping procedure will provide an intermediate level of mapping resolution between restriction endonuclease analysis and somatic cell hybridisation.

We thank Dr R. Church for carrying out procollagen assays, Ms E. Nichols, Mr D. Laska and Ms S. Pafka for technical assistance, Mrs M. Reger and Mrs M. Siniscalchi for preparation of the manuscript, and Drs R. Fournier, G. Scangos, P. D'Eustachio and D. Plotkin and Ms D. Slate for helpful discussions and critical reading of the manuscript. This work was supported by grant GM09966 from the NIH to F.H.R.

Received 20 February; accepted 2 July 1979.

1. McBride, O. W. & Ozer, H. L. Proc. natn. Acad. Sci. U.S.A. 70, 1258–1262 (1973).
2. Wullem, G. J., van der Horst, J. & Bootsma, D. Somatic Cell Genet. 1, 137–152 (1975).
3. Willecke, K. & Ruddle, F. H. Proc. natn. Acad. Sci. U.S.A. 72, 1792–1796 (1975).
4. Burch, J. W. & McBride, O. W. Proc. natn. Acad. Sci. U.S.A. 72, 1797–1801 (1975).
5. Willecke, K., Lange, R., Krüger, A. & Reber, T. Proc. natn. Acad. Sci. U.S.A. 73, 1274–1278 (1976).
6. McBride, O. W., Burch, J. W. & Ruddle, F. H. Proc. natn. Acad. Sci. U.S.A. 75, 914–918 (1978).
7. Degnen, G. E., Miller, I. L., Adelberg, E. A. & Eisenstadt, J. M. Proc. natn. Acad. Sci. U.S.A. 73, 2838–2842 (1976).
8. Willecke, K. Theor. appl. Genet. 52, 97–104 (1978).
9. McBride, O. W. & Athwal, R. S. In vitro 12, 777–786 (1976).
10. Elsevier, S. M. et al. Nature 251, 633–636 (1974).
11. Sundar Raj, C. V., Church, R. L., Klobutcher, L. A. & Ruddle, F. H. Proc. natn. Acad. Sci. U.S.A. 74, 4444–4448 (1977).
12. Friend, K. K., Dorman, B. P., Kucherlapati, R. S. & Ruddle, F. H. Expl Cell Res. 99, 31–36 (1976).
13. Kozak, C. A., Lawrence, J. B. & Ruddle, F. H. Expl Cell Res. 105, 109–117 (1977).
14. Miller, C. L. & Ruddle, F. H. Proc. natn. Acad. Sci. U.S.A. 75, 3346–3350 (1978).
15. Nichols, E. A. & Ruddle, F. H. J. Histochem. Cytochem. 21, 1066–1081 (1973).
16. McKusick, V. A. & Ruddle, F. H. Science 196, 390–405 (1977).
17. Goodgal, S. H. J. gen. Physiol. 45, 205–228 (1961).
18. Simmons, T., Lipman, M. & Hodge, L. D. Somatic Cell Genet. 4, 55–76 (1978).
19. Miller, O. J., Miller, D. A., Allderdice, P. W., Dev, V. G. & Grewal, M. S. Cytogenetics 10, 338–346 (1971).
20. Athwal, R. S. & McBride, W. O. Proc. natn. Acad. Sci. U.S.A. 74, 2943–2947 (1977).
21. Willecke, K., Mierau, R., Kruger, A. & Lange, R. Cytogenet. Cell Genet. 16, 405–408 (1976).
22. Willecke, K., Mierau, R., Kruger, A. & Lange, R. Molec. gen. Genet. 161, 49–57 (1978).
23. Fournier, R. E. K. & Ruddle, F. H. Proc. natn. Acad. Sci. U.S.A. 74, 3937–3941 (1977).
24. Littlefield, J. W. Science 145, 709–710 (1964).
25. Barile, M. F., Bodey, G. P., Snyder, J., Riggs, B. D. & Grabowski, M. W. J. natn. Cancer Inst. 36, 155–159 (1966).
26. Nichols, E. A., Elsevier, S. M. & Ruddle, F. H. Cytogenet. Cell Genet. 13, 275–278 (1974).

Reprinted from
Proc. Natl. Acad. Sci. USA
Vol. 76, No. 11, pp. 5714–5718, November 1979
Cell Biology

Passage of phenotypes of chemically transformed cells via transfection of DNA and chromatin

(chemical carcinogenesis/transformation alleles/Southern blotting)

Chiaho Shih, Ben-Zion Shilo, Mitchell P. Goldfarb, Ann Dannenberg, and Robert A. Weinberg

Center for Cancer Research and Department of Biology, Massachusetts Institute of Technology, Cambridge, Massachusetts 02139

Communicated by David Baltimore, August 10, 1979

ABSTRACT   DNA was prepared from 15 different mouse and rat cell lines transformed by chemical carcinogens *in vitro* and *in vivo*. These DNAs were applied to NIH3T3 mouse fibroblast cultures by using the calcium phosphate transfection technique. DNAs of five donor lines were able to induce foci on the recipient monolayers. Ten other donor DNAs yielded few or no foci. DNAs from control, nontransformed parental cell lines induced few or no foci. Chromosomes were transfected from one donor whose naked DNA was unable to induce foci, and morphologic transformation of recipients was observed. These experiments prove that in five of these cell lines the chemically induced phenotype is encoded in DNA, and the sequences specifying the transformed phenotype behave as a dominant allele in the NIH3T3 recipient cells. The sequences encoding the transformation are likely found on a single fragment of DNA.

The molecular mechanisms of chemical carcinogenesis are poorly understood. Work of Ames and others (1–4) has demonstrated a strong correlation between the mutagenicity and carcinogenicity of a large series of compounds, suggesting that DNA is the ultimate target of the carcinogens. Experiments of others (5–7) have shown that the rate of focal transformation elicited by chemical carcinogens on monolayer cultures *in vitro* occurs with an efficiency within an order of magnitude of the efficiency of mutagenesis of a specific marker gene carried by these cells. Taken together these experiments might suggest that the mutation of one of several target genes in these cells leads to transformation and ultimately to tumorigenicity. Nevertheless, there has been no direct proof that the chemically induced transformation phenotype is encoded within the DNA and that the phenotype is specified by a discrete segment of genetic information.

We report here experiments designed to investigate the transmissibility of the chemically transformed phenotype from cell to cell via purified DNA. This demonstration depends upon the transfection technique of Graham and van der Eb (8) in which DNA extracted from donor cells is introduced into recipients as a coprecipitate with calcium phosphate. Previous work in this laboratory utilized this technique to demonstrate the infectivity of several forms of murine leukemia virus DNA (9, 10). More recently, this technique was applied to demonstrate the biological activity of *in vitro* synthesized, subgenomic fragments of murine sarcoma virus (MSV) DNA (11) and of several forms of *in vivo* synthesized Harvey MSV (unpublished results). In addition, work of others has demonstrated the transmissibility of other viral and cellular genes via this technique (12, 13).

The transfection of these sarcoma virus DNAs led to the observation of foci of transformed cells whose behavior was

indistinguishable in many cases from that of virus-infected cells. Some of these transfections utilized donor cellular DNA in which the transforming genome was present in single copy number per haploid cell DNA complement. We reasoned that nonviral transforming genes, if present in unique copy number, might also be transferable via DNA transfection. Specifically, we attempted to demonstrate the existence of genes in the DNA of chemically transformed cells whose introduction into normal recipients would result in focal transformation of the recipient monolayer.

MATERIALS AND METHODS

Cell lines used here are described in Table 1. DNA transfection procedures were as described (11). DNA was prepared from tumors or cell lines as described (23).

Chromatin transfection procedures were as described (24) with the following exceptions: (*a*) recipients were not pretreated with mixtures of colchicine, Colcemid, and cytochalasin D before transfection, (*b*) gentamicin was not used in these experiments, (*c*) instead of counting the chromosome number under a microscope, quantitation of chromosomes was done by spectrophotometric absorbance, and (*d*) 10% dimethyl sulfoxide posttransfectional treatment was not always included.

Southern gel-filter transfer was performed as described (12). The soft agar assay was done by pouring 0.3% soft agar (Difco) containing 3000 cells over a 0.6% agar layer in a 6-cm dish. Colonies were scored 14 days later. Transforming virus rescue assays were done as described (11).

RESULTS

A series of 15 different cell lines (Table 1) were collected from various sources. These cell lines were all of murine origin and most were transformed *in vitro* by various carcinogens commonly used for *in vivo* and *in vitro* chemical carcinogenesis (25, 26). DNA from all these lines was prepared and transfected in a fashion identical to that used in the transfection of retrovirus DNAs. The recipient cells used for monitoring the biological activity of these DNAs were a subline of NIH3T3 cells which fulfills two requirements for these experiments. First, these cells take up DNA in a biologically active form at high efficiency compared with most other mouse cell lines that we have tested. Second, these cells are contact inhibited, and the monolayers they form allow relatively easy visualization of transformed foci.

After transfection of NIH3T3 cultures, the cells were reseeded and scored for foci 14–20 days after transfection. The scored foci were examined individually and were counted only if the constituent cells were hyperrefractile, grew in a crisscrossed pattern, and formed a colonial morphology that we feel

Abbreviation: MSV, murine sarcoma virus.

5714

Table 1. Origins of cell lines used in transfections

Cell line or *in vivo* dissected tumor (ref.)	Induced by[a]	Parental cell line or animal
MC5-5 [Y. Ikawa (14)]	3-MC	BALB 3T3
DMBA-BALB 3T3 (15)	DMBA	BALB 3T3
MCA16 (16)	3-MC	C3H10T1/2
MCA5 (16)	3-MC	C3H10T1/2
MB66 MCA ad 36[b]	3-MC	C3H10T1/2
MB66 MCA ACL 6[b]	3-MC	C3H10T1/2
MB66 MCA ACL 13[b]	3-MC	C3H10T1/2
MB66 MCA ACL 14[b]	3-MC	C3H10T1/2
MC-1 (17)	3-MC	C3H10T1/2
TU-2 (17)	UV	C3H10T1/2
F-17 (18)	X-ray	C3H10T1/2
F-2407-NQO c11W (19)	NQO	F-2407
BP-1 fibrosarcoma[c]	BP	C57BL × C3H/HeJ
BP-2 fibrosarcoma[c]	BP	C57BL × C3H/HeJ
BP-3 fibrosarcoma[c]	BP	C57BL × C3H/HeJ
C3H10T1/2 (20)		
NIH3T3 (21)		
BALB 3T3 (22)		

[a] 3-MC, 3-methylcholanthrene; DMBA, 7,12–dimethylbenzanthracene; NQO, 4-nitroquinolene-1-oxide; BP, benzo[a]pyrene.
[b] Gifts from U. Rapp.
[c] Animals with tumors were provided by P. Donahue and G. N. Wogan.

is representative of a true transformed colony (Fig. 1). Nevertheless, control nontransfected monolayers and monolayers transfected with control nontransformed donor DNAs occasionally exhibited spontaneous foci, a few of which were not readily distinguishable from true transformants. Therefore, unless otherwise indicated, all focus counts presented here were the results of double-blind experiments. After preparation of DNAs from transformed and nontransformed control cultures, the DNA samples were encoded before transfection. Several days before final evaluation of foci, each of the culture dishes was encoded a second time and the experimental and control cultures were randomized. After the foci in the dishes were counted, the identities of the cultures were decoded and the data were tabulated. We believe that this procedure would neutralize the effects of subjective evaluations of focal morphologies.

Transfection of DNA of Chemically Transformed Clones. It was soon apparent that the donor DNAs could be grouped

into two classes. The first class consisted of cells whose DNAs yielded none or a few (1 or 2) distinctive foci after transfection of 75 μg of DNA onto 1.5×10^6 cells (Table 3). Although this small number of foci seen upon transfection was quite distinctive and differentiable from spontaneous overgrowths, the number was so small and irreproducible that we do not presently regard these results as credible.

A second group of donor DNAs reproducibly yielded foci with high efficiency. Representative double-blind experiments to evaluate one of these high-efficiency donor DNAs are summarized in Table 2. This group consists of five lines derived from independently transformed foci of C3H10T1/2 cells and another line of BALB 3T3 origin. These high-efficiency DNAs yielded foci at a rate of 0.1–0.2 focus per μg of transfecting cellular DNA, a transfection efficiency comparable to that observed upon transfection of integrated retrovirus genomes present in low copy number in cellular DNA. In order to assure identity of two of these cell lines, independent aliquots of each were received from U. Rapp and C. Heidelberger 6–12 months after receipt of initial samples. The subsequently received cultures yielded DNA that behaved identically to their previously characterized counterparts. Further controls showed that the transmissibility of the transforming alleles is resistant to ribonuclease treatment and is destroyed by some but not by all site-specific DNA endonucleases (unpublished results).

Transfection with Chromatin of Chemically Transformed Cells. The failure to rigorously demonstrate a transmissible transforming gene in the DNAs of some cell lines listed in Table 1 could be attributable to the absence of a discrete, transmissible allele in these cells. Alternatively, the transforming alleles of these cells might well be in a configuration that allows them to exhibit only relatively low transfection efficiency. This latter possibility was plausible because our previous transfection of retrovirus DNA indicated that the presence of certain linked sequences could affect the transfection efficiency of the Moloney MSV transforming gene by as much as two orders of magnitude (11).

Work of others (27–30) had demonstrated that the transfection of chromosomes rather than of naked DNA allowed a considerable enhancement in transfection efficiency (24). Therefore, we attempted transfection of metaphase chromosomes from a dimethylbenzanthracene-transformed BALB 3T3 cell line. The DNA of these cells had previously yielded a small, irreproducible number of foci (0–2 per 75 μg of DNA). As a control, we transfected chromosomes from a cell line (MC5-5)

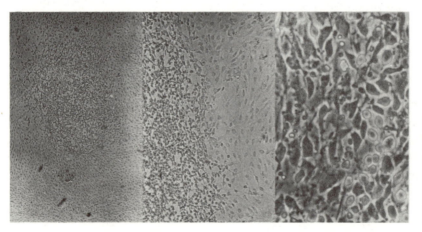

FIG. 1. Foci of transfectants at different magnifications. (*Left*) DNA of MCA16 yielded focus MCA16-5 whose DNA was in turn transfected to yield MCA16-5-1, pictured here. (*Center*) DNA of MCA5 was used to derive focus MCA5-1, pictured here (the focus is seen to the right of the frame). (*Right*) DNA of MC5-5 was used to derive focus MC5-5-4, pictured here.

Table 2.   Double-blind evaluations of foci after DNA transfections[a]

Experiment	Donor cells	Foci in individual culture dishes	Total foci per experiment
I	MC5-5-0	6, 3, 4, 3, 6, 9, 5, 2, 4, 6[b]	48
	NIH3T3	0, 0, 0, 0, 0, 0, 0, ≤1, 0, 0, 0, 0	≤1
II	MCA16	2, 2, 0, 0, 0, 1, 0, 0, 0, 0, 0, 0	5
	MB66 MCA ad 36	0, 1, 2, 0, 2, 0, 0, 1, 0, 0, 1, 1	8
	MB66 MCA ACL 6	0, 0, 0, 0, 0, 0, 0, 0, 0, 0, 0, 0	0
	MB66 MCA ACL 13	0, 0, 0, 0, 0, 0, 0, 0, 0, 0, 0, 0	0
	C3H10T1/2	0, 0, 0, 0, 0, 0, 0, 0, 0, 0, 0, 0	0

[a] DNA (75 µg) was transfected onto 1.5×10^6 NIH3T3 cells, which were reseeded into 12 100-mm dishes 4–6 hr posttransfection. The culture dishes were encoded and randomized and foci were counted 14–18 days later.
[b] Two cultures were lost due to contamination.

whose naked DNA had previously yielded foci with reasonable efficiency. As seen in Table 3, transfection of these chromatin preparations yielded, via double-blind experiments, a significant level of foci, whereas transfection of chromatin of the BALB 3T3 parent cell line yielded a low background level of foci.

Serial Passaging of Transformed Allele. In order to determine whether these alleles could be passaged serially, one focus (termed MCA16-5) was picked after transfection of DNA from the MCA16 donor. A second focus (termed MC5-5-6) was picked after transfection of the chromosomes of the MC5-5 line. These cells were subjected to single-cell cloning. Their DNA was prepared and further tested for biological activity. As seen in Table 4, both DNAs demonstrated high levels of biological activity (10–375 foci per 75 µg of DNA). Therefore, the

transforming element of transformed C3H10T1/2 and BALB 3T3 cells is passageable from donor to recipient over two cycles of transfection. More recent work demonstrates the transmissibility of several of these alleles through a third serial cycle of transfection.

Involvement of Retrovirus Genomes in Transformation. It was possible that the transforming genes detected here reflected adventitious laboratory contamination of cultures by Harvey or Moloney MSVs, both of which are used in this laboratory. Alternatively, an endogenous retrovirus genome might have been activated in these cells (16, 31, 32). We attempted to minimize this possibility by demonstrating the absence of transmissible type C retrovirus transforming genomes. Although these transforming viral genomes are normally replication defective, they can be transmitted by superinfection of virus-

Table 3.   Characterization of donor and transformants used in transfection

Donor DNA[a] or chromosomes prepared from	Transfection efficiency	Growth in agar[b] of		Rescue of transforming virus[g] from	
		Donor	Derived transformants	Donor	Transformants
Naked DNA transfections[c]					
BALB 3T3	1, 1, 0	−		−	
MC5-5	6[d], 6[d], 2	+++; >25%		−	
C3H10T1/2	1, 1, 0			−	
MCA 5	9–14[d], 8, 6, 5	++; 70%	+; 1–10%	−	−
MCA16	12[d], 7–10[d], 5, 3		++; 25%	−	−
MB66 MCA ad 36	8, 15	ND	ND	ND	ND
MC-1	10	ND	ND	ND	ND
Other controls					
NIH3T3	0, 0[d]	−		−	
NSF mouse liver	2[d], 1, 0				
MSV-Transformed NIH3T3	10[d], 100[d]	+	+	+	+
Mock[e]	2, 0				
Chromosome transfections[f]					
BALB 3T3	1, 1–2	−		−	
MC-5	20[d], 30	+++; >25%	+; 2–15%	−	−
DMBA-BALB 3T3	15[d]	±; <0.1%	−	−	−

The following lines gave 0–2 foci per 76 µg of DNA: MB66 MCA ACL 6, MB66 MCA ACL 13, MB66, MCA ACL 14, TU-2, F-17, BP-1, BP-2, BP-3, F-2407-NQO c11W, DMBA-BALB 3T3. ND, not done.
[a] These are described further in Table 1.
[b] Symbols: +++ and ++, macroscopic-sized colony; +, regular size colony; ±, small colonies; −, no colonies seen. Percentages represent plating efficiencies in soft agar.
[c] Number of foci per 75 µg of DNA per 1.5×10^6 transfected NIH3T3 cells.
[d] Not performed in double-blind experiment.
[e] Mock, no DNA added to calcium phosphate precipitate.
[f] Number of foci after transfection of chromosomes containing 75 µg of DNA applied to 1.5×10^6 NIH3T3 cells.
[g] Symbols: +, $>10^5$ focus-forming units of transforming virus rescued per ml; −, no transforming virus rescued; ND, not done. All murine leukemia virus-infected cells released $>10^6$ plaque-forming units of murine leukemia virus per ml.

593

Cell Biology: Shih *et al.*

Proc. Natl. Acad. Sci. USA 76 (1979)   5717

Table 4.   Serial passage of transformed phenotype

Primary transfection	Secondary transfection

MC5-5 $\xrightarrow[\text{(chromosomes)}]{\text{20 foci/75 μg DNA}}$ MC5-5-6 $\xrightarrow[\text{(naked DNA)}]{\text{28 foci/75 μg DNA}}$ MC5-5-6-1

MCA16 $\xrightarrow[\text{(naked DNA)}]{\substack{\text{5, 12}^{\text{a}}\text{ foci/75 μg} \\ \text{DNA}}}$ MCA16-5 $\xrightarrow[\text{(naked DNA)}]{\substack{\text{10, 375}^{\text{a}}\text{ foci/75 μg} \\ \text{DNA}}}$ MCA16-5-1

[a] The two numbers represent two independent DNA preparations.

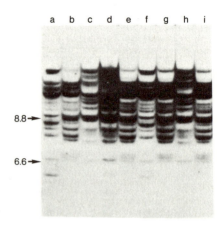

FIG. 2.   Southern gel filter hybridizations of *Eco*RI-cleaved DNAs from different cell lines. DNAs of the three reference mouse strains and the derivative transformed and transfected cell lines were cleaved with endonuclease *Eco*RI and analyzed as described (12). The DNAs analyzed here are as follows. Lanes: a, C3H10T1/2; b, NIH3T3; c, BALB 3T3; d, MC5-5; e, focus derived from MC5-5 DNA; f, MCA5; g, focus derived from MCA5 DNA; h, DMBA-BALB 3T3; i, focus derived from DMBA-BALB 3T3 DNA. Numbers are in kilobases.

transformed cells with a replication-competent, nontransforming murine leukemia virus whose genome allows pseudotyping and transmission of the transforming MSV genome. As seen in Table 3, whereas murine leukemia virus superinfection of an MSV-transformed cell line results in release of >10^5 focus-forming units of MSV per ml and >10^6 plaque-forming units of murine leukemia virus per ml, superinfection of a series of donor and derived transformant cell lines yielded high levels of the superinfecting murine leukemia virus but no rescue of retrovirus transforming genomes. Although we cannot presently exclude the intervention of various other viral genomes in this transformation, transmissible type-C retrovirus genomes do not appear to be responsible for the observed phenomena.

Genetic Background of Donor and Recipient Cell Lines. Additional control experiments were designed to rule out the possibility that the foci of transformation were the result of contamination of recipient cultures by small numbers of donor cells. This artifact is unlikely in the instance of DNA transfection because no donor cells could survive the deproteinization accompanying DNA preparation. In the case of chromatin transfection, inadvertent passage of viable cells together with chromatin was conceivable although still not likely in view of the 1% nonionic detergent used during chromosome preparation. Therefore, we wished to control the genetic origin of the donor and recipient cell lines. In the experiments described here, the donor cells were of C3H/He and BALB/c origin whereas the recipients were of NIH3T3 origin.

Southern blot analysis of cellular DNAs of these cell lines reveals a spectrum of proviruses of endogenous murine type C retroviruses (14). When *Eco*RI-cleaved cell DNA is probed with AKR virus cDNA, a characteristic and unique pattern of fragments is detected for each of the above-mentioned cell lines (unpublished results). We have used this spectrum as a characteristic signature of the genetic origin of a cell line under investigation. DNAs from the donor cell lines and from foci resulting from DNA and chromatin transfection were analyzed by this procedure to confirm their genetic origin. Prior to electrophoresis and Southern analysis, the DNA samples were encoded, and the identities were regenerated only after evaluation of the Southern blots. An example of this analysis is shown in Fig. 2. All transfected foci were found by this procedure to be of NIH3T3 origin, whereas the donor cell lines were found to be of the expected C3H/He and BALB/c origin. This analysis precludes the potential donor cell contamination hypothesized above. However, this assay is not sensitive enough to determine how much of the donor DNA was established in the transfectant.

An Additional Transformed Phenotype of Transfectant Foci. The refractile foci induced among the recipient cells

exhibit a distinct, readily distinguishable phenotype which is normally associated with transformation. To demonstrate transformation by a second criterion, transformed foci were isolated from monolayer cultures and tested together with their parental donor lines for their ability to form colonies in soft agar. Growth in this medium is usually associated with tumorigenicity and is widely used as a criterion of transformation (33, 34). As seen in Table 3, the normal NIH3T3 recipient cells did not form colonies, whereas most transformants and their respective parental donors grew to the 50- to 100-cell stage. Thus, by the criteria of morphology and anchorage independence, these transfected cells are transformed.

An exception to the transmitted anchorage independence was seen when examining the DMBA-BALB 3T3 cells used for chromatin transfection in which neither the donors nor the transfectants grew well in agar. Because neither the donor nor the transfected cells grew well in agar, this would not appear to represent an artifact of chromatin transfection. Rather it appears to reflect a weakly transforming phenotype present originally in this donor line, this phenotype being transmitted faithfully to the recipient cells upon chromosome transfection.

DISCUSSION

Studies of chemical carcinogenesis by Ames and others have suggested strongly that the carcinogenic event is a mutagenic event that alters the DNA of a target cell (1–5). Other work on *in vitro* carcinogenesis suggests that certain established cell lines can be transformed by carcinogens at rates consistent with one-hit kinetics (6, 7). These data suggested to us that a discrete, dominant allele may be present in certain chemically transformed cells whose introduction into a nontransformed counterpart would elicit transformation. The allele(s) studied here is capable of inducing transformation in NIH3T3 cells, an established cell line derived originally from outbred NIH/Swiss mice (21).

The high-efficiency donor cell lines studied here contain DNA which induces foci at the high efficiencies observed previously upon transfection of retrovirus DNA (9). Our attempts at eliciting retroviruses from these cells or their derivatives have been negative and we tentatively conclude that the transforming alleles present in these cells are of cellular origin. The copy number of the transforming alleles in the DNA of these cells is unknown.

Although not directly demonstrated here, we consider it highly unlikely that the observed transformations depend upon successful introduction of two or more unlinked genetic elements into the same recipient cell. Given the low efficiencies of transfection [*ca.* 10^{-5} events per competent transfecting molecule (35, 36)] and the $1:10^6$ dilution at which single-copy genes are found in a haploid mouse genome, we consider it almost certain that the transforming trait is localized on a single fragment of DNA. The size of this fragment is probably less than the 30-kilobase pair size to which transfecting DNA has been sheared prior to transfection.

The present experiments have concentrated on the DNAs of high-efficiency donors whose transfection readily yields foci. These high-efficiency donors represent less than half of the transformed mouse lines that we have examined. The remaining low-efficiency donors may contain alleles that are transmissible via chromosome transfection. One such chromosome-mediated transmission was reported here. Interpretation of these chromosome transfection experiments is less clear because such transmissions do not prove that the allele is encoded solely by a discrete allele present in the DNA.

The ability to transmit the transformed phenotypes of five different cell lines suggests that these cells contain alleles that act dominantly in the NIH3T3 genetic background. These alleles may represent cellular genes whose alteration by the carcinogens resulted in their activation as transforming elements. The present studies show that two different carcinogens are able to induce these alleles in the DNAs of at least two strains of mouse cells. The transmission of these alleles via naked DNA would support the notion that DNA is the target of carcinogenesis and the carrier of the transformed trait. This transmissibility should make possible the isolation of the sequences encoding these alleles.

Note Added in Proof: Recent work by L. C. Padhy and R. A. Weinberg has shown that cells transformed by DNA transfection are tumorigenic in newborn mice.

We acknowledge those who generously provided us with the cell lines and tumors studied here: Drs. U. Rapp, C. Heidelberger, J. B. Little, T. Benjamin, G. diMayorca, G. N. Wogan, and J. A. DiPaolo. We thank David Fisher who helped in the early stages of this work and Stephanie Bird and David Steffen who helped with the Southern blotting. This work was supported by the Rita Allen Foundation, of which R.A.W. is a Fellow; by National Institutes of Health Grant CA17537, and by Core Grant CA14051 to S. Luria.

1. McCann, J., Choi, E., Yamasaki, E. & Ames, B. N. (1975) *Proc. Natl. Acad. Sci USA* **72**, 5135–5139.
2. McCann, J. & Ames, B. N. (1976) *Proc. Natl. Acad. Sci. USA* **73**, 950–954.
3. Bridges, B. A. (1976) *Nature (London)* **261**, 195–200.
4. Bouck, N. & diMayorca, G. (1976) *Nature (London)* **264**, 722–727.
5. Barrett, J. C., Tsutsui, T. & Ts'o, P. O. P. (1978) *Nature (London)* **274**, 229–232.
6. Landolph, J. R. & Heidelberger, C. (1979) *Proc. Natl. Acad. Sci. USA* **76**, 930–934.
7. Huberman, E., Mager, R. & Sachs, L. (1976) *Nature (London)* **264**, 360–361.
8. Graham, F. L. & van der Eb, A. J. (1973) *Virology* **52**, 456–467.
9. Smotkin, D., Gianni, A. M., Rozenblat, S. & Weinberg, R. A. (1975) *Proc. Natl. Acad. Sci. USA* **72**, 4910–4913.
10. Rothenberg, E., Smotkin, D., Baltimore, D. & Weinberg, R. A. (1977) *Nature (London)* **269**, 122–126.
11. Andersson, P., Goldfarb, M. P. & Weinberg, R. A. (1979) *Cell* **16**, 63–75.
12. Wigler, M., Pellicer, A., Silverstein, S. & Axel, R. (1978) *Cell* **14**, 725–731.
13. Wigler, M., Pellicer, A., Silverstein, S., Axel, R., Urlaub, G. & Chasin, L. (1979) *Proc. Natl. Acad. Sci. USA* **76**, 1373–1376.
14. Benveniste, R. E., Todaro, G. J., Scolnick, E. M. & Parks, W. P. (1973) *J. Virol.* **12**, 711–720.
15. Oshiro, Y. & DiPaolo, J. A. (1972) *J. Cell. Physiol.* **81**, 133–138.
16. Rapp, U. R., Nowinski, R. C., Reznikoff, C. A. & Heidelberger, C. (1975) *Virology* **65**, 392–409.
17. Chan, G. L. & Little, J. B. (1976) *Nature (London)* **264**, 442–444.
18. Terzaghi, M. & Little, J. B. (1976) *Cancer Res.* **36**, 1367–1374.
19. Waters, R., Mishra, N., Bouck, N., DiMayorca, G. & Regan, J. D. (1977) *Proc. Natl. Acad. Sci. USA* **74**, 238–242.
20. Reznikoff, C. A., Brankow, D. W. & Heidelberger, C. (1973) *Cancer Res.* **33**, 3231–3238.
21. Todaro, G. J. & Green, H. (1963) *J. Cell Biol.* **17**, 299.
22. Aaronson, S. A. & Todaro, G. J. (1968) *Science* **162**, 1024–1026.
23. Steffen, D. L. & Weinberg, R. A. (1978) *Cell* **15**, 1003–1010.
24. Miller, C. L. & Ruddle, F. H. (1978) *Proc. Natl. Acad. Sci. USA* **75**, 3346–3350.
25. Heidelberger, C. (1975) *Annu. Rev. Biochem.* **44**, 79–121.
26. DiPaolo, J. A. & Casto, B. C. (1978) *Natl. Cancer Inst. Monogr.* **48**, 245–257.
27. McBride, W. O. & Ozer, H. L. (1973) *Proc. Natl. Acad. Sci. USA* **70**, 1258–1262.
28. Willecke, K. (1978) *Theor. Appl. Genet.* **52**, 97–104.
29. Spandidos, D. A. & Siminovitch, L. (1977) *Cell* **12**, 675–682.
30. Willecke, K. & Ruddle, F. H. (1975) *Proc. Natl. Acad. Sci USA* **72**, 1792–1796.
31. Tereba, A. (1979) *Virology* **93**, 340–347.
32. Getz, M. J., Elder, P. K. & Moses, H. L. (1978) *Cancer Res.* **38**, 566–569.
33. Macpherson, I. & Montagnier, L. (1964) *Virology* **23**, 291–294.
34. Shin, S., Freedman, V. H., Risser, R. & Pollack, R. (1975) *Proc. Natl. Acad. Sci. USA* **72**, 4435–4439.
35. Graessmann, A., Graessmann, M. & Mueller, C. (1976) *J. Virol.* **17**, 854–858.
36. Rigby, P. W. & Berg, P. (1978) *J. Virol.* **28**, 475–489.

Reprinted from Nature, Vol. 284, No. 5755, pp. 418-421, April 3 1980
© Macmillan Journals Ltd., 1980

Transforming activity of DNA of chemically transformed and normal cells

Geoffrey M. Cooper, Sharon Okenquist & Lauren Silverman

Sidney Farber Cancer Institute and Department of Pathology, Harvard Medical School, Boston, Massachusetts 02115

DNA fragments of chemically transformed and normal avian and murine cells induce transformation of NIH 3T3 mouse cells with low efficiencies. High molecular weight DNAs of cells transformed by DNA fragments induce transformation with high efficiencies in secondary transfection assays. It thus seems that endogenous transforming genes of uninfected cells can be activated and efficiently transmitted by transfection. These results are consistent with the hypothesis that normal cells contain genes that are capable of inducing transformation if expressed at abnormal levels.

SUBSTANTIAL evidence indicates that highly oncogenic retroviruses are recombinants between non-transforming viruses and normal cell genes. Examples of transforming viruses that have apparently originated by recombination with different cell genes include avian sarcoma viruses[1-3], avian myelocytomatosis virus[4,5], avian myeloblastosis virus[5], avian erythroblastosis virus[5], Moloney sarcoma virus[6,7], Abelson leukaemia virus[8], Kirsten and Harvey sarcoma viruses[9] and feline sarcoma viruses[10]. In the case of avian sarcoma viruses, it has been demonstrated that genetic information related to the viral transforming gene (*src*) is encoded in the genomes of many vertebrate species[3]. In addition, uninfected avian and mammalian cells contain a normal cell protein (p60^{sarc}) that is closely related to the protein encoded by the avian sarcoma virus *src* gene (p60^{src})[11-13]. The amount of p60^{src} present in avian sarcoma virus-transformed cells seems to be at least 100-fold higher than the amount of p60^{sarc} present in uninfected cells[11-13]. These observations suggest that viral transformation may result from overproduction of normal cell proteins as a consequence of the insertion of normal cell genes into a viral genome in a manner permitting their efficient expression. If this is the case, it also seems plausible that the transforming genes of

retroviruses represent only a subset of the normal cell genes that are potentially capable of inducing transformation if expressed at higher than normal levels. The present experiments indicate that DNAs of both chemically transformed and normal uninfected cells are capable of inducing transformation on transfection. The results thus provide direct support for the hypothesis that potential transforming genes are encoded in the genomes of normal avian and mammalian cells.

Transformation by sonicated DNA fragments of uninfected cells

The transforming activity of uninfected cell DNAs was assayed by transfection of NIH 3T3 mouse cells. These cells are transformable by direct integration of DNAs of avian sarcoma viruses[14], avian acute leukaemia viruses[15] and murine sarcoma viruses[7,16]. In addition, NIH 3T3 cells are transformable by subgenomic fragments of murine[7] and avian[15] sarcoma virus DNAs. The efficiency of transformation by subgenomic fragments of avian sarcoma virus DNA, which contain *src* but apparently lack the normal viral transcriptional promoter, is 100–1,000-fold lower than the efficiency of transformation by intact virus DNA[15]. This reduced efficiency is thought to represent the probability of integration of these fragments at a site adjacent to a transcriptionally active promoter in the DNA of the recipient cells. A similar efficiency of transformation by fragments of normal cell DNAs might therefore be expected if transformation could occur by integration of normal cell genes at a site in recipient cell DNA leading to their increased expression.

The donor DNAs used in these experiments were extracted from methylcholanthrene-transformed quail cells (QT-6 cells)[17], methylcholanthrene-transformed BALB 3T3 mouse cells (MC5-5 cells)[18], normal chicken embryo fibroblasts and non-transformed BALB 3T3 and NIH 3T3 mouse cells. Both high molecular weight (MW) and sonicated DNAs were used in transfection assays to investigate the possibility that sonication might increase the transforming activity of uninfected cell DNAs by dissociating potential transforming genes from flanking sequences that might regulate their expression. The MWs of unsonicated DNAs were greater than 20×10^6 whereas those of sonicated DNA fragments were $0.3–3 \times 10^6$ (Fig. 1). Salmon sperm DNA (MW range $0.3–20 \times 10^6$, Fig. 1) and sonicated fragments of *Escherichia coli* DNA were included as controls to estimate the frequency of spontaneous transformation of the recipient cells.

DNAs were assayed by transfection of NIH 3T3 cells as previously described[14], except that DNA-treated cells were transferred into soft agarose medium 7 days after transfection and colonies of transformed cells were counted after 2–3 weeks of further incubation. This procedure was found to reduce the

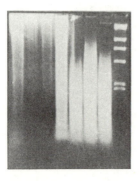

Fig. 1 Molecular weights of DNAs used in transfection assays. DNAs were extracted as described in Table 1 legend and were electrophoresed in 0.8% agarose horizontal slab gels in Tris-acetate buffer, stained with ethidium bromide and photographed under UV light[14]. High-MW DNAs of chicken embryo fibroblasts, NIH 3T3 cells and MC5-5 cells were electrophoresed in lanes *a–c*, respectively. Salmon sperm DNA was electrophoresed in lane *d*. DNAs of chicken embryo fibroblasts, BALB 3T3 cells and MC5-5 cells were sonicated as described in Table 1 legend and were electrophoresed in lanes *e–g*, respectively. The MWs of marker fragments of *Hind*III-digested λ DNA (lane *h*) are indicated ($\times 10^6$).

Table 1 Transformation by uninfected cell DNAs

Donor DNA	Fraction positive cultures Individual	Total	Transformants per μg DNA (total)
Controls			
Salmon sperm			
(MW 0.3–20×10^6)	3/396	3/513	3×10^{-4}
Sonicated *E. coli*			
(MW 0.3–3×10^6)	0/117		
High MW ($>20 \times 10^6$)			
QT-6	1/48		
MC5-5	0/48		
		1/182	3×10^{-4}
CEF*	0/46		
NIH 3T3	0/40		
Sonicated fragments			
(MW 0.3–3×10^6)			
QT-6	5/70		
MC5-5	3/48		
CEF*	3/56	16/294	3×10^{-3}
NIH 3T3	3/80		
BALB 3T3	2/40		

High MW DNAs of avian and mammalian cells were extracted essentially as previously described[22]. Cells were lysed with 0.5% SDS and extracts were deproteinised by digestion with pronase, extraction with phenol and extraction with chloroform–isoamyl alcohol. DNAs were precipitated with ethanol, dissolved in SSC (0.15 M NaCl–0.015 M sodium citrate, pH 7.0), digested with RNase A, re-digested with pronase, extracted with phenol, extracted with chloroform–isoamyl alcohol, precipitated with ethanol and dissolved in SSC. *E. coli* DNA was extracted by the same procedure after lysis of the cells by treatment with lysozyme. Salmon sperm DNA (Sigma) was dissolved in SSC, extracted with phenol, extracted with chloroform–isoamyl alcohol, precipitated with ethanol and dissolved in SSC. Fragments of DNA were prepared by sonication for 10 s using the microtip of a Branson sonifier with an output of 60 W. Recipient cultures of NIH 3T3 cells were transfected with 20 μg DNA per culture as previously described[14]. DNA-treated cells were transferred into semi-solid medium containing 0.25% agarose[14] 7 days after exposure to DNA. Colonies of transformed cells were counted after 2–3 weeks of further incubation. Data are presented as the fraction of recipient cultures that contained transformed cell colonies (generally 1–5 colonies per positive culture). Each positive culture was considered an independent event for calculation of the transformation frequency.
* CEF DNA was extracted from normal chicken embryo fibroblasts after four to six passages in culture.

background of spontaneous transformation of the recipient cells compared with that observed in assays of focus formation on the original DNA-treated plates. Control experiments indicated that the number of colonies induced by DNA of avian sarcoma virus-infected cells in this assay was similar to the number of foci induced in parallel cultures that were maintained in liquid medium (0.1–1 transforming units per μg DNA).

Results of transfection assays of uninfected cell DNAs are summarised in Table 1. Transfection by high MW DNAs of uninfected cells did not increase the frequency of transformation of recipient cells compared with the background of spontaneous transformation in control cultures exposed to salmon sperm or sonicated *E. coli* DNAs. In contrast, transfection by sonicated fragments of DNAs of both transformed and non-transformed cells resulted in a ~10-fold increase in transformation frequency. The increased transforming activity of sonicated DNA fragments of avian and mammalian cells, compared with salmon sperm and sonicated *E. coli* DNAs, was statistically significant at the $P < 0.001$ level (χ^2 test). The frequency of transformation by sonicated DNA fragments of uninfected avian and mammalian cells was similar to the frequency of transformation by subgenomic DNA fragments of avian sarcoma virus-infected cells[15].

Transformation by high molecular weight DNAs of cells transformed by DNA fragments

Individual colonies of transformed cells were picked and grown to populations of 10^8–10^9 cells for further study. These cells had morphologies typical of transformed mouse cells and grew to high cell densities. Cells transformed by fragments of uninfected cell DNAs did not produce retrovirus particles (assayed by sedimentable DNA polymerase activity of culture fluids[19]) or infectious transforming virus. In addition, transforming virus could not be rescued from these cells by superinfection with non-transforming Moloney murine leukaemia virus.

Previous studies indicated that high MW DNAs extracted from nine spontaneous transformants of NIH 3T3 cells did not induce transformation of NIH 3T3 cells in secondary transfection assays (<0.001 transforming units per μg DNA)[14]. In contrast, high MW DNAs of cells transformed by subgenomic fragments of avian sarcoma virus DNA induced transformation of NIH 3T3 cells with efficiencies of ~0.5 transforming units per μg DNA[15]. We therefore assayed the transforming activity of high MW DNAs of NIH 3T3 cells that were transformed by fragments of uninfected cell DNAs to determine whether DNAs of these cells were similarly capable of inducing transformation with high efficiencies.

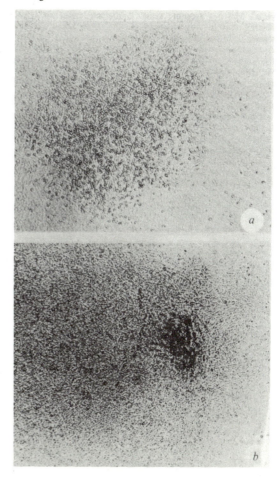

Fig. 2 Morphologies of transformed cells. Representative foci of NIH 3T3 cells transformed by DNAs of NIH(QT-6 DNA) cells (*a*) and of NIH(CEF DNA)cl 1 cells (*b*) as described in Table 2 legend were photographed with a magnification of ×38.

Table 2 Transformation by DNAs of NIH 3T3 cells transformed by fragments of uninfected cell DNAs

Donor DNA	Foci per μg DNA	Colonies per μg DNA
Salmon sperm	<0.002	<0.002
NIH 3T3	<0.002	<0.002
NIH (salmon sperm DNA)	≤0.005	<0.002
NIH(QT-6 DNA)	0.09	0.3
NIH(MC5-5 DNA)cl 1	0.16	1.5
NIH(MC5-5 DNA)cl 2	0.15	0.3
NIH(MC5-5 DNA)cl 3	0.08	0.2
NIH(NIH DNA)	ND	0.1
NIH(CEF DNA)cl 1	0.03	0.2
NIH(CEF DNA)cl 2	ND	0.8
NIH(CEF DNA)cl 3	ND	1.1

Salmon sperm DNA, high MW DNA of non-transformed NIH 3T3 cells and high MW DNAs of transformed NIH 3T3 cells isolated after exposure to salmon sperm DNA [NIH (salmon sperm DNA)] and to sonicated DNAs of QT-6 cells [NIH(QT-6 DNA)], MC5-5 cells (three independent isolates) [NIH(MC5-5 DNA)cl 1,2,3], NIH 3T3 cells [NIH(NIH DNA)] and normal chicken embryo fibroblasts (three independent isolates) [NIH(CEF DNA)cl 1,2,3] in the experiments described in Table 1 were assayed by transfection of NIH 3T3 cells. Some cultures were maintained under liquid medium and foci of transformed cells were counted 14–17 days after transfection. Other recipient cultures were transferred into soft agarose medium 7 days after transfection and colonies of transformed cells were counted after 2–3 weeks of further incubation. ND, not done.

High MW DNAs of NIH 3T3 cells and of spontaneously transformed NIH 3T3 cells isolated after exposure to salmon sperm DNA did not induce transformation in secondary transfection assays (Table 2). In contrast, high MW DNAs of NIH 3T3 cells transformed by sonicated fragments of DNAs of QT-6 cells, MC5-5 cells, NIH 3T3 cells and normal chicken embryo fibroblasts induced efficient transformation of recipient NIH 3T3 cells as assayed by either focus formation or colony formation in soft agarose (Table 2). The morphologies of representative foci of transformed cells are illustrated in Fig. 2. The efficiencies of transformation by DNAs of cells transformed by DNA fragments were ~100-fold higher than the original efficiencies of transformation by sonicated DNA fragments. Thus, NIH 3T3 cells transformed by DNA fragments of uninfected avian and mammalian cells apparently contained transforming genes that were transmitted to new recipient cells by transfection of high MW DNA with relatively high efficiencies.

The transforming activity of restriction endonuclease-digested DNAs of three independent clones of NIH 3T3 cells transformed by fragments of chicken embryo fibroblast DNA was investigated in experiments presented in Table 3. The transforming activity of DNA of NIH(CEF DNA) clone 1 cells was abolished by digestion with HindIII but not by digestion with EcoRI or BamHI. In contrast, the transforming activities of DNAs of NIH(CEF DNA) clone 2 and clone 3 cells were abolished by digestion with BamHI but not by digestion with EcoRI or HindIII. These results indicated that transformation in secondary transfection assays was mediated by specific segments of donor DNA. In addition, it seemed that the transforming activity of NIH(CEF DNA) clone 1 cells differed from the transforming activities of NIH(CEF DNA) clone 2 and clone 3 cells, although no conclusion can be drawn concerning the relationship between the transforming activities of NIH(CEF DNA) clone 2 and clone 3 cells.

Endogenous transforming genes of uninfected cells

The findings that DNA fragments of uninfected normal and chemically transformed cells induce transformation with low efficiencies and that high MW DNAs of cells transformed by

DNA fragments induce transformation with high efficiencies indicate that endogenous transforming genes of uninfected cells can be activated and efficiently transmitted by transfection. These observations are consistent with the hypothesis that normal cells contain genes that are capable of inducing transformation if expressed at abnormally high levels. A model which accounts for our results in terms of this hypothesis is presented in Fig. 3. The low efficiency of transformation by sonicated DNA fragments of uninfected cells could correspond to the probability of integration of endogenous transforming genes derived from the donor DNA in the vicinity of a transcriptionally active promoter in the recipient cell genome, thereby permitting abnormally high expression of the donor transforming gene. Note that the efficiency of transformation by sonicated fragments of uninfected cell DNAs (3×10^{-3} transformants per μg DNA) is similar to that by subgenomic fragments of avian sarcoma virus-infected cell DNAs ($\sim 1 \times 10^{-2}$ transformants per μg DNA)[15]. The transforming activity of sonicated DNA fragments, but not high MW DNAs, may indicate that the biologically active DNA fragments do not include flanking sequences that might regulate the expression of endogenous transforming genes in normal cells. The high efficiency of transformation by high MW DNAs of cells transformed by DNA fragments is consistent with transformation of the secondary recipient cells by DNAs containing promoter regions derived from the primary recipient cell genome in tandem with transforming genes derived from the original DNA fragments. Similar high efficiencies of transformation are observed in transfection assays of DNAs of cells transformed by subgenomic fragments of avian sarcoma virus DNA[15] and by fragments of herpes simplex virus DNA containing the viral thymidine kinase gene[20]. The transforming activity of avian cell DNAs in NIH 3T3 mouse cells is consistent with the transforming activity of DNAs of avian sarcoma[14] and acute leukaemia[15] viruses in these cells and with the biological activity of the chicken thymidine kinase gene in transfection of Ltk⁻ mouse cells[20].

An alternative possibility is that transformation by sonicated DNA fragments results from mutagenic activation of endogenous transforming genes of the recipient cells as a consequence of integration of donor DNA fragments into regulatory sequences that normally control expression of these genes. In this case also, the high efficiency of transformation by high MW DNAs of cells transformed by DNA fragments indicates that these cells contain transmissible activated endogenous transforming genes.

We are now investigating the possible relationships between the endogenous transforming genes identified here and the transforming genes of retroviruses. Molecular cloning of the transforming genes of cells transformed by fragments of uninfected cell DNAs should permit studies of these endogenous transforming genes at the molecular level and facilitate investigations of the biological activities of these genes in normal and transformed cells.

Table 3 Restriction endonuclease digestion of DNAs of transformed NIH 3T3 cells

Donor DNA	Colonies per μg DNA			
	Undigested	EcoRI	HindIII	BamHI
NIH(CEF DNA)cl 1	0.3	0.5	<0.03	0.5
NIH(CEF DNA)cl 2	0.8	0.3	0.3	<0.03
NIH(CEF DNA)cl 3	1.1	0.9	0.5	<0.03

DNAs of three independent clones of NIH 3T3 cells transformed by sonicated fragments of uninfected chicken embryo fibroblast DNA were digested to completion with EcoRI, HindIII and BamHI. The extent of digestion was determined by inclusion of λ DNA in an aliquot of each reaction mixture and analysis of the cleavage products by electrophoresis in agarose gels[14]. Transforming activity of undigested and digested DNAs was assayed by transfection of NIH 3T3 cells. Colonies of transformed cells in soft agarose medium were counted as described in Table 2 legend.

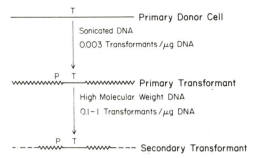

Fig. 3 Model for transfection by endogenous transforming genes. Uninfected cells used as donors of DNA contain genes that are potentially capable of inducing transformation if expressed at abnormally high levels (endogenous transforming genes). Transformation by sonicated DNAs results from integration of endogenous transforming genes (T) in the vicinity of a promoter site (P) in recipient cell DNA, resulting in abnormally high expression of the donor gene. Efficient transformation by high MW DNAs of the primary transformed cells results from transfection of secondary recipient cells by DNAs containing promoter regions (P) derived from the primary recipient genome in tandem with transforming genes (T) derived from the original donor DNA.

Implications for chemical carcinogenesis

The efficiency of transformation by high MW or sonicated DNAs of the two methylcholanthrene-transformed cell lines (QT-6 and MC5-5 cells) used as DNA donors in the present experiments did not differ from the efficiency of transformation by DNAs of non-transformed mouse cell lines (BALB 3T3 and NIH 3T3) or by DNAs of normal chicken embryo fibroblasts. Therefore, it seems that the two chemically transformed cell lines studied did not contain transforming genes that were efficiently transmitted by transfection of high MW DNA. These results suggest that the chemically induced transformation events that resulted in establishment of the QT-6 and MC5-5 cell lines were not transmissible by transfection. Mutational

inactivation of a gene encoding a *trans*-acting regulator of the expression of an endogenous transforming gene may be an example of such a transforming lesion. Analysis of the transforming activity of DNAs of a variety of spontaneous and chemically induced neoplasms will be required to determine whether some tumours contain transforming genes that are transmissible with high efficiencies by transfection, as might result from mutational alteration of *cis*-acting regulatory sequences that control expression of endogenous transforming genes.

Since submission of this manuscript, Shih *et al.*[21] have reported transformation of NIH 3T3 cells by high MW DNAs of 5 out of 15 chemically transformed mouse cell lines.

We thank D. M. Livingston and M. A. Lane for helpful comments on the manuscript. This research was supported by NIH grant CA 18689 from the NCI.

Received 31 December 1979; accepted 6 February 1980.

1. Stehelin, D., Varmus, H. E., Bishop, J. M. & Vogt, P. K. *Nature* **260**, 170–173 (1976).
2. Hanfusa, H., Halpern, C. C., Buckhagen, D. L. & Kawai, S. *J. exp. Med.* **146**, 1735–1747 (1977).
3. Spector, D. H., Varmus, H. E. & Bishop, J. M. *Proc. natn. Acad. Sci. U.S.A.* **75**, 4102–4106 (1978).
4. Sheiness, D. & Bishop, J. M. *J. Virol.* **31**, 514–521 (1979).
5. Stehelin, D., Saule, S., Roussel, M., Lagrou, C. & Rommens, C. *Cold Spring Harb. Symp. quant. Biol.* **44** (in the press).
6. Frankel, A. E. & Fischinger, P. J. *Proc. natn. Acad. Sci. U.S.A.* **73**, 3705–3709 (1976).
7. Andersson, P., Goldfarb, M. P. & Weinberg, R. A. *Cell* **16**, 63–75 (1979).
8. Baltimore, D., Shields, A. Otto, G. & Goff, S. *Cold Spring Harb. Symp. quant. Biol.* **44** (in the press).
9. Scolnick, E. M., Rands, E., Williams, D. & Parks, W. P. *J. Virol.* **12**, 458–463 (1973).
10. Frankel, A. E., Gilbert, J. H., Prozig, K. J., Scolnick, E. M. & Aaronson, S. A. *J. Virol.* **30**, 821–827 (1979).
11. Oppermann, H. D., Levinson, A., Varmus, H. E., Levintow, L. & Bishop, J. M. *Proc. natn. Acad. Sci. U.S.A.* **76**, 1804–1808 (1979).
12. Karess, R. E., Hayward, W. S. & Hanafusa, H. *Proc. natn. Acad. Sci. U.S.A.* **76**, 3154–3158 (1979).
13. Collett, M. S., Erikson, E., Purchio, A. F., Brugge, J. S. & Erikson, R. L. *Proc. natn. Acad. Sci. U.S.A.* **76**, 3159–3163 (1979).
14. Copeland, N. G., Zelenetz, A. D. & Cooper, G. M. *Cell* **17**, 993–1002 (1979).
15. Cooper, G. M., Copeland, N. G., Zelenetz, A. D. & Krontiris, T. *Cold Spring Harb. Symp. quant. Biol.* **44** (in the press).
16. Lowy, D. R., Rands, E. & Scolnick, E. M. *J. Virol.* **26**, 291–298 (1978).
17. Moscovici, C. *et al. Cell* **11**, 95–103 (1977).
18. Lieber, M. M., Livingston, D. M. & Todaro, G. J. *Science* **181**, 443–444 (1973).
19. Copeland, N. G. & Cooper, G. M. *Cell* **16**, 347–356 (1979).
20. Wigler, M., Pellicer, A., Silverstein, S. & Axel, R. *Cell* **14**, 725–731 (1978).
21. Shih, C., Shilo, B.-Z., Goldfarb, M. P., Dannenberg, A. & Weinberg, R. A. *Proc. natn. Acad. Sci. U.S.A.* **76**, 5714–5718 (1979).
22. Cooper, G. M. & Temin, H. M. *J. Virol.* **14**, 1132–1141 (1974).

Supplementary Readings

1971–1980

Andersson, P., M. Goldfarb, and R. Weinberg. 1979. A defined subgenomic fragment of in vitro synthesized Moloney sarcoma virus DNA can induce cell transformation upon translation. *Cell 16:* 63–75.

Camacho, A. and P.G. Spear. 1978. Transformation of hamster embryo fibroblasts by a specific fragment of the herpes simplex virus genome. *Cell 15:* 993–1002.

Houweling, A., P.J. Van den Elsen, and A.J. Van der Eb. 1980. Partial transformation of primary rat cells by the leftmost 4.5% fragment of adenovirus 5 DNA. *Virology 105:* 537–550.

Jaenisch, R. 1980. Retroviruses and embryogenesis: Microinjection of Moloney leukemia virus into midgestation mouse embryos. *Cell 19:* 181–188.

Kucherlapati, R., S.P. Hwang, N. Shimizu, J.K. McDougall, and M.R. Botchan. 1978. Another chromosomal assignment for a simian virus 40 integration site in human cells. *Proc. Natl. Acad. Sci. 75:* 4460–4464.

Lester, S.C., S.K. LeVan, C. Steglich, and R. DeMars. 1980. Expression of human genes for adenine phosphoribosyltransferase and hypoxanthineguanine phosphoribosyltransferase after genetic transformation of mouse cells with purified human DNA. *Somatic Cell Genet. 6:* 241–259.

Minson, A.C., P. Wildy, A. Buchan, and G. Darby. 1978. Introduction of the herpes simplex virus thymidine kinase gene into mouse cells using virus DNA or transformed cell DNA. *Cell 13:* 581–587.

Park, M., D.M. Lonsdale, M.C. Timbury, J.H. Subak-Sharpe, and J.C.M. Macnab. 1980. Genetic retrieval of viral genome sequences from herpes simplex virus transformed cells. *Nature 285:* 412–415.

Schaffner, W. 1980. Direct transfer of cloned genes from bacteria to mammalian cells. *Proc. Natl. Acad. Sci. 77:* 2163–2167.

Shiroki, K., H. Handa, H. Shimojo, S. Yano, S. Ojima, and K. Fujinaga. 1979. Establishment and characterization of rat cell lines transformed by restriction endonuclease fragments of adenovirus 12 DNA. *Virology 82:* 462–471.

Sutter, D., M. Westphal, and W. Doerfler. 1978. Patterns of integration of viral DNA sequences in the genomes of adenovirus type 12-transformed hamster cells. *Cell 14:* 569–585.

Chapter 10   Cell Membrane Organization and Function

Frye, L.D. and M. Edidin. 1970. The rapid intermixing of cell surface antigens after formation of mouse-human heterokaryons. *J. Cell Sci.* 7: 319–335.

Eckhart, W., R. Dulbecco, and M.M. Burger. 1971. Temperature-dependent surface changes in cells infected or transformed by a thermosensitive mutant of polyoma virus. *Proc. Natl. Acad. Sci.* 68: 283–286.

Hynes, R.O. 1973. Alteration of cell-surface proteins by viral transformation and by proteolysis. *Proc. Natl. Acad. Sci.* 70: 3170–3174.

Rose, B. and W.R. Loewenstein. 1975. Permeability of cell junction depends on local cytoplasmic calcium activity. *Nature* 254: 250–252.

Flanagan, J. and G.L.E. Koch. 1978. Cross-linked surface Ig attaches to actin. *Nature* 273: 278–281.

Kahn, C.R., K.L. Baird, D.B. Jarrett, and J.S. Flier. 1978. Direct demonstration that receptor cross-linking or aggregation is important in insulin action. *Proc. Natl. Acad. Sci.* 75: 4209–4213.

Ben-Ze'ev, A., A. Duerr, F. Solomon, and S. Penman. 1979. The outer boundary of the cytoskeleton: A lamina derived from plasma membrane proteins. *Cell* 17: 859–865.

The limiting membrane of a cultured cell is a complex organelle. All communication with the rest of the world passes through it, and so viability, homeostasis, and proper responses to regulatory signals all depend in turn on properties of the molecules comprising it. These molecules are of two major classes: lipids and proteins. Both are often found to be conjugated to sugar molecules as well. Hydrophobic effects organize the lipids into a bilayer which gives the membrane its first-order properties. Proteins with stretches of hydrophobic residues reside in this bilayer. Such proteins orient in the membrane asymmetrically. Sometimes the amino terminal end is free in the environment (see, for example, the immunoglobulin [IgM] heavy chain on memory B-cell membranes, Chapter 7), while the carboxy-terminal end may sometimes be imbedded in the cell membrane. Some membrane proteins have portions fixed both in the environment and in the underlying cytoplasm, like rivets. In the papers in this chapter, cultured cells are used to examine the roles of the membrane's lipids and proteins in such events as agglutination, ion permeability, lymphoid cell cap formation, and response of cells to polypeptide hormones.

Certain cell surface proteins are characteristically different in each individual of a species. When tissues are transplanted between non-related individuals, these molecules, called histocompatibility antigens, induce a strong immune response. **Frye** and **Edidin** visualize a histocompatibility antigen of mouse cells by mixing live cells with antibody to the antigen and then with fluorescein-labeled antibody to mouse immunoglobulin. This sandwich makes the mouse cells glow green under blue light. Similar treatment of a line of human cells with rabbit antibody directed against its surface antigens followed by rhodamine-labeled anti-rabbit immunoglobulin makes human cell surfaces glow red. Sendai virus fuses the human and mouse cells into heterokaryons (Chapter 6). When stained serially with the proper mix of antibodies, heterokaryons at first appear as harlequin spheres, with one hemisphere red and the other green. But after less than an hour, the surface colors mix to give a smooth display of both red and green over the entire heterokaryon surface. This result means that these two surface proteins must have rapidly moved in the bilayer of the heterokaryon to redistribute evenly over the whole surface. The figure showing this was in color in the original paper, but it is reproduced here on page 618 in black and white. In the original, one-half of the cell in frame C is green and the other half red.

By extension, such proteins also ought to be able to move in the membranes of any ordinary cell. Lateral movement of membrane proteins can lead to two sorts of redistribution: diffusion or clustering. Frye and Edidin neatly demonstrated diffusion. Clustering occurs as well, and can also be easily detected. Certain plant proteins agglutinate red blood cells (Chapter 1). Agglutination requires the redistribution of lectin-binding molecules in patches on the cell surface.

Eckhart, Dulbecco, and **Burger** use these molecules, called lectins, to agglutinate cultured cells. In this paper they show that transformed cells are far more easily agglutinated than normal cells and that ease of agglutination is temperature-dependent in cells transformed by temperature-sensitive transformation mutants of polyoma virus (Chapter 3). Since we know the *sarc* gene product (src) is localized under the cell membrane (Chapter 4), it is reasonable to hypothesize that a polyoma-virus-encoded transforming protein may be in the same place, also facilitating lateral mobility of transmembrane proteins, perhaps by generating aberrations in the specific phosphorylation of these proteins.

Hynes directly examines the proteins on the outside of the cell membrane. In the presence of H_2O_2, radioactive iodine ion may be coupled to tyrosine residues of a protein. With a large enzyme (lactoperoxidase) to do the coupling, iodination will be restricted to the outer surface of a live cell. Hynes shows that a major surface protein of normal cells is diminished or absent from a variety of transformed cells. This protein, now known to be the molecule fibronectin, also is exquisitely sensitive to proteolysis. Indeed, proteolysis sufficient to permit agglutination of normal cells by lectins is also sufficient to remove fibronectin. Although causality remains to be worked out, it seems clear that transformation by a single viral gene product can coordinately change the cell membrane in many ways, which are seen as increased lateral mobility of transmembrane proteins (TMPs), loss of a protease-sensitive major cell surface protein, and secretion of at least one new protease (Chapter 4).

Certain cells are coupled to each other by gap junctions that permit passage between cells of small molecules, such as ions, sugars, amino acids, and bases. Although epithelial cells establish gap junctions most readily, they can also be demonstrated in endothelial cells and even in some fibroblasts. The junctional channel is made from highly organized membrane proteins apposed to one another on the surfaces of two adjacent cells. **Rose** and **Loewenstein** examine the role of divalent calcium ion (Ca^{++}) in regulating the effective pore size of these channels.

In the cytoplasm, free Ca^{++} is present in micromolar concentrations. Aequorin is a molecule that emits light in the presence of free Ca^{++}. By coupling an image intensifier, photomultiplier, and TV camera to the microscope of a microinjection apparatus, Rose and Loewenstein were able to inject Ca^{++} and aequorin and to monitor directly local Ca^{++} concentration in a single cell by measuring emission of light by aequorin. They then impaled this cell with still other electrodes to apply a voltage across its membrane, while monitoring the current flowing through the membrane of an adjacent cell. Current flow in the adjacent cell measured the number of open ion channels connecting the two cells. By aequorin emission, they found that injected Ca^{++} was rapidly sequestered at the point of the injecting needle, possibly by the endoplasmic reticulum, by cytoplasmic calmodulin (Chapter 4), or by nuclear and mitochondrial membrane pumps. Puffs of cytoplasmic high Ca^{++} near the needle point had no effect on the currents through ion channels between the two adjacent cells. However, when aequorin showed the puff to have reached a region in the cytoplasm near a boundary with the adjacent cell, ionic coupling to that cell was rapidly and completely lost. These results suggest that local concentrations of intracellular

Ca^{++} may vary rapidly from place to place in the cell and that locally high Ca^{++} can modify transmembrane functions. The molecules that actually transduce changes in cytoplasmic free Ca^{++} concentration into changes in membrane junctional channel organization have not yet been isolated.

Studies on the molecules linking membrane function to cytoplasmic organization are central to our current understanding of the immune system. Memory B cells carry about a million molecules of IgM on their surfaces (Chapter 7). Antibody to IgM causes these surface molecules to redistribute into patches, which eventually coalesce into one large surface cap. Iodinated antibody to mouse IgM reveals radioactive patches on myeloma (B-cell tumor) cells and on normal spleen B lymphocytes. **Flanagan** and **Koch** follow the radioactivity of an iodinated anti-IgM antibody to show a surprisingly tight interaction of surface and cytoplasmic molecules in the caps.

Monovalent fragments of iodinated anti-IgM bind to surface IgM on B cells but do not cause patches or caps. When the cell membrane lipid was dissolved with a detergent, radioactive monovalent anti-IgM fell off an unpatched lymphocyte. However, after patching with radioactive divalent antibody, the radioactivity remained tightly associated with cytoplasmic actin (see Chapter 11), even after detergent treatment. Drugs such as azide permit patching of the labeled antibody but prevent capping. Up to 50% of iodinated anti-IgM was associated with actin even after patch formation, so capping was not necessary to bring actin and surface IgM into close association. This suggests that some sort of linkage to cytoplasmic actin may in fact be a necessary step in the redistribution of surface IgM molecules into antigen-initiated patches.

A novel form of diabetes arises in individuals who make an autoantibody to their insulin receptors. This antibody can be used to study the role of receptor movement in insulin activity. On fat cells (Chapters 1 and 2), surface receptors for insulin will redistribute into patches in the presence of bivalent receptor antibody. **Kahn, Baird, Jarrett,** and **Flier** show that antibody to insulin receptor can also modulate the response of rat adipocytes to insulin. The monovalent form of antibody to insulin interferes with insulin's ability to increase glucose intake and oxidation, presumably by blocking receptor sites. However, bivalent antibody, or monovalent antibody followed by bivalent second antibody, enhances the effects of suboptimal concentrations of insulin. The simplest explanation of these unexpected results is that insulin action in all cases depends on redistribution and clustering of receptor molecules and that external agents that enhance such clustering mimic insulin, whereas agents that interfere, inhibit its action. It is not known yet how receptor redistribution occurs when insulin itself initiates a cellular response, nor how the cell redistributes insulin receptor molecules. It is clear, though, that the cross-linking of the external portion of hormone receptor molecules is a sufficient signal to mimic the hormone. This suggests that the hormone itself need never be preserved in an active form inside the signaled cell.

The last paper shows that protein-protein interactions can be the predominant force connecting molecules that see the extracellular environment to molecules in the cytoplasm. **Ben-Ze'ev, Duerr, Solomon,** and **Penman** dissolve the membranes of live anchored cells in detergent. Most of the cell protein goes into solution, but some proteins remain anchored as a meshlike structure, which retains the nucleus. By phase microscopy, this mesh bears a remarkable resemblance to the live cell. Using iodination before extraction and assaying for virus receptors, Ben-Ze'ev et al. show that the mesh retains many surface proteins in native form. Thus, even the external face proteins of the plasma membrane may be thought of as part of a macromolecular array extending through the cytoplasm. This possibility is examined from the opposite side of the cell in the next chapter.

J. Cell Sci. **7**, 319–335 (1970)

Printed in Great Britain

THE RAPID INTERMIXING OF CELL SURFACE ANTIGENS AFTER FORMATION OF MOUSE-HUMAN HETEROKARYONS

L. D. FRYE* AND M. EDIDIN†

Department of Biology, The Johns Hopkins University,
Baltimore, Maryland 21218, U.S.A.

SUMMARY

Cells from established tissue culture lines of mouse (*c11D*) and human (*VA-2*) origin were fused together with Sendai virus, producing heterokaryons bearing both mouse and human surface antigens which were then followed by the indirect fluorescent antibody method. Within 40 min following fusion, total mixing of both parental antigens occurred in over 90% of the heterokaryons.

Mouse H-2 (histocompatibility) and human surface antigens were visualized by successive treatment of the heterokaryons with a mixture of mouse alloantiserum and rabbit anti-*VA-2* antiserum, followed by a mixture of fluorescein-labelled goat anti-mouse IgG and tetramethyl-rhodamine-labelled goat anti-rabbit IgG(Fc).

The *c11D* × *VA-2* fusions were carried out in suspension and maintained at 37 °C in a shaking water bath; aliquots were removed at various intervals and stained with the above reagents. The heterokaryon population was observed to change from an initial one (5-min post-fusion) of non-mosaics (unmixed cell surfaces of red and green fluorescence) to one of over 90% mosaics (total intermixing of the 2 fluorochromes) by 40 min after fusion. Mouse–human hybrid lines, derived from similar fusions, gave fluorescence patterns identical to those of the mosaic heterokaryons.

Four possible mechanisms would yield such results: (i) a very rapid metabolic turnover of the antigens; (ii) integration of units into the membrane from a cytoplasmic precursor pool; (iii) movement, or 'diffusion' of antigen in the plane of the membrane; or (iv) movement of existing antigen from one membrane site into the cytoplasm and its emergence at a new position on the membrane.

In an effort to distinguish among these possibilities, the following inhibitor treatments were carried out: (1) both short- and long-term (6-h pre-treatment) inhibition of protein synthesis by puromycin, cycloheximide, and chloramphenicol; (2) short-term inhibition of ATP formation by dinitrophenol (DNP) and NaF; (3) short- and long-term inhibition of glutamine-dependent pathways with the glutamine analogue 6-diazo-5-oxonorleucine; and (4) general metabolic suppression by lowered temperature.

The only treatment found effective in preventing the mosaicism was lowered temperature, from which resulted a sigmoidal curve for per cent mosaics versus incubation temperature. These results would be consistent with mechanisms iii and/or iv but appear to rule out i and ii.

From the speed with which the antigen markers can be seen to propagate across the cell membrane, and from the fact that the treatment of parent cells with a variety of metabolic inhibitors does not inhibit antigen spreading, it appears that the cell surface of heterokaryons is not a rigid structure, but is 'fluid' enough to allow free 'diffusion' of surface antigens resulting in their intermingling within minutes after the initiation of fusion.

---

* Present address: Immunochemistry Unit, Princess Margaret Hospital for Children, Subiaco, W.A. 6008, Australia.

† To whom requests for reprints should be addressed.

INTRODUCTION

The surface membranes of animal cells rapidly change shape as the cells move, form pseudopods, or ingest material from their environment. These rapid changes in shape suggest that the plasma membrane itself is fluid, rather than rigid in character, and that at least some of its component macromolecules are free to move relative to one another within the fluid. We have attempted to demonstrate such freedom of movement using specific antigen markers of 2 unlike cell surfaces. Our experiments show that marker antigens on surface membranes spread rapidly when unlike cells are fused. The speed of antigen spread and its insensitivity to a number of metabolic inhibitors offer some support for the notion of a fluid membrane.

We have approached the problem of mixing unlike, and hence readily differentiated, cell surface membranes by using Sendai virus to fuse tissue culture cells of mouse and human origin (Harris & Watkins, 1965). The antigens of the parent cell lines and of progeny heterokaryons have been visualized by indirect immunofluorescence, using heteroantiserum to whole human cells, and alloantiserum to mouse histocompatibility antigens. Both sera were cytotoxic for intact cells in the presence of complement; alloantisera have previously been shown, by both immunofluorescence and immunoferritin techniques, to bind only to the surface of intact cells (Möller, 1961; Cerottini & Brunner, 1967; Drysdale, Merchant, Shreffler & Parker, 1967; Davis & Silverman, 1968; Hammerling et al. 1968).

The surface antigens of heterokaryons between hen erythrocytes and HeLa cells and between Ehrlich ascites and HeLa cells have previously been studied using mixed agglutination techniques for antigen localization (Watkins & Grace, 1967; Harris, Sidebottom, Grace & Bramwell, 1969). In these studies intermixing of surface antigens was demonstrable within an hour or two of heterokaryon formation. However, the antigens could not readily be localized, since the marker particles used were several microns in diameter; also, observations were not made of the earliest time at which mixing occurred. We have been able to examine heterokaryons within 5 min of their formation and to show that antigen spread and intermixing occurs within minutes after membrane fusion. Studies on cells poisoned with a variety of metabolic inhibitors strongly suggest that antigen spread and intermixing   quires neither *de novo* protein synthesis nor insertion of previously synthesized subunits into surface membranes.

MATERIALS AND METHODS

Cell lines

c11D. A thymidine-kinase negative (TK-) subline of the mouse 'L' cell, isolated by Dubbs & Kitt (1964), and kindly provided by Dr H. G. Coon.

VA-2. An 8-azaguanine-resistant subclone, isolated by Weiss, Ephrussi & Scaletta (1968), obtained from *W-18-VA-2*, an SV40-transformed human line which has been free of infective virus for several years (Ponten, Jensen & Koprowski, 1963).

SaI. An ascites tumour (designated as Sarcoma I), provided by Dr A. A. Kandutsch, The Jackson Laboratory, and carried in *A/J* mice; it was used as a convenient source of mouse cells for absorption of antiglobulin reagents.

Tissue culture

The $ciiD$ and $VA-2$ lines were routinely grown in a modified F-12 medium containing 5 % foetal calf serum (FCS) (Coon & Weiss, 1969) or in Minimal Essential Medium with 5 % FCS, 5 % Fungizone and 100 units penicillin/ml. The cultures were maintained at 37 °C in a water-jacketed CO_2 incubator, 98 % humidity, 5 % CO_2.

For experiments or routine passages, cells were harvested with 2·5 % heat-inactivated chicken serum, 0·2 % trypsin and 0·002 % purified collagenase (Worthington CSL) in Moscona's (1961) solution, which is referred to as 'CTC'.

Sensitizing antibodies

Mouse alloantiserum (FAS-2). Preparation: antibodies primarily directed against the H-2^k histocompatibility antigens were obtained by a series of intraperitoneal injections of $CBA/\mathcal{J}$ (H-2^k) mouse mesenteric lymph node and spleen cells into $BALB/c\mathcal{J}$ (H-2^d) mice (4-recipients: 1 donor). Six injections were given twice weekly, followed by a booster 2 weeks after the last injection. The animals were bled from the retro-orbital sinus 4 and 5 days post-booster.

Specificity. Reaction with mouse cells ($ciiD$): Aliquots of $2·5 \times 10^5$ $ciiD$ cells were treated in suspension with 0·1 ml of two-fold dilutions of FAS-2 from 1/10 to 1/80. The cells were agitated periodically for 15 min at room temperature at which time they were washed twice in phosphate-buffered saline (PBS). They were then resuspended in 0·05 ml of fluorescein-labelled rabbit antimouse IgG, incubated, and washed as above. The cells were then put on to Vaseline-ringed slides, covered and observed in the fluorescence microscope. Ring reactions as reported by Möller (1961) were observed with decreasing brightness upon increasing dilutions of the FAS-2. As maximum brightness was desired, the 1/10 dilution was chosen for all subsequent staining reactions.

Reaction with human cells ($VA-2$): No fluorescence was observed when analagous staining reactions were carried out with human cells.

Reaction with Sendai virus: It was discovered that $VA-2$ cells pre-treated with Sendai virus became positive for the FAS-2 sensitization. Normal mouse sera from $BALB/c\mathcal{J}$, $CBA/\mathcal{J}$, $DBA/2\mathcal{J}$ and $A/\mathcal{J}$ strains were also shown to exhibit this anti-Sendai activity. This activity in FAS-2 was easily absorbed by treating a 1/5 dilution of the antiserum with 333-666 haemag-glutinating units (HAU)/ml of virus for 30 min at room temperature and overnight at 4 °C. The absorbing virus was then removed by centrifugation.

Rabbit anti-VA-2 antiserum (RaVA-2) preparation: $VA-2$ cells were grown in Falcon plastic Petri dishes, harvested with CTC, and washed 3 times in Hanks's balanced salts solution, BSS (HEPES-buffered) to remove the foetal calf serum. 2×10^5 cells were emulsified with Freund's complete adjuvant (cells : adjuvant = 1:2) and injected intradermally (flanks and footpads) into a New Zealand white rabbit. One week later 10^6 washed cells were given intradermally (flanks only; 10^5 cells/site). The rabbit was bled from the ear vein 1 and 2 weeks following this second injection. The sera were heat-inactivated at 56 °C for 30 min, aliquoted and stored at -30 °C.

Specificity of RaVA-2: Reaction with $VA-2$: $VA-2$ cells were seeded on to coverslips ($2·5 \times 10^5$/coverslip) and allowed to adhere and spread. The coverslips were then washed with Hanks's and 0·1 ml of 2-fold dilutions of the RaVA-2 from 1/2 to 1/256 were added. After incubation in a moist chamber for 15 min at room temperature, the coverslips were washed twice in Hanks's BSS and similarly treated with tetramethylrhodamine (TMR)-labelled goat anti-rabbit IgG (anti-Fc). The cells gave strong fluorescent ring reactions at the lower dilutions of the sensitizing antibody; the 1/4 dilution was chosen for all subsequent staining reactions.

Reaction with $ciiD$: When analogous staining reactions were carried out with the mouse cells, a very weak fluorescence was seen in the lower dilutions of the RaVA-2. Consequently, the serum was routinely absorbed with 5×10 $ciiD$/ml of a 1/2 diluted serum (30 min at room temperature).

Reaction with Sendai virus: The $ciiD$ cells, when pre-treated with Sendai virus, gave weak positive staining with the mouse-absorbed RaVA-2. Therefore, the RaVA-2 was also absorbed with 333 HAU/ml (30 min at room temperature and overnight at 4° C). This doubly absorbed RaVA-2 then gave a negligible background on the $ciiD$ cells.

21-2

Fluorescent antibodies

Goat anti-mouse IgG. Preparation of mouse IgG: A 16% Na_2SO_4 cut of 26 ml of normal *BALB/c* serum was dissolved in 8 ml of 0·2 M NaCl, 0·1M phosphate buffer, pH 8·0 and then dialysed against this buffer in the cold prior to chromatography on a 2·5 × 100 cm column of Sephadex G-200, in the same buffer. Included fractions comprising the second protein peak off the column were pooled and dialysed against 0·01 M phosphate buffer, pH 7·5. The dialysed material was applied to a 500-ml column of DEAE-cellulose, equilibrated with 0·01 M phosphate buffer. Material eluting stepwise from the column in 0·025 and 0·05 M phosphate buffer was pooled and concentrated in an Amicon ultrafilter. The purified material showed only an IgG arc upon immunoelectrophoresis on agarose and reaction with rabbit anti-whole mouse serum.

Immunization of goat: 10 mg of immunogen was emulsified with Freund's complete adjuvant (immunogen: Freund's = 1:2) and injected intramuscularly into a 6-month-old goat. Three and one-half weeks later, 600 ml of blood were collected from the jugular vein and the serum tested by immunoelectrophoresis. Even though the immunogen had shown no contaminants as judged by the rabbit anti-whole mouse serum, a trace amount of a more negative protein was present. Though the goat anti-mouse IgG was not monospecific, non-specificity was not observed in the indirect fluorescent antibody technique described under sensitizing antibodies.

Conjugation to FITC: The isothiocyanate derivative of fluorescein (FITC) was used for conjugation to partially purified goat antibodies, employing the method of Wood, Thompson & Goldstein (1965). The fluorescein-labelled antibodies were eluted stepwise from DEAE-cellulose with 0·05, 0·1, 0·2 and 0·3 M phosphate buffers, pH 7·5. The 0·1 M phosphate buffer cut, having an O.D. 280/495 = 2·0 and a protein concentration of 1·3 mg/ml, was used in all our experiments.

Goat anti-rabbit IgG (Fc). Source: Goat anti-rabbit IgG (Fc), prepared against the Fc portion of the gamma heavy chain, was kindly provided by Dr J. J. Cebra.

TMRITC conjugation: The preparation of tetramethylrhodamine (TMR)-labelled antibodies was carried out under the same conditions as for the fluorescein conjugation. An 0·1 M phosphate buffer cut from DEAE-cellulose had an O.D. 280/515 = 1·7 and a protein concentration of 1 mg/ml; it was used in all studies described here.

Sendai virus

The Sendai virus used in the experiments described in this report, was kindly provided by Dr H. G. Coon. Its preparation was as published (Coon & Weiss, 1969) except that the virus was inactivated with β-proprio-lactone, rather than by ultraviolet irradiation.

Formation of heterokaryons

Heterokaryons were produced by the suspension fusion technique originally described by Okada (1962) for homokaryons. The parental ratios were $cIID/VA-2$ = 2–4; 3×10^6 cells were resuspended in 0·1 ml of cold Sendai (100–250 HAU/ml) and shaken at 0–4 °C for 10 min and then at 37 °C for 5–10 min. Culture medium was then added for a 10-fold dilution of the cells.

Formation of hybrid cell lines

Mouse-human hybrid cell lines were produced by viral fusion as for the heterokaryons.

Following fusion, the cells were plated at 3×10^5/ml in normal medium; 24 h later this medium was replaced by 'HAT' (Littlefield, 1964), which was used for all subsequent feedings of these plates and the resulting hybrid lines.

Fluorescent staining of cells

Cells from fusion experiments or hybrid cells were washed in Hanks's BSS and resuspended in a mixture of sensitizing antibodies: FAS-2 and RaVA-2 (1/10 and 1/4 final dilutions, respectively), 0·1 ml mixture/3–6 × 10^5 cells; incubation at room temperature for 15 min. Finally, the cells were washed twice in Hanks's BSS, resuspended in a small volume of the same, placed on a Vaseline-ringed slide, and observed in the fluorescence microscope.

Table 1. *Filter combinations used for excitation of fluorescein and tetramethyl-rhodamine conjugates*

	For observation of	
Filter type	Fluorescein	Tetramethylrhodamine
Excitation	Blue interference filter	Green interference filter:
	Type: PAL	Type: PAL
	No.: 100·105	No.: 10157·14
	λ max.: 437 nm	λ max.: 545 nm
	T max.: 43 %	T max.: 63 %
	HW = 21 nm	HW = 21 nm
	Source: Schott and	Source: Schott and
	Gen., Mainz	Gen., Mainz
Barrier or Window	Kodak Wratten gelatin	Kodak Wratten gelatin
	Filter no. 58 (green)	Filter no. 23 A (red)
	or	and
	Kodak Wratten gelatin	RG–1 filter (Red;
	Filter no. 8 K 2 (yellow)	2 × 17 mm)
		(Schott and Gen., Mainz)

Fluorescence microscopy

All observations of fluorescent cells were made with a Leitz Ortholux Microscope, using darkfield condenser D 1·20 and an Osram HBO 200-W high-pressure mercury lamp as the light source. The exciting light was first passed through a Corning BG-38 heat filter and then through a combination of interference and barrier filters, depending upon the type of fluorescence to be maximized (see Table 1). The interference filters were patterned after those reported by Ploem (1967) to give maximum brightness for fluorescein- and tetramethylrhodamine-labelled antibodies.

Photography

Pictures of cells stained with fluorescent-labelled antibodies were taken using a Leica camera and exposing Anscochrome 200 daylight film (ASA 200) for 3–4 min or Hi-Speed Ektachrome (ASA 160) for 4–6 min.

RESULTS

Staining of *c11D* and *VA-2*

When *c11D* or *VA-2* populations were stained for either H-2 or human antigens by the protocol given in Materials and Methods, 2 basic fluorescent patterns were observed: (1) The majority of the cells gave a full ring reaction, (Möller, 1961) with intensities varying from cell to cell; an occasional cell gave no fluorescence except for a weak blue-green autofluorescence which was easily distinguishable from the FL or TMR fluorescence. On the cells giving ring reactions, distinct, tiny patches of fluorescence could be seen by focusing on the upper or lower cell surfaces. When a cell was in focus for the ring reaction (that is, at the cell equator), these patches were no longer visible. (2) Some cells gave only a partial ring reaction, and upon focusing on the upper or lower surfaces, the patches of fluorescence were in highest concentration on the cell half giving the partial ring reaction.

Staining of $c_{11}D \times VA$-2 hybrid cell lines

Hybrids between $c_{11}D$ and VA-2 were produced as described in Materials and Methods. Colonies of hybrid cells first appeared 13 days following fusion. Six of these were isolated and stained to provide positive controls for doubly antigenic cells (Fig. 3 A, B). Cells of all lines fluoresced green, indicating a content of mouse H-2 antigens, and the intensities were comparable to those given by the mouse parent, $c_{11}D$. The intensity of the red fluorescence, marking human antigens, varied from line to line. If the human parent, VA-2, is given an arbitrary intensity of $++$ units, then the hybrid lines gave the following: MH-1, $++$; MH-5, $+$; MH-2, $\pm$; and MH-6, $-$.

Time-course of staining of heterokaryons

Fusions and fluorescent antibody staining of $c_{11}D \times VA$-2 were carried out as described in Materials and Methods. Crosses of $c_{11}D \times c_{11}D$ and VA-2 $\times VA$-2 were also made as controls for antibody specificity. Following the fusion reaction (5–10 min at 37 °C), the cells were diluted 10-fold with medium and shaken at 37 °C. Aliquots removed at various times were stained immediately, and the stained cells were kept at 0 °C until observation.

Though cell fusion indices were not measured, the degree of fusion in the system was not great, as evidenced by the low number of double-staining cells obtained; the scanning of several fields was required to count adequate numbers of the heterokaryons.

The control slides of $c_{11}D \times c_{11}D$ and VA-2 $\times VA$-2 showed cells with ring reactions of only one colour; green for $c_{11}D$ and red for VA-2. No double-staining cells were seen.

Four basic types of double-staining cells were observed in the $c_{11}D \times VA$-2 crosses: (1) $M\frac{1}{2}$–$H\frac{1}{2}$, a heterokaryon showing unmixed partial ring reactions for each fluorochrome; (2) $M\frac{1}{2}$-H1, in which the heterokaryon showed a complete ring reaction for the human antigens, but only a partial one for the mouse H–2 antigenic markers; (3) M1-$H\frac{1}{2}$, the reverse of the pattern seen in $M\frac{1}{2}$-H1; this type was much rarer than $M\frac{1}{2}$-H1, and (4) M1-H1, which we term mosaics, showing complete ring reactions for both fluorochromes (Fig. 3 C–G). In addition to these 4 types, a large number of parent cells were seen which had not fused; these provided another control for reagent specificity. Homokaryons, though undoubtedly present, would not be detected since only one type of fluorescence would be seen on such cells. Also present were weakly staining cells which could not be categorized, as well as damaged cells; the latter were characterized by a diffuse fluorescence of both fluorochromes throughout the cell.

Table 2 shows the results obtained up to 2 h following the initial fusion reaction. There is a definite trend from an initial population of non-mosaics to one of over 90% mosaics (as percentage of double-stained cell population) by 40 min. Fig. 1 shows a bar graph of the population shift over time.

Table 2. *Time course of antigen spread*

Incubation time at 37 °C (min)	Double-staining category				Total	Mosaics %
	$M\frac{1}{2}$–$H\frac{1}{2}$	$M\frac{1}{2}$–$H1$	$M1$–$H\frac{1}{2}$	$M1$–$H1$		
5	26	3	0	0	29	0
5	25	1	0	0	26	0
10	6	24	0	1	31	3
10	9	4	0	0	13	0
25	2	7	0	6	15	40
25	1	20	0	25	46	54
40	0	4	0	26	30	87
40	0	2	0	14	16	88
40	0	1	0	14	15	93
40	0	0	0	37	37	100
120	0	0	0	31	31	100

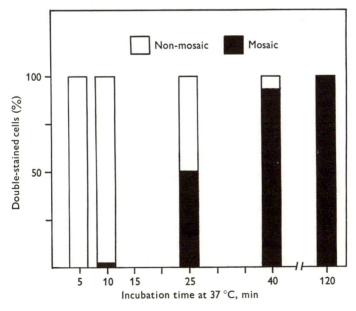

Fig. 1. Appearance of completely double-staining (mosaic) cells in the population of fused cells.

In initial studies in which the fused cells were placed on coverslips in Petri dishes and allowed to adhere and spread for up to 12 h before staining, all double-staining cells observed were mosaic in appearance.

Inhibitor studies

The rapid spread of antigens across the surfaces of heterokaryons could be due either to movement of antigens across the cell surface ('diffusion') or to new antigen synthesis. In an attempt to distinguish between these possibilities, cells were treated with various metabolic inhibitors before and during fusion. Since the time-course studies showed that over 90% of the double-stained cells were mosaic by 40 min at 37 °C, this time period was chosen for testing the effects of the inhibitors. The experiments fall into 4 categories: inhibition of protein synthesis, of ATP formation, of glutamine-dependent synthetic pathways, and generalized metabolic inhibition by lowered temperature.

Table 3. *Effect of inhibition of protein synthesis on mosaicism*

| | Double-staining category | | | | | |
Treatment	$M\frac{1}{2}-H\frac{1}{2}$	$M\frac{1}{2}-H1$	$M1-H\frac{1}{2}$	$M1-H1$	Total	Mosaics %
Puromycin, short term	0	1	0	35	36	97
Puromycin, long term	0	9	0	24	33	73
	0	1	1	43	45	95
Cycloheximide, short term	0	0	0	12	12	100
Cycloheximide, long term	0	7	0	34	41	83
	0	2	0	39	41	95
Chloramphenicol, short term	0	1	0	40	41	98
Chloramphenicol, long term	0	6	1	43	50	86

Inhibition of protein synthesis. Cycloheximide and puromycin were tested for potency in terms of the inhibition of [³H]leucine incorporation into TCA-precipitable material. Cycloheximide was found to give over 95% inhibition at 5 μg/ml after 30 min incubation, whereas comparable inhibition for our batch of puromycin required 80 μg/ml as measured after 3 h.

For the initial experiments the parent cells were suspended in medium containing either puromycin (80 μg/ml) or cycloheximide (5 μg/ml) and incubated for 15 min at 37 °C. The cells were then pelleted and fusion carried out as usual. After the fusion reaction, the cells were diluted 10-fold with inhibitor-containing medium and shaken for 30 min at 37 °C before staining. Table 3 shows the results of these and other inhibitor experiments. Neither inhibitor had any effect on mosaic formation.

If membrane or antigen subunits were synthesized some time prior to being released to the surface, short-term inhibition of protein synthesis might be without effect on such subunits, or their insertion into membrane. Therefore, the parent cells were treated with inhibitors for 6 h prior to their fusion. Confluent plates of *c11D* and *VA-2* were treated with inhibitor in the concentrations used in short-term experiments, but for 6 h; after 3 h incubation, the medium was replaced with fresh inhibitor-medium to ensure continued inhibitor potency. (Both inhibitors had been tested with [³H]-leucine for prolonged inhibition, and they did retain their potency during a 3-h incubation.) Two and one-half hours later, the cells were harvested with CTC-inhibitor, counted and checked for viability (nigrosin dye exclusion test). Viability was greater than 95%. The experiment was then continued, using the same procedures as for the short-term inhibition experiments. Table 3 gives the results of these experiments. In the initial long-term experiment, a slight effect for both inhibitors was noted; however, a repeat experiment gave results equal to control values.

Table 4. *Effect of inhibition of glutamine-dependent pathways on mosaicism*

Treatment	Double-staining category				Total	Mosaics %
	$M\frac{1}{2}-H\frac{1}{2}$	$M\frac{1}{2}-H1$	$M1-H\frac{1}{2}$	$M1-H1$		
DON, short term	0	1	0	44	45	98
DON, long term	0	2	0	40	42	95

Analogous short and long-term inhibition experiments were also carried out with chloramphenicol, at a concentration of 200 μg/ml (Table 3). Again negligible inhibition of mosaic formation was noted.

Inhibition of ATP formation. When cells are exposed to 2.5×10^{-3} M DNP $+ 2 \times 10^{-3}$ M NaF following initiation of fusion for 10 min at 37 °C, their content of ATP, as measured in terms of the light flash produced in a luciferin/luciferase system, (Denburg, Lee & McElroy, 1969) falls to 20% of that of control cells 5 min after addition of the uncouplers and to 17% at 10 min. After shaking for a further 30 min at 37 °C, the cells were stained and observed. DNP+NaF did not inhibit mosaic formation.

Inhibition of glutamine-dependent synthetic pathways. Fusion experiments were performed using cells treated with the glutamine analogue, DON (6-diazo-5-oxonorleucine), at a concentration of 250 μg/ml in the medium which contained 292 μg/ml of L-glutamine. In the short-term inhibition experiment, the parent cells were incubated in the presence of DON for 30 min at 37 °C prior to the fusion step. Following a fusion step of 5 min, the cells were diluted 10-fold in DON-containing medium, shaken for 35 min at 37 °C, stained and observed. Table 4 gives the results of this experiment, in which no effect was seen.

A long-term inhibition experiment was also performed, differing from the former only in that the cells, in culture, had been treated with DON 6 h prior to the fusion reaction. After 3 h in culture, the DON-containing medium was replaced with fresh inhibitor-medium to insure continued inhibition. Before fusion the cells were tested for viability (nigrosin dye exclusion); viability was greater than 95%. The results of this experiment are given in Table 4. Again, mosaic development proceeded as normal.

Temperature studies

To see the effects of lowered temperature upon mosaic formation, aliquots of cells were maintained for 30 min at 0, 15, 20 and 26 °C after an initial 10-min fusion step. The results are tabulated in Table 5 and Fig. 2 shows a plot of temperature *v.* per cent mosaicism in the double-stained cell population. Lowered temperature does appear to inhibit antigen spread and mixing.

In all the inhibition experiments described, the *degree* of cell fusion did not appear to be significantly reduced by any of the inhibitors.

Table 5. *Effect of temperature on antigen spread and mosaicism*

Incubation temperature (°C)	Double-staining category				Total	Mosaics %
	$M\frac{1}{2}-H\frac{1}{2}$	$M\frac{1}{2}-H1$	$M1-H\frac{1}{2}$	$M1-H1$		
0	7	4	0	0	11	0
0	18	10	0	2	30	7
0*	28	2	0	0	30	0
15	16	11	6	3	36	8
15*	22	8	0	0	30	0
20	7	18	3	20	48	42
20*	1	20	2	22	45	49
26	1	6	0	23	30	77
26*	0	10	0	32	42	76

*5-min fusion at 37 °C.

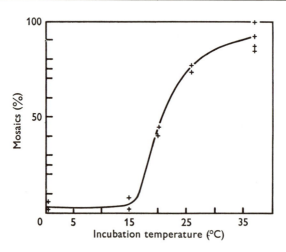

Fig. 2. Effect of temperature on the appearance of mosaic cells within 40 min of cell fusion.

DISCUSSION

Surface antigens of $c_{II}D$ and VA-2 cells and of their stable hybrids

Fluorescent antibody staining of cells of lines $c_{II}D$ and VA-2 showed surface antigens homogeneously distributed, in tiny patches, over the entire surface of most of the cells observed. The appearance of a 'ring reaction' (Möller, 1961), seems to be due to surface membrane curvature, resulting in a greater number of fluorochromes observed per unit area when focusing in the plane of a cell's equator than when focusing on its upper or lower hemispheres. The reason why some cells exhibit a partial ring reaction at the equator remains unclear; similar results were noted by Möller (1961).

The discrete patches of fluorescence seen away from the cell equator indicate that surface antigens are localized or clustered, rather than spread through the entire surface. Others have reported similar observations, using both light and electron microscopy. Cerottini & Brunner (1967), using an indirect fluorescent antibody method to detect mouse H–2 alloantigens on various tumour and normal cells, observed large patches of fluorescence, especially in ethanol-fixed cells. Two groups using ferritin-labelled antibodies have also shown H-2 antigens to be clustered in the membrane (Davis & Silverman, 1968; Hammerling *et al.* 1968).

Staining of hybrid lines between $c_{II}D$ and VA-2 provided a control for the appearance, when stained, of a cell whose surface is a mosaic of both human and mouse antigens, and is presumably synthesizing these antigens. Those lines which did stain positive for both sets of antigens were indistinguishable from the mosaic heterokaryons found 40 min after initiation of cell fusion.

Surface antigens of heterokaryons

The time-course study of antigens on the surface of newly formed heterokaryons showed that over 90% of double-staining cells are completely mosaic by 40 min after initiation of fusion. Though for technical reasons, a single cell cannot be followed through the various stages of mosaic formation, the population study would indicate a progression from $M\frac{1}{2}-H\frac{1}{2} \rightarrow M\frac{1}{2}-H_I \rightarrow M_I-H_I$. The observation that the mouse and human antigens do not spread at the same rate thus giving rise to the $M\frac{1}{2}-H_I$ class, remains unexplained. One possible reason for the apparently faster rate of human antigen mixing could be due to a concentration effect; that is, the human marker represents all VA-2 surface components antigenic for rabbits (for whole VA-2 cells were used in the rabbit immunization) whereas the mouse marker represents only H-2 alloantigens (which could presumably be in lesser amount per unit area than total human antigens per unit area on the VA-2 membrane). Those rare cells showing the reverse pattern, $M_I-H\frac{1}{2}$, could possibly be explained on the basis of the fusion of one VA-2 cell with several $c_{II}D$; cell nuclei were difficult to see under dark-field illumination for verification of multicellular fusion.

Four processes might account for the observed development of a mosaic pattern of antigen distribution: (i) a rapid synthesis of additional antigens, or a rapid metabolic turnover of existing antigens; (ii) integration of subunits, previously synthesized within the cell, into the membrane; (iii) movement or 'diffusion' of antigen within the plane of the membrane; or (iv) movement of existing antigen from one membrane site into the cytoplasm and its emergence at a new position on the membrane. Mechanisms iii and iv are difficult to distinguish operationally from each other, while it ought to be possible to distinguish i or ii from iii and iv.

If (i), rapid metabolic turnover, or synthesis of additional antigen molecules is responsible for the antigen redistribution observed, then one should be able to block the process with a suitable inhibitor. Unfortunately, lack of information on the chemistry of the antigens detected in our system precludes a definitive statement that antigen synthesis or replacement from a cytoplasmic precursor pool is inhibited by any of the inhibitors tested.

Short-term inhibition of protein synthesis, using puromycin, cycloheximide and chloramphenicol had no effect upon mosaicism. Three different inhibitors were used in an effort to block as many different sites of protein synthesis as possible. Unless one postulates that membrane proteins are synthesized in a metabolic compartment protected from all the inhibitors used, it would appear that rapid *de novo* protein synthesis is not a requirement for mosaic formation; indeed, other workers have shown that the inhibitors we used are effective in blocking the synthesis of membrane-associated molecules. Kraemer (1966, 1967) found puromycin effective in preventing the re-appearance of sialic acid on Chinese hamster cells which had previously been treated with sialidase. Warren & Glick (1968) also found that puromycin blocked the turnover of [^{14}C]glucose in L cell membranes; cycloheximide retards incorporation of [^{14}C]leucine in chloroplast membrane protein of *Chlamydomonas* (Hoober, Siekevitz & Palade, 1969).

The glutamine analogue, DON was also without effect on antigen re-arrangement, both in short- and long-term inhibition experiments. DON has been shown to be an effective inhibitor of glutamine-utilizing amino transferase reactions involved in purine and pyrimidine synthesis and in amino sugar synthesis (summarized by Meister, 1962). In each of the systems, the inhibitor has been shown to be effective in the presence of a large excess of glutamine. We feel that in our experiments its failure to inhibit appears to rule out *de novo* synthesis of oligosaccharides and their attachment to pre-existing protein chains. The presence of amino sugars and sialic acid on the cell surface is well documented (see Cook, 1968) and indeed, DON has been found to inhibit another process involving cell surfaces, glutamine-dependent reaggregation of dissociated mouse embryoid body cells, when used in the presence of glutamine (Oppenheimer, Edidin, Orr & Roseman, 1969).

The utilization of a precursor pool of membrane subunits appears to be ruled out by several of our experiments, especially that involving long-term inhibition of protein synthesis. Recent *in vivo* experiments on membrane synthesis in rat liver indicate the existence of a precursor pool of membrane which is utilized over the course of 3 h following cycloheximide injection (Ray, Lieberman & Lansing, 1968). Our 6-h pretreatment of cells should have been adequate to deplete similar precursor pools if they were present in *c11D* and *VA-2* cells. Furthermore, it might be expected that high energy phosphate bonds might be required for integration of any subunits into the surface, as indeed they seem to be required for initial membrane fusion in heterokaryon formation (Okada, Murayama & Yamada, 1966). However, no inhibition of antigen spread was produced by ATP inhibitors, which in combination quickly reduced cell ATP to less than 20% of control value. This observation also tells against, though it does not rule out, mechanism iii for antigen spread. If molecules left the surface and moved through the cytoplasm before re-emerging at a new point on the surface it might be thought that their re-integration would require energy from ATP.

In arguing the last point, it may be objected that a cell containing 15–20% of its normal ATP content may well still be capable of sustaining ATP-requiring synthetic reactions. Indeed, the degree of inhibition of total ATP generation observed is proportionally far greater than the degree of inhibition of amino acid incorporation by puromycin and cycloheximide. However, this lower degree of inhibition might be expected from the work of Atkinson (1965) showing that as ADP and AMP levels rise in a cell, and ATP levels fall, substrates tend to be shifted to the Krebs cycle, generating more ATP, rather than to synthetic pathways. Despite this possibility, we must concede that inhibitor studies with DNP and NaF can only suggest, but not strongly support, the absence of a need for ATP in antigen movement.

One treatment of fused cells did inhibit the spreading and intermixing of antigens on their surfaces; this was subjection to lowered temperature. The curve of per cent mosaics formed *v.* temperature is that expected if spread were due to diffusion of antigen-bearing molecules in a solvent, such as lipid, whose viscosity changes markedly with temperature. The rate of spread of antigens is, assuming a value of 10 nm for the radius of an antigen molecule, consistent with a membrane viscosity of 100–200 cP($1-2 \times 10^5$ Ns m^{-2}), about that of many oils (V. A. Parsegian, personal communication). Furthermore, the curve of per cent mosaics at 40 min *v.* temperature is quite similar to some of those given by De Gier, Mandersloot & Van Deenen (1968) for the penetration of glycerol into liquid crystal 'liposomes' *v.* temperature. In each instance, the system appears to have a distinct melting temperature, above which various processes may occur.

Other natural membranes have also been shown to contain at least some fluid areas. Hubbel & McConnell (1969) used spin-labelled steroids to show that fluid regions exist in myelin and another group has recently used calorimetric techniques to indicate the liquid crystalline state of mycoplasma membranes (Steim *et al.* 1969). Blasie & Worthington (1969 *a, b*) made low angle X-ray scattering measurements of the arrangement of photopigment molecules in frog retinal receptor disk membranes, showing that the nearest neighbour frequency for those molecules was altered by change in temperature, or by the addition of antirhodopsin antibody to the system. Their interpretation of the data supports the notion that the pigment molecules 'float' in a liquid-like environment.

Our observations and calculations, and the scattered examples from the literature all call attention to the possibility that elements of many biological membranes are not rigidly held in place, but are free to re-orient relative to one another. This aspect of membrane structure has not been considered in current membrane models (cf. Stoeckenius & Engleman, 1969), though modification or extension of several of these would be sufficient to account for our results.

We are deeply grateful to Dr Hayden Coon, now of the Department of Zoology, University of Indiana, for expert advice on tissue culture techniques and cell fusion, to Dr John Cebra, for instruction in the preparation of fluorescent antibody reagents, and to Dr Jeffrey Denburg for performing the ATP assay.

This work was submitted by L. D. F. in partial fulfilment of the requirements for the Ph.D. degree in the The Johns Hopkins University. It was supported by a National Institutes of Health Predoctoral fellowship to L. D. F., by NIH grant AM 11202 to M. E. and by a NIH training grant in developmental biology, in the Department of Biology.

REFERENCES

ATKINSON, D. E. (1965). Biological feedback control at the molecular level. *Science, N.Y.* **150**, 851–857.

BLASIE, J. K. & WORTHINGTON, C. R. (1969a). Molecular localisation of frog retinal receptor photopigment by electron microscopy and low-angle X-ray diffraction. *J. molec. Biol.* **39**, 407–416.

BLASIE, J. K. & WORTHINGTON, C. R. (1969b). Planar liquid-like arrangement of photopigment molecules in frog retinal receptor disk membranes. *J. molec. Biol.* **39**, 417–439.

CEROTTINI, J. C. & BRUNNER, K. T. (1967). Localization of mouse isoantigens on the cell surface as revealed by immunofluorescence. *Immunology* **13**, 395–403.

COOK, G. M. W. (1968). Chemistry of membranes. *Br. med. Bull.* **24**, 118–123.

COON, H. G. & WEISS, M. C. (1969). A quantitative comparison of formation of spontaneous and virus produced viable hybrids. *Proc. natn. Acad. Sci. U.S.A.* **62**, 852–859.

DAVIS, W. C. & SILVERMAN, L. (1968). Localization of mouse H-2 histocompatibility antigen with ferritin-labelled antibody. *Transplantation* **6**, 535–543.

DE GIER, J., MANDERSLOOT, J. G. & VAN DEENEN, L. L. M. (1968). Lipid composition and permeability of liposomes. *Biochim. biophys. Acta* **150**, 666–675.

DENBURG, J. L., LEE, R. T. & McELROY, W. D. (1969). Substrate-binding properties of firefly luciferase I. Luciferin-binding site. *Archs Biochem. Biophys.* **134**, 381–394.

DRYSDALE, R. G., MERCHANT, D. J., SHREFFLER, D. C. & PARKER, F. R. (1967). Distribution of H-2 specificities within the LM mouse cell line and derived lines. *Proc. Soc. exp. Biol. Med.* **124**, 413–418.

DUBBS, D. R. & KIT, S. (1964). Effect of halogenated pyrimidines and thymidine on growth of L-cells and a subline lacking thymidine kinase. *Expl Cell Res.* **33**, 19–28.

HAMMERLING, U., AOKI, T., DEHARVEN, E., BOYSE, E. A. & OLD, L. J. (1968). Use of hybrid antibody with anti-γG and anti-ferritin specificities in locating cell surface antigens by electron microscopy. *J. exp. Med.* **128**, 1461–1473.

HARRIS, H. & WATKINS, J. F. (1965). Hybrid cells derived from mouse and man: artificial heterokaryons of mammalian cells from different species. *Nature, Lond.* **205**, 640–646.

HARRIS, H., SIDEBOTTOM, E., GRACE, D. M. & BRAMWELL, M. E. (1969). The expression of genetic information: A study with hybrid animal cells. *J. Cell Sci.* **4**, 499–525.

HOOBER, J. K., SIEKEVITZ, P. & PALADE, G. E. (1969). Formation of chloroplast membranes in *Chlamydomonas reinhardi y-1*. *J. biol. Chem.* **244**, 2621–2631.

HUBBELL, W. L. & McCONNELL, H. M. (1969). Motion of steroid spin labels in membranes. *Proc. natn. Acad. Sci. U.S.A.* **63**, 16–22.

KRAEMER, P. M. (1966). Regeneration of sialic acid on the surface of Chinese hamster cells in culture. I. General characteristics of the replacement process. *J. cell. Physiol.* **68**, 85–90.

KRAEMER, P. M. (1967). Regeneration of sialic acid on the surface of Chinese hamster cells in culture. II. Incorporation of radioactivity from glucosamine-1-^{14}C. *J. cell. Physiol.* **69**, 199–208.

LITTLEFIELD, J. W. (1964). Selection of hybrids from matings of fibroblasts *in vitro* and their presumed recombinants. *Science, N.Y.* **164**, 709–710.

MEISTER, A. (1962). Amide nitrogen transfer. In *The Enzymes* (ed. P. Boyer, H. Lardy & K. Myrback), pp. 247–266. New York: Academic Press.

MÖLLER, G. (1961). Demonstration of mouse isoantigens at the cellular level by the fluorescent antibody technique. *J. exp. Med.* **114**, 415–433.

Moscona, A. A. (1961). Rotation-mediated histogenetic aggregation of dissociated cells. *Expl Cell Res.* **22**, 455–475.

Okada, Y. (1962). Analysis of giant polynuclear cell formation caused by HVJ virus from Ehrlich's ascites tumor cells. *Expl Cell Res.* **26**, 98–128.

Okada, Y., Murayama, F. & Yamada, K. (1966). Requirement of energy for the cell fusion reaction of Ehrlich ascites tumor cells by HVJ. *Virology* **28**, 115–130.

Oppenheimer, S., Edidin, M., Orr, C. & Roseman, S. (1969). An L-glutamine requirement for intercellular adhesion. *Proc. natn. Acad. Sci. U.S.A.* **63**, 1395–1402.

Ploem, J. S. (1967). The use of a vertical illuminator with interchangeable dichroic mirrors for fluorescence microscopy with incident light. *Z. wiss. Mikrosk.* **68**, 129–142.

Ponten, J., Jensen, F. C. & Koprowski, H. (1963). Morphological and virological investigation of human tissue cultures transformed with SV_{40}. *J. cell. comp. Physiol.* **61**, 145–163.

Ray, T. K., Lieberman, I. & Lansing, A. L. (1968). Synthesis of the plasma membrane of the liver cell. *Biochem. biophys. Res. Commun.* **31**, 54–58.

Steim, J. M., Tourtellotte, R. J. C., McElhaney, R. N. & Rader, R. L. (1969). Calorimetric evidence for the liquid-crystalline state of lipids in a biomembrane. *Proc. natn. Acad. Sci. U.S.A.* **63**, 104–109.

Stoeckenius, W. & Engelman, D. M. (1969). Current models for the structure of biological membranes. *J. Cell Biol.* **42**, 613–646.

Warren, L. & Glick, M. (1968). The metabolic turnover of the surface membrane of the L cell. In *Biological Properties of the Mammalian Surface Membrane* (ed. L. Manson), pp. 3–15. Philadelphia: The Wistar Institute Press.

Watkins, J. F. & Grace, D. M. (1967). Studies on the surface antigens of interspecific mammalian cell heterokaryons. *J. Cell Sci.* **2**, 193–204.

Weiss, M. C., Ephrussi, B. & Scaletta, L. J. (1968). Loss of T-antigen from somatic hybrids between mouse cells and SV_{40}-transformed human cells. *Proc. natn. Acad. Sci. U.S.A.* **59**, 1132–1135.

Wood, B. T., Thompson, S. H. & Goldstein, G. (1965). Fluorescent antibody staining, III. Preparation of fluorescein-isothiocyanate-labeled antibodies. *J. Immun.* **95**, 225–229.

(Received 3 February 1970)

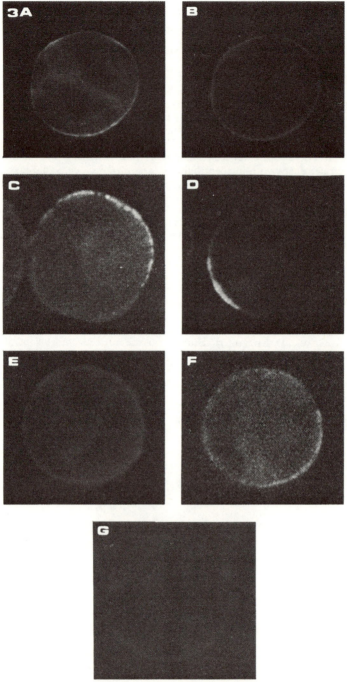

Fig. 3. All cells are doubly stained with fluorescent antibodies visualizing both mouse H-2 antigens (green fluorescence) and human surface antigens (red fluorescence). Photographs were taken through filters allowing only red or green light to reach the camera. C was doubly exposed to record both colours in a single frame. For further details see the text. × 3000 approx. A. Stable somatic hybrid line MH-1. Mouse antigens are shown. B. The same. Human antigens are shown. C. $M\frac{1}{2}$–$H\frac{1}{2}$. Double exposure, for both mouse and human antigens. D. $M\frac{1}{2}$–H1. Mouse antigens are shown. E. $M\frac{1}{2}$–H1. Human antigens are shown. F. M1–H1. Mouse antigens are shown. G. M1–H1. Human antigens are shown. The plate is made from colour prints, not from original slides.

This figure was in color in the original paper. In the original, one-half of the cell in frame C is green and the other half is red.

Proceedings of the National Academy of Sciences
Vol. 68, No. 2, pp. 283–286, February 1971

Temperature-Dependent Surface Changes in Cells Infected or Transformed by a Thermosensitive Mutant of Polyoma Virus

WALTER ECKHART, RENATO DULBECCO, AND MAX M. BURGER*

The Armand Hammer Center for Cancer Biology, The Salk Institute for Biological Studies, San Diego, California 92112; and *Department of Biochemical Sciences, Princeton University, Princeton, New Jersey 08540

Communicated November 9, 1970

ABSTRACT   Infection of BALB/3T3 cells by polyoma virus causes an alteration in the cell surface, characterized by enhanced agglutination of the cells by wheat germ agglutinin or concanavalin A. Infection by the thermosensitive mutant of polyoma, ts-3, causes the cell surface alteration at the permissive temperature, but not at the nonpermissive temperature. The cell surface alteration requires cellular DNA synthesis, but not viral DNA synthesis. BHK cells transformed by ts-3 show the surface alteration when grown at the permissive temperature, but not when grown at the nonpermissive temperature. It is concluded that the surface alteration in transformed cells is under the control of a viral gene.

Cell surface alterations that accompany transformation can be detected by enhanced agglutination of the transformed cells by wheat germ agglutinin (agglutinin) or concanavalin A (Con A) (1–4). Agglutination of normal cells, or of variants of SV40-transformed cells that do not grow to high cell densities in culture, occurs only at much higher concentrations of agglutinin than the concentrations required for agglutination of transformed cells (5, 6). A site in the cell surface containing carbohydrate is responsible for the binding of wheat germ agglutinin (1). This site is exposed in transformed cells, and can be exposed in normal cells by mild protease treatment (7). Release of 3T3 cells from inhibition of growth in crowded cultures occurs concomitantly with exposure of the agglutinin site by protease (8).

Infection of 3T3 cells by polyoma virus (Py) causes a surface alteration that is similar to the alteration in transformed cells resulting in enhanced agglutination of the infected cells by agglutinin (9). The surface alteration does not take place in the presence of inhibitors of DNA synthesis (9). A host-range mutant of Py that is unable to cause cell transformation does not cause the surface alteration after infection of nonpermissive cells (9).

Conditional lethal mutants of Py have been isolated to determine what properties of transformed cells may be controlled by viral genes (10). A thermosensitive mutant of Py, ts-3, has been shown to be defective in the induction of cellular DNA synthesis and movement in BALB/3T3 cells infected at the nonpermissive temperature, and to render one property of transformed cells, inhibition of cellular DNA synthesis by topographical factors (topoinhibition) tempera-

ture-dependent, implying that topoinhibition of transformed cells can be controlled by a viral gene (11).

In order to investigate the relationship between surface alterations and growth properties of transformed cells, we have studied the agglutination of cells infected or transformed by ts-3 by agglutinin. Enhanced agglutination of BALB/3T3 cells does not occur after infection by ts-3 at the nonpermissive temperature. BHK cells transformed by ts-3, which show temperature-dependent topoinhibition, also show temperature-dependent surface alterations; enhanced agglutination by agglutinin or Con A occurs with cells grown at permissive temperature, but not with cells grown at nonpermissive temperature. Therefore, surface alterations of transformed cells, as well as the growth property, topoinhibition, can be under the control of a viral gene.

MATERIALS AND METHODS

Cell cultures were maintained in reinforced Eagle's medium supplemented with 10% calf serum. BALB/3T3 cells were kindly provided by Dr. Stuart Aaronson. BHK cells were an early passage provided by Dr. Michael Stoker in 1962, and were kept frozen until shortly before they were used for transformation.

Virus was the wild type polyoma large plaque (LP) and ts-3, a thermosensitive mutant derived from LP by Dr. Marguerite Vogt. The properties of ts-3 will be described in detail elsewhere.

Infection was carried out using subconfluent BALB/3T3 cultures in 9-cm plastic tissue culture dishes (NUNC, Denmark). The cultures were washed once with Tris-buffered saline and infected with 0.5 ml of a virus suspension of 0.5–2 × 10^8 plaque-forming units/ml. After adsorption, the layers were covered with medium supplemented with 10% calf serum. Mock infection was carried out with buffer containing serum. In some experiments, hydroxyurea (HU) was added to the medium at a concentration of 3 × 10^{-3} M. This concentration of HU reduced incorporation of thymidine to less than 1% of that in untreated cultures. The effect of HU on thymidine incorporation was rapidly reversed upon removal of HU after treatments as long as 40 hr. The HU was recrystallized prior to use.

Transformation of BHK cells by LP or by ts-3 was carried out by methods described previously (11). Two clones of BHK cells transformed independently by ts-3 were used. Ts-3 Cl 7C had been cultured at 32°C for 3 months before being used in

Abbreviations: Con A, concanavalin A; HU, hydroxyurea; WSR, wound serum requirement; Py, polyoma virus; LP, large plaque (strain of Py); agglutinin, wheat germ agglutinin.

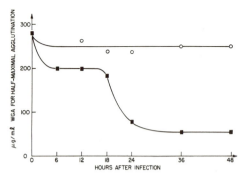

Fig. 1. Time course of appearance of surface alteration. O- - -O, mock-infected; ■- - -■ infected with wild type polyoma virus. WGA, wheat germ agglutinin.

these experiments. Ts-3 Cl 1 and WT Cl 1-A (BHK transformed by LP) were frozen shortly after being isolated from agar, and were grown at 39°C for 10 days prior to the experiments.

Wheat germ agglutination assay

Infected and transformed cells were assayed for their ability to be agglutinated by agglutinin. Subconfluent cultures were used in all experiments. Cultures were rinsed several times with saline at 37°C to remove serum. The cells were then rinsed once or twice quickly with Tris-buffered saline, pH 7.4, containing 5×10^{-4} M EDTA, which was prewarmed to 45°C so that its temperature did not fall below 35°C when layered over the cultures. When the cells loosened from the dish, they were suspended in the EDTA–saline solution and dispersed by gentle pipetting. The cells were washed twice with saline by very gentle centrifugation at about 500 rpm in the swinging bucket rotor of an International Model HN table centrifuge. Care was taken to avoid damage to the cells that might result in the release of proteolytic enzymes. The agglutination tests were performed by placing 0.1 ml of a cell suspension, at 1–2 $\times 10^6$ cells/ml, in the well of a hemagglutination tray. 0.01 ml of a pure agglutinin solution in saline was placed at the edge of the well, and cells and agglutinin were mixed in the

well. After one minute, a small drop of the suspension was placed in the well of a concavity slide, and the slide was inverted on a rocking platform with a temperature control block at 23°C to form a hanging-drop suspension. After 5 min, the suspension was examined on the concavity slide with an inverted microscope, and the proportions of single cells and clumped cells were counted in several fields. Cells not treated with agglutinin were tested at the same time. Scores of 0, +, ++, +++, and ++++ correspond to 0, 50, 75, 90, and over 97% of cells present in clumps. The concentration of agglutinin required for half-maximal agglutination (++) was determined by interpolation of scores obtained using 3–4 different concentrations of agglutinin.

RESULTS

Agglutination of cells after infection

BALB/3T3 cells were infected with wild type Py as described in *Methods*, and assayed at various times after infection for agglutination by agglutinin. The results of this experiment are shown in Fig. 1. Enhanced agglutinability of the infected cells begins to appear about 18 hr after infection, and reaches a maximum 36 hr after infection.

BALB/3T3 cells were infected with ts-3 and incubated at 39 and 32°C. Agglutination of the cells was tested at various times after infection. The results of this experiment are shown in Fig. 2. Enhanced agglutination does not occur with cells infected by ts-3 at 39°C, but does occur with cells infected at 32°C. Enhanced agglutinability appears more slowly in cells infected by wild type virus at 32°C than at 39°C, presumably because the infection proceeds more slowly at 32°C.

The lack of enhanced agglutination with cells infected by ts-3 at 39°C is not the result of failure of the virus to penetrate BALB/3T3 cells, because ts-3 behaves as thermosensitive after infection with viral DNA (W. Eckhart, unpublished observations). Lack of enhanced agglutination is not simply the result of lack of cytopathic effect at 39°C because other mutants which fail to grow at 39°C, and which do not cause a cytopathic effect at 39°C, nevertheless produced enhanced agglutinability (R. Dulbecco, manuscript in preparation).

TABLE 1. *Agglutination of transformed cells*

| Cells | Temperature, °C | Concentration of agglutinin required for half-maximal agglutination μg/ml | |
		Wheat germ	Concanavalin A
WT Cl 1-A	32	45	100
ts-3 Cl-1	32	53	
ts-3 Cl-7C	32	40	80
WT Cl 1-A	39	37	
ts-3 Cl-1	39	160	
	39 + trypsin	60	
ts-3 Cl-7C	39	180	240
	39 + trypsin	36	

Cells were grown at the temperatures indicated for several days and then tested for agglutination by wheat germ agglutinin or concanavalin A. Trypsin treatment was with 0.0003% trypsin, 5 min incubation at 37°C.

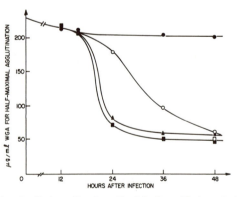

Fig. 2. Surface alteration after infection with ts mutants. ●- - -●, ts-3 at 39°C; ■- - -■, wild type at 39°C; O- - -O, ts-3 at 32°C; □- - -□, wild type at 32°C; ▲- - -▲, ts 616 at 39°C.

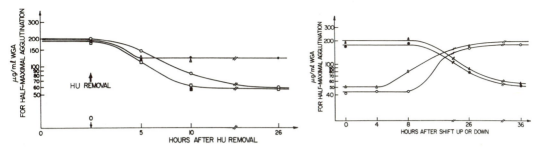

(*Left*) FIG. 3.   Requirement for DNA synthesis for surface alterations. Cells were infected at 39 and 32°C in the presence of 3×10^{-3} M hydroxyurea.   Samples were tested for agglutination before and at various times after removal of HU 24 or 36 hr after infection.   Same symbols used as in Fig. 2 except for △- - -△, mock-infected and treated with HU at 39°C for 24 hr.

(*Right*) FIG. 4.   Appearance and disappearance of surface alteration after shift in temperature of ts-3-transformed BHK cells.

▲- -▲, clone 7-C ⎫
●- - -●, clone 1 ⎬ Shifted from 39 to 32°C
△- -△, clone 1 ⎫
O- - -O, clone 7-C ⎭ Shifted from 32 to 39°C

Requirement for DNA synthesis for surface alteration after infection

BALB/3T3 cells were infected with wild type Py and incubated in the presence of HU after infection as described in *Methods.* HU was removed from the cultures at 24 hr after infection at 39°C, and at 36 hr after infection at 32°C. Agglutination of the cells by agglutinin was tested at various times after the removal of HU. The results of this experiment are shown in Fig. 3. Enhanced agglutination does not occur with cells incubated in the presence of HU. After removal of HU, enhanced agglutinability appears rapidly, and is complete by 10 hr after removal of HU. HU acts to inhibit DNA synthesis by blocking the action of ribonucleoside diphosphate reductase, preventing the formation of deoxynucleotide precursors of DNA (12). Therefore, the lack of enhanced agglutination of cells after infection in the presence of HU argues strongly that DNA synthesis is required in order for the surface alteration to take place.

In order to test whether viral DNA synthesis is required for the surface alteration to occur, we infected BALB/3T3 cells with another thermosensitive mutant, ts 616, which is defective in viral DNA synthesis at 39°C. The cell surface alteration does occur after infection by ts 616 at 39°C (Fig. 2), implying that the requirement for DNA synthesis in order for the surface alteration to occur is a requirement for cellular DNA synthesis, rather than for viral DNA synthesis.

Agglutination of transformed cells

Two clones of ts 3-transformed BHK cells, and one clone of wild type transformed BHK cells were tested for agglutination by agglutinin and Con A after being grown at 39 and 32°C. The results are shown in Table 1. Wild type transformed cells show enhanced agglutinability when grown at 39°C or at 32°C. Ts-3-transformed cells show enhanced agglutinability when grown at 32°C, but not when grown at 39°C. The surface properties of ts-3-transformed cells grown at 39°C are similar to those of normal cells, in that enhanced agglutinability can be produced by mild protease treatment (Table 1).

Appearance and disappearance of surface alterations in ts-3-transformed cells

In order to determine the length of time necessary for the surface alteration to appear and disappear in ts-3-transformed cells grown at different temperatures, we shifted ts-3-BHK cells grown at 39 to 32°C, and cells grown at 32 to 39°C. Agglutination was tested at various times after the shift in temperature. The results of this experiment are shown in Fig. 4. Masking of the surface alteration begins to be detectable between 4 and 8 hr after a shift of the cells from low to high temperature, and is essentially complete by 24 hr after the shift. Appearance of the surface alteration after shift from high to low temperature is not detectable by 12 hr after the shift, but is substantial by 24 hr, and is essentially complete by 36 hr after the shift.

DISCUSSION

It has been reported that infection of 3T3 cells by Py leads to enhanced agglutinability of the infected cells by agglutinin and that this surface alteration does not occur in the presence of inhibitors of DNA synthesis (9). We have repeated these observations, using BALB/3T3 cells.

The ts-3-mutant is defective in causing a surface alteration after infection at the nonpermissive temperature. In ts-3-transformed BHK cells, expression of the surface alteration is temperature dependent. Therefore, surface changes in infected and transformed cells can be under the control of a viral gene.

In a previous communication, two of us reported that the ts-3 mutant is partially defective in induction of cellular DNA synthesis and movement after infection of BALB/3T3 cells at the nonpermissive temperature and that BHK cells transformed by ts-3 show temperature-dependent expression of one attribute of transformed cells, inhibition of DNA synthesis by topographical factors (topoinhibition) (11). Other attributes of transformed cells, growth in agar and serum requirement for initiation of DNA synthesis in a wound (WSR), are not temperature dependent in ts-3-BHK.

The attributes of infected or transformed cells that can be controlled by the ts-3 gene may have a common origin in the surface changes in infected or transformed cells. Alteration of the cell surface by the ts-3 gene product could result in release of restraints on cellular DNA synthesis and growth such as has been observed after protease treatment of 3T3 cells and chick cells (8, 13). Topoinhibition could be mediated by surface alterations that are temperature dependent in ts-3-transformed cells.

The relation between surface changes in transformed cells, and the properties of growth in agar and WSR, which are not temperature-dependent in ts-3-BHK, is not clear. If the ts-3 gene product has a pleiotropic effect, and if different properties of transformed cells require different amounts of the gene product for expression, it is possible that the reduced amount of functional gene product present in ts-3-transformed cells at the nonpermissive temperature is sufficient to allow expression of growth in agar and WSR, but not sufficient to allow expression of enhanced agglutinability or loss of topoinhibition. Alternatively, growth in agar and decreased WSR could be unrelated to the surface changes detected by agglutination.

Characterization of ts-3 and other viral mutants that affect the properties of transformed cells should clarify some of the relationships between cell surface alterations and cell growth properties.

This work was supported by Public Health Service grants 10151 and 1-K4-Ca-16,765, Public Health Service grant CA-07592, Grant P-450 from the American Cancer Society, and National Cancer Institute Contract 67-1147. Max M. Burger thanks the Anita Mestres Fund for a travel grant to the Salk Institute.

1. Burger, M. M., and A. R. Goldberg, *Proc. Nat. Acad. Sci. USA*, **57**, 359 (1967).

2. Aub, J. C., C. Tieslau, and A. Lankester, *Proc. Nat. Acad. Sci. USA*, **50**, 613 (1963).

3. Burger, M. M., *Nature*, **219**, 499 (1968).

4. Inbar, M., and L. Sachs, *Proc. Nat. Acad. Sci. USA*, **63**, 1418 (1969).

5. Pollack, R. E., and M. M. Burger, *Proc. Nat. Acad. Sci. USA*, **62**, 1074 (1969).

6. Inbar, M., Z. Rabinowitz, and L. Sachs, *Int. J. Cancer*, **4**, 690 (1969).

7. Burger, M. M., *Proc. Nat. Acad. Sci., USA*, **62**, 994 (1969).

8. Burger, M. M., *Nature*, **227**, 170 (1970).

9. Benjamin, T. L., and M. M. Burger, *Proc. Nat. Acad. Sci. USA*, **67**, 929 (1970).

10. Eckhart, W., *Virology*, **38**, 120 (1969); diMayorca, G., J. Callander, G. Marin, and R. Giordano, *Virology*, **38**, 126 (1969); Benjamin, T. L., *Proc. Nat. Acad. Sci. USA*, **67**, 394 (1970).

11. Dulbecco, R., and W. Eckhart, *Proc. Nat. Acad. Sci. USA*, **67**, 1775 (1970).

12. Krakoff, I. H., N. C. Brown, and P. Reichard, *Cancer Res.*, **28**, 1559 (1968).

13. Sefton, B. M., and H. Rubin, *Nature*, **227**, 843 (1970).

Reprinted from

Proc. Nat. Acad. Sci. USA
Vol. 70, No. 11, pp. 3170–3174, November 1973

Alteration of Cell-Surface Proteins by Viral Transformation and by Proteolysis

(fibroblasts/lactoperoxidase/iodination/polyacrylamide gels)

RICHARD O. HYNES

Imperial Cancer Research Fund, Lincoln's Inn Fields, London, WC2A 3PX, England

Communicated by Renato Dulbecco, July 9, 1973

ABSTRACT Putative cell-surface proteins of tissue-culture cells were identified by lactoperoxidase-catalyzed iodination, a technique that attaches label only to proteins outside the cell membrane. Evidence is presented that these proteins are cell derived, not contaminating serum proteins. On "normal" cells, which exhibit density dependence of growth, one protein of high molecular weight was particularly readily iodinated. This protein was easily removed by mild proteolytic digestion, and, in virus-transformed cells, was either absent or unavailable for iodination. The possible relevance of these observations to the control of growth in cell culture is discussed.

The idea that the surfaces of virus-transformed cells differ from those of their normal counterparts has received support from various lines of research. There is immunological evidence for antigenic differences between normal and transformed cells (1, 2) and biochemical evidence for changes in surface components on transformation (3, 4). The different cell agglutinability by plant lectins, such as concanavalin A and wheat-germ agglutinin (5, 6), also suggests that there are differences in surface architecture between normal and transformed cells.

The fact that growth properties of normal cells can be influenced by mild proteolytic treatments (7, 8) implicates the surface directly in growth control, and the observation that transformed cells in culture produce proteolytic enzymes (9–12) suggests a possible mechanism for alteration of their growth properties. It would be desirable to assay directly for surface changes, in order to investigate further the roles of transformation and proteolytic enzymes and their possible involvement in growth control.

Techniques for identifying and characterizing proteins and glycoproteins on the outsides of cells have recently been developed. I wish to report some studies on normal and transformed cultured cells using lactoperoxidase-catalyzed iodination. When applied to erythrocytes or platelets, this procedure labels only external membrane proteins (13, 14). In the present work, a modification of this technique was used, in which hydrogen peroxide was generated by glucose oxidase plus glucose. The method is similar to that reported recently by Hubbard and Cohn (15).

Using this method I have been able to identify several exterior proteins on cultured animals cells. One of these is particularly heavily iodinated on normal cells and is iodinated weakly or not at all in their virus-transformed derivatives. This protein is also very sensitive to proteolytic digestion.

MATERIALS AND METHODS

Cells used were as follows. Two clones of the hamster fibroblast cell line NIL.2E (16, 17), NIL.1 and NIL.8, which show

"normal" behavior in culture, and various clonal derivatives transformed by hamster sarcoma virus (HSV) or polyoma virus (Py) were obtained from Dr. I. A. Macpherson. Eight subclones of NIL.8 were freshly isolated by plating cells at 0.5 cells per well in plastic multiple well trays (Linbro) and growing up those which initially appeared to be single colonies. All eight showed well-orientated growth patterns in culture and a low saturation density similar to that of the parent line. LX cells were kindly provided by Dr. M. Shodell (18).

NIL cells and their transformants were cultured in Dulbecco's modification of Eagle's medium with 10% calf serum. LX cells were grown in Waymouth's medium. Cells were checked periodically for contamination by *Mycoplasma* both by staining and by culture. None of the cells used ever gave a positive result.

Iodination was usually done in monolayer cultures, but in a few experiments cells were labeled in suspension after removal from dishes by trypsin–EDTA or EDTA alone. Cells were washed three times with phosphate-buffered saline (pH 7.2) to remove serum. Phosphate-buffered saline + 5 mM glucose was then added, followed by carrier-free $Na^{125}I$, usually to a final concentration of 400 $\mu Ci/ml$. The reaction was initiated by addition of lactoperoxidase (Calbiochem; EC 1.11.1.7) and glucose oxidase (Worthington; EC 1.1.3.4) to final concentrations of 20 $\mu g/ml$ and 0.1 units/ml, respectively. Reaction was allowed to continue for 10 min at room temperature with occasional swirling, and labeling was stopped by addition of phosphate-buffered iodide (phosphate-buffered saline with the NaCl replaced by NaI). This solution usually contained 2 mM phenyl methyl sulfonylfluoride to inhibit proteases. The medium was removed and the cells were washed twice more with phosphate-buffered iodide + phenyl methyl sulfonylfluoride. The cells were then scraped into phosphate-buffered iodide and phenyl methyl sulfonylfluoride, centrifuged, and dissolved in buffer containing 2% Na dodecyl sulfate and phenyl methyl sulfonylfluoride for Na dodecyl sulfate–polyacrylamide electrophoresis.

Na Dodecyl Sulfate–Polyacrylamide Gel Electrophoresis was done by the methods of Laemmli (19) as described by Studier (20). Electrophoresis was usually in slabs where 12 samples could be run in parallel on the same gel. The slabs, after they were stained with Coomassie blue if required, were dried down onto paper, and autoradiographs made on Kodirex x-ray film. Before electrophoresis, samples were reduced by addition of dithiothreitol to 0.1 M and boiling for at least 2 min. Electrophoresis was done at 50 V for about 1 hr, until the bromphenol blue marker entered the separation gel, and then at 100 V (about 10 V/cm) until the marker reached the bottom.

Enzymes. Lactoperoxidase was from Calbiochem; glucose oxidase was from Worthington Biochemicals; trypsin, twice crystallized, and soyabean trypsin inhibitor were from Sigma, London.

Radiochemicals. [^{125}I]Sodium iodide, carrier-free, and [^{14}C]-leucine 342 Ci/mol, were obtained from the Radiochemical Centre, Amersham, Bucks.

RESULTS

Monolayer cultures of the normal hamster fibroblast cell line NIL.8 were iodinated, washed, harvested by scraping, concentrated, and run on Na dodecyl sulfate–polyacrylamide gels. Fig. 1 shows an autoradiograph of a 7.5% slab gel on which several samples were run in parallel. The iodination is selective (gels *b* and *i*): the major cell proteins are not labeled and only a limited number of proteins is iodinated. Fig. 1*d* shows the result obtained when the enzymes were omitted from the iodination mixture. Omission of either of the enzymes or of glucose eliminates labeling (Table 1). Thus, iodination depends on the presence of both enzymes. Since these are macromolecules, which presumably do not enter the cell, this is suggestive evidence that the labeled proteins are external.

If the cells were removed from the monolayer *before* iodination, lysed with distilled water, and then labeled, the result shown in Fig. 1*g* was obtained. In this case, all the major cell proteins were iodinated (compare Fig. 1*g* with 1*b* and 1*i*). Thus, all or most cell proteins can be iodinated but, in the intact cell, only a few are available to the iodide–lactoperoxidase complex.

Further independent evidence for the external location of most of the iodinated proteins was obtained by tryptic digestion. Fig. 1*e* shows the result of very mild treatment with trypsin after iodination. This trypsin treatment did not cause rounding up or detachment of any cells from the dish. However, all but one of the major iodinated proteins were removed. Removal of labeled proteins was blocked by soyabean trypsin inhibitor (Fig. 1*f*) and was, therefore, due to proteolytic digestion. Identical results were obtained when trypsinization was done *before* iodination (see Fig. 4) or if the trypsin treatment was more extensive and the cells were completely detached from the dish (data not shown). One of the iodinated proteins was therefore resistant to trypsin whereas the others were removed without leaving iodine-labeled cores attached to the cells.

Table 1.　*Dependence of iodination on enzymes*

Complete mixture	Trichloroacetic acid precipitable cpm
Complete mixture	24,283
Complete mixture	22,408
Omit glucose	102
Omit glucose oxidase	111
Omit lactoperoxidase	418
Omit both enzymes	105

Cells were iodinated in suspension. The reaction was stopped by addition of Na dodecyl sulfate to 1%. Macromolecular material was assayed by trichloracetic acid precipitation

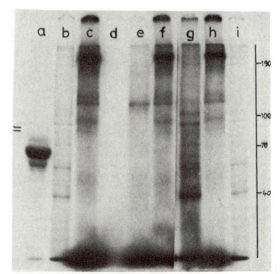

Fig. 1.　Iodination of NIL.8 cells. Autoradiogram of 7.5% gel. (*a*) Iodinated calf serum; (*b* and *i*) cells labeled with [^{14}C]-leucine for 72 hr; (*c* and *h*) cells iodinated in monolayer; (*d*) cells iodinated as for *c* except that the enzyme solution was omitted: (*e*) cells iodinated and then treated for 10 min with 10 μg/ml of trypsin in phosphate-buffered saline; (*f*) cells iodinated and then treated with 10 μg/ml of trypsin + 10 μg/ml of soyabean trypsin inhibitor; (*g*) cells lysed with distilled water and then iodinated. All iodinated cell samples contained equal quantities of cell protein. Molecular weight × 10^{-3} is marked on *right* and positions of lactoperoxidase (*top line*) and glucose oxidase (*bottom line*) on *left*.

Comparison of the pattern of iodination of NIL.8 cells (Fig. 1*c* and *h*) with that of iodinated serum (Fig. 1*a*) shows little or no correspondence between them, suggesting that the labeled proteins are not bound serum proteins. Further, when a dish that had contained only complete medium but no cells, was iodinated, washed, and then rinsed with 2% Na dodecyl sulfate to collect labeled proteins, only major serum proteins could be detected. This finding argues against selective binding of minor serum proteins to the dish. The prominent iodinated bands observed on iodination of cells were not seen (data not shown). Experiments in which prelabeled serum was incubated with cells again failed to detect selective binding of minor serum components; again only the major serum bands could be detected (data not shown). Finally, evidence that the iodinated bands are not due to serum contamination was provided by experiments with a mouse cell line, LX, (18) which grows in the absence of serum. Replicate cultures were grown for 3 days in 0, 1, or 2% fetal-calf serum, iodinated, and run in parallel on Na dodecyl sulfate–polyacrylamide gels (Fig. 2). The three profiles of radioactivity were almost identical; only one labeled band appeared in cells grown in serum and not in the controls (*arrow*, Fig. 2). The iodination pattern for the LX cells grown without serum, while broadly similar to that observed for NIL.8 cells grown in 10% serum, showed reproducible differences in its details. Both cells have the heavily labeled high-molecular-weight band, but these differ slightly in mobility on 5% gels, the one from LX cells being somewhat slower.

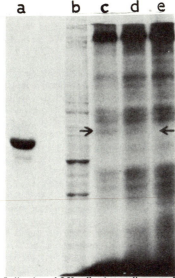

FIG. 2. Iodination of LX cells. Autoradiogram of 7.5% gel. (a) Iodinated fetal-calf serum; (b) cells labeled with [^{14}C]leucine; (c) cells grown in 2% fetal-calf serum and then iodinated; (d) cells grown in 1% fetal-calf serum and then iodinated; (e) cells grown without serum and then iodinated. Cell samples contained equal quantities of cell protein. Arrows mark serum band adhering to cells.

The migration positions of the two enzymes used for labeling are marked on Fig. 1. It was shown in separate experiments that prelabeled enzymes can bind to the cells, although

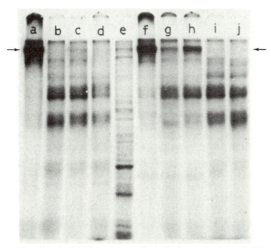

FIG. 3. Iodination of normal and virus-transformed NIL cells. Autoradiogram of 7.5% gel. (a) NIL.8; (b) NIL.8.HSV6; (c) NIL.8.HSV9; (d) NIL.8.HSV11; (e) NIL.8 labeled with [^{14}C]leucine; (f) NIL.1; (g) NIL.1.Py1; (h) NIL.1.Py8; (i) NIL.1.HSV1; (j) NIL.1.HSV3. Iodinated samples contained equal amounts of radioactivity. Arrow, location of band 1.

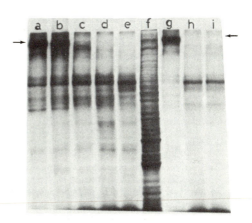

FIG. 4. Effects of tryptic digestion on the pattern of iodination of NIL.8 cells. Autoradiogram of 7.5% gel. (a–e) NIL.8 cells iodinated and subsequently treated before harvest with: (a) 10 μg/ml of trypsin plus 10 μg/ml of soyabean trypsin inhibitor, 10 min; (b) 1 μg/ml of trypsin for 1 min; (c) 1 μg/ml of trypsin for 5 min; (d) 1 μg/ml of trypsin for 10 min; (e) 10 μg/ml of trypsin for 10 min. All were at room temperature in phosphate-buffered saline; (f) [^{14}C]Leucine-labeled NIL.8 cells; (g) cells iodinated and harvested without further treatment; (h and i) cells treated with trypsin (10 μg/ml for 10 min) before (h) or after (i) iodination. Iodinated samples contained equal amounts of radioactivity. Arrow, location of band 1.

generally at low levels relative to the other iodinated bands observed on labeling cells. However, this fact may complicate interpretation of this region of the gels (molecular weight 75–85 × 10³).

Another clone of NIL cells, NIL.1, gave results very similar to those obtained with NIL.8. Eight subclones of NIL.8 all showed similar iodination profiles. Thus, there was no evidence for clonal variations. In addition to LX cells, several other fibroblastic cell lines were examined (BHK, 3T3, 3T6) and all showed similar patterns of iodination with one prominent labeled band and several minor ones. There were reproducible differences in detail between the various cell lines (unpublished data).

When virus-transformed cells were examined, a different result was obtained (Fig. 3). Cloned lines of NIL cells trans-

TABLE 2. *Iodination of normal and virus-transformed cells*

Cell line	Specific activity (acid-precipitable ^{125}I cpm per μg of protein)
NIL.8	2210
NIL.8.HSV6	632
NIL.8.HSV9	786
NIL.8.HSV11	729
NIL.1	1350
NIL.1.Py1	430
NIL.1.Py8	599
NIL.1.HSV1	738
NIL.1.HSV3	619

Cells were iodinated at confluence in 5-cm petri dishes, and processed as described in *Methods*.

formed by hamster sarcoma virus (HSV), initially isolated by their ability to grow in suspension in agar, showed a typical transformed growth pattern on dishes, i.e., they reached higher densities than their normal parents and showed irregular cell orientation. In the iodination pattern of all these transformed cells, the iodinated band with the highest molecular weight (band 1) was absent or much reduced (Fig. 3). A similar result was observed with polyoma virus-transformed NIL.1 cells (Fig. 3), although the disappearance of band 1 was not always complete. In the experiment of Fig. 3, equal amounts of radioactivity were applied to the gels. Therefore, more material from transformed cells was used, since they were less heavily labeled (Table 2), presumably because of absence of label in band 1. If the labeling of band 1 of the polyoma-transformed cells is considered relative to that of the other bands on the same track, it is clearly reduced. When equal amounts of cell protein were applied to gels, the iodinated bands of lower molecular weights seen on normal cells were unaffected by transformation. A group of fainter bands of variable intensity ahead of band 1, which appeared to be increased in certain transformed lines, could conceivably be breakdown products of band 1.

Similar results were obtained in a comparison of chicken-embryo fibroblasts and chicken-embryo fibroblasts transformed by Rous sarcoma virus Prague strain A (unpublished data). As with polyoma-transformed NIL cells, Rous sarcoma virus-transformed chicken-embryo fibroblasts generally showed a trace amount of band 1.

In view of the increased proteolytic activity observed in transformed cells (9–12), and the evidence that mild proteolytic digestion alters the surfaces of normal cells (21–25) and stimulates growth (7, 8), I tested the sensitivity of the iodinated proteins to very mild trypsin treatments (Fig. 4). 1 μg/ml of crystalline trypsin degrades band-1 protein partially within 1 min, leading to a band running ahead of band 1 (not clearly seen in Fig. 4). Band 1 was largely removed by 5 min and removed completely by 10 min of digestion. The other iodinated bands were less sensitive than band 1 but were progressively removed by increasing trypsin treatments, with the exception of one major trypsin-resistant band and several minor ones. After very light digestion, new bands running ahead of band 1 could be seen (e.g., Fig. 4) similar to those observed in transformed cells (Fig. 3).

The experiments described so far were done with approximately confluent cell cultures. The possibility arises that the differences observed between normal and transformed cells were due to the fact that the latter were dividing more rapidly. However, in other experiments with normal cells, band 1 was detected whether the cells were in exponential growth or had stopped dividing after forming a confluent sheet (unpublished data).

DISCUSSION

A major exterior cell protein (band 1) of normal fibroblasts, which is detected by lactoperoxidase-catalyzed iodination, is detected in small quantities or not at all on virus-transformed derivatives. This is true for hamster cells transformed by hamster sarcoma or polyoma virus and for chicken-embryo cells transformed by Rous sarcoma virus and is not a clonal variation unrelated to transformation.

Several possible explanations for this observation exist: either transformed cells synthesize band-1 polypeptide in reduced quantities or not at all, or else they synthesize it at normal rates but it is not available for iodination, either because it is masked or because it is continually removed from the cell surface. It is not possible to decide which of these explanations applies.

The observations that transformed cells produce proteases (9–12) and that mild trypsin treatment removes band 1 from normal cells (Fig. 4) are consistent with the idea that failure to iodinate band-1 polypeptide on transformed cells is due to its removal by proteolytic digestion, but do not prove it. The level of proteolytic digestion that brings about removal of band 1 is of the same order as that which causes normal cells to react with lectins like transformed cells (21, 23, 25). Mild treatments with proteases also stimulate quiescent normal cells into growth (7, 8) and reduce their cyclic AMP levels to those found in transformed cells (26, 27). These similarities may well be coincidental, but they encourage the speculation that transformation leads to production of proteolytic enzymes which, in turn, lead to the changes mentioned and allow transformed cells to escape normal growth controls. If this sequence of events were a necessary part of transformation, treatment of transformed cells with inhibitors of proteolytic enzymes should tend to render their surface characteristics and growth patterns normal. This has been shown to be so for density-dependent growth inhibition (28), morphology (12), and ability to grow in agar (12).

It should be noted that although the evidence presented demonstrates that the iodinatable proteins are located outside the cell membrane and are cell-derived rather than serum contaminants, it has not been shown whether they are part of the cell membrane or laid down as extracellular materials. This distinction may be a semantic one. Several of the iodinatable proteins, including band 1, appear to be glycoproteins, and evidence suggests that they are not collagen nor do they contain sulfated mucopolysaccharides (unpublished data).

After completion of the work described, I learned that similar results were obtained independently using chicken cells transformed by RSV (29) and 3T3 and its viral transformants (N. M. Hogg, personal communication).

I thank Jacqueline Bye for her very competent and conscientious technical assistance and Drs. I. A. Macpherson and R. Dulbecco for their helpful comments on the manuscript.

1. Klein, G. (1968) *Cancer Res.* **28**, 625–635.
2. Collins, J. J. & Black, P. H. (1973) *Curr. Top. Microbiol. Immunol.* **63**, in press.
3. Warren, L., Fuhrer, J. P. & Buck, C. A. (1972) *Proc. Nat. Acad. Sci. USA* **69**, 1838–1842.
4. Critchley, D. R. (1973) "Glycolipids and cancer," in *Membrane Mediated Information*, ed. Kent, P. W. (Medical Technical Publications Ltd.), in press.
5. Burger, M. M. & Goldberg, A. R. (1967) *Proc. Nat. Acad. Sci. USA* **57**, 359–366.
6. Inbar, M. & Sachs, L. (1969) *Proc. Nat. Acad. Sci. USA* **63**, 1418–1425.
7. Burger, M. M. (1970) *Nature* **227**, 170–171.
8. Sefton, B. M. & Rubin, H. (1970) *Nature* **227**, 843–845.
9. Bosmann, H. B. (1972) *Biochim. Biophys. Acta* **264**, 339–343.
10. Schnebli, H. P. (1972) *Schweiz. Med. Wochenschr.* **102**, 1194–1196.
11. Unkeless, J. C., Tobia, A., Ossowski, L., Quigley, J. P., Rifkin, E. (1973) *J. Exp. Med.* **137**, 85–111.
12. Ossowski, L., Unkeless, J. C., Tobia, A., Quigley, J. P., Rifkin, D. B. & Reich, E. (1973) *J. Exp. Med.* **137**, 112–126.
13. Phillips, D. R. & Morrison, M. (1971) *Biochemistry* **10**, 1766–1771.

14. Phillips, D. R. (1972) *Biochemistry* **11**, 4582–4588.
15. Hubbard, A. L. & Cohn, Z. A. (1972) *J. Cell Biol.* **55**, 390–405.
16. Diamond, L. (1967) *Int. J. Cancer* **2**, 143–152.
17. McAllister, R. M. & Macpherson, 1. (1968) *J. Gen. Virol.* **2**, 99–106.
18. Shodell, M. (1972) *Proc. Nat. Acad. Sci. USA* **69**, 1455–1459.
19. Laemmli, U. K. (1970) *Nature* **227**, 680–685.
20. Studier, F. W. (1972) *Science* **176**, 367–376.
21. Burger, M. M. (1969) *Proc. Nat. Acad. Sci. USA* **62**, 994–1001.
22. Inbar, M. & Sachs, L. (1969) *Nature* **223**, 710–712.
23. Ozanne, B. & Sambrook, J. (1971) *Nature New Biol.* **232**, 156–160.
24. Arndt-Jovin, D. J. & Berg, P. (1971) *J. Virol.* **8**, 716–721.
25. Nicolson, G. L. (1972) *Nature New Biol.* **239**, 193–197.
26. Sheppard, J. R. (1972) *Nature New Biol.* **236**, 14–16.
27. Burger, M. M., Bombik, B. M., Breckenridge, B. M. & Sheppard, J. R. (1972) *Nature New Biol.* **239**, 161–163.
28. Schnebli, H. P. & Burger, M. M. (1972) *Proc. Nat. Acad. Sci. USA* **69**, 3825–3827.
29. Wickus, G. G. & Robbins, P. R. (1974) *Cold Spring Harbor Symposium on Control of Proliferation in Animal Cells,* in press.

629

Reprinted from Nature, Vol. 254, No. 5497, pp. 250–252. 1975. © 1975 Macmillan Journals, Ltd.

Permeability of cell junction depends on local cytoplasmic calcium activity

MANY kinds of cells are coupled by junctions consisting of membrane channels through which molecules of a certain size range can flow freely from one cell interior to another[1].

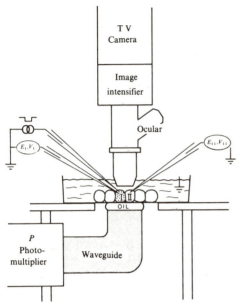

Fig. 1 Television camera coupled to image intensifier by a light guide views the aequorin luminescence of the cells through microscope (darkfield); luminescence is also measured by photomultiplier. Electrical coupling is measured with the aid of three microelectrodes, by pulsing current (*i*) between the inside of cell (I) and the outside and measuring the resulting steady-state changes (*V*) and membrane potentials (*E*) in I and adjacent cell II.

It has been proposed that the permeability of the junctional channels depends on the concentration of free ionised calcium in cytoplasm ($[Ca^{2+}]_i$) (ref. 2). This hypothesis is supported by two classes of experiments. In one, the interior of a coupled cell system is allowed to exchange freely with a known $[Ca^{2+}]$ in the exterior, through a hole in the (non-junctional) membrane; junctional conductance is reduced (uncoupling) when the $[Ca^{2+}]_i$ is above $5–8 \times 10^{-5}$ M (ref. 3). In the other class, uncoupling ensues when the (closed) cell system is treated with inhibitors of energy metabolism or with Ca^{2+} ionophores[5], or on exposure for long periods to Ca,Mg-free medium or to Li medium[6]; in these conditions a rise in $[Ca^{2+}]_i$ may be expected because of known properties of cellular Ca metabolism[7,8]. Here, we demonstrate the changes in $[Ca^{2+}]_i$ together with those in coupling in three of these conditions, using aequorin to display the distribution of Ca^{2+} in the cell. It will be shown that the uncoupling is, indeed, in each case associated with a rise in $[Ca^{2+}]_i$. Furthermore, by local injection of Ca^{2+} into the cells, it will be shown that uncoupling ensues when the rise in $[Ca^{2+}]_i$ occurs at the junction, but not when it occurs at other regions in the cell.

Chironomus salivary glands were isolated in physiological medium as described previously[4], except that the medium contained 12 mM Ca and no Mg. Purified aequorin was injected into one or two adjacent cells. This protein reacts with Ca^{2+}, emitting light. The emission, approximately proportional to $[Ca^{2+}]^2$ (ref. 9), provides a convenient $[Ca^{2+}]_i$ indicator[10-12]. The aequorin had no apparent adverse effects; membrane potentials and coupling were maintained for several hours. Light emitted inside the cells was guided by a light pipe to a photomultiplier. In addition, a television camera coupled to an image intensifier viewed the cell luminescence through a microscope (Fig. 1). This novel aspect of aequorin technique enabled us to see where the light was emitted inside the cells with a resolution of 5 μm. Ca buffered with EGTA (to stabilise the $[Ca^{2+}]$) was pressure-injected into the cell containing the current source, while electrical coupling between this cell and a contiguous cell was measured as shown in Fig. 1. Current pulses,

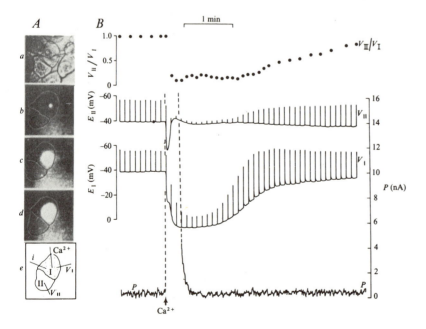

Fig. 2 Ca^{2+} injection. *A*, Dark-field television pictures (*b–d*) of aequorin luminescence in a cell, produced by three puffs of 5×10^{-5} M free Ca (buffered) of increasing magnitude, delivered to cell centre. The pictures are videotape photographs each taken at the time of maximum luminescence spread. Cell outlines are traced by superposition of brightfield video picture (*a*); cell diameter ∼ 100 μm. Puffs *b* and *c* do not reach the junctions of the cell and do not affect coupling. Puff *d* reaches one junction and causes transient uncoupling, as shown in *B*: chart records of photomultiplier current P, E_I, E_{II}, V_I, V_{II} ($i = 4 \times 10^{-8}$ A); and plot of coupling ratio V_{II}/V_I. *e*, Cell diagram showing locations of microelectrodes and of Ca injection micropipette; dotted cell preinjected with aequorin.

membrane potentials, their displacements, and photocurrent were displayed on a chart recorder and on a storage oscilloscope on to which a second television camera was focused. The two camera outputs were displayed together on a monitor and videotaped. Thus, we had continuous and simultaneous information on the electrical parameters and on the relative local Ca^{2+} activities inside the cells.

Short single injection pulses of $5–10 \times 10^{-5}$ M free Ca^{2+} into healthy cells produced aequorin glows that were confined to the immediate vicinity of the tip of the injection pipette (Fig. 2). Evidently the Ca^{2+} is prevented from diffusing freely through the cytoplasm and seems to be sequestered by intracellular elements[7,13,14]. When such an injection was made into the cell centre (cell radius, 50 μm), junctional coupling was unaffected. When it was close to a cell junction, the coupling fell (uncoupling), and this effect was confined to the junction at which the local rise in $[Ca^{2+}]$ had occurred. Longer or more frequent injections of this sort, or prolonged Ca^{2+} iontophoresis between electrodes, placed in two adjacent cells, produced glows that were diffuse over the entire cell. Then all junctions of the injected cell uncoupled. Presumably the intracellular sequestering capacity was saturated under these conditions. (When mitochondrial sequestering of Ca was blocked by ruthenium red[15], the glow was diffuse even with the shorter injection pulses.) Control injections of 0.15 M KCl did not affect coupling of $[Ca^{2+}]_i$. Massive KCl injections, however, produced transient and diffuse rise of $[Ca^{2+}]_i$ associated with transient uncoupling.

Treatment with ionophore A23187 (Lilly, 2×10^{-6} M) or X537A (Hoffman LaRoche, 1×10^{-5} M) led to enhanced Ca

influx which was detectable within 1 min as a diffuse glow that increased progressively during the next 10 min. The rise in $[Ca^{2+}]_i$ was invariably associated with uncoupling, the coupling ratio diminishing progressively with rising $[Ca^{2+}]_i$ (Fig. 3).

Similar results and equally good correlation between rising $[Ca^{2+}]_i$ and uncoupling were obtained when the cells were poisoned with sodium cyanide (5×10^{-3} M) in medium containing Ca as well as in Ca-free medium (Fig. 4).

Prolonged exposure to Ca,Mg-free medium led to a rise in $[Ca^{+2}]_i$. This rise, evidently nurtured by intracellular Ca stores (as Na accumulates inside the cells in Ca-free medium[16], the mitochondria may release Ca[17]), was asso-

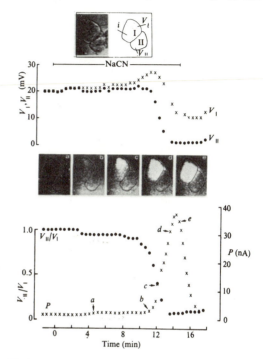

Fig. 4 Cyanide. Cells in Ca-free medium. Exposure to 5×10^{-3} M sodium cyanide (top signal). Time correspondence of television pictures a–e is marked on the P curve. $i = 4 \times 10^{-8}$ A. Cell I depolarises as P rises, reaching zero membrane potential at time 17 min.

ciated with uncoupling. These experiments were particularly instructive, because $[Ca^{2+}]_i$ and coupling fluctuated: rise and fall in $[Ca^{2+}]_i$ were associated, respectively, with fall and rise in coupling.

In cells with good membrane potentials (≥ 40 mV), the uncouplings produced by single short Ca pulses generally were spontaneously reversible (Fig. 2). Presumably the cells rid themselves of the excess Ca by sequestering it into mitochondria and by pumping it out. With more massive injections, or after ionophore treatment or prolonged exposure to Ca,Mg-free medium, the uncouplings did not reverse spontaneously. It was then possible to reverse the uncouplings by injection of Ca–EGTA solution yielding $\leq 10^{-8}$ M free Ca.

In all these conditions, the rise in $[Ca^{2+}]_i$ was accompanied by depolarisation and fall in input resistance. This effect, a consequence of increased non-junctional membrane permeability to Na^+ (refs 18 and 19), was particularly pronounced when the aequorin glow was diffuse; the

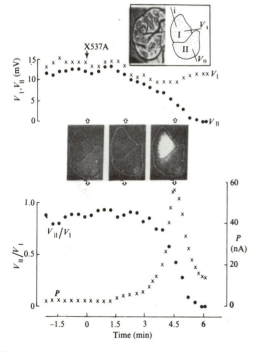

Fig. 3 Ionophore X537A. Exposure to 1×10^{-5} M X537A starts at black arrow and continues throughout the remainder of experiment. Top to bottom: V_I V_{II} ($i = 4 \times 10^{-8}$ A); darkfield television pictures of luminescence (open arrows indicate their time correspondence); coupling ratio V_{II}/V_I; photocurrent P. Cell I depolarises as P rises above background, reaching zero membrane potential at time 9 min. Peak of luminescence in cell II (not shown) occurred 5 min after peak in I. *Top inset:* bright field television picture and cell diagram; cells I and II contain aequorin.

631

depolarisation then paralleled closely the onset of un-coupling. Thus, the question presented itself of whether $[Ca^{2+}]_i$ or depolarisation is the primary cause of uncoupling. To resolve this question, we injected Ca^{2+} while clamp-ing the membrane potential at resting level with a feedback system. Uncoupling ensued none the less whenever the glow reached a junction. Depolarisation is thus clearly not necessary for uncoupling.

In a complementing series of experiments, the cells were exposed to high K (90 mM) medium. Within 2–5 min of application of K medium, the membrane potentials fell to near zero or overshot zero by 5–10 mV. The depolarised cells nevertheless stayed well coupled much longer, in some cases for several hours. Eventually, and rather abruptly, the cells uncoupled; and this was associated with an abrupt rise in $[Ca^{2+}]_i$. Depolarisation is thus also not sufficient for uncoupling.

In conclusion, the cytoplasmic free Ca concentration in the domain of the junction seems to determine the per-meability of the junction. We do not know the mechanism by which the calcium ion alters the permeability. Con-ceivably Ca^{2+} binds to the junctional membrane changing the conformation of the junctional cell-to-cell diffusion channels[2]. We favour this simple notion, because as our results show, the permeability change is readily reversed when the normal $[Ca^{2+}]_i$ is restored in cells with normal Ca-pumping and Ca-buffering ability.

We thank Dr O. Shimomura for aequorin and Drs R. Llinás and C. Nicholson for the use of a voltage-clamp system. The work was supported by research grants from the US Public Health Service and the National Science Foundation.

BIRGIT ROSE
WERNER R. LOEWENSTEIN
Department of Physiology and Biophysics,
University of Miami School of Medicine,
Miami, Florida 33136

Received December 18, 1974.

1 Loewenstein, W. R., *Ann. N. Y. Acad. Sci.*, **137**, 441–472 (1966).
2 Loewenstein, W. R., *J. Colloid Sci.*, **25**, 34–46 (1967).
3 Oliveira-Castro, G. M., and Loewenstein, W. R., *J. Membrane Biol.*, **5**, 51–77 (1971).
4 Politoff, A. L., Socolar, S. J., and Loewenstein, W. R., *J. gen. Physiol.*, **53**, 498–515 (1969).
5 Rose, B. and Loewenstein W. R., *Fedn Proc.*, **33**, 1340 (1974).
6 Rose, B., and Loewenstein, W. R., *J. Membrane Biol.*, **5**, 20–50 (1971).
7 Lehninger, A. L., Carafoli, E., and Rossi, C. J., *Adv. Enzymol.*, **29**, 259–322 (1967).
8 Baker, P. F., *Prog. Biophys. biophys. Chem.*, **24**, 177–322 (1972).
9 Shimomura, O., and Johnson, F. H., *Biochemistry*, **8**, 3991–3997 (1967).
10 Azzi, A., and Chance, B., *Biochim. biophys. Acta*, **189**, 141–151 (1969).
11 Ashley, C. C., and Ridgway, E. B., *J. Physiol., Lond.*, **209**, 105–130 (1970).
12 Baker, P. F., Hodgkin, A. L. and Ridgway, E. B., *J. Physiol., Lond.*, **218**, 709–755 (1971).
13 Harris, E. J., *Biochim. biophys. Acta*, **23**, 80–87 (1957).
14 Hodgkin, A. L., and Keynes, R. D., *J. Physiol., Lond.*, **138**, 253–281 (1957).
15 Vasington, F. D., Gazzotti, P., Tiozzo, R., and Carafoli, E., *Biochim. biophys. Acta*, **256**, 43–58 (1972).
16 Baker, P. F., Blaustein, M. P., Hodgkin, A. L., and Steinhardt, R. A., *J. Physiol., Lond.*, **200**, 431–458 (1969).
17 Carafoli, E. Tiozzo, R., Lugli, G., Crovetti, F., and Kratzing, C., *J. Mol. Cell Cardiol.*, **6**, 361–378 (1974).
18 Rose, B., and Loewenstein, W. R., *J. Membrane Biol.* (in the press).
19 Romero, P. J., and Whittam, R., *J. Physiol., Lond.*, **214**, 481–507 (1971).

Reprinted from Nature, Vol. 273, No. 5660, pp. 278–281. 1978.

Cross-linked surface Ig attaches to actin

John Flanagan & Gordon L. E. Koch

Medical Research Council, Laboratory of Molecular Biology, Hills Road, Cambridge, UK

In lymphocytes and P3 myeloma cells, cross-linking of surface Ig by the capping and patching phenomenon has been used to demonstrate the induction of a specific association between surface Ig and cellular actin.

THE redistribution of surface receptors by antibodies and lectins is thought to result from a transmembrane association of the receptors with cytoskeletal elements in the cortex of the cell[1-3]. Most of the evidence for this association is indirect, although there is some suggestion of a coordinated redistribution of the receptors and cytoskeletal elements[4-8]. However, it is not certain that this results from actual transmembrane associations, and alternative explanations cannot be dismissed. Evidence that there is actual binding of surface receptors to cytoskeletal elements would help to consolidate the transmembrane control hypothesis considerably. We describe here evidence for attachment of cross-linked surface immunoglobulin (Ig) to cellular actin.

Separation of actin

The approach used in these studies was to isolate the actin of cells and to determine whether or not the surface Ig was attached to the actin in various conditions. The surface Ig was marked by reaction with ^{125}I-labelled rabbit anti-mouse immunoglobulin antibody (RaM Ig), either as the F(ab)$_2$ or Fab fragments. The technique used to separate the actin from other cell components is referred to as the myosin affinity technique and relies on the strong and specific binding which occurs between actin and myosin filaments. Therefore, any actin-associated proteins might also be expected to bind to myosin filaments. That binding of putative actin-associated proteins has actually occurred through actin and not directly to the myosin can be confirmed by presaturating the myosin with actin, which inhibits the binding of cellular actin and its associated proteins. The technique does not distinguish between proteins which are bound directly to actin and those which bind through other proteins. The specificity of the technique for actin has been demonstrated in the preceding paper[7]. Thus, actin is the only protein or glycoprotein in a synthetic mixture to attach to myosin filaments. When NP40 lysates of lymphocytes are used, actin also binds specifically. Examination of the glycoproteins of lymphocytes shows that they have no intrinsic affinity for the myosin filaments (Fig. 1). Therefore, nonspecific binding is not likely to interfere with the application of the technique to lymphocytes.

Fig. 1 Application of the myosin affinity technique to a lysate of lymphocytes. 1×10^7 lymphocytes from the lymph nodes of a BALB/c mouse were suspended in 100 µl of Tris/NP40 buffer with 1 mM phenylmethylsulphonyl fluoride (PMSF) and the nuclei removed by centrifugation. The lysate was treated with myosin filaments as described previously and analysed by sodium dodecyl sulphate (SDS) gel electrophoresis[8]. 1 And 3, lymphocyte lysate; 2 and 4, myosin filaments treated with lymphocyte lysate. 1 And 2 were stained with Coomassie blue for protein; 3 and 4 were stained for glycoproteins with ^{125}I-concanavalin A(ref. 9). The arrow shows the position of actin.

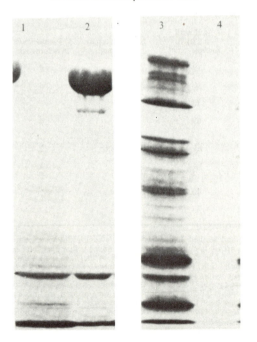

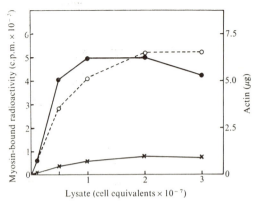

Fig. 2 Application of the myosin affinity technique to lymphocytes in capping conditions. Mouse lymphocytes were labelled with ^{125}I-RaM Ig F(ab)$_2$ or ^{125}I-RaM Ig Fab as described in Table 1. FITC-RaM Ig was added to the F(ab)$_2$-labelled cells to monitor capping. Both sets of cells were subjected to the conditions which lead to capping described in Table 1. After this treatment the cells were lysed with 1% NP40 (see Fig. 1) and increasing amounts of the lysates mixed with myosin filaments as described in Fig. 1. The radioactivity in the washed myosin filaments was measured and the pellets subjected to SDS gel electrophoresis. Densitometry was carried out on the stained gels to estimate the amount of actin in the myosin pellets. The results obtained with both F(ab)$_2$- and Fab-labelled cells were identical in this respect. ●, Actin in myosin filaments; ○, RaM Ig F(ab)$_2$-treated cells; ×, RaM Ig Fab-treated cells.

Binding of surface Ig to actin

The intrinsic affinity of surface Ig for the myosin filaments was tested by labelling lymphocytes of P3 myeloma cells with Fab fragments of RaM Ig. The binding was specific, as it could be inhibited by pre-absorbing the Fab with purified mouse immunoglobulin. Nonidet P40 lysates of the labelled cells were prepared and mixed with myosin filaments. The results show that a small proportion of the radioactivity does bind to the myosin and some of this can be eliminated by using actomyosin (Table 1). In the case of the P3 cells the amount of binding in non-cross-linking conditions is very low. However, application of capping conditions, that is, divalent ^{125}I-RaM Ig F (ab)$_2$ as labelling reagent for surface Ig and fluorescein isothiocyanate (FITC)-RaM Ig as fluorescent reagent, had a marked effect on the binding of Ig to myosin filaments. Up to half the radioactivity in the lysates was bound to the myosin filaments and most of this was eliminated by presaturation with actin, implying that surface Ig had become attached to the actin. This was supported by studies which showed that the binding to myosin of surface Ig from capped cells followed very closely the binding profile for actin (Fig. 2). This seems to be specific for surface Ig, as other glycoproteins do not bind to myosin filaments when cells are treated with RaM Ig.

Unlike the lymphoctyes, P3 cells do not undergo the capping step when treated similarly, and surface Ig redistribution is confined to patching in cross-linking conditions. However, this does not preclude binding to actin following treatment with divalent RaM Ig (Table 1), suggesting that patching is the crucial step in the attachment. This was confirmed by using lymphocytes which were prevented from capping but not patching by low temperature or by addition of sodium azide (Table 1). Both treatments effectively eliminated capping. However, attachment to actin was not prevented. This suggested that the crucial step in the attachment was the cross-linking of the surface Ig. Support for this was obtained by an experiment in which lymphocytes were labelled with ^{125}I-RaM Ig F (ab)$_2$ and then treated with increasing amounts of FITC-RaM Ig (Fig. 3). The

280

Nature Vol. 273   25 May 1978

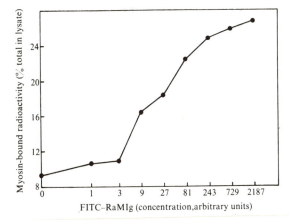

Fig. 3 Effect of cross-linking on the attachment of surface Ig to actin. P3 myeloma cells were labelled with a small amount of ^{125}I–RaM Ig F(ab)$_2$ followed by increasing amounts of FITC–RaM Ig, and incubated at 20 °C for 30 min. The patched cells were lysed with Tris/NP40/PMSF buffer and subjected to the myosin affinity technique as described in Fig. 1. The radioactivity in the myosin filaments after washing was measured to estimate the binding of the surface Ig to actin in each sample.

results show that treatment with ^{125}I–RaM Ig (ab)$_2$ causes the attachment of some radioactivity to the myosin filaments. However, as the concentration of liganded surface Ig is increased by adding cold FITC–RaM Ig, there is a further increase in the amount of radioactivity attached to the myosin filaments. Thus, increasing the degree of liganding, and presumably, therefore, of cross-linking, leads to attachment of surface Ig to the myosin filaments, suggesting that cross-linking was the crucial step in the process which leads to the attachment of surface Ig to actin. It should be noted that the possibility of the cross-linked surface Ig becoming attached to the actin after cell lysis has been examined by carrying out the lysis and myosin extraction

over a 20-fold dilution of cells. Even over such a wide range, the proportion of surface Ig which binds to the myosin filaments remains constant (data not shown), making it unlikely that the association occurs after the cells have been lysed, for in such a case the association would be expected to be dependent on the concentrations of the surface Ig and the actin. Thus, it seems that the attachment of the cross-linked surface Ig to the actin filaments takes place in the intact cell.

Interestingly the cross-linking also leads to an increase in the amount of surface Ig which associates with the nuclear pellet after detergent lysis (Table 1). However, the nuclear pellet does contain a lot of actin and this could be responsible for the phenomenon. Support for this was obtained by examining patched P3 cells which had been lysed with NP40 during examination under the microscope. The cytoplasm and membranes are dissolved away by this treatment, leaving exposed nuclei. However, the patches themselves often remain associated with a fibrillar network which envelopes the nucleus. When FITC–Fab is used the fluorescence washes away instantly. The likely explanation is that the cytoskeleton does not wash away in some cells and the attached patches remain *in situ* around the nuclei and therefore sediment with the latter.

Possible role of cross-linking in attachment

Two important questions are: is the surface Ig attached directly to the actin, and why does cross-linking lead to attachment? An answer to the former will require detailed structural analyses of the capped or patched complexes. The myosin affinity technique should prove useful in isolating this material for analysis. We would like to speculate that the cross-linking leads to attachment because of its effect on a pre-existing equilibrium direct or otherwise, between surface components such as Ig, and microfilaments. When the surface components are in a monovalent state the affinity will be low. However, an increase in the valency of one component in the equilibrium, that is, the surface component, through cross-linking, would shift the equilibrium

Table 1 Application of the myosin affinity technique to surface Ig

Cell type	Pretreatment	Incubation conditions (°C, min)	% Bound c.p.m. in lysate	% Bound c.p.m. in nuclear pellet	% Lysate c.p.m. in myosin	% Lysate c.p.m. in actomyosin
BALB/c lymphocytes	^{125}I-RaMIg Fab, 0 °C, 30 min	1.20, 30	81	19	10	5
		2.20, 30 10 mM azide	82	18	10	4
		3.0, 30	82	18	10	5
	^{125}I-RaMIg F(ab)$_2$ FITC-RaMIg 0 °C, 30 min	1. 20, 30	32	68	42	4
		2.20, 30 10 mM azide	39	61	50	3
		3.0, 30	46	54	38	3
P3 myeloma cells	^{125}I-RaMIg Fab, 0 °C, 30 min	20, 30	92	8	2	2
	^{125}I-RaMIg FITC-RaMIg 0 °C, 30 min	20, 30	60	40	54	0.7

Lymphocytes were obtained from peripheral lymph nodes of BALB/c mice and washed 3 times with Hank's balanced salts solution and 1 mM HEPES buffer, pH 7.3 (BSS/HEPES), before use. P3 Myeloma cells were grown in culture and treated as above before use. Labelling of surface Ig with ^{125}I-RaM Ig Fab or ^{125}I-RaM Ig F(ab)$_2$ was carried out on ice for 1 h in BSS/HEPES. Specificity of the RaM Ig antibody and its fragments for surface Ig was checked by preabsorbing with purified IgG from P3 cells. Redistribution of surface Ig was effected by treating cells with FITC-RaM Ig on ice for 30 min and raising the temperature to 20 °C for 30 min to induce capping. In the case of lymphocytes about 30% of the cells were labelled by the FITC-RaM Ig and over 80% of these were capped. P3 Cells were completely labelled by FITC-RaM Ig but redistribution was confined to patching. For use in the myosin affinity technique labelled cells (0.5–1 × 10^7) were pelleted and made up in 200 μl of BSS/HEPES with 1% Nonidet P40 + 1 mM phenylmethylsulphonyl fluoride on ice, and the nuclei removed by centrifugation at 400g for 10 min. The lysates were diluted 3 times with distilled water and treated with myosin filaments as described in Fig. 1 before gel electrophoresis to confirm the binding of actin to the myosin filaments. ^{125}I-radioactivity in the treated myosin filaments was measured to quantitate the binding of surface Ig. Presaturated actomyosin was prepared as described in Fig. 1 and used exactly as above.

Nature Vol. 273   25 May 1978

towards attachment and lead to the observed effect. Recent studies on the attachment of histocompatibility antigens to cytoskeletal elements also suggest that cross-linking of the surface component is the signal for attachment to the cytoskeleton[10]. Thus, it is conceivable that cross-linking of surface receptors of many types could have an important role in the transmission of transmembrane signals in general.

We thank Michelle Fink for technical assistance, John Murray for the actin and myosin and also for suggestions, Cesar Milstein for P3 myeloma cells, and Ed Lennox for discussions.

Received 25 January; accepted 4 April 1978.

1. Nicholson, G. L. *Biochim. biophys. Acta* **457**, 57–108 (1976).
2. Edelman, G. E. *Science* **192**, 218–226 (1976).
3. Berlin, R. D., Oliver, J. M., Ukena, T. E. & Yin, H. H. *Nature* **247**, 45–46 (1974).
4. Sundquist, K. G. & Ehrnst, A. *Nature* **264**, 226–231 (1976).
5. Toh, B. H. & Hard, G. C. *Nature* **269**, 695–697 (1977).
6. Gabbiani, G., Chapponnier, C., Zumbe, A. & Vassalli, P. *Nature* **269**, 697–698 (1977).
7. Koch, G. L. E. & Smith, M. J. *Nature* **273**, 271–278 (1978).
8. Laemmli, U. K. *Nature* **222**, 680–685 (1970).
9. Robinson, P. J., Bull, F. G., Anderton, B. H. & Roitt, I. M. *FEBS Lett.* **58**, 330–333 (1975).
10. Bourguignon, L. Y. W. & Singer, S. J. *Proc. natn. Acad. Sci. U.S.A.* **74**, 5031–5035 (1977).

Proc. Natl. Acad. Sci. USA
Vol. 75, No. 9, pp. 4209–4213, September 1978
Biochemistry

Direct demonstration that receptor crosslinking or aggregation is important in insulin action

(insulin receptor/receptor antibody/competitive antagonist/insulin antibody/membranes)

C. Ronald Kahn, Kathleen L. Baird, David B. Jarrett, and Jeffrey S. Flier

Diabetes Branch, National Institute of Arthritis, Metabolism, and Digestive Diseases, National Institutes of Health, Bethesda, Maryland 20014

Communicated by DeWitt Stetten, Jr., June 13, 1978

ABSTRACT   Exposure of adipocytes to antibodies to the insulin receptor results in a blockade of ^{125}I-labeled insulin binding, stimulation of glucose oxidation, and many more insulin-like effects. Allowing for differences in purity, antireceptor antibody is equipotent with insulin on a molar basis. Both the bivalent F(ab')$_2$ and monovalent Fab' fragments of the antireceptor antibody are fully active in inhibiting ^{125}I-labeled insulin binding. Bivalent F(ab')$_2$ also retains its insulin-like effects. In contrast, the monovalent Fab' loses almost all ability to stimulate glucose oxidation and acts as a competitive antagonist of insulin-stimulated glucose oxidation. Addition of anti-F(ab')$_2$ antisera, which crosslink the Fab'-receptor complexes, results in a restoration of the insulin-like activity of the antibody. Similarly, when cells are exposed to submaximal doses of insulin, addition of anti-insulin antibodies at low concentration enhances the biological activity of insulin. These data suggest that receptor occupancy by ligand is not sufficient for signal generation and that the insulin-like effects of antireceptor antibody (and perhaps insulin itself) require receptor aggregation or clustering. This aggregation, however, appears to be independent of microfilaments or microtubules because the insulin-like effects of antireceptor antibody, and in fact, of insulin itself, are unaffected by agents that are known to disrupt these structures.

The first step in insulin action is binding to a receptor site on the plasma membrane of the cell (1). Exactly how this interaction of the hormone with its receptor is transformed into a transmembrane message, however, remains unknown. Most attention has focused on the possibility that the interaction of insulin with its receptor activates some membrane-associated enzyme or transport system, which in turn generates a second intracellular messenger of hormone action, perhaps analogous to cyclic AMP (2). Recently, several investigators have presented data that insulin or one of its degradation fragments may actually enter the cell (3–5), and these workers have postulated that this entry may be important for some of insulin's biological effects.

The discovery of autoantibodies to the insulin receptor in some patients with insulin-resistant diabetes has made available a new tool for the study of insulin action (6, 7). We have shown that these antibodies will bind to the insulin receptor (8), block insulin binding (9, 10), and initiate many of insulin's biological effects (10, 11). In the present study we have prepared monovalent fragments of these antireceptor antibodies and compared their effects to those of the bivalent antibody. Like the bivalent antibody, monovalent antireceptor antibodies compete for insulin binding to the receptor. The monovalent antibodies, however, are unable to initiate a biological response, and behave as a competitive antagonist of insulin action at the receptor

level. The insulin-like activity of the monovalent antireceptor antibody can be restored by addition of a second antibody to crosslink the Fab'-receptor complexes. In addition, the activity of insulin itself is enhanced by crosslinking with anti-insulin antibody. These data provide a direct demonstration of a competitive antagonist of insulin action at the receptor level and suggest that receptor crosslinking or aggregation is important for insulin action.

MATERIALS AND METHODS

Materials. Porcine insulin (lot 7GUHSL) was purchased from Elanco Company, bovine serum albumin (Fraction V, Lot N53309) from Armour and Company, and crude collagenase (CLS45K137) and pepsin (2682 U/mg; lot PM35B735) from Worthington Biochemical Corporation. Cytochalasins B and D and colchicine were purchased from Aldrich Chemical Company, vincristine and vinblastine from Eli Lilly Company, and dinonylphthalate from Eastman Chemical Company. ^{125}I-labeled insulin (^{125}I-insulin) was prepared by a modification of the chloramine-T method (12) at specific activities of 100–200 μCi/μg.

The IgG fraction of serum from the patient with the highest concentration of antireceptor antibody activity (B-2) was prepared from the ammonium sulfate precipitate by ion exchange chromatography of DEAE-cellulose (7, 8). Bivalent F(ab')$_2$ fragment was prepared from the IgG by pepsin digestion (13) and purified by gel filtration on Sephadex G-200. To prepare monovalent Fab' fragments (14), the F(ab')$_2$ was concentrated and adjusted to pH 8.6 with 0.2 M Tris·HCl. This solution was then incubated at room temperature for 60 min with 0.01 M dithiothreitol. Iodoacetamide was added to give a final concentration of 0.022 M, and the sample was incubated for 15 min at room temperature. The sample was then dialyzed at 4°C overnight against 0.01 M potassium phosphate, pH 8.0.

Guinea pig anti-insulin serum was purchased from Peter Wright and used without further purification. Goat antibodies to human F(ab')$_2$, a generous gift of Warren Strober, were partially purified by ammonium sulfate precipitation of the antiserum prior to use.

Binding Studies and Glucose Oxidation Bioassay. Isolated adipocytes were prepared from epididymal fat pads of 100–160 gm Sprague–Dawley rats as described by Rodbell (15). Unless otherwise noted, all studies of ^{125}I-insulin binding were performed in the Krebs–Ringer buffer with albumin, pH 7.4 at 37°C as described (10, 15). Glucose oxidation was studied by measuring the conversion of [U-^{14}C]glucose to ^{14}CO$_2$ (16) with an incubation period of 30–60 min as indicated. All glucose oxidation assays were performed in triplicate.

RESULTS

Fig. 1 shows the effects of the purified IgG fraction and bivalent F(ab')$_2$ and monovalent Fab' fragments of antireceptor antibodies on both the insulin binding and glucose oxidation by isolated adipocytes. All three preparations were able to inhibit ^{125}I-insulin binding and were approximately equipotent in this effect. Significant inhibition of binding was observed with concentrations as low as 1 μg/ml. As we have previously reported (10), both the purified IgG and the F(ab')$_2$ fragment also produced insulin-like bioeffects in these cells. In both cases significant stimulation of glucose oxidation was observed at 0.3 μg/ml. Based on our earlier studies (8) which showed that approximately 1% of the total IgG are antireceptor antibodies, the antireceptor antibody is approximately equipotent with insulin on a molar basis in both binding inhibition and bioactivity.

By contrast, the monovalent Fab' has little insulin-like effect on the adipocyte. Thus at 5 μg/ml, a concentration at which bivalent F(ab')$_2$ produces full stimulation of glucose oxidation, the monovalent Fab' produces only about 5% stimulation. At higher concentrations of Fab', some stimulation of glucose oxidation does occur. This is probably due to a small contamination of the preparation with F(ab')$_2$ fragments that have not been successfully reduced. In three different preparations of monovalent Fab' fragments, the insulin-like bioactivity varied from 1% to <0.01% of that of the bivalent antibody.

The finding that the monovalent Fab' fragments block insulin binding although possessing little bioactivity suggested that the monovalent Fab' may be able to serve as a competitive antagonist of insulin at the receptor level, and this was confirmed by studying the effect of the Fab' fragment on insulin-stimulated glucose oxidation. When cells pretreated with Fab' at 10 μg/ml were subjected to further stimulation by insulin, the dose-response of glucose oxidation was clearly shifted to the right (Fig. 2). The dose of insulin producing half-maximal stimulation of the antibody-treated cells was 1.4 ng/ml, as compared with about 0.4 ng/ml for the control cells. The 3-fold shift in insulin sensitivity correlates well with the fact that pretreatment of cells with this concentration of Fab' produced about a 60–70% reduction in insulin binding (17).

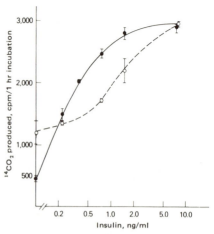

FIG. 2. Effect of monovalent antireceptor antibody on glucose oxidation by isolated adipocytes. Isolated adipocytes were incubated with buffer (●) or Fab' fragments (O) (30 μg/ml) of antireceptor antibody and the indicated concentrations of insulin. Glucose oxidation was measured. This concentration of Fab' produced a 60–70% reduction in specific insulin binding to adipocytes (see Fig. 1).

The loss of bioactivity in the monovalent Fab' fragment could be the result of the change in valency of the ligand or could be the result of chemical modifications that occur with reduction and alkylation. To explore this possibility, reconstitution of the valency was attempted by exposing the cells to the monovalent antibody and then crosslinking these by addition of second antibody. Addition of either anti-F(ab')$_2$ serum (Fig. 3) or anti-human IgG (data not shown) produced a dose-dependent

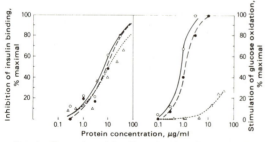

FIG. 1. Comparison of the effects of anti-insulin receptor antibody and its fragments on insulin binding and glucose oxidation by isolated adipocytes. Isolated rat adipocytes were incubated with buffer or the indicated concentrations of the IgG (●), F(ab')$_2$ (O), or Fab' (△) prepared from serum B-2 for insulin binding or glucose oxidation. The inhibitory effect of each fraction on binding and the stimulatory effect on glucose oxidation were calculated as a percentage of the maximal effect produced by insulin in each system. In the glucose oxidation experiments, the maximal effect of insulin occurred at a concentration of 2–3 ng/ml (0.3–0.5 nM) and varied between a 5- and 10-fold stimulation over the basal level. In the binding experiments, 10 μM insulin inhibited ^{125}I-insulin binding by about 90%. The data shown were obtained with antibody fractions of one sequential purification and are representative of four experiments done with three different preparations of each of the immunoglobulin fractions.

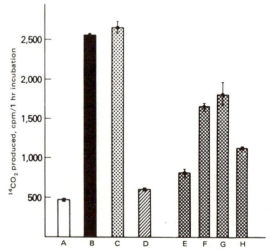

FIG. 3. Effect of crosslinking on the bioactivity of monovalent antireceptor antibodies. Isolated adipocytes were exposed to buffer, insulin (1.5 ng/ml), F(ab')$_2$ (5 μg/ml), or Fab' (5 μg/ml) for 30 min at 22°C. Anti-Fab$_2$ serum was then added at the indicated dilutions to some of the flasks containing buffer or the Fab' fragments. Glucose oxidation was then measured for 1 hr. Anti-F(ab')$_2$ serum alone had no effect on glucose oxidation at any of the concentrations used. Bars: A, basal; B, insulin at 1.5 ng/ml; C, F(ab')$_2$ at 5 μg/ml; D, Fab' at 5 μg/ml; E, F, G, and H, Fab' plus anti-F(ab')$_2$ at 1:1000, 1:200, 1:50, and 1:20, respectively.

restoration of the bioactivity of the monovalent antibody fragment. At a dilution of 1:50 of the second antibody, the effect of the monovalent Fab' reached a maximum, which was about 60% of the effect that was observed with the F(ab')$_2$. At none of the concentrations tested did anti-F(ab')$_2$ serum alone produce any insulin-like effects.

In an attempt to see if this effect of ligand crosslinking was a general factor in insulin action, experiments were performed in which the effect of anti-insulin antibodies in insulin action was studied. In these experiments, cells were exposed first to a concentration of insulin that produces a submaximal stimulation of glucose oxidation (0.2–0.3 ng/ml) and then anti-insulin antibodies were added at high dilution in hopes of crosslinking some of the insulin bound to the receptors on the cell. Consistent with the effects of anti-F(ab')$_2$ serum on Fab', low concentrations of anti-insulin serum enhanced the activity of these submaximal concentrations of insulin by about 30% (Fig. 4). Although the effect is small, it was highly reproducible and clearly significant ($p < 0.05$). The loss of the potentiating effect that occurs as the antibody concentration is increased could be simply due to the fact that the antibodies bind free insulin in solution and thus decrease its concentration or due to the fact that the antibody is more likely to act as a monovalent reagent when excess amounts are present. Taken together, these data suggest that receptor crosslinking or aggregation is important in insulin action.

A possible role of microtubules and microfilaments in this aggregation might be suggested by effects on insulin binding of the various agents known to alter these structures (18). However, pretreatment of cells with colchicine, vincristine, and vinblastine, agents which alter microtubular function (19), at concentrations of 10 μM had no effect on basal, insulin-stimulated, or antibody-stimulated glucose oxidation (Table 1). As previously described, cytochalasin B, an antimicrofilament agent (20), at 1 μg/ml lowers basal glucose oxidation (transport)

Table 1. Effect of antimicrotubular and antimicrofilament agents on stimulation of glucose oxidation by insulin and anti-insulin antibody

Addition	$^{14}CO_2$ produced, cpm/hr of incubation		
	Basal	Insulin (1.0 ng/ml)	IgG B-2 (5 μg/ml)
None	101 ± 10	745 ± 47	984 ± 17
Colchicine, 10 μM	97 ± 14	639 ± 41	860 ± 47
Vinblastine, 10 μM	98 ± 7	636 ± 63	868 ± 3
Vincristine, 10 μM	98 ± 7	670 ± 38	923 ± 66
None	302 ± 28	1732 ± 196	3019 ± 51
Cytochalasin B			
1 μg/ml	24 ± 5	143 ± 20	167 ± 17
10 μg/ml	6 ± 4	14 ± 5	13 ± 2
Cytochalasin D			
1 μg/ml	256 ± 16	1445 ± 73	2995 ± 23
10 μg/ml	228 ± 52	1377 ± 179	2806 ± 169

by over 90%. Despite this, stimulation by both insulin and IgG can still be observed, and approximately the same ratio of stimulated/basal occurs. Cytochalasin D, which also disrupts microfilaments, has little direct effect on glucose transport (21) and produced minimal inhibition of basal, insulin-stimulated or antibody-stimulated glucose oxidation.

DISCUSSION

Insulin action at the cellular level can be considered to reside at four distinct biochemical levels: (*i*) the binding of the hormone to its membrane receptor; (*ii*) transformation of this hormone-receptor interaction into some form of transmembrane signal; (*iii*) generation of an intracellular message (or messenger), and; (*iv*) subsequent chemical modification of various enzymes and transport systems all of which result in the final biological effects of insulin on carbohydrate, lipid, and protein metabolism. Although much effort has been devoted to exploring the mechanism of insulin action, most of our knowledge is limited to some understanding of the first and last steps in this process. Candidates for the role of intracellular messenger have included calcium (2), the cyclic nucleotides (2), inhibitors of protein kinase (22), and recently, several workers have suggested that insulin or one of its degradation fragments may enter the cell and act directly as a second messenger (23, 24).

Almost no clues exist as to the nature of the transmembrane signal itself. A possible role for movement of receptors in the plane of the membrane has been suggested by the finding that insulin receptors appear to be clustered (25, 26), and a temperature-dependent delay in the onset of insulin action has been noted (27). Using fluorescent derivatives of insulin, Schlessinger *et al.* (28) have shown that insulin on fibroblasts can move laterally with a diffusion coefficient (3–5 × 10^{-10} cm^2/sec) similar to that for other mobile membrane proteins.

In 1975 we discovered that sera of some patients with insulin-resistant diabetes contain autoantibodies to the insulin receptor (6, 7, 29) and this has provided new probes for the study of insulin action. These antibodies, which were initially found by their ability to inhibit insulin binding to its receptor (6), also produce insulin-like biological effects when exposed to tissues *in vitro* (10, 11). By using both adipocytes and isolated soleus muscle, we and others have shown that these antibodies stim-

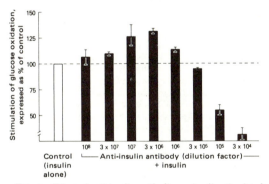

FIG. 4. Effect of anti-insulin antibodies on insulin-stimulated glucose oxidation. Isolated adipocytes were incubated with 0.2–0.3 ng of insulin/ml for 30 min at 22°C. Anti-insulin serum was then added at the indicated dilutions and glucose oxidation was measured for the next 20–60 min. Insulin alone at this concentration produced about a doubling of basal glucose oxidation. The data are expressed as a percent of the control (insulin alone) glucose oxidation and were calculated as:

$$\left(\frac{\text{experimental} - \text{basal}}{\text{control} - \text{basal}}\right) \times 100.$$

The data are presented as the mean ± SEM for three experiments, each done in triplicate. The increase in glucose oxidation observed with anti-insulin antibody at a dilution of 1:3 × 10^6 is significant at the $p < 0.05$ level. Anti-insulin antibody alone produced no systematic change in glucose oxidation (data not shown).

ulate glucose transport, glucose incorporation into glycogen and lipid, and glucose metabolism to CO_2. In addition, antireceptor antibodies stimulate amino acid transport (30) and mimic the antilipolytic effect of insulin (30, 31). In collaboration with J. Lawrence and J. Larner, we have found that these antibodies will also inhibit phosphorylase activity and activate glycogen synthase both in the presence and in the absence of glucose, two changes in activity of cytoplasmic enzymes characteristic of insulin action (unpublished data).

Although many other factors, such as lectins (32) and polyamines (33), have been shown to mimic some of insulin's actions, only the antireceptor antibody appears equipotent with insulin, and in contrast to many of these other agents, its bioeffects seem to be due to direct and specific interaction with the insulin receptor. By using ^{125}I-labeled antireceptor antibody, we have shown that these antibodies bind to cells in direct proportion to the concentration of insulin receptors and that the labeled antibody binding can be inhibited by insulin and insulin analogues in proportion to their affinity for the receptor (8). In addition, the antireceptor antibodies will specificially immunoprecipitate solubilized insulin receptors[*].

The facts that some of the compounds with insulin-like activity, such as lectins, are multivalent and that the antireceptor antibodies are bivalent suggested a possible role for receptor aggregation or crosslinking in insulin action. To test this, monovalent Fab′ fragments of antireceptor IgG were prepared. Although these retained full ability to inhibit insulin binding, the monovalent antibody fragment lost almost all bioactivity. This change in bioactivity appears to be due to the change in valence rather than to the reduction and alkylation, because activity can be restored by crosslinking the monovalent Fab′ with a second antibody (Fig. 5A). In addition, Fab fragments produced by papain digestion without alkylation also lose bioactivity while retaining their ability to inhibit binding (F. A. Karlsson, K. L. Baird, and C. R. Kahn, unpublished observation).

Although insulin in solution at concentrations at which it induces most of its biological responses is monomeric and presumably "monovalent," these observations, together with the previous findings that insulin receptors are mobile (28) or clustered (25, 26), suggested a possible role for receptor aggregation or crosslinking in the action of insulin itself. To explore this possibility, we exposed cells to a submaximal dose of insulin and then to low concentrations of anti-insulin antibody (about 0.5–1.0 mol of antibody per mol of insulin). Under these circumstances anti-insulin antibodies actually enhanced insulin's effect, consistent with an effect of the insulin antibody to crosslink the insulin-receptor complexes (Fig. 5B).

Since all insulin analogues are agonists, presumably insulin is able to either initiate the biological response in the absence of receptor aggregation or induce receptor aggregation independent of external crosslinking (Fig. 5B) (34). Interestingly, when insulin binds to the cell membrane, its local concentration may approach 10^{-6} M (J. Schlessinger, personal communication), a concentration at which insulin in solution dimerizes or aggregates (35). Schlessinger *et al.* (36) have also shown that insulin on fibroblasts can form microscopically visible patches. If this aggregation is important for activity, however, it must differ from aggregation in solution, because there are several insulins which do not dimerize but are biologically active (35, 37). Perhaps hagfish insulin, which has a lower bioactivity than affinity for the insulin receptor (38), is unable to fully induce receptor aggregation.

---

[*] Harrison, L. C., Flier, J. S., Kahn, C. R. & Roth, J. (1978) *Abstracts of the 60th Annual Meeting of the Endocrine Society,* June 14–16, p. 331.

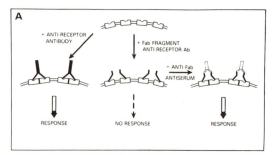

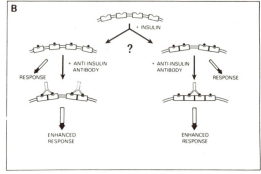

FIG. 5.   Diagrammatic representation of crosslinking of receptor by the various ligands. (*A*) Represents bivalent antireceptor antibodies or the combination of monovalent antireceptor antibodies and a second antibody. (*B*) Represents crosslinking of insulin upon addition of anti-insulin antibodies.

Proper orientation and spacing between the insulin monomers may also be important for the observed effect. Covalent insulin dimers linked by a short bridge between the A-1 phenylalanine and the B-29 lysine have an activity which is approximately an average of the activities of the two chemically modified monomers (39). Covalent dimers linked between B-1 and B-29, on the other hand, have the full biological activity of both insulins (D. Brandenberg, personal communication). Increasing chain length and flexibility could possibly lead to analogues that could bind with an affinity even higher than that of native insulin and could have increased activity.

The enhancement of insulin activity by anti-insulin antibodies may be important in one or more clinical situations. Several patients have been reported with a syndrome characterized by hypoglycemia and spontaneous development of insulin autoantibodies (40). It is interesting to speculate that perhaps these patients have developed insulin antibodies which bind to insulin in such a way that the potentiating effect predominates over the blocking activity. Such antibody modulation of insulin activity may also play a role in the "brittle" diabetic where the antibodies might enhance or inhibit insulin's action depending on the type of antibodies present and the ratio of antibody to insulin.

The data of this study also provide several important insights into the mechanism of insulin action. The first is that many, and perhaps all, of insulin's actions can be initiated by the interaction of ligands other than insulin with the insulin receptor. This suggests that the receptor, when properly triggered, contains all the biochemical attributes necessary to initiate hormone action. This argues against the theories that insulin degradation and internalization of insulin (or one of its degradation fragments) are important for all of insulin's actions. It is possible of course, that the latter are required for some of the long-term growth stimulating effects of insulin. On the other hand, it is not clear if the latter effects are mediated via the insulin receptor or one of the receptors for the insulin-like growth factors.

The second major point is that "occupancy" of the receptor is not sufficient for signal generation. Thus, monovalent Fab' fragments of antireceptor antibody can "occupy" the insulin receptor, at least measured by their ability to block insulin binding and insulin action, without generating much insulin-like effect. A similar situation exists for the IgE receptor of the basophil in which crosslinking of receptors can be accomplished by IgE plus a second antibody, antibodies to the IgE receptor, or chemically crosslinked IgE dimers (41, 42). Recently, Drachman *et al.* (43), have shown that bivalency is also required for the accelerated receptor degradation produced by antibodies to the acetylcholine receptor from patients with myasthenia gravis. Whether a hormone such as insulin is able to induce receptor aggregation without external crosslinking or exerts its signal without receptor aggregation is unclear (34); however, some enhancement of the biological activity of insulin can be obtained by crosslinking insulin-receptor complexes under some circumstances. If aggregation does occur, neither microfilaments nor microtubules appear to be important. Whether some other membrane-associated proteins are required, however, is uncertain.

We acknowledge Drs. J. Roth, L. C. Harrison, and F. A. Karlsson for their advice throughout the study, and Drs. C. Isersky and J. Schlessinger for many useful discussions. The authors also thank Drs. Isersky, Drachman, Schlessinger, and Jacobs for making available unpublished data; and Ms. Shinn and Mrs. Collins for their secretarial assistance.

1. Kahn, C. R. (1976) *J. Cell Biol.* **70**, 261–286.
2. Czech, M. P. (1977) *Annu. Rev. Biochem.* **46**, 359–384.
3. Goldfine, I. D., Smith, G. J., Wong, K. Y. & Jones, A. L. (1977) *Proc. Natl. Acad. Sci. USA* **74**, 1368–1372.
4. Carpentier, J.-L., Gorden, P., Amherdt, M., Van Obberghen, E., Kahn, C. R. & Orci, L. (1978) *J. Clin. Inv.* **61**, 1057–1070.
5. Kahn, C. R. & Baird, K. L. (1978) *J. Biol. Chem.* **253**, 4900–4906.
6. Flier, J. S., Kahn, C. R., Roth, J. & Bar, R. S. (1975) *Science* **190**, 63–65.
7. Kahn, C. R., Flier, J. S., Bar, R. S., Archer, J. A., Gorden, P., Martin, M. M. & Roth, J. (1976) *N. Engl. J. Med.* **294**, 739–745.
8. Jarrett, D. B., Roth, J., Kahn, C. R. & Flier, J. S. (1976) *Proc. Natl. Acad. Sci. USA* **73**, 4115–4119.
9. Flier, J. S., Kahn, C. R., Jarrett, D. B. & Roth, J. (1976) *J. Clin. Inv.* **58**, 1442–1449.
10. Kahn, C. R., Baird, K. L., Flier, J. S. & Jarrett, D. B. (1977) *J. Clin. Inv.* **60**, 1094–1106.
11. LeMarchand, Y., Freychet, P., Flier, J. S., Kahn, C. R. & Gorden, P. (1978) *Diabetologia* **14**, 311–318.
12. Roth, J. (1975) *Methods Enzymol.* **37**, 223–232.
13. Stanworth, D. R. & Turner, M. W. (1973) in *Immunochemistry*, ed. Weir, D. M. (Blackwell Scientific Publications, Oxford, England), pp. 10.0–10.97.
14. Nisonoff, A., Markus, G. & Wissler, F. C. (1961) *Nature (London)* **189**, 293–295.
15. Gammeltoft, S. & Gliemann, J. (1973) *Biochim. Biophys. Acta* **320**, 16–32.
16. Rodbell, M. (1964) *J. Biol. Chem.* **239**, 375–380.
17. Kahn, C. R. (1978) *Metabolism*, in press.
18. Van Obberghen, E., De Meyts, P. & Roth, J. (1976) *J. Biol. Chem.* **251**, 6844–6851.
19. Olmsted, J. B. & Borisy, G. G. (1973) *Annu. Rev. Biochem.* **42**, 507–540.
20. Kletzien, R. F., Perdue, J. F. & Springer, J. (1972) *J. Biol. Chem.* **247**, 2964–2969.
21. McDaniel, M., Roth, C., Fink, J., Fyfe, G. & Lacy, P. (1975) *Biochem. Biophys. Res. Commun.* **66**, 1089–1096.
22. Walkenbach, R. H., Hazen, R. & Larner, J. (1978) *Mol. Cell. Biochem.* **19**, 31–41.
23. Goldfine, I. D. (1977) *Diabetes* **26**, 148–155.
24. Steiner, D. F. (1977) *Diabetes* **26**, 322–340.
25. Jarrett, L. & Smith, R. M. (1974) *J. Biol. Chem.* **249**, 7024–7031.
26. Orci, L., Rufener, C., Malaisse-Lagae, F., Blondel, B., Amherdt, M., Bataille, D., Freychet, P. & Perrelet, A. (1975) *Isr. J. Med. Sci.* **11**, 639–655.
27. Ciaraldi, T. & Olefsky, J. (1978) *Clin. Res.* **26**, 127A.
28. Schlessinger, J., Schechter, Y., Cuatrecasas, P. & Pastan, I (1978) *Nature (London)*, in press.
29. Flier, J. S., Kahn, C. R., Jarrett, D. B. & Roth, J. (1977) *J. Clin. Inv.* **60**, 784–794.
30. Kasuga, M., Akanuma, Y., Tsushima, T., Suzuki, K., Kosaka, K. & Kibata, M. (1978) *J. Clin. Endocrinol. Metab.* **47**, 66–77.
31. Jacobs, S., Chang, K.-J. & Cuatrecasas, P. (1978) *Science* **200**, 1283–1284.
32. Czech, M. P., Lawrence, J. C., Jr. & Lynn, W. S. (1974) *J. Biol. Chem.* **249**, 5421–5427.
33. Livingston, J. N., Gurny, P. A. & Lockwood, D. H. (1977) *J. Biol. Chem.* **252**, 560–562.
34. Singer, S. J. (1976) in *Surface Membrane Receptors*, eds. Bradshaw, R. A., Frazier, W. A., Merrell, R. C., Gottlieb, D. I. & Hogue-Angeliti, R. A. (Plenum, New York), pp. 1–24.
35. Blundell, T., Dodson, G., Hodgkin, K. & Mercola, D. (1972) *Adv. Protein Chem.* **26**, 279–402.
36. Schlessinger, J., Schechter, Y., Willingham, M. C. & Pastan, I. (1978) *Proc. Natl. Acad. Sci. USA* **75**, 2659–2663.
37. Boesel, R. W. & Carpenter, F. H. (1972) *Fed. Proc. Fed. Am. Soc. Exp. Biol.* **31**, 255.
38. Emdin, S. O., Gammeltoft, S. & Gliemann, J. (1977) *J. Biol. Chem.* **252**, 602–607.
39. Freychet, P., Brandenburg, D. & Wollmer, A. (1975) *Diabetologia* **10**, 1–5.
40. Ohneda, A., Metsuda, K., Sato, M., Yamagata, S. & Sato, S. (1973) *Diabetes* **23**, 41–50.
41. Isersky, C., Taurog, J. D., Poy, G. & Metzger, H. (1978) *J. Immunol.*, in press.
42. Segal, D., Taurog, J. D. & Metzger, H. (1977) *Proc. Natl. Acad. Sci. USA* **74**, 2993–2997.
43. Drachman, D. B., Angus, C. W., Adams, R. N., Michelson, J. D. & Hoffman, G. J. (1978) *N. Engl. J. Med.* **298**, 1116–1122.

Cell, Vol. 17, 859–865, August 1979, Copyright © 1979 by MIT

The Outer Boundary of the Cytoskeleton: a Lamina Derived from Plasma Membrane Proteins

Avri Ben-Ze'ev,* Ann Duerr, Frank Solomon and
Sheldon Penman
Department of Biology
and Center for Cancer Research
Massachusetts Institute of Technology
Cambridge, Massachusetts 02139

Summary

We prepared the cytoskeletal framework by gently extracting cells with Triton X-100. Lipids and soluble proteins were removed, leaving a complex meshlike structure which contains the cell nucleus and is composed of the major cell filament networks as well as the microtrabeculae with attached polyribosomes. The surface sheet or lamina covering this structure contains most of the cell surface proteins by the following criteria. Intact cells are labeled externally with radioiodine and then extracted with detergent. The iodinated proteins remain almost entirely with the skeletal framework. A new major integral protein, the coat protein of Sindbis virus, is inserted into the plasma membrane of infected cells. This new protein is heavily iodinated and remains almost completely associated with the framework after extraction. Lectin binding and poliovirus binding sites are also retained after detergent extraction. Our results indicate that plasma membrane proteins form a sheet or lamina upon removal of lipids. This lamina reproduces even complex surface convolutions and appears to be supported by and intimately connected to the underlying skeleton. In this case, the surface lamina, and hence the plasma membrane of the original intact cell, might be viewed as a component of the cytoskeletal framework.

Introduction

Studies of cell structure and cytoplasmic organization reveal a subcellular protein lattice or cytoskeletal framework in cultured mammalian cells. This framework has a role in determining overall cell shape as well as detailed cell morphology, and serves to localize subcellular organelles such as polyribosomes and centrioles. The fibrous components of the framework can be observed in the cytoplasm of intact cells using high voltage electron microscopy (Wolosewick and Porter, 1976). This framework is observed even more clearly after a gentle detergent extraction that removes lipids and soluble proteins (Brown, Levinson and Spudich, 1976; Lenk et al., 1977; Small and Celis, 1978; Webster et al., 1978). An intact cagelike

* Present address: Department of Genetics, The Weizmann Institute of Science, Rehovot, Israel.

network remains, consisting of the fiber systems of microfilaments, microtubules and intermediate filaments as well as other structures including an elaborate fiberlike network which probably corresponds to the microtrabeculae of Wolosewick and Porter (1976). These presumptive microtrabeculae in extracted cytoskeletons have the majority of cell polyribosomes attached and may have a role in mRNA transport and function.

The skeletal framework also serves as a major determinant of cell structure. The preparations described in this report—structures that remain after extraction of lipids and soluble proteins—retain the morphology of the intact cell to a remarkable degree. This is true of both flat monolayer cells and spheroidal cells growing in suspension without a supporting substrate. Even more striking is the surface boundary of the skeletal framework, which appears to be a largely continuous surface that reproduces the detailed convolutions and projections of the cell surface prior to extraction.

This study examines the nature of the surface or lamina that covers the skeletal framework after extraction. Our results indicate that this lamina is derived from proteins on the cell surface and probably in the plasma membrane.

The erythrocyte has been shown to retain most plasma membrane proteins when extracted with Triton X-100 at moderate ionic strength (Yu, Fischman and Steck, 1973). We show that this result is quite general, and that the resulting membrane-derived structure resembles the normal cell surface.

Results

Morphology of the Surface Lamina

The cytoskeletal framework remains after lipids and soluble proteins are extracted from cells with a nonionic detergent. Cells are lysed in hypertonic buffer with Triton X-100, and most phospholipids and two thirds of cell protein are removed into the extraction medium (Lenk et al., 1977). Gentle physical manipulations minimize mechanical damage. An elaborate fibrous network consisting of cell filaments remains with the remnant cell nucleus as a core. This framework has many of the internal and external morphological features of the intact cell. In particular, the outer boundary of the skeletal framework is a continuous surface sheet or lamina that reproduces much of the morphology of the intact plasma membrane.

The outer boundary of a suspension-grown HeLa cell cytoskeleton can be seen in a thin-section micrograph (Figure 1). The outer edge of the cytoskeleton is bounded by a thin, sharp, densely staining line which appears continuous over most of the cell periphery. At high magnification (Figure 1, inset) this boundary appears to be 100 Å thick. A very narrow,

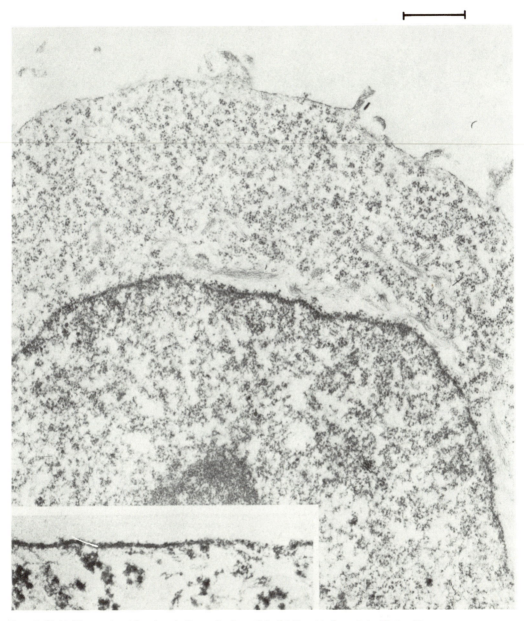

Figure 1. Skeletal Framework and Boundary of a Suspension-Grown HeLa Cell Viewed by Transmission Electron Microscopy

Suspension-grown HeLa cells were extracted, fixed and prepared for electron microscopy as described in Experimental Procedures. Bar = 0.5 μ. The inset is a high magnification view of a typical boundary or surface lamina. Magnification 118,000X.

lightly staining seam can be observed in some regions. However, the typical trilaminar pattern is gone. Internal filamentous structures often appear to terminate at the boundary. These filaments may correspond to the microfilaments of the cytoplasmic cortex (Wolosewick and Porter, 1976).

Extraction of monolayer cells such as 3T6 yields the skeletal framework, similar in essential features to that obtained from HeLa cells (Figure 2). The framework, shown in a section normal to the plane of the culture dish, has attached polyribosomes and a nucleus surrounded by the nuclear lamina; the surface

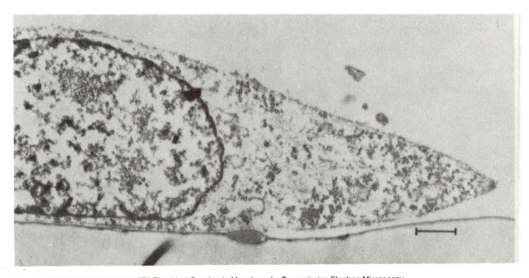

Figure 2. Skeletal Framework of a 3T6 Fibroblast Growing in Monolayer by Transmission Electron Microscopy

3T6 fibroblasts grown in monolayer cultures were lysed, fixed and prepared for electron microscopy as described in Experimental Procedures. The thin-section was cut perpendicular to the surface of the culture dish. Bar = 1 μ.

lamina is clearly visible. The efficiency of extraction is given by the data in Table 1. These preparations have lost over 90% of lipids labeled by radioactive choline, together with most tRNA and three fourths of total protein. As in the HeLa cell, the extracted structure has a relatively intact surface lamina in a configuration similar to that of the plasma membrane in whole cells.

Scanning micrographs of intact fibroblasts and the skeletal structure remaining after extraction are shown in Figure 3. The intact fibroblasts in Figure 3A display the typical morphology of these cultured cells. Figure 3B shows the cytoskeletons obtained after extraction. The extracted structure resembles the intact cell to a remarkable degree. The surface lamina covers the cell skeleton and masks the internal structure.

Some morphological features of the extracted skeleton are altered from those of the intact cell. The surface lamina is lowered and drapes over the bulge of the nucleus. Microvilli are still present but have lost their rigidity and now appear flaccid. Numerous small openings are visible at the outer periphery where the cell is thinnest. Most of the surface lamina appears continuous, however, without major breaks or discontinuities.

The surface of suspended cells is extremely convoluted, as shown by the scanning micrograph of Figure 3C. The framework remaining after detergent extraction, shown in Figure 3D, is also covered by a (somewhat flattened) lamina resembling the surface of the intact cell. The internal architecture of this

Table 1. Extraction of Choline-Labeled Lipids, tRNA and Proteins by Triton X-100

	Soluble	Skeletal	% Released
^{3}H-choline, 1 min extraction	1,859	296	86.2
^{3}H-choline, 20 min extraction	1,876	253	88.1
^{3}H-uridine in tRNA	5,275	783	88.7
^{35}S-methionine in protein	32,473	11,588	73.6

3T6 fibroblasts were labeled overnight with 1 μCi/ml ^{3}H–uridine or 2 μCi/ml ^{3}H–choline, or for 2 hr with 5 μCi/ml ^{35}S–methionine. The cells were washed and extracted with Triton X-100 as described. ^{3}H–uridine-labeled RNA was extracted from the soluble and skeletal fractions and analyzed on a 10% acrylamide gel, and the radioactivity in the 4S area was determined. The amount of ^{35}S-methionine-labeled protein in each fraction was determined by TCA precipitation. The same result was obtained for any length pulse with ^{35}S–methionine, from 10 min to 24 hr. The radioactive choline-labeled lipids were isolated from the soluble and skeletal fraction by the chloroform-methanol extraction procedure, and the radioactivity was determined.

preparation has been shown in the thin-section in Figure 1.

Origin of the Surface Lamina Proteins: Retention of Surface Iodinated Proteins

The skeletal surface lamina of these preparations retains most proteins associated with the cell surface. One method of determining the surface constituents remaining in the surface lamina uses radioiodine to label the cell surface prior to extraction (Hynes and

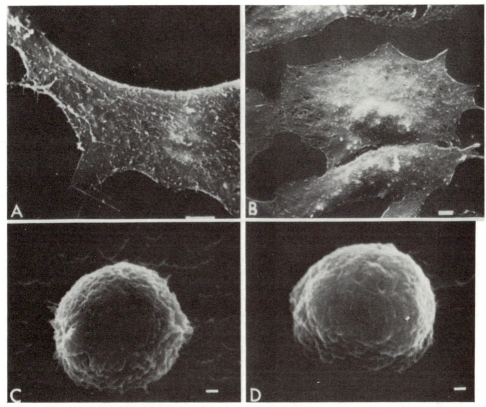

Figure 3. Intact Cells and Their Skeletal Framework by Scanning Electron Microscopy

Intact and lysed cells were prepared for scanning electron microscopy as described in Experimental Procedures. (A) Intact NIL fibroblast; (B) extracted NIL fibroblast; (C) intact HeLa cell; (D) extracted HeLa cell. In (A) and (B), bar = 5 μ; in (C) and (D), bar = 1 μ.

Bye, 1974). The iodination is specific for surface proteins accessible from the external milieu. Most of these proteins remain with the skeleton preparation after detergent extraction.

The detergent extraction removes about three fourths of the cytoplasmic protein, leaving the skeletal framework. A gel electropherogram of the ^{35}S–methionine-labeled cell proteins removed by detergent and those remaining with the skeletal framework of CHO cells is shown in Figure 4, lanes g and h. Growing CHO monolayer cultures were then iodinated. As shown in lane b of Figure 4, only three major bands are extensively extracted and a few minor iodinated bands are slightly extracted. Approximately 80% of the iodinated proteins remain with the skeleton, as shown in lane c of Figure 4. Essentially identical results are obtained with 3T6 cells. The patterns resemble those obtained with erythrocytes, where only band 3 is extensively extracted under comparable conditions (Yu et al., 1973).

The surface iodination procedure does not distinguish between integral membrane proteins and external peripheral proteins such as LETS, although the retention of a true integral membrane protein can be demonstrated. The plasma membrane is modified during lytic growth of Sindbis virus in CHO cells so that it contains a new major integral protein (Strauss and Strauss, 1977). The major glycoprotein of the virus coat is first inserted into the cellular plasma membrane. As the viral nucleocapsids bud, they are enveloped by plasma membrane containing the virus-specific glycoprotein. The CHO cells infected with Sindbis virus are labeled on their surface with radioiodine when appreciable viral glycoprotein is in the plasma membrane, but before extensive virus budding. The iodinated proteins retained in the skeletal framework from infected cells (shown in lane f of Figure 4) are similar to the spectrum from uninfected cells (lane c), except for a major new band (V). This band co-migrates with the coat protein of mature Sindbis virus. Lane e of Figure 4 shows the iodinated protein removed from the skeleton during detergent extraction. More than 80% of the virus-specific protein is retained on the skeletal framework.

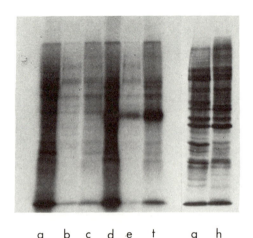

a b c d e f g h

Figure 4. Surface Iodinated Proteins Released and Retained on the Skeleton

CHO cells were labeled with radioactive iodine on their surface as described in Experimental Procedures. The cells were extracted with Triton X-100 and the soluble and skeleton fraction were analyzed on acrylamide slab gels as described in the text. (a) Uninfected cells; (b) uninfected cells, soluble fraction; (c) uninfected cells, skeleton fraction; (d) Sindbis-infected cells; (e) infected cells, soluble fraction; (f) infected cells, skeleton fraction; (g) soluble fraction, ^{35}S–methionine label; (h) skeleton fraction, ^{35}S–methionine label.

Lectin and Viral Binding Sites Retained on the Skeletal Framework

Polysaccharides on the cell surface bind plant lectins tightly, and these can serve as another marker of surface components in the skeletal framework (Lis and Sharon, 1973). Fluorescein-conjugated concanavalin A (FITC-Con A) is bound to intact cells and viewed by epifluorescence. Labeled cells are then extracted and the remaining fluorescence pattern is compared to that of the intact cell.

Con A attached to suspension-grown HeLa cells gives relatively uniform fluorescence, as shown in Figure 5A. The extracted cells show approximately the same fluorescent intensity (Figure 5B), indicating that most surface lectin-binding glycoproteins remain with the skeletal framework.

A more complex fluorescence pattern is obtained by binding Con A to well spread human diploid fibroblasts (Figure 6A); only minor changes are observed after cells are extracted (Figure 6B). In particular, small dark regions appear at the cell periphery. These probably correspond to the small holes visible in the scanning micrograph (Figure 3). Fluorescent Con A binding appears to be specific for carbohydrate since it is blocked in the presence of α–methylmannoside. The results suggest little binding of Con A to glycolipids, which are probably extracted.

Another type of cell surface component retained in these skeleton preparations is the binding sites for poliovirus. These can be measured when radioactive

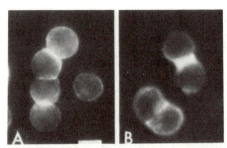

Figure 5. Surface Labeling of Suspension HeLa Cells with Fluorescent Con A

HeLa cells were labeled with FITC-Con A in suspension as described. (A) Cells were fixed and viewed by epifluorescence; (B) cells were extracted with Triton X-100 and then viewed by epifluorescence. Bar = 10 μ.

virions are bound to the virus receptors of the intact cell under conditions that do not allow virus penetration.

HeLa cells are incubated briefly with the radioactive virus at 18°C. The relatively low temperature allows passive virus attachment to the cell surface but little subsequent virus penetration or uncoating (Joklik and Darnell, 1961). Skeletons are then prepared and the amount of virus retained is measured. Table 2 shows that at least 80% of the radioactive virions remain associated with the skeletal framework.

Discussion

This report describes properties of the surface layer that forms the outer boundary of the cell skeletal framework obtained by extraction with detergent. This surface layer or lamina appears as a relatively continuous sheet, probably derived from the proteins of the plasma membrane of the intact cell. Specific components of the plasma membrane are shown to be present in these skeletal preparations; the results suggest an extensive retention of most surface proteins. These include proteins accessible to lactoperoxidase-catalyzed iodination, as well as a newly inserted integral protein, the surface glycoprotein of infecting Sindbis virus. The surface lectin binding sites are largely retained, as shown by the intensity of the fluorescent lectin patterns remaining after extraction. The same results are obtained with a wide variety of plant lectins. Finally, the receptors for poliovirus binding remain almost completely associated with the skeletal framework.

The morphology of the skeletal framework reflects that of the intact cell to a marked degree. As noted in Results, the microvilli of the monolayer fibroblasts are still visible although largely collapsed, and small openings have appeared at the cell periphery where the skeletal framework is thinnest. In general, the lamina appears lowered and the nuclear bulge exaggerated.

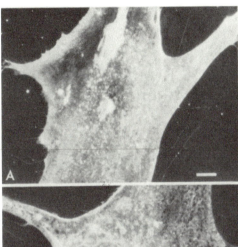

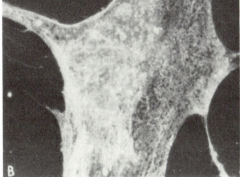

Figure 6. Cell Surface Labeling of Human Diploid Fibroblasts with Fluorescent Con A

Human diploid fibroblasts were grown on glass coverslips and labeled with rhodamine-conjugated Con A as described in Experimental Procedures. (A) Cells labeled with Rh-Con A, fixed with formaldehyde and viewed by epifluorescence; (B) cells labeled as in (A) but then extracted with Triton X-100, fixed and viewed by epifluorescence. Bar = 10 μ.

Table 2. Retention of Poliovirus on the Cytoskeleton

	Before Extraction	After Extraction	% Retained
^{3}H–uridine in poliovirus	7144	6215	86.9

Poliovirus labeled with ^{3}H–uridine in its RNA was prepared as described in Experimental Procedures. HeLa cells were incubated at 18°C with the radioactive virus for 20 min. The cells were split into two equal fractions. One fraction was washed several times, and the radioactivity was determined. The second sample was extracted with Triton X-100 as described and washed, and the radioactivity remaining associated with the cytoskeleton was determined.

The skeletal framework of the suspension HeLa cells also retains much of the detailed configuration of the intact cell surface. Here again some alterations are apparent. The blebs and ruffles of the intact cell, while still visible, are somewhat flattened.

Since it is improbable that the lamina alone could maintain this detailed morphology, an intimate association of the lamina with the underlying cytoskeleton is strongly suggested. Numerous connections are indicated in the high magnification inset in Figure 1. The formation of a relatively continuous layer at the cell surface with little change in total surface area in the absence of lipids indicates that interactions between proteins can maintain this two-dimensional structure. This observation suggests a role for protein-protein interaction in the forces that maintain the intact plasma membrane. Certainly, the formation of a lamina by a membrane after detergent extraction is not a general phenomenon, and is observed only in the case of the plasma membrane and possibly the rough endoplasmic reticulum. In contrast, the smooth endoplas-

reticulum, the inner and outer mitochondrial membranes and the outer nuclear membrane all dissolve in detergent without leaving a visible remnant.

The fluid mosaic model of Singer perhaps represents an extreme in which protein-lipid interactions are completely dominant (Singer and Nicolson, 1972). The true situation, especially in the plasma membrane, may be one in which both types of interactions are important. It is interesting to note that diffusion rates of plasma membrane components are about two orders of magnitude smaller than would be expected for integral proteins in a lipid sea. (Edidin and Wei, 1977). This would be expected if membrane proteins interacted strongly with each other and with the underlying skeletal structure.

The retention of plasma membrane proteins in a Triton-extracted skeleton has been demonstrated in the erythrocyte by Yu et al. (1973), who showed that extraction at moderate ionic strength removes phospholipids but not most surface glycoproteins. Band 3 was the only integral membrane protein extensively extracted. The results resemble those shown in Figure 6 of this paper, where one high molecular weight band is the principal surface-iodinated protein removed from the skeleton.

It is not yet clear exactly to what degree and in what manner the proteins of the cytoplasmic structures and the proteins of the lamina interact. Taken together, however, these observations do present the possibility that the surface lamina and the underlying fibrous structures constitute an important component of the cytoskeletal framework.

Experimental Procedures

Cells

NIL hamster fibroblasts, 3T6 mouse fibroblasts and human diploid fibroblasts (HDF) were grown in Dulbecco's Modified Medium containing 10% calf serum. HeLa cells (S3) were grown in spinner cultures at a density of 10^5 cells per ml in MEM containing 7% horse serum. CHO cells were grown in MEM containing 10% fetal calf serum.

Fractionation of Cells

HeLa cell fractionation was similar to the procedure described by Lenk et al. (1977), with a slight modification of the extraction buffer. Briefly, cells were washed twice with PBS; extraction buffer [0.05 M NaCl, 0.01 M HEPES (pH 7.4), 2.5 mM MgCl$_2$, 0.3 M sucrose, 2 mM PMSF] was gently pipetted onto the cell pellet; and shortly thereafter

Triton X-100 was added to a concentration of 0.5%. After mixing the detergent by gentle swirling for 30 sec in ice, the extracted cell skeletons were pelleted and the supernatant containing the soluble material was removed.

The monolayer cell cultures were extracted as follows: the cells were washed twice with PBS and once with the extraction buffer described above without Triton X-100. Extraction buffer containing 0.5% Triton X-100 was added for 2 min on ice. The plates were gently tilted from side to side during the extraction. After 2 min, the supernatant containing the soluble material was removed.

Sindbis Virus Infection

Confluent monolayers of CHO cells were infected with Sindbis virus at 10–30 pfu per cell. After 1 hr at 37°C, the unabsorbed virus was removed and the medium was replaced with fresh medium containing 1 μg/ml of actinomycin D.

Iodination of Cell Surface Proteins

5 hr after Sindbis virus infection, mock-infected and virus-infected cells were iodinated by the lactoperoxidase reaction as described by Hynes and Bye (1974). Briefly, cells were washed 3 times with PBS containing Ca^{+2} and Mg^{+2}. 350 μCi of ^{125}I were added per 35 mm dish in 500 μl of PBS containing 5 mM glucose. The reaction was started by the addition of lactoperoxidase and glucose oxidase to a final concentration of 20 μg/ml and 0.1 U/ml, respectively. After a 10 min incubation at room temperature, the reaction was stopped by the addition of cold PBS with NaI replacing NaCl (PBS$_I$). After washing twice with PBS$_I$, the cells were further extracted as described above.

Poliovirus Infection

To prepare poliovirus with radioactively labeled RNA, HeLa cells were infected with poliovirus type I at 30 pfu per cell. After 2 hr actinomycin was added to a concentration of 5 μg/ml. After 30 min the cells were labeled with 20 μCi/ml of ^{3}H-uridine. 6 hr after infection the infected cells were homogenized, layered onto a 15–30% sucrose gradient and centrifuged for 2 hr in the SW41 rotor at 40K at 4°C. The virus peak at 135S was identified by counting the radioactivity on a small aliquot of each fraction. The purified radioactive virus was used in further experiments.

Fluorescent Con A Labeling

FITC-Con A (Calbiochem) or Rhodamine-conjugated Con A (Cappell) were used at a concentration of 50–100 μg/ml. The monolayer cells grown on 18 mm coverslips were washed at room temperature with PBS and incubated with the lectin at room temperature for 20–30 min. After incubation, the cells were washed extensively with PBS and fixed in 3.5% formaldehyde for 30 min. In some cases the cells were extracted with lysis buffer (as described above) before fixation in formaldehyde. The fixed cells were washed with PBS and water before mounting on slides. The fluorescence was examined by a Zeiss microscope equipped with epifluorescent optics using a 40× objective. The photographs were taken with an exposure time of 20–30 sec using Kodak Tri-X film that was developed using Diafine.

Electron Microscopy

Extracted cells were fixed with 2% glutaraldehyde in lysis buffer for 30 min on ice, washed with cacodylate buffer and post-fixed with 0.5% osmium tetroxide buffered with cacodylate. Fixed plates or pellets were embedded in Epon-Araldite (Ladd supplies) and sectioned. For scanning electron microscopy, monolayer cells were grown on glass coverslips and HeLa cells were placed on nucleopore filters after fixation. The samples were critical point-dried after the osmium fixation and coated with gold/palladium.

Gel Electrophoresis

The iodinated extracts were analyzed on a 10% acrylamide slab gel by the Laemmli technique (Laemmli, 1970).

Acknowledgments

We are indebted to Elisabeth Beaumont for preparation of samples and for transmission electron microscopy, to Erika Hartwieg for advice and guidance on microscopy and to Leonard Sudenfield for dealing effectively with an ancient scanning instrument. This research was supported by grants from the NIH, the MIT Cancer Center, the American Cancer Society, the NSF and the Health Sciences Fund.

The costs of publication of this article were defrayed in part by the payment of page charges. This article must therefore be hereby marked *"advertisement"* in accordance with 18 U.S.C. Section 1734 solely to indicate this fact.

Received March 15, 1979; revised May 18, 1979

References

Brown, S., Levinson, W. and Spudich, J. A. (1976). Cytoskeletal elements of chick embryo fibroblasts revealed by detergent extraction. J. Supramol. Structure 5, 119–130.

Edidin, M. and Wei, T. (1977). Diffusion rates of cell surface antigens of mouse-human heterokaryons. I. Analysis of the population. J. Cell Biol. 75, 475–482.

Hynes, R. O. and Bye, J. M. (1974). Density and cell cycle dependence of cell surface proteins in hamster fibroblasts. Cell 3, 113–120.

Joklik, W. K. and Darnell, J. P. (1961). The adsorption and early fate of purified poliovirus in HeLa cells. Virology 13, 439–447.

Laemmli, U. K. (1970). Cleavage of structural proteins during the assembly of the head of bacteriophage T4. Nature New Biol. 227, 680–685.

Lenk, R., Ransom, L., Kaufman, Y. and Penman, S. (1977). A cytoskeletal structure with associated polyribosomes obtained from HeLa cells. Cell 10, 67–78.

Lis, H., and Sharon, N. (1973). The biochemistry of plant lectins (Phytohemagglutinins). Ann. Rev. Biochem. 42, 541–574.

Singer, S. J. and Nicolson, G. L. (1972). The fluid mosaic model of the structure of cell membranes. Science 175, 720–731.

Small, J. V. and Celis, J. E. (1978). Filament arrangements in negatively stained culture cells: the organization of actin. Cytobiologie 16, 308–325.

Strauss, J. H. and Strauss, E. G. (1977). Toga virus. In The Molecular Biology of Animal Viruses, 1, D. Nyak, ed. (New York: Marcel Dekker).

Webster, R. E., Henderson, D., Osborn, M. and Weber, K. (1978). Three-dimensional electron microscopic visualization of the cytoskeleton of the animal cell: immunoferritin identification of actin- and tubulin-containing structures. Proc. Nat. Acad. Sci. USA 75, 5511–5515.

Wolosewick, J. J. and Porter, K. R. (1976). Stereo high-voltage electron microscopy of whole cells of the human diploid line, W1-38. Am. J. Anat. 147, 303–324.

Yu, J., Fischman, D. A. and Steck, T. L. (1973). Selective solubilization of proteins and phospholipids from red blood cell membranes by non-ionic detergents. J. Supramol. Structure 1, 233–248.

Supplementary Readings

Before 1961

Abercrombie, M. and E. Ambrose. 1958. Interference miscroscopic studies of cell contacts in tissue culture. *Exp. Cell Res. 15:* 332–345.

Easty, G.C. and V. Mutolo. 1960. The nature of the intercellular material of adult mammalian tissues. *Exp. Cell Res. 21:* 374–385.

Moscona, A. 1957. The development in vitro of chimeric aggregates of dissociated embryonic chick and mouse cells. *Proc. Natl. Acad. Sci. 43:* 184–193.

1961–1970

Aub, J.C., B.H. Sanford, and L.H. Wang. 1965. Reactions of normal and leukemic cell surfaces to a wheat germ agglutinin. *Proc. Natl. Acad. Sci. 54:* 400–402.

Borek, C., S. Higashino, and W. Loewenstein. 1969. Intercellular communication and tissue growth. IV. Conductance of membrane junction of normal and cancerous cells in culture. *J. Membr. Biol. 1:* 274–293.

Follett, E. and R. Goldman. 1970. The occurrence of microvilli during spreading and growth of BHK21/C13 fibroblasts. *Exp. Cell Res. 59:* 124–136.

Hakomori, S.I. and W.T. Murakami. 1968. Glycolipids of hamster fibroblasts and derived malignant transformed cell lines. *Proc. Natl. Acad. Sci. 59:* 254–261.

Inbar, M. and L. Sachs. 1969. Structural difference in sites on the surface membrane of normal and transformed cells. *Nature 223:* 710–712.

Martinez-Paloma, A., C. Brailovsky, and W. Bernhard. 1969. Ultrastructural modifications of the cell surface and intercellular contacts of some transformed cell strains. *Cancer Res. 29:* 925–937.

McNutt, N.S. and R.S. Weinstein. 1969. Carcinoma of the cervix: Deficiency of nexus intercellular junctions. *Science 165:* 597–599.

Pollack, R.E. and M.M. Burger. 1969. Surface-specific characteristics of a contact-inhibited cell line containing the SV40 viral genome. *Proc. Natl. Acad. Sci. 62:* 1074–1076.

Sefton, B.M. and H. Rubin. 1970. Release from density dependent growth inhibition by proteolytic enzymes. *Nature 227:* 843–845.

1971–1980

Artzt, K., P. Dubois, D. Bennett, H. Condamine, C. Babinet, and F. Jacob. 1973. Surface antigens common to mouse cleavage embryos and primitive teratocarcinoma cells in culture. *Proc. Natl. Acad. Sci. 70:* 2988–2992.

Borek, C., M. Grob, and M.M. Burger. 1973. Surface alterations in transformed epithelial and fibroblastic cells in culture. *Exp. Cell Res. 77:* 207–215.

Campisi, J. and C.J. Scandella. 1980. Calcium-induced decrease in membrane fluidity of sea urchin egg cortex after fertilization. *Nature 286:* 185–186.

Chen, L.B. and J.M. Buchanan. 1975. Mitogenic activity of blood components. I. Thrombin and prothrombin. *Proc. Natl. Acad. Sci. 72:* 131–135.

Critchley, D.R. and I. Macpherson. 1973. Cell density dependent glycolipids in NIL2 hamster cells, derived malignant and transformed cell lines. *Biochim. Biophys. Acta 296:* 145–159.

Culp, L.A. 1974. Substrate-attached glycoproteins mediating adhesion of normal and virus-transformed mouse fibroblasts. *J. Cell Biol. 63:* 71–83.

Edidin, M. and A. Weiss. 1972. Antigen cap formations in cultured fibroblasts. *Proc. Natl. Acad. Sci. 69:* 2456–2459.

Glick, M.C., Z. Rabinowitz, and L. Sachs. 1973. Surface membrane glycopeptides correlated with tumorigenesis. *Biochemistry 12:* 4864–4867.

Glynn, R.D., C.R. Thrash, and D.D. Cunningham. 1973. Maximal concanavalin A-specific agglutinability without loss of density-dependent growth control. *Proc. Natl. Acad. Sci. 70:* 2076–2077.

Gonzalez-Ros, J.M., A. Paraschos, and M. Martinez-Carrion. 1980. Reconstitution of functional membrane-bound acetylcholine receptor from isolated *Torpedo californica* receptor protein and electroplax lipids. *Proc. Natl. Acad. Sci. 77:* 1797–1800.

Grimes, W.J. 1973. Glycosyltransferase and sialic acid levels of normal and transformed cells. *Biochemistry 12:* 990–996.

Guerin, C., A. Zachowski, B. Prigent, A. Paraf, I. Dunia, M.A. Diawarea, and E.L. Benedetti. 1974. Correlation between the mobility of inner plasma membrane structure. *Proc. Natl. Acad. Sci. 71:* 114–117.

Hogg, N.M. 1974. A comparison of membrane proteins of normal and transformed cells by lactoperoxidase labeling. *Proc. Natl. Acad. Sci. 71:* 489–492.

Horwitz, A.F., M.E. Hatten, and M.M. Burger. 1974. Membrane fatty acid replacements and their effect on growth and lectin-induced agglutinability. *Proc. Natl. Acad. Sci. 71:* 3115–3119.

Inbar, M. and M. Shinitzky. 1974. Increase of cholesterol level in the surface membrane of lymphoma cells and its inhibitory effect on ascites tumor development. *Proc. Natl. Acad. Sci. 71:* 2128–2130.

Ji, T.H. and G.L. Nicolson. 1974. Lectin binding and perturbation of the outer surface of the cell membrane induces a transmembrane organizational alteration at the inner surface. *Proc. Natl. Acad. Sci. 71:* 2212–2216.

Kiefer, H., A.J. Blume, and R.H. Kaback. 1980. Membrane potential changes during mitogenic stimulation of mouse spleen lymphocytes. *Proc. Natl. Acad. Sci. 77:* 2200–2204.

Marchmont, R.J. and M.D. Houslay. 1980. Insulin triggers cyclic AMP-dependent activation and phosphorylation of a plasma membrane cyclic AMP phosphodiesterase. *Nature 286:* 904–906.

McNutt, N.S. 1976. Ultrastructural comparison of the interface between epithelium and stroma in basal cell carcinoma and control human skin. *Lab. Invest. 335:* 132–142.

Nemanic, M.K., D.P. Carter, D.R. Pitelka, and L. Wofsy. 1975. Hapten-sandwich labeling. *J. Cell Biol. 64:* 311–321.

Nicolson, G.L. and R. Yanagimachi. 1974. Mobility and the restriction of mobility of plasma membrane lectin-binding components. *Science 184:* 1294–1296.

Porter, K.R., V. Fonte, and G. Weiss. 1974. A scanning microscope study of the topography of HeLa cells. *Cancer Res. 34:* 1385–1394.

Prives, J.M. and B.M. Paterson. 1974. Differentiation of cell membranes in cultures of embryonic chick breast muscle. *Proc. Natl. Acad. Sci. 71:* 3208–3211.

Raftery, M.A., M.W. Hunkapiller, C.D. Strader, and L.E. Hood. 1980. Acetylcholine receptor: Complex of homologous subunits. *Science 208:* 1454–1457.

Revel, J.P. and K. Wolken. 1973. Electron microscope investigations of the underside of cells in culture. *Exp. Cell Res. 78:* 1–14.

Rodbell, M. 1980. The role of hormone receptors and GTP-regulatory proteins in membrane transduction. *Nature 284:* 17–22.

Roth, S. and D. White. 1972. Intercellular contact and cell-surface galactosyl transferase activity. *Proc. Natl. Acad. Sci. 69:* 485–489.

Ruoslahti, E. and A. Vaheri. 1974. Novel human serum protein from fibroblast plasma membrane. *Nature 248:* 789–791.

Schindler, M. and M.J. Osborn. 1980. Lateral mobility in reconstituted membranes—comparisons with diffusion in polymers. *Nature 283:* 346–350.

Schlessinger, J., E. Van Obberghen, and C.R. Kahn. 1980. Insulin and antibodies against insulin receptor cap on the membrane of cultured human lymphocytes. *Nature 286:* 729–731.

Shay, J.W., K.R. Porter, and D.M. Prescott. 1974. The surface morphology and fine structure of CHO cells following enucleation. *Proc. Natl. Acad. Sci. 71:* 3059–3063.

Shields, R. and K. Pollock. 1974. The adhesion of BHK and PyBHK cells to the substratum. *Cell 3:* 31–38.

Stone, K.R., R.E. Smith, and W.K. Jokik. 1974. Changes in membrane polypeptides that occur when chick embryo fibroblasts and NRK cells are transformed with avian sarcoma virus. *Virology 58:* 86–100.

Teng, N.N.H., and L.B. Chen. 1975. The role of surface proteins in cell proliferation as studied with thrombin and other proteases. *Proc. Natl. Acad. Sci. 72:* 413–417.

Wartiovaara, J., E. Linder, E. Ruoslahti, and A. Vaheri. 1974. Distribution of fibroblast surface antigen. *J. Exp. Med. 140:* 1522–1533.

Weber, J. 1973. Relationship between cytoagglutination and saturation density of cell growth. *J. Cell. Physiol. 81:* 49–54.

Whittenberger, B., D. Raben, M. Lieberman, and L. Glaser. 1978. Inhibition of growth of 3T3 cells by extract of surface membranes. *Proc. Natl. Acad. Sci. 75:* 5457–5461.

Willingham, M.C. and I. Pastan. 1974. Cyclic AMP mediates the concanavalin A agglutinability of mouse fibroblasts. *J. Cell Biol. 63:* 288–294.

Wolf, D.E., M. Edidin, and P.R. Dragsten. 1980. Effect of bleaching light on measurements of lateral diffusion in cell membranes by the fluorescence photo-bleaching recovery method. *Proc. Natl. Acad. Sci. 77:* 2043–2045.

Yamada, K.M. and J.A. Weston. 1974. Isolation of a major cell surface glycoprotein from fibroblasts. *Proc. Natl. Acad. Sci. 71:* 3492–3496.

Chapter 11 Cytoplasmic Organization

Goldman, R.D. and E.A.C. Follett. 1969. The structure of the major cell processes of isolated BHK21 fibroblasts. *Exp. Cell Res. 57:* 263–276.

Lazarides, E. and K. Weber. 1974. Actin antibody: The specific visualization of actin filaments in non-muscle cells. *Proc. Natl. Acad. Sci. 71:* 2268–2272.

Fuller, G.M., B.R. Brinkley, and J.M. Boughter. 1975. Immunofluorescence of mitotic spindles by using monospecific antibody against bovine brain tubulin. *Science 187:* 948–950.

Heggeness, M.H., K. Wang, and S.J. Singer. 1977. Intracellular distributions of mechanochemical proteins in cultured fibroblasts. *Proc. Natl. Acad. Sci. 74:* 3883–3887.

Albrecht-Buehler, G. 1977. Phagokinetic tracks of 3T3 cells: Parallels between the orientation of track segments and of cellular structures which contain actin or tubulin. *Cell 12:* 333–339.

Heuser, J.E. and M.W. Kirschner. 1980. Filament organization revealed in platinum replicas of freeze-dried cytoskeletons. *J. Cell Biol. 86:* 212–234.

Verderame, M., D. Alcorta, M. Egnor, K. Smith, and R. Pollack. 1980. Cytoskeletal F-actin patterns quantitated with fluorescein isothiocyanate-phalloidin in normal and transformed cells. *Proc. Natl. Acad. Sci. 77:* 6624–6628.

Most cells in suspension or in mitosis are spherical. Yet, the cytoplasm of a cultured cell in suspension is not just a deformable lipid sphere filled with proteins and smaller molecules in solution: The linkage of membrane proteins to cytoplasmic mesh (Chapter 10) shows that ordered polymers as well as simple proteins are present. Now consider the consequence of the spherical normal cell settling on a wettable surface; it flattens, but not at all after the fashion of a simple droplet. Rather, the spreading cell becomes converted into a sheet whose perimeter is highly irregular.

Active, organized deformations of the cell are mediated by the copolymerization of a set of proteins in the cytoplasm. In their more polymerized states, these proteins can be seen by electron or even by light microscopy as sets of filaments. Papers in this chapter trace the development of means to visualize these cytoplasmic structures, to study their functions, and to determine the molecules from which the cell constructs them.

After spreading, cells of the BHK line of hamster fibroblasts extend a fanlike leading edge as they migrate about, pausing only upon contact with other cells (Chapter 4). **Goldman** and **Follett** look at the cytoplasmic processes of migratory BHK cells, using many different sorts of microscopes. By interference or phase-contrast microscopy, live cells showed large processes running the length of the cell. These were birefringent, suggesting a high degree of order among parallel subunits. Replicas cast from fixed cells revealed that the cell was filled with bundles running parallel to the major cell process. Electron microscopy of thin sections cut parallel and perpendicular to cell processes showed three different filamentous components oriented along the axis of the major process: microfilaments, microtubules, and intermediate filaments. Microtubules are long, curving, hollow tubes about 75 nm in diameter. Microfilaments are solid rods about 5 nm in diameter. Intermediate filaments are similar but thicker, about 10 nm. Goldman and Follett found microfilaments in a large net under the cell membrane as well as in structures aligned with the major axis of the cell. Their basic topology of the cytoskeleton remains essentially correct.

As the next three papers show, immunofluorescence microscopy (Chapter 9) can be used to identify the molecules making up major cytoskeletal components. First, a set of antibodies had to be prepared that recognized only one each of the many proteins making up the cytoskeleton. This was not easy, because actin binds well to many proteins. To obtain a pure antigen, **Lazarides** and **Weber** turned to sodium dodecyl sulfate (SDS)-denatured actin recovered from a gel electrophoresis band. The resulting antibody to actin stained fibers of spread fibroblasts. These fibers were largely in the plane of adhesion of the spread cell, where Goldman and Follett had reported a preponderance of microfilaments. Indeed, we now know that both the microfilament mesh and bundles seen in nonmuscle cells are rich in actin and that microfilaments can even be assembled in vitro by polymerization of actin.

Tubulin is the major subunit of microtubules. **Fuller, Brinkley,** and **Boughter** use antibody to brain tubulin to show how microtubules redistribute during the cell cycle. In interphase, they are part of the cytoskeleton. In mitosis, they break down to reassemble in the mitotic spindle.

The cytoskeleton is not solely a set of filaments, but also a meshwork that runs to and into the cell membrane (Chapter 10). Filamin is a protein of fibroblast cells that copolymerizes with actin in solution to form a meshwork much like the one under the membrane of a cell. The heavy head fragment of myosin (HMM) binds tightly to filaments of actin. When conjugating to fluorescein, HMM stains the same structures seen by anti-actin. **Heggeness, Wang,** and **Singer** use fluorescein-conjugated HMM and rhodamine-conjugated antibody to one of the other cytoskeletal molecules to visualize the locations of two different molecules in the same fixed cell. In this way, they surveyed the relative locations of filamin, actin, myosin, and tubulin in spread normal rat kidney fibroblasts. Ruffles and

microspikes move rapidly until cell-cell contact. Upon contact, they convert to flat, immobile sheets.

Actin was present in cables, ruffles, microspikes, and regions of cell-cell contact. Myosin was present in cables, but not in ruffles, microspikes, or regions of cell-cell contact. Tubulin was found as an independent set of microtubules. Filamin was present in cables and also in microspike bases, ruffles, and regions of cell-cell contact. Heggeness et al. suggest that both the movement and the conversion from ruffles or spikes to an immobilized sheet are consequences of actin-filamin interaction and that both go on without actin-myosin interaction. It is interesting to note that ruffles of transformed cells do not cease to move upon contact and that the src gene product has been localized under the cell membrane at regions of cell-substrate contact (Chapter 4).

Microtubules in the cytoplasm of a spread cell show no apparent preferential orientation. They are, however, often accompanied by organelles containing tubulin, the basal body, and rudimentary cilium. In a masterpiece of careful observation, **Albrecht-Buehler** shows that the most recent direction of migration of a cell is parallel to the direction of the rudimentary cilium and also parallel to the direction of the heaviest microfilament bundles. He proposes that microfilament bundles are "rails" that steady the cell and permit it to translocate smoothly over distances comparable to a cell diameter. Whether the rudimentary cilium is a guide to the direction of movement or whether it is oriented by movement remains to be seen, but the surprising linkage of Albrecht-Buehler's phagokinetic paths to cytoskeletal structures is unlikely to be merely coincidence.

The last two papers describe new ways to visualize cytoplasmic structures. Fixation for electron microscopy can introduce artifacts in the final image. To examine structures at high resolution without introducing any chemical fixatives, **Heuser** and **Kirschner** dropped the temperature of the cells to that of liquid helium very rapidly, so that large ice crystals did not form. The water from frozen cells was then sublimed in a vacuum, and the freeze-dried skeletons were viewed on an electron microscopy grid, either with or without metal shadowing. Preparations were so clean that the subunit molecules of microfilaments, filaments, and microtubules could be seen, thus permitting direct in situ identification of the three species of filaments. The lattice of 5-nm microfilaments which fills the cell was well preserved here too. These pictures are magnificent, and they are graced by experimental controls too often lacking in electron microscopy studies.

Transformation is accompanied by a loss from the cytoplasm of large actin-containing cables. Heuser and Kirschner show that this transition is accomplished by a shift of actin filaments in the cytoskeleton from bundles to a crudely interwoven structure. Apparently, then, transformation reweaves actin filaments. Phalloidin, a compound derived from poisonous mushrooms, binds very tightly to filamentous actin. To quantitate the pattern changes in actin filament organization in actin at the level of light microscopy, **Verderame, Alcorta, Egnor, Smith,** and **Pollack** stained cells with fluorescent phalloidin. They found that transformation was accompanied by a drop in actin cable size and number and that persistant SV40 early gene expression was needed to disrupt the parallel actin bundles of normal cells.

The mechanism of transformation remains unknown. The preponderance of current work assumes a central role for changes in the nucleus. From the papers in Chapters 10 and 11, it is reasonable to guess that a full understanding of transformation will also depend on a clear picture of how the surface and cytoplasm of a cell cooperate to maintain normal communication with other cells and thereby with the organism of which it is a part.

Experimental Cell Research 57 (1969) 263–276

THE STRUCTURE OF THE MAJOR CELL PROCESSES
OF ISOLATED BHK21 FIBROBLASTS

R. D. GOLDMAN[1] and E. A. C. FOLLETT

Institute of Virology, Church Street, Glasgow, W. 1., Scotland

SUMMARY

The structure of the major cell processes of BHK21/C13 fibroblasts was examined by phase contrast, differential interference and polarization optics and by the electron microscopy of thin sections and whole cell replicas. In living cells, the optical studies revealed a birefringent streak extending along the length of the processes. In fixed cells, in addition to endoplasmic reticulum and filamentous mitochondria, microtubules and smaller tubular filaments were seen within the cytoplasm and bundles of microfilaments just under the cell membrane. The latter structures could be correlated with oriented, fibrous material seen in the whole cell replicas. The microtubules and filaments were frequently associated with each other. Possible functions for these filamentous structures in producing and maintaining cell processes and in intra- and extracellular motility are proposed and the relationship of these structures to the birefringence seen in living cells is discussed.

The fully spread fibroblast-like cell in culture normally possesses one or more major cell processes. From the rigid spike-like appearance of these processes when viewed with the light microscope, it would be expected that some organized structure would be involved in their formation and maintenance.

In this paper, we report on a search for organized structure in major cell processes, utilizing BHK21/C13 fibroblast-like cells as a model system.

MATERIALS AND METHODS

Cell culture

The cells studied were the baby hamster kidney line BHK21/C13 [22]. These cells were cultured in Eagle's medium containing 10 % calf serum and 10 % tryptose

phosphate broth. All the cells used in the various experiments were subcultured from one original stock which had been stored in ampoules in 10 % glycerol at −160°C. Cells were recovered, passaged for 2 weeks, and then discarded. For light optical and surface replica studies, $3–5 \times 10^5$ cells were allowed to spread on coverslips in 60 mm petri dishes. For the examination of thin sections, cells were grown on Millipore filters according to the technique of McCombs, Benyesh–Melnick & Brunschivig [15].

Fixation

For electron microscopic studies, cells growing on glass or Millipore filters were fixed for 15 min at room temperature in 1 % glutaraldehyde in 0.05 M KH_2PO_4–NaOH buffer, pH 7.2, containing 0.1 M sucrose, 0.003 M $MgCl_2$ and 0.003 M $CaCl_2$. After rinsing in buffer the cells were postfixed for 15 min in 1 % osmium tetroxide dissolved in the same buffer and maintained at room temperature.

Light microscopy

Fully spread cells on coverslips were observed with a Reichert Zetopan microscope equipped with phase contrast, anoptral, polarized light and differential interference optics. For polarized light observations a high pressure mercury-arc light source, filtered with a 560 mμ interference filter was utilized. To obtain

---

[1] Present address: Department of Biology, Case Western Reserve University, Cleveland, Ohio, USA.

Exptl Cell Res 57

optimum contrast in polarized light, a Zeiss $\lambda/30$ Brace-Koehler rotating compensator was inserted below the analyzer.

Light micrographs were taken on Adox KB17 35 mm film and were developed with Diafine two stage developer.

Preparation of whole cell replicas for electron microscopy

Fixed cells on cover glasses were washed five times with distilled water to remove any loosely adhering material which could obscure structural detail on the cell surface, and frozen by immersion for 30 sec in liquid Arcton 22 (chlorodifluoromethane, m. pt $-162°C$) maintained at its melting point. Frozen cells on coverslips were stored in liquid nitrogen at $-180°C$.

The cells were freeze-dried by rapidly transferring cover glasses from liquid nitrogen to the surface of a brass block previously cooled in liquid nitrogen and mounted below carbon rods in a high vacuum system. Evacuation of the system to as low a pressure as possible for 2–3 h completely removed all traces of ice from the system without the appearance of a liquid phase. The temperature of the system was allowed to rise slowly overnight and a thick layer of carbon (~ 500 Å) was evaporated onto the freeze-dried cells which were then shadowed with Ni/Pd at an angle of 30°.

The cell contents were extracted by immersion of the carbon-coated, shadowed cells in 50 % KOH. Residual KOH was neutralized by transferring to 5 % acetic acid. Finally the replica was separated from the glass by flotation on water, and areas of the film were picked up on electron microscope specimen grids.

Thin sections

Fixed cells on Millipore filters were rinsed three times in buffer, dehydrated in a graded series of alcohols and embedded in Epon 812 according to the method of Luft [14]. Thin sections were cut on an LKB Ultrotome I, mounted on uncoated 400 mesh grids and stained with uranyl acetate [21] and lead citrate [19].

Electron microscopy

All specimens were examined in a Siemens Elmiskop IA at magnifications ranging from × 2000 to × 16,000. The kilovoltage was varied from 40 to 80 depending on the contrast required in a particular specimen.

RESULTS

Light microscopy

When observed with phase contrast, anoptral or differential interference optics fully spread cells possessed at least one, and usually two or more major cell processes (figs 1, 4). These processes reached lengths of 500 μ from the centre (nuclear region) of the cell, and were seen to taper towards their ends, often spreading out slightly at the tips of the process. With these optical systems, very little structure was observed within the cell projections, however in most cells filamentous mitochondria and granules were seen (fig. 2). In addition, a clear region (which did not contain large cytoplasmic particles) was often seen adjacent to the nucleus and extending out towards the tip of the process (fig. 3).

In polarized light, positive birefringence (with respect to the long axis of the cell process) was seen extending along the axis of the process (figs 5, 6). The birefringence appeared as a "streak" running continuously from near the tip of the process back towards the nucleus. In cells with one major process, the "streak" often terminated as a "cap" of birefringence (fig. 5) adjacent to the nucleus. This birefringent "cap" coincided exactly with the clear region observed with phase contrast optics (fig. 3). In cells with two or more major processes, the birefringent "streaks" of the processes were continuous alongside the nucleus (fig. 6).

Cells were examined with the light microscope after fixation with the glutaraldehyde fixative described. There were no obvious changes in fixed cells and they were virtually

---

Fig. 1. A living BHK21/C13 cell observed with anoptral optics. This micrograph shows a fully spread cell with two major processes (*P*). × 500.

Fig. 2. A higher magnification light micrograph of a cell similar to that observed in fig. 1. Note filamentous mitochondria (*M*) and granules (*G*) located within the major cell process. Observed with anoptral optics. × 1300.

Fig. 3. An anoptral micrograph demonstrating a "clear" region (*CR*) adjacent to the nucleus. × 1300.

Exptl Cell Res 57

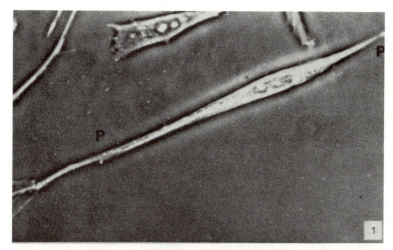

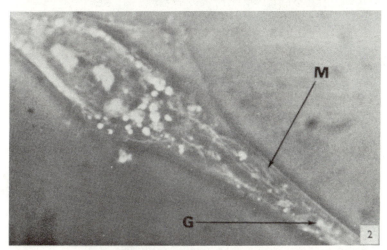

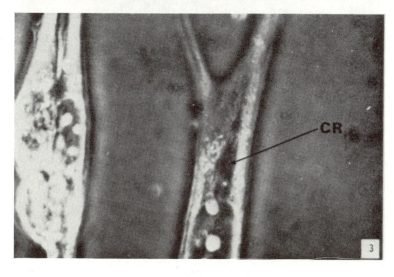

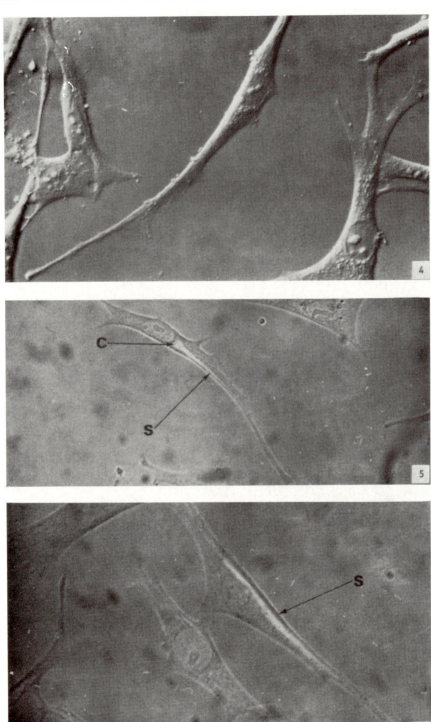

Fig. 4. A differential interference micrograph of a fully spread cell. × 500.
Figs 5, 6. Polarized light micrographs of living cells showing a bright birefringent streak (*S*). Fig. 5 is a micrograph of a cell having one major process and a cap of birefringence (*C*). Fig. 6 shows a cell with two major processes possessing a "continuous" birefringent streak. × 500.

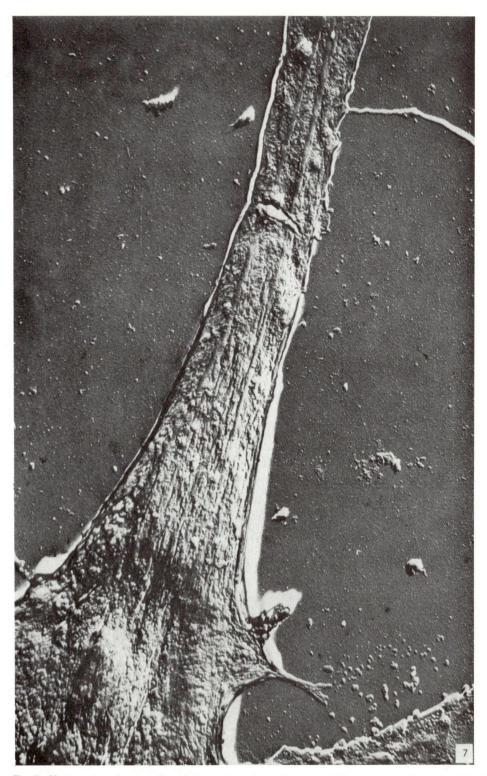

Fig. 7. Shadowed carbon replica of the surface of a cell process. Numerous long fibres extending from the nuclear region out along the process. × 14,000.

identical in structure to living preparations. Fixed cells were observed to retain their birefringent streaks when examined with polarization optics.

Observation of whole cell replicas with the electron microscope

The image of the fixed, fully-spread cell surface provided by the replica technique compared favourably with that of living and of fixed cells observed optically. The cell shape was retained and, in addition many features below the light optical resolution limit were apparent. Two distortions were noted. Cracks had developed in some replicas due to differential expansion between the glass and the frozen cell and in others, the nuclear area appeared to have partially collapsed (fig. 7).

As well as delineating the cell shape the replica revealed an organisation of material at or just under the cell surface. Long fibrous components were apparent being particularly conspicuous along the major cell processes of all fully-spread cells (fig. 7). Two types of fibrous components were noted. One was thick (1000–1500 Å), was present in the majority of cells, and decreased in number with progression along a cell process away from the nucleus (fig. 7). The other was much thinner (400–700 Å), was interspersed with the thick fibres, and was usually best observed where the thick fibrous components were few in number (figs 8 a, b). Where a cell had two major processes in close proximity, fibrous components could be seen extending along both processes (fig. 9). Fibres could be seen to extend back to the area of the nucleus (fig. 7), and occasionally to traverse it and continue out along another cell process.

Where a cell had more than two processes at considerable angles to each other, sets of fibres could be seen to cross each other within the main cell body.

These fibres were oriented approximately parallel to the long axis of the process. No regular organised structural network was observed on very thin regions at the edges of cells (fig. 9). In these regions the replicas appeared to have a random granular structure.

Observation of thin sections with the electron microscope

Longitudinal sections: Longitudinal thin sections through major processes were identified by their overall shape and dimensions when observed at low magnification with the electron microscope. This type of section demonstrated the presence of three filamentous components oriented longitudinally along the long axis of the process (fig. 10). These three components consisted of microtubules, microfilaments and a third type which we have termed filaments.

The microtubules were normally found scattered through the cytoplasm of the process and were rarely seen in regions close to the plasma membrane. They ranged in diameter from 200 to 250 Å. The walls of the microtubules were 40–60 Å thick, and most of them were quite smooth, although occasionally kinked walls were found (fig. 10).

In general, the filaments followed a path through the cytoplasm similar to that of the microtubules, although in some instances they appeared tortuous (fig. 11). In the majority of longitudinal sections they seemed to be of uniform electron density, with a diameter of 80–100 Å.

Fig. 8 (*a*) Shadowed carbon replica of the surface of a cell process showing several thick fibres interspersed with thinner fibres. ×9000. (*b*) Area of fig. 8*a* at higher magnification. One thick fibre and several thin fibres can be seen to traverse the area. ×28,000.

Fig. 9. Shadowed carbon replica of a cell with two major processes. Fibrous material can be seen extending along both processes. ×9000.

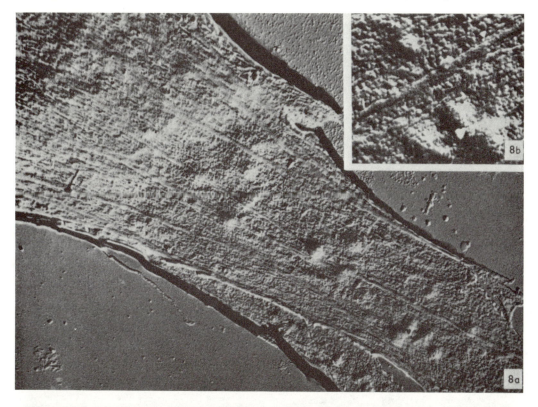

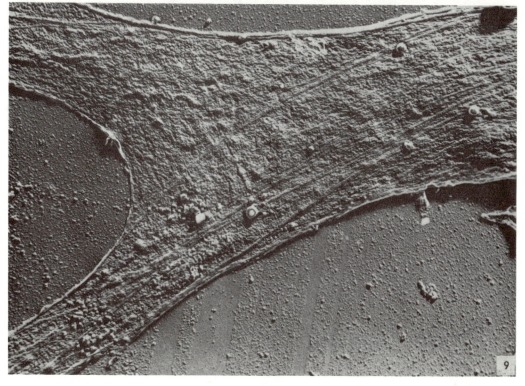

Exptl Cell Res 57

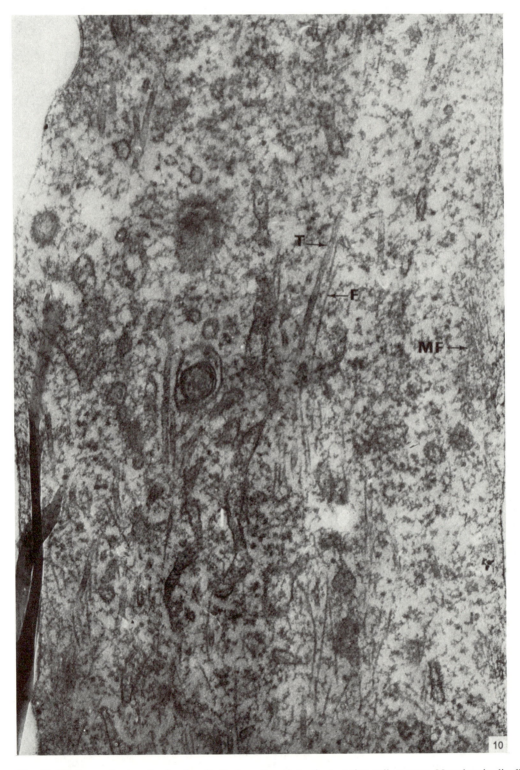

Fig. 10. Electron micrograph of a longitudinal section through a major cell process. Note longitudinally oriented microtubules (*T*) filaments (*F*), microfilaments (*MF*). × 72,000.

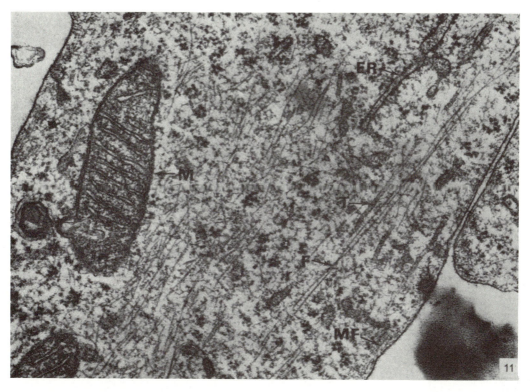

Fig. 11. Electron micrograph of a longitudinal section showing oriented elements of the endoplasmic reticulum (*ER*), portions of filamentous mitochondria (*M*), microtubules (*T*), filaments (*F*) and microfilaments (*MF*). × 44,000.

When present, microfilaments were arranged in a layer located just under the cell membrane (fig. 12). The diameter of individual microfilaments was 40–60 Å, and the layer could extend to a depth of 1500 Å below the cell membrane (fig. 12). These microfilaments were not present in all sections.

Cross sections: In cross sections the microtubules appeared as electron-dense rings, and in the majority of sections the filaments as electron-dense discs (figs 13, 14). In the thinnest sections the filaments were seen to have a less dense core (fig. 15b). On the other hand microfilaments never appeared tubular in any cross-sections.

It also became apparent from these cross-sections that many of the microtubules were in close proximity to the filaments (fig. 15,

15a. Usually, there were several filaments associated with one microtubule and the filaments were located outside an electron-transparent region which surrounded the microtubules (Fig. 15, 15a). In some cross-sections, amorphous material extended from the microtubules to the periphery of the clear region (fig. 15a). The cross-sections of microfilaments were seen just under the cell surface. They occurred either in discrete bundles (figs 13, 14) or as a continuous layer under regions of the plasma membrane (fig. 14). In cross-sections through the nuclear region of a cell, all three components could be found in their characteristic array (fig. 14).

In addition to the three oriented filamentous components, long channels of endoplasmic reticulum and portions of filamen-

18 – 691806

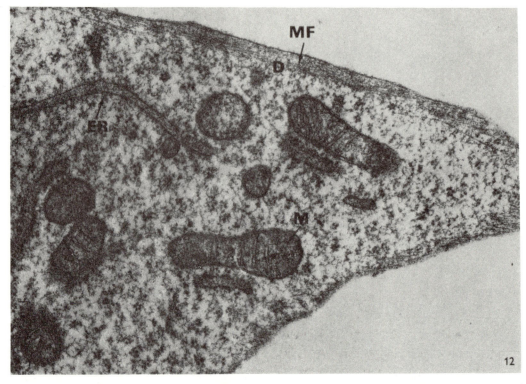

Fig. 12. Electron micrograph of longitudinal section through a major cell process showing a layer of microfilaments just under the plasma membrane (*MF*). Note also oriented portions of filamentous mitochondria (*M*) and endoplasmic reticulum (*ER*). × 40,000.

tous mitochondria were also encountered running longitudinally along a process (figs 11, 12).

DISCUSSION

The evidence provided by the various techniques used in this study supports the idea that there is organized structure in the form of filamentous sub-cellular components extending longitudinally along the long axis of the major cell processes of BHK21 cells in culture.

Both the replicas and thin sections observed with the electron microscope complement each other by demonstrating the presence of similarly oriented fibrous components distributed along the length of the processes. The fibres present in the replicas probably represent bundles of microfilaments located just under the cell surface, and indicate that the microfilaments are not distributed evenly under the entire cell surface. The fact that they are visible at all indicates that there must have been some collapse of adjacent weaker areas of the cell surface during the preparation procedure. In living cells, the existence of a clear, birefringent zone adjacent to the nucleus and extending out into the major cell process, also suggests the presence of oriented components.

At the present time, it is impossible to determine which of the oriented components (including oriented filamentous mitochondria elements of the e.r., and clear zones surrounding the microtubules) contribute to the birefringent streak observed in the processes of

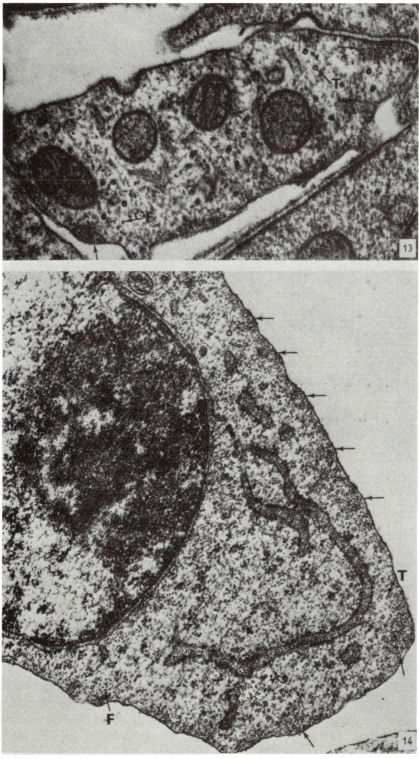

Figs 13, 14. Electron micrographs of cross sections through a major cell process, (fig. 13) and the nuclear region of a cell (fig. 14). Note bundles of microfilaments (*arrows*), cross sections through microtubules (*T*), and filaments (*F*).

665

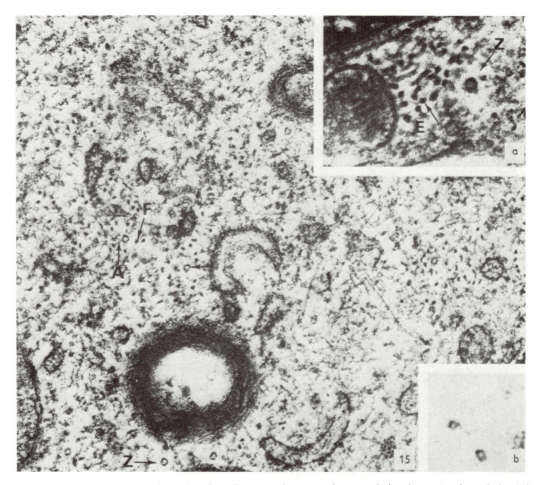

Fig. 15. (*a*, *b*) Cross section through major cell process demonstrating association between microtubules (*T*) and filaments (*F*). (*a*) An electron transparent zone (*Z*) is seen to surround most microtubules. Amorphous material (*A*) is seen in some of these electron transparent zones. (*b*) At higher magnification the tubular nature of the filaments is evident. Fig. 15, × 80,000; (*a*), × 152,000; (*b*), × 250,000.

living cells. It is likely that several or perhaps even all of these structures contribute in varying amounts to the birefringence. Preliminary evidence from observations of spreading cells suggests that the filaments are a major source of birefringence in these cells (Goldman & Follett, in preparation).

Taylor [23] has described the presence of three similar fibrous structures in cultured cells. Although essentially similar in type, the fibrous structures found by Taylor differed in size, distribution and relationship to one

another, from those seen in BHK21 cells. These differences may reflect physiological variations from one cell type to another.

Possible functions of the fibrous components

Microtubules: The microtubules located along the length of the major cell processes are normally straight and relatively rigid in appearance, indicating that they may provide a cytoskeletal system needed for the support and maintenance of these cellular extensions.

Microtubules have been seen in similar projections from a variety of cell types, including the axopods of *Actinosphaerium* [24] and the elongated processes of melanocytes [5]. Their role as a support system involved in various types of cell motility has also been proposed in other instances such as cytoplasmic streaming in plants [20], saltatory particle movements [18], mitosis [9] and the extension and retraction of Heliozoan axopods [25].

Preliminary experiments demonstrate that the major cell processes of BHK21 cells are not formed (although the cells still spread on glass) in the presence of colchicine (Goldman & Follett, in preparation). As colchicine appears to result in the disruption of microtubules [6, 7], these experiments lend support to the hypothesis that microtubules are involved in the production, as well as the maintenance and support of the major cell processes.

Filaments: The association between the filaments and the microtubules may represent a multicomponent system involved in intracellular particle movements. The electron-transparent zones surrounding the microtubules may also represent a functional component. Similar clear zones have been seen surrounding the intranuclear microtubules of crane fly spermatocytes [3, 4], although no filaments were seen in these cells. These filaments may represent one of two or more components in a complex of microtubules and filaments involved in intracellular particle movements. This system may well be involved with the directed movements of particles and mitochondria along the length of the major cell processes. Similar multi-component systems, involved in various aspects of cell motility, have been proposed for such intracellular movements as chromosome movement during mitosis [4, 8, 9] and in saltatory particle movements [18].

Microfilaments: The bundles of microfila-ments localized just under the cell membrane probably represent a contractile layer involved in membrane movements, such as the ruffled membrane associated with fibroblast movement [1, 2] phagocytosis and cytokinesis [10]. This layer may also be responsible for the formation of microvilli often seen on the surface of cultured cells. The facts that similar microfilaments observed in glycerinated fibroblasts bind heavy meromyosin giving the characteristic arrow-head structure of actin and HMM complexes [11] and that an actomyosin-like protein can be isolated from cell membrane fractions [12, 17], support the hypothesis that these bundles of microfilaments represent a contractile layer in the cortical region of the major cell processes.

Similar bundles of microfilaments have been associated with motive force production in protoplasmic streaming in the alga *Nitella* [16], in the acellular slime mould *Physarum* [13] and in the shelled amoeba *Difflugia* [26].

The findings of this study indicate that fibroblast-like cells in culture have a rather complex array of fibrous elements which may be involved in various aspects of intracellular, and extracellular motility, as well as in the formation and maintenance of the major cell processes. Studies are now in progress to elucidate the function of these structures during various stages of the cell cycle.

The authors are indebted to Miss Susan Harris and Mr A. Munro for technical assistance.

This work was supported in part by a grant to R. D. Goldman from the American Cancer Society (PF-376A). E. A. C. Follett is a member of the Medical Research Council Experimental Virus Research Unit.

REFERENCES

1. Abercrombie, M, Exptl cell res Suppl 8 (1961) 188.
2. Ambrose, E J, Exptl cell res Suppl 8 (1961) 54.
3. Behnke, O & Forer, A, Compt rend trav lab Carlsberg 35 (1966) 437.
4. — Science 153 (1966) 1536.

5. Bickle, D, Tilney, L G & Porter, K R, Protoplasma 61 (1966) 322.
6. Borisy, G G & Taylor, E W, J cell biol 34 (1967) 525.
7. — Ibid 34 (1967) 535.
8. Forer, A, Chromosoma 19 (1966) 44.
9. Goldman, R D, & Rebhun, L I, J cell sci 4 (1969) 179.
10. Harris, P, Exptl cell res 52 (1968) 677.
11. Ishikawa, H, J cell biol 39 (1968) 65 a.
12. Jones, B M, Nature 212 (1966) 362.
13. Kamiya, N, SEB Symp. XXII. (1967) 199.
14. Luft, J, J biophys biochem cytol 9 (1961) 231.
15. McCombs, R M, Benyesh-Melnick, M & Brunschivig, J P, J cell biol 36 (1968) 231.
16. Nagai, R & Rebhun, L I, J ultrastruct res 17 (1966) 571.
17. Neifakh, S A, Avramov, J A, Gaitskhoki, V S, Kazakova, T B, Monakhov, N K, Repin, V S, Turovski, V S & Vassiletz, I M, Biochim biophys acta 100 (1965) 329.
18. Rebhun, L I, Proc NY heart assoc, p. 223. Little, Brown & Co., Boston (1967).
19. Reynolds, R, J cell biol 17 (1963) 208.
20. Sabnis, D D & Jacobs, W P, J cell sci 2 (1967) 465.
21. Stempak, J & Ward, R, J cell biol 22 (1964) 697.
22. Stoker, M & Macpherson, I, Nature 203 (1964) 1355.
23. Taylor, A, J cell biol 28 (1966) 155.
24. Tilney, L G & Porter, K R, Protoplasma 60 (1965) 317.
25. Tilney, L G, Hiramoto, Y & Marsland, D, J cell biol 29 (1966) 77.
26. Wohlman, A & Allen, R D, J cell sci 3 (1968) 105.

Received March 21, 1969

Proc. Nat. Acad. Sci. USA
Vol. 71, No. 6, pp. 2268–2272, June 1974

Actin Antibody: The Specific Visualization of Actin Filaments in Non-Muscle Cells

(immunofluorescence/microfilaments/sodium dodecyl sulfate gel electrophoresis)

ELIAS LAZARIDES AND KLAUS WEBER

Cold Spring Harbor Laboratory, Cold Spring Harbor, New York 11724

Communicated by J. D. Watson, March 11, 1974

ABSTRACT   **Actin purified from mouse fibroblasts by sodium dodecyl sulfate gel electrophoresis was used as antigen to obtain an antibody in rabbits. The elicited antibody was shown to be specific for actin as judged by immunodiffusion and complement fixation against partially purified mouse fibroblast actin and highly purified chicken muscle actin. The antibody was used in indirect immunofluorescence to demonstrate by fluorescence light microscopy the distribution and pattern of actin-containing filaments in a variety of cell types. Actin filaments were shown to span the cell length or to concentrate in "focal points" in patterns characteristic for each individual cell.**

Eucaryotic cells contain three basic fibrous structures: filaments, microfilaments, and microtubules. These three structures are thought to be intimately involved in the maintenance of cell shape, in cell movement, and in other important cellular functions (1). Microfilaments are thought to contain actin. This assumption is based on the observation that these structures can be selectively decorated with heavy meromyosin, a specific proteolytic fragment of muscle myosin known to interact with muscle actin (2). Furthermore, actin is now now known to exist as a major protein component of a variety of non-muscle cellular types and in each case it has properties markedly similar to those of its muscle counterpart (3–7)*. The major protein of the microtubular system, tubulin, has been isolated and well characterized (9). The basic protein subunit of the filament structure, however, has not so far been identified. Presumptive muscle proteins like myosin (10–13) and tropomyosin (14) have been found in some non-muscle cells. However, their exact distribution within the cell, as well as their specific localization in one of these fibrous structures, is as yet undetermined.

We have developed a relatively simple way of selectively visualizing filamentous structures in the cell by using antibodies made against different structural proteins. The problem of purifying each antigen separately was circumvented by using sodium dodecyl sulfate (SDS) gel electrophoresis. The denatured proteins are antigenic, and the antibody obtained cross reacts with the native protein. Once specificity has been demonstrated, the antiserum obtained can be used in indirect immunofluorescence to visualize the structures in the cell with which the protein is associated.

In this paper we have used mouse fibroblast actin to test this approach. The actin, purified by SDS gel electrophoresis,

was used to obtain an antibody in rabbits. The antiserum obtained was shown by immunodiffusion and complement fixation to be specific for actin. This antibody was then used in indirect immunofluorescence to show that microfilaments are polymers of actin. This technique also enabled us to demonstrate the complex network of actin filaments in a variety of cell types.

MATERIALS AND METHODS

Growth of Cells. Actin was isolated from the cell line SV101, a clone of mouse fibroblast 3T3 cells transformed by Simian virus 40. This transformed cell line was chosen because it grows to a higher saturation density than the parent 3T3 cell line (15). The cells were grown in roller bottles (Vitro Corp.) in Dulbecco's modified Eagle's medium containing 10% calf serum and 50 μg/ml of gentamycin. At confluency, the medium was removed and the cells were washed with phosphate-buffered saline (PBS). The cells were then scraped off the bottles, collected by low speed centrifugation, and stored at $-70°$.

Preparation of Actin. The cells were thawed and homogenized in 20 volumes of 95% ethanol. The precipitate was collected by low speed centrifugation, washed immediately with ether, and air dried. The yield from 10 bottles was approximately 1.2 g of ethanol-ether powder. The ethanol-ether powder was stirred at 4° in 0.01 M sodium phosphate buffer (pH 6.8), 10 mM $MgCl_2$, and 1 mM dithiothreitol (15 ml of buffer per g of powder) for 3–5 hr. The supernatant was made 30% in ammonium sulfate by adding 0.17 g of ammonium sulfate per ml of extract. After stirring for 30 min at 4°, the precipitate was collected by centrifugation and dissolved in and dialyzed against 0.01 M Tris·HCl (pH 7.5), 10^{-4} M $CaCl_2$, 1 mM dithiothreitol, and 10^{-4} M ATP. The actin could be further purified by a second precipitation at 30% ammonium sulfate saturation. Under these conditions, 1 g of ethanol-ether powder yields approximately 1 mg of actin.

Highly purified chicken muscle actin was a generous gift from Dr. Susan Lowey.

Antibody Preparation. The actin used as an antigen in rabbits was purified through SDS slab gel electrophoresis from the high speed supernatant (100,000 × *g*) of a mouse fibroblast cell homogenate (see *Results*). Approximately 400 μg of the antigen was injected in complete Freund's adjuvant and 2 weeks later the rabbits were boosted with an additional 400 μg. Blood was collected 6 weeks after the last injection and the serum was clarified by centrifugation at 10,000 rpm. The gamma globulin fraction was partially purified using precipitation with half saturated ammonium sulfate. It was

---

Abbreviations: SDS, sodium dodecyl sulfate; PBS, phosphate-buffered saline.

* The authors apologize for not referring to all the contributors in this field. The reader is referred to a recent detailed review of actin and myosin in non-muscle cells for complete references (8).

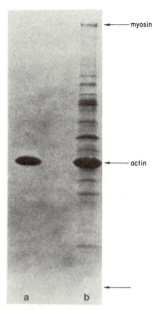

FIG. 1. SDS gel electrophoresis of actin. SDS polyacrylamide slab gel electrophoresis was performed in 12.5% slabs according to Studier (19). (a) Highly purified chicken muscle actin (10 μg). (b) A 30% ammonium sulfate cut of a low-salt extract from an ethanol powder of mouse fibroblasts (15 μg). A small amount of myosin copurifies with actin under the extraction conditions used (see *Materials and Methods*). The *arrow* at the bottom of the figure shows the dye front of the gel.

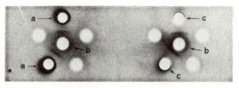

mouse fibroblast actin　　　　chicken muscle actin

FIG. 2. Immunodiffusion of actin antiserum. (a) 4 μg (upper hole) and 2 μg (lower hole) of a 30% ammonium sulfate cut from mouse fibroblasts. The heavy precipitate around the holes is due to aggregated actin. (b) 20 μl of partially purified rabbit antiserum. (c) 5 μg (upper) and 2 μg (lower) of purified chicken muscle actin. Immunodiffusion was performed at 37° for 24 hr. The plates were washed for 24–36 hr at room temperature in 0.15 M NaCl, 0.01 M Tris·HCl (pH 7.5). They were subsequently stained in 0.25% Coomassie brilliant blue–50% methanol–7.5% acetic acid for 1 hr, destained in 7.5% methanol–7.5% acetic acid, and photographed.

then dialyzed into 0.15 M NaCl, 0.01 M Tris·HCl (pH 7.5), and stored at −20° at a concentration of approximately 30 mg/ml.

Indirect Immunofluorescence. Cells were grown on glass coverslips in the appropriate medium (see figure legends). The coverslips were washed briefly in PBS to remove excess medium and fixed in PBS containing 3.5% formaldehyde for 20 min at room temperature. They were subsequently washed thoroughly in PBS, treated with absolute acetone at −10° for 7 min, and air dried. An appropriate dilution (1:20 in PBS) of the rabbit antibody was applied to the cells. After incubation in a humid atmosphere at 37° for 1 hr, the coverslips were washed 3 times in PBS and incubated for 1 hr with a 1:10 dilution of goat anti-rabbit globulins coupled to fluorescein made in PBS (Miles). The coverslips were washed 3 times in PBS and once in distilled water and mounted in Elvanol on a glass slide. Coverslips were viewed in a Zeiss PM III microscope with ultraviolet optics. Photographs were taken using Plus-X film (Kodak).

RESULTS

A protein purified from an ethanol powder of SV101 cells has several properties which identify it as mouse fibroblast actin: (*i*) it coelectrophoreses on SDS gels with highly purified chick muscle actin with a molecular weight of 45,000 (Fig. 1); (*ii*) it precipitates characteristically at a low ammonium sulfate saturation (12) (see below); (*iii*) it binds ATP on millipore filters; (*iv*) it is precipitable by 3 mM vinblastine sulfate

(16, 17) and by 50 mM Mg^{+2} ions (18); and (*v*) it undergoes monomer to polymer transformations. Similar properties characterize skeletal muscle actin and the protein extensively studied and identified as actin-like (13, 16–18) in a variety of nonmuscle cellular types including cultured chick embryo fibroblasts (3) and human platelets (6). Mouse fibroblast actin purified as described in *Materials and Methods* is approximately 70% pure and shows one major polypeptide on SDS polyacrylamide gels and on polyacrylamide gels at pH 4.5 run in the presence of 8 M urea. Fig. 1b shows the purity of mouse fibroblast actin used in the experiments below. The gel is purposely overloaded to reveal all minor contaminants.

Approximately 10% of the total cellular actin precipitates at the 30% ammonium sulfate cut of a low-salt extract of an ethanol powder. Another 50% is recovered in the 30–60% ammonium sulfate cut. The remainder can be extracted as an actomyosin-like complex from the powder in the presence of 0.6 M NaCl. This differential fractionation appears to depend in part on the state of the polymerization of the actin under the extraction conditions used. In order to obtain sufficient actin to use as antigen, we therefore decided to purify this protein by SDS gel electrophoresis from the high-speed supernatant of homogenized SV101 cells.

SDS polyacrylamide gels separate proteins according to the molecular weights of their polypeptide chains (27) and therefore all actin will move with a uniform molecular weight on the gel regardless of its original state of polymerization. This method has the further advantage that SDS denatured proteins often make good antigens and allow one to use many different proteins as antigens from the same high-speed supernatant. The procedure is of general applicability and will be described in detail elsewhere (E.L. and K.W., manuscript in preparation).

The proteins moving as a major band at molecular weight 45,000 on SDS slab gels were recovered from the gels by elution. Approximately 800 μg of protein were obtained and this material was used as an antigen. Previous experiments involving ion exchange chromatography of the high-speed supernatant had shown that actin constituted more than 85% of the total protein moving in this molecular-weight range. The remaining 15% was shown to have different chromatographic properties than actin and showed a small number of minor species on polyacrylamide gels at pH 4.5 in 8 M urea.

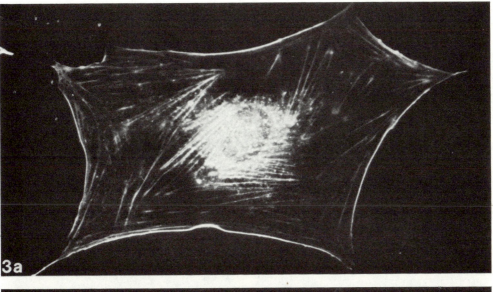

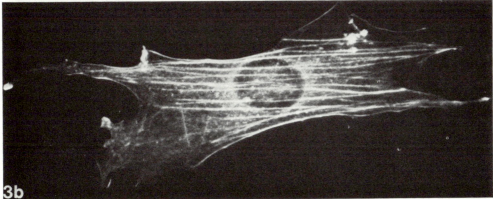

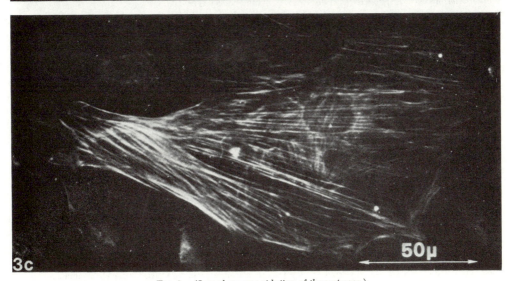

Fig. 3.   (*Legend appears at bottom of the next page.*)

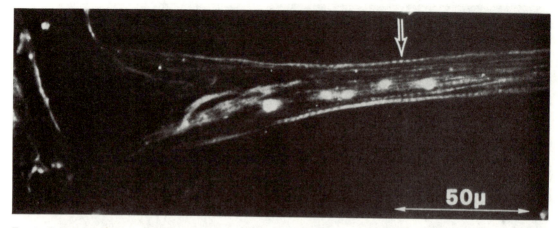

Fɪɢ. 4.   Indirect immunofluorescence of actin antibody with primary chick embryo myoblasts. The *arrow* indicates the actin striations characteristic of myofibrils. The nuclear staining seen in this figure seems to be nonspecific. Purification of the antibody used in this experiment by 50% ammonium sulfate fractionation removes the nuclear fluorescence. The culture of primary chick embryo myoblasts was prepared for us by Dr. C. M. Chang using 5% chick embryo extract, 7% horse serum, and 3% fetal-calf serum in F-12 medium.

The antiserum is specific for actin as judged both by immunodiffusion and complement fixation, using the purified mouse fibroblast actin preparation (Fig. 1b) and the homogeneous preparation of chicken muscle actin (Fig. 1a). The results of the immunodiffusion assay are shown in Fig. 2. Both mouse fibroblast actin and chicken actin show a single precipitin line when diffused against the antiserum. In complement fixation the antibody shows complement fixing ability at a dilution of 1:150 both with the mouse fibroblast actin and the chicken actin. Complement fixation was performed by a modification of the micro method of Sever (20) (Osborn and Weber, in preparation).

Since the antibody appeared to be specific for actin, we attempted to use it for indirect immunofluorescence (21) in the hope of revealing the intracellular distribution of actin. Cells grown on coverslips were fixed in formaldehyde for optimal preservation of fibrous material and were then stained with antibody using the indirect immunofluorescent technique (see *Materials and Methods*). Fig. 3 shows the fluorescence pattern of the mouse fibroblast 3T3 cell line and a baby hamster kidney cell line (BHK) stained with the actin antibody. The fluorescent staining reveals a multitude of actin-containing filaments marking clearly the cell periphery and spanning the interior of the cell frequently parallel to each other. A multitude of patterns have been observed with varying degrees of complexity and each cell appears to have its own individual way of portraying its actin filamentous network. However, after observing hundreds of cells, two major patterns appear to prevail in the cell types tested so far. One is that shown in Fig. 3b and c where the fibers run parallel to each other along the long axis of the cell. The other is that shown in Fig. 3a where actin filaments seem to converge at what we have named "focal points." Control experiments with rabbit antiserum obtained before immunization reveal no fluorescent fibers. Furthermore, antibodies obtained against

other cellular structural proteins reveal a very different fluorescent staining pattern.

As shown above (Fig. 2), the antibody cross reacts with chicken muscle actin. We therefore studied the staining pattern of cultured primary chick embryo myoblasts in early myogenesis (Fig. 4). Besides the usual actin-containing filaments, the fluorescence reveals the characteristic banding striations of actin in newly formed myofibrils.

DISCUSSION

It is known that actin is a major component of eucaryotic cells. It accounts for some 10% of the cells' proteins. Mouse fibroblast actin coelectrophoreses with chick muscle actin and has a molecular weight of 45,000. Not surprisingly, therefore, actin purified from SV101 fibroblasts shows the same properties as the protein extensively studied from a variety of different cell types.

SDS-denatured actin obtained by preparative gel electrophoresis of a cell extract after high speed centrifugation has proven to be a good antigen in rabbits. The antibody obtained reacts with native actin from SV101 cells and with highly purified chick muscle actin both by immunodiffusion and complement fixation. The antibody was directed against a specific class of proteins coelectrophoresing with actin and having a molecular weight of 45,000. While the possibility of obtaining antibodies against other minor proteins in the same molecular-weight range is likely, the possibility of obtaining antibodies to proteins with different polypeptide molecular weights is excluded. Thus the previous difficulty of obtaining an actin specific antibody due to the presence of contaminating tropomyosin is circumvented (22). Furthermore, actin is the only major component constituting more than 85% of the proteins migrating with a molecular weight of 45,000. We therefore believe that although the original antigen was not completely homogeneous, the small amounts of con-

Fɪɢ. 3 (*on preceding page*).   Indirect immunofluorescence using actin antibody. (*a, c*) A sparse mouse fibroblast cell line (3T3). (*b*) A sparse hamster established cell line (BHK). Cells were grown on coverslips in Dulbecco's modified Eagle's medium containing 10% calf serum.

taminating proteins would not interfere with the final analysis.

Immunofluorescence clearly shows the presence of actin filamentous structures. Although limited by resolution, it has the advantage over electron microscopy in revealing the two dimensional mosaic of actin filaments of a whole cell. It also demonstrates that these fibers frequently span the whole length of the cell or converge to characteristic "focal points." Furthermore, the actin filament pattern observed with immunofluorescence corresponds to the microfilament pattern observed by electron microscopy. These latter structures exist as well organized bundles running in close association with and parallel to the plasma membrane, both in baby hamster kidney cells (23) and in 3T3 cells. In 3T3 cells, microfilament bundles are also seen frequently to converge together in patterns very similar to the "focal points" observed by immunofluorescence (unpublished observations with R. Goldman). Since the fluorescence is seen also in close association with the plasma membrane, the actin filaments observed with immunofluorescence correspond to the microfilaments observed by electron microscopy. This conclusion is further substantiated by the finding that only microfilaments are decorated with heavy meromyosin (2).

The technique of immunofluorescence is convenient and fast, and allows the screening of a large number of cells under a variety of experimental conditions. The availability of antibodies to other major structural proteins will enable us to use this technique to study the intracellular localization of these proteins. This experimental approach has been previously used successfully to localize myosin (24), the light chains of myosin (25), and troponin (26) in myofibrils.

The immunofluorescent demonstration that actin exists in filamentous structures gives us a tool to compare structural differences between normal and transformed cells during various stages of their cell cycle. We hope that the convenience of this technique will aid in answering major questions of cellular structure and movement.

We thank Dr. F. Miller at Stonybrook University for the use of his animal facilities and his help in the preparation of the antibodies. We thank Dr. R. Pollack for the use of his laboratory; Drs. R. Goldman and C.-M. Chang and Art Vogel for their help with the fluorescent microscopy; and Drs. M. Osborn, C. W. Anderson, R. Goldman, and R. Pollack for their comments on the manuscript. We also thank Dr. Susan Lowey for a generous gift of purified chicken muscle actin. This investigation was supported by the National Institutes of Health (Research Grant CA-13106 from the National Cancer Institute).

1. Goldman, R. D. & Knipe, D. M. (1972) *Cold Spring Harbor Symp. Quant. Biol.* **37**, 523–534.
2. Ishikawa, H., Bischoff, R. & Holtzer, H. (1969) *J. Cell Biol.* **43**, 312–328.
3. Yang, Y. & Perdue, J. G. (1972) *J. Biol. Chem.* **247**, 4503–4509.
4. Tilney, L. G. & Mooseker, M. (1971) *Proc. Nat. Acad. Sci. USA* **68**, 2611–2615.
5. Bettex-Galland, M. & Lüscher, E. F. (1965) *Advan. Protein Chem.* **20**, 1–35.
6. Probst, E. & Luscher, R. (1972) *Biochim. Biophys. Acta* **278**, 577–584.
7. Puskin, S. & Berl, S. (1972) *Biochim. Biophys. Acta* **276**, 695–709.
8. Pollard, T. D. & Weihing, R. R. (1973) "Cytoplasmic actin and myosin and cell movement," in *Critical Reviews in Biochemistry*, Vol. 2, pp. 1–65.
9. Weisenberg, R., Borisy, G. G. & Taylor, E. W. (1968) *Biochemistry* **7**, 4466–4478.
10. Adelstein, R. S., Pollard, T. D. & Kuehl, W. M. (1971) *Proc. Nat. Acad. Sci. USA* **68**, 2703–2707.
11. Adelstein, R. S., Conti, M. A., Johnson, G. S., Pastan, I. & Pollard, T. D. (1972) *Proc. Nat. Acad. Sci. USA* **69**, 3693–3697.
12. Stossel, T. P. & Pollard, T. D. (1973) *J. Biol. Chem.* **248**, 8288–8294.
13. Adelstein, R. S. & Conti, M. A. (1971) *Cold Spring Harbor Symp. Quant. Biol.* **37**, 598–605.
14. Cohen, I. & Cohen, C. (1972) *J. Mol. Biol.* **68**, 383–387.
15. Todaro, G. J., Green, H. & Goldberg, B. D. (1964) *Proc. Nat. Acad. Sci. USA* **51**, 66–73.
16. Bray, D. (1972) *Cold Spring Harbor Symp. Quant. Biol.* **37**, 567–571.
17. Wilson, L., Bryan, J., Ruby, A. & Mazia, D. (1970) *Proc. Nat. Acad. Sci. USA* **66**, 807–814.
18. Spudich, A. J. (1972) *Cold Spring Harbor Symp. Quant. Biol.* **37**, 585–593.
19. Studier, F. W. (1972) *Science* **176**, 367–376.
20. Sever, J. L. (1962) *J. Immunol.* **88**, 320–329.
21. Coons, A. H. & Kaplan, M. H. (1950) *J. Exp. Med.* **91**, 1–13.
22. Holtzer, H., Sanger, J. W., Ishikawa, H. & Strahs, K. (1972) *Cold Spring Harbor Symp. Quant. Biol.* **37**, 549–566.
23. Goldman, R. D. (1971) *J. Cell Biol.* **51**, 752–762.
24. Pepe, F. A. (1972) *Cold Spring Harbor Symp. Quant. Biol.* **37**, 97–108.
25. Lowey, S. & Holt, J. C. (1972) *Cold Spring Harbor Symp. Quant. Biol.* **37**, 19–28.
26. Perry, S. V., Cole, H. A., Head, J. F. & Wilson, F. J. (1972) *Cold Spring Harbor Symp. Quant. Biol.* **37**, 251–262.
27. Weber, K. and Osborn, M. (1969) *J. Biol. Chem.* **244**, 4406–4412.

Reprinted from Science, Vol. 187, pp. 948–950. 1975. © 1975 AAAS

Immunofluorescence of Mitotic Spindles by Using
Monospecific Antibody against Bovine Brain Tubulin

Abstract. *Monospecific antibody directed against bovine brain tubulin has been purified by affinity chromatography and tested against soluble tubulin and intact microtubules of brain and mitotic apparatus. Binding of the tubulin antibody to the mitotic spindle of rat kangaroo cells was demonstrated in all stages of mitosis by indirect immunofluorescence.*

Within the past 5 years, considerable interest has been directed toward elucidating the molecular structure, mode of assembly, and drug-binding capability of the protein tubulin (*1–3*). Revised methods of purification (*4, 5*) have permitted the isolation of a relatively large amount of tubulin which is nearly electrophoretically homogeneous and retains the capacity to assemble into microtubules. Production of antibody directed against tubulin has been reported (*6, 7*); however, the monospecificity of the antiserum (homogeneity of the antigen tubulin) has not been confirmed. Previous methods of tubulin purification for antibody production involved the formation of paracrystals of tubulin by using high concentrations of vinblastine (*7*). Although a large amount of the crystal-line material was undoubtedly tubulin, the presence of other proteins was not ruled out in the preparation, since it was demonstrated that the resolubilized paracrystals possessed an appreciable amount of adenosine triphosphatase activity. To date, there has been no confirmation of such activity in purified 6S brain tubulin. A radioimmunoassay for outer doublet tubulin from *Naeglaria gruberi* has been reported (*8*).

In this report we present evidence for the production of monospecific antibody directed against pure bovine brain 6S tubulin. The antibody is relatively simple to produce and purify, and it has the ability to bind with microtubules as well as soluble tubulin. Studies using the indirect immunofluorescence technique demonstrate specific fluorescence of mitotic spindles during mitosis. The antibody can become a useful molecular probe for understanding more fully the architecture of this ubiquitous and important protein, in addition to providing a means of following the attachment and fate of spindle fibers during cell division.

Bovine brain tubulin was prepared by a modification of the procedure of Shelanski et al. (5). Fresh bovine brain tissue was washed and homogenized at 4°C in reassembly buffer: 0.1M 2-(N-morpholino)ethanesulfonic acid, 1 mM ethylenebis(oxyethylenenitrilo)-tetraacetic acid, 0.1 mM guanosine triphosphate, and 0.5 mM MgCl₂ (pH 6.85). After centrifugation at 100,000g for 1 hour at 4°C, an equal volume of reassembly buffer containing 8M glycerol was added to the supernatant, which was then divided into portions and frozen at −70°C until needed. Pure tubulin was obtained from the supernatants by allowing polymerization to proceed for 30 minutes at 37°C, followed by centrifugation at 100,000g for 30 minutes at 25°C. The pellet, containing polymerized tubulin, was resuspended and resolubilized in reassembly buffer at 0°C for 1 hour. Insoluble aggregated tubulin was removed from solution by a 30-minute centrifugation at 4°C and 100,000g. To the soluble tubulin was added an equal volume of 8M glycerol–reassembly buffer, and the tubules were again allowed to form at 37°C. After another 100,000g centrifugation at 25°C for 30 minutes, the tubules were solubilized and made 4M with respect to glycerol, as previously described. The twice repolymerized tubulin was then chromatographed on a Sepharose 4B (Pharmacia) column (2.5 by 60 cm) at 4°C in order to separate a small amount of higher molecular weight material from the 6S tubulin. The pure tubulin was stored in small portions at −20°C. The purity of the tubulin protein was determined by polyacrylamide gel electrophoresis on 6 percent acrylamide gels made up in 8M urea and 0.1 percent sodium dodecyl sulfate, as described by Weisenberg et al. (2). The gels were stained with 0.05 percent Coomassie brilliant blue in 25 percent isopropanol and 10 percent acetic acid, then destained in 25 percent isopropanol and 10 percent acetic acid.

Sepharose 4B was activated with cyanogen bromide by the procedure of Cuatrecasas et al. (9). Cyanogen bromide (3.5 g) was added to 20 ml of Sepharose 4B at 0°C. The pH of

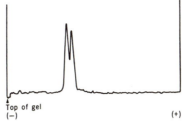

Fig. 1. Densitometric scan of a polyacrylamide gel containing 0.1 percent sodium dodecyl sulfate and 8M urea. Approximately 30 μg of purified tubulin was applied to the gel. Following electrophoresis the gel was stained with Coomassie brilliant blue, destained, and scanned at 550 nm.

Top of gel
(−)   (+)

the solution was maintained at 11.0 by the addition of 4N NaOH; the reaction was allowed to proceed for 15 minutes at 0°C with constant stirring. Following activation, the Sepharose was immediately washed in a sintered glass funnel at 4°C, first with 50 ml of distilled water and then twice with 50 ml of a solution containing 100 mM boric acid, 25 mM sodium borate, and 75 mM NaCl (pH 8.4). The Sepharose was then allowed to react with 40 to 50 mg of tubulin in 30 ml of the borate-saline buffer, while being stirred gently at 4°C for 24

hours. The coupled Sepharose-tubulin material was washed extensively in a column (2.5 by 5 cm) with 100 mM HCl to remove unbound protein, and was then equilibrated with borate-saline buffer (pH 8.4).

Purified tubulin containing only the 6S molecule was cross-linked with glutaraldehyde to form an insoluble complex. To 10 mg of tubulin in 2.0 ml of buffer was added an equal volume of 2 percent glutaraldehyde at 0°C. The precipitated tubulin was collected by centrifugation and washed twice with 0.15M NaCl. Freund's complete adjuvant (5 ml) was added and the mixture was homogenized. The rabbits were given an injection of 3.5 mg of protein, followed by two booster injections of 1.5 to 2.0 mg at 10-day intervals. One week after the last injection the rabbits were bled via cardiac puncture. The serum was precipitated by adding an equal volume of saturated ammonium sulfate. The precipitate was collected by centrifugation and dialyzed against borate-saline buffer.

Monospecific antibody against tubulin was isolated by passing the rabbit gamma-globulin fraction previously dialyzed against the borate-saline buffer, through the immunosorbent column. After all the unbound gamma globulin was eluted, the elution buffer

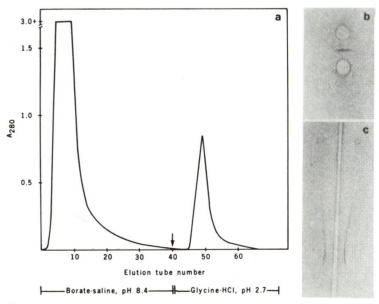

Fig. 2. (a) Absorbancy (A) at 280 nm of a chromatographic column of the gamma-globulin fraction from the immunosorbent column to which tubulin was covalently bound. The arrow indicates the point at which 200 mM glycine-HCl buffer (pH 2.7) is added to disrupt the antigen-antibody complex. (b) Double immunodiffusion of pure tubulin (6S) against antibody eluted from the immunosorbent column. (c) Immunoelectrophoresis in barbital buffer (pH 8.2).

was changed to 200 mM glycine-HCl (pH 2.7), at which point the bound tubulin antibody was released. This monospecific antibody was concentrated by ammonium sulfate precipitation and dialyzed extensively against 0.15M NaCl before being stored at −20°C.

Double immunodiffusion was carried out in 1 percent Bacto agar (Difco) in 0.15M NaCl with 0.1 percent NaN$_3$ (10). Immunoelectrophoresis was performed with 2 percent Bacto agar in 0.05M barbital buffer (pH 8.2), according to the method of Scheidegger and Roulet (11). Electrophoresis was carried out for 75 minutes at 4°C with a constant voltage of 30 volts across the agar.

Tubulin prepared from bovine brain by the in vitro reassembly procedure was contaminated with a small amount of higher molecular weight material. The additional step using gel filtration on Sepharose 4B separated the higher molecular weight proteins from 6S tubulin as judged by polyacrylamide gel electrophoresis (Fig. 1).

All six rabbits injected with tubulin produced an antibody against the antigen, although the titers, as determined by hemagglutination and hemagglutination inhibition assays, were low. Antibody directed against tubulin was isolated and concentrated by passing the gamma-globulin fraction over an affinity column to which tubulin had been covalently bound (Fig. 2a). The specificity of the tubulin antibody was determined by double immunodiffusion and immunoelectrophoresis (Fig. 2, b and c).

The binding of antibody against tubulin to spindle microtubules was tested by using an indirect immunofluorescence procedure (12). Rat kangaroo cells (strain Pt K1) were grown to confluence on glass cover slips. After being rinsed in phosphate-buffered saline (PBS), the cells were fixed for 20 minutes in PBS containing 3.5 percent formaldehyde at 22°C. The cover slips were rinsed in PBS and subsequently fixed in absolute acetone at −10°C for 7 minutes. The cover slips were air-dried and then incubated with the antibody (0.20 mg/ml in borate-saline buffer) at 37°C for 1 hour. The preparations were again washed in PBS and incubated in a 1 : 1.5 dilution of fluorescein-tagged goat antiserum against rabbit immunoglobulin G (Meloy Laboratories, Springfield, Virginia) for 1 hour at 37°C. The cover slips were rinsed in PBS followed by distilled H$_2$O and mounted on glass slides in a drop of PBS-glycerol (1 : 1) for viewing in

a Leitz microscope adapted for dark-field ultraviolet microscopy. Photographs were recorded on Tri-X Pan film (Kodak).

The mitotic spindle was clearly delineated by fluorescent stain and both chromosomal (kinetochore to pole) and interpolar (pole to pole) fibers were apparent (Fig. 3). In addition, a weak cytoplasmic fluorescence was apparent throughout mitosis. At prophase (Fig. 3a), a faint background fluorescence was apparent in the vicinity of the nucleus and around the condensing chromosomes. During various stages of prometaphase, the fluorescence pattern progressed from a central spot surrounded by chromosomes (Fig. 3b) to a more spindle-shaped pattern with chromosomes distributed throughout (Fig. 3, c and d). At metaphase, the spindle was fully formed and the chromosomes were positioned at the cell equator (Fig. 3e). On separation of the daughter chromosomes at ana-

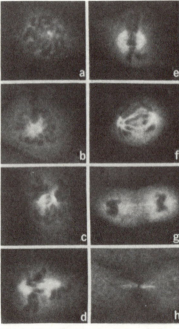

Fig. 3. Fluorescence of mitotic spindle of dividing rat kangaroo cells. (a) At prophase, a weak background fluorescence is seen around condensing chromosomes. (b) At a later stage a single bright fluorescent spot is surrounded by chromosomes. (c and d) During subsequent states of prometaphase the fluorescence becomes more spindle-shaped. (e) Metaphase. (f) Anaphase, showing both chromosomal and interpolar fibers. (g) Telophase; note the weak interpolar fibers and increased cytoplasmic fluorescence. (h) Midbody between two daughter cells.

phase, both chromosomal and interpolar fibers were apparent (Fig. 3f). At telophase, interpolar fibers were faintly fluorescent, but cytoplasmic fluorescence increased sharply (Fig. 3g), possibly due to increased free tubulin derived from spindle microtubule disassembly. Except for a brightly fluorescent midbody between daughter cells (Fig. 3h), fluorescence was greatly diminished at late telophase or the early G$_1$ phase. Electron microscopic studies of rat kangaroo cells (13) showed that microtubules were present in all the brightly fluorescent regions indicated in Fig. 3.

The procedure for preparing antibody to purified brain tubulin is relatively simple. The antibody is able to bind not only to soluble 6S tubulin but also to intact microtubules of another cell type in a different species. This observation supports and extends the concept of the ubiquitous nature of microtubule protein (7, 14). Moreover, it supports the idea that the gene for tubulin is highly conserved. Antibody to purified 6S tubulin should be a useful molecular probe for studies of the function of this important molecule.

GERALD M. FULLER
B. R. BRINKLEY
J. MARK BOUGHTER
Department of Human Biological Chemistry and Genetics, Graduate School of Biomedical Sciences, University of Texas Medical Branch, Galveston 77550

References and Notes

1. K. R. Porter, in *Principles of Biomolecular Organization*, G. E. Wolstenholme and M. O'Connor, Eds. (Little, Brown, Boston, 1966), p. 308.
2. R. C. Weisenberg, G. C. Borisy, E. W. Taylor, *Biochemistry* **7**, 4466 (1968).
3. J. B. Olmsted and G. G. Borisy, *ibid.* **12**, 4282 (1973); J. Bryan, *Fed. Proc.* **33**, 152 (1974); L. Wilson, J. R. Bamburg, S. B. Mizel, L. M. Grisham, K. M. Creswell, *ibid.*, p. 158.
4. R. C. Weisenberg, *Science* **177**, 1104 (1972).
5. M. L. Shelanski, F. Gaskin, C. Cantor, *Proc. Natl. Acad. Sci. U.S.A.* **70**, 765 (1973).
6. A. Nagayama and S. Dales, *ibid.* **66**, 464 (1970); C. Fulton, R. E. Kane, R. E. Stephens, *J. Cell Biol.* **50**, 762 (1971).
7. S. Dales, *J. Cell Biol.* **52**, 748 (1972).
8. J. D. Kowit and C. Fulton, *J. Biol. Chem.* **249**, 3638 (1974).
9. P. Cuatrecasas, M. Wilcheck, C. B. Anderson, *Proc. Natl. Acad. Sci. U.S.A.* **61**, 636 (1968).
10. O. Ouchterlony, *Prog. Allergy* **5**, 11 (1958).
11. J. J. Scheidegger and H. Roulet, *Praxis* **44**, 73 (1955).
12. E. Lazarides and K. Weber, *Proc. Natl. Acad. Sci. U.S.A.* **71**, 2268 (1974).
13. B. R. Brinkley and J. Cartwright, Jr., *J. Cell Biol.* **50**, 416 (1971); U. P. Roos, *Chromosoma* **40**, 43 (1973).
14. J. B. Olmsted and G. G. Borisy, *Annu. Rev. Biochem.* **42**, 507 (1973).
15. Supported by PHS grant CA 14675. We thank D. Highfield for help in preparing the immunofluorescent slides and W. J. Mandy for helpful advice in preparing the antibody to tubulin.

19 August 1974; revised 18 October 1974

Proc. Natl. Acad. Sci. USA
Vol. 74, No. 9, pp. 3883–3887, September 1977
Cell Biology

Intracellular distributions of mechanochemical proteins in cultured fibroblasts

(microfilament organization/cell–cell contact/cell motility/filamin)

MICHAEL H. HEGGENESS, KUAN WANG*, AND S. J. SINGER

Department of Biology, University of California at San Diego, La Jolla, California 92093

Contributed by S. J. Singer, June 24, 1977

ABSTRACT   We have used methods that have allowed simultaneous fluorescent staining of intracellular actin together with either myosin, filamin, or tubulin in normal rat kidney fibroblasts in monolayer culture. In the main portions of the cell body, the actin, myosin, and filamin are all present in two structures: in one, the three proteins are present in the same fiber bundles (stress fibers); in the other, there is a diffuse distribution of the three proteins. On portions of the cell periphery however—in the basal regions of microspikes, in ruffles, and in regions of cell–cell contact—actin and filamin are present, but myosin is severely depleted or absent. Microtubules are present in the cell body in a distribution independent of the stress fibers and are mostly absent from the cell periphery. Microspikes and ruffles are highly dynamic structures on the cell surface, and regions of cell–cell contact generally result from the association of ruffles on the two contacting cells. Therefore, the presence of filamin and actin but not myosin in these specialized regions on the cell surface, together with the recent demonstration [Wang, K. & Singer, S. J. (1977) Proc. Natl. Acad. Sci. USA 74, 2021–2025] that pure filamin interacts with individual F-actin filaments in solution to form fiber bundles and sheet-like structures, suggest that in vivo filamin–actin interactions play an important role in the control of actin filament structure, in cell motility, and in the stabilization of cell–cell contacts.

It is well known that eukaryotic nonmuscle cells contain contractile proteins such as actin and myosin similar to those found in muscle cells (for review, see ref. 1). With fibroblasts in monolayer culture, it has been observed that these proteins are, in part at least, organized into extended filaments inside the cell (2, 3). At present, however, detailed interactions among these proteins in forming such filaments, and the relationship of these structures to phenomena such as cell motility and cell–cell contact, are not understood. As a step towards the elucidation of such problems, we have carried out experiments to localize two specific mechanochemical proteins simultaneously in the same normal rat kidney (NRK) fibroblast. Our first studies have been done at the light microscopic level of resolution, by using specific fluorescence staining techniques for actin, myosin, filamin (3, 4), and tubulin. The actin was stained by a modification (5) of the fluorescein-labeled heavy meromyosin technique (6), whereas the other proteins were stained one at a time by specific rhodamine immunofluorescence methods. These experiments have revealed some interesting differential distributions of the four proteins within fibroblasts which are presented and discussed in this paper.

MATERIALS AND METHODS

Cell Culture. The cell line NRK (7) was maintained at 37° in Coons' modified F-12 medium supplemented with 10% fetal calf serum and antibiotics in an atmosphere of 90% air/10% CO₂. Cells to be stained were plated on glass coverslips at densities of 1 to 2×10^3 cells per cm² and allowed to grow for from 24 to 72 hr before fixation and staining, at which time the cell density was between 2 and 10×10^3 cells per cm².

Antibodies and Staining Reagents. Rabbit antibodies were used as primary reagents. Rabbit antibodies specific for human uterine myosin (which crossreacts with NRK myosin) (3, 8) and chicken gizzard filamin (ref. 3; K. Wang and S. J. Singer, unpublished data), have been described. Rabbit antibodies prepared against highly purified tubulin from 12-day-old chick embryo brains (9) were the gift of Melvin Simon. Goat antibodies against rabbit IgG were used for the indirect staining of the rabbit antibodies. The IgG fraction of the goat antiserum was derivatized with Lissamine rhodamine B sulfonyl chloride and was fractionated by ion exchange chromatography on DE-52 cellulose (10). The conjugates used in this study had rhodamine/IgG molar ratios of between 1.5 and 2.7.

Actin was localized in these cells using the reagents biotin-conjugated heavy meromyosin and fluorescein-conjugated avidin as described (5).

Staining of Cells. Formaldehyde (3%) in phosphate-buffered saline (PBS), pH 7.4, was warmed to 37° and applied to the cells on coverslips for 20–45 min at room temperature. The cells were then rinsed with PBS, incubated for 10 min in PBS containing 0.1 M glycine or 0.05 M NH₄Cl to quench any remaining aldehyde functions, and rinsed again in PBS. The fixed cells were then rendered permeable to protein reagents by a 2 min exposure to 0.1% Triton X-100 in PBS. This treatment resulted in better structural preservation than either acetone treatment or freezing–thawing. The cells were then washed thoroughly in PBS and were treated with a mixture of biotin-conjugated heavy meromyosin (0.2–0.7 mg/ml) and a rabbit antibody IgG (0.1–0.5 mg/ml) to either myosin, filamin, or tubulin for 20 min at room temperature. Purified IgG was always used in this step, as serum interfered with actin staining by this method. After thorough washing in PBS, the cells were then treated with a mixture of fluorescein-conjugated avidin (0.05 mg/ml) and rhodamine-conjugated goat anti-rabbit-IgG (0.05–0.2 mg/ml) in PBS for 20 min. Following a further thorough washing, the cover slip was inverted on a drop of 90% glycerol/10% PBS and the cells were observed with a Zeiss Photoscope III using epi-illumination. The filter combinations used were CZ 487710 and CZ 487714 for fluorescein and rhodamine observation, respectively. The two fluorescences were always photographed without changing focus. Specimens were photographed using Kodak Plus X or Panatomic X film.

As controls for the specificity of the staining reactions, it was

Abbreviations: NRK, normal rat kidney; PBS, phosphate-buffered saline.
* Present address: Department of Chemistry and Clayton Foundation Biochemical Institute, University of Texas at Austin, Austin, TX 78712.

3883

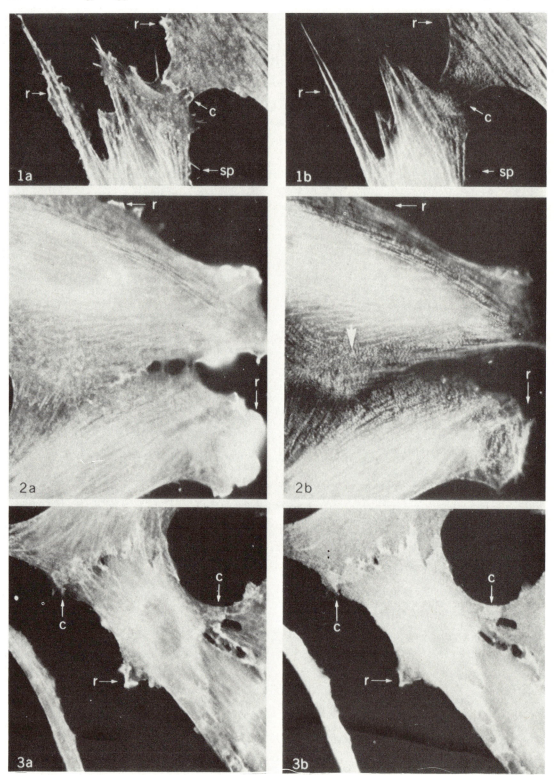

Legend to Figs. 1–5 on following page.

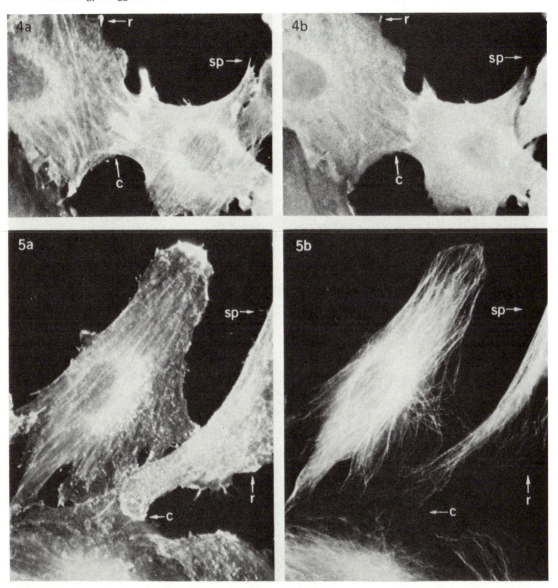

Figs. 1–3 on preceding page.

FIGS. 1–5.　Double fluorescence staining pictures of mechanochemical proteins inside NRK cells in monolayer culture. Each figure shows the same cell stained in *a* for actin and in *b* for either myosin (1*b* and 2*b*), filamin (3*b* and 4*b*), or tubulin (5*b*). The symbols represent: *sp*, microspike; *r*, ruffle; *c*, region of cell–cell contact. Note the staining for actin in *sp*, *r*, and *c* (1*a*–5*a*), and the absence of staining for myosin (1*b* and 2*b*) in these same structures. Note also the staining for filamin in *r* and *c* (3*b* and 4*b*), and in the basal portion of *sp* (4*b*). The large arrow in 2*b* points to a reticular pattern of myosin staining. (×950.)

found that myosin, filamin, and tubulin staining could be eliminated by preabsorption of the respective rabbit antibodies by the appropriate antigen. The specificity of actin staining was established by its elimination by Mg pyrophosphate (5 mM) or by free biotin (0.1 mg/ml) (5).

RESULTS

Double staining for the intracellular actin and myosin components of NRK cells gave results such as those shown in Figs.

1 and 2. The staining observed was specific, because the controls were negative. All the staining was intracellular, as demonstrated by the fact that cells were not stained if the Triton X-100 treatment was omitted. The actin and myosin were found in part to be organized into extended filaments, the so-called stress fibers that have been observed previously (2, 3). As is particularly clear in Fig. 1, the same filaments were stained with both actin and myosin reagents, and hence both proteins were present in the same fiber bundles. In addition to the stress fibers, however, there was a more diffuse staining of actin in the cell

body, as is especially evident in Figs. 2*a* and 3*a*. The myosin present outside of the stress fibers, by contrast, often exhibited a more reticular pattern than the actin (compare Fig. 2 *a* and *b*, large arrow).

As illustrated by Figs. 1 and 2, marked differences between the extents of actin and myosin staining were always observed in certain regions of the cell periphery, at microspikes (*sp*) and at ruffles (*r*) on isolated cell surfaces, as well as at regions of contact between two cells (*c*). In each of these regions, actin staining was quite intense, often more intense than in the interior of the cell, but little or no myosin staining was visible.

When double staining for actin and filamin was performed, results such as those shown in Figs. 3 and 4 were obtained. In the interior of the cell, filamin was organized on stress fibers (Fig. 4*b*, upper cell) (see also ref. 3), as well as more diffusely spread. Especially interesting was the coincident staining of filamin and actin in ruffles and in regions of cell–cell contact. Filamin staining was also always observed in the basal portions of microspikes (Fig. 4*b*; contrast with Fig. 1*b*) but not in the extremities where actin staining persisted.

Double staining for actin and tubulin (Fig. 5) showed that the stress fibers and microtubules in the cell interior, although distinctly separate structures, were often in roughly parallel alignment along the long axis of a cell. Microtubules were present in much lower density near the cell periphery than in the interior and were generally depleted or absent from microspikes and ruffles. Whereas some microtubules extended into the regions of cell–cell contact, there was no regularity to their distribution in these regions.

DISCUSSION

The intracellular distributions of the mechanochemical proteins studied in this paper, and of others as well, have previously been individually investigated with fluorescence staining methods by several investigators (2, 3, 11–13). Such single staining experiments have yielded important results, but some of the conclusions drawn from our experiments could not have been derived in the absence of double staining. The double staining was facilitated by the development of a sensitive nonantibody method for the specific staining of actin (5), which allowed us to use an indirect immunofluorescence technique for the second stain.

Considering the three proteins, actin, myosin, and filamin, in NRK fibroblasts, we have observed several different kinds of distributions, which will be discussed in more detail elsewhere. For the purposes of the present paper, however, we wish only to distinguish between the distributions found in certain specialized regions of the cell periphery and those found in the rest of the cell body. Whereas all three proteins are present in most of the cell body (in part, on the same stress fibers), by contrast in basal regions of microspikes, in ruffles, and in regions of cell–cell contact, substantial amounts of actin and filamin are present, but myosin is greatly diminished or absent. We have obtained very similar results with human WI-38 fibroblasts. Consistent with the absence of myosin from these specialized peripheral regions, Lazarides (13) has recently shown that tropomyosin is also diminished or absent in the cell ruffles of cultured myoblasts and fibroblasts. Microtubules do not appear to play any role in these specialized peripheral regions.

These results are particularly interesting because of the functional relationships that have been proposed between microspikes and ruffles and between ruffles and regions of cell–cell contact.

Microspikes and ruffles are both highly dynamic structures, forming and retracting at different sites on the cell surface. Time-lapse photography of living fibroblasts (14) has suggested that the rapid extension and retraction of several microspikes generally precede the formation of a ruffle on the same region of the cell surface. Ruffles are pancake-like structures that lift off the surface of the substrate to which the cell is attached and appear to be the principal motile apparatus of the cell. When a cell is isolated from contact with other cells, such ruffles perform oscillatory motions until they retract or become anchored to the substrate. But when two cells make contact by their respective ruffles, ruffling stops at the contact region (contact inhibition of motility) (15) and the cells remain so attached for some time. These considerations suggest therefore that microspikes, ruffles, and regions of cell–cell contact all arise sequentially at the same areas on the cell surface. The absence of myosin from all three may therefore reflect this spatial and functional relationship among them.

When the ruffles of two normal cells make contact, leading to an inhibition of motility, it has been shown by electron microscopy (16, 17) that fiber bundles rapidly appear within the two ruffle regions. Our studies suggest that these fiber bundles differ from the stress fibers found in the cell interior in that they do not contain myosin. The presence of actin and filamin in ruffles, however, is especially interesting in view of our recent demonstration (18) that filamin and F-actin interact specifically in solution. Mixtures of the two pure proteins rapidly form aggregates that contain bundles of fibers and sheet-like structures made up of F-actin filaments crosslinked by filamin molecules. The interaction of these two proteins may therefore be responsible for the formation of fiber bundles *in vivo* in ruffles that have made contact (and perhaps also in other regions of the cell body). If such were the case, then the contact between ruffles must rapidly convey a signal for the filamin and actin present within the ruffles to interact with one another. This is of interest in connection with malignant transformation, because cancer cells that make contact via their ruffles do not show contact inhibition of motility nor do they form any fiber bundles within the contacting ruffles (19).

If indeed filamin–F-actin interactions are implicated in these cell contact phenomena, the detailed molecular mechanism of the interactions and the signals required to initiate them remain to be discovered.

We are grateful to Dr. J. F. Ash for substantial contributions to the development of these techniques and to Dr. Immo Scheffler for providing cell culture facilities. We thank Ms. Donna Luong and Mr. George Anders for excellent technical assistance. S.J.S. is an American Cancer Society Research Professor, and K.W. was a U.S. Public Health Service Postdoctoral Fellow, 1974–1976. This work was supported by U.S. Public Health Service Grant GM-15971 to S.J.S.

1. Pollard, T. D. & Weihing, R. R. (1974) *C.R.C. Crit. Rev. Biochem.* **2,** 1–65.
2. Lazarides, E. & Weber, K. (1974) *Proc. Natl. Acad. Sci. USA* **71,** 2268–2272.
3. Wang, K., Ash, J. F. & Singer, S. J. (1975) *Proc. Natl. Acad. Sci. USA* **72,** 4483–4486.
4. Wang, K. (1977) *Biochemistry* **16,** 1857–1865.
5. Heggeness, M. H. & Ash, J. F. (1977) *J. Cell Biol.* **73,** 783–788.
6. Sanger, J. W. (1975) *Proc. Natl. Acad. Sci. USA* **72,** 1913–1916.
7. Duc-Nguyen, H., Rosenblum, E. N. & Zeigel, R. F. (1966) *J. Bacteriol.* **92,** 1133–1140.
8. Sheetz, M., Painter, R. G. & Singer, S. J. (1976) *Biochemistry* **15,** 4486–4492.

Cell Biology: Heggeness *et al.*

Proc. Natl. Acad. Sci USA 74 (1977)   3887

9. Shelanski, M. L., Gaskin, F. & Cantor, C. R. (1973) *Proc. Natl. Acad. Sci. USA* **70,** 765–768.
10. Brandtzaeg, P. (1973) *Scand. J. Immunol.* **2,** 273–279.
11. Brinkley, B. R., Fuller, G. M. & Highfield, D. P. (1975) *Proc. Natl. Acad. Sci. USA* **72,** 4981–4985.
12. Pollard, T. D., Fujiwara, K., Niederman, R. & Maupin-Szamier, P. (1976) in *Cell Motility,* eds. Goldman, R. D., Pollard, T. & Rosenbaum, J. (Cold Spring Harbor Laboratory, Cold Spring Harbor, NY), Vol. B, pp. 689–724.
13. Lazarides, E. (1977) *J. Supramol. Struct.,* **5,** 531–563.
14. Albrecht-Buehler, G. (1976) in *Cell Motility,* eds. Goldman, R. D., Pollard, T. & Rosenbaum, J. (Cold Spring Harbor Laboratory,

Cold Spring , Harbor, NY), Vol. A, pp. 247–264.
15. Abercrombie, M. (1961) *Exp. Cell Res. Suppl.* **8,** 188–198.
16. Heaysman, J. E. M. & Pegrum, S. M. (1973) *Exp. Cell Res.* **78,** 71–78.
17. Goldman, R. D., Schloss, J. A. & Starger, J. M. (1976) in *Cell Motility,* eds. Goldman, R. D., Pollard, T. & Rosenbaum, J. (Cold Spring Harbor Laboratory, Cold Spring Harbor, NY), Vol. A, pp. 217–245.
18. Wang, K. & Singer, S. J. (1977) *Proc. Natl. Acad. Sci. USA* **74,** 2021–2025.
19. Heaysman, J. E. M. & Pegrum, S. M. (1973) *Exp. Cell Res.* **78,** 479–481.

Cell, Vol. 12, 333–339, October 1977, Copyright © 1977 by MIT

Phagokinetic Tracks of 3T3 Cells: Parallels between the Orientation of Track Segments and of Cellular Structures Which Contain Actin or Tubulin

Guenter Albrecht-Buehler
Cold Spring Harbor Laboratory
P.O. Box 100
Cold Spring Harbor, New York 11724

Summary

Phagokinetic tracks were used to determine the current direction of migration in 3T3 cells. Comparing this direction with the orientation of actin- or tubulin-containing cellular structures by indirect immunofluorescence, the following results were obtained.

First, the main actin-containing bundles were located at the bottom and tail end of 3T3 cells and ran parallel to the current or preceding direction of migration.

Second, the 3 μm long rod-like structure (primary cilium), which contains tubulin and which has been observed by other investigators in transmission electron microscopy (Barnes, 1961; Sorokin, 1962; Wheatley, 1969) and in indirect immunofluorescence (Osborn and Weber, 1976), was oriented predominantly parallel to the substrate and to the current movement direction.

It seems possible that the primary cilium has a role in the directional control of a migrating 3T3 cell, and that the main actin containing bundles act as substrate-attached rails along which the nucleus and bulk cytoplasm slide during displacement of the cells.

Introduction

Phagokinetic tracks are observed if cultured animal cells migrate on a densely gold particle-coated solid substrate and clear the particles out of their way (Albrecht-Buehler, 1977a, 1977b). Studying such tracks of 3T3 cells, I observed a number of surprising aspects of cellular movement.

In the majority of cases, the track of one daughter cell appeared as the mirror image or identical copy of the other daughter cell's track (Albrecht-Buehler, 1977a). In several cases, details of the track of an ancestor cell were found to be repeated by descendant cells up to three generations. Different cell lines produced characteristically different track patterns (Albrecht-Buehler, 1977b).

These observations suggested a detailed predetermination of the displacements in migrating 3T3 cells and raised the suspicion that there are cellular structures able to store movement instructions and to control their timely execution by the motile apparatus of the cell.

In search of such structures, I assumed that one might approach them by looking for cellular structures which are conspicuously oriented to the current or past direction of a migrating cell. I therefore examined by indirect immunofluorescence the intracellular distributions of various antigens in individual 3T3 cells and compared them with the track patterns produced by the same cells. It appeared that the actin and tubulin patterns (Lazarides and Weber, 1974; Osborn and Weber, 1976; Fuller, Brinkley and Boughter, 1975) contained linear structures, which were oriented parallel to the track segment which the cell was following at the time of fixation. These relationships are described in this paper.

Results

Comparison between 3T3 Cells on Plain Glass and on Gold Particle-Coated Glass Substrates

More than 99% of the 3T3 cells on the test substrates produced tracks of 200–300 μm length during 1 day in the presence of culture medium (DME) containing 10% calf serum. I used low inocula of 200–400 cells per cm^2 to reduce the number of intersecting tracks. Below, I describe the actin and tubulin patterns of such cultured 3T3 cells fixed and processed 1 day after plating. Comparing the actin and tubulin patterns of such 3T3 cells (Figures 1 and 2) with the patterns of 3T3 cells on plain glass which were published earlier (Lazarides and Weber, 1974; Goldman et al., 1975; Osborn and Weber, 1976; Brinkley, Fuller and Highfield, 1976), the reader will notice that they look less complex. This difference appears to be due to the sparse inocula and short culturing time: if 3T3 cells were plated on plain glass and treated similarly, the detected fluorescent patterns of both antigens were similar to the patterns shown in Figures 1 and 2. The cells plated on gold particle-coated substrates, however, generally showed an increase of diffuse fluorescence near the nucleus compared to cells on plain glass which showed only faint rings of diffuse fluorescence in this area. The gold particles which were picked up and internalized during track formation by the cells are accumulated around the nucleus. I therefore believe that the diffuse fluorescence reflects organizations or disorganizations of the respective antigens which are involved in phagocytosis of the particles or in maintaining them around the nucleus. I will ignore this diffuse nuclear fluorescence and describe only other features of the detected fluorescent patterns.

Actin Patterns and Phagokinetic Tracks

The front and rear ends of a cell can be determined by the track direction (they are indicated as "f" or "r" in the figures). According to this definition, the major actin-containing bundles stretched between

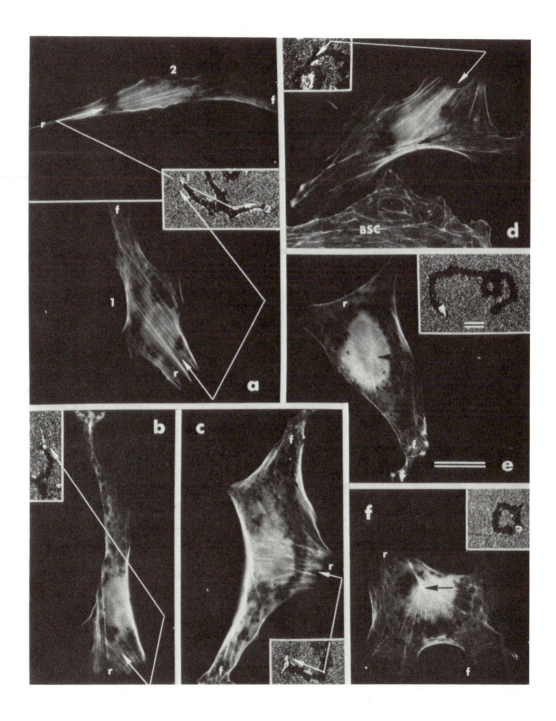

the rear end of the cell and the nucleus (Figure 1). The front end only rarely showed bundles.

There was always one most pronounced bundle stretching from the rear end along the cell bottom to the front, which also passed underneath the nucleus (Figures 1a–1d, arrows). The single fibers of this bundle could run parallel to each other or converge toward a point at or behind the rear end of the cell. In view of the "tail retraction" of migrating fibroblasts (Goldman, Schloss and Starger, 1976), I believe that the converging bundles are in the process of forming a tail (for example, Figure 1a, cell 2) or are remainders of an already retracted tail (for example, Figure 1d).

The small number of different bundle directions in the majority of 3T3 cells allowed one to compare them rather easily with the similar few directions of approximately straight track segments produced by the same cells during 1 day. Below, I will distinguish between a track segment whose front end is still occupied by the cell (current segment) and the immediately preceding segment, separated from the current segment by a major directional change in the track (previous segment).

I examined the actin patterns and corresponding tracks of 113 individual 3T3 cells. In 61% of the cases, the symmetry axis of the main bundle stretched parallel to the current segment (Figures 1a, 1c and 1d) or previous segment (Figure 1b) of the corresponding track. In 39% of the cases, I could not determine any relationship between bundles and track segments because the segments were too short or irregular. Sometimes the cells had too many different bundle directions or were rounding up.

The following three observations may be of special interest. First, quite frequently, 3T3 cells turn sideways, thus producing wide "thorns" along their tracks (Albrecht-Buehler, 1977b). In such cases, I found that the major bundle still pointed parallel to the previous track segment. Figure 1d shows such a case of a sideways extending cell as judged by the ruffles at the areas marked "f." It

therefore seems that the extension to the sides is only a transient motile action of a 3T3 cell, possibly a probing from time to time of the environment to the sides of its path. Second, if a 3T3 cell collides with a (nonmigrating) BSC-1 cell as a target cell, the 3T3 cell develops an outgoing track from the collision area, as if it were elastically reflected at the target (Albrecht-Buehler, 1977b). Figure 1d shows such a collision in progress. In the insert of Figure 1d, the nonmigrating BSC-1 cell is located inside the particle-free ring which was cleaned by the cell's surface projection after the cell settled down on the particle-coated substrate (Albrecht-Buehler, 1977b). The BSC-1 cell remains inside this ring. In contrast, one sees the track of the impacting 3T3 cell approaching from the upper right. The main bundle of the impacting cell stretched parallel to the current track segment. A bundle in the front end of the 3T3 cell, however, ran approximately tangential to the circumference of the BSC-1 cell. I collected twenty such cases. A tangential bundle could be seen in eleven of them. It remains to be seen whether such tangential bundles are constructed by the cell upon collision with another, and whether they are the structural basis of the "tangential perseverence" (Albrecht-Buehler, 1977b) of colliding 3T3 cells. Third, in less than one in 1000 cases, cells crossed their own tracks, thus producing loops. The rarity of this event suggests that almost all 3T3 cells alternate the side of directional changes as they migrate. I found twelve such "waltzing" 3T3 cells in the preparations. Five of these showed a peculiar actin organization with many fibers radiating from a single center toward the cell periphery (Figures 1e and 1f). A possible meaning of such organized actin fibers is discussed below.

Tubulin Patterns and Phagokinetic Tracks
Most of the cytoplasmic microtubular patterns appeared in the front half of the cells (Figures 2a and 2b). The density of fluorescent fibers in the rear end seemed to be reduced (Figure 2f) unless diffuse

Figure 1. Parallels between the Directions of the Main Actin-Containing Bundle and the Current or Preceding Direction of Displacement in 3T3 Cells (Anti-Actin Stain)

The inserts show darkfield light micrographs of the same cells (white clusters) in their phagokinetic tracks as seen in fluorescence light microscopy. Inserts and fluorescence micrographs are shown in identical orientations relative to the observer. Bars in (e) represent 20 μm and 100 μm. The connected arrows indicate the observed parallels between the main actin-containing bundles and the respective track directions. The letters "f" and "r" designate the front and rear ends of the cells as derived from the track directions. In cells which extend sideways (for example, c), both fronts were marked "f." The black clusters in the fluorescent micrographs are internalized clusters of gold particles.

(a) Pair of daughter cells with the main bundles parallel to the current track segment. The bundles of cell 2 converge due to the tail formation.

(b) Cell with the main bundle parallel to the previous track segment, while the other bundles parallel to the current direction are formed on the left side of the cell.

(c) A sideways turning cell with the main bundle still parallel to the previous track segment.

(d) A 3T3 cell in collision with a BSC-1 cell. The main bundle is parallel to the current track segment, while a bundle seems to be formed in the front end which runs approximately tangential to the perimeter of the BSC-1 cell (black arrow).

(e and f) "Waltzing" 3T3 cells with a peculiar centralized actin bundle pattern (black arrows point to the centers).

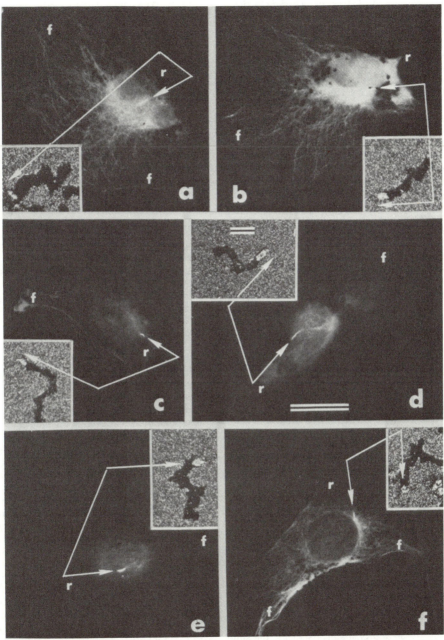

Figure 2. Parallels between the Direction of the Primary Cilium (Rod-Like Structure in Front of Arrows) and the Direction of Movement in 3T3 Cells (Anti-Tubulin Stain)

General description and magnification bars as in Figure 1.
(a and b) Cells with primary cilia close to the substrate showing the typical fine fibrous network revealed by antitubulin antibody strain. The cilia run parallel to the current orientation of the cells. The cell in (a) is in the process of extending sideways. The microtubules are slightly out of focus because the focus is raised to the level of the cilia.
(c and e) Cells with the primary cilia close to the dorsal cell surface. The focus is raised in the fluorescent micrographs, thus rendering most of the microtubular network invisible.
(f) A cell extending sideways, while the primary cilium still points in the direction of displacement of the cell.

fluorescence around the accumulated gold particles (see section 1) made it impossible to detect single fibers (Figures 2a and 2b). Due to the complexity of the cytoplasmic microtubular patterns, a relationship between them and the track directions was not obvious.

Most interesting for the search of structures which are conspicuously oriented to track segments, however, was the fluorescent-labeled rod-like structure which was described by Osborn and Weber (1976). In electron microscopical studies, Wheatley (1971, 1972) had shown earlier that 50–75% of 3T6 cells contained one single, mostly intracellular cilium growing out of a centriole ("primary cilium") whose shaft was embedded in a vacuole. The origin of primary cilia from centrioles had been observed earlier by Barnes (1961) and Sorokin (1962). Osborn and Weber (1976) argued that the fluorescent rod-like structure is located in the centrospheric region and is identical with the primary cilium described by Wheatley; I shall therefore use that term in referring to this structure.

The following results were obtained with cells between 12 and 14 passages. Screening 101 such 3T3 cells on plain glass, I found a single primary cilium in 87% of them. Invariably, it was located close to the nucleus. As far as one can tell in light microscopy using an objective lens (N.A. 1.4) with a small depth of focus (approximately $\pm$ 0.7 μm), in 61% of the cells, the cilium was straight and ran approximately parallel to the substrate. In 14% of the cells, it was bent, and in 12%, it was oriented vertically to the substrate. In one cell, I found two cilia parallel to each other.

The cilia frequency in 3T3 cells plated on gold particle-coated glass substrates was lower than on plain glass. Screening 487 single cells, I detected one single cilium in only 258 of them, corresponding to 53%. This reduced frequency may be explained by a reduced visibility of cilia on such substrates, caused by internalized gold particle clusters or a high background level of diffuse fluorescence around the nucleus. There may be other, as yet unknown, factors influencing the expression of cilia. In four different preparations, I found variations of the cilium frequency between 28 and 73%. I could not detect the source of this variation.

Comparing the orientation of the primary cilium with the direction of the phagokinetic track in 258 cells, where the structure could be detected, I found that it ran approximately parallel to the substrate and current segment in 73% of the cells (Figures 2a–2f). In the remaining 27%, it was vertical or bent or pointed in a direction which did not seem related to the track direction. Unfortunately, the resolution of the light microscope does not allow one to decide whether the cilium points in the foreward or the backward direction of the migrat-

ing cell. The cilium could be detected near the cell bottom (Figures 2a, 2b and 2f) or close to the dorsal surface (Figures 2c–2e). In cells which extended sideways, the primary cilium remained oriented parallel to the just completed segment. As judged by many 3T3 cell tracks which showed "thorns" at the places where the cells had turned sideways, the cells tend to resume their original direction afterwards. It therefore seems that in such cases, the primary cilium is also predominantly oriented parallel to the current direction of a cell's displacement.

The percentage of cells whose cilia were oriented parallel to the current track segment was independent of the cilium frequency of the whole population. In the above-mentioned four preparations with cilia frequencies between 28 and 73%, this percentage varied only between 69 and 78%.

The expression of cilia is not a sufficient condition for 3T3 cells to displace themselves. Confluent quiescent 3T3 cultures after 8–9 days without medium change showed cilia in 100% of the cells, although the cells of such treated cultures do not migrate any more. The orientation of the cilia in such cultures appeared random.

Discussion

The discussion of the above results is complicated by the difficulty in distinguishing between cause and effect in the relationships described above between the directions of the primary cilia, the main actin bundles and the track segments. Nevertheless, I should like to submit the following considerations.

The main actin-containing bundles predominantly stretched from the rear end of the cells to beneath the nucleus and ran parallel to the current or previous segments of the tracks. It therefore seems possible that the fibers of the main bundles became organized close to the substrate in parallel lines, and that the nucleus and the bulk cytoplasm moved along them as if on "rails." The bundles could conceivably lengthen further during this process. According to this model, the used parts of the rails were left behind, converged in the rear (see Figures 1a and 1d) and eventually were disassembled during the tail retraction. If they disassembled incompletely, their remainders were carried along, remaining parallel to the previous segment of the track.

This notion of cellular displacements on "rails" offers a reason why the main bundles in 3T3 cells, which are in fact microfilament bundles (Goldman et al., 1975), are located near the substrate and are often tied to the substrate by adhesion plaques (Goldman et al., 1976). Such features are implied by the term "rail" in the first place. Movement along rails would also allow the nucleus and bulk cytoplasm to move parallel to themselves, thus possibly reducing the chance of entangling the complex cytoplasmic architecture during displacement. Furthermore, it would explain the observation that 3T3 cells, if considered random walkers, move persistently during short time intervals of 2.5 hr (Gail and Boone, 1970) — that is, they have a high probability of moving in the same direction as they moved before. 2.5 hr is not enough time for 3T3 cells to displace themselves by a whole cell diameter. An observer marking the position of migrating 3T3 cells at such time intervals will find many cells continuing to move along their straight rails, thus appearing persistent in their direction of movement.

The observations of the star-like actin patterns of some "waltzing" 3T3 cells offer an additional aspect. If cells with such actin patterns follow the directions of the "spokes" sequentially, they would indeed move in loops. One may therefore speculate that the possible direction a 3T3 cell can take during interphase is predetermined intracellularly by the fibers of a special actin-containing network, which is normally not detected by immunofluorescence, unless parts of it become amplified — for example, to form the main bundle. In the exceptional cases of waltzing cells, this fiber network may be permanently detectable for reasons which are as yet unknown.

Most striking is the observed relationship between the orientation of primary cilia and current track segments. There are various possible interpretations of this relationship ranging from the assumption that the cilium shaft is passively oriented by certain cytoplasmic forces which act during cellular displacement, to the assumption that the direction of the cilium determines the direction of a migrating 3T3 cell by an as yet unknown mechanism. One could also think of the primary cilium as a "sensor" or mechano-transducer which is aligned with the "intended" direction of movement to yield to and thus detect cytoplasmic movements deviating from this direction. There is also the possibility that the cilium itself is of little relevance for the movement of cells. One may consider it simply a convenient light microscopic marker for the direction of the long axis of the centriole out of which it grew. In other words, the described parallels between the direction of primary cilia and current track segments may have to be interpreted as a geometric relationship between the direction of migration and the orientation of one centriole in migrating interphase cells. Based on the present data, however, it is impossible to decide between these alternatives. The conjecture of a sensory function of the primary cilia in cultured 3T3 cells is based on the observations that retinal rod cells

(DeRobertis, 1956) and olfactory epithelium (Reese, 1965) contain similar cilia in their sensory parts.

The cases where the primary cilia were bent or vertical to the substrate may be related to the process of turning the cilia in a new movement direction. In BHK21 cells, the primary cilia disappear at an unspecified time after S phase and reappear in late telophase (Archer and Wheatley, 1971). Thus the observed 3T3 cells without primary cilia may be approaching prophase or may not have left telophase long enough. This explanation is supported by the finding that 100% of the 3T3 cells in confluent quiescent cultures express cilia.

Wheatley (1972) noted that primary cilia are expressed in primary fibroblastic cell cultures from embryonic mice in about the same frequency as in established murine cell lines; however, they are missing or very poorly expressed in old established L-929 cells. They are also missing, but can be induced in Chinese hamster fibroblasts by colchicine treatment (Stubblefield and Brinkley, 1966). It remains to be seen whether cells with reduced cilia frequency also show reduced directionality of track formation.

Experimental Procedures

Cells, culture conditions and the method of obtaining phagokinetic tracks have been described previously (Albrecht-Buehler, 1977b). Rabbit-anti-actin antibody was a gift from Dr. K. Burridge (Cold Spring Harbor Laboratory, Cold Spring Harbor, New York), and rabbit-antitubulin antibody was a gift from Dr. G. M. Fuller (University of Texas, Galveston, Texas). Processing for indirect immunofluorescence of 3T3 cells on gold particle-coated glass substrates has been described elsewhere (Albrecht-Buehler, 1977a).

Acknowledgments

I thank Dr. James D. Watson for his continuing encouragement and support. I am grateful to Dr. Keith Burridge (Cold Spring Harbor Laboratory) and Dr. Gerald M. Fuller (University of Texas, Galveston) for their generous gifts of the antibody preparations which were used in this work, and to Dr. William Gordon (Cold Spring Harbor Laboratory) for generous permission to screen his antibody preparations. Robert M. Lancaster provided excellent experimental help, and Ms. Sallie Chait provided photographic assistence. I am grateful for Ms. Madeline Szadkowski's patient typing of the manuscript. The work was supported by a Cold Spring Harbor Laboratory Cancer Center grant from the National Cancer Institute and by grants from the NSF.

Received May 5, 1977; revised June 20, 1977

References

Albrecht-Buehler, G. (1977a). Daughter 3T3 cells. Are they mirror images of each other? J. Cell Biol. 72, 595–603.

Albrecht-Buehler, G. (1977b). The phagokinetic tracks of 3T3 cells. Cell 11, 395–404.

Archer, F. L. and D. N. Wheatley (1971). Cilia in cell-cultured fibroblasts. II. Incidence in mitotic and post-mitotic BHK21/C13 fibroblasts. J. Anat. 109, 277–292.

Barnes, B. (1961). Ciliated secretory cells in the pars distalis of the mouse hypophysis. J. Ultrastruct. Res. 5, 453–467.

Brinkley, B. R., G. M. Fuller and D. P. Highfield (1976). Tubulin antibodies as probes for microtubules in dividing and non-dividing mammalian cells. In Cell Motility, R. D. Goldman, T. Pollard and J. Rosenbaum, eds. (Cold Spring Harbor, New York: Cold Spring Harbor Laboratory), pp. 435–456.

DeRobertis, E. (1956). Electron microscope observations on the submicroscopic organization of the retinal rods. J. Biophys. Biochem. Cytol. 2, 319–329.

Fuller, G. M., B. R. Brinkley and J. M. Boughter (1975). Immunofluorescence of mitotic spindles by using monospecific antibody against bovine brain tubulin. Science 187, 948–950.

Gail, M. H. and C. W. Boone (1970). The locomotion of mouse fibroblasts in tissue culture. Biophys. J. 10, 980–993.

Goldman, R. D., J. A. Schloss and J. M. Starger (1976). Organizational changes of actinlike microfilaments during animal cell movement. In Cell Motility, R. D. Goldman, T. Pollard and J. Rosenbaum, eds. (Cold Spring Harbor, New York: Cold Spring Harbor Laboratory), pp. 217–245.

Goldman, R. D., E. Lazarides, R. Pollack and K. Weber (1975). The distribution of actin in non-muscle cells: the use of actin antibody in the localization of actin within the microfilament bundles of mouse 3T3 cells. Exp. Cell Res. 90, 333–344.

Lazarides, E. and K. Weber (1974). Actin antibody: the specific visualization of actin filaments in non-muscle cells. Proc. Nat. Acad. Sci. USA 71, 2268–2272.

Osborn, M. and K. Weber (1976). Cytoplasmic microtubules in tissue culture cells appear to grow from an organizing structure towards the plasma membrane. Proc. Nat. Acad. Sci. USA 73, 867–871.

Reese, T. S. (1965). Olfactory cilia in the frog. J. Cell Biol. 25, 209–229.

Sorokin, S. (1962). Centrioles and the formation of rudimentary cilia by fibroblasts in smooth muscle cells. J. Cell Biol. 15, 363–377.

Stubblefield, E. and Brinkley, B. R. (1966). Cilia formation in Chinese hamster fibroblasts in vitro as a response to colcemid treatment. J. Cell Biol. 30, 645–652.

Wheatley, D. N. (1969). Cilia in cell-cultured fibroblasts. I. On their occurrence and relative frequencies in primary cultures and established cell lines. J. Anat. 105, 351–362.

Wheatley, D. N. (1971). Cilia in cell-cultured fibroblasts. III. Relationship between mitotic activity and cilium frequency in mouse 3T6 fibroblasts. J. Anat. 110, 367–382.

Wheatley, D. N. (1972). Cilia in cell-cultured fibroblasts. IV. Variation within the mouse 3T6 fibroblastic cell line. J. Anat. 113, 83–93.

Filament Organization Revealed in Platinum Replicas
of Freeze-dried Cytoskeletons

J. E. HEUSER, and M. W. KIRSCHNER

Departments of Physiology, Biochemistry, and Biophysics, School of Medicine, University of California, San Francisco, California 94143, *Current address: Dept. of Physiology and Biophysics, Washington Univ., School of Medicine, St. Louis, MO 63110

ABSTRACT   This report presents the appearance of rapidly frozen, freeze-dried cytoskeletons that have been rotary replicated with platinum and viewed in the transmission electron microscope. The resolution of this method is sufficient to visualize individual filaments in the cytoskeleton and to discriminate among actin, microtubules, and intermediate filaments solely by their surface substructure. This identification has been confirmed by specific decoration with antibodies and selective extraction of individual filament types, and correlated with light microscope immunocytochemistry and gel electrophoresis patterns.

The freeze-drying preserves a remarkable degree of three-dimensionality in the organization of these cytoskeletons. They look strikingly similar to the meshwork of strands or "microtrabeculae" seen in the cytoplasm of whole cells by high voltage electron microscopy, in that the filaments form a lattice of the same configuration and with the same proportions of open area as the microtrabeculae seen in whole cells. The major differences between these two views of the structural elements of the cytoplasmic matrix can be attributed to the effects of aldehyde fixation and dehydration.

Freeze-dried cytoskeletons thus provide an opportunity to study—at high resolution and in the absence of problems caused by chemical fixation—the detailed organization of filaments in different regions of the cytoplasm and at different stages of cell development. In this report the pattern of actin and intermediate filament organization in various regions of fully spread mouse fibroblasts is described.

In recent years more and more attention has been directed toward the role of intracellular filamentous proteins in the determination of cell morphology and in the direction of cell movement. An important technical advance in isolating and visualizing these filaments has come from the discovery that extraction of living cells with nonionic detergents leaves behind a residuum that contains a large proportion of the cellular actin and intermediate filament protein (4, 20, 21). The cell's microtubules may also be preserved in this residuum under specific stabilizing conditions (3, 7, 22). This framework of filaments, which remains after detergent extraction, has been termed the cytoskeleton.

Light microscope immunocytochemistry has illustrated that the cytoskeleton contains many of the components thought to be involved in cell motility, in arrangements similar to that found in unextracted cells. For example, Osborn and Weber (21, 22) and Webster et al. (29) have demonstrated close similarities in the filament distributions in cytoskeletons vs.

whole cells, by immunofluorescence, using antibodies to actin, myosin, tropomyosin, α-actinin, tonofilaments, and tubulin. All of these antibodies bind avidly to cytoskeletons, and in fact some, including those directed against microtubules, bind to extracted cells with better definition and lower backgrounds than they do to intact cells (22, 27). These results vividly demonstrate that the cytoskeleton can serve as an excellent representation of the filaments in intact cells, at least at the level of resolution of the light microscope.

Attempts to determine how filaments in the cytoskeleton are organized on an ultrastructural level have been limited primarily by the methods used to prepare them as whole mounts for electron microscopy. In a recent protocol of Bell et al. (2) cells were cross-linked with a diimidoester, extracted with Triton X-100 (Triton) and then fixed by the standard aldehyde and osmium sequence, followed by dehydration, critical-point drying, and viewing in the scanning electron microscope. Critical-point drying prevented the collapse of the filaments and

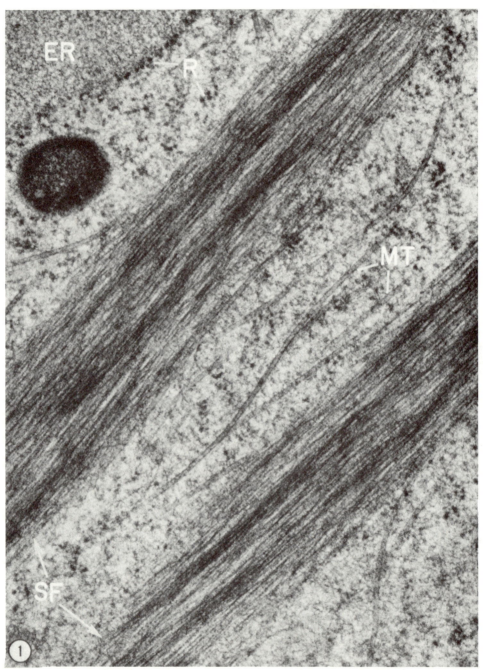

FIGURE 1 Traditional thin section of a fibroblast that was quick-frozen with a liquid helium–cooled block of copper, and then freeze-substituted in acetone-osmium at −90°C for 2 d, to illustrate the overall quality of freezing that could be achieved. Ice crystals are too small to disturb the close packing of filaments in the stress fibers (*SF*), and too small to disturb the random distribution of protein inside the cisternae of the endoplasmic reticulum (*ER*); the only hint of their presence is the small lucent area around the mitochondrion. Ribosomes (*R*) and microtubules (*MT*) look well preserved, and lie in a background matrix that looks delicately floccular. × 85,000.

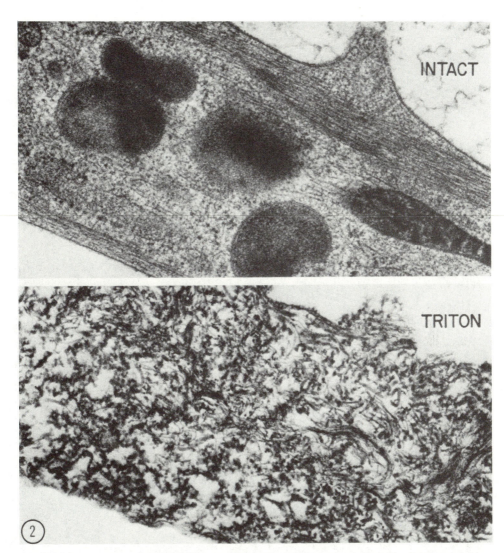

INTACT

TRITON

FIGURE 2   Cross section through an intact fibroblast prepared as in Fig. 1 (*INTACT*) compared with a section through a cytoskeleton prepared by exposing a living cell to 0.5% Triton for 30 min (*TRITON*). Membranous organelles and surface membranes are gone from the skeleton, but a rich complement of filaments is left behind. (Here the electron density of these filaments was accentuated by block-staining in 0.2% hafnium chloride in acetone for 1 h after freeze-substitution.) The pattern of filament organization cannot be easily discerned in such 0.1-µm-thick sections, but is more easily assessed in the platinum surface replicas that follow. × 70,000.

provided some sense of their three-dimensional organization. However, the filaments appeared variable in thickness as a result of the gold shadowing, and the resolution of the scanning electron microscope was not good enough to image the interactions of filaments with each other, or to identify structurally the various filament types. Nevertheless, this approach illustrated that the cytoskeletons are somewhat thinner than the whole cells, and comprise primarily the lower portions of the cells, the regions where most stress fibers occur. In many instances, the cell nucleus remains attached to this cytoplasmic foundation, but little appears to remain of the protoplasmic components that lie immediately under the plasma membrane or in the region above the nucleus.

Efforts were made originally to view cytoskeletons at high resolution employing uranyl acetate staining after aldehyde fixation, followed by air-drying. As shown by Brown et al. (4) and Small and Celis (26), air-drying causes collapse of the cytoskeletons. However, they could then be viewed flat in the transmission electron microscope at sufficient resolution to demonstrate that stress fibers and other major struts in the

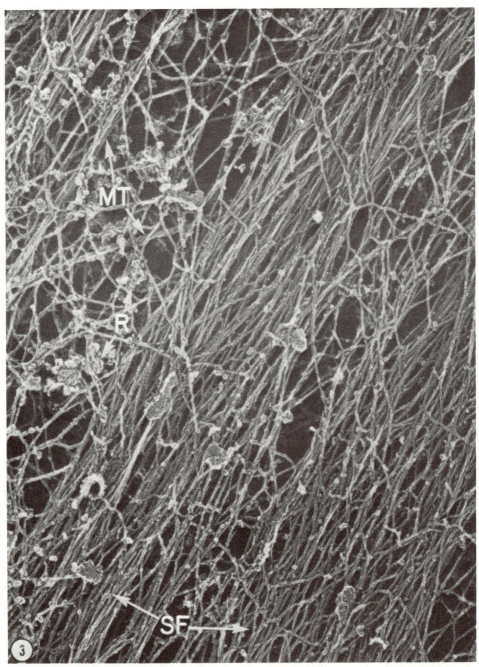

FIGURE 3   Rotary-deposited platinum replica of a freeze-dried cytoskeleton, from a fibroblast that was extracted for 30 min with 0.5% Triton. Prominent are the bundles of filaments that comprise what must have been the cells' stress fibers (*SF*). Also visible in the upper left-hand corner are two thicker structures thought to be microtubules (*MT*), and a more diffuse meshwork of filaments that is studded here and there with grapelike clusters that are the size and shape of polyribosomes (*R*). (Note: Figs. 2, 3, 7, 10, 12, 16, and 17 are all printed in the same × 70,000 magnification to facilitate comparison of the results of different experiments described herein.)

cytoskeleton are composed of bundles of 60–70 Å and 100 Å filaments. This was consistent with the biochemical and immunocytochemical analyses of the cytoskeletons, which illustrated that they contain primarily actin and intermediate filament protein (2, 4, 20, 21, 29).

More recently, Webster et al. (28) have amalgamated these preparative methods by critical-point drying cytoskeletons after aldehyde fixation and uranyl acetate staining, thus producing specimens that are beautifully three-dimensional and highly informative when viewed by stereo transmission electron microscopy. By this method, the most prevalent component of the cytoskeleton is a branched, anastomotic network, which decorates with ferritin labeled anti-actin antibodies.

A totally different approach to visualizing the polymeric elements of cytoplasm, which does not rely upon selective extraction and preparation of cytoskeletons, has been introduced by Porter and co-workers. They examined whole cells, grown on electron microscope grids, that they fixed with aldehyde and osmium tetroxide and then critical-point dried before viewing first with the standard transmission electron microscope and later with the high voltage electron microscope (5, 31, 32). In these preparations, whole cells exhibit a fine anastomotic three-dimensional network, composed of what Wolosowick and Porter call "microtrabeculae." This network looks so similar to the ones seen by Webster et al. (28) in cytoskeletons after critical-point drying, that it is tempting to conclude they are identical. However, that conclusion would imply that actin is the most prevalent component of the microtrabeculae. Yet, the microtrabeculae seen in whole cells are much more variable in shape and diameter than any known form of actin polymer. Wolosowick and Porter (32) report that they vary from <30 Å to >100 Å in diameter, whereas actin is known to polymerize into filaments of uniform diameter of ~60 Å. Moreover, microtrabeculae appear to branch or anastomose with each other at smooth, continuous junctions that do not look like simple points of filament crossover.

To determine how the images of whole cells relate to the filamentous cytoskeletons prepared by detergent extraction, we have examined both whole and Triton-extracted cells after preparing them for electron microscopy by quick-freezing at liquid helium temperature. The freezing method avoids the problems of chemical fixation and dehydration. Cytoskeletons of fibroblast cells viewed by this new protocol of extraction, rapid freezing, freeze-drying, and rotary shadowing offer new and informative images of the actin, intermediate filament, and microtubule arrays in different regions of the cell under different physiological conditions. The method has sufficient resolution to identify different filament types by their structure alone. When we compared these new images with those obtained after cells were pretreated with aldehydes, it appeared that the microtrabeculae may actually be composed of discrete filamentous components of the cytoskeleton that seem to have become cross-linked, thickened, and partly obscured by adsorption of soluble cytoplasmic proteins during the process of chemical fixation.

MATERIALS AND METHODS

Cell Culture

Fibroblast cells were isolated by trypsinization from BALB/c mouse embryos and used within four passages. They were plated onto 60-mm Falcon tissue culture dishes (Falcon Labware, Div. Becton, Dickinson & Co., Oxnard, Calif.) containing a number of 4-mm-square fragments of no. 1 Corning cover glass (Corning Glass Works, Scientific Products Div., Corning, N. Y.) and maintained

in Dulbecco's Modified Eagle's Minimal Essential Medium with 10% fetal calf serum. Cells were allowed to attach and spread on the glass coverslips for 18 h before each experiment.

Preparation of Cytoskeletons

Cytoskeletons containing actin, intermediate filaments, and microtubules were prepared in the following manner: Each dish of cells containing several glass coverslip fragments was washed for 1 min in phosphate-buffered saline at 37°C, followed by 1 min at 37°C in a buffer commonly used to polymerize and stabilize microtubules, which we will refer to as "stabilization buffer," (0.1 M PIPES at pH 6.9, 0.5 mM MgCl₂, 0.1 mM EDTA), followed by stabilization buffer plus 0.5% Triton and 4 M glycerol at 37°C. The glycerol was added simultaneously to the Triton to further stabilize microtubules as the Triton dissolved away the plasma membrane. (cf., reference 3). After 3 min an equal volume of stabilization buffer plus 4 M glycerol and 4% glutaraldehyde was added with very gentle mixing to give a final glutaraldehyde concentration of 2%. This was allowed to remain at 22°C for 30 min before washing with five changes in distilled water.

Cytoskeletons containing actin but no microtubules were prepared as above except that the glycerol was omitted during the Triton extraction. The extraction was allowed to proceed for 30–60 min, after which the cytoskeletons were washed with stabilization buffer. They were then fixed with 2% glutaraldehyde in stabilization buffer for 30 min and washed five times with distilled water. For subfragment 1 (S1) decoration of actin-containing cytoskeletons, S1 at a concentration of 3 mg/ml in stabilization buffer (provided by Dr. Roger Cooke and prepared according to the procedure of Cooke (9), was added to the cytoskeletons before the fixation step. After 10 min, the coverslips were rinsed twice with stabilization buffer and then fixed with 2% glutaraldehyde and washed with distilled water.

Actin was extracted from cytoskeletons by the following procedure. Cytoskeletons prepared in the absence of glycerol were incubated with 0.3 M potassium iodide (KI) for 3 h at 4°C. The coverslips were then washed with stabilization buffer and fixed for 30 min in stabilization buffer plus 2% glutaraldehyde.

Immunofluorescence of the Cytoskeletons

Cytoskeletons prepared by the protocols listed above were stained with antibodies, using indirect immunofluorescence. Before the glutaraldehyde fixation step, the coverslips were plunged into methanol at 20°C, washed in saline, treated with primary rabbit antiserum against tubulin, followed by reaction with fluorescein-conjugated goat anti-rabbit antibody as described previously (27). Antibody to tubulin was prepared from purified chick brain tubulin by the procedure of Connolly et al. (8). Antibody to the hamster 58,000-dalton protein (intermediate filament protein) was a kind gift of Dr. Richard Hynes (Massachusetts Institute of Technology, Cambridge, Mass.) (16). Antibody to actin was the kind gift of Dr. Robert Pollack (Columbia University, New York) and Dr. Keith Burridge (Cold Spring Harbor Laboratory, Cold Spring Harbor, N. Y.) (6).

Preparation of Microtubules

Microtubules from chicken brain were purified by three cycles of polymerization-depolymerization according to the Weingarten et al. (30) modification of the Shelanski et al. (25) procedure. They were resuspended in stabilization buffer at 37°C at a concentration of 8 mg/ml, quick frozen, fractured, and deep etched.

Quick Freezing of Samples

After washing in distilled water, the coverslips were rinsed in 10% methanol in water. We found that 10% methanol acts as a very good cryoprotectant; that is, it reduces the size of the ice crystals formed at a given rate of freezing, and yet it is volatile at the temperatures used for freeze-drying. It has no observable effect on the structures of purified actin and microtubules, nor on the overall appearance of the cytoskeleton.

Freezing was performed by mounting the cover glass on a 1-cm-diameter aluminum disk at the end of a plunger device, originally designed to catch exocytosis in isolated frog muscles. The plunger slams the sample down against a pure copper block cooled to liquid helium temperatures (13). This produced frozen cytoskeletons embedded in microcrystalline ice. The ice crystal size was smaller than 50 nm as a result of the extremely rapid freezing and the help of the volatile cryoprotectant (methanol). To avoid crushing the cytoskeleton and shattering the cover glass when the holder made forceful contact with the cold copper block, the cover glasses were mounted on a soft, spongy matrix composed of slices of aldehyde-fixed rat lung. The lungs had been cut 0.8 mm thick with a Smith-Farquhar tissue chopper and washed for weeks in distilled water before use. As they froze, the slices of lung became bonded to the surface of the aluminum disk. The disks could then be mounted on the standard rotary specimen stage of the Balzers freeze-fracture device (Balzers Corp., Hudson, N. H.).

Freeze-drying and Replication

The frozen cytoskeletons on cover glasses were kept covered with a small aluminum cap as they were transferred and clamped into the vacuum evaporator. Once a vacuum of 10^{-4} torr was achieved, the cap was removed with a special vacuum feed-through. Then the samples were warmed to $-95°C$ at a vacuum of 10^{-6} torr or better for 30–60 min, in order to sublime away all the ice from the top of the cover glass. Because there were no solutes or nonvolatile cryoprotectants in this ice, there was no scum left behind after freeze-drying. As discussed in Results, the exposed cytoskeletons did not appear to collapse very much at $-95°C$, though we found they could collapse and become distorted if warmed much above $-80°C$.

After this procedure, the freeze-dried skeletons could be viewed directly in the electron microscope, if the cells had been originally grown on Formvar-coated gold microscope grids instead of on glass. Alternatively, they could be stained with osmium vapor at ambient temperature and pressure before viewing. In either case we found that the cytoskeletons warmed and melted considerably in the 100-kV electron beam. Freeze-dried skeletons on glass were best viewed after coating with a thin layer of platinum-carbon applied in the rotary mode. In this procedure, the sample, after freeze-drying and still at $-95°C$, was exposed to evaporation of platinum-carbon from an electron beam gun mounted at 24° to the specimen plane, while the specimen was rotating at 60 rpm. The resulting platinum "replica" of the cytoskeleton was stabilized by 5 s of rotary deposition of pure carbon from a gun mounted at 75° to the sample plane. The sample and the cover glass were then removed from the vacuum evaporator and the cover glass floated off the adhering lung by placing it in 30% hydrofluoric acid, which also promptly released the replica from its surface. The replica was then floated sequentially through distilled water, household bleach, and two washes of distilled water before being picked up on a 75-mesh formvar-carbon-coated copper grid.

Examination was carried out in a JEOL 100B or 100C electron microscope operating at 100 kV. The high accelerating voltage was used to reduce heating of the replica, to minimize recrystallization, and also to reduce contrast in the final photographic negative. Stereo pairs were obtained with a high-resolution top-entry goniometer, stage tilted through ±6° for all magnifications. Micrographs were examined in negative contrast, i.e., by projecting the original electron microscopic negatives or by photographically reversing them before printing, to make platinum deposits look white and background look dark. The reversed contrast enhanced the three-dimensional appearance of the images, and made the skeletons look as if they were illuminated by diffuse "moonlight" very much the way samples look in the standard mode of scanning electron microscopy.

RESULTS

Assessment of Cell Structure after Rapid Freezing

Because the quality of structural preservation of the filament network could be no better than the initial quality of freezing, this was first evaluated by freeze-substituting whole cells that had been grown on cover glasses and rapidly frozen while alive, without prior Triton extraction. Freeze-substitution was by standard techniques. Cells thin-sectioned parallel to the culture dish (Fig. 1) displayed membranous organelles and filamentous actin bundles and microtubules typical of fibroblasts, in a natural-looking state of preservation. The background cytoplasmic matrix looked uniformly dense at lower magnifications and slightly lacy or flocculent at higher magnifications.

The effects of Triton on such tissue culture cells could be evaluated by cross-sectioning them after freeze substitution. Unextracted cells displayed the usual complement of membranous organelles and cytoplasmic filaments embedded in a vaguely wispy cytoplasmic matrix (Fig. 2a). Freezing was often good all the way down to the surface of the cover glass on which the cells were grown. In contrast, cells extracted with Triton for 3 min or longer had lost their plasma membrane and most of their internal membranous organelles. Left behind were rich tangles of filaments, whose prominence could be

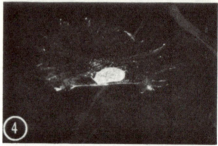

FIGURE 4 Antitubulin immunofluorescence of a mouse fibroblast cytoskeleton prepared by Triton extraction in glycerol stabilization medium (top) vs. a fibroblast cytoskeleton extracted in the absence of glycerol (bottom), which results in the loss of microtubules. × 1,000.

enhanced after freeze-substitution by block staining with tannic acid or with hafnium chloride (as was done in Fig. 2b), but whose pattern of organization could not easily be discerned in routine thin sections.[1]

Assessment of Freeze-drying

Rotary replicas of freeze-dried cytoskeletons displayed a high degree of structural preservation and detail (Fig. 3). The extracted cells appeared to be composed of myriads of long filaments that overlapped and intersected each other in complex patterns. The most obvious pattern was parallel alignment of many dozens of filaments into broad bundles that presumably represented the stress fibers seen in light microscopy. Other patterns appeared to be more isotropic in three dimensions, and thus represented webs woven out of individual filaments or small groups of filaments that crossed and inter-

---

[1] Block-staining with hafnium chloride (M. Karnovsky, personal communication) or tannic acid was employed after freeze-substitution. If either treatment was allowed to continue too long, it increased the electron density of the cytoplasmic matrix of whole cells to such a great extent that it became impossible to discern individual filamentous components. Instead, the cytoplasm of whole cells appeared to be a dense lacy spongework surrounding translucent holes, which varied from <200 Å in the best-frozen areas to >1,000 Å in the areas where membranous organelles began to look "cobblestoned" and otherwise badly frozen. Presumably, these holes represented pure ice crystals that had formed during freezing and had pushed aside the stainable components of the cytoplasm. Similar ice crystals must have formed in the cytoskeletons, but they must have been small enough not to leave permanent distortions after freeze-drying.

J. E. HEUSER AND M. W. KIRSCHNER   *Filament Organization in Cytoskeletons*   217

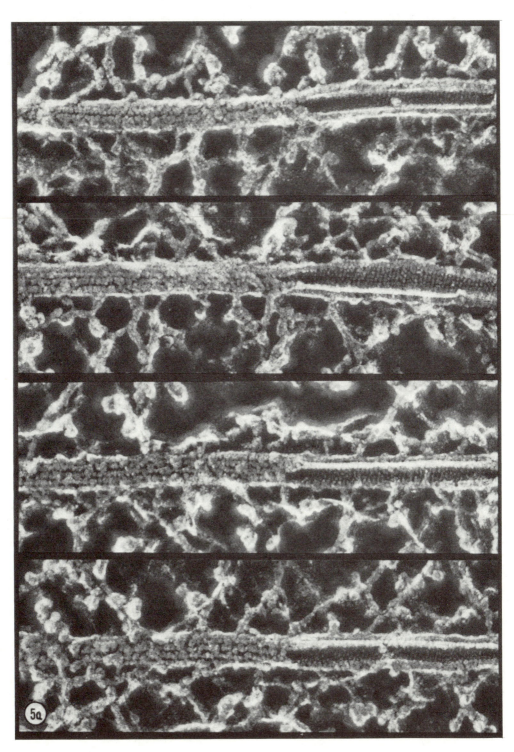

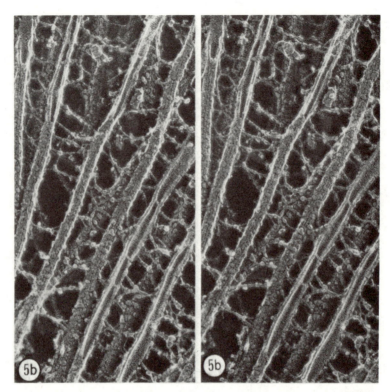

FIGURE 5 High magnification of purified microtubules that were fractured and deep etched after quick-freezing, to illustrate the resolution of the rotary replication technique. In the four examples shown in *a*, the left half of the field illustrates the outer surface of the microtubules, which display longitudinal bands of bumps spaced 55 Å apart, which may represent the microtubule's protofilaments. To the right of each figure, the microtubules are fractured open to reveal their inner luminal walls, which display characteristic oblique striations separated by 40 Å. This 3-start helix has been seen before only in optically filtered electron micrographs. (The reticulum surrounding the microtubules is thought to be unpolymerized tubulin and microtubule-associated proteins. In *b*, a similar field is shown in stereo, to illustrate more clearly the three-dimensional organization of the inner and outer substructure visible in freeze-dried microtubules. *a*, × 350,000; *b*, × 200,000.

digitated. These webs could be found between the stress fibers and above stress fibers in the areas that may have contained the organelle-rich "endoplasm" of the fibroblasts before detergent extraction.

At the magnifications that showed up the individual filaments in these skeletons, microtubules were relatively sparse and ran a relatively straight course right through the middle of the filament webs, though they did not appear to penetrate the more compact bundles of filaments. Lateral contacts between microtubules and other filaments were common, but there was no visual cue to specificity or order in this association. (Much may have been missed here, however, because these microtubules had been pre-fixed with glutaraldehyde before freeze-drying. The problem we faced was that unfixed microtubules had to be maintained in glycerol at a salt concentration of >25 mM or they would depolymerize [cf. Fig. 4], but that much nonvolatile material left behind an impossible scum when the cytoskeletons were totally freeze-dried. Only after the main body of work in this paper was done did the agent Taxol, which aided in stabilizing microtubules at low ionic strength [24], become available.)

Stereo views of the substructure of purified microtubules such as those seen in Fig. 5 illustrated the resolution of the rotary replicas. Purified microtubules were pelleted in 100 mM salt and then quick-frozen without fixation. The salt made it impossible to freeze-dry the entire pellet, but a portion could be exposed by deep-etching it for 3 min after freeze-fracture. The outside surface of these microtubules demonstrated several parallel arrays of bumps having a periodicity of ~40 Å. These arrays undoubtedly represented the subunits in the 13 protofilaments of the microtubule. Also visible in regions where a microtubule had been broken open by the fracture was a substructure on its inner surface, against its lumen. There, the subunits accentuated the 3-start helix, which had previously only been detected by optical filtering of negative stain electron microscopic images (1, 10, 12). Though the resolution of these platinum replicas was not as high as with negative stain, it could easily pick up the 40-Å repeat, and offered the advantage that inner and outer surfaces could be visualized independently. We will show next that the resolution of the technique was sufficient to distinguish differences among the various filament types in the intact cytoskeleton.

J. E. HEUSER AND M. W. KIRSCHNER *Filament Organization in Cytoskeletons* 219

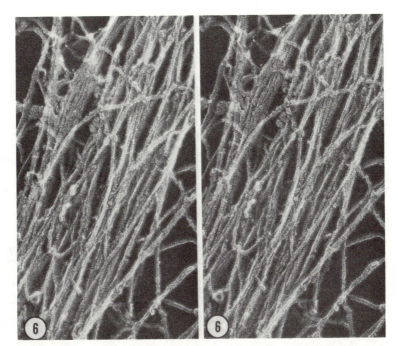

FIGURE 6 Stereo views of the bundles of filaments found in fully spread fibroblasts, which allow the observer to discern the individual filaments more clearly, to appreciate the characteristic granularity of their coat of platinum, and to discern how they are bundled together. These features are harder to see on flat views such as Fig. 3. × 200,000.

Identification of Filament Types in the Cytoskeleton

High magnification stereo views of the bundles of filaments that made up what may have been stress fibers in the whole cells, as shown in Fig. 6, allowed ready discrimination of the individual filaments. These filaments appeared to have a relatively uniform caliber of 95 ± 10 Å along their entire length. Typically, the replicas of these filaments displayed a coarse, vaguely striped pattern of platinum granularity, spaced about every 50–60 Å. Such a pattern was not observed on the surface of microtubules, nor on what turned out to be intermediate filaments, as will be discussed below. Pure freeze-dried actin showed an identical "striped" pattern (J. E. Heuser and R. Cooke, manuscript in preparation). Thus this pattern may reflect the underlying 55-Å repeat of the monomers that compose the actin filaments.

As shown in Fig. 3, similar filaments could be seen running individually between adjacent stress fibers and helter-skelter in the coarse meshwork above the stress fibers. Many of these, having the same thickness as the ones in bundles, nevertheless looked whiter and did not display the striped granular pattern of platinum deposition as clearly. Presumably this was because, being unobstructed by other structures, they received a heavier platinum deposit.

Unfortunately, the double helical pattern of the actin filament has not so far been visible in our replicas of freeze-dried pure actin, nor in the 95-Å filaments we see in cytoskeletons. Nevertheless, this helix could be brought out dramatically by decorating these filaments with S1, the head fragment of the

myosin molecule, which binds stereospecifically to actin. (This result was part of a collaborative effort, conducted with R. Cooke, that, will be reported in detail elsewhere, to visualize actin-myosin interaction.) Cytoskeletons soaked in S1 for 10 min after extraction with Triton no longer displayed any 95-Å, vaguely striped filaments. Instead, essentially all of the filaments in the stress-fiber bundles, and many of the filaments in other regions, were thickened to ~220 Å and looked clearly like double helices (Fig. 7). These helices were also vaguely striped, with foci of platinum grains spaced every 55 Å, which may have corresponded to the individual S1 molecules. An identical pattern was observed after purified skeletal muscle actin was decorated with S1 (J. E. Heuser and R. Cooke, manuscript in preparation).

Interestingly, not all the filaments in cytoskeletons such as that in Fig. 7 become decorated with S1. Some remained thin and uniform. These undecorated filaments typically occurred in between the bundles of filaments and often ran at right angles to the bundles, thus forming a "woof" in the skeletons' filamentous actin "warp". Subsequent observations described below confirmed that these were intermediate filaments, and not actin.

Further confirmation of the filament types could be obtained by selective extraction. As shown in Fig. 4, in the absence of glycerol, tubulin was extracted as judged by immunofluorescence and as shown in Fig. 8 by gel electrophoresis (cf. slots 2 and 3). Microtubules were not found ultrastructurally in the glycerol-free cytoskeletons. Similarly, we could confirm that the bundles of filaments were composed primarily of actin, by extracting actin from cytoskeletons and observing the residue

697

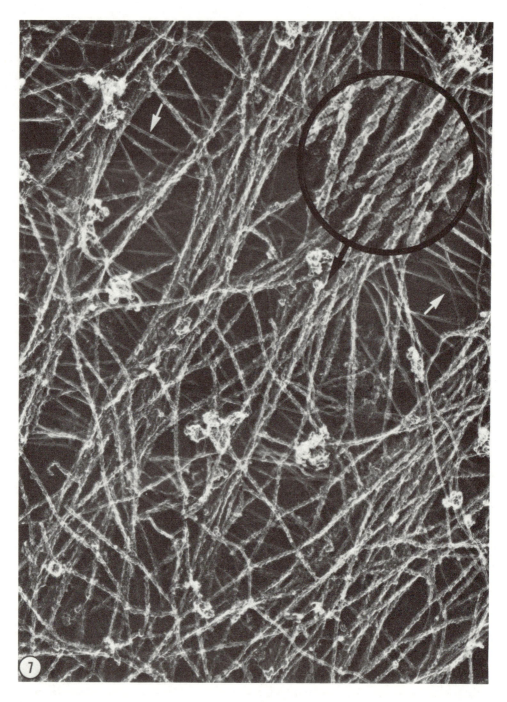

FIGURE 7   Replica of a freeze-dried cytoskeleton that was exposed to the myosin subfragment 1 (S1) before quick-freezing. Nearly all the filaments in the lengthwise bundles, and many of the intervening filaments, have been thickened and converted into ropelike double helices (see *inset*). However, some of the filaments that travel by themselves, in between the bundles, remain totally undecorated (arrows); these are presumably intermediate filaments. × 70,000; *inset*, × 200,000.

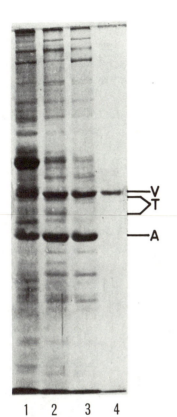

FIGURE 8 Gel electrophoresis of isolated cytoskeletons. The position of vimentin (58,000-dalton intermediate filament protein) is denoted by *V*, the tubulin dimer by *T*, and actin by *A*. Shown in slot *1* is the complete repertoire of proteins found in whole mouse embryo fibroblasts grown in tissue culture (including an intense band at 68,000, which is serum albumin from the culture medium). Shown in slot *2* are Triton cytoskeletons prepared in glycerol, which display a prominent tubulin band as well as actin and vimentin bands. Shown in slot *3* are Triton cytoskeletons prepared in the absence of glycerol, which results in loss of microtubules and leaves only actin and vimentin as the prominent bands. Shown in slot *4* is the residue that remains after 0.3 M KI extraction, which is composed almost exclusively of 58,000-dalton protein. The electrophoresis was done in an 8.5% acrylamide gel, using the procedure of Laemmli (17).

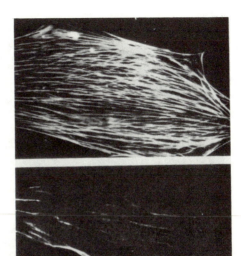

FIGURE 9 Anti-actin immunofluorescence of a mouse fibroblast cytoskeleton prepared by Triton extraction in glycerol-free stabilization medium (top) compared with the immunofluorescence of a fibroblast cytoskeleton treated with 0.3 M KI for 3 h to remove actin (bottom). × 1,250.

ultrastructurally. Actin extraction was accomplished by exposing cytoskeletons to 0.3 M KI for 3 h or exposure to pancreatic DNase 1 (14). This removed >80% of the actin, as determined by gel electrophoresis. For example, slots *3* and *4* on the gels displayed in Fig. 8 show the proportion of actin vs. vimentin (intermediate filament or 58,000-dalton protein; see for example reference 11), in skeletons before and after extraction with KI. Note that after extraction, the actin band largely disappeared, leaving the vimentin band as the principal component. This removal of actin was confirmed by staining the cells with antibody to actin using indirect immunofluorescence. Fig. 9 shows actin localization in an unextracted cytoskeleton. The actin showed its typical localization into bundles that course

the length of the cell (19). After extraction with KI, only a faint, diffuse fluorescence remained, yet the cells retained their nuclei and general contour. Similarly, cytoskeletons freeze-dried and rotary replicated after KI extraction (Fig. 10) possessed only ragged remnants of their stress fibers, further confirming that actin was one of the major components of these structures. The remaining filaments were smooth surfaced and distinctly thicker (115 ± 10 Å) than were actin filaments after shadowing. Such 115-Å filaments could also be seen in the original untreated skeletons, whether aldehyde fixed or not. They could be distinguished from the 95-Å presumptive actin filaments not only by their thickness, but also by their smooth surface and by the general course they took through the cytoskeleton. They seemed to curve more gently, bend less often at points of intersection, and form bundles much less frequently than the actin filaments, all of which suggested that they were stiffer than actin.

To substantiate that the 115-Å filaments in these replicas represented intermediate filaments coated with a thin layer of platinum, cytoskeletons were next decorated with antibody to intermediate filament protein (kindly provided by Richard Hynes, who obtained it by immunizing rabbits with the 58,000-dalton protein from hamster cells [16]). By indirect immunocytochemistry (Fig. 11), it was clear that this antibody stained a filament network different from the stress fiber–actin system or the microtubule system (compare Fig. 11 [top] with Figs. 4 [top] and 9 [top]). Fig. 11 (bottom) illustrates that this antibody staining was not altered by extraction with 0.3 M KI, a treatment that substantially removed actin-containing structures.

Examination of cytoskeletons frozen and freeze-dried after decoration with the anti-58,000-dalton antibody (Fig. 12) con-

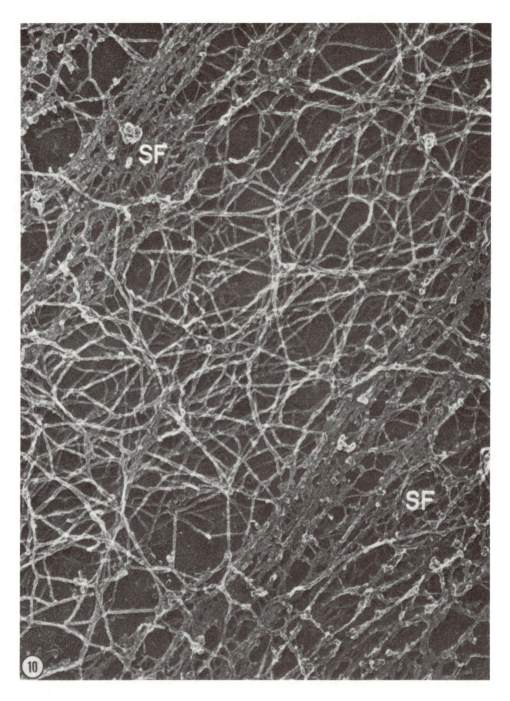

Figure 10   Replica of a freeze-dried cytoskeleton that was extracted with 0.3 M KI for 1 h before freezing. This removed ~80% of the actin, according to gel electrophoresis, and has severely disrupted the longitudinal bundles of filaments, which now resemble ragged coagulums (at *SF*). Left behind, however, is a rich network of filaments in between the damaged stress fibers, presumably equivalent to the 58,000-dalton intermediate filament protein left on gels. × 70,000.

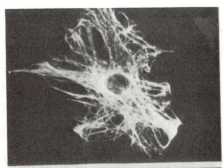

FIGURE 11 Anti-vimentin immunofluorescence of a mouse fibroblast cytoskeleton prepared by Triton extraction in glycerol-free stabilization medium (top) and a cytoskeleton extracted with 0.3 M KI (bottom), showing that KI does not remove the vimentin filaments when it removes the actin. × 1,000.

firmed that it was reacting with a type of filament different from actin. Gone were the smooth, individual filaments that were normally woven between the actin bundles; in their place were thick, coarsened strands measuring 285 Å on the average. These coarsened filaments were presumably intermediate filaments that had been coated with a knobbly layer of antibody molecules. Although antibody decoration of intermediate filaments did not look nearly as orderly as S1 decoration of actin filaments, it did unambiguously localize these filaments on the ultrastructural level. In addition, it showed that antibody-antigen association could be visualized directly by this technique of tissue preparation and replication.

In summary then, the freeze-dry rotary-replica technique appeared to be suitable for distinguishing—in addition to microtubules—the two types of filaments that remain in cytoskeletons of mouse fibroblasts after Triton extraction: actin and vimentin. Fig. 13 illustrates this distinction by displaying four fields of decorated and undecorated filaments at the same high magnification. Fig. 13 a and b are from freshly plated fibroblasts that did not have time to spread out or form stress fibers, and stained diffusely with anti-actin antibodies. They could be seen to contain an isotropic mesh of 95-Å filaments that intersected and overlapped, but did not branch or anastomose. These filaments looked coarsely granular and vaguely striped in platinum replicas and, when decorated with S1, formed distinct "ropes" (Fig. 13 b). They could therefore be identified as actin.

Fig. 13 c and d, on the other hand, were from old fibroblasts (plated at least 24 h before the experiment) whose actin was woven largely into stress fibers. In between these stress fibers

was a meshwork of 115-Å filaments, shown in Fig. 13 c, that were distinctly thicker and less coarsely granular than the actin shown in Fig. 13 a. These intervening filaments became considerably thickened and knobbly when exposed to antivimentin antibody (Fig. 13 d). These, presumably, were intermediate filaments.

Patterns of Filament Organization

With the three major filament types found in cytoskeletons thus defined by chemical extraction, specific decoration, and morphology, it became possible to recognize the overall patterns in which these fibrous components of the protoplasm were woven together, and to recognize how these patterns changed when the cells changed shape. This analysis was aided by examining the cytoskeletons in stereo. The patterns shown so far in Figs. 3, 7, 10, and 12, in which distinct actin bundles were woven together by individual actin and vimentin filaments, was typical only of the thinner regions of older cells. Occasionally, in the thicker regions of older cells, these bundles of filaments appeared to radiate out from starlike foci (Fig. 14). These were presumably the vertices of the "geodesic domes" of actin (18) discovered in such cells by immunofluorescent staining. Less focused arrays were usually found in the thicker regions of these cells, and in younger cells that had not been allowed to spread so thin. Indeed, in round cells, which stained diffusely with anti-actin antibodies, actin filaments appeared to be distributed randomly and isotropically, as shown in Fig. 13, just as they were in pure actin gels.

The concentration of the actin filaments also appeared to vary in different parts of the cell. Around the nucleus, they were relatively sparse, and formed large interstices that may have accommodated relatively large membranous organelles. In thinner regions of the cell, they were often primarily bundled, as shown in Figs. 3, 7, 10, and 12. In the periphery, especially in the lamellopodia, the actin filaments were extremely concentrated. Fig. 15 illustrates this point. The top figure shows the outer surface of the cell membrane over an intact lamellopodium from a cell that was not extracted before freezing and drying. In the center is a lamellopodium with its membrane removed by treatment with Triton for 3 min before freezing. The extraction of the membrane by Triton exposed a dense network of filamentous strands that were characteristically striped, generally 95 Å in diameter, and capable of decoration with S1 (data not shown). It is well known from the work of Lazarides and Weber (19) that "ruffles" react strongly with anti-actin antibody. However, unlike actin filaments elsewhere, these filaments did not appear to extend continuously past points of intersection. Either they were very short or else they must have bent acutely at each junction. Thus, the lamellopodia appeared to contain actin in its most concentrated and most cross-linked state.

Comparison of Cytoskeletons with Intact Cells

Fig. 15 c also shows an image of a ruffle from a cell that was fixed, while intact, with glutaraldehyde and subsequently extracted with Triton. After this sort of preparation, the cytoplasmic matrix of cultured fibroblasts showed subtle but distinct differences from that seen in cytoskeletons that were extracted first and fixed later. In thin ruffles, for example, the delicate web of actin that was retained in the cytoskeleton (Fig. 15 b) was coarsened and partly obscured by aldehyde fixation (Fig. 15 c). This appeared to be the result of lateral aggregation

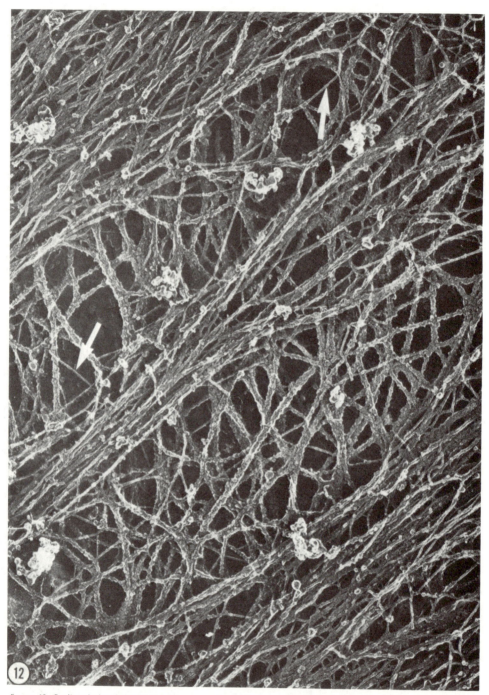

FIGURE 12 Replica of a freeze-dried cytoskeleton exposed to antibodies against intermediate filaments before quick-freezing. The filaments that form the longitudinal bundles do not appear to be altered, but the filaments that compose the intervening loose networks appear, for the most part, to be grossly thickened. A few thin, undecorated filaments persist in these areas (arrows); presumably, they were individual actin filaments. × 70,000.

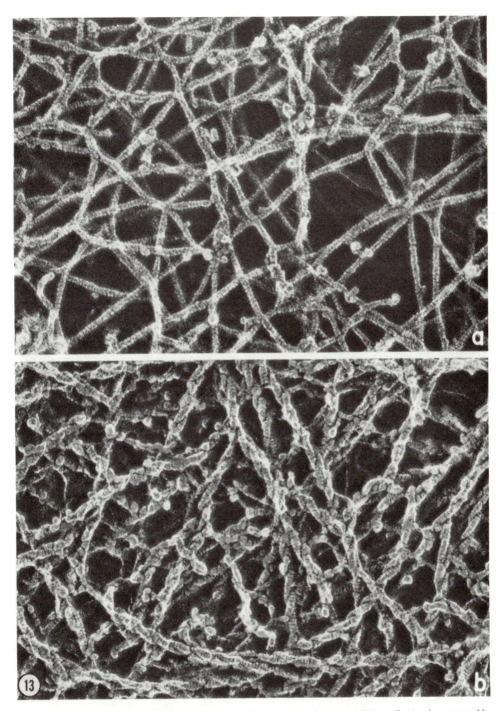

FIGURE 13 Four higher-magnification views of cytoskeletal filaments under different conditions, all printed at comparable magnification to facilitate direct comparison. *a* is a tangle of filaments from the perikaryal region of a freshly plated, somewhat round cell, which would stain diffusely for actin by light-microscope immunocytochemistry. The filaments are 95 Å in diameter and display a characteristic 55-Å surface roughness or graininess that appears to be pathognomic of actin filaments. *b* is a tangle of filaments from a region like *a*, after exposure to S1. As a result of this decoration, each individual filament appears to have

226

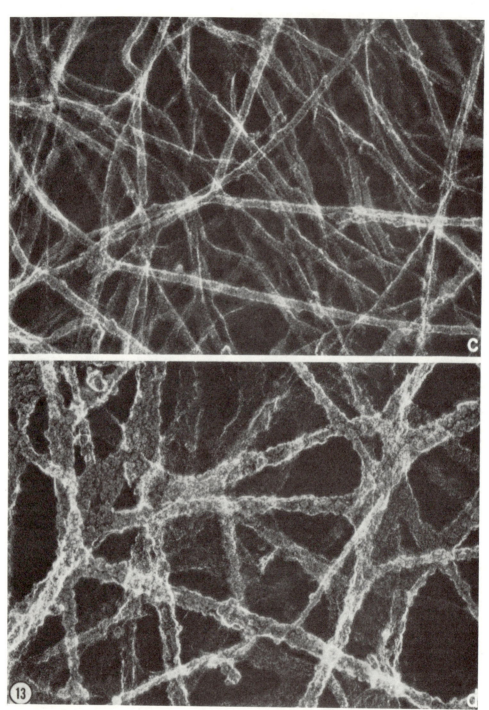

become a double-stranded "rope," in which each strand displays the 55-Å axial periodicity even more clearly than the underlying actin filament. *c* is a typical tangle of the filaments found in the regions in between stress fibers, where filaments are distinctly thicker than actin (115–120 Å vs. 95 Å for actin) and appear to be coated by a more uniform, less granular layer of platinum. *d* is an image of such "intermediate" filaments after exposure to antivimentin antibody. They become considerably thickened and knobbly, as a result of what must be an adhering layer of antibody molecules.

227

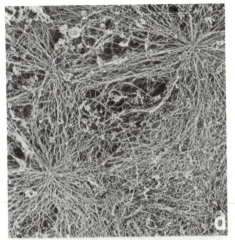

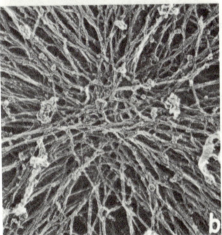

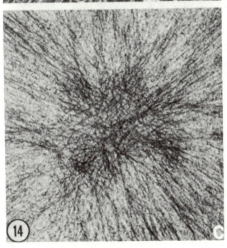

of the individual filaments and of adherence of lumpy material to the aggregated filaments.

Lower magnification survey views of the thinner parts of these fixed and subsequently extracted cells, such as that in Fig. 16 (which is displayed at the same magnification as the previous Figs. 3, 7, 10, and 12, for ready comparison), confirmed the impression that the same filamentous components were present in fixed whole cells as in the cytoskeletons; but after fixation, the various filaments were partly agglutinated and partly obscured by adhering material. This made them look more branched, more anastomotic, more variable in thickness and, in short, more like a "trabecular meshwork" than they looked in the isolated cytoskeleton.

Which image was more real, i.e., more true to nature, was not easy to decide. One might have hoped that simply fracturing whole cells open and deep-etching them, without ever fixing them, would reveal the true nature of the organization of their cytoplasm. Unfortunately, the whole cell turned out to be so loaded with nonvolatile material that it was almost like looking at a thin section, and very little depth could be seen. Unidentified granular material filled in everywhere and largely obscured the view, and little could be learned about the pattern of filament organization.

The only way we could expose a reasonable number of the filaments inside the intact cell was to swell the cell by placing it briefly in distilled water, and thus partially diluting its contents before freezing. Fig. 17 illustrates a representative region from such a swollen cell. The fracture broke partly through a filament bundle in the center of the field and, in addition, exposed a number of individual filaments on either side of the bundle. But in spite of the swelling, a large amount of granular material still obscured the filaments and prevented us from determining much about their overall organization.

DISCUSSION

The method introduced here for preparing cytoskeletons for electron microscopy achieves two major requirements for visualizing the filamentous systems in cells. First, it combines sufficient resolution to identify the filament types by their morphology alone. Second, it can be used with whole mounts of cells to give the realism and perspective of three-dimensional visualization. Using this method we have illustrated that the cytoskeleton of primary mouse embryo fibroblast cells is composed of discrete actin filaments, intermediate filaments, and microtubules woven into characteristic bundles and networks.

Identifying the Filaments

In general, filaments were identified by their specific morphology. For example, microtubules not only displayed their characteristic surface protofilament structure, but also revealed for the first time in fractured samples the structure of their interior. Their luminal wall was seen to have a prominent 3-start helical arrangement of subunits, quite distinct from the

FIGURE 14 Three views of the foci of filament convergence that may represent the vertices of the actin "geodomes" seen in fibroblasts by light-microscope immunocytochemistry (18). Superficially, the centers of these foci have filaments running into them like the Z-bands of muscles, both when viewed in freeze-dried replicas (b) and when viewed in freeze-substituted whole cells (c). (a, × 10,000; b and c, × 65,000).

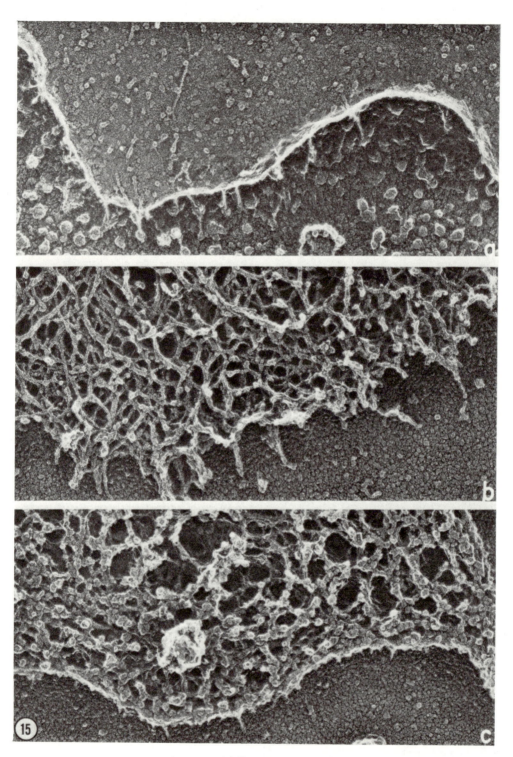

prominent 13-stranded protofilament structure of their outer surface. Similarly, actin filaments displayed a characteristic surface granularity with a repeat at approximately every 55 Å, which hinted at their basic construction from 55-Å G-actin monomers. Of course, there was no way to know whether the details visible in such platinum replicas were an accurate reflection of the real surface contours of these macromolecules or whether they were actually some sort of "decoration" phenomenon resulting from an irregular disposition of platinum on the underlying structures. In any case, these substructural details were visible on all the replicas and allowed us to distinguish one filament type from another. For example, the 55 Å granularity seen in actin filaments was quite unlike the smooth deposit of platinum that formed on the slightly thicker filaments later identified as intermediate filaments.

Secondly, the filaments were identified by selective extraction. Conditions of extraction could be altered in such a way that the cytoskeletons no longer reacted with antibodies to a particular filament as viewed in the fluorescence microscope and no longer displayed that protein by gel electrophoresis. These samples were also freeze-dried and examined in the electron microscope to see which filament type was missing. With microtubules, this was particularly easy. Simply omitting glycerol from the original Triton extraction medium caused them to disappear completely. They were only stable in the absence of glycerol when 10^{-6} M Taxol was present. Actin was a little harder to extract, but could be substantially removed, as determined by gel electrophoresis and immunocytochemistry, by exposing the cytoskeletons to 0.3 M KI for 3 h. This treatment removed the stress-fiber bundles and the ruffles from the freeze-dried cytoskeletons, but left a remarkably dense residuum of filaments. These were the filaments that looked slightly thicker and smoother than actin filaments. Gel electrophoresis illustrated that the only major cytoplasmic protein left after these treatments was the 58,000-dalton subunit of intermediate filaments recently named vimentin (11).

A third way of identifying the filaments in cytoskeletons was by specific decoration. Actin filaments were identified by exposing cytoskeletons to the myosin subfragment S1, which on binding drastically changed the appearance of the actin filaments. A characteristic "ropelike" double-stranded helix was produced that is apparently equivalent to the familiar "arrowhead" image of decorated actin seen by Huxley (15) after negative staining. (Elsewhere, we will evaluate the differences between these two images; J. E. Heuser and R. Cooke, manuscript in preparation.) Intermediate filaments were identified by exposing cytoskeletons to specific antibodies against the 58,000-dalton intermediate filament protein. These antibodies decorated a significant proportion of the filaments in the cell and specifically those that did not react with S1. Unlike the results of S1 binding to actin, the coat of antibodies was too thick and irregular to reveal any underlying order that might have existed in the intermediate filament surface lattice. Nevertheless, it was apparent that the freeze-dry replica method had

sufficient resolution to visualize antigen-antibody complexes directly, without the need for additional electron-dense tags that are usually used in electron microscope immunocytochemistry.

The Organization of Cytoskeletons

The most important conclusion that has emerged from this new preparative technique so far has been a simple one; namely, that cytoskeletons are composed almost entirely of discrete filaments. This is not a novel observation; it only confirms what has been said before on the basis of negative-stain images (4, 26). However, it does not coincide exactly with the transmission electron microscope views of cytoskeletons obtained by positive staining and critical-point drying (28). Cytoskeletal components are translucent after such preparation. When they overlap each other, their images become more difficult to interpret than images of filaments in opaque replicas or in negative stain. With the help of stereomicroscopy to sort out overlapping structures, it has become clear that individual components in critical-point-dried skeletons look both "trabecular" and filamentous. Parts of the images show structures of variable thickness, which merge at truncated intersections and look a bit more like the walls of a sponge than the strands of a web (28). Indeed, the critical-point-dried skeletons look in some respects similar to the images of critical-point-dried protoplasm seen inside whole cells viewed by stereo high-voltage electron microscopy, which Wolosewick and Porter (32) have described as a "microtrabecular meshwork."

Comparison of Cytoskeletons and "Microtrabeculae"

The appearance of the platinum replicas of freeze-dried cytoskeletons is different in important ways from the views of conventionally fixed and stained cytoskeletons and the views of unextracted cells seen by high-voltage electron microscopy. The latter two techniques and particularly high-voltage electron microscopy suggest that the cytoplasm is composed of a three-dimensional lattice of formed elements, the microtrabeculae, which support filaments and organelles and pervade every nook and cranny of the cytoplasmic space. (31, 32). These microtrabeculae appear to vary in thickness and to fuse with each other at points of intersection. They appear also to merge imperceptibly with a thin cytoplasmic "matrix" that coats microtubules and all membranes after aldehyde fixation. In contrast, the cytoskeleton seen after freeze-drying appears to be composed of nothing but discrete filaments of uniform caliber, which crisscross each other in complex patterns but do not appear to fuse. In whole freeze-fractured cells, these filaments appear to lie embedded in a granular cytoplasmic matrix but do not appear to fuse with microtubules or with the membranes of cell organelles.

These fundamental differences in appearance raise the question: Are microtrabeculae additional components superim-

---

FIGURE 15   Three moderately high powered views of ruffles or lamellopodia from fibroblasts that were fixed while whole (in *a*), were extracted with Triton before fixation (in *b*), or extracted with Triton after fixation (in *c*). In *a*, plasma membrane is intact and no internal structure can be seen. In *b*, the plasma membrane has been removed and an underlying web of "kinky" filaments revealed. In other experiments, these filaments decorate with S1, but they are much more concentrated and much more extensively interdigitated than actin in other regions of the cell. In *c*, the plasma membrane has again been removed, but only after the cell was fixed with aldehyde. The delicate meshwork of underlying filaments appears coarser after the chemical fixation. (*a*, × 140,000; *b* and *c*, × 115,000).

707

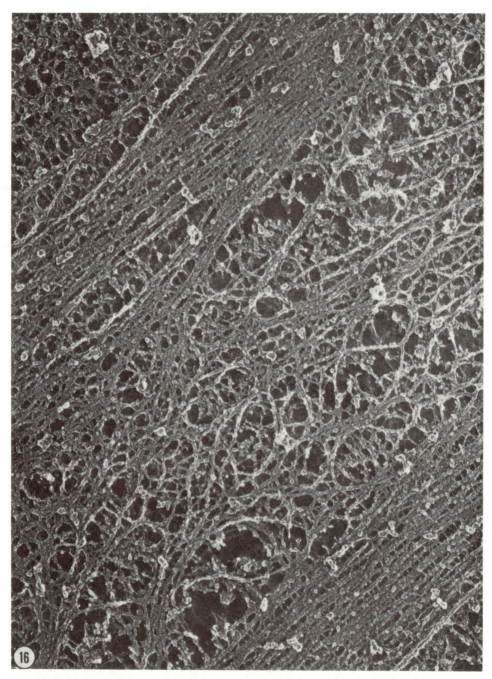

FIGURE 16   Replica of freeze-dried fibroblast that was fixed with aldehyde before extraction with Triton. Triton was still able to remove the membrane and expose the cytoplasmic matrix. It is no longer easy, however, to discern individual filaments therein, except in the obvious stress fibers (*SF*). Instead, the whole matrix looks more like a continuous spongework of curvilinear elements with smoothly varying diameters, not unlike the "microtrabecular meshwork" seen in whole cells by high-voltage electron microscopy. × 70,000.

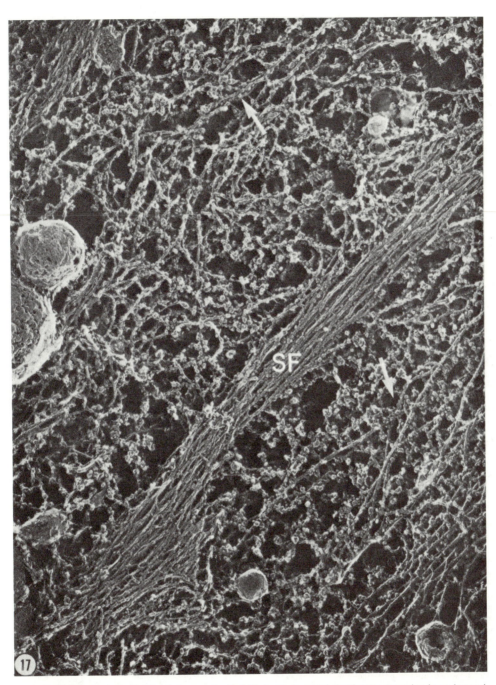

FIGURE 17 Replica of an intact fibroblast that was swollen in distilled water for 1 min before freezing. It was then freeze-fractured and deep-etched for 3 min at −100°C, rather than Triton extracted and freeze-dried. Though cytoplasmic filaments are less readily discernable in whole cells, as a result of the granular material that fills all areas, at least one stress fiber bundle (*SF*) and a number of individual filaments (arrows) can be seen. A disappointing aspect of these views was that the granular material (presumably soluble cytoplasmic protein) obscured so much that it was impossible to tell whether the whole cell might possess microtrabeculae that were not seen in the cytoskeletons. × 70,000.

posed on the cytoskeleton of actin, intermediate filaments, and microtubules, components that are extracted and therefore not visualized in cytoskeletal preparations; or is the microtrabecular lattice a different sort of image of the same cytoskeletal filaments we see in freeze-dried cells?

We favor the latter explanation. For example, in the lamellopodia where high-voltage electron microscope images reveal nothing but microtrabeculae (31), the current method resolves a dense tangle of individual actin filaments that are no less concentrated than the microtrabeculae (maximum separation, 500 Å). Nor is there any indication, at any other locus in the cell, that microtrabeculae form a finer mesh than the components of the cytoskeleton.

However, before we could conclude that the microtrabeculae are essentially cytoskeletal filaments, we would have to show how the filaments could end up looking like trabeculae after fixation, staining, and critical-point drying. To do this, we quick-froze and deep-etched whole cells after each step in such a preparative procedure. This illustrated that discrete filaments abound in the cytoplasm of unfixed cells, but that after aldehyde fixation, these filaments appear to be, by comparison with unfixed cells, partially agglutinated and decorated with irregular condensations of what may have been soluble cytoplasmic proteins. This was illustrated above by comparing Fig. 15 c with b, or Fig. 16 with Fig. 3. After fixation, organized structures such as actin bundles could still be distinguished, but individual actin fibers became impossible to recognize.

Subsequent alcohol dehydration caused further collapse and coarsening of these cross-linked cytoplasmic components. The final image looked very much like the microtrabeculae seen in high-voltage electron microscopy. Thus, from these studies, we believe that microtrabeculae are actually a different image of the individual actin and intermediate filaments that normally weave among the other formed components of the cytoplasm. It seems that after these filaments become agglutinated and decorated with other cytoplasmic proteins during aldehyde fixation, they appear as curvilinear struts in a confluent trabeculum, rather than discrete filaments woven into a complex fabric.

Future Applications

The high-resolution surface views provided by these platinum replicas have permitted us to sort out the overlapping filaments and identify them individually by their characteristic diameters and surface features. Thus, this method of viewing bridges the gap between the scanning electron microscope view of cellular shape and the negative stain view of molecular shape. It reveals detail at the macromolecular level, with a minimum of distortion, and is directly applicable to whole mounts of cells or cell organelles. We hope that in the future, with improvement of the vacuum evaporation of thin films, the technique can reach beyond the present level of resolution, and thus provide a three-dimensional surface view of molecules that would overcome some of the limitations of negative staining, such as opacity of large structures, collapse upon drying, and superimposition of the opposite sides of structures.

Nevertheless, at the present level of resolution, the surface-replica technique has already proved useful for visualizing changes in cytoskeletal organization that relate to changes in cell function. We have already learned, for example, that the loss of stress fibers and a rounding up that occurs during cell transformation is accompanied by a shift of a large proportion of the actin filaments in the cytoskeleton from a bundled state

to a randomly interwoven state. Even though the perinuclear cytoplasm of transformed cells appears, by light microscope immunocytochemistry, to stain diffusely for actin, this new technique has the resolution to show that actin is still filamentous in these areas and has only become rewoven into a diffuse, isotropic three-dimensional mesh.

This technique has not yet revealed any indication of where or how the tangled filaments are cross-linked. No specializations or discontinuities can be seen at the points of intersection, so it is impossible to know which are mere contact and which are real bonds. Thus, it remains to be seen whether the filaments that make up cytoskeletons are extensively cross-linked to each other. Because this technique can visualize antibody molecules directly without secondary labels, its use with specific antibodies to structural proteins, such as α-actinin, desmin, tropomyosin, myosin, tau, and HMW, should allow a reconstruction, at the ultrastructural level, of the organization of the fibrous systems of the cell.

We thank Louise Evans and Margaret Lopata for excellent technical help during the course of these experiments. Further, we wish to thank Roger Cooke for the gift of the S1, Robert Pollack and Keith Burridge for the gift of the anti-actin antiserum, and Richard Hynes for the gift of the antivimentin antiserum. We thank Susan Horowitz (Albert Einstein College of Medicine, Bronx, N. Y.) for the kind gift of Taxol.

Received for publication 29 January 1980, and in revised form 18 March 1980.

REFERENCES

1. Amos, L. A., and A. Klug. 1974. Arrangement of subunits in flagellar microtubules. *J. Cell Sci.* 14:533–549.
2. Bell, P. B., M. M. Miller, K. L. Carraway, and J. P. Revel. 1978. SEM-revealed changes in the distribution of the Triton-insoluble cytoskeleton on Chinese hamster ovary cells induced by dibutyryl cyclic AMP. *Scanning Electron Microsc.* II:899–906.
3. Bershadsky, A. D., V. I. Gelfand, J. M. Svitkina, and I. S. Tint. 1978. Microtubules in mouse embryo fibroblasts extracted with Triton X-100. *Cell Biol. Int. Rep.* 2:425–432.
4. Brown, S., W. Levinson, and J. A. Spudich. 1976. Cytoskeletal elements of duck embryo fibroblasts revealed by detergent extraction. *J. Supramol. Struct.* 5:119–130.
5. Buckley, I. K. 1975. Three-dimensional fine structure of cultured cells. Possible implications for subcellular motility. *Tissue & Cell.* 1:51–72.
6. Burridge, K. 1976. Changes in cellular glycoproteins after transformation: Identification of specific glycoproteins and antigens in sodium dodecyl sulfate gels. *Proc. Natl. Acad. Sci. U. S. A.* 73:4457–4461.
7. Cande, W. Z., J. Snyder, D. Smith, K. Summers, and J. R. McIntosh. 1974. A functional mitotic spindle prepared from mammalian cells in culture. *Proc. Natl. Acad. Sci. U. S. A.* 71:1559–1563.
8. Connolly, J. A., V. I. Kalnins, D. W. Cleveland, and M. W. Kirschner. 1978. Intracellular localization of the high molecular weight microtubule accessory protein by indirect immunofluorescence. *J. Cell Biol.* 76:781–786.
9. Cooke, R. 1972. A new method for producing myosin subfragment-1. *Biochem. Biophys. Res. Commun.* 49:1021–1028.
10. Erickson, H. 1974. Microtubule surface lattice and subunit structure and observations on reassembly. *J. Cell Biol.* 60:153–167.
11. Franke, W. W., E. Schmid, M. Osborn, and K. Weber. 1978. Different intermediate-sized filaments distinguished by immunofluorescence microscopy. *Proc. Natl. Acad. Sci. U. S. A.* 75:5034–5038.
12. Grimstone, A. V., and A. Klug. 1966. Observations on the substructure of flagellar fibres. *J. Cell Sci.* 1:351–362.
13. Heuser, J. E., T. S. Reese, M. J. Dennis, Y. Jan, L. Jan, and L. Evans. 1979. Synaptic vesicle exocytosis captured by quick freezing and correlated with quantal transmitter release. *J. Cell Biol.* 81:275–300.
14. Hitchock, S. E., L. Carlsson, and U. Lindberg. 1976. Depolymerization of F-actin by deoxyribonuclease I. *Cell.* 7:531–542.
15. Huxley, H. E. 1963. Electron microscope studies on the structure of natural and synthetic protein filaments from striated muscle. *J. Mol. Biol.* 7:281–308.
16. Hynes, R. O., and A. T. Destree. 1978. 10 nm filaments in normal and transformed cells. *Cell.* 13:151–163.
17. Laemmli, V. K. 1970. Cleavage of structural proteins during the assembly of the head of bacteriophage T4. *Nature (Lond.).* 227:680–685.
18. Lazarides, E. 1975. Immunofluorescence studies on the structure of actin filaments in tissue culture cells. *J. Histochem. Cytochem.* 23:507–528.
19. Lazarides, E., and K. Weber. 1974. Actin antibody: The specific visualization of actin filaments in nonmuscle cells. *Proc. Natl. Acad. Sci. U. S. A.* 71:2268–2272.
20. Lenk, R., L. Ransom, Y. Kaufmann, and S. Penman. 1977. A cytoskeletal structure with associated polyribosomes obtained from HeLa cells. *Cell.* 10:67–78.
21. Osborn, M., and K. Weber. 1977. The detergent resistant cytoskeleton of tissue culture cells includes the nucleus and the microfilament bundles. *Exp. Cell Res.* 106:339–349.
22. Osborn, M., and K. Weber. 1977. The display of microtubules in transformed cells. *Cell.* 12:561–571.

23. Raju, T. R., M. Stewart, and I. K. Buckley. 1978. Selective extraction of cytoplasmic actin-containing filaments with DNAse I. *Cytobiologie.* 17:307–311.
24. Schiff, P. B., J. Fant, and S. B. Horowitz. 1979. Promotion of microtubule assembly in vitro by Taxol. *Nature (Lond.).* 277:665–667.
25. Shelanski, M. L., F. Gaskin, and C. R. Cantor. 1973. Microtubule assembly in the absence of added nucleotides. *Proc. Natl. Acad. Sci. U. S. A.* 70:765–768.
26. Small, J. V., and J. E. Celis. 1978. Filament arrangements in negatively stained cultured cells: The organization of actin. *Cytobiologie.* 16:308–325.
27. Spiegelman, B., M. Lopata, and M. W. Kirschner. 1979. Multiple sites for the initiation of microtubule assembly in mammalian cells. *Cell.* 16:239–252.
28. Webster, R. E., D. Henderson, M. Osborn, and K. Weber. 1978. Three-dimensional electron microscopical visualization of the cytoskeleton of animal cells: Immunoferritin identification of actin and tubulin containing structures. *Proc. Natl. Acad. Sci. U. S. A.* 75:5511–5515.
29. Webster, R. E., M. Osborn, and K. Weber. 1978. Visualization of the same PtK2 cytoskeletons by both immunofluorescence and low power electron microscopy. *Exp. Cell Res.* 117:47–78.
30. Weingarten, M. D., M. M. Suter, D. R. Littman, and M. W. Kirschner. 1974. Properties of the depolymerization products of microtubules from mammalian brain. *Biochemistry.* 13:5529–5537.
31. Wolosewick, J. J., and K. R. Porter. 1976. Stereo high voltage electron microscopy of whole cells of the human diploid cell line WI-38. *Am. J. Anat.* 147:303–324.
32. Wolosewick, J. J., and K. R. Porter. 1979. Microtrabecular lattice of the cytoplasmic ground substance: Artifact or reality? *J. Cell Biol.* 82:114–139.

Proc. Natl. Acad. Sci. USA
Vol. 77, No. 11, pp. 6624–6628, November 1980
Cell Biology

Cytoskeletal F-actin patterns quantitated with fluorescein isothiocyanate-phalloidin in normal and transformed cells

(simian virus 40/tumor antigens/revertants/actin antibody/two-color fluorescence microscopy)

M. Verderame, D. Alcorta, M. Egnor, K. Smith, and R. Pollack

Department of Biological Sciences, 813 Sherman Fairchild Center, Columbia University, New York, New York 10027

Communicated by Alex B. Novikoff, July 21, 1980

ABSTRACT   Actin in cultured fibroblasts is organized into a complex set of fibers. Patterns of organization visualized with antibody to actin are similar but not identical to those visualized with fluorescein isothiocyanate-phalloidin (Fl-phalloidin), a chemical that binds to F-actin polymer with a dissociation constant of 2.7×10^{-7} M [Wulf, E., Deboben, A., Bautz, F. A., Faulstich, H. & Wieland, T. (1979) *Proc. Natl. Acad. Sci. USA* 76, 4498–4502]. Fl-phalloidin reveals that transformed cells have fewer, finer, and shorter F-actin-containing structures than do normal cells. Two-color fluorescence microscopy of single cells reveals that F-actin staining by Fl-phalloidin picks out the cytoskeletal cables more sharply than does antibody to actin, due to a reduced intracellular background fluorescence. This improved resolution permits sorting of cellular Fl-phalloidin patterns into four classes ranging in organization from 90% of the cytoplasm occupied by large cables to the absence of detectable cables. Reproducible differences in pattern distributions between normal and transformed cell lines have been quantitated. Fl-phalloidin together with rhodamine-based indirect antibody to simian virus 40 tumor antigen reveals a direct relationship between the degree of pattern change and simian virus 40 nuclear antigen expression in intermediate transformed 3T3 cell lines [Risser, R. & Pollack, R. (1974) *Virology* 59, 477–489].

Antibody to actin or transmission electron microscopy of sections cut parallel to the adherent plane reveals F-actin bundles in well-spread mammalian cells (1, 2). The bundles in normal cells are different from those in tumor cells or in many virus-transformed cell lines (3–5). In general, the difference relates to paucity and disorganization of the bundles in the abnormal cell.

A major block to quantitation of this observation has been the use of antibody to actin as the probe for actin fiber organization. Most available anti-actin antisera, even affinity-purified antisera (6), require use of a second antibody as a fluorescent probe, thereby generating an extremely complex relationship of F-actin concentration to fluorescence intensity (6). Recently, affinity-purified anti-actin has been prepared and conjugated directly to a fluorescent moiety (7). This antibody shows no nonspecific staining, and it generates patterns similar to those seen by two-step immunofluorescence. Although use of direct-conjugated antibody eliminates the problem of nonlinearity of intensity, all antisera have the drawback of intrinsic heterogeneity insofar as they are composed of a mixture of immunoglobulin molecules. In any event, quantitation of patterns with fluorescent-labeled anti-actin antibodies has depended to date upon counting the number of cells with and the number without cables (4, 5, 8). Such scores, although consistent and reproducible, totally mask the complexities of the pattern seen: two cables, however fine, make a cell as positive as a full display of thick cables filling the cytoplasm at the plane of adhesion.

Phalloidin is a small (M_r, 788), stable compound that forms a complex with F-actin with K_d 3.6×10^{-8} M (9) and promotes the formation of F-actin from dimers and trimers of G-actin (monomeric) in solution (10). Recently, phalloidin has been specifically labeled with fluorescein (Fl-phalloidin) to produce a M_r 1250 compound that retains a strong F-actin binding affinity ($K_d = 2.7 \times 10^{-7}$ M) (11). We have screened a large number of cell lines with Fl-phalloidin to determine the quantitative effects of viral transformation on F-actin distribution pattern.

METHODS

Cells and Culture Conditions. Rat postcrisis lines have been described (12, 13). They were grown in Dulbecco's modified Eagle's medium supplemented with penicillin (100 units/ml), streptomycin (100 μg/ml), and 10% fetal calf serum (Rehatuin). Mouse cell postcrisis lines have been described (14). They were grown in Dulbecco's modified Eagle's medium with penicillin (100 units/ml), streptomycin (100 μg/ml), and 10% calf serum (GIBCO).

Mouse and rat precrisis fibroblasts were obtained from embryos and used in the third to seventh passages. They were cultured in Dulbecco's modified Eagle's medium plus 10% calf and fetal calf serum, respectively. SV80 is a simian virus 40 (SV40)-transformed human cell line obtained from W. Topp (Cold Spring Harbor Laboratory).

All cells were kept at 37°C in an atmosphere of 10% CO_2/90% air and 100% humidity. Cultures were fed twice weekly and dissociated for transfer with 0.25% trypsin/0.02% EDTA.

Fixation and Staining. Cells were plated on coverslips at a density of 2.4×10^3 cells per cm^2 in 10% fetal calf serum. One day later, medium was switched to 1% serum. After an additional day, cells on coverslips were fixed in 10% formalin in phosphate-buffered (pH 7.1) saline (P_i/NaCl) for 20 min, rinsed with P_i/NaCl and extracted with 1% Nonidet P-40 in P_i/NaCl for 20 min. Subsequent steps varied with the stain desired. Phalloidin concentrations were determined in concentrated stock solutions by absorbance at 300 nm (corrected for the fluorescein absorbance by measurement at 492 nm). For Fl-phalloidin staining, 10 μl of Fl-phalloidin (1 μg/ml in P_i/NaCl; gift of T. Wieland, West Germany) was incubated with cells at 37°C for 30 min. Coverslips were then rinsed three times in P_i/NaCl and mounted on microscope slides with Aquamount.

For double staining with antibodies to actin and Fl-phalloidin, a two-step incubation was used. First, fixed cells were incubated with rabbit anti-actin (25 μg/ml in P_i/NaCl; gift of K.

Abbreviations: Fl-phalloidin, fluorescein isothiocyanate-phalloidin; P_i/NaCl, phosphate buffered saline; SV40, simian virus 40; T antigen, tumor antigen.

Burridge, Cold Spring Harbor Laboratory) (15) at 37°C for 1 hr. After three P_i/NaCl washes, cells were then incubated with a mixture of rhodamine-conjugated goat anti-rabbit IgG (0.4 mg/ml in P_i/NaCl; Cappell Laboratories, Cochranville, PA) and Fl-phalloidin (1 μg/ml). For double staining with antibody to SV40 tumor antigen (T antigen) and Fl-phalloidin, the same procedure was used but the first stain was hamster anti-T antigen (0.2 mg/ml in P_i/NaCl, National Institutes of Health) and the second stain was a mixture of Fl-phalloidin (1 μg/ml) and rabbit anti-hamster IgG (0.3 mg/ml in P_i/NaCl, National Institutes of Health). In all cases, coverslips were rinsed three times after the second stain and mounted with Aquamount.

Immunofluorescence. Stained coverslips were examined with a Leitz Orthoplan microscope. A Zeiss Planapo ×63 oil-immersion objective was coupled with a Leitz L2 exciter-barrier filter cube to reveal fluorescein or with a Leitz N2 exciter-barrier filter cube to reveal rhodamine. This apparatus completely separated the fluorescences due to rhodamine and fluorescein. Before examination, labels on all slides were blind-coded by someone other than the scorer.

Fl-phalloidin specificity for F-actin was determined in the following preincubation controls. G-actin, purified from acetone powder of chicken gizzard (refs. 16 and 17; J. Feramisco, personal communication) was converted to F-actin by the addition of KCl to a final concentration of 0.1 M. Actin preparations were mixed with Fl-phalloidin for 20 min at room temperature and then applied to coverslips of fixed 3T3 cells for 20 min at 37°C. Fl-phalloidin staining was completely blocked by F-actin at 1 mg/ml, not blocked by G-actin at 2 mg/ml, and only partially blocked by G-actin at 10 mg/ml. At this high concentration of G-actin, approximately 1% (or 0.1 mg/ml) is expected to be polymerized. Thus, staining with Fl-phalloidin was sensitive to the polymerized state of the actin. Staining by antibody to actin was completely blocked by 45-min preincubation with G-actin at 2 mg/ml.

Fl-phalloidin staining was blocked by preincubation of fixed cells with unlabeled phalloidin at 1 mg/ml (18) or the coincubation with a mixture of Fl-phalloidin (1 μg/ml) and phalloidin (1 mg/ml). Fixed cells preincubated with phalloidin at concentrations up to 100 μg/ml showed normal antibody staining with anti-actin. Thus, phalloidin blocked specific binding of Fl-phalloidin but not of actin antibody.

Photography. Photographs were taken with a Leitz Orthomat camera on Kodak Tri-X film developed in Microdol-X (1:3) at 22°C for 13 min. Prints were made on Kodak Polycontrast SC paper or Agfa TP6-WP, as contrast demanded.

RESULTS

Comparison of Fl-phalloidin and Anti-actin Patterns in Single Cells. Wulf *et al.* (11) have shown that Fl-phalloidin generates patterns that are qualitatively similar to those generated by antibody to actin. However, those studies did not include comparative examination of the two stains on the same cell populations. With two-color fluorescence, we were able to detect Fl-phalloidin and anti-actin patterns in the same cell. In cells of three types with widely varying cytoskeletons, the patterns were similar but not identical in any cell (Fig. 1). In the normal precrisis mouse embryo fibroblast cell, both compounds stained the large actin-containing stress fibers or cables (Fig. 1 *A* and *B*). However, the antibody stain also diffusely filled the perinuclear region of the cytoplasm as well as the nucleus (Fig. 1*A*); the Fl-phalloidin did not (Fig. 1*B*). As a result, Fl-phalloidin also revealed fine filaments in MEF whose intensity did not exceed the diffuse background of antibody-stained cells (Fig. 1*B*).

Patterns in the postcrisis growth-controlled cell 3T3 showed

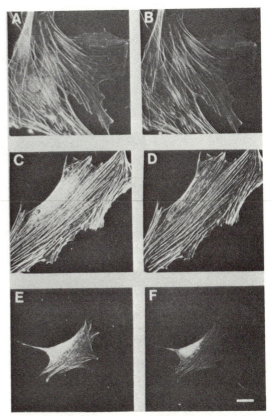

FIG. 1.   Fl-phalloidin and rhodamine anti-actin patterns in MEF cells (*A, B*), 3T3 cells (*C, D*), and SV101 cells (*E, F*). Cell cultures were fixed, stained with rabbit antibody to actin, and then exposed to a mixture of Fl-phalloidin and goat anti-rabbit IgG. Single cells were illuminated for rhodamine (*A, C, E*) or fluorescein (*B, D, F*) and photographed. See Table 1 for properties of these cells. (Bar, 20 μm.)

a similar difference (Fig. 1 *C* and *D*): the cables were in general thicker, but still Fl-phalloidin was capable of revealing finer cables than was the antibody. In the tumorigenic transformed SV101 (19), the antibody and Fl-phalloidin detected both a diffuse actin distribution and a multiplicity of short fine cables (Fig. 1 *E* and *F*) similar to those reported from an electron microscope study of these SV40-transformed 3T3 cells (20). Thus, Fl-phalloidin microscopy confirms a loss of size and length of actin-containing structures in these transformed cells and reveals that the structures can be replaced by thinner or shorter ones as well as by a diffuse distribution of actin.

States of Organization. The multiplicity of different Fl-phalloidin-generated patterns fell into four major classes (Fig. 2). In all cases the central portion of the cell was examined and scored. Class 1 cells have heavy distinct cables that cross more than 90% of the central area of the cell; 50% of which span the breadth of the cell. Class 2 have fine cables and at least two heavy distinct cables that enter the central half of the cell and extend more than half the breadth of the cell. Class 3 cells have only fine cables. Class 4 cells have no detectable cables in the central area but do have a diffuse fluorescence. Some class 4 cells have fine cables solely at their periphery. In addition, rounded cells and cells with radial polar cables have been seen in some cultures, but under standard conditions of culture these were not more than 10% of the total in any cell line.

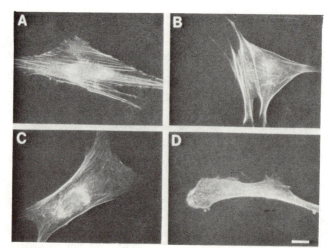

FIG. 2. Four categories of F-actin distribution used for scoring cultures stained with Fl-phalloidin. (*A*) Category 1; >90% of cell area filled with thick cables. (*B*), Category 2; at least two thick cables running under nucleus, and rest of cell area filled with fine cables. (*C*), Category 3; no thick cables, but some fine cables present. (*D*) Category 4; no cables visible in the central area of the cell. (Bar, 20 μm.)

F-actin Pattern Changes in Transformation and Reversion of Mouse Fibroblasts. Mouse cells were examined by using Fl-phalloidin and distributed into categories according to the criteria in Fig. 2. For each line, some cells fell into each category. However, the distributions were reproducibly different for different cell lines. Table 1 summarizes the properties of these cell lines and also lists two single-number measures of F-actin organization derived from the histograms. The percentage of cells in classes 1 and 2 is a single score emphasizing large cables. The percentage of cells in classes 1, 2, and 3 is a single score recording all visible cables.

The precrisis mouse embryo fibroblasts and normal cell line 3T3 had F-actin pattern distributions that were undistinguishable both by histogram (Fig. 3 *A* and *B*) and by either single threshold score (Table 1). The histogram for the transformed cell SV101 had shifted due to reduction both in cells with large cables and in cells with fine cables (Fig. 3*C*). The SV40 T-positive partial phenotypic revertant FlSV101 (19) had many cells with fine cables, but only a few cells had regained the large cables of classes 1 and 2 (Fig. 3*D*).

In this set of related lines, Fl-phalloidin showed that SV40 transformation reduced the size and number of F-actin-con-

Table 1. Properties of mouse, rat, and human cells

Cell	Pre/Post*	Virus	Clone of	Agar growth[†]		Tumors[‡]	F-actin cables, %[§]	
				RPE	CVI		1 + 2	1 + 2 + 3
Mouse:								
MEF	Pre	—	—	≤10⁻³	ND	0/7	72	94
3T3	Post	—	—	≤10⁻³	ND	0/4	76	90
SV101	Post	SV40	3T3	27	ND	5/16	11	37
FlSV101	Post	—	SV101	0.01	ND	0/2	25	82
SVR42	Post	SV40	3T3	≤10⁻³	ND	ND	33	67
SVR63	Post	SV40	3T3	≤10⁻³	ND	0/2	32	86
SVR13	Post	SV40	3T3	0.2	ND	0/2	39	76
SVR85	Post	SV40	3T3	10.6	ND	ND	8	22
SVR87	Post	SV40	3T3	38.6	ND	3/4	1	10
Rat:								
REF	Pre	—	—	<1	1.6	0/6	69	93
Rat 1	Post	—	—	7 × 10⁻³	13	2/3	16	61
14B	Post	SV40	Rat 1	15	380	3/3	1	1
1-4	Post	—	14B	3 × 10⁻³	21	0/4	4	24
3-8	Post	—	14B	4 × 10⁻³	1	0/4	26	80
MCA	Post	—	1-4	5.4	4200	3/3	2	14
Human:								
SV80	Post	SV40	—	1.6	ND	0/20[¶]	16	90

MEF, mouse embryo fibroblast; REF, rat embryo fibroblast.
* Precrisis strain or postcrisis established line.
[†] Relative plating efficiency (RPE) measured colonies >0.2 mm in diameter in agar as a percentage of the colonies on the plastic dish. Colony volume increase (CVI) measured the total increase in cell number in agar culture (14, 21–23). ND, not done.
[‡] *Nude* mice with tumors at 6 months/total animals (23, 24) injected with 10⁷ cells.
[§] Data taken from histograms of F-actin pattern distribution by Fl-phalloidin staining of adherent cells on coverslips; cells were divided into four classes (see Fig. 2).
[¶] SV80 is not tumorigenic in *nude* mice unless the animals are heavily irradiated (ref. 25; S. Shin, personal communication).

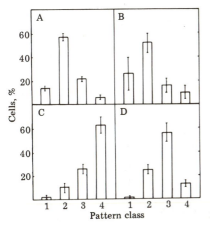

FIG. 3.   F-actin pattern distribution in mouse cell lines. Cells were examined by fluorescence microscopy and sorted into one of the four categories described in Fig. 2. For each line, 200 cells were scored, on at least two coverslips each. Results are mean ± SD for each score of each category. (*A*) Mouse embryo fibroblasts (MEF). (*B*) 3T3 normal cell line. (*C*) SV101-transformed cell lines. (*D*) FlSV101 revertant cell lines.

taining cables; partial phenotypic reversion was accompanied by an increase in number, but not a return to original cable distribution.

SV40 infection of 3T3 generates a continuum of transformed phenotypes when clones are isolated nonselectively after infection (14, 26). These phenotypes range from T-antigen-negative minimal transformants which can grow in reduced serum but not in agar, through intermediate transformants which contain small amounts of SV40 T antigens or are stable mixtures of T-antigen-positive and T-antigen-negative cells and

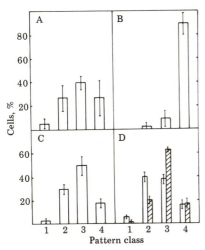

FIG. 4.   F-actin patterns in nonselectively isolated SV40-transformed cell lines. Scores were obtained as in Fig. 3. (*A*) Minimally transformed SVR42. (*B*) Average scores of two fully transformant lines, SV85 and SV87. (*C*) Average scores of two intermediate transformed lines, SVR63 and SVR13. SV40 T antigen was absent from the minimal transformant, present in about 50% of the cells of the intermediate lines, and present in all cells of the full transformant. (*D*) Resolution of intermediate lines into two patterns. Open bars, T-antigen-negative intermediate cells; hatched bars, T-antigen-positive intermediate cells.

which grow slowly and poorly in agar, to full transformants which contain large amounts of SV40 T antigen and which grow very well in agar (14). We examined representatives of these classes of transformants (Fig. 4; Table 1). The F-actin pattern of minimal transformants and intermediate transformants deviated slightly from that of 3T3 as the major category shifted from cells with at least two large cables to cells with only fine cables (Fig. 3*B*; Fig. 4 *A* and *B*). Full transformants were comparable to SV101, with most cells lacking any detectable cables (Fig. 4*B*).

F-actin Patterns and Nuclear SV40 T Antigen. The correlation of reduction in F-actin cable size and number with SV40 T antigen in these nonselected transformants suggests that T-antigen-positive cells should have fewer and finer cables than T-antigen-negative cells in the intermediate transformed lines. Using double-immunofluorescence with rhodamine-labeled antibody to T antigen and Fl-phalloidin, we examined two intermediate cell lines. The patterns of T-antigen-positive and T-antigen-negative cells within these cloned lines were significantly different, the T-antigen-positive cells having fewer and finer cables (Fig. 4*D*).

F-actin Pattern Changes in SV40 Transformation and Reversion of Rat Cells. We carried out a series of transformations and reversions of a rat cell line (Table 1). The growth-controlled cell line Rat 1 was used as the initial line (27, 28). From it, 14B was isolated after transformation with SV40 DNA (28); 14B contains SV40 large and small T antigens and one copy of integrated SV40 DNA (28). T-antigen-negative total phenotypic revertants 1-4 and 3-8 were isolated from 14B (29). 1-4 retained SV40 DNA; 3-8 lacked it. From 1-4, the phenotypically transformed subline MCA was isolated as a rare colony in agar (13); it, too, lacked SV40 T antigens.

The F-actin distributions of these lines closely mirrored those of the mouse cell lines (Table 1) in that in the more transformed lines the pattern was dominated by cells in classes 3 and 4. However, significant differences appeared upon a close comparison of the mouse and rat sets of cell lines.

Precrisis (REF and MEF) distributions were very close, but the postcrisis normal lines differ (Table 1). Rat 1 had many more cells in classes 3 and 4 than did 3T3. Transformed cells 14B essentially lacked cables. Both rat revertants had some cells with fine cables, and revertant 3-8 in addition had regained a considerable number of cells with large cables. Neither revertant had returned to the pattern of Rat 1. The spontaneous retransformant MCA was almost devoid of cells with cables but was hardly different in this regard from its parent, the revertant line 1-4. Although the revertant 1-4 lacked T antigen (29) and was anchorage-dependent (29) and nontumorigenic (23), nevertheless it had an F-actin pattern distribution quite similar to that of the tumorigenic lines 14B and MCA.

DISCUSSION

The actin-based cytoskeletal framework of cultured fibroblasts visualized by fluorescence microscopy is formidably complex in its organization. The generally accepted criteria of cytoskeletal actin organization, which up to now have been based on work with antibodies, are now redefined in terms of the patterns generated by interaction of cellular F-actin with a chemical of totally known structure and well-described properties (9, 11, 18, 30). By binding with high affinity to native F-actin (11), Fl-phalloidin can generate patterns whose reproducibility permits quantitation.

We have chosen to quantitate F-actin patterns by distributing cells into four classes (Fig. 2). These classes expand the "plus/minus" scoring system used in previous studies (3, 4, 8) in that they generate two distinct thresholds of organization (Table 1).

In general, the threshold that emphasizes large cables (classes 1 and 2 as percentage of total cells) reproduces quite well the fraction of "cable-positive" cells reported in studies with anti-actin (4, 5, 8, 23). Apparently, finer cables were not routinely detected in those studies. In light of the capacity of G-actin to completely block anti-actin but not Fl-phalloidin, it is likely that this is at least in part the consequence of antibody recognition of fixed cellular G-actin. Because affinity-purified anti-actin that has been directly conjugated to a fluorescent moiety does not show perinuclear fluorescence (7), the perinuclear fluorescence seen here may be due to nonspecific binding.

Comparisons of patterns in similar types of cells of different species reveals significant variation. For example, the normal established mouse line 3T3 has more heavy cables than the corresponding normal established rat line Rat 1 (Table 1). The SV40-transformed cell lines SV80 (human), SV101 (mouse), and 14B (rat) all have fewer cables than their respective parents but are not similar to each other, with SV80 having the most cables and 14B having essentially no cables at all.

We used two sets of cell lines descended from each other through selection for transformation or reversion, coupled with two-color immunofluorescence, to study the role of SV40 viral gene expression in changes of F-actin patterns. The results from a set of SV40-transformed 3T3 mouse cell lines suggest that, in these lines, SV40 gene expression in a given cell (i.e., nuclear T antigen) was necessary for maximal shift in F-actin pattern (Figs. 3 and 4). The results from a set of lines descended from the SV40 DNA-transformed Rat 1 rat cell line, however, show that SV40 gene products need not be present for this pattern shift (Table 1). Both the revertant 1-4 and its spontaneous re-transformed subclone MCA almost completely lack detectable F-actin cables, yet neither has any SV40 large or small T antigen. Perhaps in those cell lines the F-actin perturbation is maintained by transformation-specific molecules coded for by the host cell. We have recently shown that a 54,000-dalton protein is present in large amounts in 14B and MCA but in much smaller amounts in Rat 1 and the 14B revertants and is absent from precrisis rat fibroblasts (13).

Viral transformation is not the only perturber of F-actin pattern distribution. Our results show that establishment, the transition from precrisis cell to clonable line, can be accompanied by a significant shift: the majority of rat embryo fibroblasts are in class 2, but the majority of Rat 1 cells are in class 3 (Table 1). The cytoskeletal F-actin pattern of human fibroblasts resembles that of MEF but is changed in cells from patients with various familial colonic neoplasms (31). Such cells have patterns quite similar to those of minimal or intermediate SV40-transformed mouse 3T3 cells and an SV40-transformed human fibroblast line, SV80. Microinjection of Fl-phalloidin into live cells or very gentle treatment, such as 1 min in 0.01% Nonidet P-40, followed by 1 min in the presence of Fl-phalloidin at 1 μg/ml yields images comparable to those seen here (unpublished data). Such patterns are therefore unlikely to be the result of differential extraction during the preparation of the cells.

It is likely that the classes we have used here are arbitrary boundaries within a continuum of F-actin cable length, number, and distribution. However, proof of this will require more measurements per cell than we can make by eye. Because the emitted light from Fl-phalloidin is proportional to F-actin, we can couple fluorescence images to a vidicon-based digitalized image analyzer (32). In this way differences in cytoskeletal F-actin organization will result in signals that can be stored in our computer—for example, for prospective studies of cultured skin biopsy cells from persons at risk for cancer (31, 33).

We thank Dr. Theodor Wieland and Dr. A. Deboben for their gift of Fl-phalloidin and Dr. K. Burridge for his rabbit anti-actin antiserum. We thank Peggy Monahan for her excellent assistance and Marisa Bolognese for her help in the preparation of this manuscript. This work was supported by National Institutes of Health Grant CA-25066 and Training Grant GM-07216 and the Josiah Macy Foundation.

1. Goldman, R. D., Lazarides, E., Pollack, R. & Webster, K. (1975) *Exp. Cell Res.* **90**, 333–344.
2. Lazarides, E. (1976) *J. Supramol. Struct.* **5**, 531–563.
3. Pollack, R., Osborn, M. & Weber, K. (1975) *Proc. Natl. Acad. Sci. USA* **72**, 994–998.
4. Edelman, G. & Yahara, I. (1976) *Proc. Natl. Acad. Sci. USA* **73**, 2047–2051.
5. Tucker, R. W., Sanford, K. K. & Frankel, F. R. (1978) *Cell* **13**, 629–642.
6. Osborn, M., Born, T., Koitsch, H. J. & Weber, K. (1978) *Cell* **14**, 477–488.
7. Herman, I. M. & Pollard, T. D. (1979) *J. Cell Biol.* **80**, 509–520.
8. McClain, D. A., Maness, P. F. & Edelman, G. M. (1978) *Proc. Natl. Acad. Sci. USA* **75**, 2750–2754.
9. Faulstich, H., Schafer, A. J. & Weckauf, M. (1977) *Hoppe-Seyler's Z. Physiol. Chem.* **358**, 181–184.
10. Lengsfeld, A., Low, I., Wieland, T., Dancker, P. & Hasselbach, W. (1974) *Proc. Natl. Acad. Sci. USA* **71**, 2803–2807.
11. Wulf, E., Deboben, A., Bautz, F. A., Faulstich, H. & Wieland, T. (1979) *Proc. Natl. Acad. Sci. USA* **76**, 4498–4502.
12. Steinberg, B., Pollack, R., Topp, W. & Botchan, M. (1978) *Cell* **13**, 19–32.
13. Pollack, R., Lo, A., Steinberg, B., Smith, K., Shure, H., Blanck, G. & Verderame, M. (1980) *Cold Spring Harbor Symp. Quant. Biol.* **44**, 681–688.
14. Risser, R. & Pollack, R. (1974) *Virology* **59**, 477–489.
15. Burridge, K. (1976) *Proc. Natl. Acad. Sci. USA* **73**, 4457–4461.
16. Feramisco, J. R. & Burridge, K. (1980) *J. Biol. Chem.* **255**, 1194–1199.
17. Spudich, J. A. & Watt, S. (1971) *J. Biol. Chem.* **246**, 4866–4880.
18. Wieland, T. & Faulstich, H. (1978) *CRC Crit. Rev. Biochem.* **5**, 185–260.
19. Pollack, R., Green, H. & Todaro, G. J. (1968) *Proc. Natl. Acad. Sci. USA* **60**, 126–133.
20. Goldman, R. D., Yerna, M. & Schloss, J. (1976) *J. Supramol. Struct.* **5**, 155–183.
21. Vogel, A. & Pollack, R. (1973) *J. Cell Phys.* **82**, 189–198.
22. Steinberg, B. M. & Pollack, R. (1979) *Virology* **99**, 302–311.
23. Steinberg, B. M., Rifkin, D., Shin, S., Boone, C. & Pollack, R. (1979) *J. Supramol. Struct.* **11**, 539–546.
24. Shin, S., Freedman, V. H., Risser, R. & Pollack, R. (1975) *Proc. Natl. Acad. Sci. USA* **72**, 4435–4439.
25. Kahn, P. & Shin, S. (1979) *J. Cell Biol.* **82**, 1–16.
26. Pollack, R., Risser, R., Conlon, S., Freedman, V. H., Rifkin, D. & Shin, S. (1975) in *Proteases and Biological Controls*, eds. Reich, E. & Rifkin, D. (Cold Spring Harbor Laboratory, Cold Spring Harbor, NY), pp. 885–889.
27. Mishra, N. & Rayan, W. (1973) *Int. J. Cancer* **11**, 123–130.
28. Botchan, M., Topp, W. & Sambrook, J. (1976) *Cell* **9**, 269–287.
29. Steinberg, B., Pollack, R., Topp, W. & Botchan, M. (1978) *Cell* **13**, 19–32.
30. Wieland, T. (1977) *Naturwissenschaften* **64**, 303–309.
31. Kopelovich, L., Lipkin, M., Blattner, W. A., Fraumeni, J. F., Lynch, H. T., & Pollack, R. (1980) *Int. J. Cancer*, in press.
32. Sobel, I., Levinthal, C. & Macagno, E. (1980) *Annu. Rev. Biophys. Bioeng.* **9**, 347–362.
33. Pollack, R. & Kopelovich, L. (1979) *Methods Achiev. Exp. Pathol.* **9**, 207–230.

Supplementary Readings

Before 1961

Goldstein, L., R. Cailleau, and T.T. Crocker. 1960. Nuclear-cytoplasmic relationships in human cells in tissue culture. *Exp. Cell Res. 19:* 332–342.

1961–1970

Buckley, I.K. and K.R. Porter. 1967. Cytoplasmic fibrils in living cultured cells. *Protoplasma 64:* 349–380.

Carter, S.B. 1967. Effects of cytochalasins on mammalian cells. *Nature 213:* 261–264.

Carter, S.B. 1967. Haptotaxis and the mechanism of cell motility. *Nature 213:* 256–260.

Ishikawa, H., R. Bischoff, and H. Holtzer. 1969. Formation of arrowhead complexes with heavy meromyosin in a variety of cell types. *J. Cell Biol. 43:* 312–328.

Ridler, M. and G. Smith. 1968. The response of human cultured lymphoblasts to cytochalasin B. *J. Cell Sci. 3:* 595–602.

Weisenberg, R.C., G.G. Borisy, and E.W. Taylor. 1968. The colchicine-binding protein of mammalian brain and its relation to microtubules. *Biochemistry 7:* 4466–4479.

Wilson, L., J. Bryan, A. Ruby, and D. Mazia. 1970. Precipitation of proteins by vinblastine and calcium ions. *Proc. Natl. Acad. Sci. 66:* 807–814

1971–1980

Adelstein, R.S., M.A. Conti, G.S. Johnson, I. Pastan, and T.D. Pollard. 1972. Isolation and characterization of myosin from cloned mouse fibroblasts. *Proc. Natl. Acad. Sci. 69:* 3693–3697.

Allison, A.C., P. Davies, and S. de Petris. 1971. Role of contractile microfilaments in macrophage movement and endocytosis. *Nat. New Biol. 232:* 153–155.

Brinkley, B.R., S.S. Barham, S.C. Barranco, and G.M. Fuller. 1974. Rotenone inhibition of spindle microtubule assembly in mammalian cells. *Exp. Cell Res. 85:* 41–46.

Brouty-Boye, D. and B.R. Zetter. 1980. Inhibition of cell motility by interferon. *Science 208:* 516–518.

Burk, R.R. 1973. A factor from a transformed cell line that affects cell migration. *Proc. Natl. Acad. Sci. 70:* 369–372.

Cande, W.Z., J. Snyder, D. Smith, K. Summers, and R. McIntosh. 1974. A functional mitotic spindle prepared from mammalian cells in culture. *Proc. Natl. Acad. Sci. 71:* 1559–1563.

Dermer, G.B., J. Lue, and H.B. Neustein. 1974. Comparison of surface material, cytoplasmic filaments and intercellular junctions from untransformed and two mouse sarcoma virus transformed cell lines. *Cancer Res. 34:* 31–38.

Dirksen, E.R. 1974. Cilogenesis in the mouse oviduct. A scanning electron microscope study. *J. Cell Biol. 62:* 899–904.

Elias, E., Z. Hruban, J.B. Wade, and J.I. Boyer. 1980. Phalloidin-induced cholestasis: A microfilament-mediated change in junctional complex permeability. *Proc. Natl. Acad. Sci. 77:* 2229–2233.

Fine, R. and D. Bray. 1971. Actin in growing nerve cells. *Nature 234:* 115–118.

Fukui, Y. and H. Katsumaru. 1980. Dynamics of nuclear actin bundle induction by dimethyl sulfoxide and factors affecting its development. *J. Cell Biol. 84:* 131–140.

Gadasi, H. and E.D. Korn. 1980. Evidence for differential intracellular localization of the acanthamoeba myosin isoenzymes. *Nature 286:* 452–456.

Gard, D.L. and E. Lazarides. 1980. The synthesis and distribution of desmin and vimentin during myogenesis in vitro. *Cell 19:* 263–275.

Goldman, R.D., R. Pollack, and N.H. Hopkins. 1973. Preservation of normal behavior by enucleated cells in culture. *Proc. Natl. Acad. Sci. 70:* 750–754.

Goldman, R.D., E. Lazarides, R. Pollack, and K. Weber. 1975. The distribution of actin in non-muscle cells. *Exp. Cell Res. 90:* 333–344.

Gruenstein, E., A. Rich, and R.R. Weihing. 1975. Actin associated with membranes from 3T3 mouse fibroblast and HeLa cells. *J. Cell Biol. 64:* 223–234.

Harris, A.K., P. Wild, and D. Stopak. 1980. Silicone rubber substrata: A wrinkle in the study of cell locomotion. *Science 208:* 177–179.

Hsie, A.W. and T.T. Puck. 1971. Morphological transformation of Chinese hamster cells by dibutyryl adenosine cyclic 3':5'-monophosphate and testosterone. *Proc. Natl. Acad. Sci. 68:* 358–361.

Kelly, F. and J. Sambrook. 1973. Differential effect of cytochalasin B on normal and transformed mouse cells. *Nat. New Biol. 24:* 217–219.

Lazarides, E. 1975. Tropomyosin antibody: The specific localization of tropomyosin in nonmuscle cells. *J. Cell Biol. 65:* 549–561.

Lazarides, E. 1980. Intermediate filaments as mechanical integrators of cellular space. *Nature 283:* 249–256.

Lengsfeld, A.M., I. Low, T. Wieland, P. Dancker, and W. Hasselbach. 1974. Interaction of phalloidin with actin. *Proc. Natl. Acad. Sci. 71:* 2803–2807.

Lin, D.C., K.D. Tobin, M. Grumet, and S. Lin. 1980. Cytochalasins inhibit nuclei-induced actin polymerization by blocking filament elongation. *J. Cell Biol. 84:* 455–460.

Luduena, M.A. and N.K. Wessells. 1973. Cell locomotion, nerve elongation, and microfilaments. *Devel. Biol. 130:* 427–440.

MacLean-Fletcher, S. and T.D. Pollard. 1980. Mechanism of action of cytochalasin B on actin. *Cell 20:* 329–341.

Miller, R.A. and F.H. Ruddle. 1974. Enucleated neuroblastoma cells form neurites when treated with dibutyryl cyclic AMP. *J. Cell Biol. 63:* 295–299.

Muszbek, L. and K. Laki. 1974. Cleavage of actin by thrombin. *Proc. Natl. Acad. Sci. 71:* 2208–2211.

Nagata, K., J. Sagara, and Y. Ichikawa. 1980. Changes in contractile proteins during differentiation of myeloid leukemia cells. *J. Cell Biol. 85:* 273–282.

Ostlund, R.E., I. Pastan, and R.S. Adelstein. 1974. Myosin in cultured fibroblasts. *J. Biol. Chem. 249:* 3903–3907.

Pfeffer, L.M., E. Wang, and I. Tamm. 1980. Interferon effects on microfilament organization, cellular fibronectin distribution, and cell motility in human fibroblasts. *J. Cell Biol. 85:* 9–17.

Pollack, R., M. Osborn, and K. Weber. 1975. Patterns of organization of actin and myosin in normal and transformed cultured cells. *Proc. Natl. Acad. Sci. 72:* 994–998.

Pollack, R., R.D. Goldman, S. Conlon, and C. Chang. 1974. Properties of enucleated cells. II. Characteristic overlapping of transformed cells is reestablished by enucleates. *Cell 3:* 51–54.

Prescott, D.M., D. Myerson, and J. Wallace. 1972. Enucleation of mammalian cells with cytochalasin B. *Exp. Cell Res. 71:* 480–485.

Projan, A. and St. Tanneberger. 1973. Some findings on movement and contact of human normal and tumour cell in vitro. *Eur. J. Cancer 9:* 703–708.

Puszkin, S. and S. Berl. 1972. Actomyosin-like protein from brain. *Biochim. Biophys. Acta 246:* 695–709.

Sanger, J.M. and J.W. Sanger. 1980. Banding and polarity of actin filaments in interphase and cleaving cells. *J. Cell Biol. 86:* 568–575.

Silver, R.B., R.D. Cole, and W.Z. Cande. 1980. Isolation of mitotic apparatus containing vesicles with calcium sequestration activity. *Cell 19:* 505–516.

Spudich, J. and S. Lin. 1972. Cytochalasin B, its interaction with actin and actomyosin from muscle. *Proc. Natl. Acad. Sci. 69:* 442–446.

Stossel, T.P. and T.D. Pollard. 1973. Myosin in polymophonuclear leukocytes. *J. Biol. Chem. 248:* 8288–8294.

Tillotson, D. and A.L.F. Gorman. 1980. Non-uniform Ca^{2+} buffer distribution in a nerve cell body. *Nature 286:* 816–817.

Tilney, L.G. and M. Mooseker. 1971. Actin in the brush-border of epithelial cells of the chicken intestine. *Proc. Natl. Acad. Sci. 68:* 2611–2615.

Watterson, D.M., F. Sharief, and T.C. Vanaman. 1980. The complete amino acid sequence of the Ca^{2+}-dependent modulator protein (calmodulin) of bovine brain. *J. Biol. Chem. 255:* 962–975.

Weber, A. and J.M. Murray. 1973. Molecular control mechanisms in muscle contraction. *Physiol. Rev. 53:* 612–673.

Weber, K. and U. Croeschel-Stewart. 1974. Antibody to myosin: The specific visualization of myosin-containing filaments in non-muscle cells. *Proc. Natl. Acad. Sci. 71:* 4561–4564.

Weber, K., R. Pollack, and T. Bibring. 1975. Microtubular antibody: The specific visualization of cytoplasmic microtubules in tissue culture cells. *Proc. Natl. Acad. Sci. 72:* 459–463.

Wickus, G., E. Gruenstein, P.W. Robbins, and A. Rich. 1975. Decrease in membrane-associated actin of fibroblasts after transformation by Rous sarcoma virus. *Proc. Natl. Acad. Sci. 72:* 746–749.

Yang, Y. and J.F. Perdue. 1972. Contractile proteins of cultured cells. *J. Biol. Chem. 247:* 4503–4509.